Student, Parent, Teacher Internet Resources

Access your Student Edition on the Internet so you don't need to bring your textbook home every night. You can link to features and get additional practice with these online study tools.

Check out the following features on your **Online Learning Center:**

Study Tools

Concepts In Motion

- Interactive Tables
- Interactive Time Lines
- Animated Illustrations
- National Geographic Visualizing Animations

Study to Go
Section Self-Check Quizzes
Chapter Test Practice
Standardized Test Practice
Vocabulary Puzzlemaker
Interactive Tutor
Multilingual Science Glossary
Online Student Edition

Extensions

National Geographic Society
NASA's Picture of the Day
Prescreened Web Links
Internet GeoLabs
Science Fair Ideas
Career Links

For Teachers

Teacher Bulletin Board
Teaching Today and much more!

Safety Symbols

These safety symbols are used in laboratory and investigations in this book to indicate possible hazards. Learn the meaning of each symbol and refer to this page often. *Remember to wash your hands thoroughly after completing lab procedures.*

SAFETY SYMBOLS	HAZARD	EXAMPLES	PRECAUTION	REMEDY
DISPOSAL	Special disposal procedures need to be followed.	certain chemicals, living organisms	Do not dispose of these materials in the sink or trash can.	Dispose of wastes as directed by your teacher.
BIOLOGICAL	Organisms or other biological materials that might be harmful to humans	bacteria, fungi, blood, unpreserved tissues, plant materials	Avoid skin contact with these materials. Wear mask or gloves.	Notify your teacher if you suspect contact with material. Wash hands thoroughly.
EXTREME TEMPERATURE	Objects that can burn skin by being too cold or too hot	boiling liquids, hot plates, dry ice, liquid nitrogen	Use proper protection when handling.	Go to your teacher for first aid.
SHARP OBJECT	Use of tools or glassware that can easily puncture or slice skin	razor blades, pins, scalpels, pointed tools, dissecting probes, broken glass	Practice common-sense behavior and follow guidelines for use of the tool.	Go to your teacher for first aid.
FUME	Possible danger to respiratory tract from fumes	ammonia, acetone, nail polish remover, heated sulfur, moth balls	Make sure there is good ventilation. Never smell fumes directly. Wear a mask.	Leave foul area and notify your teacher immediately.
ELECTRICAL	Possible danger from electrical shock or burn	improper grounding, liquid spills, short circuits, exposed wires	Double-check setup with teacher. Check condition of wires and apparatus.	Do not attempt to fix electrical problems. Notify your teacher immediately.
IRRITANT	Substances that can irritate the skin or mucous membranes of the respiratory tract	pollen, moth balls, steel wool, fiberglass, potassium permanganate	Wear dust mask and gloves. Practice extra care when handling these materials.	Go to your teacher for first aid.
CHEMICAL	Chemicals that can react with and destroy tissue and other materials	bleaches such as hydrogen peroxide; acids such as sulfuric acid, hydrochloric acid; bases such as ammonia, sodium hydroxide	Wear goggles, gloves, and an apron.	Immediately flush the affected area with water and notify your teacher.
TOXIC	Substance may be poisonous if touched, inhaled, or swallowed.	mercury, many metal compounds, iodine, poinsettia plant parts	Follow your teacher's instructions.	Always wash hands thoroughly after use. Go to your teacher for first aid.
FLAMMABLE	Open flame may ignite flammable chemicals, loose clothing, or hair.	alcohol, kerosene, potassium permanganate, hair, clothing	Avoid open flames and heat when using flammable chemicals.	Notify your teacher immediately. Use fire safety equipment if applicable.
OPEN FLAME	Open flame in use, may cause fire.	hair, clothing, paper, synthetic materials	Tie back hair and loose clothing. Follow teacher's instructions on lighting and extinguishing flames.	Always wash hands thoroughly after use. Go to your teacher for first aid.

Eye Safety
Proper eye protection should be worn at all times by anyone performing or observing science activities.

Clothing Protection
This symbol appears when substances could stain or burn clothing.

Animal Safety
This symbol appears when safety of animals and students must be ensured.

Radioactivity
This symbol appears when radioactive materials are used.

Handwashing
After the lab, wash hands with soap and water before removing goggles

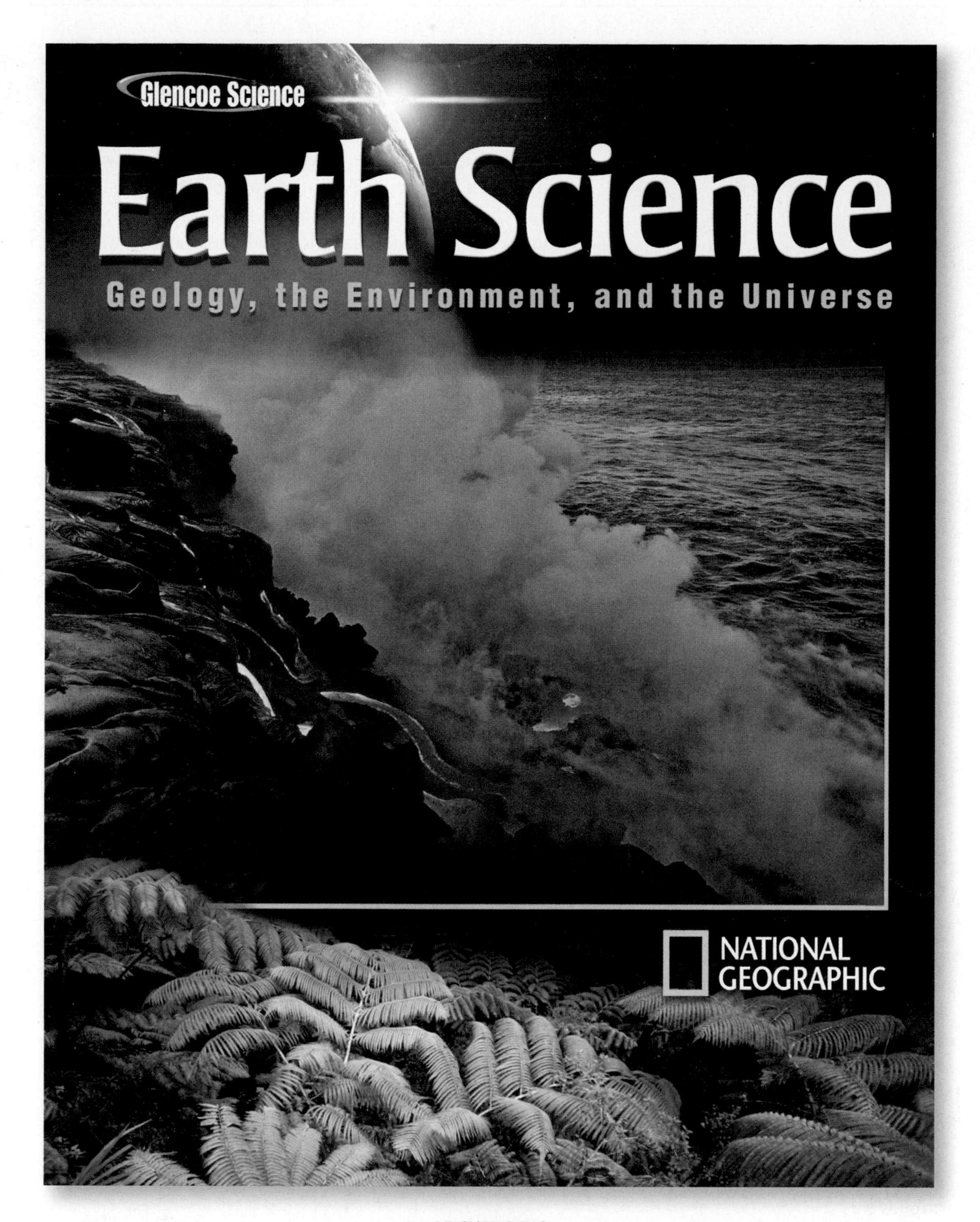

AUTHORS

Francisco Borrero • Frances Scelsi Hess • Juno Hsu
Gerhard Kunze • Stephen A. Leslie • Stephen Letro
Michael Manga • Len Sharp • Theodore Snow • Dinah Zike
National Geographic

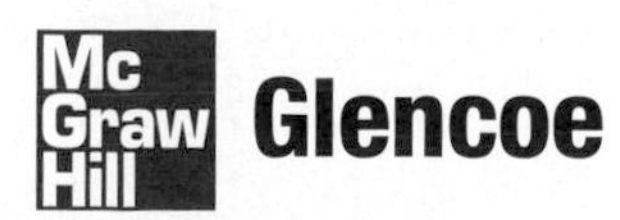

New York, New York   Columbus, Ohio   Chicago, Illinois   Woodland Hills, California

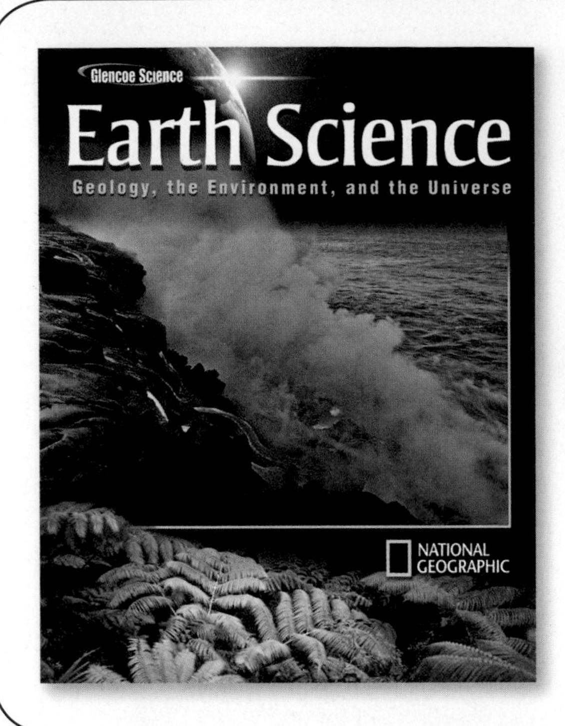

About the Photo: The lava photo on the cover was taken in Hawaii Volcanoes National Park on the big island of Hawaii. The lava in the photo is flowing from active vents on the flank of Kilauea volcano. When lava flows into the sea, sulfuric acid in the lava mixes with chlorine in the salt-water to form a mist of water vapor and hydrochloric acid.

The McGraw·Hill Companies

Mc Graw Hill Glencoe

Send all inquiries to:
Glencoe/McGraw-Hill
8787 Orion Place
Columbus, OH 43240-4027

ISBN: 978-0-07-874636-9
MHID: 0-07-874636-1

Printed in the United States of America

8 9 10 RJE/LEH 12 11

Contents in Brief

About the Authors

Dr. Francisco Borrero is a high school Earth science and Spanish teacher at Cincinnati Country Day School and a research associate and Adjunct Curator of Mollusks at Cincinnati Museum Center in Cincinnati, Ohio. He has taught Earth science and Spanish for over 20 years. Dr. Borrero holds a BS in zoology from Universidad del Valle, Colombia, and MS and PhD degrees in biological sciences from the University of South Carolina at Columbia. Dr. Borrero's research examines the relationship between physical habitat characteristics and the diversity and distribution of natural populations of mollusks.

Dr. Frances Scelsi Hess teaches Earth science at Cooperstown High School in New York. She received her BS and MS in science from the State University at Oneonta, and her EdD from Columbia University. Dr. Hess is a Fellow of the Science Teachers Association of New York State, and has received numerous teaching awards, including the Phi Delta Kappa Reed Travel Scholarship to Australia and New Zealand.

Dr. Chia Hui (Juno) Hsu currently works as a project scientist at University of California, Irvine. She holds a BS in physics and Earth science from National Taiwan Normal University, an MS in atmospheric sciences from National Taiwan University, and a PhD in atmospheric sciences from Massachusetts Institute of Technology. Before beginning her graduate work, Dr. Hsu taught ninth-grade Earth science. Her research interests include the dynamics of monsoons, climate regime shifts, and modeling global-scale atmospheric chemistry.

Dr. Gerhard Kunze is professor emeritus of geology at the University of Akron in Ohio. He has a BS in science and a PhD in geophysics from Penn State University. He was an NRC research associate at Johnson Space Center, Houston, Texas from 1973 to 1974. In 1990, Dr. Kunze was awarded a senior Fulbright scholarship to teach geophysics at the Institute of Geophysics, a department of the University of Kiel in Germany.

Dr. Stephen A. Leslie is the professor and department head of geology and environmental science at James Madison University in Harrisburg, Virginia. He was formerly an associate professor of geology at the University of Arkansas in Little Rock. His areas of research include paleontology, stratigraphy, and the evolution of early life on Earth. He has a BS in geology from Bowling Green State University, an MS in geology from the University of Idaho, and a PhD in geology from The Ohio State University.

Stephen Letro has been a meteorologist for the National Weather Service, the media, and private industry since 1971. He currently serves as the Meteorologist-in-Charge of the National Weather Service office in Jacksonville, Florida. He received his BS in meteorology from Florida State University with an emphasis on tropical meteorology. He is a member of the National Hurricane Center's Hurricane Liaison Team, and has received numerous awards, including an award for his role in restructuring the National Weather Service.

Dr. Michael Manga is a professor of Earth and planetary science at U.C. Berkeley. He has a BS in geophysics from McGill University and a PhD in Earth science from Harvard University. His areas of research include volcanology, the internal evolution and dynamics of planets, and hydrogeology. He is a MacArthur Fellow, and has received the Donath medal from the Geological Society of America and the Macelwane medal from the American Geophysical Union.

Len Sharp taught Earth science at Liverpool High School, New York, for 30 years. He has a BS in secondary education and an MS in science education from Syracuse University. Mr. Sharp was president of the Science Teachers Association of New York from 1991 to 1992, and president of the National Earth Science Teachers Association from 1992 to 1994. He was a Presidential Awardee in 1995, and received the 2005 Distinguished Teacher Award from NSTA and the 2006 NAGT—Eastern Section, Outstanding Earth Science Teacher award.

Dr. Theodore Snow is a professor of astronomy at the University of Colorado. He has a BA from Yale University, and an MS and PhD from the University of Washington. Dr. Snow is a founder and former director of the Center for Astrophysics and Space Astronomy at the University of Colorado. Dr. Snow led instrument-development programs for space-based telescopes, and is now a member of the Science Team for an ultraviolet spectrograph to be installed aboard the *Hubble Space Telescope* in early 2008.

Dinah Zike is an international curriculum consultant and inventor who has developed educational products and three-dimensional, interactive graphic organizers for over 30 years. As president and founder of Dinah-Might Adventures, L.P., Dinah is the author of more than 100 award-winning educational publications, including *The Big Book of Science.* Dinah has a BS and an MS in educational curriculum and instruction from Texas A&M University. Dinah Zike's *Foldables* are an exclusive feature of McGraw-Hill textbooks.

National Geographic, founded in 1888 for the increase and diffusion of geographic knowledge, is the world's largest nonprofit scientific and educational organization. The Children's Books and Education Division of National Geographic supports National Geographic's mission by developing innovative educational programs. National Geographic's *Visualizing* and *Expeditions* features are exclusive components of *Earth Science: Geology, the Environment, and the Universe.*

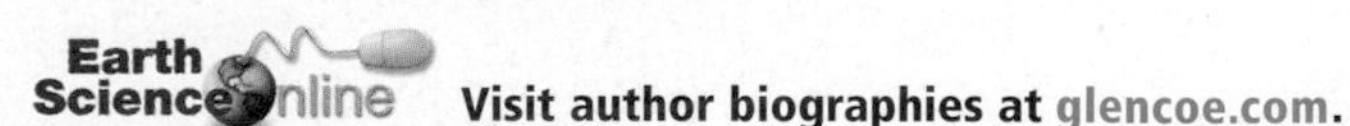

Teacher Advisory Board and Reviewers

Teacher Advisory Board

The Teacher Advisory Board gave the editorial and design team feedback on the content and design of the Student Edition. We thank these teachers for their hard work and creative suggestions.

Francisco Borrero
Cincinnati Country Day High School
Cincinnati, OH

Bill Brown
Grandview Heights High School
Columbus, OH

Carmen S. Dixon
East Knox High School
Howard, OH

Joel Heuberger
Waite High School
Toledo, OH

Jane Karabaic
Steubenville City Schools
Steubenville, OH

Terry Stephens
Edgewood High School
Trenton, OH

Reviewers

Each teacher reviewed selected chapters of *Earth Science: Geology, the Environment, and the Universe,* and provided feedback and suggestions for improving the effectiveness of the instruction.

Mark Brazo
Lincoln High School
Portland, OR

Gayle R. Dawson
Blackman High School
Murfreesboro, TN

William Dicks
Northville High School
Northville, MI

Alvin Echeverria
Del Sol High School
Las Vegas, NV

Wendy Elkins
Blue Valley Northwest High School
Overland Park, KS

Carolyn C. Elliot
South Iredell High School
Statesville, NC

Sandra Forster-Terrell
Atherton High School
Louisville, KY

Carol L. Jarocha
Northville High School
Northville, MI

Steve Kluge
Fox Lane High School
Bedford, NY

Sussan Nwabunachi Oladipo
Wells Academy High School
Chicago, IL

Michael J. Passow
White Plains Middle School
White Plains, NY

Jeremy Richardson
Lewis and Clark High School
Spokane, WA

Angela Jones Rizzo
AC Flora High School
Columbia, SC

Terry A. Stephens
Edgewood High School
Trenton, OH

Content Consultants

Content consultants each reviewed selected chapters of *Earth Science: Geology, the Environment, and the Universe* for content accuracy and clarity.

Anastasia Chopelas, PhD
Research Professor of Earth and Space Sciences
University of Washington
Seattle, WA

Diane Clayton, PhD
University of California at Santa Barbara
Santa Barbara, CA

Sarah Gille, PhD
Associate Professor
Scripps Institution of Oceanography and Department of Mechanical and Aerospace Engineering
University of California San Diego
San Diego, CA

Alan Gishlick, PhD
National Center for Science Education
Oakland, CA

Janet Herman, PhD
Professor and Director of Program of Interdisciplinary Research in Contaminant Hydrogeology
University of Virginia
Charlottesville, VA

David Ho, PhD
Storke-Doherty Lecturer & Doherty Associate Research Scientist
Lamont-Doherty Earth Observatory
Columbia University
New York, NY

Jose Miguel Hurtado, PhD
Associate Professor of Geology
University of Texas at El Paso
El Paso, TX

Monika Kress, PhD
Assistant Professor of Physics and Astronomy
San Jose State University
San Jose, CA

Amy Leventer, PhD
Associate Professor of Geology
Colgate University
Hamilton, NY

Amala Mahadevan, PhD
Associate Research Professor
Department of Earth Sciences
Boston University
Boston, MA

Nathan Niemi, PhD
Assistant Professor of Geological Sciences
University of Michigan
Ann Arbor, MI

Anne Raymond, PhD
Professor of Geology and Geophysics
Texas A&M University
College Station, TX

Contents

Your book is divided into chapters that are organized around Themes, Big Ideas, and Main Ideas of Earth Science.

THEMES are overarching concepts used throughout the entire book that help you tie what you learn together. They help you see the connections among major ideas and concepts.

BIG Ideas appear in each chapter and help you focus on topics within the themes. The Big Ideas are broken down even further into Main Ideas.

MAIN Ideas draw you into more specific details about Earth science. All the Main Ideas of a chapter add up to the chapter's Big Idea.

THEMES
Change
Structures
Geologic Time
Systems
Scientific Inquiry

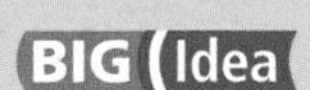

one per chapter

MAIN Idea
one per section

Contents

National Geographic Expeditions are referenced within the units and chapters at point of use, to support or extend chapter content.

Student Resources

Labs

LAUNCH Lab

Start off each chapter with a hands-on introduction to the subject matter.

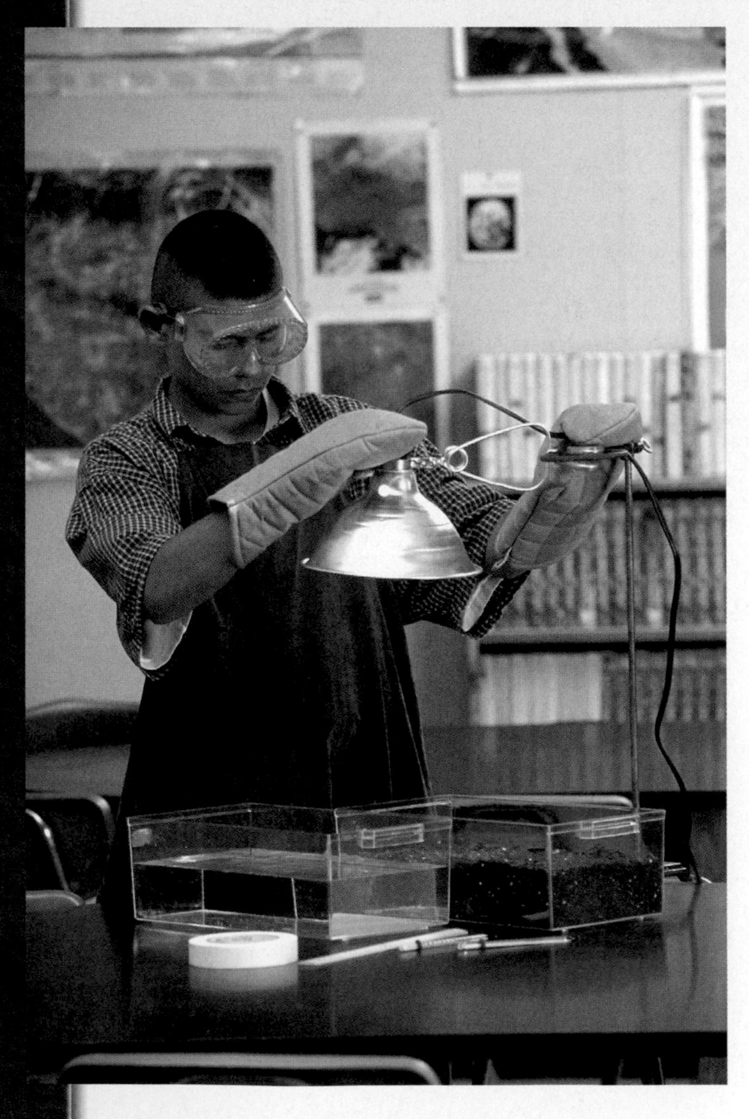

DATA ANALYSIS LAB

Build your analytical skills using actual data from real scientific sources.

PROBLEM-SOLVING LAB

Use math-based skill activities that often require data interpretation and graphing.

MiniLab

Practice scientific methods and hone your lab skills with these quick activities.

GEOLAB

Apply the skills you developed in Launch Labs, Data Analysis Labs, Problem-Solving Labs, and MiniLabs in these chapter-ending, real-world labs.

Real-World Earth Science Features

Explore today's world of Earth science. Go along on an Earth science expedition, delve into new technologies, uncover discoveries impacting the environment, and discover the hot topics in Earth science.

NATIONAL GEOGRAPHIC eXpeditions!

Get an inside look at exciting places and scientists doing real-world Earth science.

EARTH SCIENCE AND TECHNOLOGY

Discover recent advancements that have influenced Earth science.

Earth Science and the Environment

Explore environmental issues that influence Earth science.

Earth Science and Society

Examine Earth science in the news and sharpen your debating skills on complex issues in Earth science.

CAREERS IN EARTH SCIENCE

Investigate a day in the life of people working in the field of Earth science.

Concepts In Motion

History in Focus Interactive Time Line
Interactive Time Lines explore science and history through milestones in Earth science.

Concepts in Motion

Concepts In Motion — Analyze complex Earth science topics with animations of the National Geographic Visualizing pages.

Concepts In Motion — **Interactive Tables** — Check your understanding by viewing interactive versions of tables in your text.

Concepts In Motion **Animated Art** Enhance and enrich your knowledge of Earth science concepts through simple and 3D animations of visuals.

NATIONAL GEOGRAPHIC

eXpeditions!

Swim the Okavango...

Explore the African Landscape...

Dig for Dinosaurs...

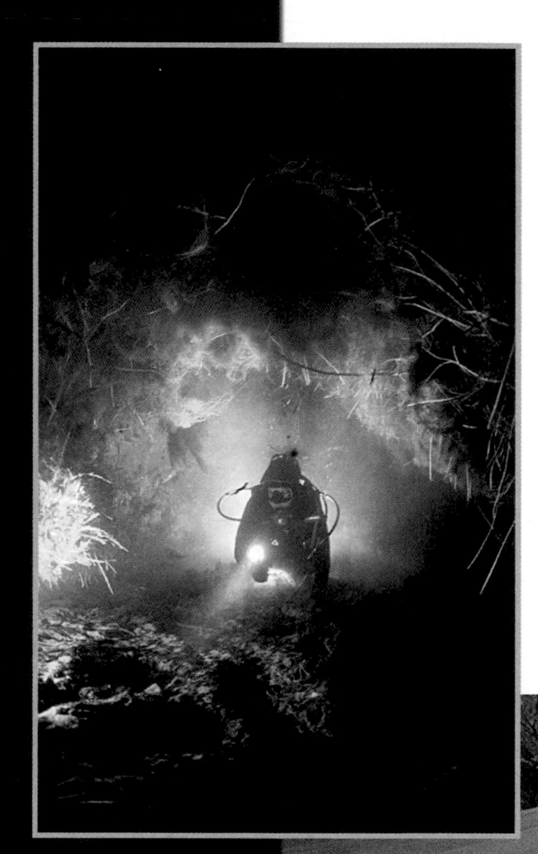

What is it like to scuba dive with crocodiles in the Okavango delta? Or fly in a bush plane over the African continent? Or dig for dinosaurs in China? The **National Geographic Expeditions** allow you to share in the excitement and adventures of explorers, scientists, and environmentalists as they venture into the unknown. Each Expedition takes you on a journey that enriches your learning about our dynamic planet.

Table of Contents

For more information on these Expeditions, visit glencoe.com. You can also link to original National Geographic articles that cover these topics and more.

Reading for Information

When you read *Earth Science: Geology, the Environment, and the Universe,* you need to read for information. Science is nonfiction writing; it describes real-life events, people, ideas, and technology. Here are some tools that this book has to help you read.

Before You Read

By reading the BIG Ideas and MAIN Ideas prior to reading the chapter or section, you will get a preview of the coming material.

Each unit preview lists the chapters in the unit. An overall BIG Idea is listed for each chapter. The Big Idea describes what you will learn in the chapter.

UNIT 5 The Dynamic Earth

Chapter 17
Plate Tectonics
BIG Idea Most geologic activity occurs at the boundaries between plates.

Chapter 18
Volcanism
BIG Idea Volcanoes develop from magma moving upward from deep within Earth.

Chapter 19
Earthquakes
BIG Idea Earthquakes are natural vibrations of the ground, some of which are caused by movement along fractures in Earth's crust.

Chapter 20
Mountain Building
BIG Idea Mountains form through dynamic processes which crumple, fold, and create faults in Earth's crust.

CAREERS IN EARTH SCIENCE
Volcanologist This volcanologist is monitoring volcanic activity to help forecast an eruption. Volcanologists spend much of their time in the field, collecting samples and measuring changes in the shape of a volcano.

WRITING in Earth Science
Visit glencoe.com to learn more about the work of volcanologists. Then write a short newspaper article about how volcanologists predicted a recent eruption.

464

Source: Unit 5, p. 464

The MAIN Ideas within a chapter support the BIG Idea of the chapter. Each section of the chapter has a Main Idea that describes the focus of the section.

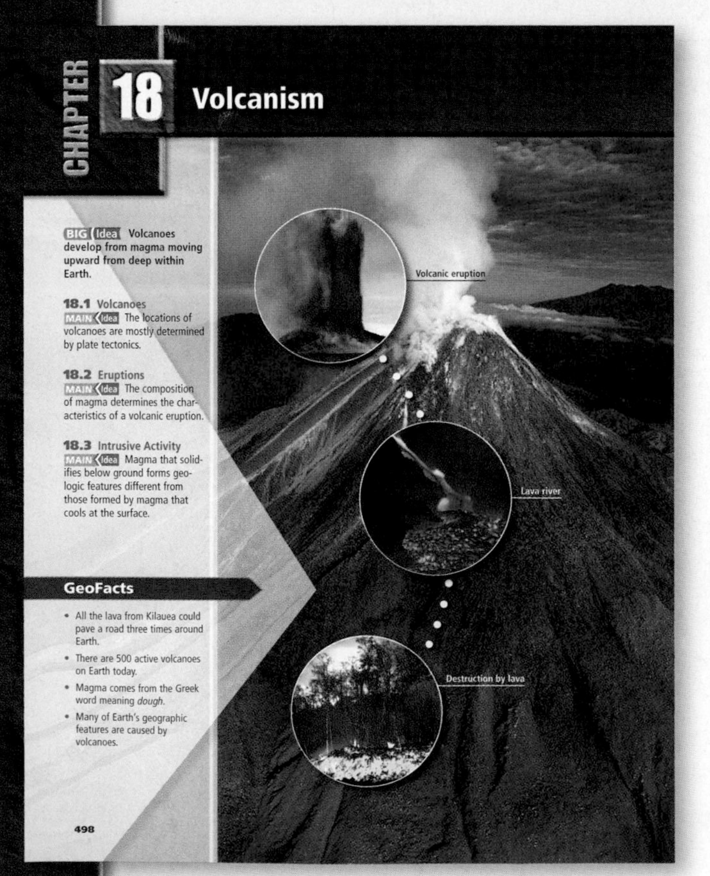
CHAPTER 18 Volcanism

BIG Idea Volcanoes develop from magma moving upward from deep within Earth.

18.1 Volcanoes
MAIN Idea The locations of volcanoes are mostly determined by plate tectonics.

18.2 Eruptions
MAIN Idea The composition of magma determines the characteristics of a volcanic eruption.

18.3 Intrusive Activity
MAIN Idea Magma that solidifies below ground forms geologic features different from those formed by magma that cools at the surface.

GeoFacts
- All the lava from Kilauea could pave a road three times around Earth.
- There are 500 active volcanoes on Earth today.
- Magma comes from the Greek word meaning *dough*.
- Many of Earth's geographic features are caused by volcanoes.

498

Source: Chapter 18, p. 498

OTHER WAYS TO PREVIEW

- **Read the chapter title to find out what the topic will be.**
- **Skim the photos, illustrations, captions, graphs, and tables.**
- **Look for vocabulary terms that are boldfaced and highlighted.**
- **Create an outline using section titles and heads.**

As You Read

Within each section you will find a tool to deepen your understanding and a tool to check your understanding.

Section 18.1

Objectives
- **Describe** how plate tectonics influences the formation of volcanoes.
- **Locate** major zones of volcanism.
- **Identify** the parts of a volcano.
- **Differentiate** between volcanic landforms.

Review Vocabulary
convergent: tending to move toward one point or to approach each other

New Vocabulary
volcanism
hot spot
flood basalt
fissure
conduit
vent
crater
caldera
shield volcano
cinder cone
composite volcano

Volcanoes

MAIN Idea **The locations of volcanoes are mostly determined by plate tectonics.**

Real-World Reading Link Road crews spread salt on icy winter roads because salt makes the ice melt at a lower temperature. At extremely high temperatures, rocks can melt. Often, if heated rocks are in contact with water, they melt more easily.

Zones of Volcanism

Volcanoes are fueled by magma. Recall from Chapter 5 that magma is a slushy mixture of molten rock, mineral crystals, and gases. As you observed in the Launch Lab, once magma forms, it rises toward Earth's surface because it is less dense than the surrounding mantle and crust. Magma that reaches Earth's surface is called lava. **Volcanism** describes all the processes associated with the discharge of magma, hot fluids, and gases.

As you read this, approximately 20 volcanoes are erupting. In a given year, volcanoes will erupt in about 60 different places on Earth. The distribution of volcanoes on Earth's surface is not random. A map of active volcanoes, shown in **Figure 18.1**, reveals striking patterns on Earth's surface. Most volcanoes form at plate boundaries. The majority form at convergent boundaries and divergent boundaries. Along these margins, magma rises toward Earth's surface. Only about 5 percent of magma erupts far from plate boundaries.

Figure 18.1 Most of Earth's active volcanoes are located along plate boundaries.

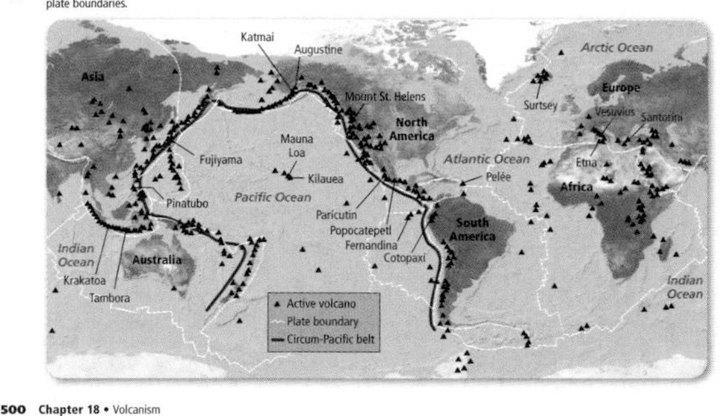

Source: Section 18.1, p. 500

❮ The **Real-World Reading Link** describes how the section's content may relate to you.

Environmental Connections point out ❯ sections and paragraphs that emphasize real-world environmental applications of Earth Science content. When you see this icon, think about how the content connects to the world around you.

Reading Checks are questions that assess your understanding.

Source: Section 23.1, p. 650

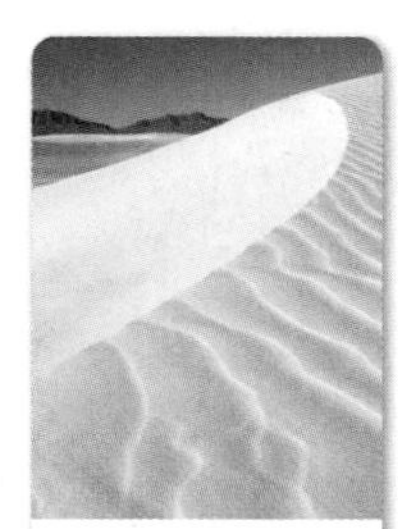

Figure 23.3 The white sands of New Mexico's White Sands National Park are made of gypsum from ancient evaporite deposits.

Evaporites Scientists can also learn about fluctuating sea level by studying evaporite deposits. Recall from Chapter 6 that evaporite deposits are rocks that have crystallized out of water that is supersaturated with dissolved minerals. Some evaporite deposits can be associated with fossilized reefs.

Reefs are made of the carbonate skeletons of tropical organisms. Reefs form in long, linear mounds parallel to a continent or island, where they absorb the energy of the waves that crash against them on their seaward side. The area behind the reef, called a lagoon, is protected from the wave's energy. Water in the lagoons evaporates in the warm, tropical sunshine, and minerals such as halite and gypsum precipitate out. Over time, cycles of evaporite deposition mark changes in water level.

Reading Check **Explain** how evaporites and reefs are related.

Mineral deposits Huge amounts of gypsum and halite evaporites were deposited in Paleozoic lagoons. The white sands of White Sands National Park, shown in **Figure 23.3**, are the remains of one such evaporite deposit. Other deposits, such as those in the Great Lakes area of North America, are mined commercially. Halite is used as road salt. Gypsum is an ingredient in plaster and drywall.

Impermeability As shown in **Figure 23.4**, reef rocks tend to have large pore spaces, allowing oil and other liquids to move through them. Evaporite rocks, in contrast, are impermeable. This means that they contain very little pore space and liquid cannot move through them. When an evaporite deposit overlies a reef rock that contains oil, it seals in the oil and prevents the oil from migrating. A good example is the Permian Basin, home to the Great Permian Reef Complex in western Texas and southeastern New Mexico. The oil in this complex rarely leaks to Earth's surface because of its tight evaporite seal.

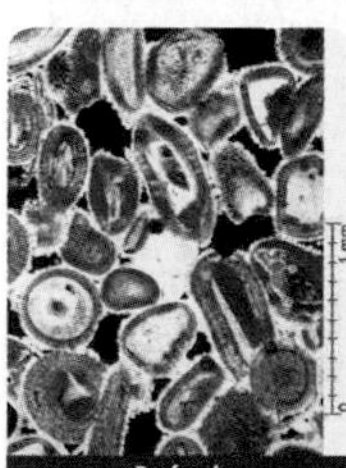

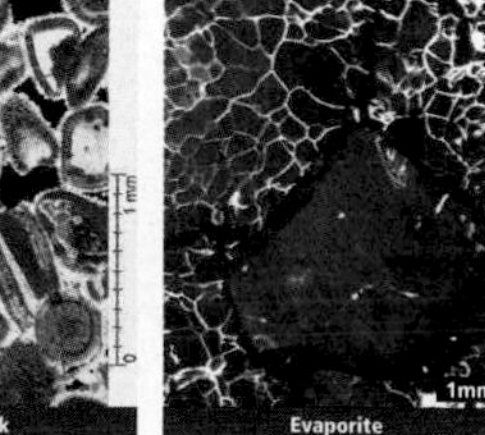

Figure 23.4 Reef rocks have large pores that can contain oil or other liquids, in contrast to evaporite rock, which is impermeable to liquids.
Infer *why ancient evaporite deposits are important to petroleum geologists.*

OTHER READING SKILLS

- Ask yourself: What is the BIG Idea? What is the MAIN Idea?
- Think about people, places, and situations that you've encountered. Are there any similarities with those mentioned in this book?
- Relate the information in this book to other areas you have studied.
- Predict events or outcomes by using clues and information that you already know.
- Change your predictions as you read and gather new information.

After You Read

Follow up your reading with a summary and assessment of the material to evaluate if you understood the text.

Each section concludes with an assessment. The assessment contains a summary and questions. The summary reviews the section's key concepts while the questions test your understanding.

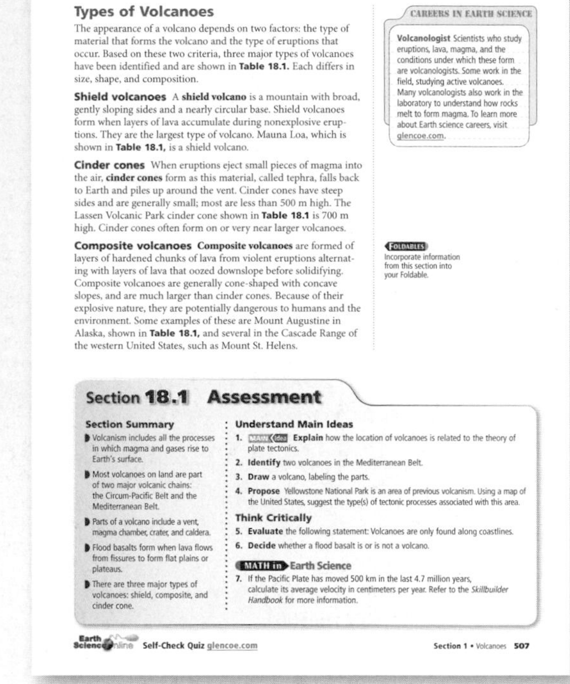

Types of Volcanoes

The appearance of a volcano depends on two factors: the type of material that forms the volcano and the type of eruptions that occur. Based on these two criteria, three major types of volcanoes have been identified and are shown in **Table 18.1.** Each differs in size, shape, and composition.

Shield volcanoes A **shield volcano** is a mountain with broad, gently sloping sides and a nearly circular base. Shield volcanoes form when layers of lava accumulate during nonexplosive eruptions. They are the largest type of volcano. Mauna Loa, which is shown in **Table 18.1,** is a shield volcano.

Cinder cones When eruptions eject small pieces of magma into the air, **cinder cones** form as this material, called tephra, falls back to Earth and piles up around the vent. Cinder cones have steep sides and are generally small; most are less than 500 m high. The Lassen Volcanic Park cinder cone shown in **Table 18.1** is 700 m high. Cinder cones often form on or very near larger volcanoes.

Composite volcanoes **Composite volcanoes** are formed of layers of hardened chunks of lava from violent eruptions alternating with layers of lava that oozed downslope before solidifying. Composite volcanoes are generally cone-shaped with concave slopes, and are much larger than cinder cones. Because of their explosive nature, they are potentially dangerous to humans and the environment. Some examples of these are Mount Augustine in Alaska, shown in **Table 18.1,** and several in the Cascade Range of the western United States, such as Mount St. Helens.

CAREERS IN EARTH SCIENCE

Volcanologist Scientists who study eruptions, lava, magma, and the conditions under which these form are volcanologists. Some work in the field, studying active volcanoes. Many volcanologists also work in the laboratory to understand how rocks melt to form magma. To learn more about Earth science careers, visit glencoe.com.

FOLDABLES Incorporate information from this section into your Foldable.

Section 18.1 Assessment

Section Summary

- Volcanism includes all the processes in which magma and gases rise to Earth's surface.
- Most volcanoes on land are part of two major volcanic chains: the Circum-Pacific Belt and the Mediterranean Belt.
- Parts of a volcano include a vent, magma chamber, crater, and caldera.
- Flood basalts form when lava flows from fissures to form flat plains or plateaus.
- There are three major types of volcanoes: shield, composite, and cinder cone.

Understand Main Ideas

1. MAIN Idea **Explain** how the location of volcanoes is related to the theory of plate tectonics.
2. **Identify** two volcanoes in the Mediterranean Belt.
3. **Draw** a volcano, labeling the parts.
4. **Propose** Yellowstone National Park is an area of previous volcanism. Using a map of the United States, suggest the type(s) of tectonic processes associated with this area.

Think Critically

5. **Evaluate** the following statement: Volcanoes are only found along coastlines.
6. **Decide** whether a flood basalt is or is not a volcano.

MATH in Earth Science

7. If the Pacific Plate has moved 500 km in the last 4.7 million years, calculate its average velocity in centimeters per year. Refer to the *Skillbuilder Handbook* for more information.

Earth Science Online Self-Check Quiz glencoe.com

Section 1 • Volcanoes 507

Source: Section 18.1, p. 507

At the end of each chapter you will find a Study Guide. The chapter's vocabulary words as well as key concepts are listed here. Use this guide for review and to check your comprehension.

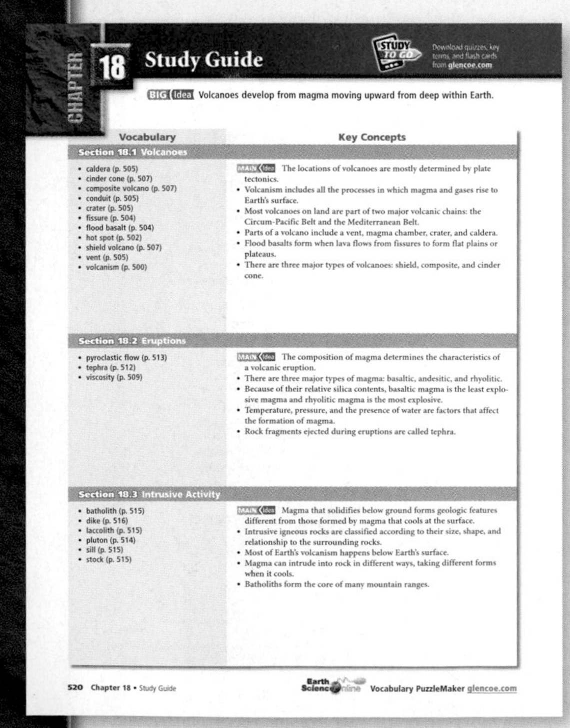

CHAPTER 18 Study Guide

STUDY TO GO Download quizzes, key terms, and flash cards from glencoe.com

BIG Idea Volcanoes develop from magma moving upward from deep within Earth.

Vocabulary	Key Concepts
Section 18.1 Volcanoes	
• caldera (p. 505) • cinder cone (p. 507) • composite volcano (p. 507) • conduit (p. 505) • crater (p. 505) • fissure (p. 504) • flood basalt (p. 504) • hot spot (p. 502) • shield volcano (p. 507) • vent (p. 505) • volcanism (p. 500)	MAIN Idea The locations of volcanoes are mostly determined by plate tectonics. • Volcanism includes all the processes in which magma and gases rise to Earth's surface. • Most volcanoes on land are part of two major volcanic chains: the Circum-Pacific Belt and the Mediterranean Belt. • Parts of a volcano include a vent, magma chamber, crater, and caldera. • Flood basalts form when lava flows from fissures to form flat plains or plateaus. • There are three major types of volcanoes: shield, composite, and cinder cone.
Section 18.2 Eruptions	
• pyroclastic flow (p. 513) • tephra (p. 512) • viscosity (p. 509)	MAIN Idea The composition of magma determines the characteristics of a volcanic eruption. • There are three major types of magma: basaltic, andesitic, and rhyolitic. • Because of their relative silica contents, basaltic magma is the least explosive magma and rhyolitic magma is the most explosive. • Temperature, pressure, and the presence of water are factors that affect the formation of magma. • Rock fragments ejected during eruptions are called tephra.
Section 18.3 Intrusive Activity	
• batholith (p. 515) • dike (p. 516) • laccolith (p. 515) • pluton (p. 514) • sill (p. 515) • stock (p. 515)	MAIN Idea Magma that solidifies below ground forms geologic features different from those formed by magma that cools at the surface. • Intrusive igneous rocks are classified according to their size, shape, and relationship to the surrounding rocks. • Most of Earth's volcanism happens below Earth's surface. • Magma can intrude into rock in different ways, taking different forms when it cools. • Batholiths form the core of many mountain ranges.

520 Chapter 18 • Study Guide

Earth Science Online Vocabulary PuzzleMaker glencoe.com

Source: Chapter 18, p. 520

OTHER WAYS TO REVIEW

- State the BIG Idea.
- Relate the MAIN Idea to the BIG Idea.
- Use your own words to explain what you read.
- Apply this new information in other school subjects or at home.
- Identify sources you could use to find out more information about the topic.

Scavenger Hunt

Earth Science: Geology, the Environment, and the Universe is full of important information and useful resources. Use the activity below to familiarize yourself with the tools and information in this book.

As you complete this scavenger hunt, either alone or with your teacher or family, you will learn quickly how this book is organized and how to get the most out of your reading and study time.

1. How many units are in this book? How many chapters?
2. On what page does the glossary begin? What glossary is online?
3. In what two areas can you find a listing of laboratory safety symbols?
4. Suppose you want to find a list of all the MiniLabs, Data Analysis Labs, and GeoLabs. Where in the front do you look?
5. How can you quickly find the pages that have information about hurricanes?
6. What is the name of the table that summarizes the key concepts of a chapter?
7. In what special feature can you find information on unit conversions? What are the page numbers?
8. On what page can you find the BIG Ideas for Unit 1? On what page can you find the MAIN Ideas for Chapter 2?
9. What feature at the start of each unit provides insight into Earth scientists in action?
10. Name four activities that are found at Earth Science Online.
11. What study tool shown at the beginning of a chapter can you make from notebook paper?
12. Where do you go to view Concepts In Motion?
13. Earth Science and Society and Earth Science and Technology are two types of end-of-chapter features. What are the other two types?

UNIT 1

Earth Science

Chapter 1
The Nature of Science
BIG Idea Earth scientists use specific methods to investigate Earth and beyond.

Chapter 2
Mapping Our World
BIG Idea Earth scientists use mapping technologies to investigate and describe the world.

CAREERS IN EARTH SCIENCE

Speleologist This **speleologist**, a scientist who studies caves, descends into a 200-m-deep sinkhole. Speleologists use scientific methods to make maps, collect samples, and make observations of incredible landforms resulting from geologic processes.

WRITING in Earth Science
Visit glencoe.com to learn more about speleologists. What would it be like to explore an undiscovered cave? Write a journal entry about leading a team of speleologists on such an adventure.

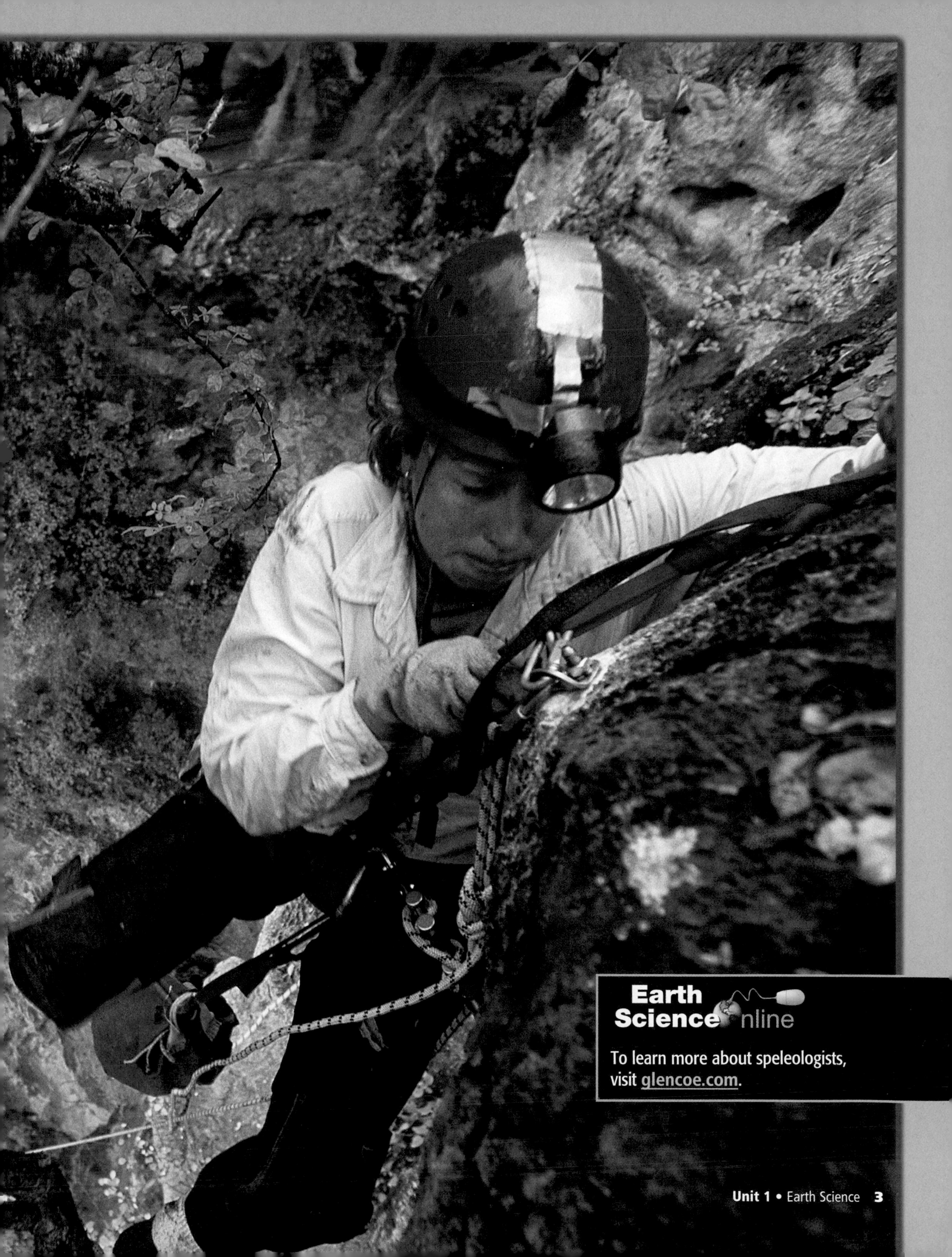
Earth Science Online
To learn more about speleologists, visit glencoe.com.

The Nature of Science

BIG Idea **Earth scientists use specific methods to investigate Earth and beyond.**

1.1 Earth Science
MAIN Idea Earth science encompasses five areas of study: astronomy, meteorology, geology, oceanography, and environmental science.

1.2 Methods of Scientists
MAIN Idea Scientists use scientific methods to structure their experiments and investigations.

1.3 Communication in Science
MAIN Idea Precise communication is crucial for scientists to share their results effectively with each other and with society.

GeoFacts

- The temperature of Earth's core is thought to be as high as 7227°C.
- It is about 6378 km to the center of Earth.
- Seventy percent of Earth's freshwater is contained in glaciers.

Start-Up Activities

LAUNCH Lab

Why is precise communication important?

Have you ever explained something to someone only later to find out that what you thought was a clear explanation was confusing, misleading, or even incorrect? Precise communication is an important skill.

Procedure

1. Read and complete the lab safety form.
2. Obtain an **object** from your teacher. Do not show it to your partner.
3. Write one sentence that accurately describes the object in detail without identifying or naming the object.
4. Give your partner the description and allow him or her a few minutes to identify your object.
5. Now use your partner's description to identify his or her object.

Analysis

1. **Identify** Were you and your partner able to identify each others' objects? Why or why not?
2. **Error Analysis** Work together to rewrite each description in your science journals to make them as accurate as possible.
3. **Compare** Trade the new descriptions with another pair of students. Did this pair of students have an easier time determining the objects than you and your partner did? Why or why not?

Earth's Systems Make this Foldable to compare Earth's four main systems.

STEP 1 Fold a sheet of paper in half lengthwise.

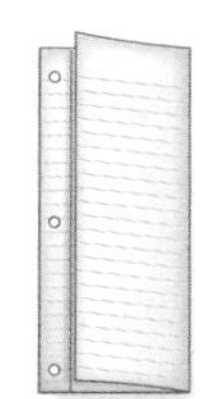

STEP 2 Fold the sheet into fourths (fold in half and half again).

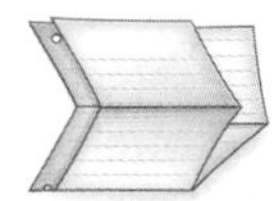

STEP 3 Unfold and cut the top flap along the fold lines to make four tabs. Label the tabs *Geosphere, Hydrosphere, Atmosphere,* and *Biosphere.*

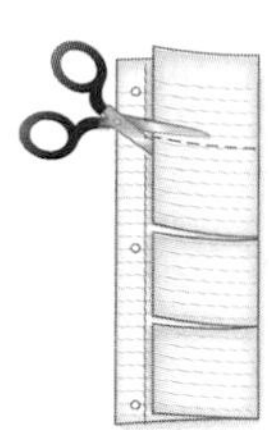

FOLDABLES Use this Foldable with Section 1.1. As you read this section, summarize Earth's systems and how they interact.

Visit glencoe.com to

- study entire chapters online;
- explore Concepts In Motion animations:
  - Interactive Time Lines
  - Interactive Figures
  - Interactive Tables
- access Web Links for more information, projects, and activities;
- review content with the Interactive Tutor and take Self-Check Quizzes.

Section 1.1

Earth Science

Objectives

- **Compare** the areas of study within Earth science.
- **Identify** Earth's systems.
- **Explain** the relationships among Earth's systems.
- **Explain** why technology is important.

Review Vocabulary

technology: the application of knowledge gained from scientific research to solve society's needs and problems

New Vocabulary

astronomy
meteorology
geology
oceanography
environmental science
geosphere
atmosphere
hydrosphere
biosphere

MAIN Idea Earth science encompasses five areas of study: astronomy, meteorology, geology, oceanography, and environmental science.

Real-World Reading Link From the maps you use when traveling, to the weather report you use when deciding whether or not to carry an umbrella, Earth science is part of your everyday life.

The Scope of Earth Science

The scope of Earth science is vast. This broad field can be broken into five major areas of specialization: astronomy, meteorology, geology, oceanography, and environmental science.

Astronomy The study of objects beyond Earth's atmosphere is called **astronomy.** Prior to the invention of sophisticated instruments, such as the telescope shown in **Figure 1.1,** many astronomers merely described the locations of objects in space in relation to each other. Today, Earth scientists study the universe and everything in it, including galaxies, stars, planets, and other bodies they have identified.

Meteorology The study of the forces and processes that cause the atmosphere to change and produce weather is **meteorology.** Meteorologists also try to forecast the weather and learn how changes in weather over time might affect Earth's climate.

Figure 1.1 The Keck I and Keck II telescopes are part of the Mauna Kea Observatories in Hawaii. One of the Keck telescopes is visible here in its protective dome.

Geology The study of the materials that make up Earth, the processes that form and change these materials, and the history of the planet and its life-forms since its origin is the branch of Earth science known as **geology.** Geologists identify rocks, study glacial movements, interpret clues to Earth's 4.6-billion-year history, and determine how forces change our planet.

Oceanography The study of Earth's oceans, which cover nearly three-fourths of the planet, is called **oceanography.** Oceanographers study the creatures that inhabit salt water, measure different physical and chemical properties of the oceans, and observe various processes in these bodies of water. When oceanographers are conducting field research, they often have to dive into the ocean to gather data, as shown in **Figure 1.2.**

Figure 1.2 Oceanographers study the life and properties of the ocean.
Investigate *What kind of training would this Earth scientist need?*

Environmental science The study of the interactions of organisms and their surroundings is called **environmental science.** Environmental scientists study how organisms impact the environment both positively and negatively. The topics an environmental scientist might study include natural resources, pollution, alternative energy sources, and the impact of humans on the atmosphere.

Subspecialties The study of our planet is a broad endeavor, and as such, each of the five major areas of Earth science consists of a variety of subspecialties, some of which are listed in **Table 1.1.**

Concepts In Motion
Interactive Table To explore more about the scope of Earth science, visit glencoe.com.

Table 1.1 Subspecialties of Earth Science

Major Area of Study	Subspecialty	Subjects Studied
Astronomy	astrophysics	physics of the universe, including the physical properties of objects found in space
	planetary science	planets of the solar system and the processes that form them
Meteorology	climatology	patterns of weather over a long period of time
	atmospheric chemistry	chemistry of Earth's atmosphere, and the atmospheres of other planets
Geology	paleontology	remains of organisms that once lived on Earth; ancient environments
	geochemistry	Earth's composition and the processes that change it
Oceanography	physical oceanography	physical characteristics of oceans, such as salinity, waves, and currents
	marine geology	geologic features of the ocean floor, including plate tectonics of the ocean
Environmental science	environmental soil science	interactions between humans and the soil, such as the impact of farming practices; effects of pollution on soil, plants, and groundwater
	environmental chemistry	chemical alterations to the environment through pollution and natural means

FOLDABLES
Incorporate information from this section into your Foldable.

Earth's Systems

Scientists who study Earth have identified four main Earth systems: the geosphere, atmosphere, hydrosphere, and biosphere. Each system is unique, yet each interacts with the others.

Geosphere The area from the surface of Earth down to its center is called the **geosphere.** The geosphere is divided into three main parts: the crust, mantle, and core. These three parts are illustrated in **Figure 1.3.**

The rigid outer shell of Earth is called the crust. There are two kinds of crust—continental crust and oceanic crust. Just below the crust is Earth's mantle. The mantle differs from the crust both in composition and behavior. The mantle ranges in temperature from 100°C to 4000°C—much warmer than the temperatures found in Earth's crust. Below the mantle is Earth's core. You will learn more about the crust, mantle, and core in Unit 5.

VOCABULARY
SCIENCE USAGE V. COMMON USAGE
Crust
Science usage: the thin, rocky, outer layer of Earth

Common usage: the hardened exterior or surface part of bread

Atmosphere The blanket of gases that surrounds our planet is called the **atmosphere.** Earth's atmosphere contains about 78 percent nitrogen and 21 percent oxygen. The remaining 1 percent of gases in the atmosphere include water vapor, argon, carbon dioxide, and other trace gases. Earth's atmosphere provides oxygen for living things, protects Earth's inhabitants from harmful radiation from the Sun, and helps to keep the planet at a temperature suitable for life. You will learn more about Earth's atmosphere and how parts of this system interact to produce weather in Unit 4.

Hydrosphere All the water on Earth, including the water in the atmosphere, makes up the **hydrosphere.** About 97 percent of Earth's water exists as salt water, while the remaining 3 percent is freshwater contained in glaciers, lakes and rivers, and beneath Earth's surface as groundwater. Only a fraction of Earth's total amount of freshwater is in lakes and rivers. You will find out more about Earth's hydrosphere in Units 3, 4, and 7.

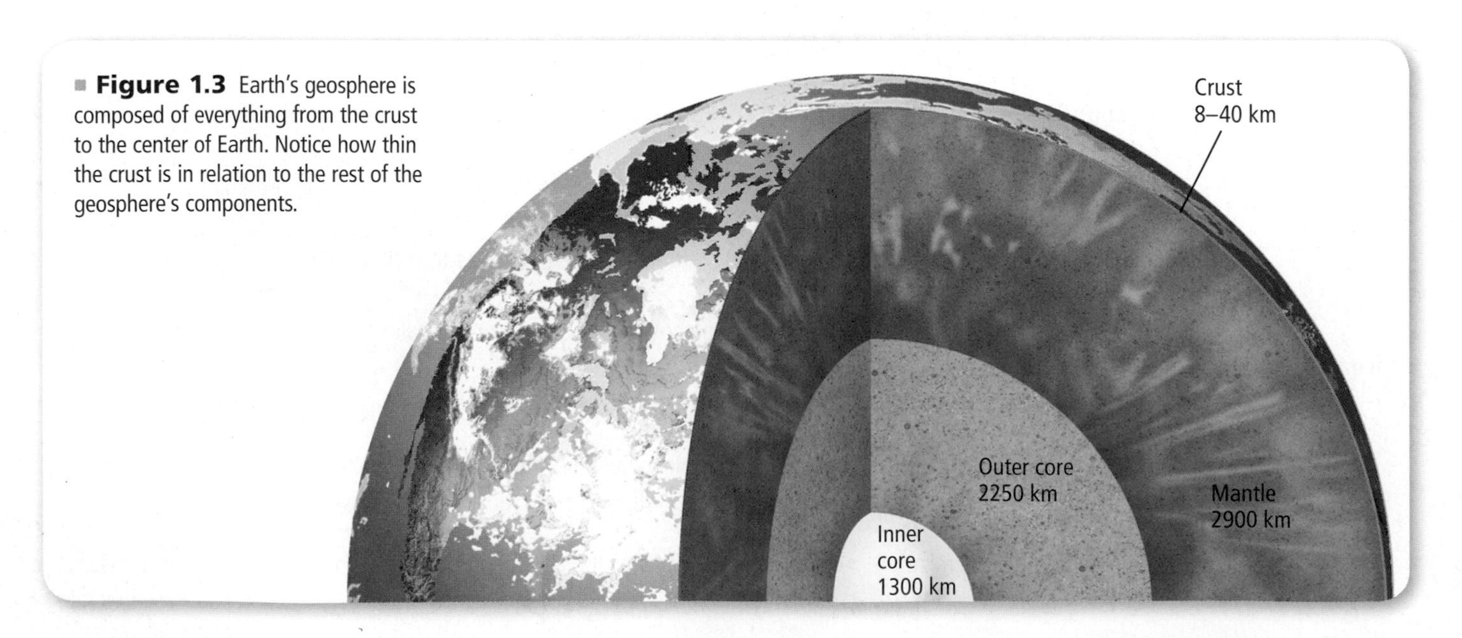

Figure 1.3 Earth's geosphere is composed of everything from the crust to the center of Earth. Notice how thin the crust is in relation to the rest of the geosphere's components.

Biosphere The **biosphere** includes all organisms on Earth as well as the environments in which they live. Most organisms live within a few meters of Earth's surface, but some exist deep beneath the ocean's surface, and others live high atop Earth's mountains. All of Earth's life-forms require interaction with at least one of the other systems for their survival.

As illustrated in **Figure 1.4,** Earth's biosphere, geosphere, hydrosphere, and atmosphere are interdependent systems. For example, Earth's present atmosphere formed millions of years ago through interactions with the geosphere, hydrosphere, and biosphere. Organisms in the biosphere, including humans, continue to change the atmosphere through their activities and natural processes. You will explore interactions among Earth's biosphere and other systems in Units 3, 4, 6, and 7.

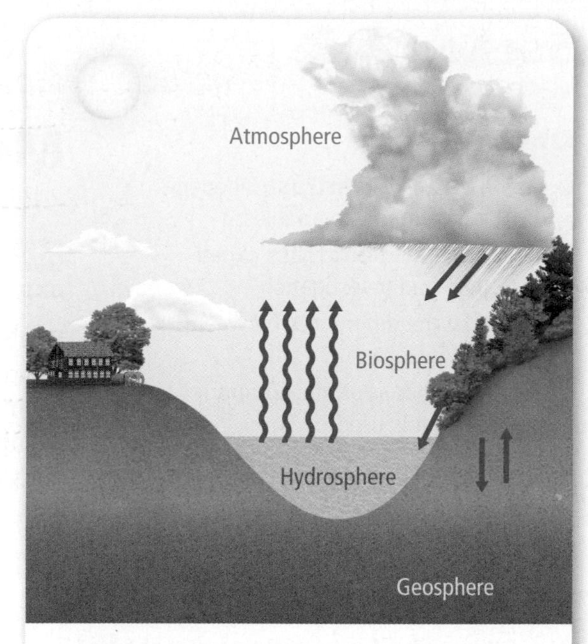

Figure 1.4 All of Earth's systems are interdependent. Notice how water from the hydrosphere enters the atmosphere, falls on the biosphere, and soaks into the geosphere.

Technology

The study of science, including Earth science, has led to many discoveries that have been applied to solve society's needs and problems. The application of scientific discoveries is called technology. Technology is transferable, which means that it can be applied to new situations. Freeze-dried foods, ski goggles, and the ultralight materials used to make many pieces of sports equipment were created from technologies used in our space program. Technology is not used only to make life easier. It can also make life safer. Most people have smoke detectors in their houses to help warn them if there is a fire. Smoke detectors were also invented as part of the space program and were adapted for use in everyday life.

Section 1.1 Assessment

Section Summary

- Earth is divided into four systems: the geosphere, hydrosphere, atmosphere, and biosphere.
- Earth systems are all interdependent.
- Identifying the interrelationships between Earth systems leads to specialties and subspecialties.
- Technology is important, not only in science, but in everyday life.
- Earth science has contributed to the development of many items used in everyday life.

Understand Main Ideas

1. MAIN Idea **Explain** why it is helpful to identify specialties and subspecialties of Earth science.
2. **Apply** What are three items you use on a daily basis that have come from research in Earth science?
3. **Compare and contrast** Earth's geology and geosphere.
4. **Hypothesize** about human impact on each of Earth's systems.
5. **Compare and contrast** the hydrosphere and biosphere.

Think Critically

6. **Predict** what would happen if the makeup of the hydrosphere chan[illegible] would happen if the atmosphere changed?

WRITING in Earth Science

7. Research a subspecialty of Earth science. Make a [illegible] this field.

Earth Science Online Self-Check Quiz glencoe.com

Section 1.2

Objectives

- **Compare and contrast** independent and dependent variables.
- **Compare and contrast** experimentation and investigation.
- **Identify** the differences between mass and weight.
- **Explain** what scientific notation is and how it is used.

Review Vocabulary

experiment: procedure performed in a controlled setting to test a hypothesis and collect precise data

New Vocabulary

scientific methods
hypothesis
independent variable
dependent variable
control
Le Système International d'Unités (SI)
scientific notation

Methods of Scientists

MAIN Idea Scientists use scientific methods to structure their experiments and investigations.

Real-World Reading Link Have you ever seen a distinct rock formation and wondered how it formed? Have you ever wondered why the soil near your home might be different from the soil in your schoolyard? If so, you have already begun to think like a scientist. Scientists often ask questions and make observations to begin their investigations.

The Nature of Scientific Investigations

Scientists work in many different places to gather data. Some work in the field, and some work in a lab, as shown in **Figure 1.5.** No matter where they work, they all use similar methods to gather data and communicate information. These methods are referred to as scientific methods. As illustrated in **Figure 1.6, scientific methods** are a series of problem-solving procedures that help scientists conduct experiments.

Whatever problem a scientist chooses to pursue, he or she must gather background information on the topic. Once the problem is defined and the background research is complete, a hypothesis is made. A **hypothesis** is a testable explanation of a situation that can be supported or disproved by careful procedures.

It is important to note that scientific methods are not rigid, step-by-step outlines to solve problems. Scientists can take many different approaches to performing a scientific investigation. In many scientific investigations, for example, scientists form a new hypothesis after observing unexpected results. A researcher might modify a procedure, or change the control mechanism. And a natural phenomenon might change the direction of the investigation.

Figure 1.5 Whether a meteorologist gathers storm data in the field or an environmental scientist analyzes microbial growth in a lab, scientific methods provide an approach to problem-solving and investigation.

Visualizing Scientific Methods

Figure 1.6 Scientific methods are used by scientists to help organize and plan their experiments and investigations. The flow chart below outlines some of the methods commonly used by scientists.

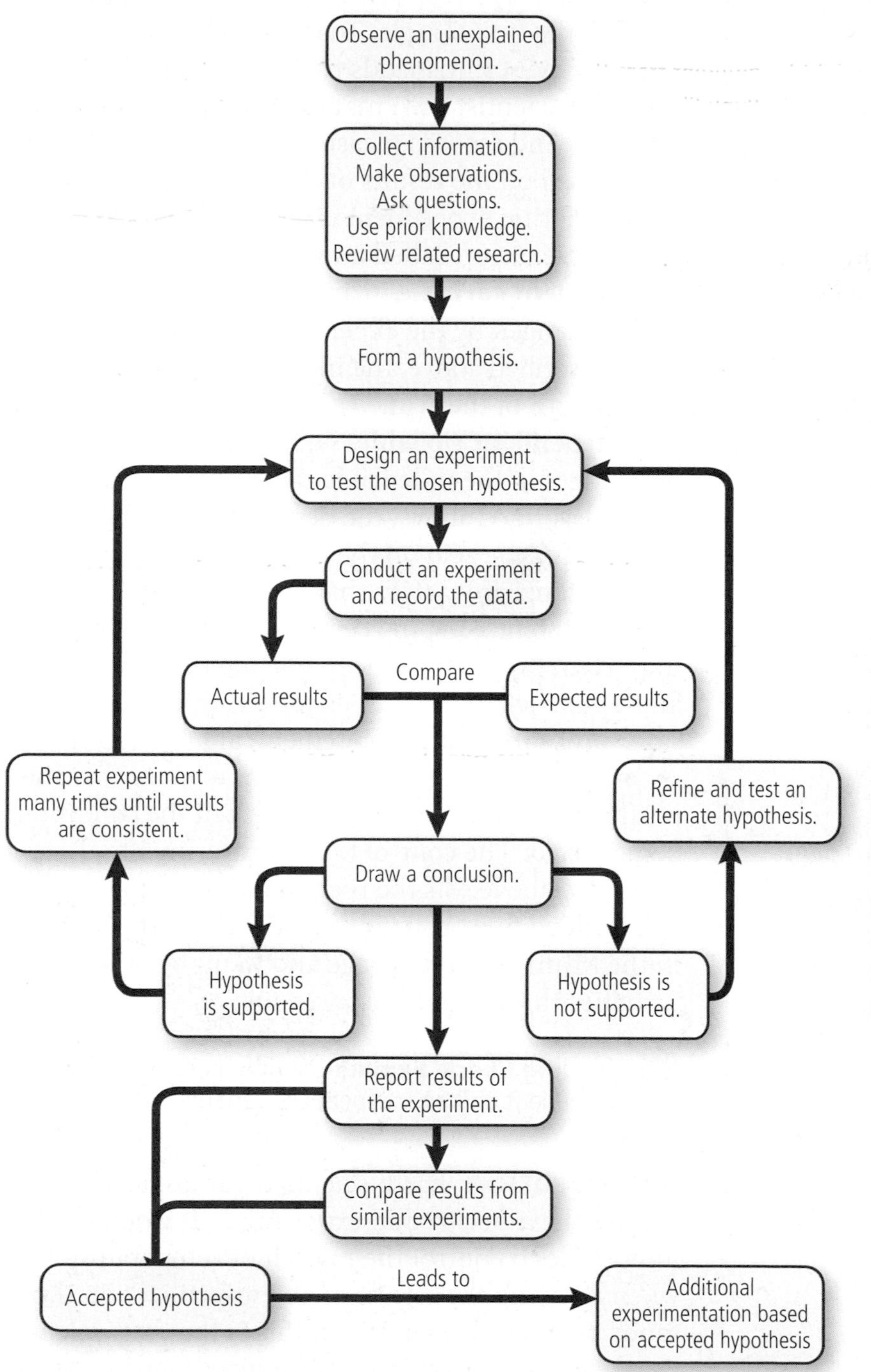

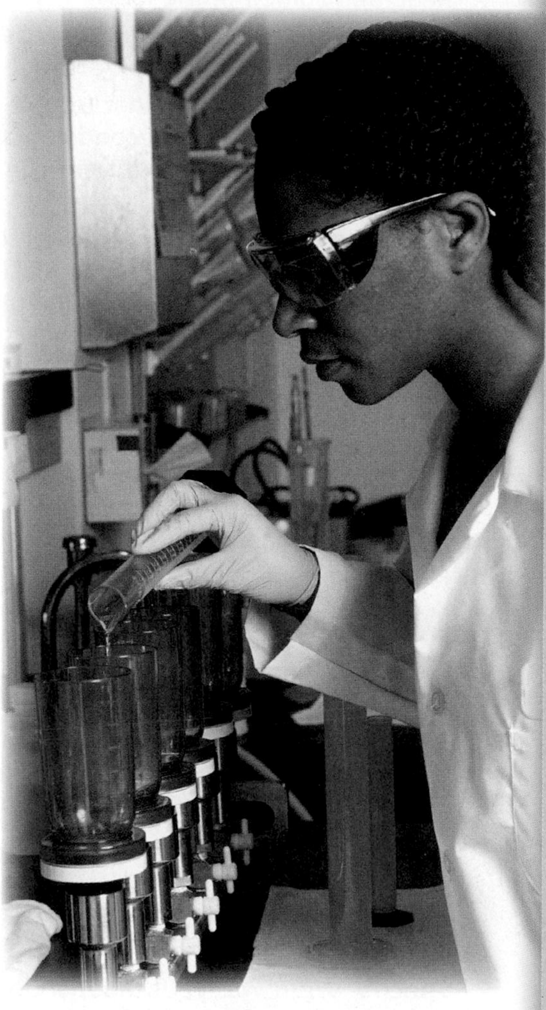

Concepts In Motion To explore more about scientific methods, visit glencoe.com.

Earth Science Online

MiniLab

Determine the Relationship Between Variables

How do the rates of heat absorption and release vary between soil and water? Different substances absorb and release heat at different rates.

Procedure

1. Read and complete the lab safety form.
2. Read the procedure and create a data table to record your temperature results.
3. Pour **soil** into **one container** until it is half full. Pour **water** into a **second container** until it is half full. Leave a **third container** empty.
4. Place **one thermometer** in the soil so that the bulb is barely covered. Use **masking tape** to secure **another thermometer** about 1 cm above the top of the soil.
5. Repeat Step 4 for the container with water.
6. In the empty container, place the bulb of one thermometer halfway into the cup and secure it with masking tape. Use the tape to secure another thermometer bulb about 2 cm higher than the first thermometer bulb.
7. Put the containers on a **sunny windowsill.** Record the temperature shown on each thermometer. Write these values in a table. Record temperature readings every 5 min for 30 min.
8. Remove the containers from the windowsill and continue to record the temperature on each thermometer every 5 min for 30 min.

Analysis

1. **Determine** Which substance absorbed heat more quickly? Which substance lost heat more quickly?
2. **Specify** What was your independent variable? What was your dependent variable?
3. **Identify** your control.

Experimentation An experiment is classified as an organized procedure that involves making observations and measurements to test a hypothesis. Collecting good qualitative and quantitative data is vital to the success of an experiment.

Imagine a scientist is conducting an experiment on the effects of acid on the weathering of rocks. In this experiment, there are three different samples of identical rock pieces. The scientist does not add anything to the first sample. To the second and third samples, the scientist adds two different strengths of acid. The scientist then makes observations (qualitative data) and records measurements (quantitative data) based on the results of the experiment.

A scientific experiment usually tests only one changeable factor, called a variable, at a time. The **independent variable** in an experiment is the factor that is changed by the experimenter. In the experiment described above, the independent variable was the strength of the acid.

A **dependent variable** is a factor that is affected by changes in the independent variable. In the experiment described above, the dependent variable was the effect of the acid on the rock samples.

Constants are factors that do not change during an experiment. Keeping certain variables constant is important to an experiment. Placing the same amount of acid on each rock tested, or using the same procedure for measurement, are two examples. A **control** is used in an experiment to show that the results of an experiment are a result of the condition being tested. The control for the experiment described above was the rock that did not have anything added to it. You will experiment with variables in the MiniLab on this page and in many other activities throughout this textbook.

Reading Check **Explain** the difference between a dependent and an independent variable.

Investigation Earth scientists cannot always control the aspects of an experiment. It would be impossible to control the rainfall or temperature when studying the effects of a new fertilizer on thousands of acres of corn. When this is the case, scientists refer to their research as an investigation. An investigation involves observation and collecting data but does not include a control. Investigations can often lead scientists to design future experiments based on the observations they have made.

Safety Many of the experiments and investigations in this book will require that you handle various materials and equipment. When conducting any scientific investigation, it is important to use all materials and equipment only as instructed. Refer to the *Reference Handbook* for additional safety information and a table of safety symbols.

Analysis and conclusions New ideas in science are carefully examined by the scientist who made the initial discovery and by other scientists in the same field. Processes, data, and conclusions must be examined to eliminate influence by expectations or beliefs, which is called bias. During a scientific experiment, all data are carefully recorded. Once an experiment is complete, graphs, tables, and charts are commonly used to display data. These data are then analyzed so that a conclusion can be drawn. Many times, a conclusion does not support the original hypothesis. In such a case, the hypothesis must be reevaluated and further research must be conducted.

VOCABULARY

ACADEMIC VOCABULARY

Bias

to influence in a particular, typically unfair, direction; prejudice

Their choice of teammates showed a bias toward their friends.

Measurement

Scientific investigations often involve making measurements. A measurement includes both a number and a unit of measure. Scientific investigations use a standard system of units called **Le Système International d'Unités** (SI), which is a modern version of the metric system. SI is based on a decimal system that uses the number 10 as the base unit. See **Table 1.2** for information on SI and metric units of measure commonly used in science.

Length The standard SI unit to measure length is the meter (m). The distance from a doorknob to the floor is about 1 m. The meter is divided into 100 equal parts called centimeters (cm). Thus, 1 cm is 1/100 of 1 m. One millimeter (mm) is smaller than 1 cm. There are 10 mm in 1 cm. Longer distances are measured in kilometers (km). There are 1000 m in 1 km.

Table 1.2 **Measurement and Units**

Measurement	SI and Metric Units Commonly Used in Science
Length	millimeter (mm), centimeter (cm), meter (m), kilometer (km)
Mass	gram (g), kilogram (kg), metric ton
Area	square meter (m^2), square centimeter (cm^2)*
Volume	cubic meter (m^3)*, milliliter (mL), liter (L) #
Density	grams per cubic centimeter (g/cm^3)*, grams per milliliter (g/mL)#, kilograms per cubic meter (kg/m^3)*
Time	second (s), hour (h)#
Temperature	kelvin (K)

* units derived from SI units   # commonly used metric units

Mass The amount of matter in an object is called mass. Mass depends on the number and types of atoms that make up the object. The mass of an object is the same no matter where the object is located in the universe. The SI unit of mass is the kilogram (kg).

Weight Weight is a measure of the gravitational force on an object. Weight is typically measured with some type of scale. Unlike mass, weight varies with location. For example, the weight of an astronaut while on the Moon is about one-sixth the astronaut's weight on Earth. This is because the gravitational force exerted by the Moon on the astronaut is one-sixth the force exerted by Earth on the astronaut. Weight is a force, and the SI unit for force is the newton (N). A 2-L bottle of soft drink with a mass of 2 kg weighs about 20 N on Earth.

 **Reading Check** **Compare** mass and weight.

Area and volume Some measurements, such as area, require a combination of SI units. Area is the amount of surface included within a set of boundaries and is expressed in square units of length, such as square meters (m^2).

The amount of space occupied by an object is the object's volume. The SI units for volume, like those for area, are derived from the SI units used to measure length. The basic SI unit of volume for a solid object is the cubic meter (m^3). Measurements for fluid volumes are usually made in milliliters (mL) or liters (L). Liters and milliliters are metric units that are commonly used to measure liquid volumes. Volume can also be expressed in cubic centimeters (cm^3)—1 cm^3 equals 1 mL.

■ **Figure 1.7**

Major Events in Earth Science

Many discoveries during the twentieth and early twenty-first centuries revolutionized our understanding of Earth and its systems.

1900 · 1920 · 1940 · 1960

1907 Scientists begin using radioactive decay to determine that Earth is billions of years old. This method will be used to develop the first accurate geological time scale.

1913 French physicists discover the ozone layer in Earth's upper atmosphere and propose that it protects Earth from the Sun's ultraviolet radiation.

1925 Cecilia Payne's analysis of the spectra of stars reveals that hydrogen and helium are the most abundant elements in the universe.

1936 Inge Lehmann proposes that Earth's center consists of a solid inner core and a liquid outer core based on her studies of seismic waves.

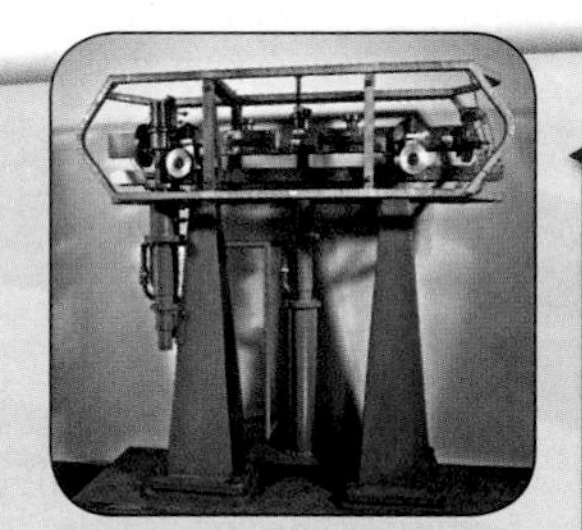

1955 Louis Essen invents a highly accurate atomic clock that tracks radiation emitted and absorbed by cesium atoms.

Density The measure of the amount of matter that occupies a given space is density. Density is calculated by dividing the mass of the matter by its volume. Density is often expressed in grams per cubic centimeter (g/cm^3), grams per milliliter (g/mL), or kilograms per cubic meter (kg/m^3).

Time The interval between two events is time. The SI unit of time is the second. In the activities in this book, you will generally measure time in seconds or minutes. Time is usually measured with a watch or clock. The atomic clock provides the most precise measure of time currently known. Known as UTC, Coordinated Universal Time is based on the atomic clock element cesium-133 and is adapted to the astronomical demarcation of day and night. See **Figure 1.7** for more information on the invention of the atomic clock and other advances in Earth science.

Temperature A measure of the average kinetic energy of the particles that make up a material is called temperature. A mass made up of particles that vibrate quickly generally has a higher temperature than a mass whose particles vibrate more slowly. Temperature is measured in degrees with a thermometer. Scientists often measure temperature using the Celsius (°C) scale. On the Celsius scale, a comfortable room temperature is about 21°C, and the normal temperature of the human body is about 37°C.

The SI unit for temperature is the kelvin (K). The coldest possible temperature, absolute zero, was established as 0 K or −273 °C. Since both temperature units are the same size, the difference between the two scales (273) is used to convert from one scale to another. For example, the temperature of the human body is 37°C, to which you would add 273 to get 310 K.

VOCABULARY

ACADEMIC VOCABULARY

Interval
space of time between two events or states
The interval for pendulum swings was three seconds.

1962 Harry Hess's seafloor spreading hypothesis, along with the discoveries made about the ocean floor, lays the foundation for plate tectonic theory.

1979–1980 *Magsat,* a NASA satellite, takes the first global measurement of Earth's magnetic field.

2004 A sediment core retrieved from the ocean floor discloses 55 million years of Earth's atmospheric and climatic history. The sample reveals that the north pole once had a warm climate.

1970 1980 1990 2000 2010

1970 George Carruthers' ultraviolet camera and spectrograph, placed on the Moon's surface, analyzes pollutants in Earth's atmosphere and detects interstellar hydrogen.

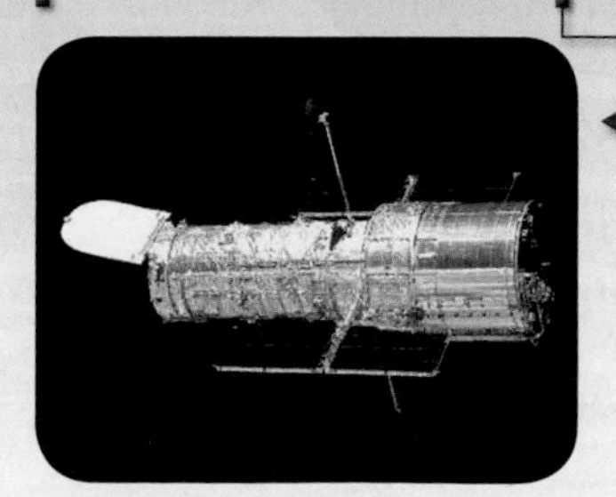

1990 The *Hubble Space Telescope* goes into orbit, exploring Earth's solar system, measuring the expansion of the universe, and providing evidence of black holes.

Concepts In Motion

Interactive Time Line To learn more about these discoveries and others, visit glencoe.com. Earth Science Online

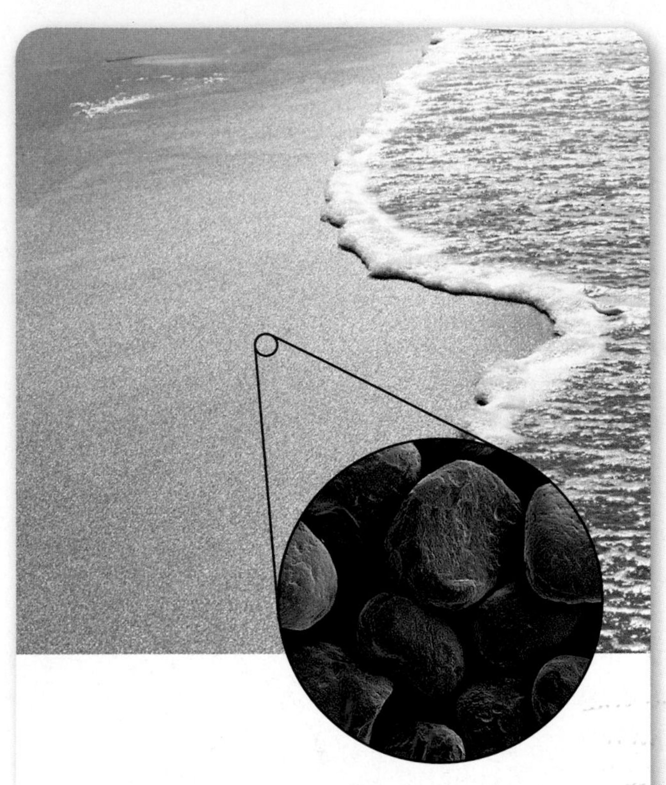

■ **Figure 1.8** On a 5-km-long beach, such as the one shown above, there might be 8×10^{15} grains of sand. The average size of a grain of sand is 0.5 mm.

Scientific Notation

In many branches of science, some numbers are very small, while others are very large. To express these numbers conveniently, scientists use a type of shorthand called **scientific notation,** in which a number is expressed as a value between 1 and 10 multiplied by a power of 10. The power of 10 is the number of places the decimal point must be shifted so that only a single digit remains to the left of the decimal point.

If the decimal point must be shifted to the left, the exponent of 10 is positive. **Figure 1.8** shows a beach covered in sand. The number of grains of sand on Earth has been estimated to be approximately 4,000,000,000,000,000,000,000. In scientific notation, this number is written as 4×10^{21}.

In astronomy, masses and distances are usually so large that writing out the numbers would be cumbersome. For example, the mass of Earth at 5,974,200,000,000,000,000,000,000 kg would be written as 5.9742×10^{24} kg in scientific notation.

If the decimal point in a number must be shifted to the right, the exponent of 10 is negative. The diameter of an atom in meters, for example, which is approximately 0.0000000001 m, is written as 1×10^{-10} m.

Section 1.2 Assessment

Section Summary

- Scientists work in many ways to gather data.
- A good scientific experiment includes an independent variable, dependent variable, and control. An investigation, however, does not include a control.
- Graphs, tables, and charts are three common ways to communicate data from an experiment.
- SI, a modern version of the metric system, is a standard form of measurement that all scientists can use.
- To express very large or very small numbers, scientists use scientific notation.

Understand Main Ideas

1. MAIN Idea **Explain** why scientific methods are important and why there is not one established way to conduct an investigation.
2. **Compare and contrast** the purpose of a control, an independent variable, and a dependent variable in an experiment.
3. **Calculate** Express 0.00049386 in scientific notation.
4. **Calculate** Convert the temperature 49°C to kelvin.
5. **Compare and contrast** volume and density.

Think Critically

6. **Construct** a plan to test the absorption of three different kinds of paper towels, including a control, dependent variable, and independent variable.
7. **Explain** which is more useful when comparing mass and weight on different planets.

MATH in Earth Science

8. If you have 20 mL of water, how many cubic centimeters of water do you have?

Self-Check Quiz glencoe.com

Section 1.3

Objectives

- **Explain** why precise communication is crucial in science.
- **Compare and contrast** scientific theories and scientific laws.
- **Identify** when it is appropriate to use a graph or a model.

Review Vocabulary

hypothesis: testable explanation of a situation

New Vocabulary

scientific model
scientific theory
scientific law

Communication in Science

MAIN Idea Precise communication is crucial for scientists to share their results effectively with each other and with society.

Real-World Reading Link If you read an advertisement for a product called "Glag" without any description, would you know whether to eat it or wear it? When a scientist does an investigation, he or she has to describe every part of it precisely so that everyone can understand his or her conclusions.

Communicating Results

There are many ways to communicate information, such as newspapers, magazines, TV, the Internet, and scientific journals. Think back to the Launch Lab from the beginning of the chapter. Although you and your lab partner both used the same form of communication, were your descriptions identical? Scientists have the responsibility to truthfully and accurately report their methods and results. To keep them ethical, a system of peer review is used in which scientists in the same field verify each other's results and examine procedures and conclusions for bias. Communicating scientific data and results, as the scientists are shown doing in **Figure 1.9,** also allows others to learn of new discoveries and conduct new investigations that build on previous investigations.

Lab reports Throughout this book, you will conduct many Earth science investigations and experiments. During and after each, you will be asked to record and analyze the information that you collected and to draw conclusions based on your data. Your written account of each lab is your lab report. This will be used by your teacher to assess your understanding. You might also be asked to compare your results with those of other students to help you find both similarities and differences among the results.

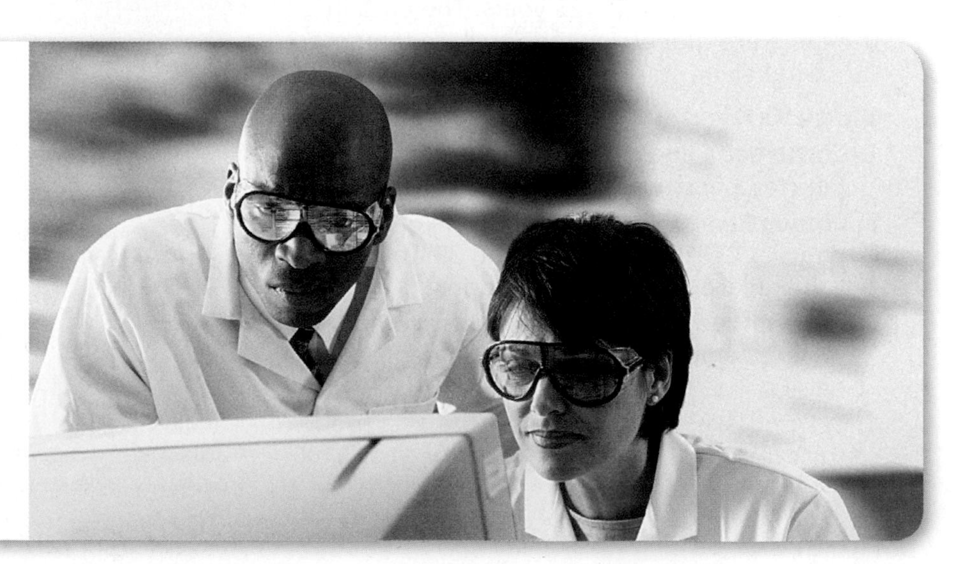

■ **Figure 1.9** Scientists, like those shown in the photo, communicate data and discoveries with each other to maintain accuracy in methods and reporting.
Infer ***what could happen if scientists did not compare results.***

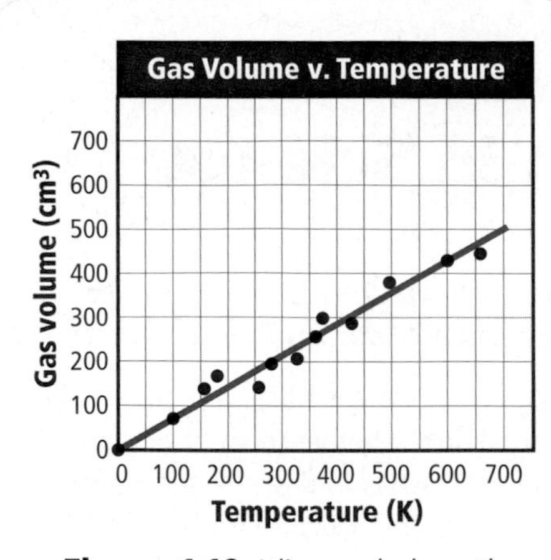

■ **Figure 1.10** A line graph shows the relationship between two variables.
Determine *Based on this graph, what is the relationship between gas volume and temperature?*

Graphs By graphing data in a variety of ways, scientists can more easily show the relationships among data sets. Graphs also allow scientists to represent trends in their data. You will be asked to graph the results of many experiments and activities in this book. There are three types of graphs you will use in this book.

Line graphs A visual display that shows how two variables are related is called a line graph. As shown in **Figure 1.10,** on a line graph, the independent variable is plotted on the horizontal (x) axis, and the dependent variable is plotted on the vertical (y) axis.

Circle graphs To show a fixed quantity, scientists often use a circle graph, also called a pie graph. The circle represents the total and the slices represent the different parts of the whole. The slices are usually presented as percentages.

Bar graphs To represent quantitative data, bar graphs use rectangular blocks called bars. The length of the bar is determined by the amount of the variable you are measuring as well as the scale of the bar graph. See the *Skillbuilder Handbook,* page 951, for examples of all the types of graphs described above.

Models In some of the investigations, you will be making and using models. A **scientific model** is an idea picture, a system, or a mathematical expression that represents the concept being explained. While a model might not have all of the components of a given idea, it should be a fairly accurate representation.

DATA ANALYSIS LAB

Based on Real Data*

Make and Use Graphs

How can graphs help interpret data? The table shows the average surface temperature of Earth over the past 125 years. The data in the table are global, average surface temperatures, in kelvin, starting in the year 1880.

Think Critically

1. **Construct** a line graph from the average surface temperatures in the data table.
2. **Convert** each temperature from kelvin to degrees Celsius by subtracting 273 from each value. Place both on your graph.
3. **Determine** from your graph the average surface temperature for 1988 in degrees Celsius.
4. **Extrapolate,** in Celsius, what the average surface temperature will be in the year 2100 if this trend continues.

Data and Observations

Average Global Surface Temperatures

Years	Average surface temperature (K)
1880–1899	286.76
1900–1919	286.77
1920–1939	286.97
1940–1959	287.02
1960–1979	286.98
1980–1999	287.33
2000–2004	287.59

*Data obtained from Goddard Institute for Space Studies, NASA Goddard Space Flight Center

Models can change when more data are gathered. As shown in **Figure 1.11,** early astronomers thought that Earth was the center of the solar system. This model was changed as the result of observations of the motions of the Sun and the planets in the night sky. The observations showed that the planets in our solar system orbit the Sun.

Theories and Laws

A **scientific theory** is an explanation based on many observations during repeated investigations. A scientific theory is valid only if it is consistent with observations, makes predictions that can be tested, and is the simplest explanation of observations. Like a scientific model, a theory can be changed or modified with the discovery of new data.

A **scientific law** is a principle that describes the behavior of a natural phenomenon. A scientific law can be thought of as a rule of nature, even though the cause of the law might not be known. The events described by a law are observed to be the same every time. An example of a scientific law is Newton's first law of motion, which states that an object at rest or in motion stays at rest or in motion unless it is acted on by an outside force. This law explains why Earth and other planets in our solar system remain in orbit around the Sun. Theories are often used to explain scientific laws.

In this book, you will communicate your observations and draw conclusions based on scientific data. You will also read that many of the models, theories, and laws used by Earth scientists to explain various processes and phenomena grow from the work of other scientists and sometimes develop from unexpected discoveries.

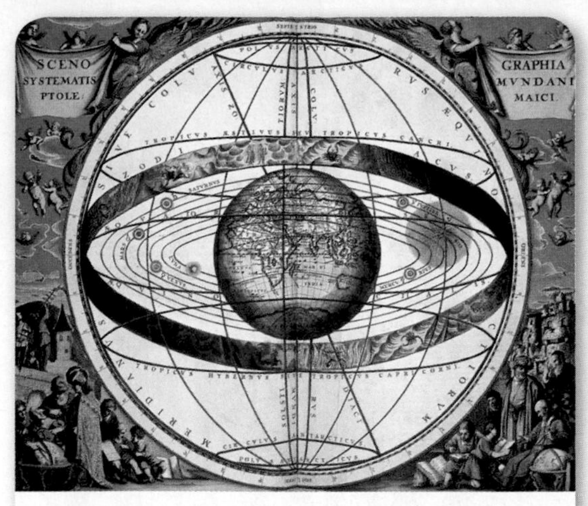

Figure 1.11 Scientific models, like this ancient one of the solar system, are used to represent a larger idea or system. As scientists gather new information, models can change or be revised. **Explain** *what is wrong with this model.*

Section 1.3 Assessment

Section Summary

- Scientists communicate data so others can learn the results, verify the results, examine conclusions for bias, and conduct new experiments.
- There are three main types of graphs scientists use to represent data: line graphs, circle graphs, and bar graphs.
- A scientific model is an accurate representation of an idea or theory.
- Scientific theories and scientific laws are sometimes discovered accidentally.

Understand Main Ideas

1. MAIN Idea **Explain** what might happen if a scientist inaccurately reported data from his or her experiment.
2. **Describe** the difference between scientific theory and scientific law.
3. **Apply** Why is it important to compare your data from a lab with that of your classmates?

Think Critically

4. **Interpret** Why would a model be important when studying the solar system?
5. **Explain** when to use a line graph, a circle graph, and a bar graph.

WRITING in **Earth Science**

6. Research scientific laws and theories, and write a concise example of each.

ON SITE: IN THE FOOTSTEPS OF DISASTER

The tsunami destroyed many homes and buildings, leaving few of the structures standing.

On December 26, 2004, a massive earthquake rattled the seafloor of the Indian Ocean. A tsunami was generated by the earthquake, which devastated the landscape and killed almost 300,000 people in 12 countries. After humanitarian efforts were underway, many Earth scientists mobilized to collect data before the area was changed by cleanup efforts.

Planning the investigation Jose Borrero, an environmental engineer at University of Southern California, wanted to determine the height of the waves associated with the tsunami, how far inland they traveled, the number of waves, and the distance between them. This information would determine where to rebuild towns and assist in the development of a warning system and a hazard plan.

Taking measurements To measure heights of the waves and the following rush of water, Borrero looked for mud or watermarks on the buildings that were left standing. He then placed a 5-m pole next to the watermark to measure the height the water reached. The closer he got to the coast, however, the less he was able to measure accurately. The water had surged up over 5 m deep, so he relied on visual estimates and photos for documentation. With each measurement, he recorded the location on the Global Positioning Satellite system (GPS).

After a six-day study of the devastation, Borrero had more than 150 data points. Upon returning to the United States, scientists used these data to determine that the waves reached 15–30 m high in Banda Aceh, and almost 3.2 km inland.

Using models It is impossible and unethical to simulate natural disasters on an actual scale, so scientists use the data collected from real incidents to create models of those events to learn more about how nature behaves. Using scientific methods and data gathered, scientists are able to provide information for model building or computer simulation. Back at the lab, Borrero applies the data to study other possible tsunami scenarios. He uses data to predict wave height and the area of inundation along the coast, should a tsunami hit the United States.

He hopes that the data collected will enable better detection and prevent widespread devastation from a natural tsunami disaster.

WRITING in Earth Science

Journal Imagine you are a geologist who is accompanying a team of scientists to a natural disaster. Describe the way you will use the scientific method to gather data for your report. Visit glencoe.com to learn more about scientific methods in the field.

GEOLAB

MEASUREMENT AND SI UNITS

Background: **Suppose someone asked you to measure the area of your classroom in square cubits. What would you use? A cubit is an ancient unit of length equal to the distance from the elbow to the tip of the middle finger. Today, SI is used as a standard system of measurement.**

Question: *Why are standard units of measure important?*

Materials

water
large graduated cylinder or beaker
graph paper
balance
pieces of string
spring scale
rock samples
ruler

Safety Precautions

Procedure

1. Read and complete the lab safety form.
2. Obtain a set of rock samples from your teacher.
3. Measure the weight and length of two rock samples using a nonstandard unit of measure. You might use your pinky, a paper clip, or anything you choose.
4. Record your measurements.
5. Working with a partner, explain your units of measure and which samples you measured. Ask your partner to measure the rocks using your units.
6. Record your partner's measurements.
7. Use the information in the *Skillbuilder Handbook* to design a data table in which to record the following measurements for each rock sample: area, volume, mass, weight, and density.
8. Carefully trace the outline of each rock onto a piece of graph paper. Determine the area of each sample and record the values in your data table.
9. Secure each rock with a piece of dry string. Place the string loop over the hook of the spring scale to determine the weight of each rock sample. Record the values in your data table.
10. Pour water into a large graduated cylinder until it is half full. Record this volume in the table. Slowly lower the sample by its string into the cylinder. Record the volume of the water. Subtract the two values to determine the volume of the rock sample.
11. Repeat Steps 9 and 10 for each rock. Make sure the original volume of water for each rock is the same as when you measured your first sample.
12. Follow your teacher's instructions about how to use the balance to determine the mass of each rock. Record the measurements in your table.

Analyze and Conclude

1. **Interpret** How did the results of your initial measurements (Step 4) compare with your lab partner's (Step 6)? If they were different, why were they?
2. **Propose** What does this tell you about the importance of standard units of measure?
3. **Compare** the area of each of your samples with the volumes determined for the same rock. Which method of measurement was more accurate? Explain.
4. **Calculate** the density of each sample using this formula: density = mass/volume. Record these values in your data table.
5. **Explain** Does mass depend on the size or shape of a rock? Explain.
6. **Identify** the variables you used to determine the volume of each sample.
7. **List** the standard units you used in this investigation and explain the standard unit advantages over your measurement units.

INQUIRY EXTENSION

Inquiry How could you find the volume of a rock, such as pumice, that floats in water? Design an investigation to test your prediction.

CHAPTER 1

Study Guide

BIG Idea Earth scientists use specific methods to investigate Earth and beyond.

Vocabulary	Key Concepts
Section 1.1 Earth Science	
• astronomy (p. 6) • atmosphere (p. 8) • biosphere (p. 9) • environmental science (p. 7) • geology (p. 7) • geosphere (p. 8) • hydrosphere (p. 8) • meteorology (p. 6) • oceanography (p. 7)	**MAIN Idea** Earth science encompasses five areas of study: astronomy, meteorology, geology, oceanography, and environmental science. • Earth is divided into four systems: the geosphere, hydrosphere, atmosphere, and biosphere. • Earth systems are all interdependent. • Identifying the interrelationships between Earth systems leads to specialties and subspecialties. • Technology is important, not only in science, but in everyday life. • Earth science has contributed to the development of many items used in everyday life.
Section 1.2 Methods of Scientists	
• control (p. 12) • dependent variable (p. 12) • hypothesis (p. 10) • independent variable (p. 12) • Le Système International d'Unités (SI) (p. 13) • scientific methods (p. 10) • scientific notation (p. 16)	**MAIN Idea** Scientists use scientific methods to structure their experiments and investigations. • Scientists work in many ways to gather data. • A good scientific experiment includes an independent variable, dependent variable, and control. An investigation, however, does not include a control. • Graphs, tables, and charts are three common ways to communicate data from an experiment. • SI, a modern version of the metric system, is a standard form of measurement that all scientists can use. • To express very large or very small numbers, scientists use scientific notation.
Section 1.3 Communication in Science	
• scientific law (p. 19) • scientific model (p. 18) • scientific theory (p. 19)	**MAIN Idea** Precise communication is crucial for scientists to share their results effectively with each other and with society. • Scientists communicate data so others can learn the results, verify the results, examine conclusions for bias, and conduct new experiments. • There are three main types of graphs scientists use to represent data: line graphs, circle graphs, and bar graphs. • A scientific model is an accurate representation of an idea or theory. • Scientific theories and scientific laws are sometimes discovered accidentally.

Earth Science Online Vocabulary PuzzleMaker glencoe.com

Assessment

Vocabulary Review

Explain the relationship between the vocabulary terms below.

1. geosphere, mantle

2. hydrosphere, atmosphere

3. oceanography, hydrosphere

4. meteorology, atmosphere

5. geology, biosphere

For Questions 6 to 9, fill in the blanks with the correct vocabulary terms from the Study Guide.

6. When conducting experiments, scientists use ________ to help guide their processes.

7. The ________ is the one factor that can be manipulated by the experimenter.

8. Scientists use a form of shorthand called ________ to express very large or very small numbers.

9. Most scientific studies and experiments use a standard system of units called ________.

Write a sentence using the following vocabulary terms.

10. scientific theory

11. scientific law

12. scientific model

Fill in the blanks with a vocabulary term from the Study Guide.

13. In the field of ________, scientists measure temperature, pressure, and humidity.

14. Their measurements come from features of the ________ and hydrosphere, and they look at how weather affects the ________ and geosphere.

15. The units of their measurements come from ________ and the metric system.

16. The numbers generally are not large, so ________ is not used.

Understand Key Concepts

17. Which one of these is NOT a specialized area of Earth science?
A. astronomy
B. environmental science
C. technology
D. oceanography

Use the figure below to answer Questions 18 and 19.

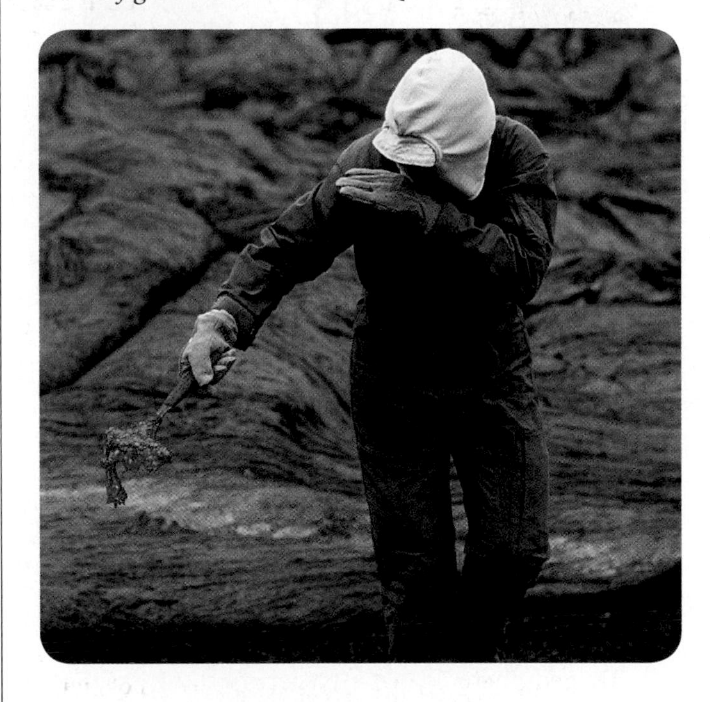

18. Which type of scientist is shown above?
A. oceanographer
B. geologist
C. astronomer
D. meteorologist

19. Which type of research is this scientist conducting?
A. field research
B. lab research
C. library research
D. biological research

20. Which is a sequence of steps a scientist might use to conduct an investigation?
A. analysis, test, question, conclude
B. test, question, conclude, analysis
C. question, test, analysis, conclude
D. conclude, test, question, analysis

Use the figure below to answer Questions 21 and 22.

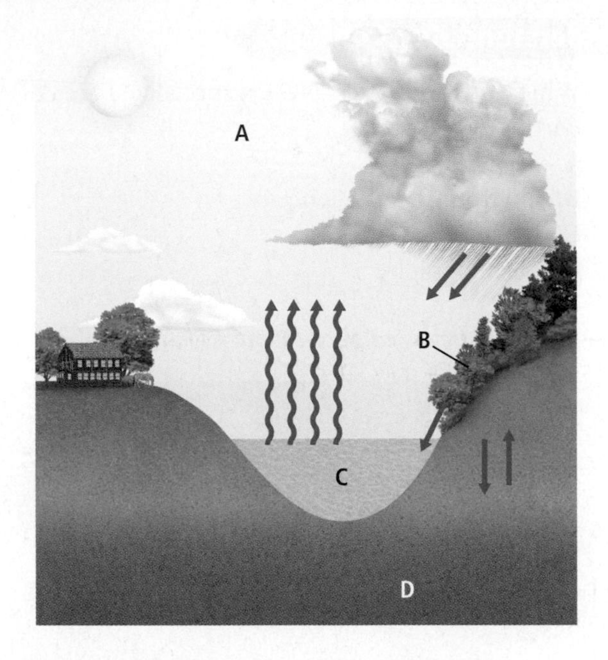

21. Identify the Earth system that is labeled *A*.
    A. atmosphere
    B. biosphere
    C. hydrosphere
    D. geosphere

22. Identify the Earth system that is labeled *B*.
    A. atmosphere
    B. biosphere
    C. hydrosphere
    D. geosphere

23. Which type makes up 97 percent of Earth's water?
    A. groundwater
    B. salt water
    C. freshwater
    D. spring water

24. Which is true of scientific models?
    A. They never change.
    B. They must be true for at least ten years.
    C. They will be modified with new observations and data.
    D. They are generally the work of one scientist.

25. Select the correct scientific notation for 150,000,000 km.
    A. 150×10^6 km
    B. 15×10^7 km
    C. 1.5×10^8 km
    D. 0.15×10^9 km

Constructed Response

26. **Explain** how technology relates to science.

Use the photo below to answer Question 27.

27. **Identify** the SI units that would be used to measure each of the above items.

28. **Summarize** each of Earth's systems and explain their relationships to each other.

29. **Compare and contrast** an investigation and an experiment.

30. **Apply** Why might a graph be more helpful in explaining data than just writing the results in words?

31. **Apply** When ice is heated above 0°C, it melts. Is this a theory or a law? Explain.

Think Critically

32. **CAREERS IN EARTH SCIENCE** Why would a meteorologist need an understanding of Earth's hydrosphere?

33. **Design an Experiment** Suppose you want to find the effect of sunlight on the temperature of a room with the shade up and the shade down. Describe how you would test this hypothesis. What would be your variables? What would you use as a control?

Earth Science Online Chapter Test glencoe.com

34. Propose An ecologist wants to study the effects of pollution on plant growth. The scientist uses two groups of plants. To the first group, a type of pollutant is added. To the second group, nothing is added. The scientist records plant growth for each plant for two weeks. What is the purpose of the second group in the scientist's study?

Use the table below to answer Question 35.

Some SI Conversions				
1 m	= ________	mm	= ________	km
1 g	= ________	mg	= ________	kg
1 cm^3	= ________	m^3	= ________	mL
3.5 km	= ________	m	= ________	cm

35. Calculate Copy the table into your notebook. Complete the table. Once you have made your conversions, express each answer in scientific notation.

Concept Mapping

36. Use the following terms to make a concept map summarizing the units used to measure each quantity discussed in the chapter: *time, density, temperature, volume, mass, weight, length, area, °C, g/mL, km, s, cm^3, m^2, kg,* and *N*. For help, refer to the *Skillbuilder Handbook.*

Challenge Question

37. Evaluate A scientist is researching a new cancer drug. Fifty patients have been diagnosed with the type of cancer the drug is designed to treat. If a control is used, the patients might not receive any medication. The patients do not know if they are receiving the placebo or the new medication. For this reason, the patients are allowed to also receive traditional treatment if they choose. How will this impact the research? How should the scientist account for this information in the results? Should the scientists be allowed to discourage patients from receiving additional treatment?

Additional Assessment

38. WRITING in Earth Science Imagine you are writing an explanation of the scientific methods for someone who has never done a scientific investigation before. Explain what the scientific methods are and why they are so important.

Document–Based Questions

Data obtained from: Annual mean sunspot numbers 1700–2002. *National Geophysical Data Center.*

Use the graphs below to answer Questions 39–41.

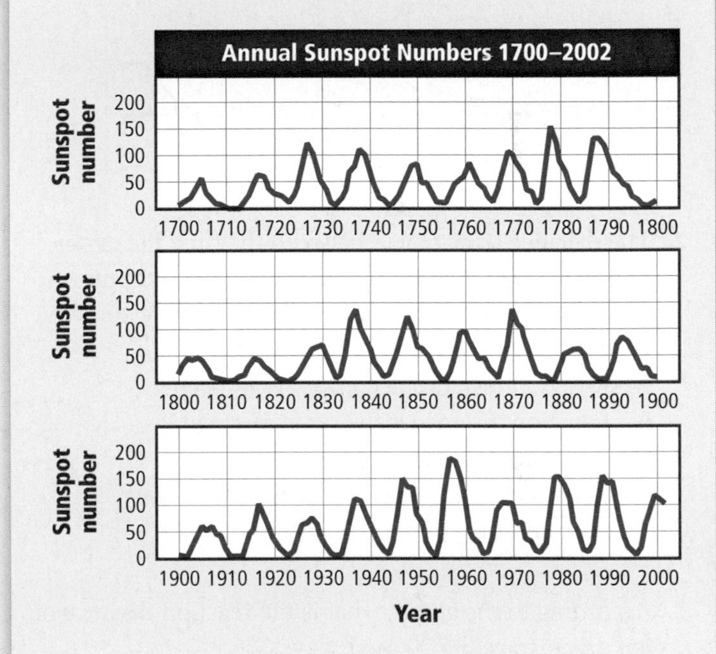

39. Is there a consistent pattern in the graphs? If so, what is the pattern showing?

40. What do the graphs express regarding the number of sunspots that have been seen and recorded since the 1700s?

41. What would you predict would be the pattern for the years 2000 to 2100?

Cumulative Review

In Chapters 2–30, Cumulative Review questions will help you review and check your understanding of concepts discussed in previous chapters.

Standardized Test Practice

Multiple Choice

1. Identify the type of Earth science that involves the study of the materials that make up Earth.
   A. astronomy
   B. meteorology
   C. geology
   D. oceanography

Use the graph below to answer Questions 2 and 3.

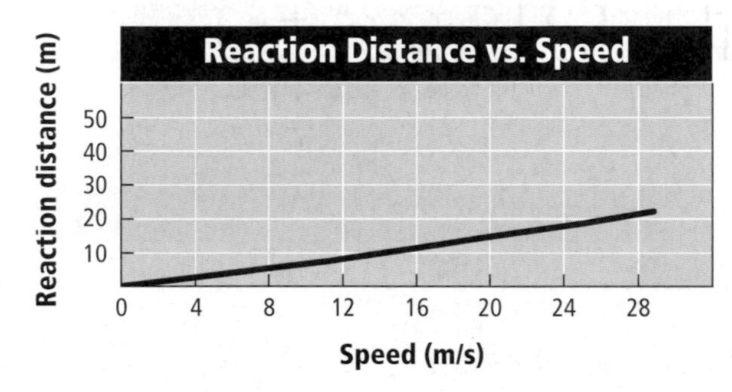

2. The distance a car travels between the time the driver decides to stop the car and the time the driver puts on the brakes is called the reaction distance. How does the reaction distance change with speed?
   A. Reaction distance decreases with speed.
   B. Reaction distance is the same as speed.
   C. Reaction distance increases with speed.
   D. There is not enough information to answer the question.

3. According to the graph, what is the reaction distance of the driver traveling 20 m/s?
   A. 3 m
   B. 15 m
   C. 20 m
   D. 28 m

4. Which lists Earth's layers from the inside out?
   A. inner core, outer core, mantle, crust
   B. crust, mantle, outer core, inner core
   C. crust, inner core, outer core, mantle
   D. mantle, outer core, inner core, crust

5. A block is 2 cm wide, 5.4 cm deep, and 3.1 cm long. The density of the block is 8.5 g/cm^3. What is the mass of the block?
   A. 33.48 g
   B. 85.10 g
   C. 399.3 g
   D. 284.58 g

6. If a conclusion is supported by data, but does not support an original hypothesis, what should a scientist do?
   A. The scientist should reevaluate the original hypothesis.
   B. The scientist should redesign the experiment.
   C. The scientist should not change anything.
   D. The scientist should modify the conclusion.

Use the illustration below to answer Questions 7 and 8.

WARNING:
Goggles and Aprons Must
Be Worn at All Times

7. This sign was found at the entrance to a chemistry laboratory. Why is this an important sign?
   A. Goggles help chemists see better.
   B. Chemicals can seriously damage eyes and skin.
   C. Accidents rarely happen in laboratories.
   D. Chemists will be fined if they do not obey the rules.

8. Why are safety rules posted, like this sign, or stated when conducting experiments?
   A. Safety rules are used to scare students.
   B. The goal of safety rules is to make an experiment boring.
   C. Safety rules are just suggestions as to how to behave during an experiment.
   D. The safety rules are given for scientists' protection.

9. What should you always do when conducting an experiment?
   A. You should clean up broken glass yourself.
   B. You should unplug cords by pulling on the cord, not the plug.
   C. You should report spills immediately.
   D. You should flush your eyes at the eyewash station.

10. Which of the following are Sir Isaac Newton's ideas on motion considered to be?
    A. scientific law
    B. scientific theory
    C. scientific model
    D. hypothesis

Short Answer

Use the graph below to answer Questions 11–13.

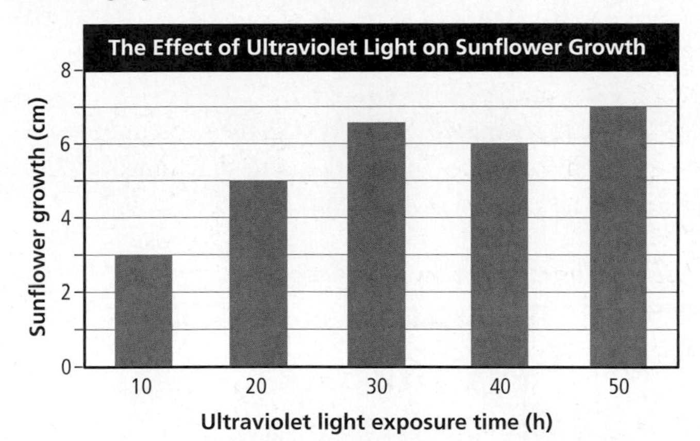

11. According to the graph, what was the greatest growth observed?

12. What type of graph is this? Why is this the best way to represent the data?

13. What are some variables that might affect the outcome of the experiment?

14. Describe the difference between the terms *astronomy* and *meteorology.*

15. Analyze the idea that technology is transferable. How is this beneficial?

16. Explain the importance of making a hypothesis before conducting an experiment.

17. Justine wants to measure how far an ant moves across a table in 1-min intervals. What would be the independent variable in this example?

Reading for Comprehension

Investigation Steps

Michael conducted an experiment to test if matter is conserved after a phase change. He filled an empty bottle with 50 mL of water and placed it in a sunny window until the liquid water changed to water vapor. The steps of the activity are listed in the table below but might not be in the correct order.

Investigation Steps	
1	Find the mass of the bottle, lid, and water vapor.
2	Pour 50 mL of water into an empty bottle.
3	Find the mass of the bottle, lid, and 50 mL of water.
4	Place a lid on the opening of the bottle to tightly seal it.

18. Which shows the investigation steps in the correct order?
A. 1, 2, 3, 4
B. 2, 4, 3, 1
C. 4, 2, 1, 3
D. 2, 4, 1, 3

19. According to the text, what scientific idea is Michael testing?
A. the rate of evaporation
B. the conservation of energy
C. the conservation of matter
D. the equilibrium of a system

20. Why does Michael put a lid on the bottle?
A. to prevent water vapor from leaving the system
B. to make the water evaporate faster
C. to prevent bacterial growth in the system
D. to establish equilibrium

NEED EXTRA HELP?																	
If You Missed Question . . .	1	2	3	4	5	6	7	8	9	10	11	12	13	14	15	16	17
Review Section . . .	1.1	1.3	1.3	1.1	1.2	1.3	1.2	1.2	1.2	1.3	1.3	1.3	1.3	1.1	1.1	1.2	1.2

CHAPTER 2

Mapping Our World

BIG Idea Earth scientists use mapping technologies to investigate and describe the world.

2.1 Latitude and Longitude
MAIN Idea Lines of latitude and longitude are used to locate places on Earth.

2.2 Types of Maps
MAIN Idea Maps are flat projections that come in many different forms.

2.3 Remote Sensing
MAIN Idea New technologies have changed the appearance and use of maps.

GeoFacts

- Maps predate written history. The earliest known map was created as a cave painting in ancient Turkey.
- China spans five international time zones; however, the entire country operates on only one standard time.
- Global Positioning System (GPS) satellites were originally designed for strategic defense and navigation purposes.

LAUNCH Lab

Can you make an accurate map?

If you have ever been asked to give someone directions, you know that it is important to include as many details as possible so that the person asking for directions will not get lost. Perhaps you drew a detailed map of the destination in question.

Procedure

1. Read and complete the lab safety form.
2. With a classmate, choose a location in your school or schoolyard.
3. Use a sheet of **graph paper** and **colored pencils** to draw a map from your classroom to the location you chose. Include landmarks such as drinking fountains and restrooms.
4. Share your map with a classmate. Compare the landmarks you chose and the path each of you chose to get to your locations. If they were different, explain why.
5. Follow your map to the location you and your partner chose. Was your map correct? Were there details you left out that might have been helpful?

Analysis

1. **Discuss** with your classmate how you could improve your maps.
2. **Examine** What details could you add?

Types of Mapping Technologies Make this Foldable to help organize information about the four major types of mapping technologies.

STEP 1 Find the middle of a horizontal sheet of paper and mark it. Fold the left and right sides of the paper to the middle and crease the folds.

STEP 2 Fold the piece of paper in half.

STEP 3 Open the last fold and cut along the fold lines to make four tabs.

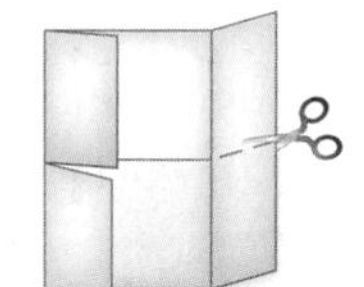

STEP 4 Label the tabs *Landsat, GPS/GIS, TOPEX/ Poseidon,* and *Sea Beam.*

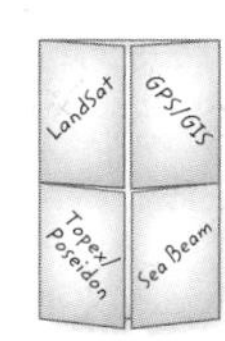

FOLDABLES **Use this Foldable with Section 2.3.** As you read this section, summarize information about the mapping technologies.

Visit glencoe.com to

- **study entire chapters online;**
- **explore Concepts In Motion animations:**
  - **Interactive Time Lines**
  - **Interactive Figures**
  - **Interactive Tables**
- **access Web Links for more information, projects, and activities;**
- **review content with the Interactive Tutor and take Self-Check Quizzes.**

Section 2.1

Objectives

- **Describe** the difference between latitude and longitude.
- **Explain** why it is important to give a city's complete coordinates when describing its location.
- **Explain** why there are different time zones from one geographic area to the next.

Review Vocabulary

time zone: a geographic region within which the same standard time is used

New Vocabulary

cartography
equator
latitude
longitude
prime meridian
International Date Line

Latitude and Longitude

MAIN Idea Lines of latitude and longitude are used to locate places on Earth.

Real-World Reading Link Imagine you were traveling from New York City, New York, to Los Angeles, California. How would you know where to go? Many people use maps to help them plan the quickest route.

Latitude

Maps are flat models of three-dimensional objects. For thousands of years people have used maps to define borders and to find places. The map at the beginning of this chapter was made in 1570. What do you notice about the size and shape of the continents? Today, more information is available to create more accurate maps. The science of mapmaking is called **cartography.**

Cartographers use an imaginary grid of parallel lines to locate exact points on Earth. In this grid, the **equator** horizontally circles Earth halfway between the north and south poles. The equator separates Earth into two equal halves called the northern hemisphere and the southern hemisphere.

Lines on a map running parallel to the equator are called lines of **latitude.** Latitude is the distance in degrees north or south of the equator as shown in **Figure 2.1.** The equator, which serves as the reference point for latitude, is numbered 0° latitude. The poles are each numbered 90° latitude. Latitude is thus measured from 0° at the equator to 90° at the poles.

Locations north of the equator are referred to by degrees north latitude (N). Locations south of the equator are referred to by degrees south latitude (S). For example, Syracuse, New York, is located at 43° N, and Christchurch, New Zealand, is located at 43° S.

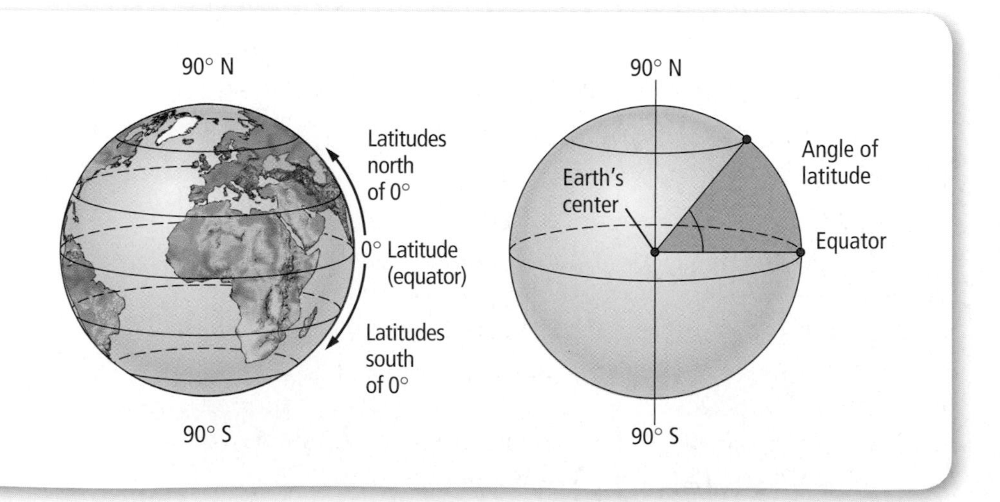

Figure 2.1 Lines of latitude are parallel to the equator. The value in degrees of each line of latitude is determined by measuring the imaginary angle created between the equator, the center of Earth, and the line of latitude as seen in the globe on the right.

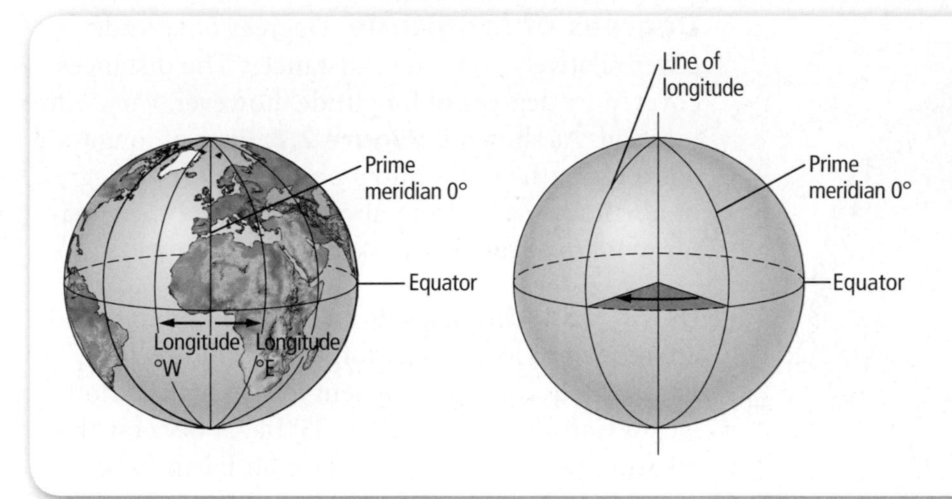

Figure 2.2 The reference line for longitude is the prime meridian. The degree value of each line of longitude is determined by measuring the imaginary angle created between the prime meridian, the center of Earth, and the line of longitude as seen on the globe on the right.

Degrees of latitude Each degree of latitude is equivalent to about 111 km on Earth's surface. How did cartographers determine this distance? Earth is a sphere and can be divided into 360°. The circumference of Earth is about 40,000 km. To find the distance of each degree of latitude, cartographers divided 40,000 km by 360°.

To locate positions on Earth more precisely, cartographers break down degrees of latitude into 60 smaller units, called minutes. The symbol for a minute is ′. The actual distance on Earth's surface of each minute of latitude is 1.85 km, which is obtained by dividing 111 km by 60′.

A minute of latitude can be further divided into seconds, which are represented by the symbol ″. Longitude is also divided into degrees, minutes, and seconds.

VOCABULARY

SCIENCE USAGE V. COMMON USAGE

Minute

Science usage: a unit used to indicate a portion of a degree of latitude

Common usage: a unit of time comprised of 60 seconds

Longitude

To locate positions in east and west directions, cartographers use lines of longitude, also known as meridians. As shown in **Figure 2.2, longitude** is the distance in degrees east or west of the prime meridian, which is the reference point for longitude.

The **prime meridian** represents 0° longitude. In 1884, astronomers decided that the prime meridian should go through Greenwich, England, home of the Royal Naval Observatory. Points west of the prime meridian are numbered from 0° to 180° west longitude (W); points east of the prime meridian are numbered from 0° to 180° east longitude (E).

Semicircles Unlike lines of latitude, lines of longitude are not parallel. Instead, they are large semicircles that extend vertically from pole to pole. For instance, the prime meridian runs from the north pole through Greenwich, England, to the south pole.

The line of longitude on the opposite side of Earth from the prime meridian is the 180° meridian. There, east lines of longitude meet west lines of longitude. This meridian is also known as the International Date Line, and will be discussed later in this section.

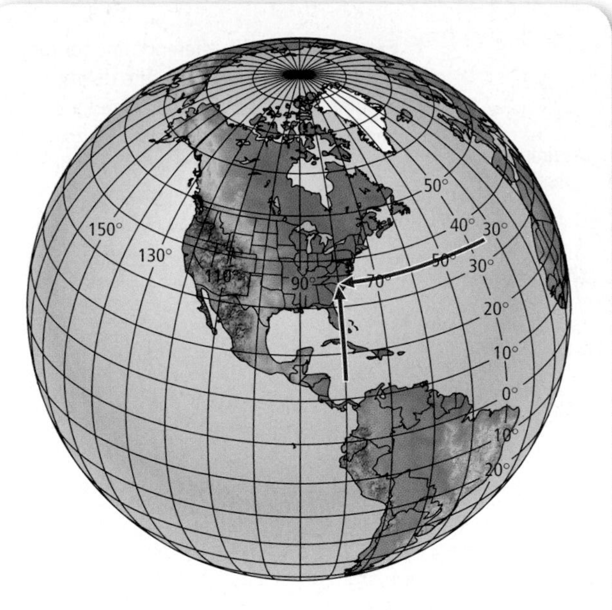

■ **Figure 2.3** The precise location of Charlotte is 35°14′N, 80°50′W. Note that latitude comes first in reference to the coordinates of a particular location.

Degrees of longitude Degrees of latitude cover relatively consistent distances. The distances covered by degrees of longitude, however, vary with location. As shown in **Figure 2.2,** lines of longitude converge at the poles into a point. Thus, one degree of longitude varies from about 111 km at the equator to 0 km at the poles.

Using coordinates Both latitude and longitude are needed to locate positions on Earth precisely. For example, it is not sufficient to say that Charlotte, North Carolina, is located at 35°14′ N because that measurement includes any place on Earth located along the 35°14′ line of north latitude.

The same is true of the longitude of Charlotte; 80°50′ W could be any point along that longitude from pole to pole. To locate Charlotte, use its complete coordinates—latitude and longitude—as shown in **Figure 2.3.**

Time zones Earth is divided into 24 time zones. Why 24? Earth takes about 24 hours to rotate once on its axis. Thus, there are 24 times zones, each representing a different hour. Because Earth is constantly spinning, time is always changing. Each time zone is 15° wide, corresponding roughly to lines of longitude. To avoid confusion, however, time zone boundaries have been adjusted in local areas so that cities and towns are not split into different time zones.

MiniLab

Locate places on Earth

How can you locate specific places on Earth with latitude and longitude?

Procedure

1. Read and complete the lab safety form.
2. Use a **world map** or **globe** to locate the prime meridian and the equator.
3. Take a few moments to become familiar with the grid system. Examine lines of latitude and longitude on the map or globe.

Analysis

1. **Locate** the following places:
   - Mount St. Helens, Washington; Niagara Falls, New York; Mount Everest, Nepal; Great Barrier Reef, Australia
2. **Locate** the following coordinates, and record the names of the places there:
   - 0°03′S, 90°30′W; 27°07′S, 109°22′W; 41°10′N, 112°30′W; 35°02′N, 111°02′W; 3°04′S, 37°22′E
3. **Analyze** How might early cartographers have located cities, mountains, or rivers without latitude and longitude lines?

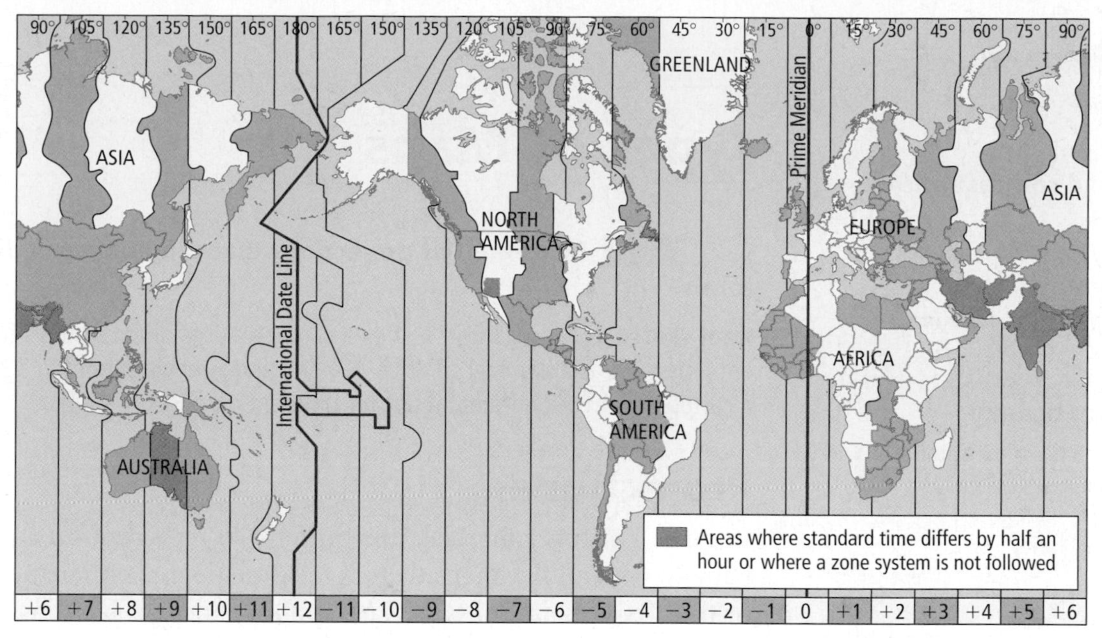

■ **Figure 2.4** In most cases, each time zone represents a different hour. However, there are some exceptions.
Identify ***two areas where the time zone is not standard.***

For example all of Morton County, North Dakota, operates within the central time zone, even though the western part of the county is within the mountain-time-zone boundary. As shown in **Figure 2.4,** there are six time zones in the United States.

International Date Line Each time you travel through a time zone, you gain or lose time until, at some point, you gain or lose an entire day. The **International Date Line,** which is 180° meridian, serves as the transition line for calendar days. If you were traveling west across the International Date Line, you would advance your calendar one day. If you were traveling east, you would move your calendar back one day.

Concepts In Motion

Interactive Figure To see an animation of time zones, visit glencoe.com.

Section 2.1 Assessment

Section Summary

- Latitude lines run parallel to the equator.
- Longitude lines run east and west of the prime meridian.
- Both latitude and longitude lines are necessary to locate exact places on Earth.
- Earth is divided into 24 time zones, each 15° wide, that help regulate daylight hours across the world.

Understand Main Ideas

1. MAIN Idea **Explain** why it is important to give both latitude and longitude when giving coordinates.
2. **Describe** how the distance of a degree of longitude varies from the equator to the poles.
3. **Estimate** the time difference between your home and places that are 60° east and west longitude of your home.

Think Critically

4. **Evaluate** If you were flying directly south from the north pole and reached 70° N, how many degrees of latitude would be between you and the south pole?

WRITING in Earth Science

5. Imagine what it would be like to fly from where you live to Paris, France. Describe what it would be like to adjust to the time difference.

Section 2.2

Objectives

- **Compare and contrast** different types of maps.
- **Explain** why different maps are used for different purposes.
- **Calculate** gradients on a topographic map.

Review Vocabulary

parallel: extending in the same direction and never intersecting

New Vocabulary

Mercator projection
conic projection
gnomonic projection
topographic map
contour line
contour interval
geologic map
map legend
map scale

Concepts In Motion

Interactive Figure To see an animation of map projections, visit glencoe.com.

■ **Figure 2.5** In a Mercator projection, points and lines on a globe are transferred onto cylinder-shaped paper. Mercator projections show true direction but distort areas near the poles.

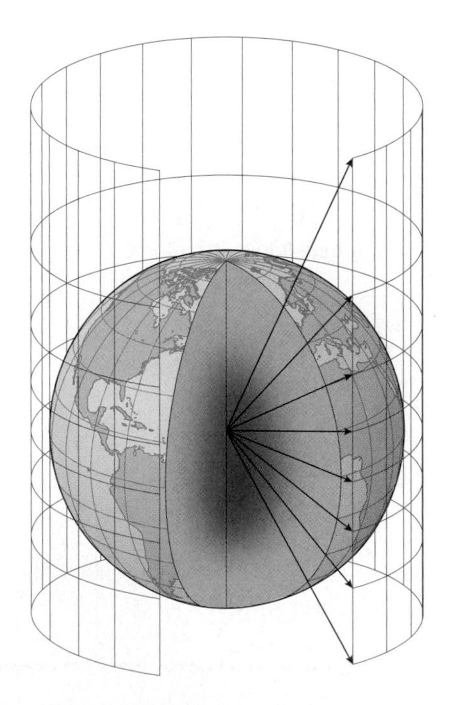

Types of Maps

MAIN Idea Maps are flat projections that come in many different forms.

Real-World Reading Link Just as a carpenter uses different tools for different jobs, such as a hammer to drive in a nail and wrench to tighten a bolt, a cartographer uses different maps for different purposes.

Projections

Because Earth is spherical, it is difficult to represent on a piece of paper. Thus, all flat maps distort to some degree either the shapes or the areas of landmasses. Cartographers use projections to make maps. A map projection is made by transferring points and lines on a globe's surface onto a sheet of paper.

Mercator projections A **Mercator projection** is a map that has parallel lines of latitude and longitude. Recall that lines of longitude meet at the poles. When lines of longitude are projected as being parallel on a map, landmasses near the poles are exaggerated. Thus, in a Mercator projection, the shapes of the landmasses are correct, but their areas are distorted.

As shown in **Figure 2.5,** Greenland appears much larger than Australia. In reality, Greenland is much smaller than Australia. Because Mercator projections show the correct shapes of landmasses and also clearly indicate direction in straight lines, they are used for the navigation of planes and ships.

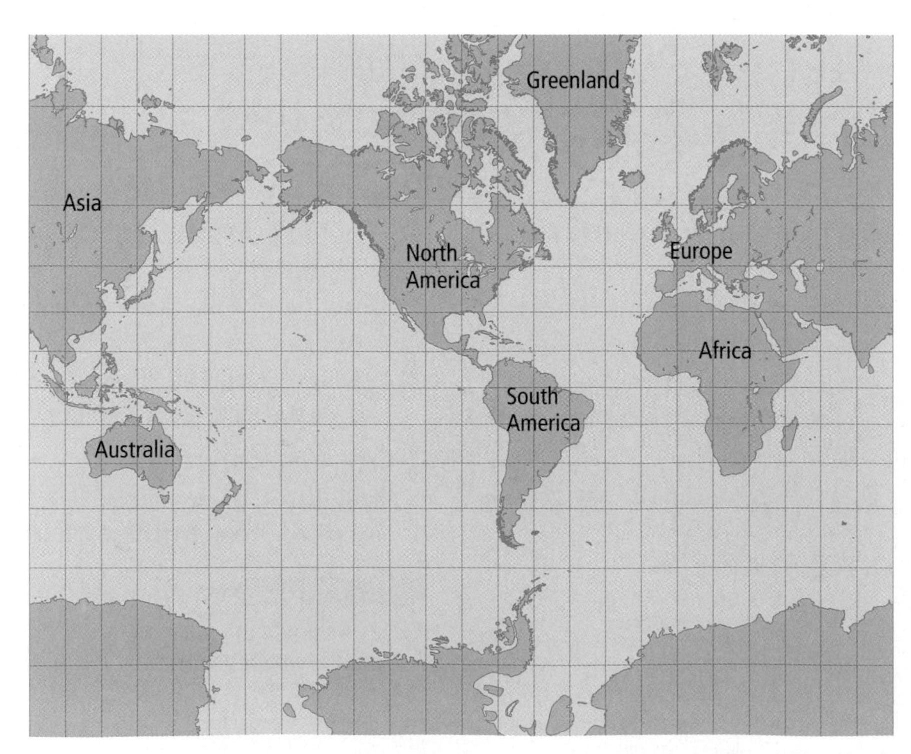

Conic projections A **conic projection** is made by projecting points and lines from a globe onto a cone, as shown in **Figure 2.6.** The cone touches the globe at a particular line of latitude. There is little distortion in the areas or shapes of landmasses that fall along this line of latitude. Distortion is evident, however, near the top and bottom of the projection. As shown in **Figure 2.6,** the landmass at the top of the map is distorted. Because conic projections have a high degree of accuracy for limited areas, they are excellent for mapping small areas. Hence, they are used to make road maps and weather maps.

Gnomonic projections A **gnomonic** (noh MAHN ihk) **projection** is made by projecting points and lines from a globe onto a piece of paper that touches the globe at a single point. At the single point where the map is projected, there is no distortion, but outside of this single point, great amounts of distortion are visible both in direction and landmass, as shown in **Figure 2.7.**

Because Earth is a sphere, it is difficult to plan long travel routes on a flat projection with great distortion, such as a conic projection. To plan such a trip, a gnomonic projection is most useful. Although the direction and landmasses on the projection are distorted, it is useful for navigation. A straight line on a gnomonic projection is the straightest route from one point to another when traveled on Earth.

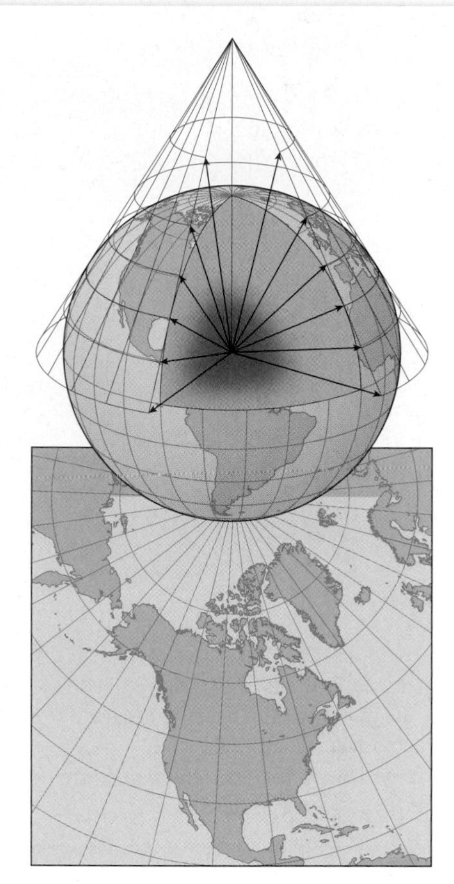

Figure 2.6 In a conic projection, points and lines on a globe are projected onto cone-shaped paper. There is little distortion along the line of latitude touched by the paper.

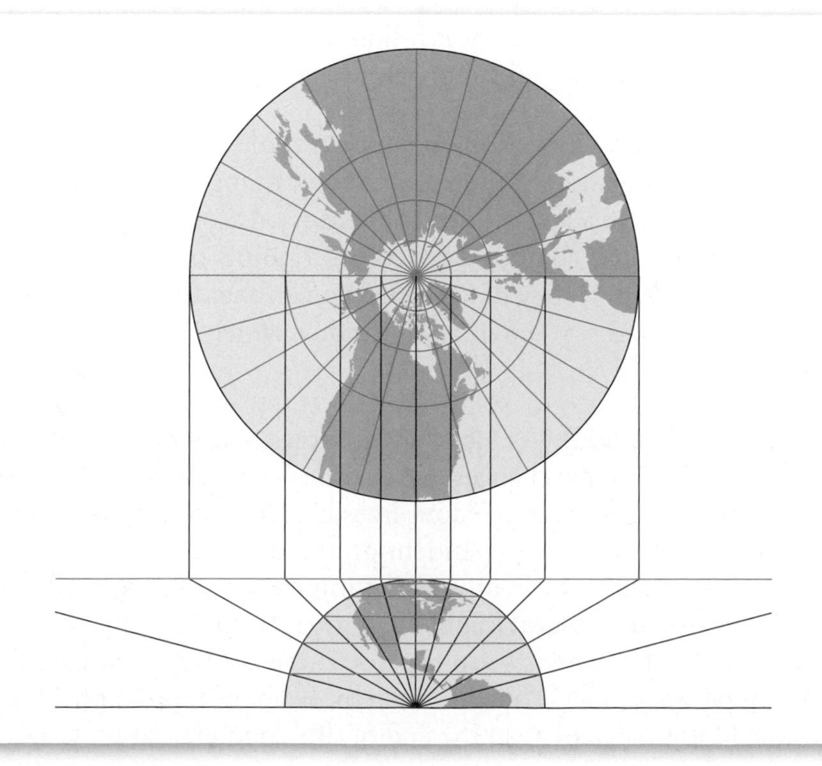

Figure 2.7 In a gnomonic projection, points and lines from a globe are projected onto paper that touches the globe at a single point.

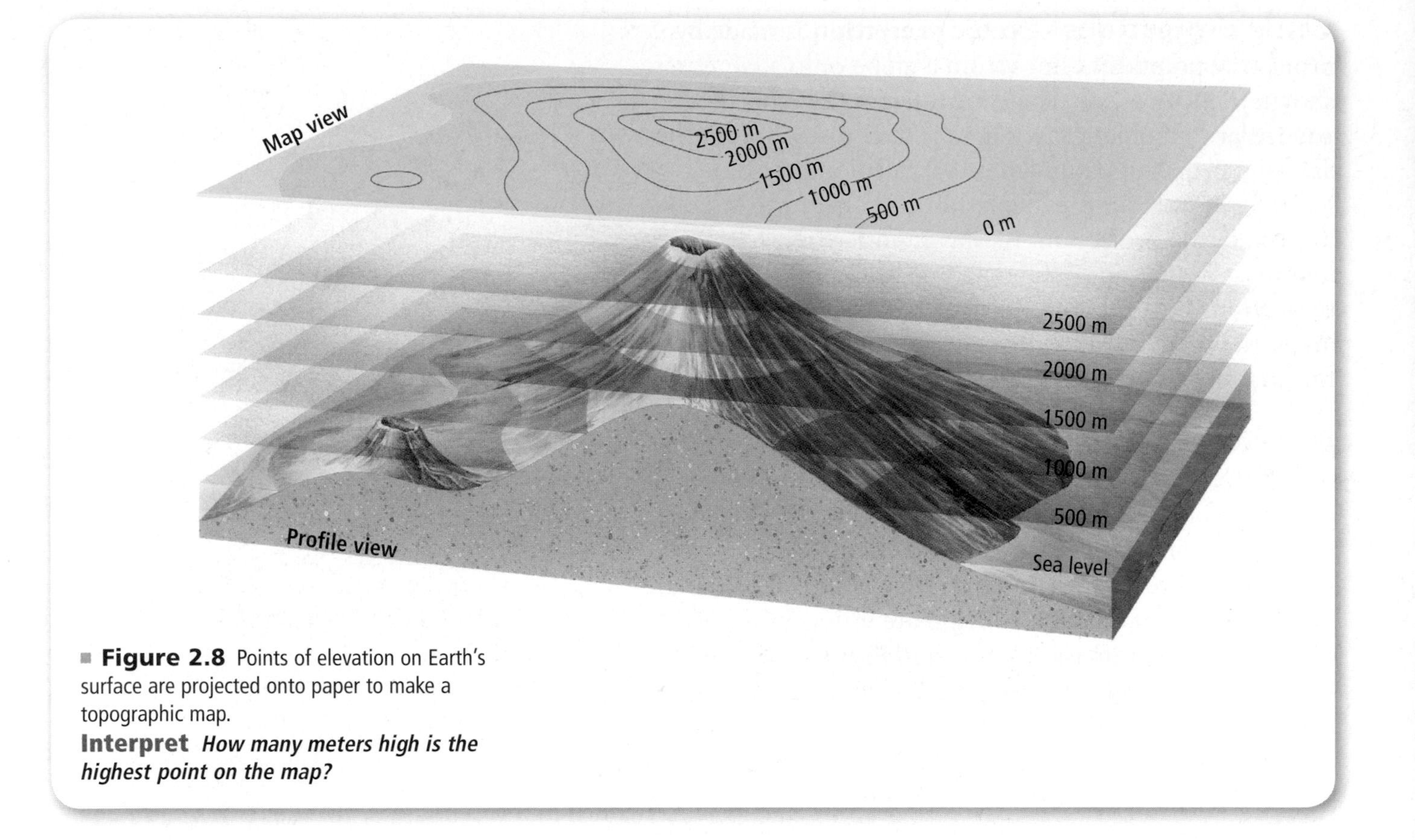

■ **Figure 2.8** Points of elevation on Earth's surface are projected onto paper to make a topographic map.
Interpret ***How many meters high is the highest point on the map?***

Topographic Maps

Detailed maps showing the hills and valleys of an area are called topographic maps. **Topographic maps** show changes in elevation of Earth's surface, as shown in **Figure 2.8.** They also show mountains, rivers, forests, and bridges, among other features. Topographic maps use lines, symbols, and colors to represent changes in elevation and features on Earth's surface.

Contour lines Elevation on a topographic map is represented by a contour line. Elevation refers to the distance of a location above or below sea level. A **contour line** connects points of equal elevation. Because contour lines connect points of equal elevation, they never cross. If they did, it would mean that the point where they crossed had two different elevations, which would be impossible.

Contour intervals As **Figure 2.8** shows, topographic maps use contour lines to show changes in elevation. The difference in elevation between two side-by-side contour lines is called the **contour interval.** The contour interval is dependent on the terrain.

For mountains, the contour lines might be very close together, and the contour interval might be as great as 100 m. This would indicate that the land is steep because there is a large change in elevation between lines. You will learn more about topographic maps in the Mapping GeoLab at the end of this chapter.

Index contours To aid in the interpretation of topographic maps, some contour lines are marked by numbers representing their elevations. These contour lines are called index contours, and they are used hand-in-hand with contour intervals to help determine elevation.

If you look at a map with a contour interval of 5 m, you can determine the elevations represented by other lines around the index contour by adding or subtracting 5 m from the elevation indicated on the index contour. Learn more about contour maps and index contours in the Problem-Solving Lab on this page.

 **Reading Check** **Analyze** If you were looking at a topographic map with a contour interval of 50 m and the contour lines were far apart, would this indicate a rapid increase or slow increase in elevation?

Depression contour lines The elevations of some features such as volcanic craters and mines are lower than that of the surrounding landscape. Depression contour lines are used to represent such features.

On a map, depression contour lines look like regular contour lines, but have hachures, or short lines at right angles to the contour line, to indicate depressions. As shown in **Figure 2.9,** the hachures point toward lower elevations.

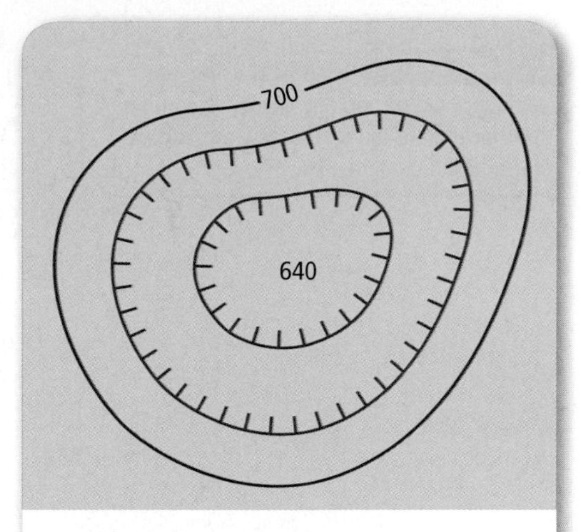

■ **Figure 2.9** The depression contour lines shown here indicate that the center of the area has a lower elevation than the outer portion of the area. The short lines pointing inward are called hachures and indicate the direction of the elevation change.

PROBLEM-SOLVING LAB

Calculate Gradients

How can you analyze changes in elevation? Gradient refers to the steepness of a slope. To measure gradient, divide the change in elevation between two points on a map by the distance between the two points. Use the map to answer the following questions, and convert your answers to SI units.

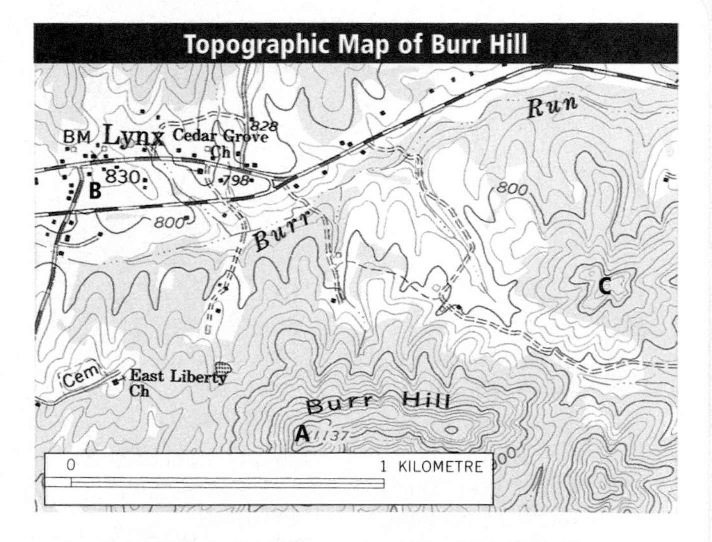

Analysis

1. **Determine** the distance from Point A to Point B using the map scale.
2. **Record** the change in elevation.
3. **Calculate** If you were to hike the distance from Point A to Point B, what would be the gradient of your climb?

Think Critically

4. **Explain** Would it be more difficult to hike from Point A to Point B, or from Point B to Point C?
5. **Calculate** Between Point A and Point C, where is the steepest part of the hike? How do you know?

To read about how one scientist is using maps and mapping technology to map the human footprint, go to the **National Geographic Expedition** on page 892.

Geologic Maps

A useful tool for a geologist is a geologic map. A **geologic map** is used to show the distribution, arrangement, and type of rocks located below the soil. A geologic map can also show features such as fault lines, bedrock, and geologic formations.

Using the information contained on a geologic map, combined with data from visible rock formations, geologists can infer how rocks might look below Earth's surface. They can also gather information about geologic trends, based on the type and distribution of rock shown on the map.

Geologic maps are most often superimposed over topographic maps and color coded by type of rock formation, as shown in **Figure 2.10.** Each color corresponds to the type of bedrock present in a given area. There are also symbols that represent mineral deposits and other structural features. Refer to **Table 2.1** on the following page to compare geologic maps to the other maps you have learned about in this chapter.

■ **Figure 2.10** Geologic maps show the distribution of surface geologic features. Notice the abundance of Older Precambrian rock formations.

Geologic Map of Grand Canyon

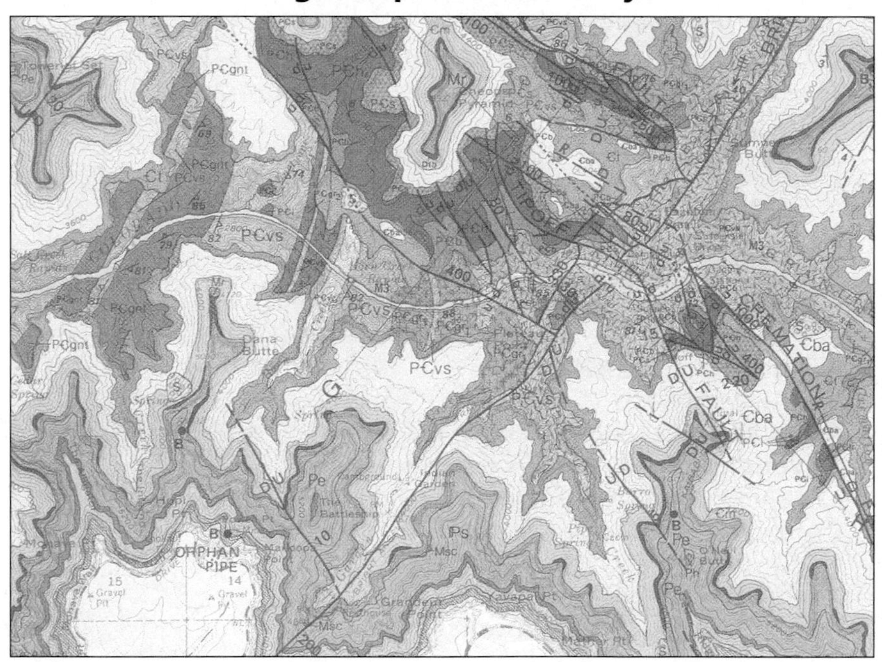

QUATERNARY
- S Landslides and rockfalls
- r River sediment

PERMIAN
- Pk Kaibab Limestone
- Pt Toroweap Formation
- Pc Coconino Sandstone
- Ph Hermit Shale
- Pe Esplanade Sandstone

PENNSYLVANIAN
- IPs Supai Formation

MISSISSIPPIAN
- Mr Redwall Limestone

DEVONIAN
- Dtb Temple Butte Limestone

CAMBRIAN
- Cm Muav Limestone
- Cba Bright Angel Shale
- Ct Tapeats Sandstone

YOUNGER PRECAMBRIAN
- PCi Diabase sills and dikes
- PCs Shinumo Quartzite
- PCh Hakatai Shale
- PCb Bass Formation

OLDER PRECAMBRIAN
- PCgr1 Zoroaster Granite
- PCgnt Trinity Gneiss
- PCvs Vishnu Schist

CONCEPTS In MOtion
Interactive Table To explore more about maps and projections, visit glencoe.com.

Table 2.1 Types of Maps and Projections

Map or Projection	Common Uses	Distortions
Mercator projection	navigation of planes and ships	The land near the poles is distorted.
Conic projection	road and weather maps	The areas at the top and bottom of the map are distorted.
Gnomonic projection	great circle routes	The direction and distance between landmasses is distorted.
Topographic map	to show elevation changes on a flat projection	It depends on the type of projections used.
Geologic map	to show the types of rocks below the surface present in a given area	It depends on the type of projection used.

Three-dimensional maps Topographic and geologic maps are two-dimensional models of Earth's surface. Sometimes, scientists need to visualize Earth three-dimensionally. To do this, scientists often rely on computers to digitize features such as rivers, mountains, valleys, and hills.

Map Legends

Most maps include both human-made and natural features located on Earth's surface. These features are represented by symbols, such as black dotted lines for trails, solid red lines for highways, and small black squares and rectangles for buildings. A **map legend,** such as the one shown in **Figure 2.11,** explains what the symbols represent. For more information about the symbols in map legends, see the *Reference Handbook.*

Reading Check **Apply** If you made a legend for a map of your neighborhood, what symbols would you include?

Map Scales

When using a map, you need to know how to measure distances. This is accomplished by using a map scale. A **map scale** is the ratio between distances on a map and actual distances on the surface of Earth. Normally, map scales are measured in SI, but as you will see on the map in the GeoLab, sometimes they are measured in different units such as miles and inches. There are three types of map scales: verbal scales, graphic scales, and fractional scales.

Figure 2.11 Map legends explain what the symbols on maps represent.

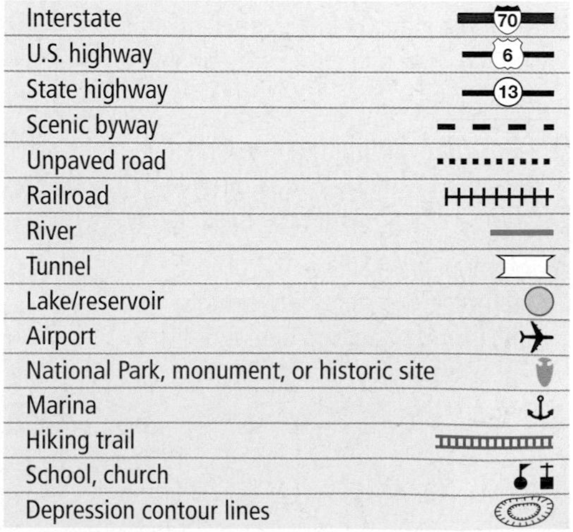

Interstate
U.S. highway
State highway
Scenic byway
Unpaved road
Railroad
River
Tunnel
Lake/reservoir
Airport
National Park, monument, or historic site
Marina
Hiking trail
School, church
Depression contour lines

Verbal scales To express distance as a statement, such as "one centimeter is equal to one kilometer," cartographers and Earth scientists use verbal scales. The verbal scale, in this example, means that one centimeter on the map represents one kilometer on Earth's surface.

Graphic scales Instead of writing the map scale out in words, graphic scales consist of a line that represents a certain distance, such as 5 km or 5 miles. The line is labeled, and then broken down into sections with hash marks, and each section represents a distance on Earth's surface. For instance, a graphic scale of 5 km might be broken down into five sections, with each section representing 1 km. Graphic scales are the most common type of map scale.

VOCABULARY

ACADEMIC VOCABULARY

Ratio
the relationship in quantity, amount, or size between two or more things
The ratio of girls to boys in the class was one to one.

Reading Check Infer why an Earth scientist might use different types of scales on different types of maps.

Fractional scales Fractional scales express distance as a ratio, such as 1:63,500. This means that one unit on the map represents 63,500 units on Earth's surface. One centimeter on a map, for instance, would be equivalent to 63,500 cm on Earth's surface. Any unit of distance can be used, but the units on each side of the ratio must always be the same.

A large ratio indicates that the map represents a large area, while a small ratio indicates that the map represents a small area. A map with a large fractional scale such as 1:100,000 km would therefore show less detail than a map with a small fractional scale such as 1:1000 km.

Section 2.2 Assessment

Section Summary

- Different types of projections are used for different purposes.
- Geologic maps help Earth scientists study patterns in subsurface geologic formations.
- Maps often contain a map legend that allows the user to determine what the symbols on the map signify.
- The map scale allows the user to determine the ratio between distances on a map and actual distances on the surface of Earth.

Understand Main Ideas

1. MAIN Idea **Explain** why distortion occurs at different places on different types of projections.
2. **Describe** how a conic projection is made. Why is this type of projection best suited for mapping small areas?
3. **Determine** On a Mercator projection, where does most of the distortion occur? Why?
4. **Compare and contrast** Mercator and gnomonic projections. What are these projections commonly used for?

Think Critically

5. **Predict** how a geologic map could help a city planner decide where to build a city park.

MATH in Earth Science

6. Determine the gradient of a slope that starts at an elevation of 55 m and ends 20 km away at an elevation of 15 m.

Earth Science Online Self-Check Quiz glencoe.com

Section 2.3

Objectives

- **Compare and contrast** different types of remote sensing.
- **Discuss** how satellites and sonar are used to map Earth's surface and its oceans.
- **Describe** the Global Positioning System and how it works.

Review Vocabulary

satellite: natural or human-made object that orbits Earth, the Moon, or other celestial body

New Vocabulary

remote sensing
Landsat satellite
TOPEX/Poseidon satellite
sonar
Global Positioning System
Geographic Information System

Remote Sensing

MAIN Idea New technologies have changed the appearance and use of maps.

Real-World Reading Link Many years ago, if you wanted a family portrait, it would be painted by an artist over many hours. Today, cameras can create a photo in seconds. Cartography has also changed. Cartographers use digital images to create maps with many more details that can be updated instantly.

Landsat Satellite

Advanced technology has changed the way maps are made. The process of gathering data about Earth using instruments mounted on satellites, airplanes, or ships is called **remote sensing.**

One form of remote sensing is detected with satellites. Features on Earth's surface, such as rivers and forests, radiate warmth at slightly different frequencies. **Landsat satellites** record reflected wavelengths of energy from Earth's surface. These include wavelengths of visible light and infrared radiation. One example of a Landsat image is shown in **Figure 2.12.**

To obtain such images, each Landsat satellite is equipped with a moving mirror that scans Earth's surface. This mirror has rows of detectors that measure the intensity of energy received from Earth. This information is then converted by computers into digital images that show landforms in great detail.

Landsat 7, launched in 1999, maps 185 km at a time and scans the entire surface of Earth in 16 days. Landsat data are also used to study the movements of Earth's plates, rivers, earthquakes, and pollution.

Figure 2.12 Notice the differences between the two Landsat photos of New Orleans.
Interpret *Which image was taken after Hurricane Katrina in 2005? Explain.*

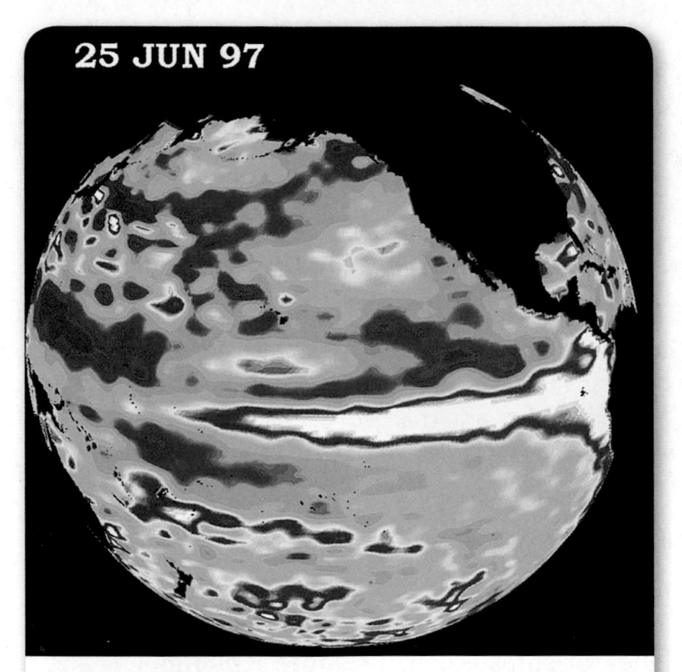

■ **Figure 2.13** This image, which focuses on the Pacific Ocean, was created with data from *TOPEX/Poseidon.* The white color in the image shows the change in ocean depth during a hurricane event relative to normal.

TOPEX/Poseidon Satellite

One satellite that uses radar to map features on the ocean floor is the ***TOPEX/Poseidon* satellite.** **TOPEX** stands for **top**ography **ex**periment and Poseidon (puh SY duhn) is the Greek god of the sea. Radar uses high-frequency signals that are transmitted from the satellite to the surface of the ocean. A receiving device then picks up the returning echo as it is reflected off the water.

The distance to the water's surface is calculated using the known speed of light and the time it takes for the signal to be reflected. Variations in time indicate the presence of certain features on the ocean floor. For instance, ocean water bulges over seafloor mountains and forms depressions over seafloor valleys.

These changes are reflected in satellite-to-sea measurements and result in images such as the one shown in **Figure 2.13,** that shows ocean depths during a hurricane. Using *TOPEX/Poseidon* data, scientists were able to estimate global sea levels with an accuracy of just a few millimeters and could repeat these calculations as often as every ten days. Scientists can also use this data and combine it with other existing data to create maps of ocean-floor features.

The *TOPEX/Poseidon* satellite also has been used to study tidal changes and global ocean currents. **Figure 2.14** below shows additional technological advances in cartography.

FOLDABLES

Incorporate information from this section into your Foldable.

■ **Figure 2.14**

Mapping Technology

Advances in mapping have relied on technological developments.

1500 B.C. — A.D. 1 — 1500

1300 B.C. An ancient Egyptian scribe draws the oldest surviving topographical map.

150 B.C. The ancient Greek scientist Ptolemy creates the first map using a coordinate grid. It depicted Earth as a sphere and included Africa, Asia, and Europe.

A.D. 1154 Arab scholar Al-Idrisi creates a world map used by European explorers for several centuries. Earlier medieval maps showed Jerusalem as the center of a flat world.

1569 Flemish geographer Gerhardus Mercator devises a way to project the globe onto a flat map using lines of longitude and latitude.

Sea Beam

Sea Beam technology is similar to the *TOPEX/Poseidon* satellite in that it is also used to map the ocean floor. However, Sea Beam is located on a ship rather than on a satellite. **Figure 2.15** shows an example of a map created with information gathered with Sea Beam technology. To map ocean-floor features, Sea Beam relies on **sonar,** which is the use of sound waves to detect and measure objects underwater.

You might have heard of sonar before. It is often used to detect other objects like ships or submarines under water. This same technology allows scientists to detect changes in elevation or calculate distances between objects.

First, to gather the information needed to map the seafloor, a sound wave is sent from a ship toward the ocean floor. A receiving device then picks up the returning echo when it bounces off the seafloor.

Computers on the ship calculate the distance from the ship to the ocean floor using the speed of sound in water and the time it takes for the sound to be reflected. Sea Beam technology is used by fishing fleets, deep-sea drilling operations, and scientists such as oceanographers, volcanologists, and archaeologists.

Reading Check **Compare and contrast** Sea Beam images with *TOPEX/Poseidon* images and how each might be used.

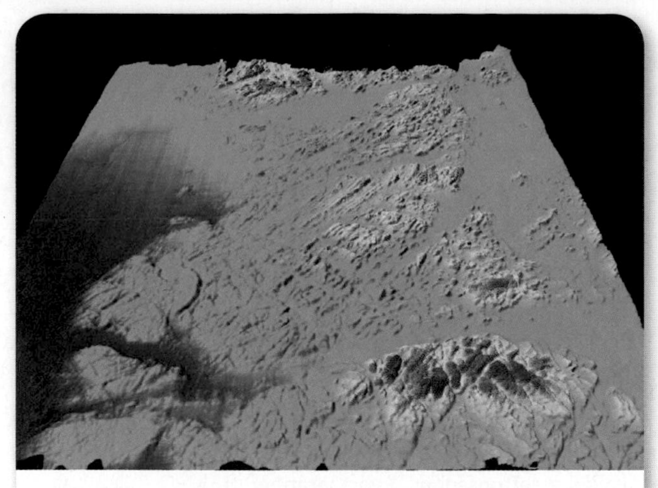

Figure 2.15 This image of Plymouth offshore was created with data from Sea Beam. The change in color indicates a change in elevation. The red-orange colors are the peaks, and the blue colors are the lowest elevations.

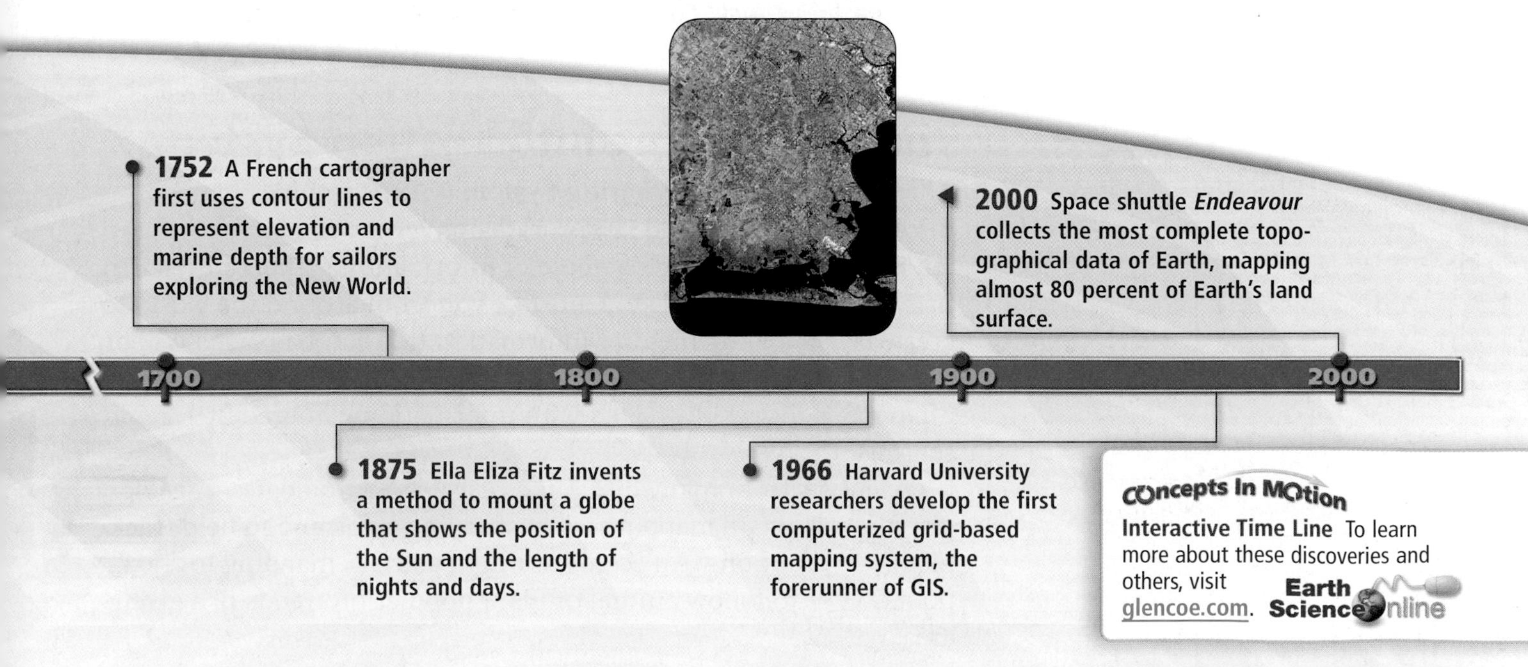

The Global Positioning System

The **Global Positioning System (GPS)** is a satellite navigation system that allows users to locate their approximate position on Earth. There are 27 satellites orbiting Earth, as shown in **Figure 2.16,** for use with GPS units. The satellites are positioned around Earth, and are constantly orbiting so that signals from at least three or four satellites can be picked up at any given moment by a GPS receiver.

To use GPS to find your location on Earth, you need a GPS receiver. The receiver calculates your approximate latitude and longitude—usually within 10 m—by processing the signals emitted by the satellites. If enough information is present, these satellites can also relay information about elevation, direction of movement, and speed. With signals from three satellites, a GPS receiver can calculate location on Earth without elevation, while four satellite signals will allow a GPS receiver to calculate elevation also. For more information on how the satellites are used to determine location, see **Figure 2.16.**

CAREERS IN EARTH SCIENCE

Cartographer An Earth scientist who works primarily with maps is called a cartographer. A cartographer might make maps, interpret maps, or research mapping techniques and procedures. To learn more about Earth science careers, visit glencoe.com.

Uses for GPS technology GPS technology is used extensively for navigation by airplanes and ships. However, as you will read later, it is also used to help detect earthquakes, create maps, and track wildlife.

GPS technology also has many applications for everyday life. Some people now have GPS receivers in their cars to help navigate to preprogrammed destinations such as restaurants, hotels, and their homes. Hikers, bikers, and other travelers often have portable, handheld GPS systems with them at all times. This allows them to find their destinations more quickly and can help them determine their location so they do not get lost. Some cell phones also contain GPS systems that can help you find your location.

 **Reading Check** **Compare** GPS satellites with *TOPEX/Poseidon.*

VOCABULARY

ACADEMIC VOCABULARY

Comprehensive
covering completely or broadly
The teacher gave the students a comprehensive study guide for the final exam.

The Geographic Information System

The **Geographic Information System (GIS)** combines many of the traditional types and styles of mapping described in this chapter. GIS mapping uses a database of information gathered by scientists, professionals, and students like you from around the world to create layers, or "themes," of information that can be placed one on top of the other to create a comprehensive map. These "themes" are often maps that were created with information gathered by remote sensing.

Scientists from many disciplines use GIS technologies. A geologist might use GIS mapping when studying a volcano to help track historical eruptions. An ecologist might use GIS mapping to track pollution or to follow animal or plant population trends of a given area.

Visualizing GPS Satellites

Figure 2.16 GPS receivers detect signals from the 27 GPS satellites orbiting Earth. Using signals from at least three satellites, the receiver can calculate location within 10 m.

First, a GPS receiver, located in New York City, receives a signal from one satellite. The distance from the satellite to the receiver is calculated. Suppose the distance is 20,000 km. This limits the possible location of the receiver to anywhere on a sphere 20,000 km from the satellite.

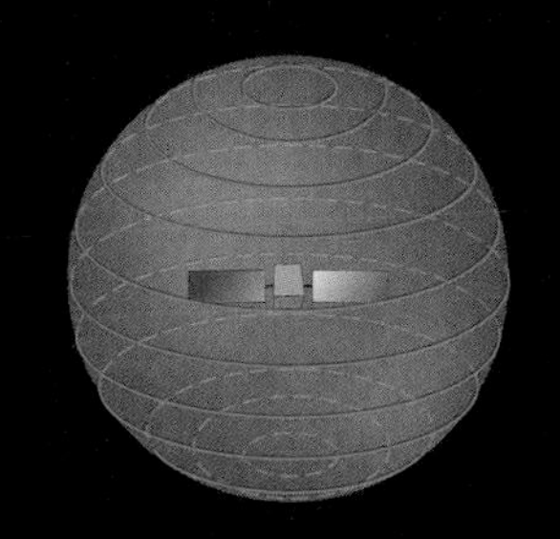

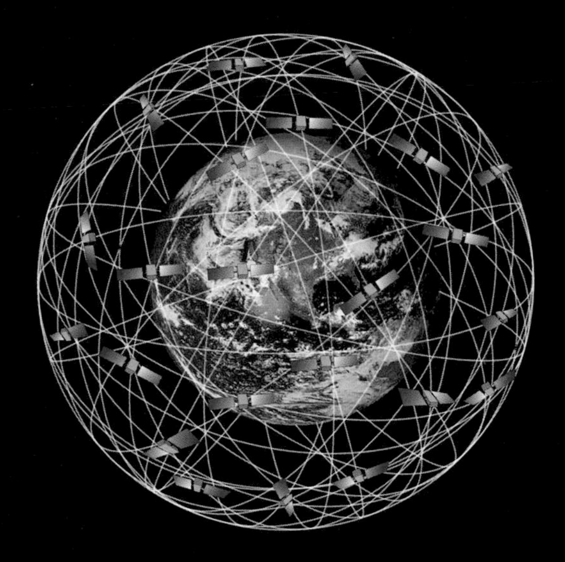

Next, the receiver measures the distance to a second satellite. Suppose this distance is calculated to be 21,000 km away. The location of the receiver has to be somewhere on the area where the two spheres intersect, shown here in yellow.

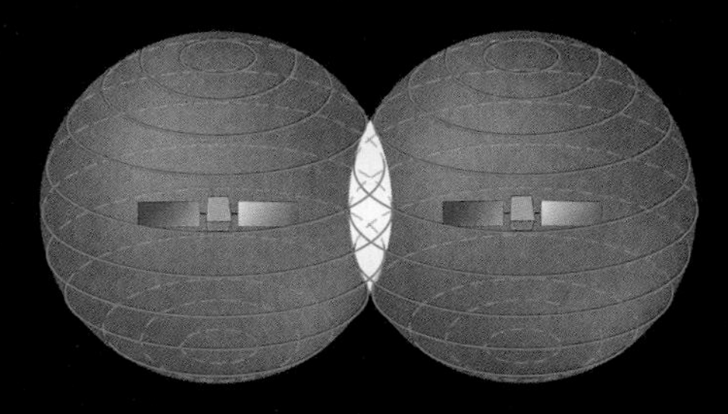

Finally, the distance to a third satellite is calculated. Using this information, the location of the receiver can be narrowed even further. By adding a third sphere, the location can be calculated to be one of two points as shown. Often one of these points can be rejected as an improbable or impossible location.

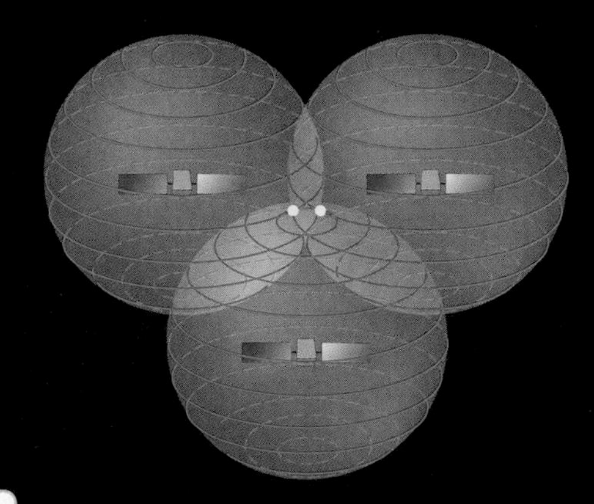

Concepts In Motion To explore more about GPS satellites, visit glencoe.com.

Earth Science Online

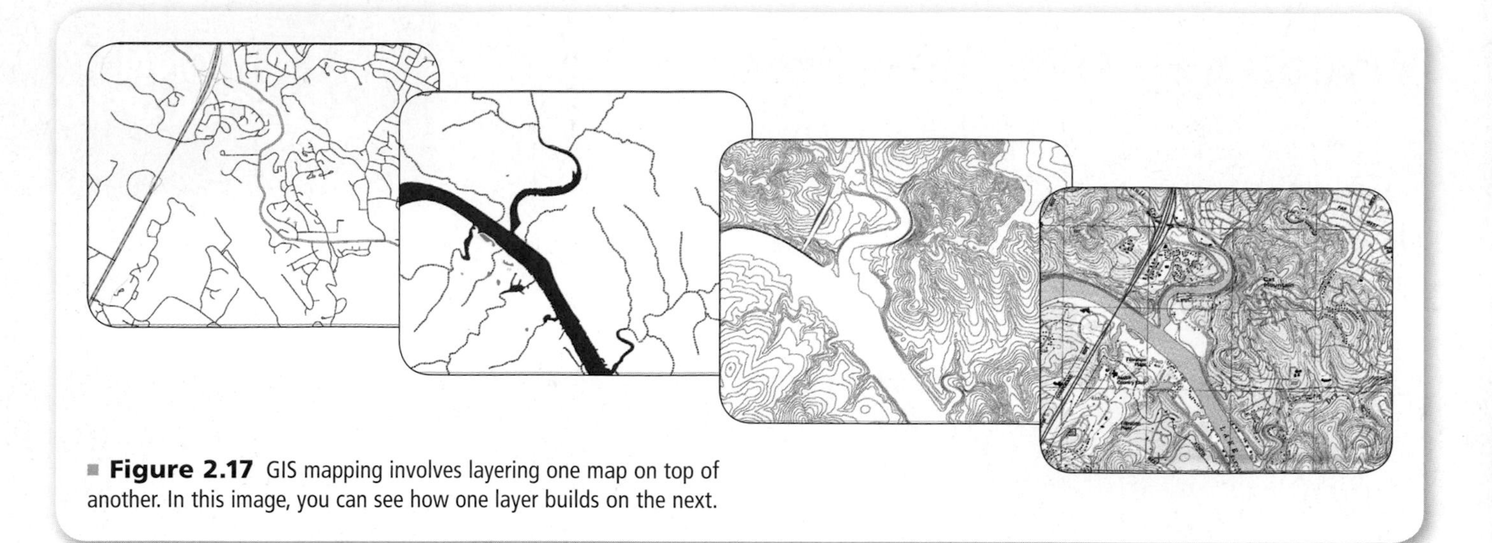

■ **Figure 2.17** GIS mapping involves layering one map on top of another. In this image, you can see how one layer builds on the next.

GIS maps might contain many layers of information compiled from several different types of maps, such as a geologic map and a topographic map. As shown in **Figure 2.17,** layers such as rivers, topography, roads, and landforms from the same geographic area can be placed on top of each other to create a comprehensive map.

One major difference between GIS mapping and traditional mapping is that a GIS map can be updated as new information is loaded into the database. Once a map is created, the layers are still linked to the original information. If this information changes, the GIS layers also change. The result is a map that is always up-to-date—a valuable resource for people who rely on current information.

Section 2.3 Assessment

Section Summary

- Remote sensing is an important part of modern cartography.
- Satellites are used to gather data about features of Earth's surface.
- Sonar is also used to gather data about features of Earth's surface.
- GPS is a navigational tool that is now used in many everyday items.

Understand Main Ideas

1. MAIN Idea **Describe** how remote sensing works and why it is important in cartography.
2. **Apply** Why is GPS navigation important to Earth scientists?
3. **Explain** the different types of information that can be gathered with satellites.
4. **Predict** why it might be important to be able to add and subtract map layers as with GIS mapping.

Think Critically

5. **Infer** How could GIS mapping be helpful in determining where to build a housing development?
6. **Explain** why it is important to have maps of the ocean floor, such as those gathered with Sea Beam technology.

WRITING in Earth Science

7. Write an article describing how GPS satellites help you locate your position on Earth.

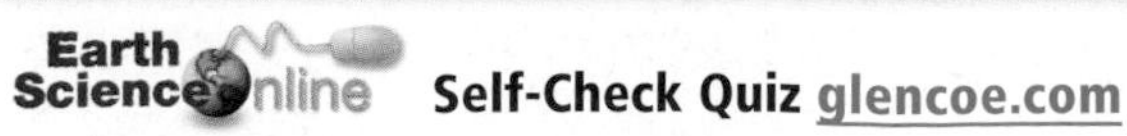

EARTH SCIENCE AND TECHNOLOGY

MAPPING DISASTER ZONES

On August 29, 2005, Hurricane Katrina hit the New Orleans area, causing $81.2 billion in damage and resulting in the deaths of nearly 2000 people. With such widespread devastation, how did relief workers reach the damaged areas? Mapping technologies helped workers to identify priority areas and create a plan to aid those affected.

GPS and disaster relief Global Positioning System (GPS) satellites send signals back to Earth telling the receiver the exact location of the user. The satellites travel at approximately 14,000 km/h, and are powered by solar energy. During Katrina, GPS signals provided up-to-the-minute information regarding destruction detail and locations of survivors and aid workers.

Using GIS Another important mapping tool used during disasters is the Geographic Information System Technology (GIS). This technology captures, stores, records, and analyzes data dependent on geography and location. As a result, many important decisions about environmental issues or relief efforts can be made using GIS data. After Katrina, GIS data provided relief workers with images of area hospitals within a small geographic area. This enabled emergency workers to get injured individuals to medical facilities quickly.

Other imaging systems Other mapping software packages provide actual pictorial images of Earth. These images show the damaged areas as well as buildings that can be appropriate for setting up relief sites.

Synthetic Aperture Radar (SAR) polarimetry is an imaging technology that is able to rapidly detect disaster zones.

This aerial image shows some of the flooding and destruction caused by Hurricane Katrina. Images like this help workers navigate through the altered landscape.

With other satellite images, views of the affected landscape can be blocked by clouds, darkness, smoke, or dust. By using radar, SAR mapping is not affected by these things, thus making the images readily available to relief workers.

Mapping areas affected by natural disasters with satellite and aerial images makes these areas accessible by relief workers. They are better able to prepare for the changes in local geography, destruction of buildings, and other physical challenges in the disaster zone. Continued improvements in mapping technologies and increased accessibility are important for continued improvement of disaster relief programs.

WRITING in Earth Science

Mapping Applications Research a recent natural disaster by visiting glencoe.com. Write a news article that describes the disaster based on the images of the disaster you find. Include several images in your news article.

GEOLAB

MAPPING: USE A TOPOGRAPHIC MAP

Background: Topographic maps show two-dimensional representations of Earth's surface. With these maps, you can determine the slope of a hill, what direction streams flow, and where mines and other features are located. In this lab, you will use the topographic map on the following page to determine elevation for several routes and to create a profile showing elevation.

Question: *How can you use a topographic map to interpret information about an area?*

Materials

ruler
string
piece of paper

Procedure

1. Read and complete the lab safety form.
2. Take a piece of paper and lay it on the map so that it intersects Point A and Point B.
3. On this piece of paper, draw a small line at each place where a contour line intersects the line from Point A to Point B. Also note the elevation at each hash mark and any rivers crossed.
4. Copy the table shown on this page into your science journal.
5. Now take your paper where you marked your lines and place it along the base of the table.
6. Mark a corresponding dot on the table for each elevation.
7. Connect the dots to create a topographic profile.
8. Use the map to answer the following questions. Be sure to check the map's scale.
9. Use the string to measure distances between two points that are not in a straight line. Lay the string along curves, and then measure the distance by laying the string along the ruler. Remember that elevations on United States Geological Survey (USGS) maps are given in feet.

Analyze and Conclude

1. **Determine** What is the contour interval?
2. **Identify** what type of map scale the map utilizes.
3. **Calculate** the stream gradient of Big Wildhorse Creek from the Gravel Pit in Section 21 to where the creek crosses the road in Section 34.
4. **Calculate** What is the highest elevation of the jeep trail? If you followed the jeep trail from the highest point to where it intersects an unimproved road, what would be your change in elevation?
5. **Apply** If you started at the bench mark (BM) on the jeep trail and hiked along the trail and the road to the Gravel Pit in section 21, how far would you hike?
6. **Analyze** What is the straight line distance between the two points in Question 4? What is the change in elevation?
7. **Predict** Does Big Wildhorse Creek flow throughout the year? Explain your answer.
8. **Calculate** What is the shortest distance along roads from the Gravel Pit in Section 21 to the secondary highway?

	820
	810
	800
	790
	780
	770
	760
	750
	740
	730
	720
	710
	700

INQUIRY EXTENSION

Make a Map Using what you have learned in this lab, create a topographic map of your hometown. For more information on topographic maps, visit glencoe.com.

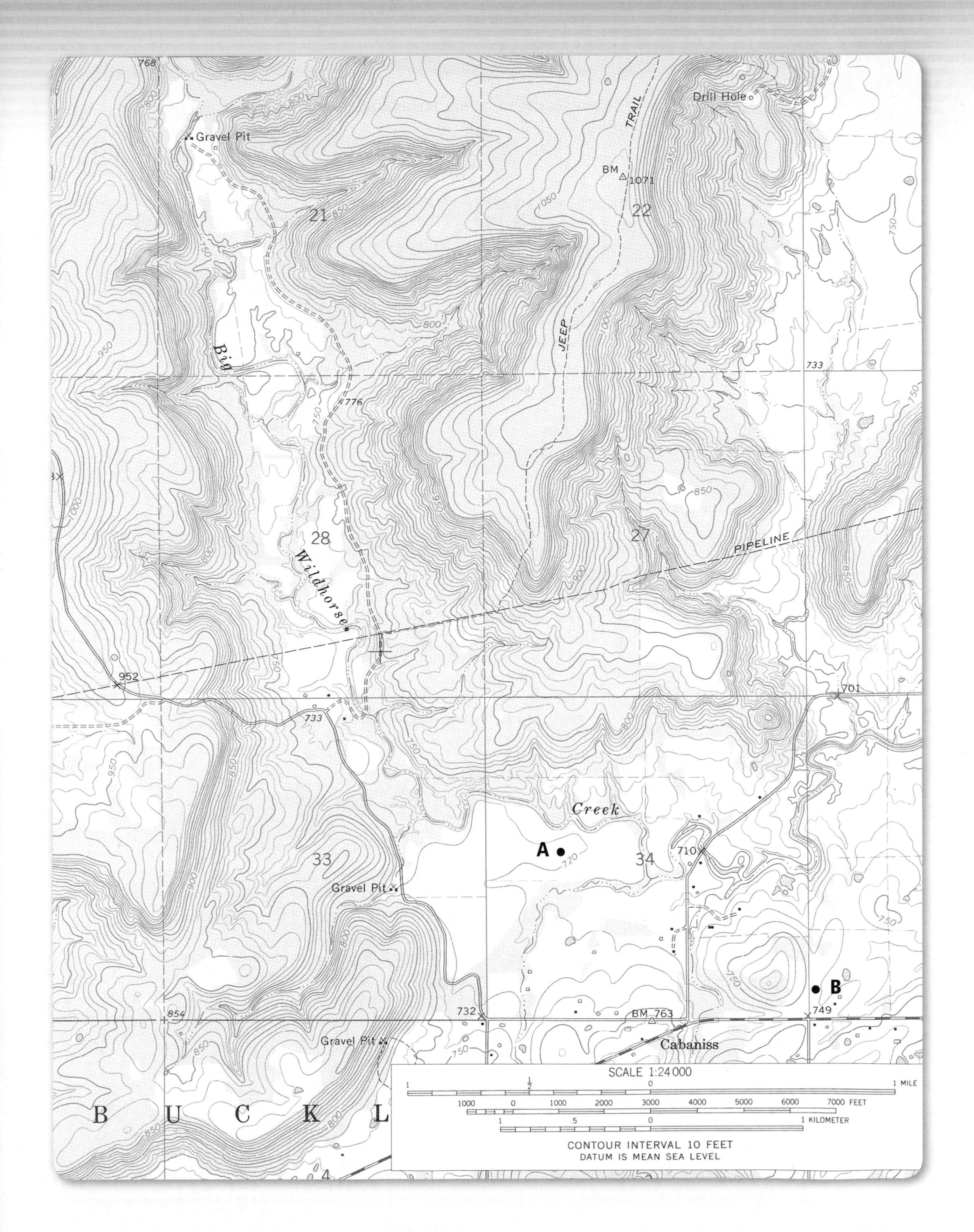
768
Gravel Pit
Drill Hole
TRAIL
BM
1071
21
22
850
800
1050
950
1000
JEEP
Big
733
776
750
28
27
Wildhorse
PIPELINE
952
701
733
Creek
A
720
34
710
33
Gravel Pit
B
854
732
BM 763
749
Gravel Pit
Cabaniss
B U C K L
4
SCALE 1:24 000
1 ½ 0 1 MILE
1000 0 1000 2000 3000 4000 5000 6000 7000 FEET
1 .5 0 1 KILOMETER
CONTOUR INTERVAL 10 FEET
DATUM IS MEAN SEA LEVEL

CHAPTER 2

Study Guide

Download quizzes, key terms, and flash cards from glencoe.com.

BIG Idea Earth scientists use mapping technologies to investigate and describe the world.

Vocabulary	Key Concepts
Section 2.1 Latitude and Longitude	
• cartography (p. 30) • equator (p. 30) • International Date Line (p. 33) • latitude (p. 30) • longitude (p. 31) • prime meridian (p. 31)	MAIN Idea Lines of latitude and longitude are used to locate places on Earth. • Latitude lines run parallel to the equator. • Longitude lines run east and west of the prime meridian. • Both latitude and longitude lines are necessary to locate exact places on Earth. • Earth is divided into 24 time zones, each 15° wide, that help regulate daylight hours across the world.
Section 2.2 Types of Maps	
• conic projection (p. 35) • contour interval (p. 36) • contour line (p. 36) • geologic map (p. 38) • gnomonic projection (p. 35) • map legend (p. 39) • map scale (p. 39) • Mercator projection (p. 34) • topographic map (p. 36)	MAIN Idea Maps are flat projections that come in many different forms. • Different types of projections are used for different purposes. • Geologic maps help Earth scientists study patterns in subsurface geologic formations. • Maps often contain a map legend that allows the user to determine what the symbols on the map signify. • The map scale allows the user to determine the ratio between distances on a map and actual distances on the surface of Earth.
Section 2.3 Remote Sensing	
• Geographic Information System (p. 44) • Global Positioning System (p. 44) • Landsat satellite (p. 41) • remote sensing (p. 41) • sonar (p. 43) • *TOPEX/Poseidon* satellite (p. 42)	MAIN Idea New technologies have changed the appearance and use of maps. • Remote sensing is an important part of modern cartography. • Satellites are used to gather data about features of Earth's surface. • Sonar is also used to gather data about features of Earth's surface. • GPS is a navigational tool that is now used in many everyday items.

CHAPTER 2

Assessment

Vocabulary Review

Each of the following sentences is false. Make each sentence true by replacing the italicized word with a vocabulary term from the Study Guide.

1. The study of mapmaking is called *topology.*

2. A *gnomonic projection* is a map that has parallel lines of latitude and longitude.

3. The process of collecting data about Earth from far above Earth's surface is called *planetology.*

4. *Landsat satellite* uses sonar waves emitted from a ship to map the ocean floor.

5. A *map scale* explains what the symbols on the map represent.

Replace the underlined words with the correct vocabulary term from the Study Guide.

6. Latitude lines run north to south and are measured from the prime meridan.

7. A map legend shows the ratio between distances on a map.

8. GPS mapping combines many traditional types of maps into one.

9. GIS technology helps determine a user's exact location.

Choose the correct vocabulary term from the Study Guide to complete the following sentences.

10. Zero longitude is known as the ________.

11. The difference in elevation between two side-by-side contour lines on a topographic map is called the ________.

12. ________ is the use of sound waves to detect and measure objects underwater.

13. The ________ serves as the transition line for calendar days.

14. A(n) ________ is used on a topographic map to indicate elevation.

Understand Key Concepts

Use the figure below to answer Questions 15 and 16.

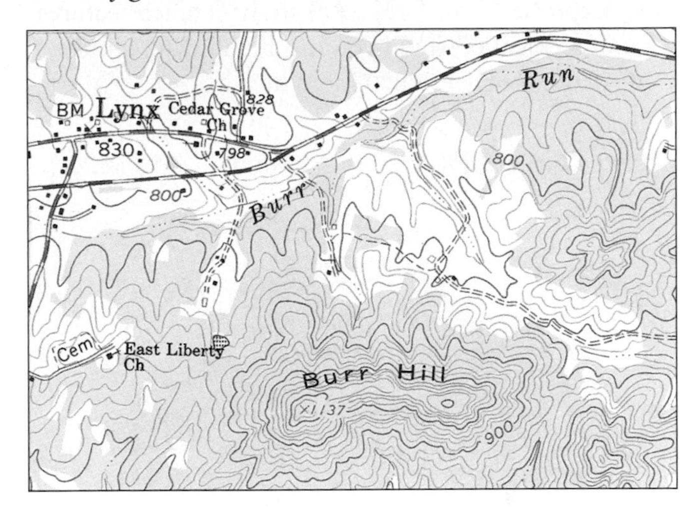

15. What is shown in this image?
A. a Landsat image
B. a topographic map
C. a gnomonic projection
D. a GIS map

16. What are the lines in the figure called?
A. hachures
B. contour lines
C. latitude lines
D. longitude lines

17. Refer to **Figure 2.4.** How many time zones are there in Australia?
A. 5
B. 1
C. 3
D. 10

18. Which is a use of Sea Beam?
A. to map continents
B. to map the ocean floor
C. to map Antarctica
D. to map mountains and valleys

19. On a topographic map, which do hachures point toward?
A. higher elevations
B. lakes
C. no change in elevation
D. lower elevations

20. Which is not usually included in map legends?
A. interstates
B. people
C. rivers
D. railroads

Constructed Response

21. **Locate** What time is it in New Orleans, LA, if it is 3 P.M. in Syracuse, NY? Refer to **Figure 2.4** for help.

22. **Explain** If you wanted to study detailed features of a volcano, would you use a map with a scale of 1:150 m or 1:150,000 m? Why?

Use the figure below to answer Questions 23 and 24.

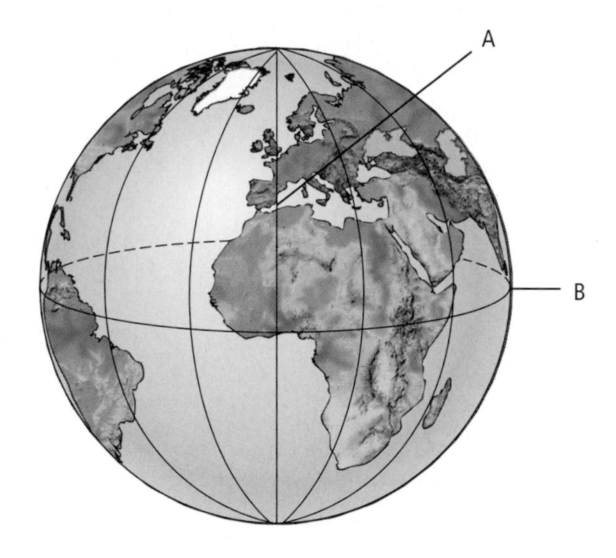

23. **Identify** What is the line labeled *A*?

24. **Identify** What is the line labeled *B*?

25. **Explain** What is the maximum potential height of a mountain if the last contour line is 2000 m and the map has a contour interval of 100 m?

26. **Describe** how radar used in the *TOPEX/Poseidon* satellite differs from the sonar used in the collection of data by Sea Beam.

27. **Infer** Based on what you have learned in this chapter, how might an astronomer map objects seen in the night sky?

28. **Practice** Think back to the Launch Lab at the beginning of the chapter. What type of map projection would be best for the map you drew? Why?

29. **Explain** how degrees of longitude are calculated.

30. **Explain** how degrees of latitude are calculated.

Use the figure below to answer Question 31.

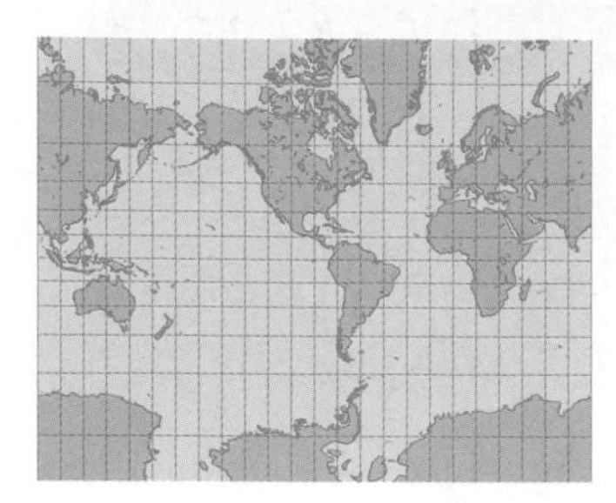

31. **Interpret** what type of projection is shown in the figure. What would this type of projection be used for?

Think Critically

32. **Apply** Would a person flying from Virginia to California have to set his or her watch backward or forward? Explain.

33. **Consider** why a large country like China might choose to follow only one time zone.

34. **CAREERS IN EARTH SCIENCE** Analyze how an architect trying to determine where to build a house and a paleontologist trying to determine where to dig for fossils might use a geologic map.

Use the figure below to answer Question 35.

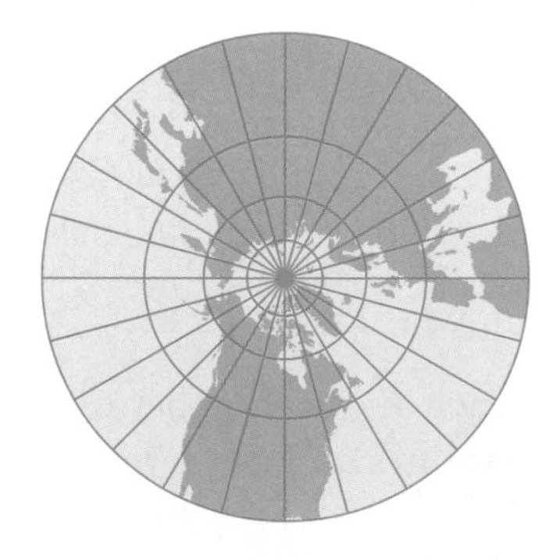

35. **Apply** What is the projection shown above? What would be two uses for this type of projection? Explain.

36. **Plan** Make a map from your school to the nearest supermarket. How will you determine the scale? What will you need to include in your legend?

37. Analyze Why isn't a conic projection used to navigate a ship or an aircraft?

38. Design an experiment to test the accuracy of several types of GPS receivers. Make sure you include your control, dependent, and independent variables.

39. Evaluate Briefly describe GIS and how it can be used by your community to develop an emergency plan for a severe storm, earthquake, blizzard, drought, or another potential local disaster.

40. Explain why it is necessary to have three satellite signals to determine elevation when using a GPS receiver.

Concept Mapping

41. Use the following to complete a concept map about remote sensing: *remote sensing*, TOPEX/ Poseidon *satellite, Landsat satellite, GPS, uses radar to map ocean floor, uses visible light and infrared radiation to map Earth's surface*, and *uses microwaves to determine location of user*. For more help, refer to the *Skillbuilder Handbook*.

Challenge Problem

Use the figure below to answer Question 42.

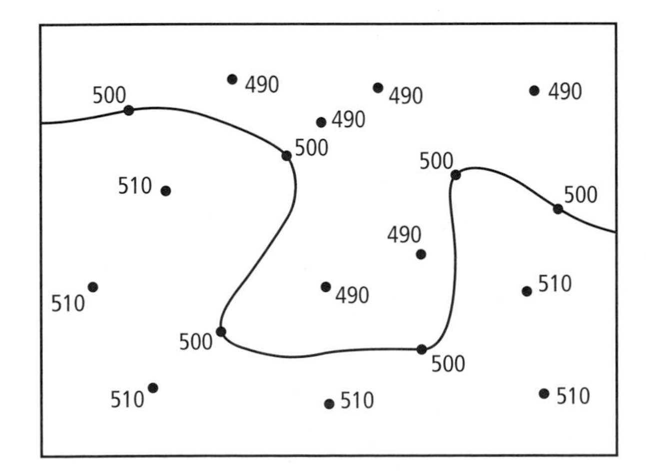

42. Assess Trace the following image to create a topographic map. Connect the elevation measurements to create contour intervals.

Additional Assessment

43. WRITING in **Earth Science** Write a journal entry for an explorer traveling across America before an accurate map was made.

DBQ Document–Based Questions

Data obtained from: NASA, *CALIPSO* satellite image.

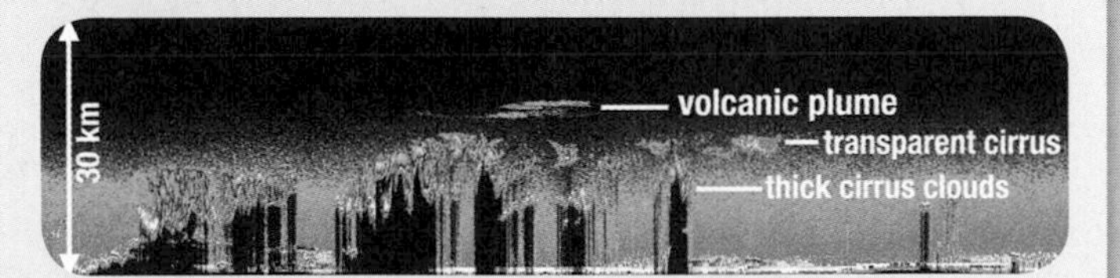

This is a satellite image from NASA's CALIPSO *satellite.* CALIPSO *is similar to the other remote sensing technologies you learned about in this chapter, but instead of radar or sonar, it uses something called lidar that sends pulses of light and measures the time it takes for the light to reflect back to the satellite. Based on the amount of time it takes for the light to reflect,* CALIPSO *and scientists can determine what is located below the satellite. The image above was collected on June 7, 2006, across the Indian Ocean. The navy blue areas indicate that no data was detected.*

44. At approximately how many kilometers is the volcanic plume located?

45. At approximately how many kilometers is the thick cirrus cloud located?

46. Why do you think the volcanic plume is higher than the thick cirrus clouds?

47. Why do you think there was no data detected below the thick cirrus clouds?

Cumulative Review

48. Why are graphs, charts, and maps useful? **(Chapter 1)**

49. Why is good communication important in the field of science? **(Chapter 1)**

Standardized Test Practice

Multiple Choice

Use the map to answer Questions 1 and 2.

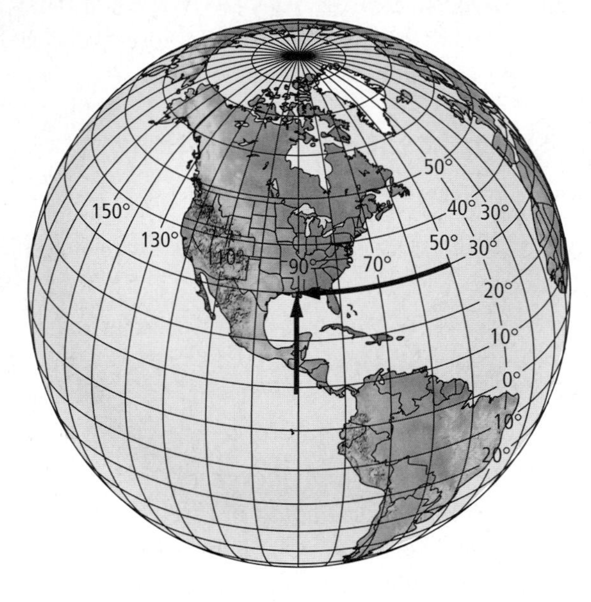

1. What is the latitude and longitude of the location pointed out by the arrows?
   A. 30° N, 100° W
   B. 45° N, 105° W
   C. 30° N, 90° W
   D. 10° N, 90° W

2. Roughly how many degrees of latitude does the United States cover?
   A. 10°
   B. 15°
   C. 20°
   D. 25°

3. Which would be most useful if you were lost in the Sahara desert?
   A. Landsat satellite
   B. *TOPEX/Poseidon* satellite
   C. Global Positioning System
   D. topographic map of Africa

4. What is the reference point for lines of longitude?
   A. the equator
   B. the prime meridian
   C. the International Date Line
   D. the 360th meridian

5. Why do cartographers break down degrees of longitude and latitude into minutes and seconds
   A. to get a better time frame of how long it takes to get from one place to the next
   B. to help travelers with planning trips
   C. to locate positions on Earth more precisely
   D. to make cartography easier to understand

Use the map below to answer Questions 6 and 7.

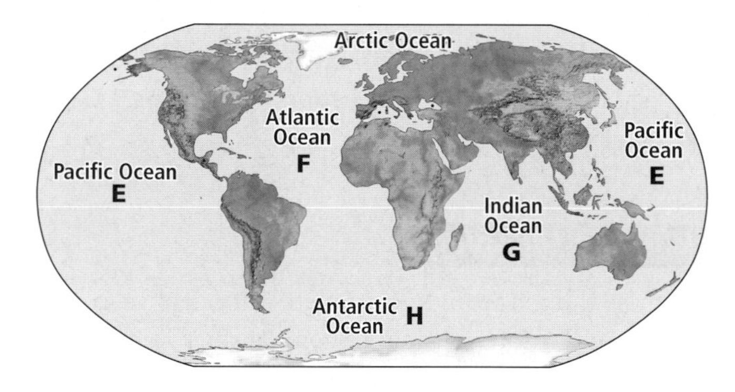

6. What problem do cartographers encounter when creating maps such as the one shown above?
   A. placing all of the continents in the correct position
   B. transferring a three-dimensional Earth onto a flat piece of paper
   C. naming all of the important locations on the map
   D. placing lines of latitude and longitude at the correct locations

7. What improvements could be made to make this map more helpful to sailors?
   A. Distort the size of the continents.
   B. Show only the water locations and not the locations of the land.
   C. Label the various continents.
   D. Add lines of latitude and longitude for navigation.

8. For what purpose are conic projection maps typically used?
   A. road and weather maps
   B. showing changes in elevation
   C. plotting long distance trips
   D. showing one specific point on Earth

Short Answer

Use the map below to answer Questions 9–11.

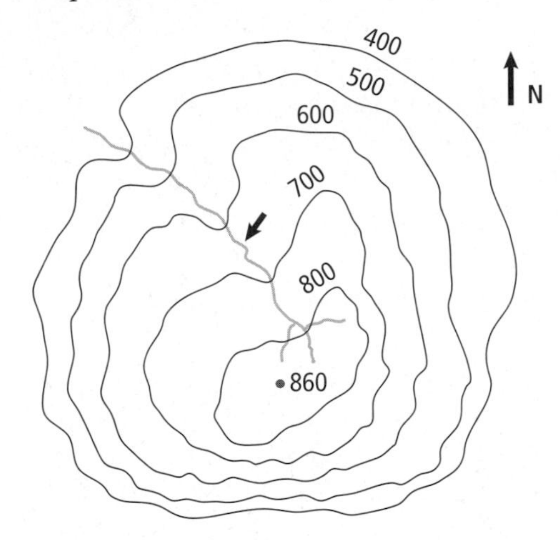

9. What is the map above showing?

10. What do the numbers on the map represent?

11. How might a hiker use this map in creating a route to get to the top?

12. Why would a ship find Sea Beam technology beneficial?

13. The distance from Earth to the Sun is 149,500,000 km. Rewrite this number using scientific notation.

14. Why is it important to include legends on a map?

15. Jenna measured the temperature of solutions before, during, and after an exothermic reaction. Which type of display would show the changes in temperature throughout the reaction most clearly and why?

Reading for Comprehension

Map Likely Fake, Experts Say

Recently, a Chinese map, including North America, Antarctica, and Australia, was unveiled. This map purported to show that a Chinese explorer discovered America in 1418, but has been met with skepticism from cartographers and historians alike. Antiquities collector Liu Gang, who unveiled the map in Beijing, says it proves that Chinese seafarer Zheng discovered America more than 70 years before Christopher Columbus set foot in the New World. But experts have dismissed the map as a fake. They say the map resembles a French seventeenth-century world map with its depiction of California as an island. That China is not shown in the center also suggests the Chinese did not make the map, one expert says.

Article obatined from: Lovgren, S. "Chinese Columbus" map likely fake, experts say. *National Geographic News.* January 23, 2006.

16. Why might a seventeenth-century map show California as an island?
    A. California really was an island back then.
    B. America had not been explored well enough to know that California was actually connected.
    C. California was so different from the rest of America that they assumed it was an island.
    D. A river was mistakenly drawn to look like part of the ocean.

17. What can be inferred from this passage?
    A. China should be put in the center of every map drawn.
    B. The map is an exact copy of the seventeenth-century world map.
    C. Liu Gang wants people to believe that the Chinese first discovered America.
    D. Liu Gang drew the map himself.

NEED EXTRA HELP?															
If You Missed Question . . .	1	2	3	4	5	6	7	8	9	10	11	12	13	14	15
Review Section . . .	2.1	2.1	2.1	2.3	2.1	2.2	2.2	2.2	2.2	2.2	2.2	2.3	1.2	2.2	1.3

UNIT 2 Composition of Earth

Chapter 3
Matter and Change
BIG Idea The variety of substances on Earth results from the way that atoms are arranged and combined.

Chapter 4
Minerals
BIG Idea Minerals are an integral part of daily life.

Chapter 5
Igneous Rocks
BIG Idea Igneous rocks were the first rocks to form as Earth cooled from a molten mass to the crystalline rocks of the early crust.

Chapter 6
Sedimentary and Metamorphic Rocks
BIG Idea Most rocks are formed from preexisting rocks through external and internal geologic processes.

CAREERS IN EARTH SCIENCE

Geologist: This **geologist** is exploring the internal structures of this giant cave. Geologists like this one might collect samples of the rocks and minerals to help describe the origins of the geologic features within the cave.

WRITING in Earth Science
Visit glencoe.com to learn more about geologists. Then write a short magazine article about how a geologist is studying a newly discovered cave.

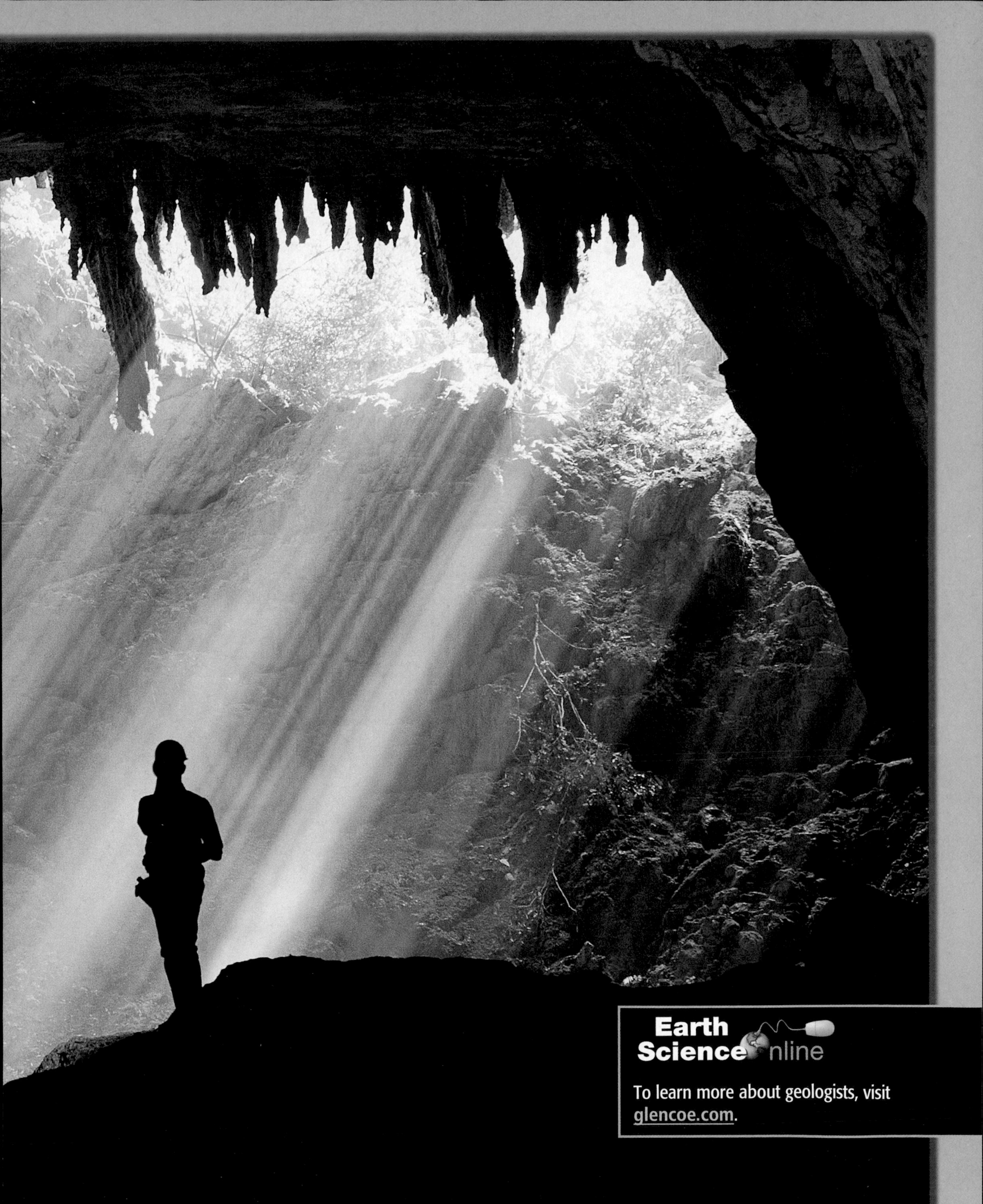

Earth Science Online

To learn more about geologists, visit glencoe.com.

CHAPTER 3

Matter and Change

BIG Idea The variety of substances on Earth results from the way that atoms are arranged and combined.

3.1 Matter
MAIN Idea Atoms are the basic building blocks of all matter.

3.2 Combining Matter
MAIN Idea Atoms combine through electric forces, forming molecules and compounds.

3.3 States of Matter
MAIN Idea All matter on Earth and in the universe occurs in the form of a solid, a liquid, a gas, or plasma.

GeoFacts

- Only atmospheres that contain oxygen and water cause iron-bearing objects to rust. Therefore, the equipment that has been left on the Moon will never rust.
- Ocher, a red pigment used as a coloring agent, is made from the iron-bearing mineral hematite.
- Mars is red because of abundant iron oxide, also known as rust, in the soil.

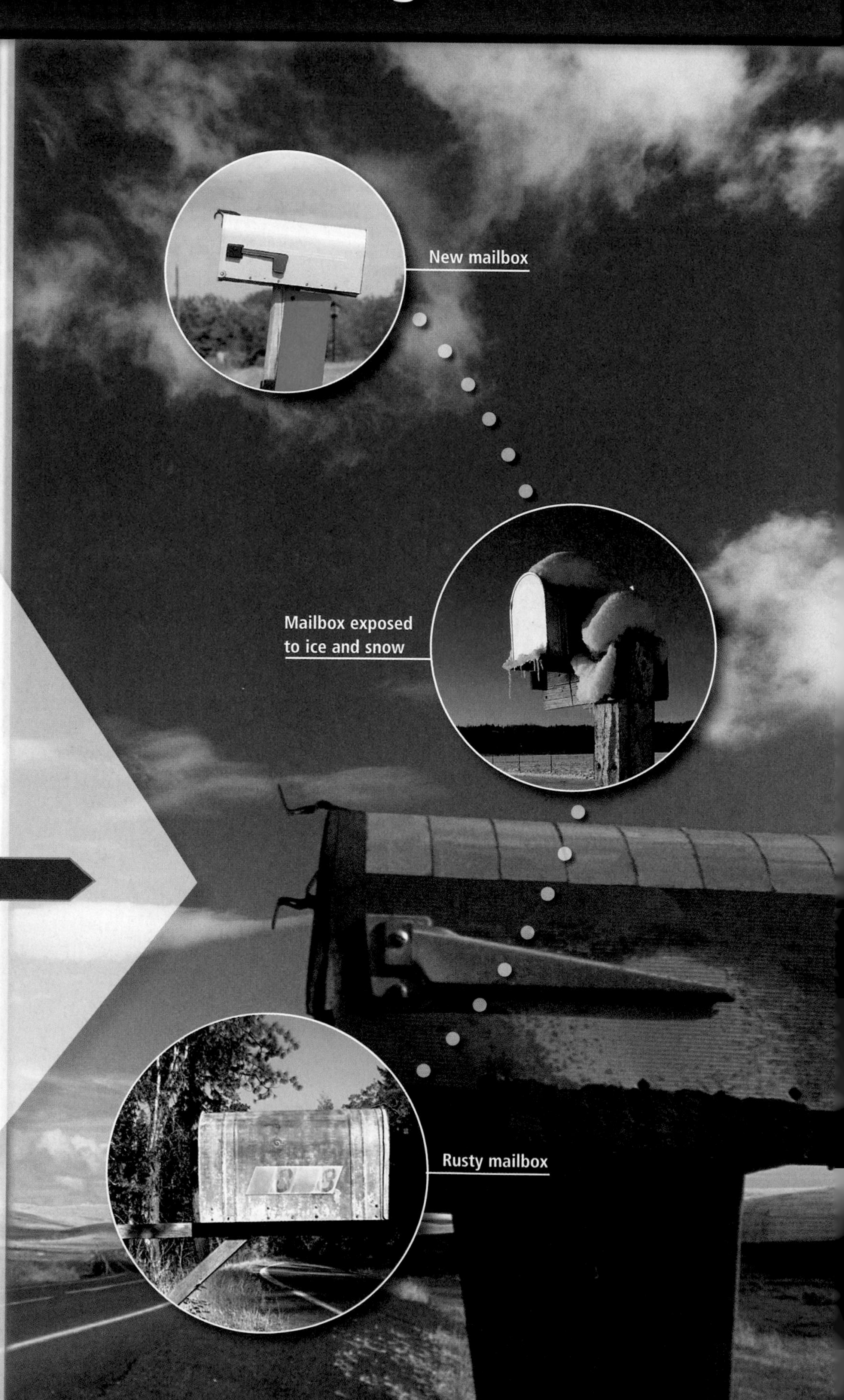

LAUNCH Lab

What do fortified cereals contain?

Everything is made up of matter; different types of matter have different properties. Some metals, such as iron, cobalt, and nickel, are attracted to magnets.

Procedure

1. Read and complete the lab safety form.
2. Tape a **small, strong magnet** to the eraser end of a **pencil.**
3. Pour 250 g of **dry, fortified cereal** into a **small, plastic bag.** Smooth the bag as you close it to release excess air.
4. Using a **rolling pin,** thoroughly crush the cereal in the plastic bag.
5. Pour the crushed cereal into a **250-mL glass beaker.** Add 150 mL of **tap water** to the beaker.
6. Using the pencil-magnet as a stirrer, stir the cereal/water mixture for 10 min, stirring slowly for the last minute.
7. Remove the stirrer from the mixture and examine the magnet end of the stirrer with a **magnifying lens.**

Analysis

1. **Describe** what you see on the magnet.
2. **Determine** Study the cereal box to determine what the substance on the magnet might be.

States of Matter Make the following Foldable to organize information about the four states of matter on Earth.

STEP 1 Fold a sheet of paper in half lengthwise, and then fold it in half twice more.

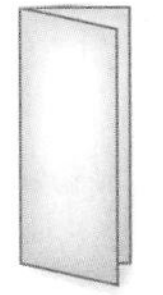

STEP 2 Unfold and cut along the folds of the top flap to make four tabs.

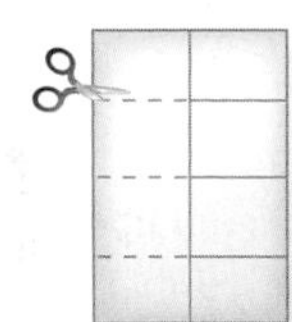

STEP 3 Label the tabs as follows: *Solids, Liquids, Gases,* and *Plasma.*

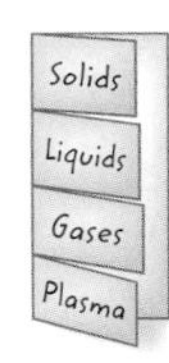

FOLDABLES **Use this Foldable with Section 3.3**
As you read this section, summarize what you learn about the states of matter.

Visit glencoe.com to
- study entire chapters online;
- explore Concepts In Motion animations:
  - Interactive Time Lines
  - Interactive Figures
  - Interactive Tables
- access Web Links for more information, projects, and activities;
- review content with the Interactive Tutor and take Self-Check Quizzes.

Section 3.1

Objectives

- **Describe** an atom and its components.
- **Relate** energy levels of atoms to the chemical properties of elements.
- **Define** the concept of isotopes.

Review Vocabulary

atom: the smallest particle of an element that retains all the properties of that element

New Vocabulary

matter
element
nucleus
proton
neutron
electron
atomic number
mass number
isotope
ion

Matter

MAIN Idea Atoms are the basic building blocks of all matter.

Real-World Reading Link Gold, which is often used in jewelry, is so soft that it can be molded, hammered, sculpted, or drawn into wire. Whatever its size or shape, the gold is still gold. Gold is a type of matter.

Atoms

Matter is anything that has volume and mass. Everything in the physical world that surrounds you is composed of matter. On Earth, matter usually occurs as a solid, a liquid, or a gas. All matter is made of substances called elements. An **element** is a substance that cannot be broken down into simpler substances by physical or chemical means. For example, gold is still gold whether it is a gold brick, coins, or a statue.

Each element has distinct characteristics. You have learned some of the characteristics of the element gold. Although aluminum has different characteristics than gold, both aluminum and gold are elements that are made up of atoms. All atoms consist of even smaller particles—protons, neutrons, and electrons. **Figure 3.1** shows one method of representing an atom. The center of an atom is called the nucleus (NEW klee us) (plural, nuclei). The **nucleus** of an atom is made up of protons and neutrons. A **proton** is a tiny particle that has mass and a positive electric charge. A **neutron** is a particle with approximately the same mass as a proton, but it is electrically neutral; that is, it has no electric charge. All atomic nuclei have a positive charge because they are composed of protons with positive electric charges and neutrons with no electric charges.

Figure 3.1 In this representation of an atom, the fuzzy area surrounding the nucleus is referred to as an electron cloud.

Concepts In Motion

Interactive Figure To see an animation of the electron cloud, visit glencoe.com.

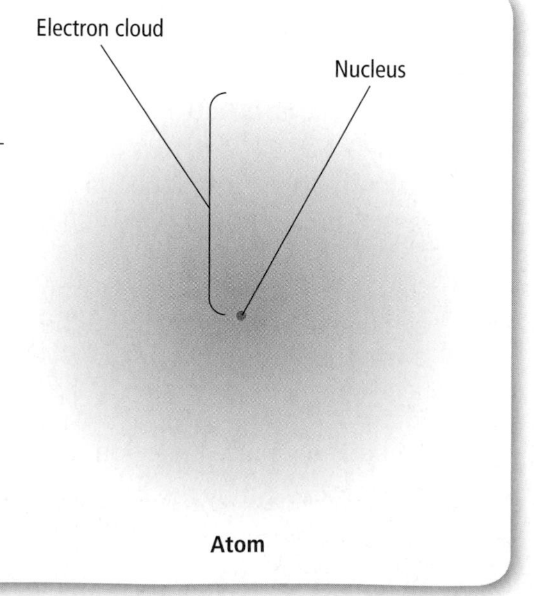

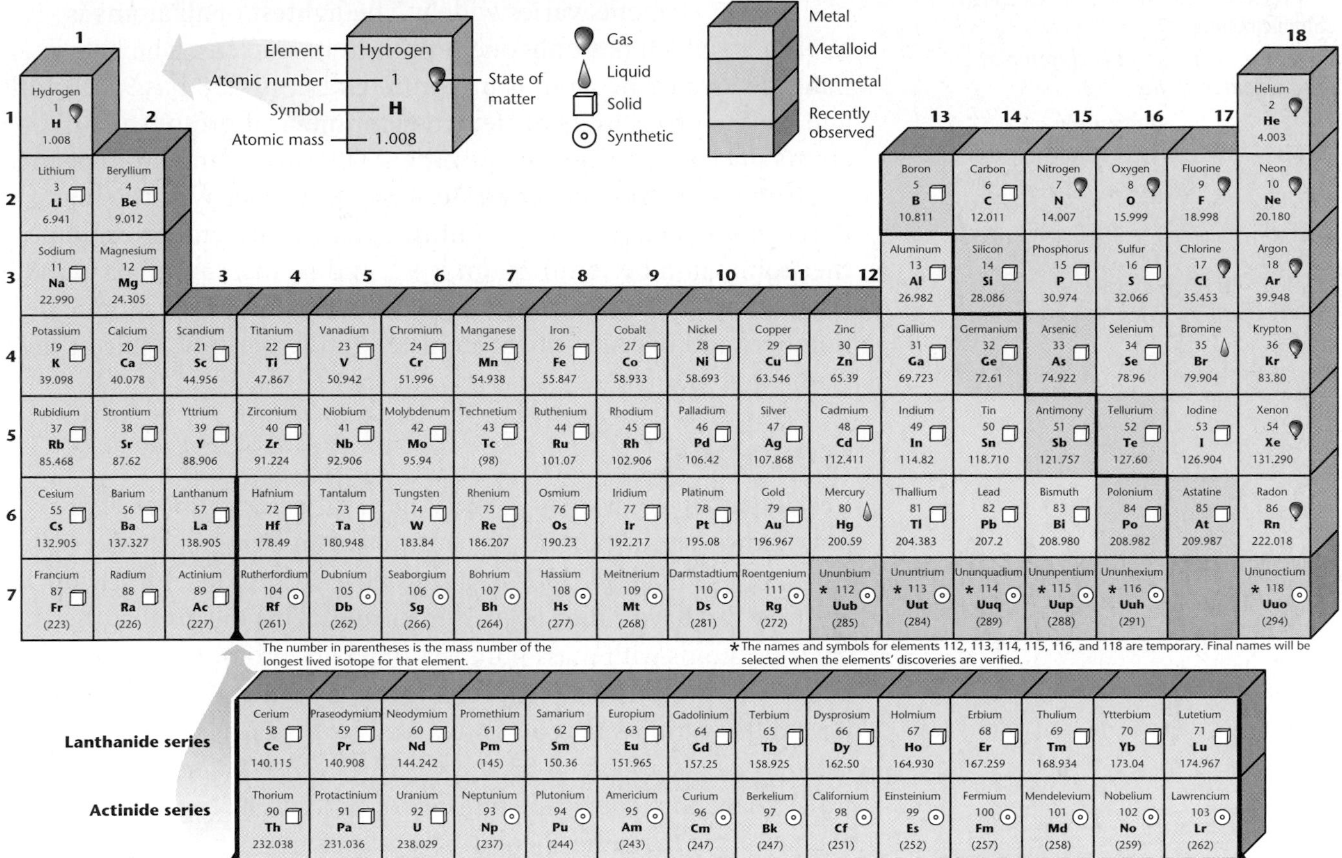

Figure 3.2 The periodic table of the elements is arranged so that a great deal of information about all of the known elements is provided in a small space.

Surrounding the nucleus of an atom are smaller particles called electrons. An **electron** (e^-) has little mass, but it has a negative electric charge that is exactly the same magnitude as the positive charge of a proton. An atom has an equal number of protons and electrons; thus, the electric charge of an electron cancels the positive charge of a proton to produce an atom that has no overall charge. Notice that the electrons in **Figure 3.1** are shown as a cloudlike region surrounding the nucleus. This is because electrons are in constant motion around an atom's nucleus, and their exact positions at any given moment cannot be determined.

Symbols for elements There are 92 elements that occur naturally on Earth and in the stars. Other elements have been produced in laboratory experiments. Generally, each element is identified by a one-, two-, or three-letter abbreviation known as a chemical symbol. For example, the symbol H represents the element hydrogen, C represents carbon, and O represents oxygen. Elements identified in ancient times, such as gold and mercury, have symbols of Latin origin. For example, gold is identified by the symbol Au for its Latin name, *aurum.* All elements are classified and arranged according to their chemical properties in the periodic table of the elements, shown in **Figure 3.2.**

Concepts In Motion

Interactive Figure To see an animation of the periodic table of elements, visit glencoe.com.

■ **Figure 3.3** The element chlorine is atomic number 17.
Infer ***In what state is chlorine at room temperature?***

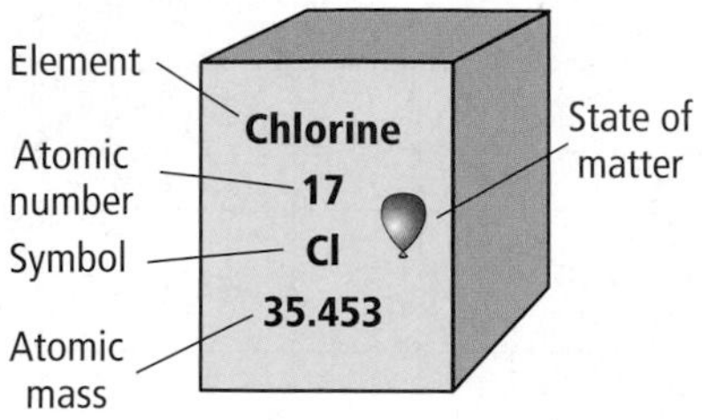

Mass number The number of protons and neutrons in atoms of different elements varies widely. The lightest of all atoms is hydrogen, which has only one proton in its nucleus. The heaviest naturally occurring atom is uranium. Uranium-238 has 92 protons and 146 neutrons in its nucleus. The number of protons in an atom's nucleus is its **atomic number.** The sum of the protons and neutrons is its **mass number.** Because electrons have little mass, they are not included in determining mass number. For example, the atomic number of uranium is 92, and its mass number is 238 (92 protons + 146 neutrons). **Figure 3.3** explains how atomic numbers and mass numbers are listed in the periodic table of the elements.

Isotopes

Recall that all atoms of an element have the same number of protons. However, the number of neutrons of an element's atoms can vary. For example, all chlorine atoms have 17 protons in their nuclei, but they can have either 18 or 20 neutrons. This means that there are chlorine atoms with mass numbers of 35 (17 protons + 18 neutrons) and 37 (17 protons + 20 neutrons). Atoms of the same element that have different mass numbers are called **isotopes.** The element chlorine has two isotopes: Cl-35 and Cl-37. Because the number of electrons in an atom equals the number of protons, isotopes of an element have the same chemical properties.

Look again at the periodic table in **Figure 3.2.** Scientists have measured the mass of atoms of elements. The atomic mass of an element is the average of the mass numbers of the isotopes of an element. Most elements are mixtures of isotopes. For example, notice in **Figure 3.2** that the atomic mass of chlorine is 35.453. This number is the average of the mass numbers of the naturally occurring isotopes of chlorine-35 and chlorine-37.

MiniLab

Identify Elements

What elements are in your classroom? Most substances on Earth occur in the form of chemical compounds. Around your classroom, there are numerous objects or substances that consist mostly of a single element.

Procedure

1. Read and complete the lab safety form.
2. Create a data table with the following column headings: Article, Element, Atomic Number, Properties.
3. Name three objects in your classroom and the three different elements of which they are made.
4. List the atomic numbers of these elements and describe some of their properties.

Analysis

1. **Categorize** List two examples of a solid, a liquid, and a gaseous object or substance.
2. **Compare and contrast** liquids, solids, and gases.

Radioactive isotopes The nuclei of some isotopes are unstable and tend to break down. When this happens, the isotope also emits energy in the form of radiation. Radioactive decay is the spontaneous process through which unstable nuclei emit radiation. In the process of radioactive decay, a nucleus can lose protons and neutrons, change a proton to a neutron, or change a neutron to a proton. Because the number of protons in a nucleus identifies an element, decay changes the identity of an element. For example, the isotope polonium-218 decays at a steady rate over time into bismuth-214. The polonium originally present in a rock is gradually replaced by bismuth. You will learn about the use of radioactive decay to calculate the ages of rocks in Chapter 21.

Electrons in Energy Levels

Although the exact position of an electron cannot be determined, scientists have discovered that electrons occupy areas called energy levels. Look again at **Figure 3.1.** The volume of an atom is mostly empty space. However, the size of an atom depends on the number and arrangement of its electrons.

Filling energy levels **Figure 3.4** presents a model to help you visualize the position of atomic particles. Note that electrons are distributed over one or more energy levels in a predictable pattern. Keep in mind that the electrons are not sitting still in one place. Each energy level can hold only a limited number of electrons. For example, the smallest, innermost energy level can hold only two electrons, as illustrated by the oxygen atom in **Figure 3.4**. The second energy level is larger, and it can hold up to eight electrons. The third energy level can hold up to 18 electrons and the fourth energy level can hold up to 32 electrons. Depending on the element, an atom might have electrons in as many as seven energy levels surrounding its nucleus.

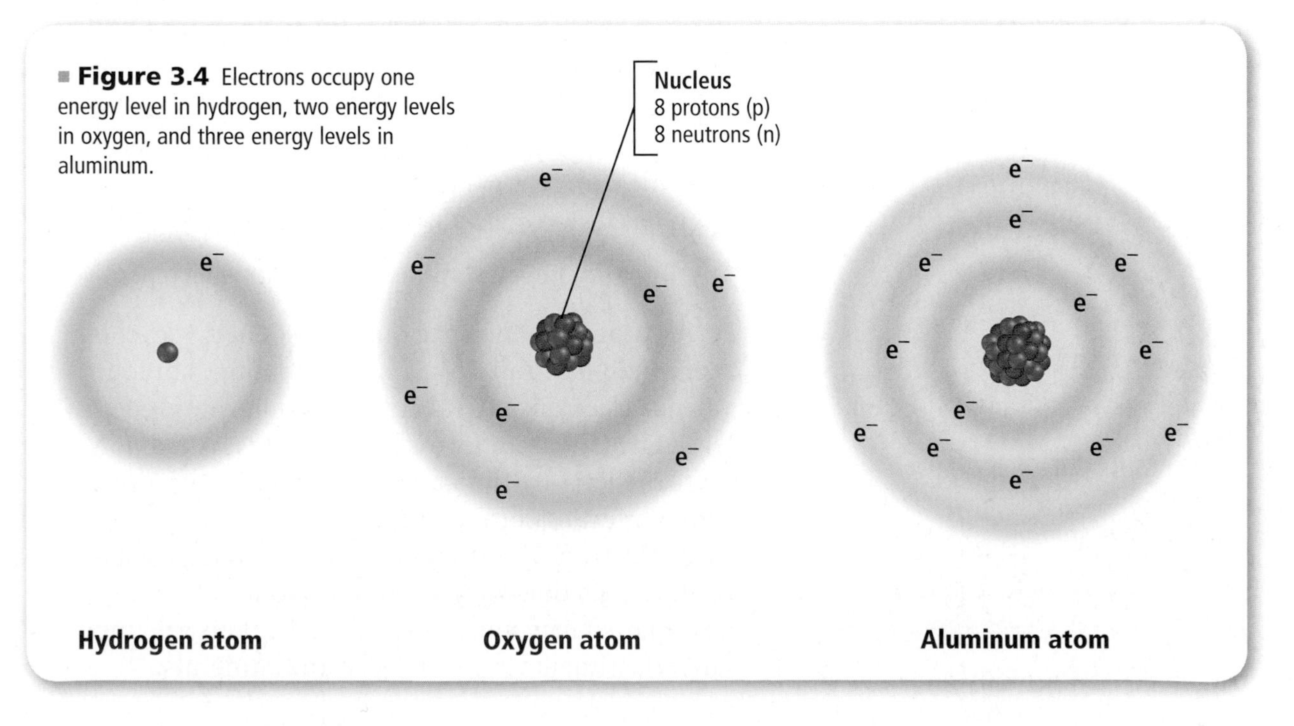

Figure 3.4 Electrons occupy one energy level in hydrogen, two energy levels in oxygen, and three energy levels in aluminum.

■ **Figure 3.5** The inert nature of argon makes it an ideal gas to use inside an incandescent light bulb because it does not react with the extremely hot filament.

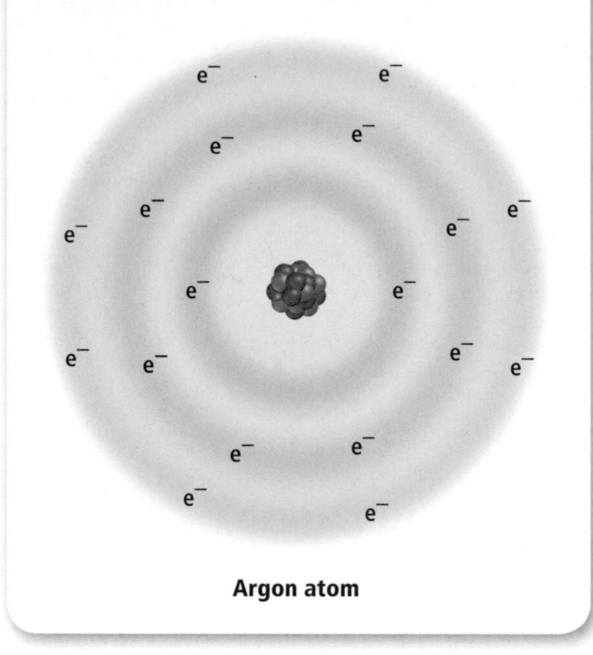

Valence electrons The electrons in the outermost energy level determine the chemical behavior of the different elements. These outermost electrons are called valence electrons. Elements with the same number of valence electrons have similar chemical properties. For example, both a sodium atom, with the atomic number 11, and a potassium atom, with the atomic number 19, have one valence electron. Thus both sodium and potassium exhibit similar chemical behavior. These elements are highly reactive metals, which means that they combine easily with many other elements.

Elements such as helium and argon have full outermost energy levels. For example, an argon atom, shown in **Figure 3.5,** has 18 electrons, with two electrons in the first energy level and eight electrons in the second and third (outermost) energy levels. Elements that have full outermost energy levels are highly unreactive. The gases helium, neon, argon, krypton, xenon, and radon have full outer energy levels.

Ions

Sometimes atoms gain or lose electrons from their outermost energy levels. Recall that atoms are electrically neutral because the number of electrons, which have negative charges, balances the number of protons, which have positive charges. An atom that gains or loses an electron has a net electric charge and is called an **ion.** In general, an atom in which the outermost energy level is less than half-full—that is, it has fewer than four valence electrons—tends to lose its valence electrons. When an atom loses valence electrons, it becomes positively charged. In chemistry, a positive ion is indicated by a superscript plus sign. For example, a sodium ion is represented by Na^{+}. If more than one electron is lost, that number is placed before the plus sign. For example, a magnesium ion, which forms when a magnesium atom has lost two electrons, is represented by Mg^{2+}.

Reading Check **Explain** what makes an ion positive.

An atom in which the outermost energy level is more than half-full—that is, it has more than four valence electrons—tends to fill its outermost energy level. Such an atom forms a negatively charged ion. Negative ions are indicated by a superscript minus sign. For example, a nitrogen atom that has gained three electrons is represented by N^{3-}. Some substances contain ions that are made up of groups of atoms—for example, silicate ions. These complex ions are important constituents of most rocks and minerals.

VOCABULARY

ACADEMIC VOCABULARY

Exhibit

to show or display outwardly

The dog exhibited aggression by baring its teeth and growling.

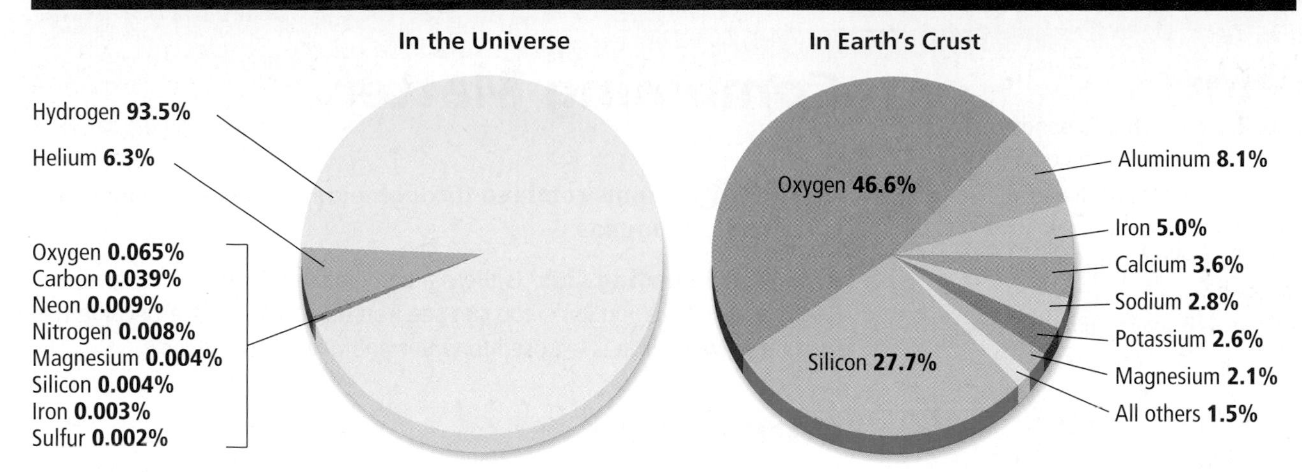

Figure 3.6 The most abundant elements in the universe are greatly different from the most abundant elements on Earth.
Hypothesize ***Where might most of the hydrogen and helium in the universe be found?***

What elements are most abundant?

Astronomers have identified the two most abundant elements in the universe as hydrogen and helium. All other elements account for less than 1 percent of all atoms in the universe, as shown in **Figure 3.6.** Analyses of the composition of rocks and minerals on Earth indicate that the percentages of elements in Earth's crust differ from the percentages in the universe. As shown in **Figure 3.6,** 98.5 percent of Earth's crust is made up of only eight elements. Two of these elements, oxygen and silicon, account for almost 75 percent of the crust's composition. This means that most of the rocks and minerals on Earth's crust contain oxygen and silicon. You will learn more about these elements and the minerals they form in Chapter 4.

Section 3.1 Assessment

Section Summary

- Atoms consist of protons, neutrons, and electrons.
- An element consists of atoms that have a specific number of protons in their nuclei.
- Isotopes of an element differ by the number of neutrons in their nuclei.
- Elements with full outermost energy levels are highly unreactive.
- Ions are electrically charged atoms or groups of atoms.

Understand Main Ideas

1. MAIN Idea **Differentiate** among the three parts of an atom in terms of their location, charge, and mass.
2. **Explain** why the elements magnesium and calcium have similar properties.
3. **Illustrate** how a neutral atom becomes an ion.
4. **Compare and contrast** these isotopes: uranium-239, uranium-238, and uranium-235.

Think Critically

5. **Illustrate** a model of a calcium atom, including the number and position of protons, neutrons, and electrons in the atom.
6. **Interpret** the representation of magnesium in the periodic table. Explain why the atomic mass of magnesium is not a whole number.

MATH in Earth Science

7. As the radioactive isotope radium-226 decays, it emits two protons and two neutrons. How many protons and neutrons are now left in the nucleus? What is the atom's new atomic number? What is the name of this element?

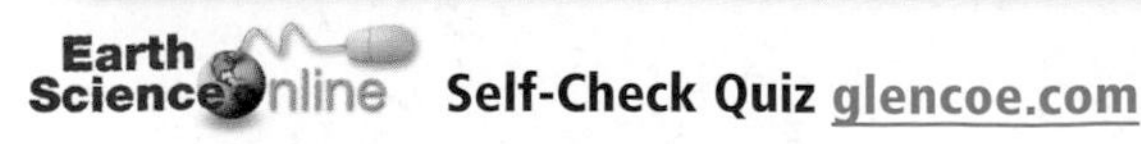

Self-Check Quiz glencoe.com

Section 3.2

Objectives

- **Describe** the chemical bonds that unite atoms to form compounds.
- **Relate** the nature of chemical bonds that hold compounds together to the physical structures of compounds.
- **Distinguish** among different types of mixtures and solutions.

Review Vocabulary

ion: an electrically charged atom

New Vocabulary

compound
chemical bond
covalent bond
molecule
ionic bond
metallic bond
chemical reaction
solution
acid
base

Combining Matter

MAIN Idea Atoms combine through electric forces, forming molecules and compounds.

Real-World Reading Link Is there a rusty mailbox or bicycle on your street? Nearly everywhere you look, you can see iron objects that have become rusty. Rust forms when iron is exposed to water and oxygen in the air.

Compounds

Can you identify the materials in **Figure 3.7?** The greenish gas in the flask is the element chlorine, which is poisonous. The solid, silvery metal is the element sodium, which is highly reactive. These two elements combine chemically to form the third material in the photograph—table salt. How can two dangerous elements combine to form a material that you sprinkle on your popcorn?

Table salt is a compound, not an element. A **compound** is a substance that is composed of atoms of two or more different elements that are chemically combined. Water is another example of a compound because it is composed of two elements—hydrogen and oxygen. Most compounds have different properties from the elements of which they are composed. For example, both oxygen and hydrogen are highly flammable gases at room temperature, but in combination they form water—a liquid.

Chemical formulas Compounds are represented by chemical formulas. These formulas include the symbol for each element followed by a subscript number that stands for the number of atoms of that element in the compound. If there is only one atom of an element, no subscript number follows the symbol. Thus, the chemical formula for table salt is $NaCl$. The chemical formula for water is H_2O.

Figure 3.7 Sodium is a silvery metal that is soft enough to cut with a knife. Chlorine is a green, poisonous gas. When they react, they produce sodium chloride, a white solid.

Covalent Bonds

Recall that an atom is chemically stable when its outermost energy level is full. A state of stability is achieved by some elements by forming chemical bonds. A **chemical bond** is the force that holds together the elements in a compound. One way in which atoms fill their outermost energy levels is by sharing electrons. For example, individual atoms of hydrogen each have just one electron. Each atom becomes more stable when it shares its electron with another hydrogen atom so that each atom has two electrons in its outermost energy level. **Figure 3.8** shows an example of this bond. How do these two atoms stay together? The nucleus of each atom has one proton with a positive charge, and the two positively charged protons attract the two negatively charged electrons. This attraction of two atoms for a shared pair of electrons that holds the atoms together is called a **covalent bond.**

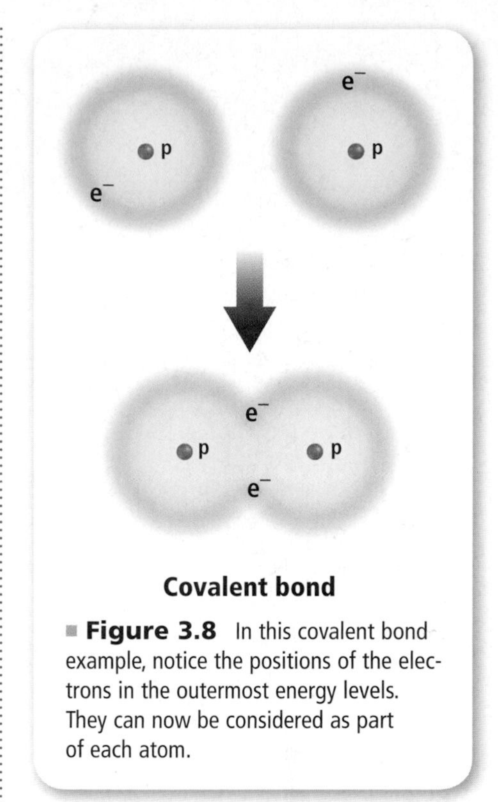

Figure 3.8 In this covalent bond example, notice the positions of the electrons in the outermost energy levels. They can now be considered as part of each atom.

Molecules A **molecule** is composed of two or more atoms held together by covalent bonds. Molecules have no overall electric charge because the total number of electrons equals the total number of protons. Water is an example of a compound whose atoms are held together by covalent bonds, as illustrated in **Figure 3.9.** The chemical formula for a water molecule is H_2O because, in this molecule, two atoms of hydrogen, each of which need to gain an electron to become stable, are combined with one atom of oxygen, which needs to gain two electrons to become stable. A compound comprised of molecules is called a molecular compound.

Polar molecules Although water molecules are held together by covalent bonds, the atoms do not share the electrons equally. As shown in **Figure 3.9,** the shared electrons in a water molecule are attracted more strongly by the oxygen atom than by the hydrogen atoms. As a result, the electrons spend more time near the oxygen atom than they do near the hydrogen atoms. This unequal sharing of electrons results in polar molecules. A polar molecule has a slightly positive end and a slightly negative end.

VOCABULARY
SCIENCE USAGE V. COMMON USAGE
Polar
Science usage: the unequal sharing of electrons

Common usage: locations of or near the north or south pole, or the ends of a magnet

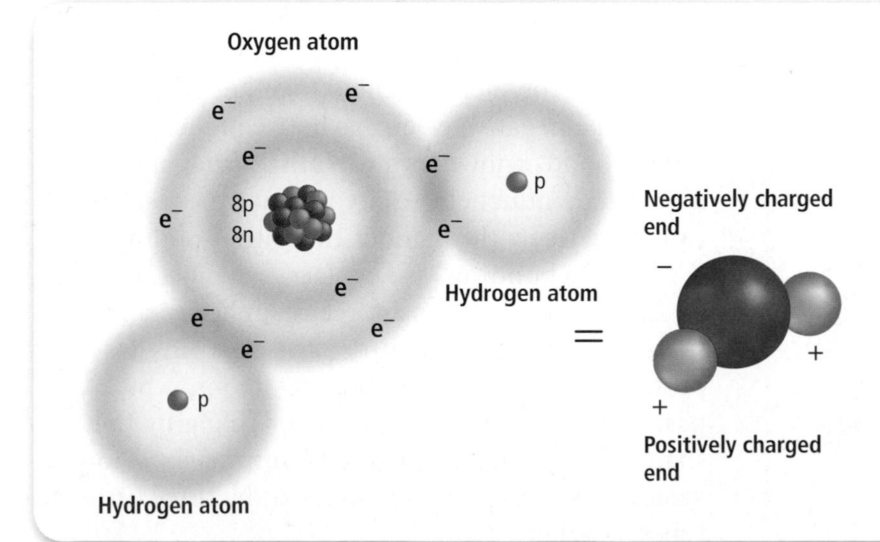

Figure 3.9 Polar molecules are similar to bar magnets. At one end of a water molecule, the hydrogen atoms have a positive charge, while at the opposite end, the oxygen atom has a negative charge.

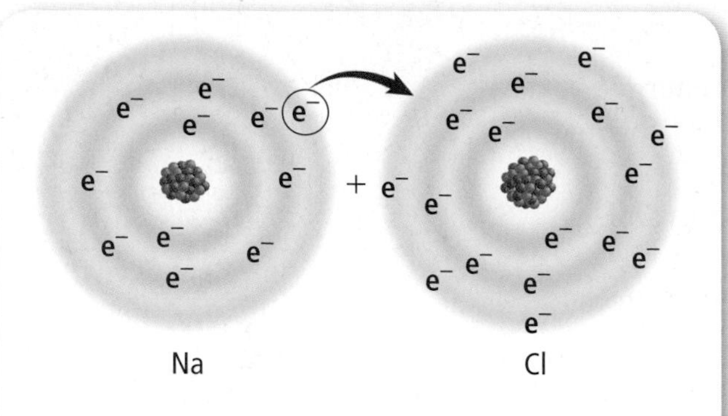

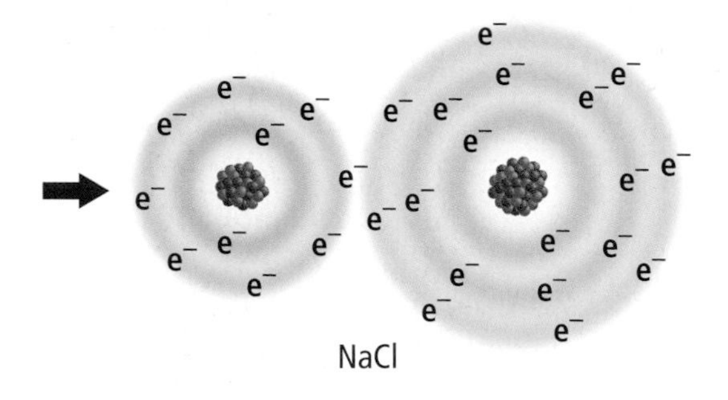

Figure 3.10 The single valence electron in a sodium atom is used to form an ionic bond with a chlorine atom. Once an ionic bond is formed, the negatively charged ion is slightly larger than the positively charged ion.

Concepts In Motion

Interactive Figure To see an animation of ionic bonds, visit glencoe.com.

Figure 3.11 Metallic bonds are formed when valence electrons are shared equally among all the positively charged atoms. Because the electrons flow freely among the positively charged ions, you can visualize electricity flowing through electrical wires.

Concepts In Motion

Interactive Figure To see an animation of electron flow, visit glencoe.com.

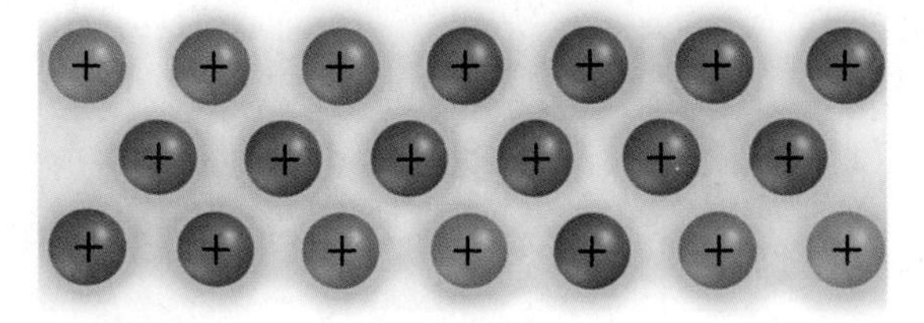

Metallic bond

Ionic Bonds

As you might expect, positive and negative ions attract each other. An **ionic bond** is the attractive force between two ions of opposite charge. **Figure 3.10** illustrates an ionic bond between a positive ion of sodium and a negative ion of chlorine called chloride. The chemical formula for common table salt is NaCl, which consists of equal numbers of sodium ions (Na^+) and chloride ions (Cl^-). Note that positive ions are always written first in chemical formulas.

Within the compound NaCl, there are as many positive ions as negative ions; therefore, the positive charge on the sodium ion equals the negative charge on the chloride ion, and the net electric charge of the compound NaCl is zero. Magnesium and oxygen ions combine in a similar manner to form the compound magnesium oxide (MgO)—one of the most common compounds on Earth. Compounds formed by ionic bonding are called ionic compounds. Other ionic compounds have different proportions of ions. For example, oxygen and sodium ions combine in the ratio shown by the chemical formula for sodium oxide (Na_2O), in which there are two sodium ions to each oxygen ion.

Reading Check **Describe** how ionic bonds form.

Metallic Bonding

Most compounds on Earth are held together by ionic or covalent bonds, or by a combination of these bonds. Another type of bond is shown in **Figure 3.11.** In metals, the valence electrons are shared by all the atoms, not just by adjacent atoms as they are in covalent compounds. You could think of a metal as a group of positive ions surrounded by a sea of freely moving negative electrons. The positive ions of the metal are held together by the attraction to the negative electrons between them. This type of bond, known as a **metallic bond,** allows metals to conduct electricity because the electrons can move freely throughout the entire solid metal.

Metallic bonding also explains why metals are so easily deformed. When a force is applied to a metal, such as the blow of a hammer, the electrons are pushed aside. This allows the metal ions to move past each other, thus deforming or changing the shape of the metal. **Figure 3.12** summarizes how valence electrons are used to form the three different types of bonds.

Visualizing Bonds

Figure 3.12 Atoms gain stability by sharing, gaining, or losing electrons to form ions and molecules. The properties of metals can be explained by metallic bonds.

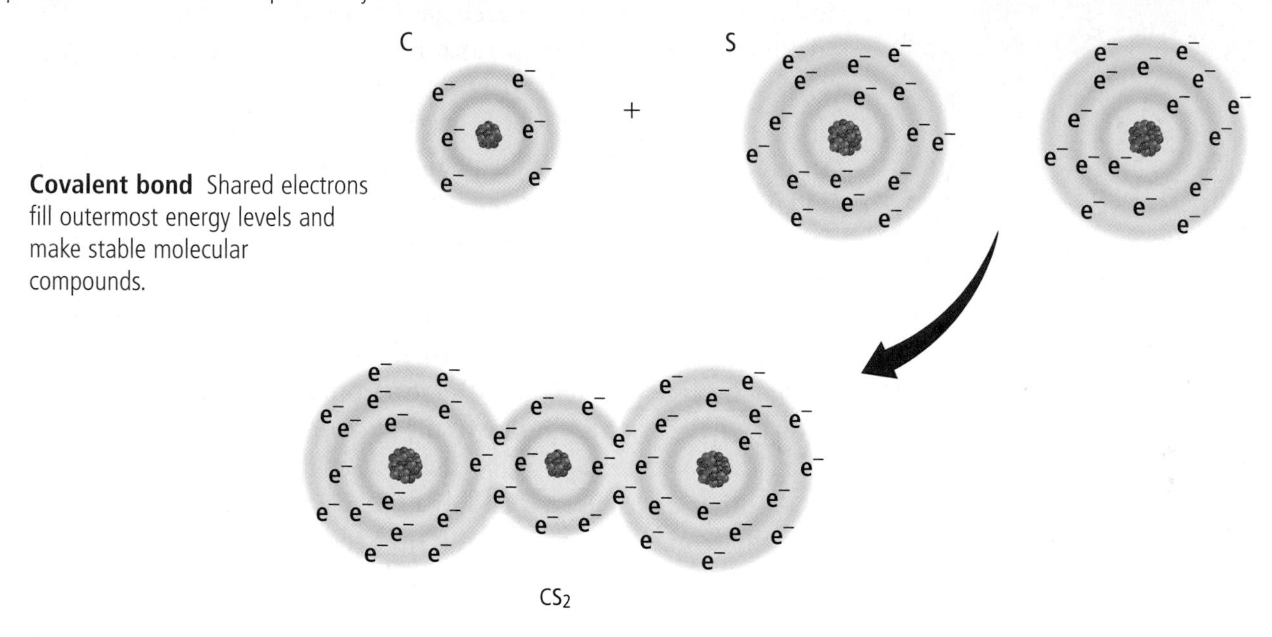

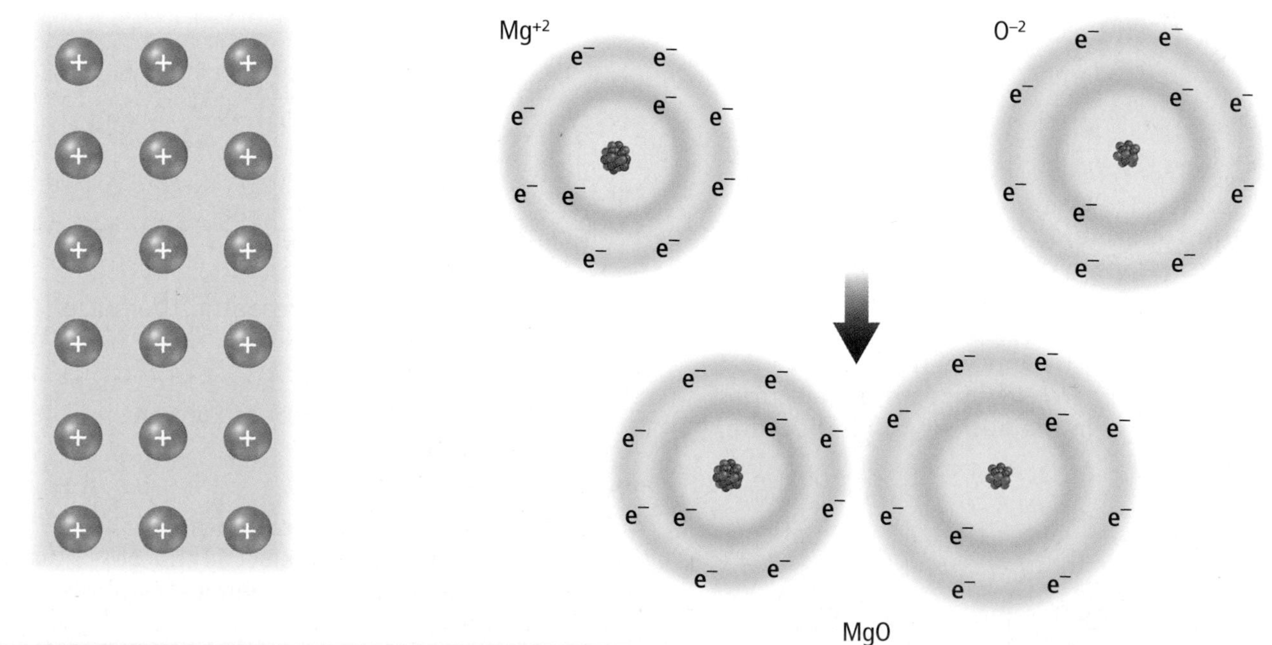

Concepts In Motion To explore more about chemical bonding, visit glencoe.com.

Earth Science Online

■ **Figure 3.13** When a copper wire is placed in the solution of silver nitrate in the beaker, a chemical reaction occurs in which silver replaces copper in the wire and an aqua-colored copper nitrate solution forms.

Chemical Reactions

You have learned that atoms gain, lose, or share electrons to become more stable and that these atoms form compounds. Sometimes, compounds break down into simpler substances. The change of one or more substances into other substances, such as those in **Figure 3.13,** is called a **chemical reaction.** Chemical reactions are described by chemical equations. For example, water (H_2O) is formed by the chemical reaction between hydrogen gas (H_2) and oxygen gas (O_2). The formation of water can be described by the following chemical equation.

$$2H_2 + O_2 \rightarrow 2H_2O$$

You can read this chemical equation as "two molecules of hydrogen and one molecule of oxygen react to yield two molecules of water." In this reaction, hydrogen and oxygen are the reactants and water is the product. When you write a chemical equation, you must balance the equation by showing an equal number of atoms for each element on each side of the equation. Therefore, the same amount of matter is present both before and after the reaction. Note that there are four hydrogen atoms on each side of the above equation ($2 \times 2 = 4$). There are also two oxygen atoms on each side of the equation.

Another example of a chemical reaction, one that takes place between iron (Fe) and oxygen (O), is represented by the following chemical equation.

$$4Fe + 3O_2 \rightarrow 2Fe_2O_3$$

You will examine how compounds form in the Problem-Solving Lab on this page.

PROBLEM-SOLVING LAB

Interpret Scientific Illustrations

How do compounds form? Many atoms gain or lose electrons in order to have eight electrons in the outermost energy level. In the diagram, energy levels are indicated by the circles around the nucleus of each element. The colored spheres in the energy levels represent electrons, and the spheres in the nucleus represent protons and neutrons.

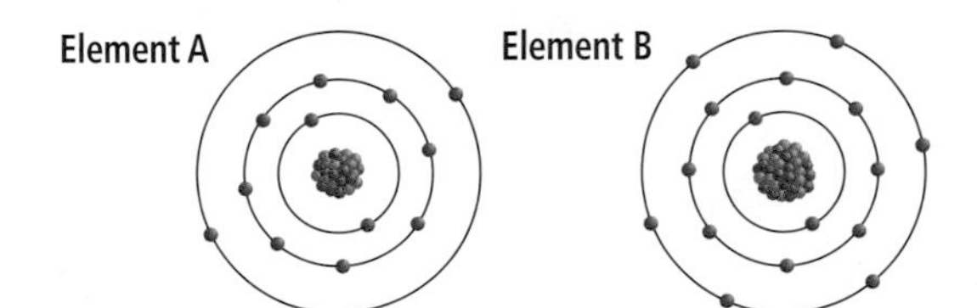

Analysis

1. How many electrons are present in atoms of Element A? Element B?
2. How many protons are present in the nuclei of these atoms?
3. Use the periodic table on page 61 to determine the name and symbol of Element A and Element B.

Think Critically

4. **Decide** if these elements can form ions. If so, what would be the electric charges (magnitude and sign) and chemical symbols of these ions?
5. **Formulate** a compound from these two elements. What is the chemical formula of the compound?

Mixtures and Solutions

Unlike a compound, in which the atoms combine and lose their identities, a mixture is a combination of two or more components that retain their identities. When a mixture's components are easily recognizable, it is called a heterogeneous mixture. For example, beach sand, shown in **Figure 3.14,** is a heterogeneous mixture because its components are still recognizable—shells, small pieces of broken shells, grains of minerals, and so on. In a homogeneous mixture, which is also called a **solution,** the component particles cannot be distinguished, even though they still retain their original properties.

A solution can be liquid, gaseous, or solid. Seawater is a solution consisting of water molecules and ions of many elements that exist on Earth. Molten rock is also a liquid solution; it is composed of ions representing all atoms that were present in the crystals of the rock before it melted. Air is a solution of gases, mostly nitrogen and oxygen molecules together with other atoms and molecules. Metal alloys, such as bronze and brass, are also solutions. Bronze is a homogeneous mixture of copper and tin atoms; brass is a similar mixture of copper and zinc atoms. Such solid homogeneous mixtures are called solid solutions. You will learn more about solid solutions in Chapters 4 and 5.

Reading Check **Describe** three examples of solutions.

Figure 3.14 Not all mixtures of beach sand and shells are alike. Mixtures from the Atlantic Ocean will contain components that are different from mixtures that form in the Pacific Ocean.

Acids Many chemical reactions that occur on Earth involve solutions called acids and bases. An **acid** is a solution containing a substance that produces hydrogen ions (H^+) in water. Recall that a hydrogen atom consists of one proton and one electron. When a hydrogen atom loses its electron, it becomes a hydrogen ion (H^+). The pH scale, shown in **Figure 3.15,** is based on the amount of hydrogen ions in a solution. This amount is referred to as the concentration. A value of 7 is considered neutral. A solution with a pH reading below 7 is considered to be acidic. The lower the number, the more acidic the solution.

Concepts In Motion

Interactive Figure To see an animation of the pH scale, visit glencoe.com.

Figure 3.15 The pH scale is not only reserved for science class. All substances have a pH value, as you can see by the common household substances shown here.

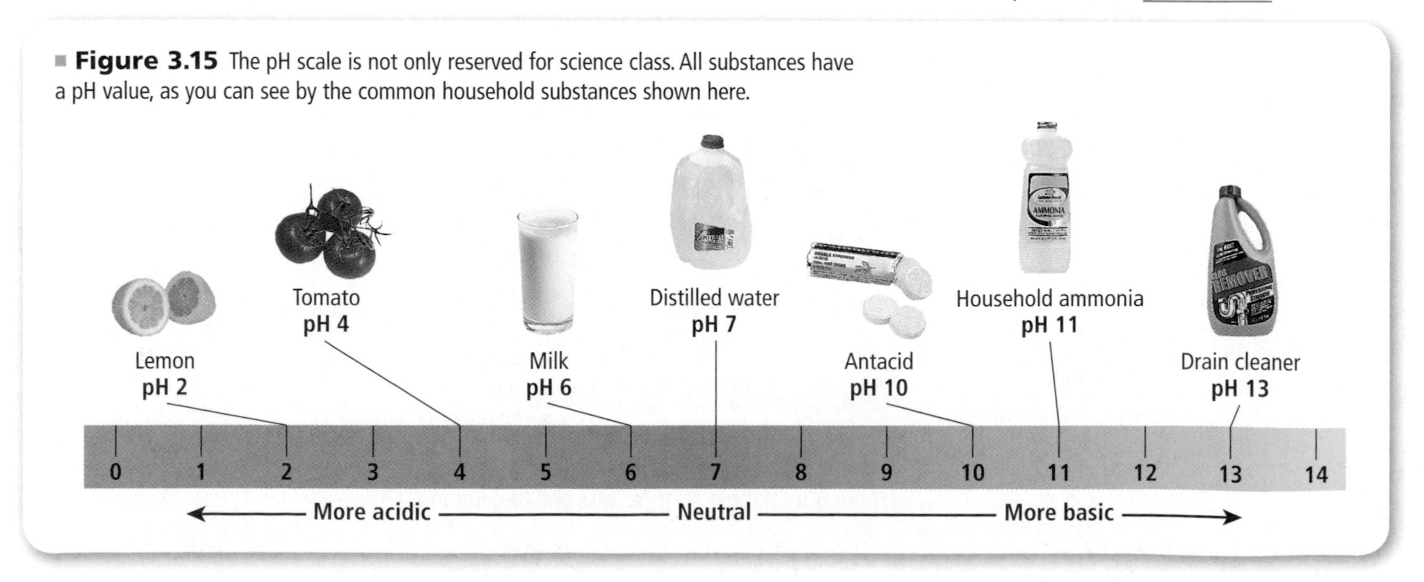

CAREERS IN EARTH SCIENCE

Geochemistry Some geochemists study the interaction of rocks, minerals and the environment. They can help mining companies reduce the amount of contamination from waste piles by understanding how the rocks and minerals break down and how toxic the byproducts might be. For more information on Earth science careers, visit glencoe.com.

The most common acid in Earth's environment is carbonic acid (H_2CO_3), which is produced when carbon dioxide (CO_2) is dissolved in water (H_2O) by the following reaction.

$$H_2O + CO_2 \rightarrow H_2CO_3$$

Some of the carbonic acid (H_2CO_3) in the water ionizes, or breaks apart, into hydrogen ions (H^+) and bicarbonate ions (HCO_3^-), as represented by the following equation.

$$H_2CO_3 \rightarrow H^+ + HCO_3^-$$

These two equations play a major role in the dissolution and precipitation of limestone and the formation of caves, discussed in Chapter 10. Many of the reaction rates involved in geological processes are very slow. For example, it might take thousands of years for enough carbonic acid in limestone to dissolve in groundwater and produce a cave.

Bases A **base** is a substance that produces hydroxide ions (OH^-) in water. A base can neutralize an acid because hydrogen ions (H^+) from the acid react with the hydroxide ions (OH^-) from the base to form water through the following reaction.

$$H^+ + OH^- \rightarrow H_2O$$

Refer again to **Figure 3.15.** A solution with a reading above 7 is considered to be basic. The higher the number, the more basic the solution. Distilled water usually has a pH of 7, but rainwater is slightly acidic, with a pH of 5.0 to 5.6. The pH values of some common substances are shown in **Figure 3.15.**

Section 3.2 Assessment

Section Summary

- Atoms of different elements combine to form compounds.
- Covalent bonds form from shared electrons between atoms.
- Ionic compounds form from the attraction of positive and negative ions.
- There are two types of mixtures—heterogeneous and homogeneous.
- Acids are solutions containing hydrogen ions. Bases are solutions containing hydroxide ions.

Understand Main Ideas

1. MAIN Idea **Explain** why molecules do not have electric charges.
2. **Differentiate** between molecules and compounds.
3. **Calculate** the number of atoms needed to balance the following equation: $CaCO + HCl \rightarrow CO_2 + H_2O + CaCl$
4. **Diagram** how an acid can be neutralized.
5. **Compare and contrast** mixtures and solutions by using specific examples of each.

Think Critically

6. **Design** a procedure to demonstrate whether whole milk, which consists of microscopic fat globules suspended in a solution of nutrients, is a homogeneous or heterogeneous mixture.
7. **Predict** what kind of chemical bond forms between nitrogen and hydrogen atoms in ammonia (NH_3). Sketch this molecule.

WRITING in Earth Science

8. Antacids are used to relieve indigestion and upset stomachs. Write an advertisement for a new antacid product. Explain how the product works in terms that people who are not taking a science class will understand.

Earth Science Online Self-Check Quiz glencoe.com

Section 3.3

Objectives

- **Describe** the states of matter on Earth.
- **Explain** the reasons that matter exists in these states.
- **Relate** the role of thermal energy to changes in state of matter.

Review Vocabulary

chemical reaction: the change of one or more substances into another substance

New Vocabulary

crystalline structure
glass
evaporation
plasma
condensation
sublimation

States of Matter

MAIN Idea All matter on Earth and in the universe occurs in the form of a solid, a liquid, a gas, or plasma.

Real-World Reading Link When your skin is wet, even on a hot day, it usually feels cool—especially if it is windy. How can warm air feel cold? When the water evaporates, it absorbs heat from your skin. The harder the wind blows, the more water evaporates and the colder your skin becomes.

Solids

Solids are substances with densely packed particles, which can be ions, atoms, or molecules. Most solids are **crystalline structures** because the particles of a solid are arranged in regular geometric patterns. Examples of crystals are shown in **Figure 3.16.** Because of their crystalline structures, solids have both a definite shape and volume.

Perfectly formed crystals are rare. When many crystals form in the same space at the same time, crowding prevents the formation of perfect crystals with smooth boundaries. The result is a mass of intergrown crystals called a polycrystalline solid. Most solid substances on Earth, including rocks, are polycrystalline solids. **Figure 3.16** shows the polycrystalline nature of the rock granite.

Some solid materials have no regular internal patterns. **Glass** is a solid that consists of densely packed atoms arranged randomly. Glasses form when molten material is chilled so rapidly that atoms do not have enough time to arrange themselves in a regular pattern. These solids do not form crystals, or their crystals are so small that they cannot be seen. Window glass consists mostly of disordered silicon and oxygen (SiO_2).

Figure 3.16 This granite is composed of mineral crystals that fit together like interlocking puzzle pieces. The minerals that make up the rock are composed of individual atoms and molecules that are aligned in a crystalline structure.

■ **Figure 3.17** Each of these containers has the same volume of liquid in it.
Explain *why the liquids are not all at the same level in the containers.*

Liquids

At any temperature above absolute zero (−273°C), the atoms in a solid vibrate. Because these vibrations increase with increasing temperature, they are called thermal vibrations. At the melting point of the material, these vibrations become vigorous enough to break the forces holding the solid together. The particles can then slide past each other, and the substance becomes liquid. Liquids take the shape of the container they are placed in, as you can see in **Figure 3.17.** However, liquids do have definite volume.

 **Reading Check** **Explain** the effect that increasing temperature has on the atoms in solids.

Gases

The particles in liquids vibrate vigorously. As a result, some particles can gain sufficient energy to escape the liquid. This process of change from a liquid to a gas at temperatures below the boiling point is called **evaporation**. When any liquid reaches its boiling point, it vaporizes quickly as a gas.

In gases, the particles are separated by relatively large distances and they travel at high speeds in one direction until they bump into another gas particle or the walls of a container. Gases, like liquids, have no definite shape. Gases also have no definite volume unless they are restrained by a container or a force such as gravity. For example, Earth's gravity keeps gases in the atmosphere from escaping into space.

FOLDABLES
Incorporate information from this section into your Foldable.

Plasma

When matter is heated to a temperature greater than 5000°C, the collisions between particles are so violent that electrons are knocked away from atoms. Such extremely high temperatures exist in stars and, as a result, the gases of stars consist entirely of positive ions and free electrons. These hot, highly ionized, electrically conducting gases are called **plasmas. Figure 3.18** shows the plasma that forms the Sun's corona. You have seen matter in the plasma state if you have ever seen lightning or a neon sign. Both lightning and the matter inside a neon tube are in the plasma state.

■ **Figure 3.18** The Sun's temperature is often expressed in kelvins; 273 K is equal to 0°C. The Sun's corona, which is a plasma, has a temperature of about 15,000,000 K.
Compare *the temperature of the corona to lightning, which is 30,000 K.*

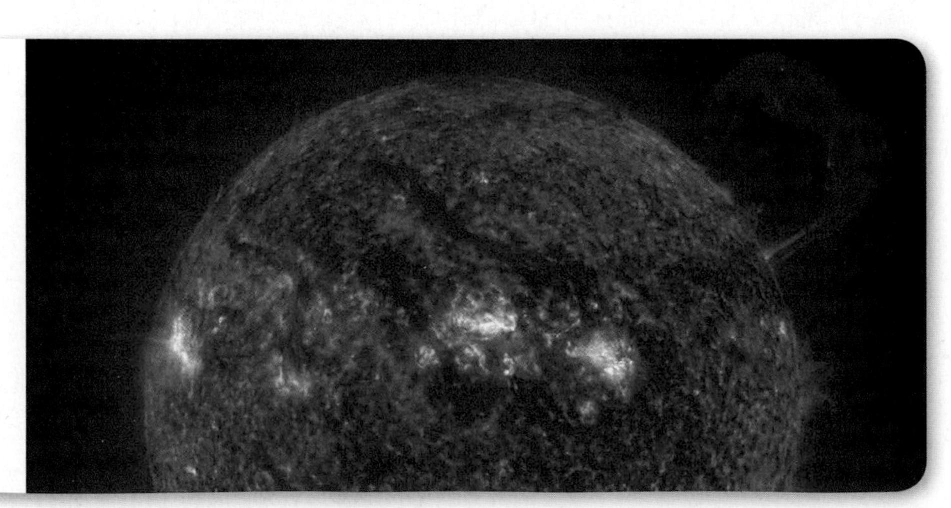

Changes of State

Solids melt when they absorb enough thermal energy to cause their orderly internal crystalline arrangement to break down. This happens at the melting point. When liquids are cooled, they solidify at that same temperature and release thermal energy. The temperature at which liquids solidify is called the freezing point.

When a liquid is heated to the boiling point and absorbs enough thermal energy, vaporization occurs and it becomes a gas. When a gas is cooled to the boiling point it becomes a liquid in a process called **condensation,** shown in **Figure 3.19.** Energy that was absorbed during vaporization is released upon condensation.

Evaporation can occur below the boiling point when thermal vibrations enable individual atoms or molecules to escape from a solid. You might have noticed that even on winter days with temperatures below freezing, snow gradually disappears. This slow change of state from a solid (ice crystals) to a gas (water vapor) without an intermediate liquid state is called **sublimation.**

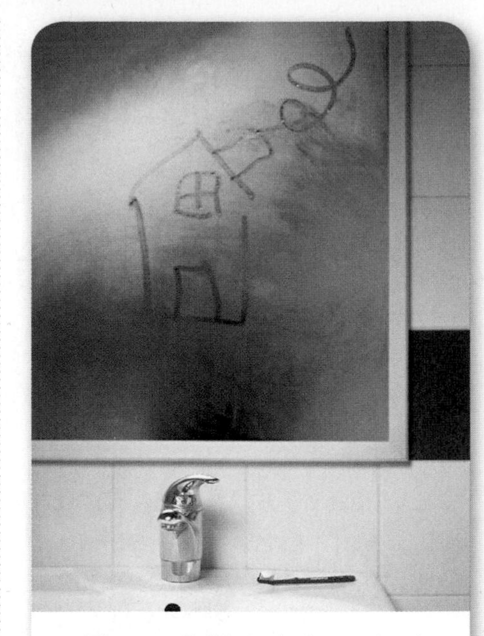

Figure 3.19 As the hot, moist air from the shower encounters the cool glass of the mirror, the water vapor in the air condenses on the glass.
Predict ***What would happen if the glass were the same temperature as the air?***

Conservation of Energy

The identity of matter can be changed through chemical reactions and nuclear processes, and its state can be changed under different thermal conditions. You have learned that a chemical equation must be balanced because matter cannot be created or destroyed. This fundamental fact is called the law of conservation of matter. Like matter, energy cannot be created or destroyed, but it can be changed from one form to another. For example, electric energy might be converted into light energy. This law, called the conservation of energy, is also known as the first law of thermodynamics.

Section 3.3 Assessment

Section Summary

- Changes of state involve thermal energy.
- The law of conservation of matter states that matter cannot be created or destroyed.
- The law of conservation of energy states that energy is neither created nor destroyed.

Understand Main Ideas

1. MAIN Idea **Explain** how thermal energy is involved in changes of state.
2. **Evaluate** the nature of the thermal vibrations in each of the four states of matter.
3. **Apply** what you know about thermal energy to compare evaporation and condensation.

Think Critically

4. **Infer** how the boiling point of water (100°C) would change if water molecules were not polar molecules.
5. **Consider** glass and diamond—two clear, colorless solids. Why does glass shatter more easily than diamond?

MATH in Earth Science

6. Refer to **Figure 3.18.** Calculate the corona's temperature in degrees. Remember that 273 K is equal to 0°C.

EARTH SCIENCE AND TECHNOLOGY

LIQUID CRYSTAL DISPLAYS

You wake up in the morning, glance at your alarm clock, and get ready for school. You microwave your breakfast, grab your music player and dash out the door, checking your wristwatch as you go. Once at school, you pull out your calculator and get ready for the big math exam. Did you know you have used liquid crystal display (LCD) technology five times already? LCD is common display technology, used often because it is thin, lightweight, and energy efficient.

Digital watch displays are made possible through LCD technology. The inset photograph shows a polarized light micrograph of a LCD.

What is a liquid crystal? You know that liquids and crystals are two states of matter; but how is it possible to be both a liquid and a crystal? Recall that particles in a liquid can slide past each other in a container, while particles in a solid are packed together and cannot move separately. Liquid crystals are long molecules that keep their orientation—if they were oriented side-to-side in a thin layer on a glass plate, they would keep that side-to-side orientation. Because of their liquid property, the crystals can move around almost like a school of fish. Therefore, they share characteristics with both solids and liquids. This unique property makes them useful for a variety of electronic applications.

How do LCDs work? Consider a digital watch, for example. If you look closely at it, you can see the numbers, even when they are not darkened. These are the tracks that are engraved in the middle layer of a display "sandwich." Two plates of glass make up the outer portion of this sandwich. The inner portion of the sandwich, the tracks, contains liquid crystals that are in their natural, "relaxed" state. In the relaxed state, light passes through the plates of glass, and is reflected out.

If an electric current is applied across a track of liquid crystals, the crystals lose their original orientation. As long as a small current passes through them, light entering the plates of glass will not be reflected. In other words, that track will appear black.

Seems simple enough, right? That is why LCD displays are becoming more and more popular. They can be all black or color. There are, however, some flaws with LCD technology that need to be corrected. For example, it has a narrow viewing angle; if you tilt your watch slightly you can no longer see the numbers as clearly, if at all. With further research, however, LCD might just become the vision of the future.

WRITING in Earth Science

Diagram Visit glencoe.com to research the different layers of an LCD. Create a drawing showing all the different layers and how they fit together.

GEOLAB

PRECIPITATE SALTS

Background: Many rocks on Earth form from salts precipitated from seawater. Salts precipitate when a salt solution becomes saturated. Solubility is the ability of a substance to dissolve in a solution. When a solution is saturated, no more of that substance can be dissolved.

Question: *Under what conditions do salt solutions become saturated, and under what conditions does salt precipitate out of solution?*

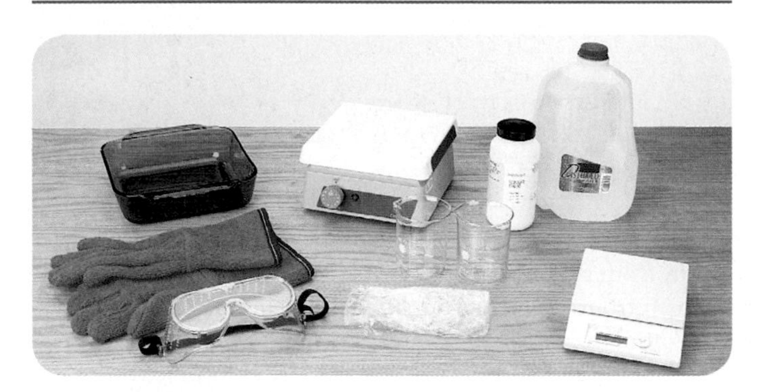

Suggested materials

Materials

halite (sodium chloride)
250-mL glass beakers (2)
distilled water
plastic wrap
laboratory scale
hot plate
shallow glass baking dish
refrigerator
glass stirring rod

Safety Precautions

Procedure

1. Read and complete the lab safety form.
2. Make a data table to record your observations.
3. Pour 150 mL of distilled water into a 250-mL glass beaker. Add 54 g of sodium chloride and stir until only a few grains remain on the bottom of the beaker.
4. Place the beaker on the hot plate, and turn on the hot plate. Stir the solution until the last few grains of sodium chloride dissolve. The salt solution will then be saturated.
5. Pour 50 mL of the warm, saturated solution into the second 250-mL glass beaker, and cover it with plastic wrap so that it forms a seal. Put this beaker in the refrigerator.
6. Pour 50 mL of the saturated solution into the glass baking dish. Place the dish on the hot plate and heat the salt solution until all the liquid evaporates. **WARNING: *The baking dish will be hot. Handle with care.***
7. Place the original beaker with 50 mL of the remaining solution on a shelf or windowsill. Do not cover the beaker.
8. Observe both beakers one day later. If crystals have not formed, wait another day.
9. Once crystals have formed in all three containers, observe the size and shape of the crystals. Write your observations in your data table.

Analyze and Conclude

1. **Describe** the shape of the precipitated crystals in the three containers. Does the shape of the crystals alone identify them as sodium chloride?
2. **Infer** how heating the salt solution affected the solubility of the sodium chloride.
3. **Interpret** what effect cooling has on the solubility of salt. What effect does evaporation have on the solubility of salt?
4. **Evaluate** the relationship between rate of cooling and crystal size.

INQUIRY EXTENSION

Use Other Substances Design an experiment to investigate other soluble substances. Test to see how much of the substance can be dissolved in a given amount of water, how long it takes for the solution to evaporate, and what crystal shapes form. Prepare a short report to share with your class.

CHAPTER 3

Study Guide

Download quizzes, key terms, and flash cards from glencoe.com.

BIG Idea The variety of substances on Earth results from the way that atoms are arranged and combined.

Vocabulary	Key Concepts
Section 3.1 Matter	
• atomic number (p. 62) • electron (p. 61) • element (p. 60) • ion (p. 64) • isotope (p. 62) • mass number (p. 62) • matter (p. 60) • neutron (p. 60) • nucleus (p. 60) • proton (p. 60)	**MAIN Idea** Atoms are the basic building blocks of all matter. • Atoms consist of protons, neutrons, and electrons. • An element consists of atoms that have a specific number of protons in their nuclei. • Isotopes of an element differ by the number of neutrons in their nuclei. • Elements with full outermost energy levels are highly unreactive. • Ions are electrically charged atoms or groups of atoms.
Section 3.2 Combining Matter	
• acid (p. 71) • base (p. 72) • chemical bond (p. 67) • chemical reaction (p. 70) • compound (p. 66) • covalent bond (p. 67) • ionic bond (p. 68) • metallic bond (p. 68) • molecule (p. 67) • solution (p. 71)	**MAIN Idea** Atoms combine through electric forces, forming molecules and compounds. • Atoms of different elements combine to form compounds. • Covalent bonds form from shared electrons between atoms. • Ionic compounds form from the attraction of positive and negative ions. • There are two types of mixtures—heterogeneous and homogeneous. • Acids are solutions containing hydrogen ions. Bases are solutions containing hydroxide ions.
Section 3.3 State of Matter	
• condensation (p. 75) • crystalline structure (p. 73) • evaporation (p. 74) • glass (p. 73) • plasma (p. 74) • sublimation (p. 75)	**MAIN Idea** All matter on Earth and in the universe occurs in the form of a solid, a liquid, a gas, or plasma. • Changes of state involve thermal energy. • The law of conservation of matter states that matter cannot be created or destroyed. • The law of conservation of energy states that energy is neither created nor destroyed.

Vocabulary PuzzleMaker glencoe.com

CHAPTER 3

Assessment

Vocabulary Review

Fill in the blank with the correct vocabulary term from the Study Guide.

1. The electrically neutral particles in the nucleus of an atom are called ________.
2. The ________ of an element is equal to the number of ________ in the nucleus of its atoms.
3. Atoms of an element that differ by their mass numbers are called ________.

Explain how both terms in each set below are related.

4. ionic, covalent
5. homogeneous mixture, solution
6. acid, base

Arrange each set of vocabulary terms into a meaningful and true sentence.

7. solid, glass
8. molecules, ions, plasma, gas
9. evaporation, condensation
10. electrons, metallic bond

Understand Key Concepts

Use the figure below to answer Questions 11 to 13.

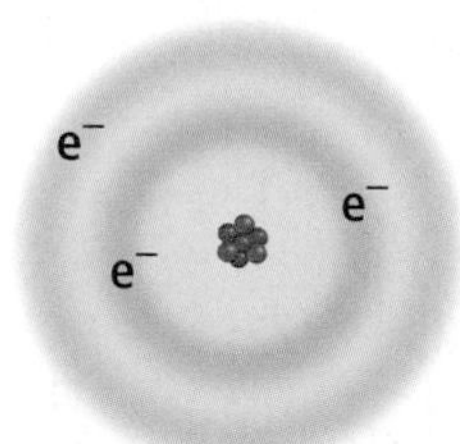

11. What is the atomic number of this atom?
    A. 3
    B. 4
    C. 5
    D. 6

12. How many valence electrons does this atom have?
    A. 1
    B. 2
    C. 3
    D. 4

13. Which element does this atom represent? (Refer to the periodic table of the elements in **Figure 3.2.**)
    A. helium
    B. beryllium
    C. lithium
    D. nitrogen

14. What ionic compound is formed by the ions Al^{3+} and O^{2-}?
    A. Al_3O_2
    B. Al_2O
    C. Al_2O_3
    D. AlO

Use the figure below to answer Question 15.

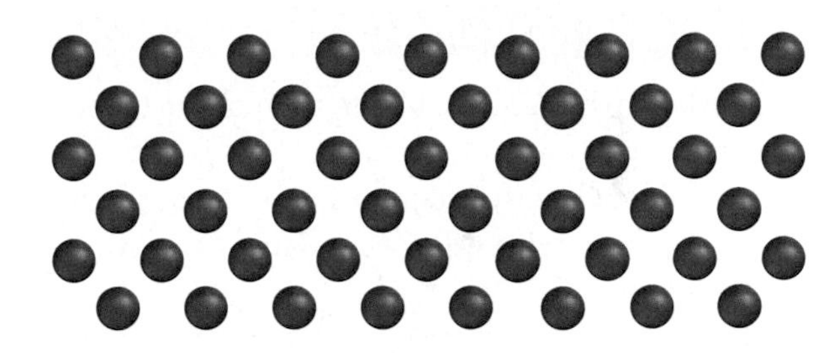

15. The figure shows the arrangement of atoms in a substance. What is this substance?
    A. gas
    B. glass
    C. liquid
    D. solid

16. Which is an example of a heterogeneous mixture?
    A. coffee
    B. soil
    C. gelatin
    D. air

17. During the process of sublimation, into what is ice converted?
    A. hydrogen ions and hydroxide ions
    B. hydrogen
    C. water
    D. water vapor

18. Many musical instruments are made of brass, which is a mixture of copper and zinc atoms. What is brass an example of?
    A solid solution
    B. ionic compound
    C. chemical reaction
    D. base

19. What happens to a gas when it condenses and forms a liquid?
    A. It releases thermal energy.
    B. It absorbs thermal energy.
    C. It increases in temperature.
    D. It decreases in temperature.

20. What kind of ion is characteristic of an acid?
    A. oxygen ion
    B. negative ion
    C. hydroxide ion
    D. hydrogen ion

Constructed Response

21. **Explain** why table salt does not conduct electricity.

22. **Explain** why gases such as neon and argon do not readily react with other elements.

23. **Illustrate** a model atom of potassium (K), indicating the positive charge of the nucleus and the idealized positions of the electrons in the various energy levels. Is potassium a metal or nonmetal? Refer to the periodic table of the elements in **Figure 3.2.**

Use the figure below to answer Questions 24 and 25.

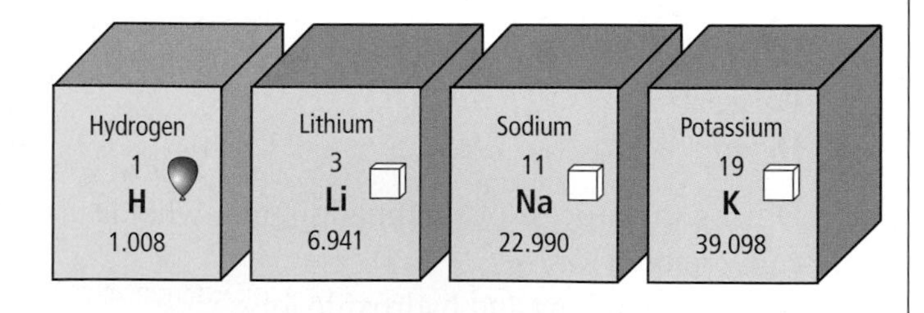

24. **Detect** What do these elements have in common?

25. **Explain** why the atomic masses of these elements are not whole numbers.

26. **Distinguish** which kind of chemical bond produces a solid that readily conducts heat and electricity.

27. **Compare and contrast** the physical properties of the elements helium and neon.

Use the figure below to answer Questions 28 and 29.

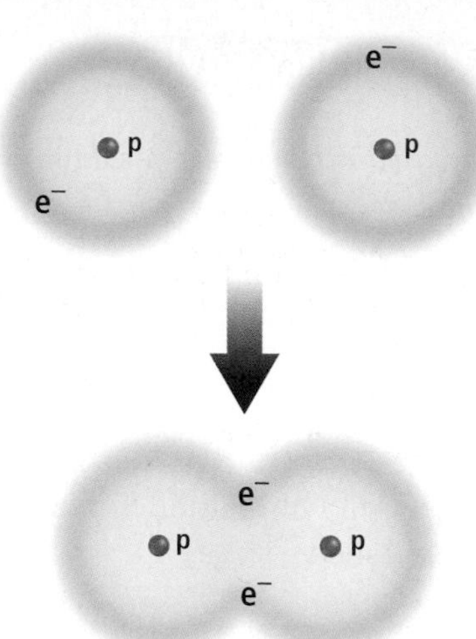

28. **Identify** the type of bond shown in the figure. Explain your reasoning.

29. **Compare** this bond to a metallic bond. Use an illustration to clarify your answer.

30. **Deduce** what the difference would be between water molecules containing deuterium and those containing ordinary hydrogen atoms. *(Hint: Deuterium is an isotope of hydrogen with mass number two. It forms the same chemical compounds as other hydrogen atoms, including water.)*

31. **Evaluate** the statement: Plasma is usually hotter than gas.

Think Critically

Use the figure below to answer Question 32.

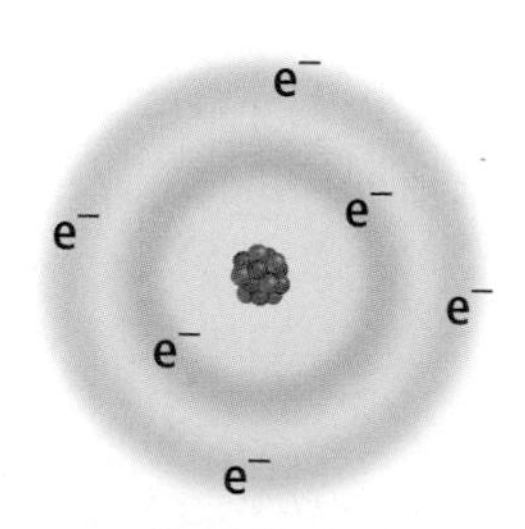

32. **Deduce** The figure shows an atom of carbon-14. This radioactive isotope decays by converting one of its neutrons to a proton. What element and isotope is produced by the radioactive decay of carbon-14?

Earth Science Online Chapter Test glencoe.com

33. Illustrate Use an illustration to show why water is effective in dissolving ionic solids such as table salt.

34. Group and list some of the properties that all metals have in common.

35. Assess the correctness of the following statement: Snow that covers the ground can disappear on cold days even when the temperature remains below 0°C.

36. Arrange When hydrochloric acid (HCl) is added to the sedimentary rock limestone ($CaCO_3$), carbon dioxide (CO_2), calcium chloride ($CaCl_2$) and water (H_2O) are given off. Arrange the chemical compounds listed above into a balanced equation that shows this chemical reaction.

37. Infer Earth's upper atmosphere—the ionosphere—conducts electricity. Infer about the state of matter in the ionosphere.

38. CAREERS IN EARTH SCIENCE Assess the importance of understanding chemical reactions in order to interpret the conditions of rocks and minerals that are present on other planets.

39. Estimate Air at sea level has a density of 0.13 g/L. Estimate how much air, in kilograms, fills your classroom. *(Hint: to start, multiply the length, width, and height of your classroom to calculate the room's volume.)*

Concept Mapping

40. Create a concept map using the following terms or phrases: *ionic bond, covalent bond, metallic bond, shared electrons, gain or lose electrons, a sea of electrons, molecule,* and *compound.*

Challenge Question

41. An atom is mostly empty space. A typical atom has a diameter of 10^{-10} m with a nucleus of diameter 10^{-15} m. To visualize this, enlarge this atom by a factor of 10^{13} (10 trillion) so that its nucleus has the size of a marble (1 cm). What would be the diameter of this enlarged atom? Would this atom fit into a football field?

Additional Assessment

42. WRITING in **Earth Science** Prepare a news release reporting on the discovery of a new chemical element. The element has 121 protons in its nucleus. Be sure to include the characteristics of this element and its location in the periodic table.

DBQ Document–Based Questions

Data obtained from: Kramer, D. A. 2005. Mineral resource of the month: Magnesium. *Geotimes* 50 (November): 57.

Magnesium is lightweight and has a high strength-to-weight ratio. It constitutes about 2 percent of Earth's crust and its concentration in seawater is 0.13 percent. Magnesium is present in more than 60 minerals and is produced from magnesium-bearing ores, seawater, and brines.

Magnesium is made into an alloy with aluminum to increase strength and corrosion resistance, especially in beverage cans. Its light weight makes it useful in aircrafts, cars, chain saws, lawn mowers, and other machine parts. Annual world magnesium production is 584,000 metric tons. China produces the most at 426,000 metric tons. Yearly production in the United States is 43,000 metric tons and U.S. consumption is 140,000 metric tons per year. Canada, China, Israel, and Russia supply 92 percent of U.S. magnesium imports. Recycling covers about 15 percent of U.S. magnesium consumption.

43. Determine the amount of magnesium in 1 m^3 of seawater. Express your answer in kilograms. (Hint: The density of seawater is 1025 kg/m^3.)

44. Analyze and explain the role of magnesium in the manufacture of cars and beverage cans.

45. Compare U.S. production of magnesium to that of other countries. How does U.S. dependence on imports affect these other countries?

Cumulative Review

46. Why is the concept of time and scale in the study of Earth science difficult to understand? **(Chapter 1)**

47. In the collection of data, measurements must follow what general guidelines? **(Chapter 1)**

Standardized Test Practice

Multiple Choice

Use the table below to answer Questions 1–3.

Atomic Structure		
Element	**Atomic Number**	**Atomic Mass**
Beryllium	4	9.01
Calcium	20	40.08
Silicon	14	28.09
Scandium	21	44.96
Titanium	22	47.88
Zirconium	40	91.22

1. If titanium has 22 protons in its nucleus, how many neutrons are present in the nucleus of its most common isotope?
A. 26
B. 28
C. 48
D. 60

2. If the most common isotope of scandium has 24 neutrons in its nucleus, how many protons does scandium have?
A. 13
B. 21
C. 45
D. 66

3. If calcium's most common isotope has 20 neutrons in its nucleus, how many neutrons can be found in another naturally occurring isotope of calcium?
A. 21
B. 30
C. 41
D. 60

4. Determine the number of valence electrons that oxygen has.
A. 2
B. 4
C. 6
D. 9

5. How should a city that is located between two time zones establish a time?
A. Have two different times within the same city.
B. Allow the people to choose what time zone they want to go by.
C. Move the time zone split outside of the city.
D. Divide the time in half to split the difference of the two times.

Use the illustration below to answer Questions 6–8.

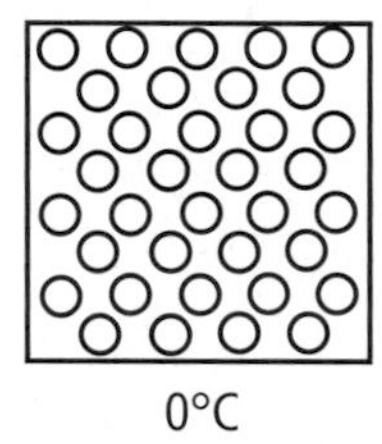
0°C

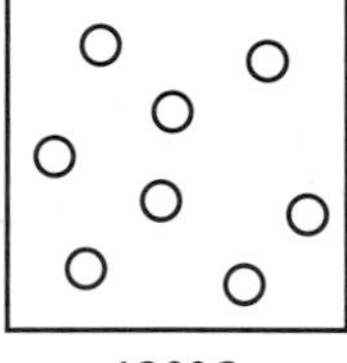
120°C

6. In a cup, an ice cube melts in liquid water. Which is true at the moment the ice melts?
A. The water has less thermal energy than the ice.
B. The water has more thermal energy than the ice.
C. The water is at a higher temperature than the ice.
D. The water is at the same temperature as the ice.

7. According to the illustration, what happens to water molecules when water is heated?
A. Their energy levels decrease.
B. They move farther apart.
C. They move more slowly.
D. They stop moving.

8. In order for a liquid to change to a gaseous state, what must it reach?
A. its freezing point
B. its condensation point
C. its melting point
D. its boiling point

9. Which is the most acidic?
A. banana (pH 4.7)
B. celery (pH 5.9)
C. grape (pH 3.0)
D. lettuce (pH 6.9)

Short Answer

Use the graph below to answer Questions 10–12.

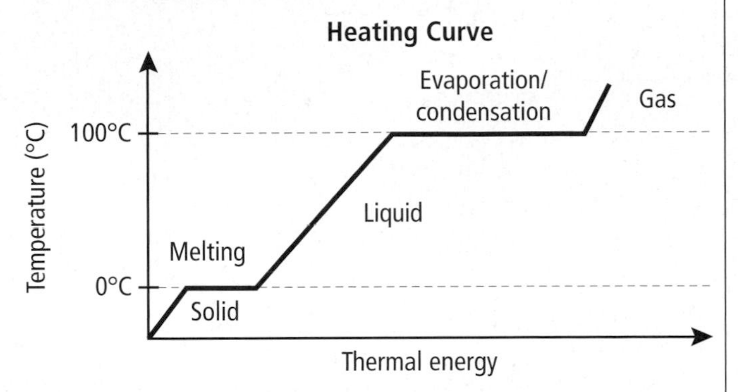

10. The diagram represents a sample of water. What does it demonstrate?

11. At which point does the water have the least amount of thermal energy?

12. What do the level lines in the diagram represent?

13. Differentiate between a theory and a hypothesis.

14. Why would a geologic map be important to a scientist studying earthquakes?

15. Does silicon have any isotopes? Explain your answer.

16. Ethical scientific researchers accurately report the data on which they base their conclusions. Why is this important?

Reading for Comprehension

Anthocyanin Pigments

Red cabbage contains a pigment molecule called flavin (an anthocyanin). This water-soluble pigment is also found in apple skin, plums, poppies, cornflowers, and grapes. Acidic solutions will turn anthocyanin a red color. Neutral solutions result in a purplish color. Basic solutions appear in greenish-yellow. Therefore, it is possible to determine the pH of a solution based on the color it turns the anthocyanin pigments in red cabbage juice.

How to make red cabbage pH indicator. *About: Chemistry.* (Online resource accessed February 12, 2007.)

17. How does red cabbage act as an acid/base indicator?
- **A.** Its pigment changes color based on the acid or base with which it comes in contact.
- **B.** Its pigment will not change when it comes into contact with a neutral solution.
- **C.** It always stays red.
- **D.** Its pigment releases water when it comes in contact with an acid or base.

18. What can be inferred from this passage?
- **A.** Red cabbage is the only food that can act as an indicator of acids and bases.
- **B.** Anthocyanin pigments in red cabbage juice change color when exposed to acids or bases.
- **C.** It is safe to eat the cabbage after using it for an acid/base experiment.
- **D.** The change in color does not indicate an acid or base.

19. What color does the cabbage become when exposed to a base?
- **A.** purplish color
- **B.** red color
- **C.** blue color
- **D.** greenish-yellow color

NEED EXTRA HELP?																
If You Missed Question . . .	1	2	3	4	5	6	7	8	9	10	11	12	13	14	15	16
Review Section . . .	3.1	3.1	3.1	3.1	2.1	3.3	3.3	3.3	3.2	3.3	3.3	3.3	1.3	2.2	3.1	1.3

CHAPTER 4

Minerals

BIG Idea Minerals are an integral part of daily life.

4.1 What is a mineral?
MAIN Idea Minerals are naturally occurring, solid, inorganic compounds or elements.

4.2 Types of Minerals
MAIN Idea Minerals are classified based on their chemical properties and characteristics.

GeoFacts

- Stalactites and other cave formations take thousands of years to form. One estimate is that a stalactite will grow only 10 cm in 1000 years. That is equal to 0.1 mm each year!
- The diameter of a soda straw is equal to the droplets of water that form them.
- The longest soda straws discovered measure more than 9 m long.

LAUNCH Lab

What shapes do minerals form?

Although there are thousands of minerals in Earth's crust, each type of mineral has unique characteristics. These characteristics are clues to a mineral's composition and to the way it formed. Physical properties can also be used to distinguish one type of mineral from another.

Procedure

1. Read and complete the lab safety form.
2. Place a few grains of **table salt** (the mineral halite) on a **microscope slide.** Place the slide on the **microscope** stage. Or, observe the grains with a **magnifying lens.**
3. Focus on one grain at a time. Count the number of sides of each grain. Make sketches of the grains.
4. Next, examine a **quartz crystal** with the microscope or magnifying lens. Count the number of sides of the quartz crystal. Sketch the shape of the quartz crystal.

Analysis

1. **Compare and contrast** the shapes of the samples of halite and quartz.
2. **Describe** some other properties of your mineral samples.
3. **Infer** what might account for the differences you observed.

Mineral Identification Make the following Foldable to explain the tests used to identify minerals.

STEP 1 Collect four sheets of paper and layer them 2 cm apart vertically. Keep the left and right edges even.

STEP 2 Fold up the bottom edges of the sheets to form seven equal tabs. Crease the fold to hold the tabs in place.

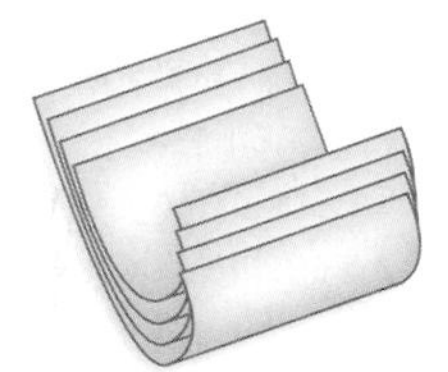

STEP 3 Staple along the fold. Label the tabs with the names of the tests used to identify minerals.

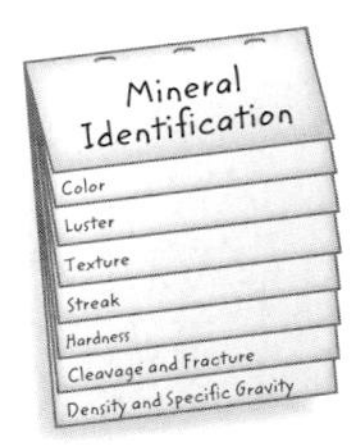

FOLDABLES Use this Foldable with Section 4.1. As you read this section, describe the chemical or physical properties of minerals that are used in each test.

Visit glencoe.com to

- **study entire chapters online;**
- **explore Concepts In Motion animations:**
  - **Interactive Time Lines**
  - **Interactive Figures**
  - **Interactive Tables**
- **access Web Links for more information, projects, and activities;**
- **review content with the Interactive Tutor and take Self-Check Quizzes.**

Section 4.1

Objectives

- **Define** a mineral.
- **Describe** how minerals form.
- **Classify** minerals according to their physical and chemical properties.

Review Vocabulary

element: a pure substance that cannot be broken down into simpler substances by chemical or physical means

New Vocabulary

mineral
crystal
luster
hardness
cleavage
fracture
streak
specific gravity

What is a mineral?

MAIN Idea **Minerals are naturally occurring, solid, inorganic compounds or elements.**

Real-World Reading Link Look around your classroom. The metal in your desk, the graphite in your pencil, and the glass in the windows are just three examples of how modern humans use products made from minerals.

Mineral Characteristics

Earth's crust is composed of about 3000 minerals. Minerals play important roles in forming rocks and in shaping Earth's surface. A select few have helped shape civilization. For example, great progress in prehistory was made when early humans began making tools from iron.

A **mineral** is a naturally occurring, inorganic solid, with a specific chemical composition and a definite crystalline structure. This crystalline structure is often exhibited by the crystal shape itself. Examples of mineral crystal shapes are shown in **Figure 4.1.**

Naturally occurring and inorganic Minerals are naturally occurring, meaning that they are formed by natural processes. Such processes will be discussed later in this section. Thus, synthetic diamonds and other substances developed in labs are not minerals. All minerals are inorganic. They are not alive and never were alive. Based on these criteria, salt is a mineral, but sugar, which is harvested from plants, is not. What about coal? According to the scientific definition of minerals, coal is not a mineral because millions of years ago, it formed from organic materials.

Figure 4.1 The shapes of these mineral crystals reflect the internal arrangement of their atoms.

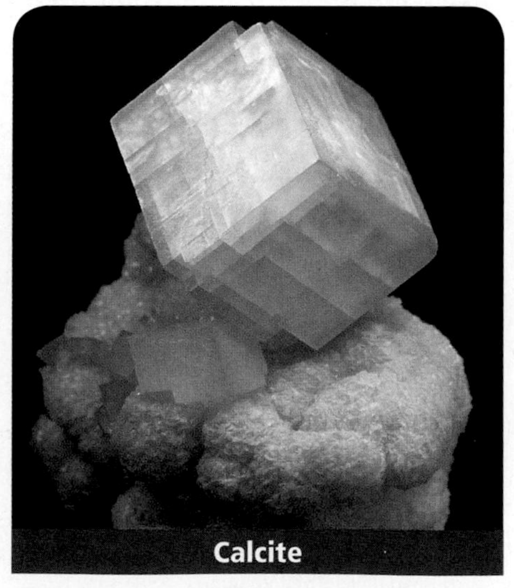

Definite crystalline structure The atoms in minerals are arranged in regular geometric patterns that are repeated. This regular pattern results in the formation of a crystal. A **crystal** is a solid in which the atoms are arranged in repeating patterns. Sometimes, a mineral will form in an open space and grow into one large crystal. The well-defined crystal shapes shown in **Figure 4.1** are rare. More commonly, the internal atomic arrangement of a mineral is not apparent because the mineral formed in a restricted space. **Figure 4.2** shows a sample of quartz that grew in a restricted space.

 **Reading Check** **Describe** the atomic arrangement of a crystal.

■ **Figure 4.2** This piece of quartz most likely formed in a restricted space, such as within a crack in a rock.

Solids with specific compositions The fourth characteristic of minerals is that they are solids. Recall from Chapter 3 that solids have definite shapes and volumes, while liquids and gases do not. Therefore, no gas or liquid can be considered a mineral.

Each type of mineral has a chemical composition unique to that mineral. This composition might be specific, or it might vary within a set range of compositions. A few minerals, such as copper, silver, and sulfur, are composed of single elements. The vast majority, however, are made from compounds. The mineral quartz (SiO_2), for example, is a combination of two atoms of oxygen and one atom of silicon. Although other minerals might contain silicon and oxygen, the arrangement and proportion of these elements in quartz are unique to quartz.

VOCABULARY

ACADEMIC VOCABULARY

Restricted

small space; to have limits

The room was so small that it felt very restricted.

Quartz

Aquamarine

■ **Figure 4.3** The range in composition and resulting appearance is specific enough to identify numerous feldspar varieties accurately.

Variations in Composition In some minerals, chemical composition can vary slightly depending on the temperature at which the mineral crystallizes. The plagioclase feldspar, shown in **Figure 4.3,** ranges from sodium-rich albite (AHL bite) at low temperatures to calcium-rich anorthite (uh NOR thite) at high temperatures. The difference in the mineral's appearance is due to a slight change in the chemical composition and a difference in growth pattern as the temperature changes. At intermediate temperatures, both calcium and sodium are incorporated in the crystal structure, building up alternating layers that allow light to refract or scatter, producing a range of colors, as show in the labradorite in **Figure 4.3.**

Rock-Forming Minerals

Although about 3000 minerals occur in Earth's crust, only about 30 of these are common. Eight to ten of these minerals are referred to as rock-forming minerals because they make up most of the rocks in Earth's crust. They are primarily composed of the eight most common elements in Earth's crust. This is illustrated in **Table 4.1.**

Table 4.1 Most Common Rock-Forming Minerals

Quartz	Feldspar	Mica	Pyroxene*
SiO_2	$NaAlSi_3O_8 - CaAl_2Si_2O_8$ & $KAlSi_3O_8$	$K(Mg,Fe)_3(AlSi_3O_{10})(OH)_2$ $KAl_2(AlSi_3O_{10})(OH)_2$	$MgSiO_3$ $CaMgSi_2O_6$ $NaAlSi_2O_6$
Amphibole*	**Olivine**	**Garnet***	**Calcite**
$Ca_2(Mg,Fe)_5Si_8O_{22}(OH)_2$ $Fe_7Si_8O_{22}(OH)_2$	$(Mg,Fe)_2SiO_4$	$Mg_3Al_2Si_3O_{12}$ $Fe_3Al_2Si_3O_{12}$ $Ca_3Al_2Si_3O_{12}$	$CaCO_3$

O 46.6% | Si 27.7% | Al 8.1% | Fe 5% | Ca 3.6% | S 2.8% | K 2.6% | Mg 2.1% | Other 1.5%

*representative mineral compositions

Minerals from magma Molten material that forms and accumulates below Earth's surface is called magma. Magma is less dense than the surrounding solid rock, so it can rise upward into cooler layers of Earth's interior. Here, the magma cools and crystallizes. The type and number of elements present in the magma determine which minerals will form. The rate at which the magma cools determines the size of the mineral crystals. If the magma cools slowly within Earth's heated interior, the atoms have time to arrange themselves into large crystals. If the magma reaches Earth's surface, comes in contact with air or water, and cools quickly, the atoms do not have time to arrange themselves into large crystals. Thus, small crystals form from rapidly cooling magma, and large crystals form from slowly cooling magma. The mineral crystals in the granite shown in **Figure 4.4** are the result of cooling magma. You will learn more about crystal size in Chapter 5.

Reading Check **Explain** how contact with water affects crystal size.

Figure 4.4 The crystals in these two samples formed in different ways. **Describe** *the differences you see in these rock samples.*

Minerals from solutions Minerals are often dissolved in water. For example, the salts that are dissolved in ocean water make it salty. When a liquid becomes full of a dissolved substance and it can dissolve no more of that substance, the liquid is saturated. If the solution then becomes overfilled, it is called supersaturated and conditions are right for minerals to form. At this point, individual atoms bond together and mineral crystals precipitate, which means that they form into solids from the solution.

Minerals also crystallize when the solution in which they are dissolved evaporates. You might have experienced this if you have ever gone swimming in the ocean. As the water evaporated off your skin, the salts were left behind as mineral crystals. Minerals that form from the evaporation of liquid are called evaporites. The rock salt in **Figure 4.4** was formed from evaporation. **Figure 4.5** shows Mammoth Hot Springs, a large evaporite complex in Yellowstone National Park.

Figure 4.5 This large complex of evaporite minerals is in Yellowstone National Park. The variation in color is the result of the variety of elements that are dissolved in the water.

Identifying Minerals

Geologists rely on several simple tests to identify minerals. These tests are based on a mineral's physical and chemical properties, which are crystal form, luster, hardness, cleavage, fracture, streak, color, texture, density, specific gravity, and special properties. As you will learn in the GeoLab at the end of this chapter, it is usually best to use a combination of tests instead of just one to identify minerals.

Crystal form Some minerals form such distinct crystal shapes that they are immediately recognizable. Halite—common table salt—always forms perfect cubes. Quartz crystals, with their double-pointed ends and six-sided crystals, are also readily recognized. However, as you learned earlier in this section, perfect crystals are not always formed, so identification based only on crystal form is rare.

Luster The way that a mineral reflects light from its surface is called **luster.** There are two types of luster—metallic luster and nonmetallic luster. Silver, gold, copper, and galena have shiny surfaces that reflect light, like the chrome trim on cars. Thus, they are said to have a metallic luster. Not all metallic minerals are metals. If their surfaces have shiny appearances like metals, they are considered to have a metallic luster. Sphalerite, for example, is a mineral with a metallic luster that is not a metal.

Minerals with nonmetallic lusters, such as calcite, gypsum, sulfur, and quartz, do not shine like metals. Nonmetallic lusters might be described as dull, pearly, waxy, silky, or earthy. Differences in luster, shown in **Figure 4.6,** are caused by differences in the chemical compositions of minerals. Describing the luster of nonmetallic minerals is a subjective process. For example, a mineral that appears waxy to one person might not appear waxy to another. Using luster to identify a mineral should usually be used in combination with other physical characteristics.

Reading Check **Define** the term *luster.*

CAREERS IN EARTH SCIENCE

Lapidary A lapidary is someone who cuts, polishes, and engraves precious stones. He or she studies minerals and their properties in order to know which minerals are the best for certain projects. To learn more about Earth science careers, visit glencoe.com.

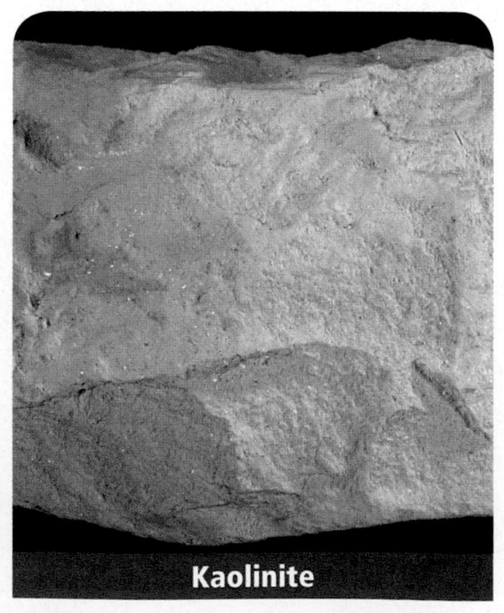

Figure 4.6 The flaky and shiny nature of talc gives it a pearly luster. Another white mineral, kaolinite, contrasts sharply with its dull, earthy luster.

Table 4.2 Mohs Scale of Hardness

Concepts In Motion **Interactive Table** To explore more about Mohs scale of hardness, visit glencoe.com.

Mineral	Hardness	Hardness of Common Objects
Diamond	10	
Corundum	9	
Topaz	8	
Quartz	7	streak plate = 7
Feldspar	6	steel file = 6.5
Apatite	5	glass = 5.5
Fluorite	4	iron nail = 4.5
Calcite	3	piece of copper = 3.5
Gypsum	2	fingernail = 2.5
Talc	1	

Hardness One of the most useful and reliable tests for identifying minerals is hardness. **Hardness** is a measure of how easily a mineral can be scratched. German geologist Friedrich Mohs developed a scale by which an unknown mineral's hardness can be compared to the known hardness of ten minerals. The minerals in the Mohs scale of mineral hardness were selected because they are easily recognized and, with the exception of diamond, readily found in nature.

Reading Check **Explain** what hardness measures.

Talc is one of the softest minerals and can be scratched by a fingernail; therefore, talc represents 1 on the Mohs scale of hardness. In contrast, diamond is so hard that it can be used as a sharpener and cutting tool, so diamond represents 10 on the Mohs scale of hardness. The scale, shown in **Table 4.2,** is used in the following way: a mineral that can be scratched by your fingernail has a hardness equal to or less than 2. A mineral that cannot be scratched by your fingernail and cannot scratch glass has a hardness value between 5.5 and 2.5. Finally, a mineral that scratches glass has a hardness greater than 5.5. Using other common objects, such as those listed in the table, can help you determine a more precise hardness and provide you with more information with which to identify an unknown mineral. Sometimes more than one mineral is present in a sample. If this is the case, it is a good idea to test more than one area of the sample. This way, you can be sure that you are testing the hardness of the mineral you are studying. **Figure 4.7** shows two minerals that have different hardness values.

Figure 4.7 The mineral on top can be scratched with a fingernail. The mineral on the bottom easily scratches glass.
Determine *Which mineral has greater hardness?*

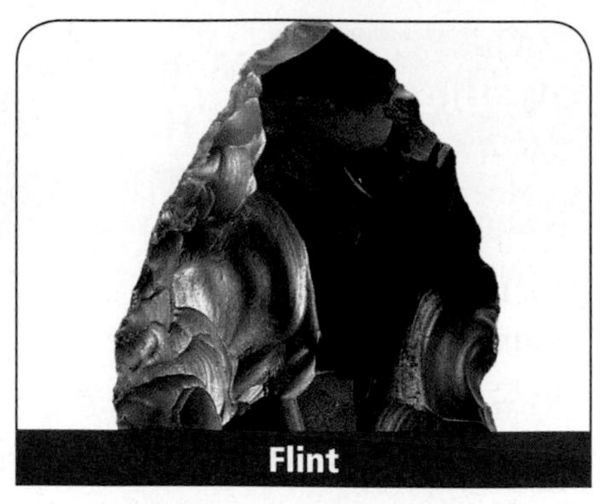

■ **Figure 4.8** Halite has perfect cleavage in three directions; it breaks apart into pieces that have 90° angles. The strong bonds in quartz prevent cleavage from forming. Conchoidal fractures are characteristic of microcrystalline minerals such as flint.

Cleavage and fracture Atomic arrangement also determines how a mineral will break. Minerals break along planes where atomic bonding is weak. A mineral that splits relatively easily and evenly along one or more flat planes is said to have **cleavage.** To identify a mineral according to its cleavage, geologists count the number of cleaved planes and study the angle or angles between them. For example, mica has perfect cleavage in one direction. It breaks in sheets because of weak atomic bonds. Halite, shown in **Figure 4.8,** has cubic cleavage, which means that it breaks in three directions along planes of weak atomic attraction.

MiniLab

Recognize Cleavage and Fracture

How is cleavage used? Cleavage forms when a mineral breaks along a plane of weakly bonded atoms. If a mineral has no cleavage, it exhibits fracture. Recognizing the presence or absence of cleavage and determining the number of cleavage planes is a reliable method of identifying minerals.

Procedure

Part 1

1. Read and complete the lab safety form.
2. Obtain five **mineral samples** from your teacher. Separate them into two sets—those with cleavage and those without cleavage.
3. Arrange the minerals that have cleavage in order from fewest to most cleavage planes. How many cleavage planes does each sample have? Identify these minerals if you can.
4. Examine the samples that have no cleavage. Describe their surfaces. Identify these minerals if you can.

Part 2

5. Obtain two more samples from your teacher. Are these the same mineral? How can you tell?
6. Use a **protractor** to measure the cleavage plane angles of both minerals. Record your measurements.

Analysis

1. **Record** the number of cleavage planes or presence of fracture for all seven samples.
2. **Compare** the cleavage plane angles for Samples 6 and 7. What do they tell you about the mineral samples?
3. **Predict** the shape each mineral would exhibit if you were to hit each one with a hammer.

Quartz, shown in **Figure 4.8,** breaks unevenly along jagged edges because of its tightly bonded atoms. Minerals that break with rough or jagged edges are said to have **fracture.** Flint, jasper, and chalcedony (kal SEH duh nee) (microcrystalline forms of quartz) exhibit a unique fracture with arclike patterns resembling clamshells, also shown in **Figure 4.8.** This fracture is called conchoidal (kahn KOY duhl) fracture and is diagnostic in identifying the rocks and minerals that exhibit it.

Streak A mineral rubbed across an unglazed porcelain plate will sometimes leave a colored powdered streak on the surface of the plate. **Streak** is the color of a mineral when it is broken up and powdered. The streak of a nonmetallic mineral is usually white. Streak is most useful in identifying metallic minerals.

Sometimes, a metallic mineral's streak does not match its external color, as shown in **Figure 4.9.** For example, the mineral hematite occurs in two different forms, resulting in two distinctly different appearances. Hematite that forms from weathering and exposure to air and water is a rusty red color and has an earthy feel. Hematite that forms from crystallization of magma is silver and metallic in appearance. However, both forms make a reddish-brown streak when tested. The streak test can be used only on minerals that are softer than a porcelain plate. This is another reason why streak cannot be used to identify all minerals.

Figure 4.9 Despite the fact that these pieces of hematite appear remarkably different, their chemical compositions are the same. Thus, the streak that each makes is the same color.

Reading Check **Explain** which type of mineral can be identified using streak.

Color One of the most noticeable characteristics of a mineral is its color. Color is sometimes caused by the presence of trace elements or compounds within a mineral. For example, quartz occurs in a variety of colors, as shown in **Figure 4.10.** These different colors are the result of different trace elements in the quartz samples. Red jasper, purple amethyst, and orange citrine contain different amounts and forms of iron. Rose quartz contains manganese or titanium. However, the appearance of milky quartz is caused by the numerous bubbles of gas and liquid trapped within the crystal. In general, color is one of the least reliable clues of a mineral's identity.

FOLDABLES
Incorporate information from this section into your Foldable.

Figure 4.10 These varieties of quartz all contain silicon and oxygen. Trace elements determine their colors.

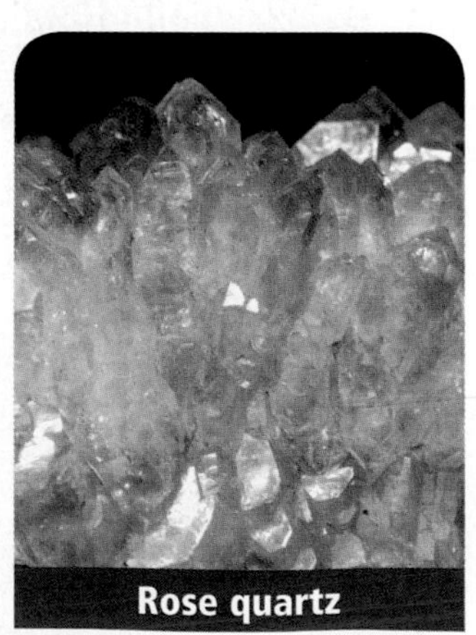

Concepts In Motion
Interactive Table To explore more about the special properties of minerals, visit glencoe.com.

Table 4.3	Special Properties of Minerals				
Property	**Double refraction** occurs when a ray of light passes through the mineral and is split into two rays.	**Effervescence** occurs when reaction with hydrochloric acid causes calcite to fizz.	**Magnetism** occurs between minerals that contain iron; only magnetite and pyrrhotite are strongly magnetic.	**Iridescence**—a play of colors, caused by the bending of light rays.	**Fluorescence** occurs when some minerals are exposed to ultraviolet light, which causes them to glow in the dark.
Mineral	Calcite—Variety Iceland Spar	Calcite	Magnetite Pyrrhotite	Labradorite	Fluorite Calcite
Example					

Special properties Several special properties of minerals can also be used for identification purposes. Some of these properties are magnetism, striations, double refraction, effervescence with hydrochloric acid, and fluorescence, shown in **Table 4.3**. For example, Iceland spar is a form of calcite that exhibits double refraction. The arrangement of atoms in this type of calcite causes light to be bent in two directions when it passes through the mineral. The refraction of the single ray of light into two rays creates the appearance of two images.

DATA ANALYSIS LAB

Based on Real Data*

Make and Use a Table

What information should you include in a mineral identification chart?

Mineral Identification Chart			
Mineral Color	Streak	Hardness	Breakage Pattern
copper red		3	hackly, fracture
	red or reddish brown	6	irregular fracture
pale to golden yellow	yellow		
	colorless	7.5	conchoidal fracture
gray, green or white			two cleavage planes

Analysis

1. Copy the data table and use the *Reference Handbook* to complete the table.
2. Expand the table to include the names of the minerals, other properties, and uses.

Think Critically

3. **Determine** which of these minerals will scratch glass? Explain.
4. **Identify** which of these minerals might be present in both a painting and your desk.
5. **Identify** any other information you could include in the table.

*Data obtained from: Klein, C. 2002. *The Manual of Mineral Science.*

Texture Texture describes how a mineral feels to the touch. This, like luster, is subjective. Therefore, texture is often used in combination with other tests to identify a mineral. The texture of a mineral might be described as smooth, rough, ragged, greasy, or soapy. For example, fluorite, shown in **Figure 4.11,** has a smooth texture, while the texture of talc, shown in **Figure 4.6,** is greasy.

■ **Figure 4.11** Textures are interpreted differently by different people. The texture of this fluorite is usually described as smooth.

Density and specific gravity Sometimes, two minerals of the same size have different weights. Differences in weight are the result of differences in density, which is defined as mass per unit of volume. Density is expressed as follows.

$$D = \frac{M}{V}$$

In this equation, D = density, M = mass and V = volume. For example, pyrite has a density of 5.2 g/cm^3, and gold has a density of 19.3 g/cm^3. If you had a sample of gold and a sample of pyrite of the same size, the gold would have greater weight because it is denser.

Density reflects the atomic mass and structure of a mineral. Because density is not dependent on the size or shape of a mineral, it is a useful identification tool. Often, however, differences in density are too small to be distinguished by lifting different minerals. Thus, for accurate mineral identification, density must be measured. The most common measure of density used by geologists is **specific gravity,** which is the ratio of the mass of a substance to the mass of an equal volume of water at 4°C. For example, the specific gravity of pyrite is 5.2. The specific gravity of pure gold is 19.3.

Section 4.1 Assessment

Section Summary

- A mineral is a naturally occurring, inorganic solid with a specific chemical composition and a definite crystalline structure.
- A crystal is a solid in which the atoms are arranged in repeating patterns.
- Minerals form from magma or from supersaturated solutions.
- Minerals can be identified based on their physical and chemical properties.
- The most reliable way to identify a mineral is by using a combination of several tests.

Understand Main Ideas

1. MAIN Idea **List** two reasons why petroleum is not a mineral.
2. **Define** *naturally occurring* in terms of mineral formation.
3. **Contrast** the formation of minerals from magma and their formation from solution.
4. **Differentiate** between subjective and objective mineral properties.

Think Critically

5. **Develop** a plan to test the hardness of a sample of feldspar using the following items: glass plate, copper penny, and streak plate.
6. **Predict** the success of a lab test in which students plan to compare the streak colors of fluorite, quartz, and feldspar.

MATH in Earth Science

7. Calculate the volume of a 5-g sample of pure gold.

Section 4.2

Objectives

- **Identify** different groups of minerals.
- **Illustrate** the silica tetrahedron.
- **Discuss** how minerals are used.

Review Vocabulary

chemical bond: the force that holds two atoms together

New Vocabulary

silicate
tetrahedron
ore
gem

Types of Minerals

MAIN Idea **Minerals are classified based on their chemical properties and characteristics.**

Real-World Reading Link Everything on Earth is classified into various categories. Food, animals, and music are all classified according to certain properties or features. Minerals are no different; they, too, are classified into groups.

Mineral Groups

You have learned that elements combine in many different ways and proportions. One result is the thousands of different minerals present on Earth. In order to study these minerals and understand their properties, geologists have classified them into groups. Each group has a distinct chemical nature and specific characteristics.

Silicates Oxygen is the most abundant element in Earth's crust, followed by silicon. Minerals that contain silicon and oxygen, and usually one or more other elements, are known as **silicates.** Silicates make up approximately 96 percent of the minerals present in Earth's crust. The two most common minerals, feldspar and quartz, are silicates. The basic building block of the silicates is the silica tetrahedron, shown in **Figure 4.12.** A **tetrahedron** (plural, tetrahedra) is a geometric solid having four sides that are equilateral triangles, resembling a pyramid. Recall from Chapter 3 that the electrons in the outermost energy level of an atom are called valence electrons. The number of valence electrons determines the type and number of chemical bonds an atom will form. Because silicon atoms have four valence electrons, silicon has the ability to bond with four oxygen atoms. As shown in **Figure 4.13,** silica tetrahedra can share oxygen atoms. This structure allows tetrahedra to combine in a number of ways, which accounts for the large diversity of structures and properties of silicate minerals.

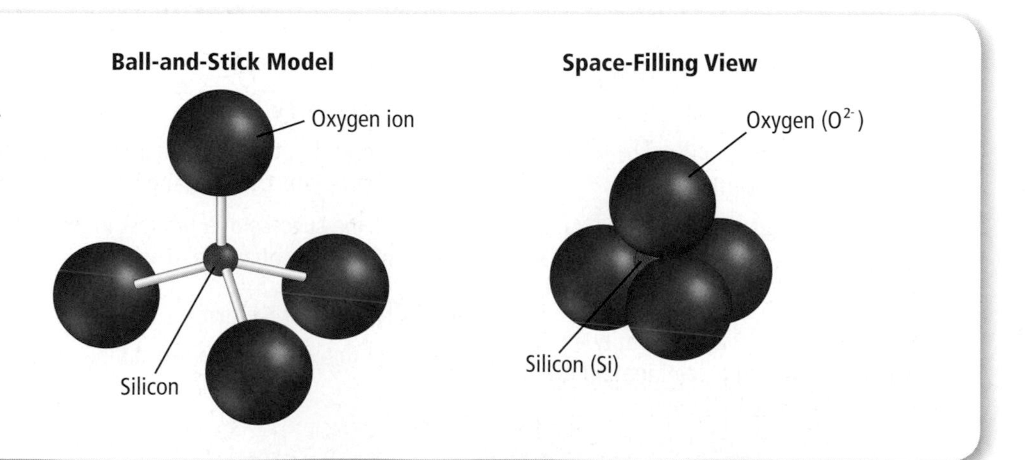

Figure 4.12 The silicate polyatomic ion SiO_4^{2-} forms a tetrahedron in which a central silicon atom is covalently bonded to oxygen ions.
Specify *How many atoms are in one tetrahedron?*

Visualizing the Silica Tetrahedron

Figure 4.13 The tetrahedron formed by silicates contains four oxygen ions bonded to a central silicon atom. Chains, sheets, and complex structures form as the tetrahedra bond with other tetrahedra. These structures become the numerous silicate minerals that are present on Earth.

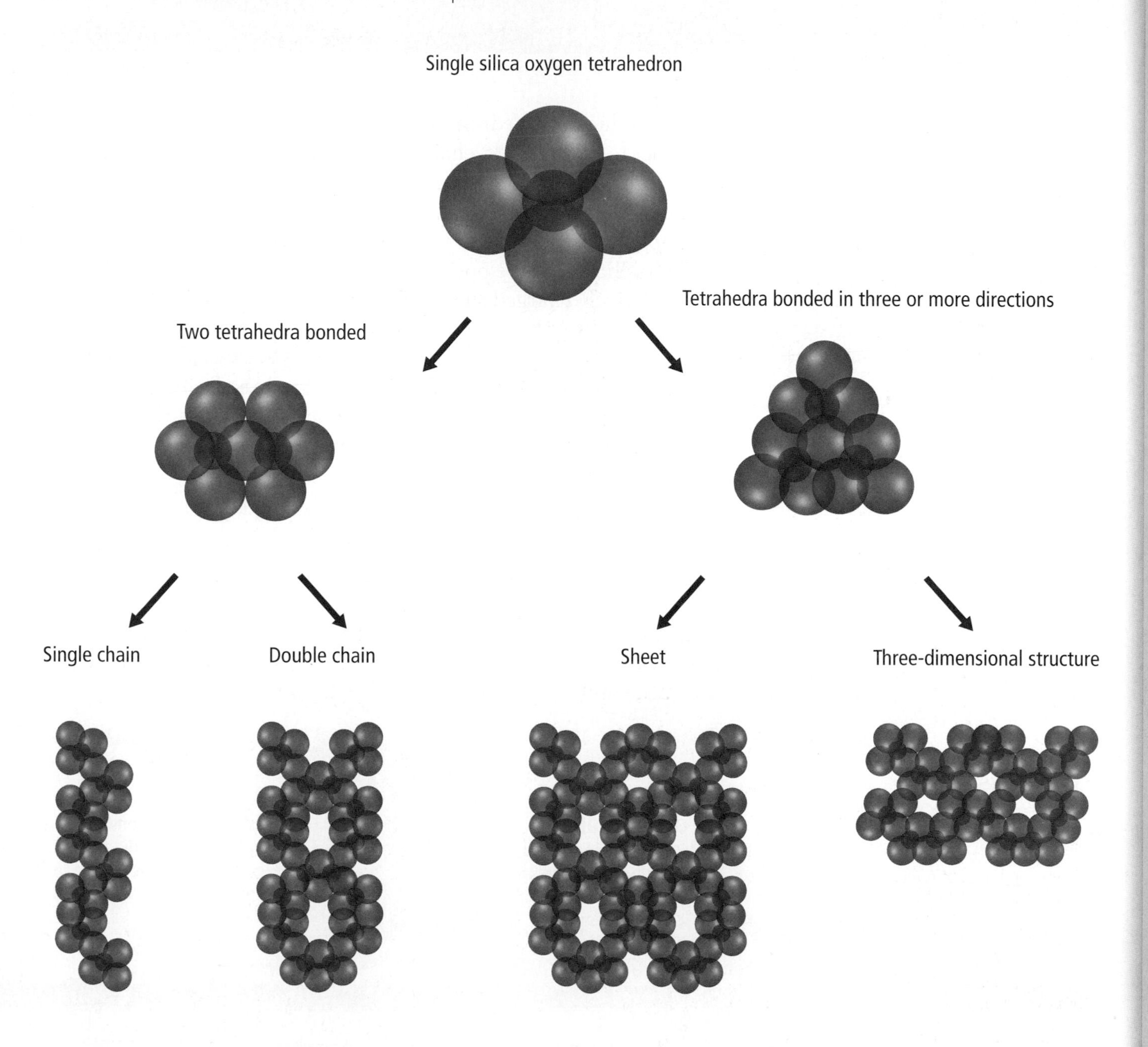

Concepts In Motion To explore more about the bonding behavior of the silica tetrahedron, visit glencoe.com.

Earth Science Online

■ **Figure 4.14** The differences in silicate minerals are due to the differences in the arrangement of their silica tetrahedra. Certain types of asbestos consist of weakly bonded double chains of tetrahedra, while mica consists of weakly bonded sheets of tetrahedra.

Asbestos

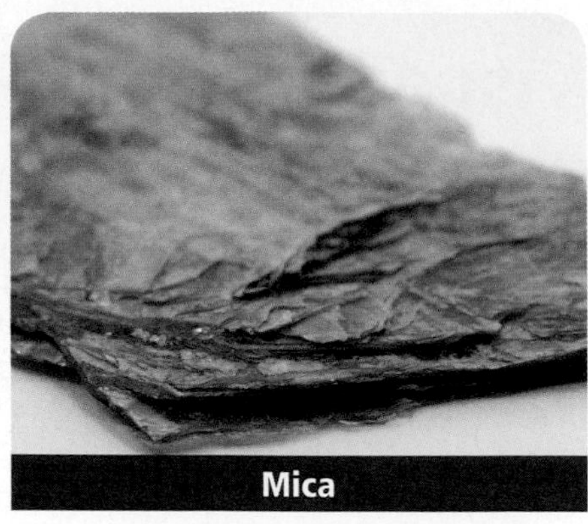
Mica

VOCABULARY
SCIENCE USAGE V. COMMON USAGE
Phyllo
Science usage: the sheets of silica tetrahedra

Common usage: sheets of dough used to make pastries and pies

Individual tetrahedron ions are strongly bonded. They can bond together to form sheets, chains, and complex three-dimensional structures. The bonds between the atoms help determine several mineral properties, including a mineral's cleavage or fracture. For example, mica, shown in **Figure 4.14,** is a sheet silicate, also called a phyllosilicate, where positive potassium or aluminum ions bond the negatively charged sheets of tetrahedra together. Mica separates easily into sheets because the attraction between the tetrahedra and the aluminum or potassium ions is weak. Asbestos, also shown in **Figure 4.14,** consists of double chains of tetrahedra that are weakly bonded together. This results in the fibrous nature shown in **Figure 4.14.**

Carbonates Oxygen combines easily with almost all other elements, and forms other mineral groups, such as carbonates. Carbonates are minerals composed of one or more metallic elements and the carbonate ion CO_3^{2-}. Examples of carbonates are calcite, dolomite, and rhodochrosite. Carbonates are the primary minerals found in rocks such as limestone and marble. Some carbonates have distinctive colorations, such as the colorful varieties of calcite and the pink of rhodochrosite shown in **Figure 4.16.**

■ **Figure 4.15**
Mineral Use Through Time

The value and uses of minerals have changed over time.

10,000 B.C. — 3000 B.C. — 500 B.C.

12,000–9000 B.C. The demand for flint—a hard volcanic glass used for tools—produces the first known long-distance trade route.

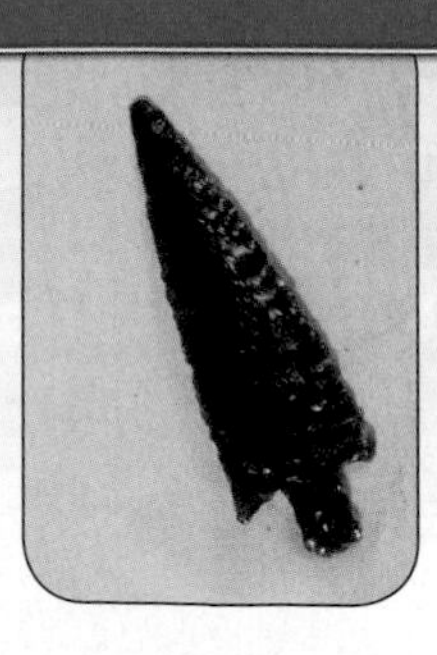

3300–3000 B.C. Bronze weapons and tools become common in the Near East as large cities and powerful empires arise.

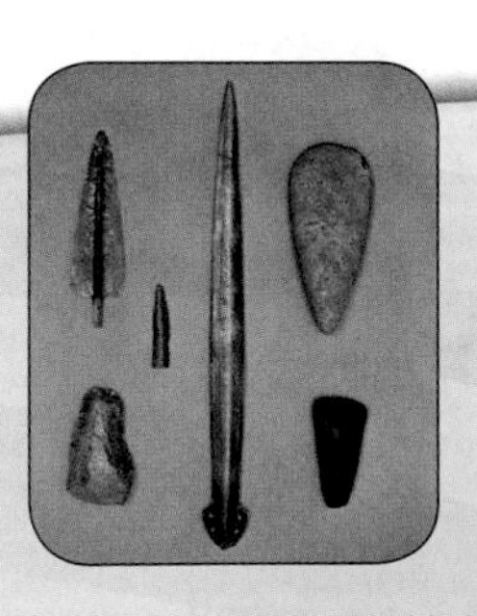

1200–1000 B.C. In the Near East, bronze becomes scarce and is replaced by iron in tools and weapons.

800 B.C. Diamond use spreads from India to other parts of the world to be used for cutting, engraving, and in ceremonies.

506 B.C. Rome takes over the salt industry at Ostia. The word *salary* comes from *salarium argentums,* the salt rations paid to Roman soldiers.

Figure 4.16 Carbonates such as calcite and rhodochrosite occur in distinct colors due to trace elements found in them.

Oxides Oxides are compounds of oxygen and a metal. Hematite (Fe_2O_3) and magnetite (Fe_3O_4) are common iron oxides and good sources of iron. The mineral uraninite (UO_2) is valuable because it is the major source of uranium, which is used to generate nuclear power.

Other groups Other major mineral groups are sulfides, sulfates, halides, and native elements. Sulfides, such as pyrite (FeS_2), are compounds of sulfur and one or more elements. Sulfates, such as anhydrite ($CaSO_4$), are composed of elements with the sulfate ion SO_4^{2-}. Halides, such as halite (NaCl), are made up of chloride or fluoride along with calcium, sodium, or potassium. A native element such as silver (Ag) or copper (Cu), is made up of one element only.

Economic Minerals

Minerals are virtually everywhere. They are used to make computers, cars, televisions, desks, roads, buildings, jewelry, beds, paints, sports equipment, and medicines, in addition to many other things. You can learn about the uses of minerals throughout history by examining **Figure 4.15.**

800–900 Chinese alchemists combine saltpeter with sulfur and carbon to make gunpowder, which is first used for fireworks and later used for weapons.

1546 South American silver mines help establish Spain as a global trading power, supplying silver needed for coinage.

2006 There are 242 uranium-fueled nuclear power plants in operation worldwide with a net capacity of 369.566 GW(e).

A.D. 500 — 1500 — 2000

A.D. 200–400 Iron farming tools and weapons allow people to migrate across Africa clearing and cultivating land for agricultural settlement and driving out hunter-gatherer societies.

1927 The first quartz clock improves timekeeping accuracy. The properties of quartz make it instrumental to the development of radio, radar, and computers.

Concepts In Motion

Interactive Time Line To learn more about these discoveries and others, visit glencoe.com. Earth Science Online

Concepts In Motion

Interactive Table To explore more about major mineral groups, visit glencoe.com.

Table 4.4 Major Mineral Groups

Group	Examples	Economic Use
Silicates	mica (biotite)	furnace windows
	olivine (Mg_2SiO_4)	gem (as peridot)
	quartz (SiO_2)	timepieces
	vermiculite	potting soil additive; swells when wet
Sulfides	pyrite (FeS_2)	used to make sulfuric acid; often mistaken for gold (fool's gold)
	marcasite (FeS_2)	jewelry
	galena (PbS)	lead ore
	sphalerite (ZnS)	zinc ore
Oxides	hematite (Fe_2O_3)	iron ore; red pigment
	corundum (Al_2O_3)	abrasive, gem (as in ruby or sapphire)
	uraninite (UO_2)	uranium source
	ilmenite ($FeTiO_3$)	titanium source; pigment; replaced lead in paint
	chromite ($FeCr_2O_4$)	chromium source, plumbing fixtures, auto accessories
Sulfates	gypsum ($CaSO_4 \cdot 2H_2O$)	plaster, drywall; slows drying in cement
	anhydrite ($CaSO_4$)	plaster; name indicates absence of water
Halides	halite (NaCl)	table salt, stock feed, weed killer, food preparation and preservative
	fluorite (CaF_2)	steel manufacturing, enameling cookware
	sylvite (KCl)	fertilizer
Carbonates	calcite ($CaCO_3$)	Portland cement, lime, chalk
	dolomite ($CaMg(CO_3)_2$)	Portland cement, lime; source of calcium and magnesium in vitamin supplements
Native elements	gold (Au)	monetary standard, jewelry
	copper (Cu)	coinage, electrical wiring, jewelry
	silver (Ag)	coinage, jewelry, photography
	sulfur (S)	sulfa drugs and chemicals; match heads; fireworks
	graphite (C)	pencil lead, dry lubricant

Figure 4.17 Parts of this athlete's wheelchair are made of titanium. Its light weight and extreme strength makes it an ideal metal to use.

Ores Many of the items just mentioned are made from ores. A mineral is an **ore** if it contains a valuable substance that can be mined at a profit. Hematite, for instance, is an ore that contains the element iron. Consider your classroom. If any items are made of iron, their original source might have been the mineral hematite. If there are items in the room made of aluminum, their original source was the ore bauxite. A common use of the metal titanium, obtained from the mineral ilmenite, is shown in **Figure 4.17. Table 4.4** summarizes the mineral groups and their major uses.

The classification of a mineral as an ore can also change if the supply of or demand for that mineral changes. Consider a mineral that is used to make computers. Engineers might develop a more efficient design or a less costly alternative material. In either of these cases, the mineral would no longer be used in computers. Demand for the mineral would drop. It would not be profitable to mine. The mineral would no longer be considered an ore.

Mines Ores that are located deep within Earth's crust are removed by underground mining. Ores that are near Earth's surface are obtained from large, open-pit mines. When a mine is excavated, unwanted rock and dirt, known as gangue, are dug up along with the valuable ore. The overburden must be separated from the ore before the ore can be used. Removing the overburden can be expensive and, in some cases, harmful to the environment, as you will learn in Chapters 24 and 26. If the cost of removing the overburden becomes higher than the value of the ore itself, the mineral will no longer be classified as an ore. It would no longer be economical to mine.

Gems What makes a ruby more valuable than mica? Rubies are rarer and more visually pleasing than mica. Rubies are thus considered gems. **Gems** are valuable minerals that are prized for their rarity and beauty. They are very hard and scratch resistant. Gems such as rubies, emeralds, and diamonds are cut, polished, and used for jewelry. Because of their rareness, rubies and emeralds are more valuable than diamonds. **Figure 4.18** shows a rough diamond and a polished diamond.

In some cases, the presence of trace elements can make one variety of a mineral more colorful and more prized than other varieties of the same mineral. Amethyst, for instance, is the gem form of quartz. Amethyst contains traces of iron, which gives the gem a purple color. The mineral corundum, which is often used as an abrasive, also occurs as rubies and sapphires. Rubies contain trace amounts of chromium, while sapphires contain trace amounts of cobalt or titanium.

Figure 4.18 The real beauty of gemstones is revealed once they are cut and polished.

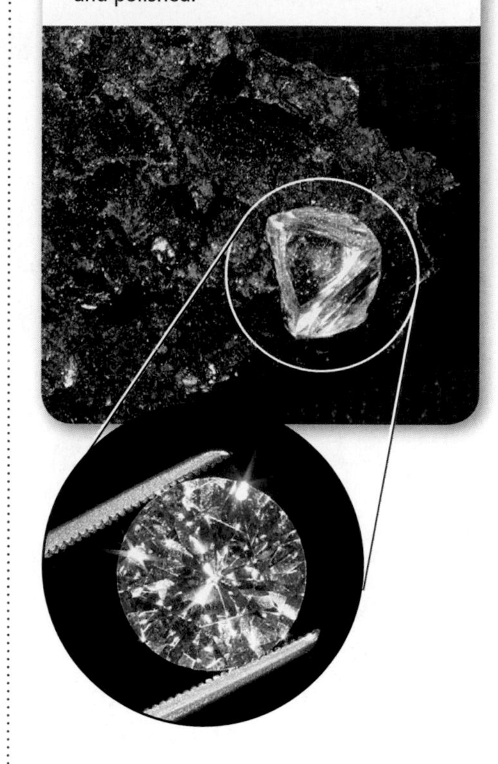

Section 4.2 Assessment

Section Summary

- In silicates, one silicon atom bonds with four oxygen ions to form a tetrahedron.
- Major mineral groups include silicates, carbonates, oxides, sulfides, sulfates, halides, and native elements.
- An ore contains a valuable substance that can be mined at a profit.
- Gems are valuable minerals that are prized for their rarity and beauty.

Understand Main Ideas

1. MAIN Idea **Formulate** a statement that explains the relationship between chemical elements and mineral properties.
2. **List** the two most abundant elements in Earth's crust. What mineral group do these elements form?
3. **Hypothesize** what some environmental consequences of mining ores might be.

Think Critically

4. **Hypothesize** why the mineral opal is often referred to as a mineraloid.
5. **Evaluate** which of the following metals is better to use in sporting equipment and medical implants: titanium—specific gravity = 4.5, contains only Ti; or steel—specific gravity = 7.7, contains Fe, O, Cr.

WRITING in Earth Science

6. Design a flyer advertising the sale of a mineral of your choice. You might choose a gem or industrially important mineral. Include any information that you think will help your mineral sell.

ON SITE: CRYSTALS AT LARGE IN MEXICO

Cave of Crystals, part of Naica Cave in Chihuahua, Mexico is known for its large crystals.

Eloy and Javier Delgado walk slowly into the Naica Cave in Chihuahua, Mexico. The cave is very hot, making it difficult for them to breathe. They enter a room in the cave and before them are huge 4.5-m crystals that are clear and brilliant. How did these crystals grow this large? What kinds of conditions make these crystals possible?

The climate inside the cave The large gypsum minerals are present in the Cave of Crystals, a room in Naica Cave, located 300 m below Earth's surface. Temperatures there hover around 58°C. The air here has a relative humidity of 100 percent. These extreme conditions mean that anyone entering the cave can remain only for a few minutes at a time.

Crystal formations in the cave The crystals in the Naica Cave are a crystalline form of gypsum called selenite. The crystals in this cave grow into three distinct shapes. Crystals that grow from the floor of the cave are plantlike in appearance. They are grayish in color from the mud that seeps into them as they grow. Swordlike crystals cover the walls of the cave. These crystals grow to lengths of 0.5 m to 1 m and are opaque white in color. Within the main room of the cave, there are crystals with masses of up to 27 kg and up to 8.25 m long and 1 m wide.

How did these crystals form? Crystals need several things in order to form. First, they need a space—in this case, a cave. Caves form as a result of water circulating along weak planes in a rock. Over time, the rock dissolves and a cave is formed. Second, crystals need a source of water that is rich in dissolved minerals. Crystal formation also depends on factors such as pressure, temperature, level of water in the cave, and the chemistry of the mineral-rich water.

In 2006, geologists determined that the crystals' massive sizes resulted from the steady temperature of about 58°C while the cave was full of mineral-rich water. As long as the crystals remained in this environment, they continued to grow. Because the crystals are so large, scientists think that the Cave of Crystals had these conditions for thousands of years.

WRITING in Earth Science

Research Visit glencoe.com to conduct research about the processes that form crystals in a cave. Pick a cave and make a brochure describing and illustrating the types of crystals found there.

GEOLAB

DESIGN YOUR OWN: MAKE A FIELD GUIDE FOR MINERALS

Background: Have you ever used a field guide to identify a bird, flower, rock, or insect? If so, you know that field guides include more than photographs. A typical field guide for minerals might include background information about minerals in general and specific information about the formation, properties, and uses of each mineral.

Question: *Which mineral properties should be included in a field guide to help identify unknown minerals?*

Materials

Choose materials that would be appropriate for this lab.

mineral samples
magnifying lens
glass plate
streak plate
the Mohs scale of mineral hardness
steel file or nail
piece of copper
paper clip
magnet
dilute hydrochloric acid
dropper
Reference Handbook

Safety Precautions

Procedure

1. Read and complete the lab safety form.
2. As a group, list the steps that you will take to create your field guide. Keep the available materials in mind as you plan your procedure.
3. Should you test any of the properties more than once for any of the minerals? How will you determine whether certain properties indicate a specific mineral?
4. Design a data table to summarize your results. Be sure to include a column to record whether or not a particular test will be included in the guide. You can use this table as the basis for your field guide.
5. Read over your entire plan to make sure that all steps are in a logical order.
6. Have you included a step for additional research? You might have to use the library or glencoe.com to gather all the necessary information for your field guide.
7. What additional information will be included in the field guide? Possible data include how each mineral formed, its uses, its chemical formula, and a labeled photograph or drawing of the mineral.
8. Make sure your teacher approves your plan before you proceed.

Analyze and Conclude

1. **Interpret** Which properties were most reliable for identifying minerals? Which properties were least reliable? Discuss reasons that one property is more useful than others.
2. **Observe and Infer** What mineral reacted with the hydrochloric acid? Why did the mineral bubble? Write the balanced equation that describes the chemical reaction that took place between the mineral and the acid.
3. **Summarize** What information did you include in the field guide? What resources did you use to gather your data? Describe the layout of your field guide.
4. **Evaluate** the advantages and disadvantages of field guides.
5. **Conclude** Based on your results, is there any one definitive test that can always be used to identify a mineral? Explain your answer.

WRITING in Earth Science

Peer Review Trade field guides with another group and test them out by using them to identify a new mineral. Provide feedback to the authors of the guide that you use.

CHAPTER 4

Study Guide

Download quizzes, key terms, and flash cards from glencoe.com.

BIG Idea Minerals are an integral part of daily life.

Vocabulary	Key Concepts
Section 4.1 What is a mineral?	
• cleavage (p. 92) • crystal (p. 87) • fracture (p. 93) • hardness (p. 91) • luster (p. 90) • mineral (p. 86) • specific gravity (p. 95) • streak (p. 93)	MAIN Idea Minerals are naturally occurring, solid, inorganic compounds or elements. • A mineral is a naturally occurring, inorganic solid with a specific chemical composition and a definite crystalline structure. • A crystal is a solid in which the atoms are arranged in repeating patterns. • Minerals form from magma or from supersaturated solutions. • Minerals can be identified based on their physical and chemical properties. • The most reliable way to identify a mineral is by using a combination of several tests. 
Section 4.2 Types of Minerals	
• gem (p. 101) • ore (p. 100) • silicate (p. 96) • tetrahedron (p. 96)	MAIN Idea Minerals are classified based on their chemical properties and characteristics. • In silicates, one silicon atom bonds with four oxygen ions to form a tetrahedron. • Major mineral groups include silicates, carbonates, oxides, sulfides, sulfates, halides, and native elements. • An ore contains a valuable substance that can be mined at a profit. • Gems are valuable minerals that are prized for their rarity and beauty. 

CHAPTER 4 Assessment

Vocabulary Review

Use what you know about the vocabulary terms listed on the Study Guide to answer the following questions.

1. What is a naturally occurring, solid, inorganic compound or element?
2. What term refers to the regular, geometric shapes that occur in many minerals?
3. What is the term for minerals containing silicon and oxygen?

Explain the relationship between the vocabulary terms in each pair.

4. ore, gem
5. silicate, tetrahedron

Complete the sentences below using vocabulary terms from the Study Guide.

6. Minerals that break randomly exhibit ________.
7. The ________ test determines what materials a mineral will scratch.

Understand Key Concepts

Use the photo below to answer Question 8.

8. Which mineral property is being tested?
   A. texture
   B. hardness
   C. cleavage
   D. streak
9. Which property causes the mineral galena to break into tiny cubes?
   A. density
   B. crystal structure
   C. hardness
   D. luster
10. What characteristic is used for classifying minerals into individual groups?
    A. internal atomic structure
    B. presence or absence of silica tetrahedrons
    C. chemical composition
    D. density and hardness
11. A mineral has a mass of 100 g and a volume of 50 cm^3. What is its density?
    A. 5000 g/cm^3
    B. 2 g/cm^3
    C. 5 g/cm^3
    D. 150 g/cm^3
12. What is the correct chemical formula for a silica tetrahedron?
    A. SiO_2
    B. $Si_2O_2^{+4}$
    C. SiO_4^{-4}
    D. Si_2O_2

Use the diagram below to answer Questions 13 and 14.

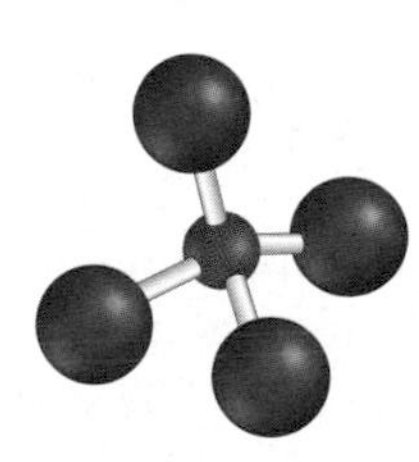

13. Where do the tetrahedra bond to each other?
    A. the center of the silicon atom
    B. at any oxygen atom
    C. only the top oxygen atom
    D. only the bottom oxygen atoms
14. What group of minerals is composed mainly of these tetrahedra?
    A. silicates
    B. oxides
    C. carbonates
    D. sulfates
15. Which is an example of a mineral whose streak cannot be determined with a porcelain streak plate?
    A. hematite
    B. gold
    C. feldspar
    D. magnetite

16. Which is one of the three most common elements in Earth's crust?
    A. sodium
    B. silicon
    C. iron
    D. carbon

Use the table below to answer Question 17.

Mineral Formulas	
Name	**Formula**
Quartz	SiO_2
Feldspar	$NaAlSi_3O_8$— $CaAl_2Si_2O_8$ & $KalSi_3O_8$
Amphibole	$Ca_2(Mg,Fe)_5Si_8O_{22}(OH)_2$ $Fe_7Si_8O_{22}(OH)_2$
Olivine	$(Mg,Fe)_2SiO_4$

17. What is the main factor that determines the formation of the minerals listed in the table?
    A. rate of magma cooling
    B. temperature of the magma
    C. presence or absence of water
    D. changes in pressure

18. Calcite is the dominant mineral in the rock limestone. In which mineral group does it belong?
    A. silicates
    B. oxides
    C. carbonates
    D. sulfates

19. What mineral fizzes when it comes in contact with hydrochloric acid?
    A. quartz
    B. gypsum
    C. calcite
    D. fluorite

20. *Dull, silky, waxy, pearly,* and *earthy* are terms that best describe which property of minerals?
    A. luster
    B. color
    C. streak
    D. cleavage

21. For a mineral to be considered an ore, which requirement must it meet?
    A. It must be a common mineral.
    B. Its production must not generate pollution.
    C. It must be naturally occurring.
    D. Its production must generate a profit.

Constructed Response

22. **Explain** why rubies and sapphires, which are both forms of the mineral corundum, are different colors.

23. **Describe** the visual effect of placing a piece of clear, Iceland spar on top of the word *geology* in a textbook.

24. **Summarize** the process of sugar crystals forming in a glass of sugar-sweetened hot tea.

25. **Hypothesize** which mineral properties are the direct result of the arrangement of atoms or ions in a crystal. Explain your answer.

26. **Compare and Contrast** Diamond and graphite have the same chemical composition. Compare and contrast these two to explain why diamond is a gem and graphite is not.

Think Critically

27. **Describe** the differences that might be exhibited by the garnets listed in **Table 4.1.**

Use the figure below to answer Question 28.

28. **Illustrate** what the atomic structure might be if the crystal shape is an external reflection of it.

29. **Recommend** which minerals, other than diamond, would be best for making sandpaper. Explain your answer. Refer to **Table 4.2.**

Earth Science Online Chapter Test glencoe.com

30. **Decide** which of the following materials are not minerals, and explain why: petroleum, wood, coal, steel, concrete, and glass.

31. **Infer** how early prospectors used density to determine whether they had found gold or pyrite in a mine.

32. **Assess** Imagine that a new gem is discovered that is more beautiful than the most stunning diamond or ruby. Assess the factors that will determine its cost compared to other known gems.

Use the figure below to answer Questions 33–34.

33. **Infer** Mica is a mineral with a sheet silicate structure. The atomic arrangement is shown above. Infer what is holding these sheets, which consist of negatively charged silicon-oxygen tetrahedra, together.

34. **Describe** the type of cleavage that occurs in minerals with the atomic arrangement shown.

Concept Mapping

35. Create a concept map using the following terms: *silicates, oxides, halides, sulfates, sulfides, native elements,* and *carbonates.* Add any other terms that are helpful. For more help, refer to the *Skillbuilder Handbook.*

Challenge Question

36. **Arrange** In addition to sheet silicates, there are chain silicates, tectosilicates, and cyclosilicates. Arrange six silica tetrahedra in a cyclosilicate form. Be sure to bond the oxygen atoms correctly.

Additional Assessment

37. *WRITING in* **Earth Science** Imagine that you are planning a camping trip. What tools should you pack if you want to identify interesting minerals? How would you use these tools?

DBQ Document–Based Questions

Data obtained from: Plunkert, P.A. 2005. Mineral resource of the month: Aluminum. *Geotimes* 50:57.

Aluminum is an abundant metallic element in Earth's crust. It is lightweight, ductile [bendable], corrosion resistant, and a good conductor of electricity. It is used most often in the manufacture of cars, buses, trailers, ships, aircraft, railway and subway cars. Other uses include beverage cans, aluminum foil, machinery, and electrical equipment.

Aluminum is produced from bauxite (hydrated aluminum-oxide) deposits, located mostly in Guinea, Australia, and South America. The United States does not have bauxite deposits; it imports it from Brazil, Guinea, and Jamaica. Total world aluminum production is approximately 30 million metric tons per year. U.S. aluminum production is less than U.S. aluminum consumption. Leading aluminum producers are China and Russia. A major part (3 million metric tons per year) of the U.S. aluminum supply comes from recycling.

38. Interpret the relationship between aluminum's resistance to corrosion and its use in transportation vehicles.

39. Propose a plan for how the United States can increase aluminum production without increasing the amount it imports.

40. Predict the possible effects an increase in U.S. production would have on Guinea, Jamaica, and China.

Cumulative Review

41. How do different isotopes of an element differ from each other? **(Chapter 3)**

42. Why is an understanding of the study of Earth science important to us as residents of Earth? **(Chapter 1)**

Standardized Test Practice

Multiple Choice

1. What is the second most abundant element in Earth's crust?
   A. nitrogen
   B. oxygen
   C. silicon
   D. carbon

Use the table below to answer Questions 2 and 3.

Mineral Characteristics			
Mineral	**Hardness**	**Specific Gravity**	**Luster/Color**
Feldspar	6–6.5	2.5–2.8	nonmetallic/colorless or white
Fluorite	4	3–3.3	nonmetallic/yellow, blue, purple, rose, green, or brown
Galena	2.5–2.75	7.4–7.6	metallic/grayish black
Quartz	7	2.65	nonmetallic/colorless in pure form

2. What is the hardest mineral in the table?
   A. feldspar
   B. fluorite
   C. galena
   D. quartz

3. Which mineral most likely has a shiny appearance?
   A. feldspar
   B. fluorite
   C. galena
   D. quartz

4. What can be inferred about an isotope that releases radiation?
   A. It has unstable nuclei.
   B. It has stable nuclei.
   C. It has the same mass number as another element.
   D. It is not undergoing decay.

5. How do electrons typically fill energy levels?
   A. from lowest to highest
   B. from highest to lowest
   C. in no predictable pattern
   D. all in one energy level

6. What is the most reliable clue to a mineral's identity?
   A. color
   B. streak
   C. hardness
   D. luster

Use the table below to answer Questions 7 and 8.

Mineral	Hardness
Talc	1
Gypsum	2
Calcite	3
Fluorite	4
Apatite	5
Feldspar	6
Quartz	7
Topaz	8
Corundum	9
Diamond	10

7. Which mineral will scratch feldspar but not topaz?
   A. quartz
   B. calcite
   C. apatite
   D. diamond

8. What can be implied about diamond based on the table?
   A. It is the heaviest mineral.
   B. It is the slowest mineral to form.
   C. It has the most defined crystalline structure.
   D. It cannot be scratched by any other mineral.

9. A well-planned experiment must have all of the following EXCEPT
   A. technology
   B. a control
   C. a hypothesis
   D. collectible data

10. What name is given to the imaginary line circling Earth halfway between the north and south poles?
   A. prime meridian
   B. equator
   C. latitude
   D. longitude

Short Answer

Use the conversion factor and table below to answer Questions 11–13.

1.0 carat = 0.2 grams

Diamond	Carats	Grams
Uncle Sam: largest diamond found in United States	40.4	?
Punch Jones: second largest; named after boy who discovered it	?	6.89
Theresa: discovered in Wisconsin in 1888	21.5	4.3
2001 diamond production from western Australia	21,679,930	?

11. List the three diamonds from least to greatest according to carats, and list the carats.

12. How many kilograms of diamonds were produced in western Australia in 2001?

13. Why would a diamond excavator want to convert the diamond measurement from carats to grams?

14. Why are map scales important parts of a map?

15. Discuss how a scientist might use a Landsat satellite image to determine the amount of pollution being produced by a city.

16. Why might a mineral no longer be classified as an ore?

Reading for Comprehension

Silicon Valley

Silicon (Si) is the second most abundant element in Earth's crust, but we didn't hear much about it until Silicon Valley. It is present in measurable amounts in nearly every rock, in all natural waters, as dust in the air, in the skeletons of many plants and some animals, and even in the stars. Silicon is never found in the free state like gold or silver, but is always with oxygen (O), aluminum (Al), magnesium (Mg), calcium (Ca), sodium (Na), potassium (K), iron (Fe), or other elements in combinations called the silicates. Silicates are the largest and most complicated group of minerals. Silicon is dull gray in appearance and has a specific gravity of 2.42. It has valence electrons like carbon (C) and can form a vast array of chemical compounds like silicon carbide abrasive, silicon rubber and caulking, oils and paints. Pure silicon is used in semiconductors, as solar panels to generate electricity from light, and in microchips for transistors.

Information obtained from: Ellison, B. Si and SiO_2...or what a difference a little O makes. (online resource accessed October 2006.)

17. According to the text, what is the most challenging aspect of silicon?
A. It has valence electrons.
B. It is dull gray in appearance.
C. It is never found in its free state.
D. It is present in many places.

18. Which is NOT a use of silicon as a chemical compound given in this passage?
A. silicon rubber and caulking
B. silicon carbide abrasive
C. microchips for transistors
D. oils and paints

19. Why was silicon not widely known until Silicon Valley?

NEED EXTRA HELP?																
If You Missed Question . . .	1	2	3	4	5	6	7	8	9	10	11	12	13	14	15	16
Review Section . . .	4.2	4.1	4.1	3.1	3.6	4.1	4.1	4.1	1.2	2.1	4.2	4.2	1.2	2.2	2.3	3.2

CHAPTER 5

Igneous Rocks

BIG Idea **Igneous rocks were the first rocks to form as Earth cooled from a molten mass to the crystalline rocks of the early crust.**

5.1 What are igneous rocks?
MAIN Idea Igneous rocks are the rocks that form when molten material cools and crystallizes.

5.2 Classification of Igneous Rocks
MAIN Idea Classification of igneous rocks is based on mineral composition, crystal size, and texture.

GeoFacts

- In the monument pictured here, Crazy Horse's head is over 26 m tall.
- The monument, located in South Dakota, was started in 1948 and is still a work in progress. The next component created will be his arm, which will measure more than 70 m.
- When completed, the monument will be more than 170 m tall and 195 m long. Nearly 10,000,000 metric tons of rock have already been blasted away.

LAUNCH Lab

How are minerals identified?

Igneous rocks are composed of different types of minerals. It is often possible to identify the different minerals in a sample of rock.

Procedure

1. Read and complete the lab safety form.
2. Examine a **sample of granite** from a distance of about 1 m. Record your observations.
3. Use a **magnifying lens** or **microscope** to observe the granite sample. Record your observations.

Analysis

1. **Illustrate** what you saw through the magnifying glass or microscope. Include a scale for your drawing.
2. **List** the different minerals that you observed in your sample.
3. **Describe** the sizes and shapes of the mineral crystals.
4. **Describe** any evidence that suggests that these crystals formed from molten rock.

Types of Igneous Rocks Make this Foldable to compare intrusive and extrusive igneous rock.

STEP 1 Fold the bottom of a horizontal sheet of paper up about 3 cm.

STEP 2 Fold in half.

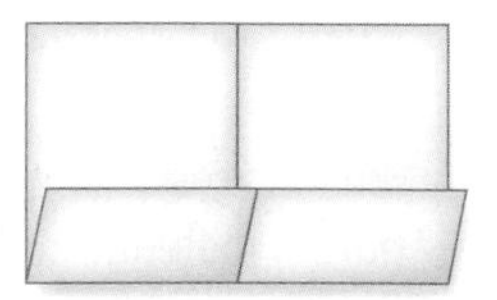

STEP 3 Unfold once, and dot with glue or staple to make two pockets. Label as shown.

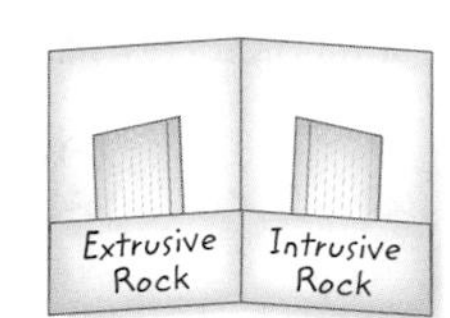

FOLDABLES **Use this Foldable with Section 5.1.** As you read this section, use index cards or quarter sheets of paper to summarize how each type of rock forms and give examples.

Visit glencoe.com to

- **study entire chapters online;**
- **explore Concepts In Motion animations:**
  - **Interactive Time Lines**
  - **Interactive Figures**
  - **Interactive Tables**
- **access Web Links for more information, projects, and activities;**
- **review content with the Interactive Tutor and take Self-Check Quizzes.**

Section 5.1

Objectives

- **Summarize** igneous rock formation.
- **Describe** the composition of magma.
- **Identify** the factors that affect how rocks melt and crystallize.

Review Vocabulary

silicate: mineral that contains silicon and oxygen, and usually one or more other elements

New Vocabulary

lava
igneous rock
partial melting
Bowen's reaction series
fractional crystallization

What are igneous rocks?

MAIN Idea Igneous rocks are the rocks that form when molten material cools and crystallizes.

Real-World Reading Link At any given point in time, igneous rocks are forming somewhere on Earth. The location and the conditions that are present determine the types of igneous rocks that form.

Igneous Rock Formation

If you live near an active volcano, you can literally watch igneous rocks form. A hot, molten mass of rock can solidify into solid rock overnight. As you read in Chapter 4, magma is molten rock below Earth's surface. **Lava** is magma that flows out onto Earth's surface. **Igneous rocks** form when lava or magma cools and minerals crystallize.

In the laboratory, most rocks must be heated to temperatures of 800°C to 1200°C before they melt. In nature, these temperatures are present in the upper mantle and lower crust. Where does this heat come from? Scientists theorize that the remaining energy from Earth's molten formation and the heat generated from the decay of radioactive elements are the sources of Earth's thermal energy.

Composition of magma The type of igneous rock that forms depends on the composition of the magma. Magma is often a slushy mix of molten rock, dissolved gases, and mineral crystals. The common elements present in magma are the same major elements that are in Earth's crust: oxygen (O), silicon (Si), aluminum (Al), iron (Fe), magnesium (Mg), calcium (Ca), potassium (K), and sodium (Na). Of all the compounds present in magma, silica is the most abundant and has the greatest effect on magma characteristics. As summarized in **Table 5.1,** magma is classified as basaltic, andesitic, or rhyolitic, based on the amount of silica it contains. Silica content affects melting temperature and impacts how quickly magma flows.

Concepts In Motion Interactive Table To explore more about magma composition, visit glencoe.com.

Table 5.1 Types of Magma

Group	Silica Content	Example Location
Basaltic	42–52%	Hawaiian Islands
Andesitic	52–66%	Cascade Mountains, Andes Mountains
Rhyolitic	more than 66%	Yellowstone National Park

Once magma is free of the overlying pressure of the rock layers around it, dissolved gases are able to escape into the atmosphere. Thus, the chemical composition of lava is slightly different from the chemical composition of the magma from which it developed.

Magma formation Magma can be formed either by melting of Earth's crust or by melting within the mantle. The four main factors involved in the formation of magma are temperature, pressure, water content, and the mineral content of the crust or mantle. Temperature generally increases with depth in Earth's crust. This temperature increase, known as the geothermal gradient, is plotted in **Figure 5.1.** Oil-well drillers and miners have firsthand experience with the geothermal gradient. Drill bits, such as the one shown in **Figure 5.2,** can encounter temperatures in excess of 200°C when drilling deep oil wells.

Pressure also increases with depth. This is a result of the weight of overlying rock. Laboratory experiments show that as pressure on a rock increases, its melting point also increases. Thus, a rock that melts at 1100°C at Earth's surface will melt at 1400°C at a depth of 100 km.

The third factor that affects the formation of magma is water content. Rocks and minerals often contain small percentages of water, which changes the melting point of the rocks. As water content increases, the melting point decreases.

Reading Check **List** the main factors involved in magma formation.

Mineral content In order to better understand how the types of elements and compounds present give magma its overall character, it is helpful to discuss this fourth factor in more detail. Different minerals have different melting points. For example, rocks such as basalt, which are formed of olivine, calcium feldspar, and pyroxene (pi RAHK seen), melt at higher temperatures than rocks such as granite, which contain quartz and potassium feldspar. Granite has a melting point that is lower than basalt's melting point because granite contains more water and minerals that melt at lower temperatures. In general, rocks that are rich in iron and magnesium melt at higher temperatures than rocks that contain higher levels of silicon.

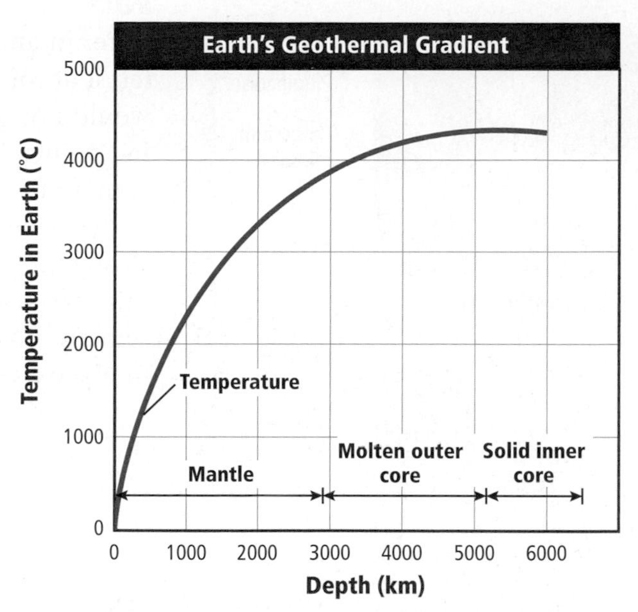

Figure 5.1 The average geothermal gradient in the crust is about 25°C/km, but scientists think that it drops sharply in the mantle to as low as 1°C/km.

Figure 5.2 The temperature of Earth's upper crust increases with depth by about 30°C for each 1 km. At a depth of 3 km, this drill bit will encounter rock that is close to the temperature of boiling water.

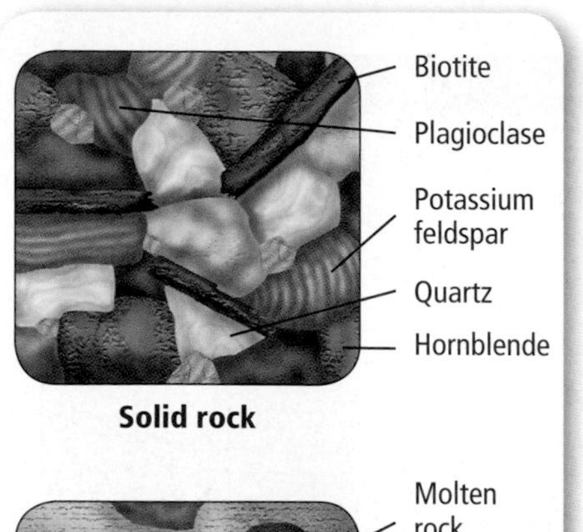

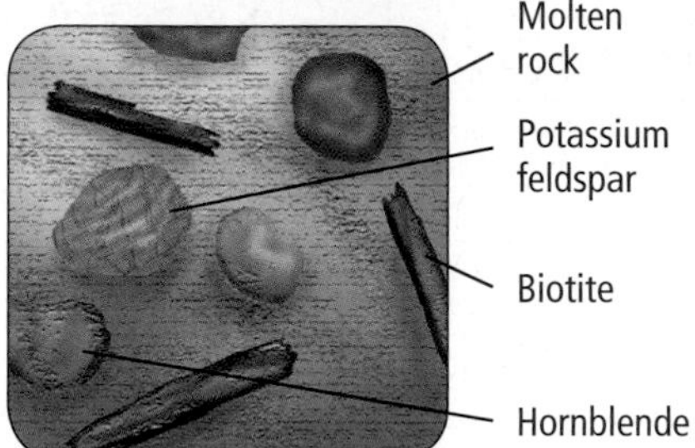

■ **Figure 5.3** As the temperature increases in an area, minerals begin to melt.
Determine ***What can you suggest about the melting temperature of quartz based on this diagram?***

Partial melting Suppose you froze melted candle wax and water in an ice cube tray. If you took the tray out of the freezer and left it at room temperature, the ice would melt, but the candle wax would not. This is because the two substances have different melting points. Rocks melt in a similar way because the minerals they contain have different melting points. Not all parts of a rock melt at the same temperature. This explains why magma is often a slushy mix of crystals and molten rock. The process whereby some minerals melt at relatively low temperatures while other minerals remain solid is called **partial melting.** Partial melting is illustrated in **Figure 5.3.** As each group of minerals melts, different elements are added to the magma mixture thereby changing its composition. If temperatures are not high enough to melt the entire rock, the resulting magma will have a different composition than that of the original rock. This is one way in which different types of igneous rocks form.

 **Reading Check** **Summarize** the formation of magma that has a different chemical composition from the original rock.

Bowen's Reaction Series

In the early 1900s, Canadian geologist N. L. Bowen demonstrated that as magma cools and crystallizes, minerals form in predictable patterns in a process now known as the **Bowen's reaction series.** **Figure 5.4** illustrates the relationship between cooling magma and the formation of minerals that make up igneous rock. Bowen discovered two main patterns, or branches, of crystallization. The right-hand branch is characterized by a continuous, gradual change of mineral compositions in the feldspar group. An abrupt change of mineral type in the iron-magnesium groups characterizes the left-hand branch.

■ **Figure 5.4** On the left side of Bowen's reaction series, minerals rich in iron and magnesium change abruptly as the temperature of the magma decreases.
Compare ***How does this compare to the feldspars on the right side of the diagram?***

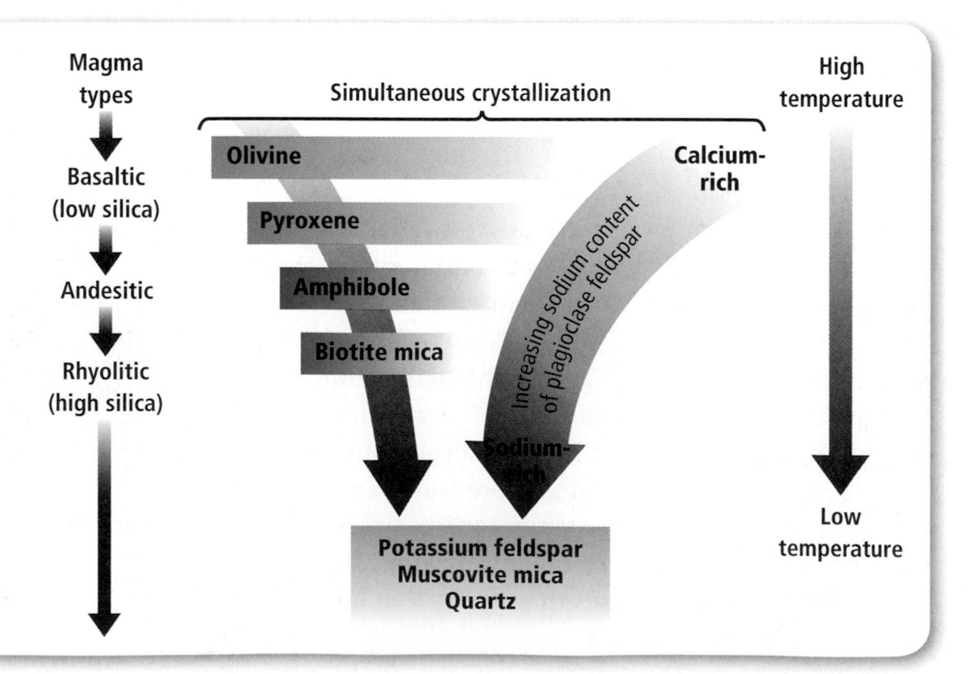

Iron-rich minerals The left branch of Bowen's reaction series represents the iron-rich minerals. These minerals undergo abrupt changes as magma cools and crystallizes. For example, olivine is the first mineral to crystallize when magma that is rich in iron and magnesium begins to cool. When the temperature decreases enough for a completely new mineral, pyroxene, to form, the olivine that previously formed reacts with the magma and is converted to pyroxene. As the temperature decreases further, similar reactions produce the minerals amphibole and biotite mica.

Feldspars In Bowen's reaction series, the right branch represents the plagioclase feldspars, which undergo a continuous change of composition. As magma cools, the first feldspars to form are rich in calcium. As cooling continues, these feldspars react with magma, and their calcium-rich compositions change to sodium-rich compositions. In some instances, such as when magma cools rapidly, the calcium-rich cores are unable to react completely with the magma. The result is a zoned crystal, as shown in **Figure 5.5.**

Figure 5.5 When magma cools quickly, a feldspar crystal might not have time to react completely with the magma and might retain a calcium-rich core. The result is a crystal with distinct calcium-rich and sodium-rich zones.

FOLDABLES
Incorporate information from this section into your Foldable.

Fractional Crystallization

When magma cools, it crystallizes in the reverse order of partial melting. That is, the first minerals that crystallize from magma are the last minerals that melted during partial melting. This process, called **fractional crystallization,** is similar to partial melting in that the composition of magma can change. In this case, however, early formed crystals are removed from the magma and cannot react with it. As minerals form and their elements are removed from the remaining magma, it becomes concentrated in silica.

MiniLab

Compare Igneous Rocks

How do igneous rocks differ? Igneous rocks have many different characteristics. Color and crystal size are some of the features that differentiate igneous rocks.

Procedure

1. Read and complete the lab safety form.
2. Obtain a set of **igneous rock samples** from your teacher.
3. Carefully observe the following characteristics of each rock: overall color, crystal size, and, if possible, mineral composition.
4. Design a data table to record your observations.

Analysis

1. **Classify** your samples as either basaltic, andesitic, or rhyolitic. [Hint: The more silica in the rock, the lighter it is in color.]
2. **Compare and contrast** your samples using the data from the data table. How do they differ? What characteristics do each of the groups share?
3. **Speculate** in which order the samples crystallized. [Hint: Use Bowen's reaction series as a guide.]

Visualizing Fractional Crystallization and Crystal Settling

Figure 5.6 The Palisades Sill in the Hudson River valley of New York and New Jersey is a classic example of fractional crystallization and crystal settling. In the basaltic intrusion, small crystals formed in the chill zone as the outer areas of the intrusion cooled more quickly than the interior.

As magma in an intrusion begins to cool, crystals form and settle to the bottom. This layering of crystals is fractional crystallization.

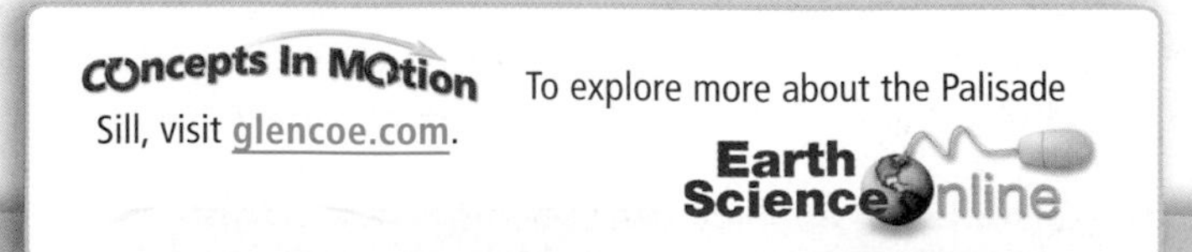

Concepts In Motion To explore more about the Palisade Sill, visit glencoe.com.

As is often the case with scientific inquiry, the discovery of Bowen's reaction series led to more questions. For example, if olivine converts to pyroxene during cooling, why is olivine found in rock? Geologists hypothesize that, under certain conditions, newly formed crystals are separated from magma, and the chemical reactions between the magma and the minerals stop. This can occur when crystals settle to the bottom of the magma body, and when liquid magma is squeezed from the crystal mush to form two distinct igneous bodies with different compositions. **Figure 5.6** illustrates this process and the concept of fractional crystallization with an example from the Hudson River valley in New York and New Jersey. This is one way in which the magmas listed in **Table 5.1** are formed.

As fractional crystallization continues and more magma is separated from the crystals, the magma becomes more concentrated in silica, aluminum, and potassium. This is why the last two minerals to form are potassium feldspar and quartz. Potassium feldspar is one of the most common feldspars in Earth's crust. Quartz often occurs in veins, as shown in **Figure 5.7,** because it crystallizes while the last liquid portion of magma is squeezed into rock fractures.

Figure 5.7 These quartz veins represent the last remnants of a magma body that cooled and crystallized.

Section 5.1 Assessment

Section Summary

- Magma consists of molten rock, dissolved gases, and mineral crystals.
- Magma is classified as basaltic, andesitic, or rhyolitic, based on the amount of silica it contains.
- Different minerals melt and crystallize at different temperatures.
- Bowen's reaction series defines the order in which minerals crystallize from magma.

Understand Main Ideas

1. MAIN Idea **Predict** the appearance of an igneous rock that formed as magma cooled quickly and then more slowly.
2. **List** the eight major elements present in most magmas. Include the chemical symbol of each element.
3. **Summarize** the factors that affect the formation of magma.
4. **Compare and contrast** magma and lava.

Think Critically

5. **Predict** If the temperature increases toward the center of Earth, why does the inner core become solid?
6. **Infer** the silica content of magma derived from partial melting of an igneous rock. Would it be higher, lower, or about the same as the rock itself? Explain.

WRITING in Earth Science

7. A local rock collector claims that she has found the first example of pyroxene and sodium-rich feldspar in the same rock. Write a commentary about her claim for publication in a rock collector society newsletter.

Earth Science Online **Self-Check Quiz** glencoe.com

Section 5.2

Classification of Igneous Rocks

Objectives

- **Classify** different types and textures of igneous rocks.
- **Recognize** the effects of cooling rates on the grain sizes in igneous rocks.
- **Describe** some uses of igneous rocks.

Review Vocabulary

fractional crystallization: a sequential process during which early formed crystals are removed from the melt and do not react with the remaining magma.

New Vocabulary

intrusive rock
extrusive rock
basaltic rock
granitic rock
texture
porphyritic texture
vesicular texture
pegmatite
kimberlite

MAIN Idea Classification of igneous rocks is based on mineral composition, crystal size, and texture.

Real-World Reading Link Many statues, floors, buildings, and countertops have something in common. Many of them are made of the popular rock type granite—one of the most abundant rocks in Earth's crust.

Mineral Composition of Igneous Rocks

Igneous rocks are broadly classified as intrusive or extrusive. When magma cools and crystallizes below Earth's surface, **intrusive rocks** form. If the magma is injected into the surrounding rock, it is called an igneous intrusion. Crystals of intrusive rocks are generally large enough to see without magnification. Magma that cools and crystallizes on Earth's surface forms **extrusive rocks.** These are sometimes referred to as lava flows or flood basalts. The crystals that form in these rocks are small and difficult to see without magnification. Geologists classify these rocks by their mineral compositions. In addition, physical properties such as grain size and texture serve as clues for the identification of various igneous rocks.

Igneous rocks are classified according to their mineral compositions. **Basaltic rocks,** such as gabbro, are dark-colored, have lower silica contents, and contain mostly plagioclase and pyroxene. **Granitic rocks,** such as granite, are light-colored, have high silica contents, and contain mostly quartz, potassium feldspar, and plagioclase feldspar. Rocks that have a composition of minerals that is somewhere in between basaltic and granitic are called intermediate rocks. They consist mostly of plagioclase feldspar and hornblende. Diorite is a good example of an intermediate rock. **Figure 5.8** shows examples from these three main compositional groups of igneous rocks. A fourth category, called ultrabasic, contains the rock peridotite. These rocks contain only iron-rich minerals such as olivine and pyroxene and are always dark. **Figure 5.9** summarizes igneous rock identification.

■ **Figure 5.8** Differences in magma composition can be observed in the rocks that form when the magma cools and crystallizes.
Observe *Describe the differences you see in these rocks.*

Gabbro

Granite

Diorite

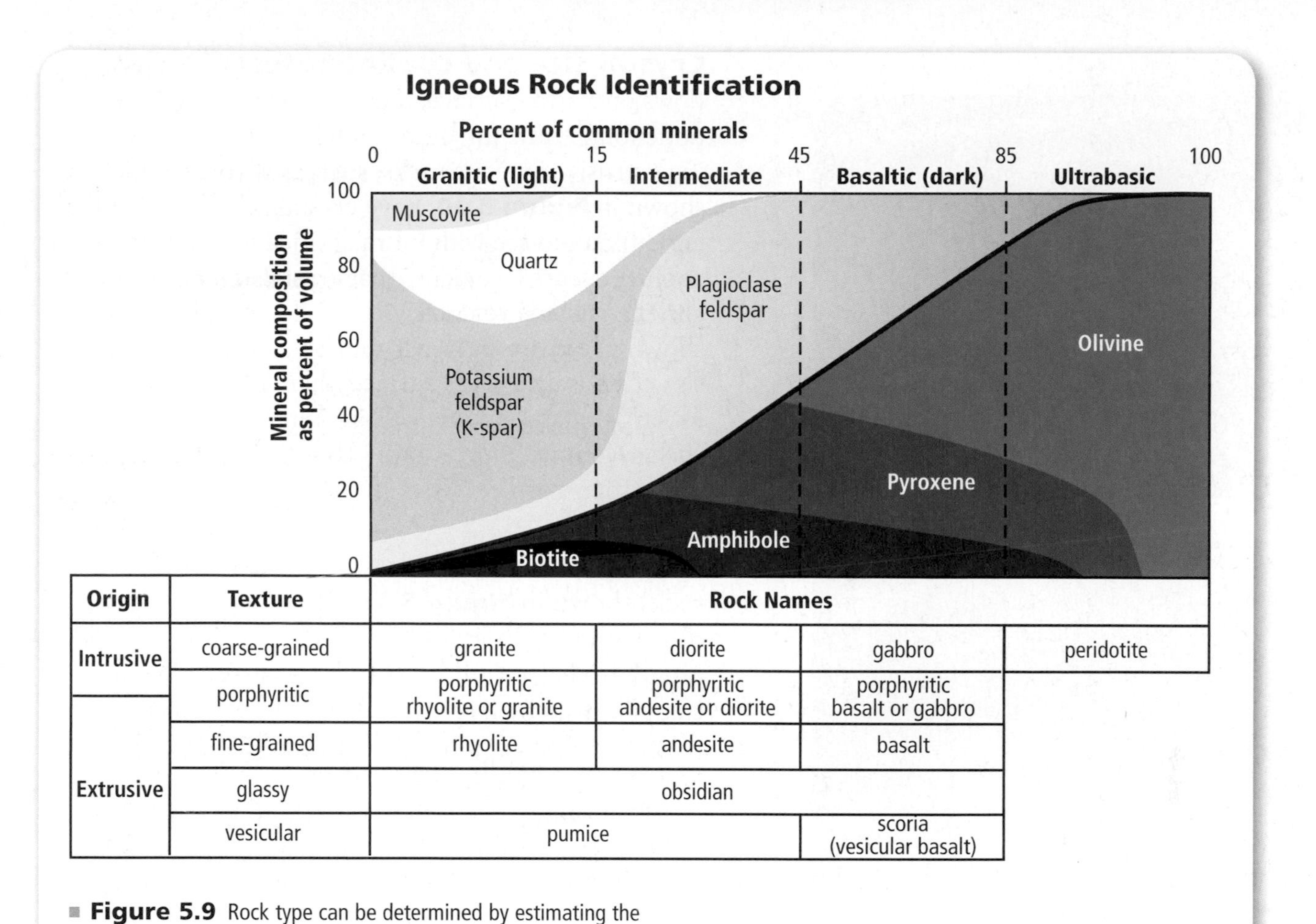

Origin	Texture	Rock Names			
Intrusive	coarse-grained	granite	diorite	gabbro	peridotite
	porphyritic	porphyritic rhyolite or granite	porphyritic andesite or diorite	porphyritic basalt or gabbro	
Extrusive	fine-grained	rhyolite	andesite	basalt	
	glassy	obsidian			
	vesicular	pumice		scoria (vesicular basalt)	

■ **Figure 5.9** Rock type can be determined by estimating the relative percentages of minerals in the rocks.

Texture

In addition to differences in their mineral compositions, igneous rocks differ in the sizes of their grains or crystals. **Texture** refers to the size, shape, and distribution of the crystals or grains that make up a rock. For example, as shown in **Figure 5.10,** the texture of rhyolite can be described as fine-grained, while granite can be described as coarse-grained. The difference in crystal size can be explained by the fact that one rock is extrusive and the other is intrusive.

■ **Figure 5.10** Rhyolite, granite, and obsidian have different textures because they formed in different ways.

Rhyolite

Granite

Obsidian

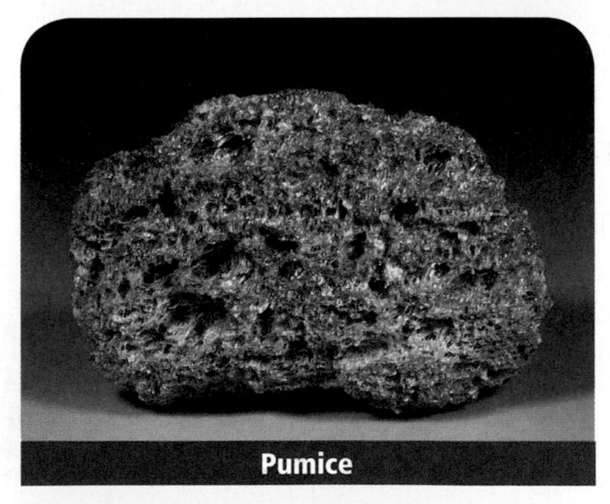

■ **Figure 5.11** Rock textures provide information about a rock's formation. Evidence of the rate of cooling and the presence or absence of dissolved gases is preserved in the rocks shown here.

Crystal size and cooling rates When lava flows on Earth's surface, it cools quickly and there is not enough time for large crystals to form. The resulting extrusive igneous rocks, such as rhyolite, which is shown in **Figure 5.10,** have crystals so small that they are difficult to see without magnification. Sometimes, cooling occurs so quickly that crystals do not form at all. The result is volcanic glass, called obsidian, also shown in **Figure 5.10.** In contrast, when magma cools slowly beneath Earth's surface, there is sufficient time for large crystals to form. Thus, intrusive igneous rocks, such as granite, diorite, and gabbro, can have crystals larger than 1 cm.

Porphyritic rocks Look at the textures of the rocks shown in **Figure 5.11.** The top photo shows a rock with two different crystal sizes. This rock has a **porphyritic** (por fuh RIH tihk) **texture,** which is characterized by large, well-formed crystals surrounded by finer-grained crystals of the same mineral or different minerals.

What causes minerals to form both large and small crystals in the same rock? Porphyritic textures indicate a complex cooling history during which a slowly cooling magma suddenly began cooling rapidly. Imagine a magma body cooling slowly, deep in Earth's crust. As it cools, the first crystals to form grow large. If this magma were to be suddenly moved higher in the crust, or if it erupted onto Earth's surface, the remaining magma would cool quickly and form smaller crystals.

Vesicular rocks Magma contains dissolved gases that escape when the pressure on the magma lessens. If the lava is thick enough to prevent the gas bubbles from escaping, holes called vesicles are left behind. The rock that forms looks spongy. This spongy appearance is called **vesicular texture.** Pumice and vesicular basalt are examples shown in **Figure 5.11.**

Reading Check **Explain** what causes holes to form in igneous rocks.

Thin Sections

It is usually easier to observe the sizes of mineral grains than it is to identify the mineral. To identify minerals, geologists examine samples that are called thin sections. A thin section is a slice of rock, generally 2 cm × 4 cm and only 0.03 mm thick. Because it is so thin, light is able to pass through it.

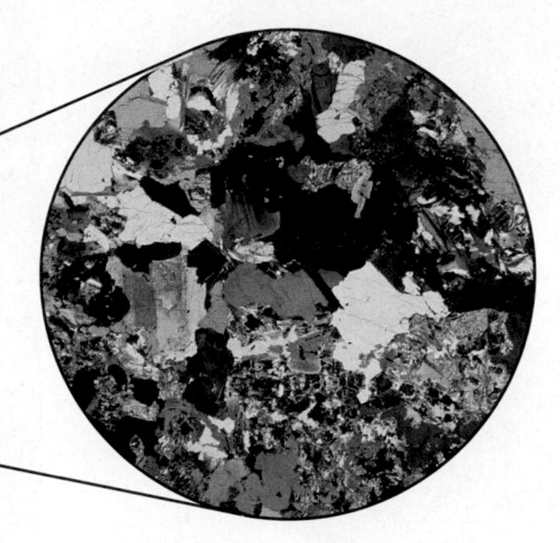

■ **Figure 5.12** The minerals that make up this piece of granite can be identified in a thin section.

When viewed through a special microscope, called a petrographic microscope, mineral grains exhibit distinct properties. These properties allow geologists to identify the minerals present in the rock. For example, feldspar grains often show a distinct banding called twinning. Quartz grains might appear wavy as the microscope stage is rotated. Calcite crystals become dark, or extinguish, as the stage is rotated. **Figure 5.12** shows the appearance of a thin section of granite under a petrographic microscope.

Igneous Rocks as Resources

The cooling and crystallization history of igneous rocks sometimes results in the formation of unusual but useful minerals. These minerals can be used in many fields, including construction, energy production, and jewelry making. Some of these uses are described in the following paragraphs.

Veins As you learned in Chapter 4, ores are minerals that contain a useful material that can be mined for a profit. Valuable ore deposits often occur within igneous intrusions. At other times, ore minerals are found in the rocks surrounding intrusions. These types of deposits sometimes occur as veins. Recall from Bowen's reaction series that the fluid left during magma crystallization contains high levels of silica and water. This fluid also contains any leftover elements that were not incorporated into the common igneous minerals. Some important metallic elements that are not included in common minerals are gold, silver, lead, and copper. These elements, along with the dissolved silica, are released at the end of magma crystallization in a hot, mineral-rich fluid that fills cracks and voids in the surrounding rock. This fluid solidifies to form metal-rich quartz veins, such as the gold-bearing veins in the Sierra Nevada. An example of gold formed in a quartz vein is shown in **Figure 5.13.**

Reading Check **Explain** why veins have high amounts of quartz.

■ **Figure 5.13** Gold and quartz are extracted from mines together. The two are later separated.
Infer ***What can you determine from this photo about the melting temperature of gold?***

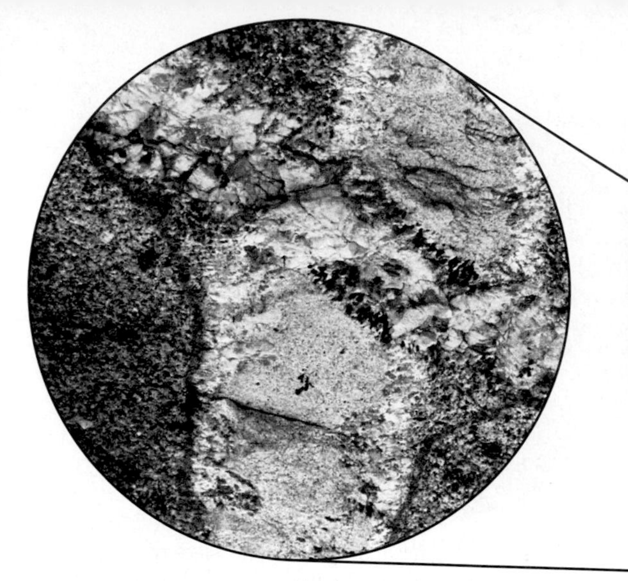

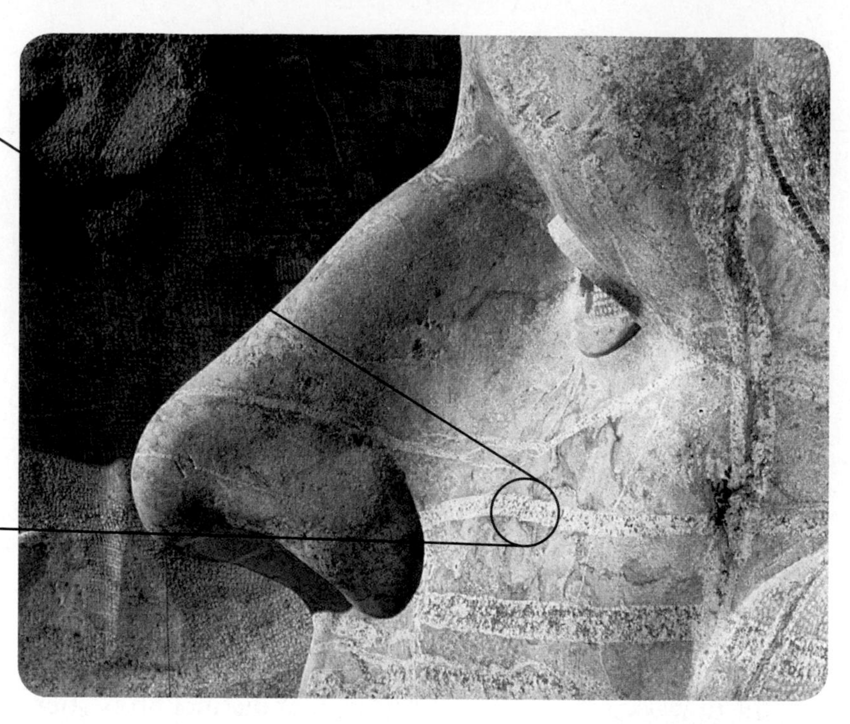

■ **Figure 5.14** Pegmatite veins cut through much of the rock from which Mount Rushmore National Memorial is carved. You can see the veins running across Thomas Jefferson's face.

Pegmatites Vein deposits can contain other valuable resources in addition to metals. Veins of extremely large-grained minerals are called **pegmatites.** Ores of rare elements, such as lithium (Li) and beryllium (Be), form in pegmatites. In addition to ores, pegmatites can produce beautiful crystals. Because these veins fill cavities and fractures in rock, minerals grow into voids and retain their shapes. Some of the world's most beautiful minerals have been found in pegmatites. A famous pegmatite is the rock source for the Mount Rushmore National Memorial located near Keystone, South Dakota. A close-up view of President Thomas Jefferson, shown in **Figure 5.14,** reveals the huge mineral veins that run through the rock.

PROBLEM-SOLVING LAB

Interpret Scientific Illustrations

How do you estimate mineral composition? Igneous rocks are classified by their mineral compositions. In this activity, you will use the thin section in **Figure 5.12** to estimate the different percentages of minerals in the sample.

Analysis

1. Design a method to estimate the percentages of the minerals in the rock sample shown in **Figure 5.12.**
2. Make a data table that lists the minerals and their estimated percentages.

Think Critically

3. **Interpret Figure 5.9** to determine where in the chart this rock sample fits.
4. **Compare** your estimates of the percentages of minerals in the rock with those of your classmates. Why do the estimates vary? What are some possible sources of error?
5. **Propose** a method to improve the accuracy of your estimate.

Kimberlites Diamond is a valuable mineral found in rare, ultrabasic rocks known as **kimberlites,** named after Kimberly, South Africa, where the intrusions were first identified. These unusual rocks are a variety of peridotite. They most likely form deep in the crust or in the mantle at depths of 150 to 300 km, because diamond and other minerals present in kimberlites can form only under very high pressure.

Geologists hypothesize that kimberlite magma is intruded rapidly upward toward Earth's surface, forming long, narrow, pipelike structures. These structures extend many kilometers into the crust, but they are only 100 to 300 m in diameter. Most of the world's diamonds come from South African mines, such as the one shown in **Figure 5.15.** Many kimberlites have been discovered in the United States, but diamonds have been found only in Arkansas and Colorado. The diamond mine in Colorado is the only diamond mine currently in operation in the United States.

Igneous rocks in construction Igneous rocks have several characteristics that make them especially useful as building materials. The interlocking grain textures of igneous rocks make them strong. In addition, many of the minerals present in igneous rocks are resistant to weathering. Granite is among the most durable of igneous rocks. You have probably seen many items, such as countertops, floors, and statues, made from the wide variety of granite that has formed on Earth.

Figure 5.15 Diamonds are mined from kimberlite in mines like this one in Richtersveld, Northern Cape, South Africa.

Section 5.2 Assessment

Section Summary

- Classification of igneous rocks is based on three main characteristics.
- The rate of cooling determines crystal size.
- Ores often occur in pegmatites. Diamonds occur in kimberlites.
- Some igneous rocks are used as building materials because of their strength, durability, and beauty.

Understand Main Ideas

1. MAIN Idea **Infer** why obsidian, which is black or red in color, can have a granitic composition.
2. **Describe** the three major compositional groups of igneous rocks.
3. **Apply** what you know about cooling rates to explain differences in crystal sizes.
4. **Distinguish** between andesite and diorite using two physical properties of igneous rocks.

Think Critically

5. **Speculate** why there are almost no extrusive ultrabasic rocks in Earth's crust.
6. **Determine** whether quartz or plagioclase feldspar is more likely to form a well-shaped crystal in an igneous rock. Explain.

MATH in Earth Science

7. A granite slab has a density of 2.7 g/cm^3. What is the mass of a 2-cm-thick countertop that is 0.6 m $\times$ 2.5 m? How many grams is this?

Earth Science and the Environment

Moon Rocks

During each of the six Apollo missions, lunar rocks were collected with the hope of providing information about the Moon's origin, history, and environment. How do moon rocks compare with rocks on Earth?

Moon rock types Between 1969 and 1972, astronauts collected approximately 380 kg of lunar rocks. The 2415 individual pieces range in size from a grain of sand to a basketball.

Generally, moon rocks vary in color from gray to black to white to green. Some rocks are glassy, some are hard, and others are fragile. Analysis of the rocks has revealed at least three different rock types on the Moon. Basaltic rocks formed from lava flows and volcanic ash that reached the surface through cracks and fissures caused by meteorite impacts. Breccias formed when meteorites shattered rocks and then fused the pieces together with the heat generated by the impact. Pristine rock is rock that has not been hit by meteorites. Pristine rock is commonly composed of calcium-rich plagioclase feldspar and is gray in color.

Moon rock composition Moon rocks are unique in two ways. First, they contain no water and are not oxidized. Considering how much iron is contained in the rocks, this is a sharp contrast to weathered and rusty iron-bearing rocks on Earth. Second, the surfaces of some moon rocks are covered with tiny pockmarks called zap pits. These are caused by micrometeoroids that impact the rocks on the Moon's surface. Zap pits do not occur on Earth rocks because friction from Earth's atmosphere causes tiny meteoroids to burn up long before they reach Earth's surface.

Moon rock classification Scientists use the same categories for classifying lunar rocks as they use for igneous rocks on Earth.

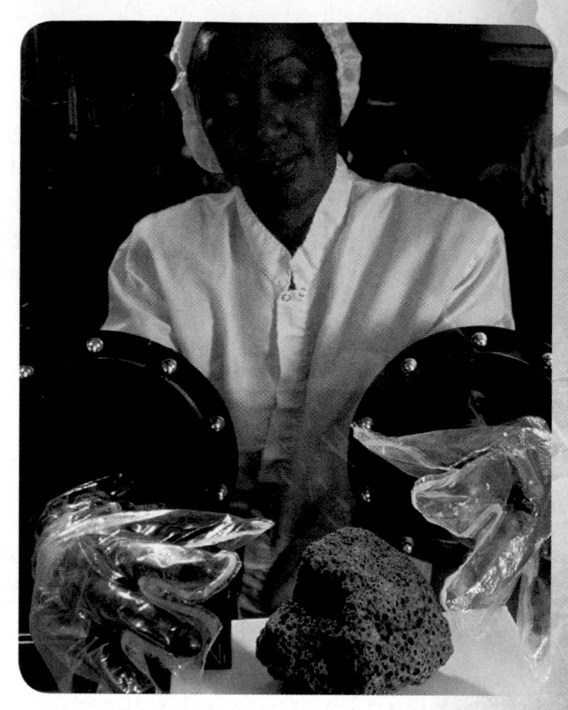

This scientist is studying a piece of basalt that was collected from the lunar surface during the *Apollo 15* mission.

Based on mineral composition, scientists named a new class of moon rocks called KREEP rocks. These contain high amounts of potassium (K), rare Earth elements (REE), and phosphorus (P). These rocks are more radioactive and higher in thorium than Earth rocks.

Moon rock research Lunar rock research continues at the Johnson Space Center in Houston, Texas. The rocks are protected in stainless steel vaults in a dry nitrogen atmosphere to keep them moisture- and rust-free. Scientists continue to pose questions about these rocks as they study the Moon's origin and history.

WRITING in Earth Science

Lunar Rock Game Use resources to design a game that involves the collection and analysis of lunar rocks by scientists. Trade games with classmates to increase your understanding of lunar rocks.

GEOLAB

DESIGN YOUR OWN: MODEL CRYSTAL FORMATION

Background: The rate at which magma cools affects the grain size of the resulting igneous rock. Observing the crystallization of magma is difficult because molten rock is very hot and the crystallization process is sometimes very slow. Other materials, however, crystallize at lower temperatures. These materials can be used to model crystal formation.

Question: *How do minerals crystallize from magma?*

Materials

clean, plastic petri dishes
saturated alum solution
200-mL glass beaker
magnifying lens
dark-colored construction paper
thermometer
paper towels
water
hot plate

Safety Precautions

WARNING: ***The alum solution can cause skin irritation and will be hot when it is first poured into the petri dishes. If splattering occurs, wash skin with cold water.***

Procedure

1. Read and complete the lab safety form.
2. As a group, plan how you will change the cooling rate of a hot solution poured into a petri dish. Each group member should choose a petri dish in a predetermined location to observe during the investigation. Make sure your teacher approves your plan before you begin.
3. Place a piece of dark-colored construction paper on a level surface where it will not be disturbed. Be sure to put the paper in all of the predetermined locations. Place the petri dishes on top of the paper.
4. Using the glass beaker, obtain about 150 mL of saturated alum solution from your teacher. The temperature should be about 95°C to 98°C, just below boiling temperature.
5. Carefully pour some of the solution into each petri dish so that it is half full. Use caution when pouring the hot liquid to avoid splatters and burns.

6. Every 5 min for 30 min, record your observations of your petri dish. Make drawings of any crystals that begin to form.

Analyze and Conclude

1. **Compare** your methods of cooling with those of other groups. Did some methods appear to work better than others? Explain.
2. **Examine** your alum crystals. What do the crystals look like? Are they all the same size? Do all the crystals have the same shape?
3. **Draw** the most common crystal shape in your science journal. Compare your drawings with those of other groups. Describe any patterns that you see.
4. **Deduce** what factors affected the size of the crystals in the different petri dishes. How do you know?
5. **Infer** why the crystals changed shape as they grew.
6. **Compare and contrast** this experiment with magma crystallization.
7. **Evaluate** the relationship between cooling rate and crystal formation.

SHARE YOUR DATA

Peer Review Visit glencoe.com and post a summary of your data. Compare and contrast your results with those of other students who have completed this lab.

CHAPTER 5

Study Guide

Download quizzes, key terms, and flash cards from glencoe.com.

BIG Idea Igneous rocks were the first rocks to form as Earth cooled from a molten mass to the crystalline rocks of the early crust.

Vocabulary	Key Concepts

Section 5.1 What are igneous rocks?

Vocabulary

- Bowen's reaction series (p. 114)
- fractional crystallization (p. 115)
- igneous rock (p. 112)
- lava (p. 112)
- partial melting (p. 114)

Key Concepts

MAIN Idea Igneous rocks are the rocks that form when molten material cools and crystallizes.

- Magma consists of molten rock, dissolved gases, and mineral crystals.
- Magma is classified as basaltic, andesitic, or rhyolitic, based on the amount of silica it contains.
- Different minerals melt and crystallize at different temperatures.
- Bowen's reaction series defines the order in which minerals crystallize from magma.

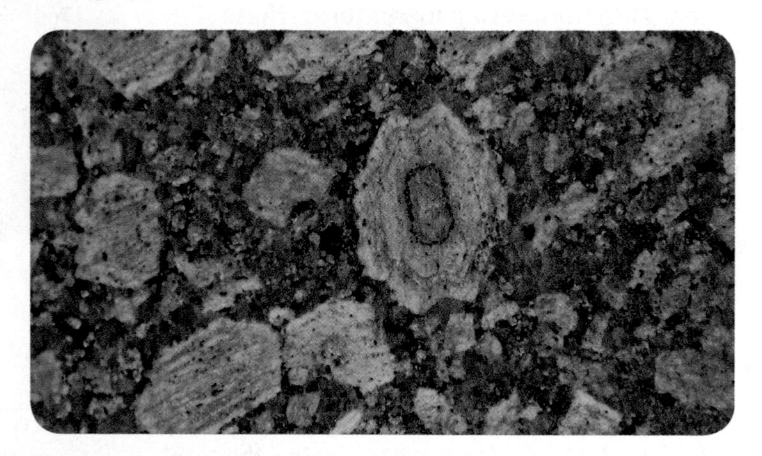

Section 5.2 Classification of Igneous Rocks

Vocabulary

- basaltic rock (p. 118)
- extrusive rock (p. 118)
- granitic rock (p. 118)
- intrusive rock (p. 118)
- kimberlite (p. 123)
- pegmatite (p. 122)
- porphyritic texture (p. 120)
- texture (p. 119)
- vesicular texture (p. 120)

Key Concepts

MAIN Idea Classification of igneous rocks is based on mineral composition, crystal size, and texture.

- Classification of igneous rocks is based on three main characteristics.
- The rate of cooling determines crystal size.
- Ores often occur in pegmatites. Diamonds occur in kimberlites.
- Some igneous rocks are used as building materials because of their strength, durability, and beauty.

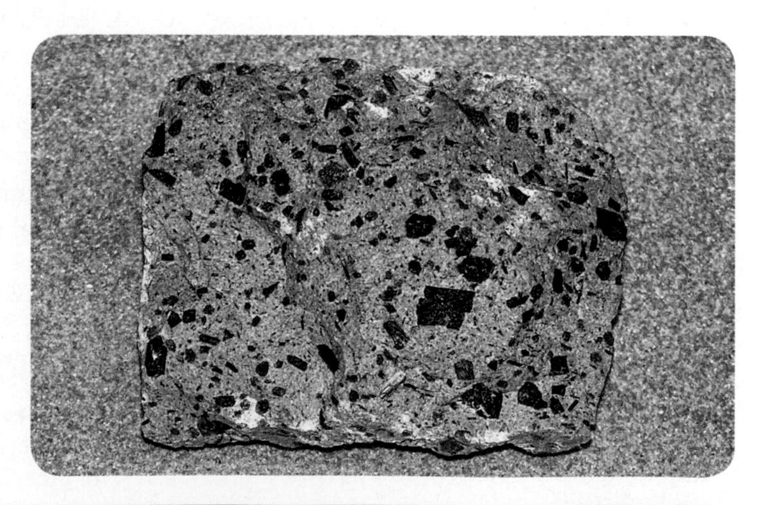

CHAPTER 5 Assessment

Vocabulary Review

The sentences below are incorrect. Make each sentence correct by replacing the italicized word or phrase with a vocabulary term from the Study Guide.

1. Gases escape from *magma* as it flows out onto Earth's surface.
2. *Mohs scale of hardness* describes the order in which minerals crystallize.
3. *Lava* forms deep beneath Earth's crust.

Complete the sentences by filling in the blank with the correct vocabulary term from the Study Guide.

4. An igneous texture characterized by large crystals embedded in a fine-grained background is called a ________.
5. Igneous rocks that form under conditions of fast cooling are said to be ________.
6. Light-colored rocks with large crystals are said to be ________.

Understand Key Concepts

7. Which is the first mineral to form in cooling magma?
   - **A.** quartz
   - **B.** mica
   - **C.** potassium feldspar
   - **D.** olivine

Use the diagram below to answer Question 8.

8. Which process is occurring in the diagram?
   - **A.** fractional separation
   - **B.** crystal separation
   - **C.** fractional crystallization
   - **D.** partial melting

9. Which minerals are associated with the right-hand branch of Bowen's reaction series?
   - **A.** olivine and pyroxene
   - **B.** feldspars
   - **C.** mica and feldspars
   - **D.** quartz and biotite

10. Which magma type contains the greatest amount of silica?
    - **A.** basaltic
    - **B.** andesitic
    - **C.** rhyolitic
    - **D.** peridotic

11. Which does not affect the formation of magma?
    - **A.** volume
    - **B.** temperature
    - **C.** pressure
    - **D.** mineral composition

12. Which extrusive rock has the same composition as andesite?
    - **A.** granite
    - **B.** basalt
    - **C.** obsidian
    - **D.** diorite

Use the figure below to answer Question 13.

13. Which process formed this rock?
    - **A.** slow cooling
    - **B.** fast cooling
    - **C.** very fast cooling
    - **D.** slow, then fast cooling

14. Which type of ultrabasic rock sometimes contains diamonds?
    - **A.** pegmatite
    - **B.** kimberlite
    - **C.** granite
    - **D.** rhyolite

15. What effect does a fast cooling rate have on grain size in igneous rocks?
    A. It forms fine-grained crystals.
    B. It forms large-grained crystals.
    C. It forms light crystals.
    D. It forms dark crystals.

16. What term describes igneous rocks that crystallize inside Earth?
    A. magma
    B. intrusive
    C. lava
    D. extrusive

17. Which minerals are most common in granite?
    A. quartz and feldspar
    B. plagioclase feldspar and amphibole
    C. olivine and pyroxene
    D. quartz and olivine

Constructed Response

18. **List** some uses of igneous rocks in the construction industry.

19. **Explain** how and why the plagioclase feldspar in basaltic rocks differs from that in granitic rocks.

Use the photos below to answer Questions 20 and 21.

20. **Draw** a flowchart documenting the formation of the holes in this sample of vesicular basalt.

21. **Speculate** on the reasons that samples of pumice are able to float in water.

22. **Illustrate** how fractional crystallization changes the composition of magma, using the formation of iron-rich olivine to illustrate the point.

23. **Apply** the concepts of temperature and crystallization to explain why magma is often described as a slushy mixture of crystals and molten rock.

Use the table below to answer Questions 24 and 25.

Rock Composition

Mineral	Mineral Percentage			
	Rock 1	Rock 2	Rock 3	Rock 4
Quartz	5	35	0	0
Potassium feldspar	0	15	0	0
Plagioclase feldspar	55	25	0	55
Biotite	15	15	0	10
Amphibole	25	10	0	30
Pyroxene	0	0	40	5
Olivine	0	0	60	0

24. **Analyze** the data in the table, and explain which rock is most likely granite.

25. **Incorporate** Use the data for Rock 4 and the fact that it is fine-grained to determine the name of Rock 4.

Think Critically

26. **Compare** obsidian and granite to explain why granite is more easily carved into statues and monuments.

27. **Evaluate** this statement: It is possible for magma to have a higher silica content than the rock that forms from it.

28. **Apply** what you know about mineral hardness to explain why stainless steel knives do not harm granite cutting boards.

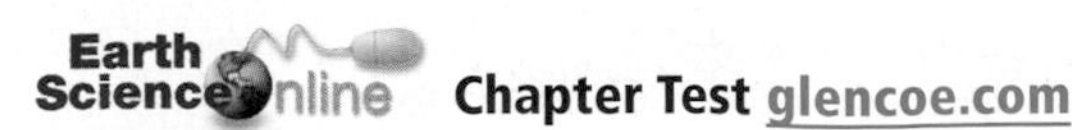

29. **Infer** Kimberlites are the source of most diamonds. Infer why scientists study kimberlites to learn more about Earth's mantle.

30. **Assess** Rocks generally consist of minerals. When molten rock is chilled rapidly, it becomes a glass. Volcanic glass is an extrusive igneous rock. Assess whether this rock contains minerals. Explain your answer. [Hint: Recall the definition of a mineral from Chapter 4.]

31. **Infer** why rocks that are composed of minerals that crystallize first according to Bowen's reaction series are unstable and break down quickly at Earth's surface.

32. **Hypothesize** what the Palisades Sill would look like if the magma that formed it was granitic in composition.

Concept Mapping

33. Use the following terms to create a concept map showing the relationship among position in Earth's crust and mantle, crystal size, and rock type: *fast, slow, slowest, intrusive, extrusive, magma, lava, granite, rhyolite, basalt, gabbro, obsidian,* and *pumice.*

Challenge Question

Use the diagram below to answer Question 34.

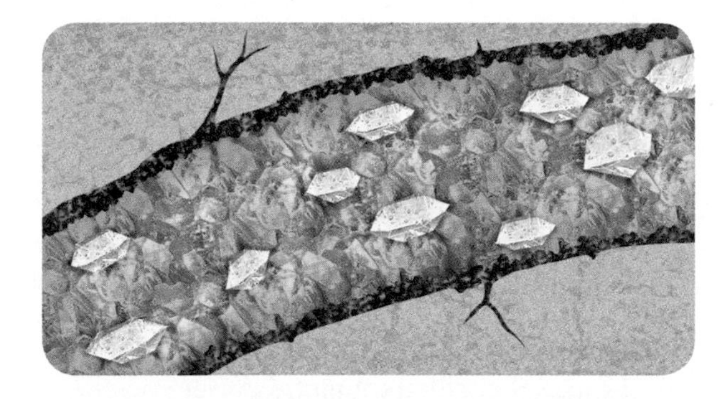

34. **Determine** The diagram shows a cross section of the Leopard Lode, an igneous rock unit in Wyoming. Determine the formation history of this rock unit.

Additional Assessment

35. WRITING in Earth Science Building stone is expensive. Suppose you are selling kitchen countertops that look like granite, but consist of a less-expensive synthetic material. List the specific characteristics of granite that your customers would look for in the imitation granite.

DBQ Document–Based Questions

Data obtained from: Gerya, T.V., et al. 2003. Cold fingers in a hot magma: numerical modeling of country-rock diapirs in the Bushveld Complex, South Africa. *Geology* 31 (9): 753.

The Bushveld Complex is the world's largest layered intrusion. It was injected as a hot, dense basaltic magma between overlying volcanic and underlying sedimentary rocks. Modeling of this event indicates that finger-shaped bodies of heated, metamorphosed sedimentary rocks subsequently intruded the overlying igneous layers. The model assumed the igneous rock properties shown in the table.

Igneous Rock Properties

Rock Type	Density (kg/m^3)	T Solidus (°C)	T Liquidus (°C)
Granitic	2700 (solid) 2400 (molten)	675	925
Basaltic/ ultrabasic	3000 (solid) 2900 (molten)	950	1100

36. Compare and contrast the density of solid and molten rocks in this model.

37. Speculate about why the overlying rhyolitic rocks could not penetrate, or sink into, the basaltic magma.

38. Infer the meaning of the terms *liquidus* and *solidus.* At what temperature do the first crystals in granitic rocks melt?

Cumulative Review

39. What is a molecule? **(Chapter 3)**

40. Name a gemstone that consists of corundum. **(Chapter 4)**

Standardized Test Practice

Multiple Choice

Use the table below to answer Questions 1 and 2.

Characteristics of Rocks			
	Color	Silica Content	Composition
Rock A	light	high	quartz and feldspars
Rock B	dark	low	iron and magnesium

1. Rock A is most likely what kind of rock?
A. granitic
B. basaltic
C. ultrabasic
D. adesitic

2. Which type of rock is Rock B?
A. granite
B. diorite
C. gabbro
D. pegmatite

3. Which is most abundant in magma and has the greatest effect on its characteristics?
A. O
B. Ca
C. Al
D. SiO_2

4. Which process describes how minerals form in predictable sequences?
A. partial melting
B. fractional crystallization
C. Bowen's reaction series
D. geothermal gradient

5. Which is NOT a feature used for identifying minerals?
A. hardness
B. color
C. density
D. volume

6. Which is distorted on a Mercator projection map?
A. shapes of the landmasses
B. areas of the landmasses
C. latitude lines
D. longitude lines

Use the graph below to answer Questions 7 and 8.

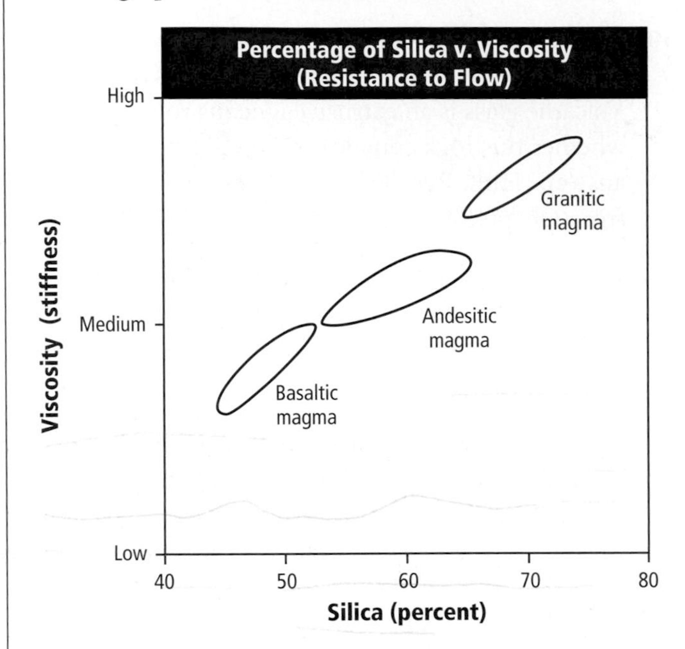

7. What relationship can be inferred from the graph?
A. Magmas that have more silica are more viscous.
B. Magmas that have less silica are more viscous.
C. Magmas always have low viscosity.
D. There is no relationship between silica content and viscosity (resistance to flow).

8. Which is a true statement about rhyolitic magma?
A. Rhyolitic magma is heavier than the other two types of magma.
B. Rhyolitic magma is lighter than the other two types of magma.
C. Rhyolitic magma flows more quickly than the other two types of magma.
D. Rhyolitic magma flows more slowly than the other two types of magma.

9. Which is a combination of two or more components that retain their identities?
A. chemical
B. solution
C. mixture
D. element

10. Which is the lightest of all atoms?
A. uranium atom
B. oxygen atom
C. carbon atom
D. hydrogen atom

Standardized Test Practice glencoe.com

Short Answer

Use the picture below to answer Questions 11–13.

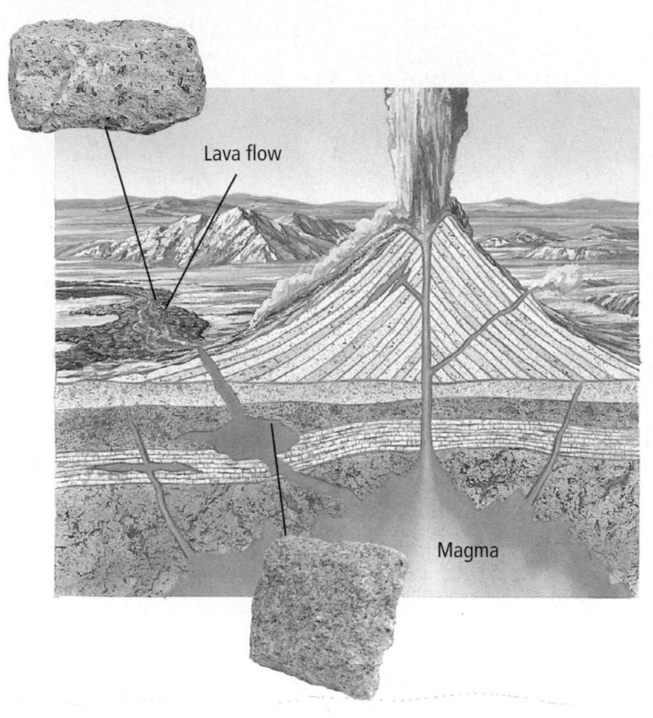

11. Name the type of igneous rock located at the bottom of the picture, and state a common example of that type of rock and explain how this rock is formed.

12. Name the type of igneous rock located at the top of the picture and state a common example of that type of rock and explain how this type of rock is formed.

13. Contrast the formation of the two types of igneous rock.

14. What does it mean to say that minerals are naturally occurring and inorganic?

15. Why are some minerals classified as gems?

16. Why are both latitude and longitude lines necessary when identifying a location?

Reading for Comprehension

Marianas Island Research

Billowing ash plumes, molten sulfur droplets, feisty shrimp feasting on fish killed by noxious gases, and red lava jetting from a vent are all part of the action recently filmed at an underwater volcano in the western Pacific Ocean. The images are the first ever direct observations of an active, submarine-arc volcano. Unlike volcanic activity at mid-ocean ridges, island-arc volcanoes can remain fixed over their magma sources for thousands of years, allowing them to sometimes grow above water level and become islands. The new studies at the Marianas Islands are giving scientists a firsthand look into this formation process. The volcano has been going through nearly constant low-level eruptions since at least 2004, when it was first observed, Embley says. It could potentially keep erupting for decades, giving scientists the opportunity to monitor its growth.

Article obtained from: Roach, J. "Deep-Sea Volcano Erupts on Film—A First" *National Geographic News.* 24 May 2006.

17. What are the benefits of the new studies at the Marianas Islands?
    - **A.** The studies give scientists a firsthand look into the formation process.
    - **B.** The studies reveal that the volcano could potentially keep erupting for decades.
    - **C.** The studies show life near the vent.
    - **D.** The studies are the first ever direct observations of an active submarine arc-volcano.

18. What can you infer from this passage?
    - **A.** Volcanoes constantly erupt at some level of intensity.
    - **B.** Volcanic activity occurs only at mid-ocean ridges.
    - **C.** Shrimp only eat fish killed by noxious gasses.
    - **D.** There are many active submarine volcanoes.

NEED EXTRA HELP?																
If You Missed Question . . .	1	2	3	4	5	6	7	8	9	10	11	12	13	14	15	16
Review Section . . .	5.2	5.2	5.1	5.1	4.1	2.2	5.1	5.1	3.2	3.1	5.2	5.2	5.2	4.1	4.2	2.1

CHAPTER 6

Sedimentary and Metamorphic Rocks

BIG Idea **Most rocks are formed from preexisting rocks through external and internal geologic processes.**

6.1 Formation of Sedimentary Rocks
MAIN Idea Sediments produced by weathering and erosion form sedimentary rocks through the process of lithification.

6.2 Types of Sedimentary Rocks
MAIN Idea Sedimentary rocks are classified by their mode of formation.

6.3 Metamorphic Rocks
MAIN Idea Metamorphic rocks form when preexisting rocks are exposed to increases in temperature and pressure and to hydrothermal solutions.

GeoFacts

- The exterior of the Empire State Building is made of limestone, marble, granite, and metal.
- 5663 m^3 of Indiana limestone and granite, 929 m^2 of Rose Famosa and Estrallante marble, and 27,870 m^2 of Hauteville and Rocheron marble were used in the building's construction.
- Overall, the Empire State Building weighs 331,122.43 metric tons.

LAUNCH Lab

What happened here?

Fossils are the remains of once-living plants and animals. In this activity, you will interpret animal activity from the pattern of fossil footprints.

Procedure

1. Read and complete the lab safety form.
2. Study the **photograph of a set of footprints that have been preserved in sedimentary rock.**
3. Write a description of how these tracks might have been made.
4. Draw your own diagram of a set of fossilized footprints that records the interactions of organisms in the environment.
5. Give your diagram to another student and have him or her interpret what happened.

Analysis

1. **Determine** the number of animals that made these tracks.
2. **Infer** types of information that can be obtained by studying fossil footprints.
3. **Interpret** another group's diagram. Is your answer the same as theirs? What might have caused any differences?

The Rock Cycle Make the following Foldable to show possible paths of rock formation.

STEP 1 Mark the middle of a vertical sheet of paper. Fold the top and bottom to the middle to form two flaps.

STEP 2 Fold into thirds.

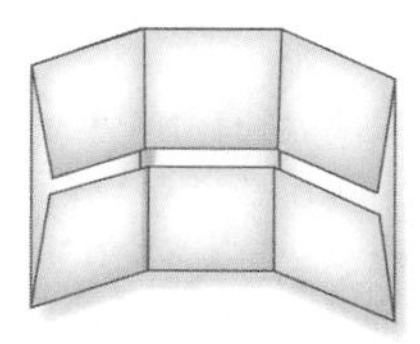

STEP 3 Unfold the paper and cut the flaps along the fold lines as shown.

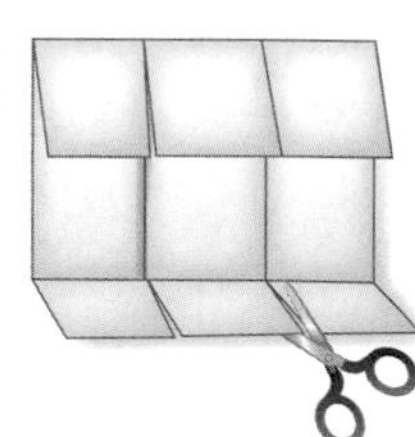

STEP 4 Label the tabs as shown in the diagram to the right.

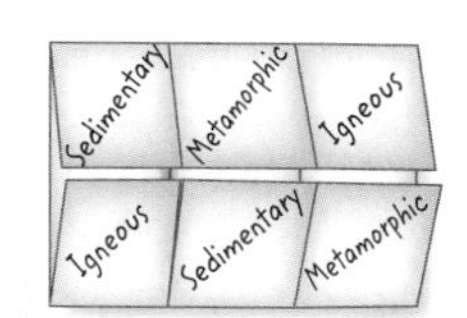

FOLDABLES **Use this Foldable throughout the chapter.** Record under each tab the processes rocks might undergo as they change into the type of rock on an adjoining tab of the Foldable.

Visit glencoe.com to

- **study entire chapters online;**
- **explore Concepts In Motion animations:**
  - **Interactive Time Lines**
  - **Interactive Figures**
  - **Interactive Tables**
- **access Web Links for more information, projects, and activities;**
- **review content with the Interactive Tutor and take Self-Check Quizzes.**

Section 6.1

Objectives

- **Sequence** the formation of sedimentary rocks.
- **Explain** the process of lithification.
- **Describe** features of sedimentary rocks.

Review Vocabulary

texture: the physical appearance or feel of a rock

New Vocabulary

sediment
lithification
cementation
bedding
graded bedding
cross-bedding

Formation of Sedimentary Rocks

MAIN Idea Sediments produced by weathering and erosion form sedimentary rocks through the process of lithification.

Real-World Reading Link Whenever you are outside, you might see pieces of broken rock, sand, and soil on the ground. What happens to this material? With one heavy rain, these pieces of broken rock, sand, and soil could be on their way to becoming part of a sedimentary rock.

Weathering and Erosion

Wherever rock is exposed at Earth's surface, it is continuously being broken down by weathering—a set of physical and chemical processes that breaks rock into smaller pieces. **Sediments** are small pieces of rock that are moved and deposited by water, wind, glaciers, and gravity. When sediments become glued together, they form sedimentary rocks. The formation of sedimentary rocks begins when weathering and erosion produce sediments.

Weathering Weathering produces rock and mineral fragments known as sediments. These sediments range in size from huge boulders to microscopic particles. Chemical weathering occurs when the minerals in a rock are dissolved or otherwise chemically changed. What happens to more-resistant minerals during weathering? While the less-stable minerals are chemically broken down, the more-resistant grains are broken off of the rock as smaller grains. During physical weathering, however, minerals remain chemically unchanged. Rock fragments break off of the solid rock along fractures or grain boundaries. The rock in **Figure 6.1** has been chemically and physically weathered.

■ **Figure 6.1** When exposed to both chemical and physical weathering, granite eventually breaks apart and might look like the decomposed granite shown here.
Explain *which of the three common minerals —quartz, feldspar and mica—will be most resistant to weathering.*

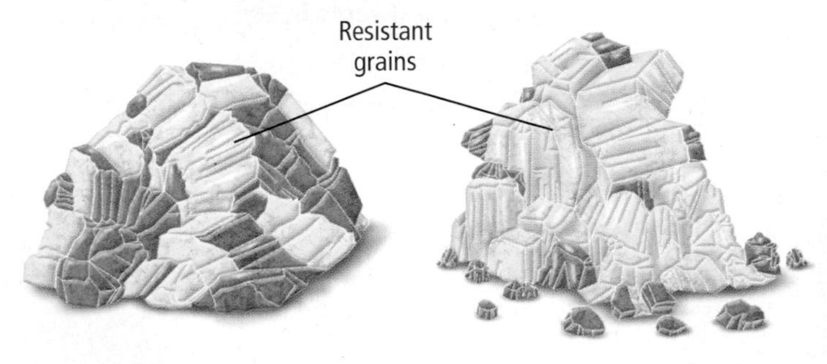

Erosion The removal and transport of sediment is called erosion. **Figure 6.2** shows the four main agents of erosion: wind, moving water, gravity, and glaciers. Glaciers are large masses of ice that move across land. Visible signs of erosion are all around you. For example, water in streams becomes muddy after a storm because eroded silt and clay-sized particles have been mixed in it. You can observe erosion in action when a gust of wind blows soil across the infield at a baseball park. The force of the wind removes the soil and carries it away.

After rock fragments and sediments have been weathered out of the rock, they often are transported to new locations through the process of erosion. Eroded material is almost always carried downhill. Although wind can sometimes carry fine sand and dust to higher elevations, particles transported by water are almost always moved downhill. Eventually, even windblown dust and fine sand are pulled downhill by gravity. You will learn more about weathering and erosion in Chapter 7.

Reading Check Summarize what occurs during erosion.

■ **Figure 6.2** Rocks and sediment are eroded and transported by the main agents of erosion—wind, moving water, gravity, and glaciers.

Wind

Moving water

Gravity

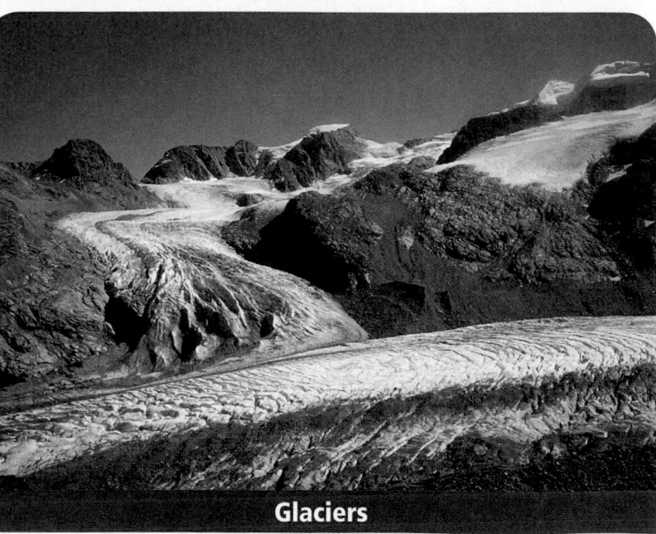
Glaciers

MiniLab

Model Sediment Layering

How do layers form in sedimentary rocks? Sedimentary rocks are usually found in layers. In this activity, you will investigate how layers form from particles that settle in water.

Procedure

1. Read and complete the lab safety form.
2. Obtain 100 mL of **sediment** from a location specified by your teacher.
3. Place the sediment in a **200 mL jar with a lid.**
4. Add **water** to the jar until it is three-fourths full.
5. Place the lid on the jar securely.
6. Pick up the jar with both hands and turn it upside down several times to mix the water and sediment. Hesitate briefly with the jar upside down before tipping it up for the last time. Place the jar on a flat surface.
7. Let the jar sit for about 5 min.
8. Observe the settling process.

Analysis

1. **Illustrate** what you observed in a diagram.
2. **Describe** what type of particles settle out first.
3. **Describe** what type of particles form the topmost layers.

Deposition When transported sediments are deposited on the ground or sink to the bottom of a body of water, deposition occurs. During the MiniLab, what happened when you stopped turning the jar full of sediment and water? The sediment sank to the bottom and was deposited in layers with the largest grains at the bottom and the smallest grains at the top. Similarly, sediments in nature are deposited when transport stops. Perhaps the wind stops blowing or a river enters a quiet lake or an ocean. In each case, the particles being carried will settle out, forming layers of sediment with the largest grains at the bottom.

Energy of transporting agents Fast-moving water can transport larger particles better than slow-moving water. As water slows down, the largest particles settle out first, then the next largest, and so on, so that different-sized particles are sorted into layers. Such deposits are characteristic of sediment transported by water and wind. Wind, however, can move only small grains. For this reason, sand dunes are commonly made of fine, well-sorted sand, as shown in **Figure 6.3.** Not all sediment deposits are sorted. Glaciers, for example, move all materials with equal ease. Large boulders, sand, and mud are all carried along by the ice and dumped in an unsorted pile as the glacier melts. Landslides create similar deposits when sediment moves downhill in a jumbled mass.

Lithification

Most sediments are ultimately deposited on Earth in low areas such as valleys and ocean basins. As more sediment is deposited in an area, the bottom layers are subjected to increasing pressure and temperature. These conditions cause **lithification,** the physical and chemical processes that transform sediments into sedimentary rocks. *Lithify* comes from the Greek word *lithos*, which means *stone.*

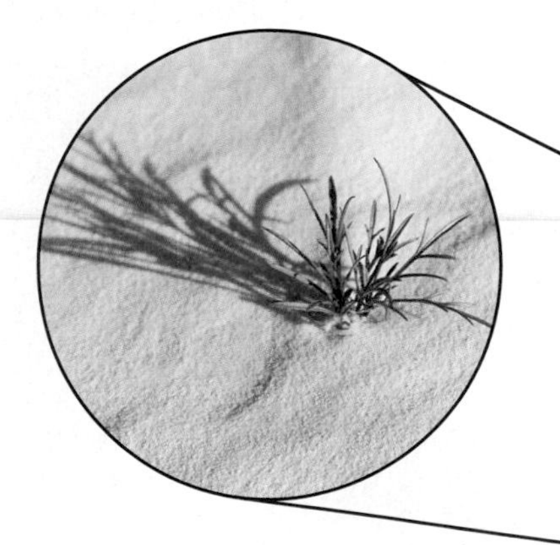

Figure 6.3 These sand dunes at White Sands National Monument in New Mexico were formed by wind-blown sand that has been transported and redeposited. Notice the uniform size of the sand grains.

Compaction Lithification begins with compaction. The weight of overlying sediments forces the sediment grains closer together, causing the physical changes shown in **Figure 6.4.** Layers of mud can contain up to 60 percent water, and these shrink as excess water is squeezed out. Sand does not compact as much as mud during burial. One reason is that individual sand grains, usually composed of quartz, do not deform under normal burial conditions. Grain-to-grain contacts in sand form a supporting framework that helps maintain open spaces between the grains. Groundwater, oil, and natural gas are commonly found in these spaces in sedimentary rocks.

Cementation Compaction is not the only force that binds the grains together. **Cementation** occurs when mineral growth glues sediment grains together into solid rock. This occurs when a new mineral, such as calcite ($CaCO_3$) or iron oxide (Fe_2O_3), grows between sediment grains as dissolved minerals precipitate out of groundwater. This process is illustrated in **Figure 6.5.**

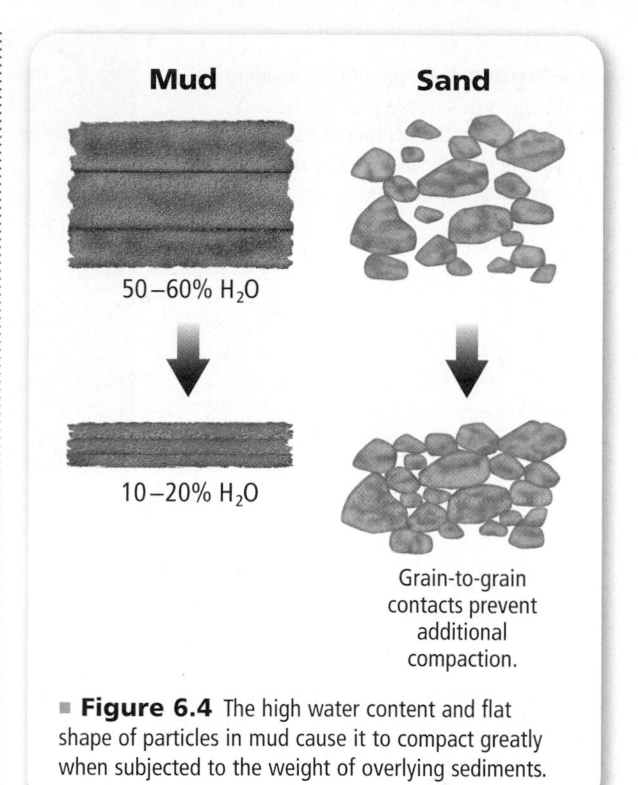

Figure 6.4 The high water content and flat shape of particles in mud cause it to compact greatly when subjected to the weight of overlying sediments.

Sedimentary Features

Just as igneous rocks contain information about the history of their formation, sedimentary rocks also have features and characteristics that help geologists interpret how they formed and the history of the area in which they formed.

FOLDABLES Incorporate information from this section into your Foldable.

Bedding The primary feature of sedimentary rocks is horizontal layering called **bedding.** This feature results from the way sediment settles out of water or wind. Individual beds can range in thickness from a few millimeters to several meters. There are two different types of bedding, each dependent upon the method of transport. However, the size of the grains and the material within the bedding depend upon many other factors.

Figure 6.5 Minerals precipitate out of water as it flows through pore spaces in the sediment. These minerals form the cement that glues the sediments together.

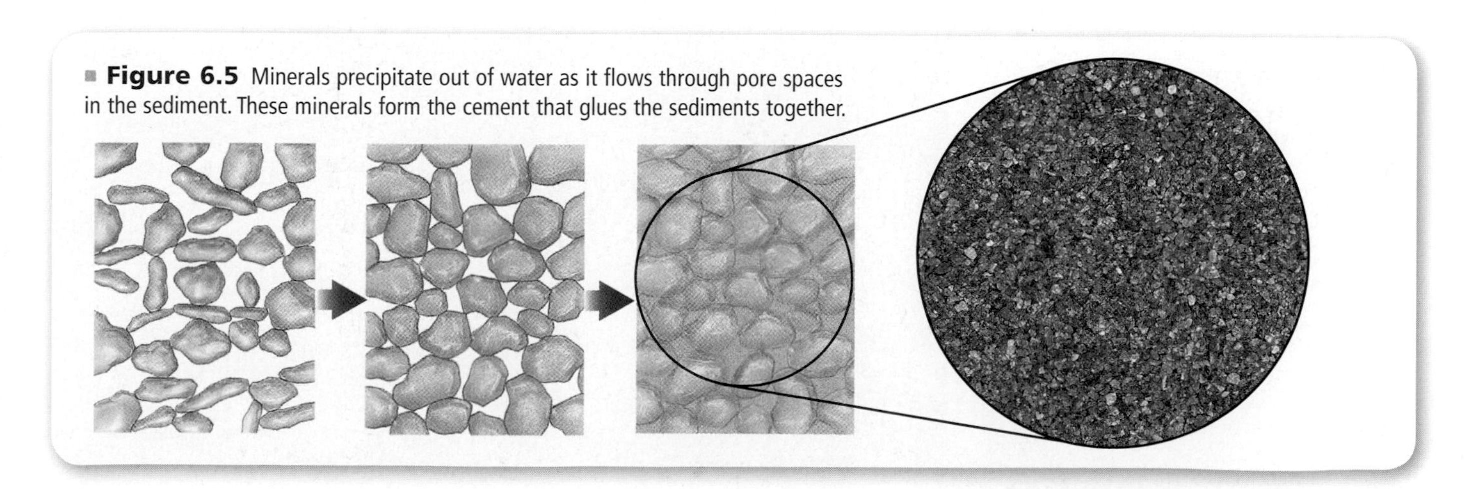

■ **Figure 6.6** The graded bedding shown in this close-up of the Navajo Sandstone in Zion National Park records an episode of deposition during which the water slowed and lost energy.

CAREERS IN EARTH SCIENCE

Sedimentologist Studying the origin and deposition of sediments and their conversion to sedimentary rocks is the job of a sedimentologist. Sedimentologists are often involved in searching for and finding oil, natural gas, and economically important minerals. To learn more about Earth science careers, visit glencoe.com.

Graded bedding Bedding in which the particle sizes become progressively heavier and coarser toward the bottom layers is called **graded bedding.** Graded bedding is often observed in marine sedimentary rocks that were deposited by underwater landslides. As the sliding material slowly came to rest underwater, the largest and heaviest material settled out first and was followed by progressively finer material. An example of graded bedding is shown in **Figure 6.6.**

Cross-bedding Another characteristic feature of sedimentary rocks is cross-bedding. **Cross-bedding,** such as that shown in **Figure 6.7,** is formed as inclined layers of sediment are deposited across a horizontal surface. When these deposits become lithified, the cross-beds are preserved in the rock. This process is illustrated in **Figure 6.8.** Small-scale cross-bedding forms on sandy beaches and along sandbars in streams and rivers. Most large-scale cross-bedding is formed by migrating sand dunes.

Ripple marks When sediment is moved into small ridges by wind or wave action or by a river current, ripple marks form. The back-and-forth movement of waves forms ripples that are symmetrical, while a current flowing in one direction, such as in a river or stream, produces asymmetrical ripples. If a rippled surface is buried gently by more sediment without being disturbed, it might later be preserved in solid rock. The formation of ripple marks is illustrated in **Figure 6.8.**

■ **Figure 6.7** The large-scale cross-beds in these ancient dunes at Zion National Park were deposited by wind.

Visualizing Cross-Bedding and Ripple Marks

Figure 6.8 Moving water and loose sediment result in the formation of sedimentary structures such as cross-bedding and ripple marks.

Cross-Bedding

A Wind direction

Sand particles

B Wind direction

Sand carried by wind gets deposited on the downwind side of a dune. As the wind changes direction, cross-bedding is formed that records this change in direction.

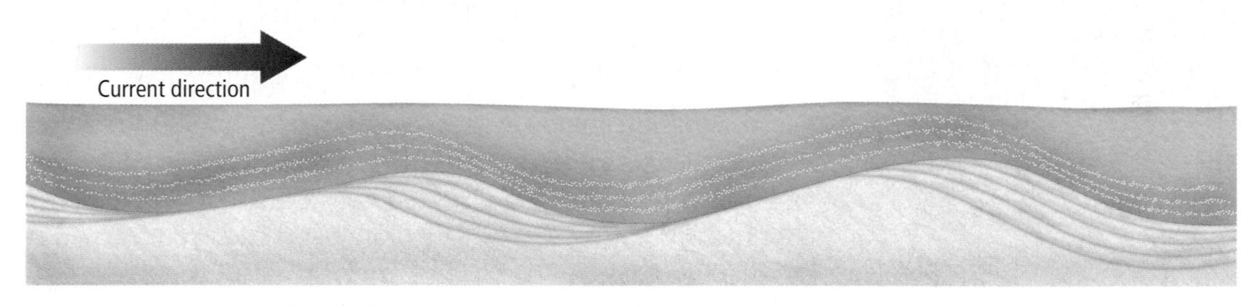

Sediment on the river bottom gets pushed into small hills and ripples by the current. Additional sediment gets deposited at an angle on the downcurrent side of these hills forming cross-beds. Eventually, it levels out or new hills form and the process begins again.

Symmetrical Ripple Marks

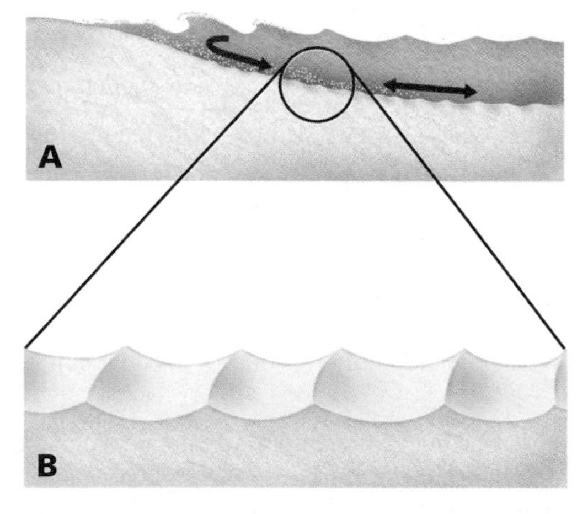

The back-and-forth wave action on a shore pushes the sand on the bottom into symmetrical ripple marks. Grain size is evenly distributed.

Asymmetrical Ripple Marks

Current direction

River channel

River bed

A

Current direction

B

Current that flows in one direction, such as that of a river, pushes sediment on the bottom into asymmetrical ripple marks. They are steeper upstream and contain coarser sediment on the upstream side.

Concepts In Motion To explore more about cross-bedding and ripple marks, visit glencoe.com.

Earth Science Online

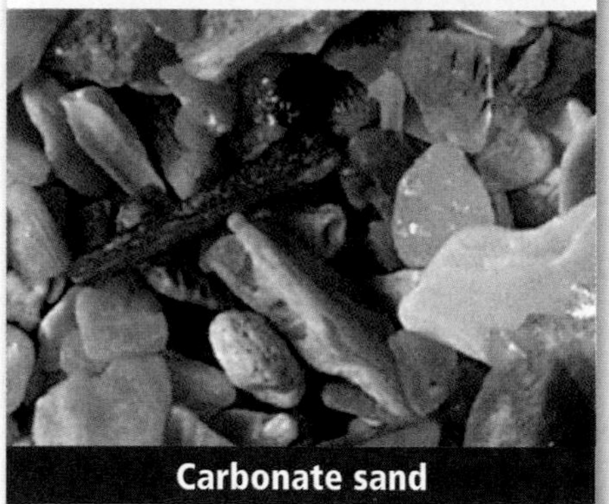

Figure 6.9 The carbonate sand has sharp, jagged pieces and is not as rounded and smooth as the quartz sand.

Sorting and rounding Close examination of individual sediment grains reveals that some have jagged edges and some are rounded. When a rock breaks apart, the pieces are angular in shape. As the sediment is transported, individual pieces knock into each other. The edges are broken off and, over time, the pieces become rounded. The amount of rounding is influenced by how far the sediment has traveled. Additionally, the harder the mineral, the better chance it has of becoming rounded before it breaks apart and becomes microscopic in size. As shown in **Figure 6.9,** quartz sand on beaches is nearly round while carbonate sand, which is made up of softer seashells and calcite, is usually more angular because it is deposited closer to the source of the sediment.

Evidence of past life Probably the best-known features of sedimentary rocks are fossils. Fossils are the preserved remains, impressions, or any other evidence of once-living organisms. When an organism dies, it sometimes is buried before it decomposes. If its remains are buried without being disturbed, it might be preserved as a fossil. During lithification, parts of the organism can be replaced by minerals and turned into rock, such as shells that have been mineralized. Fossils are of great interest to Earth scientists because fossils provide evidence of the types of organisms that lived in the distant past, the environments that existed in the past, and how organisms have changed over time. You will learn more about fossils and how they form in Chapter 21. You learned firsthand how fossils can be used to interpret past events when you completed the Launch Lab at the beginning of this chapter.

Section 6.1 Assessment

Section Summary

- The processes of weathering, erosion, deposition, and lithification form sedimentary rocks.
- Sediments are lithified into rock by the processes of compaction and cementation.
- Fossils are the remains or other evidence of once-living organisms that are preserved in sedimentary rocks.
- Sedimentary rocks might contain features such as horizontal bedding, cross-bedding, and ripple marks.

Understand Main Ideas

1. MAIN Idea **Describe** how sediments are produced by weathering and erosion.
2. **Sequence** Use a flowchart to show why sediment deposits tend to form layers.
3. **Illustrate** the formation of graded bedding.
4. **Compare** temperature and pressure conditions at Earth's surface and below Earth's surface, and relate them to the process of lithification.

Think Critically

5. **Evaluate** this statement: It is possible for a layer of rock to show both cross-bedding and graded bedding.
6. **Determine** whether you are walking upstream or downstream along a dry mountain stream if you notice that the shape of the sediment is getting more angular as you continue walking. Explain.

WRITING in Earth Science

7. Imagine you are designing a display for a museum based on a sedimentary rock that contains fossils of corals and other ocean-dwelling animals. Draw a picture of what this environment might have looked like, and write the accompanying description that will be posted next to the display.

Section 6.2

Objectives

- **Describe** the types of clastic sedimentary rocks.
- **Explain** how chemical sedimentary rocks form.
- **Describe** biochemical sedimentary rocks.

Review Vocabulary

saturated: the maximum possible content of dissolved minerals in solution

New Vocabulary

clastic sedimentary rock
clastic
porosity
evaporite

Types of Sedimentary Rocks

MAIN Idea Sedimentary rocks are classified by their mode of formation.

Real-World Reading Link If you have ever walked along the beach or along a riverbank, you might have noticed different sizes of sediments. The grain size of the sediment determines what type of sedimentary rock it can become.

Clastic Sedimentary Rocks

The most common sedimentary rocks, **clastic sedimentary rocks,** are formed from the abundant deposits of loose sediments that accumulate on Earth's surface. The word **clastic** comes from the Greek word *klastos,* meaning *broken*. These rocks are further classified according to the sizes of their particles. As you read about each rock type, refer to **Table 6.1** on the next page, which summarizes the classification of sedimentary rocks based on grain size, mode of formation, and mineral content.

Coarse-grained rocks Sedimentary rocks consisting of gravel-sized rock and mineral fragments are classified as coarse-grained rocks, samples of which are shown in **Figure 6.10.** Conglomerates have rounded, gravel-sized particles. Because of its relatively large mass, gravel is transported by high-energy flows of water, such as those generated by mountain streams, flooding rivers, some ocean waves, and glacial meltwater. During transport, gravel becomes abraded and rounded as the particles scrape against one another. This is why beach and river gravels are often well rounded. Lithification turns these sediments into conglomerates.

In contrast, breccias are composed of angular, gravel-sized particles. The angularity indicates that the sediments from which they formed did not have time to become rounded. This suggests that the particles were transported only a short distance and deposited close to their source. Refer to **Table 6.1** to see how these rocks are named.

■ **Figure 6.10** Conglomerates and breccias are made of coarse sediments that have been transported by high-energy water.
Infer *the circumstances that might cause the types of transport necessary for each to form.*

Conglomerate

Breccia

Concepts In Motion

Interactive Table To explore more about sedimentary rock formation, visit glencoe.com.

Table 6.1 Classification of Sedimentary Rocks

Classification	Texture/Grain Size	Composition	Rock Name
Clastic	coarse (> 2 mm)	Fragments of any rock type—quartz, chert and quartzite common } rounded / angular	conglomerate breccia
	medium (1/16 mm to 2 mm)	quartz and rock fragments quartz, potassium feldspar and rock fragments	sandstone arkose
	fine (1/256 mm–1/16 mm)	quartz and clay	siltstone
	very fine ($< 1/256$ mm)	quartz and clay	shale
Biochemical	microcrystalline with conchoidal fracture	calcite ($CaCO_3$)	micrite
	abundant fossils in micrite matrix	calcite ($CaCO_3$)	fossiliferous limestone
	oolites (small spheres of calcium carbonate)	calcite ($CaCO_3$)	oolitic limestone
	shells and shell fragments loosely cemented	calcite ($CaCO_3$)	coquina
	microscopic shells and clay	calcite ($CaCO_3$)	chalk
	variously sized fragments	highly altered plant remains, some plant fossils	coal
Chemical	fine to coarsely crystalline	calcite ($CaCO_3$)	crystalline limestone
	fine to coarsely crystalline	dolomite $(Ca,Mg)CO_3$ (will effervesce if powdered)	dolostone
	very finely crystalline	quartz (SiO_2)—light colored —dark colored	chert flint
	fine to coarsely crystalline	gypsum ($CaSO_4 \cdot 2H_2O$)	rock gypsum
	fine to coarsely crystalline	halite (NaCl)	rock salt

Medium-grained rocks Stream and river channels, beaches, and deserts often contain abundant sand-sized sediments. Sedimentary rocks that contain sand-sized rock and mineral fragments are classified as medium-grained clastic rocks. Refer to **Table 6.1** for a listing of rocks with sand-sized particles. Sandstone usually contains several features of interest to scientists. For example, because ripple marks and cross-bedding indicate the direction of current flow, geologists use sandstone layers to map ancient stream and river channels.

Another important feature of sandstone is its relatively high porosity. **Porosity** is the percentage of open spaces between grains in a rock. Loose sand can have a porosity of up to 40 percent. Some of these open spaces are maintained during the formation of sandstone, often resulting in porosities as high as 30 percent. When pore spaces are connected to one another, fluids can move through sandstone. This feature makes sandstone layers valuable as underground reservoirs of oil, natural gas, and groundwater.

VOCABULARY

ACADEMIC VOCABULARY

Reservoir

a subsurface area of rock that has enough porosity to allow for the accumulation of oil, natural gas, or water

The newly discovered reservoir contained large amounts of natural gas and oil.

Fine-grained rocks Sedimentary rocks consisting of silt- and clay-sized particles are called fine-grained rocks. Siltstone and shale are fine-grained clastic rocks. These rocks represent environments such as swamps and ponds which have still or slow-moving waters. In the absence of strong currents and wave action, these sediments settle to the bottom where they accumulate in thin horizontal layers. Shale often breaks along thin layers, as shown in **Figure 6.11.** Unlike sandstone, fine-grained sedimentary rock has low porosity and often forms barriers that hinder the movement of groundwater and oil. **Table 6.1** shows how these rocks are named.

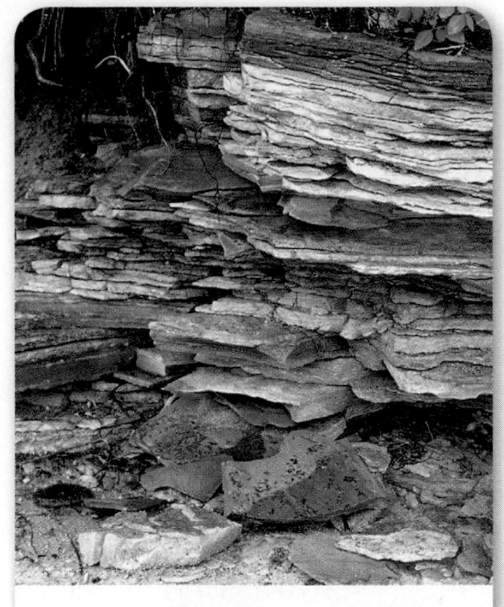

Figure 6.11 The very fine-grained sediment that formed this shale was deposited in thin layers in still waters.

Reading Check Identify the types of environments in which fine-grained rocks form.

Chemical and Biochemical Sedimentary Rocks

The formation of chemical and biochemical rocks involves the processes of evaporation and precipitation of minerals. During weathering, minerals can be dissolved and carried into lakes and oceans. As water evaporates from the lakes and oceans, the dissolved minerals are left behind. In arid regions, high evaporation rates can increase the concentration of dissolved minerals in bodies of water. The Great Salt Lake, shown in **Figure 6.12,** is an example of a lake that has high concentrations of dissolved minerals.

Chemical sedimentary rocks When the concentration of dissolved minerals in a body of water reaches saturation, crystal grains precipitate out of solution and settle to the bottom. As a result, layers of chemical sedimentary rocks form, most of which are called **evaporites.** Evaporites most commonly form in arid regions and in drainage basins on continents that have low water flow. Because little freshwater flows into these areas, the concentration of dissolved minerals remains high. Even as more dissolved minerals are carried into the basins, evaporation continues to remove freshwater and maintain high mineral concentrations. Over time, thick layers of evaporite minerals can accumulate on the basin floor, as illustrated in **Figure 6.12.**

Figure 6.12 The constant evaporation from a body of salt water results in precipitation of large amounts of salt. This process has been occurring in the Great Salt Lake in Utah for approximately 18,000 years.

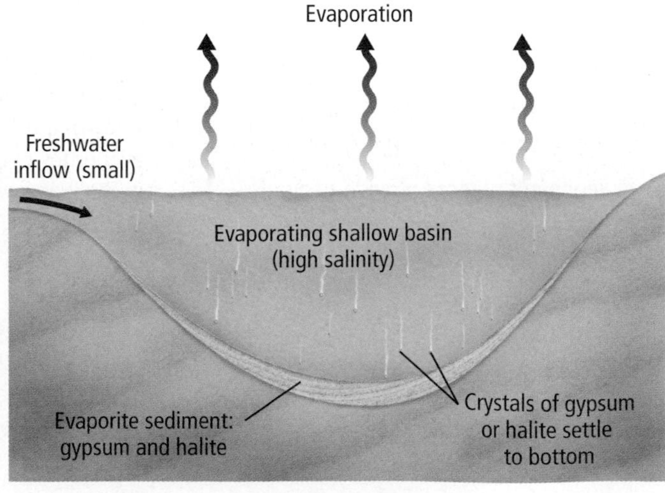

Biochemical sedimentary rocks Biochemical sedimentary rocks are formed from the remains of once-living organisms. The most abundant of these rocks is limestone, which is composed primarily of calcite. Some organisms that live in the ocean use the calcium carbonate that is dissolved in seawater to make their shells. When these organisms die, their shells settle to the bottom of the ocean and can form thick layers of carbonate sediment. During burial and lithification, calcium carbonate precipitates out of the water, crystallizes between the grains of carbonate sediment, and forms limestone.

Limestone is common in shallow water environments, such as those in the Bahamas, where coral reefs thrive in 15 to 20 m of water just offshore. The skeletal and shell materials that are currently accumulating there will someday become limestone as well. Many types of limestone contain evidence of their biological origin in the form of abundant fossils. As shown in **Figure 6.13,** these fossils can range from large-shelled organisms to microscopic, unicellular organisms. Not all limestone contains fossils. Some limestone has a crystalline texture, some consists of tiny spheres of carbonate sand, and some is composed of fine-grained carbonate mud. These are listed in **Table 6.1.**

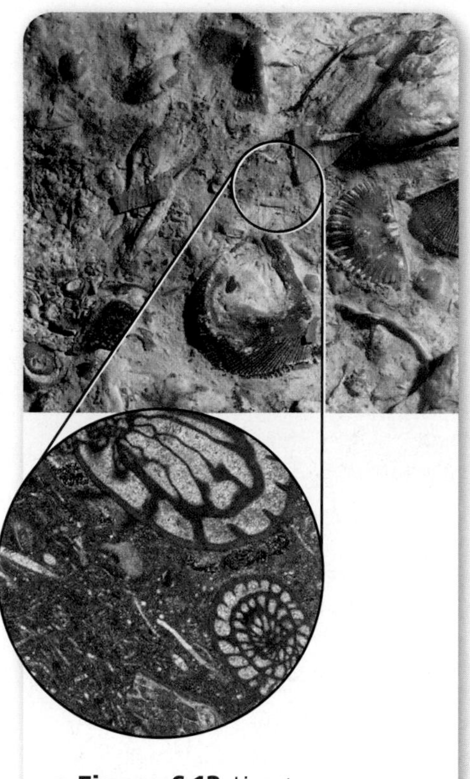

■ **Figure 6.13** Limestone can contain many different fossil organisms. Geologists can interpret where and when the limestone formed by studying the fossils within the rock.

Other organisms use silica to make their shells. These shells form sediment that is often referred to as siliceous ooze because it is rich in silica. Siliceous ooze becomes lithified into the sedimentary rock chert, which is also listed in **Table 6.1.**

Section 6.2 Assessment

Section Summary

- Sedimentary rocks can be clastic, chemical, or biochemical.
- Clastic rocks form from sediments and are classified by particle size and shape.
- Chemical rocks form primarily from minerals precipitated from water.
- Biochemical rocks form from the remains of once-living organisms.
- Sedimentary rocks provide geologists with information about surface conditions that existed in Earth's past.

Understand Main Ideas

1. MAIN Idea **State** the type of sedimentary rock that is formed from the erosion and transport of rocks and sediments.
2. **Explain** why coal is a biochemical sedimentary rock.
3. **Calculate** the factor by which grain size increases with each texture category.
4. **Analyze** the environmental conditions to explain why most chemical sedimentary rocks form mainly in areas that have high rates of evaporation.

Think Critically

5. **Propose** a scenario to explain how it is possible to form additional layers of evaporites in a body of seawater when the original amount of dissolved minerals in the water was enough to form only a thin evaporite.
6. **Examine** the layers of shale in **Figure 6.11** and explain why shale contains no cross-bedding or ripple marks.

MATH in Earth Science

7. Assume that the volume of a layer of mud will decrease by 35 percent during deposition and compaction. If the original sediment layer is 30 cm thick, what will be the thickness of the shale layer after compaction?

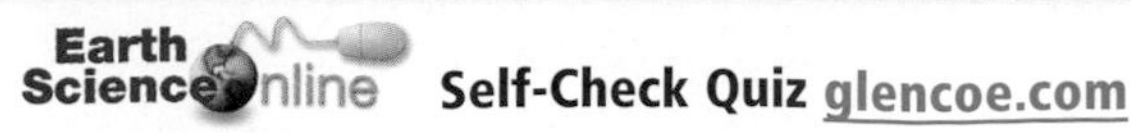

Section 6.3

Objectives

- **Compare and contrast** the different types and causes of metamorphism.
- **Distinguish** among metamorphic textures.
- **Explain** how mineral and compositional changes occur during metamorphism.
- **Apply** the rock cycle to explain how rocks are classified.

Review Vocabulary

intrusive: rocks that form from magma that cooled and crystallized slowly beneath Earth's surface

New Vocabulary

foliated
nonfoliated
regional metamorphism
contact metamorphism
hydrothermal metamorphism
rock cycle

Metamorphic Rocks

MAIN Idea **Metamorphic rocks form when preexisting rocks are exposed to increases in temperature and pressure and to hydrothermal solutions.**

Real-World Reading Link When you make a cake, all of the individual ingredients that you put into the pan change into something new. When rocks are exposed to high temperatures, their individual characteristics also change into something new and form a completely different rock.

Recognizing Metamorphic Rock

The rock layers shown in **Figure 6.14** have been metamorphosed (meh tuh MOR fohzd)—this means that they have been changed. How do geologists know that this has happened? Pressure and temperature increase with depth. When temperature or pressure becomes high enough, rocks melt and form magma. But what happens if the rocks do not reach the melting point? When high temperature and pressure combine and change the texture, mineral composition, or chemical composition of a rock without melting it, a metamorphic rock forms. The word *metamorphism* is derived from the Greek words *meta,* meaning *change*, and *morphé,* meaning *form.* During metamorphism, a rock changes form while remaining solid.

The high temperatures required for metamorphism are ultimately derived from Earth's internal heat, either through deep burial or from nearby igneous intrusions. The high pressures required for metamorphism come from deep burial or from compression during mountain building.

Figure 6.14 Strong forces were required to bend these rock layers into the shape they are today.
***Hypothesize** the changes that occurred to the sediments after they were deposited.*

Figure 6.15 Metamorphic minerals, such as mica, staurolite, garnet, and talc (shown above, clockwise from top left), occur in many colors, shapes, and crystal sizes. Colors can be dark or bright and crystal form can be unique.

Metamorphic minerals How do minerals change without melting? Think back to the concept of fractional crystallization, discussed in Chapter 5. Bowen's reaction series shows that all minerals are stable at certain temperatures and they crystallize from magma along a range of different temperatures. Scientists have discovered that these stability ranges also apply to minerals in solid rock. During metamorphism, the minerals in a rock change into new minerals that are stable under the new temperature and pressure conditions. Minerals that change in this way are said to undergo solid-state alterations. Scientists have conducted experiments to identify the metamorphic conditions that create specific minerals. When the same minerals are identified in rocks, scientists are able to interpret the conditions inside the crust during the rocks' metamorphism. **Figure 6.15** shows some common metamorphic minerals.

Reading Check **Explain** what metamorphic minerals are.

Metamorphic textures Metamorphic rocks are classified into two textural groups: foliated and nonfoliated. Geologists use metamorphic textures and mineral composition to identify metamorphic rocks. **Figure 6.16** shows how these two characteristics are used in the classification of metamorphic rocks.

Foliated rocks Layers and bands of minerals characterize **foliated** metamorphic rocks. High pressure during metamorphism causes minerals with flat or needlelike crystals to form with their long axes perpendicular to the pressure, as shown in **Figure 6.17.** This parallel alignment of minerals creates the layers observed in foliated metamorphic rocks.

Figure 6.16 Increasing grain size parallels changes in composition and development of foliation. Grain size is not a factor in nonfoliated rocks.

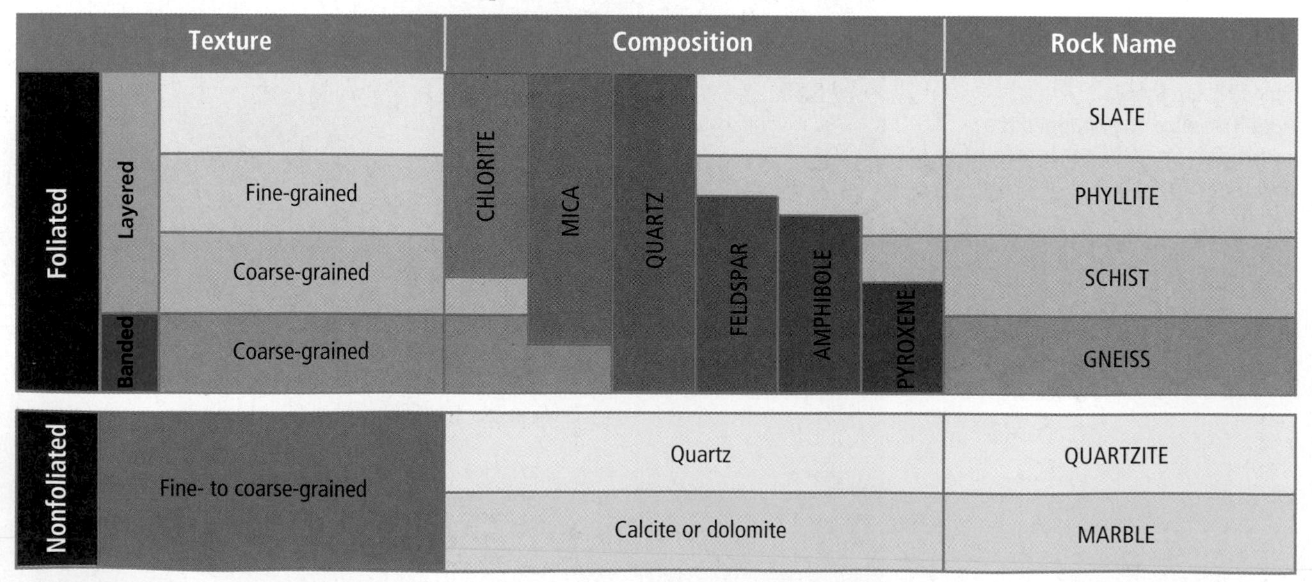

Metamorphic Rock Identification Chart

Texture			Composition	Rock Name
Foliated	Layered		CHLORITE, MICA, QUARTZ	SLATE
Foliated	Layered	Fine-grained	CHLORITE, MICA, QUARTZ, FELDSPAR, AMPHIBOLE	PHYLLITE
Foliated	Layered	Coarse-grained	CHLORITE, MICA, QUARTZ, FELDSPAR, AMPHIBOLE, PYROXENE	SCHIST
Foliated	Banded	Coarse-grained	MICA, QUARTZ, FELDSPAR, AMPHIBOLE, PYROXENE	GNEISS
Nonfoliated		Fine- to coarse-grained	Quartz	QUARTZITE
Nonfoliated		Fine- to coarse-grained	Calcite or dolomite	MARBLE

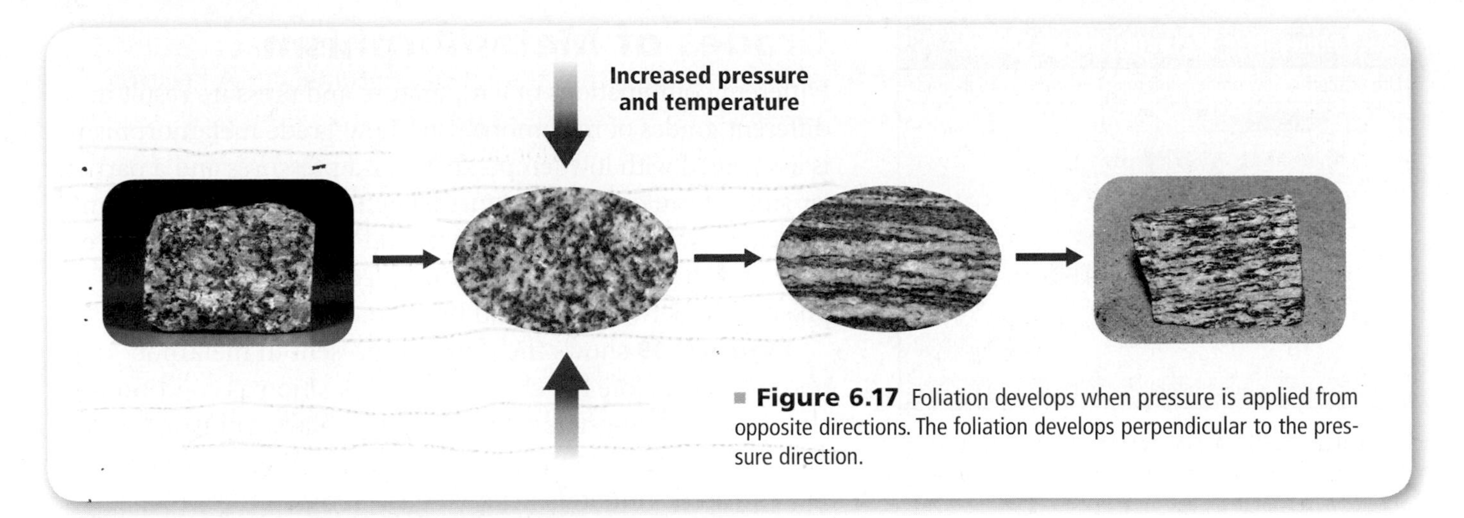

■ **Figure 6.17** Foliation develops when pressure is applied from opposite directions. The foliation develops perpendicular to the pressure direction.

Nonfoliated rocks Unlike foliated rocks, **nonfoliated** metamorphic rocks are composed mainly of minerals that form with blocky crystal shapes. Two common examples of nonfoliated rocks, shown in **Figure 6.18,** are quartzite and marble. Quartzite is a hard, often light-colored rock formed by the metamorphism of quartz-rich sandstone. Marble is formed by the metamorphism of limestone. Some marbles have smooth textures that are formed by interlocking grains of calcite. These marbles are often used in sculptures. Fossils are rarely preserved in metamorphic rocks.

Under certain conditions, new metamorphic minerals can grow large while the surrounding minerals remain small. The large crystals, which can range in size from a few millimeters to a few centimeters, are called porphyroblasts. Although these crystals resemble the very large crystals that form in pegmatite granite, they are not the same. Instead of forming from magma, they form in solid rock through the reorganization of atoms during metamorphism. Garnet, shown in **Figure 6.18,** is, a mineral that commonly forms porphyroblasts.

■ **Figure 6.18** As a result of the extreme heat and pressure during metamorphism, marble rarely contains fossils. Metamorphism does not, however, always destroy cross-bedding and ripple marks, which can be seen in some quartzites. Garnet porphyroblasts can grow to be quite large in some rocks.

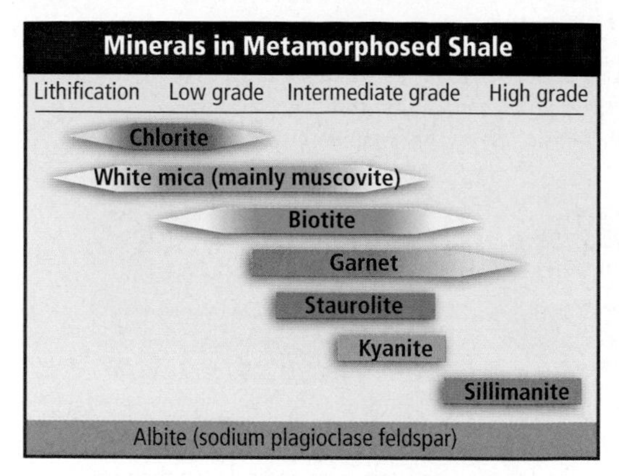

■ **Figure 6.19** Metamorphism of shale results in the formation of minerals that provide the wide variety of color observed in slate.

Grades of Metamorphism

Different combinations of temperature and pressure result in different grades of metamorphism. Low-grade metamorphism is associated with low temperatures and pressures and a particular suite of minerals and textures. High-grade metamorphism is associated with high temperatures and pressures and a different suite of minerals and textures. Intermediate-grade metamorphism is in between low- and high-grade metamorphism.

Figure 6.19 shows the minerals present in metamorphosed shale. Note the change in composition as conditions change from low-grade to high-grade metamorphism. Geologists can create metamorphic maps by plotting the location of metamorphic minerals. Knowing the temperatures that certain areas experienced when rocks were forming helps geologists locate valuable metamorphic minerals such as garnet and talc. Studying the distribution of metamorphic minerals helps geologists to interpret the metamorphic history of an area.

Types of Metamorphism

The effects of metamorphism can be the result of contact metamorphism, regional metamorphism, or hydrothermal metamorphism. The minerals that form and the degree of change in the rocks provide information as to the type and grade of metamorphism that occurred.

PROBLEM-SOLVING LAB

Interpret Scientific Illustrations

Which metamorphic minerals will form? The minerals that form in metamorphic rocks depend on the metamorphic grade and composition of the original rock. The figure below and **Figure 6.19** show the mineral groups that form under different metamorphic conditions.

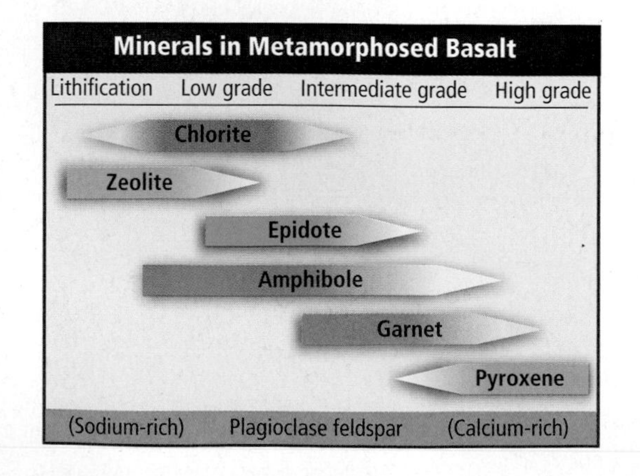

Analysis

1. What mineral is formed when shale and basalt are exposed to low-grade metamorphism?
2. Under high-grade metamorphism, what mineral is formed in shale but not in basalt?

Think Critically

3. **Compare** the mineral groups that you would expect to form from intermediate-grade metamorphism of shale and basalt.
4. **Describe** the major compositional differences between shale and basalt. How are these differences reflected in the minerals formed during metamorphism?
5. **Explain** When limestone is metamorphosed, there is little change in mineral composition. Calcite is still the dominant mineral. Explain why this happens.

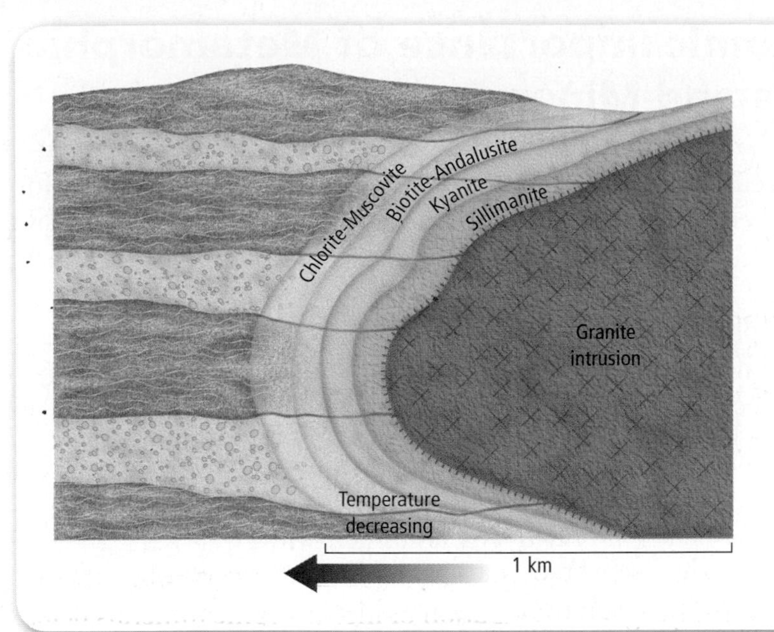

■ **Figure 6.20** Contact metamorphism from the intrusion of this granite batholith has caused zones of metamorphic minerals to form.

Apply ***what you know about contact metamorphism to determine the type of rock that is now present along the edge of the intrusion.***

Regional metamorphism When high temperature and pressure affect large regions of Earth's crust, they produce large belts of **regional metamorphism.** The metamorphism can range in grade from low to high grade. Results of regional metamorphism include changes in minerals and rock types, plus folding and deforming of the rock layers that make up the area. The folded rock layers shown in **Figure 6.14** experienced regional metamorphism.

Contact metamorphism When molten material, such as that in an igneous intrusion, comes in contact with solid rock, a local effect called **contact metamorphism** occurs. High temperature and moderate-to-low pressure form mineral assemblages that are characteristic of contact metamorphism. **Figure 6.20** shows zones of different minerals surrounding an intrusion. Because temperature decreases with distance from an intrusion, metamorphic effects also decrease with distance. Recall from Chapter 5 that minerals crystallize at specific temperatures. Metamorphic minerals that form at high temperatures occur closest to the intrusion, where it is hottest. Because lava cools too quickly for the heat to penetrate far into surface rocks, contact metamorphism from extrusive igneous rocks is limited to thin zones.

Hydrothermal metamorphism When very hot water reacts with rock and alters its chemical and mineral composition, **hydrothermal metamorphism** occurs. The word *hydrothermal* is derived from the Greek words *hydro,* meaning *water,* and *thermal,* meaning *heat.* As hot fluids migrate in and out of the rock during metamorphism, the original mineral composition and texture of the rock can change. Chemical changes are common during contact metamorphism near igneous intrusions and active volcanoes. Valuable ore deposits of gold, copper, zinc, tungsten, and lead are formed in this manner. The gold deposited in the quartz shown in **Figure 6.21** is the result of hydrothermal metamorphism.

VOCABULARY
SCIENCE USAGE V. COMMON USAGE

Intrusion

Science usage: the placement of a body of magma into preexisting rock

Common usage: joining or coming into without being invited

■ **Figure 6.21** When the hydrothermal solution in the quartz cooled, gold veins formed.

Economic Importance of Metamorphic Rocks and Minerals

The modern way of life is made possible by a great number of naturally occurring Earth materials. We need salt for cooking, gold for trade, other metals for construction and industrial purposes, fossil fuels for energy, and rocks and various minerals for construction, cosmetics, and more. **Figure 6.22** shows two examples of how metamorphic rocks are used in construction. Many of these economic mineral resources are produced by metamorphic processes. Among these are the metals gold, silver, copper, and lead, as well as many significant nonmetallic resources.

Metallic mineral resources Metallic resources occur mostly in the form of metal ores, although deposits of pure metals are occasionally discovered, many metallic deposits are precipitated from hydrothermal solutions and are either concentrated in veins or spread throughout the rock mass. Native gold, silver, and copper deposits tend to occur in hydrothermal quartz veins near igneous intrusions or in contact metamorphic zones. However, most hydrothermal metal deposits are in the form of metal sulfides such as galena (PbS) or pyrite (FeS_2). The iron ores magnetite and hematite are oxide minerals often formed by precipitation from iron-bearing hydrothermal solutions.

Reading Check **State** what resources hydrothermal metamorphism produces.

Nonmetallic mineral resources Metamorphism of ultrabasic igneous rocks produces the minerals talc and asbestos. Talc, with a hardness of 1, is used as a dusting powder, as a lubricant, and to provide texture in paints. Because it is not combustible and has low thermal and electric conductivity, asbestos has been used in fireproof and insulating materials. Prior to the recognition of its cancer-causing properties, it was also widely utilized in the construction industry. Many older buildings still have asbestos-containing materials. Graphite, the main ingredient of the lead in pencils, may be formed by the metamorphism of coal.

Figure 6.22 Marble and slate are metamorphic rocks that have been used in construction for centuries.

The Rock Cycle

Metamorphic rocks form when other rocks change. The three types of rock—igneous, sedimentary, and metamorphic—are grouped according to how they form. Igneous rocks crystallize from magma; sedimentary rocks form from cemented or precipitated sediments; and metamorphic rocks form from changes in temperature and pressure.

Once a rock forms, does it remain the same type of rock always? Possibly, but it most likely will not. Heat and pressure can change an igneous rock into a metamorphic rock. A metamorphic rock can be changed into another metamorphic rock or melted to form an igneous rock. Alternately, the metamorphic rock can be weathered and eroded into sediments that might become cemented into a sedimentary rock. In fact, any rock can be changed into any other type of rock. The continuous changing and remaking of rocks is called the **rock cycle.** The rock cycle is summarized in **Figure 6.23.** The arrows represent the different processes that change rocks into different types.

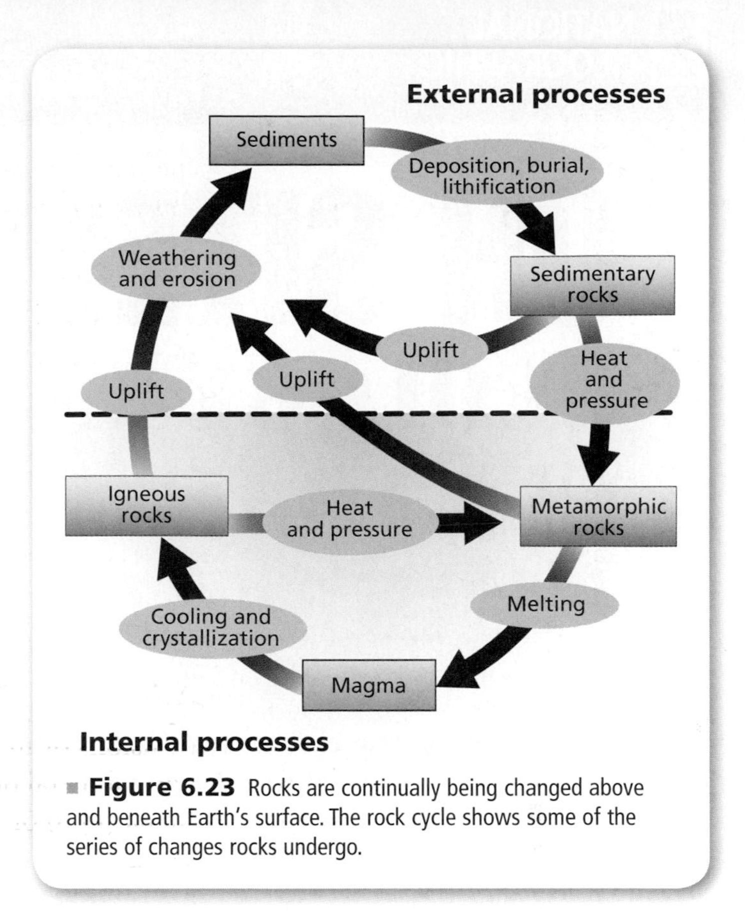

Figure 6.23 Rocks are continually being changed above and beneath Earth's surface. The rock cycle shows some of the series of changes rocks undergo.

Section 6.3 Assessment

Section Summary

- The three main types of metamorphism are regional, contact, and hydrothermal.
- The texture of metamorphic rocks can be foliated or nonfoliated.
- During metamorphism, new minerals form that are stable under the increased temperature and pressure conditions.
- The rock cycle is the set of processes through which rocks continuously change into other types of rocks.

Understand Main Ideas

1. MAIN Idea **Summarize** how temperature increases can cause metamorphism.
2. **Summarize** what causes foliated metamorphic textures to form.
3. **Apply** the concept of the rock cycle to explain how the three main types of rocks are classified.
4. **Compare and contrast** the factors that cause the three main types of metamorphism.

Think Critically

5. **Infer** which steps in the rock cycle are skipped when granite metamorphoses to gneiss.
6. **Predict** the location of an igneous intrusion based on the following mineral data. Muscovite and chlorite were collected in the northern portion of the area of study; garnet and staurolite were collected in the southern portion of the area.

MATH in Earth Science

7. Gemstones often form as porphyroblasts. Gemstones are described in terms of carat weight. A carat is equal to 0.2 g or 200 mg. A large garnet discovered in New York in 1885 weighs 4.4 kg and is 15 cm in diameter. What is the carat weight of this gemstone?

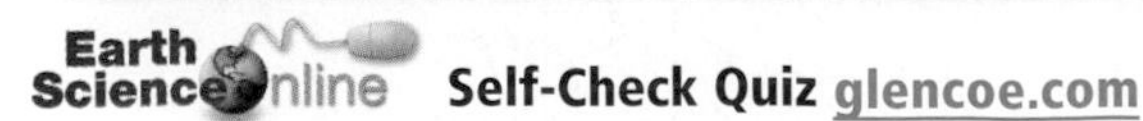

ON SITE: GEOLOGY IN CENTRAL PARK

Some people travel to remote locations of the world to see different types of rock. However, examples of rocks often can easily be found in urban areas. Central Park in New York City is an excellent place to find examples of igneous, sedimentary, and metamorphic rock, both naturally occurring and used for sculptures, monuments, and bridges.

The Obelisk Weighing 221 tons and standing 21 m high, Cleopatra's Needle is the oldest human-made object in Central Park. The granite was quarried in Egypt more than 3000 years ago in 1475 B.C. The sculpture remained in Egypt until 1879, when it was moved to the United States. Granite is more resistant to weathering than other types of rock, and engravings made in granite can be read for hundreds of years, making it an excellent rock for the construction of monuments.

Cleopatra's Needle

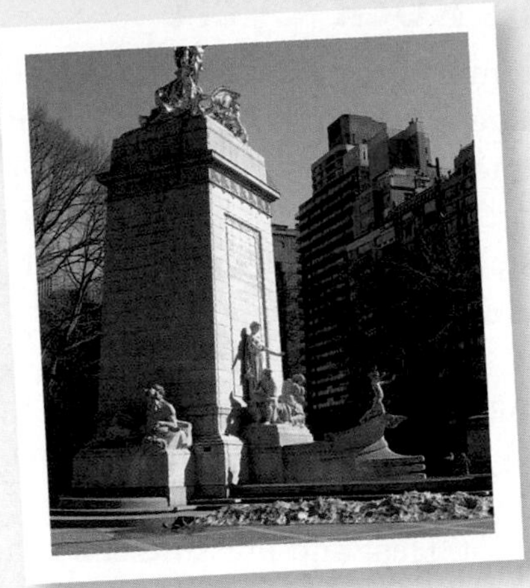

Maine Monument

The Maine Monument Located at the main entrance to Central Park, the Maine Monument is an immense structure made of marble, limestone, and bronze. The massive bow of a ship that makes up the base of the monument was sculpted out of marble, a type of metamorphic rock. A bronze statue sits atop a 15-m limestone pylon.

Schist and gneiss These two types of metamorphic rock occur naturally in Central Park. Outcroppings of these rocks, formed from sedimentary or igneous rock under intense heat and pressure, can be found throughout the park. The Gapstow Bridge was constructed using the local bedrock.

Gapstow Bridge

WRITING in Earth Science

Promotional Brochure Research more information about the type of rock used to build structures and that occur naturally in your area. Create a promotional brochure that describes a tour focused on local geology. To learn more about the different types of rock, visit glencoe.com.

GEOLAB

INTERPRET CHANGES IN ROCKS

Background: As the rock cycle continues and rocks change from one type to another, more changes occur than meet the eye. Color, grain size, texture, and mineral composition are easily observed and described visually. Yet, with mineral changes come changes in crystal structure and density. How can these be accounted for and described? Studying pairs of rocks can show you how.

Question: *How do the characteristics of sedimentary or igneous rocks compare to metamorphic rocks?*

Materials

samples of sandstone, shale, limestone, granite, quartzite, slate, marble, and gneiss
magnifying lens
paper
beam balance
100-mL graduated cylinder or beaker that is large enough to hold the rock samples
water

Safety Precautions

Procedure

1. Read and complete the lab safety form.
2. Prepare a data table similar to the one at the right. Adjust the width of the columns as needed.
3. Observe each rock sample. Record your observations in the data table.
4. Recall that density = mass/volume. Make a plan that will allow you to measure the mass and volume of a rock sample.
5. Determine the density of each rock sample, and record this information in the data table.

Sample Data Table

Sample Number	1	2	3	4
Rock type				
Specific characteristics				
Mass				
Volume				
Density				

Analyze and Conclude

1. **Compare and contrast** sandstone and quartzite.
2. **Describe** how the grain size of sandstone changes during metamorphism.
3. **Describe** the textural differences you observe between shale and slate.
4. **Infer** Compare your calculated densities to those calculated by other students. Infer why yours might differ.
5. **Explain** why the color of a sedimentary rock may change during metamorphism.
6. **Evaluate** the changes in density between shale and slate, sandstone and quartzite, limestone and marble, and granite and gneiss. Does density always change? Explain your results.

SHARE YOUR DATA

Peer Review Discuss your results with other groups in your class. Speculate on the reasons for variations in mass, volume, and density.

CHAPTER 6

Study Guide

Download quizzes, key terms, and flash cards from glencoe.com.

BIG Idea Most rocks are formed from preexisting rocks through external and internal geologic processes.

Vocabulary	Key Concepts
Section 6.1 Formation of Sedimentary Rocks	
• bedding (p. 137) • cementation (p. 137) • cross-bedding (p. 138) • graded bedding (p. 138) • lithification (p. 136) • sediment (p. 134)	**MAIN Idea** Sediments produced by weathering and erosion form sedimentary rocks through the process of lithification. • The processes of weathering, erosion, deposition, and lithification form sedimentary rocks. • Sediments are lithified into rock by the processes of compaction and cementation. • Fossils are the remains or other evidence of once-living organisms that are preserved in sedimentary rocks. • Sedimentary rocks might contain features such as horizontal bedding, cross-bedding, and ripple marks.
Section 6.2 Types of Sedimentary Rocks	
• clastic (p. 141) • clastic sedimentary rock (p. 141) • evaporite (p. 143) • porosity (p. 142)	**MAIN Idea** Sedimentary rocks are classified by their mode of formation. • Sedimentary rocks can be clastic, chemical, or biochemical. • Clastic rocks form from sediments and are classified by particle size and shape. • Chemical rocks form primarily from minerals precipitated from water. • Biochemical rocks form from the remains of once-living organisms. • Sedimentary rocks provide geologists with information about surface conditions that existed in Earth's past.
Section 6.3 Metamorphic Rocks	
• contact metamorphism (p. 149) • foliated (p. 146) • hydrothermal metamorphism (p. 149) • nonfoliated (p. 147) • regional metamorphism (p. 149) • rock cycle (p. 151)	**MAIN Idea** Metamorphic rocks form when preexisting rocks are exposed to increases in temperature and pressure and to hydrothermal solutions. • The three main types of metamorphism are regional, contact, and hydrothermal. • The texture of metamorphic rocks can be foliated or nonfoliated. • During metamorphism, new minerals form that are stable under the increased temperature and pressure conditions. • The rock cycle is the set of processes through which rocks continuously change into other types of rocks.

CHAPTER 6 Assessment

Vocabulary Review

Complete the sentences below using vocabulary terms from the Study Guide.

1. Compaction and cementation of clastic sediments result in ________.
2. Sedimentary layers that are deposited on an angle are called ________.
3. Cooling and crystallization, igneous rocks, uplift, and weathering and erosion describe a path along the ________.
4. Hot fluids that come in contact with solid rock result in ________.

Replace the italicized word with the correct vocabulary term from the Study Guide.

5. *Cementation* occurs when sediment gets deposited as the energy of the water decreases.
6. *Foliated* rocks have square, blocky crystals.

Write a sentence using each pair of words.

7. contact metamorphism, regional metamorphism
8. porosity, clastic sedimentary rock
9. sediment, bedding
10. clastic, evaporite

Understand Key Concepts

11. Which clastic sediment has the smallest grain size?
    A. sand
    B. clay
    C. pebbles
    D. silt

12. Which is a coarse-grained clastic rock that contains angular fragments?
    A. limestone
    B. conglomerate
    C. sandstone
    D. breccia

13. Which is a biochemical rock that contains fossils?
    A. chert
    B. limestone
    C. sandstone
    D. breccia

14. Which process forms salt beds?
    A. deposition
    B. cementation
    C. evaporation
    D. lithification

15. Which does not cause metamorphism?
    A. lithification
    B. hydrothermal solutions
    C. heat
    D. pressure

Use the diagram below to answers Questions 16 and 17.

16. Which term best describes this rock's texture?
    A. crystalline
    B. nonfoliated
    C. foliated
    D. clastic

17. From what igneous rock does this sample usually form?
    A. rhyolite
    B. basalt
    C. granite
    D. gabbro

18. Which agent of erosion can usually move only sand-sized or smaller particles?
    A. landslides
    B. glaciers
    C. water
    D. wind

19. Which would you expect to have the greatest porosity?
    A. sandstone
    B. gneiss
    C. shale
    D. quartzite

20. By what process are surface materials removed and transported from one location to another?
    A. weathering
    B. erosion
    C. deposition
    D. cementation

Constructed Response

Use the diagram to answer Question 21.

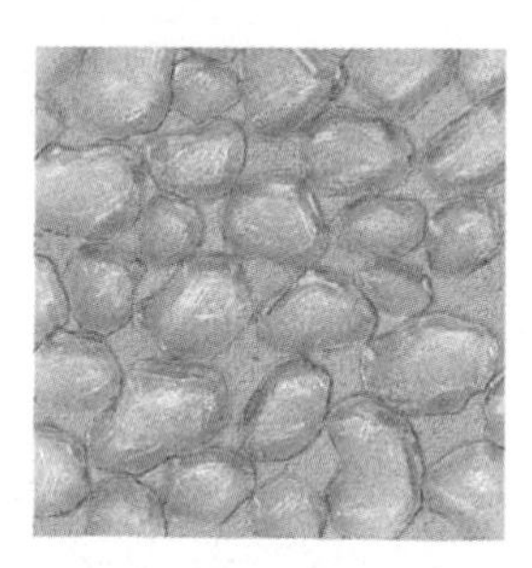

21. **Describe** how the grains in the diagram become glued together.

22. **Summarize** the main difference between coquina and fossiliferous limestone. Use **Table 6.1** for help.

23. **Calculate** A sandstone block has a volume of 1 m^3 and a porosity of 30 percent. How many liters of water can this block hold?

24. **Illustrate** the two conditions necessary to form a foliated metamorphic rock.

25. **Compare and contrast** the modes of lithification for sand and mud.

26. **Classify** the following types of sediments as either poorly sorted or well sorted: dune sand, landslide material, glacial deposits, and beach sand.

27. **Analyze** the effect that precipitation of calcite or iron oxide minerals has on clastic sediments.

28. **Compare and contrast** the character and formation of breccia and conglomerate.

Use the diagram below to answer Question 29.

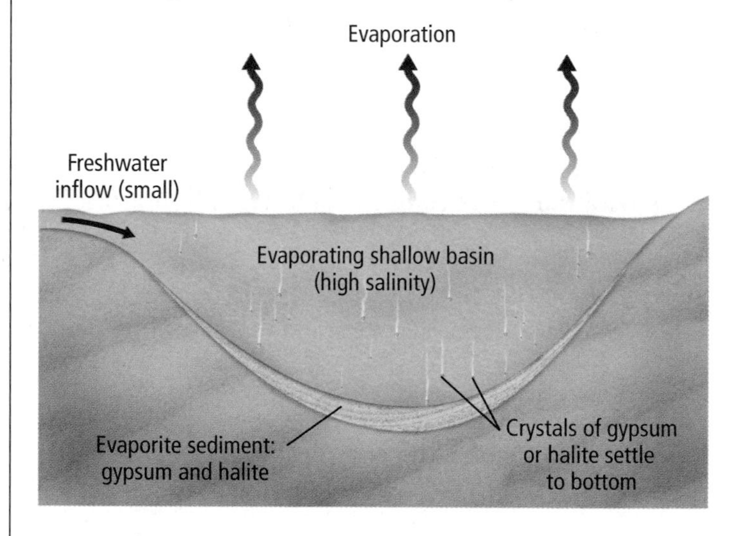

29. **Evaluate** the effect that an opening to the ocean would have on this environment.

Think Critically

30. **Incorporate** what you know about crystal form to explain why marble, even if formed under high pressure, does not show foliation.

31. **Compose** a statement to explain why the sedimentary rock coal does not meet the standard definition of a rock—an aggregate of minerals.

32. CAREERS IN EARTH SCIENCE Some sedimentologists work in sand and gravel pits where they analyze the material to best decide where and how it should be used. Infer why it is important for the sedimentologists to understand what would happen to the porosity of sand if finer-grained sediment were mixed in with the sand.

33. **Illustrate** an oil reservoir made up of layers of sandstone and shale. Indicate the position of the oil within the rocks.

34. **Assess** whether ripple marks and animal footprints preserved in sandstone are fossils. Explain your reasoning.

Use the figure below to answer Questions 35 and 36.

35. **Evaluate** the sediment in the layers in the figure. What type of bedding is this, and how well is it sorted? Explain.

36. **Infer** Look at **Figure 6.2** and explain which agents of erosion can produce the layers shown.

37. **Deduce** why glass on a quartz sand beach becomes rounded and frosted, while glass on a carbonate sand beach stays sharp and glassy.

Concept Mapping

38. Use the following terms to create a concept map that organizes sedimentary features: *ripple marks, graded bedding, horizontal bedding, asymmetrical, symmetrical, river current, wave action, wind deposited,* and *water deposited.* Some terms can be used more than once.

Challenge Question

39. **Hypothesize** At an approximate ocean depth of 4000 km, the carbonate compensation depth occurs. Below this depth, no calcium carbonate precipitates and no shells accumulate on the ocean floor. Hypothesize why this condition exists.

Additional Assessment

40. WRITING in **Earth Science** Imagine that you are planning a geologic walking tour of your community. Create a brochure highlighting the various natural building stones that are used in homes and buildings in your town or neighborhood.

DBQ Document–Based Questions

Data obtained from: Mineral Commodity Summaries. January 2006. *United States Geological Survey.*

Dimension stone is natural rock material used in construction, for monuments, and home interiors, such as kitchen countertops and floors. The principal rock types used are granite, limestone, marble, sandstone, and slate. Global resources of dimension stone are virtually limitless. Production of dimension stone in the United States and elsewhere has been steadily increasing.

Dimension Stone Production	U.S. Sold or Used (tonnage)	U.S. Sales or Uses (by value)
Limestone	39%	34%
Granite	29%	39%
Sandstone	14%	9%
Misc. stone	10%	7%
Marble	7%	6%
Slate	1%	5%

41. Construct a graph comparing the amount of dimension stone used by the value of the types of dimension stone.

42. Propose an explanation for why the value of granite is the highest of the dimension stones listed.

Cumulative Review

43. Compare and contrast the terms *science* and *technology.* **(Chapter 1)**

44. What is the formula of the ionic compound magnesium chloride? **(Chapter 3)**

45. Explain the concepts of partial melting and fractional crystallization. **(Chapter 5)**

Standardized Test Practice

Multiple Choice

Use the illustration below to answer Questions 1 and 2.

1. Which rocks are most likely to metamorphose from the lava flow?
   A. only the rocks in the crater of the volcano, where the lava is hottest
   B. rocks in the crater and rocks along the top half of the mountain
   C. all the rocks on the mountain
   D. all the rocks reached by the lava flow

2. As the lava cools and crystallizes, what type of rock will form?
   A. sedimentary
   B. metamorphic
   C. extrusive igneous
   D. intrusive igneous

3. What is NaCl commonly known as?
   A. table salt **C.** water
   B. sugar **D.** natural chlorine

4. What initiates the process that changes sediments into sedimentary rocks?
   A. bedding **C.** cementation
   B. burial **D.** compaction

5. Identify the unit that is NOT an example of the Le Système International D'Unités (SI).
   A. metric ton **C.** ampere
   B. kilogram **D.** Fahrenheit

6. Which rocks are composed of minerals that form with blocky crystal shapes?
   A. foliated **C.** porphyroblasts
   B. nonfoliated **D.** phenocrysts

Use the diagram below to answer Questions 7 and 8.

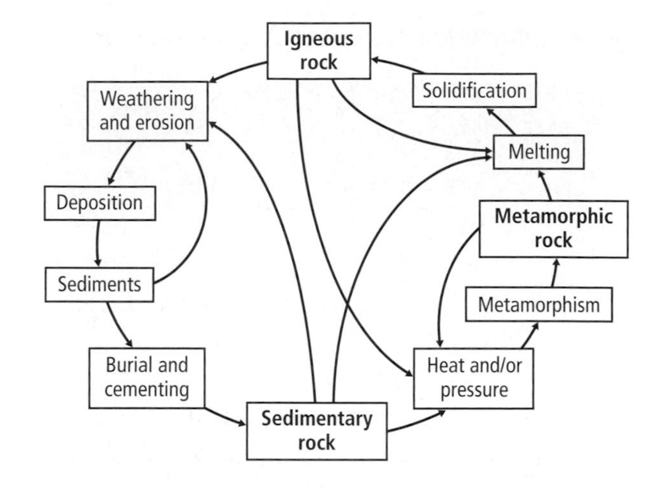

7. Based on the diagram, which is the most reasonable hypothesis?
   A. Igneous rocks have layers caused by deposition.
   B. Sedimentary rocks contain grains of other rocks.
   C. Metamorphic rocks never have layers.
   D. Sedimentary rocks are always the same color.

8. According to the rock cycle shown above, what most likely happens after the deposition of sediment?
   A. Weathering forms more sediment.
   B. Magma cools and forms igneous rock.
   C. Heat and pressure cause the sediment to melt.
   D. Cementation occurs and forms sedimentary rock.

9. Where are valence electrons located?
   A. every energy level
   B. middle energy levels
   C. the outermost energy level
   D. the innermost energy level

10. Which sedimentary rock is used to make cement for the construction industry?
    A. shale
    B. sandstone
    C. phosphate
    D. limestone

Short Answer

Use the illustration below to answer Questions 11 and 12.

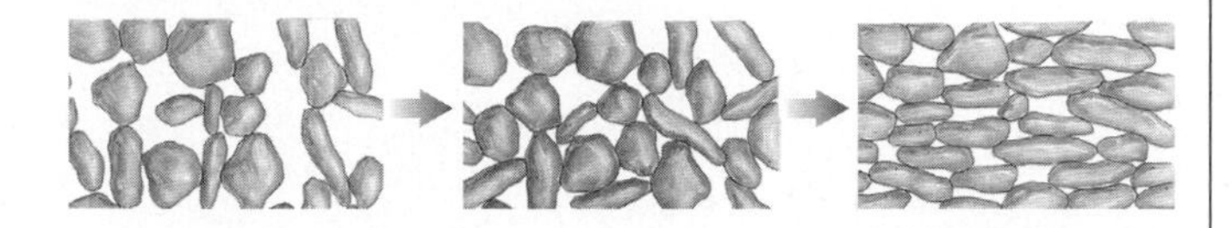

11. What do you notice about the formation of sedimentary rock above?

12. Does this process represent compaction or cementation? Describe the difference between the two.

13. The results of an experiment show that as temperature increases, enzyme activity decreases. Describe what a line graph made from this data would look like.

14. Define *luster.* Why is it difficult to use luster to identify minerals?

15. What process does Bowen's reaction series illustrate?

16. Boron has an atomic number of 5. Describe an atom of boron with a mass number of 10 and an atom of boron with a mass number of 11 in terms of their atomic particles. What is unique about these two atoms of boron?

17. Briefly describe the process by which magma becomes igneous rock.

18. How does studying sedimentary rock layers and understanding how they form help paleontologists learn about Earth's history?

Reading for Comprehension

Sedimentary Rock Layers

Paleontologists wanted to study the sedimentary rock layers and their contents of a particular area. The diagram shows a cross section of the rock layers they studied. The table shows the data the scientists were able to collect.

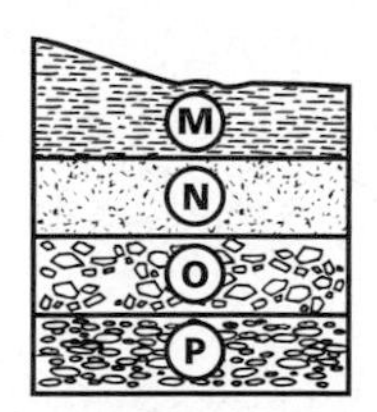

Age of Sedimentary Rock Layers			
Layer	**Composition**	**Estimated Age (years)**	**Depth (meters)**
M	sedimentary rock	100,000	0–4
N	sedimentary rock	Unknown	5–7
O	sedimentary rock	6 million	8–9
P	sedimentary rock	6.1 million	9–10

19. What could the paleontologists have recorded to improve their study?
    A. time of year
    B. age of layer N
    C. location of the work site
    D. mass of the sedimentary rocks

20. If fossils of a species were found in Layers O and P, but not M and N, which could you conclude?
    A. The species does not exist anywhere on Earth today.
    B. The species evolved into a completely different species.
    C. The species became extinct less than 100,000 years ago.
    D. The species disappeared from the area around 6 mya.

NEED EXTRA HELP?																		
If You Missed Question . . .	1	2	3	4	5	6	7	8	9	10	11	12	13	14	15	16	17	18
Review Section . . .	6.3	5.1	3.2	6.1	1.2	6.3	6.3	6.3	3.1	6.2	6.1	6.1	1.3	4.1	5.1	3.1	5.1	6.1

Surface Processes on Earth

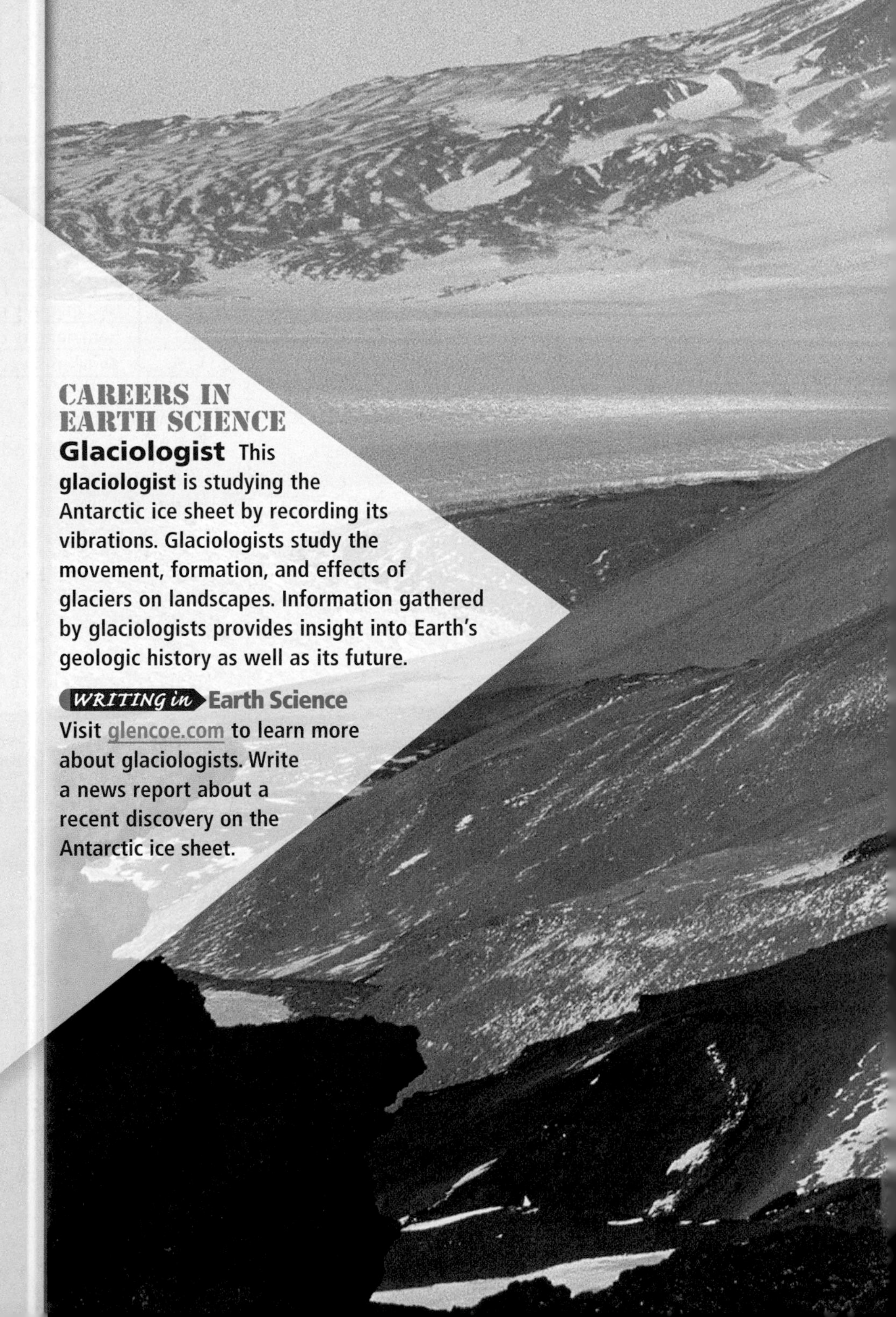

CAREERS IN EARTH SCIENCE

Glaciologist This **glaciologist** is studying the Antarctic ice sheet by recording its vibrations. Glaciologists study the movement, formation, and effects of glaciers on landscapes. Information gathered by glaciologists provides insight into Earth's geologic history as well as its future.

WRITING in **Earth Science**
Visit glencoe.com to learn more about glaciologists. Write a news report about a recent discovery on the Antarctic ice sheet.

Earth Science Online

To learn more about glaciologists, visit glencoe.com.

CHAPTER 7

Weathering, Erosion, and Soil

BIG Idea Weathering and erosion are agents of change on Earth's surface.

7.1 Weathering
MAIN Idea Weathering breaks down materials on or near Earth's surface.

7.2 Erosion and Deposition
MAIN Idea Erosion transports weathered materials across Earth's surface until they are deposited.

7.3 Soil
MAIN Idea Soil forms slowly as a result of mechanical and chemical processes.

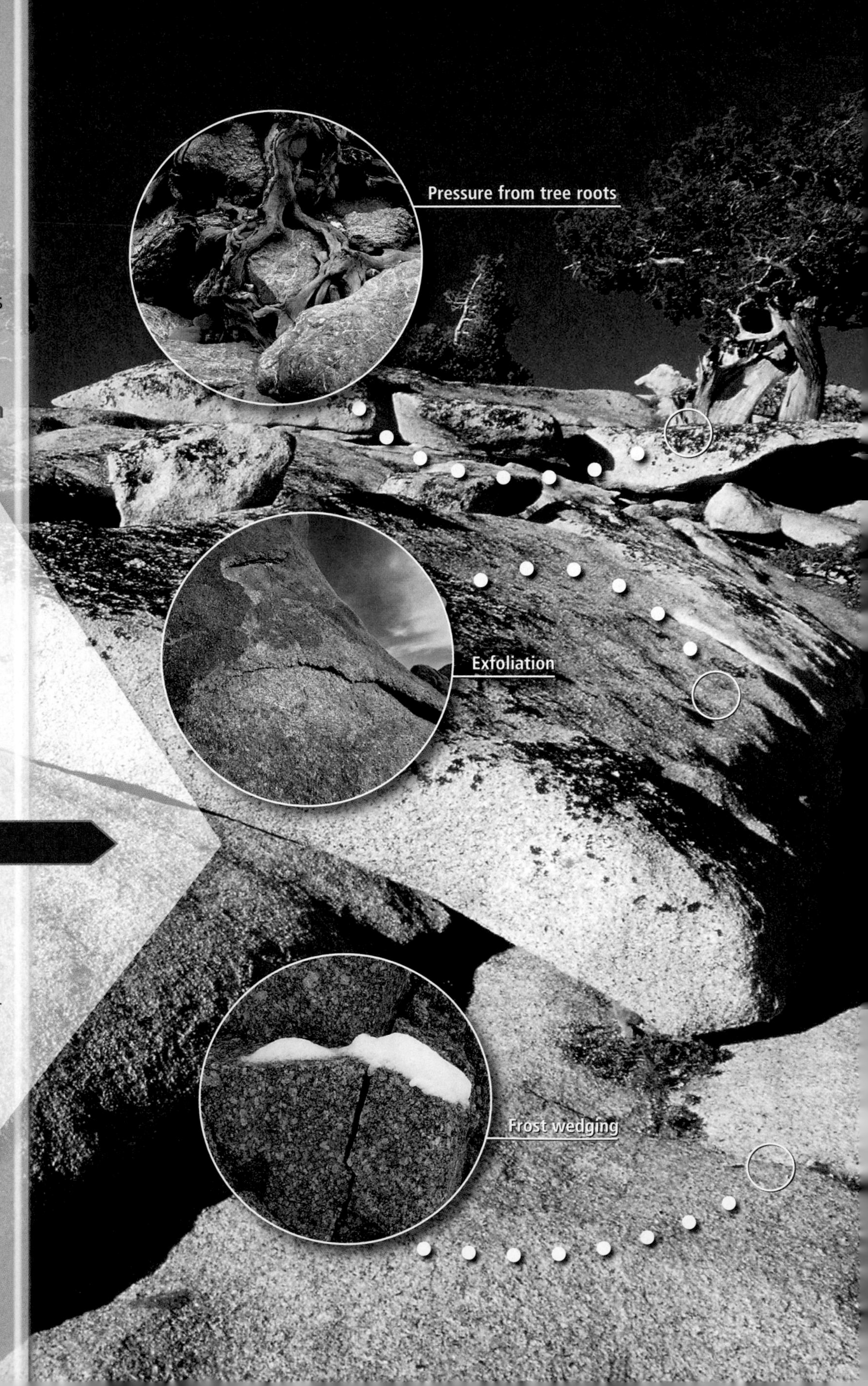

GeoFacts

- When plants sprout as seedlings in cracks in rocks, their growing roots can split rocks in two.
- Exfoliated rock weathers in layers, much like the layers of an onion.
- When water in the cracks of rocks freezes, it increases in volume, which can cause rocks to split.

LAUNCH Lab

How does change relate to surface area?

Surface area is a measure of the interface between an object and its environment. An object having more surface area can be affected more rapidly by its surroundings.

Procedure

1. Read and complete the lab safety form.
2. Fill two **250-mL beakers** with **water** at room temperature.
3. Drop a **sugar cube** in one beaker and 5 mL of **granulated sugar** in the other beaker at the same time. Record the time.
4. Slowly and continuously use a **stirring rod** to stir the solution in each beaker.
5. Observe the sugar in both beakers. Using a **stopwatch**, record the amount of time it takes for the sugar to completely dissolve in each beaker of water.

Analysis

1. **Describe** what happened to the sugar cube and the granulated sugar.
2. **Explain** why one form of sugar dissolved faster than the other.
3. **Infer** how you could decrease the time required for the slower-dissolving form of sugar.

Types of Weathering Make this Foldable to explain the types of weathering and what affects the rate of weathering.

STEP 1 Fold a sheet of paper in half vertically.

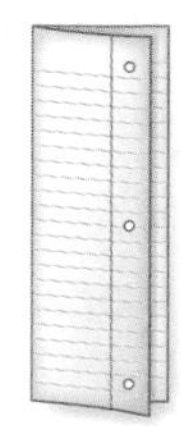

STEP 2 Make a 3-cm fold at the top and crease.

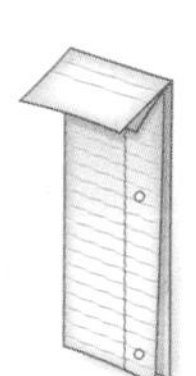

STEP 3 Unfold the paper and draw lines along the fold lines. Label the columns *Mechanical Weathering* and *Chemical Weathering.*

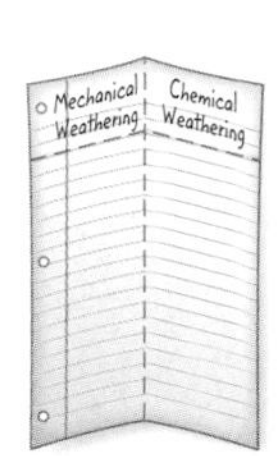

FOLDABLES **Use this Foldable with Section 7.1.** As you read this section, explain the types of weathering and the variables in the processes.

Visit glencoe.com to

- **study entire chapters online;**
- **explore Concepts In Motion animations:**
  - **Interactive Time Lines**
  - **Interactive Figures**
  - **Interactive Tables**
- **access Web Links for more information, projects, and activities;**
- **review content with the Interactive Tutor and take Self-Check Quizzes.**

Section 7.1

Objectives

- **Distinguish** between mechanical and chemical weathering.
- **Describe** the different factors that affect mechanical and chemical weathering.
- **Identify** variables that affect the rate of weathering.

Review Vocabulary

acid: solution that contains hydrogen ions

New Vocabulary

weathering
mechanical weathering
frost wedging
exfoliation
chemical weathering
oxidation

Weathering

MAIN Idea Weathering breaks down materials on or near Earth's surface.

Real-World Reading Link While walking on a sidewalk, you might notice that it has been pushed upward in places. In areas where trees are close to sidewalks, tree roots can cause the sidewalk to rise, buckle, and break.

Mechanical Weathering

Weathering is the process in which materials on or near Earth's surface break down and change. **Mechanical weathering** is a type of weathering in which rocks and minerals break down into smaller pieces. This process is also called physical weathering. Mechanical weathering does not involve any change in a rock's composition, only changes in the size and shape of the rock. A variety of factors are involved in mechanical weathering, including changes in temperature and pressure.

Effect of temperature Temperature plays a role in mechanical weathering. When water freezes, it expands and increases in volume by 9 percent. You have observed this increase in volume if you have ever frozen water in an ice cube tray. In many places on Earth's surface, water collects in the cracks of rocks and rock layers. If the temperature drops to the freezing point, water freezes, expands, exerts pressure on the rocks, and can cause the cracks to widen slightly, as shown in **Figure 7.1.** When the temperature increases, the ice melts in the cracks of rocks and rock layers. The freeze-thaw cycles of water in the cracks of rocks is called **frost wedging.** Frost wedging is responsible for the formation of potholes in many roads in the northern United States where winter temperatures vary frequently between freezing and thawing.

Figure 7.1 Frost wedging begins in hairline fractures of a rock. Repeated cycles of freeze and thaw cause the crack to expand over time.
Predict *the results of additional frost wedging on this boulder.*

Concepts In Motion
Interactive Figure To see an animation of frost wedging, visit glencoe.com.

Effect of pressure Another factor involved in mechanical weathering is pressure. Roots of trees and other plants can exert pressure on rocks when they wedge themselves into the cracks in rocks. As the roots grow and expand, they exert increasing amounts of pressure which often causes the rocks to split, as shown in **Figure 7.2.**

On a much larger scale, pressure also functions within Earth. Bedrock at great depths is under tremendous pressure from the overlying rock layers. A large mass of rock, such as a batholith, may originally form under great pressure from the weight of several kilometers of rock above it. When the overlying rock layers are removed by processes such as erosion or even mining, the pressure on the bedrock is reduced. The bedrock surface that was buried expands, and long, curved cracks can form. These cracks, also known as joints, occur parallel to the surface of the rocks. Reduction of pressure also allows existing cracks in the bedrock to widen. For example, when several layers of overlying rocks are removed from a deep mine, the sudden decrease of pressure can cause large pieces of rocks to explode off the walls of the mine tunnels.

Over time, the outer layers of rock can be stripped away in succession, similar to the way an onion's layers can be peeled. The process by which outer rock layers are stripped away is called **exfoliation.** Exfoliation often results in dome-shaped formations, such as Moxham Mountain in New York and Half Dome in Yosemite National Park in California, shown in **Figure 7.3.**

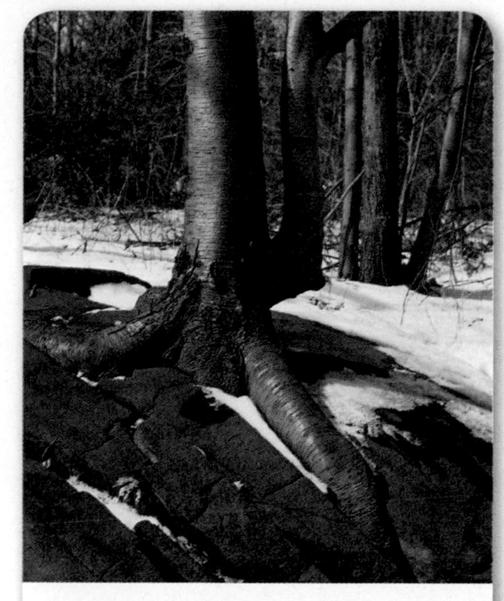

Figure 7.2 Tree roots can grow within the cracks and joints in rocks and eventually cause the rocks to split.

FOLDABLES
Incorporate information from this section into your Foldable.

Figure 7.3 The rock that makes up Half Dome in Yosemite National Park fractures along its outer surface in a process called exfoliation. Over time this has resulted in the dome shape of the outcrop.

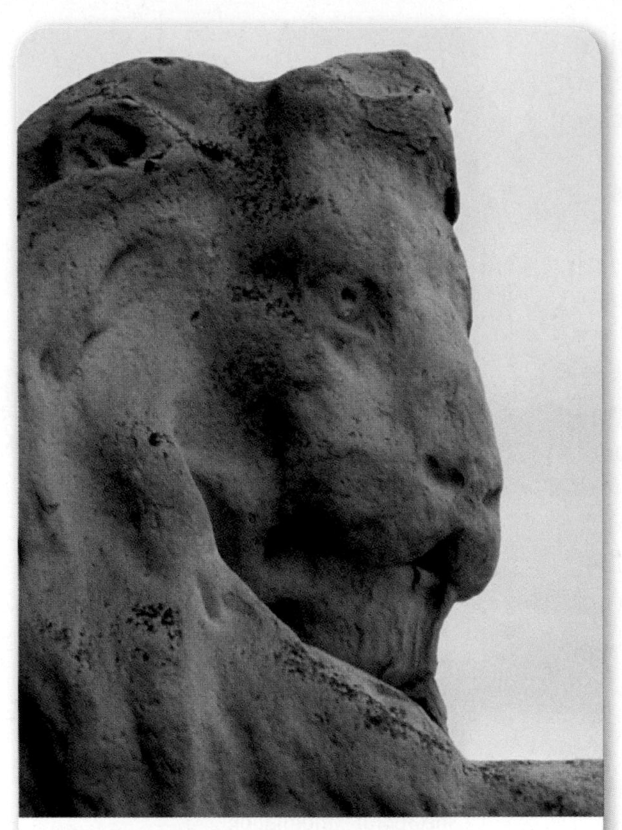

■ **Figure 7.4** This statue has been chemically weathered by acidic water and atmospheric pollutants.

Chemical Weathering

Chemical weathering is the process by which rocks and minerals undergo changes in their composition. Agents of chemical weathering include water, oxygen, carbon dioxide, and acid precipitation. The interaction of these agents with rock can cause some substances to dissolve, and some new minerals to form. The new minerals have properties different than those that were in the original rock. For example, iron often combines with oxygen to form iron oxide, such as in hematite.

Reading Check **Express** in your own words the effect that chemical weathering has on rocks.

The composition of a rock determines the effects that chemical weathering will have on it. Some minerals, such as calcite, which is composed of calcium carbonate, can decompose completely in acidic water. Limestone and marble are made almost entirely from calcite, and are therefore greatly affected by chemical weathering. Buildings and monuments made of these rocks usually show signs of wear as a result of chemical weathering. The statue in **Figure 7.4** is an example of chemical weathering from acid precipitation.

Temperature is another significant factor in chemical weathering because it influences the rate at which chemical interactions occur. Chemical reaction rates increase as temperature increases. With all other factors being equal, the rate of chemical weathering reactions doubles with each 10°C increase in temperature.

Effect of water Water is an important agent in chemical weathering because it can dissolve many kinds of minerals and rocks. Water also plays an active role in many reactions by serving as a medium in which the reactions can occur. Water can also react directly with minerals in a chemical reaction. In one common reaction with water, large molecules of the mineral break down into smaller molecules. This reaction decomposes and transforms many silicate minerals. For example, potassium feldspar decomposes into kaolinite, a fine-grained clay mineral common in soils.

Effect of oxygen An important element in chemical weathering is oxygen. The chemical reaction of oxygen with other substances is called **oxidation.** Approximately 21 percent of Earth's atmosphere is oxygen gas. Iron in rocks and minerals combines with this atmospheric oxygen to form minerals with the oxidized form of iron. A common mineral that contains the oxidized form of iron is hematite.

Effect of carbon dioxide Another atmospheric gas that contributes to the chemical weathering process is carbon dioxide. Carbon dioxide is a gas that occurs naturally in the atmosphere as a product of living organisms. When carbon dioxide combines with water in the atmosphere, it forms a very weak acid called carbonic acid that falls to Earth's surface as precipitation.

Precipitation includes rain, snow, sleet, and fog. Natural precipitation has a pH of 5.6. The slight acidity of precipitation causes it to dissolve certain rocks, such as limestone.

Decaying organic matter and respiration produce high levels of carbon dioxide. When slightly acidic water from precipitation seeps into the ground and combines with carbon dioxide in the soil, carbonic acid becomes an agent in the chemical weathering process. Carbonic acid slowly reacts with minerals such as calcite in limestone and marble to dissolve rocks. After many years, limestone caverns can form where the carbonic acid flowed through cracks in limestone rocks and reacted with calcite.

VOCABULARY

ACADEMIC VOCABULARY

Process
a natural phenomenon marked by gradual changes that lead toward a particular result
The process of growth changes a seedling into a tree.

Effect of acid precipitation Another agent of chemical weathering is acid precipitation, which is caused by sulfur dioxide, carbon dioxide, and nitrogen oxides. These compounds are released into the atmosphere, often by human activities. Sulfur dioxide and carbon dioxide are primarily the product of burning fossil fuels. Motor vehicle exhaust contributes to the emissions of nitrogen oxides. These three gases combine with oxygen and water in the atmosphere and form strong sulfuric, nitric, and carbonic acids.

The acidity of a solution is described using the pH scale, as you learned in Chapter 3. Acid precipitation is precipitation that has a pH value below 5.6—the pH of normal rainfall. Because strong acids can be harmful to many organisms and destructive to human-made structures, acid precipitation often creates problems. Many plant and animal populations cannot survive even slight changes in acidity. Acid precipitation is a serious issue in New York, as shown in **Figure 7.5,** and in West Virginia and much of Pennsylvania.

Figure 7.5 The forests of the Adirondack Mountains have been damaged by the effects of acid precipitation. Acid precipitation can make forests more vulnerable to disease and damage by insects.

Rate of Weathering

The natural weathering of Earth materials occurs slowly. For example, it can take 2000 years to weather 1 cm of limestone, and most rocks weather at even slower rates. Certain conditions and interactions can accelerate or slow the weathering process, as demonstrated in the GeoLab at the end of this chapter.

Effects of climate on weathering Climate is the major influence on the rate of weathering of Earth materials. Precipitation, temperature, and evaporation are factors of climate. The interaction between temperature and precipitation in a given climate determines the rate of weathering in a region.

Reading Check **Explain** why different climates have different rates of weathering.

Rates of chemical weathering Chemical weathering is rapid in climates with warm temperatures, abundant rainfall, and lush vegetation. These climatic conditions produce soils that are rich in organic matter. Water from heavy rainfalls combines with the carbon dioxide in soil organic matter and produces high levels of carbonic acid. The resulting carbonic acid accelerates the weathering process. Chemical weathering has the greatest effects along the equator, where rainfall is plentiful and the temperature tends to be high, as shown in **Figure 7.6.**

Figure 7.6 The impact of chemical weathering is related to a region's climate. Warm, lush areas such as the tropics experience the fastest chemical weathering.
Infer *what parts of the world experience less chemical weathering.*

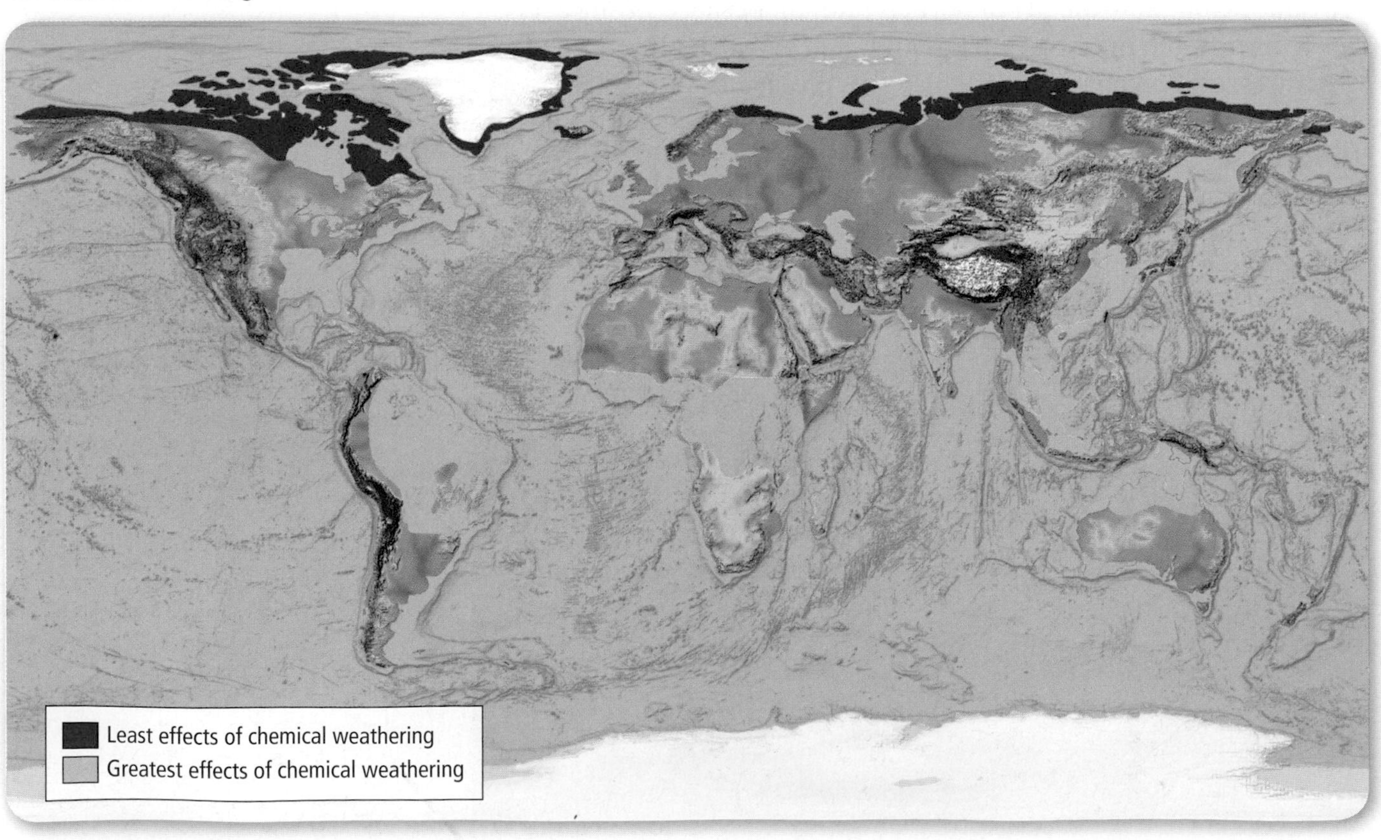

Rates of physical weathering Conversely, physical weathering can break down rocks more rapidly in cool climates. Physical weathering rates are highest in areas where water in cracks within the rocks undergoes repeated freezing and thawing. Conditions in such climates do not favor chemical weathering because cool temperatures slow or inhibit chemical reactions. Little or no chemical weathering occurs in areas that are frigid year-round.

The different rates of weathering caused by different climatic conditions can be emphasized by a comparison of Asheville, North Carolina, and Phoenix, Arizona. Phoenix has dry, warm, conditions; temperatures do not drop below the freezing point of water, and humidity is low. In Asheville, temperatures frequently drop below freezing during the winter months, and Asheville has more monthly rainfall and higher levels of humidity than Phoenix. Because of these differences in their climates, rocks and man-made structures in Asheville experience higher rates of mechanical and chemical weathering than those in Phoenix.

Figure 7.7 shows how rates of weathering are dependent on climate. Both Egyptian obelisks were carved from granite more than one thousand years ago. For more than a thousand years, they stood in Egypt's dry climate, showing few effects of weathering. In 1881, Cleopatra's Needle was transported from Egypt to New York City. In the time that has passed since then, the acid precipitation and the repeated cycles of freezing and thawing in New York City accelerated the processes of chemical and physical weathering. In comparison, the obelisk that remains in Egypt appears unchanged.

Rock type and composition. Not all the rocks in the same climate weather at the same rate. The effects of climate on the weathering of rock also depends on the rock type and composition. For example, rocks containing mostly calcite, such as limestone and marble, are more easily weathered than rocks containing mostly quartz, such as granite and quartzite.

NATIONAL GEOGRAPHIC To read about desert landscapes formed by weathering and erosion, go to the **National Geographic Expedition** on page 898.

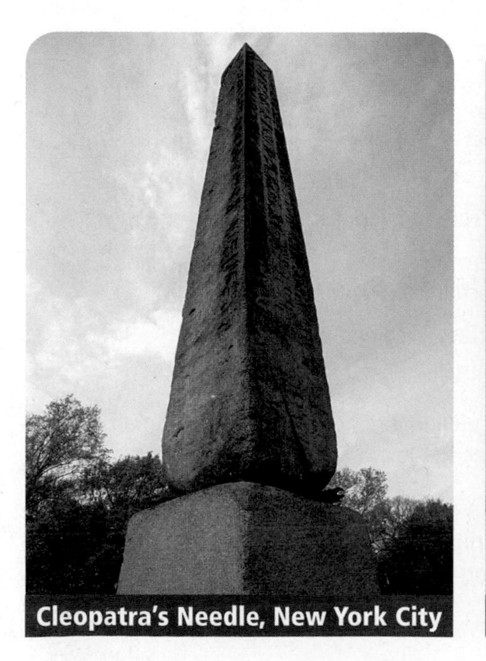

Cleopatra's Needle, New York City

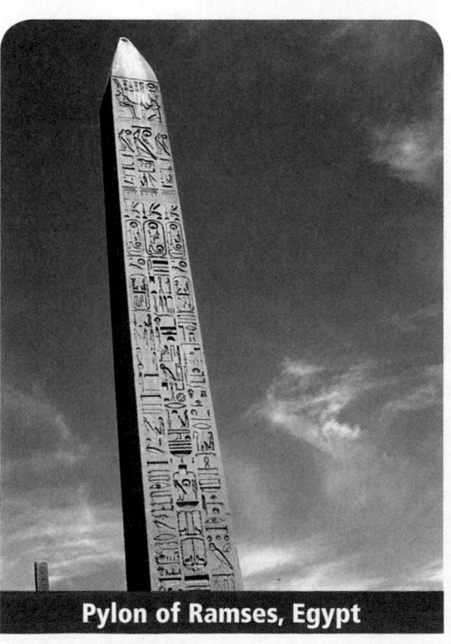

Pylon of Ramses, Egypt

■ **Figure 7.7** The climate of New York City caused the obelisk on the left to weather rapidly. The obelisk on the right has been preserved by Egypt's dry, warm climate.

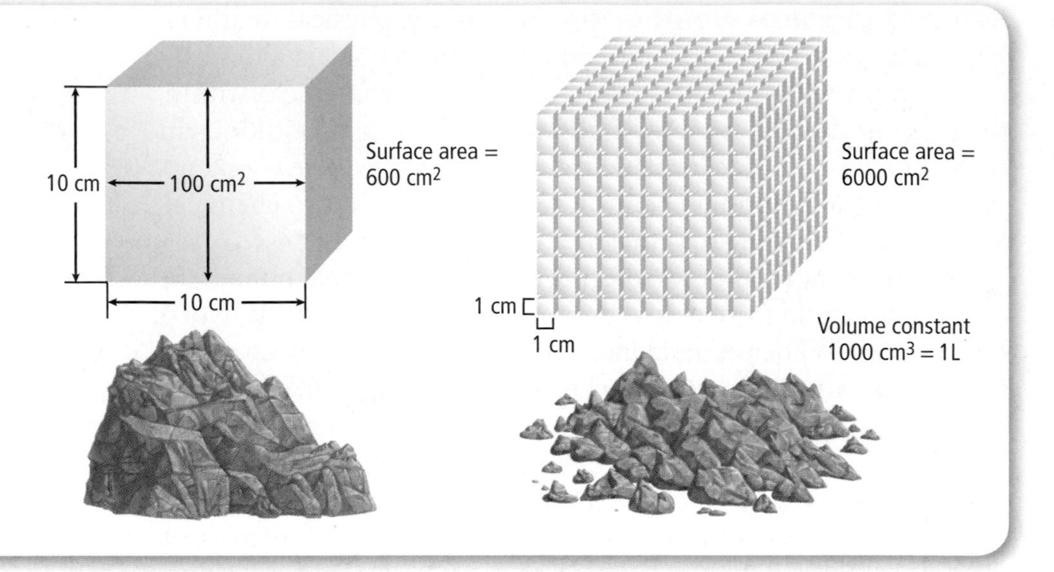

■ **Figure 7.8** When the same object is broken into two or more pieces, the surface area increases. The large cube has a volume of 1000 cm^3. When it is broken into 1000 pieces, the volume is unchanged, but the surface area is increased one thousand times.

Surface area The rate of weathering also depends on the surface area that is exposed. Mechanical weathering breaks rocks into smaller pieces. As the pieces get smaller, their surface area increases, as illustrated in **Figure 7.8.** When this happens, there is more total surface area available for chemical weathering. The result is that weathering has more of an effect on smaller particles, as you learned in the Launch Lab.

Topography The slope of a landscape also determines the rate of weathering. Rocks on level areas are likely to remain in place over time, whereas the same rocks on slopes tend to move as a result of gravity. Steep slopes therefore promote erosion and continually expose less-weathered material.

Section 7.1 Assessment

Section Summary

- Mechanical weathering changes a rock's size and shape.
- Frost wedging and exfoliation are forms of mechanical weathering.
- Chemical weathering changes the composition of a rock.
- The rate of chemical weathering depends on the climate, rock type, surface area, and topography.

Understand Main Ideas

1. MAIN Idea **Distinguish** between the characteristics of an unweathered rock and those of a highly weathered rock.
2. **Describe** the factors that control the rate of chemical weathering and those that control the rate of physical weathering.
3. **Compare** chemical weathering to mechanical weathering.
4. **Analyze** the relationship between surface area and weathering.

Think Critically

5. **Infer** which would last longer, the engraving in a headstone made of marble, or an identical engraving in a headstone made of granite.

MATH in Earth Science

6. Infer the relationship between weathering and surface area by graphing the relationship between the rate of weathering and the surface area of a material.

Earth Science Online Self-Check Quiz glencoe.com

Section 7.2

Objectives

- **Describe** the relationship of gravity to all agents of erosion.
- **Contrast** the features left from different types of erosion.
- **Analyze** the impact of living and nonliving things on the processes of weathering and erosion.

Review Vocabulary

gravity: a force of attraction between objects due to their masses

New Vocabulary

erosion
deposition
rill erosion
gully erosion

Erosion and Deposition

MAIN Idea Erosion transports weathered materials across Earth's surface until they are deposited.

Real-World Reading Link Have you ever noticed the mud that collects on sidewalks and streets after a heavy rainfall? Water carries sediment to the sidewalks and streets and deposits it as mud.

Gravity's Role

Recall that the process of weathering breaks rock and soil into smaller pieces, but never moves it. The removal of weathered rock and soil from its original location is a process called **erosion.** Erosion can remove material through a number of different agents, including running water, glaciers, wind, ocean currents, and waves. These agents of erosion can carry rock and soil thousands of kilometers away from their source. After the materials are transported, they are dropped in another location in a process known as **deposition.**

Gravity is associated with many erosional agents because the force of gravity tends to pull all materials downslope. Without gravity, neither streams nor glaciers would flow. In the process of erosion, gravity pulls loose rock downslope. **Figure 7.9** shows the effects of gravity on the landscape of Watkins Glen State Park in New York. The effects of gravity on erosion by running water can often produce dramatic landscapes with steep valleys.

Figure 7.9 Within about 3000 m, the stream descends 120 m at Watkins Glen State Park in New York.
Calculate *the average descent of the stream per meter along the river.*

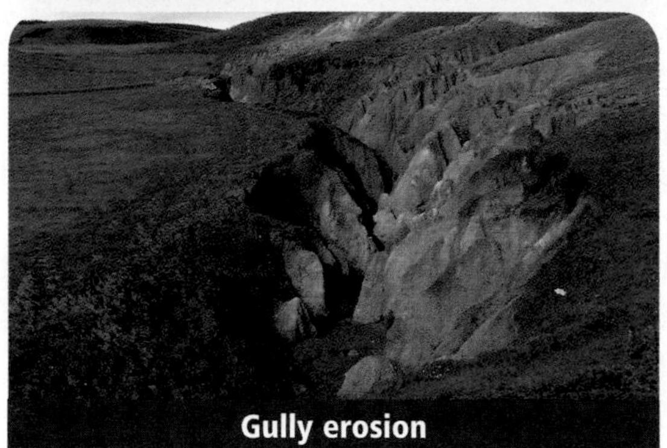

■ **Figure 7.10** Rill erosion can occur in an agricultural field. Gully erosion often develops from rills.
Suggest *land management practices that can slow or prevent the development of gully erosion.*

Erosion by Water

Moving water is perhaps the most powerful agent of erosion. Stream erosion can reshape entire landscapes. Stream erosion is greatest when a large volume of water is moving rapidly, such as during spring thaws and torrential downpours. Water flowing down steep slopes has additional erosive potential resulting from gravity, causing it to cut downward into the slopes, carving steep valleys and carrying away rock and soil. Water that flows swiftly or in large volumes can independently carry more material. The Mississippi River carries over 400,000 metric tons of sediment each day from thousands of kilometers away due to the volume of water in the river.

 **Reading Check** **Predict** what time of year water has the most potential for erosion.

Erosion by water can have destructive results. For example, water flowing downslope can carry away fertile agricultural soil. **Rill erosion** develops when running water cuts small channels into the side of a slope, as shown in **Figure 7.10.** When a channel becomes deep and wide, rill erosion evolves into **gully erosion,** also shown in **Figure 7.10.** The channels formed in gully erosion can transport much more water, and consequently more soil, than rills. Gullies can be more than 3 m deep and can cause major problems in farming and grazing areas.

MiniLab

Model Erosion

How do rocks erode? When rocks are weathered by their surrounding environment, particles can be carried away by erosion.

Procedure

1. Read and complete the lab safety form.
2. Carve your name deeply into a **bar of soap** with a **toothpick.** Measure the mass of the soap.
3. Measure and record the depth of the letters carved into the soap.
4. Place the bar of soap on its edge in a **catch basin.**
5. Slowly pour **water** over the bar of soap until a change occurs in the depth of the carved letters.
6. Measure and record the depth of the carved letters.

Analysis

1. **Describe** how the depth of the letters carved into the bar of soap changed.
2. **Infer** whether the shape, size, or mass of the bar of soap changed.
3. **Consider** what additional procedure you could follow to determine whether any soap wore away.

Rivers and streams Each year, streams carry billions of metric tons of sediments and weathered material to coastal areas. Once a river enters the ocean, the current slows down, which reduces the potential of the stream to carry sediment. As a result, streams deposit large amounts of sediments in the region where they enter the ocean. The buildup of sediments over time forms deltas, such as the Colorado River Delta, shown in **Figure 7.11.** The volume of river flow and the action of tides determines the shapes of deltas, most of which contain fertile soil. The Colorado River Delta shows the classic fan shape associated with many deltas.

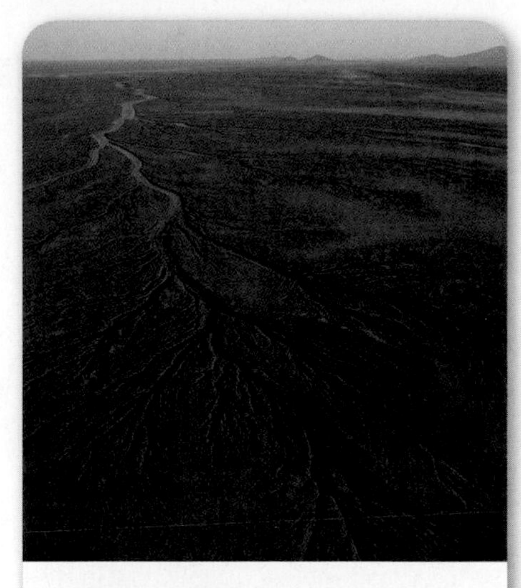

Figure 7.11 Streams slow down when they meet the ocean. In these regions, sediments are deposited by the river, resulting in the development of a delta.

Wave action Erosion of materials also occurs along the ocean floor and at continental and island shorelines. The work of ocean currents, waves, and tides carves out cliffs, arches, and other features along the continents' edges. In addition, sand particles accumulate on shorelines and form dunes and beaches. The constant movement of water and the availability of accumulated weathered material result in a continuous erosional process, especially along ocean shorelines. Sand along a shoreline is repeatedly picked up, moved, and deposited by ocean currents. As a result, sandbars form from offshore sand deposits. If the sandbars continue to be built up with sediments, they can develop into barrier islands. Many barrier islands, such as the Outer Banks of North Carolina shown in **Figure 7.12,** have formed along both the Gulf and Atlantic Coasts of the United States.

Just as shorelines are built by the process of deposition in some areas, they are reduced by the process of coastal erosion in other areas. Changing tides and conditions associated with coastal storms can also have a great impact on coastal erosion. Human development and population growth along shorelines have led to attempts to control the erosion of sand. However, efforts to keep the sand on one beachfront disrupt the natural migration of sand along the shore, depleting sand from another area. You will learn more about ocean and shoreline features in Chapters 15 and 16.

Figure 7.12 The Outer Banks of North Carolina have been built over time by deposition of sand and sediments.

■ **Figure 7.13** Iceberg Lake in Glacier National Park, Montana, was formed by glaciers.

Glacial Erosion

Although glaciers currently cover less than 10 percent of Earth's surface, they have covered over 30 percent of Earth's surface in the past. Glaciers left their mark on much of the landscape, and their erosional effects are large-scale and dramatic. Glaciers scrape and gouge out large sections of Earth's landscape. Because they can move as dense, enormous rivers of slowly flowing ice, glaciers have the capacity to carry huge rocks and piles of debris over great distances and grind the rocks beneath them into flour-sized particles. Glacial movements scratch and grind surfaces. The features left in the wake of glacial movements include steep U-shaped valleys and lakes, such as the one shown in **Figure 7.13.**

The effects of glaciers on the landscape also include deposition. For example, soils in the northern parts of the United States are formed from material that was transported and deposited by glaciers. Although the most recent ice age ended 15,000 years ago, glaciers continue to affect erosional processes on Earth.

Wind Erosion

Wind can be a major erosional agent, especially in arid and coastal regions. Such regions tend to have little vegetation to hold soil in place. Wind can easily pick up and move fine, dry particles. The effects of wind erosion can be both dramatic and devastating. The abrasive action of windblown particles can damage both natural features and human-made structures. Winds can blow against the force of gravity and easily move fine-grained sediments and sand uphill.

Wind barriers One farming method that can reduce the effects of wind erosion is the planting of wind barriers, also called windbreaks, shown in **Figure 7.14.** Windbreaks are trees or other vegetation planted perpendicular to the direction of the wind. A wind barrier might be a row of trees along the edge of a field. In addition to reducing erosion, wind barriers can trap blowing snow, conserve moisture, and protect crops from the effects of the wind.

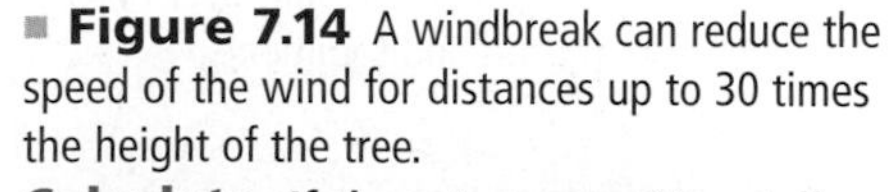

■ **Figure 7.14** A windbreak can reduce the speed of the wind for distances up to 30 times the height of the tree.
Calculate ***If these trees are 10 m tall, what is the distance over which they can serve as a windbreak?***

■ **Figure 7.15** In this construction project, the landscape was considerably altered. **Analyze** ***the results of this alteration of the landscape.***

Erosion by Living Things

Plants and animals also play a role in erosion. As plants and animals carry out their life processes, they move Earth's surface materials from one place to another. For example, Earth materials are moved when animals burrow into soil. Humans excavate large areas and move soil from one location to another, as shown in **Figure 7.15.** Planting a garden, developing a new athletic field, and building a highway are all examples of human activities that result in the moving of Earth materials from one place to another. You will learn more about how human activity impacts erosion in Chapter 26.

Section 7.2 Assessment

Section Summary

- The processes of erosion and deposition have shaped Earth's landscape in many ways.
- Gravity is the driving force behind major agents of erosion.
- Agents of erosion include running water, waves, glaciers, wind, and living things.

Understand Main Ideas

1. MAIN Idea **Discuss** how weathering and erosion are related.
2. **Describe** how gravity is associated with many erosional agents.
3. **Classify** the type of erosion that could move sand along a shoreline.
4. **Compare and contrast** rill erosion and gully erosion.

Think Critically

5. **Generalize** about which type of erosion is most significant in your area.
6. **Diagram** a design for a wind barrier to prevent wind erosion.

WRITING in Earth Science

7. Research how a development in your area has alleviated or contributed to erosion. Present your results to the class, including which type of erosion occurred, and where the eroded materials will eventually be deposited.

Section 7.3

Objectives

- **Describe** how soil forms.
- **Recognize** soil horizons in a soil profile.
- **Differentiate** among the factors of soil formation.

Review Vocabulary

organism: anything that has or once had all the characteristics of life

New Vocabulary

soil
residual soil
transported soil
soil profile
soil horizon

Soil

MAIN Idea Soil forms slowly as a result of mechanical and chemical processes.

Real-World Reading Link What color is soil? Soils can be many different colors—dark brown, light brown, red, or almost white. Soils develop through the interaction of a number of factors, which determine the color of soil.

Soil Formation

What is soil? It is found almost everywhere on Earth's surface. Weathered rock alone is not soil. **Soil** is the loose covering of weathered rock particles and decaying organic matter, called humus, overlying the bedrock of Earth's surface, and serves as a medium for the growth of plants. Soil is the product of thousands of years of chemical and mechanical weathering and biological activity.

Soil development The soil-development process often begins when weathering breaks solid bedrock into smaller pieces. These pieces of rock continue to undergo weathering and break down into smaller pieces. Worms and other organisms help break down organic matter and add nutrients to the soil as well as creating passages for air and water, as shown in **Figure 7.16.**

As nutrients are added to the soil, its texture changes, and the soil's capacity to hold water increases. While all soil contains some organic matter in various states of decay, the amount varies widely among different types of soil. For example, as much as 5 percent of the volume of prairie soils is organic matter, while most desert soils have almost no organic matter.

■ **Figure 7.16** Organisms in the soil change the soil's structure over time by adding nutrients and passages for air.
Infer *how animals also alter the soil by adding organic material.*

Soil Layers

During the process of its formation, soil develops layers. Most of the volume of soil is formed from the weathered products of a source rock, called the parent material. The parent material of a soil is often the bedrock. As the parent material weathers, the weathering products rest on top of the parent material. Over time, a layer of the smallest pieces of weathered rock develops above the parent material. Eventually, living organisms such as plants and animals become established, and use nutrients and shelter available in the material. Rainwater seeps through this top layer of materials and dissolves soluble minerals, carrying them into the lower layers of the soil.

A soil whose parent material is the local bedrock is called **residual soil.** Kentucky's bluegrass soil is an example of residual soil, as are the red soils in Georgia. Not all soil develops from local bedrock. **Transported soil,** shown in the valley in **Figure 7.17,** is soil that develops from parent material that has been moved far from its original location. Agents of erosion transport parent material from its place of origin to new locations. For example, glaciers have transported sediments from Canada to many parts of the United States. Streams and rivers, especially during times of flooding, also transport sediments downstream to floodplains. Winds also carry sediment to new locations. Over time, processes of soil formation transform these deposits into mature soil layers.

Reading Check **Explain** how residual soils are different from transported soils.

CAREERS IN EARTH SCIENCE

Landscaper A landscaper uses his or her knowledge of soils and performs tests to evaluate soils at different sites. Landscapers use the information they gather to choose plants that are appropriate to the soil conditions. To learn more about Earth science careers, visit glencoe.com.

Figure 7.17 In a stream valley, transported soils are often found in the flood plain. Residual soils are often found in the higher, mountainous regions.

■ **Figure 7.18** An undeveloped soil has few, if any, distinct layers, while mature soils are characterized by several soil horizons that have developed over time.

Soil profiles Digging a deep hole in the ground will reveal a soil profile. A **soil profile** is a vertical sequence of soil layers. Some soil profiles have more distinct layers than others. Relatively new soils that have not yet developed distinct layers are called undeveloped soils, shown in **Figure 7.18.** It can take tens of thousands of years for distinct layers to form in a soil. Those soils are called mature. An example is shown in **Figure 7.18.**

Reading Check **Explain** the difference between a mature and an undeveloped soil.

Soil horizons A distinct layer within a soil profile is called a **soil horizon.** There are typically four major soil horizons in mature soils, O, A, B, and C. The O-horizon is the top layer of organic material, which is made of humus and leaf litter. Below that, the A-horizon is a layer of weathered rock combined with a rich concentration of dark brown organic material. The B-horizon, also called the zone of accumulation, is a red or brown layer that has been enriched over time by clay and minerals deposited by water flowing from the layers above, or percolating upward from layers below. Usually the clay gives a blocky structure to the B-horizon. Accumulations of certain minerals can result in a hard layer called hardpan. Hardpan can be so dense that it allows little or no water to pass through it. The C-horizon contains little or no organic matter, and is often made of broken-down bedrock. The development of each horizon depends on the factors of soil formation.

Factors of Soil Formation

Five factors influence soil formation: climate, topography, parent material, biological organisms, and time. These factors combine to produce different types of soil, called soil orders, from region to region. Soil taxonomy (tak SAH nuh mee) is the system that scientists use to classify soils into orders and other categories. The five factors of soil formation result in 12 different soil orders.

Climate Climate is the most significant factor controlling the development of soils. Temperature, wind, and the amount of rainfall determines the type of soil that can develop.

Recall from Section 7.1 that rocks tend to weather rapidly under humid, temperate conditions, such as those found in climates along the eastern United States. Weathering results in soils that are rich in aluminum and iron oxides. Water from abundant rainfall moves downward, carrying dissolved minerals into the B-horizon. In contrast, the soils of arid regions are so dry that water from below ground moves up through evaporation, and leaves an accumulation of white calcium carbonate in the B-horizon. Tropical areas experience high temperatures and heavy rainfall. These conditions lead to the development of intensely weathered soils where all but the most insoluble minerals have been flushed out.

Topography Topography, which includes the slope and orientation of the land, affects the type of soil that forms. On steep slopes, weathered rock is carried downhill by agents of erosion. As a result, hillsides tend to have shallow soils, while valleys and flat areas develop thicker soils with more organic material. The orientation of slopes also affects soil formation. In the northern hemisphere, slopes that face south receive more sunlight than other slopes. The extra sunlight allows more vegetation to grow. Slopes without vegetation tend to lose more soil to erosion. **Figure 7.19** shows how the orientation and slope of a landscape can affect the formation of soil.

Figure 7.19 The slope on the right side faces south, and the slope on the left side faces north.
Interpret *why one slope has more vegetation than the other.*

Parent material Recall that a soil can be either residual or transported. If the soil is residual, it will have the same chemical composition as the local bedrock. For example, in regions near volcanoes, the soils form from weathered products of lava and ash. Volcanic soils tend to be rich in the minerals that were present in the lava. If the soil is transported, the minerals in the soil are likely to be different from those in the local bedrock.

Biological organisms Organisms including fungi and bacteria, as well as plants and animals, interact with soil. Microorganisms decompose dead plants and animals. Plant roots can open channels, and when they decompose, they add organic material to the soil. Different types of biological organisms in a soil can result in different soil orders. Mollisols (MAH lih sawlz), which are called prairie soils, and alfisols (AL fuh sawlz), also called woodland soils, both develop from the same climate, topography, and parent material. The different sets of organisms result in two soils with entirely different characteristics. For example, the activity of prairie organisms in mollisols produces a thick A-horizon, rich in organic matter. Some of the most fertile agricultural lands in the Great Plains region are mollisols.

Reading Check **Describe** how microorganisms affect soil formation.

Time The effects of time alone can determine the characteristics of a soil. New soils, such as entisols (EN tih sawlz), are often found along rivers, where sediment is deposited by periodic flooding. This type of soil is shown as a light blue color in **Figure 7.20.** These soils have had little time to weather and develop soil horizons. The effects of time on soil can be easy to recognize. After tens of thousands of years of weathering, most of the original minerals in a soil are changed or washed away. Minerals containing aluminum and iron remain, which can give older soils, such as ultisols (UL tih sawlz), a red color. **Figure 7.21** shows the locations of the 12 soil orders in the United States.

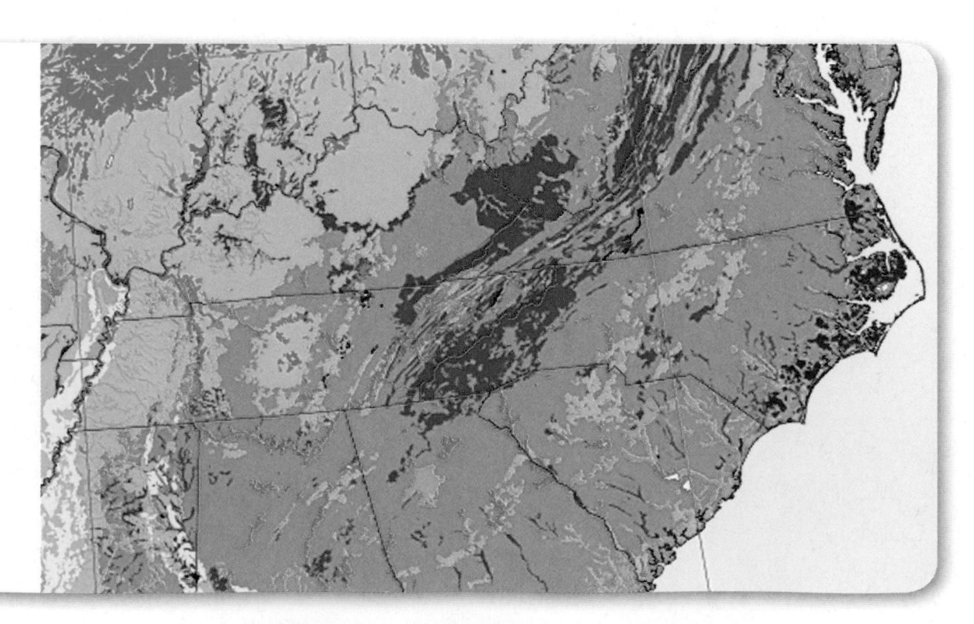

Figure 7.20 Soil types vary widely from one area to the next, depending on the local climate, topography, parent material, organisms, and age of the soil. Entisols are shown in light blue and ultisols are shown in orange on this map.
Infer *how differences in topography have affected the types of soils in North Carolina.*

Visualizing Soil Orders

Figure 7.21 The five factors of soil formation determine how the soil orders are distributed across the United States. Soil profiles of three soil orders from different parts of the country are shown. Each soil profile has soil horizons expressed differently.

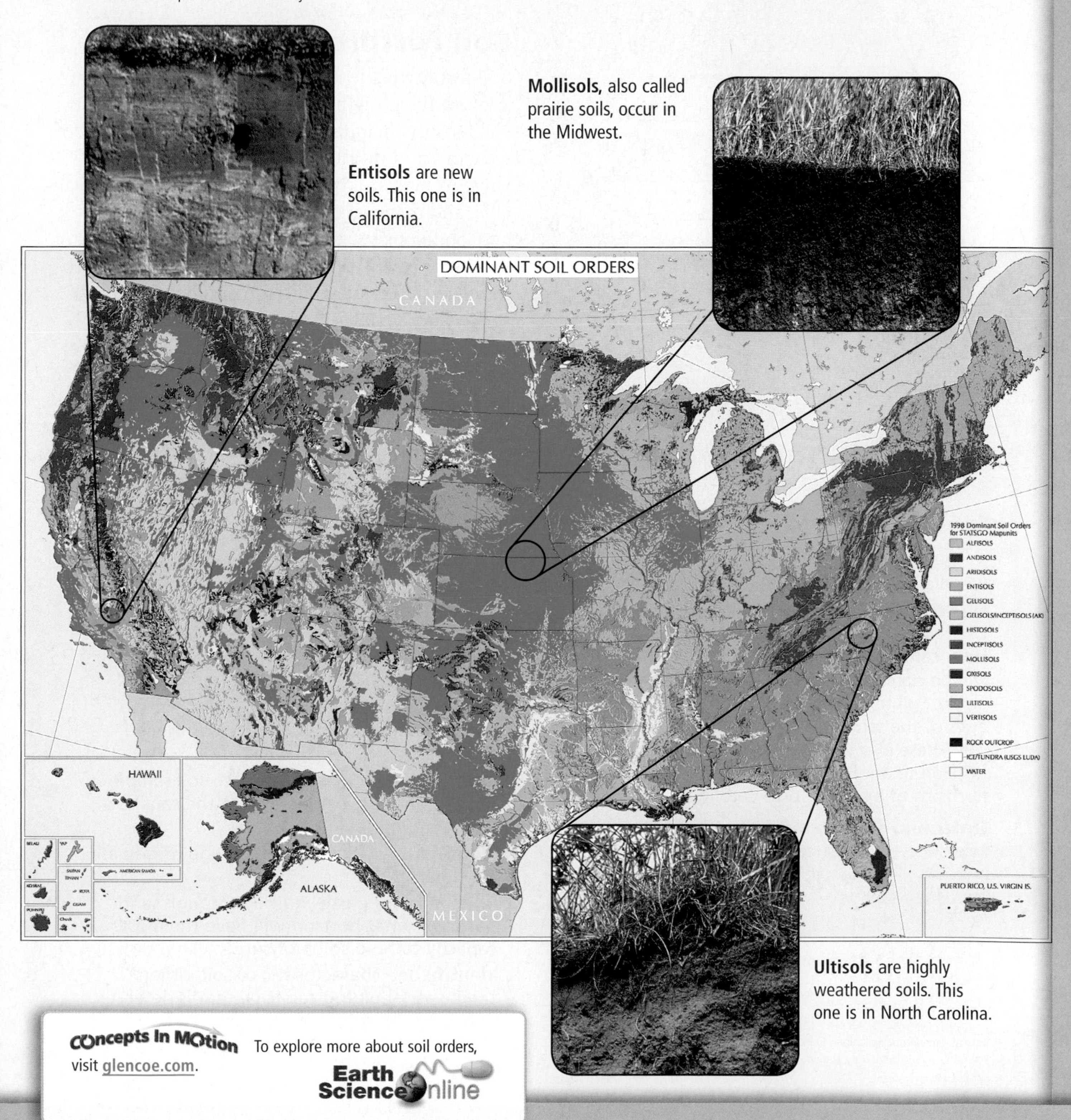

Concepts In Motion To explore more about soil orders, visit glencoe.com.

Earth Science Online

USDA Soil Classification

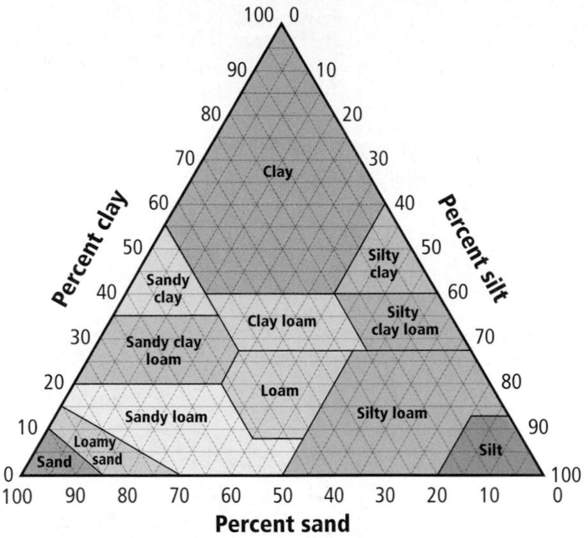

■ **Figure 7.22** A soil textural triangle is used to determine a soil's texture.

Soil Texture

Particles of soil are classified according to size as clay, silt, or sand, with clay being the smallest and sand being the largest. The relative proportions of particle sizes determine a soil's texture, as shown in **Figure 7.22.** Soil texture affects its capacity to retain moisture and therefore its ability to support plant growth. Soil texture also varies with depth.

Soil Fertility

Soil fertility is the measure of how well a soil can support the growth of plants. Factors that affect soil fertility include the topography, availability of minerals and nutrients, the number of microorganisms present, the amount of precipitation available, and the level of acidity.

Conditions necessary for growth vary with plant species. Farmers use natural and commercially produced fertilizers to replace minerals and maintain soil fertility. Commercial fertilizers add nitrates, potassium, and phosphorus to soil. The planting of legumes, such as beans and clover, allows bacteria to grow on plant roots and replace nitrates in the soil. Pulverized limestone is often added to soil to reduce acidity and enhance crop growth.

DATA ANALYSIS LAB

Based on Real Data*

Interpret the Data

How can you determine a soil's texture? Soils can be classified with the use of a soil textural triangle. Soil texture is determined by the percentages of the sand, silt, and clay that make up the soil. These also vary with depth, from one soil horizon to another. Below are data from three horizons of a soil in North Carolina.

Data and Observations

Soil Sample	Percent Clay	Percent Silt	Percent Sand	Texture
A	11	48		Loam
B	67		5	
C		53	38	

Data obtained from: Soil Survey Staff. 2006. National Soil Survey Characterization Data. Soil Survey Laboratory. National Soil Survey Center. (November 9) USDA-NRCS-Lincoln, NE

Think Critically

1. **Examine** the soil texture triangle shown in **Figure 7.22** to complete the data table. Record the percentages of particle sizes in the soil samples and the names of their textures.
2. **Infer** from the data table which soil sample has the greatest percentage of the smallest-sized particles.
3. **Identify** the maximum percentage of clay in clay loam.
4. **Infer,** if water passes quickly through sand particles, what horizon will have the most capacity to hold soil moisture.
5. **Identify** one characteristic of soil, other than water-holding capacity, that is determined by the soil's particle size.

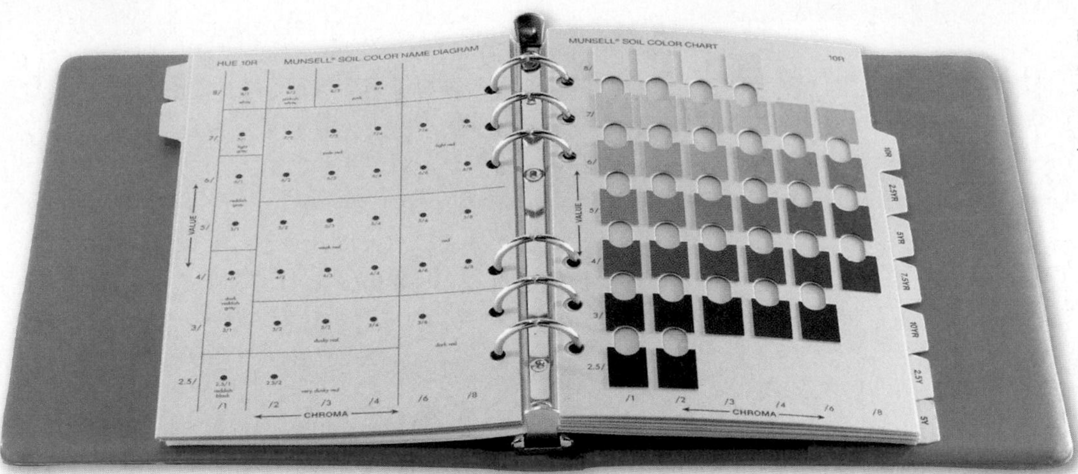

■ **Figure 7.23** Hue, value, and chroma can be found using the Munsell System of Color Notation.

Soil Color

The minerals, organic matter, and moisture in each soil horizon determine its color. An examination of the color of a soil can reveal many of its properties. For example, the layers that compose the O-horizon and A-horizon are usually dark-colored because they are rich in humus. Red and yellow soils might be the result of oxidation of iron minerals. Yellow soils are usually poorly drained and are often associated with environmental problems. Grayish or bluish soils are common in poorly drained regions where soils are constantly wet and lack oxygen.

Scientists use the Munsell System of Color Notation, shown in **Figure 7.23,** to describe soil color. This system consists of three parts: hue (color), value (lightness or darkness), and chroma (intensity). Each color is shown on a chip from a soil book. Using the components of hue, value, and chroma, a soil's color can be precisely described.

Section 7.3 Assessment

Section Summary

- Soil consists of weathered rock and humus.
- Soil is either residual or transported.
- A typical soil profile has O-horizon, A-horizon, B-horizon, and C-horizon.
- Five factors influence soil formation: climate, topography, parent material, biological organisms, and time.
- Characteristics of soil include texture, fertility, and color.

Understand Main Ideas

1. MAIN Idea **Describe** how soil forms.
2. **Summarize** the features of each horizon of soil.
3. **Classify** a soil profile based on whether it is mature or immature.
4. **Generalize** the effect that topography has on soil formation.

Think Critically

5. **Infer** Soil scientists discover that a soil in a valley has a C-horizon of sand that is 1 km deep. Is this a transported soil or a residual soil? Justify your answer.
6. **Hypothesize** what type of soil exists in your area, and describe how you would determine whether your hypothesis is correct.

WRITING in Earth Science

7. Soil in a portion of a garden is found to be claylike and acidic. Design a plan for improving the fertility of this soil.

EARTH SCIENCE AND TECHNOLOGY

SPACE-AGE TECHNOLOGY SHAPES MODERN FARMING

Many years ago, farmers planted and plowed with their hands, a few tools, and sometimes large animals, such as horses. Since then, new technology has revolutionized the work of farmers. In the United States, agriculture is a multi-billion dollar industry, in part because of something called precision farming.

Precision farming Precision farming, which is also called site-specific farming, is a method of farming that involves giving special attention to certain areas of a field.

The fields across a farm can vary greatly. Soil fertility might differ from one area to the next, some areas might retain water more easily than others, and the topography might vary. In the past, a farmer would have made decisions about planting, fertilizing, irrigation, and pesticide applications based on the average characteristics of a field. So some areas of the field would then receive too much fertilizer, while other areas of the field would not receive enough. Precision farming allows farmers to account for the differences across the field, which can increase crop yields, reduce waste, and protect natural resources. Precision farming relies on tools called the geographic information systems (GIS) and global positioning system (GPS).

GIS mapping The GIS helps farmers plot many types of information onto a computerized map of their fields. Farmers can record areas on a field that are prone to pest infestations, or areas where there is a change in elevations. Images of the field taken from satellites can be combined with observations made by the farmer. A computer program incorporates all of the information that is added, and creates GIS map layers.

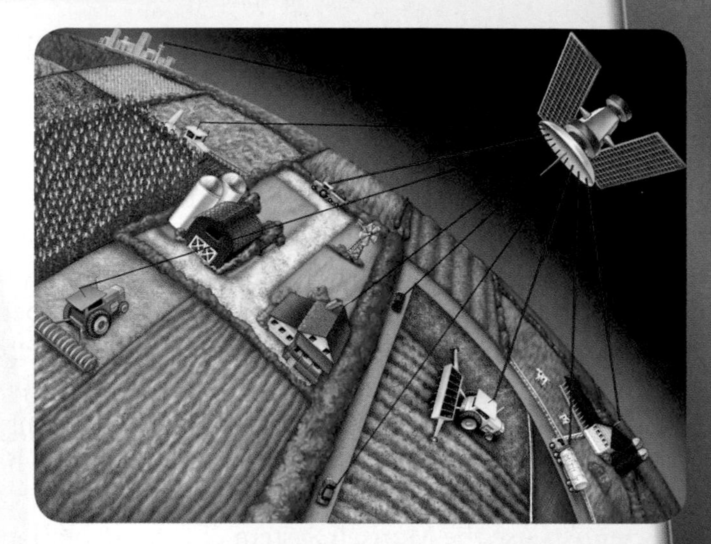

Satellites give information to farmers about their exact locations.

These layers are used to create detailed maps of the farm which can be used to plan for future crops, and to help plan where fertilizer or herbicides should be applied.

GPS navigation A system of satellites in orbit around Earth constantly relay their signals to Earth's surface. Specialized devices called GPS receivers can pick up the signals from these satellites, and use them to instantly calculate their exact location on a GIS map. This technology is used in many ways, including helping farmers find their location within a few centimeters' accuracy. Using GPS, farmers can program their tractors to plow rows that are perfectly straight, and know exactly how much fertilizer to apply to the soil.

WRITING in Earth Science

Write a journal entry about what it would be like to run a farm where all the tractors were operated remotely. For more information on precision farming, visit www.glencoe.com.

GEOLAB

MODEL MINERAL WEATHERING

Background: Many factors affect the rate of weathering of Earth materials. Two major factors that affect the rate at which a rock weathers include the length of time it is exposed to a weathering agent and the composition of the rock.

Question: *What is the relationship between exposure time and weathering?*

Materials

plastic jar with lid
water (300 mL)
halite chips (100 g)
balance
timer
paper towels

Safety Precautions

Procedure

1. Read and complete the lab safety form.
2. Soak 100 g of halite chips in water overnight.
3. As a class, decide on a uniform method of shaking the jars.
4. Pour off the water, and use paper towels to gently dry the halite chips. Divide them into four piles on the paper towel.
5. Use a balance to find the starting mass of one pile of the chips.
6. Place the halite chips in the plastic jar.
7. Add 300 mL of water to the jar.
8. Secure the lid on the jar, and shake the jar for the assigned period of time.
9. Pour the water from the jar.
10. Use paper towels to gently dry the halite chips.
11. Use a balance to find the final mass of the chips. Record your measurement in a data table similar to the one provided.
12. Subtract the final mass from the starting mass to calculate the change in mass of the halite chips.
13. Repeat Steps 4 to 12 using a fresh pile of halite chips for each period of time.

Weathering Data

Shaking Time (min)	Starting Mass of Chips (g)	Final Mass of Chips (g)	Change in Mass of Chips (g)
2			
4			
6			
8			

Analyze and Conclude

1. **State** What real-world process did you model in this investigation?
2. **Infer** Why did you need to soak the halite chips before conducting the experiment?
3. **Compare** the lab procedure with actual weathering processes. What did the halite represent? What process did shaking the jar represent?
4. **Deduce** How would acid precipitation affect this process in the real world?
5. **Conclude** How would the results of your investigation be affected if you used pieces of quartz instead of halite?

INQUIRY EXTENSION

Design an Experiment This lab demonstrated the relationship between exposure time and weathering. Consider other factors that affect weathering. Design an experiment to measure the effects of those factors.

CHAPTER 7

Study Guide

Download quizzes, key terms, and flash cards from glencoe.com.

BIG Idea Weathering and erosion are agents of change on Earth's surface.

Vocabulary	Key Concepts
Section 7.1 Weathering	
• chemical weathering (p. 166) • exfoliation (p. 165) • frost wedging (p. 164) • mechanical weathering (p. 164) • oxidation (p. 166) • weathering (p. 164)	**MAIN Idea** Weathering breaks down materials on or near Earth's surface. • Mechanical weathering changes a rock's size and shape. • Frost wedging and exfoliation are forms of mechanical weathering. • Chemical weathering changes the composition of a rock. • The rate of chemical weathering depends on the climate, rock type, surface area, and topography.
Section 7.2 Erosion and Deposition	
• deposition (p. 171) • erosion (p. 171) • gully erosion (p. 172) • rill erosion (p. 172)	**MAIN Idea** Erosion transports weathered materials across Earth's surface until they are deposited. • The processes of erosion and deposition have shaped Earth's landscape in many ways. • Gravity is the driving force behind major agents of erosion. • Agents of erosion include running water, waves, glaciers, wind, and living things.
Section 7.3 Soil	
• residual soil (p. 177) • soil (p. 176) • soil horizon (p. 178) • soil profile (p. 178) • transported soil (p. 177)	**MAIN Idea** Soil forms slowly as a result of mechanical and chemical processes. • Soil consists of weathered rock and humus. • Soil is either residual or transported. • A typical soil profile has O-horizon, A-horizon, B-horizon, and C-horizon. • Five factors influence soil formation: climate, topography, parent material, biological organisms, and time. • Characteristics of soil include texture, fertility, and color.

Earth Science Online Vocabulary PuzzleMaker glencoe.com

Assessment

Vocabulary Review

Match the correct vocabulary term from the Study Guide to the following definitions.

1. the process of breaking down and changing rocks on or near Earth's surface

2. the removal of weathered materials from a location by running water, wind, ice, or waves

3. the fracturing of rock along curved lines that results when pressure is removed from bedrock

Each of the following sentences is false. Make each sentence true by replacing the italicized words with the correct vocabulary term from the Study Guide.

4. *Weathering* is caused by water flowing down the side of a slope.

5. The process by which eroded materials are left at a new location is called *physical weathering.*

6. *Mechanical weathering* is the process during which smaller eroded channels become deep and wide.

7. A *soil horizon* is formed from parent material that was moved away from its original source by water, wind, or a glacier.

8. A *soil profile* is a distinct layer or zone within a cross section of Earth's surface.

9. *Humus* is the loose covering of broken rock particles and decaying organic matter overlying the bedrock of Earth's surface.

Distinguish between the vocabulary terms in each pair.

10. weathering, erosion

11. chemical weathering, mechanical weathering

12. gully erosion, rill erosion

13. soil horizon, soil profile

14. erosion, deposition

15. residual soil, transported soil

Understand Key Concepts

16. Approximately what percent of Earth's surface is presently covered by glaciers?
A. 5 percent
B. 10 percent
C. 20 percent
D. 50 percent

17. In which horizon is humus most concentrated?
A. A-horizon
B. B-horizon
C. C-horizon
D. O-horizon

18. Which is usually the primary factor that affects the rate of weathering?
A. topography
B. volume
C. climate
D. biological organisms

Use the figure below to answer Questions 19 and 20.

19. Which process most likely produced the present appearance of this feature found in Arches National Park, Utah?
A. chemical weathering
B. mechanical weathering
C. earthquake activity
D. acid precipitation

20. Which erosional agent was most likely responsible for the appearance of this feature?
A. water
B. wind
C. glaciers
D. biological organisms

21. Frost wedging primarily relies on which process(es)?
   A. freezing and thawing
   B. gravity
   C. oxidation
   D. depth

22. Which is not considered a factor of soil formation?
   A. topography
   B. parent material
   C. time
   D. chemistry

23. Which does not contribute to the rate of weathering?
   A. rock type
   B. rock composition
   C. climate
   D. fossils

Use the photo below to answer Questions 24 and 25.

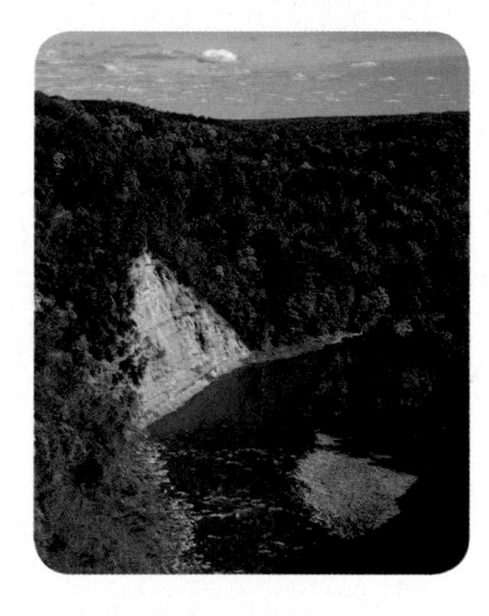

24. Which agent of erosion is shown in the picture of Letchworth State Park, known as the Grand Canyon of the East, in central New York?
   A. glaciers
   B. wind
   C. running water
   D. earthquakes

25. Which is the underlying force behind the agent of erosion in Letchworth State Park?
   A. pressure
   B. gravity
   C. temperature
   D. light

26. Which describes a residual soil?
   A. soil from sediment deposited by glaciers
   B. sand that has collected in a floodplain
   C. fine-grained sediment that was deposited by wind
   D. layers of material that weathered from bedrock below

27. Which soil horizon is a zone of accumulation consisting of soluble minerals that have been carried by water from above?
   A. A-horizon
   B. B-horizon
   C. C-horizon
   D. O-horizon

28. A mature soil most likely possesses which characteristic?
   A. thin B-horizons
   B. thick B-horizons
   C. fertility
   D. dark color

Constructed Response

29. **Analyze** the relationship between surface area and rate of mechanical weathering of a rock.

30. **Classify** how different climates affect the way rocks weather.

Use the figure below to answer Question 31.

31. **Design** a method that would have prevented the erosion occurring at this location.

32. **List** the factors that control the formation of soil, and give an example of the effects of each.

33. **CAREERS IN EARTH SCIENCE** A soil scientist stated that the soil in your area is acidic. Suggest a solution for the local gardeners.

Think Critically

34. **Examine** how the processes of erosion and deposition cause barrier islands to migrate.

Use the figure below to answer Question 35.

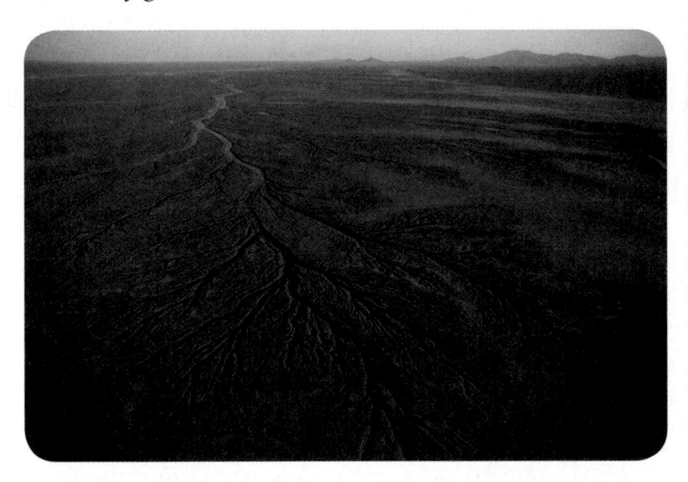

35. **Summarize** how the processes of erosion and deposition have resulted in this landscape feature.

36. **Create** a poster that illustrates the effects of erosion and deposition in your community.

37. **Draw and label** a soil profile of a mature soil containing an O-horizon, A-horizon, B-horizon, C-horizon, and bedrock. Describe how each layer of the soil was developed.

Concept Mapping

38. Create a concept map using the following terms: *weathering, erosion, deposition, chemical weathering, mechanical weathering, gully erosion,* and *rill erosion.* Refer to the *Skillbuilder Handbook* for more information.

Challenge Question

39. **Critique** this statement: Weathering, erosion and deposition are all parts of the same process.

Additional Assessment

40. *WRITING in* **Earth Science** Imagine that you are a soil scientist studying a sample in the lab. Write a journal entry describing the soil sample. Include information about what you can infer from the soil sample.

DBQ Document–Based Questions

Data obtained from: United States Department of Agriculture, Natural Resources Conservation Service. Honeoye—New York State Soil, 2006.

Honeoye [HON ee yah] soils are exceptionally fertile soils that occur in New York. The word Honeoye *is from the Iroquois* Hay-e-a-yeah.

41. Using the photograph, create an illustration of the Honeoye soil and label the following layers: *A-horizon, B-horizon,* and *C-horizon.*

42. Describe the soil profile.

43. Is the soil pictured above undeveloped or mature? How can you tell?

Cumulative Review

44. What is the difference between latitude and longitude? **(Chapter 2)**

45. What is a mineral? **(Chapter 4)**

46. Which common chemical sedimentary rock consists of calcite? **(Chapter 5)**

Standardized Test Practice

Multiple Choice

1. Which farming method is used to reduce wind erosion?
A. planting different crops
B. planting wind barriers
C. building earth mounds
D. building stone walls

Use the figure below to answer Questions 2–4.

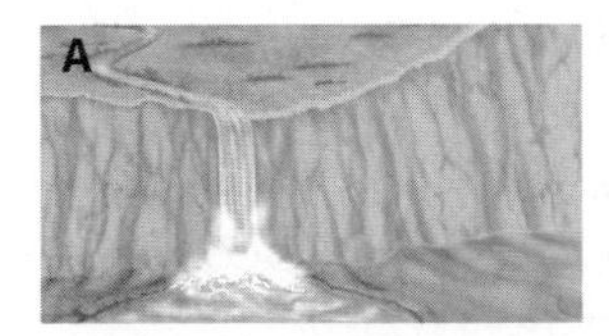

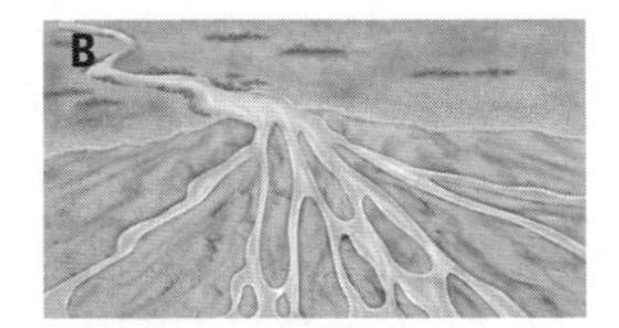

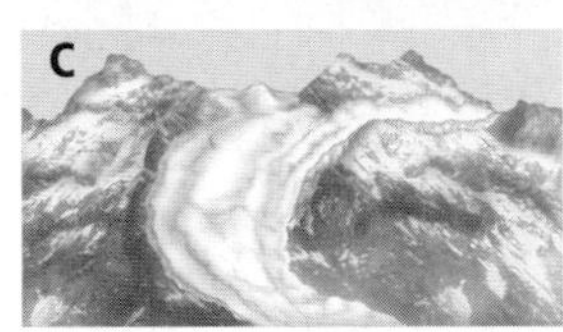

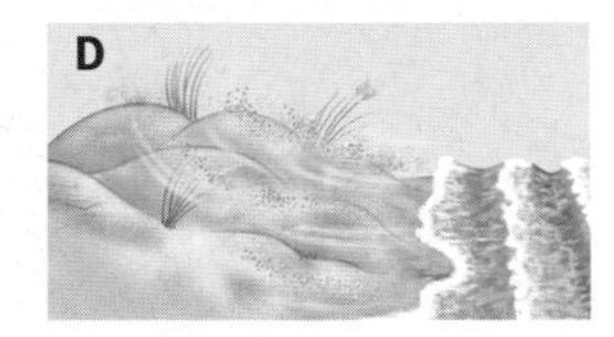

2. Which image shows the erosional agent that was responsible for leaving behind U-shaped valleys, hanging valleys, lakes, and deposits of sediment in New England and New York State?
A. A **C.** C
B. B **D.** D

3. Which image shows the erosional agent responsible for dunes formed along the Gulf and Atlantic coasts of the United States?
A. A **C.** C
B. B **D.** D

4. What common factor is responsible for three of the four erosional processes pictured?
A. wind
B. heat
C. human intervention
D. gravity

5. What is the best-known feature of sedimentary rocks?
A. ripple marks
B. fossils
C. graded bedding
D. cross-bedding

6. How does granite differ from gabbro in coloring and silica content?
A. Granite is lighter colored with higher silica content.
B. Granite is darker colored with lower silica content.
C. Granite is darker colored with higher silica content.
D. Granite is lighter colored with lower silica content.

7. Which is NOT an agent of chemical weathering?
A. water **C.** carbon dioxide
B. oxygen **D.** wind

Use the map below to answer Questions 8 and 9.

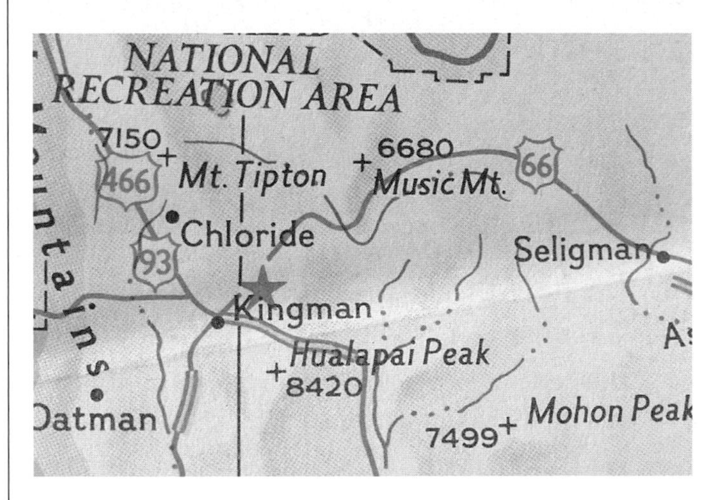

8. What can you infer about the location of the area shown in the road map?
A. It is largely uninhabited.
B. It is in a major city.
C. It is mostly impassable terrain.
D. It is a mountainous area.

9. Pikes Peak in Colorado is about 14,100 ft high. If a hiker wanted to climb an equivalent distance in the area located on this map, what two mountains should he climb?
A. Music Mountain and Mohon Peak
B. Music Mountain and Mount Tipton
C. Hualapai Peak and Mohon Peak
D. Mount Tipton and Mohon Peak

Short Answer

Use the graph below to answer Questions 10–12.

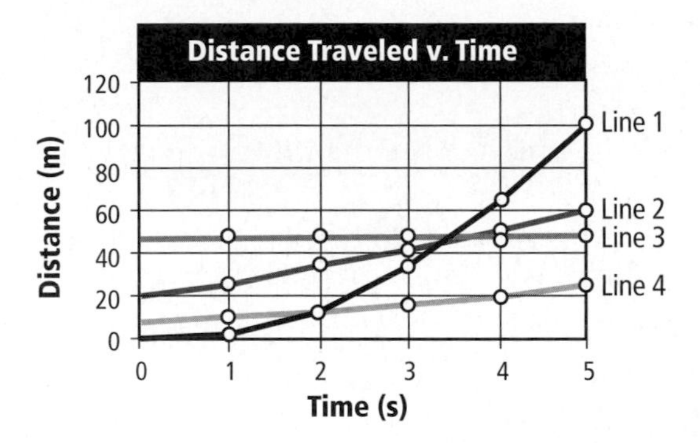

10. What was the average speed of the object represented by Line 1 during the 5 s its time was recorded?

11. Explain why Line 3 is horizontal.

12. What are the independent and dependent variables in this graph?

13. Describe how limestone forms.

14. Explain whether or not coal is a mineral.

15. When does regional metamorphism occur and what are its results?

Reading for Comprehension

Agricultural Land Use

Food production takes up almost half of Earth's land surface and threatens to consume the fertile land that still remains. The global impact of farming on the environment is revealed in new maps, which show that 40 percent of Earth's land is used for agriculture. Navin Ramankutty, a land-use researcher with Wisconsin-Madison's Center for Sustainability and the Global Environment (SAGE), posed the following question: "How can we continue to produce food from the land while preventing negative environmental consequences, such as deforestation, water pollution, and soil erosion?" One potential solution could be "precision farming." This model uses new technology to improve productivity while reducing the use of water and the application of fertilizer and other potentially harmful chemicals. The precision system, currently being developed by NASA geoscientists, would use satellite data to help farmers decide how to use their resources with pinpoint accuracy based on the requirements of different areas of each field.

Article obtained from: Owen, J. Farming claims almost half of Earth's land, new maps show. *National Geographic News.* December 9, 2005.

16. According to this passage, which statement is false?
A. Farming can harm Earth.
B. Satellite data can improve farming.
C. Farming does not cause pollution.
D. "Precision farming" is a solution.

17. Which one is not a negative environmental consequence of farming listed in the passage?
A. deforestation
B. air pollution
C. water pollution
D. soil erosion

18. What can be inferred from this text?
A. There are solutions to improve farming and its effects on the land.
B. Wisconsin is the only state with farming problems.
C. People need to eat less so that less land is needed for food.
D. There is no fertile land left to cultivate.

NEED EXTRA HELP?															
If You Missed Question . . .	1	2	3	4	5	6	7	8	9	10	11	12	13	14	15
Review Section . . .	7.2	7.1	7.1	7.2	6.1	5.2	7.2	2.2	2.2	1.2	1.3	1.2	6.2	4.1	6.3

CHAPTER 8

Mass Movements, Wind, and Glaciers

BIG Idea Movements due to gravity, winds, and glaciers shape and change Earth's surface.

8.1 Mass Movements
MAIN Idea Mass movements alter Earth's surface over time due to gravity moving sediment and rocks downslope.

8.2 Wind
MAIN Idea Wind modifies landscapes in all areas of the world by transporting sediment.

8.3 Glaciers
MAIN Idea Glaciers modify landscapes by eroding and depositing rocks.

GeoFacts

- More than 100,000 glaciers exist in Alaska, but ice covers only 5 percent of the state.
- Glaciers form when more snow falls in an area than melts in the same area.
- Layers of snow on the glacier create pressure that changes the snow underneath to ice.

LAUNCH Lab

How does water affect sediments on slopes?

Water has a significant effect on sediments on slopes. In this activity, you will demonstrate how the addition of water affects how sediments are held together.

Procedure

1. Read and complete the lab safety form.
2. Place 225 mL of **sand** in each of three separate **containers,** such as aluminum pie plates.
3. Add 20 mL of **water** to the first container of sand, and mix well. Add 100 mL of water to the second container of sand, and mix well. Add 200 mL of water to the third container of sand, and mix well.
4. Tilt each pan to test the effect of slopes. Start with a slight tilt and increase until the sand begins to move.
5. Test each mixture for its ability to be molded and retain its shape. Compare your results for the three samples.

WARNING: ***Wipe up any spilled water.***

Analysis

1. **Describe** how the addition of water affected the sand's ability to be molded in the three samples.
2. **Explain** why one mixture was better able to maintain its shape than the others.
3. **Explain** how water affects sediment on slopes.

External Processes that Shape Earth Make this Foldable to explain different processes that shape Earth's surface.

STEP 1 Fold the bottom of a horizontal sheet of paper up about 3 cm.

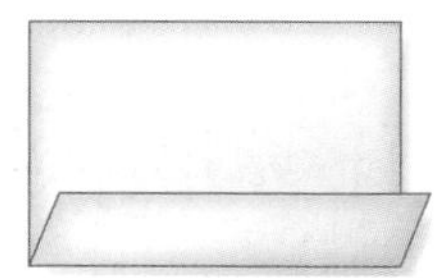

STEP 2 Fold in thirds.

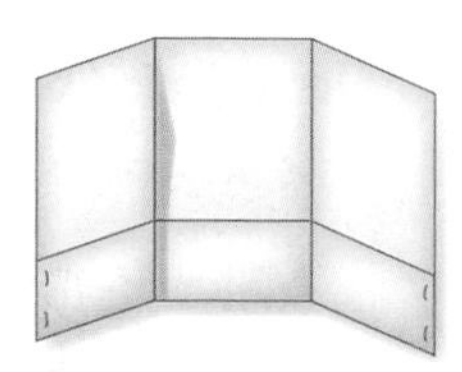

STEP 3 Unfold and dot with glue or staple to make three pockets. Label as shown.

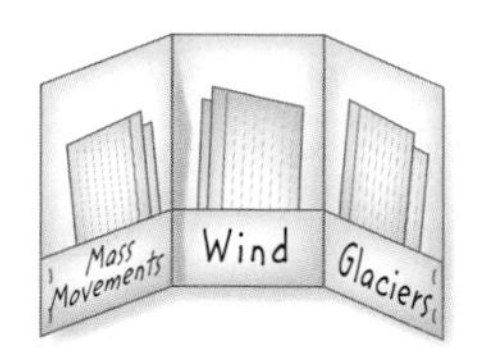

FOLDABLES **Use this Foldable with Sections 8.1, 8.2, and 8.3.** As you read, use index cards to summarize information in your own words and place them in the appropriate pockets.

Visit glencoe.com to

- **study entire chapters online;**
- **explore Concepts In Motion animations:**
  - **Interactive Time Lines**
  - **Interactive Figures**
  - **Interactive Tables**
- **access Web Links for more information, projects, and activities;**
- **review content with the Interactive Tutor and take Self-Check Quizzes.**

Section 8.1

Mass Movements

Objectives

- **Analyze** the relationship between gravity and mass movements.
- **Identify** factors that affect mass movements.
- **Distinguish** between types of mass movements.
- **Relate** how mass movements affect people.

Review Vocabulary

gravity: the force every object exerts on every other object due to their masses

New Vocabulary

mass movement
creep
mudflow
landslide
slump
avalanche

MAIN Idea Mass movements alter Earth's surface over time due to gravity moving sediment and rocks downslope.

Real-World Reading Link How fast can you travel on a waterslide? A number of factors might come into play, including the angle of the slide, the amount of water on the slide, the material of the slide, friction, and your own mass. These factors also affect mass movements on Earth's surface.

Mass Movements

How do landforms, such as mountains, hills, and plateaus, wear down and change? Landforms can change through processes involving wind, ice, and water, and sometimes through the force of gravity alone. The downslope movement of soil and weathered rock resulting from the force of gravity is called **mass movement.** Recall from Chapter 7 that weathering processes weaken and break rock into smaller pieces. Mass movements often carry the weathered debris downslope. Because climate has a major effect on the weathering activities that occur in a particular area, climatic conditions determine the extent of mass movement.

All mass movements, such as the one shown in **Figure 8.1,** occur on slopes. Because few places on Earth are completely flat, almost all of Earth's surface undergoes mass movement. Mass movements range from motions that are barely detectable to sudden slides, falls, and flows. The Earth materials that are moved range in size from fine-grained mud to large boulders.

 **Reading Check** **Describe** how gravity causes a mass movement.

Figure 8.1 Mass movements can cause tree trunks to curve in order to continue growing opposite the pull of gravity, which is toward the center of Earth.

Factors that Influence Mass Movements

Several factors influence the mass movements of Earth's material. One factor is the material's weight, which works to pull the material downslope. A second factor is the material's resistance to sliding or flowing, which depends on the amount of friction, how cohesive the material is, and whether it is anchored to the bedrock. A third factor is a trigger, such as an earthquake, that shakes material loose. Mass movement occurs when the forces pulling material downslope are stronger than the material's resistance to sliding, flowing, or falling.

Water is a fourth variable that influences mass movements. The landslide shown in **Figure 8.2** occurred after days of heavy rains. Saturation by water greatly increases the weight of soils and sediments. In addition, as the water fills the tiny open spaces between grains, it acts as a lubricant between the grains, reducing the friction between them.

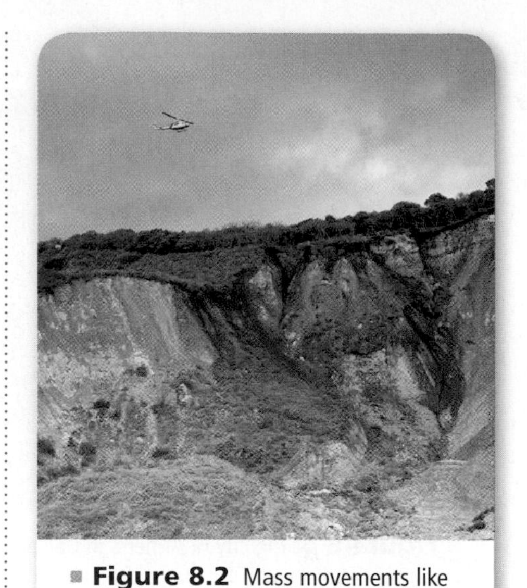

Figure 8.2 Mass movements like the one shown here can significantly alter landscapes.
Summarize *the factors that might have been involved in this mass movement.*

Types of Mass Movements

Mass movements are classified as creep, flows, slides, and rockfalls. Mass movements move different types of materials in various ways.

Creep The slow, steady, downhill flow of loose, weathered Earth materials, especially soils, is called **creep.** Because movement might be as little as a few centimeters per year, the effects of creep are usually noticeable only over long periods of time. One way to tell whether creep has occurred is to observe the positions of structures and objects. As illustrated in **Figure 8.3,** creep can cause once-vertical utility poles and fences to tilt, and trees and walls to break. Loose materials on almost all slopes undergo creep.

Creep that usually occurs in regions of permafrost, or permanently frozen soil, is called solifluction (SOH luh fluk shun). The material moved in solifluction is a mudlike liquid that is produced when water is released from melting permafrost during the warm season. The water saturates the surface layer of soil and is unable to move downward. As a result, the surface layer can slide slowly downslope.

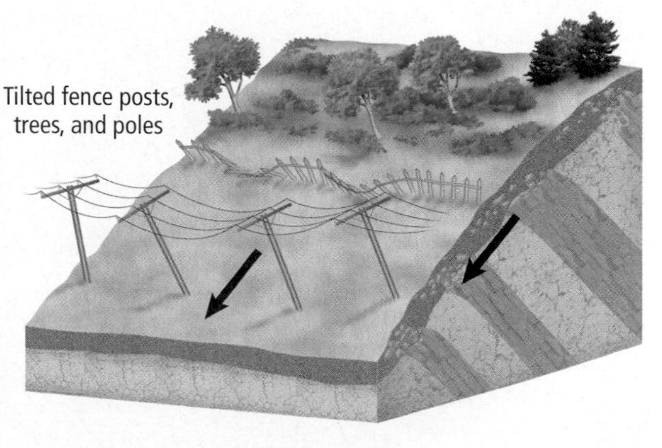

Figure 8.3 All slopes undergo creep to some extent. Tilting of vertical objects is often the result.

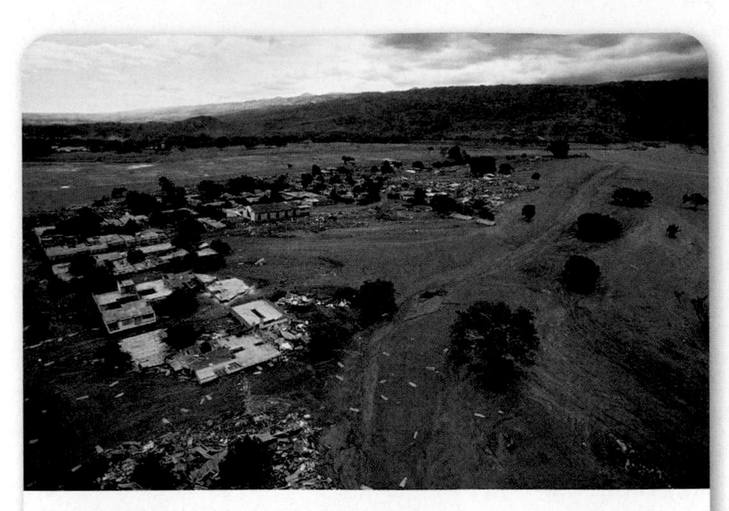

■ **Figure 8.4** The city of Armero, in Colombia, was covered in mud and debris by a lahar that contained snowmelt and volcanic material.
Describe *the effect of the lahar on the city shown above.*

Flows In some mass movements, Earth materials flow as if they were a thick liquid. The materials might move as slowly as a few centimeters per year or as rapidly as hundreds of kilometers per hour. Earth flows are moderately slow movements of soils, whereas **mudflows** are swiftly moving mixtures of mud and water. Mudflows can be triggered by earthquakes or similar vibrations and are common in volcanic regions where the heat from a volcano melts snow on nearby slopes that have fine sediment and little vegetation. The meltwater fills the spaces between the small particles of sediment and allows them to slide readily over one another and move downslope.

A lahar (LAH har) is a type of mudflow that occurs after a volcanic eruption. Often a lahar results when a snow-topped volcanic mountain erupts and melts the snow on top of a mountain. The melted snow mixes with ash and flows downslope. **Figure 8.4** shows how a lahar that originated from Nevado del Ruiz, one of the volcanic mountains in the Andes, devastated a town. The Nevado del Ruiz is 5389 m high and covered with 25 km^2 of snow and ice, which melted when it erupted. Four hours after Nevado del Ruiz erupted, lahars had traveled more than 100 km downslope. As a result of these lahars, which occurred in 1985, approximately 23,000 people were killed, 5000 were injured, and 5000 homes were destroyed.

Reading Check **Determine** what triggers a lahar.

Mudflows are also common in sloped, semi-arid regions that experience intense, short-lived rainstorms. The Los Angeles Basin in Southern California is an example of an area where mudflows are common. In such areas, periods of drought and forest fires leave the slopes with little protective vegetation. When heavy rains eventually fall in these areas, they can cause massive, destructive mudflows because there is little vegetation to anchor the soil. Mudflows are especially destructive in areas where urban development has spread to the bases of mountainous areas. These mudflows can bury homes, as shown in **Figure 8.5.**

■ **Figure 8.5** Mudflows can be extremely destructive and can result in severe property damage, road closures, and power outages.

■ **Figure 8.6** Typical of landslides, this soil moved in a large block.

Slides A rapid, downslope movement of Earth materials that occurs when a relatively thin block of soil, rock, and debris separates from the underlying bedrock is called a **landslide,** shown in **Figure 8.6.** The material rapidly slides downslope as one block, with little internal mixing. A landslide mass eventually stops and becomes a pile of debris at the bottom of a slope, sometimes damming rivers and causing flooding. Landslides are common on steep slopes, especially when soils and weathered bedrock are fully saturated by water. This destructive form of mass movement causes damage costing almost 2 billion dollars and 25 to 50 associated deaths per year in the United States alone. You will explore the movement of a landslide in the GeoLab at the end of this chapter.

A rockslide is a type of landslide that occurs when a sheet of rock moves downhill on a sliding surface. During a rockslide, some blocks of rock are broken into smaller blocks as they move downslope, as shown in **Figure 8.7.** Often triggered by earthquakes, rockslides can move large amounts of material.

■ **Figure 8.7** During this rockslide, blocks of rock were broken into smaller blocks as they moved downslope.

Concepts In Motion

Interactive Figure To see an animation of a rockslide, visit glencoe.com.

■ **Figure 8.8** Slumps leave distinct crescent-shaped scars on hillsides as the soil rotates downward.

Slumps When the mass of material in a landslide moves along a curved surface, a **slump** results. Material at the top of the slump moves downhill, and slightly inward, while the material at the bottom of the slump moves outward. Slumps can occur in areas that have thick soils on moderate-to-steep slopes. Sometimes, slumps occur along highways where the slopes of soils are extremely steep. Slumps are common after rains, when water reduces the frictional contact between grains of soil and acts as a lubricant between surface materials and underlying layers. The weight of the additional water pulls material downhill. As with other types of mass movement, slumps can be triggered by earthquakes. Slumps leave crescent-shaped scars on slopes, as shown in **Figure 8.8.**

 **Reading Check** **Describe** what conditions can cause a slump.

Avalanches Landslides that occur in mountainous areas with thick accumulations of snow are called **avalanches.** About 10,000 avalanches occur each year in the mountains of the western United States. Radiation from the Sun can melt surface snow, which then refreezes at night into an icy crust. Snow that falls on top of this crust can eventually build up, become heavy, slip off, and slide downslope as an avalanche. Avalanches can happen in early winter when snow accumulates on the warm ground. The snow in contact with the warm ground melts, then refreezes into a layer of jagged, slippery snow crystals.

Avalanches of dangerous size, like the one shown in **Figure 8.9,** occur on slope angles between 30° and 45°. When the angle of a slope is greater than 45°, enough snow cannot accumulate to create a large avalanche. At angles less than 30°, the slope is not steep enough for snow to begin sliding. A vibrating trigger, even from a single skier, can send this unstable layer sliding down a mountainside. Avalanches pose significant risks in places such as Switzerland, where more than 50 percent of the population lives in avalanche terrain.

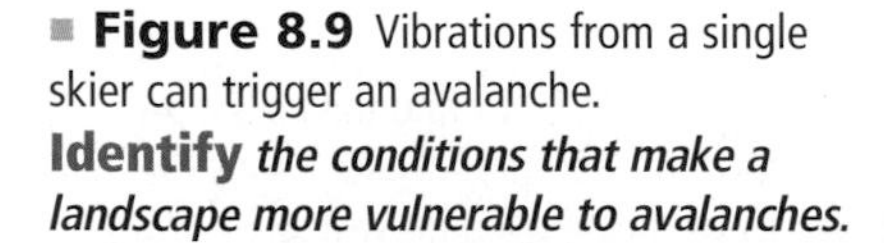

■ **Figure 8.9** Vibrations from a single skier can trigger an avalanche.
Identify *the conditions that make a landscape more vulnerable to avalanches.*

■ **Figure 8.10** This rockfall in Topanga Canyon, California, was unusual in that it involved mainly one large rock.

Rockfalls On high cliffs, rocks are loosened by physical weathering processes, such as freezing and thawing, and by plant growth. As rocks break up and fall directly downward, they can bounce and roll, ultimately producing a cone-shaped pile of coarse debris, called talus, at the base of the slope. Rockfalls, such as the one shown in **Figure 8.10,** commonly occur at high elevations, in steep road cuts, and on rocky shorelines. Rockfalls are less likely to occur in humid regions where the rock is typically covered by a thick layer of soil, vegetation, and loose materials. On human-made rock walls, such as road cuts, rockfalls are particularly common.

Mass Movements Affect People

While mass movements are natural processes, human activities often contribute to the factors that cause mass movements. Activities such as the construction of buildings, roads, and other structures can make slopes unstable. In addition, poor maintenance of septic systems, which often leak, can trigger slides. In the Philippines, mudslides, shown in **Figure 8.11,** were triggered after ten days of torrential rains delivered 200 cm of precipitation. A village estimated to have 3000 residents was totally destroyed.

■ **Figure 8.11** The mudflow on the island of Luzon occurred after days of rain.

■ **Figure 8.12** Covering hillsides with steel nets can reduce risks of mass movements and harm to humans.
Identify *the type of mass movement that these steel nets help prevent.*

Reducing the risks Catastrophic mass movements are most common on slopes greater than 25° that experience annual rainfall of over 90 cm. Risk increases if that rainfall tends to occur in a short period of time. Humans can minimize the destruction caused by mass movements by not building structures on or near the base of steep and unstable slopes.

Although preventing mass-movement disasters is not easy, some actions can help reduce the risks. For example, a series of trenches can be dug to divert running water around a slope and control its drainage. Landslides and rockfalls can be controlled by covering steep slopes with materials such as steel nets, shown in **Figure 8.12,** and constructing fences along highways in areas where mass movements are common. Other approaches involve the installation of retaining walls to support the bases of weakened slopes. Most of these efforts at slope stabilization and mass-movement prevention are only temporarily successful.

The best way to reduce the number of disasters related to mass movements is to educate people about the problems of building on steep slopes. For example, The United States Geological Survey (USGS) collects data about landslides in an effort to learn more about where and when landslides will occur. This information helps people decide where they can safely build homes or businesses.

Section 8.1 Assessment

Section Summary

- Mass movements are classified in part by how rapidly they occur.
- Factors involved in the mass movement of Earth materials include the material's weight, its resistance to sliding, the trigger, and the presence of water.
- Mass movements are natural processes that can affect human life and activities.
- Human activities can increase the potential for the occurrence of mass movements.

Understand Main Ideas

1. MAIN Idea **Organize** the following types of mass movements in order of increasing speed: slides, creep, flows, and rockfalls.
2. **Identify** the underlying force behind all forms of mass movement.
3. **Analyze** how water affects mass movements by using two examples of mass movement.
4. **Appraise** the effects of one type of mass movement on humans.

Think Critically

5. **Generalize** in which regions of the world mudflows are more common.
6. **Evaluate** how one particular human activity can increase the risk of mass movement and suggest a solution to the problem.

WRITING in Earth Science

7. Make a poster that compares and contrasts solifluction and a slump. Consider the way soil moves and the role of water.

Earth Science Online **Self-Check Quiz** glencoe.com

Section 8.2

Objectives

- **Describe** conditions that contribute to the likelihood that an area will experience wind erosion.
- **Identify** wind-formed landscape features.
- **Describe** how dunes form and migrate.

Review Vocabulary

velocity: the speed of an object and its direction of motion

New Vocabulary

deflation
abrasion
ventifact
dune
loess

Wind

MAIN Idea **Wind modifies landscapes in all areas of the world by transporting sediment.**

Real-World Reading Link If you have ever been on a beach on a windy day, you might have felt the stinging of sand on your face. Sand travels in the wind if the wind is fast enough.

Wind Erosion and Transport

A current of rapidly moving air can pick up and carry sediment in the same way that water does. However, except for the extreme winds of hurricanes, tornadoes, and other strong storms, winds cannot generally carry particles as large as those transported by moving water. Regardless, wind is a powerful agent of erosion.

Winds transport materials by causing their particles to move in different ways. For example, wind can move sand on the ground in a rolling motion. A method of transport by which strong winds cause small particles to stay airborne for long distances is called suspension. Another method of wind transport, called saltation, causes a bouncing motion of larger particles. Saltation accounts for most sand transport by wind. Limited precipitation leads to an increase in the amount of wind erosion because precipitation holds down sediments and allows plants to grow. Thus, wind transport and erosion primarily occur in areas with little vegetative cover, such as deserts, semiarid areas, seashores, and some lakeshores. Wind erosion is a problem in many parts of the United States, as shown in **Figure 8.13.**

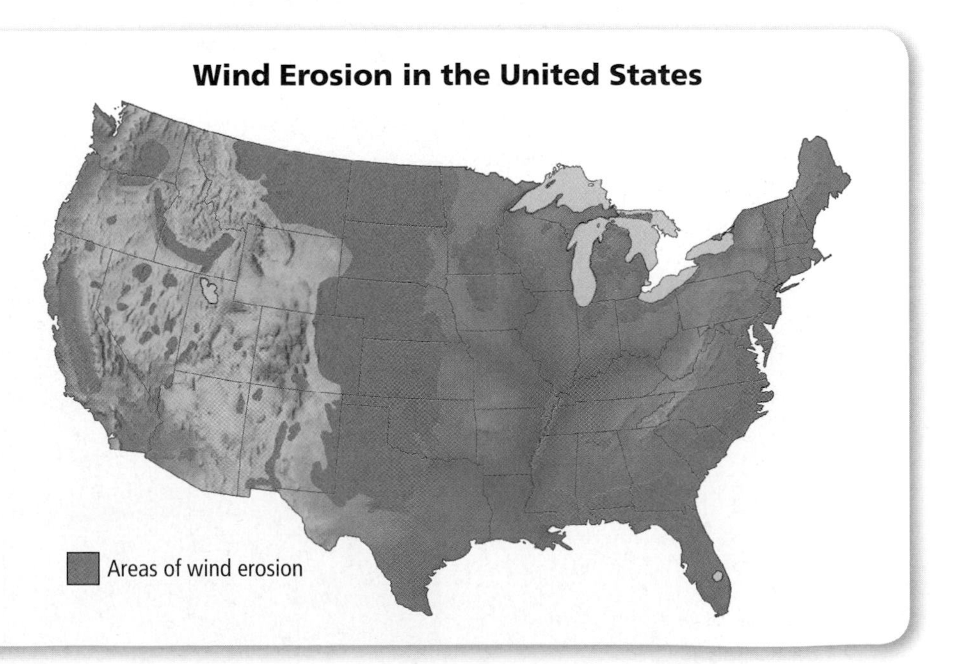

■ **Figure 8.13** Wind erosion does not affect all areas of the United States equally.
Observe *which areas are subject to wind erosion.*

■ **Figure 8.14** Through deflation, the wind can create a bowl-shaped blowout.

Deflation The lowering of the land surface that results from the wind's removal of surface particles is called **deflation.** During the 1930s, portions of the Great Plains region, which stretches from Montana to Texas, experienced severe drought. The area was already suffering from the effects of poor agricultural practices, in which large areas of natural vegetation were removed to clear the land for farming. Strong winds readily picked up the dry surface particles, which lacked any protective vegetation. Severe dust storms resulted in daytime skies that were often darkened, and the region became known as the Dust Bowl.

Today, the Great Plains are characterized by thousands of shallow depressions known as deflation blowouts. Many are the result of the removal of surface sediment by wind erosion during the 1930s. The depressions range in size from a few meters to hundreds of meters in diameter. Deflation blowouts are also found in other areas that have sandy soil, as shown in **Figure 8.14.** Wind erosion continues today throughout the world, as shown by the duststorm in **Figure 8.15.**

Reading Check **Explain** how deflation removes surface particles.

■ **Figure 8.15** A duststorm in a desert region fills the air with dust.

Deflation is a major problem in many agricultural areas of the world as well as in deserts, where wind has been consistently strong for thousands of years. In areas of intense wind erosion, coarse gravel and pebbles are usually left behind as the finer surface material is removed by winds. The coarse surface left behind is called desert pavement.

Abrasion Another process of erosion, called **abrasion,** occurs when particles such as sand rub against the surface of rocks or other materials. Abrasion occurs as part of the erosional activities of winds, streams, and glaciers. In wind abrasion, wind picks up materials such as sand particles and blows them against anything in their path. Because sand is often made of quartz, a hard mineral, wind abrasion can be an effective agent of erosion—windblown sand particles eventually wear away rocks. Structures, such as telephone poles, can also be worn away or undermined by wind abrasion, and paint and glass on homes and vehicles can be damaged by windblown sand.

Materials that are exposed to wind abrasion show unique characteristics. For example, windblown sand causes rocks to become pitted and grooved. With continued abrasion, rocks become polished on the windward side and develop smooth surfaces with sharp edges. In areas of shifting winds, abrasion patterns correspond to wind shifts, and different sides of rocks become polished and smooth. Rocks shaped by windblown sediments, such as those shown in **Figure 8.16,** are called **ventifacts.** Ventifacts are found in various shapes and sizes, and include arches and pillars.

Reading Check **Identify** the unique characteristics of materials shaped by abrasion.

NATIONAL GEOGRAPHIC To read about how wind has shaped desert landscapes, read the **National Geographic Expedition** on page 898.

Figure 8.16 Ventifacts form in different types of environments but most commonly in arid climates where wind can be a dominant erosional force.

■ **Figure 8.17** Great Sand Dunes National Monument, in southern Colorado, contains North America's highest sand dunes of more than 228.6 m.
Identify *the dominant direction of wind in the figure.*

Wind Deposition

Wind deposition occurs in areas where wind velocity decreases. As the wind velocity slows down, some of the windblown sand and other materials cannot stay airborne, and they drop out of the air stream to form a deposit on the ground.

Dunes In windblown environments, sand particles tend to accumulate where an object, such as a rock, landform, or piece of vegetation, blocks the forward movement of the particles. Sand continues to be deposited as long as winds blow in one general direction. Over time, the pile of windblown sand develops into a **dune,** as shown in **Figure 8.17.** All dunes have a characteristic profile. The gentler slope of a dune, located on the side from which the wind blows, is called the windward side. The steeper slope, on the side protected from the wind, is called the leeward side. The conditions under which a dune forms determine its shape. These conditions include the availability of sand, wind velocity, wind direction, and the amount of vegetation present. The different types of dunes are shown in **Table 8.1.**

VOCABULARY
ACADEMIC VOCABULARY
migrate
to move from one location to another
Dunes migrate as wind blows over sand.

Dune migration As long as winds continue to blow, dunes will migrate. As shown in **Figure 8.18,** dune migration is caused when prevailing winds continue to move sand from the windward side of a dune to its leeward side, causing the dune to move slowly over time.

Concepts In Motion
Interactive Figure To see an animation of dune migration, visit glencoe.com.

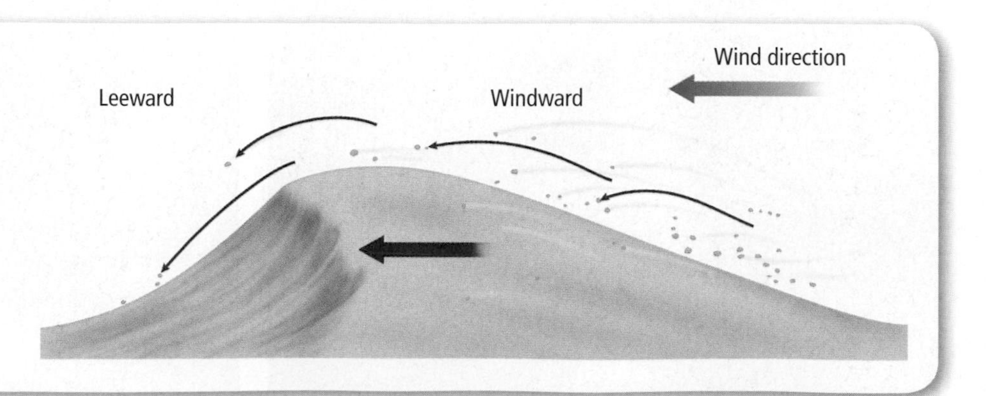

■ **Figure 8.18** Dune migration is caused by wind.

Concepts In Motion

Interactive Table To explore more about sand dunes, visit glencoe.com.

Table 8.1 Types of Dunes

Example of Dune	Description
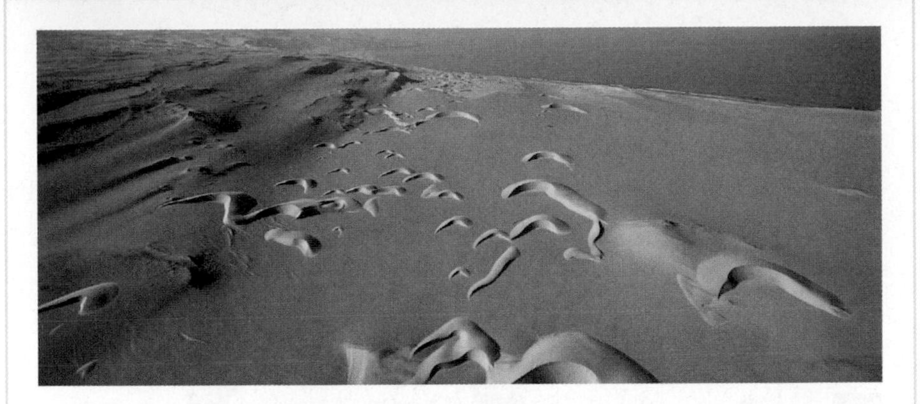	**Barchan Dunes** • form solitary, crescent shapes • form from a small amount of sand • covered by minimal or no vegetation • form in flat areas of constant wind direction • crests point downwind • reach maximum size of 30 m
	**Transverse Dunes** • form series of ridge shapes • form from a large amount of sand • covered by minimal or no vegetation • form in ridges that are perpendicular to the direction of the strong wind • reach maximum size of 25 m
	**Parabolic Dunes** • form U-shapes • form from a large amount of sand • covered by minimal vegetation • form in humid areas with moderate winds • crests point upwind • reach maximum size of 30 m
	**Longitudinal Dunes** • form series of ridge shapes • form from small or large amounts of sand • covered by minimal or no vegetation • form parallel to variable wind direction • reach maximum height of 300 m

■ **Figure 8.19** This map shows the location of loess deposits in the continental United States.

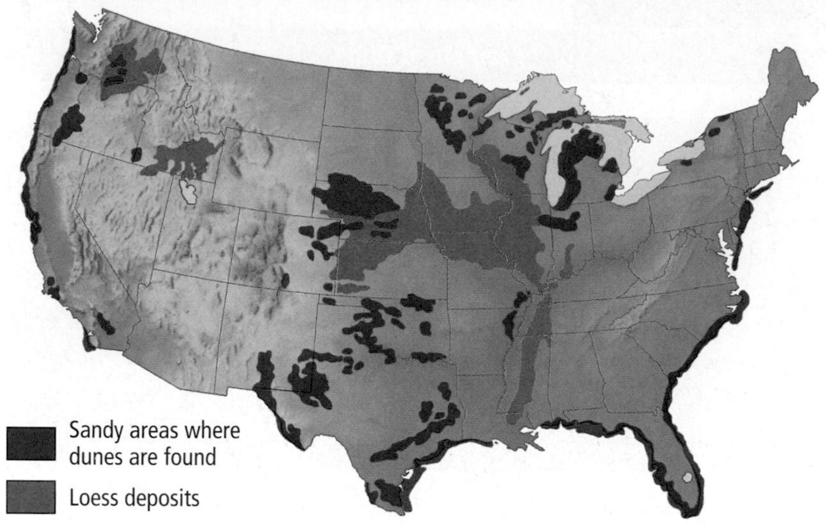

Distribution of Loess Deposits in the United States

Sandy areas where dunes are found

Loess deposits

Loess Wind can carry fine, lightweight particles such as silt and clay in great quantities and for long distances. Many parts of Earth's surface are covered by thick layers of windblown silt, which are thought to have accumulated as a result of thousands of years of dust storms. The source of these silt deposits might have been the fine sediments that were exposed when glaciers melted after the last ice age, more than 10,000 years ago. These thick, windblown silt deposits are known as **loess** (LES). Loess soils are some of the most fertile soils because they contain abundant minerals and nutrients. **Figure 8.19** shows the states where agriculture has benefited from the loess deposits.

Section 8.2 Assessment

Section Summary

- Wind is a powerful agent of erosion.
- Wind can transport sediment in several ways, including suspension and saltation.
- Dunes form when wind velocity slows down and windblown sand is deposited.
- Dunes migrate as long as winds continue to blow.

Understand Main Ideas

1. MAIN Idea **Distinguish** the various types of landforms formed by wind and how these landforms are created.
2. **Identify** conditions that can contribute to an increase in wind erosion.
3. **Examine** why loess can travel much greater distances than sand.
4. **Classify** the four types of dunes as they are related to wind, vegetation, and amount of sand available.

Think Critically

5. **Infer** how the movement of sand grains by saltation affects the overall movement of dunes.
6. **Evaluate** why wind erosion is an effective agent of erosion.

WRITING in Earth Science

7. Predict how human activities directly affect wind erosion on coastlines.

Section 8.3

Glaciers

Objectives

- **Explain** how glaciers form.
- **Compare and contrast** the conditions that produce valley glaciers with those that produce continental glaciers.
- **Describe** how glaciers modify landscapes.
- **Recognize** glacial features.

Review Vocabulary

latitude: distance in degrees north and south of the equator

New Vocabulary

glacier
valley glacier
continental glacier
cirque
moraine
outwash plain
drumlin
esker
kame
kettle

MAIN Idea **Glaciers modify landscapes by eroding and depositing rocks.**

Real-World Reading Link Have you ever wondered what formed the landscape around you? Glaciers might have left deposits of sediment as well as carved features in rock that you see every day.

Moving Masses of Ice

A large, moving mass of ice is called a **glacier.** Glaciers form near Earth's poles and in mountainous areas at high elevations. They currently cover about 10 percent of Earth's surface, as shown in **Figure 8.20.** In the past, glaciers were more widespread than they are today. During the last ice age, which began about 1.6 mya and ended more than 10,000 years ago, ice covered about 30 percent of Earth.

Areas at extreme northern and southern latitude, such as Greenland and Antarctica, and areas of high elevations, such as the Alps, have temperatures near 0°C year-round. Cold temperatures keep fallen snow from completely melting, and each year the snow that has not melted accumulates in an area called a snowfield. Thus, the total thickness of the snow layer increases as the years pass. The accumulated snow develops into a glacier. The weight of the top layers of snow eventually exerts enough downward pressure to force the accumulated snow below to recrystallize into ice. A glacier can develop in any location that provides the necessary conditions. Glaciers can be classified as one of two types—valley glaciers or continental glaciers.

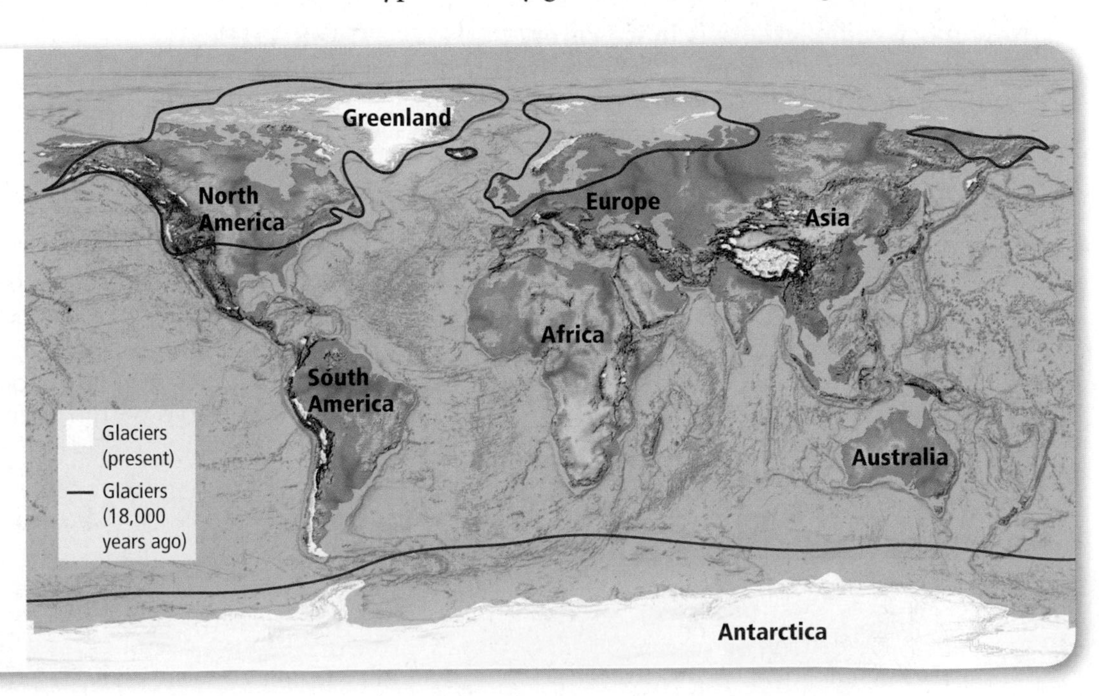

Figure 8.20 Glaciers around the world have changed in distribution throughout geologic time.
Infer *what changes have occurred in the distribution of glaciers around the world.*

Concepts In Motion
Interactive Figure To see an animation of glacier formation, visit glencoe.com.

Valley glaciers Glaciers that form in valleys in high, mountainous areas are called **valley glaciers.** The movement of a valley glacier occurs when the growing ice mass becomes so heavy that the ice maintains its rigid shape and begins to flow, much like toothpaste. For most valley glaciers, flow begins when the accumulation of snow and ice exceeds 20 m in thickness. As a valley glacier moves, deep cracks in the surface of the ice, called crevasses, can form.

The speed of a valley glacier's movement is affected by the slope of the valley floor, the temperature and thickness of the ice, and the shape of the valley walls. The sides and bottom of a valley glacier move more slowly than the middle because friction slows down the sides and bottom where the glacier comes in contact with the ground. Movement downslope is usually slow—less than a few millimeters per day. Over time, as valley glaciers flow downslope, their powerful carving action transitions V-shaped stream valleys into U-shaped glacial valleys.

FOLDABLES Incorporate information from this section into your Foldable.

Reading Check **Describe** how V-shaped valleys become U-shaped.

Continental glaciers Glaciers that cover broad, continent-sized areas are called **continental glaciers.** These glaciers form in cold climates where snow accumulates over many years. A continental glacier is thickest at its center. The weight of the center forces the rest of the glacier to flatten in all directions. In the past, when Earth experienced colder average temperatures than it does today, continental glaciers covered huge portions of Earth's surface. Today, they are confined to Greenland and Antarctica.

DATA ANALYSIS LAB

Based on Real Data*

Interpret the Data

How much radioactivity is in ice cores? Glaciologists have found that ice cores taken from the arctic region contain preserved radioactive fallout. Data collected from the study of these ice cores have been plotted on the graph.

Data and Observations

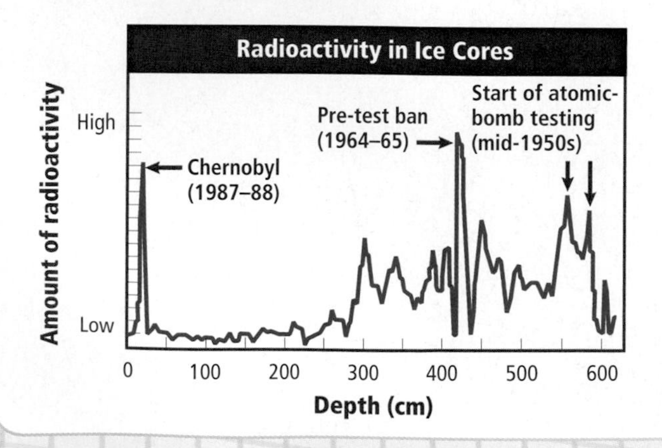

Think Critically

1. **Determine** the depth in the ice cores where the highest and lowest amounts of radioactivity were found.
2. **Describe** what happened to the amount of radioactivity in the ice cores between the pretest ban and Chernobyl.
3. **Infer** what happened to the amount of radioactivity in the ice cores after Chernobyl.
4. **Explain** what information or material other than radioactive fallout you think ice cores might preserve within them.

*Data obtained from: Mayewski, et al. 1990. Beta radiation from snow. *Nature* 345:25.

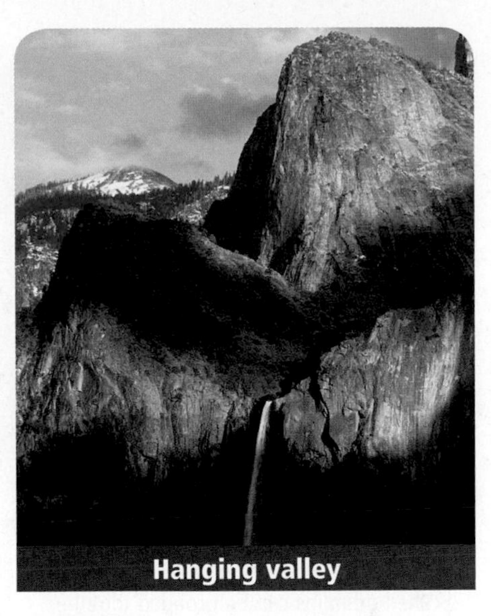

■ **Figure 8.21** Glacial erosion by valley glaciers creates features such as cirques, horns, and hanging valleys.

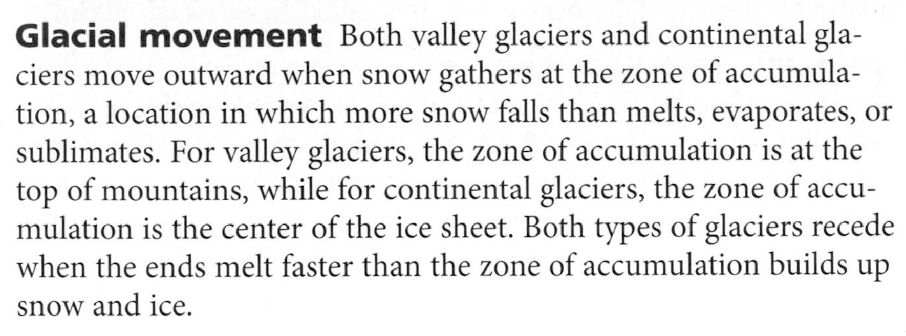

Glacial movement Both valley glaciers and continental glaciers move outward when snow gathers at the zone of accumulation, a location in which more snow falls than melts, evaporates, or sublimates. For valley glaciers, the zone of accumulation is at the top of mountains, while for continental glaciers, the zone of accumulation is the center of the ice sheet. Both types of glaciers recede when the ends melt faster than the zone of accumulation builds up snow and ice.

Glacial Erosion

Of all the erosional agents, glaciers are the most powerful because of their great size, weight, and density. When a valley glacier moves, it breaks off pieces of rock through a process called plucking. When glaciers with embedded rocks move over bedrock, they act like the grains on a piece of sandpaper, grinding parallel scratches into the bedrock. Small scratches are called striations, and larger ones are called grooves. Striations and grooves provide evidence of a glacier's history and indicate its direction of movement.

Glacial erosion by valley glaciers can create features like those shown in **Figure 8.21.** At the high elevations where snow accumulates, valley glaciers also scoop out deep depressions, called **cirques.** Where two cirques on opposite sides of a valley meet, they form a sharp, steep ridge called an arête. When there are glaciers on three or more sides of a mountaintop, the carving action creates a steep, pyramid-shaped peak. This is known as a horn. The most famous example of this feature is Switzerland's Matterhorn.

Valley glaciers can also leave hanging valleys in the glaciated landscape. Hanging valleys are formed when tributary glaciers converge with the primary glaciers and later retreat. The primary glacier is so thick that it meets the height of the smaller tributary glacier. When the glaciers melt, the valley is left hanging high above what is now a river in the primary valley floor. Hanging valleys today are often characterized by waterfalls where the tributary glacier used to be.

■ **Figure 8.22** Elongated landforms called drumlins can be grouped together as a drumlin field in areas once covered by continental glaciers.
Describe *how you could identify a drumlin on a topographic map.*

Glacial Deposition

Glacial till is the unsorted rock, gravel, sand, and clay that glaciers carry embedded in their ice and on their tops, sides, and front edges. Glacial till is formed from the grinding action of the glacier on underlying rock. Glaciers deposit unsorted ridges of till called **moraines** when the glacier melts. Terminal moraines are found along the edge where the retreating glacier melts, and lateral moraines are located parallel to the direction of a valley glacier flow.

Outwash When the farthest ends of a glacier melt and the glacier begins to recede, meltwater floods the valley below. Meltwater contains gravel, sand, and fine silt. When this sediment is deposited by meltwater carried away from the glacier, it is called outwash. Because of the way water transports sediment, outwash is always sorted by particle size. The area at the leading edge of the glacier where the meltwater flows and deposits outwash is called an **outwash plain.**

Drumlins, eskers, and kames Continental glaciers that move over older moraines form the material into elongated landforms called **drumlins,** shown in **Figure 8.22.** A drumlin's steeper slope faces the direction from which the glacier came. Streams flowing under melting glaciers leave long, winding ridges of layered sediments called **eskers,** shown in **Figure 8.23.** A **kame** is a mound of layered sediment deposited at the retreating glacier face and is conical in shape. Kames are also shown in **Figure 8.23.**

MiniLab

Model Glacial Deposition

How do glaciers deposit different types of rocks and sediments? Glaciers are powerful forces of erosion. As they move across the land, they pick up rocks and sediments, and carry them to new locations. When a glacier melts, these materials are left behind and deposits form in different shapes.

Procedure

1. Read and complete the lab safety form.
2. Work with a group of 2 to 3 other students. One student should obtain four **glaciers** from your teacher.
3. Place the glaciers on a **baking pan.** In front of each glacier, place a **popsicle stick** (to prevent the glacier from sliding down the pan).
4. Place a **textbook** under one end of the baking pan (your glaciers should be toward the elevated end of the pan).
5. Observe what happens as the glaciers melt. Record your observations in your science journal.
6. Dispose of your materials as your teacher instructs.

Analysis

1. **Discuss** Did the materials differ in the way they were deposited by the melting ice cubes? Were your results similar to those of your classmates? Explain.
2. **Explain** how this activity modeled the formation of meltwater.
3. **Apply** Which materials in this activity modeled glacial till?
4. **Apply** How did this activity model glacial deposition and the formation of a moraine?

Visualizing Continental Glacial Features

Figure 8.23 Continental glaciers carve out vast regions of landscape, leaving behind distinctive features such as kames, eskers, drumlins, and moraines.

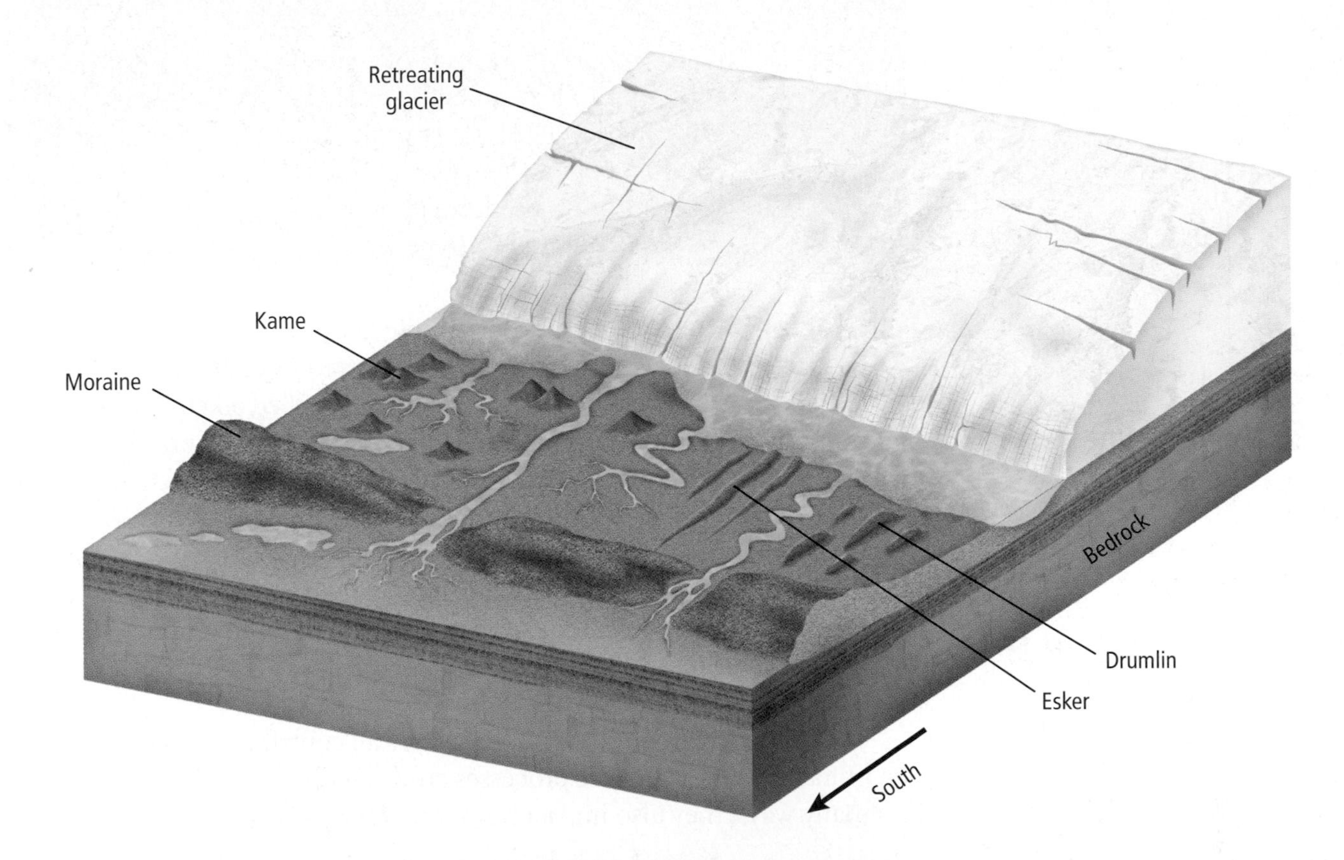

Kames are short cone-shaped mounds of sorted deposits. They are shaped from outwash left as glaciers recede.

Eskers are long ridges of sorted deposits. They are shaped from outwash deposited by water flowing through tunnels in the glacier.

Drumlins are shaped as the glacier moves over old moraines. They are unsorted.

Concepts In Motion To explore more about glacial features, visit glencoe.com.

Earth Science Online

■ **Figure 8.24** These kettle lakes in North Dakota are a result of glacial retreat. **Describe** *how you might be able to locate kettles on a topographic map.*

Glacial lakes Sometimes, a large block of ice breaks off a continental glacier and the surrounding area is covered by sediment. When the ice block melts, it leaves behind a depression called a kettle hole. After the ice block melts, the kettle hole fills with water from precipitation and runoff to form a kettle lake. **Kettles** or kettle lakes, such as those shown in **Figure 8.24,** are common in New England, New York, and Wisconsin. With valley glaciers, cirques can also fill with water, and they become cirque lakes. When a terminal moraine blocks off a valley, the valley fills with water to form a lake. Moraine-dammed lakes include the Great Lakes and the Finger Lakes of northern New York, which are long and narrow.

Mass movements, wind, and glaciers all contribute to the changing of Earth's surface. These processes erode landforms constantly, and in many ways, they also impact human populations and activities.

VOCABULARY
SCIENCE USAGE V. COMMON USAGE
Kettle
Science usage: a steep-sided depression formed by a glacier

Common usage: a metallic pot used for cooking

Section 8.3 Assessment

Section Summary

- Glaciers are large moving masses of ice that form near Earth's poles and in mountain areas.
- Glaciers can be classified as valley glaciers or continental glaciers.
- Glaciers modify the landscape by erosion and deposition.
- Features formed by glaciers include U-shaped valleys, hanging valleys, moraines, drumlins, and kettles.

Understand Main Ideas

1. MAIN Idea **Describe** two examples of how glaciers modify landscapes.
2. **Explain** how glaciers form.
3. **Compare and contrast** the characteristics of valley glaciers and continental glaciers.
4. **Differentiate** among different glacial depositional features.

Think Critically

5. **Evaluate** the evidence of past glaciers that can be found on Earth today.
6. **Infer** whether valley glaciers or continental glaciers have shaped more of the landscape of the United States.

WRITING in Earth Science

7. Deduce how you might distinguish a lake formed in a cirque and a lake formed in a kettle.

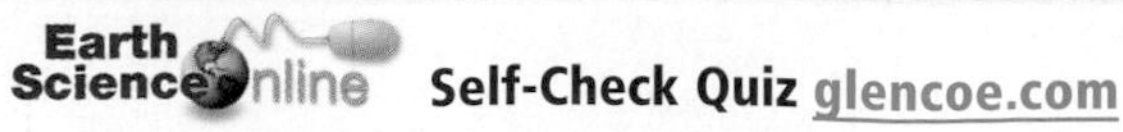

Earth Science & Society

Slipping Away

On the morning of January 10, 2005, the residents of La Conchita, California, awoke to find the highway out of town closed in both directions, due to landslides. Around 12:30 P.M. many residents heard an ominous roar as the bluff above the town unleashed 600,000 metric tons of dirt and mud, covering four blocks in 10 m of debris. Scientists went to the scene to discover exactly what had caused this enormous landslide and whether one could happen again.

The mass movement at La Conchita, California, in 2005 killed ten people.

The setting La Conchita is built on a narrow swatch of land between the highway and a huge bluff. The bluff is held together weakly, so it is susceptible to being loosened by heavy water content, such as a prolonged, heavy rain. The slope is further weakened from the effects of regular landslides, as well as being on a fault line.

In the two weeks prior to the landslide, the area had received a record amount of rain—about 35 cm—the amount it normally receives in a year! The excess water caused the earth to literally slide off the face of the mountain.

A history of landslides This event was not, however, the first landslide to hit the area. In fact, the mountain bluff is scarred with the evidence of many landslides. Ten years earlier, in March of 1995, two devastating landslides hit the area in the span of a week. These landslides were also caused by a large amount of rain, but the movement of the earth was relatively slow, so residents were able to get away. The 2005 landslide was a continuation of the 1995 slide—the soil that was deposited by the earlier slide was loosened by the rainwater and slipped down the slope. After the 1995 slide, the state government erected a retaining wall to keep the landslides at bay. However, soil, mud, and debris from the 2005 disaster passed right over parts of the wall.

The debate Could the 2005 landslide have been detected and the people warned in time to prevent loss of life? Most likely, yes. In fact, some of the residents of the town are suing the government for failure to protect their citizens, as well as failure to adequately notify them of the impending danger.

Are governments responsible for providing warnings and protection to citizens who move into areas that are prone to natural disasters? Or, does the responsibility lie with the citizens that might not have understood the dangers of living in the area? These questions and more are sure to be considered by the residents and government of La Conchita, as well as cities and local governments of disaster- prone areas throughout the United States for years to come.

WRITING in Earth Science

Debate Research information about a natural disaster that has occurred near your location. Hold a classroom debate on the topic of why people should, or should not, live in an area where natural disasters have occurred. To learn more about natural disasters, visit glencoe.com.

GEOLAB

MAPPING: MAP A LANDSLIDE

This image shows the Tully Valley landslide three days after it occurred. The Tully Farms Road is covered up to 5 m deep with clay.

Background: **Around midday on April 27, 1993, in a normally quiet, rural area of New York, the landscape dramatically changed. Unexpectedly, almost 1 million m^3 of earth debris slid 300 m down the lower slope of Bare Mountain and into Tully Valley. The debris flowed over the road and buried nearby homes. The people who lived there had no knowledge of any prior landslides occurring in the area, yet this landslide was the largest to occur in New York in more than 75 years.**

Question: *How can you use a drawing based on a topographic map to infer how the Tully Valley Landslide occurred?*

Materials

metric ruler

Procedure

Imagine that you work for the United States Geological Survey (USGS) specializing in mass movements. You have just been asked to evaluate the Tully Valley Landslide.

1. Read and complete the lab safety form.
2. Check the map's scale.
3. Measure the length and width of the Tully Valley in kilometers. Double-check your results.

Analyze and Conclude

1. **Interpret Data** What does the shape of the valley tell you about how it formed?
2. **Determine** In what direction did the landslide flow?
3. **Determine** In what direction does the Onondaga Creek flow?
4. **Infer** from the map which side of Tully Valley has the steepest valley walls.
5. **Deduce** What conditions must have been present for the landslide to occur?
6. **Infer** At the time of the Tully Valley Landslide, the trees were bare. How could this have affected the conditions that caused the landslide?

WRITING in Earth Science

Explain why the mass movement event you examined in this GeoLab is classified as a landslide. Differentiate a landslide from a creep, slump, flow, avalanche, and rockfall.

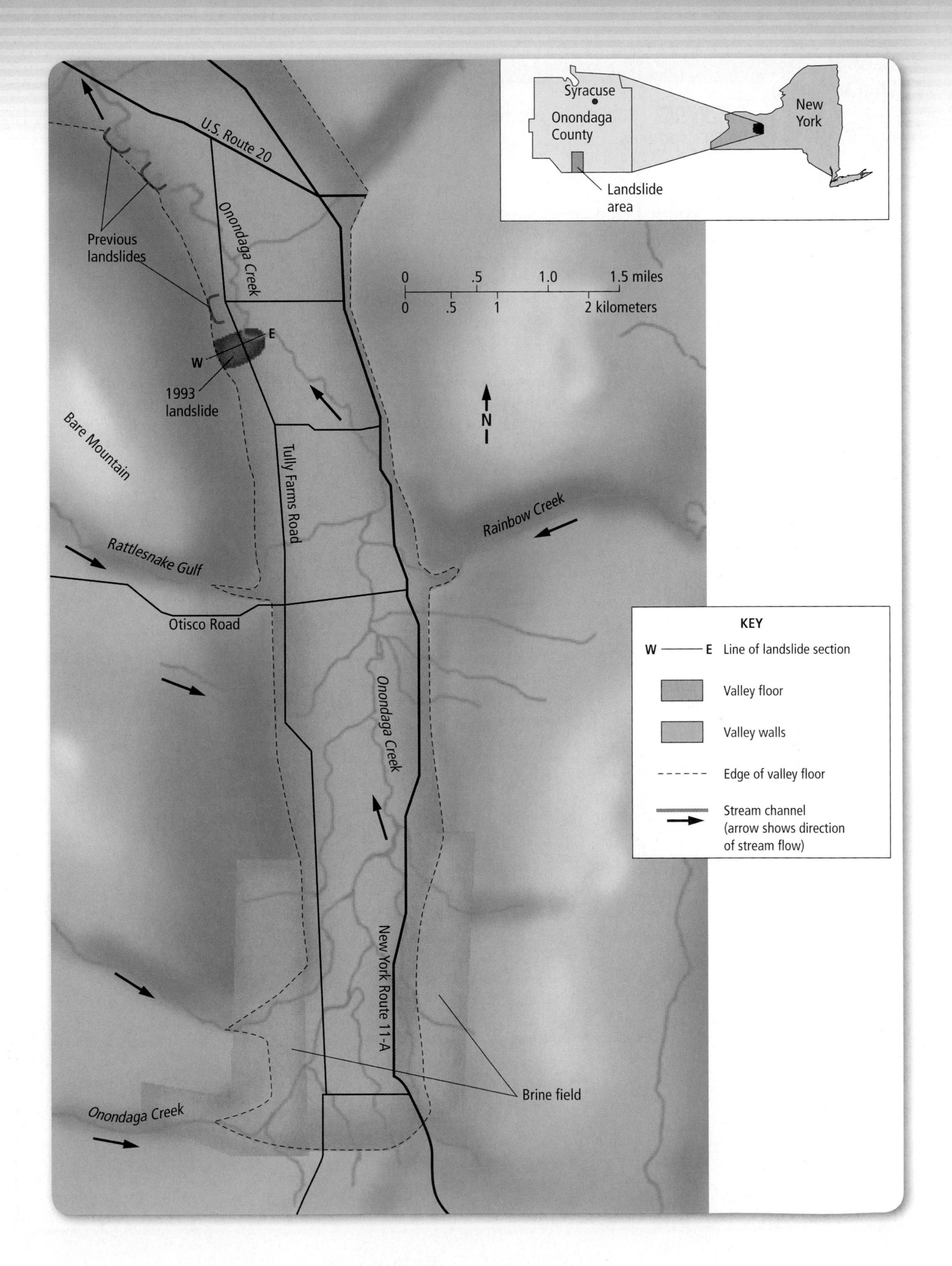
Syracuse
Onondaga County
New York
Landslide area
U.S. Route 20
Previous landslides
Onondaga Creek
0 .5 1.0 1.5 miles
0 .5 1 2 kilometers
E
W
1993 landslide
N
Bare Mountain
Tully Farms Road
Rainbow Creek
Rattlesnake Gulf
Otisco Road
KEY
W E Line of landslide section
Valley floor
Valley walls
Edge of valley floor
Stream channel (arrow shows direction of stream flow)
Onondaga Creek
New York Route 11-A
Brine field
Onondaga Creek

CHAPTER 8

Study Guide

Download quizzes, key terms, and flash cards from glencoe.com.

BIG Idea Movements due to gravity, winds, and glaciers shape and change Earth's surface.

Vocabulary	Key Concepts
Section 8.1 Mass Movements	
• avalanche (p. 198) • creep (p. 195) • landslide (p. 197) • mass movement (p. 194) • mudflow (p. 196) • slump (p. 198)	**MAIN Idea** Mass movements alter Earth's surface over time due to gravity moving sediment and rocks downslope. • Mass movements are classified in part by how rapidly they occur. • Factors involved in the mass movement of Earth materials include the material's weight, its resistance to sliding, the trigger, and the presence of water. • Mass movements are natural processes that can affect human life and activities. • Human activities can increase the potential for the occurrence of mass movements.
Section 8.2 Wind	
• abrasion (p. 203) • deflation (p. 202) • dune (p. 204) • loess (p. 206) • ventifact (p. 203)	**MAIN Idea** Wind modifies landscapes in all areas of the world by transporting sediment. • Wind is a powerful agent of erosion. • Wind can transport sediment in several ways, including suspension and saltation. • Dunes form when wind velocity slows down and windblown sand is deposited. • Dunes migrate as long as winds continue to blow.
Section 8.3 Glaciers	
• cirque (p. 209) • continental glacier (p. 208) • drumlin (p. 210) • esker (p. 210) • glacier (p. 207) • kame (p. 210) • kettle (p. 212) • moraine (p. 210) • outwash plain (p. 210) • valley glacier (p. 208)	**MAIN Idea** Glaciers modify landscapes by eroding and depositing rocks. • Glaciers are large moving masses of ice that form near Earth's poles and in mountain areas. • Glaciers can be classified as valley glaciers or continental glaciers. • Glaciers modify the landscape by erosion and deposition. • Features formed by glaciers include U-shaped valleys, hanging valleys, moraines, drumlins, and kettles.

CHAPTER 8 Assessment

Vocabulary Review

Match the correct vocabulary term from the Study Guide to the following definitions.

1. rapid downslope movement of a mass of loose sediment
2. rapidly flowing, often destructive mixtures of mud and water
3. slow, steady downhill movement of loose, weathered Earth materials

Replace each underlined word with the correct vocabulary term from the Study Guide.

4. Barchans are rock structures shaped by windblown sediments.
5. Thick, windblown, fertile deposits of silt that contain high levels of nutrients and minerals are known as desert pavement.
6. Deflation occurs when particles such as sand rub against the surface of rocks.

Explain the differences between the vocabulary terms in the following sets.

7. valley glacier, continental glacier
8. esker, kame
9. moraine, outwash plain

Understand Key Concepts

10. What are elongated landforms made of older moraines over which glaciers move?
    - **A** drumlins
    - **B.** kettle lakes
    - **C.** eskers
    - **D.** outwash plains
11. Which particles can wind move most easily?
    - **A.** sand
    - **B.** pebbles
    - **C.** silt
    - **D.** gravel
12. Which is the underlying force that causes all forms of mass movement?
    - **A.** friction
    - **B.** gravity
    - **C.** magnetism
    - **D.** the Coriolis Effect
13. The last continental ice age covered approximately what percent of Earth's surface?
    - **A.** 10 percent
    - **B.** 20 percent
    - **C.** 30 percent
    - **D.** 50 percent
14. Where are large deposits of glacial loess primarily found?
    - **A.** eastern United States
    - **B.** southeastern United States
    - **C.** southwestern United States
    - **D.** midwestern United States
15. Which has the fastest movement?
    - **A.** solifluction
    - **B.** creep
    - **C.** mudflow
    - **D.** avalanche

Use the photo below to answer Questions 16 and 17.

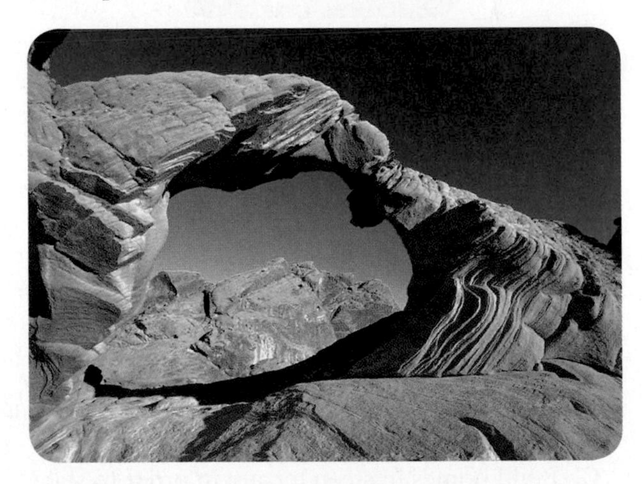

16. Which formed the structure in the photo?
    - **A.** ice
    - **B.** water
    - **C.** wind
    - **D.** organisms
17. Which process formed the structure in the photo?
    - **A.** abrasion
    - **B.** deflation
    - **C.** deposition
    - **D.** migration

18. Which range of slope angles is associated with producing an avalanche?
   A. 10 to 20 degrees
   B. 20 to 35 degrees
   C. 30 to 45 degrees
   D. 45 to 60 degrees

19. Which statement best describes sediments deposited by glaciers and rivers?
   A. Both glacial and river deposits are sorted.
   B. Glacial deposits are sorted, and river deposits are unsorted.
   C. Glacial deposits are unsorted, and river deposits are sorted.
   D. Both glacial and river deposits are unsorted deposits.

Use the photo below to answer Question 20.

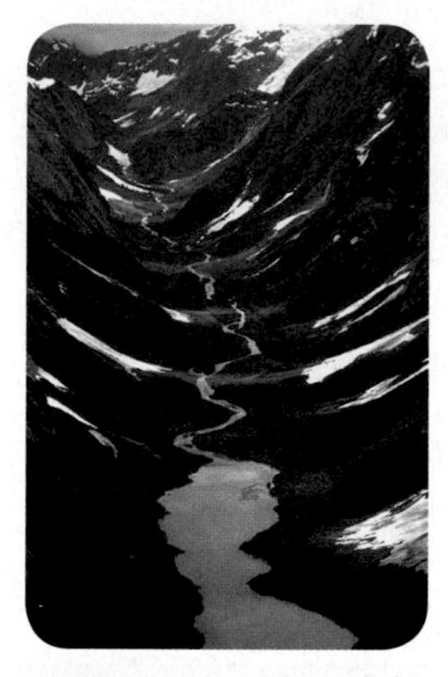

20. Which most likely created the valley?
   A. running water
   B. glacial ice
   C. landslide
   D. strong prevailing winds

21. Which is a way to reduce the risk of mass movements?
   A. Develop hillsides with roads so they become stable.
   B. Allow septic systems to run unmaintained so that they provide a source of nutrients for the soil.
   C. Build homes in steep terrain in order to stabilize the slope.
   D. Avoid construction and structures on vulnerable slopes.

22. Which property is used to classify dunes?
   A. size
   B. composition
   C. shape
   D. density

23. Which does NOT affect the speed of a valley glacier's movement?
   A. slope of the valley floor
   B. shape of the valley wall
   C. temperature and thickness of the ice
   D. internal chemistry of the glacier

Constructed Response

24. **Compare and contrast** suspension and saltation as they relate to transport of materials by wind.

25. **Infer** What happens to sand particles as the sand becomes saturated with water?

Use the figure below to answer Question 26.

26. **Identify** the features of the glacial landscape.

27. **Diagram** and label a migrating sand dune. Indicate the prevailing wind direction.

28. **Contrast** a slide, flow, and rockfall.

29. **Describe** the relationship between permafrost and solifluction.

30. **Identify** which mass movements are dependent on the addition of water.

31. **Describe** how particles eroded by wind differ from particles eroded by water.

32. **Infer** how human activities could affect the formation and migration of sand dunes in coastal areas.

33. **Compare and contrast** the formation, shape, and size of particles of a sand dune and a drumlin. How are these features used to indicate the direction of wind and glacial movement?

Think Critically

Use the figure below to answer Question 34.

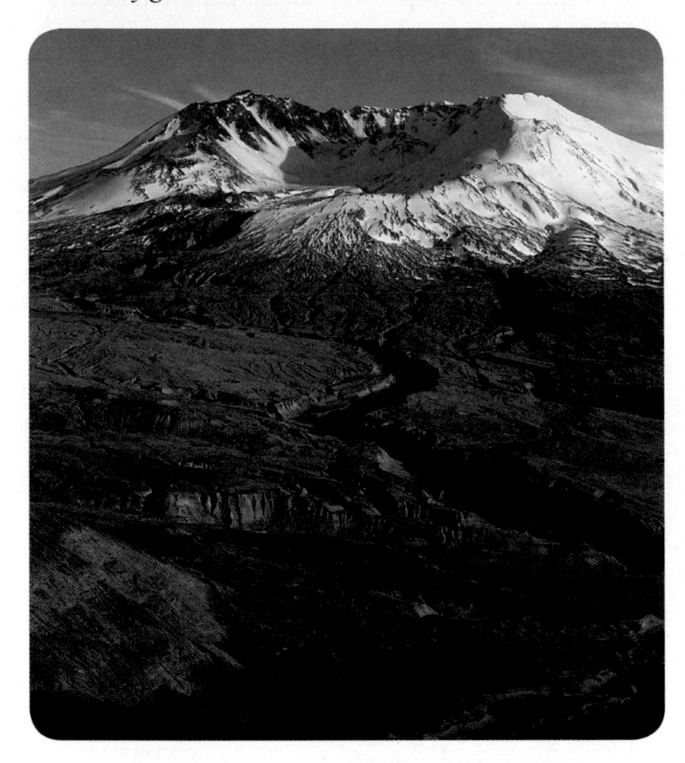

34. **Hypothesize** what kind of mass movements might have occurred after the eruption of Mount St. Helens.

35. **Analyze** the conditions that contribute to the likelihood that an area will experience wind erosion, and identify at least three areas in the United States that are prone to wind erosion.

36. **Infer** why wind abrasion is such an effective agent of erosion.

37. **Predict** the shape of a lake formed by a valley glacier.

Concept Mapping

38. Create a concept map to compare the terms *drumlin*, *esker*, and *kame*. For more help, refer to the *Skillbuilder Handbook*.

Challenge Question

39. **Hypothesize** how the Dust Bowl of the 1930s might have been avoided.

Additional Assessment

40. *WRITING in* **Earth Science** Write an editorial for a newspaper explaining why laws are needed to prevent developers from building homes on relatively steep and loosely consolidated hillside areas.

DBQ Document–Based Questions

Data obtained from: Natural Hazards-Landslides Information Sheet. 2006. *USGS*.

The photo below shows the potential for landslides across the continental United States.

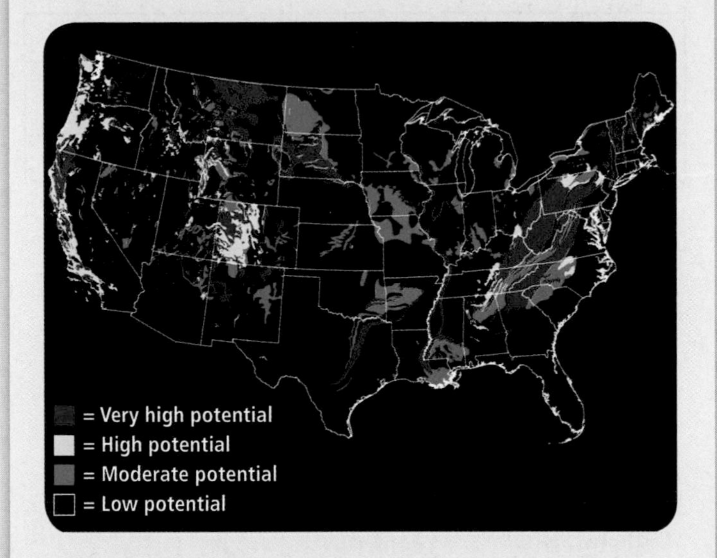

41. Identify landscapes or possible triggers for two areas that have very high potential for landslides.

42. Infer why the potential for landslides occurring in Florida is low.

43. What can be done to reduce the number of deaths due to landslide?

Cumulative Review

44. How many valence electrons does oxygen (atomic number 8) have? **(Chapter 3)**

45. Which compositional type of igneous rock has the lowest silica content? **(Chapter 5)**

46. What are fossils? **(Chapter 6)**

Standardized Test Practice

Multiple Choice

1. What is the strongest factor that controls the development of soils?
   A. parent material
   B. topography
   C. climate
   D. time

Use the table below to answer Questions 2 and 3.

Region	Characteristics
A	semiarid; experiences intense but brief rainstorms
B	permafrost; much loose, waterlogged material
C	mountainous; thick accumulations of snow
D	thick soils on semi-steep and steep slopes; occasional earthquake activity
E	arid; high cliffs and rocky shorelines

2. Which mass movement is most likely to occur in Region A?
   A. mudflow
   B. avalanche
   C. slump
   D. rockfall

3. Which mass movement is most likely to occur in Region B?
   A. solifluction
   B. mudflow
   C. avalanche
   D. slump

4. Which branch of science studies humans' interactions with the environment?
   A. planetary science
   B. environmental science
   C. oceanography
   D. geology

5. When do minerals form from a solution?
   A. when the solution is saturated
   B. when the solution is supersaturated
   C. when the solution is unsaturated
   D. when the solution is ultrasaturated

6. Why are the 24 time zones located approximately 15° apart?
   A. to line up with the equator
   B. to roughly match lines of latitude
   C. to roughly match lines of longitude
   D. to line up with the prime meridian

7. Identify the term used to describe wind transportation of materials by a bouncing motion of particles.
   A. suspension
   B. deflation
   C. saltation
   D. abrasion

Use the geologic cross section below to answer Questions 8 and 9.

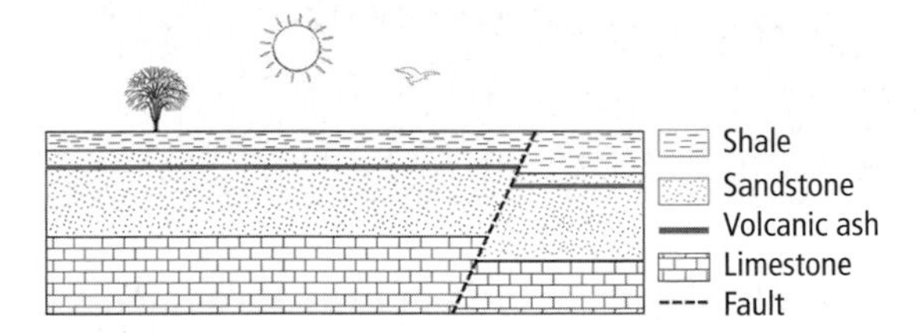

8. Assuming the rock layers shown are in the same orientation that they were deposited, which layer is the oldest?
   A. shale
   B. sandstone
   C. volcanic ash
   D. limestone

9. Which layer was probably created by sediments deposited by slow-moving water?
   A. shale
   B. sandstone
   C. volcanic ash
   D. limestone

10. Which is NOT a feature of valley glaciers?
    A. cirque
    B. loess
    C. moraine
    D. arête

Short Answer

Use the table below to answer Questions 11 and 12.

Liquid	Final Color of Litmus Paper
ammonia	blue
lemon juice	red
tea	red
vinegar	red

11. What conclusions can you draw from the results of the litmus paper tests on the liquids as shown in the table?

12. If an unknown liquid did not change the color of litmus paper, what could you infer?

13. Describe the formation of soil.

14. Evaluate the negative impact of building in coastal-dune areas.

15. Distinguish between weathering and erosion.

16. What is one reason granite is commonly used in construction?

17. What are some benefits of communicating scientific results?

18. What are isotopes of an element?

Reading for Comprehension

Arctic Ice Levels

The amount of sea ice in the Arctic shrank dramatically this summer and is now smaller than it has been in a century of record-keeping, new research reveals. Scientists say rising temperatures brought on by human-made global warming is probably to blame for the melting trend. Most scientists attribute this warming to human activities such as burning fossil fuels. The shift could lead to increased coastal erosion and shrinking of habitat for animals like polar bears. Melting sea ice may lead to greater coastal erosion, because Arctic storms could produce much larger waves on the open ocean. As the sea ice continues to melt, polar habitat also continues to shrink. If the decline in sea ice continued, summers in the Arctic could become completely ice-free before the end of this century, scientists warn.

Article obtained from Lovgren, S. Arctic ice levels at record low, may keep melting, study warns. *National Geographic News.* October 3, 2005.

19. How could melting sea ice possibly lead to greater coastal erosion?

A. Polar bears would use more of the land.
B. Humans would use the exposed land for fossil fuels.
C. It would increase global temperatures, ruining the land.
D. Arctic storms could produce larger waves to erode the shoreline.

20. What is causing the sea ice to melt?

A. increased temperatures
B. arctic storms
C. polar bears
D. time

21. What can be inferred from this text?

NEED EXTRA HELP?																		
If You Missed Question . . .	1	2	3	4	5	6	7	8	9	10	11	12	13	14	15	16	17	18
Review Section . . .	7.3	8.1	8.1	1.1	4.1	2.1	8.2	6.1	6.1	8.3	3.2	3.2	7.3	8.2	7.1	5.2	1.3	3.1

CHAPTER 9 Surface Water

BIG Idea Surface water moves materials produced by weathering and shapes the surface of Earth.

9.1 Surface Water Movement
MAIN Idea Running water is an agent of erosion, carrying sediments in streams and rivers and depositing them downstream.

9.2 Stream Development
MAIN Idea Streams erode paths through sediment and rock, forming V-shaped stream valleys.

9.3 Lakes and Freshwater Wetlands
MAIN Idea As the amount of water changes and the amount of sediments increases, lakes can be transformed into wetlands and eventually into dry land.

GeoFacts

- The United States has approximately 5,600,000 km of rivers.
- The Missouri River is about 4087 km long, making it the longest river in North America.
- The Mississippi River Basin drains 41 percent of the United States.

LAUNCH Lab

How does water infiltrate?

When water soaks into the ground, it moves at different rates through the different materials that make up Earth's surface.

Procedure

1. Read and complete the lab safety form.
2. Place a **small plastic window screen** on each of two **clear plastic shoe boxes.**
3. Place an 8-cm × 16-cm **clump of grass or sod** on one screen.
4. Place an 8-cm × 16-cm **clump of barren soil** on the other screen.
5. Lightly sprinkle 500 mL of **water** on each clump.
6. Observe the clumps for 5 min.
7. Measure the amount of water in each box.

Analysis

1. **Describe** what happens to the water after 5 min.
2. **Infer** the reason for any differences in the amount of water collected in each box.

Visit glencoe.com to

- study entire chapters online;
- explore Concepts In Motion animations:
  - Interactive Time Lines
  - Interactive Figures
  - Interactive Tables
- access Web Links for more information, projects, and activities;
- review content with the Interactive Tutor and take Self-Check Quizzes.

Stream Development Make this Foldable that features the steps in stream development.

STEP 1 Fold three sheets of notebook paper in half horizontally to find the middle. Holding two of the sheets together, make a 3-cm cut at the fold line on each side of the paper.

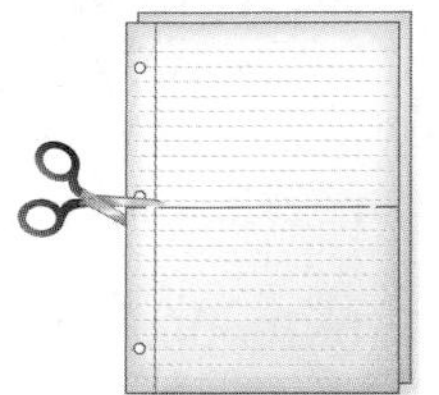

STEP 2 On the third sheet, cut along the fold line to within 3-cm of each edge.

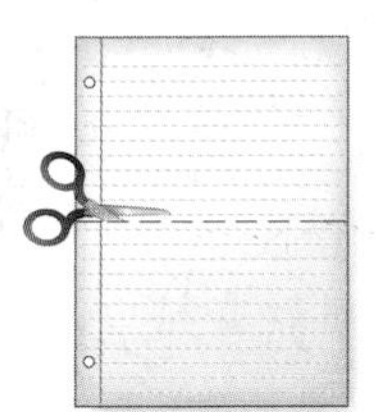

STEP 3 Slip the first two sheets through the cut in the third sheet to make a six-page book.

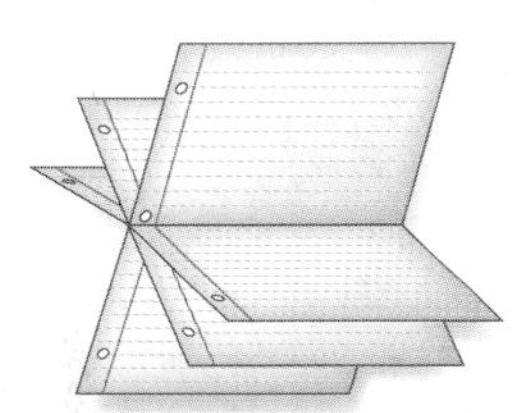

STEP 4 Label your book *Stream Development.*

FOLDABLES **Use this Foldable with Section 9.2.** As you read this section, use the pages of your Foldable to describe and illustrate the steps in stream development.

Section 9.1

Objectives

- **Describe** how surface water can move weathered materials.
- **Explain** how a stream carries its load.
- **Describe** how a floodplain develops.

Review Vocabulary

solution: a homogeneous mixture in which the component particles cannot be distinguished

New Vocabulary

runoff
watershed
divide
suspension
bed load
discharge
flood
floodplain

Surface Water Movement

MAIN Idea Running water is an agent of erosion, carrying sediments in streams and rivers and depositing them downstream.

Real-World Reading Link Have you ever noticed that sometimes a river is muddy but other times it is clear? In floods, rivers can carry greater amounts of materials, which makes them muddy. Under normal conditions, they often carry less sediment, which makes them clearer.

The Water Cycle

Earth's water supply is recycled in a continuous process called the water cycle, shown in **Figure 9.1.** Water molecules move continuously through the water cycle following many pathways: they evaporate from a body of water or the surface of Earth, condense into cloud droplets, fall as precipitation back to Earth's surface, and infiltrate the ground. As part of a continuous cycle, the water molecules eventually evaporate back to the atmosphere, form clouds, fall as precipitation, and the cycle repeats. Understanding the mechanics of the water cycle will help you understand the reasons for variations in the amount of water that is available throughout the world.

Often, a water molecule's pathway involves time spent within a living organism or as part of a snowfield, glacier, lake, or ocean. Although water molecules might follow a number of different pathways, the overall process is one of repeated evaporation and condensation powered by the Sun's energy.

Reading Check Explain What happens once water reaches Earth's surface?

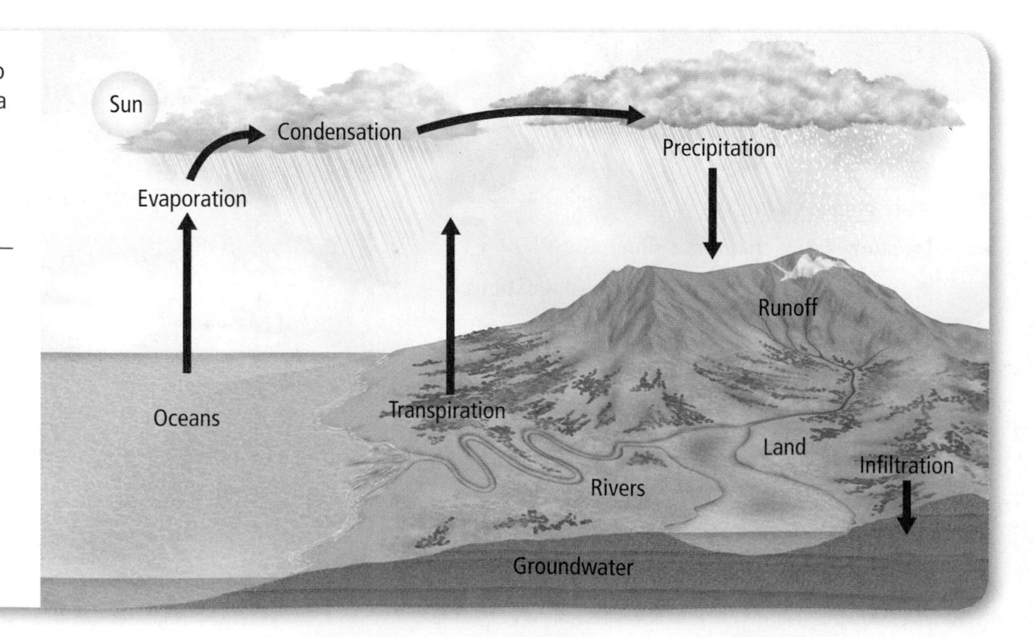

Figure 9.1 The water cycle, also referred to as the hydrologic cycle, is a never-ending, natural circulation of water through Earth's systems.
Identify *the driving force for the water cycle.*

Concepts In Motion
Interactive Figure To see an animation of the water cycle, visit glencoe.com.

Runoff

Water flowing downslope along Earth's surface is called **runoff.** Runoff might reach a stream, river, or lake, it might evaporate, or it might accumulate as puddles in small depressions and infiltrate the ground. During and after heavy rains, you can observe these processes in your yard or local park. Water that infiltrates Earth's surface becomes groundwater.

A number of conditions determine whether water on Earth's surface will infiltrate the ground or become runoff. For water to enter the ground, there must be large enough pores or spaces in the soil and rock to accommodate the water's volume, as in the loose soil illustrated in **Figure 9.2.** If the pores already contain water, the newly fallen precipitation will either remain in puddles on top of the ground or, if the area has a slope, run downhill. Water standing on the surface of Earth eventually evaporates, flows away, or slowly enters the groundwater.

Soil composition The physical and chemical composition of soil affects its water-holding capacity. Soil consists of decayed organic matter, called humus, and minerals. Humus creates pores in the soil, thereby increasing a soil's ability to retain water. The minerals in soil have different particle sizes, which are classified as sand, silt, or clay. As you learned in Chapter 7, the percentages of particles of each size vary from soil to soil. Soil with a high percentage of coarse particles, such as sand, has relatively large pores between its particles that allow water to enter and pass through the soil quickly. In contrast, soil with a high percentage of fine particles, such as clay, clumps together and has few or no spaces between the particles. Small pores restrict both the amount of water that can enter the ground and the ease of movement of water through the soil.

Rate of precipitation Light, gentle precipitation can infiltrate dry ground. However, the rate of precipitation might temporarily exceed the rate of infiltration. For example, during heavy precipitation, water falls too quickly to infiltrate the ground and becomes runoff. Thus, a gentle, long-lasting rainfall is more beneficial to plants and causes less erosion by runoff than a torrential downpour. If you have a garden, remember that more water will enter the ground if you water your plants slowly and gently.

VOCABULARY

ACADEMIC VOCABULARY

Accommodate
to hold without crowding or inconvenience
The teacher said she could accommodate three more students in her classroom.

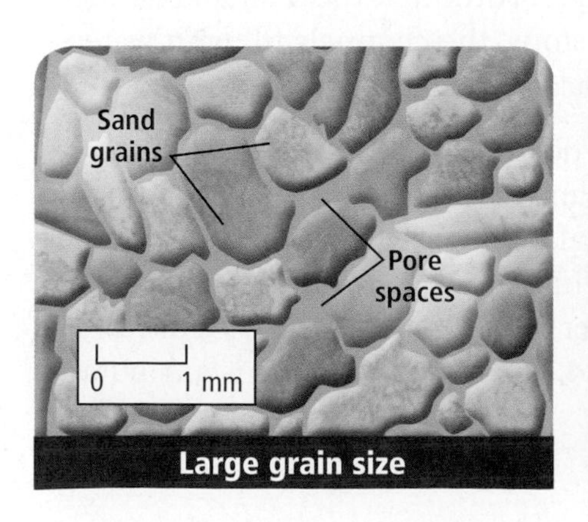

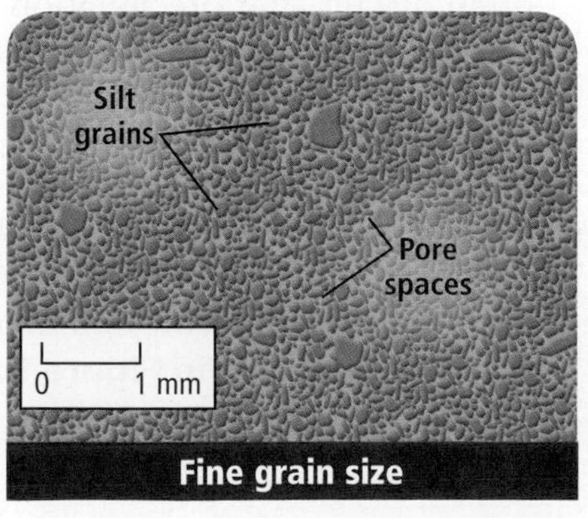

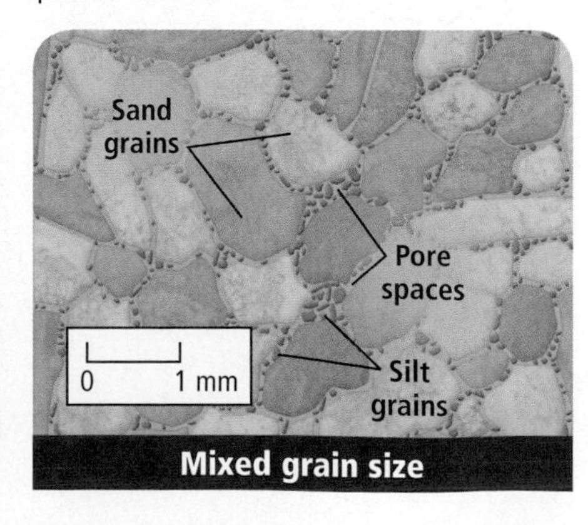

■ **Figure 9.2** Soil that has open surface pores allows water to infiltrate. The particle size that makes up a soil helps determine the pore space of the soil.

Figure 9.3 Vegetation can slow the rate of runoff of surface water. Raindrops are slowed when they strike the leaves of trees or blades of grass, and they trickle down slowly.

Vegetation Soils that contain grasses or other vegetation allow more water to enter the ground than do soils with no vegetation. Precipitation falling on vegetation slowly flows down leaves and branches and eventually drops gently to the ground, where the plants' root systems help maintain the pore space needed to hold water, as shown in **Figure 9.3.** In contrast, precipitation falls with far more force onto barren land. In such areas, soil particles clump together and form dense aggregates with little space between them. The force of falling rain can then push the soil clumps together, thereby closing pores and allowing less water to enter.

Slope The slope of a land area plays a significant role in determining the ability of water to enter the ground. Water from precipitation falling on slopes flows to areas of lower elevation. The steeper the slope, the faster the water flows. There is also greater potential for erosion on steep slopes. In areas with steep slopes, much of the precipitation is carried away as runoff.

Stream Systems

Precipitation that does not enter the ground usually runs off the surface quickly. Some surface water flows in thin sheets and eventually collects in small channels, which are the physical areas where streams flow. As the amount of runoff increases, the channels widen, deepen, and become longer. Although these small channels often dry up after precipitation stops, the channels fill with water each time it rains and become larger and longer.

Tributaries All streams flow downslope to lower elevations. However, the path of a stream can vary considerably, depending on the slope and the type of material through which the stream flows. Some streams flow into lakes, while others flow directly into the ocean. Rivers that flow into other streams are called tributaries. For example, as shown in **Figure 9.4,** the Missouri River is a tributary of the Mississippi River.

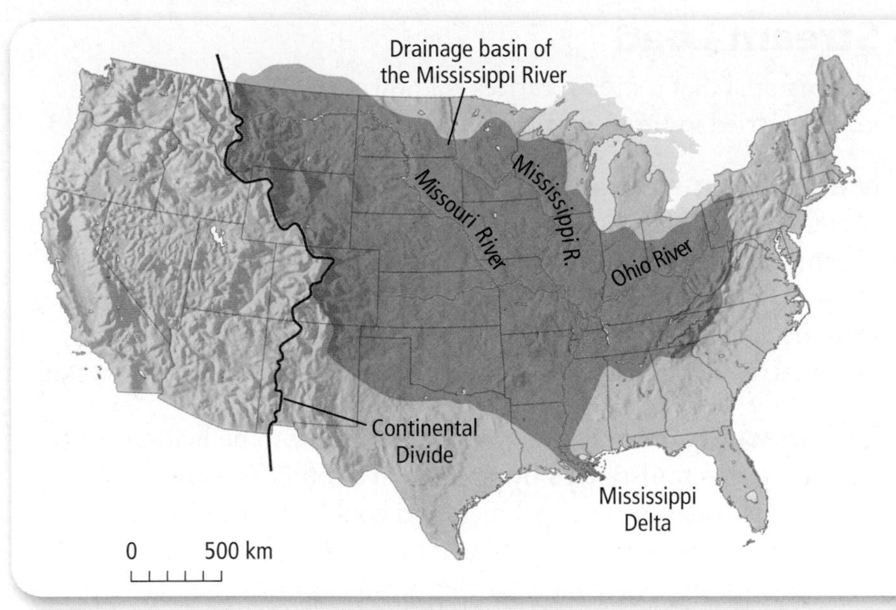

■ **Figure 9.4** The watershed of the Mississippi River includes many stream systems, including the Mississippi, Missouri and Ohio Rivers. The Continental Divide marks the western boundary of the watershed.
Identify ***what portion of the continental United States eventually drains into the Mississippi River.***

Watersheds and divides All of the land area whose water drains into a stream system is called the system's **watershed.** Watersheds can be relatively small or extremely large in area. A **divide** is a high land area that separates one watershed from another. In a watershed, the water flows away from the divide, as this is the high point of the watershed.

Each tributary in a stream system has its own watershed and divides, but they are all part of the larger stream system to which the tributary belongs. The watershed of the Mississippi River, shown in **Figure 9.4,** is the largest in North America.

Reading Check **Describe** what a divide is and what role it plays in a watershed.

PROBLEM-SOLVING LAB

Interpret the Graph

How do sediments move in a stream? The critical velocity of water determines the size of particles that can be moved. The higher the stream velocity, the larger the particles that can be transported.

Think Critically

1. **Identify** at what velocity flowing water would pick up a pebble.
2. **Identify** at what range of velocities flowing water would carry a pebble.
3. **Infer** which object would not fall into the same size range as a pebble: an egg, a baseball, a golf ball, a table tennis ball, a volleyball, and a pea. How would you test your conclusions?

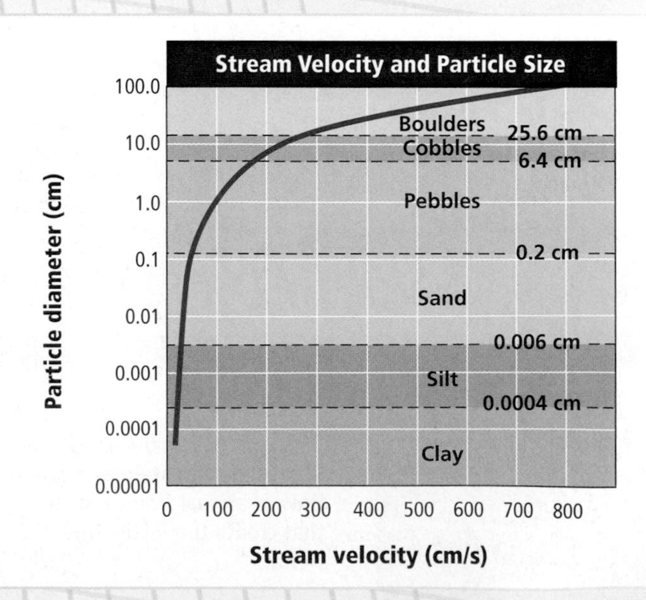

■ **Figure 9.5** Particles rub, scrape, and grind against one another in a streambed, which can create potholes.

Stream Load

The material that a stream carries is known as stream load. Stream load is carried in three ways.

Materials in suspension **Suspension** is the method of transport for all particles small enough to be held up by the turbulence of a stream's moving water. Particles such as silt, clay, and sand are part of a stream's suspended load. The amount of material in suspension varies with the volume and velocity of the stream water. Rapidly moving water carries larger particles in suspension than slowly moving water.

Bed load Sediment that is too large or heavy to be held up by turbulent water is transported by streams in another manner. A stream's **bed load** consists of sand, pebbles, and cobbles that the stream's water can roll or push along the bed of the stream. The faster the water moves, the larger the particles it can carry. As the particles move, they rub against one another or the solid rock of the streambed, which can erode the surface of the streambed, as shown in **Figure 9.5.**

Materials in solution Solution is the method of transport for materials that are dissolved in a stream's water. When water runs through or over rocks with soluble minerals, it dissolves small amounts of the minerals and carries them away in the solution. Groundwater adds the majority of the dissolved load to streams. The amount of dissolved material that water carries is often expressed in parts per million (ppm). For example, a measurement of 10 ppm means that there are 10 parts of dissolved material for every 1 million parts of water. The total concentration of materials in solution in streams averages 115–120 ppm, although some streams carry as little dissolved material as 10 ppm. Values greater than 10,000 ppm have been observed for streams draining desert basins.

■ **Figure 9.6**
Floods in Focus
Floods have shaped the landscape and affected human lives.

1900 — **1925** — **1950**

1902 In Egypt, the Aswan Dam is built to stabilize the flow of annual flood waters that create the fertile Nile Delta.

1927 Heavy rains flood the Mississippi River from Illinois to Louisiana leaving more than 600,000 people homeless.

1931 China's Yellow River floods when heavy rain causes the river's large silt deposits to shift and block the channel.

1958 Following a flood that claimed almost 2000 lives, Holland begins creating a vast network of dams, dikes, and barriers, shortening its coastline by 700 km.

Stream Carrying Capacity

The ability of a stream to transport material, referred to as its carrying capacity, depends on both the velocity and the amount of water moving in the stream. The channel's slope, depth, and width all affect the speed and direction the water moves within it. A stream's water moves more quickly where there is less friction; consequently, smooth-sided channels with great slope and depth allow water to move the most rapidly. The total volume of moving water also affects a stream's carrying capacity. **Discharge,** shown in **Figure 9.7,** is the measure of the volume of stream water that flows past a particular location within a given period of time. Discharge is commonly expressed in cubic meters per second (m^3/s). The following formula is used to calculate the discharge of a stream.

$$\underset{(m^3/s)}{\text{discharge}} = \underset{(m)}{\text{average width}} \times \underset{(m)}{\text{average depth}} \times \underset{(m/s)}{\text{average velocity}}$$

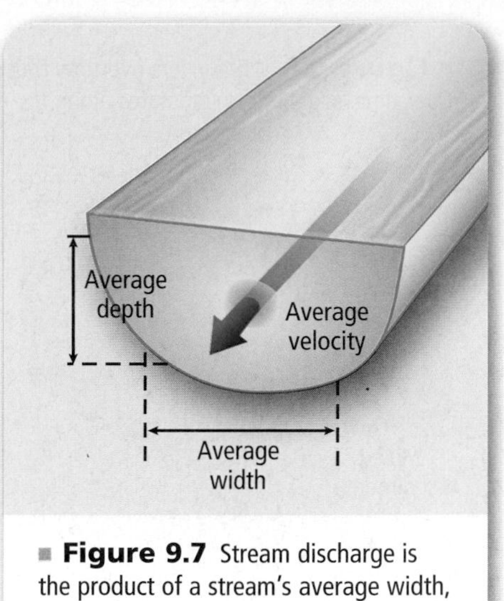

■ **Figure 9.7** Stream discharge is the product of a stream's average width, average depth, and the velocity of the water.

The largest river in North America, the Mississippi River, has a huge average discharge of about 17,000 m^3/s. The Amazon River, the largest river in the world, has a discharge of about ten times that amount. The discharge from the Amazon River over a two-hour period would supply New York City's water needs for an entire year!

As a stream's discharge increases, its capacity also increases. Both water velocity and volume increase during times of heavy precipitation, rapid melting of snow, and flooding. In addition to increasing a stream's carrying capacity, these conditions heighten a stream's ability to erode the land over which it passes. As a result of an increase in erosional power, a streambed can widen and deepen, adding to the stream's carrying capacity. Streams shape the landscape both during periods of normal flow and during floods, as highlighted in **Figure 9.6.**

1970 — 1985 — 2000

1974 The United Kingdom begins building the Thames Barrier to protect London from rising tide levels as the city sinks and sea levels rise.

1988 Monsoon rains in Bangladesh flood two-thirds of the country, affecting 45 million people.

1996 Volcanic eruptions in Iceland release meltwater from under the Vatnajökull glacier that washes away power lines, major roads, and bridges.

2005 Category 5 Hurricane Katrina slams into Louisiana, Mississippi, and Alabama, devastating New Orleans.

Concepts In Motion

Interactive Time Line To learn more about these discoveries and others, visit glencoe.com. **Earth Science Online**

■ **Figure 9.8** When rivers overflow their banks, the floodwater deposits sediment. Over time, sediment accumulates along the edges of a river, resulting in natural levees.

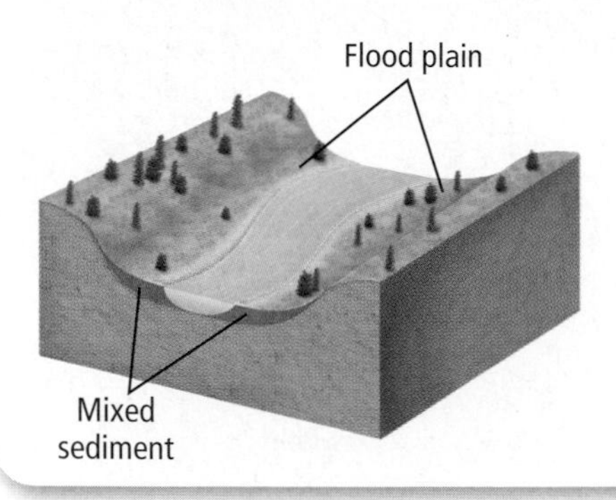

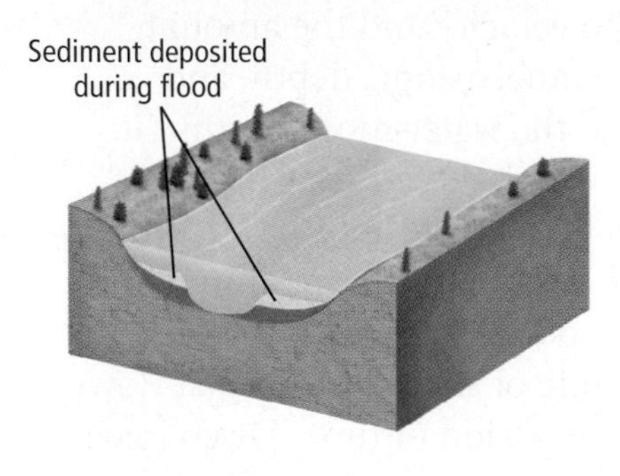

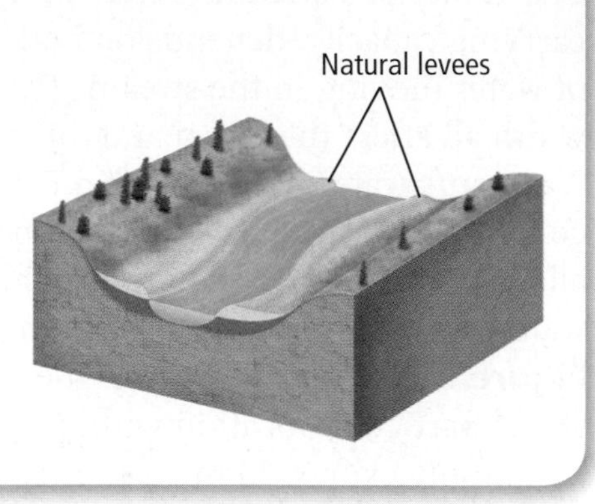

Floods

The amount of water being transported in a particular stream at any given time varies with weather conditions. Sometimes, more water pours into a stream than the banks of the stream channel can hold. A **flood** occurs when water spills over the sides of a stream's banks onto the adjacent land. The broad, flat area that extends out from a stream's bank and is covered by excess water during times of flooding is known as the stream's **floodplain.**

Floodwater carries along with it a great amount of sediment eroded from Earth's surface and the sides of the stream channel. As floodwater recedes and its volume and speed decrease, the water drops its sediment load onto the stream's floodplain. After repeated floods over time, sediments deposited by floods tend to accumulate along the banks of the stream. These develop into continuous ridges along the sides of a river, called natural levees, as shown in **Figure 9.8.** Floodplains develop highly fertile soils as more sediment is deposited with each subsequent flood. The fertile soils of floodplains make some of the best croplands in the world.

 **Reading Check** **Describe** what happens when floodwaters recede.

■ **Figure 9.9** This flood was caused by heavy rainfall upstream. Notice the farm fields that have been covered in floodwater.
Analyze ***What long-term effects might this flood have on the crops grown in this area?***

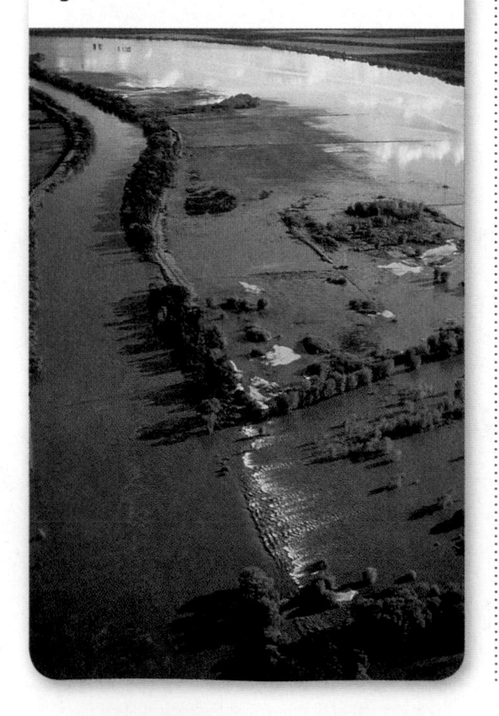

Flood stages Floods are a natural occurrence. After a rain event or snowmelt, it takes time for runoff water to reach the streams. As water enters the streams, the water level continues to rise and might reach its highest point, called its crest, days after precipitation ends. When the water level in a stream rises higher than its banks, the river is said to be at flood stage. The resulting flooding might occur over localized areas or across large regions. The flooding of a small area is known as an upstream flood.

Heavy accumulation of excess water from large regional drainage systems results in downstream floods. Such floods occur during or after long-lasting, intense storms or spring thaws of large snowpacks. The tremendous volume of water involved in a downstream flood can result in extensive damage. The effects of flooding on the landscape are shown in **Figure 9.9.**

■ **Figure 9.10** Gaging stations, like this one, can send data to meteorologic stations. There, scientists can process the information and alert the public to potential floods.

Flood Monitoring and Warning Systems

In order to provide warnings for people at risk, government agencies, such as the National Weather Service, monitor potential flood conditions. Earth-orbiting weather satellites photograph Earth and collect and transmit information about weather conditions, storms, and streams. In addition, the U.S. Geological Survey (USGS) has established approximately 7300 gaging stations in the United States to provide a continuous record of the water level in each stream as shown in **Figure 9.10.** These gaging systems often transmit data to satellites and telephone lines where the information is then sent to the local monitoring office.

In areas that are prone to severe flooding, warning systems are the first step in implementing emergency management plans. Flood warnings and emergency plans often allow people to safely evacuate an area in advance of a flood.

Section 9.1 Assessment

Section Summary

- Infiltration of water into the ground depends on the number of open pores.
- All the land area that drains into a stream system is the system's watershed.
- Elevated land areas called divides separate one watershed from another.
- A stream's load is the material the stream carries.
- Flooding occurs in small, localized areas as upstream floods or in large downstream floods.

Understand Main Ideas

1. MAIN Idea **Analyze** ways in which moving water can carve a landscape.
2. **Describe** the three ways in which a stream carries its load.
3. **Analyze** the relationship between the carrying capacity of a stream and its discharge and velocity.
4. **Determine** why little water from runoff infiltrates the ground in areas with steep slopes.

Think Critically

5. **Determine** how a floodplain forms and why people live on floodplains.
6. **Analyze** how levees form.

MATH in Earth Science

7. Design a data table that compares how silt, clay, sand, and large pebbles settle to the bottom of a stream as the velocity of water decreases.

Section 9.2

Objectives

- **Describe** some of the physical features of stream development.
- **Describe** the relationship between meanders and stream flow.
- **Explain** the process of rejuvenation in the stream development.

Review Vocabulary

abrasion: process of erosion in which windblown or waterborne particles, such as sand, scrape against rock surfaces or other materials and wear them away

New Vocabulary

stream channel
stream bank
base level
meander
delta
rejuvenation

Stream Development

MAIN Idea Streams erode paths through sediment and rock, forming V-shaped stream valleys.

Real-World Reading Link When was the last time you saw water flow uphill? Water in all rivers travels downslope to the lowest point. This allows geologists to predict the path of the river based on the features of an area.

Supply of Water

Stream formation relies on an adequate water supply. Precipitation provides water for the beginnings of stream formation. Streams can also be fed by underground deposits of water. As a stream develops, it changes width and size, and shapes the land over which it flows.

Stream channels The region where water first accumulates to supply a stream is called the headwaters. It is common for a stream's headwaters to be high in the mountains. Falling precipitation accumulates in small gullies at these higher elevations and forms briskly moving streams. As surface water begins its flow, its path might not be well defined. In time, the moving water carves a narrow pathway into the sediment or rock called the **stream channel.** The channel widens and deepens as more water accumulates and cuts into Earth's surface. **Stream banks** hold the moving water within them.

When small streams erode away the rock or soil at the head of a stream, it is known as headward erosion. These streams move swiftly over rough terrain and often form waterfalls and rapids as they flow over steep inclines. Sometimes a stream erodes the high area separating two drainage basins, joining another stream. It then draws water away from the other stream in a process called stream capture, as shown in **Figure 9.11.**

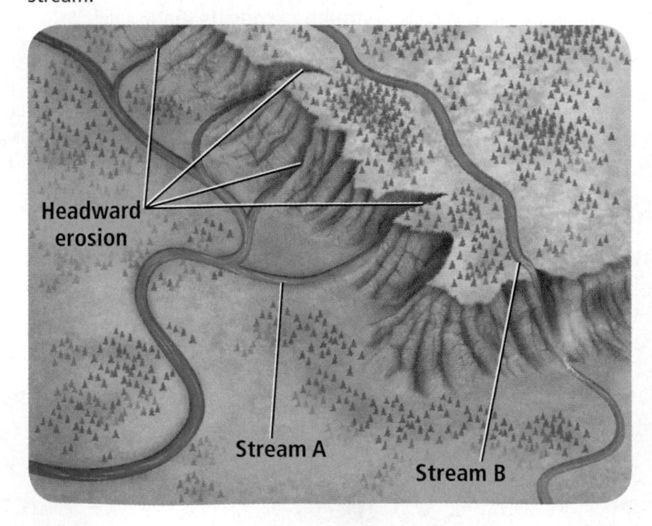

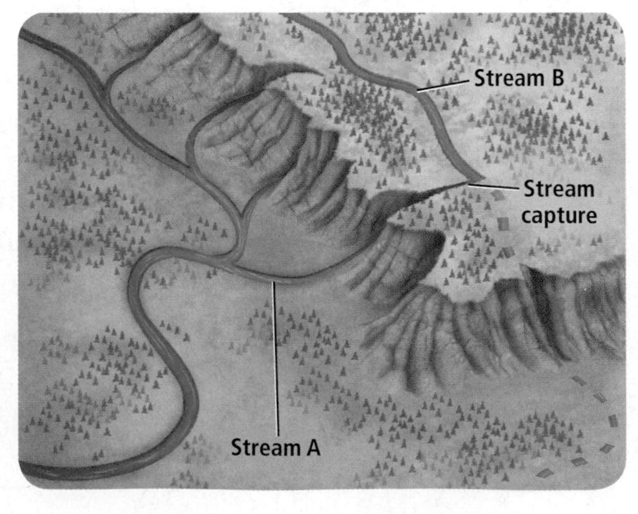

Figure 9.11 The headward erosion of Stream A cuts into Stream B and draws its water away into one stream.

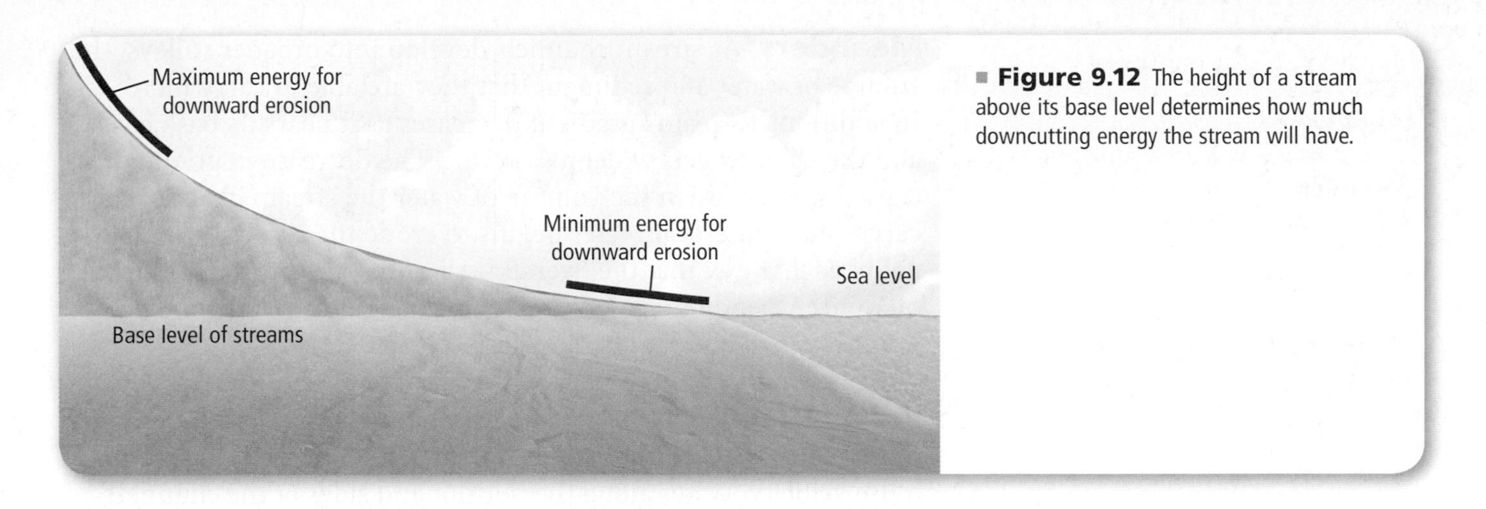

■ **Figure 9.12** The height of a stream above its base level determines how much downcutting energy the stream will have.

Formation of Stream Valleys

The driving force of a stream is the force of gravity on water. This means that the energy of a stream comes from the movement of water down a slope called a stream gradient. When the gradient of a stream is steep, water in the stream moves downhill rapidly, cutting steep valleys. The gradient of the stream depends on its **base level,** which is the elevation at which it enters another stream or body of water. The lowest base level possible for any stream is sea level, the point at which the stream enters the ocean, as shown in **Figure 9.12.**

Far from its base level, a stream actively erodes a path through the sediment or rock, and a V-shaped channel develops. V-shaped channels have steep sides and sometimes form canyons or gorges. The Yellowstone River in Wyoming flows through an impressive example of this type of narrow, deep gorge carved by a stream. **Figure 9.13** shows the classic V-shaped valley. As a stream approaches its base level, it has less energy for downward erosion. Instead, streams that are near their base level tend to erode at the sides of the stream channel, and over time result in broader valleys with gentle slopes, as shown in **Figure 9.13.**

FOLDABLES
Incorporate information from this section into your Foldable.

■ **Figure 9.13** A V-shaped valley is formed by the downcutting of a stream. A wide, broad valley is a result of stream erosion over a long period of time.
Identify *which river is closer to its base level.*

VOCABULARY

SCIENCE USAGE V. COMMON USAGE

Meander

Science usage: a bend or curve in a stream channel caused by moving water

Common usage: a winding path or course

Meanders As stream channels develop into broader valleys, the volume of water and sediment that they are able to carry increases. In addition, a stream's gradient decreases as it nears its base level, and the channel gets wider as a result. The decrease in gradient causes an increase in the volume of water the stream channel can carry. Sometimes, the water begins to erode the sides of the channel in such a way that the overall path of the stream starts to bend or wind. A bend or curve in a stream channel caused by moving water is called a **meander,** shown in **Figure 9.14.**

Water in the straight parts of a stream flows at different velocities, depending on its location in the channel. In a straight length of a stream, water in the center of the channel flows at the maximum velocity. Water along the bottom and sides of the channel flows more slowly because it experiences friction as it moves against the land.

In contrast, the water moving along the outside of a meander curve experiences the greatest velocity within the meander. The water that flows along this outside part of the curve continues to erode away the sides of the streambed, thus making the meander larger. Along the inside of the meander, the water moves more slowly and deposition is dominant. These differences in the velocity within meanders cause the meanders to become more accentuated over time. This process is illustrated in **Figure 9.15.**

Oxbow lakes Stream meanders continue to develop and become larger and wider over time. After enough winding, however, it is common for a stream to cut off a meander and once again flow along a straighter path. The stream then deposits material along the adjoining meander and eventually blocks off its water supply, as shown in **Figure 9.14.** The blocked-off meander becomes an oxbow lake, which eventually dries up.

As a stream approaches a larger body of water or its endpoint, the ocean, the streambed's gradient flattens out and its channel becomes very wide. The area of the stream that leads into the ocean or another large body of water is called the mouth.

■ **Figure 9.14** As the path of the stream bends and winds, it creates meanders and eventually oxbow lakes.

Concepts In Motion

Interactive Figure To see an animation of meander formation, visit glencoe.com.

Visualizing Erosion and Deposition in a Meander

Figure 9.15 As the water travels down a meander, the area of maximum velocity changes. As shown in cross-section A, when the meander is straight, the maximum velocity is located near the center. When the meander curves, the maximum velocity shifts to the outside of the curve, as shown in cross-section B. As the meander travels around to cross-section C, the maximum velocity shifts again to the outside of the curve. Erosion occurs around curves in the meander in areas of high velocity. The high velocity of the water carries the sediment downstream and deposits it where the velocity decreases, on the inside of a curve. The area where the erosion occurs is called a cutbank and the area where the deposition occurs is called a point bar.

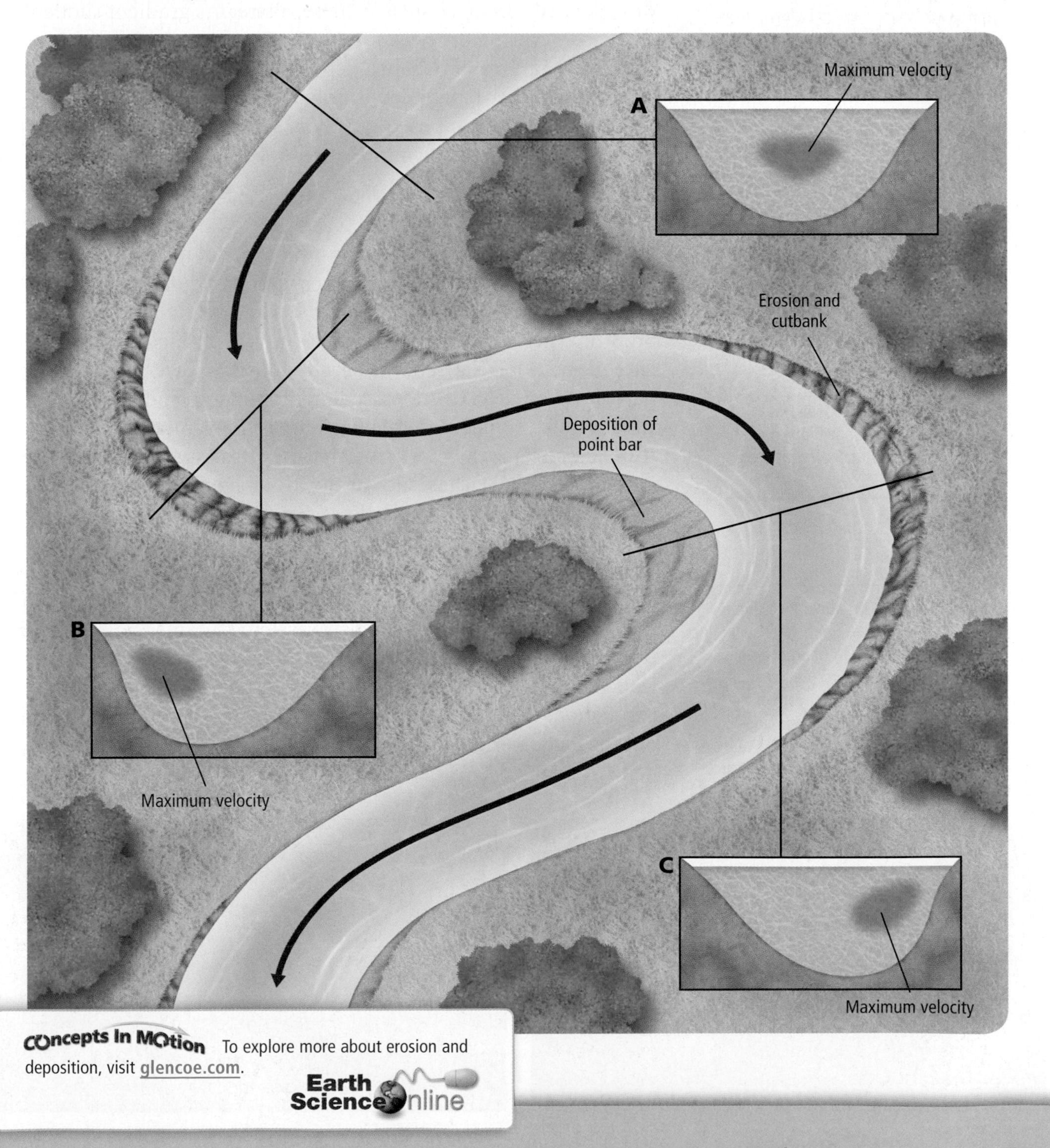

Concepts In Motion To explore more about erosion and deposition, visit glencoe.com.

Earth Science Online

■ **Figure 9.16** An alluvial fan is a fan-shaped depositional feature.

Deposition of Sediment

The velocity of a stream determines how much sediment it can transport. Rapidly flowing streams have the energy to transport sediment as large as gravel. When streams lose velocity, they lose some of the energy needed to transport sediment, and deposition of sediment occurs.

Alluvial fans A stream's velocity lessens and its sediment load drops when its gradient abruptly decreases. In dry regions such as the North American Southwest, mountain streams flow intermittently down steep, rocky slopes and then flatten out onto expansive dry lake beds. In areas such as these, a stream's gradient suddenly decreases, causing the stream to drop its sediment at the base of the mountain in a fan-shaped deposit called an alluvial fan. Alluvial fans are sloping depositional features formed at the bases of slopes and are composed mostly of sand and gravel. An example of an alluvial fan is shown in **Figure 9.16.**

Reading Check **Describe** how an alluvial fan is formed.

Deltas Streams also lose velocity and some of their capacity to carry sediment when they join larger bodies of quiet water. The often triangular deposit that forms where a stream enters a large body of water is called a **delta,** named for the triangle-shaped Greek letter delta (Δ). Delta deposits usually consist of layers of silt and clay particles. As a delta develops, sediments build up and slow the stream water, sometimes even blocking its movement. Smaller distributary streams then form to carry the stream water through the developing delta. Deltas, such as the Mississippi River Delta, are normally areas where the stream flow changes direction frequently.

NATIONAL GEOGRAPHIC To read about the Okavango delta, go to the **National Geographic Expedition** on page 904.

Over the course of thousands of years, the Mississippi River Delta has changed frequently. Today, any small change in the drainage channels can result in catastrophic flooding for local communities. To prevent floods, an extensive system of dams and levees is in place to protect people and economic activities. A consequence of flood control is the decrease in the regular deposition of sediment throughout the delta. In the absence of regular deposition throughout the delta, normal processes of coastal erosion have caused the delta to shrink over time, as shown in **Figure 9.17.**

■ **Figure 9.17** The Mississippi River Delta was formed from the deposition of river sediments. The area in the top left of both images is a marshland used for both recreation and business. Since 1973, waters upstream of the Mississippi River have been dammed, reducing the sediment flow. Over the course of 30 years, the area of the marshland has decreased without the sediment from upstream.

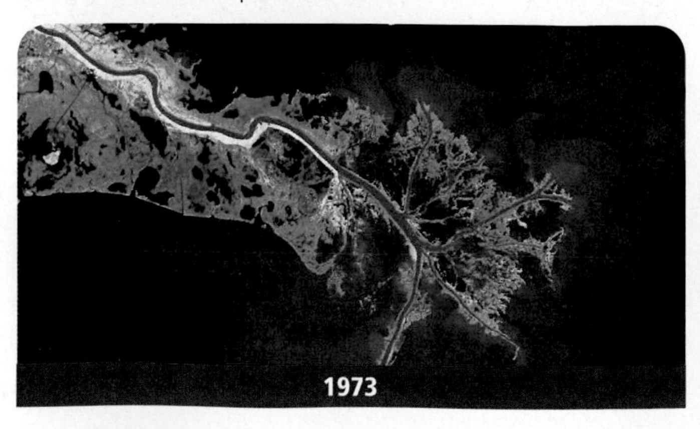

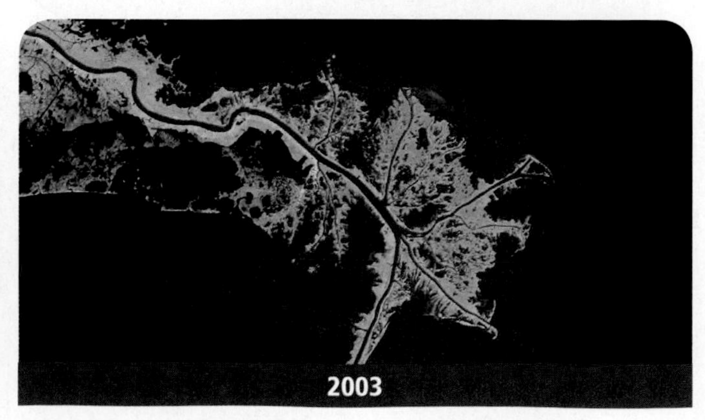

Rejuvenation

During the process of stream formation, downcutting can occur. Downcutting is the wearing away of the streambed and is a major erosional process that influences the stream until it reaches its base level. If the base level drops as a result of geologic processes, the stream undergoes rejuvenation.

Rejuvenation means *to make young again.* During **rejuvenation**, a stream actively resumes the process of downcutting toward its base level. This causes an increase in the stream's velocity and the stream's channel once again cuts downward into the existing meanders. Rejuvenation can cause deep-sided canyons to form. A well-known example of rejuvenation is the Grand Canyon, shown in **Figure 9.18.**

Millions of years ago, the Colorado River was near its base level, like much of the Mississippi River today. Then the land was uplifted compared to the level of the ocean, which caused the base level of the Colorado River to drop. This caused the process of rejuvenation, in which the river began cutting downward into the existing meanders. The result is the 1.6-km-deep canyons, which attract millions of visitors each year from all over the world.

Figure 9.18 Rejuvenation shaped the Grand Canyon when the base level of the Colorado River changed and the river began downcutting into existing meanders.

Section 9.2 Assessment

Section Summary

- Water from precipitation gathers in gullies at a stream's headwaters.
- Stream water flows in channels confined by the stream's banks.
- Alluvial fans and deltas form when stream velocity decreases and sediment is deposited.
- Alluvial fans are fan-shaped and form where water flows down steep slopes onto flat plains.
- Deltas are often triangular and form when streams enter wide, relatively quiet bodies of water.

Understand Main Ideas

1. MAIN Idea **Describe** how a V-shaped valley is formed.
2. **Identify** four changes that a stream undergoes before it reaches the ocean.
3. **Compare** the velocity on the inside of a meander curve with that on the outside of the curve.

Think Scientifically

4. **Analyze** how the type of bedrock over which a stream flows affects the time it takes for the stream to reach its base level.
5. **Infer** how you can tell that rejuvenation has modified the landscape.

MATH in Earth Science

6. Create a line graph that plots the direction of change in a hypothetical stream's rate of flow at the stream's headwaters, at midstream, and at its mouth.

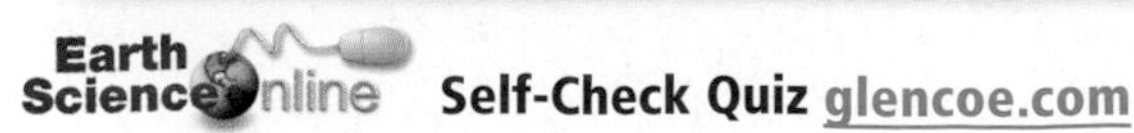

Section 9.3

Objectives

- **Explain** the formation of freshwater lakes and wetlands.
- **Describe** the process of eutrophication.
- **Recognize** the effects of human activity on lake development.

Review Vocabulary

kettle: a depression resulting from the melting of an ice block left behind by a glacier

New Vocabulary

lake
eutrophication
wetland

Lakes and Freshwater Wetlands

MAIN Idea As the amount of water changes and the amount of sediments increases, lakes can be transformed into wetlands and eventually into dry land.

Real-World Reading Link Have you ever felt the bottom of a lake with your feet? It was probably soft and squishy from deposits of fine sediments. Lakes and ponds receive materials that are carried by rivers from upland areas. Over time, accumulation of these sediments changes the characteristics of the lake.

Origins of Lakes

Natural **lakes**, bodies of water surrounded by land, form in different ways in surface depressions and in low areas. As you learned in Section 9.2, oxbow lakes form when streams cut off meanders and leave isolated channels of water. Lakes also form when stream flow becomes blocked by sediment from landslides or other sources. Still other lakes have glacial origins, as you learned in Chapter 8. The basins of these lakes formed as glaciers gouged out the land during the ice ages. Most of the lakes in Europe and North America are in recently glaciated areas. Glacial moraines originally dammed some of these depressions and restricted the outward flow of water. The lakes that formed as a result are known as moraine-dammed lakes. In another process, cirques carved high in the mountains by valley glaciers filled with water to form cirque lakes. Other lakes formed as blocks of ice left on the outwash plain ahead of melting glaciers eventually melted, leaving depressions called kettles. When these depressions filled with water, they formed kettle lakes such as those shown in **Figure 9.19.**

Figure 9.19 Lakes such as these in Minnesota were formed from blocks of ice that melted after glaciers retreated.

Lakes Undergo Change

Water from precipitation, runoff, and underground sources can maintain a lake's water supply. Some lakes contain water only during times of heavy rain or excessive runoff from spring thaws. A depression that receives more water than it loses to evaporation or use by humans will exist as a lake for a long period of time. However, most lakes are temporary water-holding areas; over hundreds of thousands of years, lakes usually fill in with sediment and become part of a new landscape.

Eutrophication Through the process of photosynthesis, plants such as green algae add oxygen to lake water. Animals that live in a lake need oxygen in the water. Throughout their life cycle, the animals add waste products to the water. Oxygen is also consumed during the decay process that occurs after plants and animals living in the body of water die. Scientists use the amount of dissolved oxygen present in a body of water to assess the overall quality of the water. Dissolved oxygen is one quality a body of water must have to support life.

The process by which the surrounding watershed enriches bodies of water with nutrients that stimulate excessive plant growth is called **eutrophication.** Although eutrophication is a natural process, it can be sped up with the addition of nutrients, such as fertilizers, that contain nitrogen and phosphorus. Other major sources of nutrients that concentrate in lakes are animal wastes and phosphate detergents.

When eutrophication occurs, the animal and plant communities in the lake can change rapidly. Algae growing at the surface of the water can suddenly multiply very quickly. The excessive algae growth in a lake or pond appears as a green blanket, as shown in **Figure 9.20.** Other organisms that eat the algae can multiply in numbers as well. In addition, the population of algae on the surface can block sunlight from penetrating to the bottom of the lake, causing sunlight-dependent plants and other organisms below the surface to die. The resulting overpopulation and, later, the decay of a large number of plants and animals depletes the water's oxygen supply. Fish and other sensitive organisms might die as a result of the lack of oxygen in the water. In some cases, the algae can also release toxins into the water that are harmful to the other organisms.

Reading Check **Identify** the effects of eutrophication on the aquatic animals in an affected lake system.

Figure 9.20 Eutrophication is a natural process that can be accelerated with the addition of nitrogen and phosphorus to a body of water. Once the process begins, it can cause rapid changes in the plant and animal communities in the affected body of water.

Freshwater wetlands A **wetland** is any land area that is covered with water for a part of the year. Wetlands include environments commonly known as bogs, marshes, and swamps. They have certain soil types and support specific plant species. Their soil types depend on the degree of water saturation.

CAREERS IN EARTH SCIENCE

Geochemist Technician Some geochemist technicians take core samples from lakes to analyze the pollutants in lake sediments. To learn more about Earth science careers, visit glencoe.com.

Bogs Bogs are not stream-fed but instead receive their water from precipitation. The waterlogged soil tends to be rich in *Sphagnum*, also called peat moss. The breakdown of peat moss produces acids, thereby contributing to the soil's acidity. The waterlogged, acidic soil supports unusual plant species, including insect-eating pitcher plants such as sundew and Venus flytrap.

Reading Check **Identify** how a bog receives water.

Marshes Freshwater marshes frequently form along the mouths of streams and in areas with extensive deltas. The constant supply of water and nutrients allows for the lush growth of marsh grasses. The shallow roots of the grasses anchor deposits of silt and mud on the delta, thereby slowing the water and expanding the marsh area. Grasses, reeds, sedges, and rushes, along with abundant wildlife, are common in marsh areas.

Swamps Swamps are low-lying areas often located near streams. Swamps can develop from marshes that have filled in sufficiently to support the growth of shrubs and trees. As these larger plants grow and begin to shade the marsh plants, the marsh plants die. Swamps that existed about 200 mya developed into present-day coal reserves that are common in Pennsylvania and many other locations in the United States and around the world.

MiniLab

Model Lake Formation

How do surface materials determine where lakes form? Lakes form when depressions or low areas fill with water. Different Earth materials allow lakes to form in different places.

Procedure 

1. Read and complete the lab safety form.
2. Use three **clear plastic shoe boxes.** Half fill each one with Earth materials: **clay, sand,** and **gravel.**
3. Slightly compress the material in each shoe box. Make a shallow depression in each surface.
4. Slowly pour 500 mL of **water** into each of the depressions.

Analysis

1. **Describe** what happened to the 500 mL of water that was added to each shoe box.
2. **Compare** this activity to what happens on Earth's surface when a lake forms.
3. **Infer** in which Earth materials lakes most commonly form.

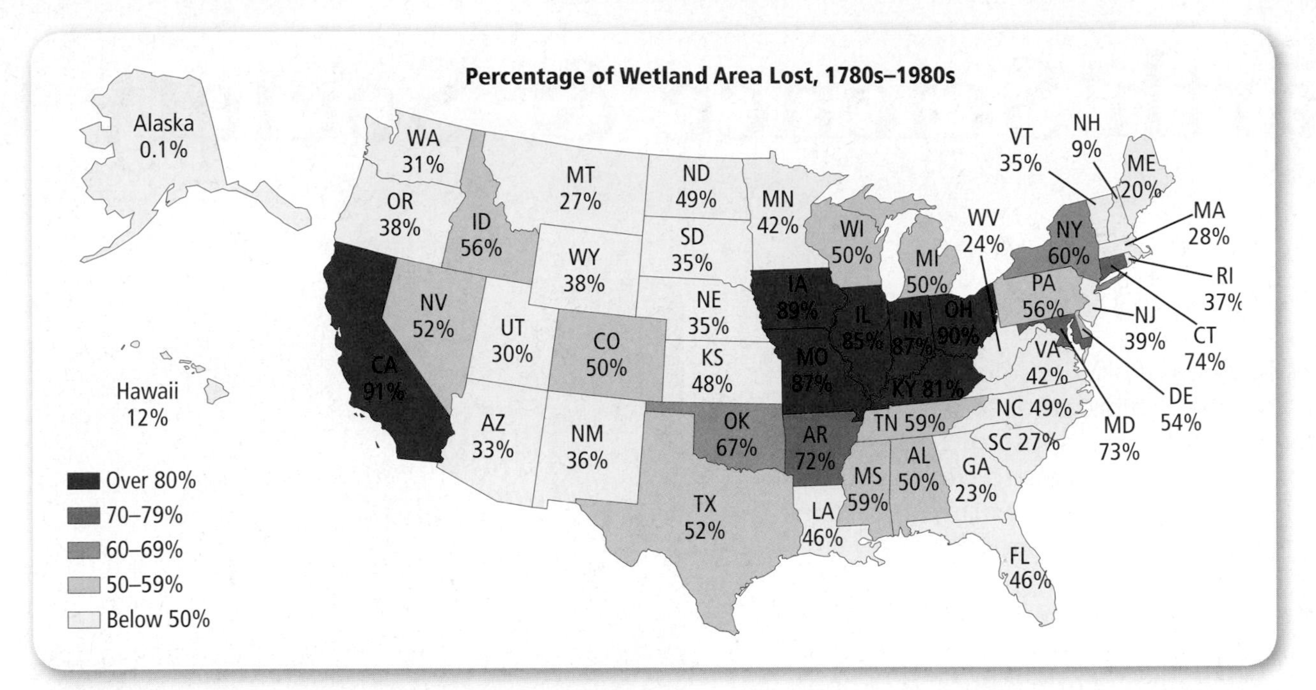

Figure 9.21 The area of wetlands in the United States was drastically reduced until the 1980s. Since then, efforts have been made to preserve wetlands.

Wetlands and water quality Wetlands play a valuable role in improving water quality. They serve as a filtering system that traps pollutants, sediments, and pathogenic bacteria contained in water sources. Wetlands also provide vital habitats for migratory waterbirds and homes for an abundance of other wildlife. In the past, it was common for wetland areas to be filled in to create more land on which to build. Government data reveal that from the late 1700s to the mid-1980s, the continental United States lost 50 percent of its wetlands, as shown in **Figure 9.21.** By 1985, it was estimated that 50 percent of the wetlands in Europe were drained. Now, however, the preservation of wetland areas has become a global concern.

Section 9.3 Assessment

Section Summary

- Lakes form in a variety of ways when depressions on land fill with water.
- Eutrophication is a natural nutrient-enrichment process that can be accelerated when nutrients from fertilizers, detergents, or sewage are added.
- Wetlands are low-lying areas that are periodically saturated with water and support specific plant species.

Understand Main Ideas

1. MAIN Idea **Explain** the transformation process that a lake might undergo as it changes to dry land.
2. **Describe** the conditions necessary for the formation of a natural lake.
3. **Identify** human activities that might affect the process of eutrophication in a lake near you.

Think Critically

4. **Organize** a data table to compare various types of lakes and their origins.
5. **Analyze** a situation in which protection of wetlands might conflict with human plans for land use.

WRITING in **Earth Science**

6. Write an essay explaining the role wetlands play in improving water quality.

Earth Science & Society

The World of Water

Humans have basic physiological needs. These include the need to breathe, to eat, to regulate body temperature, to dispose of bodily wastes, to sleep, and to have access to clean water. Humans need clean water to drink, for cleaning, cooking, and waste disposal.

A global problem Almost every continent has areas that lack safe drinking water. Rural areas of developing countries and overpopulated urban areas often have inadequate supplies of safe drinking water. Even though adequate supplies of this natural resource may exist globally, it is not distributed evenly. In addition, naturally occurring contaminants and pollution from human impact can make a water supply unhealthy.

Safe water The World Health Organization (WHO) defines safe drinking water as water from a source that is less than 1 km away from where it is used; that at least 20 L of water per member of the household per day can be obtained reliably; and that meets the national standards for microbial and contaminant levels.

Health concerns In developing countries, children are at the greatest risk for water-related diseases. Worldwide, more than 5000 children under the age of five die each day from water-related diseases. The most common health concerns from contaminated water are diarrhea and intestinal worms.

Diarrhea is a common condition caused by bacteria often found in unsafe drinking water. Without proper treatment, diarrhea can lead to severe dehydration and death, especially in children. In developed countries, children suffering from diarrhea often receive the necessary treatment. However, in developing countries, diarrhea accounts for the death of nearly 2 million children each year.

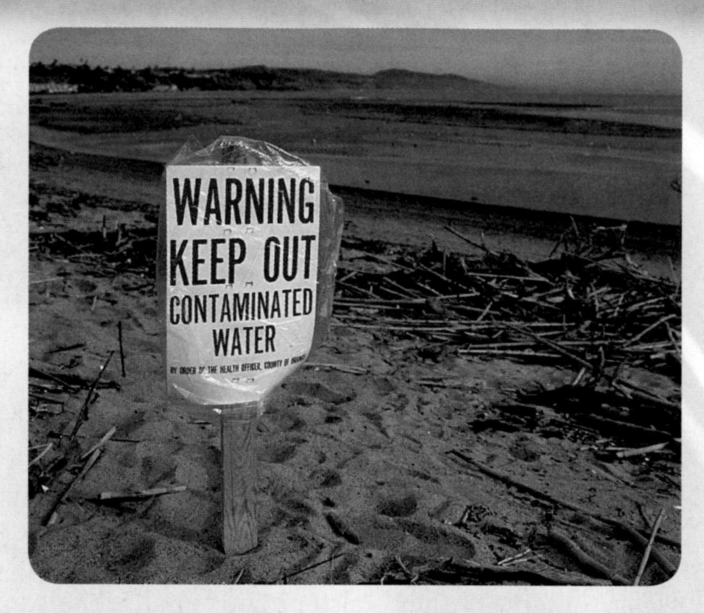

Contaminated water can be a problem in developed countries as well as developing countries. This beach is closed because of unsafe water.

Another danger from contaminated water, especially for children, is intestinal parasites. Parasites that live in the intestines of the host, humans in this case, can cause malnutrition, anemia, and other illnesses.

A global solution The inability to adequately supply this basic human need has been acknowledged by the United Nations as one of the greatest failures of the twentieth century. The United Nations has created an international task force to help fund the creation of sanitation systems and water purifiers. In the future, with effort and global cooperation, every human being might have access to safe drinking water and proper sanitation.

WRITING in Earth Science

Brochure March 22 is World Water Day. Create a brochure explaining the need for such an event and why more people should participate. For more information on World Water Day, visit glencoe.com.

GEOLAB

PREDICT THE VELOCITY OF A STREAM

Background: Water in streams flows from areas of high elevation to areas of low elevation. Stream flow is measured by recording the water's velocity. The velocity varies from one stream to another and also in different areas of the same stream. Many components of the stream affect the velocity, including sediment, slope, and rainfall.

Question *How does slope affect velocity?*

Materials

1-m length of vinyl gutter pipe
ring stand and clamp
water source with hose
protractor with plumb bob
sink or container to catch water
stopwatch
grease pencil
meterstick
paper
three-hole punch

Safety Precautions

Procedure

1. Read and complete the lab safety form.
2. Work in groups of three to four.
3. Use a three-hole punch to make 10 to 15 paper circles to be used as floating markers.
4. Use the illustration as a guide to set up the protractor with the plumb bob.
5. Use the grease pencil to mark two lines across the inside of the gutter pipe at a distance of 40 cm apart.
6. Use the ring stand and clamp to hold the gutter pipe at an angle of 10°. Place the end of the pipe in a sink or basin to collect the discharged flow of water.
7. Attach a long hose to a water faucet in the sink.
8. Keep the hose in the sink until you are ready to use it. Turn on the water and adjust the flow until the water moves quickly enough to provide a steady flow.
9. Bend the hose to block the water flow until the hose is positioned at least 5 cm above the top line marked on the pipe. Allow the water to flow. Allow the water to flow at the same rate for all slope angles.

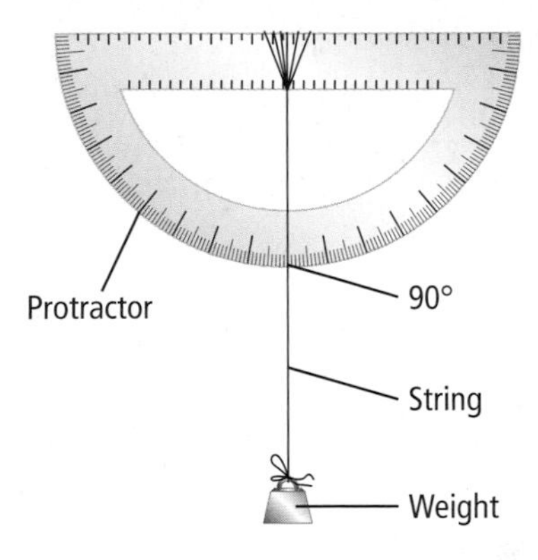

10. Drop a floating marker approximately 4 cm above the top line on the pipe into the flowing water.
11. Measure the time it takes for the floating marker to move from the top line to the bottom line. Record the time in your science journal.
12. Repeat Step 9 two more times.
13. Repeat Steps 9 and 10, but change the slope to 20°, 30°, and then 40°.
14. Make a line graph of the average velocity.

Analyze and Conclude

1. **Interpret Data** What is the relationship between the velocity and the angle of the slope?
2. **Apply** Describe one reason that a stream's slope might change.
3. **Infer** Where would you expect to find streams with the highest velocity?
4. **Predict** Using your graph, predict the velocity for a 35° slope.

INQUIRY EXTENSION

Design Your Own As discussed in the chapter, the texture of the streambed can affect the rate of stream flow. Design an experiment to test this variable.

CHAPTER 9 Study Guide

Download quizzes, key terms, and flash cards from glencoe.com.

BIG Idea Surface water moves materials produced by weathering and shapes the surface of Earth.

Vocabulary	Key Concepts
Section 9.1 Surface Water Movement	
• bed load (p. 228) • discharge (p. 229) • divide (p. 227) • flood (p. 230) • floodplain (p. 230) • runoff (p. 225) • suspension (p. 228) • watershed (p. 227)	**MAIN Idea** Running water is an agent of erosion, carrying sediments in streams and rivers and depositing them downstream. • Infiltration of water into the ground depends on the number of open pores. • All the land area that drains into a stream system is the system's watershed. • Elevated land areas called divides separate one watershed from another. • A stream's load is the material the stream carries. • Flooding occurs in small, localized areas as upstream floods or in large downstream floods.
Section 9.2 Stream Development	
• base level (p. 233) • delta (p. 236) • meander (p. 234) • rejuvenation (p. 237) • stream bank (p. 232) • stream channel (p. 232)	**MAIN Idea** Streams erode paths through sediment and rock, forming V-shaped stream valleys. • Water from precipitation gathers in gullies at a stream's headwaters. • Stream water flows in channels confined by the stream's banks. • Alluvial fans and deltas form when stream velocity decreases and sediment is deposited. • Alluvial fans are fan-shaped and form where water flows down steep slopes onto flat plains. • Deltas are triangular and form when streams enter wide, relatively quiet bodies of water.
Section 9.3 Lakes and Freshwater Wetlands	
• eutrophication (p. 239) • lake (p. 238) • wetland (p. 240)	**MAIN Idea** As the amount of water changes and the amount of sediments increases, lakes can be transformed into wetlands and eventually into dry land. • Lakes form in a variety of ways when depressions on land fill with water. • Eutrophication is a natural nutrient-enrichment process that can be accelerated when nutrients from fertilizers, detergents, or sewage are added. • Wetlands are low-lying areas that are periodically saturated with water and support specific plant species.

Assessment

Vocabulary Review

Choose the vocabulary term from the Study Guide that best describes each phrase.

1. influenced by vegetation, precipitation, soil composition, and slope
2. the land area whose water drains into a stream
3. sediments that are transported by streams but are too large to be held up in suspension
4. the measure of the volume of stream water that flows over a particular location within a given period of time
5. the triangular deposit of sediment that forms where a stream enters a large body of water such as a lake or ocean
6. the narrow channel carved over time by the water of a stream into sediment or rock layers
7. the process by which a stream resumes downcutting toward its base level

The sentences below include terms that have been used incorrectly. Make the sentences true by replacing each italicized word with a vocabulary term from the Study Guide.

8. A depression that receives more water than is removed will exist as a *stream* for a long period of time.
9. *Enrichment* is the process by which lakes become rich in nutrients, resulting in a change in the kinds of organisms in the lake.
10. Marshes, swamps, and bogs are all types of *inundated* areas.
11. *Solution* is the method of transport for all particles small enough to be held up by the turbulence of a stream's moving water.
12. The area that extends from a stream's bank and is covered by excess water during times of flooding is known as a *wetland.*

Understand Key Concepts

13. Which scenario most likely formed most large lakes in North America and Europe?
    A. Lakes formed from stream meanders that were cut off.
    B. Landslides blocked the flow of streams, creating lakes.
    C. Glacial activity on continental masses scoured the landscape and left depressions behind.
    D. Reservoirs were created for the purpose of storing water for communities.
14. What is the driving force of a stream?
    A. velocity
    B. gravity
    C. discharge
    D. downcutting

Use the figure below to answer Question 15.

15. Which part of a river is shown?
    A. the headwater
    B. the main channel
    C. the streambed
    D. the mouth
16. When does downcutting of the streambed stop during stream formation?
    A. when the water reaches a porous layer
    B. when the water reaches a base level
    C. when the water reaches a new level
    D. when the water reaches a impermeable layer

Use the figure below to answer Question 17.

17. How did this terrain feature form?
A. It formed by a meander of a stream that has been cut off.
B. It formed from a glacier that scoured out the land.
C. It is the result of a flood.
D. It is the result of eutrophication.

18. What type of streams form V-shaped valleys?
A. streams that carry a lot of sediment
B. streams that are far from ultimate base level
C. streams that meander
D. streams that carry no bed load

19. Which is not a way in which lakes are typically formed?
A. from cutoff meanders of streams
B. from asteroid craters
C. from landslides that block rivers
D. from glacial carving

20. Which substance plays a major role in the eutrophication process?
A. iron
B. phosphorus
C. ozone
D. salt

21. Which factors determine the discharge of a stream?
A. width, length, depth
B. width, length, velocity
C. width, depth, velocity
D. length, depth, runoff

22. Which process would result in rejuvenation of a stream?
A. lifting of existing base level
B. lowering of existing base level
C. lowering of the land
D. deposition on the stream banks

23. Which characteristic of the soil in a depression is most important in allowing the formation of a lake?
A. high content of organic material
B. high content of mostly of inorganic material
C. gravelly soil
D. relatively nonporous, clay-rich layer of soil

24. If a stream is carrying sand, large boulders, clay, and small pebbles, which type of particle is deposited last as the stream begins to slow down?
A. clay
B. sand
C. large boulders
D. small pebbles

Constructed Response

25. Compare and contrast the formation of a river delta to an alluvial fan.

Use the following aerial view of a stream to answer Questions 26 to 28.

26. Identify the location at which the stream has the greatest velocity.

27. Identify the location at which deposition most actively occurs.

28. Identify the location at which erosion most actively occurs.

29. Calculate the discharge of a stream that has a velocity of 300 m/s and is 25 m wide and 3 m deep.

Think Critically

30. Analyze how water in a stream is related to water in the atmosphere.

Use the figure below to answer Question 31.

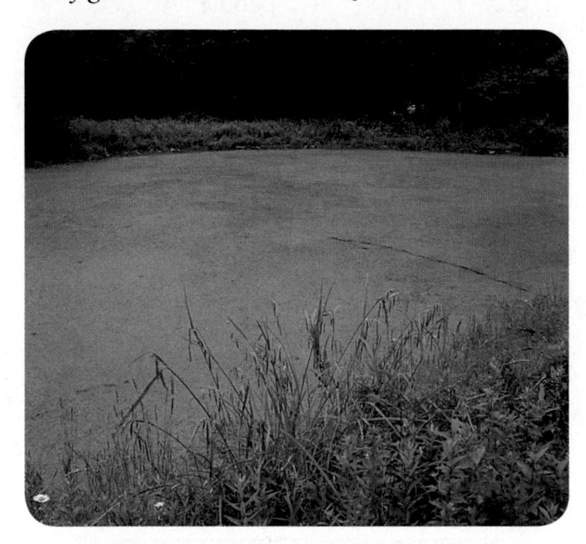

31. Interpret the features in this pond in terms of the processes that might have formed them.

32. Hypothesize how an increase in a river's turbulence could decrease its bed load.

33. Analyze how specific soil characteristics determine how much water from precipitation infiltrates or runs off.

34. Infer why upstream tributaries often have relatively small yet turbulent flow, whereas downstream portions usually have larger but smooth-flowing discharges.

35. Discuss which areas of a stream are most likely to contain fertile soil.

Concept Mapping

36. Construct a concept map of the parts of the water cycle and illustrate the relationships among them.

Challenge Question

37. Recommend measures that a town whose wastewater runs into a large lake should take to ensure the long-term quality of its lake water.

Additional Assessment

38. WRITING in **Earth Science** Write a newspaper article to explain recent flood destruction in your town. In the article, explain how erosion and transportation of sediment might be different during a flood.

DBQ Document–Based Questions

Data obtained from: Varis, O. 2005. Are floods growing? *Ambio* 34 (August): 478–480.

Dramatic changes in river flow have taken place with water flow to the sea in China's Huang He River, spanning a period of almost 30 years.

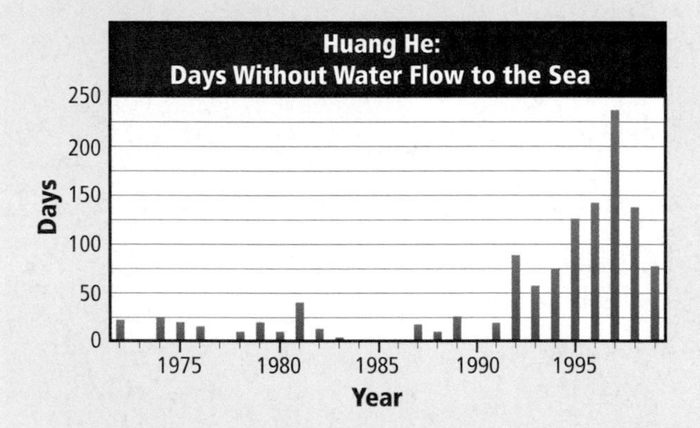

39. Based on the graph, beginning in approximately what year were the first signs evident of a pattern of progressively decreasing river flow?

40. In the worst year of records presented, approximately how many days was there a water flow from the Huang He River to the sea?

41. What kinds of conditions might cause such dramatic reductions in water flow?

Cumulative Review

42. Describe a topographic map and its uses. **(Chapter 2)**

43. Which of the following is NOT a mineral: quartz, ice, coal, or native gold? **(Chapter 4)**

44. Explain how water affects the process of mass movement. **(Chapter 8)**

Standardized Test Practice

Multiple Choice

1. Which condition would create the most runoff?
A. land covered with vegetation
B. plants in densely packed soil
C. light precipitation
D. soil with a high percentage of sand

Use the photo below to answer Questions 2 and 3.

2. What most likely caused the odd shape of the boulder?
A. a rock slide
B. glacier erosion
C. wind deflation
D. wind abrasion

3. What clue would help scientists determine the method of erosion for this boulder?
A. The boulder has smooth surfaces with smooth edges.
B. The boulder has a coarse surface.
C. The boulder is polished on the windward side.
D. The boulder has a rough surface and rough edges.

4. If you were creating a model of rock formation, you would represent the different layers of rock. In this model, which type of rock would represent particles that have been compressed and hardened?
A. sedimentary
B. volcanic
C. igneous
D. intrusive

5. As the velocity of a stream decreases, which transported particle size would settle to the stream's bottom first?
A. clay
B. silt
C. pebble
D. sand

6. Which condition helps determine the quality of lake water?
A. the amount of nitrogen
B. the amount of dissolved calcium carbonate
C. the amount of potassium
D. the amount of dissolved oxygen

7. Which state of matter are all minerals?
A. solids
B. liquids
C. gases
D. plasma

Use the table below to answer Questions 8–10.

Texture Data for a Soil Profile

Horizon	Percent		
	Sand	Silt	Clay
A	16.2	54.4	29.4
B	10.5	50.2	39.3
C	31.4	48.4	20.2
R (bedrock)	31.7	50.1	18.2

8. What inferences can scientists make from this soil profile?
A. The soil is newly layered.
B. The soil is well developed and mature.
C. The soil is poorly developed.
D. The soil profile came from the West.

9. Which horizon most likely contains the hard material known as hardpan?
A. A-horizon
B. B-horizon
C. C-horizon
D. R-horizon

10. O-horizon was not listed on this table. What could be the reason?
A. It was too deep to be studied.
B. It only contained sand and not silt or clay.
C. It was insignificant to the study containing only humus and leaf litter.
D. It contained only clay and not sand or silt.

Short Answer

Use the table below to answer Questions 11 and 12.

Mineral Characteristics				
Mineral	**Color**	**Streak**	**Hardness**	**Specific Gravity**
Sulfur	yellow	yellow	2	2.1
Schorl	black	white	7	3.2
Topaz	blue	colorless	9	3.6
Zinc	white	light gray	2	6.9

11. How is this table organized?

12. From this table, what can you infer about hardness and specific gravity?

13. According to the water cycle, what happens after water molecules evaporate and condense into cloud droplets?

14. How does limestone form?

15. What causes dune migration?

16. What are some possible strategies a road construction crew could use to protect highways located at the bottom of steep slopes susceptible to landslides?

Reading for Comprehension

Inland Flooding

According to the National Weather Service, inland flooding is one of the deadliest effects of hurricanes. Below is a list of steps to help reduce your risk of being caught in inland flooding when you hear about a potential hurricane and live in a potential flood zone.

- If advised to evacuate, do so immediately. Move to a safe area before access is cut off by flood water.
- Keep abreast of road conditions through the news media.
- Do not attempt to cross flowing water. As little as six inches of water might cause you to lose control of your vehicle—two feet of water will carry most cars away.
- Develop a flood emergency action plan with your community leaders.

Article obtained from: Hurricane flooding: a deadly danger. *NOAA's National Weather Service.* March 2001. (Online resource accessed October 2006)

17. What is important to know about water?
A. It is safe to drive on a road with flowing water.
B. Flowing water is safer than standing water.
C. Six inches of water will do no harm.
D. Two feet of water can carry most cars away.

18. According to the text, which is not a step to take to ensure your safety from inland flooding?
A. Do not attempt to cross flowing water.
B. Move to the highest level of your house.
C. Evacuate when advised to do so.
D. Develop a flood emergency action plan.

19. What is the goal of the National Weather Service in distributing this list?
A. to discuss hurricanes
B. to offer advice for people to read and use if they want
C. to inform people of the hazards of inland flooding and offer steps to protect themselves
D. to scare people

20. Suppose you live in a flood zone that could possibly be affected by inland flooding. Develop a strategy that you would follow to stay safe.

NEED EXTRA HELP?																
If You Missed Question . . .	1	2	3	4	5	6	7	8	9	10	11	12	13	14	15	16
Review Section . . .	9.1	2.2	2.2	4.1	9.2	9.3	4.1	7.3	7.3	7.3	4.1	4.1	9.1	8.2	8.2	8.1

CHAPTER 10

Groundwater

BIG Idea **Precipitation and infiltration contribute to groundwater, which is stored in underground reservoirs until it surfaces as a spring or is drawn from a well.**

10.1 Movement and Storage of Groundwater
MAIN Idea Groundwater reservoirs provide water to streams and wetlands wherever the water table intersects the surface of the ground.

10.2 Groundwater Weathering and Deposition
MAIN Idea Chemical weathering of limestone by water causes the characteristic topography of karst areas.

10.3 Groundwater Supply
MAIN Idea Water is not always available in the quantities and in the locations where it is needed and might be compromised by pollution.

GeoFacts

- The Strokkur geyser in Iceland erupts every 5 to 10 minutes.
- The eruptions of Strokkur geyser reach heights of more than 30 m.
- There are around 1000 geysers in the world.

LAUNCH Lab

How is water stored underground?

Beneath your feet, there are vast amounts of water. This water fills in the pore spaces and fractures in rock and unconsolidated sediment. In this activity, you will model groundwater storage.

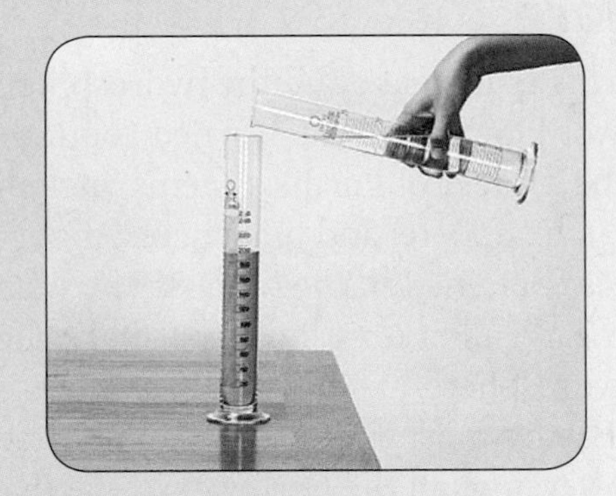

Procedure

1. Read and complete the lab safety form.
2. Fill a **250-mL graduated cylinder** with **fine, dry sand.**
3. Fill **another 250-mL graduated cylinder** with **water.**
4. Pour water from the second cylinder into the sand-filled cylinder until the water level is flush with the surface of the sand. Measure and record the volume of saturated sand in the cylinder.
5. Measure and record how much water is left in the second cylinder.
6. Repeat the experiment twice using **coarse sand** and **clay.**

Analysis

1. **Describe** how much water is present in the saturated fine sand, coarse sand, and clay.
2. **Calculate** the ratio of water volume to the volume of fine sand, coarse sand, and clay, and express the value as a percentage.
3. **Infer** how many liters of water could be stored in a cubic meter of each sediment.

FOLDABLES™ Study Organizer

Threats to the Water Supply Make this Foldable to summarize the major problems that threaten groundwater supplies.

STEP 1 Fold a sheet of paper in half lengthwise.

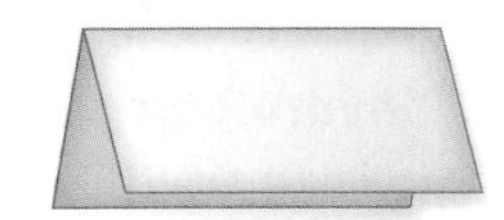

STEP 2 Fold the sheet in half and then into thirds.

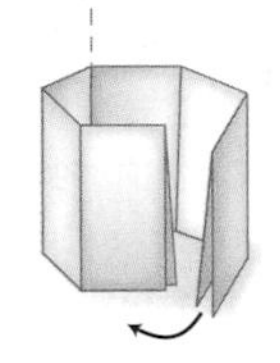

STEP 3 Unfold and cut along the fold lines of the top flap to make six tabs.

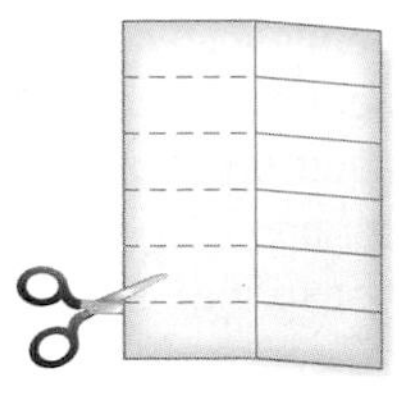

STEP 4 Label the tabs as you read.

FOLDABLES Use this Foldable with Section 10.3. As you read this section, summarize the problems that can threaten groundwater.

Visit glencoe.com to

- **study entire chapters online;**
- **explore Concepts In Motion animations:**
  - **Interactive Time Lines**
  - **Interactive Figures**
  - **Interactive Tables**
- **access Web Links for more information, projects, and activities;**
- **review content with the Interactive Tutor and take Self-Check Quizzes.**

Section 10.1

Objectives

- **Describe** how groundwater storage and underground movement relate to the water cycle.
- **Illustrate** an aquifer and an aquiclude.
- **Relate** the components of aquifers with the presence of springs.

Review Vocabulary

hydrologic cycle: a never-ending natural circulation of water through Earth's systems

New Vocabulary

infiltration
zone of saturation
water table
zone of aeration
permeability
aquifer
aquiclude
spring
hot spring
geyser

Movement and Storage of Groundwater

MAIN Idea **Groundwater reservoirs provide water to streams and wetlands wherever the water table intersects the surface of the ground.**

Real-World Reading Link Have you ever noticed that a stream flows even when it has not rained in a long time? Rainfall contributes to the flow in a stream, but much of the water comes from beneath the ground.

The Hydrosphere

The water on and in Earth's crust makes up the hydrosphere, named after *hydros,* the Greek word for *water*. You learned about the hydrosphere in Chapter 1 in the context of Earth's systems, including the geosphere, hydrosphere, atmosphere, and biosphere. About 97 percent of the hydrosphere is contained in the oceans. The water contained by landmasses—nearly all of it freshwater—makes up only about 3 percent of the hydrosphere.

Freshwater is one of Earth's most abundant and important renewable resources. However, of all the freshwater, between 70 and 80 percent is held in polar ice caps and glaciers. All the rivers, streams, and lakes on Earth represent only a small fraction of Earth's liquid freshwater, as shown in **Table 10.1.** Recall from Chapter 9 that water in the hydrosphere moves through the water cycle.

Table 10.1 World's Water Supply

Concepts In Motion
Interactive Table To explore more about Earth's water supply, visit glencoe.com.

Location	Percentage of Total Water	Water Volume (km^3)	Estimated Average Residence Time of Water
Oceans	97.2	1,230,000,000	thousands of years
Ice caps and glaciers	2.15	28,600,000	tens of thousands of years and longer
Groundwater	0.31	4,000,000	hundreds to many thousands of years
Lakes	0.009	123,000	tens of years
Atmosphere	0.001	12,700	nine days
Rivers and streams	0.0001	1200	two weeks

Groundwater and Precipitation

The ultimate source of all water on land is the oceans. Evaporation of seawater cycles water into the atmosphere in the form of invisible water vapor and visible clouds. Winds and weather systems move this atmospheric moisture all over Earth, with much of it concentrated over the continents. Precipitation brings atmospheric moisture back to Earth's surface. Some of this precipitation falls directly into the oceans and some falls on land.

Infiltration is the process by which precipitation that has fallen on land trickles into the ground and becomes groundwater. Only a small portion of precipitation becomes runoff and is returned directly to the oceans through streams and rivers. Groundwater slowly moves through the ground, eventually returns to the surface through springs and seepage into wetlands and streams, and then flows back to the oceans.

 **Reading Check** **Identify** the ultimate source of all water on land.

Groundwater Storage

Puddles of water that are left after it rains quickly disappear, partly by infiltrating the ground. On sandy soils, rain soaks into the ground almost immediately. Where does that water go? The water seeps into small openings within the ground. Although Earth's crust appears solid, it is composed of soil, sediment, and rock that contain countless small openings, called pore spaces.

Pore spaces make up large portions of some of these materials. The amount of pore space in a material is its porosity. The greater the porosity, the easier water can flow through the material. Subsurface materials have porosities ranging from 2 percent to more than 50 percent. For example, the porosity of well-sorted sand is 30 percent; however, in poorly sorted sediment, smaller particles occupy some of the pore spaces and reduce the overall porosity of the sediment, as shown in **Figure 10.1**. Similarly, the cement that binds the grains of sedimentary rocks together reduces the rocks' porosity. Because of the enormous volume of sediment and rock beneath Earth's surface, enormous quantities of groundwater are stored in the pore spaces.

Figure 10.1 Porosity depends on the size and variety of particles in a material.
Compare *the porosities shown in each sample.*

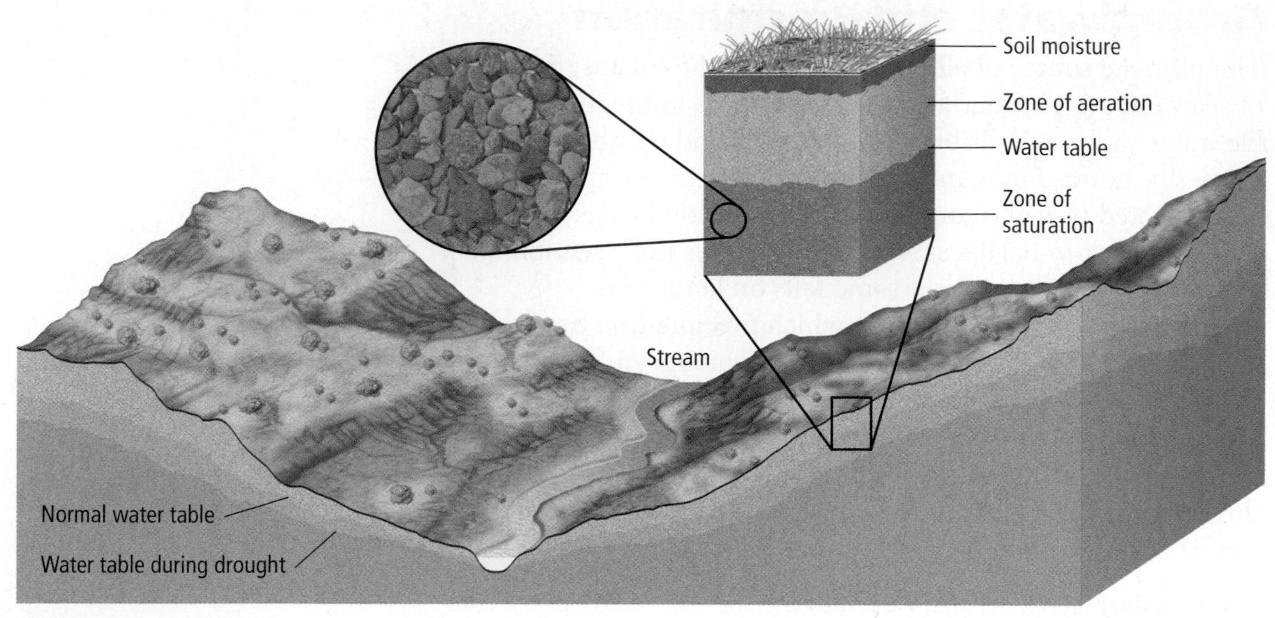

■ **Figure 10.2** The zone of saturation is where groundwater completely fills all the pores of a material below Earth's surface. **Describe** *what is above the zone of saturation.*

The Zone of Saturation

The region below Earth's surface in which groundwater completely fills all the pores of a material is called the **zone of saturation.** The upper boundary of the zone of saturation is the **water table,** shown in **Figure 10.2.** Strictly speaking, only the water in the zone of saturation is called groundwater. In the **zone of aeration,** which is above the water table, materials are moist, but because they are not saturated with water, air occupies much of the pores.

Water movement Water in the zone of saturation and zone of aeration can be classified as either gravitational water or capillary water. Gravitational water is water that trickles downward as a result of gravity. Capillary water is water that is drawn upward through capillary action above the water table and is held in the pore spaces of rocks and sediment because of surface tension. Capillary action can be seen when the tip of a paper towel is dipped into water and the water seems to climb up through the fibers of the paper towel.

The water table The depth of the water table often varies depending on local conditions. For example, in stream valleys, groundwater is relatively close to Earth's surface, and thus the water table can be only a few meters deep. In swampy areas, the water table is at Earth's surface, whereas on hilltops or in arid regions, the water table can be tens to hundreds of meters or more beneath the surface. As shown in **Figure 10.2,** the topography of the water table generally follows the topography of the land above it. For example, the slope of the water table corresponds to the shape of valleys and hills on the surface above.

Because of its dependence on precipitation, the water table fluctuates with seasonal and other weather conditions. It rises during wet seasons, usually in spring, and drops during dry seasons, often in late summer.

Groundwater Movement

Groundwater flows downhill in the direction of the slope of the water table. Usually, this downhill movement is slow because the water has to flow through numerous tiny pores in the subsurface material. The tendency of a material to let water pass through it is its **permeability.** Materials with large, connected pores, such as sand and gravel, have high permeability and permit relatively high flow velocities up to hundreds of meters per hour. Other permeable subsurface materials include highly fractured bedrock, sandstone, and limestone.

Permeability Groundwater flows through permeable sediment and rock, called **aquifers,** such as the one shown in **Figure 10.3.** In aquifers, the pore spaces are large and connected. Fine-grained materials have low permeabilities because their pores are small. These materials are said to be impermeable. Groundwater flows so slowly through impermeable materials that the flow is often measured in millimeters per day. Some examples of impermeable materials include silt, clay, and shale. Clay is so impermeable that a clay-lined depression will hold water. For this reason, clay is often used to line artificial ponds and landfills. Impermeable layers, called **aquicludes,** are barriers to groundwater flow.

Flow velocity The flow velocity of groundwater depends on the slope of the water table and the permeability of the material through which the groundwater is moving. The force of gravity pulling the water downward is greater when the slope of the water table surface is steeper. Water also flows faster through a large opening than through a small opening. The flow velocity of groundwater is proportional to both the slope of the water table and the permeability of the material through which the water flows.

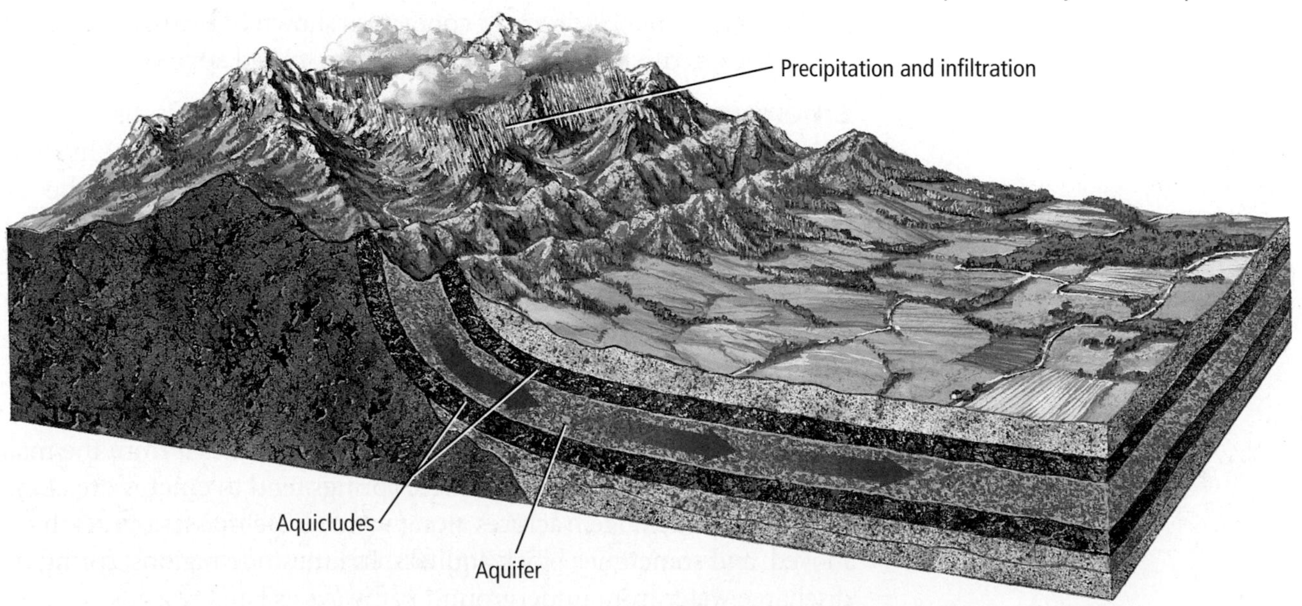

Figure 10.3 An aquifer is a layer of permeable subsurface material that is saturated with water. This aquifer is located between two impermeable layers called aquicludes.

■ **Figure 10.4** Springs occur when the groundwater emerges at points where the water table intersects Earth's surface.

Springs

Groundwater moves slowly but continuously through aquifers and eventually returns to Earth's surface. In most cases, groundwater emerges wherever the water table intersects Earth's surface. Such intersections commonly occur in areas that have sloping surface topography. The exact places where groundwater emerges depend on the arrangement of aquifers and aquicludes in an area.

 **Reading Check** **Explain** how the slope of the land can affect where groundwater emerges.

As you learned on the previous page, aquifers are permeable underground layers through which groundwater flows easily, and aquicludes are impermeable layers. Aquifers are commonly composed of layers of sand and gravel, sandstone, and limestone. In contrast, aquicludes, such as layers of clay or shale, block groundwater movement. As a result, groundwater tends to discharge at Earth's surface where an aquifer and an aquiclude are in contact, as shown in **Figure 10.4.** These natural discharges of groundwater are called **springs.**

Emergence of springs The volume of water that is discharged by a spring might be a mere trickle or it might form a stream. In some regions called karst regions, an entire river might emerge from the ground. Such a superspring is called a karst spring. Karst springs occur in limestone regions where springs discharge water from underground pathways. In regions of nearly horizontal sedimentary rocks, springs often emerge on the sides of valleys at about the same elevation, at the bases of aquifers, as shown in **Figure 10.5.** Springs might also emerge at the edges of perched water tables. In a perched water table, a zone of saturation that overlies an aquiclude separates it from the main water table below. Other areas where springs tend to emerge are along faults, which are huge fractures along which large masses of rock have moved, and sometimes block aquifers. In limestone regions, springs discharge water from underground pathways as karst springs.

Visualizing Springs

Figure 10.5 A spring is the result of groundwater that emerges at Earth's surface. Springs can be caused by a variety of situations.
Compare and contrast ***the origin of the four types of springs.***

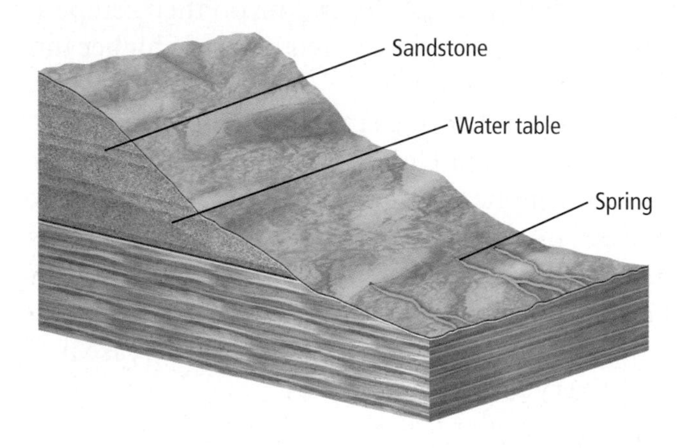

A spring forms where a permeable layer and impermeable layer come together.

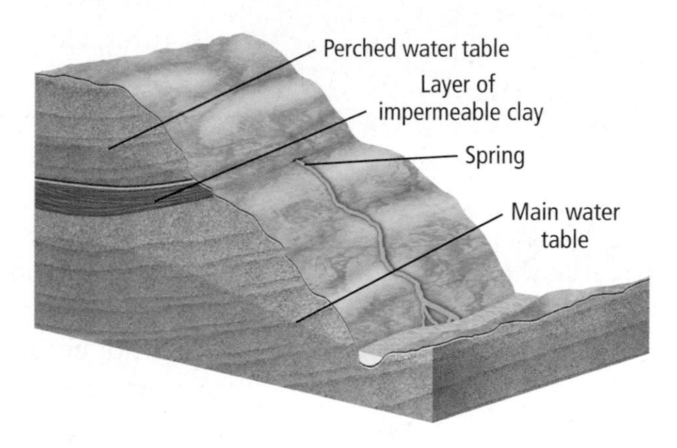

A layer of impermeable rock or clay can create a perched water table. Springs can result where groundwater emerges from a perched water table.

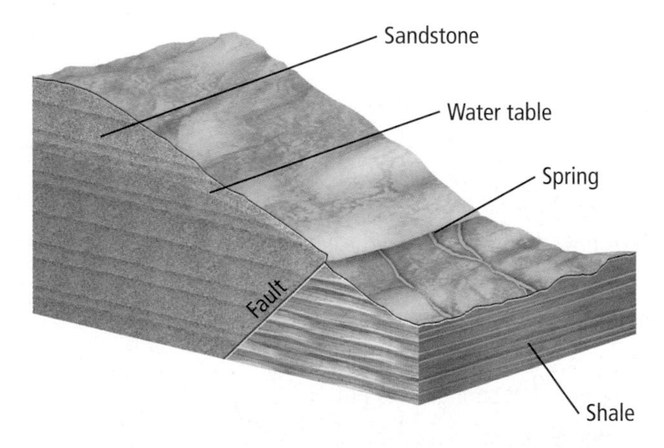

Some springs form where a fault has brought together two different types of bedrock, such as a porous rock and a non-porous rock.

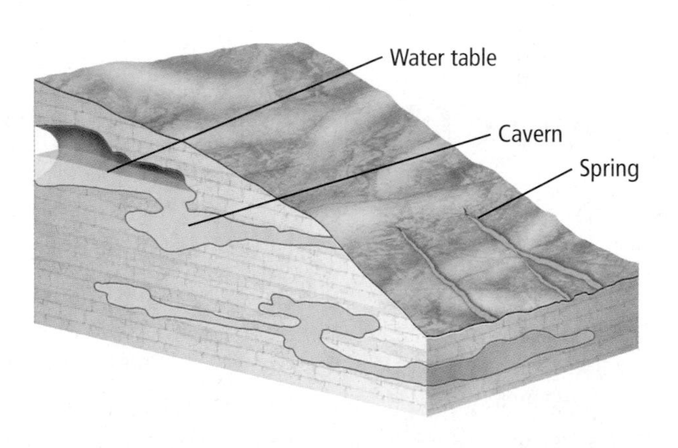

Karst springs form where groundwater weathers through limestone bedrock, and water in the underground caverns emerges at Earth's surface.

Concepts In Motion To explore more about springs, visit glencoe.com.

Earth Science Online

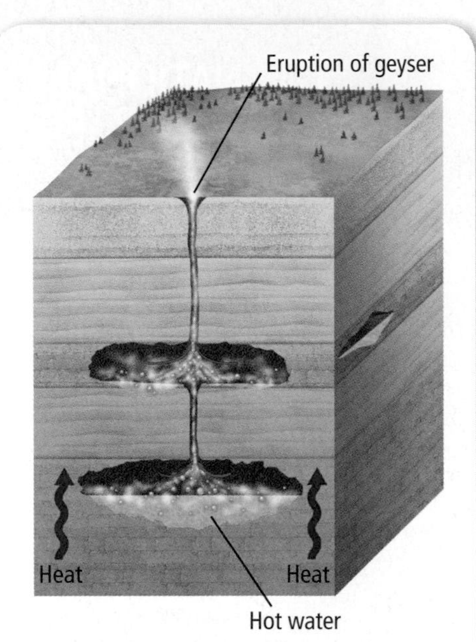

■ **Figure 10.6** A geyser is a type of hot spring from which very hot water and vapor erupt at the surface.
Identify *the origin of a geyser.*

Temperature of springs People usually think of spring water as being cool and refreshing. But the temperature of groundwater that is discharged through a spring is generally the average annual temperature of the region in which it is located. Thus, springs in New England have year-round temperatures of about 10°C, while further south, springs in the Gulf states have temperatures of about 20°C.

Compared to air temperatures, groundwater is generally colder in the summer and warmer in the winter. However, in some regions around the world, springs discharge water that is much warmer than the average annual temperature. These springs are called warm springs or **hot springs,** depending on their temperatures. Hot springs are springs that have a temperature higher than that of the human body, which is 37°C.

There are thousands of hot springs in the United States. Most of them are located in the western United States in areas where the subsurface is still hot from nearby igneous activity. A number of hot springs also occur in some eastern states. These hot springs emerge from aquifers that descend to tremendous depths in Earth's crust and through which deep, hot water rises. The water is hot because temperatures in Earth's upper crust increase by an average of 25°C for every km of depth.

Among the most spectacular features produced by Earth's underground thermal energy in volcanic regions are geysers, shown in **Figure 10.6**. **Geysers** are explosive hot springs. In a geyser, water is heated past its boiling point, causing it to vaporize. The resulting water vapor builds up tremendous pressure. This pressure is what fuels the eruptions. One of the world's most famous geysers, Old Faithful, is located in Yellowstone National Park, Wyoming.

Section 10.1 Assessment

Section Summary

- Some precipitation infiltrates the ground to become groundwater.
- Groundwater is stored below the water table in pore spaces of rocks and sediment.
- Groundwater moves through permeable layers called aquifers and is trapped by impermeable layers called aquicludes.
- Groundwater emerges from the ground where the water table intersects Earth's surface.

Understand Main Ideas

1. MAIN Idea **Explain** how the movement of groundwater is related to the water cycle.
2. **Illustrate** how the relative positions of an aquifer and aquiclude can result in the presence of a spring.
3. **Describe** how the water in hot springs gets hot.
4. **Analyze** the factors that determine flow velocity.

Think Critically

5. **Differentiate** between porosity and permeability in subsurface materials.
6. **Infer** why it is beneficial for a community to have an aquiclude located beneath the aquifer from which it draw its water supply.

WRITING in Earth Science

7. Develop a set of guidelines in which you describe where you would be most likely to find groundwater.

Earth Science Online Self-Check Quiz glencoe.com

Section 10.2

Objectives

- **Explain** how groundwater dissolves and deposits rocks and minerals.
- **Illustrate** how caves form.
- **Describe** how the features of karst topography shape the landscape.

Review Vocabulary

hydrolysis: chemical reaction of water with other substances

New Vocabulary

cave
sinkhole
karst topography
stalactite
stalagmite

Groundwater Weathering and Deposition

MAIN Idea Chemical weathering of limestone by water causes the characteristic topography of karst areas.

Real-World Reading Link You might have seen an old gravestone, statue, or sculpture that has been weathered by acidic water. Similar processes form limestone caves underground.

Carbonic Acid

Acids are aqueous solutions that contain hydrogen ions. Most groundwater is slightly acidic due to carbonic acid. Carbonic acid forms when carbon dioxide gas dissolves in water and combines with water molecules. This happens when precipitation falls through the atmosphere and interacts with carbon dioxide gas or when groundwater infiltrates the products of decaying organic matter in soil. As a result of these processes, groundwater is usually slightly acidic and attacks carbonate rocks, especially limestone. Limestone mostly consists of calcite, also called calcium carbonate, which reacts with any kind of acid. The results of this reaction over time are shown in **Figure 10.7.** This process occurs above ground and below ground.

■ **Figure 10.7** Carbonic acid has dissolved large portions of this limestone. This resulting formation is the Stone Forest in China.

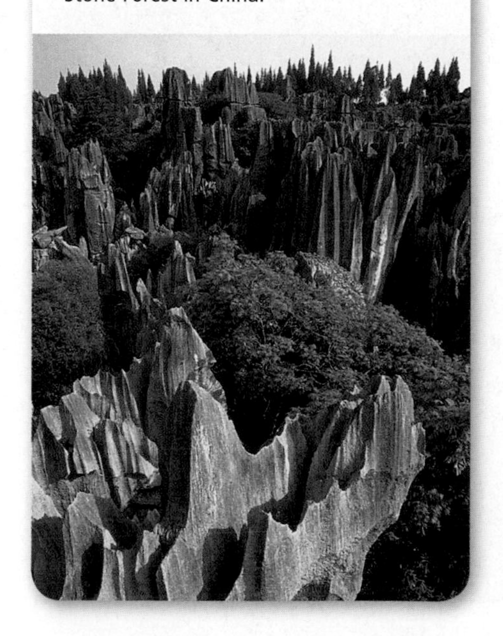

Dissolution by Groundwater

The process by which carbonic acid forms and dissolves calcite, can be described by three simple chemical reactions.

In the first reaction, carbon dioxide (CO_2) and water (H_2O) combine to form carbonic acid (H_2CO_3), as represented by the following equation.

$$CO_2 + H_2O \rightarrow H_2CO_3$$

In the second reaction, carbonic acid splits into hydrogen ions (H^+) and bicarbonate ions (HCO_3^-). This process is represented by the following equation.

$$H_2CO_3 \rightarrow H^+ + HCO_3^-$$

In the third reaction, the hydrogen ions (H^+) react with calcite ($CaCO_3$) and form calcium ions (Ca^{2+}) and bicarbonate ions (HCO_3^-).

$$CaCO_3 + H^+ \rightarrow Ca^{2+} + HCO_3^-$$

The resulting calcium ions (Ca^{2+}) and bicarbonate ions are then carried away in the groundwater. Eventually, they precipitate, which means they crystallize out of the solution, somewhere else. Precipitation occurs when the groundwater evaporates or when the carbon dioxide gas leaves the water. The processes of dissolving, called dissolution, and precipitation of calcite both play a major role in the formation of limestone caves, such as those shown in **Figure 10.8.**

Caves A natural underground opening with a connection to Earth's surface is called a **cave** or a cavern. Some caves form three-dimensional mazes of passages, shafts, and chambers that stretch for many kilometers. Some caves are dry, while some contain underground streams or lakes. Others are totally flooded and can be explored only by cave divers. Mammoth Cave in Kentucky, shown in **Figure 10.8,** is composed of a series of connected underground passages.

Most caves are formed when groundwater dissolves limestone. The development of most caves begins in the zone of saturation just below the water table. As groundwater infiltrates the cracks and joints of limestone formations, it gradually dissolves the adjacent rock and enlarges these passages to form an interconnected network of openings. As the water table is lowered, the cave system becomes filled with air. New caves then form beneath the lowered water table. If the water table continues to drop, the thick limestone formations eventually become honeycombed with caves. This is a common occurrence in limestone regions that have been uplifted by tectonic forces.

 **Reading Check** **Explain** how most caves form.

■ **Figure 10.8** Groundwater dissolution and precipitation result in a variety of features in caves.
Identify *which chemical reactions might be at work.*

Carlsbad Caverns, New Mexico

Mammoth Cave, Kentucky

Figure 10.9 Karst topography is characterized by a landscape of sinkholes formed by dissolution of limestone.
Identify *what controls the rate of dissolution of bedrock in karst topography.*

Karst topography **Figure 10.9** shows some of the surface features produced by the dissolution of limestone bedrock. One of the main features is a **sinkhole**—a depression in the ground caused by the collapse of a cave or by the direct dissolution of limestone by acidic water. Another type of feature, called a disappearing stream, forms when a surface stream drains into a cave system and continues flowing underground, leaving a dry valley above. Disappearing streams sometimes reemerge on Earth's surface as karst springs.

Limestone regions that have sinkholes and disappearing streams are said to have **karst topography.** The word *karst* comes from the name of a region in Croatia where these features are especially well developed. Prominent karst regions in the United States are located in Kentucky, Indiana, Florida, and Missouri. The Mammoth Cave region in Kentucky has karst topography that contains tens of thousands of sinkholes.

In karst areas, sinkholes proliferate, grow, and eventually join to form wide valleys. Most of the original surface has been dissolved, with the exception of scattered mesas and small buttes. The rate of the dissolution process varies greatly among locations, depending on factors such as humidity and soil composition. In humid areas, where there is more precipitation, more water infiltrates areas of porous soil, and dissolves the limestone in the subsurface.

Groundwater Deposits

Calcium ions eventually precipitate from groundwater and form new calcite minerals. These minerals create spectacular natural features.

Dripstones The most remarkable features produced by groundwater are the rock formations called dripstone that decorate many caves above the water table, as shown in **Figure 10.10.** These formations are built over time as water drips through caves. Each drop of water hanging on the ceiling of a cave loses some carbon dioxide and precipitates some calcite. A form of dripstone, called a **stalactite,** hangs from the cave's ceiling like icicles and forms gradually. As the water drips to the floor of the cave, it may also slowly build mound-shaped dripstone called **stalagmites.**

Figure 10.10 Stalactites are dripstones produced by a buildup of minerals precipitated from groundwater.

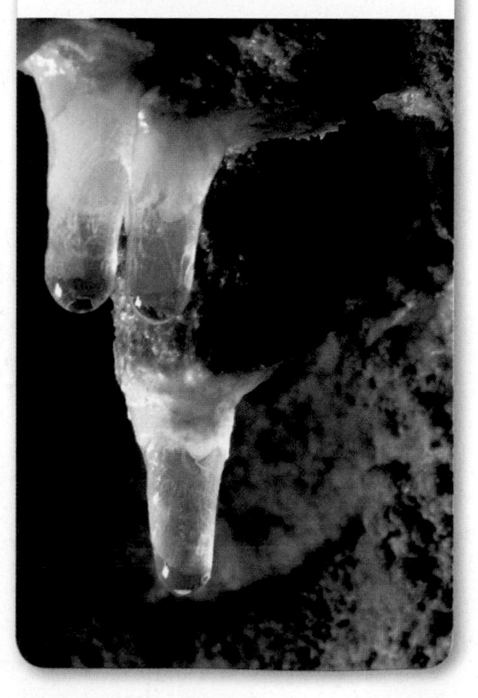

■ **Figure 10.11** Hard water contains high concentrations of minerals., which leave precipitates in household water pipes such as this one.
Identify ***one of the likely precipitates in this pipe.***

Sometimes stalactites and stalagmites grow together to form dripstone columns. Increasingly, researchers are finding abundant and varied microorganisms associated with dripstone formations in caves. It is possible that these organisms play an important role in the deposition of at least some of the materials found in caves.

Hard water You are probably aware that tap water contains various dissolved solids. While some of these materials are added by water treatment facilities, others come from the dissolution of minerals in soils and subsurface rock and sediment. Water that contains high concentrations of calcium, magnesium, or iron is called hard water. Hard water is common in areas where the subsurface rock is limestone. Because limestone is made of mostly calcite, the groundwater in these areas contain significant amounts of dissolved calcite. Hard water used in households can sometimes cause problems. Just as calcite precipitates in caves, it can also precipitate in water pipes, as shown in **Figure 10.11,** and on the heating elements of appliances. Over time, deposits of calcite can clog water pipes and render some electrical appliances useless.

Section 10.2 Assessment

Section Summary

- Groundwater dissolves limestone and forms underground caves.
- Sinkholes form at Earth's surface when bedrock is dissolved or when caves collapse.
- Irregular topography caused by groundwater dissolution is called karst topography.
- The precipitation of dissolved calcite forms stalactites and stalagmites in caves.

Understand Main Ideas

1. MAIN Idea **Analyze** how limestone is weathered, and identify the features that are formed as a result of this dissolution.
2. **Identify** the acid that is most common in groundwater.
3. **Illustrate** in a series of pictures how caves are formed.
4. **Examine** Why is hard water more common in some areas than others?

Think Critically

5. **Compare and contrast** the formation of stalactites and stalagmites.
6. **Analyze** how you might be able to tell an area of karst topography on a topographic map.

WRITING in Earth Science

7. Explain how subsurface limestone is related to karst topography.

Section 10.3

Objectives

- **Explain** how groundwater is withdrawn from aquifers by wells.
- **Describe** the major problems that threaten groundwater supplies.

Review Vocabulary

runoff: water flowing downslope along Earth's surface

New Vocabulary

well
drawdown
recharge
artesian well

Groundwater Supply

MAIN Idea Water is not always available in the quantities and in the locations where it is needed and might be compromised by pollution.

Real-World Reading Link If you have a bank account, can you withdraw as much money as you want? Of course not. Like a bank account, groundwater can be withdrawn, but only in the amount that has been deposited there.

Wells

Wells are holes dug or drilled into the ground to reach an aquifer. There are two main types of wells: ordinary wells and artesian wells.

Ordinary wells The simplest wells are those that are dug or drilled below the water table, into what is called a water-table aquifer, as shown in **Figure 10.12.** In a water-table aquifer, the level of the water in the well is the same as the level of the surrounding water-table. As water is drawn out of a well, it is replaced by surrounding water in the aquifer.

Overpumping occurs when water is drawn out of the well at a rate that is faster than that at which it is replaced. Overpumping of the well lowers the local water level and results in a cone of depression around the well, as shown in **Figure 10.12.** The difference between the original water-table level and the water level in the pumped well is called the **drawdown.** If many wells withdraw water from a water-table aquifer, the cones of depression can overlap and cause an overall lowering of the water table, causing shallow wells to become dry. Water from precipitation replenishes the water content of an aquifer in the process of **recharge.** Groundwater recharge from precipitation and runoff sometimes replaces the water withdrawn from wells. However, if withdrawal of groundwater exceeds the aquifer's recharge rate, the drawdown increases until all wells in the area become dry.

Figure 10.12 Overpumping from one well or multiple wells can result in a cone of depression and a general lowering of the water table.

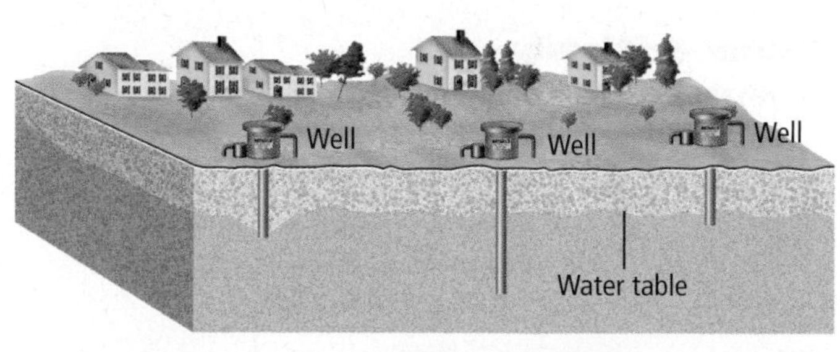

Before heavy pumping

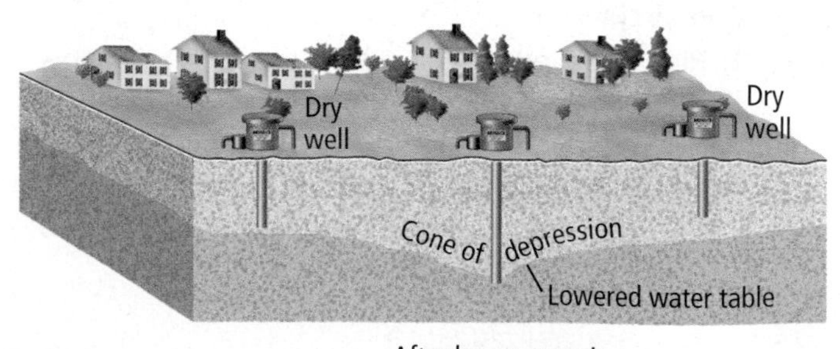

After heavy pumping

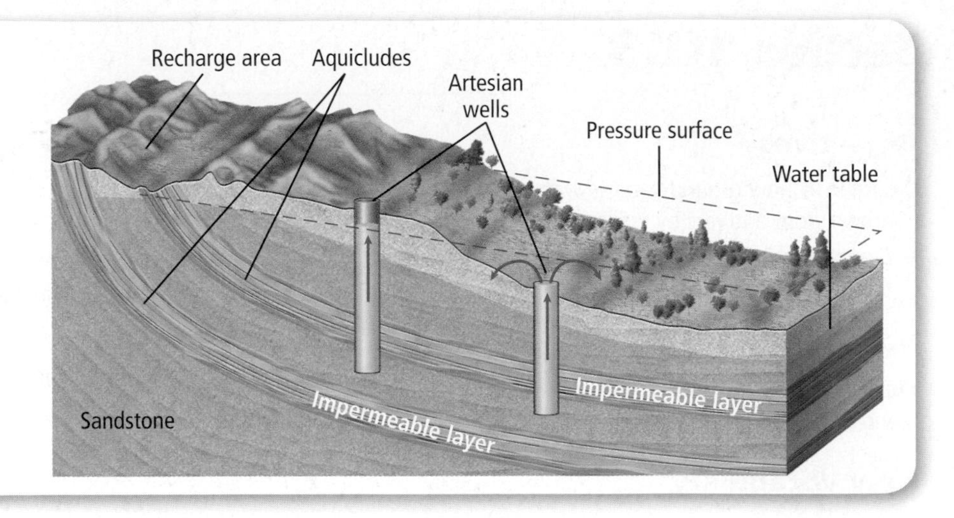

■ **Figure 10.13** An artesian aquifer contains water under pressure.
Identify ***the features that cause the primary difference between an ordinary well and an artesian well.***

Artesian wells An aquifer's area of recharge is often at a higher elevation than the rest of the aquifer. An aquifer located between aquicludes, called a confined aquifer, can contain water that is under pressure. This is because the water at the top of the slope exerts gravitational force on the water downslope, as you will learn in the Problem-Solving Lab on this page. An aquifer that contains water under pressure is called an artesian aquifer. When the rate of recharge is high enough, the pressurized water in a well drilled into an artesian aquifer can spurt above the land surface in the form of a fountain known as an **artesian well.** The level to which water in an open well can rise is called its pressure surface, as shown in **Figure 10.13.** Similarly, a spring that discharges pressurized water is called an artesian spring. The name *artesian* is derived from the French province of Artois, where such wells were first drilled almost 900 years ago.

PROBLEM-SOLVING LAB

Make a Topographic Profile

How does water level vary in an artesian well? Artesian aquifers contain water under pressure. The table provides data about an artesian aquifer for three sites that are spaced 100 m apart along a survey line. It shows the following: elevations of the land surface, the water table, the upper surface of the aquiclude on top of the artesian aquifer, and artesian pressure surface.

Analysis

1. Plot the elevation data on a graph with the sites on the *x*-axis and the elevations on the *y*-axis.
2. Make a topographic profile of the survey line from Site 1 to Site 3. Use a heavy line to indicate land surface.

Aquifer Data

Site	Surface Elevation (m)	Water Table Elevation (m)	Aquiclude Elevation (m)	Pressure Surface (m)
1	396	392	388	394
2	394	390	386	393
3	390	388	381	392

Think Critically

3. **Analyze** How close to the surface will water rise in a well drilled at each site?
4. **Evaluate** what would happen if a well were drilled into the confined aquifer at Site 3.
5. **Consider** how drilling an artesian well at one of the sites would affect the other wells.

Threats to Our Water Supply

Freshwater is Earth's most precious natural resource. Human demands for freshwater are enormous, because it is essential for life. Water is also used extensively in agriculture and industry. **Figure 10.14** shows freshwater usage in the United States. Groundwater supplies much of this water.

 **Reading Check** **Summarize** why freshwater is Earth's most precious natural resource.

Estimates of water supplies are the result of a dynamic equilibrium between various factors. These factors include the amounts of precipitation and infiltration, the surface drainage, the porosity and permeability of subsurface rock or sediment, and the volume of groundwater naturally discharged back to the surface. Several of these factors vary naturally over time, and several can be affected by human activities. Changes to groundwater supplies can lead to environmental issues such as a lowered water table, subsidence, and pollution.

An important aquifer in the United States is the Ogallala Aquifer, which underlies the Great Plains. This aquifer delivers water to a huge area stretching from South Dakota to Texas. The recharge areas of the Ogallala Aquifer are located in the Black Hills and the Rocky Mountains.

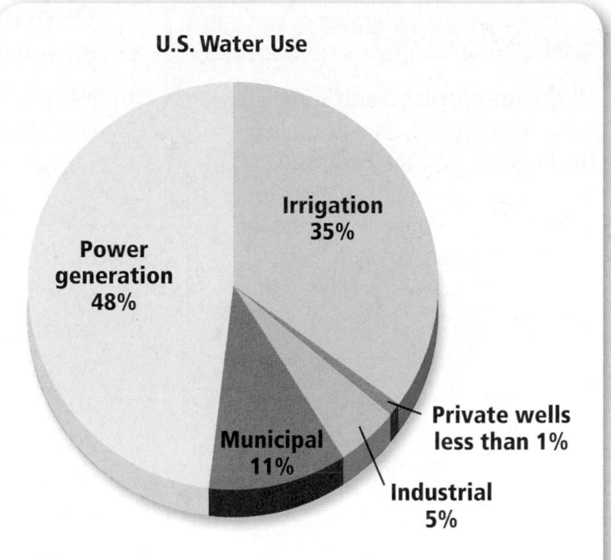

Figure 10.14 Municipal water and private wells supply your daily water needs.
Identify *how you are involved with water use in each of the other areas.*

MiniLab

Model an Artesian Well

How does an artesian well form? What causes the water to rise above the ground surface?

Procedure

1. Read and complete the lab safety form.
2. Half fill a **plastic shoe box** or other container with **sand.** Add enough **water** to saturate the sand. Cover the sand completely with a 1- or 2-cm layer of **clay** or a similar impermeable material.
3. Tilt the box at an angle of about 10°. Use a **book** for a prop.
4. Using a **straw**, punch three holes through the clay, one near the low end, one near the middle, and one near the high end of the box. Insert a **clear straw** through each hole into the sand below. Seal the holes around the straws.

Analysis

1. **Observe** the water levels in the straws. Where is the water level the highest? The lowest?
2. **Identify** the water table in the box.
3. **Analyze** Where is the water under greatest pressure? Explain.
4. **Predict** what will happen to the water table and the surface if the water flows from one of the straws.

CAREERS IN EARTH SCIENCE

Hydrogeologist Earth scientists who map groundwater are called hydrogeologists. They use field methods, maps, and aerial photographs to determine where groundwater is located. To learn more about Earth science careers, visit glencoe.com.

Overuse Groundwater supplies can be depleted. If groundwater is pumped out at a rate greater than the recharge rate, the groundwater supply will decrease and the water table will drop. This is happening to the Ogallala Aquifer. Its water, used mostly for irrigation, is being withdrawn at a rate much higher than the recharge rate.

Subsidence Another problem caused by the excessive withdrawal of groundwater is ground subsidence—the sinking of land. The volume of water underground helps support the weight of the soil, sediment, and rock above. When the height of the water table drops, the weight of the overlying material is increasingly transferred to the aquifer's mineral grains, which then squeeze together more tightly. As a result, the land surface above the aquifer sinks.

A dramatic example of subsidence can be seen along parts of the Gulf coast of Texas, where heavy usage of groundwater over many decades resulted in a wide-scale drop in the ground level. In a region of 12,000 km^2, the average subsidence was 15 cm, while some areas dropped by as much as 3 m. This has presented flooding hazards for much of the coastal region.

Pollution in groundwater In general, the most easily polluted groundwater reservoirs are water-table aquifers, which lack a confining layer above them. Confined aquifers are affected less frequently by local pollution because they are protected by impermeable barriers. When the recharge areas of confined aquifers are polluted, however, those aquifers can also become contaminated.

Reading Check **Identify** which kind of aquifer is more vulnerable to pollution.

Sources of groundwater pollution include sewage from faulty septic tanks and farms, landfills, and other waste disposal sites. Pollutants usually enter the ground above the water table, but they eventually infiltrate to the water table. In highly permeable aquifers, pollutants can spread quickly in a specific direction, such as toward the wells shown in **Figure 10.15.**

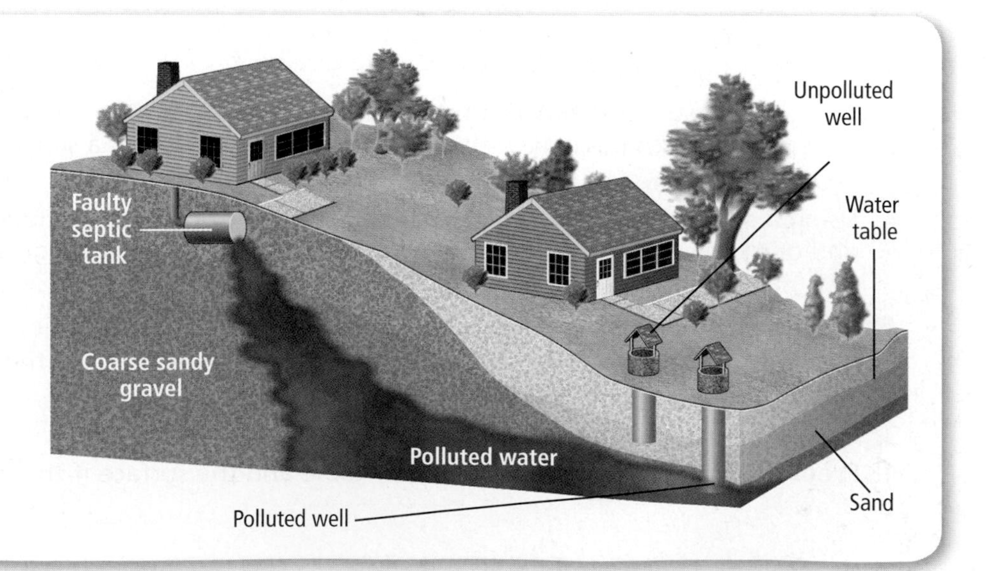

Figure 10.15 Pollutants can spread rapidly through a highly permeable aquifer. Note how the polluted well has drawn the pollution toward it as it has withdrawn water from the water table.

Chemicals Because chemicals dissolved and transported with groundwater are submicroscopic in size, they can travel through the smallest pores of fine-grained sediment. For this reason, chemicals such as arsenic can contaminate any type of aquifer. The chemicals generally move downslope from a source in the form of a pollution plume, which is a mass of contaminants that spreads through the aquifer. Once chemical contaminants have entered groundwater, they cannot be easily removed. In the GeoLab at the end of the chapter, you will learn more about how geologists predict the risks of chemical contamination of groundwater based on a region's topography.

Reading Check **Explain** why chemicals such as arsenic can contaminate any kind of aquifer.

Sewage, landfills, and other waste disposal sites can include a variety of contaminants. Chemical contaminants can be leached, meaning dissolved by infiltrating groundwater. When chemical and biological contaminants enter the groundwater, they flow through the aquifer at the same rate as the rest of the groundwater. Over time, an entire aquifer can become contaminated and toxic to humans. Aquifers are particularly vulnerable to pollution in humid areas where the water table is shallow and can more easily come in contact with waste.

Salt Not all pollutants are toxic or unhealthful in and of themselves. For example, ordinary table salt is used to season food, but water is undrinkable when its salt content is too high. In like manner, groundwater is unusable after the intrusion of salt water. Salt pollution is one of the major threats to groundwater supplies, especially in coastal areas, where the intrusion of salt water into groundwater is a major problem. In coastal areas, salty seawater, which is denser than freshwater, underlies the groundwater near Earth's surface, as shown in **Figure 10.16.** The overpumping of wells can cause the underlying salt water to rise into the wells and contaminate the freshwater aquifer.

VOCABULARY

ACADEMIC VOCABULARY

Transport
to move from one place to another
Airplanes transport packages across the country.

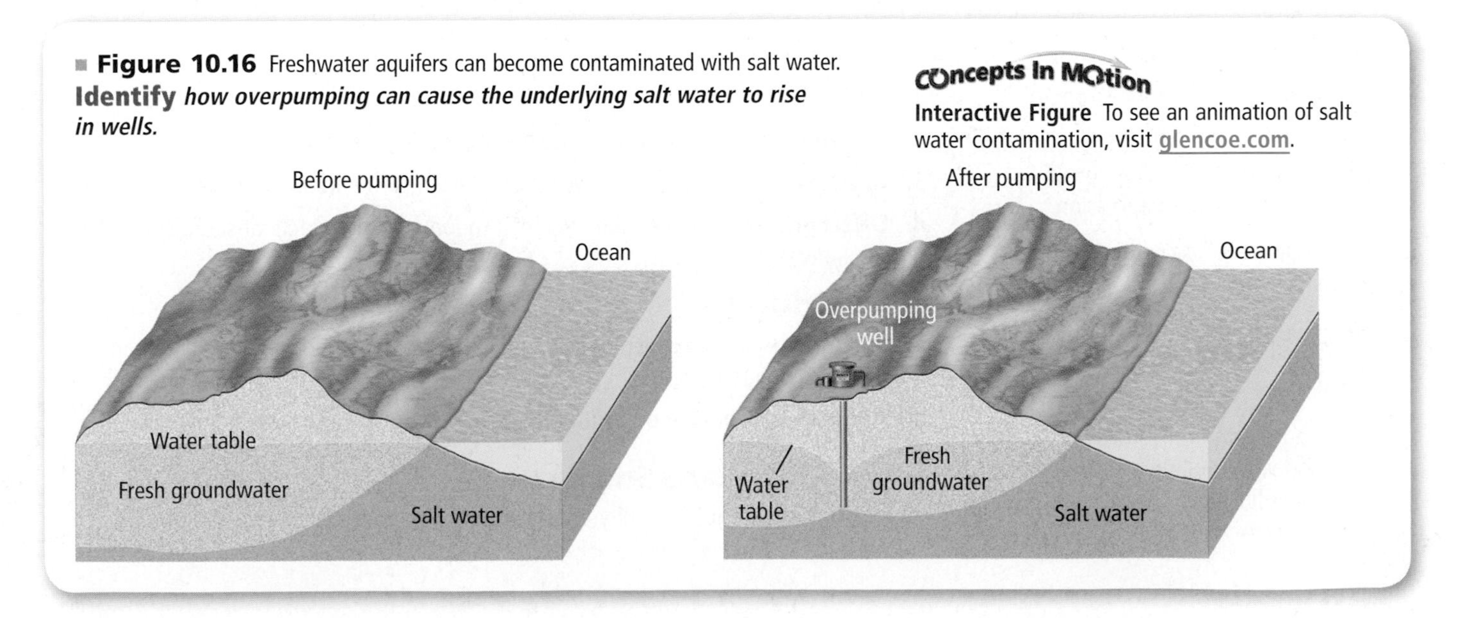

■ **Figure 10.16** Freshwater aquifers can become contaminated with salt water.
Identify *how overpumping can cause the underlying salt water to rise in wells.*

Concepts In Motion
Interactive Figure To see an animation of salt water contamination, visit glencoe.com.

Table 10.2	Groundwater Pollution Sources
Infiltration from fertilizers	
Leaks from storage tanks	
Drainage of acid from mines	
Seepage from faulty septic tanks	
Saltwater intrusion into aquifers near shorelines	
Leaks from waste disposal sites	
Radon	

Radon Another source of natural groundwater pollution is radon gas, which is one of the leading causes of cancer in the United States. Radon found in groundwater is one of the products of the radioactive decay of uranium in rocks and sediment, and it usually occurs in very low concentrations in all groundwater. However, some rocks, especially granite and shale, contain more uranium than others. Therefore the groundwater in areas where these rocks are present contains higher levels of radon. Some radon can seep into houses, and, because it is heavier than air, it can accumulate in poorly ventilated basements. The United States Environmental Protection Agency (EPA) advises homeowners in radon-prone regions to have their homes tested regularly for radon gas.

Protecting Our Water Supply

There are a number of ways by which groundwater resources can be protected and restored. First, major pollution sources, many of which are listed in **Table 10.2,** need to be identified and eliminated. Pollution plumes that already exist can be monitored with observation wells and other techniques. Most pollution plumes spread slowly providing adequate time for alternate water supplies to be found. In some cases, pollution plumes can be stopped by building impermeable underground barriers around the polluted area. Sometimes, polluted groundwater can be pumped out for chemical treatment on the surface.

While these measures can have limited success, they alone cannot save Earth's water supply. Humans must be aware of how their activities impact the groundwater system so that they can protect the water supply.

Section 10.3 Assessment

Section Summary

- Wells are drilled into the zone of saturation to provide water.
- Overpumping of shallow wells produces cones of depression.
- Artesian wells tap confined aquifers in which water is under pressure.
- When groundwater withdrawal exceeds recharge, it lowers the water table
- The most common sources of groundwater pollution include sewage, landfills, and other waste disposal sites.

Understand Main Ideas

1. MAIN Idea **Evaluate** the problems associated with overpumping wells.
2. **Explain** why artesian wells contain water under pressure.
3. **Illustrate** the difference between an artesian well and an ordinary well.
4. **Differentiate** between the effects of radon and the effects of salt dissolved in groundwater.

Think Critically

5. **Formulate** an experiment which would test if there were impermeable barriers around a polluted area.
6. **Analyze** how best to prevent groundwater pollution in a residential area.

WRITING in Earth Science

7. Predict how the permeability of an aquifer can affect the spread of pollutants.

Earth Science Online Self-Check Quiz glencoe.com

Earth Science and the Environment

Watcher of the Water

Safe drinking water is something that many people take for granted. Most of the water that is used for human consumption comes from groundwater. Who ensures that groundwater sources remain safe?

Hydrogeologists A groundwater scientist, called a hydrogeologist, is responsible for finding and monitoring groundwater sources to ensure the water supply is free of contaminants and is not used faster than it is replaced. What does a typical day in the life of a hydrogeologist look like? One day might be spent in the field conducting tests on the water levels. The next day might be spent evaluating the data in the office. The day after that might involve looking for trouble in the water-supply line of a house.

Aquifer case study Suppose a farmer wants to install an irrigation system, which involves digging a new well. First, the water level in the area's aquifer must be checked to ensure that a new well will not cause shortages for other users. The hydrogeologist finds an active well nearby and hooks it up to a pump that continuously draws water for 24 hours. Periodic checks of other wells in the area determine the changes in the water level and quality. From the data gathered, he or she computes how much water the aquifer contains and determines the amount of water available for a new well.

Suppose that, after the farm starts using the irrigation system, a house down the road loses its water supply. The hydrogeologist goes to the house and checks for technical problems, such as a hole in the well casing. If the cause is not technical, he or she will reassess the irrigation system by rechecking the water supply in the aquifer.

These hydrogeologists collect water from a well to determine whether or not it has been contaminated.

Quality assurance Hydrogeologists are also responsible for checking water quality. If the water from a particular aquifer develops a strange taste and odor, the residents would want to ensure the water is safe to drink. The hydrogeologist gathers samples and sends them to a lab to test for various contaminants, such as sewage, pesticides, dissolved metals, or organic material. If a contaminant is found, the hydrogeologist will advise the residents not to drink the water until the source is discovered and the problem is resolved. The hydrogeologist will then begin investigating the problem and searching for clues to find and stop the contamination.

WRITING in Earth Science

Journal Research more about what a hydrogeologist does at glencoe.com. Then, imagine you are accompanying a groundwater scientist on a day on the job. Describe what you saw, what you did, and what you learned about aquifers.

GEOLAB

MAPPING: TRACK GROUNDWATER POLLUTION

Background: You can use a topographic map to estimate the direction of groundwater flow. Groundwater pollution spreads out from its source and follows the flow of groundwater. The spread and movement of the pollution resembles a plume that stems from its source.

Question: *How can you determine the movement of a pollution plume?*

Materials

U. S. Geological Survey topographic map of Forest City, Florida
transparent paper
ruler
graph paper
calculator

Procedure

Imagine that Jim's Gas Station has discovered a major gasoline leak from one of its underground tanks. As the local hydrogeologist, you are asked to determine the path that the gasoline will take through the groundwater, and to notify the residents of the areas that might be affected by the contamination.

1. Read and complete the lab safety form.
2. Identify the lakes and swamps in the southwest corner of the map and list their names and elevations in a data table. (Note: *The elevations are given or can be estimated from the contour lines. The elevation of the water table in each area can be estimated from the elevations of nearby bodies of water*).
3. Note the location of Jim's gas station on Forest City Rd., about 1400 feet north of the Seminole County line (at the 96-foot elevation mark).
4. Take out a piece of paper to construct a cross section of the surface topography and the water table. Lay the paper on the map from Lake Lotus to Lake Lucien (through Jim's Gas Station).
5. On this piece of paper, mark the location of Jim's gas station.
6. Draw a small line at each place where a contour line intersects the line from Lake Lotus to Lake Lucien. Also note the elevation at each hash mark and any rivers crossed.
7. Draw a table to use for your topographic profile, using the width representing the distance between Lake Lotus to Lake Lucien. For the y-axis, use the elevations 60, 70, 80, 90, and 100 ft.
8. Now take your paper where you marked your lines and place it along the base of the table.
9. Mark a corresponding dot on the table for each elevation, and mark the position of Jim's gas station.
10. Connect the dots to create a topographic profile.
11. Note the elevations of the nearby bodies of water to approximate the distance from the ground surface to the water table. Use dots to indicate those distances on the topographic profile. Connect the dots to draw the water table on the topographic profile.

Analyze and Conclude

1. **Calculate** the slope of the ground surface on either side of Jim's Gas Station.
2. **Estimate** the slope of the water table at Jim's Gas Station.
3. **Infer** the direction toward which the pollution plume will move.
4. **Identify** the houses and bodies of water that are threatened by this pollution plume.
5. **Conclude** Prepare a written statement to present to the local community. Explain the path the plume is predicted to take, and how this was determined.

APPLY YOUR SKILL

Design Using what you have learned in this lab and in the chapter, develop a plan for stopping the pollution plume. Make a map showing where your plan will be implemented. Indicate the sites where water quality will be monitored regularly.

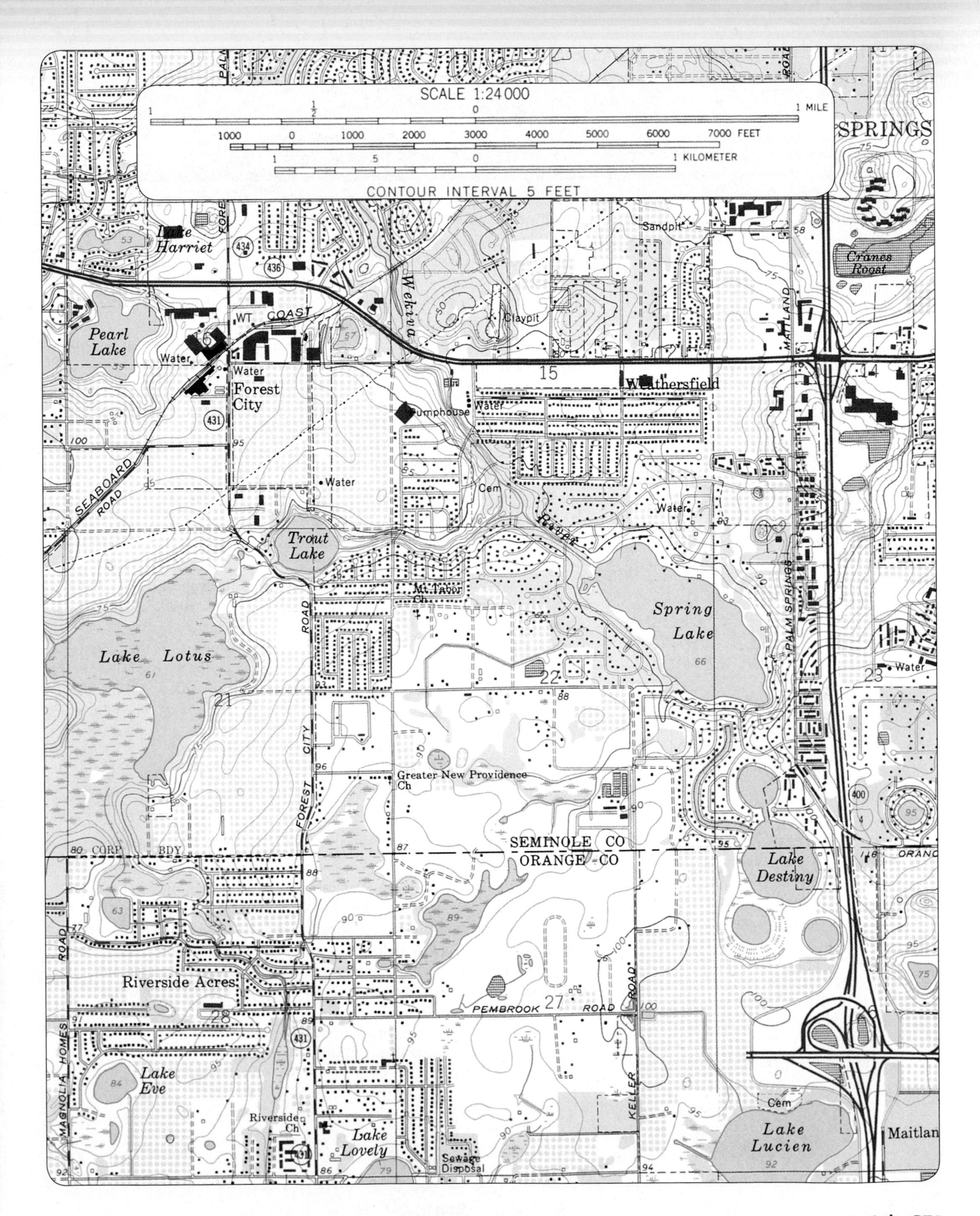
SCALE 1:24 000
1 MILE
1000 0 1000 2000 3000 4000 5000 6000 7000 FEET
1 KILOMETER
CONTOUR INTERVAL 5 FEET
SPRINGS
Lake Harriet
Cranes Roost
Pearl Lake
Wekiva
Claypit
Sandpit
WT COAST
Water
Forest City
Weathersfield
Pumphouse
SEABOARD ROAD
Cem
Trout Lake
Mt. Tabor Ch
Spring Lake
PALM SPRINGS
MAITLAND
Lake Lotus
FOREST CITY ROAD
Greater New Providence Ch
SEMINOLE CO
ORANGE CO
CORP BDY
Lake Destiny
Riverside Acres
PEMBROOK ROAD
KELLER ROAD
MAGNOLIA HOMES
Lake Eve
Riverside Ch
Lake Lovely
Sewage Disposal
Lake Lucien
Maitlan

CHAPTER 10

Study Guide

Download quizzes, key terms, and flash cards from glencoe.com.

BIG Idea Precipitation and infiltration contribute to groundwater, which is stored in underground reservoirs until it surfaces as a spring or is drawn from a well.

Vocabulary	Key Concepts
Section 10.1 Movement and Storage of Groundwater	
• aquiclude (p. 255) • aquifer (p. 255) • geyser (p. 258) • hot spring (p. 258) • infiltration (p. 253) • permeability (p. 255) • spring (p. 256) • water table (p. 254) • zone of aeration (p. 254) • zone of saturation (p. 254)	**MAIN Idea** Groundwater reservoirs provide water to streams and wetlands wherever the water table intersects the surface of the ground. • Some precipitation infiltrates the ground to become groundwater. • Groundwater is stored below the water table in pore spaces of rocks and sediment. • Groundwater moves through permeable layers called aquifers and is trapped by impermeable layers called aquicludes. • Groundwater emerges from the ground where the water table intersects Earth's surface.
Section 10.2 Groundwater Weathering and Deposition	
• cave (p. 260) • karst topography (p. 261) • sinkhole (p. 261) • stalactite (p. 261) • stalagmite (p. 261)	**MAIN Idea** Chemical weathering of limestone by water causes the characteristic topography of karst areas. • Groundwater dissolves limestone and forms underground caves. • Sinkholes form at Earth's surface when bedrock is dissolved or when caves collapse. • Irregular topography caused by groundwater dissolution is called karst topography. • The precipitation of dissolved calcite forms stalactites and stalagmites in caves.
Section 10.3 Groundwater Supply	
• artesian well (p. 264) • drawdown (p. 263) • recharge (p. 263) • well (p. 263)	**MAIN Idea** Water is not always available in the quantities and in the locations where it is needed and might be compromised by pollution. • Wells are drilled into the zone of saturation to provide water. • Overpumping of shallow wells produces cones of depression. • Artesian wells tap confined aquifers in which water is under pressure. • When groundwater withdrawal exceeds recharge, it lowers the water table. • The most common sources of groundwater pollution include sewage, landfills, and other waste disposal sites.

CHAPTER 10 Assessment

Vocabulary Review

Match each phrase with a vocabulary term from the Study Guide.

1. the depth below Earth's surface at which all pores in layers of soil are filled with water
2. the vertical movement of water through ground layers
3. all of the permeable layers at a location
4. the replacement of water content in an aquifer

Each of the following sentences is false. Make each sentence true by replacing the italicized words with a vocabulary term from the Study Guide.

5. *Drawdown* is produced in limestone regions that have sinkholes and sinking streams.
6. *Stalagmites* are icicle-shaped deposits hanging from the ceiling of caves.
7. Collapsing caves or dissolution of bedrock at the surface produce *systems of caves.*

Use what you know about the vocabulary terms found on the Study Guide to answer the following questions.

8. What two features are most often associated with the formation of springs?
9. What is the main difference between regular springs and artesian springs?
10. What are explosive hot springs that develop in volcanic areas?

Understand Key Concepts

11. Which single source of freshwater represents the largest volume of freshwater worldwide readily available for use by humans?
   A. ice caps and glaciers
   B. freshwater lakes
   C. rivers and streams
   D. groundwater deposits

12. Sinkholes may eventually join to form
   A. wide valleys
   B. zone of aeration
   C. dripstones
   D. aquifer

13. What is the name of a layer of sediment or rock that does not allow water to pass through it?
   A. a permeable layer
   B. an aquiclude
   C. an aquifer
   D. a nonaqueous layer

Use the diagram below to answer Questions 14 and 15.

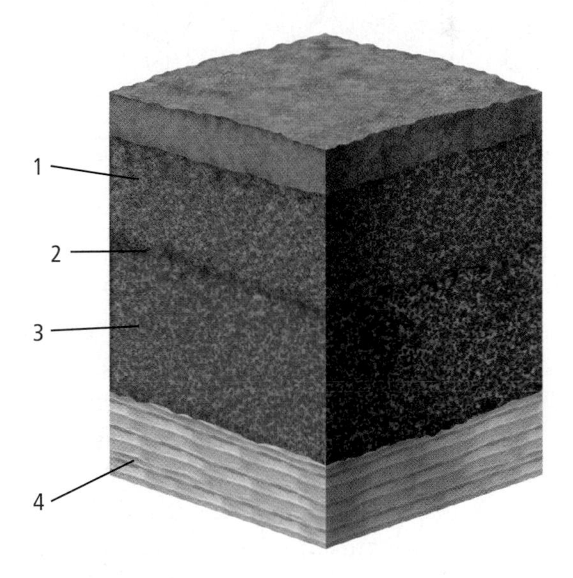

14. Which sequence of terms correctly labels the features shown in the diagram?
   A. 2: water table, 3: impermeable layer
   B. 3: surface zone, 4: impermeable layers
   C. 1: zone of aeration, 3: zone of saturation
   D. 1: zone of saturation, 3: zone of aeration

15. In which layer do the pores contain mostly air, although the materials are moist?
   A. layer 1
   B. layer 2
   C. layer 3
   D. layer 4

16. Which characteristics do most areas with karst topography share?
   A. they are dry areas; limestone bedrock
   B. they are humid areas; granite bedrock
   C. they are humid areas; limestone bedrock
   D. they are dry areas; granite bedrock

Use the graph below from a single well in North Carolina to answer Questions 17 and 18.

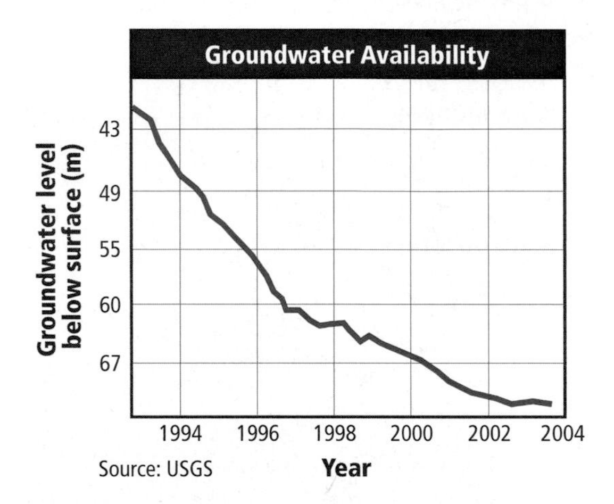

17. Which statement is a logical conclusion that can be drawn from information in the graph?
   A. From 1993 through 2003, groundwater availability at this well has increased.
   B. From 2002 through 2003, the water table has fallen faster than from 1993 through 1994.
   C. From 1993 through 1994, the water table has fallen less than from 2002 through 2003.
   D. From 1993 through 2003, groundwater availability at this well has declined.

18. What year was the groundwater level the highest?
   A. 2004
   B. 2003
   C. 1996
   D. 1993

19. What forms when carbon dioxide dissolves in water?
   A. calcite
   B. acid rain
   C. carbonic acid
   D. hydrogen ions

20. What characteristic must porous rocks have for them to be permeable?
   A. They must be above the water table.
   B. Their pores must be large.
   C. Their pores must be interconnected.
   D. They must be below the water table.

Use the diagram below to answer Question 21.

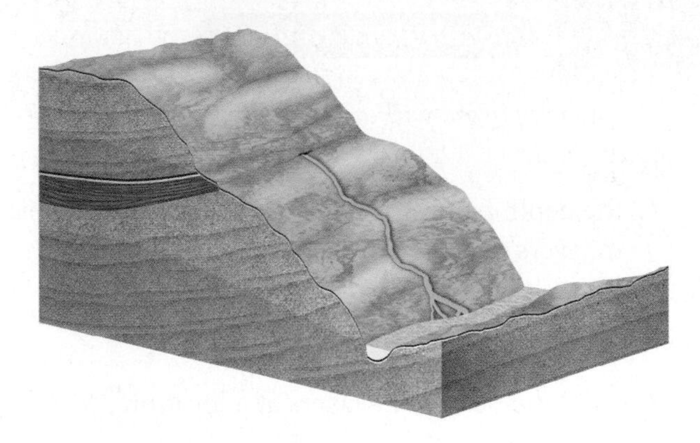

21. What conditions are required for the formation of the spring?
   A. defined areas of aeration, saturation, and an impermeable layer
   B. an aquiclude holding water above defined areas of aeration and saturation
   C. an aquiclude holding water above the main water table, and recharged from above
   D. an aquiclude defining a main water table, and recharged from above

Constructed Response

22. **Classify** where the water table is located in a lake or wetland as opposed to a region with no standing water.

23. **Identify** the two features an aquifer must have to be a source of artesian water.

24. **Compare and contrast** how the water table differs between humid and arid regions.

25. **Examine** how the cement that binds the grains of sedimentary rocks affects the porosity and permeability of the rock.

26. **Predict** how a small aquifer will be affected by a multiyear drought.

27. **Generalize** whether caves are more likely to develop in a region containing limestone bedrock or sandstone bedrock. Justify your answer.

28. **Explain** why disposal of toxic waste into a sinkhole can pose serious hazards for local drinking water.

Earth Science Online Chapter Test glencoe.com

Think Critically

29. Formulate an explanation for why stalactites have a tapering shape whereas stalagmites usually have less regular shapes and broader bases.

30. Hypothesize the effect that a severely lowered water table would have on the emergence of springs on a hillside.

31. Infer why caves often include dry chambers although most caves develop in the zone of saturation just below the water table.

Use the photo below to answer Question 32.

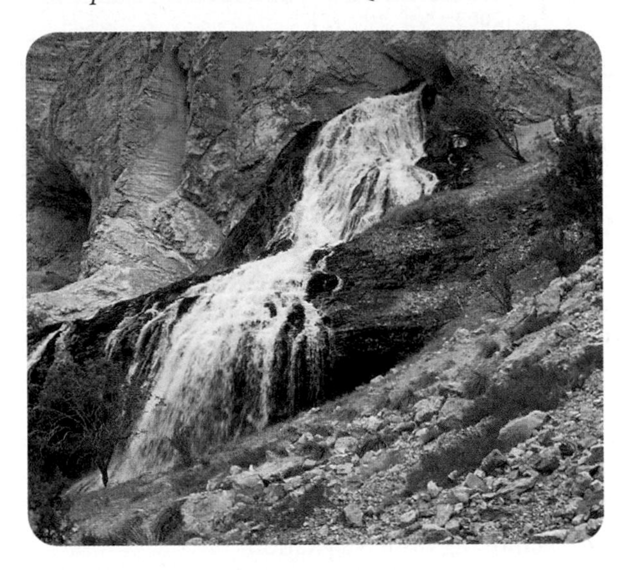

32. Consider the water pouring from the cliffside. Diagram a scenario that would explain the role of groundwater in the photo.

33. Assess what would be an important consequence of sea level rise on groundwater supplies in coastal areas.

Concept Mapping

34. Make a concept map using the following terms: *ordinary well, artesian well, aquiclude, confined, unconfined,* and *water-table aquifer.*

Challenge Question

35. Infer the effect that increased atmospheric CO_2 concentration might have on structures made of calcite and the development of karst topography.

Additional Assessment

36. WRITING in Earth Science Write a short story to demonstrate how shared groundwater resources could cause conflict between neighboring states or countries.

Document–Based Questions

Data obtained from: Lerch, R.N., C.M Wicks, and P.L. Moss. 2006. Hydrological characterization of two karst recharge areas in Boone County, Missouri. *Journal of Cave and Karst Studies* 67 (3): 158–173.

In the graphs, monthly precipitation (bar graph) and monthly discharge (line graph) are shown for Devil's Icebox cave streams in Missouri.

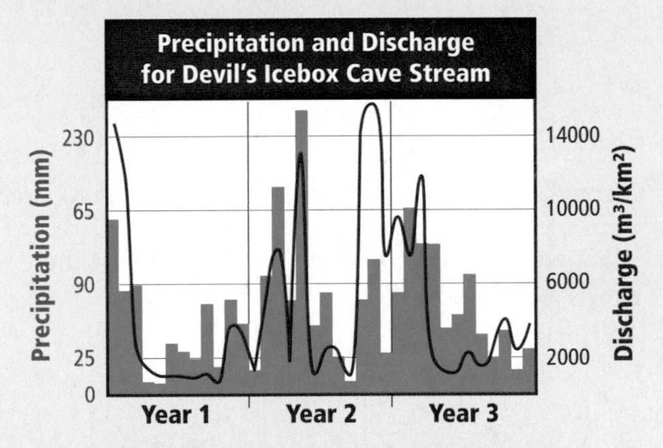

37. Which year had the most precipitation?

38. What is the general relationship between precipitation and discharge according to the graph?

39. Identify an exception to the general relationship described in Question 38. Suggest a possible explanation for this exception.

Cumulative Review

40. Compare and contrast the ionic bond and the covalent bond. **(Chapter 3)**

41. What is the significance of the rock cycle? **(Chapter 6)**

42. List factors that affect the rate of weathering. Underline the factor you think is the most important in chemical weathering and explain why. **(Chapter 8)**

Standardized Test Practice

Multiple Choice

1. Which materials would be best suited for lining a pond?
   A. gravel
   B. limestone
   C. clay
   D. sand

Use the concept map to answer Questions 2 and 3.

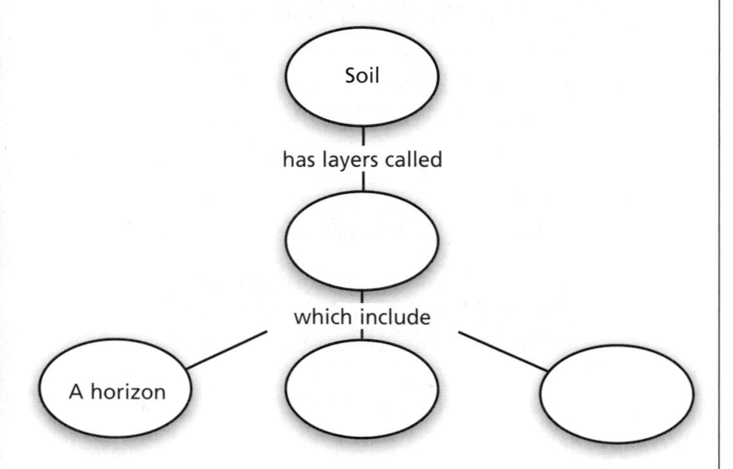

2. What word would complete the circle below the words *has layers called?*
   A. horizons
   B. profiles
   C. levels
   D. humus

3. The bottom two circles should be filled in with ________ and ________.
   A. O horizon; R horizon
   B. B horizon; C horizon
   C. A level; A profile
   D. humus; litter

4. Which is NOT a value of wetlands?
   A. feeding lakes and deltas with nutrients and oxygen-rich water
   B. filtering water by trapping pollutants, sediments, and pathogenic bacteria
   C. providing habitats for migratory birds and other wildlife
   D. preserving fossils due to the anaerobic and acidic conditions

5. What is the reaction of water with other substances known as?
   A. aquification
   B. oxidation
   C. hydrolysis
   D. carbonation

6. Which water sources are the most easily polluted?
   A. water-table aquifers
   B. confined aquifers
   C. artesian wells
   D. hot springs

Use the table below to answer Questions 7 and 8.

Year	Erosion (meters)
2001	2
2002	18
2003	7
2004	5

7. Which could have caused the unusual level of sand loss in 2002 as shown in the table?
   A. lower than usual tides
   B. higher than usual tides
   C. lower than usual storm activity
   D. higher than usual storm activity

8. What human intervention could have caused the drop in erosion from 2002 to 2003?
   A. dune building
   B. removing dune vegetation
   C. constructing buildings along the coast
   D. building fences along the coast

9. What are natural structures hanging from a cave's ceiling?
   A. geyserites
   B. travertines
   C. stalagmites
   D. stalactites

10. In which part of a meander does the water travel the fastest?
    A. along the inside curve of the meander
    B. along the bottom of the meander
    C. along the outside curve of a meander
    D. all parts of the meander are equal

Short Answer

Use the illustration below to answer Questions 11–13.

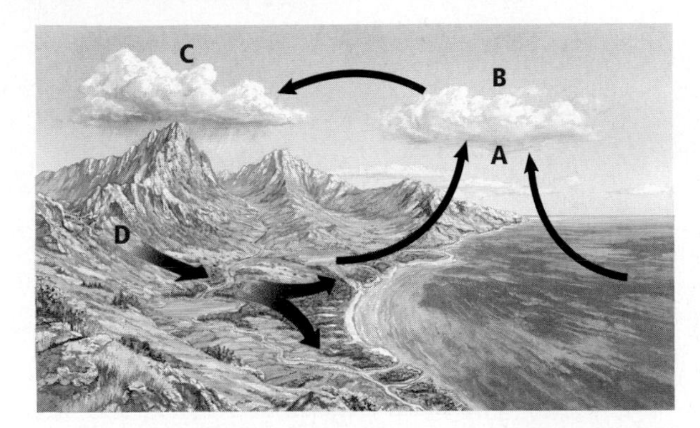

11. Explain the process being illustrated. Be sure to include the name of the process.

12. Why are there two arrows rising during the stage labeled with the letter *A*?

13. What process is occurring from Step C to Step D?

14. How does the silica tetrahedron benefit silicate minerals?

15. What is the risk with overpumping a well?

16. How can human activities trigger mass movements?

Reading for Comprehension

Living Caves

Cool air billowed from a crack in Arizona's desert and lured cave hunters underground. There, formations of rock sprouted from the ground and hung from the ceiling. The explorers discovered a so-called living cave. Tufts and Tenen were the first known to set foot in the Kartchner Caverns, which are among the world's top show caves. The caverns are *living,* a term used to describe active caves. "The cave formations still have water on them; they're still continuing to grow," said Rick Toomey, a staff scientist at the Kartchner Caverns in Benson. Rainwater from the surface seeps through the ground, absorbing calcium carbonate along the way. Inside the cave, the mixture drips from the ceiling. As it hardens, it forms the icicle-like stalactites on the ceiling and sproutlike stalagmites on the floor. With the exception of small cracks, the caverns are closed off to the outside world. The isolation allows the caverns to maintain an average temperature of 20°Celsius and near 99 percent humidity.

Article obtained from: Roach, J. Arizona tried tourism to save 'living cave.' *National Geographic News.* April 19, 2005.

17. According to the passage, what makes a cavern living?
    A. There are animals in the cave.
    B. People can go into the cave.
    C. There are plants in the cave.
    D. Cave formations continue to grow.

18. What is the first ingredient in the process that allows the stalactites and stalagmites to grow as listed in the text?
    A. soil
    B. rainwater
    C. calcium carbonate
    D. rocks

19. What can be inferred from this passage?
    A. Arizona is the only state that has living caves.
    B. The temperature in the caves is quite cold.
    C. Being isolated from the world has protected the cave.
    D. Tufts and Tenen were the first people to discover a cave.

NEED EXTRA HELP?

If You Missed Question . . .	1	2	3	4	5	6	7	8	9	10	11	12	13	14	15	16
Review Section . . .	10.1	7.3	7.3	9.3	7.1	10.3	7.2	8.2	10.2	9.2	9.1	9.1	9.1	4.2	10.2	8.1

UNIT 4

The Atmosphere and the Oceans

Chapter 11
Atmosphere
BIG Idea The composition, structure, and properties of Earth's atmosphere form the basis of Earth's weather and climate.

Chapter 12
Meteorology
BIG Idea Weather patterns can be observed, analyzed, and predicted.

Chapter 13
The Nature of Storms
BIG Idea The exchange of thermal energy in the atmosphere sometimes occurs with great violence that varies in form, size, and duration.

Chapter 14
Climate
BIG Idea The different climates on Earth are influenced by natural factors as well as human activities.

Chapter 15
Earth's Oceans
BIG Idea Studying oceans helps scientists learn about global climate and Earth's history.

Chapter 16
The Marine Environment
BIG Idea The marine environment is geologically diverse and contains a wealth of natural resources.

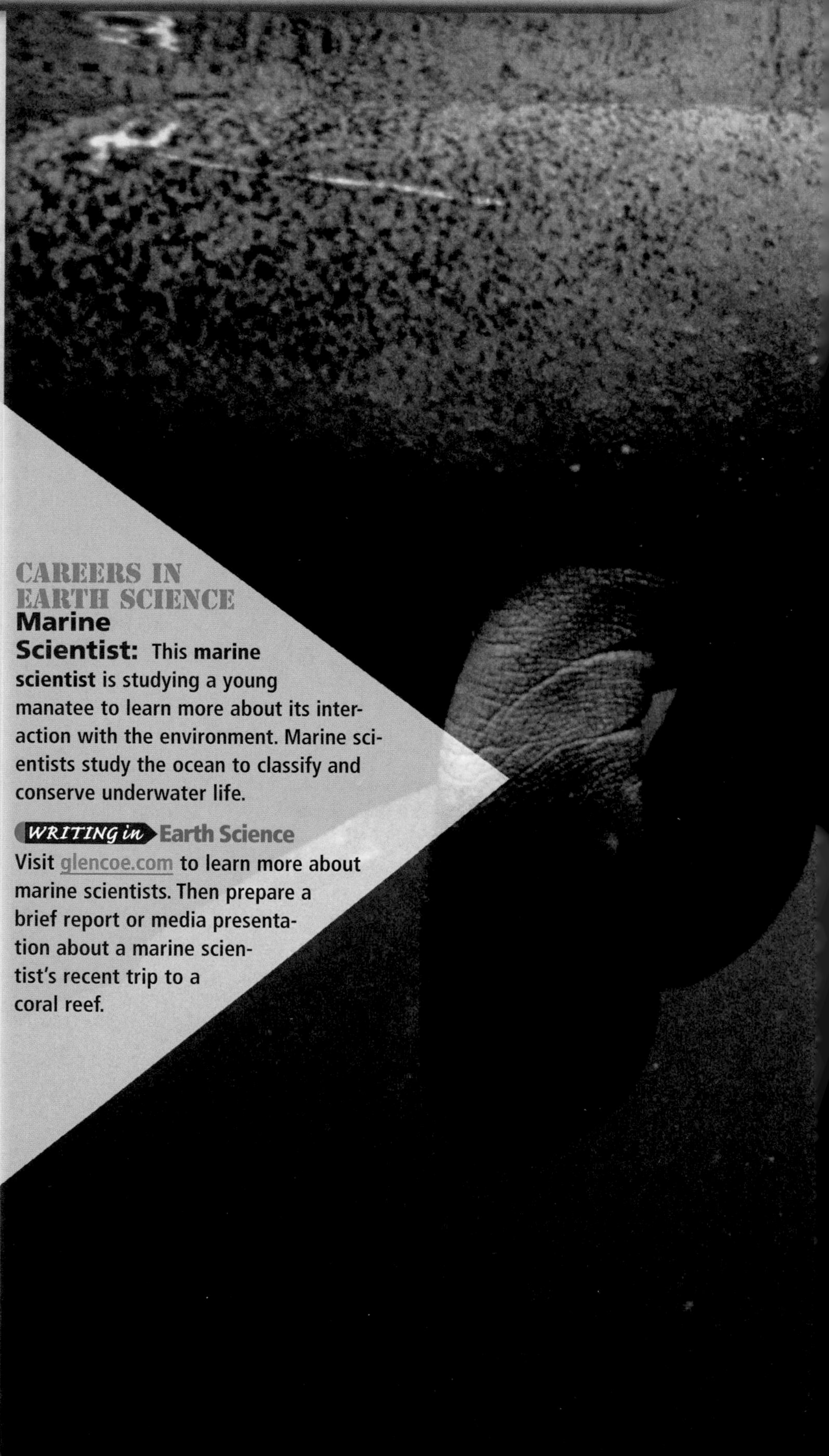

CAREERS IN EARTH SCIENCE

Marine Scientist: This marine scientist is studying a young manatee to learn more about its interaction with the environment. Marine scientists study the ocean to classify and conserve underwater life.

WRITING in Earth Science
Visit glencoe.com to learn more about marine scientists. Then prepare a brief report or media presentation about a marine scientist's recent trip to a coral reef.

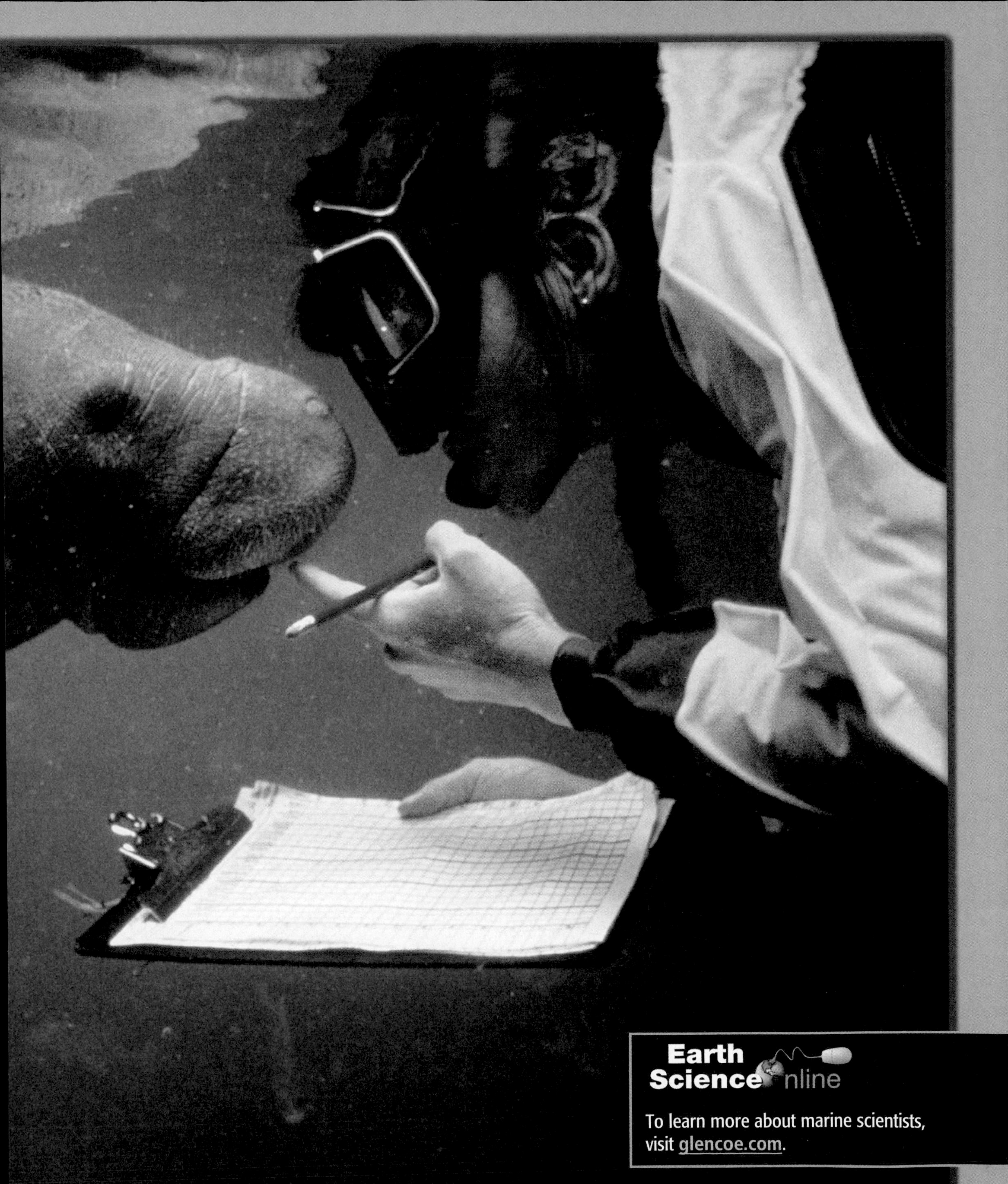

Earth Science Online

To learn more about marine scientists, visit glencoe.com.

CHAPTER 11

Atmosphere

BIG Idea **The composition, structure, and properties of Earth's atmosphere form the basis of Earth's weather and climate.**

11.1 Atmospheric Basics
MAIN Idea Energy is transferred throughout Earth's atmosphere and surface.

11.2 Properties of the Atmosphere
MAIN Idea Atmospheric properties, such as temperature, air pressure, and humidity describe weather conditions.

11.3 Clouds and Precipitation
MAIN Idea Clouds vary in shape, size, height of formation, and type of precipitation.

GeoFacts

- Cirrus clouds are named for the Latin word meaning *curl of hair* because they often appear wispy and hairlike.
- High cirrus clouds are often pushed along by the jet stream and can move at speeds exceeding 160 km/h.
- Clouds can appear gray or even black if they are high enough in the atmosphere, or dense enough that light cannot penetrate them.

LAUNCH Lab

What causes cloud formation?

Clouds form when water vapor in the air condenses into water droplets or ice. These clouds might produce rain, snow, hail, sleet, or freezing rain.

Procedure

1. Read and complete the lab safety form.
2. Pour about 125 mL of **warm water** into a **clear, plastic bowl.**
3. Loosely cover the top of the bowl with **plastic wrap.** Overlap the edges of the bowl by about 5 cm.
4. Fill a **self-sealing plastic bag** with **ice cubes,** seal it, and place it in the center of the plastic wrap on top of the bowl. Push the bag of ice down so that the plastic wrap sags in the center but does not touch the surface of the water.
5. Use **tape** to seal the plastic wrap around the bowl.
6. Observe the surface of the plastic wrap directly under the ice cubes every 10 min for 30 min, or until the ice melts.

Analysis

1. **Infer** What formed on the underside of the wrap? Why did this happen?
2. **Relate** your observations to processes in the atmosphere.
3. **Predict** what would happen if you repeated this activity with hot water in the bowl.

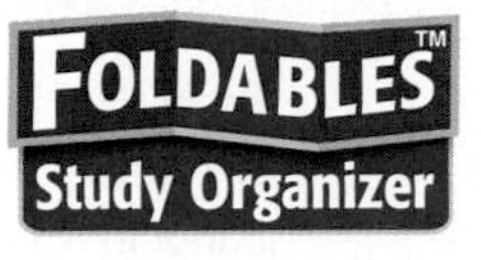

Layers of the Atmosphere Make the following Foldable to organize information about the layers of Earth's atmosphere.

STEP 1 Collect three sheets of paper, and layer them about 2 cm apart vertically.

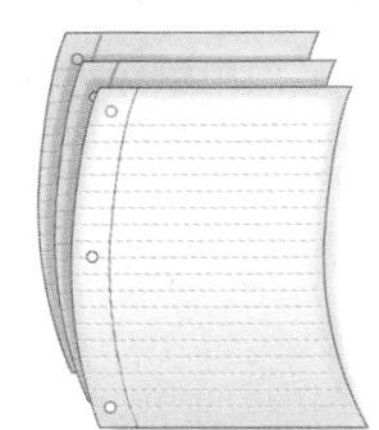

STEP 2 Fold up the bottom edges of the sheets to form five equal tabs. Crease the fold to hold the tabs in place.

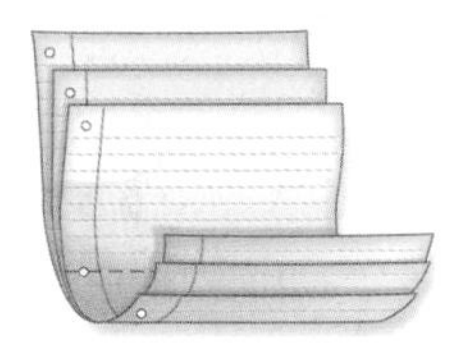

STEP 3 Staple along the fold. Label the tabs *Exosphere, Thermosphere, Mesosphere, Stratosphere,* and *Troposphere.*

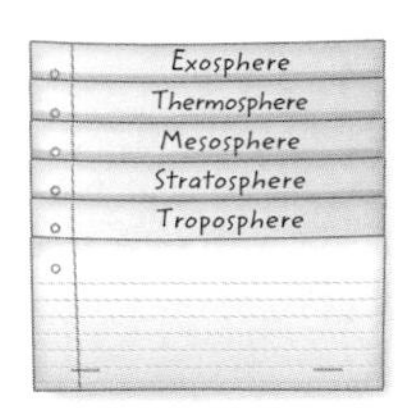

FOLDABLES Use this Foldable with Section 11.1. Sketch the layers on the first tab and summarize information about each layer on the appropriate tabs.

Visit glencoe.com to

- **study entire chapters online;**
- **explore Concepts In Motion animations:**
  - **Interactive Time Lines**
  - **Interactive Figures**
  - **Interactive Tables**
- **access Web Links for more information, projects, and activities;**
- **review content with the Interactive Tutor and take Self-Check Quizzes.**

Section 11.1

Objectives

- **Describe** the gas and particle composition of the atmosphere.
- **Compare and contrast** the five layers of the atmosphere.
- **Identify** three ways energy is transferred in the atmosphere.

Review Vocabulary

atmosphere: the layer of gases that surrounds Earth

New Vocabulary

troposphere
stratosphere
mesosphere
thermosphere
exosphere
radiation
conduction
convection

Atmospheric Basics

MAIN Idea Energy is transferred throughout Earth's atmosphere.

Real-World Reading Link If you touch something made of metal, it will probably feel cool. Metals feel cool because they conduct thermal energy away from your hand. In a similar way, energy is transferred directly from the warmed air near Earth's surface to the air in the lowest layer of the atmosphere.

Atmospheric Composition

The ancient Greeks thought that air was one of the four fundamental elements from which all other substances were made. In fact, air is a combination of gases, such as nitrogen and oxygen, and particles, such as dust, water droplets, and ice crystals. These gases and particles form Earth's atmosphere, which surrounds Earth and extends from Earth's surface to outer space.

Permanent atmospheric gases About 99 percent of the atmosphere is composed of nitrogen (N_2) and oxygen (O_2). The remaining 1 percent consists of argon (Ar), carbon dioxide (CO_2), water vapor (H_2O), and other trace gases, as shown in **Figure 11.1.** The amounts of nitrogen and oxygen in the atmosphere are fairly constant over recent time. However, over Earth's history, the composition of the atmosphere has changed greatly. For example, Earth's early atmosphere probably contained mostly helium (He), hydrogen (H_2), methane (CH_4), and ammonia (NH_3). Today, oxygen and nitrogen are continually being recycled between the atmosphere, living organisms, the oceans, and Earth's crust.

Variable atmospheric gases The concentrations of some atmospheric gases are not as constant over time as the concentrations of nitrogen and oxygen. Gases such as water vapor and ozone (O_3) can vary significantly from place to place. The concentrations of some of these gases, such as water vapor and carbon dioxide, play an important role in regulating the amount of energy the atmosphere absorbs and emits back to Earth's surface.

Water vapor Water vapor is the invisible, gaseous form of water. The amount of water vapor in the atmosphere can vary greatly over time and from one place to another. At a given place and time, the concentration of water vapor can be as much as 4 percent or as little as nearly zero. The concentration varies with the seasons, with the altitude of a particular mass of air, and with the properties of the surface beneath the air. Air over deserts, for instance, contains much less water vapor than the air over oceans.

Figure 11.1 Earth's atmosphere consists mainly of nitrogen (78 percent) and oxygen (21 percent).

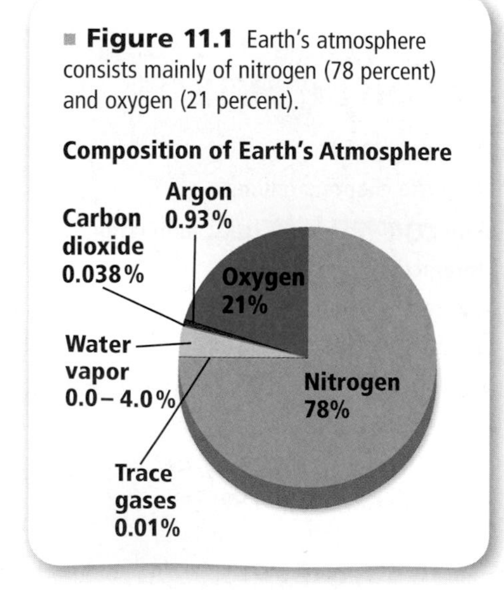

Carbon dioxide Carbon dioxide, another variable gas, currently makes up about 0.039 percent of the atmosphere. During the past 150 years, measurements have shown that the concentration of atmospheric carbon dioxide has increased from about 0.028 percent to its present value. Carbon dioxide is also cycled between the atmosphere, the oceans, living organisms, and Earth's rocks.

The recent increase in atmospheric carbon dioxide is due primarily to the burning of fossil fuels, such as oil, coal, and natural gas. These fuels are burned to heat buildings, produce electricity, and power vehicles. Burning fossil fuels can also produce other gases, such as sulfur dioxide and nitrous oxides, that can cause various respiratory illnesses, as well as other environmental problems.

Ozone Molecules of ozone are formed by the addition of an oxygen atom to an oxygen molecule, as shown in **Figure 11.2.** Most atmospheric ozone is found in the ozone layer, 20 km to 50 km above Earth's surface, as shown in **Figure 11.3.** The maximum concentration of ozone in this layer—9.8×10^{12} molecules/cm^3—is only about 0.0012 percent of the atmosphere.

The ozone concentration in the ozone layer varies seasonally at higher latitudes, reaching a minimum in the spring. The greatest seasonal changes occur over Antarctica. During the past several decades, measured ozone levels over Antarctica in the spring have dropped significantly. This decrease is due to the presence of chemicals called chlorofluorocarbons (CFCs) that react with ozone and break it down in the atmosphere.

Atmospheric particles Earth's atmosphere also contains variable amounts of solids in the form of tiny particles, such as dust, salt, and ice. Fine particles of dust and soil are carried into the atmosphere by wind. Winds also pick up salt particles from ocean spray. Airborne microorganisms, such as fungi and bacteria, can also be found attached to microscopic dust particles in the atmosphere.

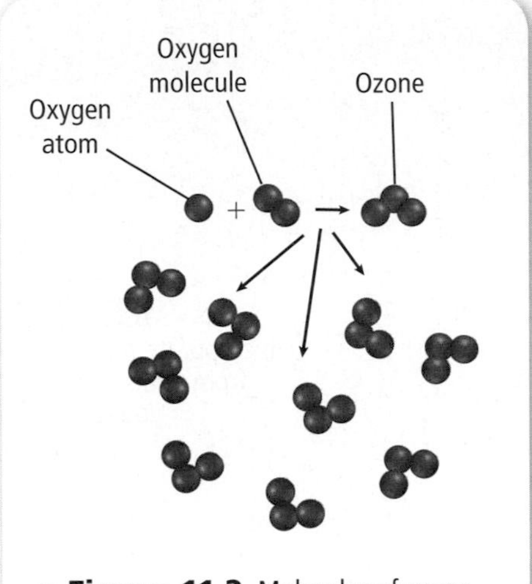

Figure 11.2 Molecules of ozone are formed by the addition of an oxygen atom to an oxygen molecule.

NATIONAL GEOGRAPHIC For more information on the ozone layer and the atmosphere, go to the **National Geographic Expedition** on page 910.

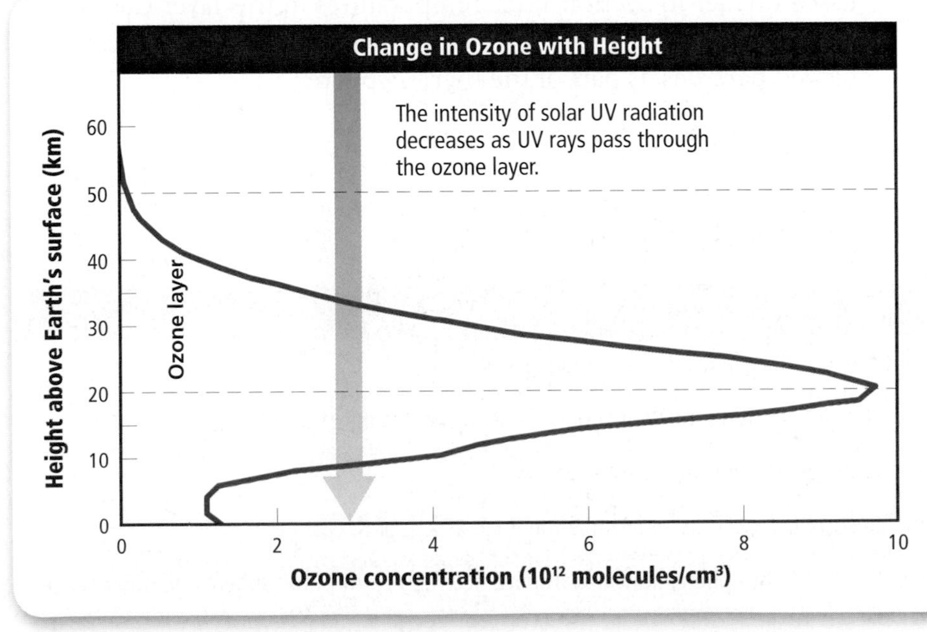

Figure 11.3 The ozone layer blocks harmful ultraviolet rays from reaching Earth's surface. Ozone concentration is highest at about 20 km above Earth's surface, in the ozone layer.

Atmospheric Layers

The atmosphere is classified into five different layers, as shown in **Table 11.1** and **Figure 11.4.** These layers are the troposphere, stratosphere, mesosphere, thermosphere, and exosphere. Each layer differs in composition and temperature profile.

FOLDABLES
Incorporate information from this section into your Foldable.

Troposphere The layer closest to Earth's surface, the **troposphere,** contains most of the mass of the atmosphere. Weather occurs in the troposphere. In the troposphere, air temperature decreases as altitude increases. The altitude at which the temperature stops decreasing is called the tropopause. The height of the tropopause varies from about 16 km above Earth's surface in the tropics to about 9 km above it at the poles. Temperatures at the tropopause can be as low as –60°C.

Stratosphere Above the tropopause is the **stratosphere,** a layer in which the air temperature mainly increases with altitude and contains the ozone layer. In the lower stratosphere below the ozone layer, the temperature stays constant with altitude. However, starting at the bottom of the ozone layer, the temperature in the stratosphere increases as altitude increases. This heating is caused by ozone molecules, which absorb ultraviolet radiation from the Sun. At the stratopause, air temperature stops increasing with altitude. The stratopause is about 48 km above Earth's surface. About 99.9 percent of the mass of Earth's atmosphere is below the stratopause.

Mesosphere Above the stratopause is the **mesosphere,** which is about 50 km to 100 km above Earth's surface. In the mesosphere, air temperature decreases with altitude, as shown in **Figure 11.4.** This temperature decrease occurs because very little solar radiation is absorbed in this layer. The top of the mesosphere, where temperatures stop decreasing with altitude, is called the mesopause.

Thermosphere The **thermosphere** is the layer between about 100 km and 500 km above Earth's surface. In this layer, the extremely low density of air causes the temperature to rise. This will be discussed further in Section 11.2. Temperatures in this layer can be more than 1000°C. The ionosphere, which is made of electrically charged particles, is part of the thermosphere.

Table 11.1 Components of the Atmosphere

Concepts In Motion
Interactive Table To explore more about layers of the atmosphere, visit glencoe.com.

Atmospheric Layer	Components
Troposphere	layer closest to Earth's surface, ends at the tropopause
Stratosphere	layer above the troposphere, contains the ozone layer, and ends at the stratopause
Mesosphere	layer above the stratosphere, ends at the mesopause
Thermosphere	layer above the mesosphere, absorbs solar radiation
Exosphere	outermost layer of Earth's atmosphere, transitional space between Earth's atmosphere and outer space

Visualizing the Layers of the Atmosphere

Figure 11.4 Earth's atmosphere is made up of five layers. Each layer is unique in composition and temperature. As shown, air temperature changes with altitude. When you fly in a plane, you might be flying at the top of the troposphere, or you might enter into the stratosphere.

In the exosphere, gas molecules can be exchanged between the atmosphere and space.

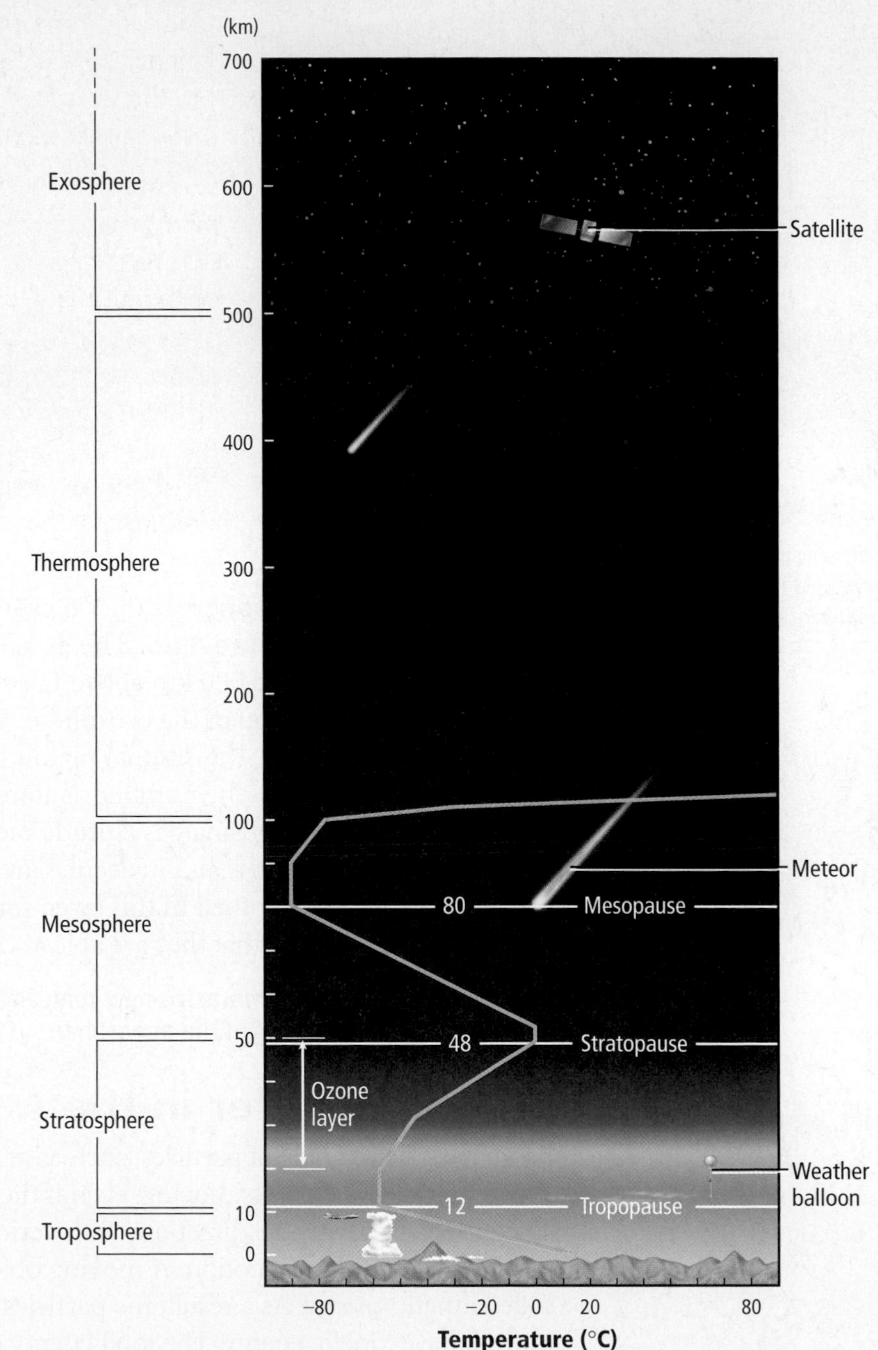

Noctilucent clouds are shiny clouds that can be seen in the twilight in the summer around 50°–60° latitude in the northern and southern hemispheres. These are the only clouds that form in the mesosphere.

Concepts In Motion To explore more about the layers of the atmosphere, visit glencoe.com. **Earth Science Online**

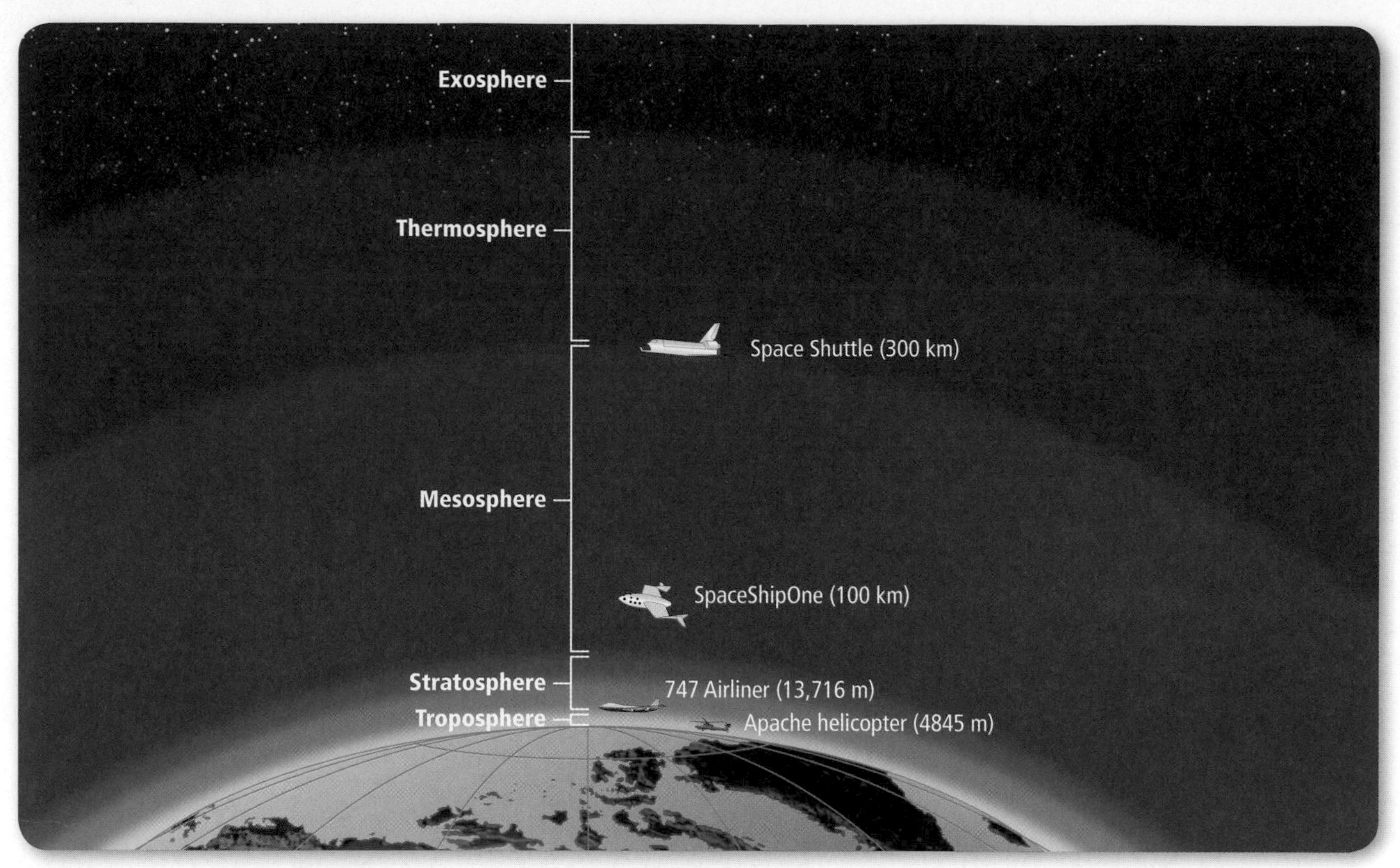

■ **Figure 11.5** Different spacecraft can traverse the various layers of the atmosphere.
Compare *the number of atmospheric layers each spacecraft can reach in its flight path.*

Exosphere The **exosphere** is the outermost layer of Earth's atmosphere, as shown in **Figure 11.5.** The exosphere extends from about 500 km to more than 10,000 km above Earth's surface. There is no clear boundary at the top of the exosphere. Instead, the exosphere can be thought of as the transitional region between Earth's atmosphere and outer space. The number of atoms and molecules in the exosphere becomes very small as altitude increases.

In the exosphere, atoms and molecules are so far apart that they rarely collide with each other. In this layer, some atoms and molecules are moving fast enough that they are able to escape into outer space.

Reading Check **Summarize** how temperature varies with altitude in the four lowest layers of the atmosphere.

Energy Transfer in the Atmosphere

All materials are made of particles, such as atoms and molecules. These particles are always moving, even if the object is not moving. The particles move in all directions with various speeds—a type of motion called random motion. A moving object has a form of energy called kinetic energy. As a result, the particles moving in random motion have kinetic energy. The total energy of the particles in an object due to their random motion is called thermal energy.

Heat is the transfer of thermal energy from a region of higher temperature to a region of lower temperature. In the atmosphere, thermal energy can be transferred by radiation, conduction, and convection.

Radiation Light from the Sun heats some portions of Earth's surface at all times, just as the heat lamp in **Figure 11.6** uses the process of radiation to warm food. **Radiation** is the transfer of thermal energy by electromagnetic waves. The heat lamp emits visible light and infrared waves that travel from the lamp and are absorbed by the food. The thermal energy carried by these waves causes the temperature of the food to increase. In the same way, thermal energy is transferred from the Sun to Earth by radiation. The solar energy that reaches Earth is absorbed and reflected by Earth's atmosphere and Earth's surface.

■ **Figure 11.6** A heat lamp transfers thermal energy by radiation. Here, the thermal energy helps to keep the french fries hot.

Absorption and reflection Most of the solar energy that reaches Earth is in the form of visible light waves and infrared waves. Almost all of the visible light waves pass through the atmosphere and strike Earth's surface. Most of these waves are absorbed by Earth's surface. As the surface absorbs these visible light waves, it also emits infrared waves. The atmosphere absorbs some infrared waves from the Sun and emits infrared waves with different wavelengths, as shown in **Figure 11.7.**

About 30 percent of solar radiation is reflected into space by Earth's surface, the atmosphere, or clouds. Another 20 percent is absorbed by the atmosphere and clouds. About 50 percent of solar radiation is absorbed directly or indirectly by Earth's surface and keeps Earth's surface warm.

Rate of absorption The rate of absorption for any particular area varies depending on the physical characteristics of the area and the amount of solar radiation it receives. Different areas absorb energy and heat at different rates. For example, water heats and cools more slowly than land. Also, as a general rule, darker objects absorb energy faster than light-colored objects. For instance, a black asphalt driveway heats faster on a sunny day than a light-colored concrete driveway.

■ **Figure 11.7** Incoming solar radiation is either reflected back into space or absorbed by Earth's atmosphere or its surface.
Trace ***the pathways by which solar radiation is absorbed and reflected.***

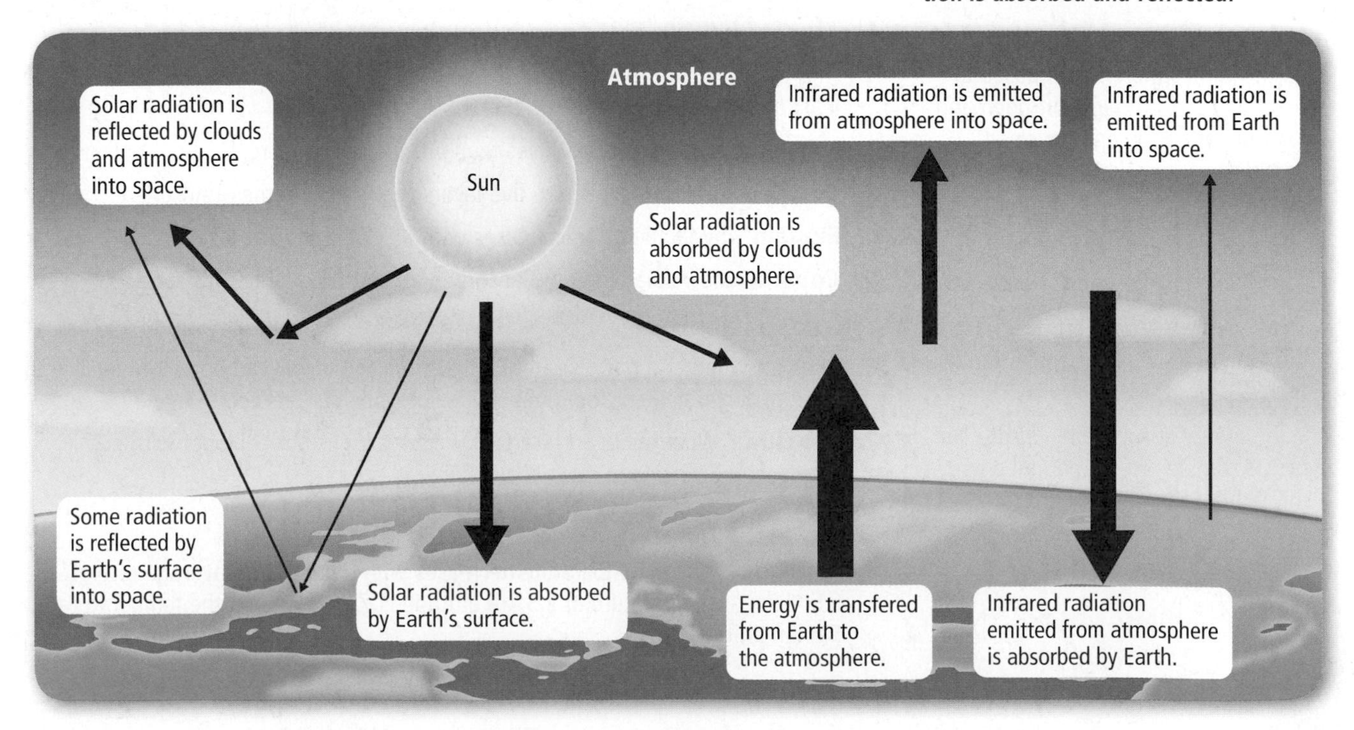

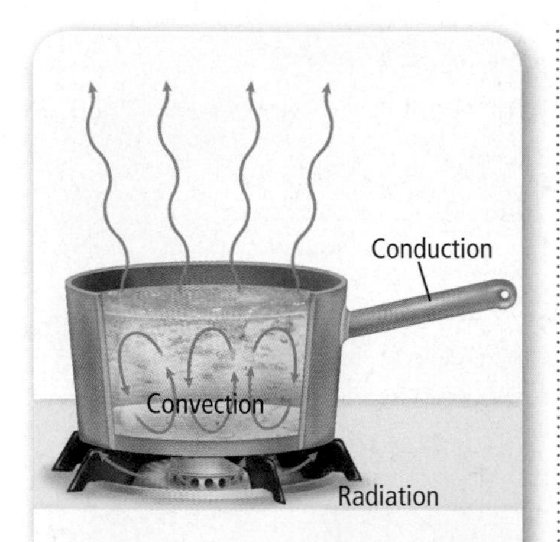

Figure 11.8 Thermal energy is transferred to the burner from the heat source by radiation. The burner transfers the energy to the atoms in the bottom of the pan, which collide with neighboring atoms. As these collisions occur, thermal energy is transferred by conduction to other parts of the pan, including the handle.

Concepts In Motion

Interactive Figure To see an animation of conduction, convection, and radiation, visit glencoe.com.

Conduction Another process of energy transfer can occur when two objects at different temperatures are in contact. **Conduction** is the transfer of thermal energy between objects when their atoms or molecules collide, as shown in **Figure 11.8.** Conduction can occur more easily in solids and liquids, where particles are close together, than in gases, where particles are farther apart. Because air is a mixture of gases, it is a poor conductor of thermal energy. In the atmosphere, conduction occurs between Earth's surface and the lowest part of the atmosphere.

Convection Throughout much of the atmosphere, thermal energy is transferred by a process called convection. The process of convection occurs mainly in liquids and gases. **Convection** is the transfer of thermal energy by the movement of heated material from one place to another. **Figure 11.8** illustrates the process of convection in a pan of water. As water at the bottom of the pan is heated, it expands and becomes less dense than the water around it. Because it is less dense, it is forced upward. As it rises, it transfers thermal energy to the cooler water around it, and cools. It then becomes denser than the water around it and sinks to the bottom of the pan, where it is reheated.

A similar process occurs in the atmosphere. Parcels of air near Earth's surface are heated, become less dense than the surrounding air, and rise. As the warm air rises, it cools and its density increases. When it cools below the temperature of the surrounding air, the air parcel becomes denser than the air around it and sinks. As it sinks, it warms again, and the process repeats. Convection currents, as these movements of air are called, are the main mechanism for energy transfer in the atmosphere.

Section 11.1 Assessment

Section Summary

- Earth's atmosphere is composed of several gases, primarily nitrogen and oxygen, and also contains small particles.
- Earth's atmosphere consists of five layers that differ in their compositions and temperatures.
- Solar energy reaches Earth's surface in the form of visible light and infrared waves.
- Solar energy absorbed by Earth's surface is transferred as thermal energy throughout the atmosphere.

Understand Main Ideas

1. MAIN Idea **Rank** the gases in the atmosphere in order from most abundant to least abundant.
2. **Name** the four types of particles found in the atmosphere.
3. **Compare and contrast** the five layers that make up the atmosphere.
4. **Explain** why temperature increases with height in the stratosphere.
5. **Compare** how solar energy is absorbed and emitted by Earth's surface.

Think Critically

6. **Predict** whether a pot of water heated from the top would boil more quickly than a pot of water heated from the bottom. Explain your answer.
7. **Conclude** What might surface temperatures be like on a planet with no atmosphere?

MATH in Earth Science

8. In the troposphere, temperature decreases with height at an average rate of 6.5°C/km. If temperature at 2.5 km altitude is 7.0°C, what is the temperature at 5.5 km altitude?

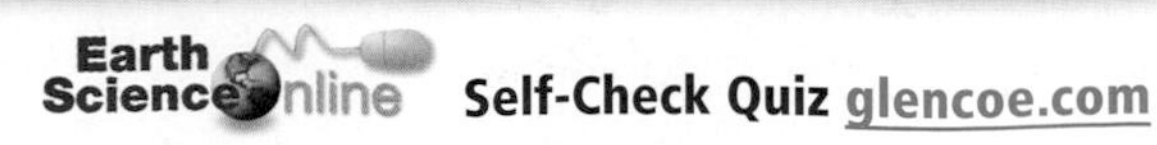

Section 11.2

Objectives

- **Identify** three properties of the atmosphere and how they interact.
- **Explain** why atmospheric properties change with changes in altitude.

Review Vocabulary

density: the mass per unit volume of a material

New Vocabulary

temperature inversion
humidity
saturation
relative humidity
dew point
latent heat

Properties of the Atmosphere

MAIN Idea Atmospheric properties, such as temperature, air pressure, and humidity describe weather conditions.

Real-World Reading Link Have you noticed the weather today? Maybe it is hot or cold, humid or dry, or even windy. These properties are always interacting and changing, and you can observe those changes every time you step outside.

Temperature

When you turn on the burner beneath a pot of water, thermal energy is transferred to the water and the temperature increases. Recall that particles in any material are in random motion. Temperature is a measure of the average kinetic energy of the particles in a material. Particles have more kinetic energy when they are moving faster, so the higher the temperature of a material, the faster the particles are moving.

Measuring temperature Temperature is usually measured using one of two common temperature scales. These scales are the Fahrenheit (°F) scale, used mainly in the United States, and the Celsius (°C) scale. The SI temperature scale used in science is the Kelvin (K) scale. **Figure 11.9** shows the differences among these temperature scales. The Fahrenheit and Celsius scales are based on the freezing point and boiling point of water. The zero point of the Kelvin scale is absolute zero—the lowest temperature that any substance can have.

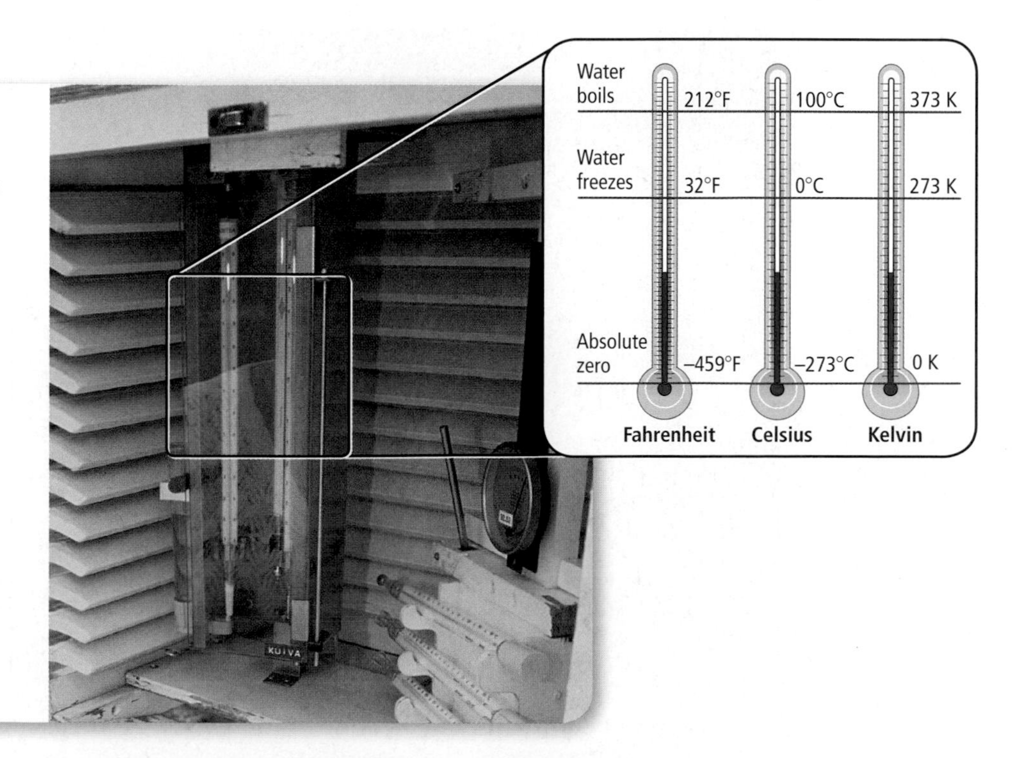

Figure 11.9 Temperature can be measured in degrees Fahrenheit, degrees Celsius, or in kelvin. The Kelvin scale starts at 0 K, which corresponds to –273°C and –459°F.

VOCABULARY
SCIENCE USAGE V. COMMON USAGE

Force

Science usage: an influence that might cause a body to accelerate

Common usage: violence, compulsion, or strength exerted upon or against a person or thing

Air Pressure

If you hold your hand out in front of you, Earth's atmosphere exerts a downward force on your hand due to the weight of the atmosphere above it. The force exerted on your hand divided by its area is the pressure exerted on your hand. Air pressure is the pressure exerted on a surface by the weight of the atmosphere above the surface.

Because pressure is equal to force divided by area, the units for pressure are N/m^2. Air pressure is often measured in units of millibars (mb), where 1 mb equals 100 N/m^2. At sea level, the atmosphere exerts a pressure of about 100,000 N/m^2, or 1000 mb. As you go higher in the atmosphere, air pressure decreases as the mass of the air above you decreases. **Figure 11.10** shows how pressure in the atmosphere changes with altitude.

 **Reading Check** **Deduce** why air pressure does not crush a human.

Density of air The density of a material is the mass of material in a unit volume, such as 1 m^3. Atoms and molecules become farther apart in the atmosphere as altitude increases. This means that the density of air decreases with increasing altitude, as shown in **Figure 11.10.** Near sea level, the density of air is about 1.2 kg/m^3. At the average altitude of the tropopause, or about 12 km above Earth's surface, the density of air is about 25 percent of its sea-level value. At the stratopause, or about 48 km above Earth's surface, air density has decreased to only about 0.2 percent of the air density at sea level.

■ **Figure 11.10** The density and pressure of the atmosphere decrease as altitude increases.

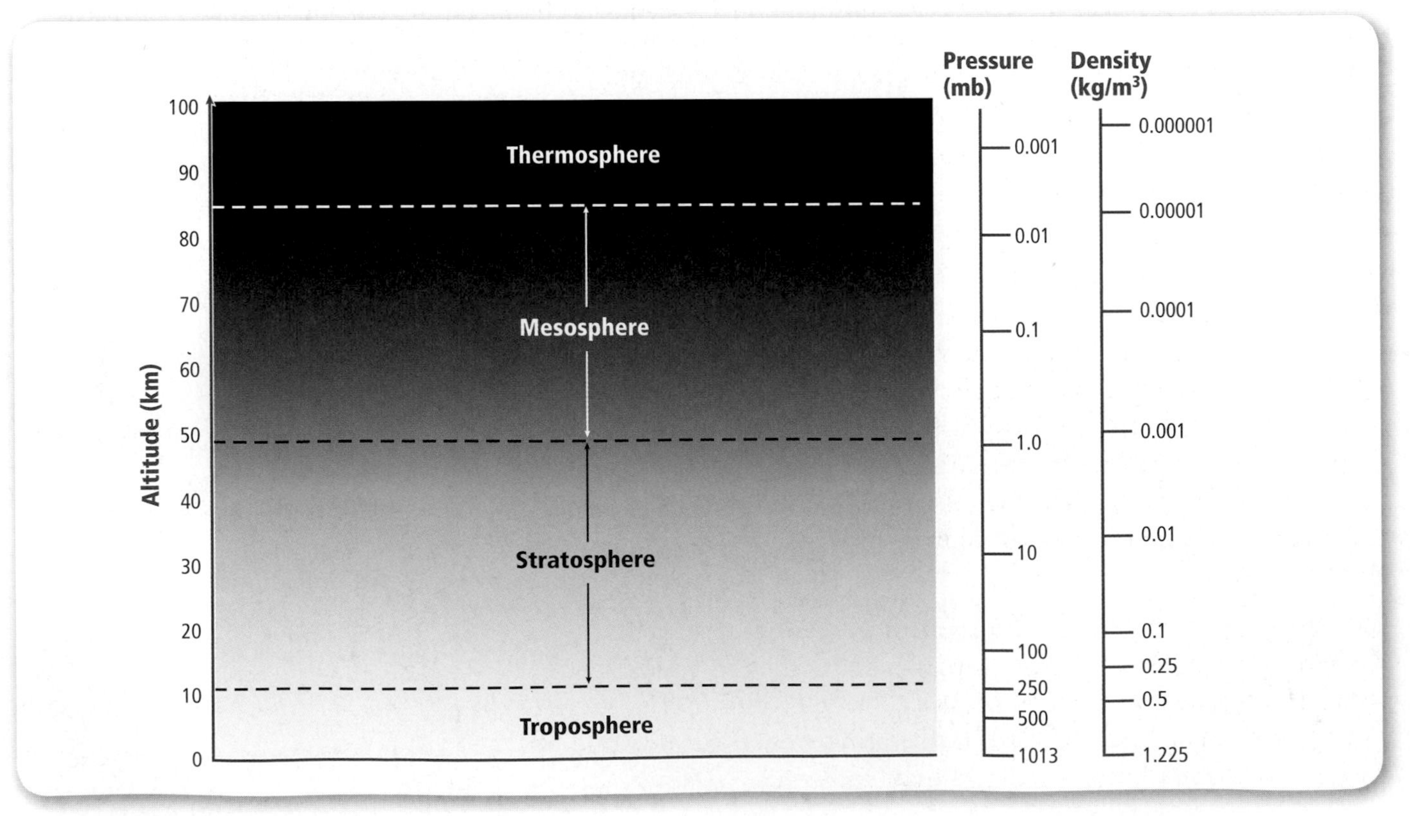

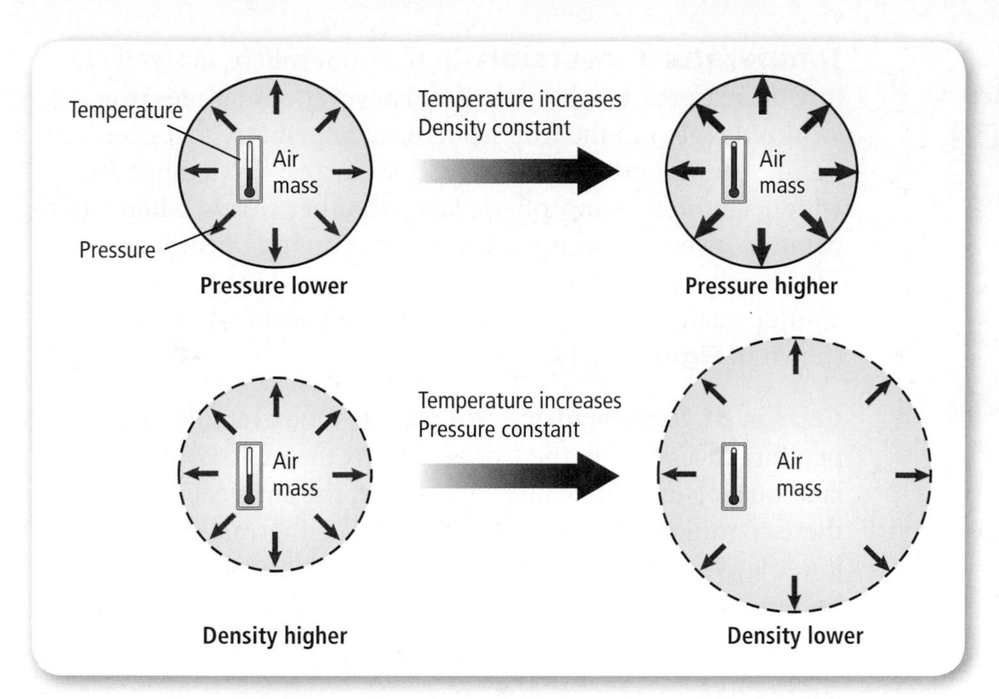

Figure 11.11 Temperature, pressure, and density are all related to one another. If temperature increases, but density is constant, the pressure increases. If the temperature increases and the pressure is constant, the density decreases.

Pressure-temperature-density relationship In the atmosphere, the temperature, pressure, and density of air are related to each other, as shown in **Figure 11.11.** Imagine a sealed container containing only air. The pressure exerted by the air inside the container is related to the air temperature inside the container and the air density. How does the pressure change if the air temperature or density changes?

Air pressure and temperature The pressure exerted by the air in the container is due to the collisions of the gas particles in the air with the sides of the container. When these particles move faster due to an increase in temperature, they exert a greater force when they collide with the sides of the container. The air pressure inside the container increases. This means that for air with the same density, warmer air is at a higher pressure than cooler air.

Air pressure and density Imagine that the temperature of the air does not change, but that more air is pumped into the container. Now there are more gas particles in the container, and therefore, the mass of the air in the container has increased. Because the volume has not changed, the density of the air has increased. Now there are more gas particles colliding with the walls of the container, and so more force is being exerted by the particles on the walls. This means that at the same temperature, air with a higher density exerts more pressure than air with a lower density.

Temperature and density Heating a balloon causes the air inside to move faster, causing the balloon to expand and increase in volume. As a result, the air density inside the balloon decreases. The same is true for air masses in the atmosphere. At the same pressure, warmer air is less dense than cooler air.

VOCABULARY

ACADEMIC VOCABULARY

Exert

to put forth (as strength)

Susan exerted a lot of energy playing basketball.

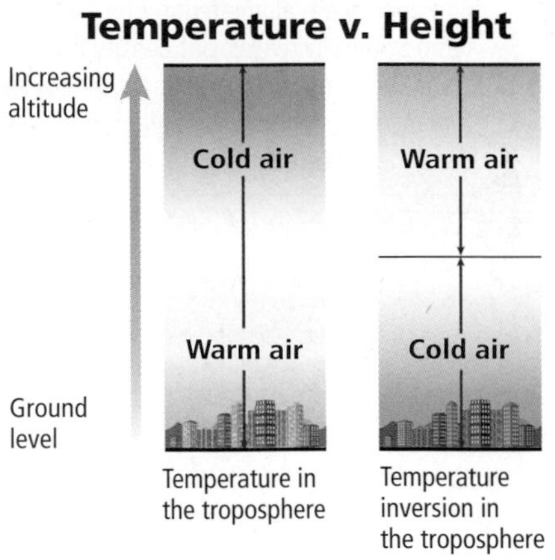

Temperature v. Height

Increasing altitude

Ground level

Cold air

Warm air

Temperature in the troposphere

Warm air

Cold air

Temperature inversion in the troposphere

■ **Figure 11.12** In a temperature inversion, the warm air is located on top of the cooler air.

Temperature inversion In the troposphere, air temperature decreases as height increases. However, sometimes over a localized region in the troposphere, a temperature inversion can occur. A **temperature inversion** is an increase in temperature with height in an atmospheric layer. In other words, when a temperature inversion occurs, warmer air is on top of cooler air. This is called a temperature inversion because the temperature-altitude relationship is inverted, or turned upside down, as shown in **Figure 11.12.**

Causes of temperature inversion One example of a temperature inversion on the troposphere is the rapid cooling of land on a cold, clear, winter night when the air is calm. Under these conditions, the land does not radiate thermal energy to the lower layers of the atmosphere. As a result, the lower layers of air become cooler than the air above them, so that temperature increases with height and forms a temperature inversion.

Effects of temperature inversion If the sky is very hazy, there is probably an inversion somewhere in the lower atmosphere. A temperature inversion can lead to fog or low-level clouds. Fog is a significant factor in lowering visibility in many coastal cities, such as San Francisco. In some cities, such as the one shown in **Figure 11.13,** a temperature inversion can worsen air-pollution problems. The heated air rises as long as it is warmer than the air above it and then it stops rising, acting like a lid to trap pollution under the inversion layer. Pollutants are consequently unable to be lifted from Earth's surface. Temperature inversions that remain over an industrial area for a long time usually result in episodes of severe smog—a combination of smoke and fog—that can cause respiratory problems.

■ **Figure 11.13** A temperature inversion in New York City traps air pollution above the city.
Describe ***the effect of temperature inversion on air quality in metropolitan areas.***

Wind Imagine you are entering a large, air-conditioned building on a hot summer day. As you open the door, you feel cool air rushing past you out of the building. This sudden rush of cool air occurs because the warm air outside the building is less dense and at a lower pressure than the cooler air inside the building. When the door opens, the difference in pressure causes the cool, dense air to rush out of the building. The movement of air is commonly known as wind.

Wind and pressure differences In the example above, the air in the building moves from a region of higher density to a region of lower density. In the lower atmosphere, air also generally moves from regions of higher density to regions of lower density. These density differences are produced by the unequal heating and cooling of different regions of Earth's surface. In the atmosphere, air pressure generally increases as density increases, so regions of high and low density are also regions of high and low air pressure respectively. As a result, air moves from a region of high pressure to a region of low pressure.

Wind speed and altitude Wind speed and direction change with height in the atmosphere. Near Earth's surface, wind is constantly slowed by the friction that results from contact with surfaces including trees, buildings, and hills, as shown **Figure 11.14.** Even the surface of water affects air motion. Higher up from Earth's surface, air encounters less friction and wind speeds increase. Wind speed is usually measured in miles per hour (mph) or kilometers per hour (km/h). Ships at sea usually measure wind in knots. One knot is equal to 1.85 km/h.

Figure 11.14 When wind blows over a forested area by a coast, it encounters more friction than when it blows over flatter terrain. This occurs because the wind encounters friction from the mountains, trees, and then the water, slowing the wind's speed.

Humidity

The distribution and movement of water vapor in the atmosphere play an important role in determining the weather of any region. **Humidity** is the amount of water vapor in the atmosphere at a given location on Earth's surface. Two ways of expressing the water vapor content of the atmosphere are relative humidity and dew point.

Relative humidity Consider a flask containing water. Some water molecules evaporate, leaving the liquid and becoming part of the water vapor in the flask. At the same time, other water molecules condense, returning from the vapor to become part of the liquid. Just as the amount of water vapor in the flask might vary, so does the amount of water vapor in the atmosphere. Water on Earth's surface evaporates and enters the atmosphere and condenses to form clouds and precipitation.

In the example of the flask, if the rate of evaporation is greater than the rate of condensation, the amount of water vapor in the flask increases. **Saturation** occurs when the amount of water vapor in a volume of air has reached the maximum amount. A saturated solution cannot hold any more of the substance that is being added to it. When a volume of air is saturated, it cannot hold any more water.

The amount of water vapor in a volume of air relative to the amount of water vapor needed for that volume of air to reach saturation is called **relative humidity.** Relative humidity is expressed as a percentage. When a certain volume of air is saturated, its relative humidity is 100 percent. If you hear a weather forecaster say that the relative humidity is 50 percent, it means that the air contains 50 percent of the water vapor needed for the air to be saturated.

PROBLEM-SOLVING LAB

Interpret the Graph

How do you calculate relative humidity? Relative humidity is the ratio of the actual amount of water vapor in a volume of air relative to the maximum amount of water vapor needed for that volume of air to reach saturation. Use the graph at the right to answer the following questions.

Think Critically

1. **Compare** the maximum amount of water vapor 1 m^3 of air could hold at 15°C and 25°C.
2. **Calculate** the relative humidity of 1 m^3 of air containing 10 g/m^3 at 20°C.
3. **Analyze** Can relative humidity be more than 100 percent? Explain your answer.

Data and Observations

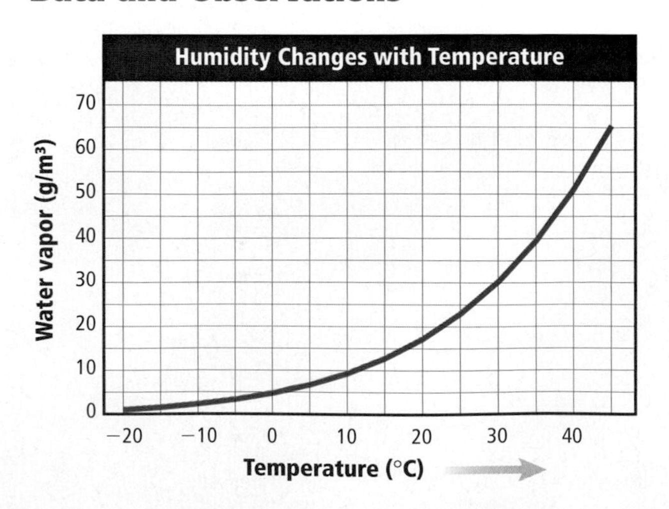

Dew point Another common way of describing the moisture content of air is the dew point. The **dew point** is the temperature to which air must be cooled at constant pressure to reach saturation. The name *dew point* comes from the fact that when the temperature falls to this level, dew begins to form. If the dew point is nearly the same as the air temperature, then the relative humidity is high.

Latent heat As water vapor in the air condenses, thermal energy is released. Where does this energy come from? To change liquid water to water vapor, thermal energy is added to the water by heating it. The water vapor then contains more thermal energy than the liquid water. This is the energy that is released when condensation occurs. The extra thermal energy contained in water vapor compared to liquid water is called **latent heat.**

When condensation occurs, as in **Figure 11.15,** latent heat is released and warms the air. At any given time, the amount of water vapor present in the atmosphere is a significant source of energy because it contains latent heat. When water vapor condenses, the latent heat released can provide energy to a weather system, such as a hurricane, increasing its intensity.

Condensation level An air mass can change temperature without being heated or cooled. A process in which temperature changes without the addition or removal of thermal energy from a system is called an adiabatic process. An example of an adiabatic process is the heating of air in a bicycle pump as the air is compressed. In a similar way, an air mass heats up as it sinks and cools off as it rises. Adiabatic heating occurs when air is compressed, and adiabatic cooling occurs when air expands.

MiniLab

Investigate Dew Formation

How does dew form? Dew forms when moist air near the ground cools and the water vapor in the air condenses into water droplets.

Procedure

1. Read and complete the lab safety form.
2. Fill a **glass** about two-thirds full of **water**. Record the temperature of the room and the water.
3. Add **ice cubes** until the glass is full. Record the temperature of the water at 10-s intervals.
4. Observe the outside of the glass. Note the time and the temperature at which changes occur on the outside of the glass.
5. Repeat the investigation outside. Record the temperature of the water and the air outside.

Analysis

1. **Compare and contrast** what happened to the outside of the glass when the investigation was performed in your classroom and when it was performed outside. If there was a difference, explain.
2. **Relate** your observations to the formation of dew.

Evaporation-Condensation Equilibrium

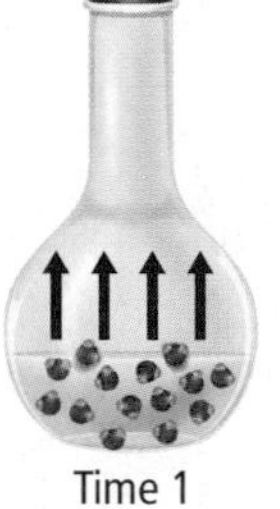

Time 1
25°C
Water molecules begin to evaporate.

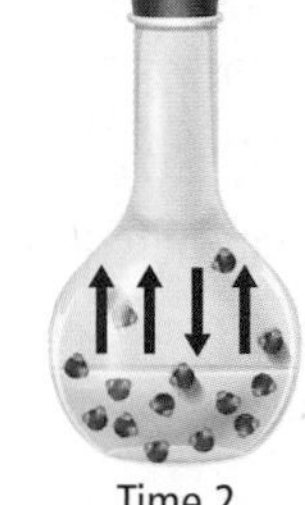

Time 2
25°C
Evaporation continues, and condensation begins.

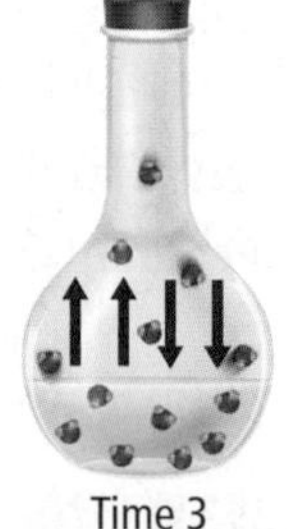

Time 3
25°C
Rate of evaporation equals rate of condensation or saturation.

Figure 11.15 During evaporation, water molecules escape from the surface of the liquid and enter the air as water vapor. During condensation, water molecules return to the liquid state. At equilibrium, evaporation and condensation continue, but the amount of water in the air and amount of water in the liquid form remain constant.

■ **Figure 11.16** Condensation occurs at the lifted condensation level (LCL). Air above the LCL is saturated and thus cools more slowly than air below the LCL.
Explain *why air above the LCL cools more slowly than air below the LCL.*

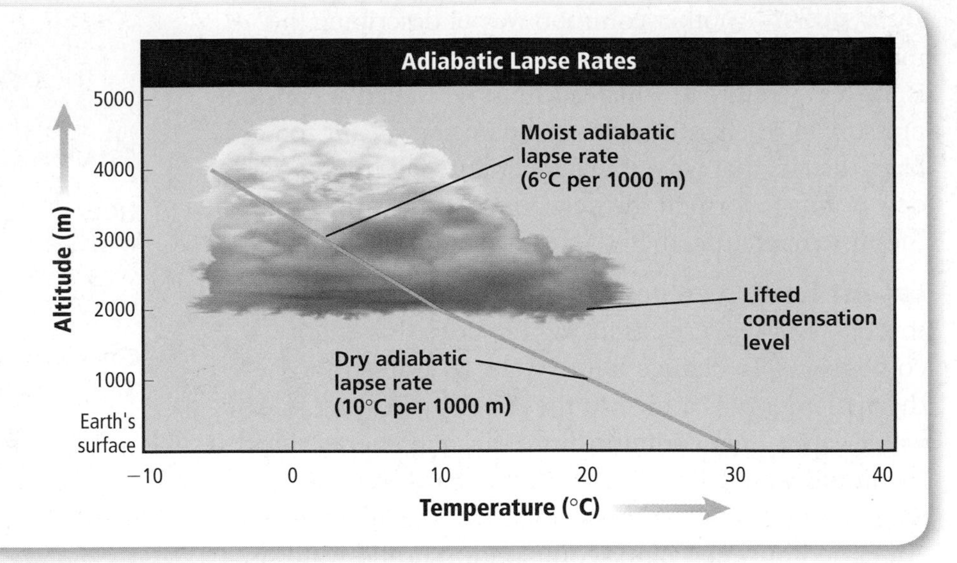

A rising mass of air cools because the air pressure around it decreases as it rises, causing the air mass to expand. A rising air mass that does not exchange thermal energy with its surroundings will cool by about 10°C for every 1000 m it rises. This is called the dry adiabatic lapse rate—the rate at which unsaturated air will cool as it rises if no thermal energy is added or removed. If the air mass continues to rise, eventually it will reach saturation. The height at which condensation occurs is called the lifted condensation level (LCL).

The rate at which saturated air cools is called the moist adiabatic lapse rate. This rate ranges from about 4°C/1000 m in very warm air to almost 9°C/1000 m in very cold air. This rate is slower than the dry adiabatic rate, as shown in **Figure 11.16,** because water vapor in the air is condensing as the air rises and is releasing latent heat.

Section 11.2 Assessment

Section Summary

- At the same pressure, warmer air is less dense than cooler air.
- Air moves from regions of high pressure to regions of low pressure.
- The dew point of air depends on the amount of water vapor the air contains.
- Latent heat is released when water vapor condenses and when water freezes.

Understand Main Ideas

1. MAIN Idea **Identify** three properties of the atmosphere and describe how they vary with height in the atmosphere.
2. **Explain** what occurs during a temperature inversion.
3. **Describe** how the motion of particles in a material changes when the temperature of the material increases.

Think Critically

4. **Predict** how the relative humidity and dew point change in a rising mass of air.
5. **Design** an experiment that shows how average wind speeds change over different types of surfaces.

MATH in Earth Science

6. If the average thickness of the troposphere is 11 km, what would be the temperature difference between the top and bottom of the troposphere if the temperature decrease is the same as the dry adiabatic lapse rate?

Earth Science Online **Self-Check Quiz** glencoe.com

Section 11.3

Objectives

- **Explain** the difference between stable and unstable air.
- **Compare and contrast** low, middle, high, and vertical development clouds.
- **Explain** how precipitation forms.

Review Vocabulary

condensation: process in which water vapor changes to a liquid

New Vocabulary

condensation nucleus
orographic lifting
cumulus
stratus
cirrus
precipitation
coalescence

Clouds and Precipitation

MAIN Idea **Clouds vary in shape, size, height of formation, and type of precipitation.**

Real-World Reading Link If you look up at the sky, you might notice differences among the clouds from day to day and hour to hour. Some clouds signal fair weather and others signal violent storms.

Cloud Formation

A cloud can form when a rising air mass cools. Recall that Earth's surface heats and cools by different amounts in different places. This uneven heating and cooling of the surface causes air masses near the surface to warm and cool. As an air mass is heated, it becomes less dense than the cooler air around it. This causes the warmer air mass to be pushed upward by the denser, cooler air.

However, as the warm air mass rises, it expands and cools adiabatically. The cooling of an air mass as it rises can cause water vapor in the air mass to condense. Recall that the lifted condensation level is the height at which condensation of water vapor occurs in an air mass. When a rising air mass reaches the lifted condensation level, water vapor condenses around condensation nuclei, as shown in **Figure 11.17.** A **condensation nucleus** is a small particle in the atmosphere around which water droplets can form. These particles are usually less than about 0.001 mm in diameter and can be made of ice, salt, dust, and other materials. The droplets that form can be liquid water or ice, depending on the surrounding temperature. When the number of these droplets is large enough, a cloud is visible.

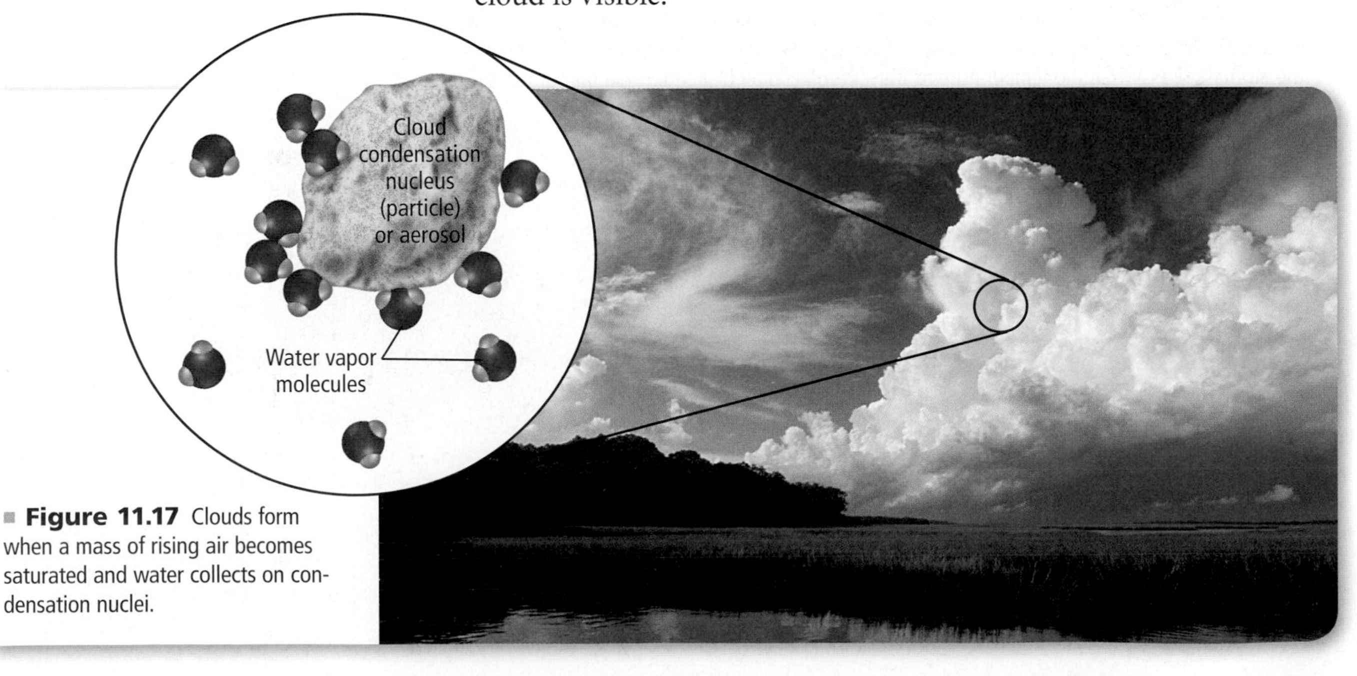

Figure 11.17 Clouds form when a mass of rising air becomes saturated and water collects on condensation nuclei.

CAREERS IN EARTH SCIENCE

Weather Observer A weather observer collects information for meteorologists about weather and sea conditions using weather equipment, radar scans, and satellite photographs. An education that includes biology, Earth science, environmental science, and geology is useful for a weather observer. To learn more about Earth science careers, visit glencoe.com.

Atmospheric stability As an air mass rises, it cools. However, the air mass will continue to rise as long as it is warmer than the surrounding air. Under some conditions, an air mass that has started to rise sinks back to its original position. When this happens, the air is considered stable because it resists rising. The stability of air masses determines the type of clouds that form and the associated weather patterns.

Stable air The stability of an air mass depends on how the temperature of the air mass changes relative to the atmosphere. The air temperature near Earth's surface decreases with altitude. As a result, the atmosphere becomes cooler as the air mass rises. At the same time, the rising air mass is also becoming cooler. Suppose that the temperature of the atmosphere decreases more slowly with increasing altitude than does the temperature of the rising air mass. Then the rising air mass will cool more quickly than the atmosphere. The air mass will finally reach an altitude at which it is colder than the atmosphere. It will then sink back to the altitude at which its density is the same as the atmosphere, as shown in **Figure 11.18.** Because the air mass stops rising and sinks downward, it is stable. Fair weather clouds form under stable conditions.

Reading Check **Describe** the factors that affect the stability of air.

Unstable air Suppose that the temperature of the surrounding air cools faster than the temperature of the rising air mass. Then the air mass will always be less dense than the surrounding air. As a result, the air mass will continue to rise, as shown in **Figure 11.18.** The atmosphere is then considered to be unstable. Unstable conditions can produce the type of clouds associated with thunderstorms.

■ **Figure 11.18** Stable air has a tendency to resist movement. Unstable air does not resist vertical displacement. When the temperature of a mass of air is greater than the temperature of the surrounding air, the air mass rises. When the temperature of the surrounding air is greater than that of the air mass, it sinks.

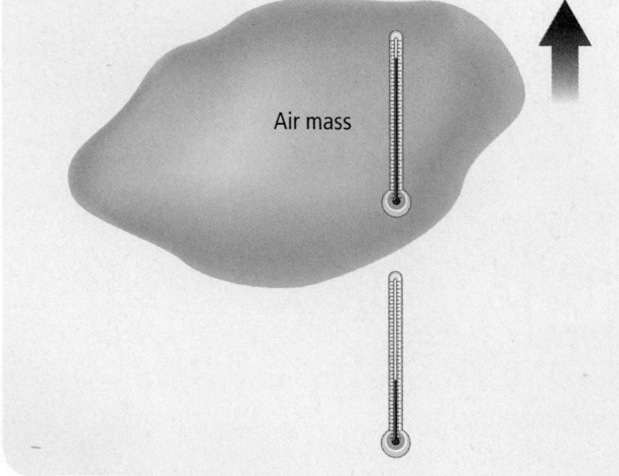

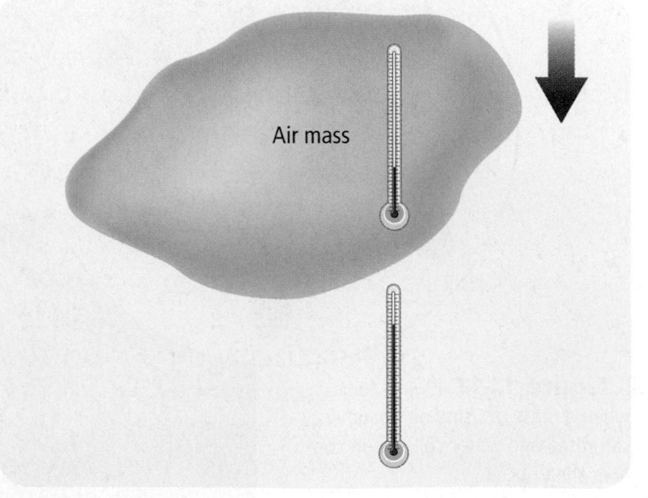

■ **Figure 11.19** Orographic lifting occurs when warm, moist air is cooled because it is forced to rise over a mountain.

Atmospheric lifting Clouds can form when moist air rises, expands, and cools. Air rises when it is heated and becomes warmer than the surrounding air. This process is known as convective lifting. Clouds can also form when air is forced upward or lifted by mechanical processes. Two of these processes are orographic lifting and convergence.

Orographic lifting Clouds can form when air is forced to rise over elevated land or other topographic barriers. This can happen, for example, when an air mass approaches a mountain range. **Orographic lifting** occurs when an air mass is forced to rise over a topographic barrier, as shown in **Figure 11.19.** The rising air mass expands and cools, with water droplets condensing when the temperature reaches the dew point. Many of the rainiest places on Earth are located on the windward sides of mountain slopes, such as the coastal side of the Sierra Nevadas. The formation of clouds and the resulting heavy precipitation along the west coast of Canada are also primarily due to orographic lifting.

Convergence Air can be lifted by convergence, which occurs when air flows into the same area from different directions. Then some of the air is forced upward. This process is even more pronounced when air masses at different temperatures collide. When a warm air mass and a cooler air mass collide, the warmer, less-dense air is forced upward over the denser, cooler air. As the warm air rises, it cools adiabatically. If the rising air cools to the dew-point temperature, then water vapor can condense on condensation nuclei and form a cloud. This cloud formation mechanism is common at middle latitudes where severe storm systems form along the cold polar front. Convergence also occurs near the equator where the trade winds meet at the intertropical convergence zone. You will read more about these topics in Chapter 12.

Types of Clouds

You have probably noticed that clouds have different shapes. Some clouds look like puffy cotton balls, while others have a thin, feathery appearance. These differences in cloud shape are due to differences in the processes that cause clouds to form. Cloud formation can also take place at different altitudes—sometimes even right at Earth's surface, in which case the cloud is known as fog.

Clouds are generally classified according to a system developed in 1803, and only minor changes have been made since it was first introduced. **Figure 11.20** shows the most common types of clouds. This system classifies clouds by the altitudes at which they form and by their shapes. There are three classes of clouds based on the altitudes at which they form: low, middle, and high. In addition, there are clouds with vertical development. Low clouds typically form below 2000 m. Middle clouds form mainly between 2000 m and 6000 m. High clouds form above 6000 m. Unlike the other three classes of clouds, those with vertical development can form at all altitudes.

■ **Figure 11.20** Clouds form at different altitudes and in different shapes.
Compare and contrast *cirrus and stratus clouds.*

Low clouds Clouds can form when warm, moist air rises, expands, and cools. If conditions are stable, the air mass stops rising at the altitude where its temperature is the same as that of the surrounding air. If a cloud has formed, it will flatten out and winds will spread it horizontally into stratocumulus or layered cumulus clouds, as shown in **Figure 11.20. Cumulus** (KYEW myuh lus) clouds are puffy, lumpy-looking clouds that usually occur below 2000 m. Another type of cloud that forms at heights below 2000 m is a **stratus** (STRAY tus), a layered sheetlike cloud that covers much or all of the sky in a given area. Stratus clouds often form when fog lifts away from Earth's surface.

Middle clouds Altocumulus and altostratus clouds form at altitudes between 2000 m and 6000 m. They are made up of ice crystals and water droplets due to the colder temperatures generally present at these altitudes. Middle clouds are usually layered. Altocumulus clouds are white or gray in color and form large, round masses or wavy rows. Altostratus clouds have a gray appearance, and they form thin sheets of clouds. Middle clouds sometimes produce mild precipitation.

High clouds High clouds, made up of ice crystals, form at heights of 6000 m where temperatures are below freezing. Some, such as **cirrus** (SIHR us) clouds, often have a wispy, indistinct appearance. Another type of cirrus cloud, called a cirrostratus, forms as a continuous layer that can cover the sky. Cirrostratus clouds vary in thickness from almost transparent to dense enough to block out the Sun or the Moon.

Reading Check **Identify** types of low, middle, and high clouds.

Vertical development clouds If the air that makes up a cumulus cloud is unstable, the cloud will be warmer than the surrounding air and will continue to grow upward. As it rises, water vapor condenses, and the air continues to increase in temperature due to the release of latent heat. The cloud can grow through middle altitudes as a towering cumulonimbus, as shown in **Figure 11.21,** and, if conditions are right, it can reach the tropopause. Its top is then composed entirely of ice crystals. Strong winds can spread the top of the cloud into an anvil shape. What began as a small mass of unstable moist air is now an atmospheric giant, capable of producing the torrential rains, strong winds, and hail characteristic of some thunderstorms.

Figure 11.21 Cumulonimbus clouds, such as the large, puffy cloud here, are associated with thunderstorms.
Describe *how a cumulonimbus cloud can form.*

Precipitation

All forms of water that fall from clouds to the ground are **precipitation.** Rain, snow, sleet, and hail are the four main types of precipitation. Clouds contain water droplets that are so small that the upward movement of air in the cloud can keep the droplets from falling. In order for these droplets to become heavy enough to fall, their size must increase by 50 to 100 times.

Coalescence One way that cloud droplets can increase in size is by coalescence. In a warm cloud, coalescence is the primary process responsible for the formation of precipitation. **Coalescence** (koh uh LEH sunts)occurs when cloud droplets collide and join together to form a larger droplet. These collisions occur as larger droplets fall and collide with smaller droplets. As the process continues, the droplets eventually become too heavy to remain suspended in the cloud and fall to Earth as precipitation. Rain is precipitation that reaches Earth's surface as a liquid. Raindrops typically have diameters between 0.5 mm and 5 mm.

Snow, sleet, and hail The type of precipitation that reaches Earth depends on the vertical variation of temperature in the atmosphere. In cold clouds where the air temperature is far below freezing, ice crystals can form that finally fall to the ground as snow. Sometimes, even if ice crystals form in a cloud, they can reach the ground as rain if they fall through air warmer than 0°C and melt.

In some cases, air currents in a cloud can cause cloud droplets to move up and down through freezing and nonfreezing air, forming ice pellets that fall to the ground as sleet. Sleet can also occur when raindrops freeze as they fall through freezing air near the surface.

If the up-and-down motion in a cloud is especially strong and occurs over large stretches of the atmosphere, large ice pellets known as hail can form. **Figure 11.22** shows a sample of hail. Most hailstones are smaller in diameter than a dime, but some stones have been found to weigh more than 0.5 kg. Larger stones are often produced during severe thunderstorms.

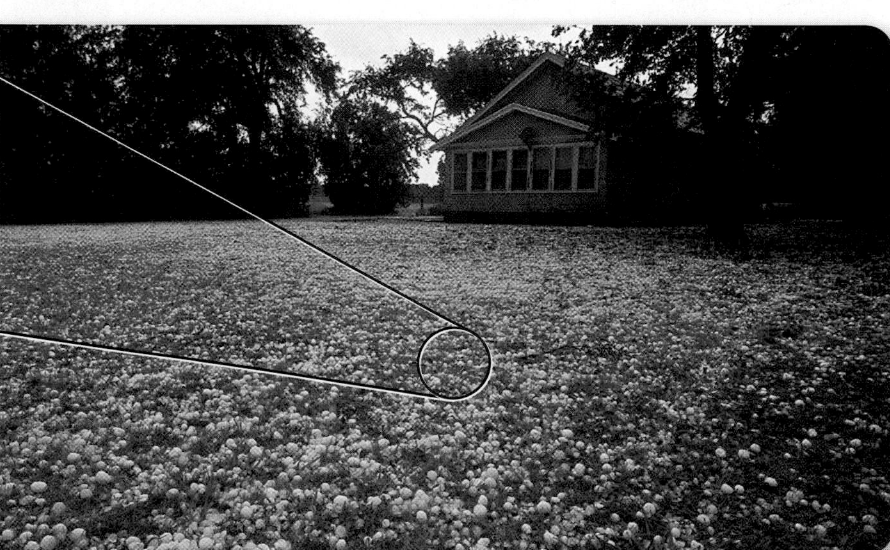

■ **Figure 11.22** Hail is precipitation in the form of balls or lumps of ice that is produced by intense thunderstorms.
Infer *How might the layers in the cross section of the hailstone form?*

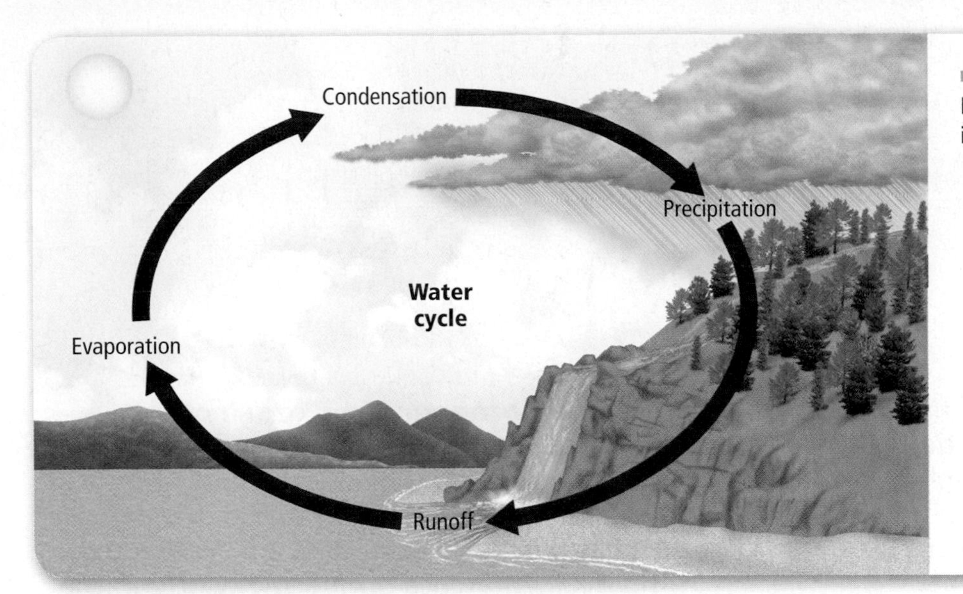

■ **Figure 11.23** Water moves from Earth to the atmosphere and back to Earth in the water cycle.

The water cycle More than 97 percent of Earth's water is in the oceans. At any one time, only a small percentage of water is present in the atmosphere. Still, this water is vitally important because, as it continually moves between the atmosphere and Earth's surface, it nourishes living things. The constant movement of water between the atmosphere and Earth's surface is known as the water cycle.

The water cycle is summarized in **Figure 11.23.** Radiation from the Sun causes liquid water to evaporate. Water evaporates from lakes, streams, and oceans and rises into Earth's atmosphere. As water vapor rises, it cools and condenses to form clouds. Water droplets combine to form larger drops that fall to Earth as precipitation. This water soaks into the ground and enters lakes, streams, and oceans, or it falls directly into bodies of water and eventually evaporates, continuing the water cycle.

Section 11.3 Assessment

Section Summary

- Clouds are formed as warm, moist air is forced upward, expands, and cools.
- An air mass is stable if it tends to return to its original height after it starts rising.
- Cloud droplets form when water vapor is cooled to the dew point and condenses on condensation nuclei.
- Clouds are classified by their shapes and the altitudes at which they form.
- Cloud droplets collide and coalesce into larger droplets that can fall to Earth as rain, snow, sleet, or hail.

Understand Main Ideas

1. MAIN Idea **Summarize** the differences between low clouds, middle clouds, and high clouds.
2. **Describe** how precipitation forms.
3. **Determine** the reason precipitation will fall as snow rather than rain.
4. **Compare** stable and unstable air.

Think Critically

5. **Evaluate** how a reduction in the number of condensation nuclei in the troposphere would affect precipitation. Explain your reasoning.

WRITING in Earth Science

6. Describe the path a drop of rain might follow throughout the water cycle.

Earth Science and the Environment

Ozone Variation

Atoms, such as chlorine and bromine, when located in the stratosphere, can destroy ozone molecules. The decline in stratospheric ozone measured since the early 1980s might be showing signs of recovery due to a decrease in stratospheric chlorine.

Variations in ozone amounts The total amount of ozone in the atmosphere over Earth's surface varies with location and also changes with time. Total ozone increases with latitude, being low at the equator and highest in the polar regions. Ozone amounts also vary seasonally, usually decreasing from winter to summer. The largest seasonal changes occur at high latitudes, particularly in the polar regions.

The Antarctic ozone hole Over Antarctica, the lowest ozone amounts occur in early spring. Since the late 1970s, total ozone amounts in the spring have greatly decreased. This decrease in springtime ozone over Antarctica is called the Antarctic ozone hole. The Antarctic ozone hole is caused by chlorofluorocarbons (CFCs) and the presence of polar stratospheric clouds (PSCs). These clouds form over Antarctica during the winter in the lower stratosphere. CFCs break down, producing molecules that contain chlorine atoms. These molecules undergo chemical reactions on ice crystals in the PSCs, producing chlorine and other compounds that destroy ozone.

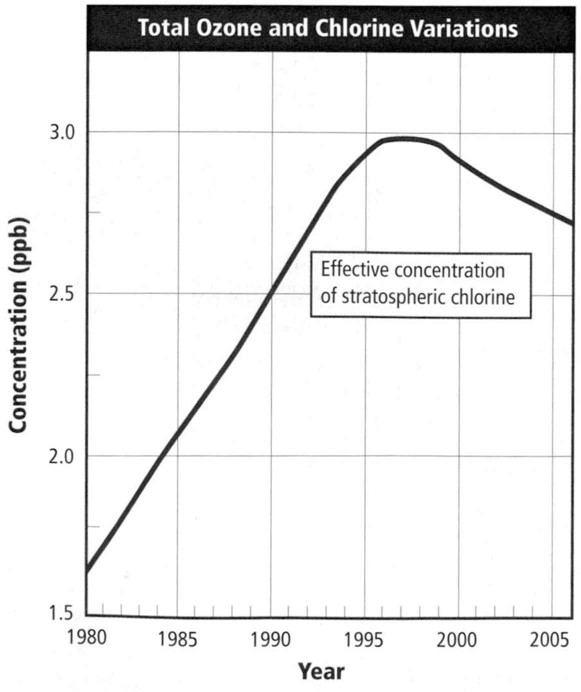

Total Ozone and Chlorine Variations

Concentration (ppb)
3.0
2.5
2.0
1.5
1980
1985
1990
1995
2000
2005
Year
Effective concentration of stratospheric chlorine

The Montreal Protocol Satellite measurements beginning in the late 1970s also showed a decrease in global ozone amounts of several percent. Concerns over decreasing ozone led to the adoption of the Montreal Protocol in 1987. This international agreement requires countries to phase out the production and use of CFCs and similar chemicals. As a result, levels of chlorine and other ozone-destroying chemicals in the stratosphere have been declining since the late 1990s, as shown in the graph.

Signs of recovery? Between 1996 and 2006, the decrease in total ozone leveled off in most regions. Part of these changes might be due to natural causes, such as solar variability, as well as to the Montreal Protocol. Measurements over several more years will be needed to determine whether the ozone layer is recovering.

WRITING in Earth Science

Magazine Article Research how natural processes, such as volcanic eruptions, solar activity, and air movements affect ozone levels in the stratosphere. Write a magazine article that reports what you found. To learn more about the ozone layer, visit glencoe.com.

GEOLAB

INTERPRET PRESSURE-TEMPERATURE RELATIONSHIPS

Background: As you go up a mountain, both temperature and air pressure decrease. Temperature decreases as you get farther away from the atmosphere's heat source—Earth's surface. Pressure decreases as you ascend the mountain because there are fewer particles in the air above you. Pressure and temperature are also related through the expansion and compression of air, regardless of height.

Question: *How does the expansion and compression of air affect temperature?*

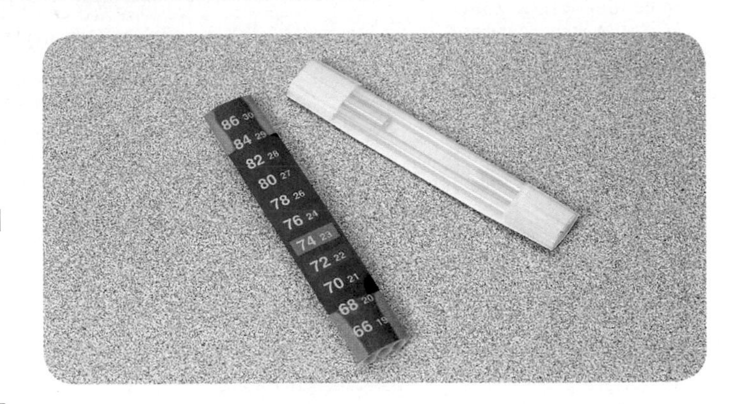

Materials

clean, clear, plastic 2-L bottle with cap
plastic straws
scissors
thin, liquid-crystal temperature strip
tape
watch or timer

Safety Precautions

Procedure

1. Read and complete the lab safety form.
2. Working with a partner, cut two pieces of straw, each the length of the temperature strip. Then cut two 2-cm pieces of straw. **WARNING:** ***Scissors pose a skin puncture or cut hazard.***
3. Lay the two long pieces on a table. Place the two shorter pieces within the space created by the longer pieces so that the four pieces form a support for the temperature strip as shown in the figure.
4. Tape the four pieces of straw together. Place the temperature strip lengthwise on the straws. Tape the strip to the straws.
5. Slide the temperature-strip-straw assembly into the clean, dry bottle. Screw the cap on tightly.
6. Place the sealed bottle on the table so that the temperature strip faces you and is easy to read. Do not handle the bottle any more than is necessary so that the temperature will not be affected by your hands.
7. Record the temperature of the air inside the bottle as indicated by the temperature strip.
8. Position the bottle so that half its length extends beyond the edge of the table. Placing one hand on each end of the bottle, push down on both ends so that the bottle bends in the middle. Hold the bottle this way for 2 min while your partner records the temperature every 15 s.
9. Release the pressure on the bottle. Observe and record the temperature every 15 s for the next 2 min.

Analyze and Conclude

1. **Interpret Data** What was the average temperature of the air inside the bottle as you applied pressure? How did this differ from the average temperature of the bottled air when you released the pressure?
2. **Graph** the temperature changes.
3. **Explain** how these temperature changes are related to changes in pressure.
4. **Predict** how the experiment would change if you took the cap off the bottle.
5. **Infer** Given your observations and what you know about the behavior of warm air, would you expect the air over an equatorial desert at midday to be characterized by high or low pressure?

WRITING in Earth Science

Research how pressure changes can affect the daily weather. Share your findings with your classmates. For more information on weather, visit glencoe.com.

CHAPTER 11 Study Guide

Download quizzes, key terms, and flash cards from glencoe.com.

BIG Idea The composition, structure, and properties of Earth's atmosphere form the basis of Earth's weather and climate.

Vocabulary	Key Concepts
Section 11.1 Atmospheric Basics	
• conduction (p. 288) • convection (p. 288) • exosphere (p. 286) • mesosphere (p. 284) • radiation (p. 287) • stratosphere (p. 284) • thermosphere (p. 284) • troposphere (p. 284)	**MAIN Idea** Energy is transferred throughout Earth's atmosphere. • Earth's atmosphere is composed of several gases, primarily nitrogen and oxygen, and also contains small particles. • Earth's atmosphere consists of five layers that differ in their compositions and temperatures. • Solar energy reaches Earth's surface in the form of visible light and infrared waves. • Solar energy absorbed by Earth's surface is transferred as thermal energy throughout the atmosphere.
Section 11.2 Properties of the Atmosphere	
• dew point (p. 295) • humidity (p. 294) • latent heat (p. 295) • relative humidity (p. 294) • saturation (p. 294) • temperature inversion (p. 292)	**MAIN Idea** Atmospheric properties, such as temperature, air pressure, and humidity describe weather conditions. • At the same pressure, warmer air is less dense than cooler air. • Air moves from regions of high pressure to regions of low pressure. • The dew point of air depends on the amount of water vapor the air contains. • Latent heat is released when water vapor condenses and when water freezes.
Section 11.3 Clouds and Precipitation	
• cirrus (p. 301) • coalescence (p. 302) • condensation nucleus (p. 297) • cumulus (p. 301) • orographic lifting (p. 299) • precipitation (p. 302) • stratus (p. 301)	**MAIN Idea** Clouds vary in shape, size, height of formation, and type of precipitation. • Clouds are formed as warm, moist air is forced upward, expands, and cools. • An air mass is stable if it tends to return to its original height after it starts rising. • Cloud droplets form when water vapor is cooled to the dew point and condenses on condensation nuclei. • Clouds are classified by their shapes and the altitudes at which they form. • Cloud droplets collide and coalesce into larger droplets that can fall to Earth as rain, snow, sleet, or hail.

CHAPTER 11 Assessment

Vocabulary Review

Match each description below with the correct vocabulary term from the Study Guide.

1. outermost layer of Earth's atmosphere
2. transfer of energy from a higher to a lower temperature by collisions between particles
3. temperature at which condensation of water vapor can occur
4. occurs when the amount of water vapor in a volume of air has reached the maximum amount
5. the amount of water vapor present in air

Complete the sentences below using vocabulary terms from the Study Guide.

6. ________ are small particles in the atmosphere around which water droplets form.
7. The atmospheric layer that is closest to Earth's surface is the ________.
8. Types of ________ include hail, sleet, and snow.

Each of the following sentences is false. Make each sentence true by replacing the italicized words with terms from the Study Guide.

9. *Convection* occurs when small cloud droplets collide to form a larger droplet.
10. *Mesosphere* is the layer of Earth's atmosphere that contains the ozone layer.
11. The transfer of energy in matter or space by electromagnetic waves is called *latent heat.*
12. When the bottom of a pan of water is heated and the water expands, becoming less dense than the surrounding water, it is forced upward. As it rises, the water cools and sinks back to the bottom of the pan. This process is called *precipitation.*
13. When *saturation* occurs, an air mass is forced to rise over a topographic barrier.

Understand Key Concepts

14. Which gas has increased in concentration by about 0.011 percent over the past 150 years?
    A. oxygen
    B. nitrogen
    C. carbon dioxide
    D. water vapor

Use the diagram below to answer Question 15.

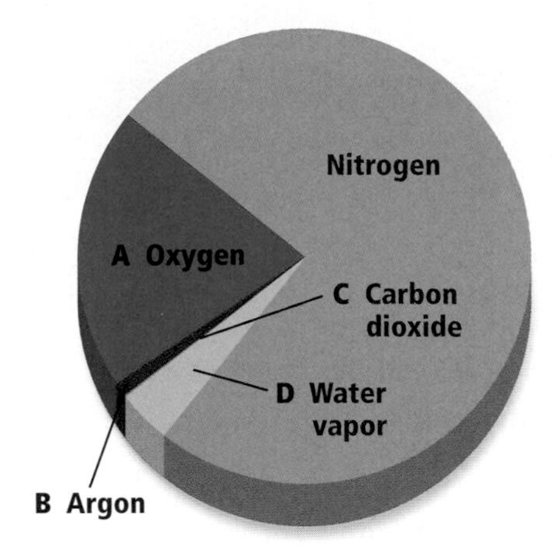

15. Which gas is least abundant in Earth's atmosphere?
    A. A
    B. B
    C. C
    D. D

16. Which is the primary cause of wind?
    A. air saturation
    B. pressure imbalances
    C. pollution
    D. movement of water

17. Which process takes up latent heat?
    A. condensation of water vapor
    B. evaporation of water vapor
    C. adiabatic heating
    D. pressure increase

18. Wind speed on Earth is reduced by which?
    A. temperature
    B. friction
    C. weather
    D. convergence

Use the diagram below to answer Question 19.

19. Which mechanical process is causing the air to rise?
A. coalescence
B. convection
C. orographic lifting
D. convergence

20. Which is a vertical development cloud?
A. cumulonimbus
B. cirrus
C. stratus
D. altocumulus

21. Almost all weather, clouds, and storms occur in which layer of the atmosphere?
A. thermosphere
B. mesosphere
C. stratosphere
D. troposphere

22. What color would be best for a home designed to absorb energy?
A. red
B. white
C. gray
D. black

23. Which temperature is coldest?
A. 32°F
B. 10°C
C. 280 K
D. 5°C

Constructed Response

24. Explain why precipitation from a cumulonimbus cloud is generally heavier than that from a nimbostratus cloud.

25. Identify the role that evaporation and condensation play in Earth's water cycle.

26. Compare what happens to latent heat in the atmosphere during evaporation to what occurs during condensation.

Use the figure below to answer Question 27.

27. Describe the process that causes the cloud type shown to reach heights of over 6000 m.

28. Determine whether the average relative humidity on a small island in the ocean would likely be higher or lower than 100 km inland on a continent.

29. Explain If clouds absorb only a small amount of solar radiation, how is Earth's atmosphere heated?

30. Distinguish between convection and conduction as methods of transferring energy in the atmosphere.

31. Compare the temperature and composition of the troposphere and the stratosphere.

32. Determine what causes precipitation to fall as rain or snow.

33. Relate dew point and saturation.

34. Describe the importance of water vapor in the atmosphere.

Think Critically

35. **CAREERS IN EARTH SCIENCE** Research information about the workday of a weather observer.

36. **Predict** how the concentration of ozone molecules would change if the concentration of oxygen molecules decreased.

37. **Infer** Using the idea that almost all weather occurs in the troposphere, infer why many airliners usually fly at altitudes of 11 km or higher.

38. **Predict** whether afternoon summertime temperatures near the beach would be warmer or cooler than temperatures farther inland. Explain.

39. **Predict** why spring is often the windiest time of the year based on your knowledge of temperature and wind.

40. **Predict** how the energy absorbed by the Arctic Ocean would change if the amount of the sea ice covering the ocean is reduced. Keep in mind that sea ice reflects more incoming solar energy than water does.

41. **Assess** which cloud type would be of most interest to a hydrologist who is concerned with possible heavy rain and flooding over large regions. Why?

42. **Analyze** why relative humidity usually decreases after the Sun rises and increases after the Sun sets.

Concept Mapping

43. Use the following terms and phrases to construct a concept map that describes the process of the water cycle: *water cycle; evaporation; condensation; precipitation; water changes from liquid to gas; water changes from gas to liquid; water falls as rain, snow, sleet, or hail.* For more help, refer to the *Skillbuilder Handbook*.

Challenge Question

44. Based on what you know about radiation and conduction, what conclusion might you make about summer temperatures in a large city compared with those in the surrounding countryside?

Additional Assessment

45. **WRITING in Earth Science** Write and illustrate a short story for elementary students that describes cumulonimbus cloud formation and the kinds of weather patterns they produce.

DBQ Document–Based Questions

Data obtained from: Climatological normals 1971–2000. *National Oceanographic and Atmospheric Administration, National Climatic Data Center.*

The graphs show the monthly variations in temperature and precipitation at three locations in the United States. Use the data to answer the questions below.

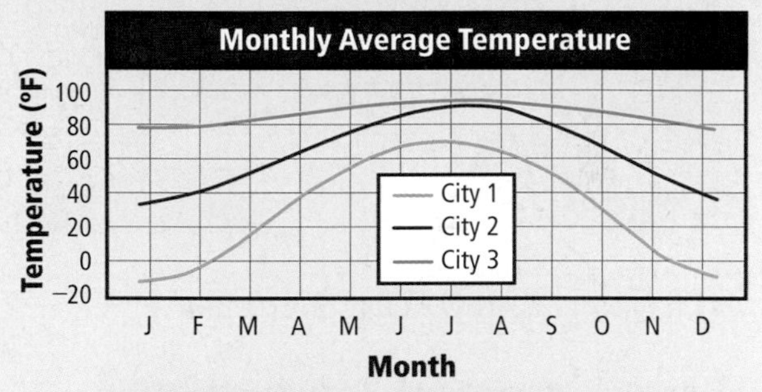

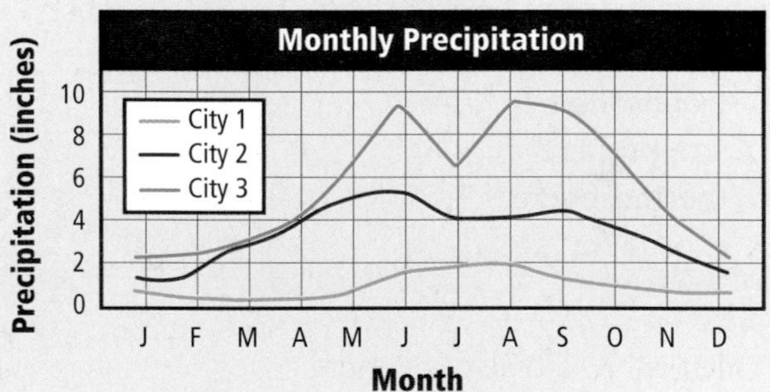

46. Estimate from the data which location probably receives the least annual solar radiation.

47. In which location would you expect heavy precipitation?

48. Deduce from the graphs which station probably receives the most annual snowfall.

Cumulative Review

49. Describe the properties of a contour line. **(Chapter 2)**

50. What process is explained by Bowen's reaction series? **(Chapter 5)**

Standardized Test Practice

Multiple Choice

1. What is the composition of dripstone formations?
   A. gravel
   B. limestone
   C. clay
   D. sand

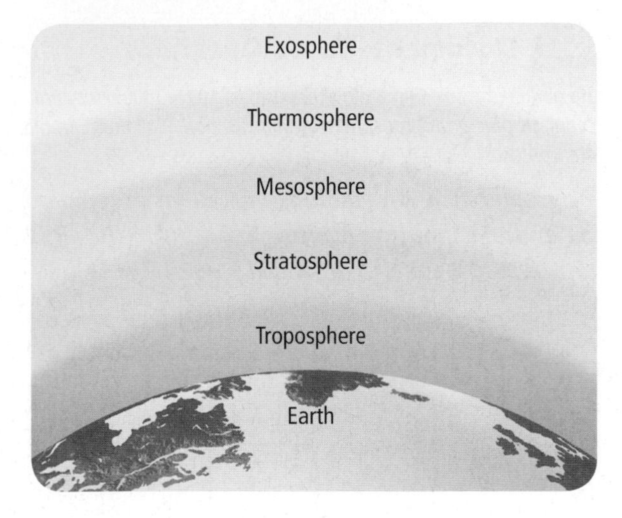

Use the diagram to answer Questions 2 and 3.

2. In which layer of Earth's atmosphere is air most likely warmed by conduction?
   A. troposphere
   B. stratosphere
   C. thermosphere
   D. exosphere

3. Which is NOT true of ozone?
   A. It absorbs ultraviolet radiation.
   B. Its concentration is decreasing.
   C. It is concentrated in the atmospheric layer called the mesosphere.
   D. It is a gas formed by the addition of one oxygen atom to an oxygen molecule.

4. Which describes the temperature of groundwater flowing through a natural spring?
   A. hotter than the region's average temperature
   B. cooler than the region's average temperature
   C. the same temperature no matter where the spring is located
   D. the same temperature as the region's average temperature

5. Why do deserts experience wind erosion?
   A. There is limited rain to allow plants to grow and hold down sediment.
   B. Saltation does not occur readily in desert areas.
   C. The increased amount of heat increases wind patterns.
   D. Wind can carry larger particles than water.

6. Which is NOT a significant agent of chemical weathering?
   A. oxygen
   B. nitrogen
   C. carbon dioxide
   D. water

Use the table below to answer Questions 7 and 8.

Population of Unknown Organisms				
	Spring	Summer	Autumn	Winter
1995	564	14,598	25,762	127
1996	750	16,422	42,511	102
1997	365	14,106	36,562	136

7. What inference can be made based upon the data?
   A. Scientists have a hard time consistently tracking the organism.
   B. The organism migrates yearly.
   C. The organism is most abundant during summer and fall.
   D. The organism should be placed on the endangered species list.

8. What would be the best graphical representation of this data?
   A. bar graph
   B. line graph
   C. circle graph
   D. model

9. Which is most likely to cause orographic lifting?
   A. a sandy beach
   B. a flowing river
   C. a rocky mountain
   D. a sand dune

10. Why are the lakes in Central Florida considered to have karst topography?
    A. They are depressions in the ground near caves.
    B. They are part of a sinking stream.
    C. They are layered with limestone.
    D. They are sinkholes.

Short Answer

Use the illustration below to answer Questions 11–13.

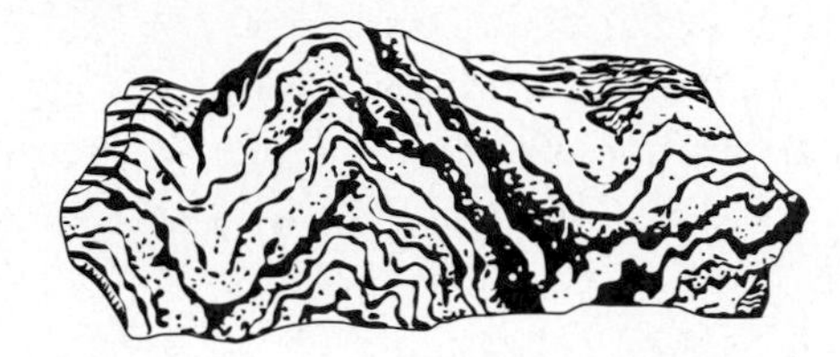

11. What type of rock is shown above? What features indicate this?

12. Hypothesize how this sample of rock formed.

13. According to the rock cycle, what changes could occur in this rock? What new type of rock would be produced?

14. Describe temperature and heat.

15. Define ion and explain how an ion is formed.

16. Describe how a flood might cause residual soil to become transported soil.

Reading for Comprehension

Ozone Layer Recovery

Damage to the ozone layer, caused by chlorofluorocarbon (CFC) chemicals and other pollutants, may be starting to reverse itself, according to data collected by NASA satellites. Ozone degradation continues despite global bans on ozone-depleting pollutants. The rate has slowed enough in the upper stratosphere that scientists think ozone could start to be replenished there within several years.

Evidence suggests that international efforts to reduce chlorofluorocarbon (CFC) pollution are working. Some predictions suggest that the ozone layer will have recovered to preindustrial levels by the late twenty-first century, though total recovery could happen within 50 years.

17. According to the passage, what is the major cause of the replenishing of the ozone layer?
    A. the ban of chlorofluorocarbons
    B. preindustrial pollution
    C. the upper stratosphere
    D. NASA satellites

18. What can be inferred from this passage?
    A. The ozone layer is recovering, but will never be fully restored.
    B. CFC pollution is no longer occurring.
    C. The upper stratosphere is the only layer with ozone depletion.
    D. Ozone depletion in the upper stratosphere has slowed down.

19. According to the text, how long could it take for a full recovery of the ozone layer?
    A. a decade
    B. until the late twenty-first century
    C. 50 years
    D. several years

20. Why is it important that the ozone layer in the upper stratosphere is replenished?

NEED EXTRA HELP?																
If You Missed Question . . .	1	2	3	4	5	6	7	8	9	10	11	12	13	14	15	16
Review Section . . .	10.2	11.1	11.1	10.1	8.2	7.1	1.2	1.3	11.3	10.2	6.3	6.3	6.3	11.2	3.1	7.3

CHAPTER 12 Meteorology

BIG Idea Weather patterns can be observed, analyzed, and predicted.

12.1 The Causes of Weather
MAIN Idea Air masses have different temperatures and amounts of moisture because of the uneven heating of Earth's surface.

12.2 Weather Systems
MAIN Idea Weather results when air masses with different pressures and temperatures move, change, and collide.

12.3 Gathering Weather Data
MAIN Idea Accurate measurements of atmospheric properties are a critical part of weather analysis and prediction.

12.4 Weather Analysis and Prediction
MAIN Idea Several methods are used to develop short-term and long-term weather forecasts.

GeoFacts

- The coldest temperature ever recorded in the United States was –62.1°C at Prospect Creek, Alaska.
- The sunniest place in the United States is Yuma, Arizona, with an average of 4133 hours of sunshine per year.

LAUNCH Lab

How does a cold air mass form?

An air mass is a large volume of air that has the characteristics of the area over which it formed.

Procedure

1. Read and complete the lab safety form.
2. Place a **full tray of ice cubes** on a table. Place a **pencil** under each end of the tray to raise it off the table.
3. Slide a **liquid-crystal temperature strip** under the ice-cube tray.
4. Place two **pencils** across the top of the tray, and **another temperature strip** across them.
5. Record the temperature of each strip at 1-min intervals for about 5 min.
6. Make a graph of the temperature changes over time for each temperature strip.

Analysis

1. **Describe** what happened to the temperatures above and below the tray.
2. **Explain** how this models a mass of cold air.

Visit glencoe.com to

- study entire chapters online;
- explore Concepts In Motion animations:
  - Interactive Time Lines
  - Interactive Figures
  - Interactive Tables
- access Web Links for more information, projects, and activities;
- review content with the Interactive Tutor and take Self-Check Quizzes.

Types of Fronts Make the following Foldable to help identify the four types of fronts.

STEP 1 Layer three sheets of paper so that the top margin or about 3 cm of each sheet can be seen.

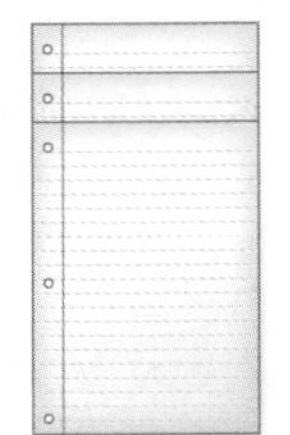

STEP 2 Make a 3-cm horizontal cut through all three sheets on about the sixth line of the top sheet.

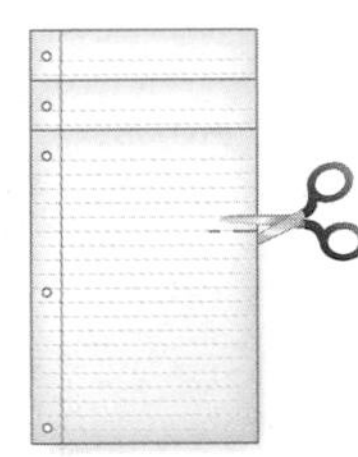

STEP 3 Make a vertical cut up from the bottom to meet the horizontal cut.

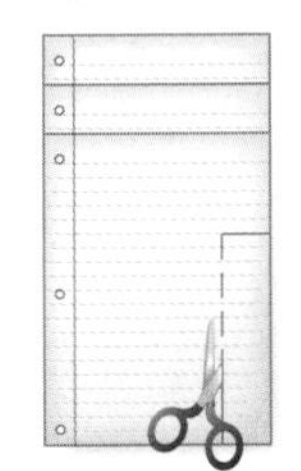

STEP 4 Place the three sheets on top of a fourth sheet and align the tops and sides of all sheets. Label the four tabs *Cold Fronts, Warm Fronts, Stationary Fronts,* and *Occluded Fronts.* The Foldable can be placed in a notebook or stapled along the left edge.

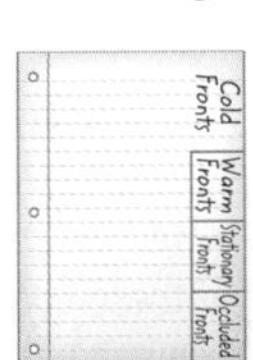

FOLDABLES **Use this Foldable with Section 12.2.** As you read this section, summarize what you learn about the different fronts. Include sketches of air movement and the weather map symbol for each type.

Section 12.1

Objectives

- **Compare and contrast** weather and climate.
- **Analyze** how imbalances in the heating of Earth's surface create weather.
- **Describe** how air masses form.
- **Identify** five types of air masses.

Review Vocabulary

heat: transfer of thermal energy from a warmer material to a cooler material

New Vocabulary

weather
climate
air mass
source region

The Causes of Weather

MAIN Idea Air masses have different temperatures and amounts of moisture because of the uneven heating of Earth's surface.

Real-World Reading Link Have you ever walked barefoot on cool grass and then stepped onto hot pavement on a sunny summer day? Around the world, the Sun heats the different surfaces on Earth to different extents. This uneven heating causes weather.

What is meteorology?

What do you enjoy doing on a summer afternoon? Do you like to watch clouds move across the sky, listen to leaves rustling in a breeze, or feel the warmth of sunlight on your skin? Clouds, breezes, and the warmth of sunlight are examples of atmospheric phenomena. Meteorology is the study of atmospheric phenomena. The root word of *meteorology* is the Greek word *meteoros,* which means *high in the air*. Anything that is high in the sky—raindrops, rainbows, dust, snowflakes, fog, and lightning—is an example of a meteor.

Atmospheric phenomena are often classified as types of meteors. Cloud droplets and precipitation—rain, snow, sleet, and hail—are types of hydrometeors (hi droh MEE tee urz). Smoke, haze, dust, and other particles suspended in the atmosphere are lithometeors (lih thuh MEE tee urz). Examples of electrometeors are thunder and lightning—signs of atmospheric electricity that you can hear or see. Meteorologists study these various meteors.

Weather versus climate Short-term variations in atmospheric phenomena that interact and affect the environment and life on Earth are called **weather.** These variations can take place over minutes, hours, days, weeks, months, or years. **Climate** is the long-term average of variations in weather for a particular area. Meteorologists use weather-data averages over 30 years to define an area's climate, such as that of the desert shown in **Figure 12.1.** You will read more about Earth's climates in Chapter 14.

Reading Check **Differentiate** between weather and climate.

Figure 12.1 A desert climate is dry with extreme variations in day and night temperatures. Only organisms adapted to these conditions, such as this ocotillo, can survive there.

Heating Earth's Surface

As you learned in Chapter 11, sunlight, which is a part of solar radiation, is always heating some portion of Earth's surface. Over the course of a year, the amount of thermal energy that Earth receives is the same as the amount that Earth radiates back to space. In meteorology, a crucial question is how solar radiation is distributed around Earth.

Imbalanced heating Why are average January temperatures warmer in Miami, Florida, than in Detroit, Michigan? Part of the explanation is that Earth's axis of rotation is tilted relative to the plane of Earth's orbit. Therefore, the number of hours of daylight and amount of solar radiation is greater in Miami during January than in Detroit.

Another factor is that Earth is a sphere and different places on Earth are at different angles to the Sun, as shown in **Figure 12.2.** For most of the year, the amount of solar radiation that reaches a given area at the equator covers a larger area at latitudes nearer the poles. The greater the area covered, the smaller amount of heat per unit of area. Because Detroit is farther from the equator than Miami is, the same amount of solar radiation that heats Miami will heat Detroit less. Investigate this relationship in the MiniLab on this page.

Thermal energy redistribution Areas around Earth maintain about the same average temperatures over time due to the constant movement of air and water among Earth's surfaces, oceans, and atmosphere. The constant movement of air redistributes thermal energy around the world. Weather—from thunderstorms to large-scale weather systems—is part of the constant redistribution of Earth's thermal energy.

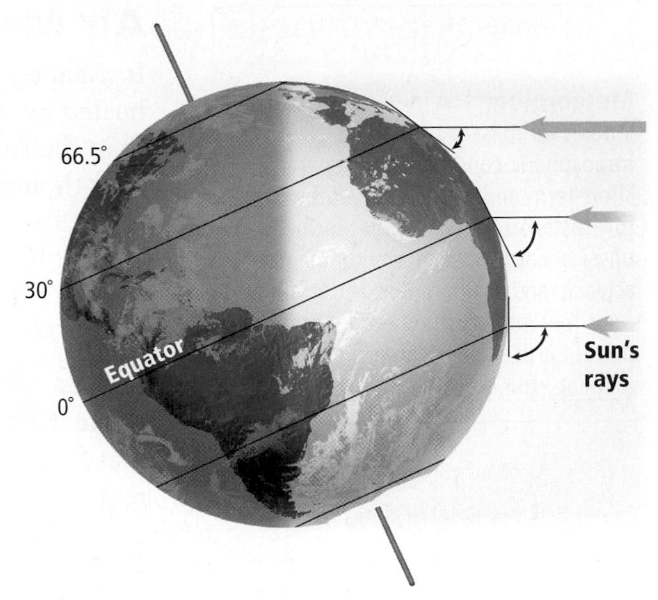

Figure 12.2 Solar radiation is unequal partly due to the changing angle of incidence of the sunlight. In this example it is perpendicular south of the equator, at the equator it is 60°, and north of the equator it is 40°.
Explain *why average temperatures decline from the equator to the poles.*

MiniLab

Compare the Angles of Sunlight to Earth

What is the relationship between the angle of sunlight and amount of heating? The angle at which sunlight reaches Earth's surface varies with latitude. This results in uneven heating of Earth.

Procedure

1. Read and complete the lab safety form.
2. Turn on a **flashlight,** and hold it 20 cm above a piece of **paper.** Point the flashlight straight down.
3. Use a pencil to trace the outer edge of the light on the paper. This models the angle of sunlight to Earth at the equator.
4. Keep the flashlight the same distance above the paper, but rotate it about 30°.
5. Trace the outer edge of the light. This is similar to the angle of sunlight to Earth at latitudes nearer the poles.

Analysis

1. **Describe** how the outline of the light differed between Step 3 and Step 5. Explain why it differed.
2. **Compare** the amount of energy per unit of area received near the equator to the amount at latitudes nearer the poles.

CAREERS IN EARTH SCIENCE

Meteorologist A meteorologist studies air masses and other atmospheric conditions to prepare short-term and long-term weather forecasts. An education that includes physics, Earth science, environmental science, and mathematics is useful for a meteorologist. To learn more about Earth science careers, visit glencoe.com.

Air Masses

In Chapter 11, you learned that air over a warm surface can be heated by conduction. This heated air rises because it is less dense than the surrounding air. On Earth, this process can take place over thousands of square kilometers for days or weeks. The result is the formation of an air mass. An **air mass** is a large volume of air that has the same characteristics, such as humidity and temperature, as its **source region**—the area over which the air mass forms. Most air masses form over tropical regions or polar regions.

Types of air masses The five types of air masses, listed in **Table 12.1,** influence weather in the United States. These air masses are all common in North America because there is a source region nearby.

Tropical air masses The origins of maritime tropical air are tropical bodies of water, listed in **Table 12.1.** In the summer, they bring hot, humid weather to the eastern two-thirds of North America. The southwestern United States and Mexico are a source region of continental tropical air, which is hot and dry, especially in summer.

Polar air masses Maritime polar air masses form over the cold waters of the North Atlantic and North Pacific. The one that forms over the North Pacific primarily affects the West Coast of the United States, occasionally bringing heavy rains in winter. Continental polar air masses form over the interior of Canada and Alaska. In winter, these air masses can carry frigid air southward. In the summer, however, cool, relatively dry, continental polar air masses bring relief from hot, humid weather.

Reading Check **Compare and contrast** tropical and polar air masses.

Table 12.1 Air Mass Characteristics

Concepts In Motion
Interactive Table To explore more about air masses, visit glencoe.com.

Air Mass Type	Weather Map Symbol	Source Region	Characteristics	
			Winter	Summer
Arctic	A	Siberia, Arctic Basin	bitter cold, dry	cold, dry
Continental polar	cP	interiors of Canada and Alaska	very cold, dry	cool, dry
Continental tropical	cT	southwest United States, Mexico	warm, dry	hot, dry
Maritime polar	mP	North Pacific Ocean	mild, humid	mild, humid
		North Atlantic Ocean	cold, humid	cool, humid
Maritime tropical	mT	Gulf of Mexico, Caribbean Sea, tropical and subtropical Atlantic Ocean and Pacific Ocean	warm, humid	hot, humid

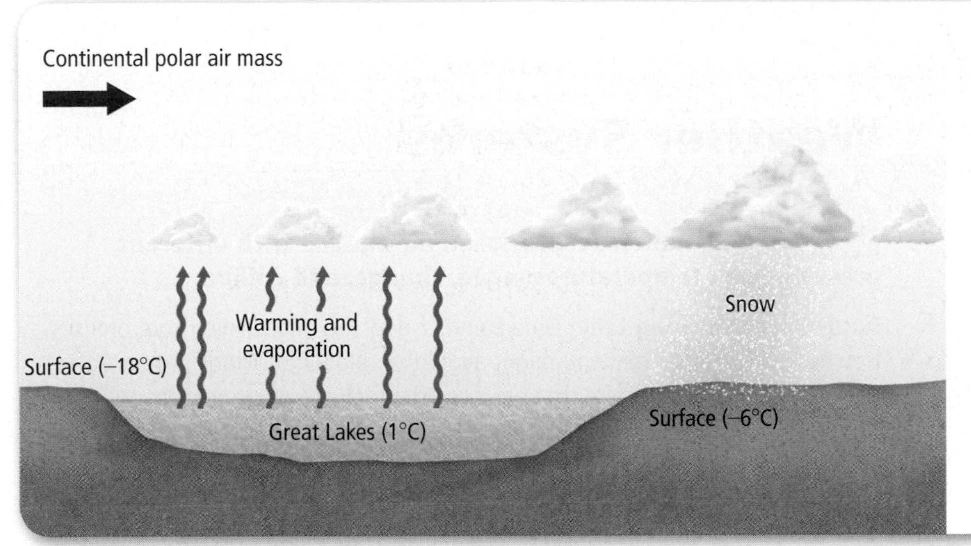

■ **Figure 12.3** As the cold, continental polar air moves over the warmer Great Lakes, the air gains thermal energy and moisture. This modified air cools as it is uplifted because of convection and topographic features, and produces lake-effect snows.

Arctic air masses Earth's ice- and snow-covered surfaces above 60° N latitude in Siberia and the Arctic Basin are the source regions of arctic air masses. During part of the winter, these areas receive almost no solar radiation but continue to radiate thermal energy. As a result, they become extremely cold and can bring the most frigid temperatures during winter.

Air mass modification Air masses do not stay in one place indefinitely. Eventually, they move, transferring thermal energy from one area to another. When an air mass travels over land or water that has characteristics different from those of its source region, the air mass can acquire some of the characteristics of that land or water, as shown in **Figure 12.3.** When this happens, the air mass undergoes modification; it exchanges thermal energy and/or moisture with the surface over which it travels.

Section 12.1 Assessment

Section Summary

- Meteorology is the study of atmospheric phenomena.
- Solar radiation is unequally distributed between Earth's equator and its poles.
- An air mass is a large body of air that takes on the moisture and temperature characteristics of the area over which it forms.
- Each type of air mass is classified by its source region.

Understand Main Ideas

1. MAIN Idea **Summarize** how an air mass forms.
2. **Explain** the process that prevents the poles from steadily cooling off and the tropics from heating up over time.
3. **Distinguish** between the causes of weather and climate.
4. **Differentiate** among the five types of air masses.

Think Critically

5. **Predict** which type of air mass you would expect to become modified more quickly: an arctic air mass moving over the Gulf of Mexico in winter or a maritime tropical air mass moving into the southeastern United States in summer.

WRITING in Earth Science

6. Describe how a maritime polar air mass formed over the North Pacific is modified as it moves west over North America.

Section 12.2

Objectives

- **Compare and contrast** the three major wind systems.
- **Identify** four types of fronts.
- **Distinguish** between high- and low-pressure systems.

Review Vocabulary

convection: the transfer of thermal energy by the flow of a heated substance

New Vocabulary

Coriolis effect
polar easterlies
prevailing westerlies
trade winds
jet stream
front

Weather Systems

MAIN Idea Weather results when air masses with different pressures and temperatures move, change, and collide.

Real-World Reading Link On a summer day, you might enjoy cool breezes. However, on a winter day, you might avoid the cold wind. Winds are part of a global air circulation system that balances thermal energy around the world.

Global Wind Systems

If Earth did not rotate on its axis, two large air convection currents would cover Earth, as shown in **Figure 12.4.** The colder and more dense air at the poles would sink to the surface and flow toward the tropics. There, the cold air would force warm, equatorial air to rise. This air would cool as it gained altitude and flowed back toward the poles. However, Earth rotates from west to east, which prevents this situation.

The directions of Earth's winds are influenced by Earth's rotation. This **Coriolis effect** results in fluids and objects moving in an apparent curved path rather than a straight line. Thus, as illustrated in **Figure 12.5,** moving air curves to the right in the northern hemisphere and curves to the left in the southern hemisphere. Together, the Coriolis effect and the heat imbalance on Earth create distinct global wind systems. They transport colder air to warmer areas near the equator and warmer air to colder areas near the poles. Global wind systems help to equalize the thermal energy on Earth.

There are three basic zones, or wind systems, at Earth's surface in each hemisphere. They are polar easterlies, prevailing westerlies, and trade winds.

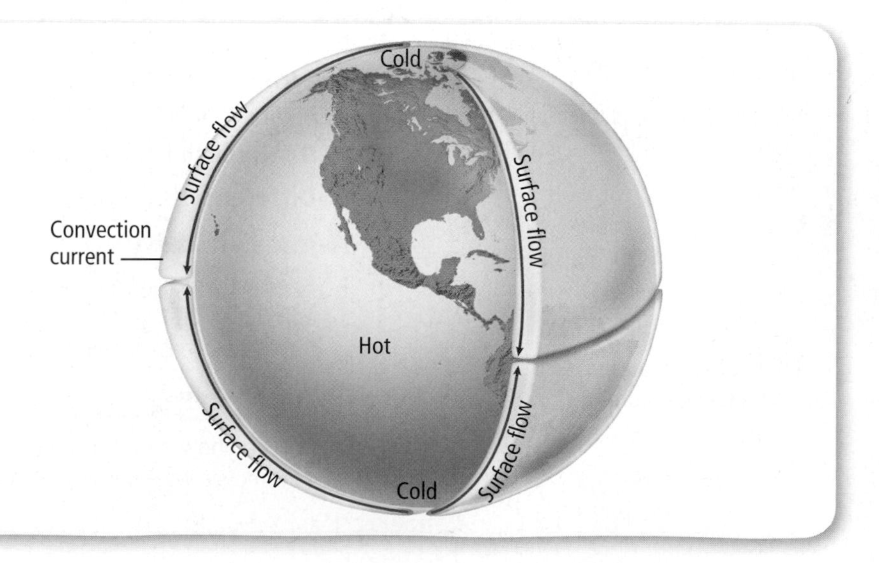

Figure 12.4 If Earth did not rotate, two large convection currents would form as denser polar air moved toward the equator. These currents would warm and rise as they approached the equator, and cool as they moved toward each pole.

Visualizing the Coriolis Effect

Figure 12.5 The Coriolis effect results in fluids and objects moving in an apparent curved path rather than a straight line.

Recall that distance divided by time equals speed. The equator has a length of about 40,000 km—Earth's circumference—and Earth rotates west to east once about every 24 hours. This means that things on the equator, including the air above it, move eastward at a speed of about 1670 km/h.

However, not every location on Earth moves eastward at this speed. Latitudes north and south of the equator have smaller circumferences than the equator. Those objects not on the equator move less distance during the same amount of time. Therefore, their eastward speeds are slower than objects on the equator.

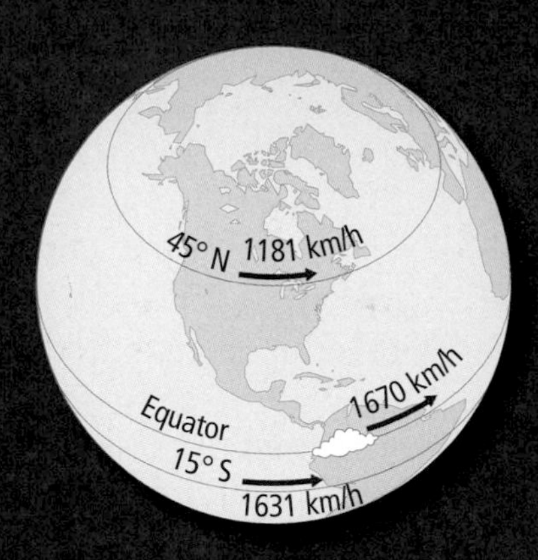

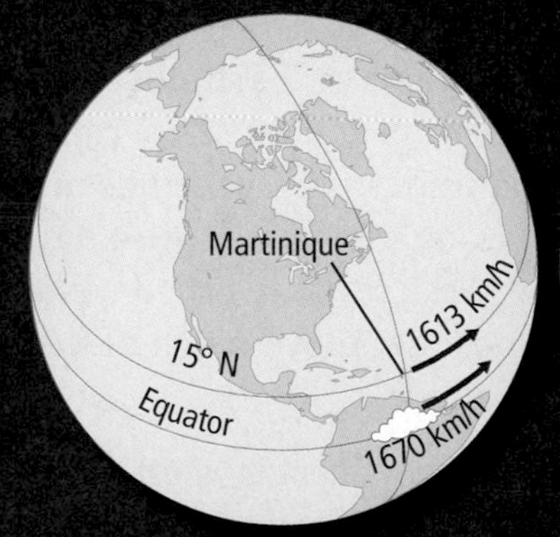

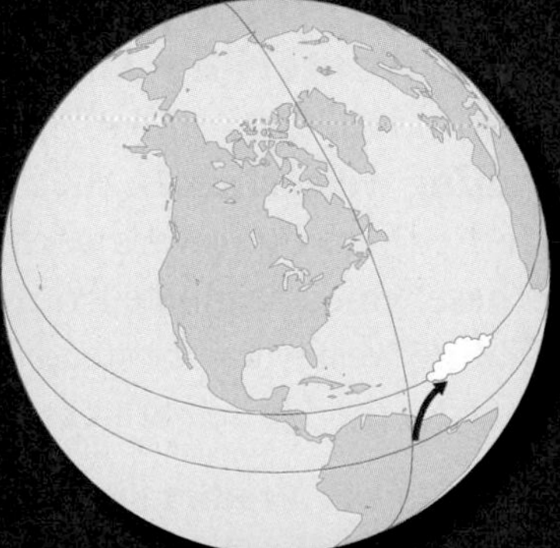

The island of Martinique is located at approximately 15°N latitude. Suppose that rising equatorial air is on the same line of longitude as Martinique. When this air arrives at 15°N latitude a day later, it will be east of Martinique because the air was moving to the east faster than the island was moving to the east.

The result is that air moving toward the poles appears to curve to the right, or east. The opposite is true for air moving from the poles to the equator because the eastward speed of polar air is slower than the eastward speed of the land over which it is moving.

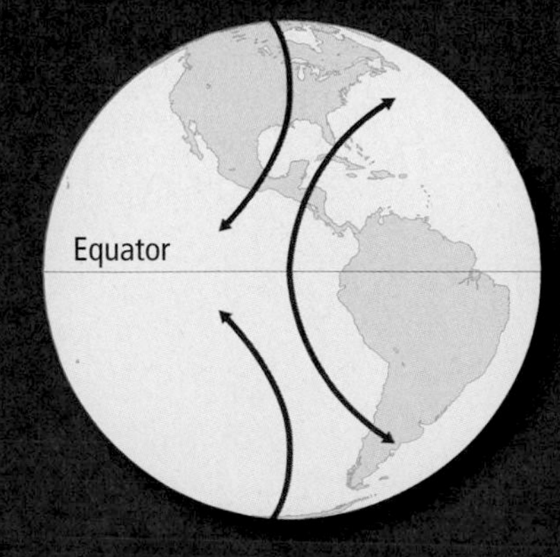

Concepts In Motion To explore more about the Coriolis effect, visit glencoe.com.

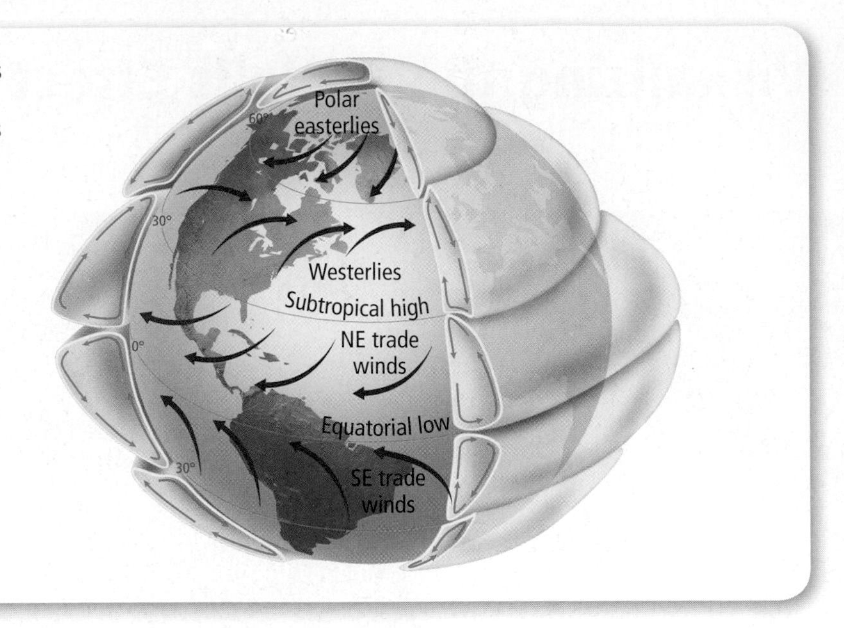

■ **Figure 12.6** The directions of Earth's wind systems, such as the polar easterlies and the trade winds, vary with the latitudes in which they occur. Note that a wind is named for the direction from which it blows. A north wind blows from the north.

Polar easterlies The wind zones between 60° N latitude and the north pole, and 60° S latitude and the south pole are called the **polar easterlies,** also shown in **Figure 12.6.** Polar easterlies begin as dense polar air that sinks. As Earth spins, this cold, descending air is deflected in an easterly direction away from each pole. The polar easterlies are typically cold winds. Unlike the prevailing westerlies, these polar easterlies are often weak and sporadic.

Between polar easterlies and prevailing westerlies is an area called a polar front. Earth has two polar fronts located near latitudes 60° N and 60° S. Polar fronts are areas of stormy weather.

Vocabulary

Science usage v. Common usage

Circulation

Science usage: movement in a circle or circuit

Common usage: condition of being passed about and widely known; distribution

Prevailing westerlies The wind systems on Earth located between latitudes 30° N and 60° N, and 30° S and 60° S are called the **prevailing westerlies.** In the midlatitudes, surface winds move in a westerly direction toward each pole, as shown in **Figure 12.6.** Because these winds originate from the West, they are called westerlies. Prevailing westerlies are steady winds that move much of the weather across the United States and Canada.

 **Reading Check Predict** the direction of movement for most tornadoes in the United States.

Trade winds Between latitudes 30° N and 30° S are two circulation belts of wind known as the **trade winds,** which are shown in **Figure 12.6.** Air in these regions sinks, warms, and moves toward the equator in an easterly direction. When the air reaches the equator, it rises and moves back toward latitudes 30° N and 30° S, where it sinks and the process repeats.

Horse latitudes Near latitudes 30° N and 30° S, the sinking air associated with the trade winds creates an area of high pressure. This results in a belt of weak surface winds called the horse latitudes. Earth's major deserts, such as the Sahara, are under these high-pressure areas.

Intertropical convergence zone Near the equator, trade winds from the North and the South meet and join, as shown in **Figure 12.6.** The air is forced upward, which creates an area of low pressure. This process, called convergence, can occur on a small or large scale. Near the equator, it occurs over a large area called the intertropical convergence zone (ITCZ). The ITCZ drifts south and north of the equator as seasons change. In general, it follows the positions of the Sun in relation to the equator. In March and September it is directly over the equator. Because the ITCZ is a region of rising air, it has bands of cloudiness and thunderstorms, which deliver moisture to many of the world's tropical rain forests.

VOCABULARY

ACADEMIC VOCABULARY

Generate (JE nuh rayt)
to bring into existence
Wind is generated as air moves from an area of high pressure to an area of low pressure.

Jet Streams

Atmospheric conditions and events that occur at the boundaries between wind zones strongly influence Earth's weather. On either side of these boundaries, both surface air and upper-level air differ greatly in temperature and pressure. Recall from Chapter 11 that warmer air has higher pressure than cooler air, and that the difference in air pressure causes wind. Wind is the movement of air from an area of high pressure to an area of low pressure.

A large temperature gradient in upper-level air combined with the Coriolis effect results in strong westerly winds called jet streams. A **jet stream,** shown in **Figure 12.7,** is a narrow band of fast wind. Its speed varies with the temperature differences between the air masses at the wind zone boundaries. A jet stream can have a speed up to 185 km/h at altitudes of 10.7 km to 12.2 km.

The position of a jet stream varies with the season. It generally is located in the region of strongest temperature differences on a line from the equator to a pole. The jet stream can move almost due south or north, instead of following its normal westerly direction. It can also split into branches and re-form later. Whatever form or position it takes, the jet stream represents the strongest core of winds.

Figure 12.7 Weather in the middle latitudes is strongly influenced by fast-moving, high-altitude jet streams.

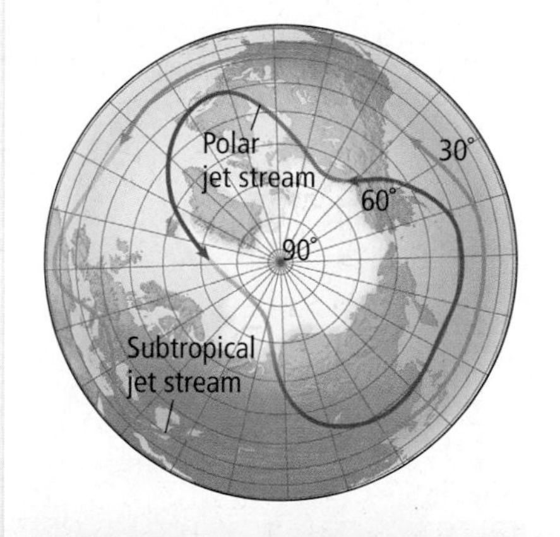

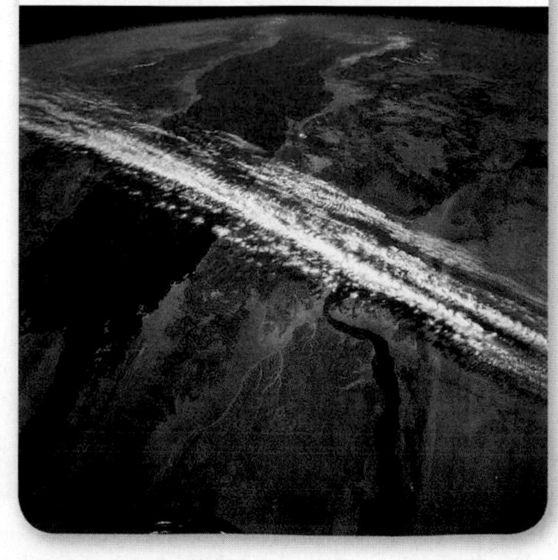

Types of jet streams The major jet streams, called the polar jet streams, separate the polar easterlies from the prevailing westerlies in the northern and southern hemispheres. The polar jet streams occur at about latitudes 40° N to 60° N and 40° S to 60° S, and move west to east. The minor jet streams are the subtropical jet streams. They occur where the trade winds meet the prevailing westerlies, at about latitudes 20° N to 30° N and 20° S to 30° S.

Jet streams and weather systems Storms form along jet streams and generate large-scale weather systems. These systems transport cold surface air toward the tropics and warm surface air toward the poles. Weather systems generally follow the path of jet streams. Jet streams also affect the intensity of weather systems by moving air of different temperatures from one region of Earth to another.

■ **Figure 12.8** The type of front formed depends on the types of air masses that collide.
Identify ***the front associated with high cirrus clouds.***

Concepts In Motion
Interactive Figure To see an animation of fronts, visit glencoe.com.

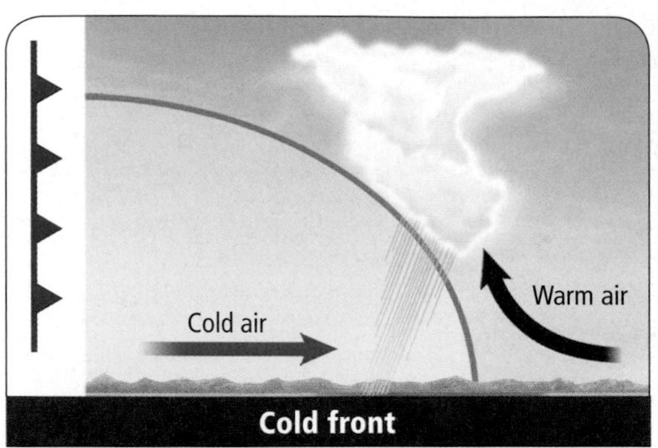

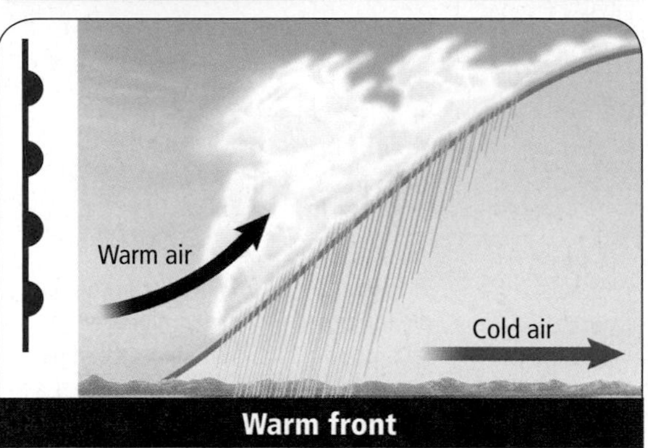

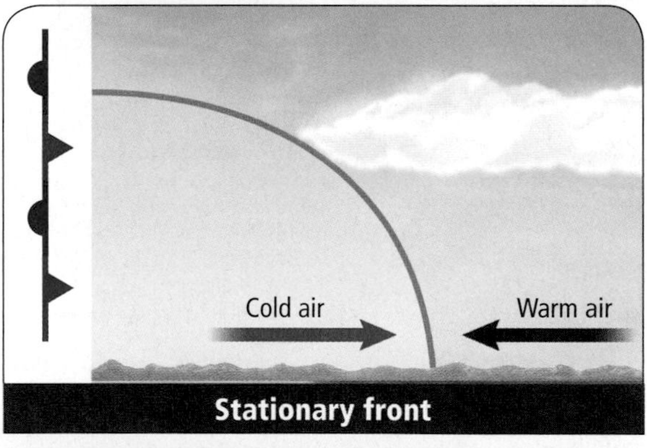

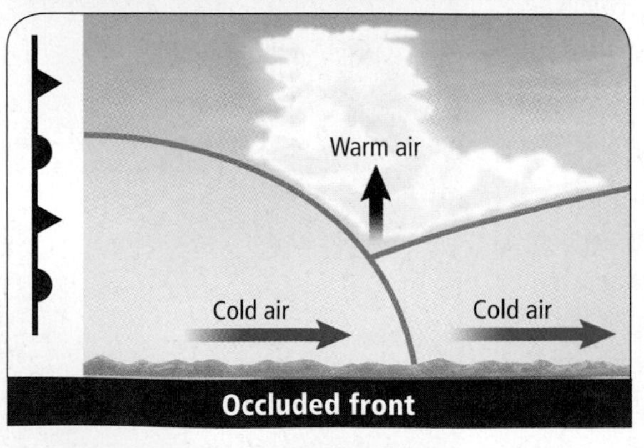

Fronts

Air masses with different characteristics can collide and result in dramatic weather changes. A collision of two air masses forms a **front**—a narrow region between two air masses of different densities. Recall that the density of an air mass results from its temperature, pressure, and humidity. Fronts can cover thousands of kilometers of Earth's surface.

Cold front When cold, dense air displaces warm air, it forces the warm air, which is less dense, up along a steep slope, as shown in **Figure 12.8.** This type of collision is called a cold front. As the warm air rises, it cools and condenses. Intense precipitation and sometimes thunderstorms are common with cold fronts. A blue line with evenly spaced blue triangles represents a cold front on a weather map. The triangles point in the direction of the front's movement.

Warm front Advancing warm air displaces cold air along a warm front. A warm front develops a gradual boundary slope, as illustrated in **Figure 12.8.** A warm front can cause widespread light precipitation. On a weather map, a red line with evenly spaced, red semicircles pointing in the direction of the front's movement indicates a warm front.

Stationary front When two air masses meet but neither advances, the boundary between them stalls. This front—a stationary front, as shown in **Figure 12.8**—frequently occurs between two modified air masses that have small temperature and pressure gradients between them. The air masses can continue moving parallel to the front. Stationary fronts sometimes have light winds and precipitation. A line of evenly spaced, alternating cold- and warm-front symbols pointing in opposite directions represents a stationary front on a weather map.

Occluded front Sometimes, a cold air mass moves so rapidly that it overtakes a warm front and forces the warm air upward, as shown in **Figure 12.8.** As the warm air is lifted, the advancing cold air mass collides with the cold air mass in front of the warm front. This is called an occluded front. Strong winds and heavy precipitation are common along an occluded front. An occluded front is shown on a weather map as a line of evenly spaced, alternating purple triangles and semicircles pointing in the direction of the occluded front's movement.

Pressure Systems

In Chapter 11, you learned that at Earth's surface, sinking air is associated with high pressure and rising air is associated with low pressure. Air always flows from an area of high pressure to an area of low pressure. Sinking or rising air, combined with the Coriolis effect, results in the formation of rotating high- and low-pressure systems in the atmosphere. Air in these systems moves in a circular motion around either a high- or low-pressure center.

Low-pressure systems In surface low-pressure systems, air rises. When air from outside the system replaces the rising air, this air spirals inward toward the center and then upward. Air in a low-pressure system in the northern hemisphere moves in a counterclockwise direction, as shown in **Figure 12.9.** The opposite occurs in the southern hemisphere for a low-pressure system. As air rises, it cools and often condenses into clouds and precipitation. Therefore, a low-pressure system, whether in the northern or southern hemisphere, is often associated with cloudy weather and precipitation.

High-pressure systems In a surface high-pressure system, sinking air moves away from the system's center when it reaches Earth's surface. The Coriolis effect causes the sinking air to move to the right, making the air circulate in a clockwise direction in the northern hemisphere and in a counter clockwise direction in the southern hemisphere. High-pressure systems are usually associated with fair weather. They dominate most of Earth's subtropical oceans and provide generally pleasant weather.

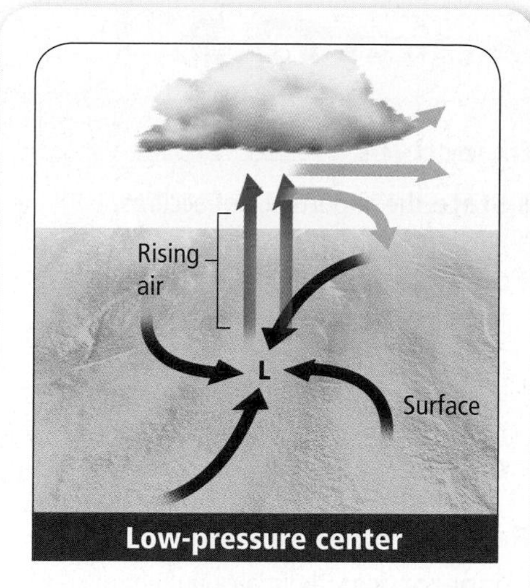

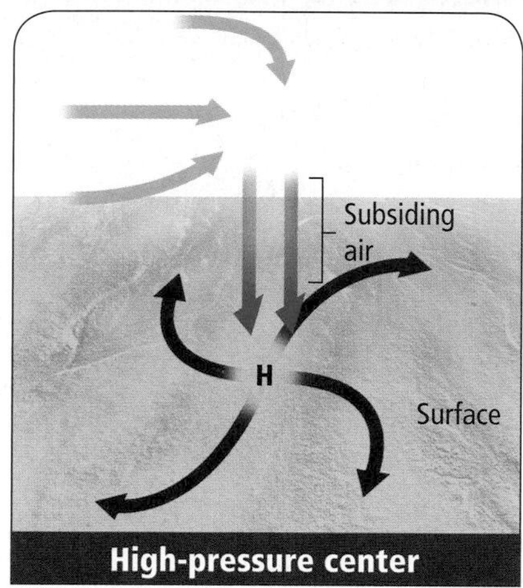

Figure 12.9 In the northern hemisphere, winds move counterclockwise around a low-pressure center, and clockwise around a high-pressure center.

Section 12.2 Assessment

Section Summary

- The three major wind systems are the polar easterlies, the prevailing westerlies, and the trade winds.
- Fast-moving, high-altitude jet streams greatly influence weather in the middle latitudes.
- The four types of fronts are cold fronts, warm fronts, occluded fronts, and stationary fronts.
- Air moves in a generally circular motion around either a high- or low-pressure center.

Understand Main Ideas

1. MAIN Idea **Summarize** information about the four types of fronts. Explain how they form and lead to changes in weather.
2. **Distinguish** among the three main wind systems.
3. **Describe** the Coriolis effect.
4. **Explain** why most tropical rain forests are located near the equator.
5. **Describe** how a jet stream affects the movement of air masses.
6. **Compare and contrast** high-pressure and low-pressure systems.

Think Critically

7. **Analyze** why most of the world's deserts are located between latitudes 10°N to 30°N and 10°S to 30°S.

WRITING in **Earth Science**

8. Write a summary about how the major wind systems form.

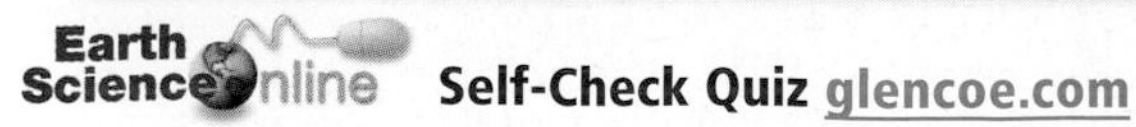

Self-Check Quiz glencoe.com

Section 12.3

Objectives

- **State** the importance of accurate weather data.
- **Summarize** the instruments used to collect weather data from Earth's surface.
- **Analyze** the strengths and weaknesses of weather radar and weather satellites.

Review Vocabulary

temperature: the measurement of how rapidly or slowly particles move

New Vocabulary

thermometer
barometer
anemometer
hygrometer
radiosonde
Doppler effect

Gathering Weather Data

MAIN Idea Accurate measurements of atmospheric properties are a critical part of weather analysis and prediction.

Real-World Reading Link Before a doctor can make a diagnosis, he or she must accurately assess the patient's state of health. This usually includes measuring body temperature and blood pressure. Similarly, in order to forecast the weather, meteorologists must have accurate measurements of the atmosphere.

Data from Earth's Surface

Meteorologists measure atmospheric conditions, such as temperature, air pressure, wind speed, and relative humidity. The quality of the data is critical for complete weather analysis and precise predictions. Two important factors in weather forecasting are the accuracy of the data and the amount of available data.

Temperature and air pressure A **thermometer,** shown in **Figure 12.10,** measures temperature using either the Fahrenheit or Celsius scale. Thermometers in most homes are liquid-in-glass or bimetallic-strip thermometers. Liquid-in-glass thermometers contain a column of either mercury or alcohol sealed in a glass tube. The liquid expands when heated, causing the column to rise, and contracts when it cools, causing the column to fall. A bimetallic-strip thermometer has a dial with a pointer. It contains a strip of metal made from two different metals that expand at different rates when heated. The strip is long and coiled into a spiral, making it more sensitive to temperature changes.

A **barometer** measures air pressure. Some barometers have a column of mercury in a glass tube. One end of the tube is submerged in an open container of mercury. Changes in air pressure change the height of the column. Another type of barometer is an aneroid barometer, shown in **Figure 12.10.** It has a sealed, metal chamber with flexible sides. Most of the air is removed, so the chamber contracts or expands with changes in air pressure. A system of levers connects the chamber to a pointer on a dial.

°F °C
120 50
110 40
100
90 30
80

Liquid-in-glass thermometer

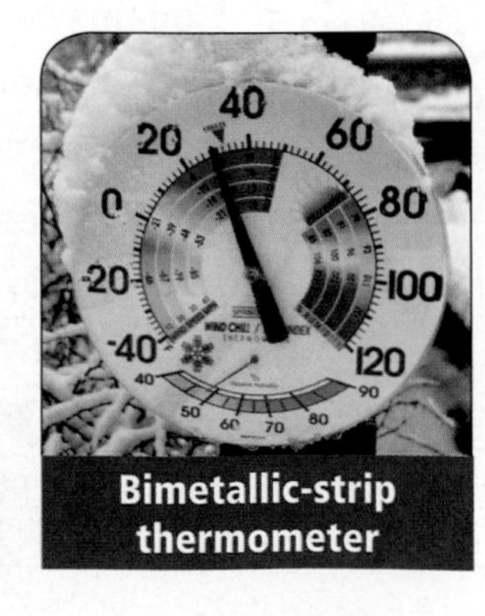

Bimetallic-strip thermometer

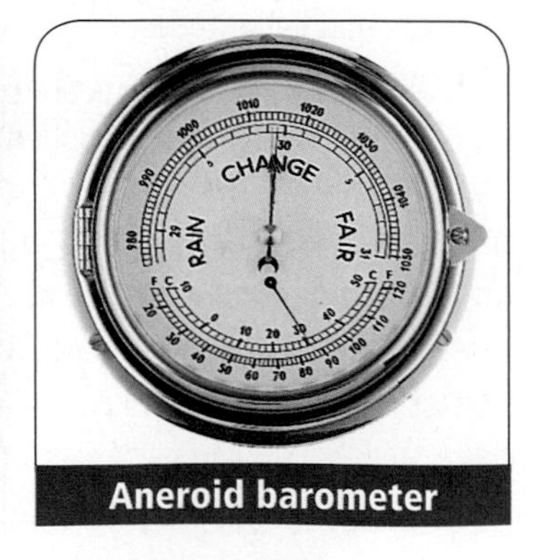

Aneroid barometer

■ **Figure 12.10** Thermometers and barometers are common weather instruments.

Wind speed and relative humidity An **anemometer** (a nuh MAH muh tur), shown in **Figure 12.11,** measures wind speed. The simplest type of anemometer has three or four cupped arms, positioned at equal angles from each other, that rotate as the wind blows. The wind's speed can be calculated using the number of revolutions of the cups over a given time. Some anemometers also have a wind vane that shows the direction of the wind.

A **hygrometer** (hi GRAH muh tur), such as the one in **Figure 12.11,** measures humidity. This type of hygrometer has wet-bulb and dry-bulb thermometers and requires a conversion table to determine relative humidity. When water evaporates from the wet bulb, the bulb cools. The temperatures of the two thermometers are read at the same time, and the difference between them is calculated. The relative humidity table lists the specific relative humidity for the difference between the thermometers.

Reading Check **Analyze** the relationship between the amount of moisture in air and the temperature of the wet bulb in a hygrometer.

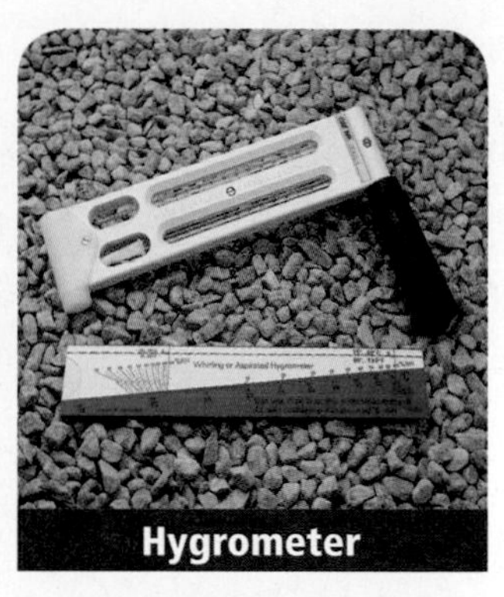

■ **Figure 12.11** Anemometers are used to measure wind speed based on the rotation of the cups as the wind blows. Hygrometers measure humidity using techniques such as finding the temperature difference between the wet bulb and the dry bulb.

Automated Surface Observing System Meteorologists need a true "snapshot" of the atmosphere at one particular moment to develop an accurate forecast. To obtain this, meteorologists analyze and interpret data gathered at the same time from weather instruments at many different locations. Coordinating the collection of this data was a complicated process until late in the twentieth century. With the development of reliable automated sensors and computer technology, instantaneously collecting and broadcasting accurate weather-related data became possible.

In the United States, the National Weather Service (NWS), the Federal Aviation Administration, and the Department of Defense jointly established a surface-weather observation network known as the Automated Surface Observing System (ASOS). It gathers data in a consistent manner, 24 hours a day, every day. It began operating in the 1990s and more than doubled the number of full-time observation sites, such as the one shown in **Figure 12.12.** ASOS provides essential weather data for aviation, weather forecasting, and weather-related research.

■ **Figure 12.12** This weather station in the United Kingdom consists of several instruments that measure atmospheric conditions.

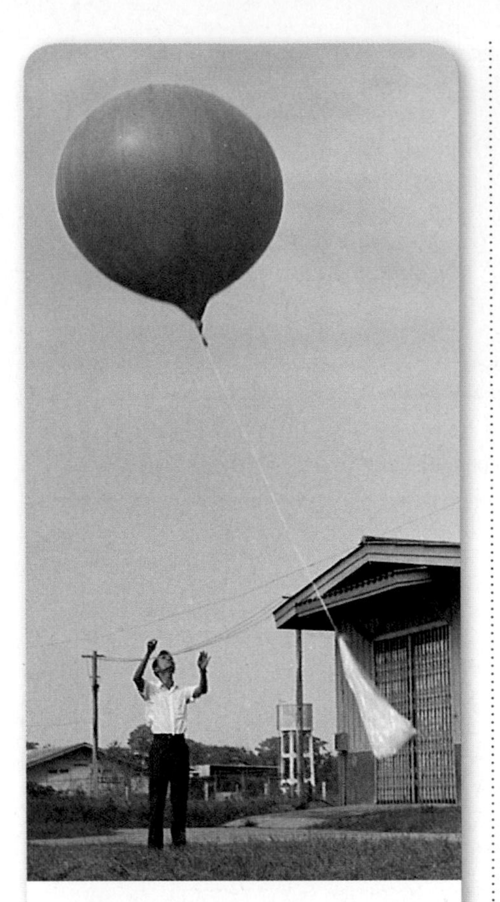

■ **Figure 12.13** Radiosondes gather upper-level weather data such as air temperature, pressure, and humidity.

Data from the Upper Atmosphere

While surface-weather data are important, the weather is largely the result of changes that take place high in the troposphere. To make accurate forecasts, meteorologists must gather data at high altitudes, up to 30,000 m. This task is more difficult than gathering surface data, and it requires sophisticated technology.

An instrument used for gathering upper-atmospheric data is a **radiosonde** (RAY dee oh sahnd), shown in **Figure 12.13.** It consists of a package of sensors and a battery-powered radio transmitter. These are suspended from a balloon that is about 2 m in diameter and filled with helium or hydrogen. A radiosonde's sensors measure the air's temperature, pressure, and humidity. Radio signals constantly transmit these data to a ground station that tracks the radiosonde's movement. If a radiosonde also measures wind direction and speed, it is called a rawinsonde (RAY wuhn sahnd), **ra**dar + **win**d + radio**sonde**.

Tracking is a crucial component of upper-level observations. The system used since the 1980s has been replaced with one that uses Global Positioning System (GPS) and the latest computer technology. Meteorologists can determine wind speed and direction by tracking how fast and in what direction a rawinsonde moves. The various data are plotted on a chart that gives meteorologists a profile of the temperature, pressure, humidity, wind speed, and wind direction of a particular part of the atmosphere. Such charts are used to forecast atmospheric changes that affect surface weather.

 **Reading Check** **Describe** the function of a radiosonde.

VOCABULARY
ACADEMIC VOCABULARY
Compute (kuhm PYEWT)
to perform mathematical operations
Jane used a calculator to compute the answers for her math homework.

Weather Observation Systems

There are many surface and upper-level observation sites across the United States. However, data from these sites cannot be used to locate exactly where precipitation falls without the additional help of data from weather radars and weather satellites.

Weather radar A weather radar system detects specific locations of precipitation. The term *radar* stands for **ra**dio **d**etection **a**nd **r**anging. How does radar work? A radar system generates radio waves and transmits them through an antenna at the speed of light. Recall that radio waves are electromagnetic waves with wavelengths greater than 10^{-3} m. The transmitter is programmed to generate waves that only reflect from particles larger than a specific size. For example, when the radio waves encounter raindrops, some of the waves scatter. Because an antenna cannot send and receive signals at the same time, radars send a pulse and wait for the return before another pulse is sent. An amplifier increases the received wave signals, and then a computer processes and displays them on a monitor. From these data, the distance to precipitation and its location relative to the receiving antenna.

Doppler weather radar You have probably noticed that the pitch produced by the horn of an approaching car gets higher as it comes closer to you and lower as it passes and moves away from you. This sound phenomenon is called the Doppler effect. The **Doppler effect** is the change in pitch or frequency that occurs due to the relative motion of a wave, such as sound or light, as it comes toward or goes away from an observer.

The NWS uses Weather Surveillance Radar-1988 Doppler (WSR-88D), shown in **Figure 12.14,** based on the Doppler effect of moving waves. Analysis of Doppler radar data can be used to determine the speed at which precipitation moves toward or away from a radar station. Because the movement of precipitation is caused by wind, Doppler radar can also provide a good estimation of the wind speeds associated with precipitation areas, including those with severe weather, such as thunderstorms and tornados. The ability to measure wind speeds gives Doppler radar a distinct advantage over conventional weather radar systems.

■ **Figure 12.14** Norman, Oklahoma, was the site of the first Doppler radar installation.
Relate ***the importance of this location to severe weather conditions.***

Weather satellites In addition to communications, one of the main uses of satellites orbiting Earth is to observe weather. Cameras mounted aboard a weather satellite take photos of Earth at regular intervals. A weather satellite can use infrared, visible-light, or water-vapor imagery to observe the atmosphere.

Infrared imagery Some weather satellites use infrared imagery to make observations at night. Objects radiate thermal energy at slightly different frequencies. Infrared imagery detects these different frequencies, which enables meteorologists to map either cloud cover or surface temperatures. Different frequencies are distinguishable in an infrared image, as shown in **Figure 12.15.**

As you learned in Chapter 11, clouds form at different altitudes and have different temperatures. Using infrared imagery, meteorologists can determine the cloud's temperature, its type, and its altitude. Infrared imagery is useful especially in detecting strong thunderstorms that develop and reach high altitudes. Consequently, they appear as very cold areas on an infrared image. Because the strength of a thunderstorm is related to the altitude that it reaches, infrared imagery can be used to establish a storm's potential to produce severe weather.

■ **Figure 12.15** This infrared image shows cloud cover across most of the United States.

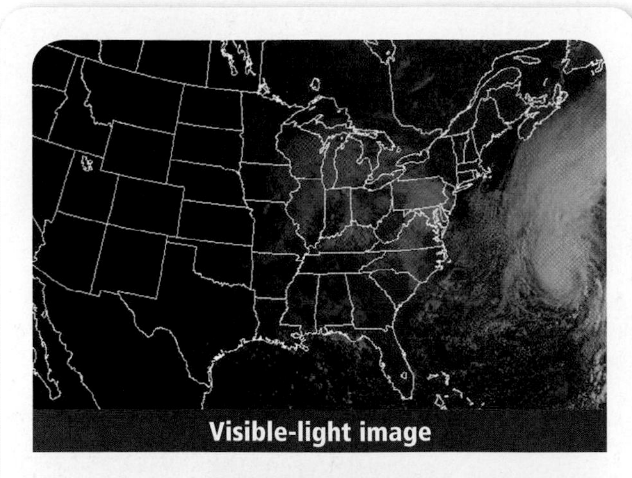

■ **Figure 12.16** These images were taken at the same time as the one in **Figure 12.15.** Each type of image shows different atmospheric characteristics. Together, they help meteorologists accurately analyze and predict weather.

Visible-light imagery Some satellites use cameras that require visible light to photograph Earth. These digital photos, like the one in **Figure 12.16,** are sent back to ground stations, and their data are plotted on maps. Unlike weather radar, which tracks precipitation but not clouds, satellites track clouds but not necessarily precipitation. By combining radar and visible imagery data, meteorologists can determine where both clouds and precipitation are occurring.

Water-vapor imagery Another type of satellite imagery that is useful in weather analysis and forecasting is called water-vapor imagery, also shown in **Figure 12.16.** Water vapor is an invisible gas and cannot be photographed directly, but it absorbs and emits infrared radiation at certain wavelengths. Many weather satellites have sensors that are able to provide a measure of the amount of water vapor present in the atmosphere.

Water-vapor imagery is a valuable tool for weather analysis and prediction because it shows moisture in the atmosphere, not just cloud patterns. Because air currents that guide weather systems are often well defined by trails of water vapor, meteorologists can closely monitor the development and change in storm systems even when clouds are not present.

Section 12.3 Assessment

Section Summary

- To make accurate weather forecasts, meteorologists analyze and interpret data gathered from Earth's surface by weather instruments.
- A radiosonde collects upper-atmospheric data.
- Doppler radar locates where precipitation occurs.
- Weather satellites use infrared, visible-light, or water-vapor imagery to observe and monitor changing weather conditions on Earth.

Understand Main Ideas

1. MAIN Idea **Identify** two important factors in collecting and analyzing weather data in the United States.
2. **Compare and contrast** methods for obtaining data from Earth's surface and Earth's upper atmosphere.
3. **State** the main advantage of Doppler radar over conventional weather radar.
4. **Summarize** the three kinds of weather satellite imagery using a graphic organizer.

Think Critically

5. **Predict** whether you would expect weather forecasts to be more accurate for the state of Kansas or a remote Caribbean island, based on what you know about weather observation systems. Explain.

WRITING in Earth Science

6. Write a newspaper article about the use of water-vapor imagery to detect water on the planet Mars.

Earth Science Online Self-Check Quiz glencoe.com

Section 12.4

Objectives

- **Analyze** a basic surface weather chart.
- **Distinguish** between digital and analog forecasting.
- **Describe** problems with long-term forecasts.

Review Vocabulary

model: an idea, system, or mathematical expression that represents an idea

New Vocabulary

station model
isobar
isotherm
digital forecast
analog forecast

Weather Analysis and Prediction

MAIN Idea Several methods are used to develop short-term and long-term weather forecasts.

Real-World Reading Link It is usually easier to predict what you will be doing later today than what you will be doing a week from now. Weather predictions also are easier for shorter time spans than for longer time spans.

Surface Weather Analysis

Newspapers, radio and television stations, and Web sites often give weather reports. These data are plotted on weather charts and maps and are often accompanied by radar and satellite imagery.

Station models After weather data are gathered, meteorologists plot the data on a map using station models for individual cities or towns. A **station model** is a record of weather data for a particular site at a particular time. Meteorological symbols, such as the ones shown in **Figure 12.17,** are used to represent weather data in a station model. A station model allows meteorologists to fit a large amount of data into a small space. It also gives meteorologists a uniform way of communicating weather data.

Plotting station model data Station models provide information for individual sites. To plot data nationwide and globally, meteorologists use lines that connect points of equal or constant values. The values represent different weather variables, such as pressure or temperature. Lines of equal pressure, for example, are called **isobars,** while lines of equal temperature are called **isotherms.** The lines themselves are similar to the contour lines—lines of equal elevation—that you studied in Chapter 2.

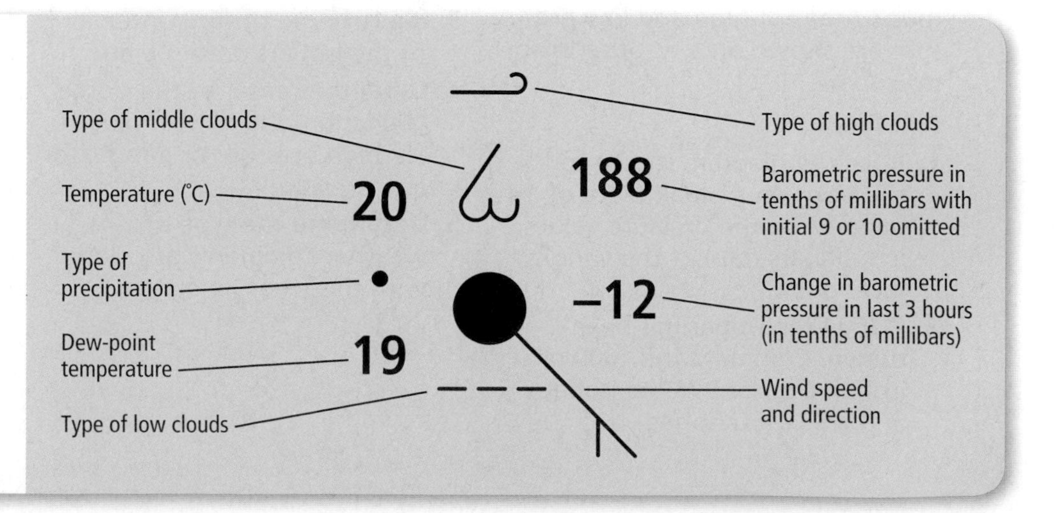

Figure 12.17 A station model shows temperature, wind direction and speed, and other weather data for a particular location at a particular time. **Explain** *the advantage of using meteorological symbols.*

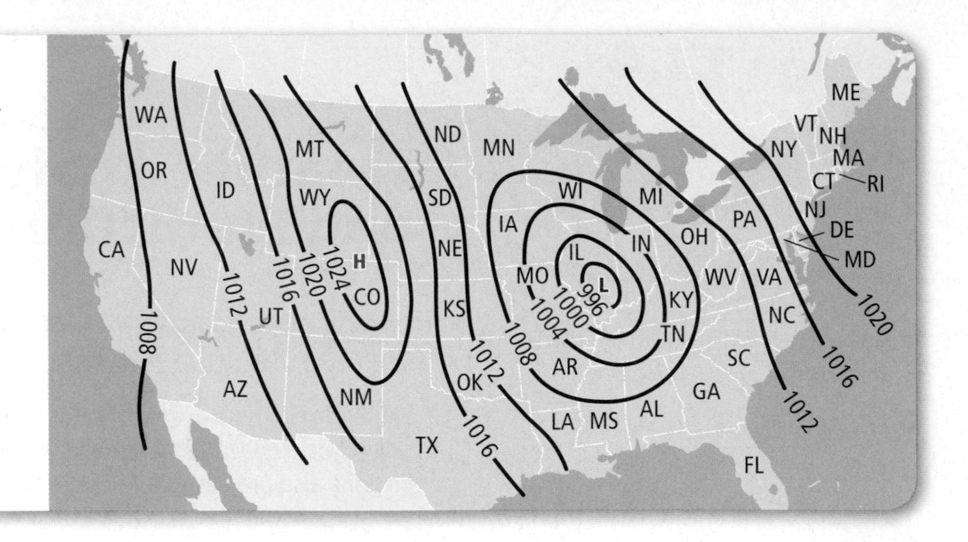

■ **Figure 12.18** The weather map shows isobars and air pressure data for the continental United States.
Determine *where on the weather map you would expect the strongest winds.*

Interpreting station model data Recall that inferences about elevation can be made by studying contour intervals on a map. Inferences about weather, such as wind speed, can be made by studying isobars and isotherms on a map. Isobars that are close together indicate a large pressure difference over a small area, which means strong winds. Isobars that are far apart indicate a small difference in pressure and light winds. As shown in **Figure 12.18,** isobars also indicate the locations of high- and low-pressure systems. Combining this information with that of isotherms helps meteorologists to identify frontal systems.

Using isobars, isotherms, and station-model data, meteorologists can analyze current weather conditions for a particular location. This is important because meteorologists must understand current weather conditions before they can forecast the weather.

PROBLEM-SOLVING LAB

Interpret a Scientific Illustration

How do you analyze a weather map? Areas of high and low pressure are shown on a weather map by isobars.

Analysis

1. Trace the diagram shown to the right on a blank piece of paper. Add the pressure values in millibars (mb) at the various locations.
2. A 1004-mb isobar has been drawn. Complete the 1000-mb isobar. Draw a 996-mb isobar and a 992-mb isobar.

Think Critically

3. **Identify** the contour interval of the isobars on this map.
4. **Label** the center of the closed 1004-mb isobar with a blue *H* for high pressure or a red *L* for low pressure.
5. **Determine** the type of weather commonly associated with this pressure system.

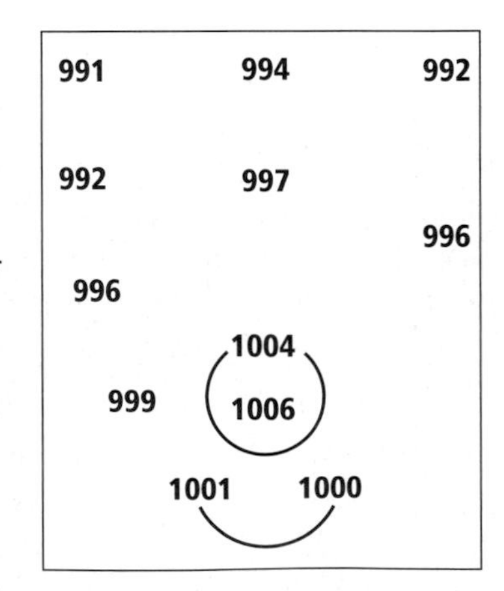

Types of Forecasts

A meteorologist, shown in **Figure 12.19,** must analyze data from different levels in the atmosphere, based on current and past weather conditions, to produce a reliable forecast. Two types of forecasts are digital forecasts and analog forecasts.

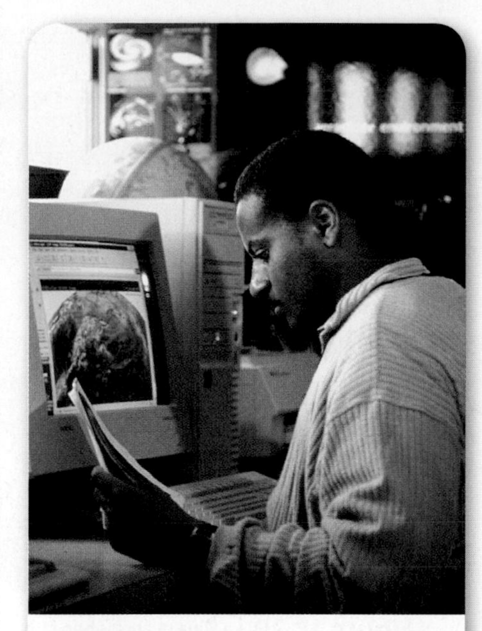

■ **Figure 12.19** This meteorologist is analyzing data from various sources to prepare a weather forecast.

Digital forecasts The atmosphere behaves like a fluid. Physical principles that apply to a fluid, such as temperature, pressure, and density, can be applied to the atmosphere and its variables. In addition, they can be expressed as mathematical equations to determine how atmospheric variables change over time.

A **digital forecast** is created by applying physical principles and mathematics to atmospheric variables and then making a prediction about how these variables will change over time. Digital forecasting relies on numerical data. Its accuracy is related directly to the amount of available data. It would take a long time for meteorologists to solve atmospheric equations on a global or national scale. Fortunately, computers can do the job quickly. Digital forecasting is the main method used by present-day meteorologists.

 **Reading Check** **State** the relationship between the accuracy of a digital forecast and the data on which it is based.

Analog forecasts Another type of forecast, an **analog forecast,** is based on a comparison of current weather patterns to similar weather patterns from the past. Meteorologists coined the term *analog forecasting* because they look for a pattern from the past that is similar, or analogous, to a current pattern. To ensure the accuracy of an analog forecast, meteorologists must find a past event that had similar atmosphere, at all levels and over a large area, to a current event.

The main disadvantage of analog forecasting is the difficulty in finding the same weather pattern in the past. Still, analog forecasting is useful for conducting monthly or seasonal forecasts, which are based mainly on the past behavior of cyclic weather patterns.

Short-Term Forecasts

The most accurate and detailed forecasts are short term because weather systems change directions, speeds, and intensities over time. For hourly forecasts, extrapolation is a reliable forecasting method because small-scale weather features that are readily observable by radar and satellites dominate current weather.

One- to three-day forecasts are no longer based on the movement of observed clouds and precipitation, which change by the hour. Instead, these forecasts are based on the behavior of larger surface and upper-level features, such as low-pressure systems. A one- to three-day forecast is usually accurate for expected temperatures, and for when and how much precipitation will occur. For this time span, however, the forecast will not be able to pinpoint an exact temperature or sky condition at a specific time.

VOCABULARY

ACADEMIC VOCABULARY

Extrapolation
(ihk stra puh LAY shun)
the act of inferring a probable value from an existing set of values
Short-term weather forecasts can be extrapolated from data collected by radar and satellites.

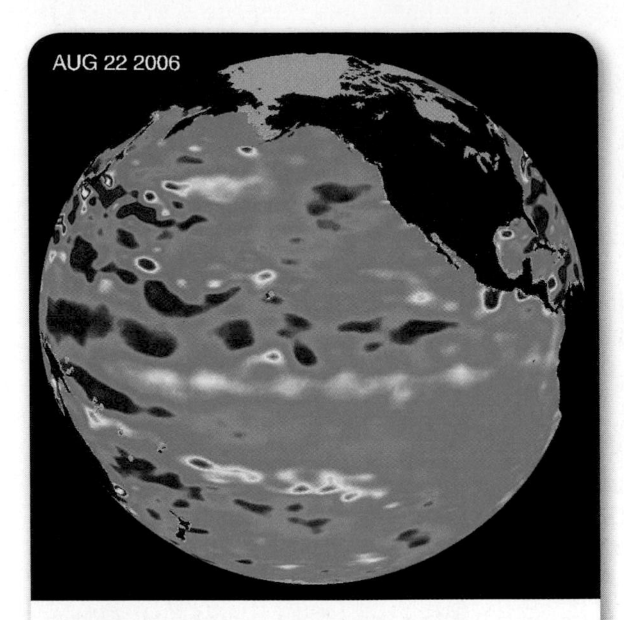

■ **Figure 12.20** La Niña occurs when stronger-than-normal trade winds carry the colder water (blue) from the coast of South America to the equatorial Pacific Ocean. This happens about every three to five years and can affect global weather patterns.

Long-Term Forecasts

Because it is impossible for computers to model every variable that affects the weather at a given time and place, all long-term forecasts are less reliable than short-term forecasts. Recall that features on Earth's surface affect the amount of thermal energy absorbed at any location. This affects the pressure at that location, which affects the wind. Wind influences cloud formation and virtually all other aspects of the weather in that location. Over time, these factors interact and create more complicated weather scenarios.

Meteorologists use changes in surface weather systems based on circulation patterns throughout the troposphere and lower stratosphere for four- to seven-day forecasts. They can estimate each day's weather but cannot pinpoint when or what specific weather conditions will occur. One- to two-week forecasts are based on changes in large-scale circulation patterns. Thus, these forecasts are vague and are based mainly on similar conditions that have occurred in the past.

Forecasts for months and seasons are based mostly on weather cycles or patterns. These cycles, such as the one shown in **Figure 12.20,** can involve changes in the atmosphere, ocean currents, and solar activity that might occur at the same time. Improvements in weather forecasts depend on identifying the influences of the cycles involved, understanding how they interact, and determining their ultimate effect on weather over longer time periods.

Section 12.4 Assessment

Section Summary

- A station model is used to plot different weather variables.
- Meteorologists plot lines on a map that connect variables of equal value to represent nationwide and global trends.
- Two kinds of forecasts are digital and analog.
- The longer the prediction period, the less reliable the weather forecast.

Understand Main Ideas

1. MAIN Idea **Describe** the methods used for illustrating weather forecasts.
2. **Identify** some of the symbols used in a station model.
3. **Model** how temperature and pressure are shown on a weather map.
4. **Compare and contrast** analog and digital forecasts.
5. **Explain** why long-term forecasts are not as accurate as short-term forecasts.

Think Critically

6. **Assess** which forecast type—digital or analog—would be more accurate for three days or less.

MATH in Earth Science

7. Using a newspaper or other media sources, find and record the high and low temperatures in your area for five days. Calculate the average high and low temperatures for the five-day period.

Earth Science Online **Self-Check Quiz** glencoe.com

Earth Science & Society

Weather Forecasting—Precision from Chaos

On a rainy evening in New Jersey, four teens went out to play soccer. They began to play, expecting the rain to clear before the game got into full swing. However, as the game progressed, the clouds darkened to a charcoal grey and thickened. When the thunder and lightning began, the teens decided to leave the field. As they walked from the field, they were struck by lightning. Two of the teens died in the hospital a few hours later. The deaths rocked the community. The storm had not been predicted in the weather forecast. Why isn't weather forecasting more predictable?

Chaos and weather systems In 1963, a meteorologist named Edward Lorenz first presented chaos theory, which states that formulated systems are dependent on initial conditions and that the precision of initial measurements has an exponential impact on the expected outcome.

Years after Lorenz published his findings in meteorology journals, other scientists recognized the importance of his work. The simplified equations Lorenz created through his studies helped form the basis of modern weather forecasting.

The beginning of a forecast Weather forecasting begins with observations. Data are collected from various sources and fed into supercomputers, which create mathematical models of the atmosphere. In the United States, the National Weather Service operates these computers and releases their data to local and regional forecasters.

Meteorologists generally agree that useful day-to-day broadcasts are limited to only five days. Most meteorologists also agree that reliable forecasts of day-to-day weather for up to six or seven days ahead are not now possible.

Weather forecasts are created from data collected from the atmosphere.

Meteorologists hope that improved measurements, computer technology, and weather models might someday predict day-to-day weather up to three weeks in advance.

Limitations of long-range forecasting Meteorologists generally find that day-to-day forecasts for more than a week in the future are unreliable. Their approach to long-range forecasting is based instead on comparisons of current and past weather patterns, as well as global ocean temperatures, to determine the probability that temperature and precipitation values will be above or below normal ranges. The National Weather Service's Climate Prediction Center, as well as other organizations, offers monthly and seasonal predictions for these values.

WRITING in Earth Science

Evaluate Use a newspaper or other local news source to obtain a weather report for the next seven days. Record the temperature and weather conditions for your city during the next week and compare the forecasted weather with the observed weather. Write a summary to share your observations with your class.

GEOLAB

MAPPING: INTERPRET A WEATHER MAP

Background: The surface weather map on the following page shows actual weather data for the United States. In this activity, you will use the station models, isobars, and pressure systems on the map to forecast the weather.

Question: *How can you use a surface weather map to interpret information about current weather and to forecast future weather?*

Materials

ruler
Reference Handbook, Weather Map Symbols, p. 959

Procedure

1. Read and complete the lab safety form.
2. The map scale is given in nautical miles. Refer to the scale when calculating distances.
3. The unit for isobars is millibars (mb). In station models, pressure readings are abbreviated. For example, 1021.9 mb is plotted on a station model as 219 but read as 1021.9.
4. Wind shafts point in the direction from which the wind is blowing. Refer to Weather Map Symbols in the table on the right and the *Reference Handbook* to learn about the symbols that indicate wind speed.
5. Each number around a city represents a different atmospheric measure. By convention, the same atmospheric measure is always in the same relative location in a station model. Refer to **Figure 12.17** and Weather Map Symbols in the *Reference Handbook* to learn what numbers represent in a station model.

Analyze and Conclude

1. **Identify** the contour interval of the isobars.
2. **Find** the highest and lowest isobars and where they are located.
3. **Describe** the winds across Texas and Louisiana.
4. **Determine** and record with their locations the coldest and warmest temperatures on the map.
5. **Infer** whether the weather in Georgia and Florida is clear or rainy. Explain.
6. **Predict** Low-pressure systems in eastern Canada and off the Oregon coast are moving east at about 24 km/h. Predict short-term weather forecasts for northern New York and Oregon.

Symbols Used in Plotting Report	
Fronts and Pressure Systems	
(H) or High **(L) or Low**	Center of high- or low-pressure systems
▲▲▲▲	Cold front
◖◖◖◖	Warm front
◖▲◖▲	Occluded front
◖▼◖▼	Stationary front

APPLY YOUR SKILL

Forecasting Find your area on the map. Based on the data shown in the map, use the extrapolation method to forecast the next day's weather for your location.

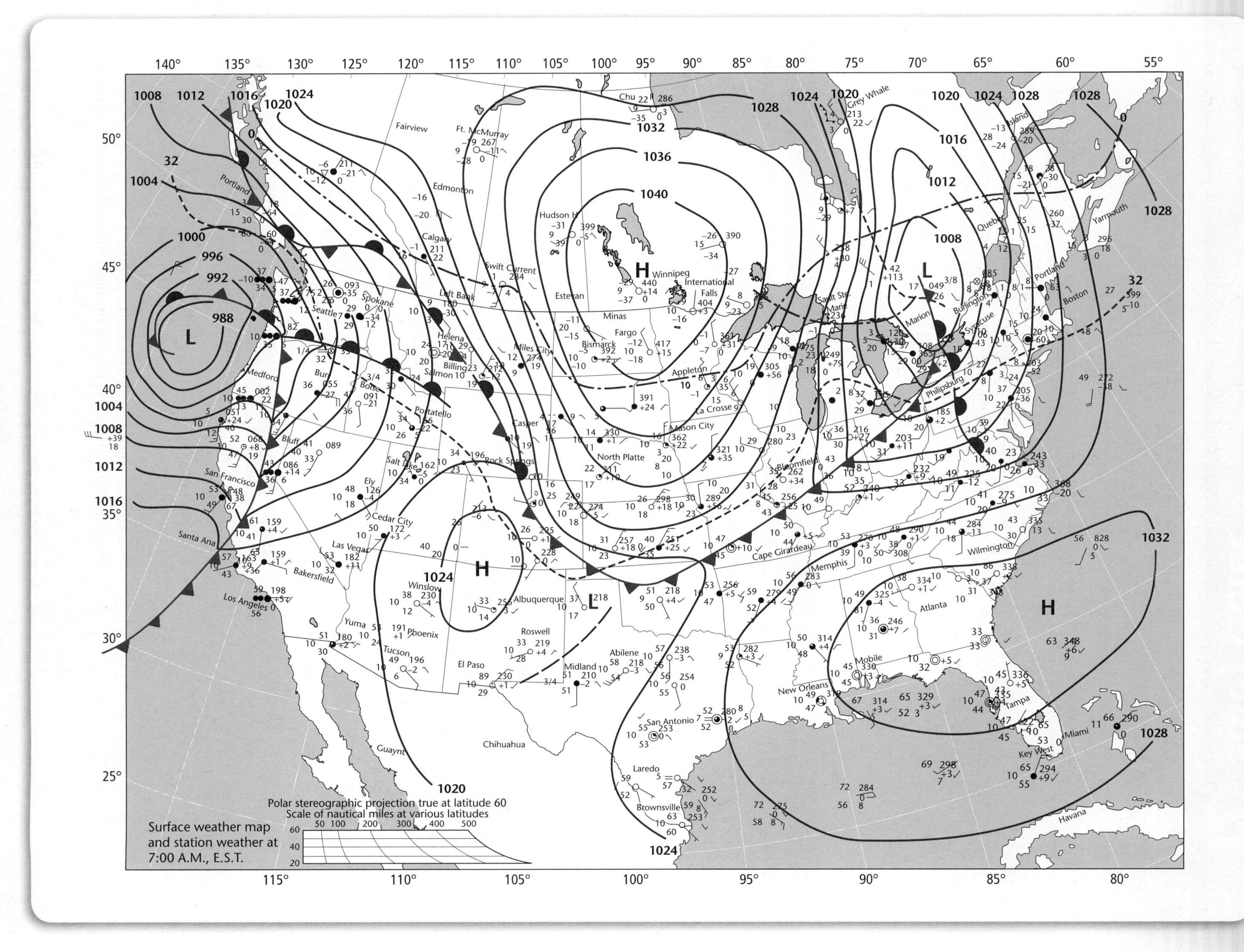

Surface weather map and station weather at 7:00 A.M., E.S.T.

CHAPTER 12

Study Guide

Download quizzes, key terms, and flash cards from glencoe.com.

BIG Idea Weather patterns can be observed, analyzed, and predicted.

Vocabulary	Key Concepts
Section 12.1 The Causes of Weather	
• air mass (p. 316) • climate (p. 314) • source region (p. 316) • weather (p. 314)	**MAIN Idea** Air masses have different temperatures and amounts of moisture because of the uneven heating of Earth's surface. • Meteorology is the study of atmospheric phenomena. • Solar radiation is unequally distributed between Earth's equator and its poles. • An air mass is a large body of air that takes on the moisture and temperature characteristics of the area over which it forms. • Each type of air mass is classified by its source region.
Section 12.2 Weather Systems	
• Coriolis effect (p. 318) • front (p. 322) • jet stream (p. 321) • polar easterlies (p. 320) • prevailing westerlies (p. 320) • trade winds (p. 320)	**MAIN Idea** Weather results when air masses with different pressures and temperatures move, change, and collide. • The three major wind systems are the polar easterlies, the prevailing westerlies, and the trade winds. • Fast-moving, high-altitude jet streams greatly influence weather in the middle latitudes. • The four types of fronts are cold fronts, warm fronts, occluded fronts, and stationary fronts. • Air moves in a generally circular motion around either a high- or low-pressure center.
Section 12.3 Gathering Weather Data	
• anemometer (p. 325) • barometer (p. 324) • Doppler effect (p. 327) • hygrometer (p. 325) • radiosonde (p. 326) • thermometer (p. 324)	**MAIN Idea** Accurate measurements of atmospheric properties are a critical part of weather analysis and prediction. • To make accurate weather forecasts, meteorologists analyze and interpret data gathered from Earth's surface by weather instruments. • A radiosonde collects upper-atmospheric data. • Doppler radar locates where precipitation occurs. • Weather satellites use infrared, visible-light, or water-vapor imagery to observe and monitor changing weather conditions on Earth.
Section 12.4 Weather Analysis and Prediction	
• analog forecast (p. 331) • digital forecast (p. 331) • isobar (p. 329) • isotherm (p. 329) • station model (p. 329)	**MAIN Idea** Several methods are used to develop short-term and long-term weather forecasts. • A station model is used to plot different weather variables. • Meteorologists plot lines on a map that connect variables of equal value to represent nationwide and global trends. • Two kinds of forecasts are digital and analog. • The longer the prediction period, the less reliable the weather forecast.

Vocabulary PuzzleMaker glencoe.com

CHAPTER 12

Assessment

Vocabulary Review

Match each description below with the correct vocabulary term from the Study Guide.

1. lines of equal pressure on a weather map
2. current state of the atmosphere
3. a forecast that relies on numerical data
4. long-term variations in weather conditions over a particular area
5. large volume of air that takes on the characteristics of the area over which it forms
6. depiction of weather data for a particular location at a particular time

Complete the sentences below using vocabulary terms from the Study Guide.

7. A ________ is used to measure relative humidity.
8. ________ describes the narrow region separating two air masses of different densities.
9. The deflection of air due to the rotation of Earth is called the ________.
10. Lines of equal temperature on a weather map are called ________.

Each of the following sentences is false. Make each sentence true by replacing the italicized words with vocabulary terms from the Study Guide.

11. The *horse latitudes* are two belts of surface winds that occur between latitudes 30° N and 60° N, and 30° S and 60° S.
12. Meteorologists use a special kind of radar, which is based on the *polar easterlies,* to plot the movement of precipitation.
13. Narrow bands of fast winds are called *trade winds.*
14. A balloon-transported package of sensors is called a *source region.*
15. An instrument that measures wind speed is called a *barometer.*

Understand Key Concepts

16. What does a large temperature gradient at high altitudes of the atmosphere cause?
    - **A.** trade winds
    - **B.** Coriolis effect
    - **C.** ITCZ
    - **D.** polar jet streams

Use the diagram below to answer Questions 17 and 18.

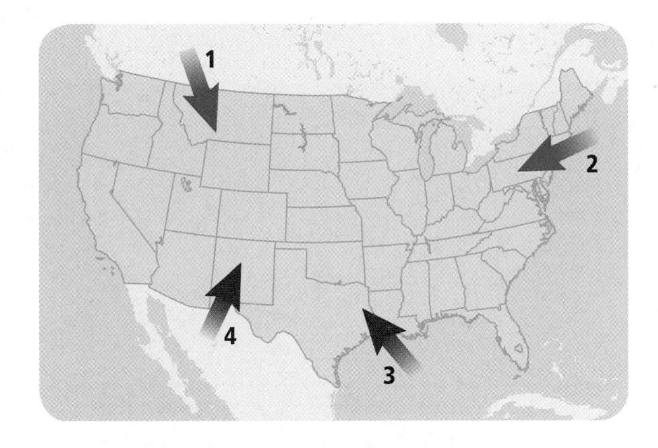

17. Which is probably the coldest air mass?
    - **A.** 1
    - **B.** 2
    - **C.** 3
    - **D.** 4
18. Which air mass is hot and dry?
    - **A.** 1
    - **B.** 2
    - **C.** 3
    - **D.** 4
19. Which is not one of Earth's three basic wind systems or zones?
    - **A.** polar easterlies
    - **B.** polar jet streams
    - **C.** trade winds
    - **D.** prevailing westerlies
20. Which location on Earth receives the most solar radiation in any given year?
    - **A.** the poles
    - **B.** the oceans
    - **C.** the tropics
    - **D.** the continents

21. Which is an example of climate?
 A. today's high temperature
 B. yesterday's rainfall
 C. tomorrow's highest wind speed
 D. average rainfall over 30 years

22. Which instrument is not used to measure surface weather?
 A. barometer
 B. hygrometer
 C. radiosonde
 D. thermometer

Use the diagram below to answer Questions 23 and 24.

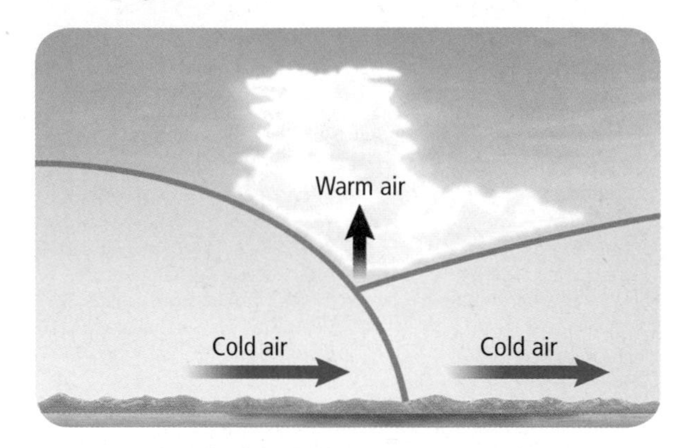

23. Which type of front is illustrated above?
 A. cold front
 B. occluded front
 C. precipitation front
 D. stationary front

24. Which weather conditions occur as a result of this type of front?
 A. warm temperatures and precipitation
 B. cool temperatures and thunderstorms
 C. light winds and precipitation
 D. strong winds and precipitation

25. Which is the most accurate forecast?
 A. long-term digital forecast
 B. short-term digital forecast
 C. long-term analog forecast
 D. short-term analog forecast

26. What is a station model used to create?
 A. digital forecast
 B. depiction of a jet stream
 C. long-term forecast
 D. surface weather map

Constructed Response

27. **Infer** why weather ahead of a warm front might be cloudier and rainier than weather ahead of a cold front.

28. **Identify** the weather feature that might be indicated by a drastic temperature change over a short distance on a surface analysis.

29. **Generalize** the problems that could result from making a weather analysis based on observations at several locations made at different times.

30. **Discuss** why light to no winds characterize the horse latitudes and why they occur at those latitudes.

31. **Compare and contrast** the temperature and moisture properties of a continental polar air mass and a maritime tropical air mass.

32. **State** the main benefit of the digital forecast method.

33. **Describe** two different weather-data methods you could use to determine if it is a rainy day at a given location.

34. **Distinguish** between isobars and isotherms.

Use the table below to answer Question 35.

Air Mass Descriptions

Air Mass	Source Region	Summer
Arctic	Siberia, Arctic Basin	cold, dry
Continental polar	interiors of Canada and Alaska	cool, dry
Continental tropical	southwest United States, Mexico	hot, dry
Maritime polar	North Pacific Ocean	mild, humid
	North Atlantic Ocean	cool, humid
Maritime tropical	Gulf of Mexico, Caribbean Sea, tropical and subtropical Atlantic Ocean and Pacific Ocean	hot, humid

35. **Choose** the summertime air mass that would most likely be associated with significant precipitation. Explain your choice.

Think Critically

36. **Propose** an ideal weather-data collection system for your school.

Use the diagram below to answer Question 37.

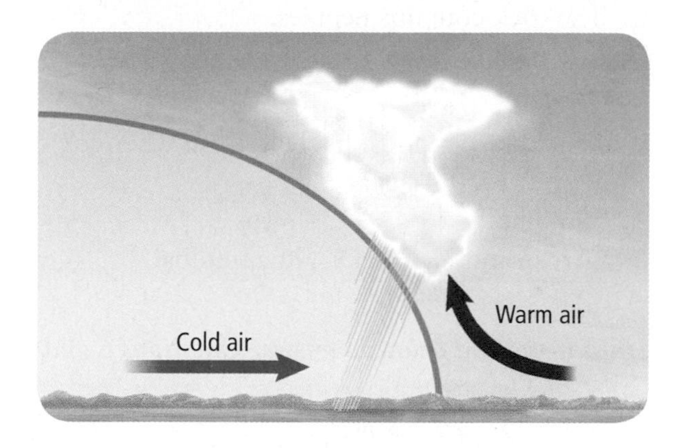

37. **Sequence** the weather changes that a person on the ground will observe for the front shown above.

38. **Compare** the challenges of forecasting weather for Seattle, Washington, with those of forecasting weather for New York City.

39. CAREERS IN EARTH SCIENCE Develop a one-day weather forecast for your city using the weather data on page 335. Role-play a television meteorologist, and give your weather report.

40. **Determine** the type of modification to an arctic air mass moving southward over the north Atlantic during the summer.

41. **Evaluate** whether temperature readings taken near an asphalt parking lot on a summer day would represent those for the entire city.

Concept Mapping

42. Create a concept map showing the relationships among the types of weather data collection.

Challenge Question

43. **Explain** why cold fronts become stationary and break down as they move through Florida.

Additional Assessment

44. WRITING in Earth Science Research a local weather-related organization, and write a short essay about the kind of analyses that it performs.

DBQ Document–Based Questions

Data obtained from: National Weather Service, National Center for Environmental Prediction. January 2006. *NOAA.*

The graphs below show the accuracy of three numerical forecast models (A, B, and C) in predicting maximum daily temperatures during January 2006 for up to a period of eight days.

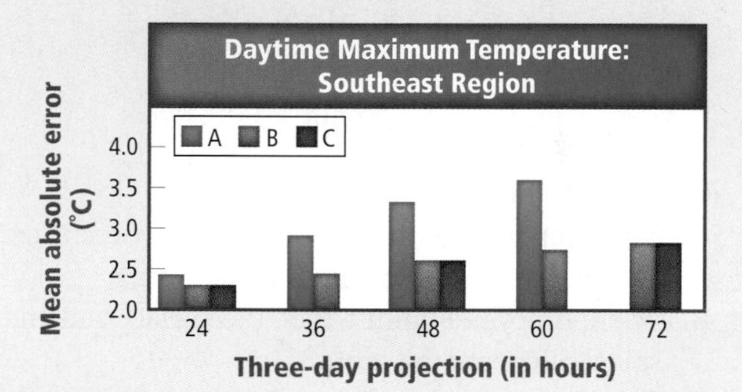

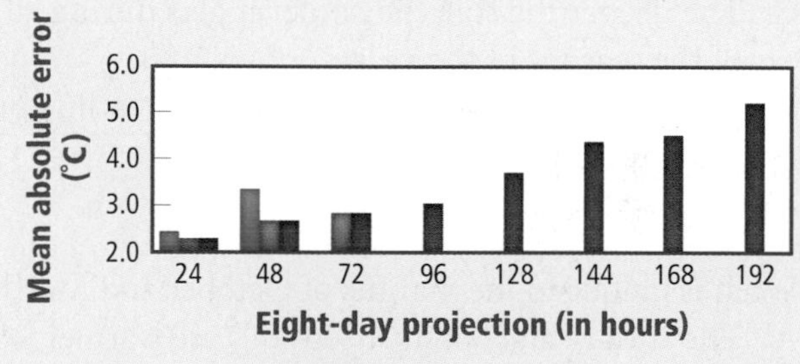

45. Which model had the greatest mean absolute error over the first 60 hours?

46. Which model could be used to find a maximum-temperature forecast for a week from now?

47. Which of the three models shown is the most valuable overall? Explain your answer.

Cumulative Review

48. How is a hand-held GPS receiver used to locate a position? **(Chapter 2)**

49. What determines the maximum height to which water from an artesian well will rise? **(Chapter 10)**

Standardized Test Practice

Multiple Choice

Use the graph to answer Questions 1–3.

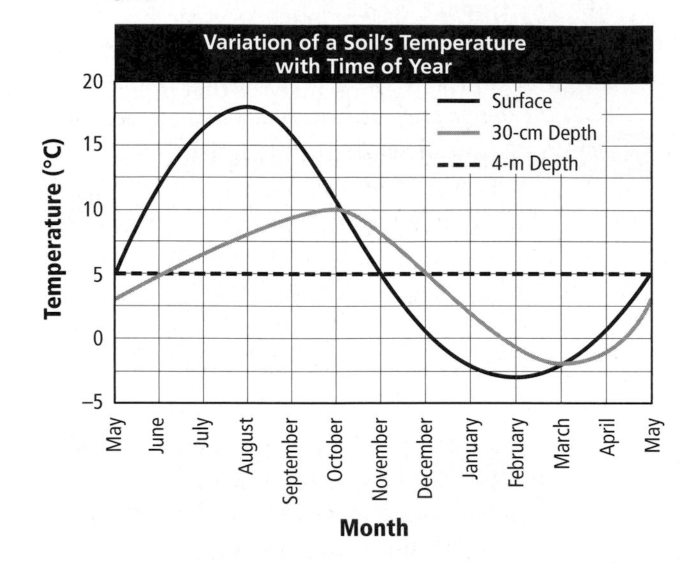

1. What can be inferred from the graph?
   - **A.** The temperature varies the greatest between surface and 30-cm-deep soil in the summer.
   - **B.** There is never a point where the surface soil and soil at 30cm are the same.
   - **C.** The deeper the soil, the cooler it gets during all of the seasons of the year.
   - **D.** All soil layers vary in temperature depending on the time of year.

2. What is unique to the months of October and March?
   - **A.** The surface and 30-cm-deep soils are warmer than the 4-m-deep soil.
   - **B.** The surface and 30-cm-deep soils have the same temperature.
   - **C.** The surface and 30-cm-deep soils are colder than the 4-m-deep soil.
   - **D.** The soil is the same temperature at all depths.

3. Why is 4m deep in the soil an important spot to record?
   - **A.** The temperature of the soil never changes there.
   - **B.** No other layer of soil ever reaches that temperature.
   - **C.** It is the only soil that is always frozen.
   - **D.** It is the deepest that soil goes below Earth's surface.

4. Which observation about a rock could lead you to identify it as igneous?
   - **A.** The rock has well defined layers.
   - **B.** The rock has a glassy texture.
   - **C.** The rock contains pebbles.
   - **D.** The rock is made of calcite.

5. Which clouds are most likely to form when fog lifts away from Earth's surface?
   - **A.** cumulus
   - **B.** cirrostratus
   - **C.** stratus
   - **D.** altocumulus

Use the illustration below to answer Questions 6 and 7.

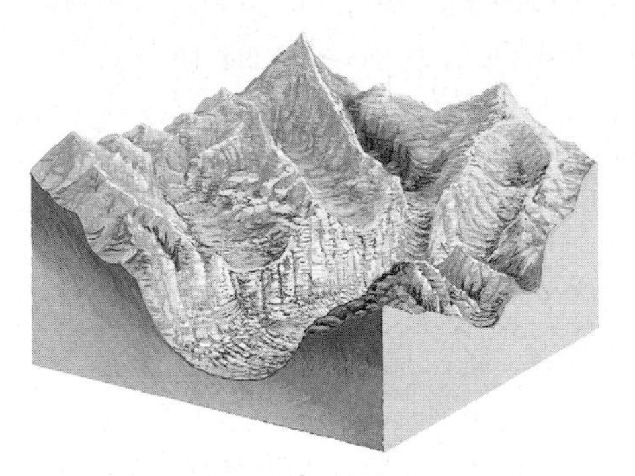

6. What event created the features shown above?
   - **A.** water erosion
   - **B.** wind erosion
   - **C.** glacial erosion
   - **D.** asteroid impact

7. What can scientists learn by studying areas similar to the illustration?
   - **A.** how rivers create U-shaped valleys
   - **B.** glacier history and its direction of movement
   - **C.** why glaciers moved
   - **D.** the impact of wind on mountainous features

8. Which statement is true about fossils found in previously undisturbed strata of sedimentary rock?
   - **A.** Fossils in the upper strata are younger than those in the lower strata.
   - **B.** Fossils in the upper strata are older than those in the lower strata.
   - **C.** Fossils in the upper strata generally are less complex than those in the lower strata.
   - **D.** There are no fossils in the upper strata that resemble those in the lower strata.

Short Answer

Use the image below to answer Questions 9 and 10.

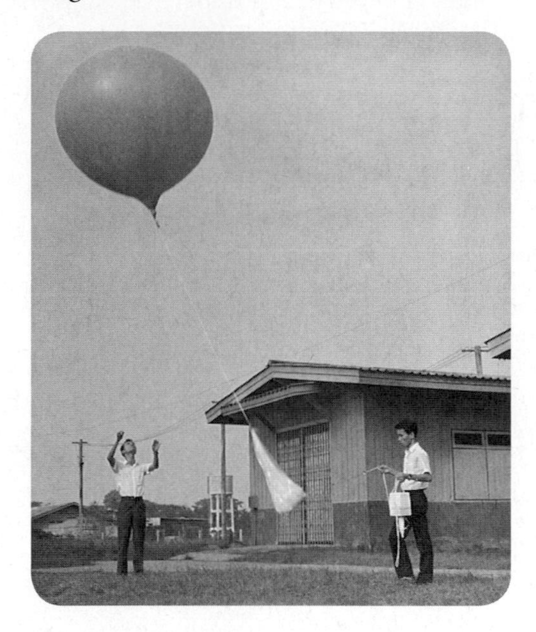

9. What weather instrument is shown in the image above? How does it work?

10. Why is it important for meteorologists to use tools like the one shown for gathering weather data?

11. Suppose surface water that is not absorbed forms a channel and quickly dries up. What can happen over time?

12. Discuss ground subsidence and its threat to our water supply.

13. Suppose a stream is 6 m wide and 3 m deep with a velocity of 12 m/s. How could you determine the stream's discharge?

14. Differentiate between the mass of a brick and the weight of the same brick.

Reading for Comprehension

Another Use for Radar

Doppler radar tracks moving objects, such as raindrops, through the atmosphere, by bouncing electromagnetic energy off them and measuring the amplitude as well as the change in frequency. Radar doesn't distinguish between bats and hailstones. To the radar, the millions of bats emerging from their caves look like a huge storm that starts at a point on the ground and spreads rapidly up and over the landscape. "It didn't take long for word to get around among bat researchers that we could view bat colonies on the new radar," recalls Jim Ward, science and operations officer at New Braunfels. "We saw bats flying as high as 10,000 feet." Since bats in other locales pursue insects close to the ground, we wondered why the free-tails were flying as high as 10,000 feet. Again, Doppler radar offered clues by detecting the billions of insects that swarm high above Texas. Since the 1980s, researchers have used radar to map the flight patterns of some of North America's most destructive agricultural pests—fall armyworms, beet armyworms, tobacco budworms, and corn earworms.

Article obtained from: McCracken G. F. and J. K Westbrook. 2002. Bat patrol. *National Geographic Magazine* (April): 1.

15. What can be inferred from this passage?
- **A.** Doppler radar is not useful for monitoring weather.
- **B.** Doppler radar can track both a major storm and a swarm of bats.
- **C.** Doppler radar should be used only for studying weather.
- **D.** Doppler radar should be used only for studying bats.

16. What important insight about technology can you gain by reading this article?

NEED EXTRA HELP?														
If You Missed Question . . .	1	2	3	4	5	6	7	8	9	10	11	12	13	14
Review Section . . .	7.3	7.3	7.3	5.1	11.3	8.3	8.3	6.1	12.3	12.3	9.1	10.3	9.1	1.2

CHAPTER 13

The Nature of Storms

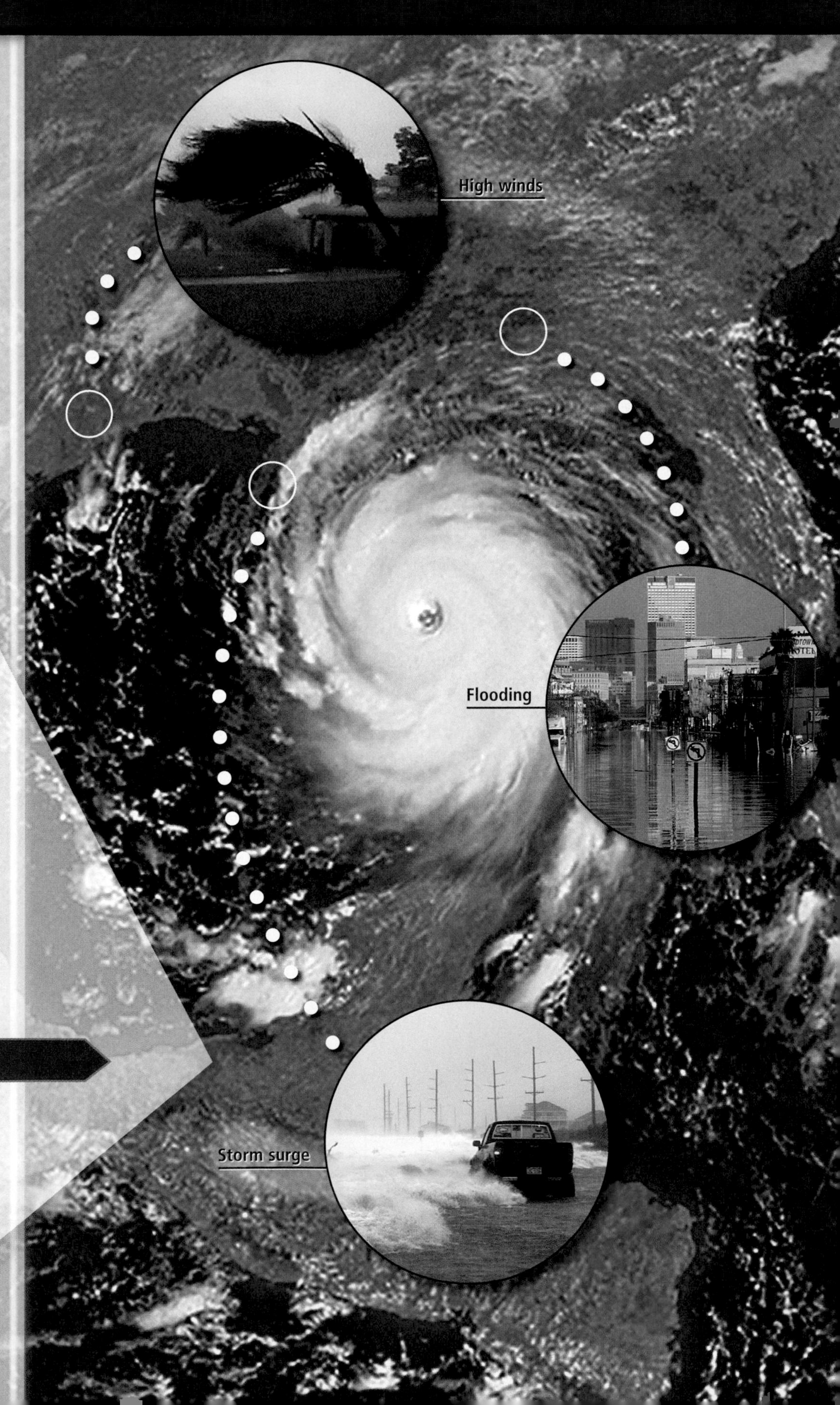

BIG Idea **The exchange of thermal energy in the atmosphere sometimes occurs with great violence that varies in form, size, and duration.**

13.1 Thunderstorms
MAIN Idea The intensity and duration of thunderstorms depend on the local conditions that create them.

13.2 Severe Weather
MAIN Idea All thunderstorms produce wind, rain, and lightning, which can have dangerous and damaging effects under certain circumstances.

13.3 Tropical Storms
MAIN Idea Normally peaceful, tropical oceans are capable of producing one of Earth's most violent weather systems—the tropical cyclone.

13.4 Recurrent Weather
MAIN Idea Even a relatively mild weather system can become destructive and dangerous if it persists for long periods of time.

GeoFacts

- Hurricanes, tornadoes, and everyday thunderstorms follow the same life cycles.
- The largest hailstone measured was nearly 18 cm in diameter.
- An F5 tornado can pack winds that will flatten a building.

LAUNCH Lab

Why does lightning form?

You have probably felt the shock of static electricity when you scuff your feet on a rug and then touch a doorknob. Your feet pick up additional electrons, which are negatively charged. These electrons are attracted to the positively charged protons of the doorknob metal, causing a small electrical current to form. The current causes you to feel a small shock.

Procedure

1. Read and complete the safety lab form.
2. With a **paper punch,** create 10 **paper circles.**
3. Place the circles in two piles of 5 on your desk.
4. Blow up a small **balloon** and mark one side with an *X*.
5. Rub the *X* side of the balloon on some **fabric.**
6. Hold the *X* side of the balloon 2 cm above one pile of paper circles.
7. Turn the balloon over, opposite the *X*, and hold it 2 cm above the other pile of paper circles.

Analysis

1. **Describe** what happened to the paper circles.
2. **Explain** what happened when you rubbed the balloon on the fabric.
3. **Infer** how the static attracting the paper is similar to the static electricity you produced on a rug.
4. **Infer** what causes lightning to jump from spot to spot.

Thunderstorm Development Make the following Foldable to summarize the stages of thunderstorm development.

STEP 1 Make a 3-cm fold along the long side of a sheet of paper and crease.

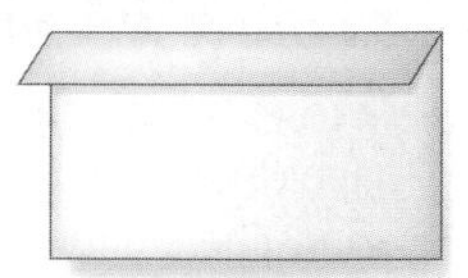

STEP 2 Fold the sheet into thirds.

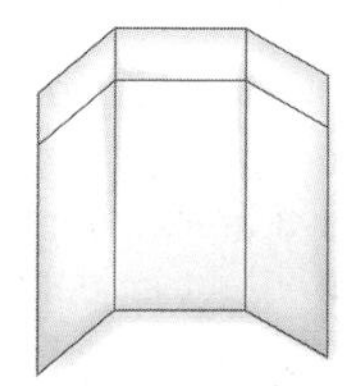

STEP 3 Unfold the paper and draw lines along the fold lines. Label the columns *Cumulus Stage, Mature Stage,* and *Dissipation Stage.*

Cumulus Stage	Mature Stage	Dissipation Stage

FOLDABLES **Use this Foldable with Section 13.1.** As you read this section, diagram the air movement, and describe the conditions at each stage.

Visit glencoe.com to

- study entire chapters online;
- explore Concepts In Motion animations:
  - Interactive Time Lines
  - Interactive Figures
  - Interactive Tables
- access Web Links for more information, projects, and activities;
- review content with the Interactive Tutor and take Self-Check Quizzes.

Section 13.1

Objectives

- **Identify** the processes that form thunderstorms.
- **Compare and contrast** different types of thunderstorms.
- **Describe** the life cycle of a thunderstorm.

Review Vocabulary

latent heat: stored energy in water vapor that is not released to warm the atmosphere until condensation occurs

New Vocabulary

air-mass thunderstorm
mountain thunderstorm
sea-breeze thunderstorm
frontal thunderstorm
stepped leader
return stroke

Thunderstorms

MAIN Idea The intensity and duration of thunderstorms depend on the local conditions that create them.

Real-World Reading Link Think about how an engine processes fuel to produce energy that powers an automobile. Thunderstorms are atmospheric engines that use heat and moisture as fuel and expend their energy in the form of clouds, rain, lightning, and wind.

Overview of Thunderstorms

At any given moment, nearly 2000 thunderstorms are in progress around the world. Most do little more than provide welcome relief on a muggy summer afternoon, or provide a spectacle of lightning. Some, however, grow into atmospheric monsters capable of producing hail the size of baseballs, swirling tornadoes, and surface winds of more than 160 km/h. These severe thunderstorms can also provide the energy for nature's most destructive storms—hurricanes. These severe thunderstorms, regardless of intensity, have certain characteristics in common. **Figure 13.1** shows which areas of the United States experience the most thunderstorms annually.

How thunderstorms form In Chapter 11, you read that the stability of the air is determined by whether or not an air mass can lift. Cooling air masses are stable and those that receive warming from the land or water below them are not. Under the right conditions, convection can cause a cumulus cloud to grow into a cumulonimbus cloud. The conditions that produce cumulonimbus clouds are the same conditions that produce thunderstorms. For a thunderstorm to form, three conditions must exist: a source of moisture, lifting of the air mass, and an unstable atmosphere.

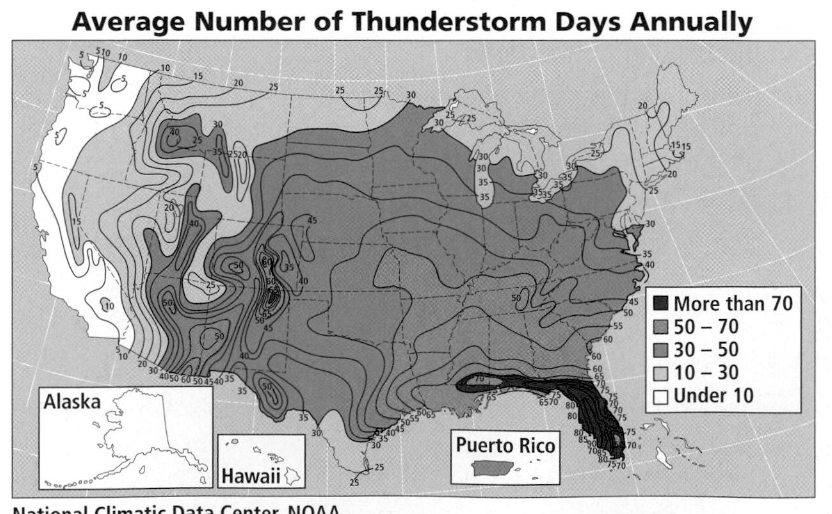

■ **Figure 13.1** Both geography and air mass movements make thunderstorms most common in the southeastern United States.
Predict *why the Pacific Coast has so few thunderstorms and Florida has so many.*

■ **Figure 13.2** This cumulus cloud is growing as a result of unstable conditions. As the cloud continues to develop into a cumulonimbus cloud, a thunderstorm might develop.

Moisture First, for a thunderstorm to form, there must be an abundant source of moisture in the lower levels of the atmosphere. Air masses that form over tropical oceans or large lakes become more humid from water evaporating from the surface below. This humid air is less dense than the surrounding dry air and is lifted. The water vapor it contains condenses into the droplets that constitute clouds. Latent heat, which is released from the water vapor during the process of condensation, warms the air causing it to rise further, cool further, and condense more of its water vapor.

Lifting Second, there must be some mechanism for condensing moisture to release its latent heat. This occurs when a warm air mass is lifted into a cooler region of the atmosphere. Dense, cold air along a cold front can push warmer air upward, just like an air mass does when moving up a mountainside. Warm land areas, heat islands such as cities, and bodies of water can also provide heat for lifting an air mass. Only when the water vapor condenses can it release latent heat and keep the cloud rising.

Stability Third, if the surrounding air remains cooler than the rising air mass, the unstable conditions can produce clouds that grow upward. This releases more latent heat and allows continued lifting. However, when the density of the rising air mass and the surrounding air are nearly the same, the cloud stops growing. **Figure 13.2** shows a cumulus cloud that is on its way to becoming a cumulonimbus cloud that can produce thunderstorms.

 **Reading Check** **Describe** the three conditions for thunderstorm growth.

Limits to thunderstorm growth The conditions that limit thunderstorm growth are the same ones that form the storm. Conditions that create lift, condense water vapor, and release latent heat keep the air mass warmer than the surrounding air. The air mass will continue to rise until it reaches a layer of equal density that it cannot overcome. Because atmospheric stability increases with height, most cumulonimbus clouds are limited to about 12,000 m. Thunderstorms are also limited by duration and size.

■ **Figure 13.3** Temperature differences exist over land and water and vary with the time of day.
Infer ***why water is warmer than the land at night.***

During the day, the temperature of land increases faster than the temperature of water. The warm air over land expands and rises, and the colder air over the sea moves inland and replaces the warm air. These conditions can produce strong updrafts that result in thunderstorms.

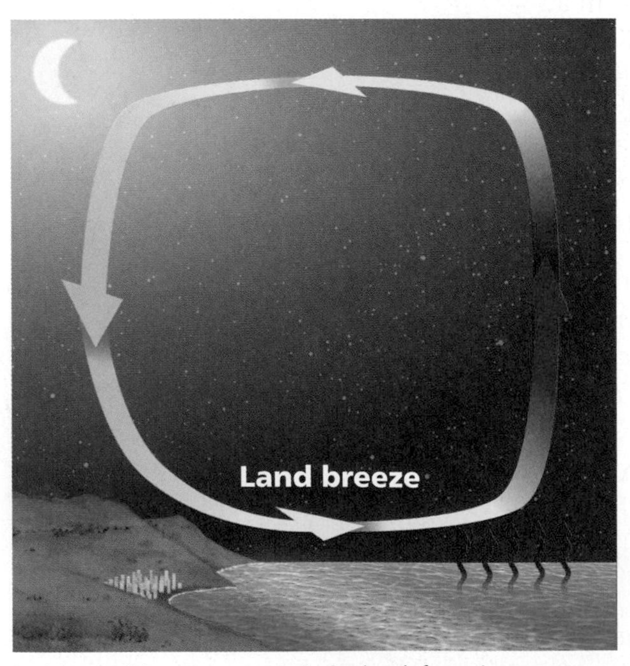

At night, conditions are reversed. The land cools faster than water, so the warmer sea air rises, and cooler air from above land moves over the water and replaces it. Nighttime conditions are considered stable.

Types of Thunderstorms

Thunderstorms are often classified according to the mechanism that causes the air mass that formed them to rise. There are two main types of thunderstorms: air-mass and frontal.

Air-mass thunderstorms When air rises because of unequal heating of Earth's surface within one air mass, the thunderstorm is called an **air-mass thunderstorm.** The unequal heating of Earth's surface reaches its maximum during mid-afternoon, so it is common for air-mass thunderstorms, also called pop-up storms, to occur.

There are two kinds of air-mass thunderstorms. **Mountain thunderstorms** occur when an air mass rises by orographic lifting, which involves air moving up the side of a mountain. **Sea-breeze thunderstorms** are local air-mass thunderstorms that occur because land and water store and release thermal energy differently. Sea-breeze thunderstorms are common along coastal areas during the summer, especially in the tropics and subtropics. Because land heats and cools faster than water, temperature differences can develop between the air over coastal land and the air over water, as shown in **Figure 13.3.**

Frontal thunderstorms The second main type is **frontal thunderstorms,** which are produced by advancing cold fronts and, more rarely, warm fronts. In a cold front, dense, cold air pushes under warm air, which is less dense, rapidly lifting it up a steep cold-front boundary. This rapid upward motion can produce a thin line of thunderstorms, sometimes hundreds of kilometers long, along the leading edge of the cold front. Cold-front thunderstorms get their initial lift from the push of the cold air. Because they are not dependent on daytime heating for their initial lift, cold-front thunderstorms can persist long into the night. Flooding from soil saturation is common with these storms. Floods are the main cause of thunderstorm-related deaths in the United States each year.

Less frequently, thunderstorms can develop along the advancing edge of a warm front. In a warm-front storm, a warm air mass slides up and over a gently sloping cold air mass. If the warm air behind the warm front is unstable and moisture levels are sufficiently high, a relatively mild thunderstorm can develop.

Thunderstorm Development

A thunderstorm usually has three stages: the cumulus stage, the mature stage, and the dissipation stage. The stages are classified according to the direction the air is moving.

Cumulus stage In the cumulus stage, air starts to rise vertically, as shown in **Figure 13.4.** The updrafts are relatively localized and cover an area of about 5–8 km. This creates updrafts, which transport water vapor to the cooler, upper regions of the cloud. The water vapor condenses into visible cloud droplets and releases latent heat. As the cloud droplets coalesce, they become larger and heavier until the updrafts can no longer sustain them and they fall to Earth as precipitation. This begins the mature stage of a thunderstorm.

Mature stage In the mature stage, updrafts and downdrafts exist side by side in the cumulonimbus cloud. Precipitation, composed of water and ice droplets that formed at high, cool levels of the atmosphere, cools the air as it falls. The newly cooled air is more dense than the surrounding air, so it sinks rapidly to the ground along with the precipitation. This creates downdrafts. As **Figure 13.4** shows, the updrafts and downdrafts form a convection cell which produces the surface winds associated with thunderstorms. The average area covered by a thunderstorm in its mature stage is 8–15 km.

Dissipation stage The convection cell can exist only if there is a steady supply of warm, moist air at Earth's surface. Once that supply is depleted, the updrafts slow down and eventually stop. In a thunderstorm, the cool downdrafts spread in all directions when they reach Earth's surface. This cools the areas from which the storm draws its energy, the updrafts cease, and clouds can no longer form. The storm is then in the dissipation stage shown in **Figure 13.4.** This stage will last until all of the previously formed precipitation has fallen.

FOLDABLES Incorporate information from this section into your Foldable.

Concepts In Motion
Interactive Figure To see an animation of thunderstorm development, visit glencoe.com.

■ **Figure 13.4** The cumulus stage of a thunderstorm is characterized mainly by updrafts. The mature stage is characterized by strong updrafts and downdrafts. The storm loses energy in the dissipation stage.

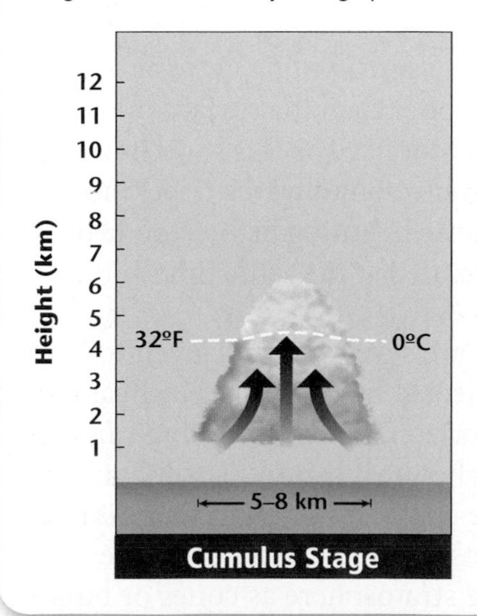

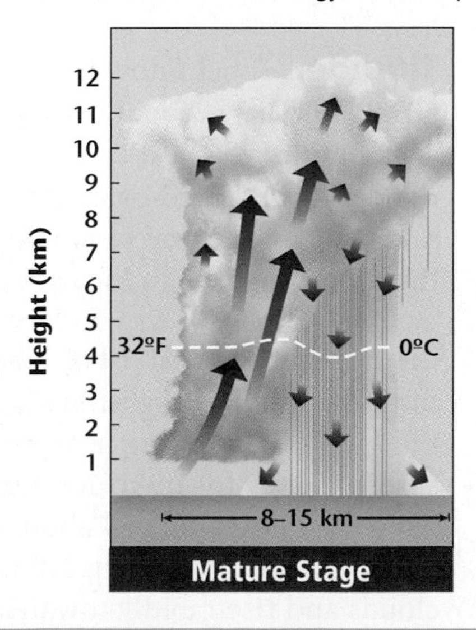

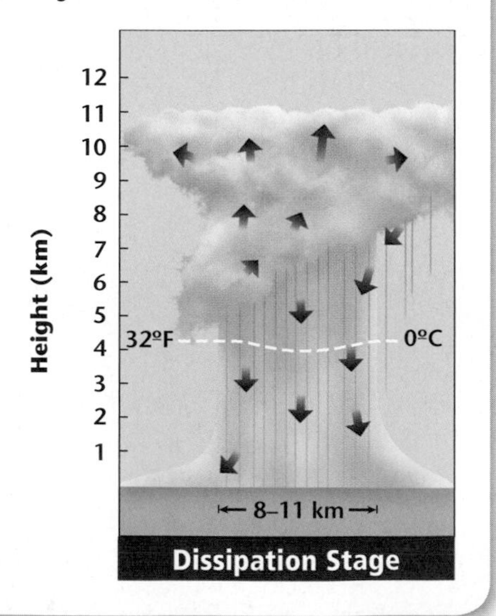

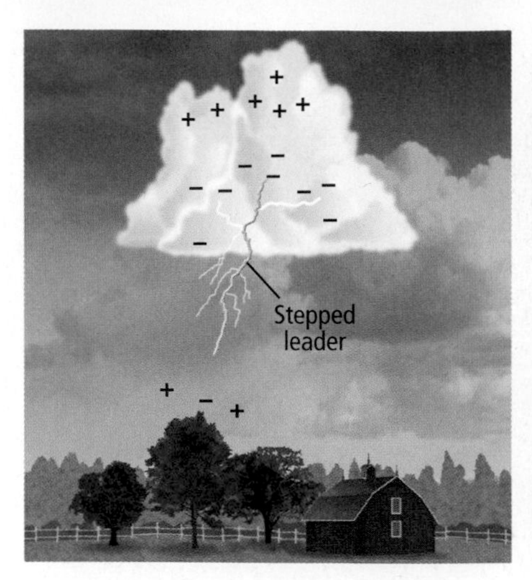

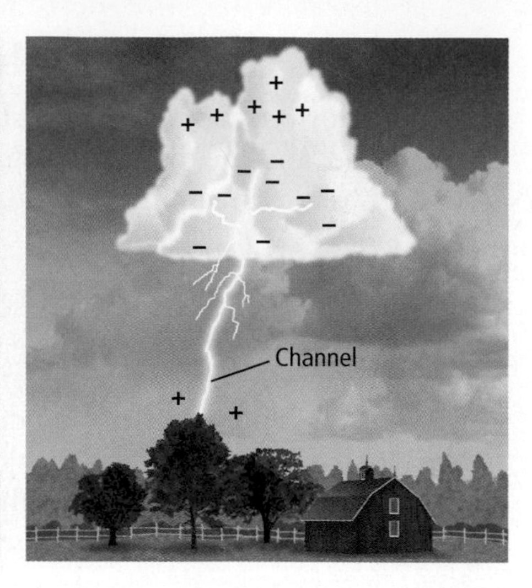

■ **Figure 13.5** When a stepped leader nears an object on the ground, a powerful surge of electricity from the ground moves upward to the cloud and lightning is produced.
Sequence ***Make an outline sequencing the steps of lightning formation.***

Lightning

Have you ever touched a metal object on a dry winter day and been zapped by a spark from static electricity? The static electricity was generated from friction, and the spark is similar to lightning. Lightning is the transfer of electricity generated by the rapid rushes of air in a cumulonimbus cloud. Clouds become charged when friction between the updrafts and downdrafts removes electrons from some of the atoms in the cloud. The atoms that lose electrons become positively charged ions. Other atoms receive the extra electrons and become negatively charged ions. As **Figure 13.5** shows, this creates regions of air with opposite charges. Eventually, the differences in charges break down, and a branched channel of partially charged air is formed between the positive and negative regions. The channel of partially charged air is called a **stepped leader,** and it generally moves from the center of the cloud toward the ground. When the stepped leader nears the ground, a branched channel of positively charged particles, called the **return stroke,** rushes upward to meet it. The return stroke surges from the ground to the cloud, illuminating the connecting channel with about 100 million volts of electricity. That illumination is the brightest part of lightning.

Thunder A lightning bolt heats the surrounding air to about 30,000°C. That is about five times hotter than the surface of the Sun. The thunder you hear is the sound produced as this superheated air rapidly expands and contracts. Because sound waves travel more slowly than light waves, you might see lightning before you hear thunder, even though they are generated at the same time.

Lightning variations There are several names given to lightning effects. Sheet lightning is reflected by clouds, while heat lightning is sheet lightning near the horizon. Spider lightning can crawl across the sky for up to 150 km. The most bizarre is ball lightning which is a hovering ball about the size of a pumpkin that disappears in a fizzle or a bang. Blue jets and red sprites originate in clouds and rise rapidly toward the stratosphere as cones or bursts.

■ **Figure 13.6** Five times hotter than the surface of the Sun, a lightning bolt can be spectacular. But when an object such as this pine tree is struck, it can be explosive.

Thunderstorm and lightning safety Each year in the United States, lightning causes about 7500 forest fires, which result in the loss of thousands of square kilometers of forest. In addition, lightning strikes in the United States cause a yearly average of 300 injuries and 93 deaths to humans. **Figure 13.6** indicates how destructive a lightning strike might be.

Avoid putting yourself in danger of being struck by lightning. If you are outdoors and feel your hair stand on end, squat low on the balls of your feet. Duck your head and make yourself the smallest target possible. Small sheds, isolated trees, and convertible automobiles are hazardous as shelters. Using electrical appliances and telephones during a lightning storm can lead to electric shock. Stay out of boats and away from water during a thunderstorm.

Section 13.1 Assessment

Section Summary

- The cumulus stage, the mature stage, and the dissipation stage comprise the life cycle of a thunderstorm.
- Clouds form as water is condensed and latent heat is released.
- Thunderstorms can be produced either within air masses or along fronts.
- From formation to dissipation, all thunderstorms go through the same stages.
- Lightning is a natural result of thunderstorm development.

Understand Main Ideas

1. MAIN Idea **List** the conditions needed for a thunderstorm's cumulus stage.
2. **Explain** how a thunderstorm is formed along a front.
3. **Differentiate** between a sea-breeze thunderstorm and a mountain thunderstorm.
4. **Identify** what causes a thunderstorm to dissipate.
5. **Compare and contrast** how a cold front and a warm front can create thunderstorms.
6. **Describe** two different types of lightning.

Think Critically

7. **Infer** Lightning occurs during which stage of thunderstorm formation?
8. **Determine** the conditions in thunderstorm formation that creates lightning.

WRITING in **Earth Science**

9. Write a setting for a movie using a storm as part of the opening scene.

Section 13.2

Objectives

- **Explain** why some thunderstorms are more severe than others.
- **Recognize** the dangers of severe weather, including lightning, hail, and high winds.
- **Describe** how tornadoes form.

Review Vocabulary

air mass: large body of air that takes on the characteristics of the area over which it forms

New Vocabulary

supercell
downburst
tornado
Fujita tornado intensity scale

Severe Weather

MAIN Idea All thunderstorms produce wind, rain, and lightning, which can have dangerous and damaging effects under certain circumstances.

Real-World Reading Link Sliding down a park slide might seem mild and safe compared to a roller coaster's wild and chaotic ride. Similarly, while a gentle rain is appreciated by many, the same weather processes can create thunderstorms on a massive atmospheric scale, resulting in disaster.

Weather Cells

All thunderstorms are not created equal. Some die out within minutes, while others flash and thunder throughout the night. What makes one thunderstorm more severe than another? The increasing instability of the air intensifies the strength of a storm's updrafts and downdrafts, which makes the storm severe.

Supercells Severe thunderstorms can produce some of the most violent weather conditions on Earth. They can develop into self-sustaining, extremely powerful storms called **supercells.** Supercells are characterized by intense, rotating updrafts taking 10 to 20 minutes to reach the top of the cloud. These furious storms can last for several hours and can have updrafts as strong as 240 km/h. It is not uncommon for a supercell to spawn long-lived tornadoes. **Figure 13.7** shows an illustration of a supercell. Notice the anvil-shaped cumulonimbus clouds associated with severe storms. The tops of the supercells are chopped off by wind shear. Of the estimated 100,000 thunderstorms that occur each year in the United States, only about 10 percent are considered to be severe, and fewer still reach supercell proportions.

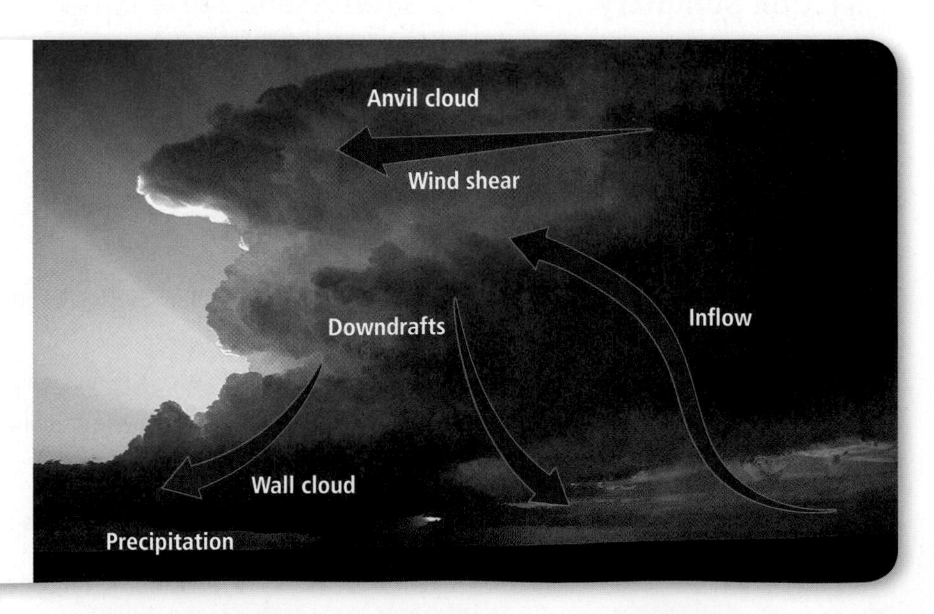

Figure 13.7 An anvil-shaped cumulonimbus cloud is characteristic of many severe thunderstorms. The most severe thunderstorms are called supercells.

Strong Winds

Recall that rain-cooled downdrafts descend to Earth's surface during a thunderstorm and spread out as they reach the ground. Sometimes, instead of dispersing that downward energy over a large area underneath the storm, the energy becomes concentrated in a local area. The resulting winds are exceptionally strong, with speeds of more than 160 km/h. Violent downdrafts that are concentrated in a local area are called **downbursts.**

Based on the size of the area they affect, downbursts are classified as either macrobursts or microbursts. Macrobursts can cause a path of destruction up to 5 km wide. They have wind speeds of more than 200 km/h and can last up to 30 minutes. Smaller in size, though deadlier in force, microbursts affect areas of less than 3 km but can have winds exceeding 250 km/h. Despite lasting fewer than 10 minutes on average, a microburst is especially deadly because its small size makes it extremely difficult to predict and detect. **Figure 13.8** shows a microburst.

Figure 13.8 A microburst, such as this one in Kansas, can be as destructive as a tornado.

Hail

Each year in the United States, almost one billion dollars in damage is caused by hail—precipitation in the form of balls or lumps of ice. Hail can do tremendous damage to crops, vehicles, and rooftops, particularly in the central United States where hail occurs most frequently. Hail is most common during the spring growing season. **Figure 13.9** shows some conditions associated with hail.

Hail forms because of two characteristics common to thunderstorms. First, water droplets enter the parts of a cumulonimbus cloud where the temperature is below freezing. When these supercooled water droplets encounter ice pellets, the water droplets freeze on contact and cause the ice pellets to grow larger. The second characteristic that allows hail to form is an abundance of strong updrafts and downdrafts existing side by side within a cloud. The growing ice pellets are caught alternately in the updrafts and downdrafts, so that they constantly encounter more supercooled water droplets. The ice pellets keep growing until they are too heavy for even the strongest updrafts to keep aloft, and they finally fall to Earth as hail.

Figure 13.9 This hail storm in Sydney, Australia, caused slippery conditions for the traffic as well as damage to property.

Tornadoes

In some parts of the world, the most feared form of severe weather is the tornado. A **tornado** is a violent, whirling column of air in contact with the ground. When a tornado does not reach the ground, it is called a funnel cloud. Tornadoes are often associated with supercells—the most severe thunderstorms. The air in a tornado is made visible by dust and debris drawn into the swirling column, sometimes called the vortex, or by the condensation of water vapor into a visible cloud.

Reading Check **Define** the term *tornado.*

VOCABULARY

ACADEMIC VOCABULARY

Phenomenon
an object or aspect known through the senses rather than by thought or intuition
Students observing the phenomenon realized later that the powerful wind was a microburst.

Development of tornadoes A tornado forms when wind speed and direction change suddenly with height, a phenomenon associated with wind shear. Current thinking suggests that tornadoes form when small pockets of cooler air are given a horizontal, rolling-pin type of rotation near Earth's surface, as shown in **Figure 13.10.** If this rotation occurs close enough to the thunderstorm's updrafts, the twisting column of wind can be tilted from a horizontal to a vertical position. As updrafts stretch the column, the rotation is accelerated. Air is removed from the center of the column, which in turn lowers the air pressure in the center. The extreme pressure gradient between the center and the outer portion of the tornado produces the violent winds associated with tornadoes. Although tornadoes rarely exceed 200 m in diameter and usually last only a few minutes, they can be extremely destructive. A tornado is classified according to its destructive force.

Concepts In Motion
Interactive Figure To see an animation of tornado formation, visit glencoe.com.

Figure 13.10 Tornado formation is associated with changes in wind speed and direction.
Infer *what would cause the updrafts.*

A change in wind direction and speed creates a horizontal rotation in the lower atmosphere.

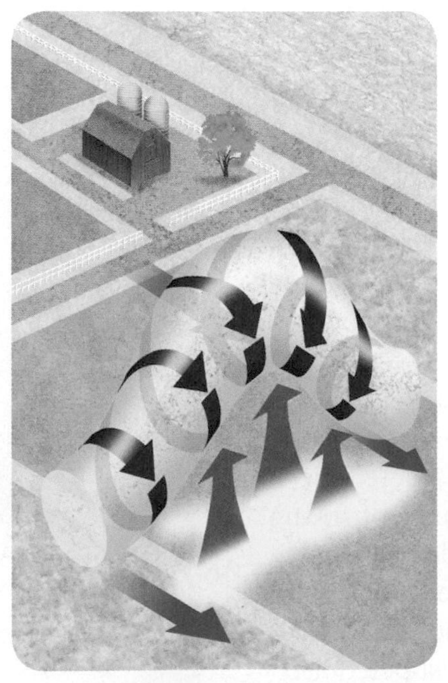

Strong updrafts tilt the rotating air from a horizontal to a vertical position.

A tornado forms within the rotating winds.

Concepts In Motion
Interactive Table To explore more about the Fujita tornado intensity scale, visit glencoe.com.

Table 13.1 Fujita Tornado Intensity Scale

Fujita scale tornadoes	**Weak (F0 and F1)** 80 percent of all tornadoes Path: up to 4 km Duration: 1–10 min Wind speed: 70–180 km/h	**Strong (F2 and F3)** 19 percent of all tornadoes Path: 24 km + Duration: 20 min + Wind speed: 181–332 km/h	**Violent (F4 and F5)** 1 percent of all tornadoes Path: 80 km + Duration: 1 h + Wind speed: 333–512+ km/h
Photo of tornado			

Tornado classification Tornadoes vary greatly in size and intensity. The **Fujita tornado intensity scale,** which ranks tornadoes according to their path of destruction, wind speed, and duration, is used to classify tornadoes. The Fujita scale was named for Japanese tornado researcher Dr. Theodore Fujita. The scale ranges from F0, which is characterized by winds of up to 118 km/h, to the incredibly violent F5, which can pack winds of more than 500 km/h. Most tornadoes do not exceed the F1 category. In fact, only about 1 percent reach F4 or F5. Those that do, however, can lift entire buildings from their foundations and toss automobiles and trucks around like toys. The Fujita scale is shown in **Table 13.1.**

Tornado distribution While tornadoes can occur at any time and at any place, there are some times and locations where they are more likely to form. Most tornadoes—especially violent ones—form in the spring during the late afternoon and evening, when the temperature contrasts between polar air and tropical air are the greatest. Large temperature contrasts occur most frequently in the central United States, where cold continental polar air collides with maritime tropical air moving northward from the Gulf of Mexico. These large temperature contrasts often spark the development of supercells, which are each capable of producing several strong tornadoes. More than 700 tornadoes touch down each year in the United States. Many of these occur in a region called "Tornado Alley," which extends from northern Texas through Oklahoma, Kansas, and Missouri.

Figure 13.11 In some areas, tornado shelters are common. If you are caught in a tornado, take shelter in the southwest corner of a basement, a small downstairs room or closet, or a tornado shelter like this one.

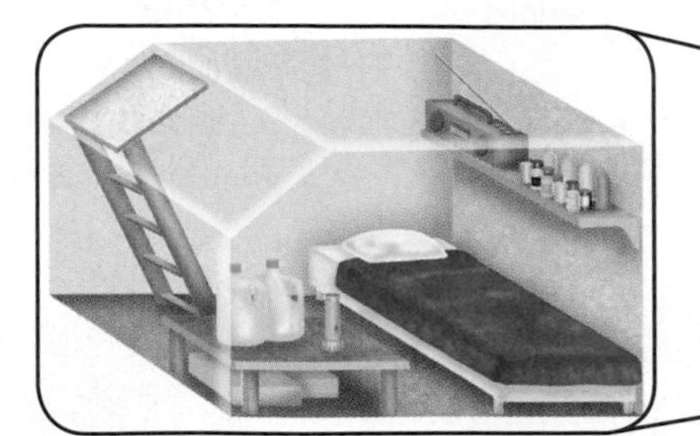

Tornado safety In the United States, an average of 80 deaths and 1500 injuries result from tornadoes each year. In an ongoing effort to reduce tornado-related fatalities, the National Weather Service issues tornado watches and warnings before a tornado strikes. These advisories are broadcast on local radio stations when tornadoes are indicated on weather radar or spotted in the region. During a severe thunderstorm, the presence of dark, greenish skies, a towering wall of clouds, large hailstones, and a loud, roaring noise similar to that of a freight train are signs of an approaching or developing tornado.

The National Weather Service stresses that despite advanced tracking systems, some tornadoes develop very quickly. In these cases, advance warnings might not be possible. However, the threat of tornado-related injury can be substantially decreased when people seek shelter, such as the one shown in **Figure 13.11,** at the first sign of threatening skies.

Section 13.2 Assessment

Section Summary

- Intense rotating updrafts are associated with supercells.
- Downbursts are strong winds that result in damage associated with thunderstorms.
- Hail is precipitation in the form of balls or lumps of ice that accompany severe storms.
- The worst storm damage comes from a vortex of high winds that moves along the ground as a tornado.

Understand Main Ideas

1. MAIN Idea **Identify** the characteristics of a severe storm.
2. **Describe** two characteristics of thunderstorms that lead to hail formation.
3. **Explain** how some hail can become baseball sized.
4. **Compare and contrast** a macroburst and a microburst.
5. **Identify** the steps that change wind shear into a tornado.
6. **Identify** the conditions that lead to high winds, hail, and lightning.

Think Critically

7. **Explain** Why are there more tornado-producing storms in flat plains than in mountainous areas?
8. **Analyze** the data of the Fujita scale, and determine why F5 tornadoes have a longer path than F1 tornadoes.

WRITING in **Earth Science**

9. Design a pamphlet about tornado safety.

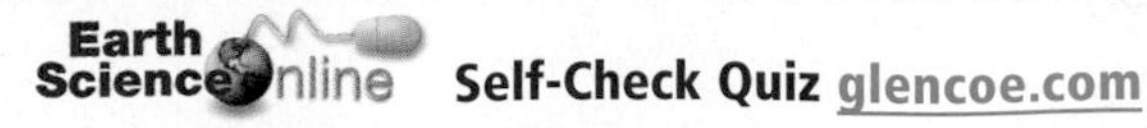

Section 13.3

Tropical Storms

Objectives

- **Identify** the conditions required for tropical cyclones to form.
- **Describe** the life cycle of a tropical cyclone.
- **Recognize** the dangers of hurricanes.

Review Vocabulary

Coriolis effect: caused by Earth's rotation, moving particles, such as air, are deflected to the right north of the equator, and to the left south of the equator

New Vocabulary

tropical cyclone
eye
eyewall
Saffir-Simpson hurricane scale
storm surge

MAIN Idea **Normally peaceful, tropical oceans are capable of producing one of Earth's most violent weather systems—the tropical cyclone.**

Real-World Reading Link If you try mixing cake batter in a shallow bowl, you might find that a low speed works well, but a high speed creates a big mess. Tropical cyclones form from processes similar to other storm systems, but their high winds can bring devastation to locations in their path.

Overview of Tropical Cyclones

During summer and fall, the tropics experience conditions ideal for the formation of large, rotating, low-pressure tropical storms called **tropical cyclones.** In different parts of the world, the largest of these storms are known as hurricanes, typhoons, and cyclones.

Cyclone location Favorable conditions for cyclone formation exist in all tropical oceans except the South Atlantic Ocean and the Pacific Ocean off the west coast of South America. The water in these areas is somewhat cooler and these areas contain regions of nearly permanently stable air. As a consequence, tropical cyclones do not normally occur in these areas. They do occur in the large expanse of warm waters in the western Pacific Ocean where they are known as typhoons. To people living near the Indian Ocean, they are known as cyclones. In the North Atlantic Ocean, the Caribbean Sea, the Gulf of Mexico, and along the western coast of Mexico, the strongest of these storms are called hurricanes. **Figure 13.12** shows where cyclones generally form.

Concepts In Motion
Interactive Figure To see an animation of tropical cyclones, visit glencoe.com.

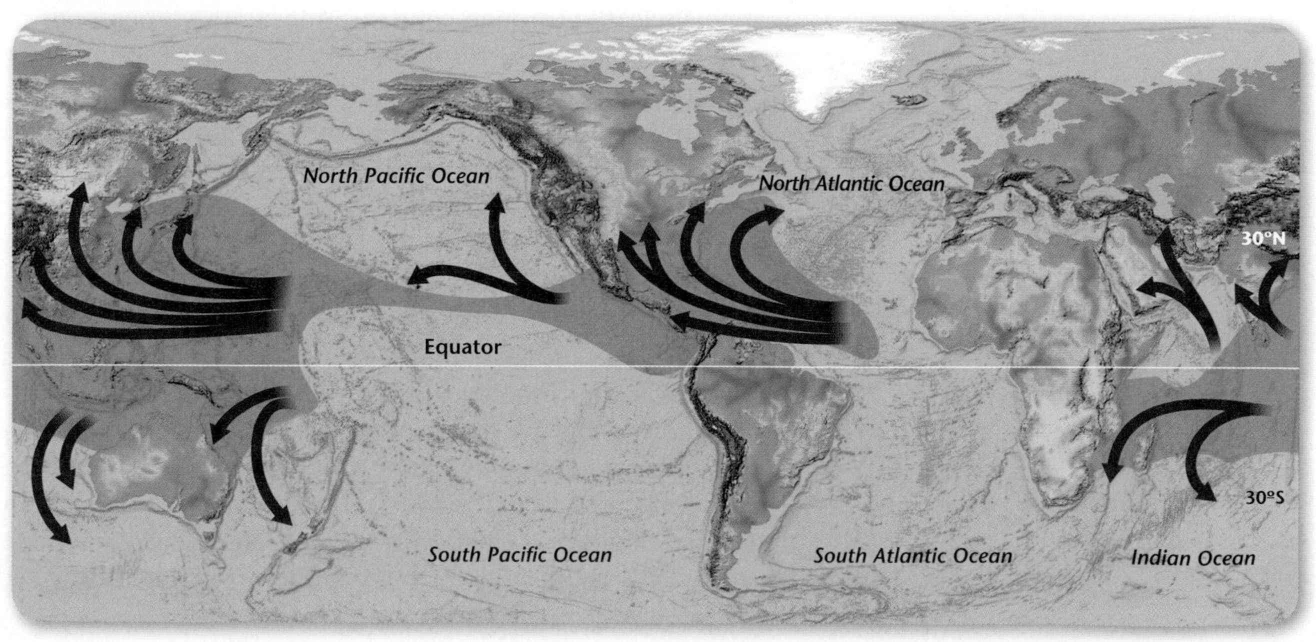

■ **Figure 13.12** Tropical cyclones are common in all of Earth's tropical oceans except in the relatively cool waters of both the South Pacific and South Atlantic Oceans.

■ **Figure 13.13** The characteristic rotating nature of cyclonic storms is evident in this tropical depression that formed over the Atlantic Ocean.

Cyclone formation Tropical cyclones require two basic conditions to form: an abundant supply of warm ocean water and some sort of mechanism to lift warm air and keep it rising. Tropical cyclones thrive on the tremendous amount of energy in warm, tropical oceans. As water evaporates from the ocean surface, latent heat is stored in water vapor. This latent heat is later released when the air rises and the water vapor condenses.

The air usually rises because of some sort of existing weather disturbance moving across the tropics. Many disturbances originate along the equator. Others are the result of weak, low-pressure systems called tropical waves. Tropical disturbances are common during the summer and early fall. Regardless of their origin, only a small percentage of tropical disturbances develop into cyclones. There are three stages in the development of a full tropical cyclone.

 **Reading Check Infer** what is produced when water vapor condenses.

Formative stage The first indications of a building tropical cyclone is a moving tropical disturbance. Less-dense, moist air is lifted, triggering rainfall and air circulation. As these disturbances produce more precipitation, more latent heat is released. In addition, the rising air creates an area of low pressure at the ocean surface. As more warm, dense air moves toward the low-pressure center to replace the air that has risen, the Coriolis effect causes the moving air to turn counterclockwise in the northern hemisphere. This produces the cyclonic (counterclockwise) rotation of a tropical cyclone, as shown in **Figure 13.13.** When a disturbance over a tropical ocean acquires a cyclonic circulation around a center of low pressure, it has reached the developmental stage and is known as a tropical depression, as illustrated in **Figure 13.14.**

Mature stage As the moving air approaches the center of the growing storm, it rises, rotates, and increases in speed as more energy is released through condensation. In the process, air pressure in the center of the system continues to decrease. As long as warm air is fed into the system at the surface and removed in the upper atmosphere, the storm will continue to build and the winds of rotation will increase as the air pressure drops.

When wind speeds around the low-pressure center of a tropical depression exceed 65 km/h, the system is called a tropical storm. If air pressure continues to fall and winds around the center reach at least 120 km/h, the storm is officially classified as a cyclone. Once winds reach these speeds, another phenomenon occurs—the development of a calm center of the storm called the **eye,** shown in **Figure 13.14.** The eye of the cyclone is a span of 30 to 60 km of calm weather and blue sky. The strongest winds in a hurricane are usually concentrated in the **eyewall**—a tall band of strong winds and dense clouds that surrounds the eye. The eyewall is visible because of the clouds that form there and mark the outward edge of the eye.

VOCABULARY

SCIENCE USAGE V. COMMON USAGE

Depression

Science usage: a pressing down or lowering, the low spot on a curved line

Common usage: a state of feeling sad

Visualizing Cyclone Formation

Figure 13.14 Like most storms, cyclones begin with warm moist air rising.

5. As the lighter air rises, moist air from the ocean takes its place, creating a wind current.

4. Condensation releases latent heat into the atmosphere, making the air less dense.

3. As the water vapor rises, the cooler upper air condenses it into liquid droplets.

2. Water vapor is lifted into the atmosphere.

1. Warm air absorbs moisture from the ocean.

Moving air starts to spin as a result of the Coriolis effect.

Tropical Depression The first indications of a building storm are a tropical depression with good circulation, thunderstorms, and sustained winds of 37–62 km/h.

Tropical Storm As winds increase to speeds of 63–117 km/h, strong thunderstorms develop and become well defined. They are now tropical storms.

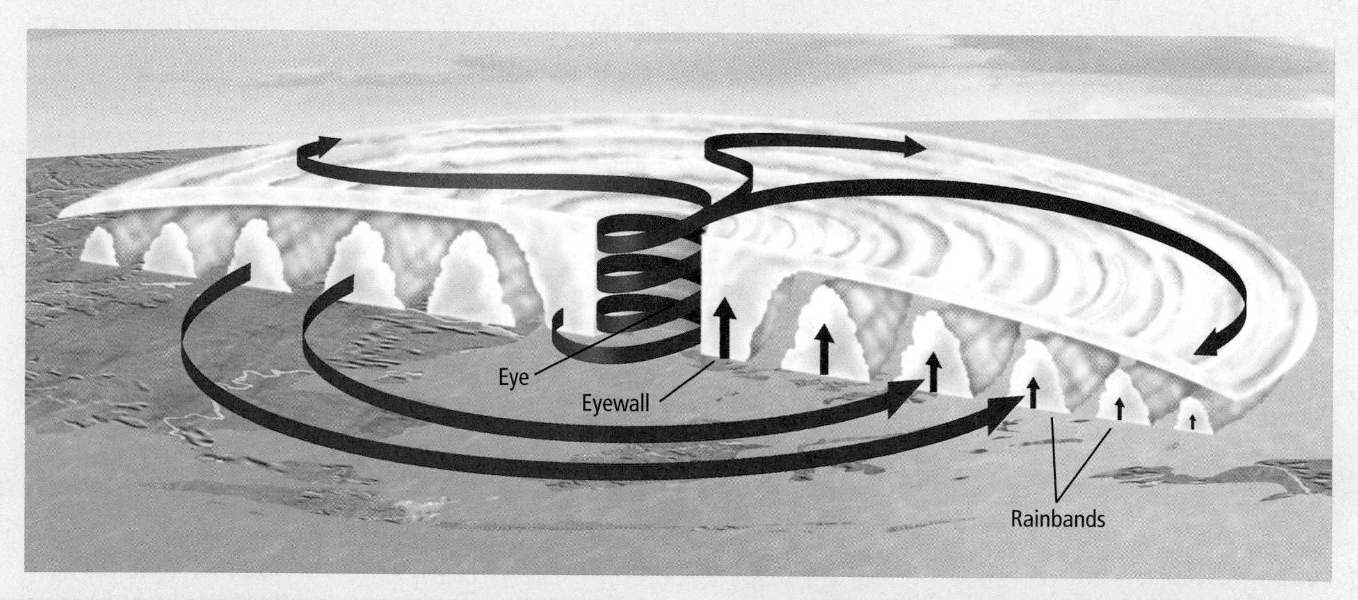

Cyclone With sustained winds of 118 km/h, an intense tropical weather system with well-defined circulation becomes a cyclone, also called a typhoon or hurricane.

Concepts In Motion To explore more about cyclone formation, visit glencoe.com.

Earth Science Online

Saffir-Simpson Hurricane Scale

Category	Winds (km/h)	Change in sea level	Damage
5	>250	>5.5 m	catastrophic
4	210–249	4.0–5.5 m	extreme
3	178–209	2.8–3.7 m	extensive
2	154–177	1.8–2.5 m	moderate
1	119–153	1.2–1.5 m	minimal

■ **Figure 13.15** The Saffir-Simpson hurricane scale classifies hurricanes according to wind speed, potential for flooding, and potential for property damage.

Dissipation stage A cyclone will last until it can no longer produce enough energy to sustain itself. This usually happens when the storm has moved either over land or over colder water. During its life cycle, a cyclone can undergo several fluctuations in intensity as it interacts with other atmospheric systems.

Tropical cyclone movement Like all large-scale storms, tropical cyclones move according to the wind currents that steer them. Recall that many of the world's oceans are home to subtropical high-pressure systems that are present to some extent throughout the year. Tropical cyclones are often caught up in the circulation of these high-pressure systems. They move steadily west, then eventually turn poleward when they reach the far edges of the high-pressure systems. There they are guided by prevailing westerlies and begin to interact with midlatitude systems. At this point, the interaction of the various wind and weather systems makes the movement of the storms unpredictable.

Hurricane Hazards

The **Saffir-Simpson hurricane scale,** as shown in **Figure 13.15,** classifies hurricanes according to wind speed, potential for flooding in terms of the effect on the height of sea level, and potential for property damage. The sea level and damage are dependent upon shore depth and the density of population and structures in the affected area.

■ **Figure 13.16**

Storm Tracking

Scientists have worked to develop weather prediction technology to protect people against the different types of storms that cause damage and loss of life.

1888 A three-day blizzard dumps 125 cm of snow on the northeast United States, creating 17-m snowdrifts burying houses and trains, killing 400 people, and sinking 200 ships.

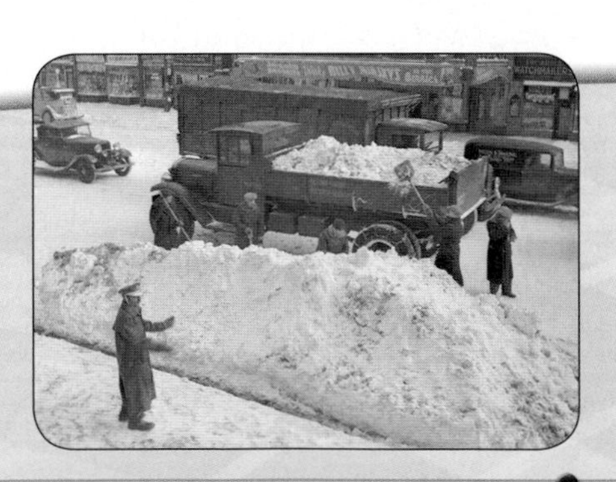

1850 — 1900 — 1925

1861 An English newspaper publishes the first daily weather forecasts based on countrywide data that is compiled via the recently invented telegraph.

1900 A Category 4 hurricane hits Texas. Five-m waves sweep over Galveston Island, killing more than 8000 people and washing away half the homes on the island.

1925 An F5 tornado rips through Missouri, Illinois, and Indiana covering 352 km in three hours.

Damage Hurricanes can cause extensive damage, particularly along coastal areas, which tend to be where human populations are the most dense. Evidence of storm damage is documented in **Figure 13.16** by a photo from a hurricane that hit Galveston, Texas, in 1900.

Winds Much of the damage caused by hurricanes is associated with violent winds. The strongest winds in a hurricane are usually located at the eyewall. Outside of the eyewall, winds taper off as distance from the center increases, although winds of more than 60 km/h can extend as far as 400 km from the center of a hurricane.

Storm surge Strong winds moving onshore in coastal areas are partly responsible for the largest hurricane threat—storm surges. A **storm surge** occurs when hurricane-force winds drive a mound of ocean water toward coastal areas where it washes over the land. Storm surges can sometimes reach 6 m above normal sea level, as shown in **Figure 13.17.** When this occurs during high tide, the surge can cause enormous damage. In the northern hemisphere, a storm surge occurs primarily on the right side of a storm relative to the direction of its forward motion. That is where the strongest onshore winds occur. This is due to the counterclockwise rotation of the storm.

Hurricanes produce great amounts of rain because of their continuous uptake of warm, moist ocean water. Thus, floods from intense rainfall are an additional hurricane hazard, particularly if the storm moves over mountainous areas, where orographic lifting enhances the upward motion of air and the resulting condensation of water vapor.

Figure 13.17 Storm surges can sometimes reach 6 m above normal sea level and cause enormous damage.

NATIONAL GEOGRAPHIC To read about increasingly strong hurricanes and the science behind them, go to the **National Geographic Expedition** on page 910.

1949 Lightning sparks a wildfire in Helena National Forest, Montana, that claims the lives of 13 firefighters and destroys 20 km^2 of land in five days.

1970 Large hailstones, more than 13 cm in diameter, fell on Coffeyville, Kansas.

2005 The Atlantic hurricane season unleashes the most hurricanes and Category 5 storms in history.

1950 — 1975 — 2000

1960 The U.S. government launches TIROS, the first weather satellite.

1990 The United States deploys its first operational Doppler radar system after more than 30 years of research.

Concepts In Motion

Interactive Time Line To learn more about these discoveries and others, visit glencoe.com. Earth Science Online

Figure 13.18 This residential area has been engulfed in debris left behind from the flood waters of Hurricane Katrina. Most of the deaths associated with a hurricane come from flooding, not high winds.

CAREERS IN EARTH SCIENCE

Hurricane Hunter A hurricane hunter flies an instrument-laden airplane into a hurricane to measure wind speed and gather weather data on the features of a hurricane. To learn more about Earth science careers, visit glencoe.com.

Hurricane advisories and safety The National Hurricane Center, which is responsible for tracking and forecasting the intensity and motion of tropical cyclones in the western hemisphere, issues a hurricane warning at least 24 hours before a hurricane is predicted to strike. The center also issues regular advisories that indicate a storm's position, strength, and movement. Using this information, people can then track a storm on a hurricane-tracking chart, such as the one you will use in the Internet GeoLab at the end of this chapter. Awareness, combined with proper safety precautions, has greatly reduced death tolls associated with hurricanes in recent years. **Figure 13.18** shows debris and destruction left by hurricane flooding; loss of life can be prevented by evacuating residents before the storm hits.

Section 13.3 Assessment

Section Summary

- Cyclones rotate counterclockwise in the northern hemisphere.
- Cyclones are also known as hurricanes and typhoons.
- Cyclones go through the same stages of formation and dissipation as other storms.
- Cyclones are moved by various wind systems after they form.
- The most dangerous part of a tropical cyclone is the storm surge.
- Hurricane alerts are given at least 24 hours before the hurricane arrives.

Understand Main Ideas

1. MAIN Idea **Identify** the three main stages of a tropical cyclone.
2. **Describe** the changing wind systems that guide a tropical cyclone as it moves from the tropics to the midlatitudes.
3. **Identify** two conditions that must exist for a tropical cyclone to form.
4. **Explain** what causes a cyclone to dissipate.

Think Critically

5. **Analyze** Imagine that you live on the eastern coast of the United States and are advised that the center of a hurricane is moving inland 70 km north of your location. Would a storm surge be a major problem in your area? Why or why not?
6. **Compare** the Saffir-Simpson scale with the Fujita scale. How are they different? Why?

MATH in Earth Science

7. Determine the average wind speed for each hurricane category shown in **Figure 13.15**.

Section 13.4

Objectives

- **Describe** recurring weather patterns and the problems they create.
- **Identify** atmospheric events that cause recurring weather patterns.
- **Distinguish** between heat waves and cold waves.

Review Vocabulary

Fahrenheit scale: a temperature scale in which water freezes at 32° and boils at 212°

New Vocabulary

drought
heat wave
cold wave
windchill index

Recurrent Weather

MAIN Idea Even a relatively mild weather system can become destructive and dangerous if it persists for long periods of time.

Real-World Reading Link Have you ever eaten so much candy you made yourself sick? Too much of any specific type of weather—cold, wet, warm, or dry—can also be unwelcome because of the serious consequences that can result from it.

Floods

An individual thunderstorm can unleash enough rain to produce floods, and hurricanes also cause torrential downpours, which result in extensive flooding. Floods can also occur, however, when weather patterns cause even mild storms to persist over the same area. For example, a storm with a rainfall rate of 1.5 cm/h is not much of a problem if it lasts only an hour or two. If this same storm were to remain over one area for 18 hours, however, the total rainfall would be 27 cm, which is enough to create flooding in most areas. In the spring of 2005, week-long storms caused flooding throughout much of New England, shown in **Figure 13.19.**

Low-lying areas are most susceptible to flooding, making coastlines particularly vulnerable to storm surges during hurricanes. Rivers in narrow-walled valleys and streambeds can rise rapidly, creating high-powered and destructive walls of water. Building in the floodplain of a river or stream can be inconvenient and potentially dangerous during a flood.

Figure 13.19 A week of prolonged rains caused this river in New York to flood.
Infer *What areas are most affected by flooding?*

■ **Figure 13.20** Cotton plants struggle to survive in dried, cracked mud during a drought.

Droughts

Too much dry weather can cause nearly as much damage as too much rainfall. **Droughts** are extended periods of well-below-average rainfall. One of the most extreme droughts in American history occurred during the 1930s in the central United States. This extended drought put countless farmers out of business, as rainfall was inadequate to grow crops.

Droughts are usually the result of shifts in global wind patterns that allow large, high-pressure systems to persist for weeks or months over continental areas. Under a dome of high pressure, air sinks on a large scale. Because the sinking air blocks moisture from rising through it, condensation cannot occur, and drought sets in until global patterns shift enough to move the high-pressure system. **Figure 13.20** shows some of the impacts of long-term drought.

Heat waves An unpleasant side effect of droughts often comes in the form of **heat waves,** which are extended periods of above-average temperatures. Heat waves can be formed by the same high-pressure systems that cause droughts. As the air under a large high-pressure system sinks, it warms by compression and causes above-average temperatures. The high-pressure system also blocks cooler air masses from moving into the area, so there is little relief from the heat. Because it is difficult for condensation to occur under the sinking air of the high-pressure system, there are few, if any, clouds to block the blazing sunshine. The jet stream, or "atmospheric railway," that weather systems normally follow is farther poleward and weaker during the summer. Thus, any upper-air currents that might guide the high-pressure system are so weak that the system barely moves.

MiniLab

Model Flood Conditions

How can mild rains cause floods? Flooding can result from repeated, slow-moving storms that drop rain over the same area for a long period of time.

Procedure

1. Read and complete the lab safety form.
2. Place an **ice cube tray** on the bottom of a **large sink or tub.**
3. Pour **water** into a clean, **plastic dishwashing-detergent bottle** until it is two-thirds full. Replace the cap on the bottle.
4. Hold the bottle upside down with the cap open about 8 cm above one end of the ice cube tray. Gently squeeze the bottle to maintain a constant flow of water into the tray.
5. Slowly move the bottle from one end of the tray to the other over the course of 30 s. Try to put approximately equal amounts of water in each ice cube compartment.
6. Measure the depth of water in each compartment. Calculate the average depth.
7. Repeat Steps 2 to 4, but move the bottle across the ice cube tray in 15 s.

Analysis

1. **Compare** How did the average depth of the water differ in Steps 5 and 7? How might you account for the difference?
2. **Infer** Based on these results, infer how the speed of a moving storm affects the amount of rain received in any one area.
3. **Deduce** How could you alter the experiment to simulate different rates of rainfall?

Heat index Increasing humidity can add to the discomfort and potential danger of a heat wave. Human bodies cool by evaporating moisture from the surface of the skin. In the process, thermal energy is removed from the body. If air is humid, the rate of evaporation is reduced, which diminishes the body's ability to regulate internal temperature. During heat waves, this can lead to serious health problems such as heatstroke, sunstroke, and even death.

Because of the dangers posed by the combination of heat and humidity, the National Weather Service (NWS) routinely reports the heat index, shown in **Table 13.2.** Note that the NWS uses the Fahrenheit scale in the heat index, as well as several other scales it produces because most United States citizens are more familiar with this scale.

The heat index assesses the effect of the body's increasing difficulty in regulating its internal temperature as relative humidity rises. This index estimates how warm the air feels to the human body. For example, an air temperature of 85°F (29°C) combined with relative humidity of 80 percent would require the body to cool itself at the same rate as if the air temperature were 97°F (36°C).

Reading Check Identify the cause of serious health problems associated with heat waves.

Concepts In Motion
Interactive Table To explore more about the heat index, visit glencoe.com.

Table 13.2 The Heat Index

Relative Humidity (%)	Air Temperature (°F)										
	70	75	80	85	90	95	100	105	110	115	120
	Apparent Temperature (°F)										
0	64	69	73	78	83	87	91	95	99	103	107
10	65	70	75	80	85	90	95	100	105	111	116
20	66	72	77	82	87	93	99	105	112	120	130
30	67	73	78	84	90	96	104	113	123	135	148
40	68	74	79	86	93	101	110	123	137	151	
50	69	75	81	88	96	107	120	135	150		
60	70	76	82	90	100	114	132	149			
70	70	77	85	93	106	124	144				
80	71	78	86	97	113	136					
90	71	79	88	102	122						
100	72	80	91	108							

Source: National Weather Service, NOAA

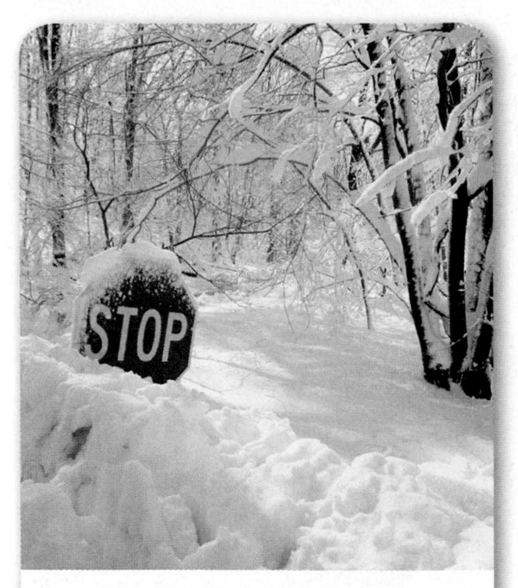

■ **Figure 13.21** Prolonged cold or recurrent cold waves can create blizzard conditions such as these that fell on Denver in 2006.

Cold Waves

The opposite of a heat wave is a **cold wave,** which is an extended period of below-average temperatures. Interestingly, cold waves are also brought on by large, high-pressure systems. However, cold waves are caused by systems of continental polar or arctic origin. During the winter, little sunlight is available to provide warmth. At the same time, the snow-covered surface is constantly reflecting the sunlight back to space. The combined effect of these two factors is the development of large pools of extremely cold air over polar continental areas. Because cold air sinks, the pressure near the surface increases, creating a strong high-pressure system.

Because of the location and the time of year in which they occur, winter high-pressure systems are much more influenced by the jet stream than are summer high-pressure systems. Moved along by the jet stream, these high-pressure systems rarely linger in any area. However, the winter location of the jet stream can remain essentially unchanged for days or even weeks. This means that several polar high-pressure systems can follow the same path and subject the same areas to continuous numbing cold. Some effects of prolonged periods of cold weather are shown in **Figure 13.21.**

Reading Check **Explain** why the Sun's energy has little effect on air temperature in the arctic.

DATA ANALYSIS LAB

Based on Real Data*

Interpret the Table

How can you calculate a heat wave? The following data represent the daily maximum and minimum temperatures for seven consecutive summer days in Chicago. A heat wave is defined as two or more days with an average temperature of 29.4°C or higher.

Analysis

1. Calculate the average temperature for each day in your table.
2. Plot the daily maximum and minimum temperatures on a graph with the days on the *x*-axis and the maximum temperatures on the *y*-axis. Using the data points, draw a curve to show how the temperatures changed over the seven-day period. Add the average temperatures.

Think Critically

3. **Determine** What day did the city heat wave begin? How long did it last?
4. **Compare** the average temperature for the days of the heat wave to the average temperature of the remaining days.

Data and Observations

Daily Temperatures			
Day	Maximum (°C)	Minimum (°C)	Average (°C)
1	32	23	
2	37	24	
3	41	27	
4	39	29	
5	37	25	
6	34	24	
7	32	23	

* Data obtained from: Klinenberg, E. 2002. *Heat Wave: A social autopsy of disaster in Chicago, IL.* Chicago: University of Chicago Press.

Windchill Chart

Temperature (°F) \ Wind (mph)	5	10	15	20	25	30	35	40	45	50	55	60
Calm 40	36	34	32	30	29	28	28	27	26	26	25	25
35	31	27	25	24	23	22	21	20	19	19	18	17
30	25	21	19	17	16	15	14	13	12	12	11	10
25	19	15	13	11	9	8	7	6	5	4	4	3
20	13	9	6	4	3	1	0	-1	-2	-3	-3	-4
15	7	3	0	-2	-4	-5	-7	-8	-9	-10	-11	-11
10	1	-4	-7	-9	-11	-12	-14	-15	-16	-17	-18	-19
5	-5	-10	-13	-15	-17	-19	-21	-22	-23	-24	-25	-26
0	-11	-16	-19	-22	-24	-26	-27	-29	-30	-31	-32	-33
-5	-16	-22	-26	-29	-31	-33	-34	-36	-37	-38	-39	-40
-10	-22	-28	-32	-35	-37	-39	-41	-43	-44	-45	-46	-48
-15	-28	-35	-39	-42	-44	-46	-48	-50	-51	-52	-54	-55
-20	-34	-41	-45	-48	-51	-53	-55	-57	-58	-60	-61	-62
-25	-40	-47	-51	-55	-58	-60	-62	-64	-65	-67	-68	-69

Frostbite times ■ 30 min ■ 10 min ■ 5 min

■ **Figure 13.22** The windchill chart was designed to show the dangers of cold and wind.
What ***wind speed and temperature is the same as 10°F on a calm day?***

Windchill index The effects of cold air on the human body are magnified by wind. Known as the windchill factor, this phenomenon is measured by the **windchill index** in **Figure 13.22.** The index estimates how cold the air feels to the human body. While the windchill index is helpful, it does not account for individual variations in sensitivity to cold, the effects of physical activity, or humidity. In 2001, the NWS revised the calculations to utilize advances in science, technology, and computer modeling. These revisions provide a more accurate, understandable, and useful index for estimating the dangers caused by winter winds and freezing temperatures.

Section 13.4 Assessment

Section Summary

- Too much heat and too little precipitation causes droughts.
- Too little heat and a stalled jet stream can cause weeks of cold weather in an area.
- Heat index estimates the effect on the human body when the air is hot and the humidity is high.
- Windchill index tells how wind and temperature affect your body in winter.
- Windchill is a factor used to warn about the effect of cold air and wind on the human body.

Understand Main Ideas

1. MAIN Idea **Explain** how everyday weather can become recurrent and dangerous.
2. **Describe** how relatively light rain could cause flooding.
3. **Compare and contrast** a cold wave and a heat wave.
4. **Explain** why one type of front would be more closely associated with flooding than another.

Think Critically

5. **Explain** why air in a winter high-pressure system is very cold despite compressional warming.
6. **Compare** the data of the heat-index scale and the windchill scale. What variables influence each scale?

MATH in Earth Science

7. A storm stalls over Virginia, dropping 0.75 cm of rain per hour. If the storm lingers for 17 hours, how much rain will accumulate?

ON SITE: STORM SPOTTERS

Figure 1: Storm chasers videotape a tornado crossing a road near Manchester, South Dakota.

When storm spotters hear that severe weather is approaching the area, do they seek the safety of their house or basement like most people do? No, they head out to the edge of town or to a high point to check on the exact wind and weather conditions.

Volunteers for the NWS Storm spotters work as volunteers for the National Weather Service (NWS) to help give NWS forecasters a clear picture of what is really happening on the ground. Although Doppler radar and other systems are sophisticated data collectors, these devices can only detect weather conditions that might produce a severe thunderstorm or a tornado. The NWS typically uses this information to issue a severe storm or tornado watch. When a watch is issued, spotters travel to key lookout points and report their observations. The observations made on the ground by storm spotters are essential to the NWS in upgrading watches to warnings.

Making Reports The NWS trains spotters to assess certain weather conditions such as wind speed, hail size, and cloud formation. For example, if large tree branches begin to sway, umbrellas are difficult to use, and the wind creates a whistling noise along telephone wires, spotters know that wind speed is between 40 and 50 km/h. If trees are uprooted or TV antennas break, wind speed is estimated to be between 85 and 115 km/h.

Spotters study the clouds to determine where hail is falling, where a tornado might develop, and in what direction the storm is headed. When they call in, they report the event, its location, its direction, and whether there is need for emergency assistance.

High Risk Mobile spotters risk their own safety in order to protect their community. The major risks they face stem from driving in bad weather and standing on a high spot where lightning might strike. Spotters always travel with a partner, so that one person can drive and the other can watch the sky. To stay safe, spotters keep watch in all directions, keep the car engine running, and have an escape plan.

The combination of technology and the work of spotters has saved many lives since the volunteer system was started by the NWS in the 1970s. The number of deaths as a result of tornadoes and other severe weather has decreased significantly since the program began.

WRITING in Earth Science

Make a Pamphlet Research more information about how to become a storm spotter and the training involved. Write and illustrate a pamphlet about storm spotting that includes this information. To learn more about storm spotting, visit glencoe.com.

GEOLAB

INTERNET: TRACK A TROPICAL CYCLONE

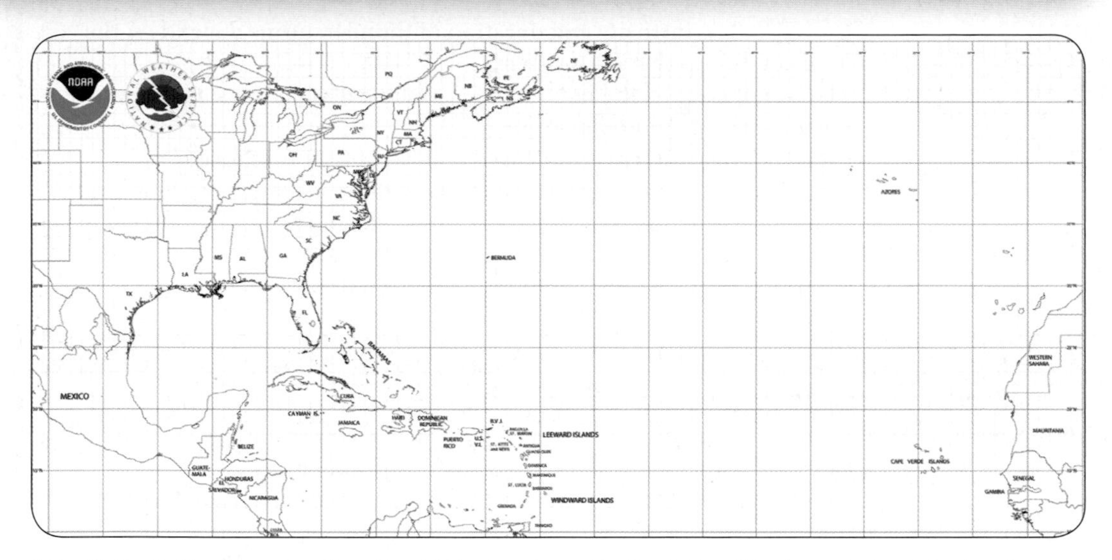

Atlantic Basin Hurricane Tracking Chart This chart is used by the National Hurricane Center to track active hurricanes in the Atlantic basin.

Background: Tropical cyclones form very violent storms. That is why it is important to have advanced warning before they hit land. By tracking the changing position of a storm on a chart and connecting these positions with a line, you can model or predict a cyclone's path.

Question: *What information can you obtain by studying the path of a tropical cyclone?*

Procedure

1. Read and complete the lab safety form.
2. Form a hypothesis about how a tropical cyclone's path can be used to predict the strength of the storm and where the most damage might be inflicted.
3. Visit glencoe.com to find links to tropical cyclone data.
4. Choose the track of a tropical cyclone that has occurred during the past five years.
5. Plot the position, air pressure, wind speed, and stage of the tropical cyclone at 6-h intervals throughout its existence.
6. Plot the changing position of the tropical cyclone on your hurricane-tracking chart.
7. Incorporate your research into a data table. Add any additional information that you think is important.

Analyze and Conclude

1. **Identify** What was the maximum wind speed in knots that the tropical cyclone reached?
2. **Calculate** Multiply the value from Question 1 by 1.85 to find the wind speed in kilometers per hour. Based on this value, how would the tropical cyclone be classified on the Saffir-Simpson hurricane scale shown in **Figure 13.15?**
3. **List** the landmasses over which the tropical cyclone passed.
4. **Identify** What was the life span of your tropical cyclone? What was the name of your cyclone?
5. **Infer** Where would you expect the storm surge to have been the greatest? Explain.
6. **Examine** How was the tropical cyclone's strength affected when its center passed over land?

SHARE YOUR DATA

Peer Review Visit glencoe.com and post a summary of your data. Compare your data to other data collected for this investigation.

CHAPTER 13 Study Guide

Download quizzes, key terms, and flash cards from glencoe.com.

BIG Idea The exchange of thermal energy in the atmosphere sometimes occurs with great violence that varies in form, size, and duration.

Section 13.1 Thunderstorms

Vocabulary

- air-mass thunderstorm (p. 346)
- frontal thunderstorm (p. 346)
- mountain thunderstorm (p. 346)
- return stroke (p. 348)
- sea-breeze thunderstorm (p. 346)
- stepped leader (p. 348)

Key Concepts

MAIN Idea The intensity and duration of thunderstorms depend on the local conditions that create them.

- The cumulus stage, the mature stage, and the dissipation stage comprise the life cycle of a thunderstorm.
- Clouds form as water is condensed and latent heat is released.
- Thunderstorms can be produced either within air masses or along fronts.
- From formation to dissipation, all thunderstorms go through the same stages.
- Lightning is a natural result of thunderstorm development.

Section 13.2 Severe Weather

Vocabulary

- downburst (p. 351)
- Fujita tornado intensity scale (p. 353)
- supercell (p. 350)
- tornado (p. 352)

Key Concepts

MAIN Idea All thunderstorms produce wind, rain, and lightning, which can have dangerous and damaging effects under certain circumstances.

- Intense rotating updrafts are associated with supercells.
- Downbursts are strong winds that result in damage associated with thunderstorms.
- Hail is precipitation in the form of balls or lumps of ice that accompany severe storms.
- The worst storm damage comes from a vortex of high winds that moves along the ground as a tornado.

Section 13.3 Tropical Storms

Vocabulary

- eye (p. 356)
- eyewall (p. 356)
- Saffir-Simpson hurricane scale (p. 358)
- storm surge (p. 359)
- tropical cyclone (p. 355)

Key Concepts

MAIN Idea Normally peaceful, tropical oceans are capable of producing one of Earth's most violent weather systems—the tropical cyclone.

- Cyclones rotate counterclockwise in the northern hemisphere.
- Cyclones are also known as hurricanes and typhoons.
- Cyclones go through the same stages of formation and dissipation as other storms.
- Cyclones are moved by various wind systems after they form.
- The most dangerous part of a tropical cyclone is the storm surge.
- Hurricane alerts are given at least 24 hours before the hurricane arrives.

Section 13.4 Recurring Weather

Vocabulary

- cold wave (p. 364)
- drought (p. 362)
- heat wave (p. 362)
- windchill index (p. 365)

Key Concepts

MAIN Idea Even a relatively mild weather system can become destructive and dangerous if it persists for long periods of time.

- Too much heat and too little precipitation causes droughts.
- Too little heat and a stalled jet stream can cause weeks of cold weather in an area.
- Heat index estimates the effect on the human body when the air is hot and the humidity is high.
- Windchill index tells how wind and temperature affect your body in winter.
- Windchill is a factor used to warn about the effect of cold air and wind on the human body.

Earth Science Online Vocabulary PuzzleMaker glencoe.com

CHAPTER 13 Assessment

Vocabulary Review

Choose the correct italicized vocabulary term to complete each sentence.

1. A(n) ________ thunderstorm is characterized by temperature differences within a mass of air. *frontal, severe, air-mass*

2. Intense, self-sustaining thunderstorms are known as ________. *downbursts, tornadoes, supercells*

Compare and contrast the following pairs of vocabulary terms.

3. cold wave, wind-chill factor

4. eye, eyewall

5. air-mass thunderstorm, frontal thunderstorm

6. stepped leader, return stroke

Replace the underlined term with the correct one from the vocabulary list on the Study Guide.

7. More lives are lost during a hurricane's heat wave than from its winds.

8. A microburst is a form of weather that is difficult to predict.

9. Another name for a typhoon is a forecast.

10. A weather map is an extended period of extreme cold in one area.

11. The Fujita tornado intensity scale can tell the possible storm surge heights from a hurricane.

Understand Key Concepts

12. Which would work against the development of a thunderstorm?
 A. rising air
 B. stable air
 C. moisture
 D. unstable air

13. Which does not describe a type of damaging thunderstorm wind?
 A. downburst
 B. microburst
 C. land breeze
 D. macroburst

Use the diagram below to answer Question 14.

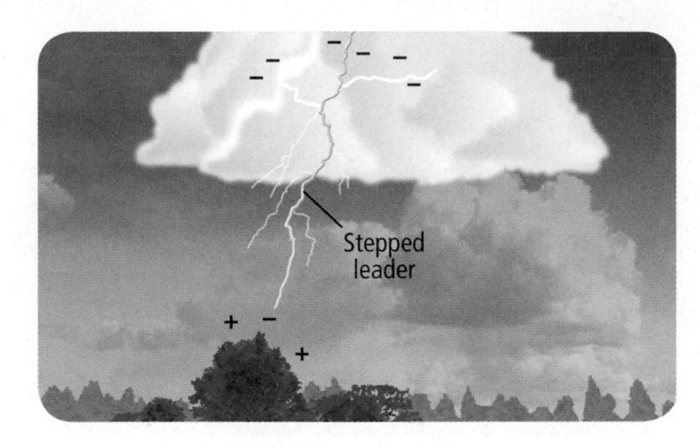

14. What phrase describes a stepped leader?
 A. return stroke
 B. partially charged air
 C. positive charge
 D. downdraft

15. Which does not play a key role in the development of hail?
 A. supercooled water
 B. freezing temperatures
 C. warm ocean water
 D. strong updrafts

16. Heat waves involve high-pressure systems that cause air to sink and warm by which process?
 A. compression
 B. conduction
 C. evaporation
 D. condensation

17. Which weather hazard involves a lack of moisture?
 A. hail
 B. drought
 C. storm surge
 D. flood

18. Flooding is most likely to take place because of rains associated with which type of storm?
    A. Category 5 hurricane moving at 25 m/s
    B. F2 tornado moving at 10 m/s
    C. A stationary tropical storm
    D. thunderstorm moving at 2 m/s

Use this photo to answer Question 19.

19. Which part of the hurricane is not visible in the photo above?
    A. eye
    B. eyewall
    C. storm surge
    D. cirrus overcast

20. What percentage of tornadoes are classified as F4 or F5 on the Fujita tornado intensity scale?
    A. 1 percent
    B. 10 percent
    C. 50 percent
    D. 75 percent

21. Which factor, if increased, would increase the chance that a severe thunderstorm would occur?
    A. upper-level temperature
    B. surface moisture
    C. strength of the jet stream
    D. conduction

22. In which ocean would you not expect to experience a hurricane?
    A. West Pacific
    B. Indian
    C. North Atlantic
    D. South Atlantic

23. What weather events are cold waves most often associated with?
    A. floods
    B. polar high-pressure systems
    C. tropical high-pressure systems
    D. droughts

Constructed Response

24. **Determine** the effect humidity has on the heat index.

25. **Compare and contrast** storm surges of Category 1 and Category 4 hurricanes using **Table 13.2.**

Use the photo below to answer Question 26.

26. **Predict** the area of landfall, the areas affected by storm surge, and evacuation routes for this hurricane.

27. **Compare and contrast** tornadoes and hurricanes.

28. **Point out** the differences between a microburst and a macroburst.

29. **Develop** a plan of safety for a family that might encounter flooding, lightning, and a tornado in the area where they live.

Think Critically

30. **Explain** how temperature and condensation are limiting factors to the growth of a thunderstorm.

31. **Distinguish** how an investigator would differentiate between a microburst and a tornado.

32. **Point out** the features of the South Atlantic that might deter the formation of hurricanes.

Earth Science Online Chapter Test glencoe.com

33. **Explain** why supercells that produce tornadoes also often produce large hailstones.

34. **Predict** why extreme cold waves are more common in the northern hemisphere than in the southern hemisphere.

35. **Describe** how cold fronts are more likely to produce severe thunderstorms than warm fronts.

Use the diagram below to answer Questions 36 and 37.

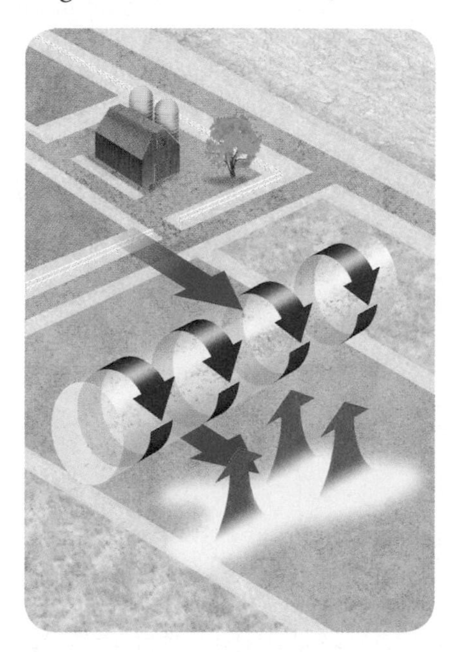

36. **Explain** how the wind shears that produce a tornado go from horizontal to vertical.

37. **Identify** what causes the lift for the vertical motion.

38. **CAREERS IN EARTH SCIENCE** Imagine you work for the National Weather Service and it is your job to write public service announcements. Write a safety plan for people who live in places where hurricanes are frequent.

Concept Mapping

39. Make a concept map to differentiate among thunderstorms, tornadoes, and tropical cyclones.

Challenge Question

40. **Appraise** the use of the wind-chill factor to determine school delays and work closings versus temperature, daylight, or icy conditions alone.

Additional Assessment

41. **WRITING in Earth Science** Imagine that you are a weather research scientist researching a way to stop hurricanes. Tell about your research that includes ways to cool the ocean's surface.

DBQ Document–Based Questions

Data obtained from: National Oceanic and Atmospheric Administration, 2004. *Hurricane Research Division.*

When a hurricane passes over the surface of the ocean, the sea surface temperature (SST) can become several degrees cooler. This is the result of cooler water being churned up to the surface from lower levels. The radial distance is the distance from the eye (TC center) measured in degrees of latitude.

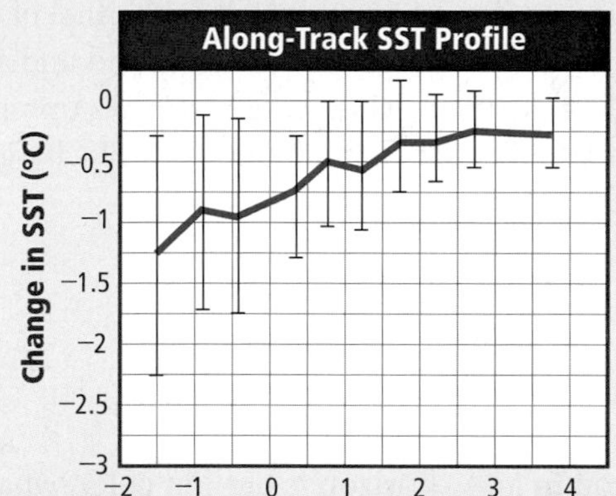

42. What is the estimated range of temperature change for each radial distance indicated?

43. A difference of 0.5°C can be the difference between a storm that intensifies and one that stops developing. At what distance from the TC center would that range be most critical?

44. What factors in a hurricane might cause the increased temperature changes to happen closer to the eye than at the edges?

Cumulative Review

45. What are the two most abundant elements in Earth's crust? **(Chapter 4)**

46. Describe why radiosonde data is important. **(Chapter 12)**

Standardized Test Practice

Multiple Choice

Use the illustration to answer Questions 1 and 2.

1. With what type of cloud is lightning associated?
   A. altocumulus
   B. stratocumulus
   C. cirrus
   D. cumulonimbus

2. Lightning occurs when an invisible channel of negatively charged air descends to the ground and a channel of positively charged ions rushes upward to meet it. What is the channel of positively charged ions called?
   A. return stroke
   B. stepped leader
   C. ground stroke
   D. electronic leader

3. If the soil in an A-horizon is dark in color, what does this imply?
   A. The layer of soil is rich in humus.
   B. The layer of soil is extremely fertile.
   C. The layer of soil is from a poorly drained area.
   D. The layer of soil is rich with iron materials.

4. Where is the majority of freshwater found?
   A. in the atmosphere
   B. underground
   C. in rivers, streams, and lakes
   D. in polar ice caps and glaciers

5. Why is the Geographic Information System (GIS) beneficial to science?
   A. It is very similar to traditional mapping.
   B. It can be used by scientists in many different disciplines.
   C. It limits maps to just one layer of information.
   D. It does not change with new information.

6. What does Doppler radar monitor?
   A. the motion of moving raindrops
   B. atmospheric pressure
   C. temperature, air pressure, and humidity
   D. the height of cloud layers

7. The data gathered by Doppler radar can be used to make a type of forecast that relies on numerical data. What is this type of forecast called?
   A. an analog forecast
   B. a digital forecast
   C. an isopleth
   D. ASOS

Use the illustration below to answer Questions 8 and 9.

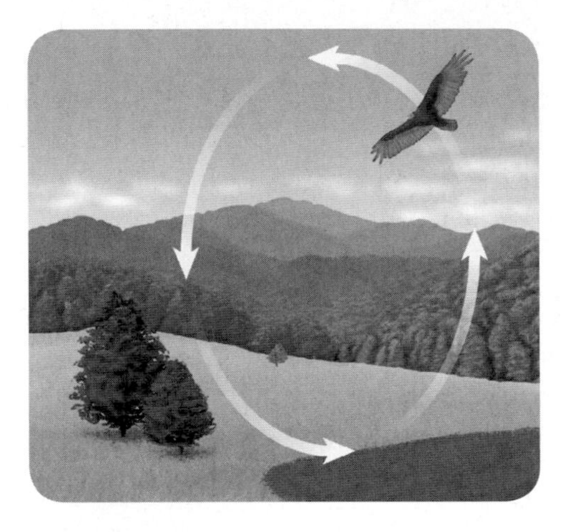

8. What process is being demonstrated by the circulating arrows?
   A. radiation
   B. a convection current
   C. a wind current
   D. a conduction current

9. Describe the process of radiation to warm Earth's surface.
   A. It is one of the methods that transfer energy from the Sun through different forms of electromagnetic waves to warm Earth.
   B. It transfers energy through the collision of molecules to warm Earth.
   C. It transfers energy through the flow of a heated substance to warm Earth.
   D. It allows the atmosphere to absorb the Sun's rays or reflect them back into space.

Short Answer

Use the diagram below to answer Questions 11–13.

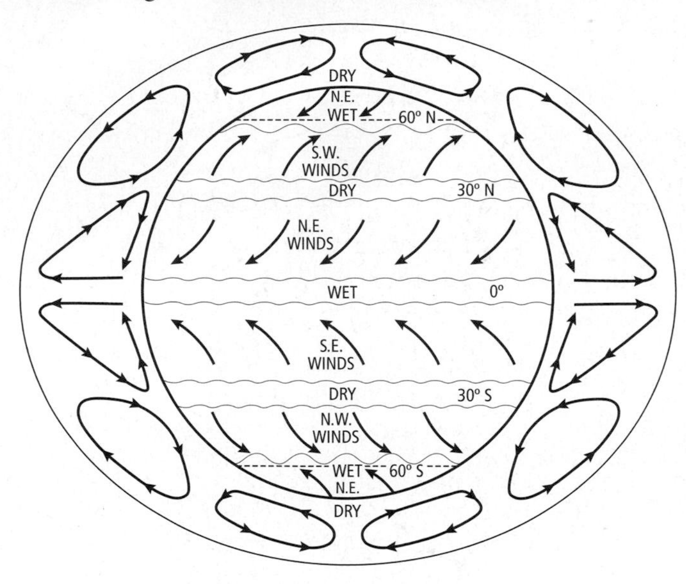

10. How does the rotation of Earth affect wind systems?

11. Analyze the wind pattern occurring between the equator and 30° north and south latitudes.

12. Why would it benefit sailors to know what type of wind system they are traveling in?

13. Why do tornadoes form in the spring during the late afternoon and evening?

14. Describe a step farmers can take to improve soil fertility.

15. Explain the properties of a geyser.

16. How does the amount of water on Earth change as a result of the water cycle?

Reading for Comprehension

Hurricane Preparedness

The Saffir-Simpson scale, used by the National Weather Service since the 1970s to classify hurricane strength, ranks storm strength from Level 1 to 5, and is used to give an estimate of the possible property damage and flooding expected when the hurricane makes landfall. A Category 5 storm maintains winds in excess of 155 mph and is capable of causing widespread damages such as destroying or causing extensive structural damage to homes. It can also trigger a storm surge more than 18 ft above normal.

Two of the most fundamental steps people can take are assembling a disaster supplies kit and making an emergency plan. They are simple measures, but can make all the difference when disasters strike.

Article obtained from: Booth, M. Hurricane Isabel strengthens to Category 5. *In the News.* September 12, 2006. (Online resource accessed October 2006.)

17. For what purpose is the Saffir-Simpson scale used?
A. estimating property damage and flooding when a hurricane makes landfall
B. estimating the height of a storm surge above normal
C. determining who should be told to evacuate before a hurricane hits
D. measuring the strength of an approaching hurricane

18. What can be inferred from this passage?
A. Only Category 5 hurricanes will cause major damage.
B. Only people living along the shoreline are in danger during hurricanes.
C. Category 5 hurricanes have the potential to cause major damage.
D. People should have a disaster supply kit or an evacuation plan in place before a hurricane hits.

NEED EXTRA HELP?																
If You Missed Question . . .	1	2	3	4	5	6	7	8	9	10	11	12	13	14	15	16
Review Section . . .	13.1	13.2	7.3	10.1	2.3	12.3	12.3	11.1	11.1	12.2	12.2	12.2	13.2	7.3	10.1	11.3

CHAPTER 14 Climate

BIG Idea **The different climates on Earth are influenced by natural factors as well as human activities.**

14.1 Defining Climate
MAIN Idea Climate is affected by several factors including latitude and elevation.

14.2 Climate Classification
MAIN Idea Climates are categorized according to the average temperatures and precipitation amounts.

14.3 Climatic Changes
MAIN Idea Earth's climate is constantly changing on many different timescales.

14.4 Impact of Human Activities
MAIN Idea Over time, human activities can alter atmospheric conditions enough to influence changes in weather and climate.

GeoFacts

- The temperate rain forests of Olympic National Park receive up to 500 cm of precipitation each year.
- Deciduous forests in the northeastern United States receive between 75 and 150 cm of precipitation each year.
- Desert areas receive less than 25 cm of precipitation each year.

LAUNCH Lab

How can you model cloud cover?

Some areas are generally cloudier than others. This affects both the temperature and the amount of precipitation that these areas receive.

Procedure

1. Read and complete the lab safety form.
2. Lay two sheets of **dark construction paper** on the grass in an open area. Place a **rock** on each sheet of paper to prevent it from blowing away.
3. Open an **umbrella** and anchor it in the ground over one of the sheets of paper.
4. During each of the next four days, observe what happens to the sheets of paper.

Analysis

1. **Describe** any differences in dew formation that you observed each day.
2. **Explain** How is the umbrella in this activity similar to clouds in the atmosphere?
3. **Infer** how temperatures during the night might differ between climates with extensive cloud cover and climates with few clouds.

Climate Classification Make this Foldable to explain the five main types of climates using the Köppen classification system.

STEP 1 Layer three sheets of paper about 2 cm apart.

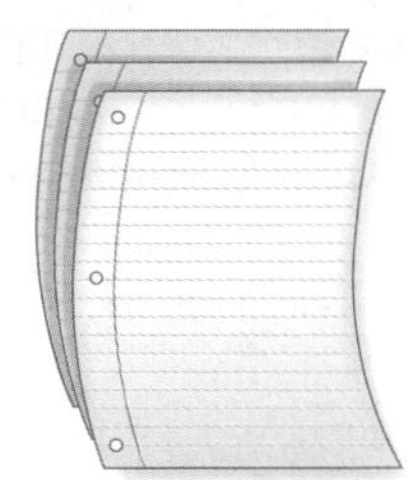

STEP 2 Fold up the bottom of the sheets to make five equal tabs. Staple along the fold.

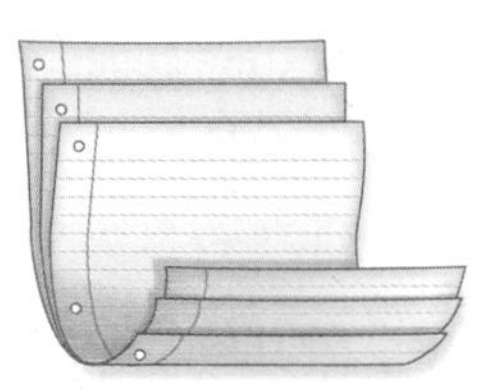

STEP 3 Label the tabs *Tropical, Dry, Mild, Continental,* and *Polar.*

FOLDABLES **Use this Foldable with Section 14.2.** As you read this section, record the major characteristics of each type of climate.

Visit glencoe.com to

- study entire chapters online;
- explore Concepts In Motion animations:
  - Interactive Time Lines
  - Interactive Figures
  - Interactive Tables
- access Web Links for more information, projects, and activities;
- review content with the Interactive Tutor and take Self-Check Quizzes.

Section 14.1

Objectives

- **Recognize** limits associated with the use of normals.
- **Explain** why climates vary.
- **Compare and contrast** temperatures in different regions on Earth.

Review Vocabulary

jet stream: a high-altitude, narrow, westerly wind band that occurs above large temperature changes

New Vocabulary

climatology
normal
tropics
temperate zones
polar zones

Defining Climate

MAIN Idea Climate is affected by several factors including latitude and elevation.

Real-World Reading Link Just because you observed someone eating a steak dinner, you probably would not assume that they ate steak for every meal. In nature, taking a one-day "snapshot" of the weather does not necessarily describe what that location experiences over the course of many days.

Annual Averages and Variations

Fifty thousand years ago, the United States had much different weather patterns than those that exist today. The average temperature was several degrees cooler, and the jet stream was probably farther south. Understanding and predicting such climatic changes are the basic goals of climatology. **Climatology** is the study of Earth's climate and the factors that affect past, present, and future climatic changes.

Climate describes the long-term weather patterns of an area. These patterns include much more than average weather conditions. Climate also describes annual variations of temperature, precipitation, wind, and other weather variables. Studies of climate show extreme fluctuations of these variables over time. For example, climatic data can indicate the warmest and coldest temperatures recorded for a location. **Figure 14.1** shows weather differences between summer and winter in Chicago, Illinois. This type of information, combined with comparisons between recent conditions and long-term averages, can be used by businesses to decide where to build new facilities and by people who have medical conditions that require them to live in certain climates.

Figure 14.1 Climate data include the warmest and coldest temperatures recorded for a location. The highest temperature on record for Chicago, IL, is 40°C, which occurred in June 1988. The lowest temperature on record for Chicago, IL, is –33°C, which occurred in January 1985.

Chicago, IL, in the summer

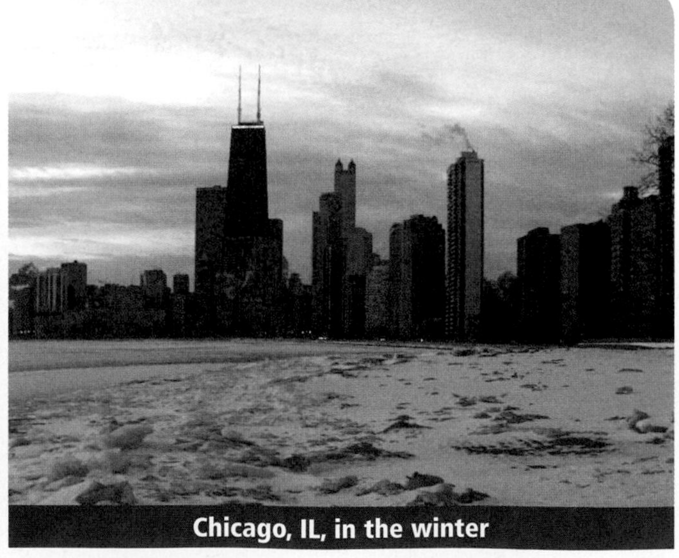

Chicago, IL, in the winter

Normals The data used to describe an area's climate are compiled from meteorological records, which are continuously gathered at thousands of locations around the world. These data include daily high and low temperatures, amounts of rainfall, wind speed and direction, humidity, and air pressure. The data are averaged on a monthly or annual basis for a period of at least 30 years to determine the **normals,** which are the standard values for a location.

Reading Check **Identify** data that can be used to calculate normals.

Limitations of normals While normals offer valuable information, they must be used with caution. Weather conditions on any given day might differ widely from normals. For instance, the normal high temperature in January for a city might be 0°C. However, it is possible that no single day in January had a high of exactly 0°C. Normals are not intended to describe usual weather conditions; they are the average values over a long period of time.

While climate describes the average weather conditions for a region, normals apply only to the specific place where the meteorological data were collected. Most meteorological data are gathered at airports, which cannot operate without up-to-date, accurate weather information. However, many airports are located outside city limits. When climatic normals are based on airport data, they might differ from actual weather conditions in nearby cities. Changes in elevation and other factors, such as proximity to large bodies of water, can cause climates to vary.

CAREERS IN EARTH SCIENCE

Climatologist Scientists who study long-term trends in climate are called climatologists. Climatologists might collect data by drilling holes in ice or sampling ocean water temperatures. To learn more about Earth science careers, visit glencoe.com.

DATA ANALYSIS LAB

Based on Real Data*

Interpret the Data

What is the temperature in Phoenix, Arizona? The table contains temperature data for Phoenix, Arizona, based on data collected from July 1, 1948, through December 31, 2005.

Analysis

1. Plot the monthly values for average maximum temperatures. Place the month on the *x*-axis and temperature on the *y*-axis.
2. Repeat Step 1 using the monthly values for the average minimum temperatures.

Think Critically

3. **Identify** the months that were warmer than the average maximum temperature.
4. **Identify** the months that were colder than the average minimum temperature.
5. **Infer** What is the climate of Phoenix, Arizona, based on average temperatures?

Data and Observations

Monthly Temperature Summary for Phoenix, AZ

Temperature (°C)	Jan	Feb	Mar	Apr	May	Jun	Jul	Aug	Sep	Oct	Nov	Dec	Average
Average maximum	19	21	24	29	34	39	41	39	37	31	24	19	29.8
Average minimum	5	7	9	13	18	22	27	26	22	16	9	6	15

*Data obtained from: Western Regional Climate Center. 2005.

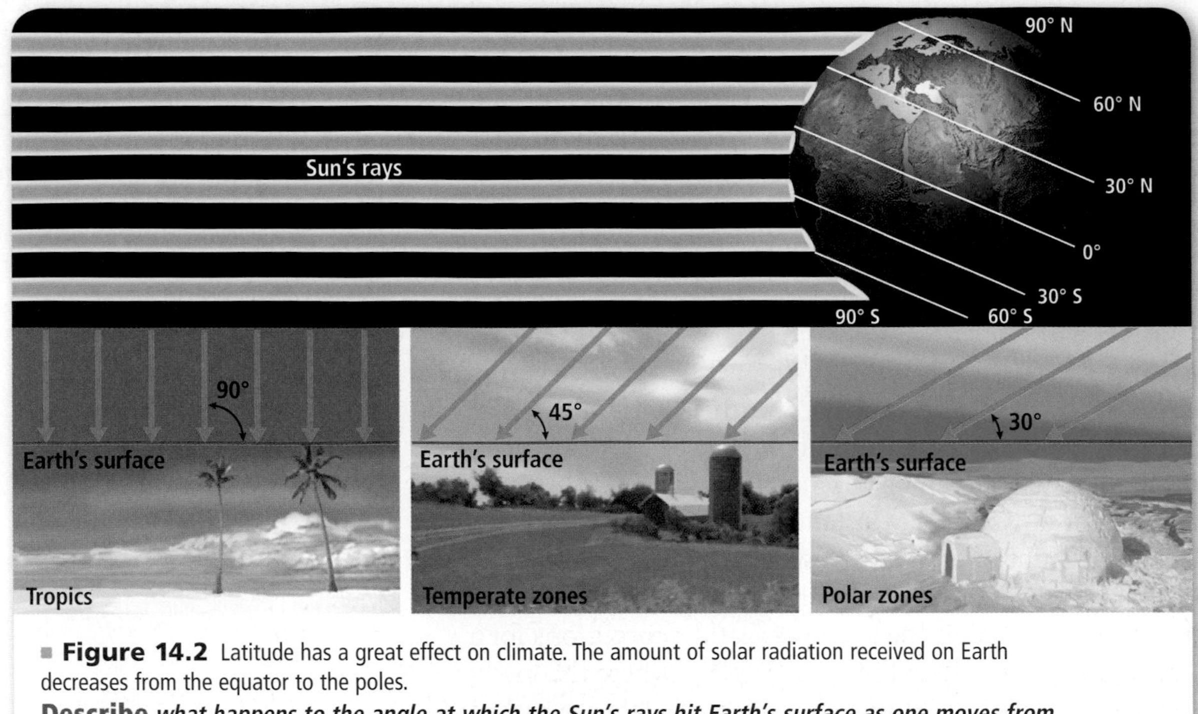

■ **Figure 14.2** Latitude has a great effect on climate. The amount of solar radiation received on Earth decreases from the equator to the poles.
Describe ***what happens to the angle at which the Sun's rays hit Earth's surface as one moves from the equator to the poles.***

Causes of Climate

You probably know from watching the weather reports that climates around the country vary greatly. For example, on average, daily temperatures are much warmer in Dallas, Texas, than in Minneapolis, Minnesota. There are several reasons for such climatic variations, including differences in latitude, topography, closeness of lakes and oceans, availability of moisture, global wind patterns, ocean currents, and air masses.

Latitude Recall that different parts of Earth receive different amounts of solar radiation. The amount of solar radiation received by any one place varies because Earth is tilted on its axis, and this affects how the Sun's rays strike Earth's surface. The area between 23.5° S and 23.5° N of the equator is known as the **tropics.** As **Figure 14.2** shows, tropical areas receive the most solar radiation because the Sun's rays are nearly perpendicular to Earth's surface. As you might expect, temperatures in the tropics are generally warm year-round. For example, Caracas, Venezuela, located at about 10° N, enjoys average maximum temperatures between 24°C and 27°C year-round. The **temperate zones** lie between 23.5° and 66.5° north and south of the equator. As their name implies, temperatures in these regions are moderate. The **polar zones** are located from 66.5° north and south of the equator to the poles. Solar radiation strikes the polar zones at a low angle. Thus, polar temperatures tend to be cold. Thule, Greenland, located at 77° N, has average maximum temperatures between −20°C and 8°C year-round.

VOCABULARY
ACADEMIC VOCABULARY
Imply
to indicate by association rather than by direct statement
The title of the movie implied that it was a love story.

Topographic effects Water heats up and cools down more slowly than land. Thus, large bodies of water affect the climates of coastal areas. Many coastal regions are warmer in the winter and cooler in the summer than inland areas at similar latitudes.

Also, temperatures in the lower atmosphere generally decrease with altitude. Thus, mountain climates are usually cooler than those at sea level. In addition, climates often differ on either side of a mountain. Air rises up one side of a mountain as a result of orographic lifting. The rising air cools, condenses, and drops its moisture, as shown in **Figure 14.3.** The climate on this side of the mountain—the windward side—is usually wet and cool. On the opposite side of the mountain—the leeward side—the air is drier, and it warms as it descends. For this reason, deserts are common on the leeward side of mountains.

Reading Check **Explain** how large bodies of water affect the climate of coastal areas.

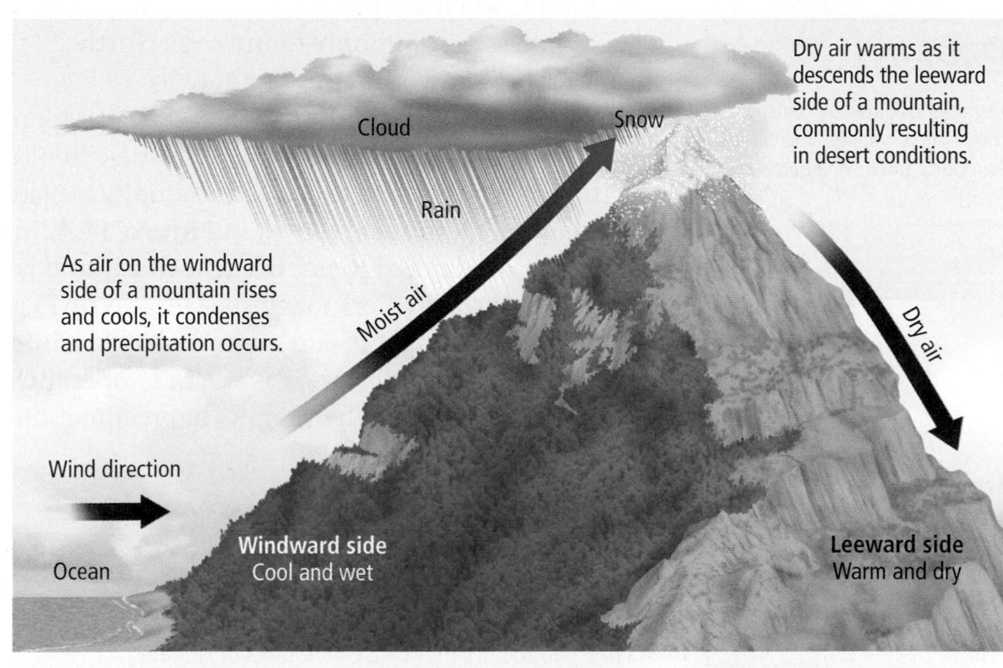

Figure 14.3 Orographic lifting leads to rain on the windward side of a mountain. The leeward side is usually dry and warm.

Major Air Masses Over North America

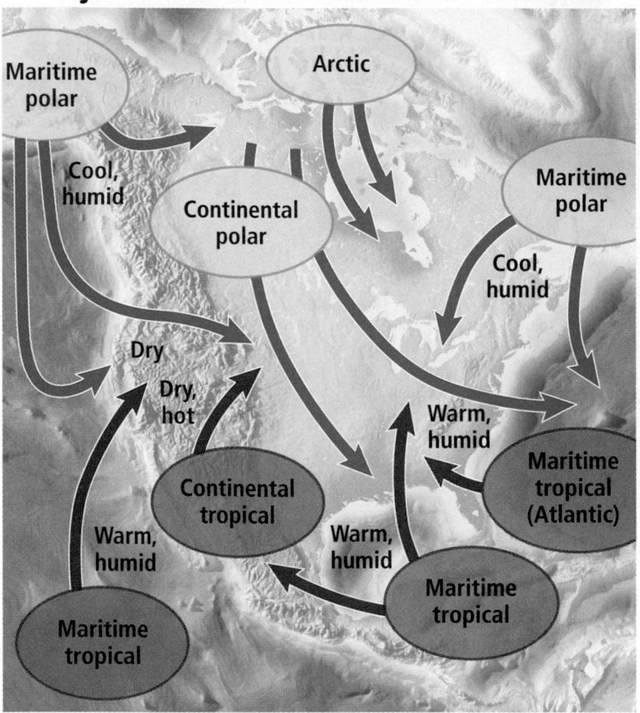

■ **Figure 14.4** Air masses affect regional climates by transporting the temperature and humidity of their source regions. The warm and humid maritime tropical air mass supports the lush vegetation on the island of Dominica.

Air masses Two of the main causes of weather are the movement and interaction of air masses. Air masses also affect climate. Recall from Chapter 12 that air masses have distinct regions of origin, caused primarily by differences in the amount of solar radiation. The properties of air masses also depend on whether they formed over land or water. The air masses commonly found over North America are shown in **Figure 14.4.**

Average weather conditions in and near regions of air-mass formation are similar to those exhibited by the air masses themselves. For example, consider the island of Dominica, shown in **Figure 14.4,** in the tropical Atlantic Ocean. Because this island is located in an area where maritime tropical (mT) air masses dominate, the island's climate has maritime tropical characteristics, such as warm temperatures, high humidity, and high amounts of precipitation.

Section 14.1 Assessment

Section Summary

- Climate describes the long-term weather patterns of an area.
- Normals are the standard climatic values for a location.
- Temperatures vary among tropical, temperate, and polar zones.
- Climate is influenced by several different factors.
- Air masses have distinct regions of origin.

Understand Main Ideas

1. MAIN Idea **Describe** two factors that cause variations in climate.
2. **Identify** What are some limits associated with the use of normals?
3. **Compare and contrast** temperatures in the tropics, temperate zones, and polar zones.
4. **Infer** how climate data can be used by farmers.

Think Critically

5. **Assess** Average daily temperatures for City A, located at 15° S, are 5°C cooler than average daily temperatures for City B, located at 30° S. What might account for the cooler temperatures in City A, even though it is closer to the equator?

WRITING in Earth Science

6. Write a hypothesis that explains why meteorological data gathered at an airport would differ from data gathered near a large lake. Assume all other factors are constant.

Earth Science Online Self-Check Quiz glencoe.com

Section 14.2

Objectives

- **Describe** the criteria used to classify climates.
- **Compare and contrast** different climates.
- **Explain** and give examples of microclimates.

Review Vocabulary

precipitation: all solid and liquid forms of water—including rain, snow, sleet, and hail—that fall from clouds

New Vocabulary

Köppen classification system
microclimate
heat island

Climate Classification

MAIN Idea Climates are categorized according to the average temperatures and precipitation amounts.

Real-World Reading Link What sort of place comes to mind when you think of a vacation in a tropical climate? A place with hot weather and a lot of rain? If so, you already know something about a tropical climate, even if you have never visited one.

Köppen Classification System

The graph on the left in **Figure 14.5** shows climate data for a desert in Reno, Nevada. The graph on the right shows climate data for a tropical rain forest in New Guinea. What criteria are used to classify the climates described in the graphs? Temperature is an obvious choice, as is amount of precipitation. The **Köppen classification system** is a classification system for climates that is based on the average monthly values of temperature and precipitation. Developed by German climatologist Wladimir Köppen, the system also takes into account the distinct vegetation found in different climates.

Köppen decided that a good way to distinguish among different climatic zones was by natural vegetation. Palm trees, for instance, are not located in polar regions, but instead are largely limited to tropical and subtropical regions. Köppen later realized that quantitative values would make his system more objective and therefore more scientific. Thus, he revised his system to include the numerical values of temperature and precipitation. A map of global climates according to a modified version of Köppen's classification system is shown in **Figure 14.6.**

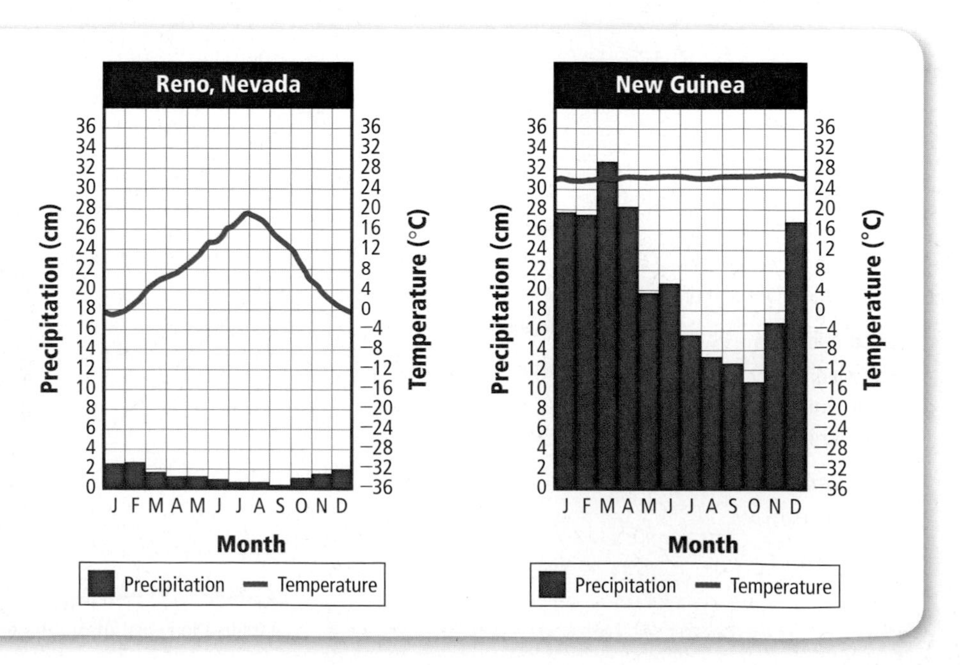

Figure 14.5 These graphs show temperature and precipitation for two different climates—a desert in Reno, Nevada, and a tropical rain forest in New Guinea.
Describe *the difference in temperature between these two climates.*

Visualizing Worldwide Climates

Figure 14.6 Köppen's classification system, shown here in a modified version, is made up of five main divisions based on temperature and precipitation.
Estimate ***Use the map to determine the approximate percentage of land covered by tropical wet climates.***

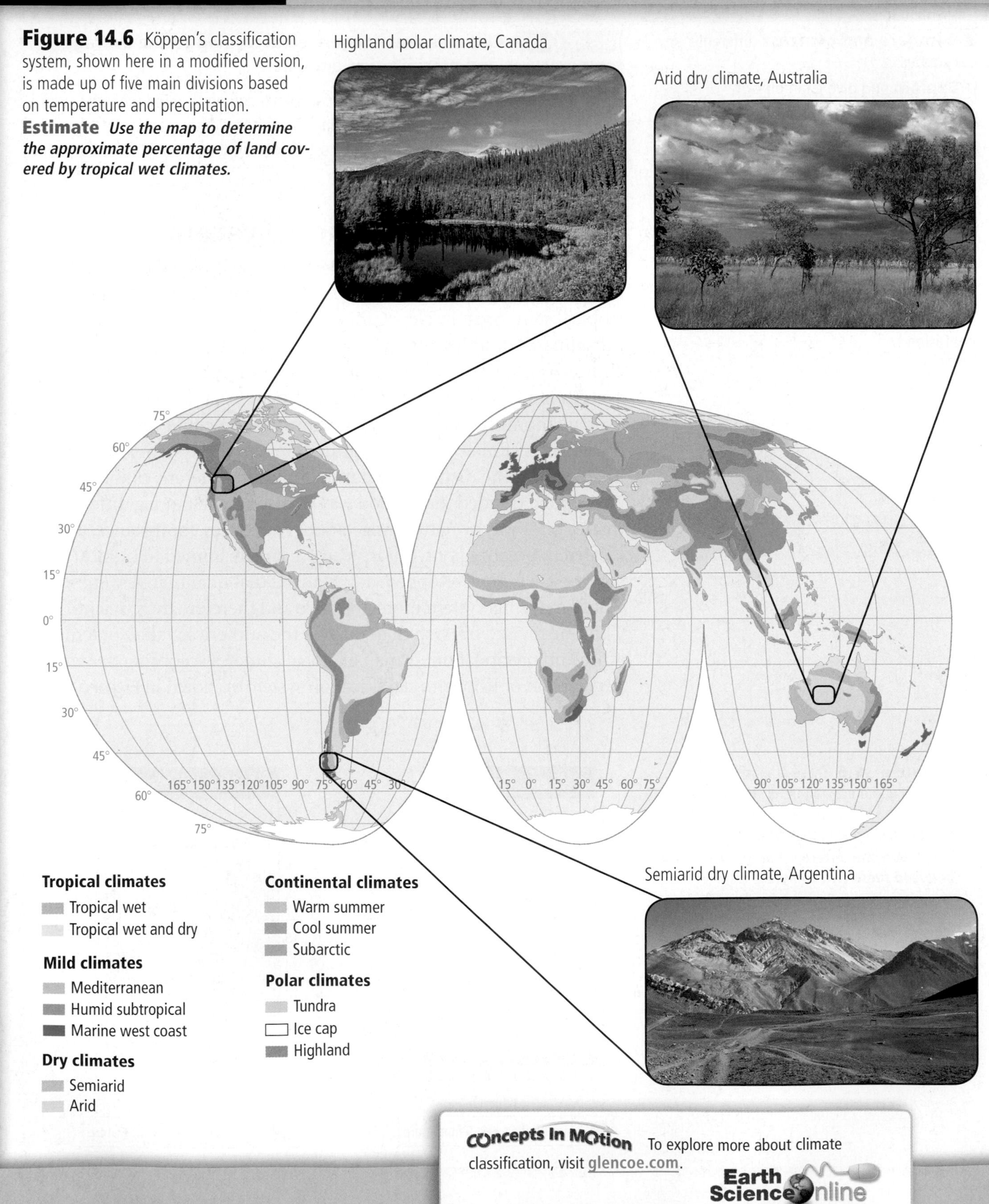

Concepts In Motion To explore more about climate classification, visit glencoe.com.

Earth Science Online

Tropical climates Year-round high temperatures characterize tropical climates. In tropical wet climates, the locations of which are shown in **Figure 14.6,** high temperatures are accompanied by up to 600 cm of rain each year. The combination of warmth and rain produces tropical rain forests, which contain some of the most dramatic vegetation on Earth. Tropical regions are almost continually under the influence of maritime tropical air.

The areas that border the rainy tropics to the north and south of the equator are transition zones, known as the tropical wet and dry zones. Tropical wet and dry zones include savannas. These tropical grasslands are found in Africa, among other places. These areas have distinct dry winter seasons as a result of the seasonal influx of dry continental air masses. **Figure 14.7** shows the average monthly temperature and precipitation readings for Normanton, Australia—a savanna in northeast Australia.

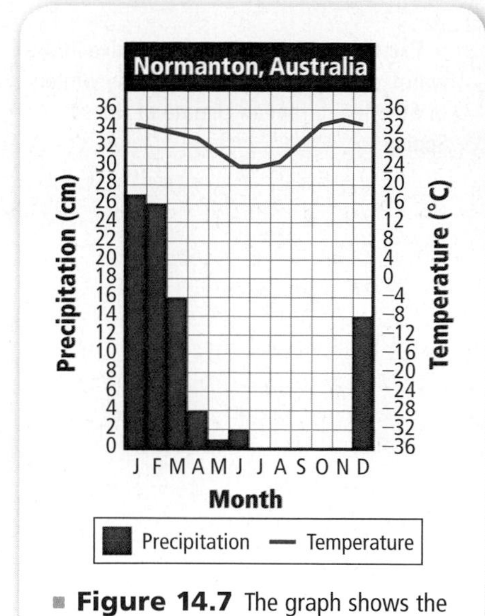

Figure 14.7 The graph shows the temperature and precipitation readings for a tropical savanna in Australia.
Analyze ***How does the rainfall in this area differ from that of a tropical rain forest?***

 **Reading Check Explain** the difference between tropical wet and tropical wet and dry climate zones.

Dry climates Dry climates, which cover about 30 percent of Earth's land area, make up the largest climatic zone. Most of the world's deserts, such as the Sahara, the Gobi, and the Australian, are classified as dry climates. In these climates, continental tropical (cT) air dominates, precipitation is low, and vegetation is scarce. Many of these areas are located near the tropics. Thus, intense solar radiation results in high rates of evaporation and few clouds. Overall, evaporation rates exceed precipitation rates. The resulting moisture deficit gives this zone its name. Within this classification, there are two subtypes: arid regions, called deserts, and semiarid regions, called semideserts. Semideserts, like the one shown in **Figure 14.8,** are usually more humid than deserts. They generally separate arid regions from bordering wet climates.

Figure 14.8 This semidesert in Kazakhstan is another example of a transition zone. It separates deserts from bordering climates that are more humid.

■ **Figure 14.9** Olive trees thrive in the warm, dry summers and cool, rainy winters of the Mediterranean climate of Huesca, Spain.

Mild climates Mild climates can be classified into three subtypes: humid subtropical climates, marine west-coast climates, and Mediterranean climates. Humid subtropical climates are influenced by the subtropical high-pressure systems that are normally found over oceans in the summer. The southeastern United States has this type of climate. There, warm, muggy weather prevails during the warmer months and dry, cool conditions predominate during the winter. The marine west-coast climates are dominated by the constant inland flow of air off the ocean, which creates mild winters and cool summers, with abundant precipitation throughout the year. Mediterranean climates, named for the climate that characterizes much of the land around the Mediterranean Sea, are also found in California and parts of South America. An example of this type of climate is shown in **Figure 14.9.** Summers in Mediterranean climates are generally warm and dry because of their nearness to the dry midlatitude climates from the south. Winters are cool and rainy as a result of the midlatitude weather systems that bring storm systems from the north.

Reading Check **Compare and contrast** humid subtropical and marine west-coast climates.

■ **Figure 14.10** Tornadoes, such as this one in Kansas, occur in continental climates.

Continental climates Continental climates are also classified into three subtypes: warm summer climates, cool summer climates, and subarctic climates. Tropical and polar air masses often form fronts as they meet in continental climates. Thus, these zones experience rapid and sometimes violent changes in weather, including severe thunderstorms or tornadoes like the one shown in **Figure 14.10.** Both summer and winter temperatures can be extreme because the influence of polar air masses is strong in winter, while warm tropical air dominates in summer. The presence of warm, moist air causes summers to be generally more wet than winters, especially in latitudes that are relatively close to the tropics.

Polar climates To the north of subarctic climate lies one of the polar climates—the tundra. Just as the tropics are known for their year-round warmth, tundra is known for its low temperatures—the mean temperature of the warmest month is less than 10°C. There are no trees in the tundra and precipitation is generally low because cold air contains less moisture than warm air. Also, the amount of heat radiated by Earth's surface is too low to produce the strong convection currents needed to release heavy precipitation. The ice-cap polar climate, found at the highest latitudes in both hemispheres, does not have a single month in which average temperatures rise above 0°C. No vegetation grows in an ice-cap climate and the land is permanently covered by ice and snow. **Figure 14.11** shows an ice-cap polar climate.

A variation of the polar climate, called a highland climate, is found at high elevations. This type of climate includes parts of the Andes Mountains in South America, which lie near the equator. The intense solar radiation found near such equatorial regions is offset by the decrease in temperature that occurs with altitude.

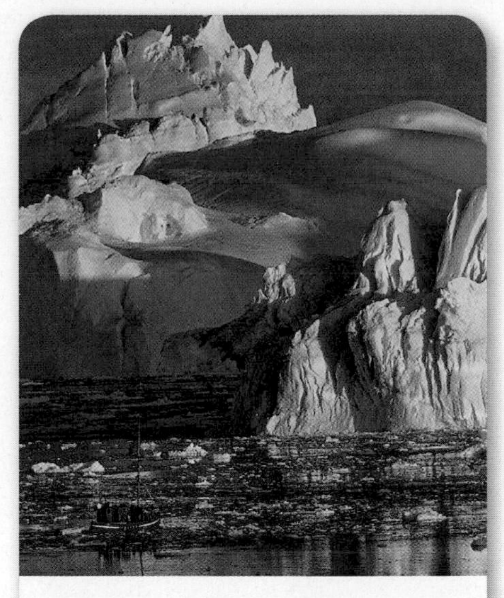

■ **Figure 14.11** Icebergs float in the sea in the ice-cap polar climate of Greenland.

Microclimates

Sometimes the climate of a small area can be much different than that of the larger area surrounding it. A localized climate that differs from the main regional climate is called a **microclimate.** If you climb to the top of a mountain, you can experience a type of microclimate; the climate becomes cooler with increasing elevation. **Figure 14.12** shows a microclimate created by the buildings and concrete in a city.

Heat islands Sometimes the presence of buildings can create a microclimate in the area immediately surrounding it. Many concrete buildings and large expanses of asphalt can create a **heat island,** where the climate is warmer than in surrounding rural areas, as shown in **Figure 14.12.** This effect was first recognized in the early nineteenth century when Londoners noted that the temperature in the city was noticeably warmer than in the surrounding countryside.

FOLDABLES Incorporate information from this section into your Foldable.

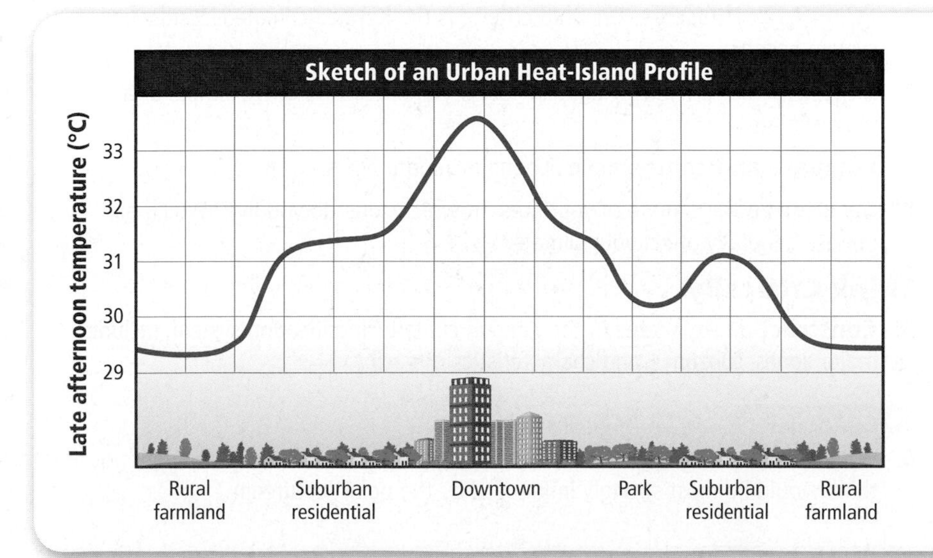

■ **Figure 14.12** This diagram shows the difference in temperature between the downtown area of a city and the surrounding suburban and rural areas.
Analyze ***How much warmer is it in the city compared to the rural areas?***

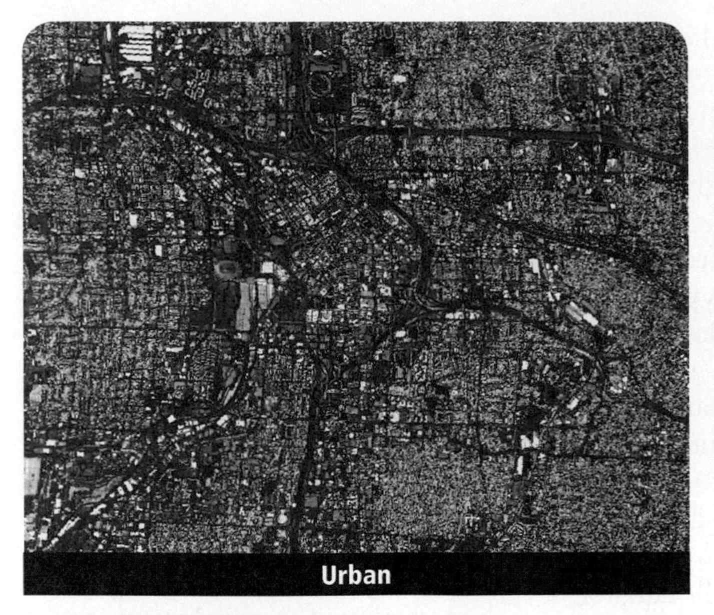

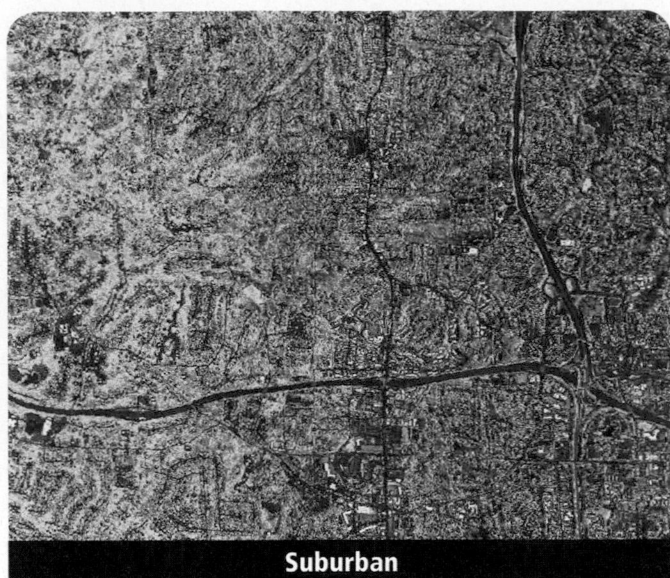

■ **Figure 14.13** These thermal images show differences in daytime temperatures between an urban area and a suburban area. The coolest temperatures are represented by blue; the warmest temperatures are represented by red.

Pavement, buildings, and roofs made of dark materials, such as asphalt, absorb more energy from the Sun than surrounding vegetation. This causes the temperature of these objects to rise, heating the air around them. This also causes mean temperatures in large cities to be significantly warmer than in surrounding areas, as shown in **Figure 14.13.** The heat-island effect also causes greater changes in temperature with altitude, which sparks strong convection currents. This, in turn, produces increased cloudiness and up to 15 percent more total precipitation in cities.

Heat islands are examples of climatic change on a small scale. In Sections 14.3 and 14.4, you will examine large-scale climatic changes caused by both natural events and human activities.

Section 14.2 Assessment

Section Summary

- German scientist Wladimir Köppen developed a climate classification system.
- There are five main climate types: tropical, dry, mild, continental, and polar.
- Microclimates can occur within cities.

Understand Main Ideas

1. MAIN Idea **Describe** On what criteria is the Köppen climate classification system based?
2. **Explain** What are microclimates? Identify and describe one example of a microclimate.
3. **Compare and contrast** the five main climate types.
4. **Categorize** the climate of your area. In which zone do you live? Which air masses generally affect your climate?

Think Critically

5. **Construct** Make a table of the Köppen climate classification system. Include major zones, subzones, and characteristics of each.

WRITING in **Earth Science**

6. Write a short paragraph that explains which of the different climate types you think would be most strongly influenced by the polar jet stream.

Earth Science Online **Self-Check Quiz** glencoe.com

Section 14.3

Objectives

- **Distinguish** between long-term and short-term climatic changes.
- **Identify** natural causes of climate change.
- **Recognize** why climatic changes occur.

Review Vocabulary

glacier: large, moving mass of ice that forms near Earth's poles and in mountainous regions at high elevations

New Vocabulary

ice age
season
El Niño
Maunder minimum

Climatic Changes

MAIN Idea Earth's climate is constantly changing on many different timescales.

Real-World Reading Link You might not notice changes in your friends' physical appearance from day to day; however, if you only see someone once a year, he or she might appear to have changed a lot. Climate changes on long timescales with differences that might not be noticed day to day.

Long-Term Climatic Changes

Some years might be warmer, cooler, wetter, or drier than others, but during the average human lifetime, climates do not appear to change significantly. However, a study of Earth's history over hundreds of thousands of years shows that climates have always been, and currently are, in a constant state of change. These changes usually take place over long time periods.

Ice ages A good example of climatic change involves glaciers, which have alternately advanced and retreated over the past 2 million years. At times, much of Earth's surface was covered by vast sheets of ice. During these periods of extensive glacial coverage, called **ice ages,** average global temperatures decreased by an estimated 5°C. Global climates became generally colder and snowfall increased, which sparked the advance of existing ice sheets. Ice ages alternate with warm periods—called interglacial intervals—and Earth is currently experiencing such an interval. The most recent ice age, shown in **Figure 14.14,** ended only about 10,000 years ago. In North America, glaciers spread from the east coast to the west coast and as far south as Indiana.

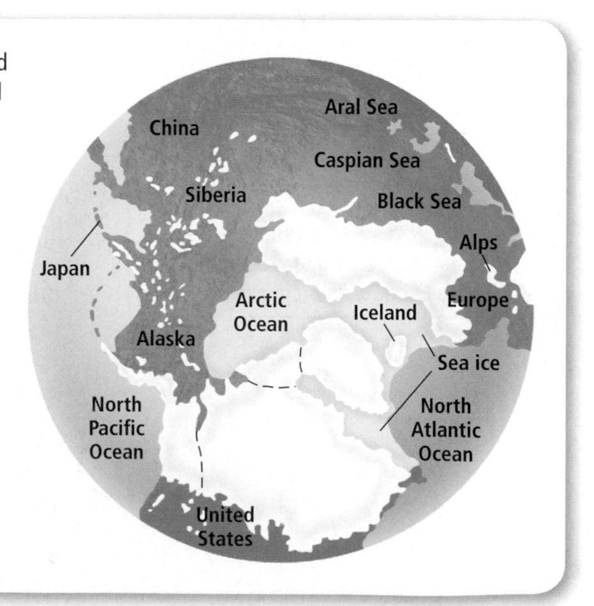

Figure 14.14 The last ice age covered large portions of North America, Europe, and Asia. Average global temperatures were roughly 5°C lower than they are today.
Explain *how decreased global temperatures can lead to an ice age.*

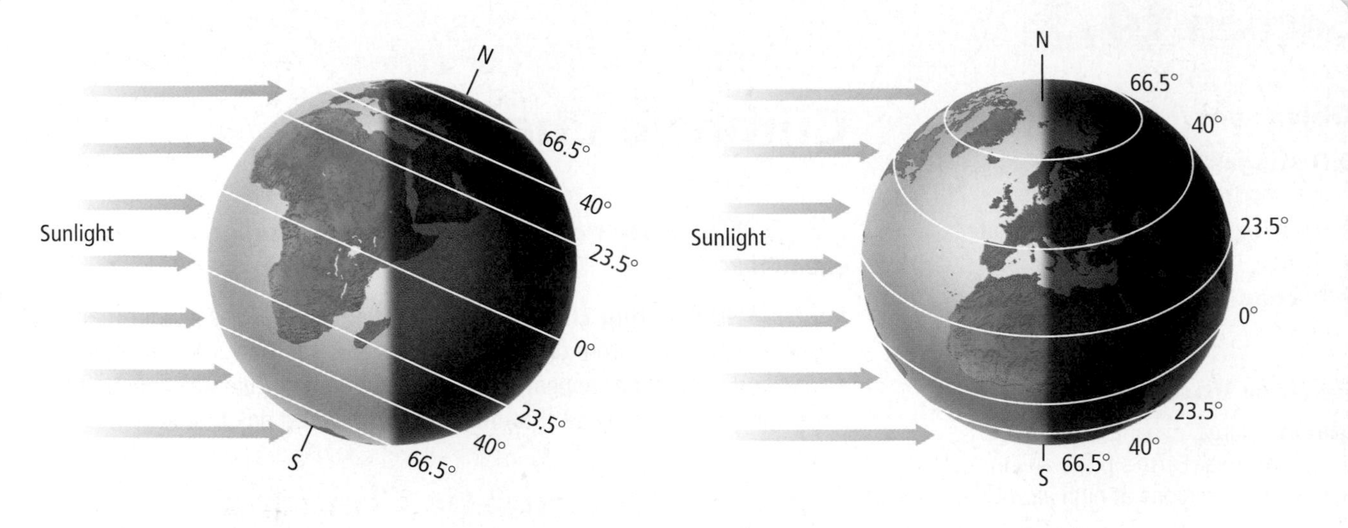

■ **Figure 14.15** When the north pole is pointed away from the Sun, the northern hemisphere experiences winter and the southern hemisphere experiences summer. During spring and fall, neither pole points toward the Sun.

Concepts In Motion
Interactive Figure To see an animation of seasons, visit glencoe.com.

Short-Term Climatic Changes

While an ice age might last for tens of thousands of years, other climatic changes occur over much shorter time periods. Climatic change can affect seasons differently. **Seasons** are short-term periods with specific weather conditions caused by regular variations in daylight, temperature, and weather patterns.

Seasons The variations that occur with seasons are the result of changes in the amount of solar radiation an area receives. As **Figure 14.15** shows, the tilt of Earth on its axis as it revolves around the Sun causes different areas of Earth to receive different amounts of solar radiation. During winter in the northern hemisphere, the north pole is tilted away from the Sun, and this hemisphere experiences long hours of darkness and cold temperatures. At the same time, it is summer in the southern hemisphere. The south pole is tilted toward the Sun, and the southern hemisphere experiences long hours of daylight and warm temperatures. Throughout the year, the seasons are reversed in the northern and southern hemispheres. During the spring and fall, neither pole points toward the Sun.

To read about how seasonal changes affect the Okavango delta, go to the **National Geographic Expedition** on page 904.

El Niño Other short-term climatic changes include those caused by **El Niño,** a band of anomalously warm ocean temperatures that occasionally develops off the western coast of South America. Under normal conditions in the southeastern Pacific Ocean, atmospheric and ocean currents along the coast of South America move north, transporting cold water from the Antarctic region.

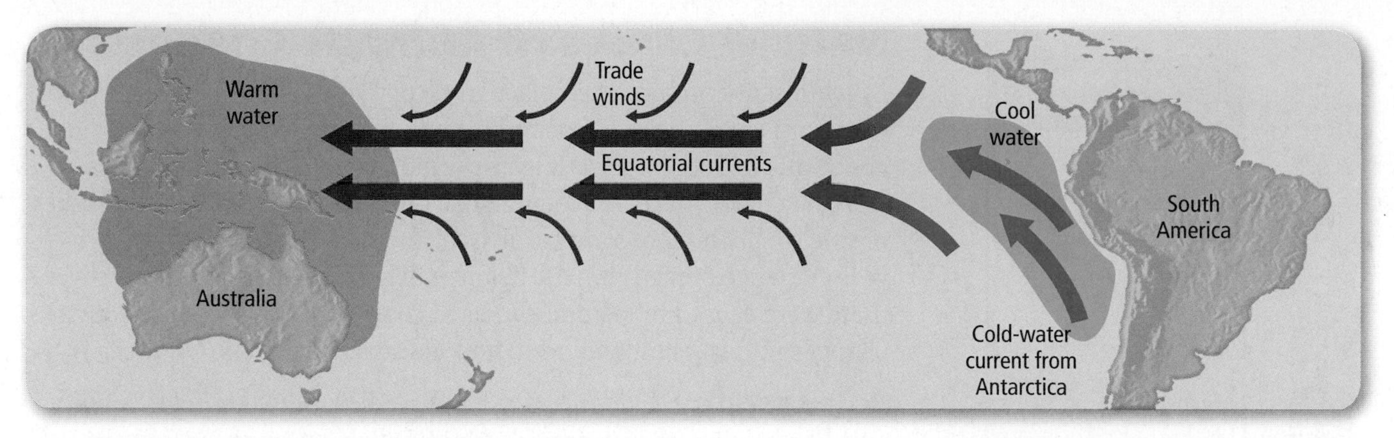

■ **Figure 14.16** Under normal conditions, trade winds and ocean currents move warm water west across the Pacific Ocean.

Meanwhile, the trade winds and ocean currents move westward across the tropics, keeping warm water in the western Pacific, as shown in **Figure 14.16.** This circulation, driven by a semipermanent high-pressure system, creates a cool, dry climate along much of the northwestern coast of South America.

Occasionally, however, for reasons that are not fully understood, this high-pressure system and its associated trade winds weaken drastically, which allows the warm water from the western Pacific to surge eastward toward the South American coast, as shown in **Figure 14.17.** These conditions are referred to as an El Niño event.

The sudden presence of this warm water heats the air near the surface of the water. Convection currents strengthen, and the normally cool and dry northwestern coast of South America becomes much warmer and wetter. The increased convection pumps large amounts of heat and moisture into the upper atmosphere, where upper-level winds transport the hot, moist air eastward across the tropics. This hot, moist air in the upper atmosphere is responsible for dramatic climate changes, including violent storms in California and the Gulf Coast, stormy weather to areas farther east that are normally dry, and drought conditions to areas that are normally wet. Eventually, the South Pacific high-pressure system becomes reestablished and El Niño weakens.

Sometimes the trade winds blow stronger than normal and warm water is pulled across the Pacific toward Australia. The coast of South America becomes unusually cold and chilly. These conditions are called La Niña.

Vocabulary

Science usage v. Common usage

Pressure

Science usage: the force that a column of air exerts on the air below it

Common usage: the burden of physical or mental distress

■ **Figure 14.17** During El Niño, warm water surges back toward South America, changing weather patterns.

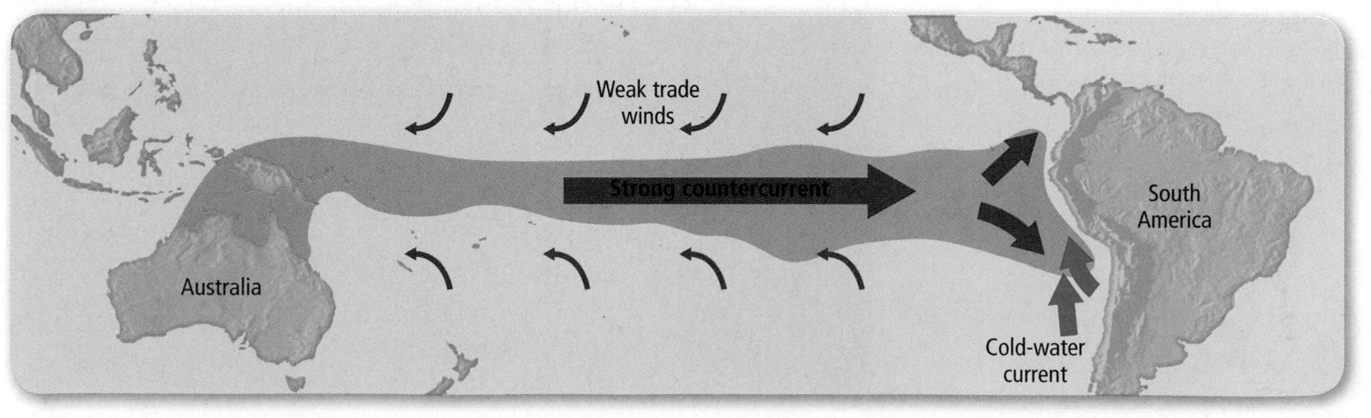

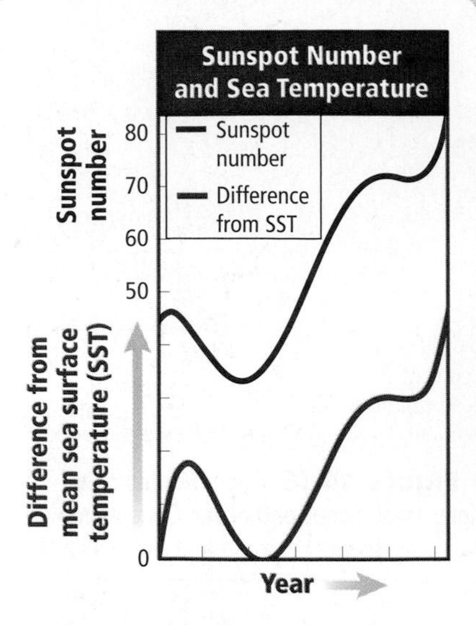

Figure 14.18 Scientists theorize that solar activity might be linked to climatic changes.
Evaluate *How is the number of sunspots related to changes in sea surface temperature?*

Natural Causes of Climatic Changes

Much discussion has taken place in recent years about whether Earth's climate is changing as a result of human activities. You will read more about this in Section 14.4. It is important to note that many cycles of climatic change occurred long before humans inhabited Earth. Studies of tree rings, ice-core samples, fossils, and radiocarbon samples provide evidence of past climatic changes. These changes in Earth's climate were caused by natural events such as variations in solar activity, changes in Earth's tilt and orbit, and volcanic eruptions.

Solar activity Evidence of a possible link between solar activity and Earth's climate was provided by English astronomer Edward Walter Maunder in 1893. The existence of sunspot cycles lasting approximately 11 years had been recognized by German scientist Samuel Heinrich Schwabe in 1843. However, Maunder found that from 1645 to 1716, the number of sunspots was scarce to nonexistent. The **Maunder minimum** is the term used to describe this period of low numbers of sunspots. This period closely corresponds to an unusually cold climatic episode called the Little Ice Age. During this time, much of Europe experienced bitterly cold winters and below-normal temperatures year-round. Residents of London are said to have ice-skated on the Thames River in June. The relationship between sea surface temperature, which is used as an indicator of climate, and periods of low sunspot numbers is illustrated in **Figure 14.18.** Studies indicate that increased solar activity coincides with warmer-than-normal sea surface temperatures, while periods of low solar activity, such as the Maunder minimum, coincide with colder sea surface temperatures.

Earth's orbit Climatic changes might also be triggered by changes in Earth's axis and orbit. The shape of Earth's elliptical orbit appears to change, becoming more elliptical, then more circular, over the course of a 100,000-year cycle. As **Figure 14.19** shows, when the orbit elongates, Earth travels for part of the year in a path closer to the Sun. As a result, temperatures become warmer than normal. When the orbit is more circular, Earth remains in an orbit that is farther from the Sun, and temperatures dip below average.

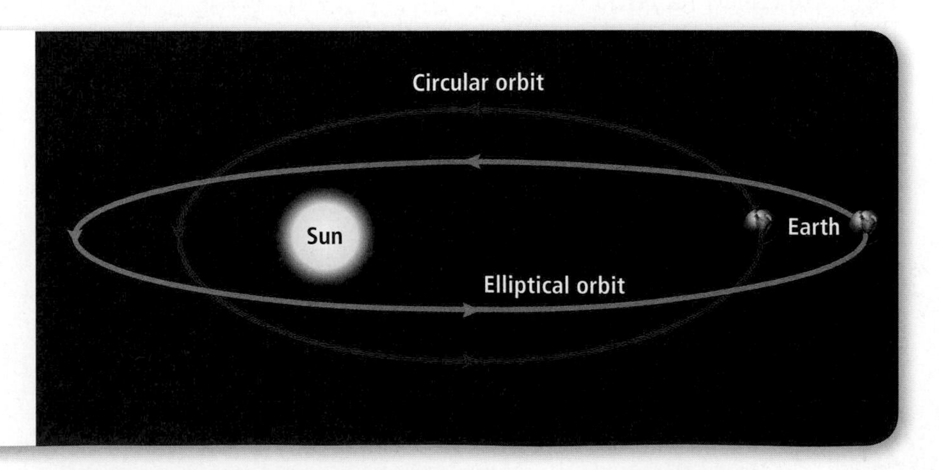

Figure 14.19 Scientists hypothesize that a more elliptical orbit around the Sun could produce significant changes in Earth's climate.

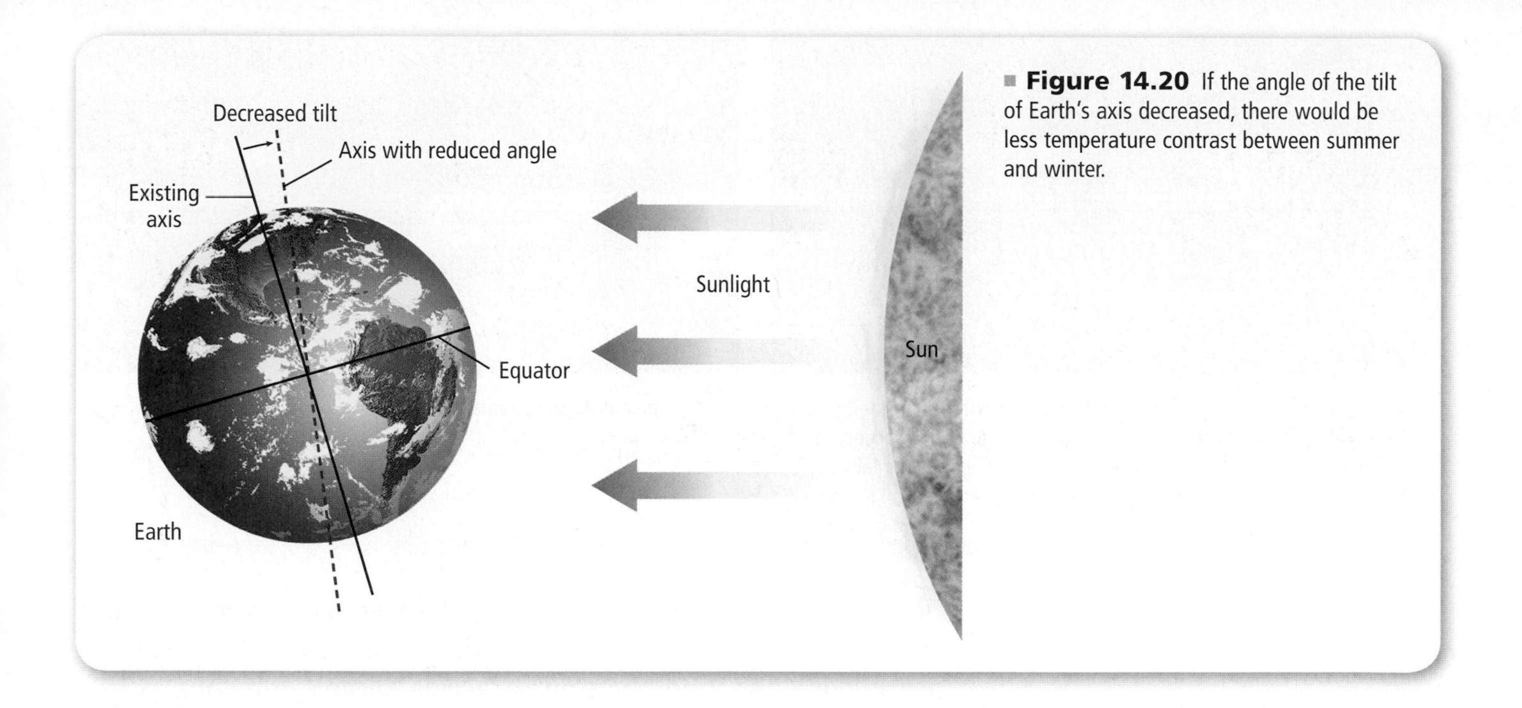

■ **Figure 14.20** If the angle of the tilt of Earth's axis decreased, there would be less temperature contrast between summer and winter.

Earth's tilt As you know, seasons are caused by the angle of the tilt of Earth's axis. At present, the angle of the tilt is 23.5°. However, the angle of tilt varies from a minimum of 22.1° to a maximum of 24.5° every 41,000 years. Scientists theorize that these changes in angle affect the differences in seasons. For example, a decrease in the angle of the tilted axis, shown in **Figure 14.20,** might cause a decrease in the temperature difference between winter and summer. Winters would be more warm and wet, and summers would be cooler. The additional snow in latitudes near the poles would not melt in summer because temperatures would be cooler than average. This could result in increased glacial formation and coverage. In fact, some scientists hypothesize that changes in the angle of Earth's tilted axis can cause ice sheets to form near the poles.

 **Reading Check Describe** how a change to the angle of Earth's tilt can lead to climate change.

Earth's wobble Another movement of Earth might be responsible for climatic changes. Over a period of about 26,000 years, Earth wobbles as it spins around on its axis. Currently, the axis points toward the North Star, Polaris, as shown in **Figure 14.21.** Because of Earth's wobbling, however, the axis will eventually rotate away from Polaris and toward another star, Vega, in about 13,000 years. Currently, winter occurs in the northern hemisphere when the direction of the tilt of Earth causes the northern hemisphere to receive more direct radiation from the Sun. However, in 13,000 years, the northern hemisphere will be tilted in the opposite direction relative to the Sun. So, during the time of year associated with winter today, the northern hemisphere will be tilted toward the Sun and will experience summer.

■ **Figure 14.21** Earth's wobble determines the timing of the seasons. When the northern hemisphere points toward the star, Vega, in 13,000 years, the northern hemisphere will experience summer during the time now associated with winter.

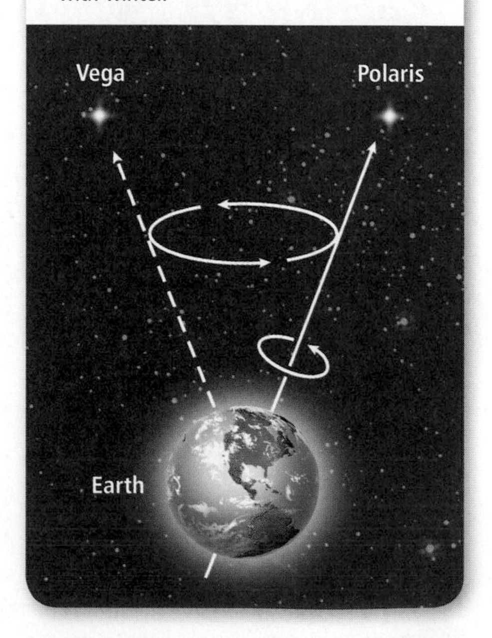

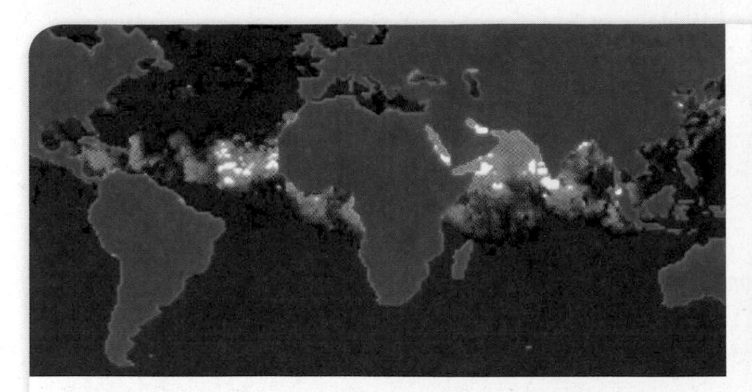
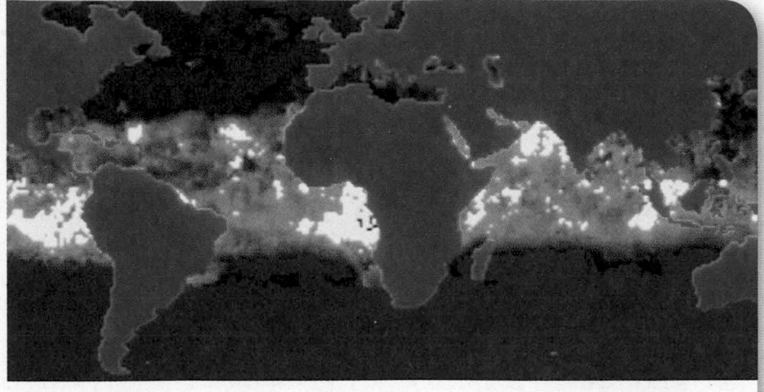

■ **Figure 14.22** After Mount Pinatubo's eruption in the Philippines, aerosol concentration increased worldwide. High concentrations appear in white and low concentrations in brown. The first image was taken immediately after the eruption and the second was taken two months later.
Infer *How did this affect global climates?*

Volcanic activity Climatic changes can also be triggered by the immense quantities of dust-sized particles, called aerosols, that are released into the atmosphere during major volcanic eruptions, as shown in **Figure 14.22.** Volcanic dust can remain suspended in the atmosphere for several years, blocking incoming solar radiation and thus lowering global temperatures. Some scientists theorize that periods of high volcanic activity cause cool climatic periods. Climatic records from the past century show that several large eruptions have been followed by below-normal global temperatures.

For example, the ash released during the 1991 eruption of Mount Pinatubo in the Philippines resulted in slightly cooler temperatures around the world the following year. Generally, volcanic eruptions appear to have only short-term effects on climate. These effects, as well as the others you have read about thus far, are a result of natural causes.

Section 14.3 Assessment

Section Summary

- Climate change can occur on a long-term or short-term scale.
- Changes in solar activity have been correlated with periods of climate change.
- Changes in Earth's orbit, tilt, and wobble are all associated with changes in climate.

Understand Main Ideas

1. MAIN Idea **Identify** and explain an example of long-term climatic change.
2. **Describe** What are seasons? What causes them?
3. **Illustrate** how El Niño might affect weather in California and along the Gulf Coast.
4. **Analyze** How does volcanic activity affect climate? Are these effects short-term or long-term climatic changes?

Think Critically

5. **Assess** What might be the effect on seasons if Earth's orbit became more elliptical and, at the same time, the angle of the tilt of Earth's axis increased?

MATH in Earth Science

6. Study **Figure 14.18.** During which period were sunspot numbers lowest? During which period were sunspot numbers highest?

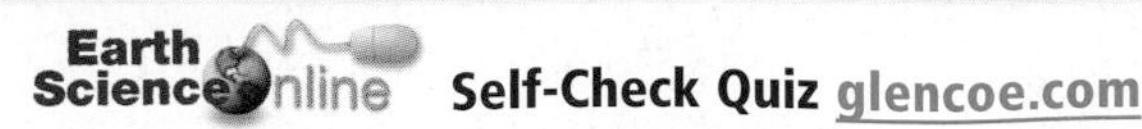
Self-Check Quiz glencoe.com

Section 14.4

Objectives

- **Explain** the greenhouse effect.
- **Describe** global warming.
- **Describe** how humans impact climate.

Review Vocabulary

radiation: transfer of thermal energy by electromagnetic waves

New Vocabulary

greenhouse effect
global warming

Impact of Human Activities

MAIN Idea Over time, human activities can alter atmospheric conditions enough to influence changes in weather and climate.

Real-World Reading Link If your computer has been affected by a virus attached to a downloaded file, the way it operates might change. If not watched closely, human activities can produce changes in Earth's natural systems.

Influence on the Atmosphere

Earth's atmosphere significantly influences its climate. Solar radiation that is not reflected by clouds passes freely through the atmosphere. It is then absorbed by Earth's surface and released as long wavelength radiation. This radiation is absorbed by atmospheric gases such as methane and carbon dioxide. Some of this absorbed energy is reradiated back to Earth's surface.

The greenhouse effect This process of the absorption and radiation of energy in the atmosphere results in the **greenhouse effect**—the natural heating of Earth's surface caused by certain atmospheric gases called greenhouse gases. The greenhouse effect, shown in **Figure 14.23,** warms Earth's surface by more than 30°C. Without the greenhouse effect, life as it currently exists on Earth would not be possible.

Scientists hypothesize that it is possible to increase or decrease the greenhouse effect by changing the amount of atmospheric greenhouse gases, particularly carbon dioxide and methane. An increase in the amount of these gases would theoretically result in increased absorption of energy in the atmosphere. Levels of atmospheric carbon dioxide and methane are increasing. This can lead to a rise in global temperatures, known as **global warming.**

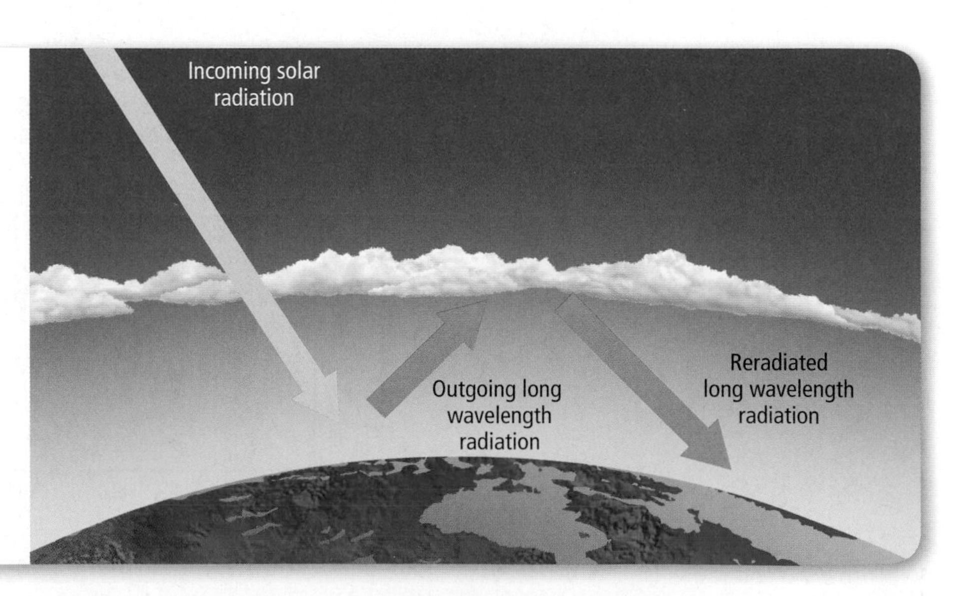

Figure 14.23 Solar radiation reaches Earth's surface where it is reradiated as long wavelength radiation. This radiation does not easily escape through the atmosphere and is mostly absorbed and rereleased by atmospheric gases. This process is called the greenhouse effect.

Concepts In Motion

Interactive Figure To see an animation of the greenhouse effect, visit glencoe.com.

MiniLab

Model the Greenhouse Effect

How does the atmosphere trap radiation? The greenhouse effect is a natural phenomenon that occurs because the atmosphere traps outgoing radiation.

Procedure

1. Read and complete the lab safety form.
2. On a clear day, place a **cardboard box** outside in a shaded area.
3. Prop two **thermometers** vertically against the box. Make sure the thermometers are not in direct sunlight.
4. Cover one thermometer with a **clean glass jar.**
5. Observe and record the temperature changes of each thermometer every 2 min over a 30-min period.

Analysis

1. **Identify** the independent variable and the dependent variable in this investigation.
2. **Construct** a graph showing how the temperatures of the two thermometers changed over time.
3. **Evaluate** Based on your graph, which thermometer experienced the greatest increase in temperature? Why?
4. **Relate** your observations to the greenhouse effect in the atmosphere.

Global Warming

Temperatures worldwide have shown an upward trend over the past 200 years, with several of the warmest years on record having occurred within the last two decades. This trend is shown in **Figure 14.24.** If the trend continues, polar ice caps and mountain glaciers might melt. This could lead to a rise in sea level and the flooding of coastal cities. Other possible consequences include the spread of deserts into fertile regions, an increase in sea surface temperature, and an increase in the frequency and severity of storms.

Based on available temperature data, many scientists agree that global warming is occurring. They disagree, however, about what is causing this warming. Some scientists hypothesize that natural cycles adequately explain the increased temperatures. Mounting evidence suggests that the rate of global temperature changes over the past 150 years is largely due to human activity.

Burning fossil fuels One of the main sources of atmospheric carbon dioxide from humans is from the burning of fossil fuels including coal, oil, and natural gas. Ninety-eight percent of these carbon dioxide emissions in the United States come from burning fossil fuels to run automobiles, heat homes and businesses, and power factories. Almost any process that involves the burning of fossil fuels results in the release of carbon dioxide. Burning fossil fuels also releases other greenhouse gases, such as methane and nitrous oxide, into the atmosphere.

Reading Check **Explain** how burning fossil fuels might contribute to global warming.

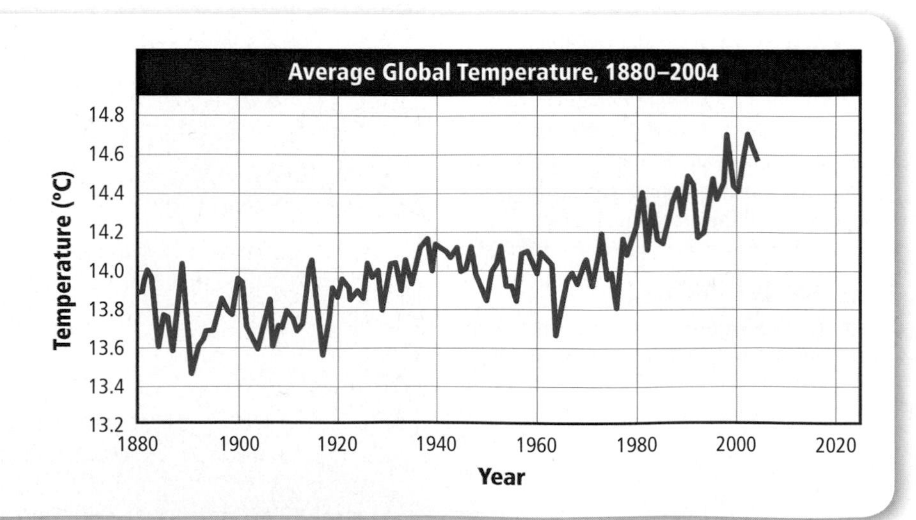

■ **Figure 14.24** The warmest years of the last century all happened within the last 20 years of the century.

■ **Figure 14.25** Deforestation, the mass removal of trees, has occurred in British Columbia, Canada.
Explain *how deforestation can lead to global warming.*

Deforestation Deforestation—the mass removal of trees—also plays a role in increasing levels of atmospheric carbon dioxide. During photosynthesis, vegetation removes carbon dioxide from the atmosphere. When trees, such as the ones shown in **Figure 14.25,** are cut down, photosynthesis is reduced, and more carbon dioxide remains in the atmosphere. Many scientists suggest that deforestation intensifies global warming trends.

Environmental efforts Individuals reduce the amount of carbon dioxide emitted to the atmosphere by conserving energy, which reduces fossil fuel consumption. Some easy ways to conserve energy include turning off electrical appliances and lights when not in use, turning down thermostats in the winter, recycling, and reducing the use of combustion engines, such as those in cars and lawn mowers. You will learn more about resources and conservation in Unit 7.

Section 14.4 Assessment

Section Summary

- The greenhouse effect influences Earth's climate.
- Worldwide temperatures have shown an upward trend over the past 200 years.
- Human activities can influence changes in weather and climate.
- Individuals can reduce their environmental impact on climate change.

Understand Main Ideas

1. MAIN Idea **Describe** some human activities that might have an impact on Earth's climate.
2. **Explain** the greenhouse effect.
3. **Apply** What is global warming? What are some possible consequences of global warming?
4. **Reason** Why do some scientists theorize that global warming might not be the result of increases in atmospheric carbon dioxide?

Think Critically

5. **Evaluate** the analogy of tropical rain forests being referred to as the "lungs" of Earth.

WRITING in Earth Science

6. Write a pamphlet that explains global warming and its possible causes. Include tips on how individuals can reduce CO_2 emissions into the atmosphere.

Earth Science & Society

Effects of Global Warming on the Arctic

Air temperatures in some areas of the Arctic have risen about 2°C in the past 30 years. As permafrost thaws and sea ice thins, houses are collapsing, roads are sinking, and flooding and erosion are increasing.

Thawing permafrost About 85 percent of the ground in Alaska lies above permafrost, which is a layer of soil that remains frozen for two or more years and has a temperature of at least 0°C. Recent data show that the temperature of permafrost across the Arctic has risen anywhere from 0.1°C to 2.8°C, resulting in thawing of the frozen soil in some areas. In areas where permafrost has thawed, the ground has dropped as much as 5 m, affecting roads, airport runways, homes, and businesses. Buildings, such as hospitals and schools, are unusable due to the sinking effect, and roads in Fairbanks, Alaska, have needed costly repairs.

Thinning sea ice People in the village of Shishmaref, on the northwestern coast of Alaska, moved houses to higher ground to avoid having them collapse into the surrounding sea. As the sea ice that helps protect the village from strong waves thins, the land is more vulnerable to erosion. The nearby village of Kivalina, Alaska, is in a similar situation. Engineers estimated that the cost of moving the village's 380 residents to more stable ground is between $100 and $400 million.

Disrupting traditions Changes in temperature also affect hunting practices of people native to the Arctic. Ice-fishing seasons used to begin in October but now do not start until December, when the sea finally freezes. Native languages have also been affected by the changing temperatures and seasonal conditions.

A house near the coast in Shishmaref, Alaska, collapsed as a result of thinning sea ice and thawing permafrost.

The Inuit word *qiqsuqqaqtug* is used to refer to the month of June. The word describes specific snow conditions that occur in June—when a thin layer of melted snow sits on the surface and refreezes at night, forming a crust. With changing temperatures, this condition now occurs in May. As a result, some Inuit think that the word no longer accurately describes the month of June.

Releasing carbon dioxide Permafrost consists of soil that contains high amounts of organic material. As thawing occurs, the organic material in the soil decomposes, releasing carbon dioxide. With more than 1 million km^2 of soil in the Arctic, scientists think that thawing could release large amounts of carbon dioxide into the atmosphere.

WRITING in Earth Science

Bulletin Board Display Research more information about the effects of climate warming on the Arctic. Prepare a display for a bulletin board that explains several examples and includes either illustrated figures or photos. To learn more about global warming and the Arctic, visit glencoe.com.

GEOLAB

DESIGN YOUR OWN: IDENTIFY A MICROCLIMATE

Background: **Microclimates can be caused by tall buildings, large bodies of water, and mountains, among other things. In this activity, you'll observe different microclimates and then attempt to determine which factors strengthen microclimates and how these factors change with distance from Earth's surface.**

Question: *Which type of surface creates the most pronounced microclimate?*

Materials

thermometer
psychrometer
paper strip or wind sock
meterstick
relative humidity chart

Safety Precautions

WARNING: ***Be careful when you handle glass thermometers, especially those that contain mercury. If the thermometer breaks, do not touch it. Have your teacher properly dispose of the glass and mercury.***

Procedure

1. Read and complete the lab safety form.
2. Working in groups of three to four, determine a hypothesis based on the question listed above.
3. Create a plan to test your hypothesis. Include how you will use your equipment to measure temperature, relative humidity, and wind speed on different surfaces and at various heights above these surfaces. Make sure you include provisions for controlling your variables.
4. Select your sites.
5. Make a map of your test sites. Design and construct data tables for recording your observations.
6. Identify your constants and variables in your plan.
7. Have your teacher approve your plan before you proceed.
8. Carry out your experiment.
9. Map your data. Color-code the areas on your map to show which surfaces have the highest and lowest temperatures, the highest and lowest relative humidity, and the greatest and least wind speed. On your map, include data for surface area only.
10. Graph your data for each site, showing differences in temperature with height. Plot temperature on the x-axis and height on the y-axis. Repeat this step for relative humidity and wind speed.

Analyze and Conclude

1. **Analyze** your maps, graphs, and data to find patterns. Which surfaces had the most pronounced microclimates?
2. **Conclude** Did height above the surface affect your data? Why or why not?
3. **Analyze** your hypothesis and the results of your experiment. Was your hypothesis supported? Explain.
4. **Infer** Why did some areas have more pronounced microclimates than others? Which factors seemed to contribute the most to the development of microclimates?
5. **Determine** which variable changed the most with height: temperature, relative humidity, or wind speed?
6. **Determine** which variable changed the least with height.
7. **Infer** why some variables changed more than others with height.

APPLY YOUR SKILL

Plan an Experiment Based on what you have learned in this lab, plan an experiment that would test for microclimates in your state. How would this large-scale experiment be different from the one you just completed?

CHAPTER 14

Study Guide

Download quizzes, key terms, and flash cards from glencoe.com.

BIG Idea The different climates on Earth are influenced by natural factors as well as human activities.

Vocabulary	Key Concepts
Section 14.1 Defining Climate	
• climatology (p. 376) • normal (p. 377) • polar zones (p. 378) • temperate zones (p. 378) • tropics (p. 378)	**MAIN Idea** Climate is affected by several factors including latitude and elevation. • Climate describes the long-term weather patterns of an area. • Normals are the standard climatic values for a location. • Temperatures vary among tropical, temperate, and polar zones. • Climate is influenced by several different factors. • Air masses have distinct regions of origin.
Section 14.2 Climate Classification	
• heat island (p. 385) • Köppen classification system (p. 381) • microclimate (p. 385)	**MAIN Idea** Climates are categorized according to the average temperatures and precipitation amounts. • German scientist Wladimir Köppen developed a climate classification system. • There are five main climate types: tropical, dry, mild, continental, and polar. • Microclimates can occur within cities.
Section 14.3 Climatic Changes	
• El Niño (p. 388) • ice age (p. 387) • Maunder minimum (p. 390) • season (p. 388)	**MAIN Idea** Earth's climate is constantly changing on many different timescales. • Climate change can occur on a long-term or short-term scale. • Changes in solar activity have been correlated with periods of climate change. • Changes in Earth's orbit, tilt, and wobble are all associated with changes in climate.
Section 14.4 Impact of Human Activities	
• global warming (p. 393) • greenhouse effect (p. 393)	**MAIN Idea** Over time, human activities can alter atmospheric conditions enough to influence changes in weather and climate. • The greenhouse effect influences Earth's climate. • Worldwide temperatures have shown an upward trend over the past 200 years. • Human activities can influence changes in weather and climate. • Individuals can reduce their environmental impact on climate change.

Vocabulary PuzzleMaker glencoe.com

CHAPTER 14 Assessment

Vocabulary Review

Write a description to correctly define the following vocabulary terms.

1. climatology

2. temperate zone

3. normal

4. polar zone

5. tropics

Fill in the blank with the correct vocabulary term from the Study Guide.

6. The differences in temperature caused by the presence of a large city is an example of a(n) ________.

7. A(n) ________ is the name given to localized climate changes such as those at the top of a mountain.

8. A climatic change occurring on a timescale of months is a(n) ________.

9. A(n) ________ is a climatic change that occurred as a result of a change in the number of sunspots.

Replace each underlined vocabulary term with the correct term from the Study Guide.

10. Increasing levels of atmospheric carbon dioxide have been suggested as a cause of <u>the greenhouse effect</u>.

11. Retention of heat in the atmosphere is a result of <u>global warming</u>.

Understand Key Concepts

12. Which is not true about climatic normals?
A. They are averaged over a 30-year period.
B. They are used to predict daily weather.
C. They describe average conditions.
D. They are gathered at one location.

Use the diagram below to answer Questions 13 and 14.

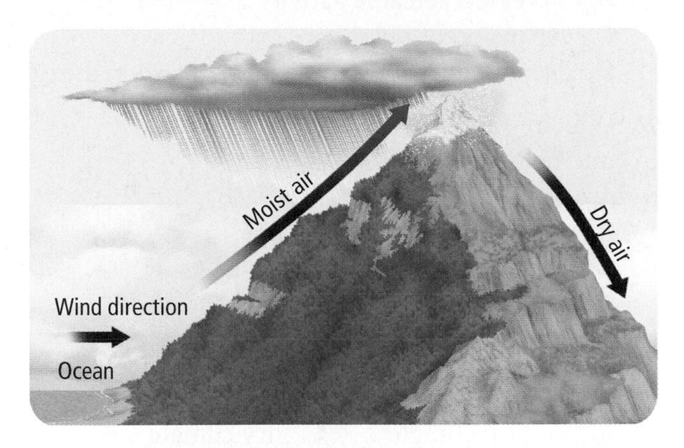

13. Which best describes the climate on the leeward side of a mountain?
A. warm and rainy
B. cool and dry
C. warm and dry
D. cool and rainy

14. What happens to air as it passes over the windward side of a mountain?
A. It sinks and gathers moisture.
B. It sinks and begins to condense.
C. It rises and begins to condense.
D. It rises and evaporates.

15. Which is a long-term climate change?
A. fall
B. an ice age
C. summer
D. El Niño

16. El Niño develops because of a weakening of what?
A. the polar front
B. the trade winds
C. the prevailing westerlies
D. the jet stream

17. Which phenomenon has not been suggested as a factor in global warming?
A. deforestation
B. El Niño
C. burning of fossil fuels
D. industrial emissions

18. Which could be a result of global warming?
    A. rising sea levels
    B. increased volcanic activity
    C. expansion of polar ice caps
    D. decreased levels of atmospheric carbon dioxide

19. Which would not be likely to produce a microclimate?
    A. an ocean shoreline
    B. a valley
    C. a flat prairie
    D. a large city

20. A heat island is an example of which type of climate?
    A. tropical climate
    B. microclimate
    C. dry climate
    D. polar climate

Use the table below to answer Questions 21 and 22.

World Climates

Location	Climate Description
New Caledonia, South Pacific	constant high temperatures, plenty of rain
Southern Israel	humid in summer, dry in winter
Gobi Desert, Mongolia	continental tropical air, low precipitation, scarce vegetation
Bogota, Colombia	mild winters, cool summers, abundant precipitation
Yukon, Canada	year-round cold, low precipitation

21. Southern Israel has which type of climate?
    A. tropical
    B. dry
    C. mediterranean
    D. continental

22. Where is a semidesert most likely to be found?
    A. New Caledonia
    B. Gobi Desert
    C. Bogota
    D. Yukon

Constructed Response

23. **Identify** two situations for which climatic normals might not be accurate to use in determining daily weather.

24. **Apply** Which of the main temperature zones probably experiences the least temperature change during the year? Why?

25. **Relate** why fog is a common characteristic of marine west-coast climates.

26. **Compare and contrast** two of the different types of climates classified as mild by Köppen.

27. **Contrast** the climatic changes produced by changes in Earth's orbit to those produced by changes in the tilt of Earth's axis.

28. **Deduce** the reason that most of the dry climates are located near the tropics.

Use the graph below to answer Question 29.

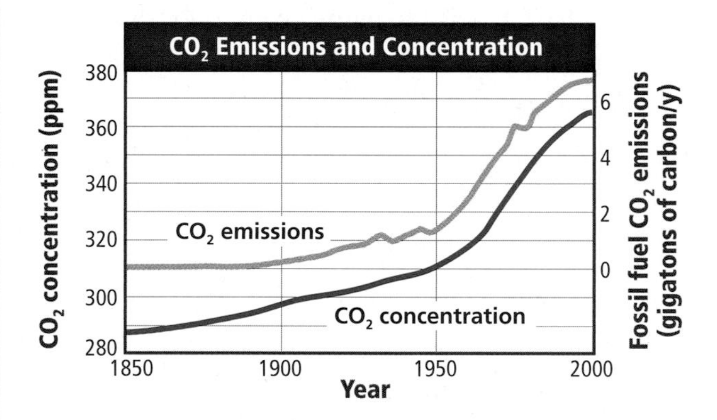

29. **Analyze** the relationship between carbon dioxide emissions and carbon dioxide concentration in the atmosphere over the last 150 years.

30. **Analyze** why the greenhouse effect is considered both essential to life on Earth and also possibly destructive.

31. **Deduce** why some scientists have proposed that global warming might affect the frequency and severity of hurricanes.

32. **Cause and Effect** How does El Niño cause short-term climatic changes?

33. **Suggest** an explanation for why scientists do not agree on the cause of global warming.

Think Critically

34. Formulate a reason why a change in temperature as little as 5°C could be responsible for a climatic change as dramatic as an ice age.

35. Predict Would you expect temperatures during the night to drop more sharply in marine climates or continental climates? Why?

36. Suggest The ironwood tree grows in the desert. Temperatures beneath an ironwood tree can be up to 8°C cooler than temperatures a few feet away. What situation does this describe and how might it affect organisms living in the desert?

Use the diagram below to answer Question 37.

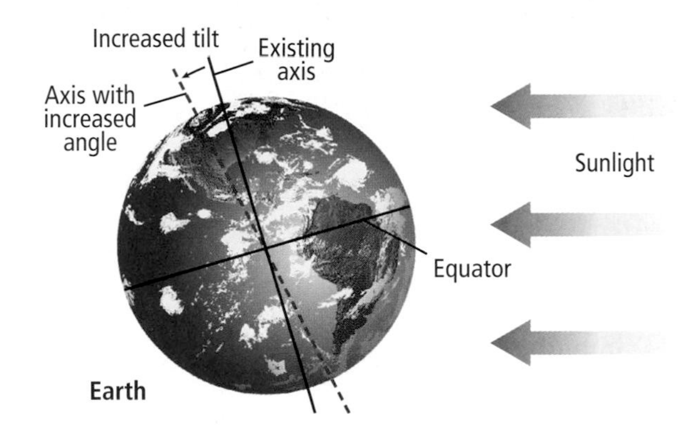

37. Predict the effect on Earth's climate if the tilt of Earth's axis increased to 25°.

38. Assess If you wanted to build a home that was solar heated, would you build it on the leeward or windward side of a mountain? Explain.

Concept Mapping

39. Draw a concept map that organizes information about El Niño.

Challenge Question

40. Analyze Use a world map and **Figure 14.5** to determine the climate classifications of the following cities: Paris, France; Athens, Greece; London, England; and Sydney, Australia.

Additional Assessment

41. WRITING in **Earth Science** Suppose a friend makes the statement that humans must be to blame for climate change. Write a paragraph describing evidence for and against the assertion.

DBQ Document–Based Questions

Data obtained from: The National Snow and Ice Data Center. 2006.

The graph below shows the area of oceans covered by sea ice each year in January in the northern hemisphere for a 27-year period. Scientists use the area of sea ice as an indicator of global climate change.

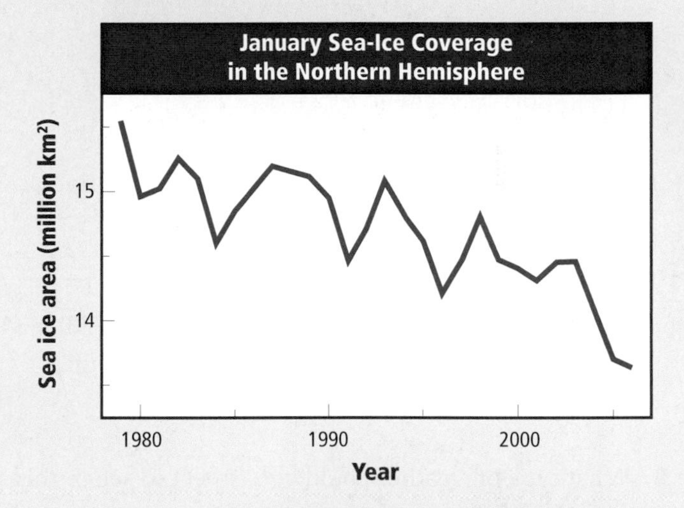

42. Describe the trend of sea-ice coverage over the period shown in the graph. Which year experienced the least sea-ice coverage?

43. What inferences about global climate can be drawn from the changes in sea-ice coverage in the last 27 years?

44. The mean sea-ice coverage from 1979 to 2000 was 14.8 million km^2. How much lower than average was the sea-ice coverage in January 2006?

Cumulative Review

45. Compare and contrast the terms *magma* and *lava*. **(Chapter 5)**

46. List the three primary mechanisms of heat transfer within the atmosphere. **(Chapter 11)**

Standardized Test Practice

Multiple Choice

1. Which type of air masses are most likely to form over land near the equator?
   A. mP
   C. cP
   B. mT
   D. cT

Use the figure below to answer Questions 2 and 3.

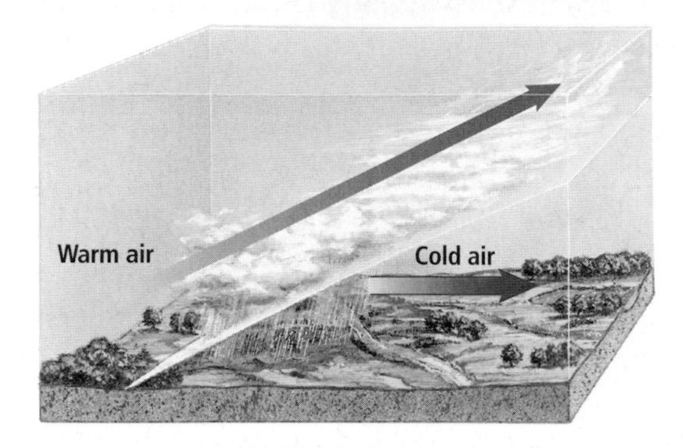

2. The front shown above is a warm front. How does it occur?
   A. Warm air gradually slides over colder air.
   B. Warm air steeply climbs over cold air.
   C. Warm air is quickly forced upward over cold air.
   D. Warm air collides with cold air and does not advance.

3. What type of weather could you expect to see in this situation?
   A. light precipitation over a wide band
   B. extensive precipitation over a narrow band
   C. extensive precipitation over a wide band
   D. light precipitation over a narrow band

4. About 99 percent of the Earth's atmosphere is made up of what two gases?
   A. carbon dioxide and oxygen
   B. nitrogen and oxygen
   C. carbon dioxide and nitrogen
   D. water vapor and oxygen

5. What occurs when winds of at least 120 km/h drive a mound of ocean water toward coastal areas?
   A. downburst
   B. cold wave
   C. storm surge
   D. tsunami

6. Which factor is NOT associated with a heat wave?
   A. a high-pressure system
   B. a weakened jet stream
   C. above-normal temperatures
   D. increased cloud cover

Use the table below to answer Questions 7 and 8.

Location	Climate Description
New Caledonia, South Pacific	constant high temperatures, plenty of rain
South Carolina	humid in summer, dry in winter
Gobi Desert, Mongolia	continental tropical air, low precipitation, scarce vegetation
Bogotá, Columbia	mild winters, cool summers, abundant precipitation
Yukon, Canada	year-round cold, low precipitation

7. According to the modified Köppen classification system, South Carolina has what kind of climate?
   A. tropical
   B. dry
   C. humid subtropical
   D. continental

8. Where is a semiarid region most likely to be found?
   A. New Caledonia
   B. Gobi Desert
   C. Bogotá
   D. Yukon

9. Why does conduction affect only the atmospheric layer near Earth's surface?
   A. It is the only layer that comes in direct contact with Earth.
   B. It is the only layer that contains particles of air.
   C. It is the coolest layer.
   D. It is the warmest layer.

10. The current state of the atmosphere is known as
    A. climate
    B. weather
    C. precipitation
    D. air mass

Short Answer

Use the graph below to answer Questions 11 and 12.

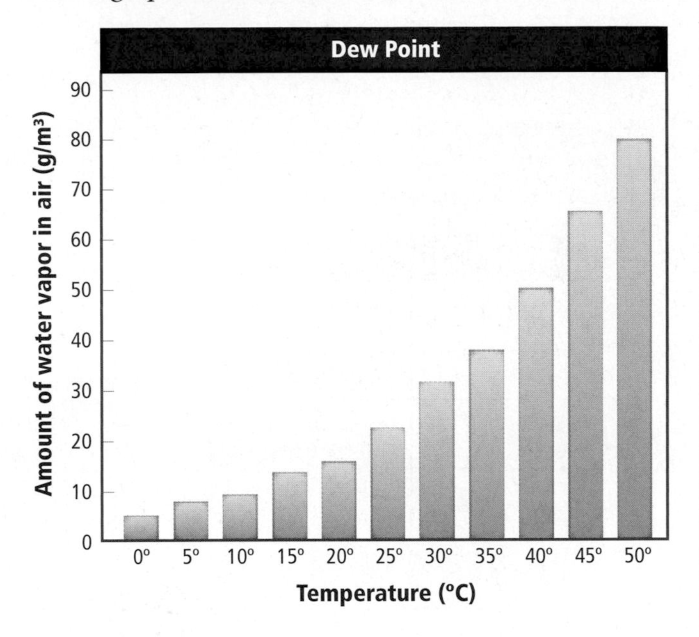

11. What can be determined about dew point according to the graph?

12. What would happen if the temperature were 35°C and the amount of water vapor present were 50 g/m³?

13. Why might the usage of normals be disadvantageous in describing an area's climate?

14. Discuss how the Geographic Information System (GIS) works.

15. Describe how the sound of thunder is produced.

16. Gold is an expensive, soft, highly malleable metal. Infer two reasons why a jeweler might choose to make a ring out of an alloy of gold and copper instead of out of pure gold.

Reading for Comprehension

Global Warming

John Harte, an ecosystem sciences professor at the University of California, Berkeley, is studying possible future outcomes of global warming. For 15 years, he has artificially heated sections of a Rocky Mountain meadow by about 3.6°F (2°C) to study the projected effects of global warming. Harte has documented dramatic changes in the meadow's plant community. Sagebrush, though at the local altitude limit of its natural range, is replacing alpine flowers. More tellingly, soils in test plots have lost about 20 percent of their natural carbon. This effect, if widespread, could dramatically increase Earth's atmospheric CO_2 levels far above even conventional worst-case models. "Soils around the world hold about 5 times more carbon than the atmosphere in the form of organic matter," Harte noted. If similar carbon loss was repeated on a global scale, it could double the amount of carbon in the atmosphere.

Article obtained from: Handwerk, B. Global warming: How hot? How soon? *National Geographic News.* July 27, 2005.

17. What can be predicted from the Rocky Mountain study?

A. The changes occurring in the Rocky Mountain study will occur in other parts of the globe.

B. These changes will only affect the Rocky Mountain area.

C. More studies are needed to determine if the changes really will happen.

D. The Rocky Mountain area is unique in its makeup and therefore is affected most by global warming.

18. What should people learn from reading this passage?

NEED EXTRA HELP?																
If You Missed Question . . .	1	2	3	4	5	6	7	8	9	10	11	12	13	14	15	16
Review Section . . .	12.1	12.2	12.2	11.1	13.3	13.4	14.2	14.2	11.1	12.1	11.2	11.2	14.1	2.3	13.2	3.2

CHAPTER 15

Earth's Oceans

BIG Idea Studying oceans helps scientists learn about global climate and Earth's history.

15.1 An Overview of Oceans
MAIN Idea The global ocean consists of one vast body of water that covers more than two-thirds of Earth's surface.

15.2 Seawater
MAIN Idea Oceans have distinct layers of water masses that are characterized by temperature and salinity.

15.3 Ocean Movements
MAIN Idea Waves and currents drive the movements of ocean water and lead to the distribution of heat, salt, and nutrients from one region of the ocean to another.

GeoFacts

- Tidal pools are formed on rocky shores when water remains on shore after the tide recedes.
- The largest tidal range in the world is found in Nova Scotia, Canada, with a 16.8-m difference between high tide and low tide.
- A sea star can extend its stomach outside of its mouth to digest prey that live in shells.

LAUNCH Lab

How much of Earth's surface is covered by water?

Earth is often referred to as the blue planet because so much of its surface is covered by water. If you study a globe or a photograph of Earth taken from space, you can see that oceans cover much more of Earth than landmasses do.

Procedure

1. Read and complete the lab safety form.
2. Stretch a **piece of string** about 1 m in length around the equator of a **globe.**
3. Use a **blue marker** to color the sections of the string that cross the oceans.
4. Using a **ruler,** measure the length of the globe's equator, then measure the length of each blue section on the string. Add the lengths of the blue sections.
5. Divide the total length of the blue sections by the length of the globe's equator.

Analysis

1. **Calculate** What percentage of the globe's equator is made up of oceans? What percentage of the globe's equator is made up of land?
2. **Observe** Study the globe again. Which hemisphere is covered with more water?

Wave Characteristics Make this Foldable to show the characteristics of waves.

STEP 1 Fold a sheet of paper in half lengthwise.

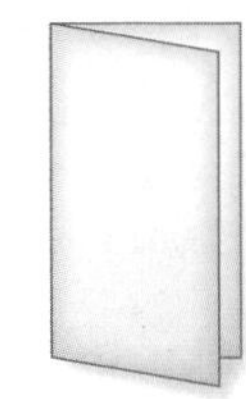

STEP 2 Fold in half and then in half again, as shown.

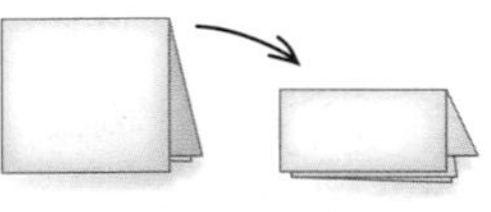

STEP 3 Unfold and cut along the fold lines of the top flap to make four tabs.

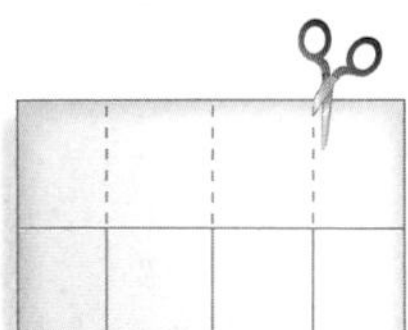

STEP 4 Label the tabs *crest, trough, wave height,* and *wavelength.*

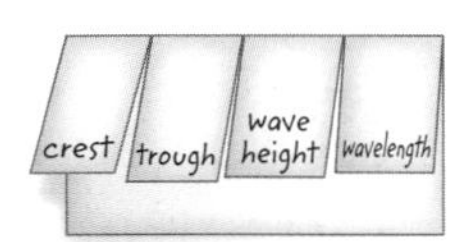

FOLDABLES **Use this Foldable with Section 15.3.** As you read this section, sketch and explain the physical properties associated with waves on the tabs. Under the tabs, illustrate and label the movement of ocean water.

Visit glencoe.com to

- **study entire chapters online;**
- **explore Concepts In Motion animations:**
  - **Interactive Time Lines**
  - **Interactive Figures**
  - **Interactive Tables**
- **access Web Links for more information, projects, and activities;**
- **review content with the Interactive Tutor and take Self-Check Quizzes.**

Section 15.1

Objectives

- **Identify** methods used by scientists to study Earth's oceans.
- **Discuss** the origin and composition of the oceans.
- **Describe** the distribution of water at Earth's surface.

Review Vocabulary

lake: natural or human-made body of water that can form when a depression on land fills with water

New Vocabulary

side-scan sonar
sea level

An Overview of Oceans

MAIN Idea The global ocean consists of one vast body of water that covers more than two-thirds of Earth's surface.

Real-World Reading Link How could you tell a person's age by only looking at him or her? The presence of wrinkles can signify age as skin changes over time. Similarly, scientists look for clues about changes in rocks formed at the bottom of the ocean to estimate the age of the ocean.

Data Collection and Analysis

Oceanography is the scientific study of Earth's oceans. In the late 1800s, the British ship *Challenger* became the first research ship to use relatively sophisticated measuring devices to study the oceans. Since then, oceanographers have been collecting data with instruments both at the surface and from the depths of the ocean floor. Technologies such as sonar, floats, satellites, submersibles, and computers have become central to the continuing exploration of the ocean. **Figure 15.1** chronicles some of the major discoveries that have been made about oceans.

At the surface Sonar, which stands for **so**und **na**vigation and **r**anging, is used by oceanographers to learn more about the topography of the ocean floor. To determine ocean depth, scientists send a sonar signal to the ocean floor and time how long it takes for the sound to reach the bottom and return to the surface as an echo. Knowing that sound travels at a constant velocity of 1500 m/s through water, scientists can determine the depth by multiplying the total time by 1500 m/s, then dividing the answer by 2.

Figure 15.1
Developments in Oceanography
Technological development has led to many new discoveries in oceanography over time.

1900 — 1925 — 1950

1872 The *Challenger* expedition marks the beginning of oceanography. Scientists measure sea depth, study the composition of the seafloor, and collect a variety of oceanic data.

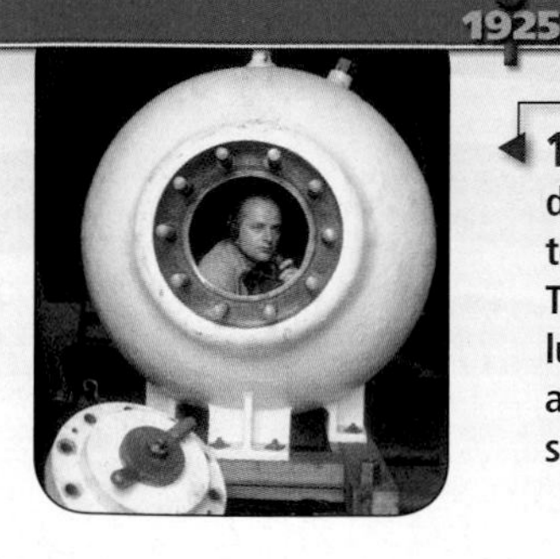

1925 The German *Meteor* expedition surveys the South Atlantic floor with sonar equipment and discovers the Mid-Atlantic Ridge.

1932–1934 The first deep-ocean dives use a tethered bathysphere. The dives uncover luminescent creatures and provide sediment samples.

1943 In France, the first diving equipment is invented from hoses, mouthpieces, air tanks, and a redesigned car regulator that supplies compressed air to divers.

1955 A survey ship detects linear magnetic stripes along the ocean floor. These magnetic patterns lead to the formulation of the theory of plate tectonics.

Large portions of the seafloor have been mapped using **side-scan sonar,** a technique that directs sound waves to the seafloor at an angle, so that the sides of underwater hills and other topographic features can be mapped.

Oceanographers use floats that contain sensors to learn more about water temperature, salinity, and the concentration of gases and nutrients in surface water. Floats can also be used to record wave motion and the speed at which currents are moving. Satellites such as the *TOPEX/Poseidon*, which you read about in Chapter 2, continually monitor the ocean's surface temperatures, currents, and wave conditions.

■ **Figure 15.2** *Alvin* is a deep-sea submersible that can hold two scientists and a pilot.

In the deep sea Submersibles, underwater vessels which can be remotely operated or carry people to the deepest areas of the ocean, have allowed scientists to explore new frontiers. *Alvin,* shown in **Figure 15.2,** is a modern submersible that can take two scientists and a pilot to depths as deep as 4500 m. *Alvin* has been used to discover geologic features such as hydrothermal vents and previously unknown sea creatures. It can also be used to bring sediments and water samples to the surface.

Reading Check **List** some discoveries made using submersibles.

Computers An integral tool in both the collection and analysis of data from the ocean is computers. Information from satellites and float sensors can be transmitted and downloaded directly to computers. Sophisticated programs use mathematical equations to analyze data and produce models. When combined with observations, ocean models provide information about subsurface currents that are not observed directly. Operating in a fashion similar to weather forecasting models, global ocean models play a role in simulating Earth's changing climate. Ocean models are also used to simulate tides, tsunamis, and the dispersion of coastal pollution.

1962 France builds the first underwater habitat where scientists live for days at a time conducting experiments.

1984 An observation system in the Pacific Ocean helps scientists predict El Niño and begin to understand the connection between oceanic events and weather.

2002 Data collected from the seafloor reveals a new ocean wave associated with earthquakes.

2006 An internet portal provides scientists around the world with access to live data collected from ocean-floor laboratories.

1960 1980 2000

1977 The submersible *Alvin* discovers hydrothermal vents and a deep-sea ecosystem including giant worms and clams that can survive without energy from the Sun.

1995 Scientists map the entire seafloor using satellite data.

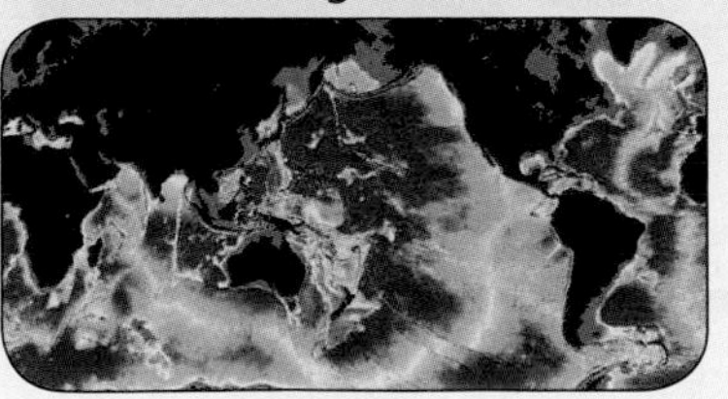

Concepts In Motion

Interactive Time Line To learn more about these discoveries and others, visit glencoe.com. Earth Science Online

Origin of the Oceans

Several geologic clues indicate that oceans have existed almost since the beginning of geologic history. Studies of radioactive isotopes indicate that Earth is about 4.56 billion years old. Scientists have found rocks nearly as old that formed from sediments deposited in water. Ancient lava flows are another clue—some of these lava flows have glassy crusts that form only when molten lava is chilled rapidly underwater. Radioactive studies and lava flows offer evidence that there has been abundant water throughout Earth's geologic history.

CAREERS IN EARTH SCIENCE

Oceanographer Scientists who investigate oceans are called oceanographers. Oceanographers might investigate water chemistry, wave action, marine organisms, or sediments. To learn more about Earth science careers, visit glencoe.com.

Reading Check **Explain** the evidence that suggests that oceans have existed almost since the beginning of Earth's geologic history.

Where did the water come from? Scientists hypothesize that Earth's water originated from either a remote source or a local source, or both. Comets and meteorites are two remote sources that could have contributed to the accumulation of water on Earth. Comets, such as the one shown in **Figure 15.3,** travel throughout the solar system and occasionally collide with Earth. These impacts release enough water over time that they could have contributed to filling the ocean basins over geologic time.

Meteorites, such as the one shown in **Figure 15.3,** are composed of the same material that might have formed the early planets. Studies indicate that meteorites contain up to 0.5 percent water. Meteorite bombardment releases water into Earth's systems.

If early Earth contained the same percentage of water as meteorites, it would have been sufficient to form early oceans. However, some mechanism must have existed to allow the water to rise from Earth's interior to its surface. Scientists theorize that this mechanism was volcanism.

■ **Figure 15.3** Comets are composed of dust and rock particles mixed with frozen water and gases. Comet impacts with Earth may have released enough water to help fill ocean basins. Meteorites contain up to 0.5 percent water.

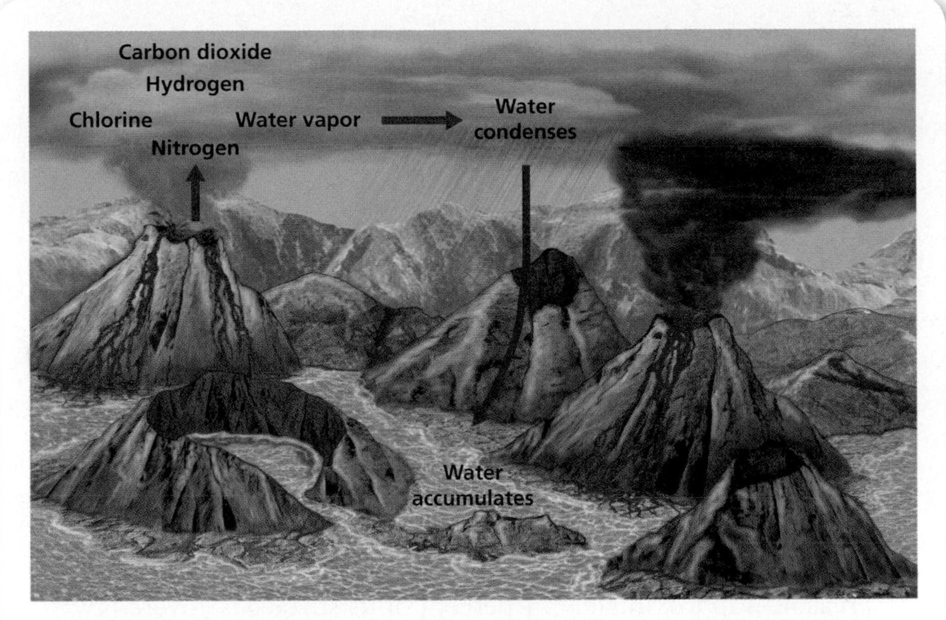

Figure 15.4 In addition to comets, water for Earth's early oceans might have come from volcanic eruptions. An intense period of volcanism occurred shortly after the planet formed. This volcanism released large quantities of water vapor and other gases into the atmosphere. The water vapor eventually condensed into oceans.

Volcanism During volcanic eruptions, significant quantities of gases are emitted. These volcanic gases consist mostly of water vapor and carbon dioxide. Shortly after the formation of Earth, when the young planet was much hotter than it is today, an episode of massive, violent volcanism took place over the course of perhaps several hundred million years. As shown in **Figure 15.4,** this volcanism released huge amounts of water vapor, carbon dioxide, and other gases, which combined to form Earth's early atmosphere. As Earth's crust cooled, the water vapor gradually condensed, fell to Earth's surface as precipitation, and accumulated to form oceans. By the time the oldest known crustal rock formed about 4 bya, Earth's oceans might have been close to their present size. Water is still being added to the hydrosphere by volcanism, but some water molecules in the atmosphere are continually being destroyed by ultraviolet radiation from the Sun. These two processes balance each other.

Distribution of Earth's Water

As shown in **Figure 15.5,** the oceans contain 97 percent of the water found on Earth. Another 3 percent is freshwater located in the frozen ice caps of Greenland and Antarctica and in rivers, lakes, and underground sources. The percentage of ice on Earth has varied over geologic time from near zero to perhaps as much as 10 percent of the hydrosphere. As you read further in this section, you will learn more about how these changes affect sea level.

Figure 15.5 About 97 percent of water on Earth is salt water found in the oceans.

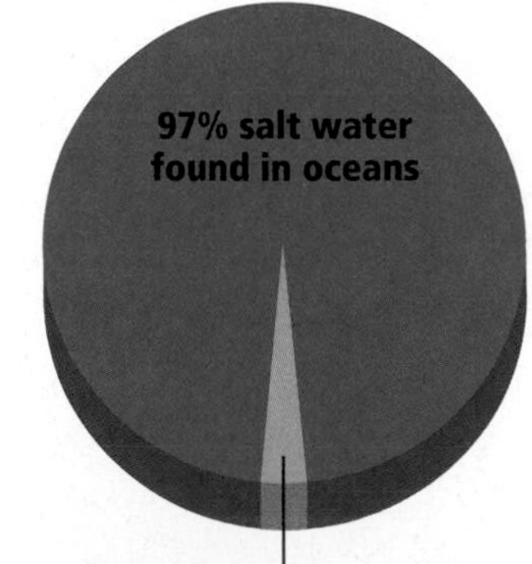

■ **Figure 15.6** The northern hemisphere is covered by slightly more water than land. The southern hemisphere, however, is almost completely covered by water.

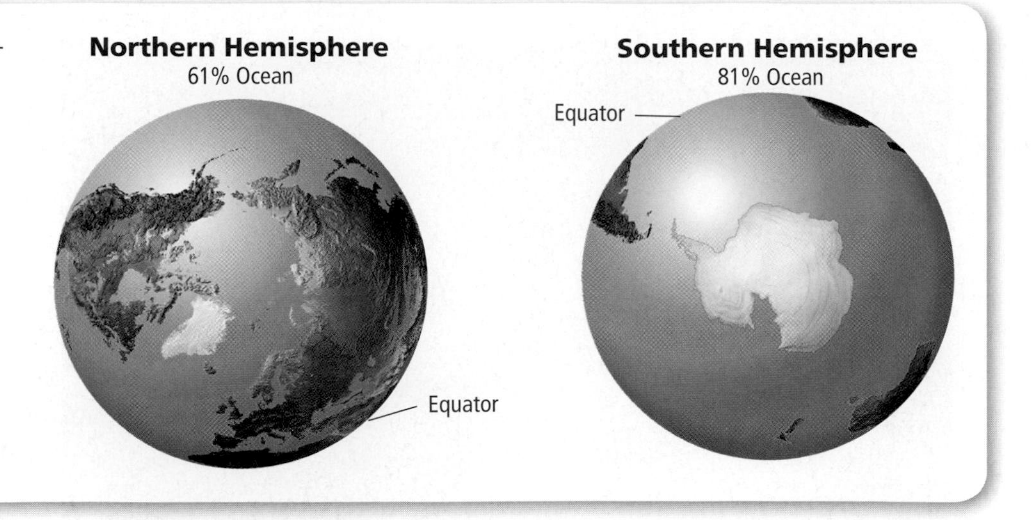

The blue planet Earth is known as the blue planet for good reason—approximately 71 percent of its surface is covered by oceans. The average depth of these oceans is 3800 m. Earth's landmasses are like huge islands, almost entirely surrounded by water. Because most landmasses are in the northern hemisphere, oceans cover only 61 percent of the surface there. However, 81 percent of the southern hemisphere is covered by water. **Figure 15.6** shows the distribution of water in the northern and southern hemispheres. Note that all the oceans are one vast, interconnected body of water. They have been divided into specific oceans and seas largely because of historic and geographic considerations.

Sea level Global **sea level,** which is the level of the oceans' surfaces, has risen and fallen by hundreds of meters in response to melting ice during warm periods and expanding glaciers during ice ages. Other processes that affect sea level are tectonic forces that lift or lower portions of Earth's crust. A rising seafloor causes a rise in sea level, while a sinking seafloor causes sea level to drop. **Figure 15.7** shows that sea level rose at a rate of about 3 mm per year between 1994 and 2004. Scientists hypothesize that this rise in sea level is related to water that has been released by the melting of glaciers and thermal expansion of the ocean due to warming.

■ **Figure 15.7** Scientists at NASA used floats and satellites to collect data on sea level changes over the period 1994 to 2004.
Infer ***What is a possible cause for rising sea level over this period?***

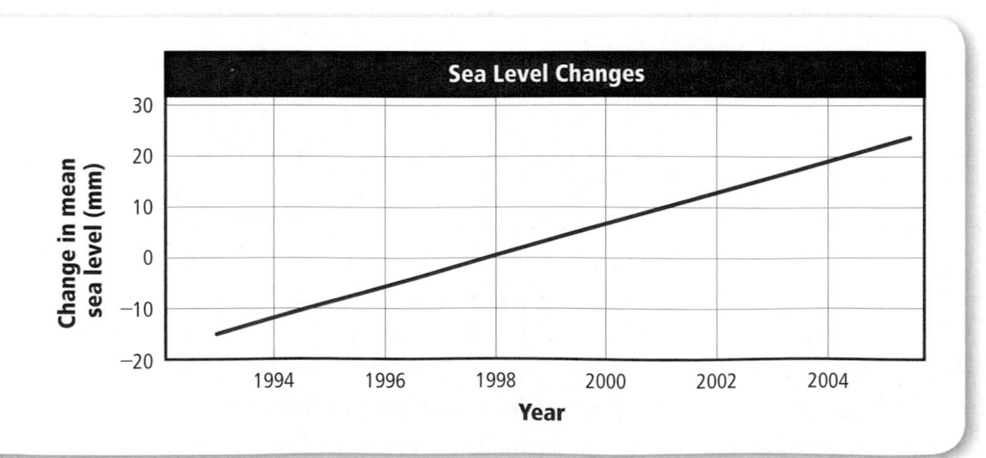

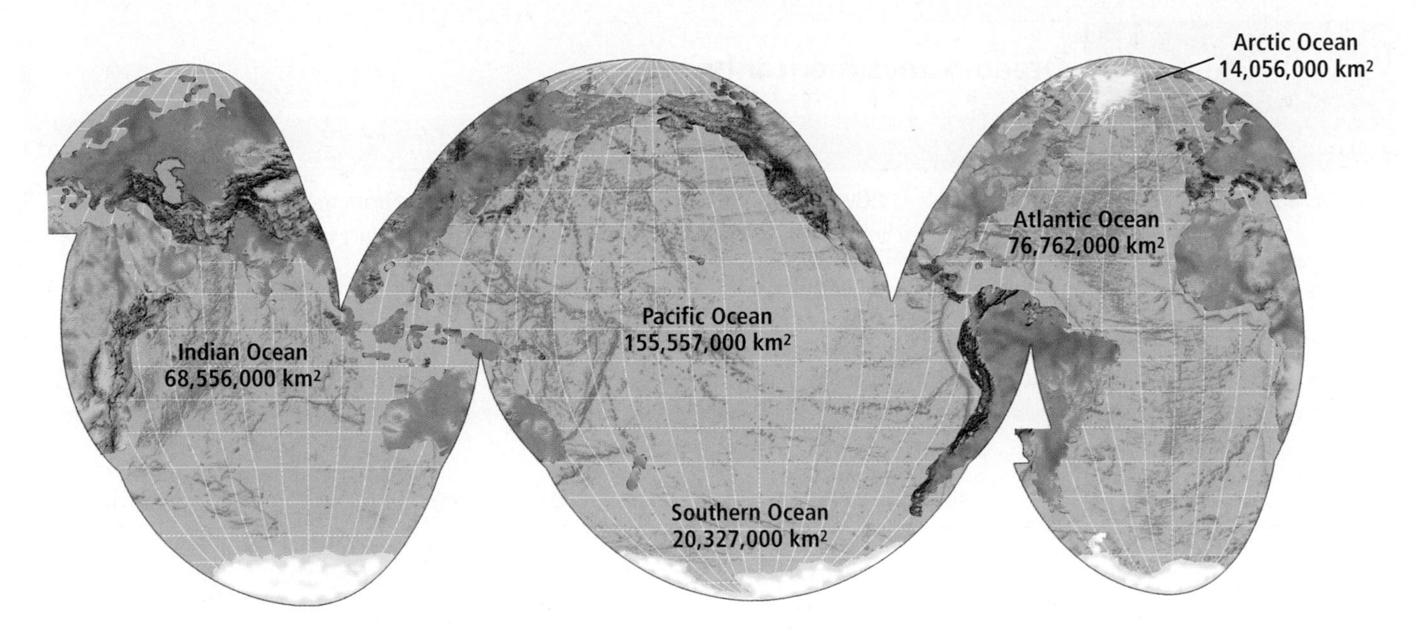

■ **Figure 15.8** The Pacific, Atlantic, and Indian Oceans stretch from Antarctica to the north. The smaller Arctic Ocean and Southern Ocean are located near the north and south poles respectively.

Major oceans As **Figure 15.8** shows, there are three major oceans: the Pacific, the Atlantic, and the Indian. The Pacific Ocean is the largest. Containing roughly half of Earth's seawater, it is larger than all of Earth's landmasses combined. The second-largest ocean, the Atlantic, extends for more than 20,000 km from Antarctica to the Arctic Circle. North of the Arctic Circle, the Atlantic Ocean is often referred to as the Arctic Ocean. The third-largest ocean, the Indian, is located mainly in the southern hemisphere. The storm-lashed region surrounding Antarctica, south of about 50° south latitude, is known as the Southern Ocean.

Reading Check **Identify** the largest ocean.

Polar oceans The Arctic and Southern oceans are covered by vast expanses of sea ice, particularly during the winter. In summer, the ice breaks up somewhat. Because ice is less dense than water, it floats. When sea-ice crystals first form, an ice-crystal slush develops at the surface of the water. The thickening ice eventually solidifies into individual round pieces called pancake ice, shown in **Figure 15.9.** Eventually, these pieces of pancake ice thicken and freeze into a continuous ice cover called pack ice. In the coldest parts of the Arctic and Southern oceans, there is no summer thaw, and the pack ice is generally several meters thick. In the winter, the pack-ice cover can be more than 1000 km wide.

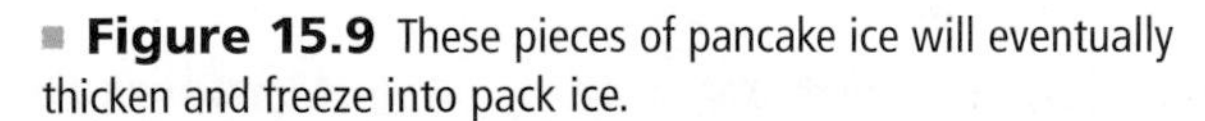

■ **Figure 15.9** These pieces of pancake ice will eventually thicken and freeze into pack ice.

Table 15.1 Ocean-Atmospheric Interactions

Example	Description
Oceans are a source of atmospheric oxygen.	Fifty percent of oxygen in the atmosphere comes from marine phytoplankton, which release oxygen into surface waters as a product of photosynthesis.
Oceans are a reservoir for carbon dioxide.	When cold, dense surface water in polar oceans sinks, dissolved carbon dioxide moves to the bottom of the ocean.
Oceans are a source of heat and moisture.	Warm ocean water in equatorial regions heats the air above it, fueling hurricanes.

Ocean and atmospheric interaction Oceans provide moisture and heat to the atmosphere and influence large-scale circulation patterns. In Chapter 13, you learned that warm ocean water energizes tropical cyclones, influences the position and strength of jet streams, and plays a role in El Niño events.

Oceans are also a vast reservoir of carbon dioxide. Dissolved carbon dioxide in surface waters sinks in water masses to the deep ocean, returning to the surface hundreds of years later. Without this natural uptake by the ocean, the accumulation of carbon dioxide in the atmosphere would be much larger than currently observed. There is also an uptake of carbon dioxide by phytoplankton during photosynthesis in the sunlit surface ocean. In the process, carbon is stored in the ocean and excess oxygen is released to the atmosphere to make Earth habitable. **Table 15.1** summarizes some of the interactions between oceans and the atmosphere.

Section 15.1 Assessment

Section Summary

- Scientists use many different instruments to collect and analyze data from oceans.
- Scientists have several ideas as to where the water in Earth's oceans originated.
- A large portion of Earth's surface is covered by ocean.
- Earth's oceans are the Pacific, the Atlantic, the Indian, the Arctic, and the Southern.

Understand Main Ideas

1. MAIN Idea **State** how much of Earth is covered by oceans. How is ocean water distributed over Earth's surface?
2. **Describe** two tools scientists use to collect data about oceans.
3. **Relate** What evidence indicates that oceans formed early in Earth's geologic history?
4. **Specify** Where did the water in Earth's early oceans originate?

Think Critically

5. **Predict** some possible consequences of rising sea level.
6. **Suggest** A recent study showed a 30 percent decrease in phytoplankton concentrations in northern oceans over the last 25 years. How might a significant decrease in marine phytoplankton affect atmospheric levels of oxygen and carbon dioxide?

MATH in Earth Science

7. Calculate the distance to the ocean floor if a sonar signal takes 6 s to return to a ship's receiver.

Section 15.2

Objectives

- **Identify** the chemical and physical properties of seawater.
- **Illustrate** ocean layering.
- **Describe** the formation of deepwater masses.

Review Vocabulary

feldspar: a rock-forming mineral that contains silicon and oxygen

New Vocabulary

salinity
estuary
temperature profile
thermocline

Seawater

MAIN Idea **Oceans have distinct layers of water masses that are characterized by temperature and salinity.**

Real-World Reading Link A person's accent can reveal a lot about his or her place of origin. Similarly, the temperature and salinity of water masses can often reveal when and where the water was first formed on the sea surface.

Chemical Properties of Seawater

Ocean water contains dissolved gases, including oxygen and carbon dioxide, and dissolved nutrients such as nitrates and phosphates. Chemical profiles of seawater vary based on both location and depth, as shown in **Figure 15.10.** Factors that influence the amount of a substance in an area of ocean water include wave action, vertical movements of water, and biological activity.

Figure 15.10 shows that oxygen levels are high at the surface in both the Atlantic and Pacific oceans. This occurs in part because oxygen is released by surface-dwelling photosynthetic organisms. Silica levels for both oceans are also shown in **Figure 15.10.** Because many organisms remove silica from ocean water and use it to make shells, silica levels near the surface are usually low. Silica levels usually increase with depth because decaying organisms sink to the ocean bottom, returning silica to the water.

Salinity The measure of the amount of dissolved salts in seawater is **salinity.** Oceanographers express salinity as grams of salt per kilogram of water, or parts per thousand (ppt). The total salt content of seawater averages 35 ppt, or 3.5 percent. The most abundant salt in seawater is sodium chloride. Other salts in seawater are chlorides and sulfates of magnesium, potassium, and calcium.

Figure 15.10 Concentrations of dissolved gases and nutrients in seawater, measured in micromolars (μM), vary by location and depth.
Examine *How do oxygen levels differ between the North Atlantic and North Pacific oceans?*

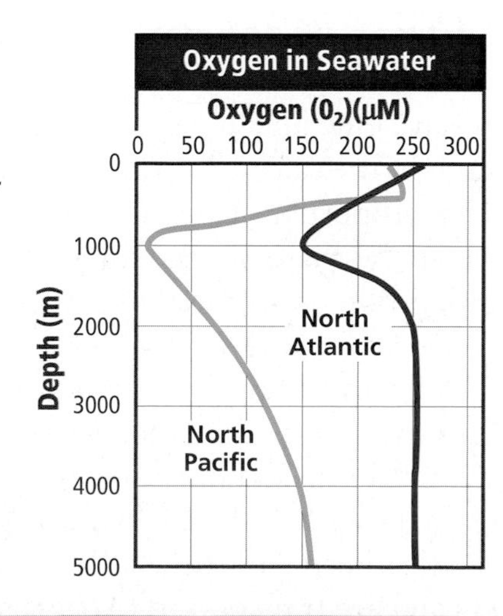

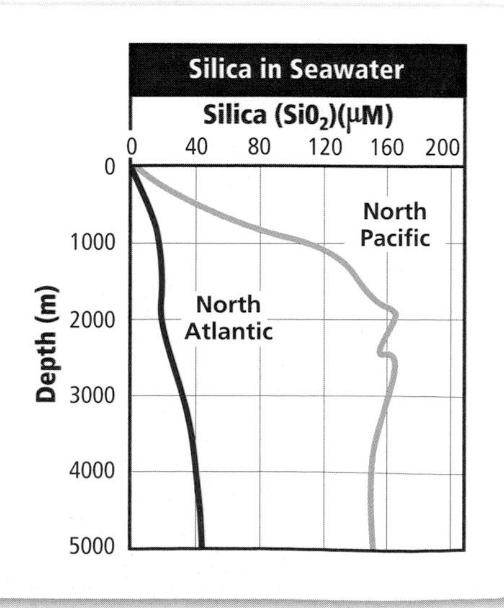

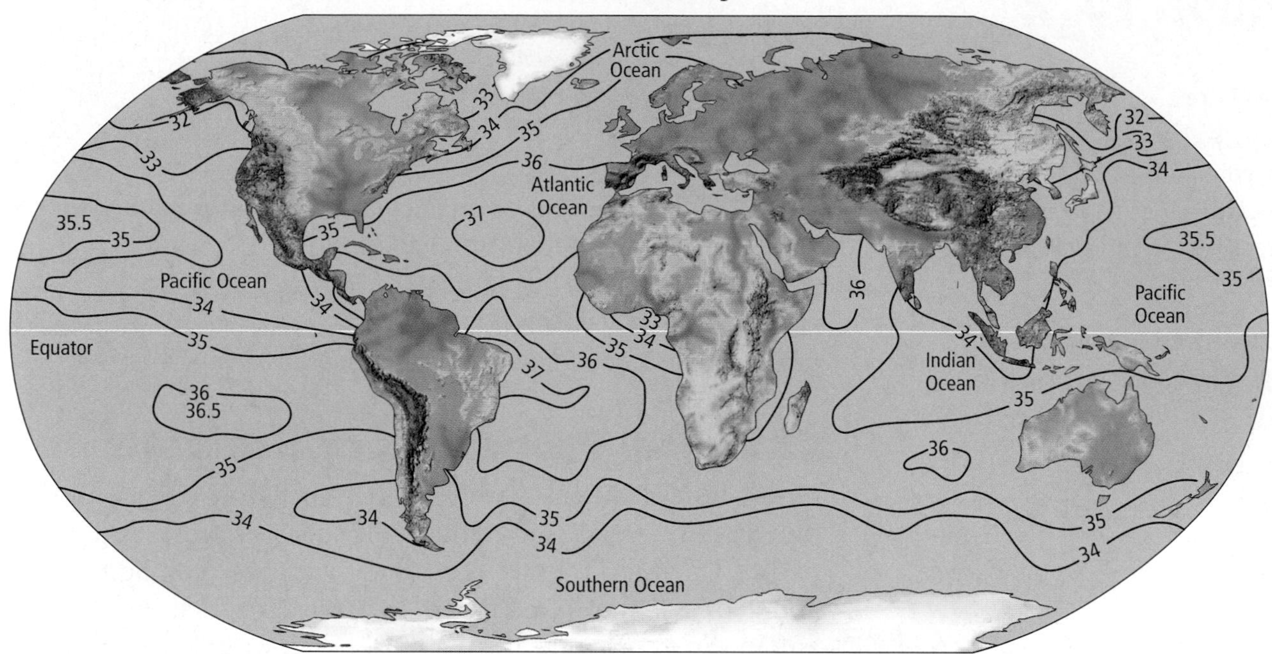

■ **Figure 15.11** Ocean salinity varies from place to place. High salinity is common in areas with high rates of evaporation. Low salinity often occurs in estuaries.

Variations in salinity Although the average salinity of the oceans is 35 ppt, actual salinity varies from place to place, as shown in **Figure 15.11.** In subtropical regions where rates of evaporation exceed those of precipitation, salt left behind by the evaporation of water molecules accumulates in the surface layers of the ocean. There, salinity can be as high as 37 ppt. In equatorial regions where precipitation is abundant, salinity is lower. Even lower salinities of 32 or 33 ppt occur in polar regions where seawater is diluted by melting sea ice. The lowest salinity often occurs where large rivers empty into the oceans, creating areas of water called **estuaries.** Even though salinity varies, the relative proportion of major types of sea salts is constant because all ocean water continually intermingles throughout Earth's oceans.

Reading Check **Describe** the factors that affect the salinity of water.

Sources of sea salt Geologic evidence indicates that the salinity of ancient seas was not much different from that of today's oceans. One line of evidence is based on the proportion of magnesium in the calcium-carbonate shells of some marine organisms. That proportion depends on the overall salinity of the water in which the shells formed. Present-day shells contain about the same proportion of magnesium as similar shells throughout geologic time.

Sources of sea salts have also stayed the same over time. Sulfur dioxide and chlorine, gases released by volcanoes, dissolve in water, forming sulfate and chlorine ions. Most of the other ions in seawater, including sodium and calcium, come from the weathering of crustal rocks, such as feldspars. Iron and magnesium come from the weathering of rocks rich in these elements. These ions enter rivers and are transported to oceans, as shown in **Figure 15.12.**

Visualizing the Salt Cycle

Figure 15.12 Salts are added to seawater by volcanic eruptions and by the weathering and erosion of rocks. Salts are removed from seawater by biological processes and the formation of evaporites. Also, wind carries salty droplets inland.

Ions, such as sodium, calcium, iron, and magnesium, enter oceans in river runoff as the weathering of rocks releases them.

Gases from volcanic eruptions contain water vapor, chloride, and sulfur dioxide. These gases dissolve in water and form the chloride and sulfate ions in seawater.

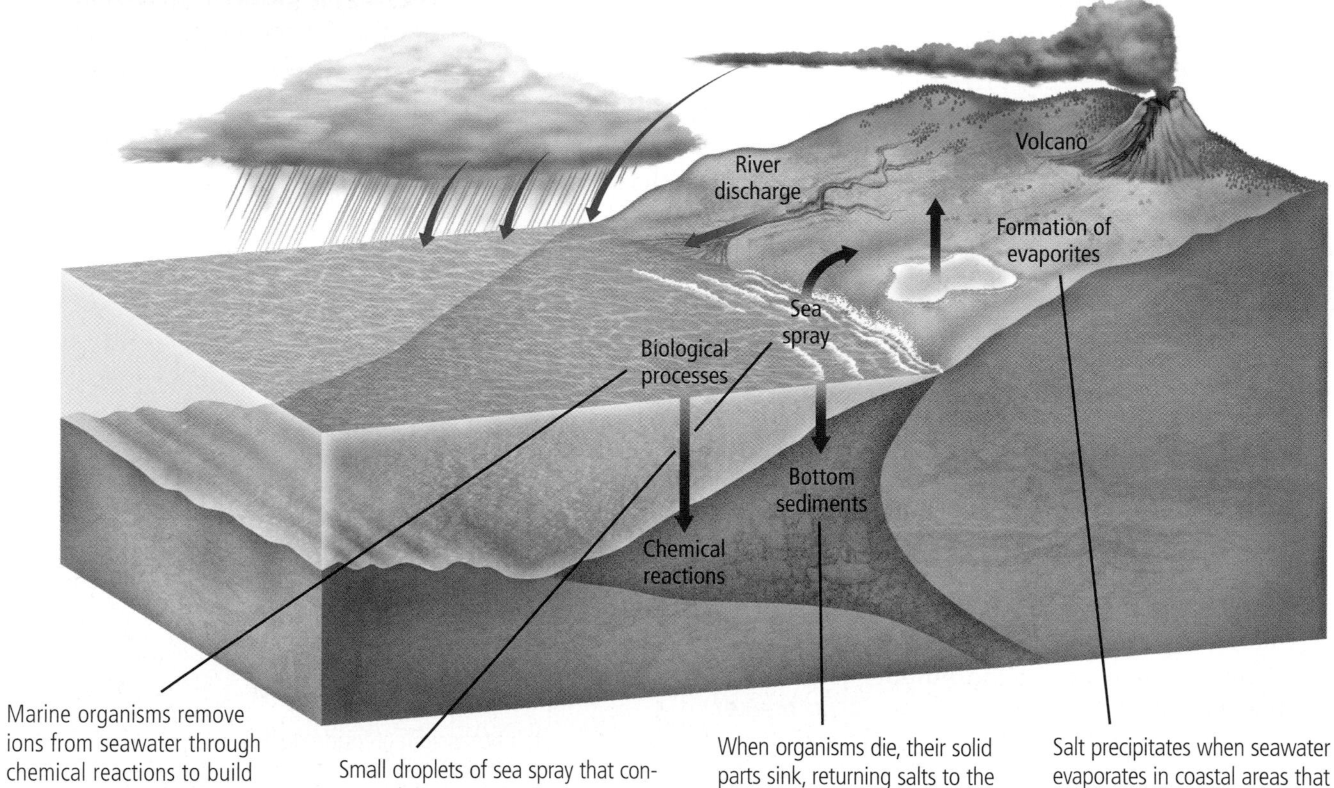

Marine organisms remove ions from seawater through chemical reactions to build their shells, bones, and teeth.

Small droplets of sea spray that contain salt are carried inland by winds.

When organisms die, their solid parts sink, returning salts to the bottom sediments.

Salt precipitates when seawater evaporates in coastal areas that are hot and dry.

Concepts In Motion To explore more about the salt cycle, visit glencoe.com.

Earth Science Online

Concepts In Motion
Interactive Table To explore more about salts in the ocean, visit glencoe.com.

Table 15.2 Removal of Sea Salts

Process	Description	Example
Evaporate formation	Solid salt is left behind when water evaporates from concentrated solutions of salt water.	
Biological activity	Organisms remove calcium ions from water to build shell, bones, and teeth.	

Removal of sea salts Although salt ions are continuously added to seawater, salinity does not increase because salts are also continuously removed. **Table 15.2** describes two processes through which sea salts are removed. Recall from Chapter 6 that evaporites form when water evaporates from concentrated solutions. In arid coastal regions, water evaporates from seawater and leaves solid salt behind. Marine organisms remove ions from seawater to build shells, bones, and teeth. As organisms die, their solid parts accumulate on the seafloor and become part of bottom sediments. Winds can also pick up salty droplets from breaking waves and deposit the salt further inland. The existing salinity of seawater represents a balance between the processes that remove salts and those that add them.

MiniLab

Model Seawater

What is the chemical composition of seawater? Determine the chemical composition of seawater using the ingredients listed in the table. The salinity of seawater is commonly measured in parts per thousand (ppt).

Procedure

1. Read and complete the lab safety form.
2. Carefully measure the **ingredients** listed in the table on the right and combine them in a **large beaker.**
3. Add 965.57 g of **distilled water** and mix.

Analysis

1. **Calculate** How many grams of solution do you have? What percentage of this solution is made up of salts?
2. **Apply** What is the salinity of your solution in ppt?
3. **Infer** how your solution differs from actual seawater.

Ingredient	Amount
Sodium chloride ($NaCl$)	23.48 g
Magnesium chloride ($MgCl_2$)	4.98 g
Sodium sulfate (Na_2SO_4)	3.92 g
Calcium chloride ($CaCl_2$)	1.10 g
Potassium chloride (KCl)	0.66 g
Sodium bicarbonate ($NaHCO_3$)	0.19 g
Potassium bromide (KBr)	0.10 g

Physical Properties of Seawater

The presence of various salts causes the physical properties of seawater to be different from those of freshwater.

Density Freshwater has a maximum density of 1.00 g/cm^3. Because salt ions add to the overall mass of the water in which they are dissolved, they increase the density of water. Seawater is therefore more dense than freshwater, and its density increases with salinity. Temperature also affects density—cold water is more dense than warm water. Because of salinity and temperature variations, the density of seawater ranges from about 1.02 g/cm^3 to 1.03 g/cm^3. These variations might seem small, but they are significant. They affect many oceanic processes, which you will learn about in Chapter 16.

 **Reading Check** **Explain** how temperature and salinity affect the density of seawater.

Freezing point Variations in salinity also cause the freezing point of seawater to be somewhat lower than that of freshwater. Freshwater freezes at 0°C. Because salt ions interfere with the formation of the crystal structure of ice, the freezing point of seawater is –2°C.

Absorption of light If you have ever swum in a lake, you might have noticed that the intensity of light decreases with depth. The water might be clear, but if the lake is deep, the bottom waters will be dark. Water absorbs light, which gives rise to another physical property of oceans—darkness. In general, light penetrates only the upper 100 m of seawater. Below that depth, all is darkness.

Figure 15.13 illustrates how light penetrates ocean water. Notice that red light does not penetrate as far as blue light. Red objects, such as the giant red shrimp shown in **Figure 15.13,** appear black below a certain depth and other reflecting objects in the water appear green or blue. Although some fading blue light can reach depths of a few hundred meters, light sufficient for photosynthesis exists only in the top 100 m of the ocean. In the darkness of the deep ocean, some organisms, including some fishes, shrimps, and crabs, are blind. Other organisms attract prey by producing light, called bioluminescence, through a chemical reaction.

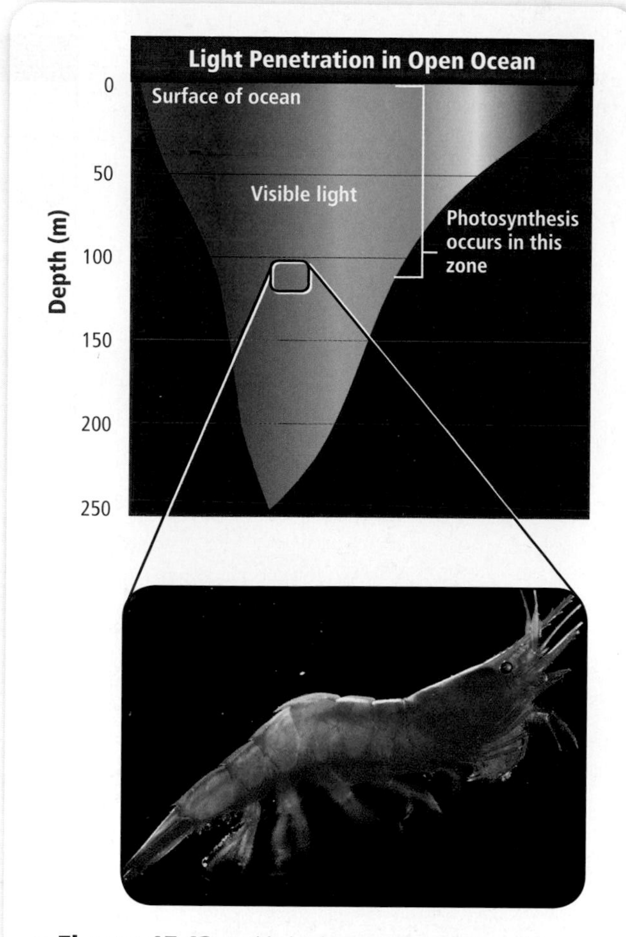

■ **Figure 15.13** Red light does not penetrate as far as blue light in the ocean. Marine organisms that are some shades of red, such as deep-sea shrimp, appear black below a depth of 10 m. This helps them escape predators.
Identify ***To what depth does blue light penetrate ocean water?***

■ **Figure 15.14** Ocean water temperatures decrease with depth. Tropical areas have warmer ocean surface temperatures than do temperate or polar areas.

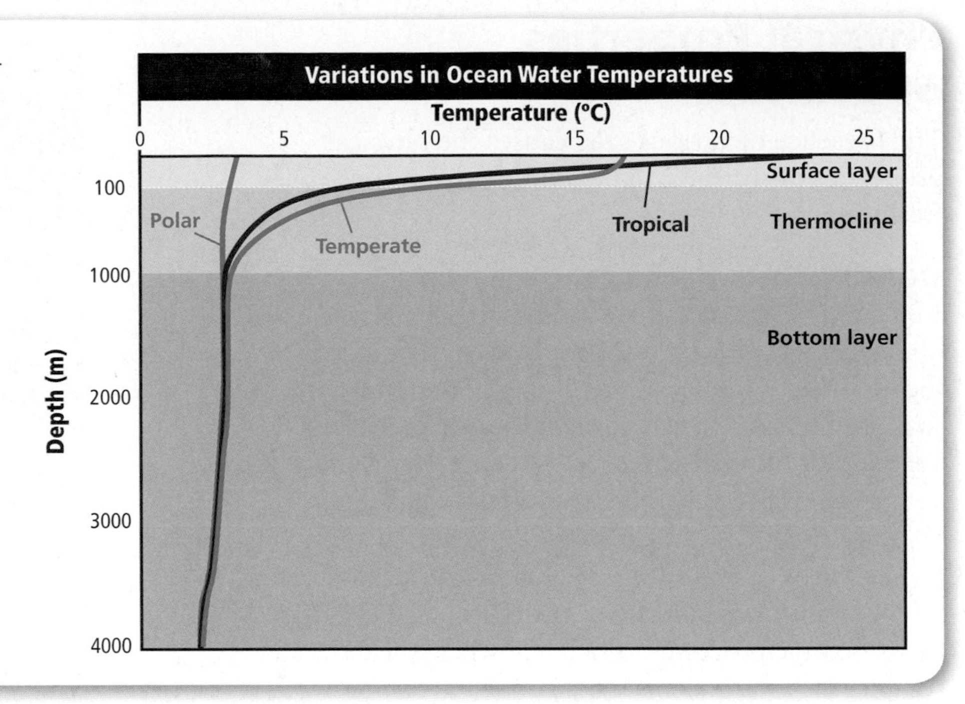

Ocean Layering

Ocean surface temperatures range from −2°C in polar waters to 30°C in equatorial regions, with the average surface temperature being 15°C. Ocean water temperatures, however, decrease significantly with depth. Thus, deep ocean water is always cold, even in tropical oceans.

Temperature profiles **Figure 15.14** shows typical ocean **temperature profiles,** which plot changing water temperatures against depth. Such profiles vary, depending on location and season. In the temperature profiles shown here, beneath roughly 100 m, temperatures decrease continuously with depth to around 4°C at 1000 m. The dark waters below 1000 m have fairly uniform temperatures of less than 4°C. Based on these temperature variations, the ocean can be divided into three layers, also shown in **Figure 15.14.** The first is a relatively warm, sunlit surface layer approximately 100 m thick. Notice that tropical areas have warmer surface temperatures than temperate or polar areas. Under the surface layer is a transitional layer known as the **thermocline,** which is characterized by rapidly decreasing temperatures with depth. The bottom layer is cold and dark with temperatures near freezing. Both the thermocline and the warm surface layer are absent in polar seas, where water temperatures are cold from top to bottom. In general, ocean layering is caused by density differences. Because cold water is more dense than warm water, cold water sinks to the bottom, while less-dense, warm water is found near the ocean's surface.

VOCABULARY

ACADEMIC VOCABULARY

Variation

the range in which a factor changes

The variation in temperature in New York was a shock for the person from California.

 **Reading Check** **Describe** the three main layers of water in oceans.

Water Masses

The temperature of the bottom layer of ocean water is near freezing. This is true even in tropical oceans, where surface temperatures are warm. Where does all this cold water come from?

Deepwater masses Cold water comes from Earth's polar seas. Recall that high salinity and cold temperatures cause seawater to become more dense. Study **Figure 15.15,** which shows how deepwater masses are formed. When seawater freezes during the arctic or antarctic winter, sea ice forms. Because salt ions are not incorporated into the growing ice crystals, they accumulate beneath the ice. Consequently, the cold water beneath the ice becomes saltier and more dense than the surrounding seawater, and this saltier water sinks. This salty water then migrates toward the equator as a cold, deepwater mass along the ocean floor. Other cold, deepwater masses form when surface currents in the ocean bring relatively salty midlatitude or subtropical waters into polar regions. In winter, these waters become colder and denser than the surrounding polar surface waters, and thus, they sink.

Three water masses account for most of the deepwater masses in the oceans—Antarctic Bottom Water, North Atlantic Deep Water, and Antarctic Intermediate Water. Antarctic Bottom Water forms when antarctic seas freeze during the winter. With temperatures below 0°C, this deepwater mass is the coldest and densest in all the oceans, as shown in **Figure 15.16** on page 420. North Atlantic Deep Water forms in a similar manner offshore from Greenland. Antarctic Bottom Water is colder and denser than North Atlantic Deep Water, so it sinks below it.

 **Reading Check** **Identify** the three water masses that make up most of the deepwater masses in the oceans.

■ **Figure 15.15** Dense polar water sinks, producing a deepwater mass.
Explain ***the relationship between the density of water and the formation of deepwater masses.***

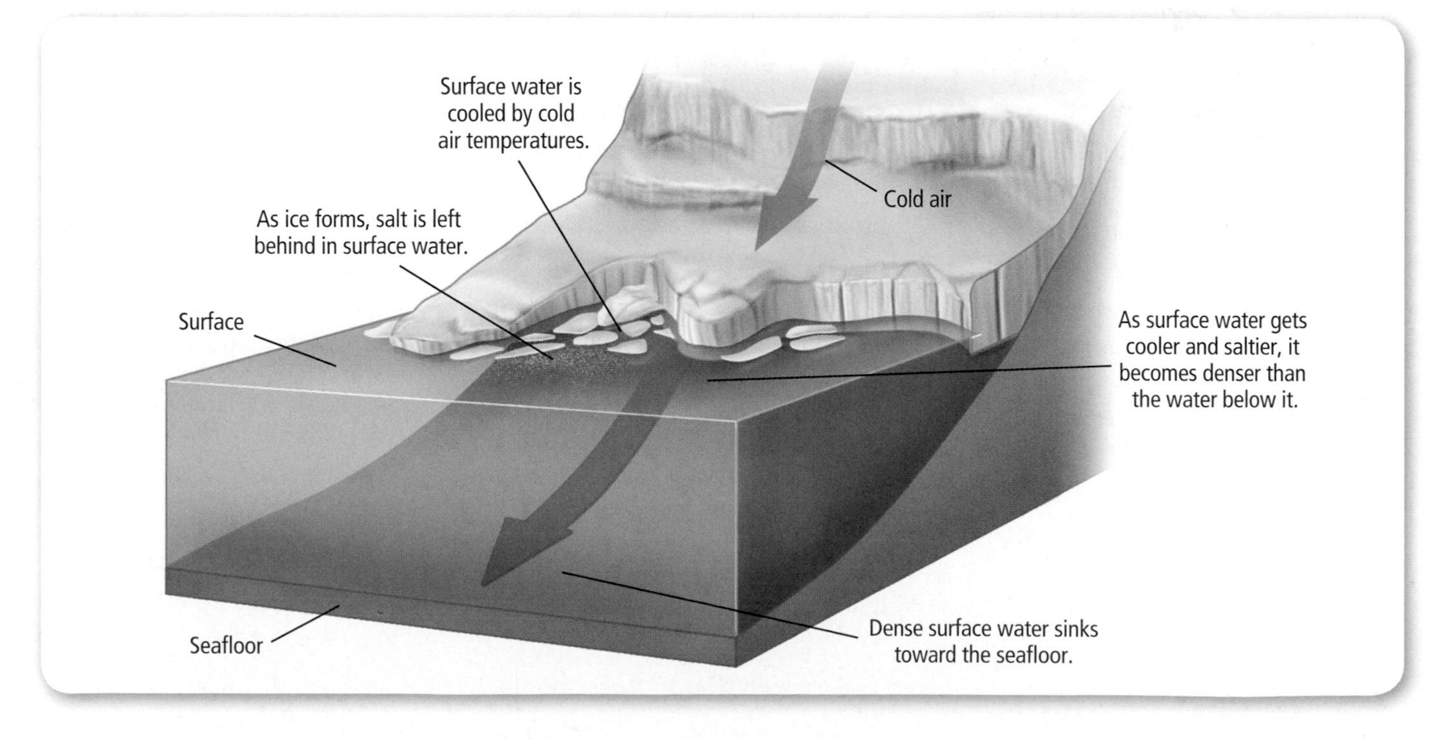

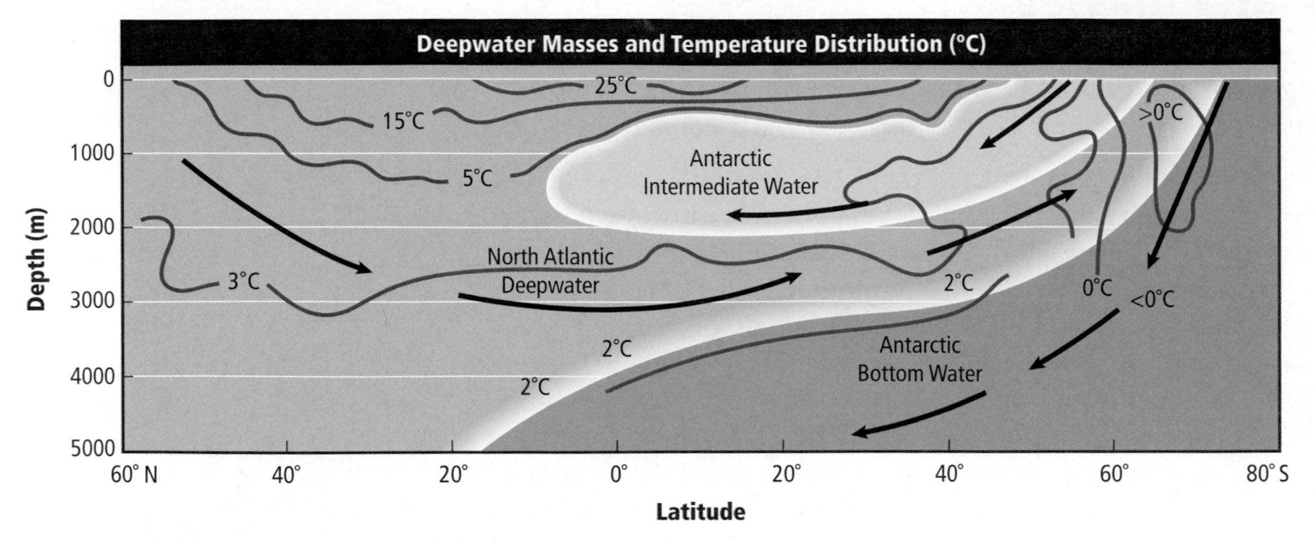

■ **Figure 15.16** Antarctic Bottom Water is the densest and coldest deepwater mass. It is overridden by the slightly warmer and less dense North Atlantic Deep Water. Antarctic Intermediate Water is still warmer and less dense, and thus it overrides the other two deepwater masses.

Intermediate water masses Antarctic Intermediate Water, shown in **Figure 15.16,** forms when the relatively salty waters near Antarctica decrease in temperature and sink during winter. Because Antarctic Intermediate Water is slightly warmer and less dense than North Atlantic Deep Water, it does not sink as deep as the other two deepwater masses. While the Atlantic Ocean contains all three major deepwater masses, the Indian and Pacific Oceans contain only the two Antarctic deepwater masses. In Section 15.3, you will learn about other water movements in the ocean.

Section 15.2 Assessment

Section Summary

- Ocean water contains dissolved gases, nutrients, and salts.
- Salts are added to and removed from oceans through natural processes.
- Properties of ocean water, including temperature and salinity, vary with location and depth.
- Many of the oceans' deepwater masses sink from the surface of polar oceans.

Understand Main Ideas

1. MAIN Idea **Compare and contrast** North Atlantic Deep Water and Antarctic Bottom Water.
2. **Identify** What factors affect the chemical properties of seawater?
3. **Illustrate** the three layers into which ocean water is divided based on temperature.
4. **Sequence** the steps involved in the formation of deepwater masses.

Think Critically

5. **Hypothesize** Which is more dense, cold freshwater or warm seawater?
6. **Predict** what color a yellow fish would appear to be in ocean water depths greater than about 50 m.

MATH in Earth Science

7. If the density of a sample of seawater is 1.02716 g/mL, calculate the mass of 4.0 mL of the sample.

Earth Science Online Self-Check Quiz glencoe.com

Objectives

- **Describe** the physical properties of waves.
- **Explain** how tides form.
- **Compare and contrast** various ocean currents.

Review Vocabulary

prevailing westerlies: global wind system located between 30°N and 60°N that moves from the west to the east toward each pole

New Vocabulary

wave
crest
trough
breaker
tide
spring tide
neap tide
surface current
upwelling
density current

Concepts In Motion
Interactive Figure To see an animation of waves, visit glencoe.com.

Ocean Movements

MAIN Idea Waves and currents drive the movements of ocean water and lead to the distribution of heat, salt, and nutrients from one region of the ocean to another.

Real-World Reading Link Think about the last time you watched a sporting event and the audience did "the wave" to cheer players by standing up and sitting down at the right time. Even though the audience does not move around the stadium, the wave does. The same idea applies to ocean waves.

Waves

Oceans are in constant motion. Their most obvious movement is that of waves. A **wave** is a rhythmic movement that carries energy through space or matter—in this case, ocean water. Ocean waves are generated mainly by wind blowing over the water's surface. In the open ocean, a typical wave has the characteristics shown in **Figure 15.17.** The highest point of a wave is the **crest,** and the lowest point is the **trough.** The vertical distance between crest and trough is the wave height, and the horizontal crest-to-crest distance is the wavelength. As energy is added, both the wavelength and speed increase. Thus, longer waves travel faster than shorter waves.

As an ocean wave passes, the water moves up and down in a circular pattern and returns to its original position, as shown in **Figure 15.17.** Only the energy moves steadily forward. The water itself moves in circles until the energy passes, but it does not move forward. The wavelength also determines the depth to which the wave disturbs the water. That depth, called the wave base, is equal to half the wavelength.

Figure 15.17 Wave characteristics include wave height, wavelength, crest, and trough. In an ocean wave, water moves in circles that decrease in size with depth. At a depth equal to half the wavelength, water movement essentially stops.

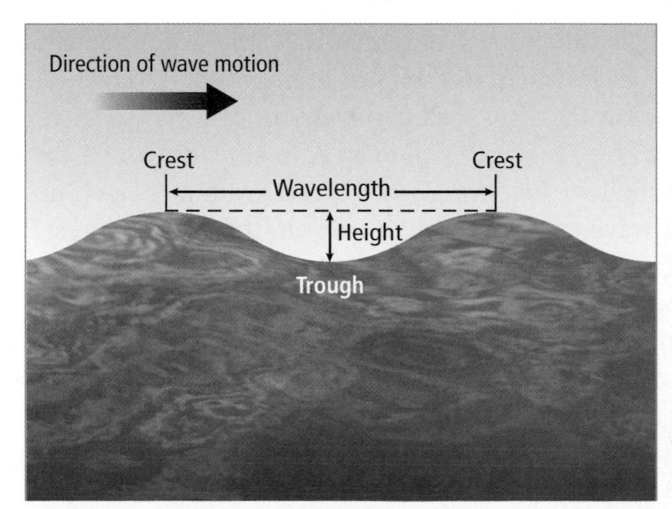

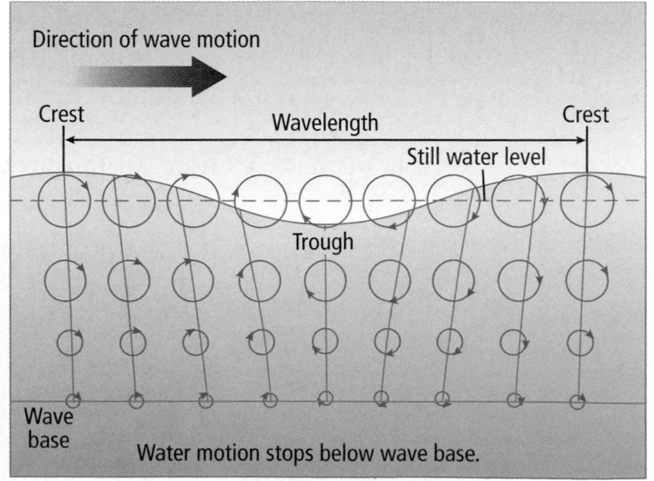

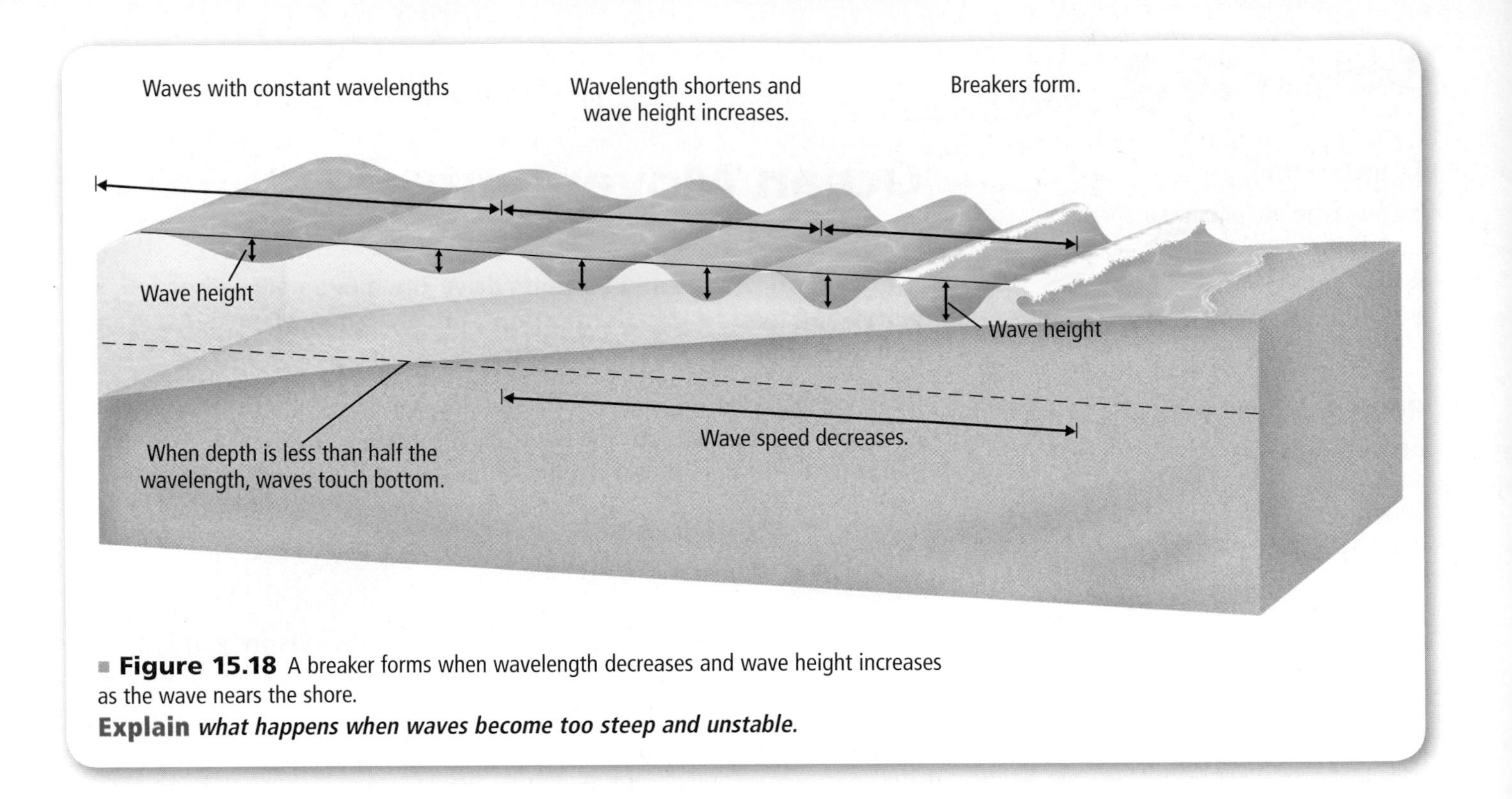

■ **Figure 15.18** A breaker forms when wavelength decreases and wave height increases as the wave nears the shore.
Explain ***what happens when waves become too steep and unstable.***

Wave height Wave height depends on three factors: fetch, wind duration, and wind speed. Fetch refers to the expanse of water that the wind blows across. The longer the wind can blow without being interrupted (wind duration) over a large area of water (fetch), the larger the waves will be. Also, the faster the wind blows (wind speed) for a longer period of time over the ocean, the larger the waves will be. The highest waves are usually found in the Southern Ocean, an area over which strong winds blow almost continuously. Waves created by large storms can also be much higher than average. For instance, hurricanes can generate waves more than 10 m high.

FOLDABLES Incorporate information from this section into your Foldable.

Reading Check **Identify** the three factors that affect the height of a wave.

Breaking waves Study **Figure 15.18.** It shows that as ocean waves reach the shallow water near shorelines, the water depth eventually becomes less than one-half of their wavelength. The shallow depth causes changes to the movement of water particles at the base of the wave. This causes the waves to slow down. As the water becomes shallow, incoming wave crests gradually catch up with the slower wave crests ahead. As a result, the crest-to-crest wavelength decreases. The incoming waves become higher, steeper, and unstable, and their crests collapse forward. Collapsing waves are called **breakers.** The formation of breakers is also influenced by the motion of wave crests, which overrun the troughs. The collapsing crests of breakers, like the one shown in **Figure 15.19,** move at high speeds toward shore and play a major role in shaping shorelines. You will learn more about breakers and shoreline processes in Chapter 16.

■ **Figure 15.19** As waves move into shallow water, breakers form.

Tides

Tides are the periodic rise and fall of sea level. The highest level to which water regularly rises is known as high tide, and the lowest level is called low tide. Because of differences in topography and latitude, the tidal range—the difference in height between high tide and low tide—varies from place to place. In the Gulf of Mexico, the tidal range is less than 1 m. In New England, it can be as high as 6 m. The greatest tidal range occurs in the Bay of Fundy between New Brunswick and Nova Scotia, Canada, where it is as much as 16.8 m. Generally, a daily cycle of high and low tides takes 24 hours and 50 minutes. Differences in topography and latitude cause three different daily tide cycles, as shown in **Figure 15.20.** Areas with semidiurnal cycles experience two high tides in about a 24-hour period. Areas with mixed cycles have one pronounced and one smaller high tide in about a 24-hour period. Areas with diurnal cycles have one high tide in about a 24-hour period.

Reading Check **Explain** the difference between semidiurnal tides and mixed tides.

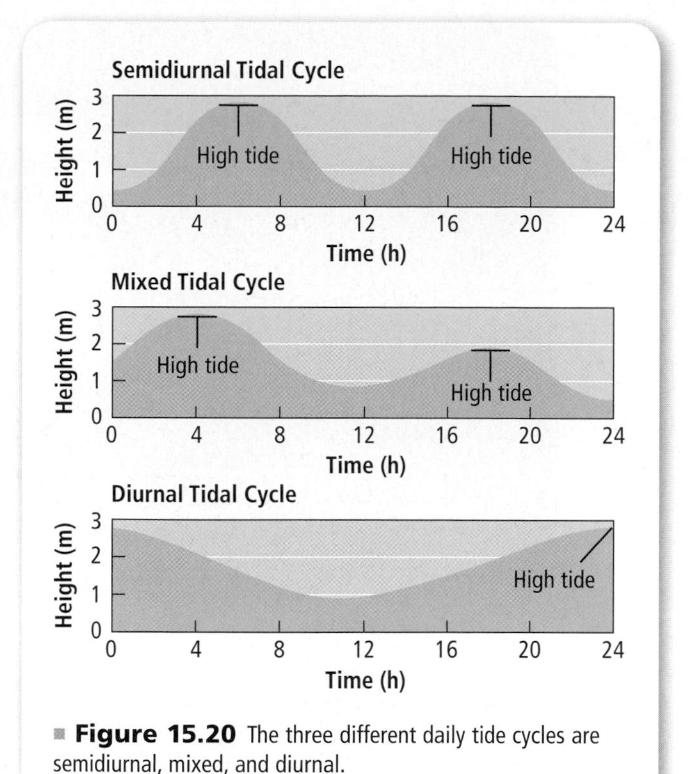

Figure 15.20 The three different daily tide cycles are semidiurnal, mixed, and diurnal.

DATA ANALYSIS LAB

Based on Real Data*

Graph Data

When does the tide come in? Tidal data is usually measured in hourly increments. The water levels shown in the data table were measured over a 24-hour period.

Data and Observations

Tidal Record			
Time (h)	Water Level (m)	Time (h)	Water Level (m)
00:00	2.11	13:00	1.70
01:00	1.79	14:00	1.37
02:00	1.33	15:00	1.02
03:00	0.80	16:00	0.68
04:00	0.36	17:00	0.48
05:00	0.10	18:00	0.50
06:00	0.03	19:00	0.69
07:00	0.20	20:00	1.11
08:00	0.55	21:00	1.58
09:00	0.99	22:00	2.02
10:00	1.45	23:00	2.27
11:00	1.74	24:00	2.30
12:00	1.80		

Think Critically

1. **Apply** Plot these water levels on a graph with time on the *x*-axis and water level on the *y*-axis.
2. **Estimate** the approximate times and water levels of high tides and low tides.
3. **Identify** the type of daily tidal cycle this area experiences.
4. **Determine** the tidal range for this area.
5. **Predict** the water level at the next high tide and estimate when it will occur.

*Data obtained from: The National Oceanic and Atmospheric Administration, Center for Operational Oceanographic Products and Services.

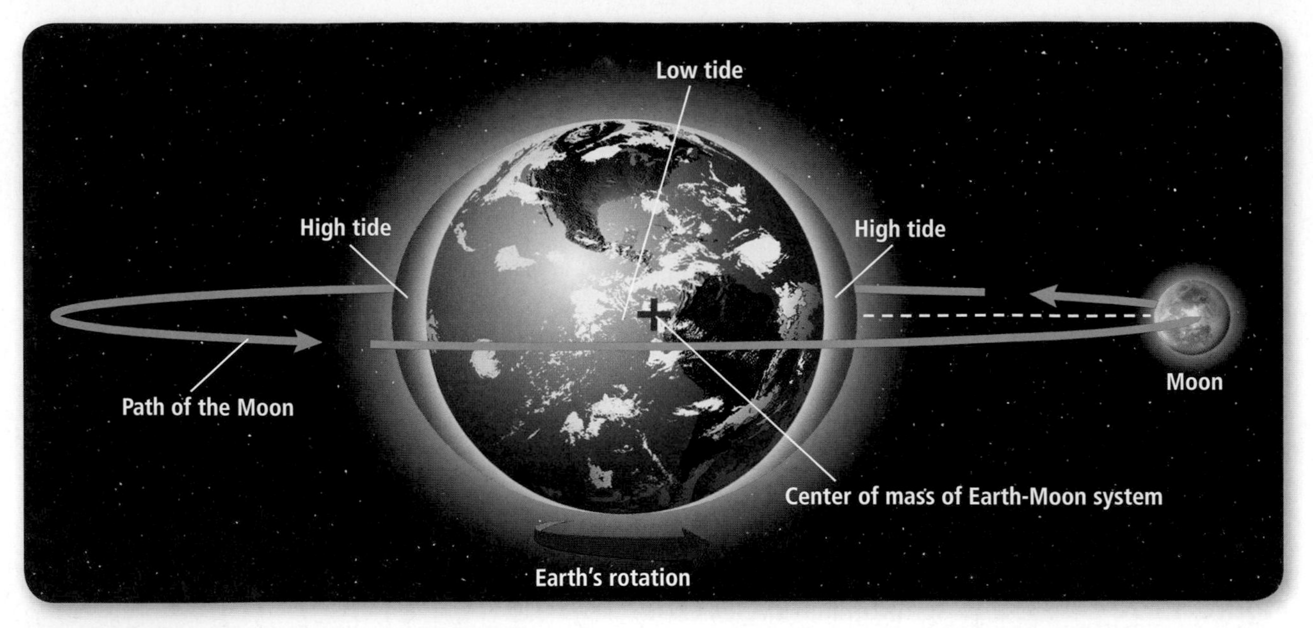

■ **Figure 15.21** The Moon and Earth revolve around a common center of gravity and experience unbalanced gravitational forces. These forces cause tidal bulges on opposite sides of Earth. (Note: *diagram is not to scale.*)

The Moon's influence The basic causes of tides are the gravitational attraction among Earth, the Moon, and the Sun, as well as the effect of Earth's rotation. Consider the Earth-Moon system. Both Earth and the Moon orbit a common center of gravity, shown as a red plus sign in **Figure 15.21.** As a result of their motions, both Earth and the Moon experience differing gravitational forces. These unbalanced forces generate tidal bulges on opposite sides of Earth. The gravitational effect of the Moon on Earth's oceans is similar to what happens to the liquid in a coffee cup inside a car as the car goes around a curve. The liquid sloshes toward the outside of the curve.

The Sun's influence The gravitational attraction between Earth and the Sun, and Earth's orbital motion around the Sun influences tides. However, even though the Moon is much smaller than the Sun, lunar tides are more than twice as high as those caused by the Sun because the Moon is much closer to Earth. Consequently, Earth's tidal bulges are aligned with the Moon.

VOCABULARY
SCIENCE USAGE V. COMMON USAGE
Attraction
Science usage: a force acting between particles of matter, tending to pull them together

Common usage: something that pulls people in by appealing to their tastes

Depending on the phases of the Moon, solar tides can either enhance or diminish lunar tides, as illustrated in **Figure 15.22.** Notice in **Figure 15.22** that during both a full and a new moon, the Sun, the Moon, and Earth are all aligned. When this occurs, solar tides enhance lunar tides, causing high tides to be higher than normal and low tides to be lower than normal. The tidal range is highest during these times. These types of tides are called **spring tides.** Spring tides have a greater tidal range during the winter in the northern hemisphere, when Earth is closest to the Sun. Study **Figure 15.22** again. Notice that when there is a first- or third-quarter moon, the Sun, the Moon, and Earth form a right angle. When this occurs, solar tides diminish lunar tides, causing high tides to be lower and low tides to be higher than normal. The tidal range is lowest during these times. These types of tides are called **neap tides.** Spring and neap tides alternate every two weeks.

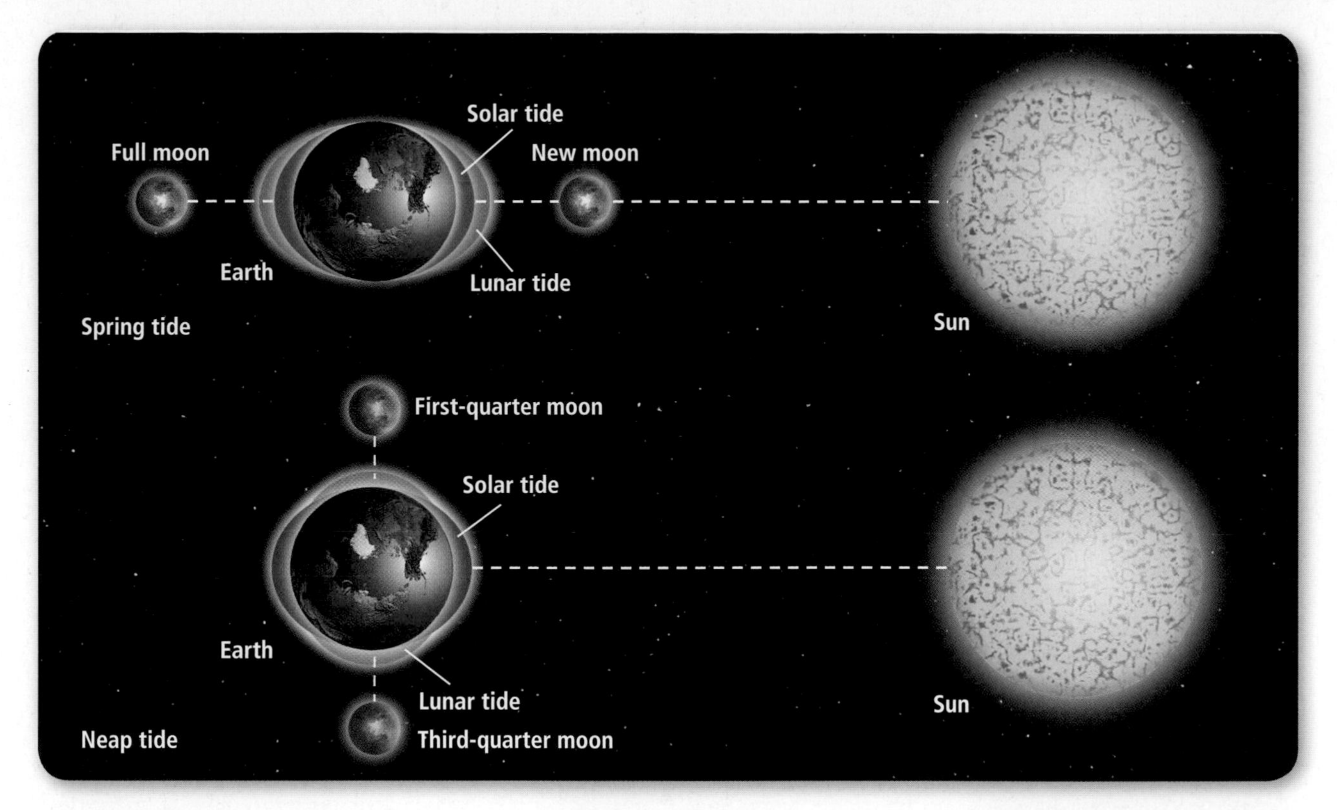

■ **Figure 15.22** Spring tides occur when the Sun, the Moon, and Earth are aligned. Neap tides occur when the Sun, the Moon, and Earth form a right angle. (Note: *diagram is not to scale.*)

Currents

Currents in the ocean can move horizontally or vertically. They can also move at the surface or deep in the ocean. Currents at the surface are usually generated by wind. Some currents are the result of tides. Deep-ocean currents usually result from differences in density between water masses.

Surface currents Mainly the top 100 to 200 m of the ocean experience **surface currents,** which can move at a velocity of about 100 km per day. Surface currents follow predictable patterns and are driven by Earth's global wind systems. Recall from Chapter 12 that, in the northern hemisphere, tropical trade winds blow from east to west. The resulting tropical ocean surface currents also flow from east to west. In northern midlatitudes, the prevailing westerlies and resulting ocean surface currents move from west to east. In northern polar regions, polar easterly winds push surface waters from east to west.

The direction of surface currents can also be affected by landforms, such as continents, as well as the Coriolis effect. Recall from Chapter 12 that the Coriolis effect deflects moving particles to the right in the northern hemisphere and to the left in the southern hemisphere.

Reading Check **Explain** how winds influence surface currents.

Gyres If Earth had no landmasses, the global ocean would have simple belts of easterly and westerly surface currents. Instead, the continents deflect ocean currents to the north and the south so that closed circular current systems, called gyres (JI urz), develop.

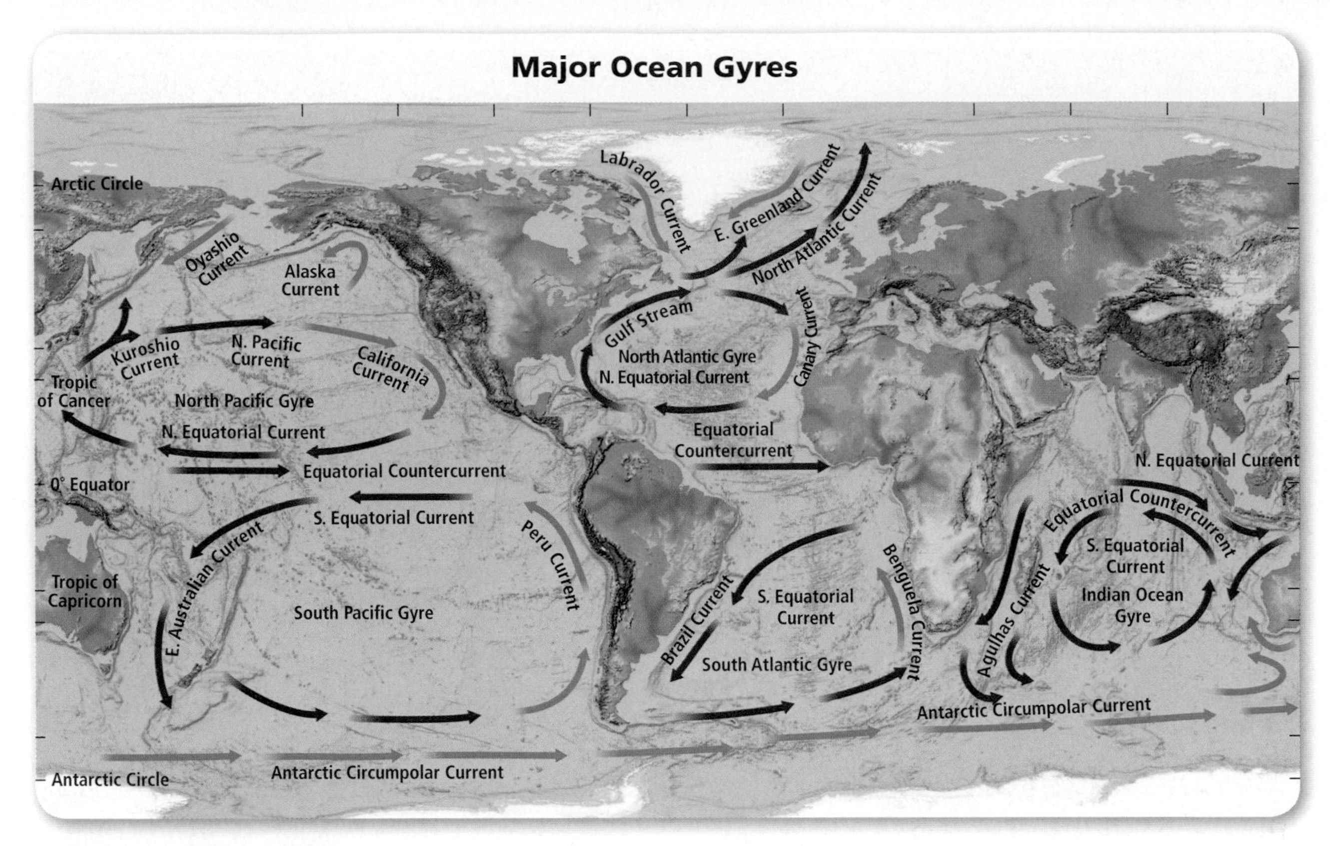

Figure 15.23 Large gyres in each ocean are formed by surface currents. Red arrows represent the movement of warm water, blue arrows represent the movement of cold water.
Identify ***the currents that make up the gyre in the South Atlantic Ocean.***

As shown in **Figure 15.23,** there are five major gyres—the North Pacific, the North Atlantic, the South Pacific, the South Atlantic, and the Indian Ocean. Because of the Coriolis effect, the gyres of the northern hemisphere circulate in a clockwise direction and those of the southern hemisphere circulate in a counterclockwise direction. The parts of all gyres closest to the equator move toward the west as equatorial currents. When these currents encounter a landmass, they are deflected toward the poles. These poleward-flowing waters carry warm, tropical water into higher, colder latitudes. An example of a warm current is the Gulf Stream Current in the North Atlantic.

After these warm waters enter polar regions, they gradually cool and, deflected by landmasses, move back toward the equator. The resulting currents then bring cold water from higher latitudes into tropical regions. An example of this kind of current is the California Current in the eastern North Pacific.

Upwelling In addition to moving horizontally, ocean water moves vertically. The upward motion of ocean water is called **upwelling.** Upwelling waters originate in deeper waters, below the thermocline, and thus are usually cold. Areas of upwelling exist mainly off the western coasts of continents in the trade-wind belts. For example, **Figure 15.24** shows what happens off the coast of California. Winds blowing from the north cause surface water to begin moving. The Coriolis effect acts on the moving water, deflecting it to the right of its direction of movement, which results in surface water being moved offshore. The surface water is then replaced by upwelling deep water.

Figure 15.24 Upwelling occurs when surface water is moved offshore and deep, colder water rises to the surface to replace it.

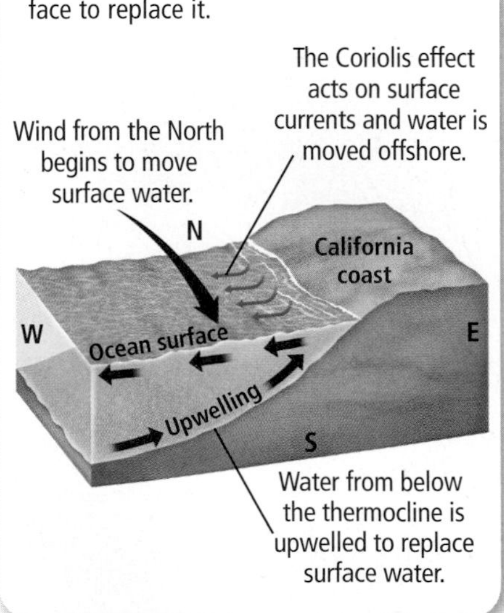

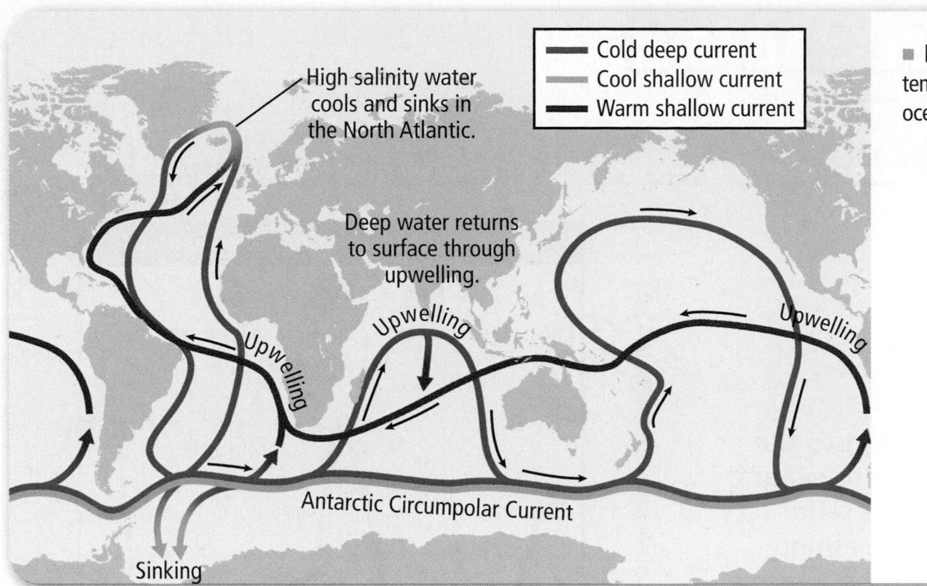

■ **Figure 15.25** Differences in salinity and temperature generate density currents in the deep ocean.

Density currents Recall the discussion of Antarctic Bottom Water in Section 15.2. The sinking of Antarctic Bottom Water is an example of an ocean current. In this case, the current is called a **density current** because it is caused by differences in the temperature and salinity of ocean water, which in turn affect density. Density currents move slowly in deep ocean waters, following a general path that is sometimes called the global conveyer belt.

The conveyor belt, a model of which is shown in **Figure 15.25,** begins when cold, dense water, including North Atlantic Deep Water and Antarctic Bottom Water, sinks at the poles. After sinking, these water masses slowly move away from the poles and circulate through the major ocean basins. After hundreds of years, the deep water eventually returns to the surface through upwelling. Once at the surface, the deep water is warmed by solar radiation.

Section 15.3 Assessment

Section Summary

- Energy moves through ocean water in the form of waves.
- Tides are influenced by both the Moon and the Sun.
- Surface currents circulate in gyres in the major ocean basins.
- Vertical currents in the ocean include density currents and upwelling.

Understand Main Ideas

1. MAIN Idea **Describe** how surface currents in gyres redistribute heat between the equator and the poles.
2. **Illustrate** a wave. Label the following characteristics: *crest, trough, wavelength, wave height,* and *wave base.*
3. **Explain** how tides form.
4. **Compare and contrast** surface currents and density currents.

Think Critically

5. **Predict** the effects on marine ecosystems if upwelling stopped.
6. **Assess** the difference between spring tides and neap tides.

WRITING in Earth Science

7. Write a step-by-step explanation of how upwelling occurs.

Earth Science and the Environment

Bacterial Counts and Full Moons

You've been looking forward to going to the beach all week. Under the hot Sun—towel and lunch in hand—you head toward the sand. You can't wait to get in the cool, refreshing water. As you near the entrance of the beach, you see a posted sign that reads "Beach Closed: High Bacterial Counts in Water."

Bacteria in the water Although most of the bacteria in seawater are harmless to humans, some types are thought to cause gastrointestinal illnesses, with symptoms that include diarrhea and vomiting, in swimmers. Water is routinely tested on many beaches for a type of bacteria called enterococci (en tur oh KAHK i), which normally live in the intestines of mammals and birds. Although enterococci are usually harmless, their presence in the water is considered a strong indicator of the presence of other, illness-causing, bacteria. If enterococci counts rise above a certain level, authorities close beaches for the safety of swimmers.

Bacterial counts and moon phases Scientists have found that higher levels of enterococci in seawater are associated with new moon and full moon phases, as shown in the graph on the right. Recall that spring tides occur during the new moon and full moon phases. During spring tides, high tides are at their highest levels and low tides are at their lowest, resulting in a large tidal range.

After compiling and analyzing data for 60 beaches along the Southern California coast, scientists found that at 50 of the 60 beaches there was a pattern of high bacterial counts during spring tides. Lower bacterial counts were associated with neap tides, which occur during first-quarter and three-quarter moon phases.

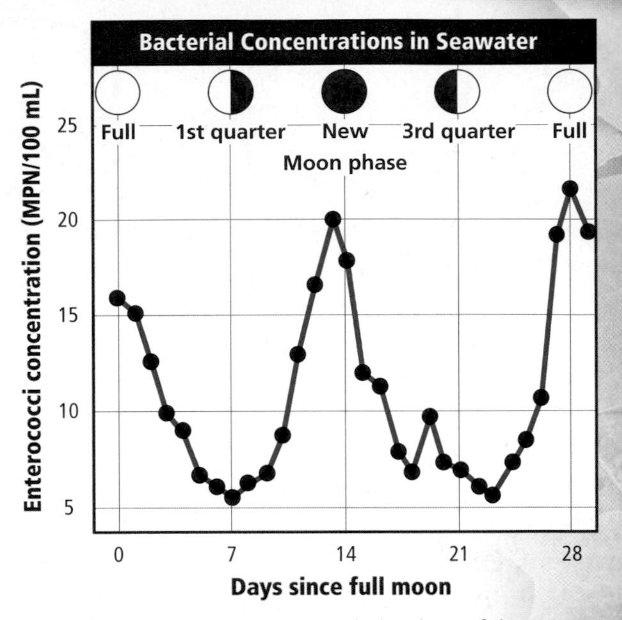

Bacterial counts in seawater vary with the phase of the Moon.

Data also showed that higher counts of bacteria were found specifically during an ebbing spring tide. The term *ebbing tide* refers to water that is receding after reaching its highest point.

Possible sources of bacteria Scientists have several hypotheses to explain possible sources for the bacteria in seawater during the spring tides. One is that the bacteria are present in high numbers in groundwater that only mixes with seawater during spring tides. Other possible sources include decaying organic material that collects on the sand, or bird droppings near the high tide line, both of which would mix with seawater during high spring tides.

WRITING in Earth Science

Newscast Suppose you are a newscaster presenting a story for the nightly news about bacterial levels at beaches. Present your story to the class, explaining results of scientific studies on patterns of bacteria and why these results are important to swimmers. To learn more about bacterial counts in seawater, visit glencoe.com.

GEOLAB

MODEL WATER MASSES

Background: Water in oceans is layered because water masses with higher densities sink below those with lower densities. The density of seawater depends on its temperature and salinity.

Question: *How do changes in salinity and temperature affect water density?*

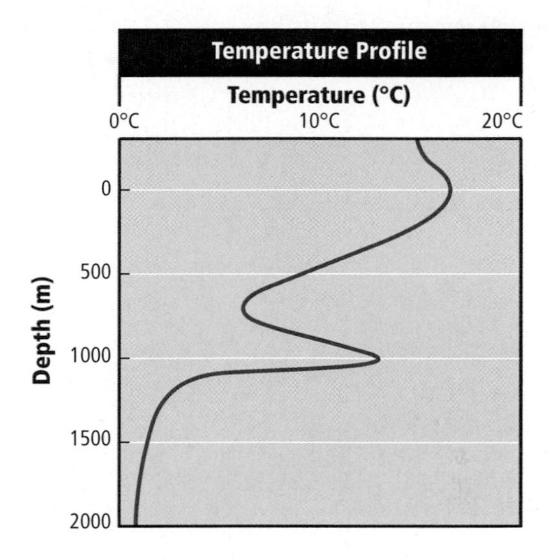

Materials

scale
graduated 500-mL cylinder
100-mL glass beakers (4)
water
red, yellow, and blue food coloring
salt
thermometer
eyedropper
graph paper
ruler
calculator

Safety Precautions

Procedure

1. Read and complete the lab safety form.
2. Mix 200 mL of water and 7.5 g of salt in the graduated cylinder. Pour equal amounts of the salt solution into two beakers. Fill each of the two other beakers with 100 mL of freshwater.
3. Put a few drops of red food coloring in one of the salt solutions. Put a few drops of yellow food coloring in the other salt solution. Put a few drops of blue food coloring in one of the freshwater beakers. Do not add food coloring to the other freshwater beaker.
4. Place the beakers with the red salt solution and the blue freshwater in the refrigerator. Refrigerate them for 30 min.
5. Measure and record the temperature of the water in all four beakers.
6. Put several drops of the cold, red salt water into the beaker with the warm, yellow salt water and observe what happens. Record your observations.
7. Put several drops of the cold, blue freshwater into the beaker with the warm, clear freshwater and observe what happens. Record your observations.
8. Put several drops of the cold, blue freshwater into the beaker with the warm, yellow salt water and observe what happens. Record your observations.

Analyze and Conclude

1. **Describe** the movement of the cold, red salt water in Step 6. Compare this to the movement of the cold, blue freshwater in Step 8. What accounts for the differences you observed?
2. **Identify** the water samples by color in order of increasing density.
3. **Explain** If you poured the four water samples into the graduated cylinder, how would they arrange themselves into layers by color, from top to bottom?
4. **Construct** Assume that four water masses in a large body of water have the same characteristics as the water in the four beakers. The warm water layers are 100 m thick, and the cold layers are 1000 m thick. Construct a graph that shows the temperature profile of the large body of water.

APPLY YOUR SKILL

Infer The temperature profile above was constructed from measurements taken in the Atlantic Ocean off the coast of Spain. Study the profile, then infer why a high-temperature layer exists beneath the thermocline. Is this layer denser than the colder water above? Explain.

CHAPTER 15

Study Guide

Download quizzes, key terms, and flash cards from glencoe.com.

BIG Idea Studying oceans helps scientists learn about global climate and Earth's history.

Vocabulary	Key Concepts
Section 15.1 An Overview of Oceans	
• sea level (p. 410) • side-scan sonar (p. 407)	**MAIN Idea** The global ocean consists of one vast body of water that covers more than two-thirds of Earth's surface. • Scientists use many different instruments to collect and analyze data from oceans. • Scientists have several ideas as to where the water in Earth's oceans originated. • A large portion of Earth's surface is covered by ocean. • Earth's oceans are the Pacific, the Atlantic, the Indian, the Arctic, and the Southern.
Section 15.2 Seawater	
• estuary (p. 414) • salinity (p. 413) • temperature profile (p. 418) • thermocline (p. 418)	**MAIN Idea** Oceans have distinct layers of water masses that are characterized by temperature and salinity. • Ocean water contains dissolved gases, nutrients, and salts. • Salts are added to and removed from oceans through natural processes. • Properties of ocean water, including temperature and salinity, vary with location and depth. • Many of the oceans' deepwater masses sink from the surface of polar oceans.
Section 15.3 Ocean Movements	
• breaker (p. 422) • crest (p. 421) • density current (p. 427) • neap tide (p. 424) • spring tide (p. 424) • surface current (p. 425) • tide (p. 423) • trough (p. 421) • upwelling (p. 426) • wave (p. 421)	**MAIN Idea** Waves and currents drive the movements of ocean water and lead to the distribution of heat, salt, and nutrients from one region of the ocean to another. • Energy moves through ocean water in the form of waves. • Tides are influenced by both the Moon and the Sun. • Surface currents circulate in gyres in the major ocean basins. • Vertical currents in the ocean include density currents and upwelling.

CHAPTER 15

Assessment

Vocabulary Review

Match each description below with the correct vocabulary term from the Study Guide.

1. the amount of dissolved salts in a fixed volume of seawater
2. a transitional layer in which temperature rapidly decreases with depth
3. plots changing water temperatures against depth

Complete the sentences below using vocabulary terms from the Study Guide.

4. The lowest point of a wave is the ________ .
5. The level of Earth's oceans is ________.
6. ________ sends sound waves to the seafloor at an angle.
7. ________ are the periodic rise and fall of sea level.
8. A ________ is caused by differences in the temperature and salinity of ocean water.
9. The upward motion of ocean water is called ________.
10. The highest point of a wave is the ________.
11. Waves that collapse near shore are called ________.

Understand Key Concepts

12. Which is used to measure ocean depth?
    A. bottom dredges
    B. nets
    C. sonar
    D. tidal patterns
13. What is the average depth of the oceans?
    A. 380 m
    B. 38 m
    C. 3800 m
    D. 3 km
14. Which are the most common gases emitted by volcanoes?
    A. hydrogen and helium
    B. oxygen and nitrogen
    C. water vapor and carbon dioxide
    D. chlorine and hydrogen

Use the graph below to answer Questions 15 and 16.

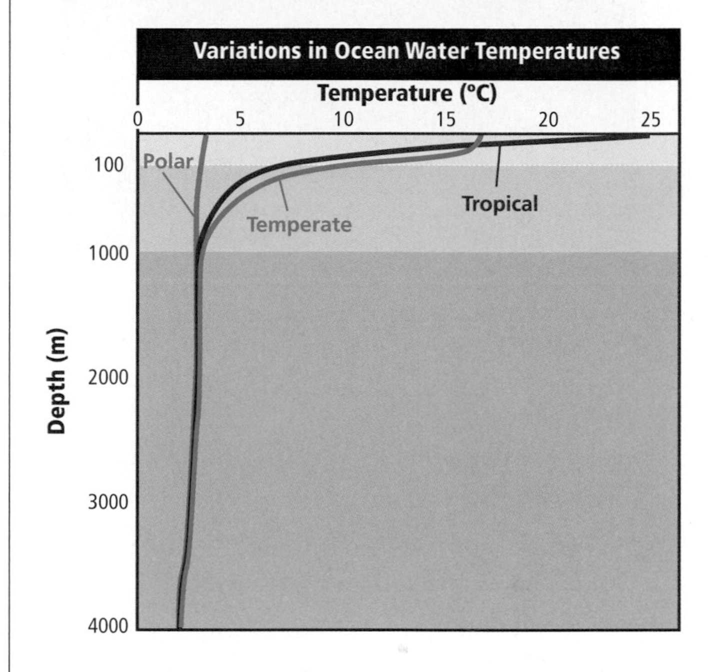

15. Between which depths is the thermocline?
    A. 0 and 100 m
    B. 100 and 1000 m
    C. 100 and 4000 m
    D. 1000 and 4000 m
16. What is the average temperature of deep water below the thermocline?
    A. 15°C
    B. more than 4°C
    C. less than 4°C
    D. 0°C
17. What basic motion does water follow during the passage of a wave?
    A. forward
    B. backward
    C. up and down
    D. circular
18. Which would have the largest impact on global ocean water density?
    A. strong winds
    B. increase in daylight hours
    C. long-term increase in air temperature
    D. thunderstorm with heavy precipitation
19. Which process does not remove salt from ocean water?
    A. precipitation of salt in dry, coastal regions
    B. evaporation of water in subtropical regions
    C. sea spray being carried inland by wind
    D. absorption by marine organisms

Use the diagram below to answer Question 20.

20. Which type of tides occur when the Sun, the Moon, and Earth are aligned as shown above?
A. spring tides
B. neap tides
C. lunar tides
D. solar tides

21. Which type of tides occur when the Sun, the Moon, and Earth form a right angle?
A. spring tides
B. neap tides
C. lunar tides
D. solar tides

Constructed Response

22. Illustrate Make a diagram that shows how the ocean acts as a sink for carbon.

23. Relate Where in the oceans are the highest values of salinity found? What processes lead to areas of high salinity water?

24. Solve What would be the depth of the wave base for a wave that is 200 m long?

25. Interview Write several questions you would ask an oceanographer about the use of computers to analyze data from the ocean.

26. Determine Which gyre would have clockwise circulation: the North Pacific, the South Pacific, the South Atlantic, or the Indian Ocean? Explain.

27. Diagram Draw a diagram that shows how the Sun influences tides.

28. Analyze Why does a wave break?

29. Explain how the Moon influences tides.

Use the diagram below to answer Question 30.

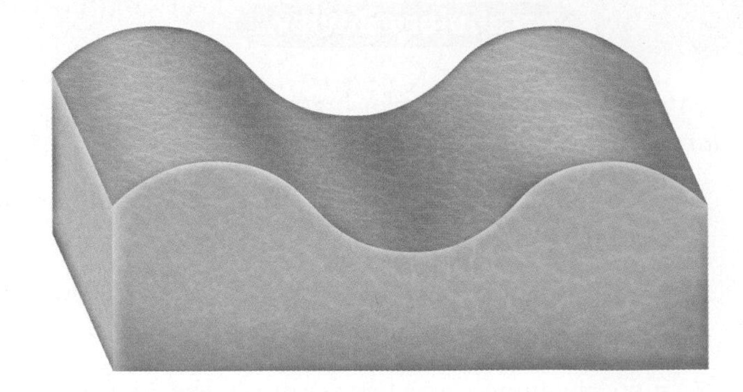

30. Diagram Copy the illustration shown above. Then use the following terms to label the characteristics of an ocean wave: *crest, trough, wave height,* and *wavelength.*

31. Cause and Effect Cold water masses are generally denser than warm water masses, yet warm water from the Mediterranean Sea sinks to a depth of more than 1000 m when it flows into the Atlantic Ocean. What causes this effect?

Think Critically

32. Predict Based on what you have learned about water density, describe the movement of freshwater from a river as it flows into a sea.

33. Hypothesize Use your knowledge of global warming to hypothesize why sea level is rising.

34. Suggest a real-world application for adding salt to lower the melting point of ice.

35. Plan Use **Figure 15.23** to plan the fastest round trip by ship from Boston, Massachusetts, to London, England. Will the return route be the same as the outbound trip? Explain.

36. CAREERS IN EARTH SCIENCE Suppose you are a lead scientist at NASA. Design an experiment that would allow you to test the hypothesis that Earth's water originated from comets.

37. Compare and contrast Antarctic Intermediate Water and North Atlantic Deep Water.

38. Explain how biological and physical processes affect levels of carbon dioxide in different areas of the ocean.

Use the diagram below to answer Question 39.

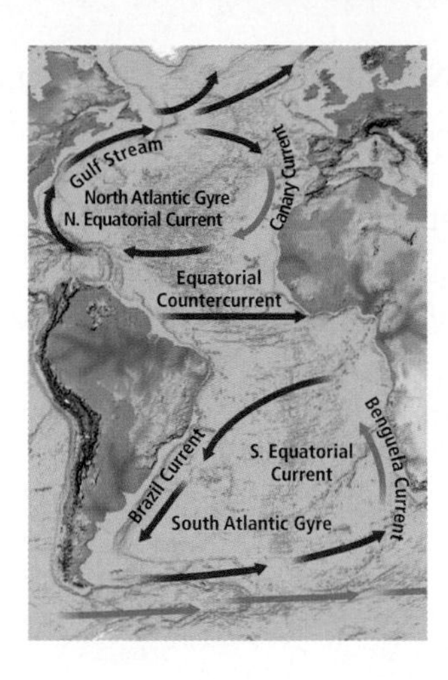

39. Assess Surface currents can affect coastal climates. Would the Gulf Stream and the Benguela Current, both of which are surface currents, have the same effect on coastal climate? Explain.

40. Predict One of the effects of El Niño, which you learned about in Chapter 14, is that the trade winds that blow across the equatorial Pacific Ocean weaken. Predict how this might affect upwelling off the coast of Peru.

41. Assess How do density currents and the conveyor belt help move dissolved gases, such as oxygen, and nutrients, such as nitrogen, from one area of the ocean to another?

Concept Mapping

42. Create a concept map using the following words or phrases that describe waves: *lowest point of a wave, wave characteristics, crest, wavelength, wave height, trough, horizontal crest-to-crest distance, highest point of a wave,* and *vertical distance between crest and trough.*

Challenge Question

43. Research the percentages of the different sources of freshwater on Earth, including glaciers, ice caps, rivers, lakes, and groundwater. Construct a new circle graph like the one shown in **Figure 15.5** that represents the new data.

Additional Assessment

44. WRITING in **Earth Science** Write a summary about majoring in oceanography for a college brochure. Include information about the prerequisites, requirements for completion, and career opportunities in your summary.

DBQ Document–Based Questions

Data obtained from: Alford, M. 2003. The redistribution of energy available for mixing by long-range propagation of internal waves. *Nature* 423:159–162.

Ocean mixing due to waves is important for pollution dispersal, marine productivity, and global climate. In the figure below, the arrows represent the direction and magnitude of waves from different data collection sites (shown as black dots). The absence of arrows from the black dots means that the waves were small enough to ignore.

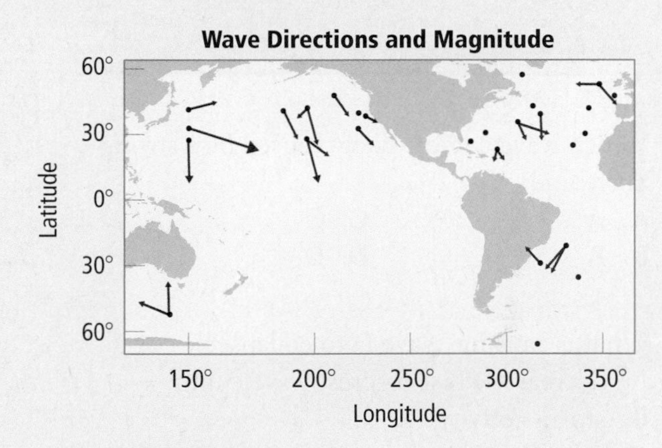

45. In which direction do most of the waves in the northern hemisphere travel? How does this differ from waves in the southern hemisphere?

46. Given the number of collection sites in the southern hemisphere, can you draw any general conclusion about waves in this area?

Cumulative Review

47. What steps are usually included in a scientific method? **(Chapter 1)**

48. Name an example of a carbonate mineral. What is its chemical composition? **(Chapter 4)**

Standardized Test Practice

Multiple Choice

1. Why is deforestation often linked to global warming?
   A. It increases the amount of dry land on Earth's surface.
   B. It releases toxic gases and pollutants into the atmosphere.
   C. It increases the amount of CO_2 released into the atmosphere.
   D. It decreases the amount of CO_2 released into the atmosphere.

Use the illustration to answer Questions 2 and 3.

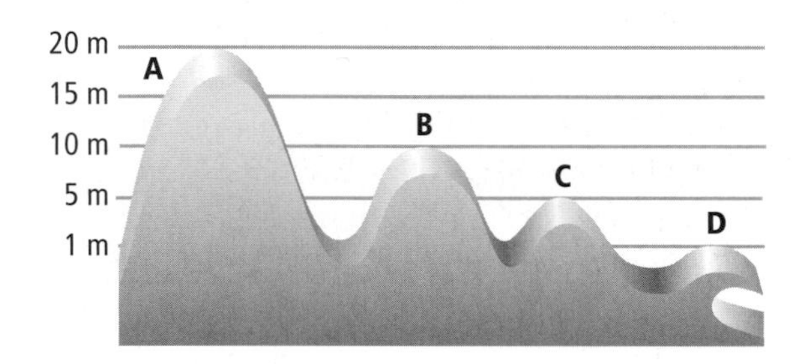

2. Which wave is most likely caused by a strong hurricane?
   A. A
   B. B
   C. C
   D. D

3. What is causing Wave D to collapse?
   A. decreased crest-to-crest wavelength
   B. storm activity
   C. increased crest-to-crest wavelength
   D. friction from the ocean floor

4. Where is a heat island most likely to be found?
   A. a farm
   B. a beach
   C. a mountain top
   D. an inner city

5. Which would be the most abundant evidence that an area had been affected by continental glaciation?
   A. thick deposits of sediment
   B. long, snakelike ridges of sand and gravel
   C. plants and animals adapted to a cold climate
   D. large, shallow lakes

6. Which is NOT a characteristic the Fujita tornado intensity scale uses to rank tornadoes?
   A. funnel length
   B. wind speed
   C. path of destruction
   D. duration

Use the map below to answer Questions 7 and 8.

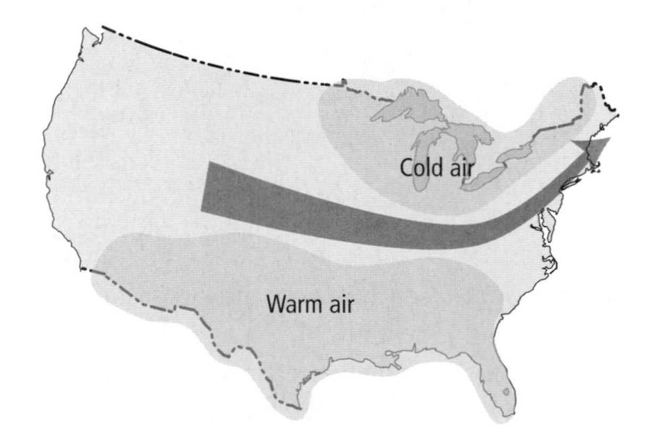

7. What air current is shown by the arrow?
   A. intertropical convergence zone
   B. prevailing westerlies
   C. prevailing easterlies
   D. trade wind

8. If the cold air were to dip down into the region of warm air, what prediction could be made about the type of weather that might result?
   A. cloudiness and widespread, mild precipitation
   B. some cloudiness with no precipitation
   C. clouds with showers and thunderstorms
   D. cool weather with no clouds

9. Which region's seawater is most likely to have the highest concentration of dissolved salts?
   A. an equatorial region
   B. a subtropical region
   C. a polar region
   D. a delta where rivers empty into oceans

10. If there is a sudden period of calm during a hurricane, what should a person assume?
    A. The hurricane is over and it is safe to go out.
    B. The eye of the hurricane is over his or her area.
    C. The hurricane has weakened in intensity.
    D. The hurricane is over, but it is not safe to go out.

Short Answer

Use the illustration below to answer Questions 11–13.

11. Describe what is being represented in the above illustration.

12. What does the dotted line represent? What is its purpose?

13. Why is the Sun important to Earth?

14. Discuss what one-to-three day forecasts can and cannot do.

15. How does the rate of solar absorption differ on water than on land?

16. What is the benefit of using water vapor imagery?

Reading for Comprehension

Cooling with Seawater

Engineers have turned to the deep ocean as a cooling source. Because of the churning action of wind, waves, and currents, ocean water must be drawn from great depths to get consistently cold temperatures. The Natural Energy Laboratory of Hawaii Authority (NELHA) runs its own deep-source cooling plant to cool buildings on the agency's campus. The plant draws 42.8°F (6°C) seawater from a depth of 610 m. "NELHA saves about [U.S.] 3000 dollars a month in electrical costs by using the cold seawater air-conditioning process," said Jan War, an operations manager. "We still use a freshwater loop to cool our buildings, since seawater is so corrosive." So far deep-source cooling is only practical for communities with numerous buildings located near large bodies of water. But many of the world's major cities, settled during the golden age of sailing ships, are close to shore—something to think about the next time a dip in the ocean takes your breath away.

Article obtained from: Smith, J. The AC of tomorrow? Tapping deep water for cooling. *National Geographic News.* September 10, 2004.

17. What can be inferred from this passage?
- **A.** Using seawater will only work in cities along the Pacific Ocean.
- **B.** Using seawater to cool buildings is a good option, but the process needs more study and improvement.
- **C.** Cooling with seawater is an expensive project.
- **D.** Cooling with seawater will eventually take over cooling with freshwater all over.

18. Why would this new technology benefit many of the world's major cities?
- **A.** Most cities have many buildings close to the shore.
- **B.** Cities have the money to fund the new technology.
- **C.** Cities are located along the Pacific Ocean.
- **D.** Cities are not close to freshwater.

19. Discuss two reasons that a small farming town in Kansas would not benefit from this saltwater technology.

NEED EXTRA HELP?																
If You Missed Question . . .	1	2	3	4	5	6	7	8	9	10	11	12	13	14	15	16
Review Section . . .	14.4	15.3	15.3	14.2	8.3	13.2	12.2	12.2	15.2	13.3	11.1	11.1	11.1	12.4	11.1	12.3

CHAPTER 16 The Marine Environment

BIG Idea The marine environment is geologically diverse and contains a wealth of natural resources.

16.1 Shoreline Features
MAIN Idea The constant erosion of the shoreline and deposition of sediments by ocean waves creates a changing coastline.

16.2 Seafloor Features
MAIN Idea The ocean floor contains features similar to those on land and is covered with sediments of several origins.

GeoFacts

- Coral reefs cover only 1 percent of the ocean floor, yet about 25 percent of all marine fish species are found living on coral reefs.
- A chemical from the Caribbean sea whip, a soft coral, is used to make skin-care products because it has anti-inflammatory properties.
- The blood of tunicates—soft-bodied organisms closely related to vertebrates—contains high levels of the rare metal vanadium.

LAUNCH Lab

Where does chalk form?

Although you might not live near a coast, parts of your environment were shaped by the ocean. For example, you might be just a few meters away from former seafloor deposits that are now part of the bedrock underground. One such seafloor deposit is chalk. How can you tell that chalk formed on the seafloor?

Procedure

1. Read and complete the lab safety form.
2. Use a **mortar and pestle** to grind a small piece of **natural chalk** into a powder. Make a **slide** of the powdered chalk.
3. Observe the chalk powder through a **microscope.**

Analysis

1. **Describe** the powder. Are the grains irregular in shape or size? Do some of the grains have patterns?
2. **Analyze** your data and hypothesize the origin of the chalk. On what evidence do you base your conclusion?

Visit glencoe.com to

- study entire chapters online;
- explore Concepts In Motion animations:
  - Interactive Time Lines
  - Interactive Figures
  - Interactive Tables
- access Web Links for more information, projects, and activities;
- review content with the Interactive Tutor and take Self-Check Quizzes.

Seafloor Features Make this Foldable to diagram the major geologic features of continental margins and ocean basins.

STEP 1 Fold a horizontal sheet of paper in half from side to side.

STEP 2 Fold into thirds. Unfold and draw lines lightly along the fold lines to indicate three tabs.

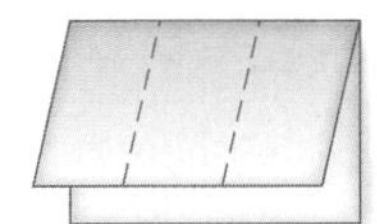

STEP 3 Label the uncut tabs *Continent, Continental Margin,* and *Ocean Basin.* Draw a diagram of these seafloor features here.

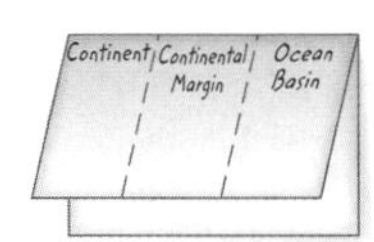

STEP 4 Cut along the fold lines to make three tabs after your diagram is complete.

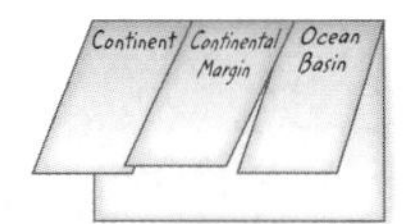

FOLDABLES **Use this Foldable with Section 16.2.** As you read this section, diagram the major features of the seafloor. When you finish your diagram, cut the tabs as described in Step 4. Summarize the features shown in your diagram under each of the tabs.

Section 16.1

Objectives

- **Explain** how shoreline features are formed and modified by marine processes.
- **Describe** the major erosional and depositional shoreline features.
- **Identify** protective structures used near shore.

Review Vocabulary

breaker: collapsing wave that forms when a wave enters shallow water

Vocabulary

beach
wave refraction
longshore bar
longshore current
barrier island

Shoreline Features

MAIN Idea The constant erosion of the shoreline and deposition of sediments by ocean waves creates a changing coastline.

Real-World Reading Link Have you ever picked up a smoothly polished pebble on the beach? A sculptor forms a masterpiece out of a chunk of wood or stone by chipping away one tiny bit at a time. Waves are nature's chisel, and beaches display the carefully carved features.

The Shore

Shown in **Figure 16.1,** the shore is the area of land between the lowest water level at low tide and the highest area of land that is affected by storm waves. Shores are places of continuous, often dramatic geologic activity—places where you can see geologic changes occurring almost daily. The shoreline is the place where the ocean meets the land. Shorelines are shaped by the action of waves, tides, and currents. The location of the shoreline constantly changes as the tide moves in and out. As waves erode some shorelines, they create some of the most impressive rock formations on Earth. In other areas, waves deposit loose material and build wide, sandy beaches.

Beaches Long stretches of U. S. coasts are lined with wide, sandy beaches. A **beach,** shown in **Figure 16.1,** is the area in which sediment is deposited along the shore. Beaches are composed of loose sediments deposited and moved about by waves along the shoreline. The size of sediment particles depends on the energy of the waves striking the coast and on the source of the sediment. Beaches pounded by large waves or formed on rocky coasts usually consist of coarse materials such as pebbles and cobbles.

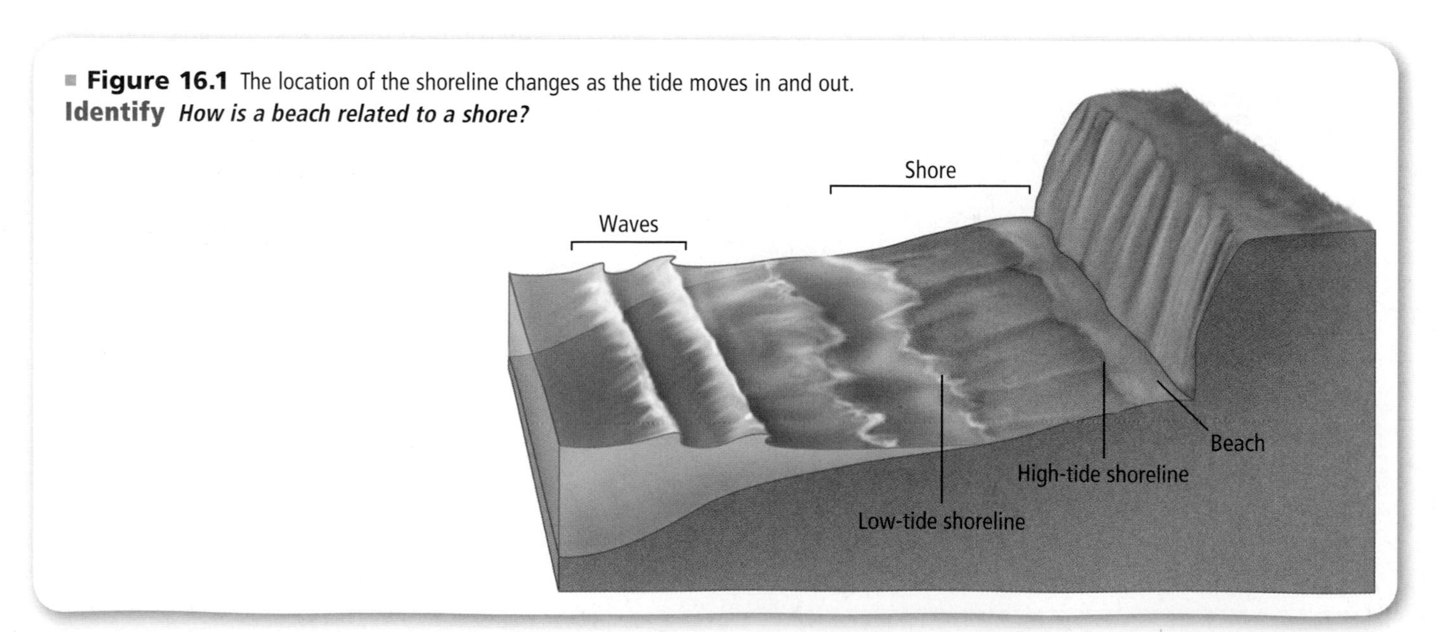

■ **Figure 16.1** The location of the shoreline changes as the tide moves in and out.
Identify *How is a beach related to a shore?*

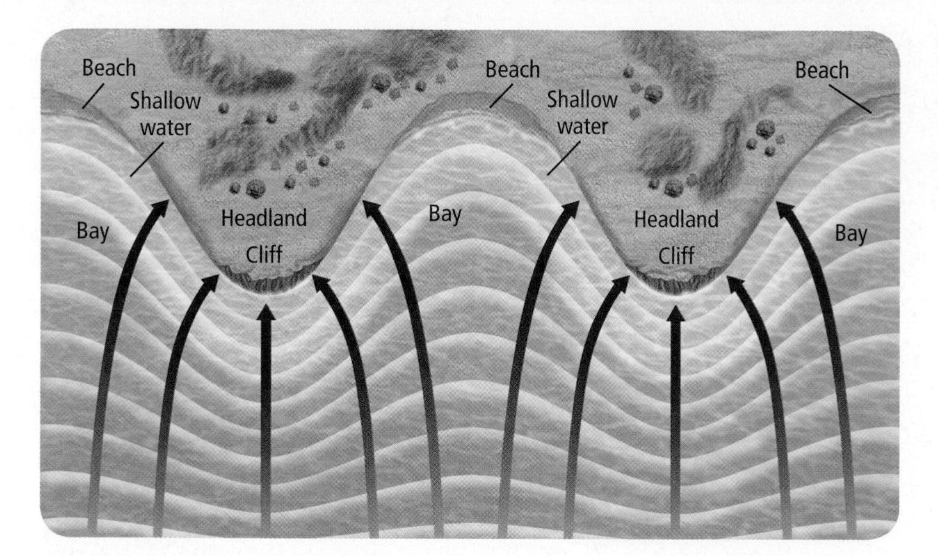

■ **Figure 16.2** Wave crests advance toward the shoreline and slow down when they encounter shallow water. This causes the wave crests to bend toward the headlands and move in the direction of the arrows.

Formation of Shoreline Features

Large breaking waves can hurl thousands of metric tons of water, along with suspended rock fragments, against a shore with such force that they are capable of eroding solid rock.

Erosional features Waves move faster in deep water than in shallow water. This difference in wave speed causes initially straight wave crests to bend when part of the crest moves into shallow water, a process known as **wave refraction,** illustrated in **Figure 16.2.** Along an irregular coast with headlands and bays, the wave crests bend toward the headlands. As a result, most of the breaker energy is concentrated along the relatively short section of the shore around the tips of the rocky headlands, while the remaining wave energy is spread out along the much longer shoreline of the bays. The headlands thus undergo severe erosion. The material eroded from the headlands is swept into the bays, where it is deposited to form crescent-shaped beaches. The headlands are worn back and the bays are filled in until the shoreline straightens.

Wave-cut platforms Many headlands have spectacular rock formations. Generally, as a headland is worn away, a flat erosional surface called a wave-cut platform is formed. The wave-cut platform terminates against a steep wave-cut cliff, as shown in **Figure 16.3.**

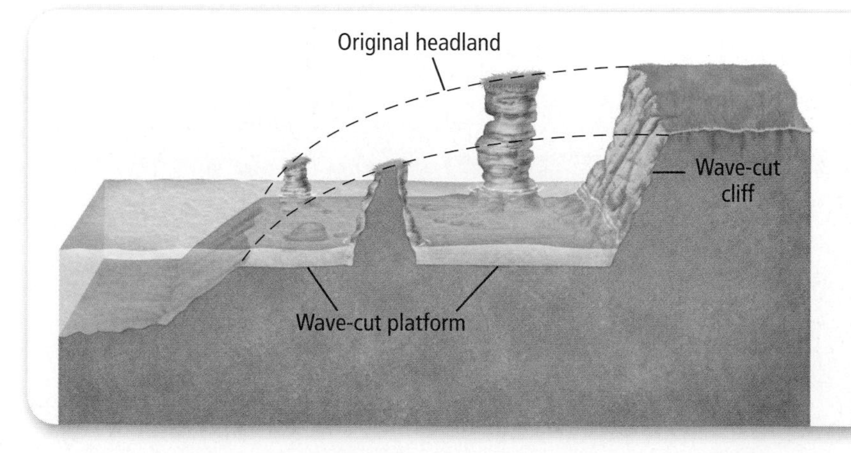

■ **Figure 16.3** A headland can be modified by wave erosion. The dotted lines indicate the original shape of the headland.

■ **Figure 16.4** The sea stack and sea arch shown here, along the coast of France, were formed by wave refraction at a rocky headland.

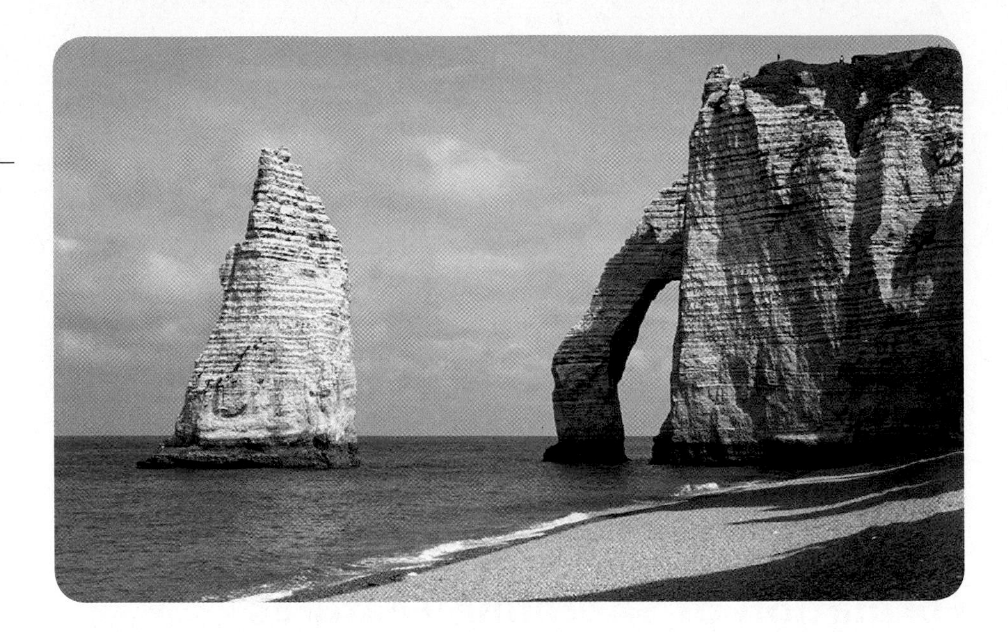

Sea stacks Differential erosion, the removal of weaker rocks or rocks near sea level, produces many of the other characteristic landforms of rocky headlands. As shown in **Figure 16.4,** a sea stack is an isolated rock tower or similar erosional remnant left on a wave-cut platform. A sea arch, also shown in **Figure 16.4,** is formed as stronger rocks are undercut by wave erosion. Sea caves are tubelike passages blasted into the headlands at sea level by the constant assault of the breakers.

Longshore currents Suppose you stood on a beach at the edge of the water and began to walk out into the ocean. As you walked, the water might get deeper for a while, but then it would become shallow again. The shallow water offshore lies above a sandbar, called a **longshore bar,** that forms in front of most beaches, as illustrated in **Figure 16.5.** Waves break on the longshore bar in the area known as the surf zone. The deeper water between the shore and longshore bar is called the longshore trough.

The waves striking the beach are almost parallel to the shoreline, although the waves seaward of the longshore bar are generally not parallel to the shore. This is another case of wave refraction. The slowing of the waves in shallow water causes the wave crests to bend toward the shore.

■ **Figure 16.5** A typical beach profile includes longshore troughs and bars within the surf zone.
Explain ***What is the difference between a longshore trough and a longshore bar?***

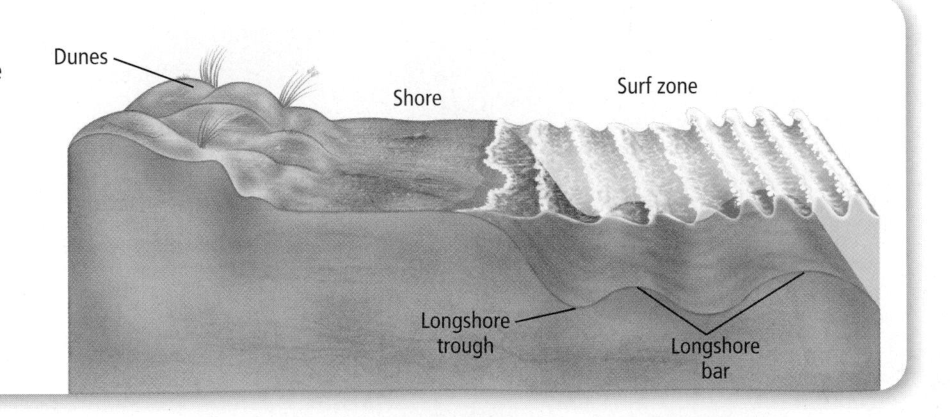

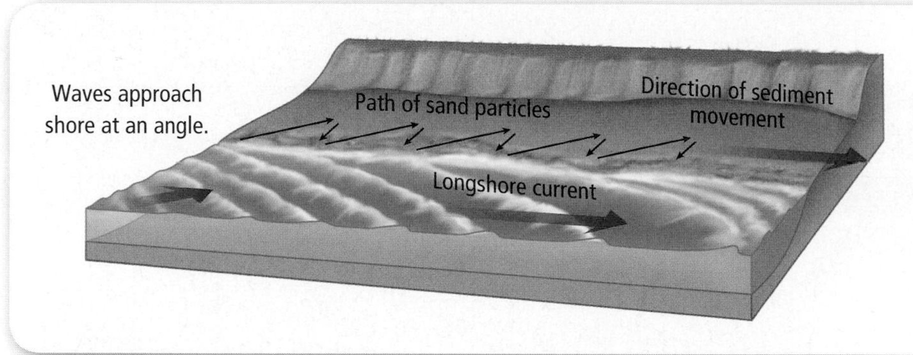

Figure 16.6 Longshore currents are driven by incoming waves.

Concepts In Motion

Interactive Figure To see an animation of longshore currents, visit glencoe.com.

As water from incoming breakers spills over the longshore bar, a current flowing parallel to the shore, called the **longshore current,** is produced. This current varies in strength and direction from day to day. Over the course of a year, because of prevailing winds and wave patterns, one direction usually dominates.

Movement of sediments Longshore currents, shown in **Figure 16.6,** move large amounts of sediment along the shore. Fine-grained material, such as sand, is suspended in the turbulent, moving water, and larger particles are pushed along the bottom by the current. Additional sediment is moved back and forth on the beach by incoming and retreating waves. Incoming waves also move sediment at an angle to the shoreline in the direction of wave motion. Overall, the transport of sediment is in the direction of the longshore current. On both the Atlantic and Pacific coasts of the United States, longshore transport moves generally toward the south.

 **Reading Check** **Explain** how the longshore current moves sediments.

Rip currents Wave action also produces rip currents, which flow out to sea through gaps in the longshore bar. Rip currents, shown in **Figure 16.7,** return the water spilled into the longshore trough to the open ocean. These dangerous currents can reach speeds of several kilometers per hour. If you are ever caught in a rip current, you should not try to swim against it, instead swim parallel to shore to get out of it.

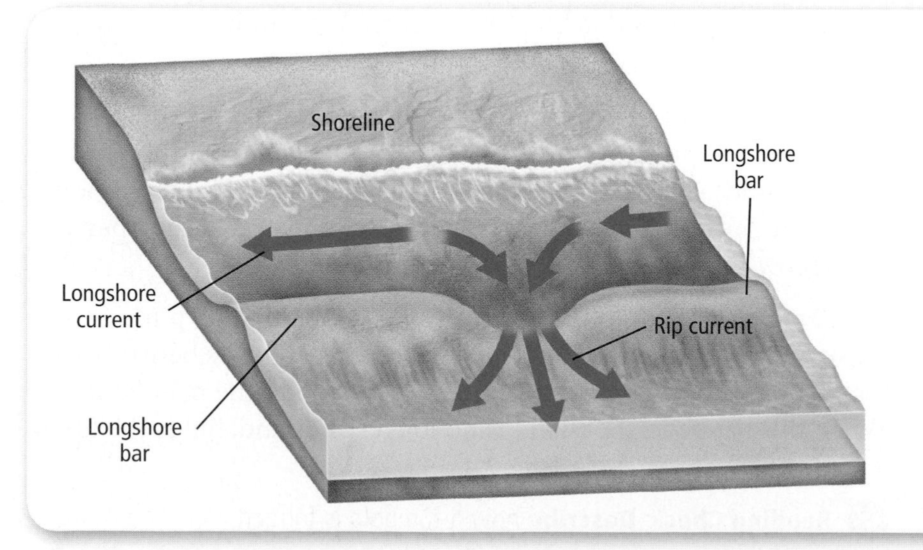

Figure 16.7 Rip currents return water through gaps in the longshore bar out to sea. Rip currents spread out and weaken beyond the longshore bar.

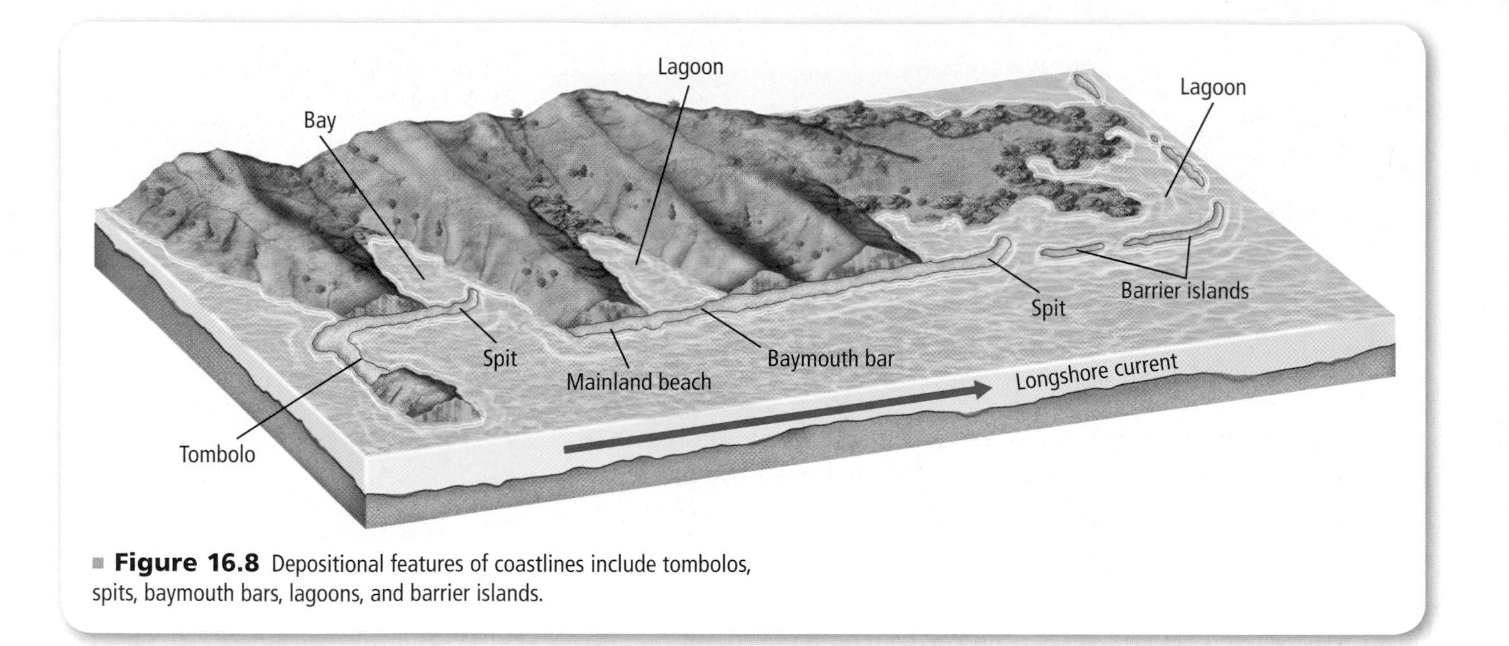

■ **Figure 16.8** Depositional features of coastlines include tombolos, spits, baymouth bars, lagoons, and barrier islands.

Depositional features As a result of wave erosion, longshore transport, and sediment deposition, most seashores are in a constant state of change. Sediments are eroded by large storm waves and deposited wherever waves and currents slow down. Sediments moved and deposited by longshore currents build various characteristic coastal landforms, such as spits and barrier islands, illustrated in **Figure 16.8.**

Spit A narrow bank of sand that projects into the water from a bend in the coastline is called a spit. A spit, which forms where a shoreline changes direction, is protected from wave action. When a growing spit crosses a bay, a baymouth bar forms.

Barrier islands Long ridges of sand or other sediment, deposited or shaped by the longshore current, that are separated from the mainland are called **barrier islands.** Barrier islands, like the ones shown in **Figure 16.9,** can be several kilometers wide and tens of kilometers long. Most of the Gulf Coast and the eastern coast south of New England is lined with an almost continuous chain of barrier islands.

■ **Figure 16.9** This barrier island off the coast of North Carolina is one of many that line the eastern coast.

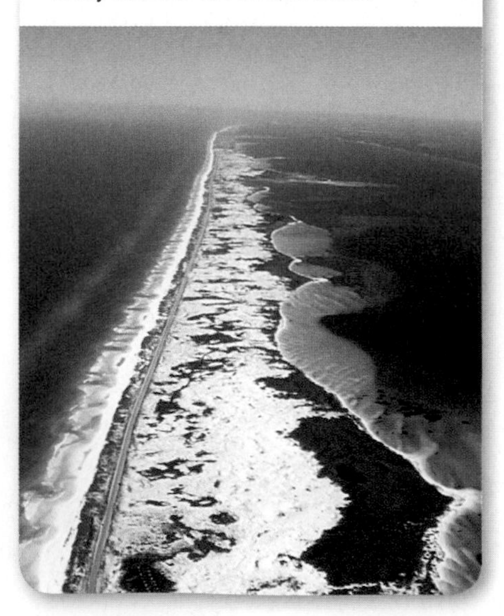

Baymouth bars A baymouth bar, shown in **Figure 16.10,** forms when a spit closes off a bay. The shallow, protected bodies of water behind baymouth bars and barrier islands are called lagoons, which are saltwater coastal lakes that are connected to the open sea by shallow, restricted outlets.

Tombolo Another coastal landform is a tombolo, shown in **Figure 16.10.** A tombolo is a ridge of sand that forms between the mainland and an island and connects the island to the mainland. When this happens, the island is no longer an island, but is the tip of a peninsula.

Reading Check **Describe** how a tombolo is formed.

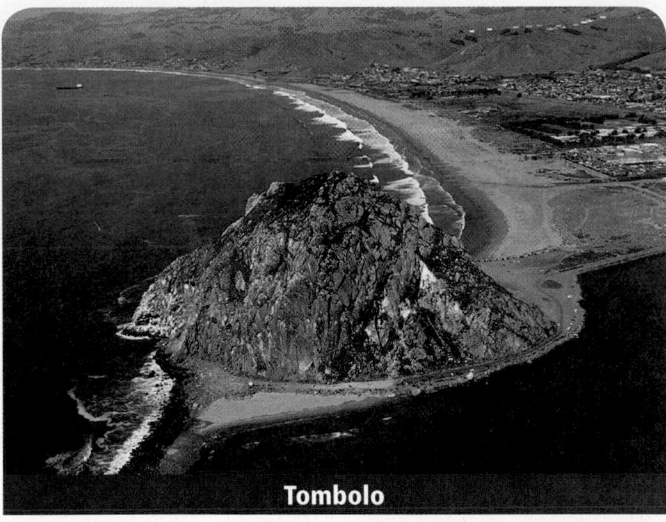

■ **Figure 16.10** Baymouth bars and tombolos are examples of features formed by the deposition of sediments.

Natural and human effects on the coast All of these depositional coastal landforms, including large barrier islands, are unstable and temporary. Occasionally, major storms sweep away entire sections of barrier islands and redeposit the material elsewhere. **Figure 16.11** shows the existence of South Gosier Island, a barrier island off the coast of Louisiana in August of 2004, a month before Hurricane Ivan passed over it. The island was completely destroyed by the strong waves generated by Hurricane Ivan. Even in the absence of storms, however, changing wave conditions can slowly erode beaches and rearrange entire shorelines. For example, the shoreline of Cape Cod, Massachusetts, is retreating by as much as 1 m per year.

People are drawn to coastal areas for their rich natural resources, mild climate, and recreation opportunities. Coastal areas in the United States make up only 17 percent of contiguous land areas, but are currently home to over half of the population in the nation. Increasing population has caused substantial coastal environmental changes, including pollution, shoreline erosion, and wetland and wildlife-habitat loss. These coastal changes in turn increase the susceptibility of the communities to natural hazards caused by hurricanes and tsunamis.

■ **Figure 16.11** The island of South Gosier, a barrier island off the coast of Louisiana, was completely washed away by Hurricane Ivan in September, 2004.

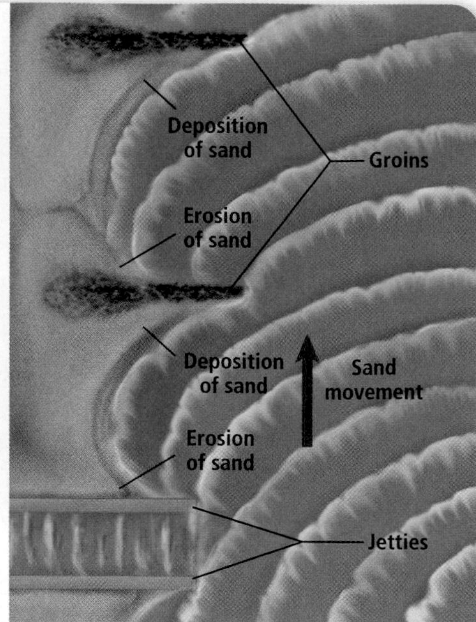

■ **Figure 16.12** Seawalls deflect the energy of waves on the beach. Groins and jetties deprive downshore beaches of sand.

Protective Structures

In many coastal areas, protective structures such as seawalls, groins, jetties, and breakwaters are built in an attempt to prevent beach erosion and destruction of oceanfront properties. However, these artificial structures interfere with natural shoreline processes and can have unexpected negative effects.

Seawalls Structures called seawalls, shown in **Figure 16.12,** are built parallel to shore, often to protect beachfront properties from powerful storm waves. Seawalls reflect the energy of such waves back toward the beach, where they worsen beach erosion. Eventually, seawalls are undercut and have to be rebuilt larger and stronger than before.

Groins and jetties Groins, shown in **Figure 16.12,** are wall-like structures built into the water perpendicular to the shoreline for the purpose of trapping beach sand. Groins interrupt natural long-shore transport and deprive beaches down the coast of sand. The result is aggravated beach erosion down the coast from groins. Similar effects are caused by jetties, which are walls of concrete built to protect a harbor entrance from drifting sand. Jetties are also shown in **Figure 16.12.**

Breakwaters Structures like the one shown in **Figure 16.13** that are built to provide anchorages for small boats or a calm beach area are called breakwaters. Breakwaters are built parallel to the shoreline. When the current slows down behind a breakwater and is no longer able to move sediment, it deposits sediment behind the breakwater. If the accumulating sediment is left alone, it can eventually fill an anchorage. To prevent this, anchorages have to be dredged regularly.

■ **Figure 16.13** A breakwater slows the movement of water against the coast, which results in sediment being deposited behind the structure.

Changes in Sea Level

At the height of the last ice age, approximately 20,000 years ago, the global sea level was about 130 m lower than it is at present. Since that time, the melting of most of the ice-age glaciers has raised the ocean to its present level. In the last 100 years, the global sea level has risen 10 to 15 cm. It continues to rise slowly; estimates suggest a rise in sea level of 3 mm/year.

Many scientists contend that this continuing rise in sea level is the result of global warming. During the last century, Earth's average surface temperature has increased by approximately 0.5°C. As Earth's surface temperature rises, seawater warms up and expands, which adds to the total volume of the seas. In addition, higher temperatures on Earth's surface cause glaciers to melt, and the meltwater flowing into the oceans increases their volume.

Effects of sea level changes If Earth's remaining polar ice sheets in Greenland and Antarctica melted completely, their meltwater would raise sea level by another 70 m. This rise would completely flood some countries, such as the Netherlands, along with some coastal cities in the United States, such as New York City, and low-lying states, such as Florida and Louisiana. Measurements from NASA's *GRACE* satellite, launched in 2002, have indicated that the Greenland ice sheet is melting at an accelerating rate. The West Antarctic Ice Sheet has also been thinning according to a decade of satellite measurements. A complete melt of the Greenland Ice Sheet would lead to a sea level rise of about 6 to 7 m.

Many of the barrier islands of the Atlantic and Gulf Coasts might be former coastal dunes that were drowned by rising sea levels. Other features produced by rising sea levels are the fjords of Norway, shown in **Figure 16.14.** Fjords (fee ORDZ) are deep coastal valleys that were scooped out by glaciers during the ice age and later flooded when the sea level rose.

VOCABULARY

ACADEMIC VOCABULARY

Estimate

a rough or approximate calculation

The estimate for the new roof was $3000.

Figure 16.14 Fjords are flooded U-shaped valleys that can be up to 1200 m deep.

■ **Figure 16.15** Elevated marine terraces are former wave-cut platforms that are now well above the current sea level. This elevated marine terrace is near Half Moon Bay, California.

Effects of tectonic forces Other processes that affect local sea levels are tectonic uplift and sinking. If a coastline sinks, there is a relative rise in sea level along that coast. A rising coastline, however, produces a relative drop in sea level. As a result of tectonic forces in the western United States, much of the West Coast is being pushed up much more quickly than the sea level is rising. Because much of the West Coast was formerly under water, it is called an emergent coast.

Emergent coasts tend to be relatively straight because the exposed seafloor topography is smoother than typical land surfaces with hills and valleys. Other signs of an emergent coast are former shoreline features such as sandy beach ridges located far inland. Among the most interesting of these features are elevated marine terraces—former wave-cut platforms that are now dry and well above current sea level. **Figure 16.15** shows a striking example of such a platform. Some old wave-cut platforms in Southern California are hundreds of meters above current sea level.

Section 16.1 Assessment

Section Summary

- Wave erosion of headlands produces wave-cut platforms and cliffs, sea stacks, sea arches, and sea caves.
- Wave action and longshore currents move sediment along the shore and build depositional features.
- Artificial protective structures interfere with longshore transport.
- Sea levels in the past were 130 m lower than at present.

Understand Main Ideas

1. MAIN Idea Shorelines have headlands and bays. Which experiences the most severe erosion by breakers? Why?
2. **Describe** What are sea stacks, and how are they formed?
3. **Apply** How do jetties and groins affect the longshore current?
4. **Analyze** What effect does a seawall have on a beach?

Think Critically

5. **Evaluate** why resort communities built on barrier islands spend thousands of dollars each year to add sand to the beaches along the shoreline.

MATH in Earth Science

6. The city of Orlando, Florida, is 32 m above sea level. If the sea level continues to rise at the highest estimated rate of 3 mm/y, in how many years might Orlando be under water?

Section 16.2

Objectives

- **Describe** the major geologic features of continental margins.
- **Identify** the major geologic features of ocean basins.
- **Describe** the different types of marine sediments and their origin.

Review Vocabulary

sediment: solid particles deposited on Earth's surface that can form sedimentary rocks by processes such as weathering, erosion, deposition, and lithification

Vocabulary

continental margin
continental shelf
continental slope
turbidity current
continental rise
abyssal plain
deep-sea trench
mid-ocean ridge
seamount
guyot

Seafloor Features

MAIN Idea The ocean floor contains features similar to those on land and is covered with sediments of several origins.

Real-World Reading Link Why do you shake a container of orange juice before pouring a glass? The juice appears thicker at the bottom of the container because the pulp sinks to the bottom. Similarly, gravity causes fine grains of sand to settle to the bottom of the ocean.

The Continental Margin

If you were asked to draw a map of the seafloor, what kind of topographic features would you include? Until recently, most people had little knowledge of the features of the ocean floor. However, modern oceanographic techniques, including satellite data, reveal that the topography of the ocean bottom is as varied as that of the continents.

The topography of the seafloor is surprisingly rough and irregular, with numerous high mountains and deep depressions. The deepest place on the seafloor, the Marianas Trench in the Pacific Ocean, is about 11 km deep.

Study **Figure 16.16.** Notice that the **continental margin** is the area where edges of continents meet the ocean. It consists of continental crust, covered with sediments, that eventually meets oceanic crust. Continental margins represent the shallowest parts of the ocean. As shown in **Figure 16.16,** a continental margin includes the continental shelf, the continental slope, and the continental rise.

Continental shelf The shallowest part of a continental margin extending seaward from the shore is the **continental shelf.** Continental shelves vary greatly in width, averaging 60 km wide. On the Pacific coast of the United States, the continental shelf is only a few kilometers wide, whereas the continental shelf of the Atlantic coast is hundreds of kilometers wide.

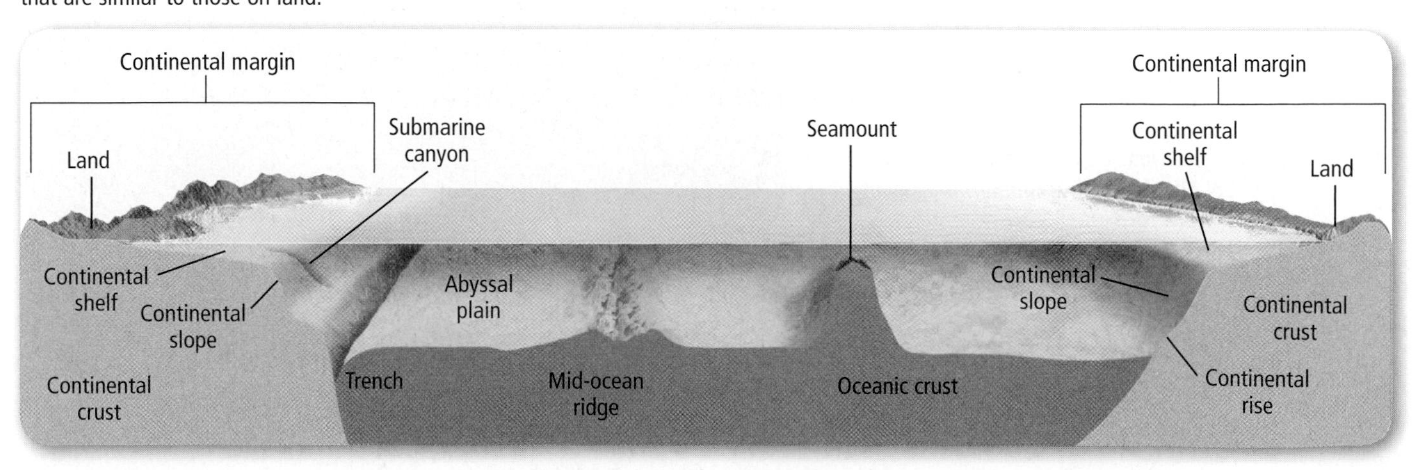

■ **Figure 16.16** A cross section of the ocean reveals a diverse topography with many features that are similar to those on land.

VOCABULARY
SCIENCE USAGE V. COMMON USAGE
Shelf
Science usage: the sloping border of a continent or island

Common usage: a flat piece of material attached to a wall on which objects are placed

The average depth of the water above continental shelves is about 130 m. Recall that sea level during the last ice age was approximately 130 m lower than at present; therefore, most of the world's continental shelves must have been above sea level at that time. As a result, present-day coastlines are radically different from the way they were during the last ice age. At that time, Siberia was attached to North America by the Bering land bridge, Great Britain was attached to Europe, and a large landmass existed where today there are only the widely scattered islands of the Bahamas.

When Earth's surface began to warm after the last ice age, and the continental ice sheets began to melt, the sea gradually covered up the continental shelves. Beaches and other coastal landforms from that time are now submerged and located far beyond the present shoreline. Commercially valuable fishes now inhabit the shallow, nutrient-rich waters of the continental shelves. In addition, the thick sedimentary deposits on the shelves are significant sources of oil and natural gas.

 **Reading Check** **List** the resources that can be found on the continental shelf.

Continental slope Beyond the continental shelves, the seafloor drops away quickly to depths of several kilometers, with slopes averaging nearly 100 m/km. These sloping regions are the **continental slopes.** To marine geologists, the continental slope is the true edge of a continent because it generally marks the edge of the continental crust. In many places, this slope is cut by deep submarine canyons, which are shown in **Figure 16.17.** Submarine canyons are similar to canyons on land and some are comparable in size to the Grand Canyon in Arizona.

These submarine canyons were cut by **turbidity currents,** which are rapidly flowing water currents along the bottom of the sea that carry heavy loads of sediments, similar to mudflows on land. Turbidity currents, shown in **Figure 16.18,** might originate as underwater landslides on the continental slope that are triggered by earthquakes, or they might originate from sediment stirred up by large storm waves on the continental shelf. Turbidity currents can reach speeds exceeding 30 km/h and effectively erode bottom sediments and bedrock.

■ **Figure 16.17** Submarine canyons are similar to canyons on land. These deep cuts in the continental slope vary in size and can be as deep as the Grand Canyon.
Explain *how submarine canyons are formed.*

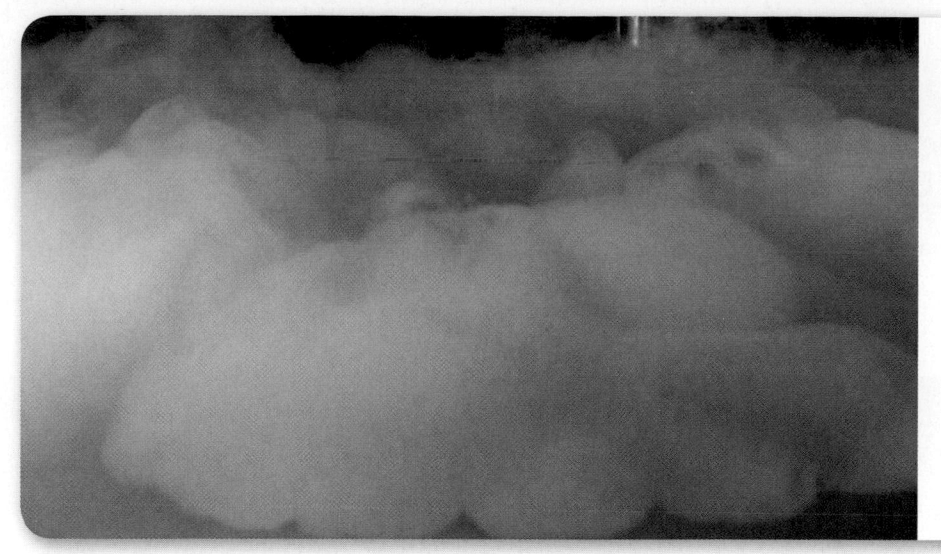

■ **Figure 16.18** Turbidity currents flow along the seafloor because the seawater-sediment mixture of the current is denser than seawater alone. Scientists study turbidity currents in the lab by simulating them in glass tanks, as shown.

Continental rise The gently sloping accumulation of deposits from turbidity currents that forms at the base of the continental slope is called a **continental rise.** A continental rise can be several kilometers thick. The rise gradually becomes thinner and eventually merges with the sediments of the seafloor beyond the continental margin. In some places, especially around the Pacific Ocean, the continental slope ends in deeper depressions in the seafloor, known as deep-sea trenches. In such places, there is no continental rise at the foot of the continental margin.

Reading Check **Differentiate** between the continental slope and the continental rise.

PROBLEM-SOLVING LAB

Interpret Graphs

How do surface elevations compare? A useful comparison of the heights of the continents to the depths of the oceans is given by the curve in the graph below.

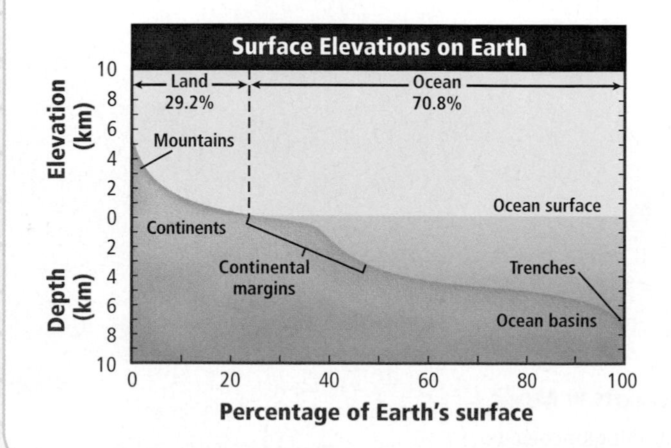

Analysis

1. Approximately how tall is the highest mountain on Earth's surface in kilometers?
2. At about what depth would you begin to find trenches on the ocean floor?
3. What percentage of Earth's surface is above current sea level?
4. What percentage of Earth's surface is represented by the continental margin?

Think Critically

5. **Calculate** The oceanic crust is the part of the crust that is at a depth of 2 km or more below sea level. What percentage of Earth's surface lies above the oceanic crust?

Visualizing the Ocean Floor

Figure 16.19 The ocean floor has topographic features, including mid-ocean ridges, trenches, abyssal plains, and seamounts.

Fracture zones run perpendicular to mid-ocean ridges.

Abyssal plains are smooth, flat areas of the deep ocean covered with fine-grained sediments.

Seamounts are submerged volcanoes that are more than 1 km high.

Mid-ocean ridges run through all ocean basins.

Trenches are the deepest areas of the ocean.

Concepts In Motion To explore more about the ocean floor, visit glencoe.com.

Earth Science Online

Deep-Ocean Basins

Beyond the continental margin are ocean basins, which represent about 60 percent of Earth's surface and contain some of Earth's most interesting topography. **Figure 16.19** shows the topography of the ocean basin beneath the Atlantic Ocean.

Abyssal plains The flattest parts of the ocean floor 5 or 6 km below sea level are called **abyssal plains.** Abyssal plains, shown in **Figure 16.19,** are plains covered with hundreds of meters of fine-grained muddy sediments and sedimentary rocks that were deposited on top of basaltic volcanic rocks.

Deep-sea trenches The deepest parts of the ocean basins are the **deep-sea trenches,** which are elongated, sometimes arc-shaped depressions in the seafloor several kilometers deeper than the adjacent abyssal plains. Many deep-sea trenches lie next to chains of volcanic islands, such as the Aleutian Islands of Alaska, and most of them are located around the margins of the Pacific Ocean. Deep-sea trenches are relatively narrow, about 100 km wide, but they can extend for thousands of kilometers. The Peru-Chile trench, shown in **Figure 16.20,** is almost 6000 km long and has an average width of 40 km.

Mid-ocean ridges The most prominent features of the ocean basins are the **mid-ocean ridges,** which run through all the ocean basins and have a total length of more than 65,000 km—a distance greater than Earth's circumference. Mid-ocean ridges have an average height of 1500 m, but they can be thousands of kilometers wide. The highest peaks in mid-ocean ridges are over 6 km tall and emerge from the ocean as volcanic islands. Mid-ocean ridges are sites of frequent volcanic eruptions and earthquake activity. The crests of these ridges often have valleys called rifts running through their centers. Rifts can be up to 2 km deep.

Mid-ocean ridges do not form continuous lines. They are broken into a series of shorter, stepped sections, which run at right angles across each mid-ocean ridge. The areas where these breaks occur are called fracture zones, shown in **Figure 16.21.** Fracture zones are about 60 km wide, and they curve gently across the seafloor, sometimes for thousands of kilometers.

Figure 16.20 The Peru-Chile trench runs along the west coast of South America.

FOLDABLES Incorporate information from this section into your Foldable.

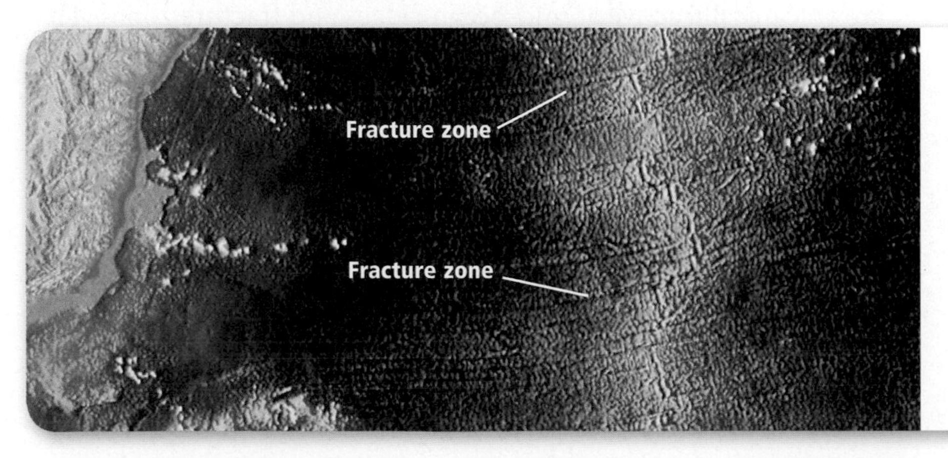

Figure 16.21 There are many fracture zones along the Mid-Atlantic Ridge.

■ **Figure 16.22** Black smokers form when metal oxides and sulfides precipitate out of fluid heated by magma. White smokers form when elements such as calcium and barium precipitate out of warm water ejected from rifts in mid-ocean ridges.

Hydrothermal vents A hydrothermal vent is a hole in the seafloor through which fluid heated by magma erupts. Most hydrothermal vents are located along the bottom of the rifts in mid-ocean ridges. When the heated fluid that erupts from these vents contains metal oxides and sulfides, they immediately precipitate out of the fluid and produce thick, black, smokelike plumes. This type of hydrothermal vent, known as a black smoker, ejects superheated water with temperatures of up to 350°C. **Figure 16.22** illustrates the black smokers found in the rift valley of a mid-ocean ridge.

A second type of vent, known as a white smoker, also shown in **Figure 16.22,** ejects warm water. Smokers are caused by seawater circulating through the hot crustal rocks in the centers of mid-ocean ridges. The fundamental cause of mid-ocean ridges and the volcanic activity associated with them is plate tectonics, which you will read about in Chapter 17.

Reading Check **Identify** where most hydrothermal vents are located.

Seamounts and guyots Satellite data have revealed that the ocean floor is dotted with tens of thousands of solitary mountains. These mountains are not located near areas of active volcanism. How, then, did they form? You have learned that the ocean basins are volcanically active at mid-ocean ridges and fracture zones. The almost total absence of earthquakes in most other areas of the seafloor suggests that volcanism in those areas must have ceased a long time ago. Thus, most of the mountains on the seafloor are probably extinct volcanoes.

Investigations of individual volcanoes on the seafloor have revealed that there are two types: seamounts and guyots (GEE ohz). **Seamounts** are submerged basaltic volcanoes more than 1 km high. Many linear chains of seamounts, such as the Emperor seamount chain, are stretched out across the Pacific Ocean basin in roughly the same direction. **Guyots** are large, extinct, basaltic volcanoes with flat, submerged tops.

Marine Sediments

The sediments that cover the ocean floor come from a variety of sources, but most come from the continents. Land-derived sediments include mud and sand washed into the oceans by rivers, as well as dust and volcanic ash blown over the ocean by winds. Much of the coarser material supplied by rivers settles out near shorelines or on beaches, but fine-grained material such as silt and clay settles so slowly through water that some tiny particles take centuries to reach the bottom.

Terrigenous sediments Ocean currents disperse fine silt, clay, and volcanic ash from land, called terrigenous sediments, throughout the ocean basins. Thus, the dominant type of sediment on the deep ocean floor is fine-grained, deep-sea mud. Deep-sea mud usually has a reddish color because the iron present in some of the sediment grains becomes oxidized during the descent to the ocean bottom. Closer to land, the sediments become mixed with coarser materials such as sand, but some sandy sediments occasionally reach the abyssal plains in particularly strong turbidity currents.

Biogenous sediments Deep-sea sediments that come from biological activity are called biogenous sediments. When some marine organisms die, such as the diatoms shown in **Figure 16.23,** their shells settle on the ocean floor. Sediments containing a large percentage of particles derived from once-living organisms are called oozes. Most of these particles are small and consist of either calcium carbonate or silica. The oozes and deep-sea mud of the deep ocean typically accumulate at a rate of only a few millimeters per thousand years.

Figure 16.23 The shells of microscopic marine organisms, such as diatoms, as well as shell fragments and hard parts from larger marine organisms make up biogenous sediments.
SEM: magnification unknown

MiniLab

Measure Sediment Settling Rates

How fast do sediment grains sink?

Procedure

1. Read and complete the lab safety form.
2. Obtain **sediment grains** with approximate diameters of 0.5 mm, 1 mm, 2 mm, 5 mm, and 10 mm.
3. Draw a data table with these headings: *Type of Particle, Diameter (mm), Distance (cm), Time (s),* and *Settling Speed (cm/s).*
4. Measure and record the diameters of each specimen using a **set of sieves.**
5. Fill a **250-mL graduated cylinder** with **cooking oil.** Measure the height of the cooking oil.
6. Drop the largest specimen into the oil. Use a **stopwatch** to measure and record the time it takes for the specimen to sink to the bottom of the cylinder.
7. Repeat Step 6 for the remaining specimens.

Analysis

1. **Calculate** the settling speed for each specimen, and fill in your data table.
2. **Plot** the settling speed (cm/s) against particle diameter (mm) on a graph.
3. **Explain** How do settling speeds change as particle sizes decrease?

■ **Figure 16.24** Metals, such as manganese, precipitate directly from seawater. These manganese nodules consist of manganese and iron and range in size from a few centimeters to 10 cm across.

Hydrogenous sediments While terrigenous sediments are derived from land and biogenous sediments are derived from biological activity, another sediment type, called hydrogenous sediments, is derived from elements in seawater. For example, manganese nodules, shown in **Figure 16.24,** consist of oxides of manganese, iron, and copper and other valuable metals that have precipitated directly from seawater. The precipitation happens slowly, therefore growth rates of manganese nodules are extremely slow. Growth rates of manganese nodules are measured in millimeters per million years.

Manganese nodules cover huge areas of the deep-sea floor. Although some have tried to mine manganese nodules for their valuable metals, the difficulty and cost involved in removing them from a depth of 5000 to 6000 m has made progress slow.

Section 16.2 Assessment

Section Summary

- The oceans cover parts of the continental crust as well as oceanic crust.
- A continental margin consists of the continental shelf, the continental slope, and the continental rise.
- Deep-ocean basins consist of abyssal plains, trenches, mid-ocean ridges, seamounts, and guyots.
- Most deep-sea sediments are fine-grained and accumulate slowly.

Understand Main Ideas

1. MAIN Idea **Describe** the features of deep-ocean basins.
2. **Identify** which sediment sinks faster—pebbles or sand grains.
3. **Summarize** the differences between deep-sea mud and oozes.
4. **Compare and contrast** the characteristics of the three major areas of the continental margin.

Think Critically

5. **Suggest** If there is little volcanic activity on abyssal plains, yet they are dotted with thousands of seamounts, from where did these extinct volcanoes come?

WRITING in Earth Science

6. Suppose you are taking side-scan sonar readings as your ship moves across the Pacific Ocean from east to west. Describe how you would interpret the sonar data according to the features you would expect to find beneath the surface.

Earth Science Online Self-Check Quiz glencoe.com

ON SITE:
SURVEYING THE DEEP OCEAN FLOOR

Alvin is a human-occupied submersible vehicle that is used to study hydrothermal vents and other deep-sea features.

There are a variety of vessels that can collect data on the deep ocean floor ranging from human-occupied submersible vehicles to free-moving autonomous robots to remotely operated vehicles. These vessels are used by oceanographers to learn more about the topography, biology, and chemistry of the deep ocean.

Submersibles *Alvin* is one of several human-occupied deep-sea submersibles used worldwide. In *Alvin,* the trip to the bottom of the ocean takes about two hours, and it takes two hours to return to the surface. Under the immense pressure of thousands of kilograms of water and in the darkness and cold of the deep ocean, scientists have about four hours of bottom time to record video and collect water, sediment, and biological samples. Using *Alvin,* scientists got their first view of hydrothermal vents in 1977, a major discovery for both geologists and biologists.

Autonomous underwater vehicles The Autonomous Benthic Explorer (ABE) has a mass of over 500 kg and is 2 m long, but, thanks to computer systems, the bulky vehicle is able to dive into narrow trenches and zip around seamounts at depths as deep as 5000 m. Autonomous underwater vehicles are able to move around independent of a pilot, ship, or submersible.

The ABE has completed over 150 dives to the deep seafloor, recording videos of the deep sea environment and taking temperature and water chemistry samples.

Remotely operated vehicles *Jason/Medea* is a remotely operated vehicle (ROV) system that can collect data at depths as deep as 6000 m. An ROV is deployed from a ship, but unlike the ABE, it needs a pilot, an engineer, and a navigator to control its operations from the ship. In 2006, *Jason/Medea* explored part of the western Pacific Ocean near the Marianas Trench, collecting data on water chemistry near active undersea volcanoes and hydrothermal vents. Each dive provided scientists with views—from glowing, red, hot lava pouring out a vent from a volcano, to black-smoker hydrothermal vents and the unique organisms that live near them—that were never disappointing.

WRITING in Earth Science

Time Line Research more information about the history of exploration of the deep sea using human-occupied vessels at glencoe.com. Create a time line that displays dates of important technological advances and discoveries in this area.

GEOLAB

MAPPING: IDENTIFY COASTAL LANDFORMS

Background: Topographic maps of coastal areas show a two-dimensional representation of coastal landforms. You can identify an emergent coast by the landforms along the coastline as well as landforms found inland.

Question: *How can you identify and describe the coastal landforms of an emergent coast on a topographic map?*

Materials

metric ruler
drafting compass
graph paper
calculator

Procedure

1. Read and complete the lab safety form.
2. Determine the map scale and the contour interval.
3. On the inset map, plot a west-east cross section of the coast just north of Coon Creek from sea level depth contour to a point 2000 ft inland. Use a horizontal scale of 1:24,000.
4. Use both maps to answer the following questions.

Analyze and Conclude

1. **Describe** What kind of coastal landform is the Morro Rock Peninsula?
2. **Explain** What kind of feature is Pillar Rock, and how was it formed?
3. **Interpret** On what coastal feature is Morro Bay State Park located? How was the feature formed?
4. **Infer** What is the direction of the longshore transport along Morro Bay?
5. **Apply** Your west-east cross section shows an elevated flat area next to the shoreline. What kind of coastal landform is this? How was it formed?

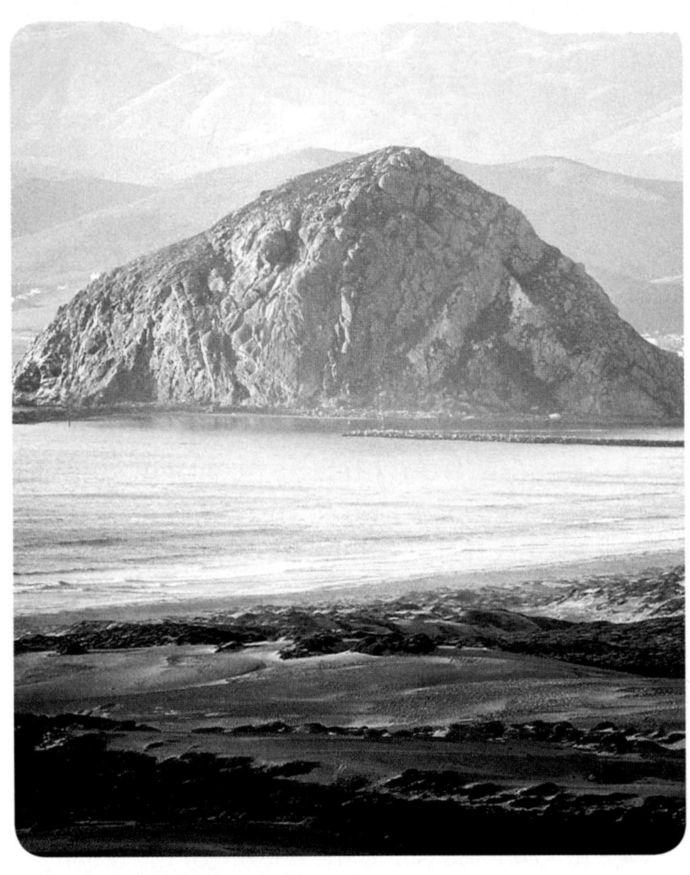

Morro Rock is located off the coast of California.

6. **Draw Conclusions** If sea level dropped 10 m, how would the shoreline change? How far would it move seaward? Would it become more regular or irregular? What would happen to Morro Bay?
7. **Suggest** three major changes that could occur to the coastal region if sea level rose 6 m.

APPLY YOUR SKILL

Compare and contrast this coastal section with a section of the Texas coast between Corpus Christi and Galveston. Which coastal features are similar? Which are different?

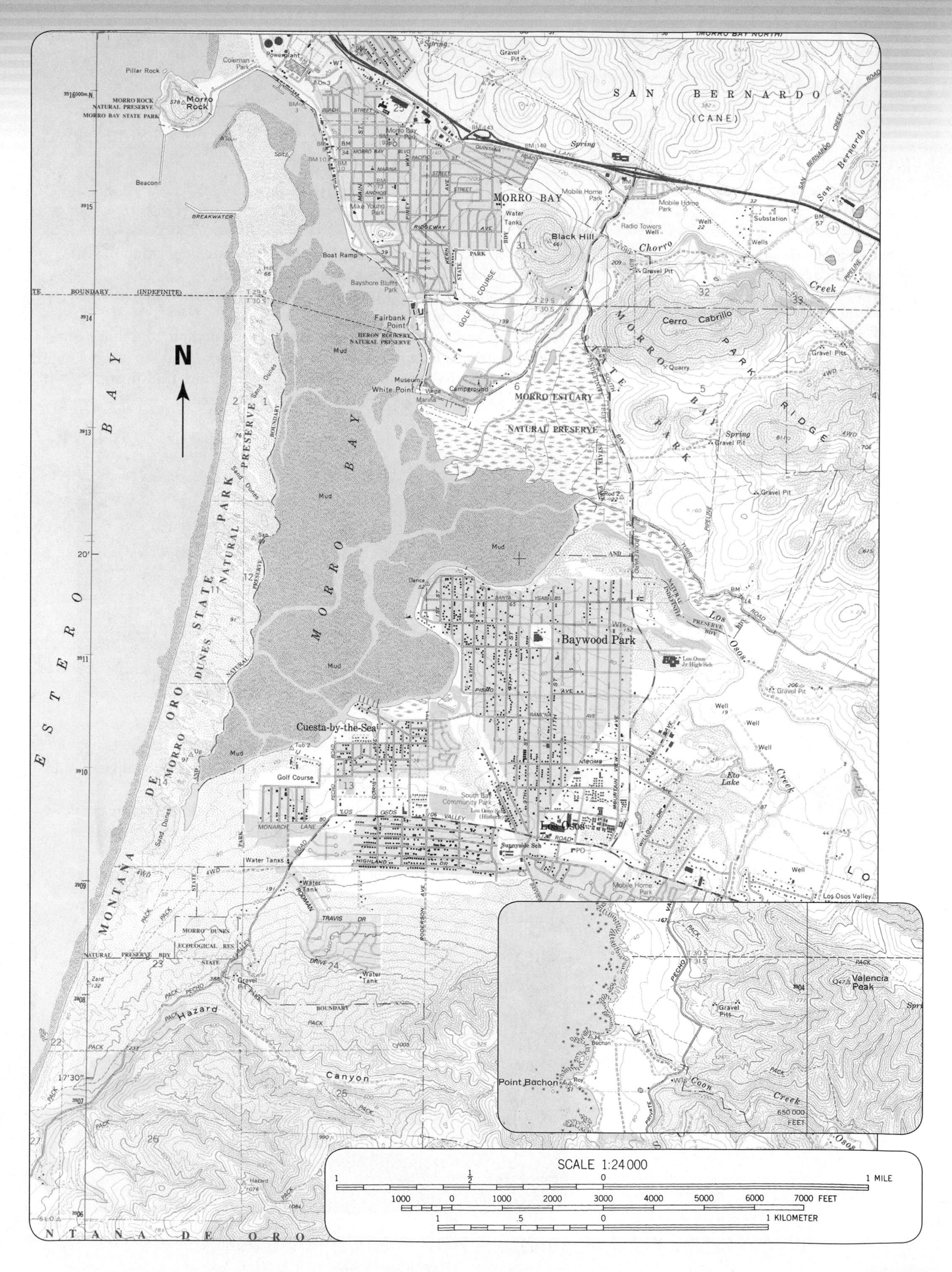
N
SCALE 1:24 000
1 MILE
1000 0 1000 2000 3000 4000 5000 6000 7000 FEET
1 .5 0 1 KILOMETER
Morro Rock
Pillar Rock
MORRO ROCK NATURAL PRESERVE
MORRO BAY STATE PARK
Coleman Park
Powerplant
Beacon
BREAKWATER
SAN BERNARDO (CANE)
MORRO BAY
Black Hill
Chorro Creek
Cerro Cabrillo
MORRO BAY STATE PARK
MORRO PARK RIDGE
MORRO ESTUARY NATURAL PRESERVE
Fairbank Point
HERON ROOKERY NATURAL PRESERVE
White Point
Campground
Museum
Marina
Boat Ramp
Bayshore Bluffs Park
GOLF COURSE
Mud
MORRO BAY
ESTERO BAY
MORRO DUNES NATURAL PRESERVE
MONTAÑA DE ORO STATE PARK
Sand Dunes
Baywood Park
Los Osos Jr High Sch
Los Osos Creek
Cuesta-by-the-Sea
Golf Course
South Bay Community Park
Los Osos
Sunnyside Sch
Eto Lake
Mobile Home Park
Los Osos Valley
MORRO DUNES ECOLOGICAL RES
Water Tanks
Water Tank
Hazard Canyon
Gravel Pit
Point Buchon
Valencia Peak
Coon Creek
650 000 FEET
NTAÑA DE ORO

CHAPTER 16 Study Guide

BIG Idea The marine environment is geologically diverse and contains a wealth of natural resources.

Vocabulary	Key Concepts

Section 16.1 Shoreline Features

Vocabulary

- barrier island (p. 442)
- beach (p. 438)
- longshore bar (p. 440)
- longshore current (p. 441)
- wave refraction (p. 439)

Key Concepts

MAIN Idea The constant erosion of the shoreline and deposition of sediments by ocean waves creates a changing coastline.

- Wave erosion of headlands produces wave-cut platforms and cliffs, sea stacks, sea arches, and sea caves.
- Wave action and longshore currents move sediment along the shore and build depositional features.
- Artificial protective structures interfere with longshore transport.
- Sea levels in the past were 130 m lower than at present.

Section 16.2 Seafloor Features

Vocabulary

- abyssal plain (p. 451)
- continental margin (p. 447)
- continental rise (p. 449)
- continental shelf (p. 447)
- continental slope (p. 448)
- deep-sea trench (p. 451)
- guyot (p. 452)
- mid-ocean ridge (p. 451)
- seamount (p. 452)
- turbidity current (p. 448)

Key Concepts

MAIN Idea The ocean floor contains features similar to those on land and is covered with sediments of several origins.

- The oceans cover parts of the continental crust as well as oceanic crust.
- A continental margin consists of the continental shelf, the continental slope, and the continental rise.
- Deep-ocean basins consist of abyssal plains, trenches, mid-ocean ridges, seamounts, and guyots.
- Most deep-sea sediments are fine-grained and accumulate slowly.

Assessment

Vocabulary Review

Write the definition for each vocabulary term listed below.

1. barrier island
2. beach
3. longshore bar
4. wave refraction

Complete the sentences below using vocabulary terms from the Study Guide.

5. ________ can create submarine canyons by cutting through bottom sediments and bedrocks of seafloors.
6. The deepest part of an ocean basin is a(n) ________.
7. The ________ is the part of a continent that is submerged under the ocean.
8. Submerged basaltic volcanoes more than 1 km high are ________.

Understand Key Concepts

9. Which coastal features are usually found in the bays along irregular coasts with headlands?
   A. sea stacks
   B. wave-cut cliffs
   C. wave-cut platforms
   D. beaches

10. Which coastal landform is not produced by longshore transport?
    A. barrier island
    B. sand spit
    C. baymouth bar
    D. sea stack

11. Which is the correct order of features on the continental margin moving from land out to sea?
    A. continental slope, continental shelf, continental rise
    B. continental rise, continental trench, continental shelf
    C. continental shelf, continental slope, continental rise
    D. continental slope, continental shelf, continental trench

12. What percentage of Earth's surface is represented by ocean basins?
    A. 10 percent
    B. 30 percent
    C. 50 percent
    D. 60 percent

13. What do the sediments of the abyssal plains mostly consist of?
    A. sand and gravel
    B. salt
    C. seashells
    D. mud and oozes

14. Where are most deep-sea trenches located?
    A. in the Atlantic Ocean
    B. in the Pacific Ocean
    C. in the Indian Ocean
    D. in the Arctic Ocean

15. Which runs through all the oceans?
    A. abyssal plains
    B. the mid-ocean ridges
    C. deep sea trenches
    D. seamounts and guyots

Use the diagram below to answer Questions 16 and 17.

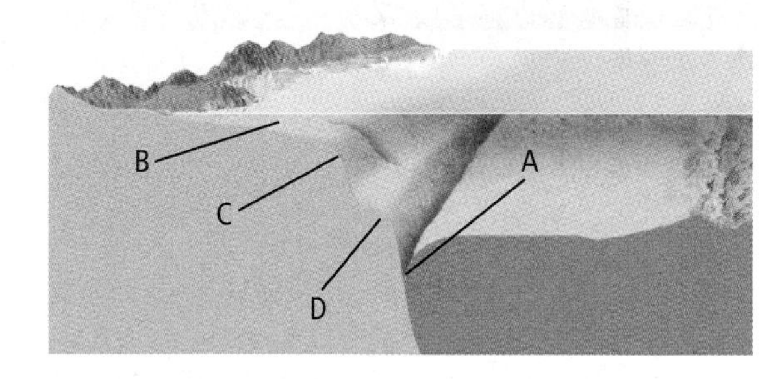

16. Which letter indicates the continental shelf?
    A. A
    B. B
    C. C
    D. D

17. Which feature is indicated by the letter A?
    A. guyot
    B. continental slope
    C. continental rise
    D. trench

18. Which marks the true edge of a continent?
   A. submarine canyon
   B. continental slope
   C. continental shelf
   D. abyssal plain

19. Which seafloor feature can be found along rifts in the mid-ocean ridges?
   A. hydrothermal vents
   B. manganese nodules
   C. deep-sea trenches
   D. seamounts

20. Which represents the flattest part of Earth's surface?
   A. deep-sea trenches
   B. continental margins
   C. abyssal plains
   D. mid-ocean ridges

Use the diagram below to answer Question 21.

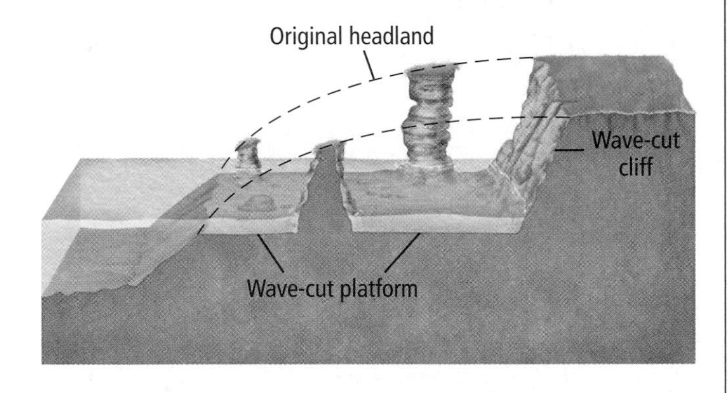

21. Which features are not caused by erosion?
   A. wave-cut platforms
   B. wave-cut cliffs
   C. original headlands
   D. sea stacks

22. Which features of the seafloor are cut by turbidity currents?
   A. longshore bars
   B. submarine canyons
   C. abyssal plains
   D. baymouth bars

23. Which are not associated with mid-ocean ridges?
   A. black smokers
   B. guyots
   C. fracture zones
   D. hydrothermal vents

Constructed Response

24. **Explain** Is global sea level currently rising, falling, or staying the same? During the last ice age, was the sea level higher, lower, or the same as at present?

25. **Explain** the relationship between oozes and the sedimentary rock known as chalk.

26. **Illustrate** the effect that wave refraction has on incoming wave crests that approach the coast at an angle.

Use the diagram below to answer Question 27.

27. **Apply** Copy the diagram above onto a sheet of paper, and label the following features: *shore, beach, low-tide shoreline, high-tide shoreline, coast,* and *waves.*

28. **Illustrate** how incoming waves along a shoreline create the longshore current.

29. **Apply** If a coast has elevated marine terraces, is it more likely to be rising or sinking? Explain.

30. **Differentiate** between a baymouth bar and a tombolo.

31. **Discuss** the processes that can cause sea level changes.

32. **Compare and contrast** seamounts and guyots. How are they formed?

33. **Analyze** Why does deep-sea mud usually have a reddish color?

34. **Compare and contrast** the origins of marine sediments.

Think Critically

35. **Infer** Form an inference to explain why seamounts are extinct volcanoes.

36. **Suppose** you are surfing at a beach in California. Describe which near-shore currents might affect your position relative to where you start out on the beach after each cycle of surfing. Will you come back to the same point?

37. **Suggest** Evidence shows that woolly mammoths moved from Siberia to North America during the last ice age. How could they have crossed the two continents?

38. **Suppose** that you are swimming in the ocean toward the shoreline, and you suddenly find that, despite all your efforts to swim ahead, you remain in the same location. Explain why this might happen, and describe the path in which you should swim to get yourself out of this situation.

39. **Assess** How is it possible to have a coast that is sinking when global sea level is falling?

Use the diagram below to answer Question 40.

40. **Evaluate** The arrow is pointing to what ocean floor feature? What is the significance of this feature?

Concept Mapping

41. Use the following terms to construct a concept map about the continental margin: *continental shelf, continental slope, continental rise, turbidity currents,* and *submarine canyons.* Refer to the *Skillbuilder Handbook* for more information.

Challenge Question

42. **Investigate** the causes and effects of a natural disaster that occurred in the past five years and resulted in coastal erosions in the United States or in any other country.

Additional Assessment

43. WRITING in **Earth Science** Write an editorial article for a newspaper that discusses some of the risks involved in living in a low-lying coastal area.

DBQ Document–Based Questions

Data obtained from: National assessment of shoreline change: part 1 historical shoreline changes and associated land loss along the U.S. Gulf of Mexico. Report 2004–1043. *U.S. Geological Survey.*

The graph below shows the average annual sea level based on tide data for a 100-year period for two cities on the Gulf Coast of the United States.

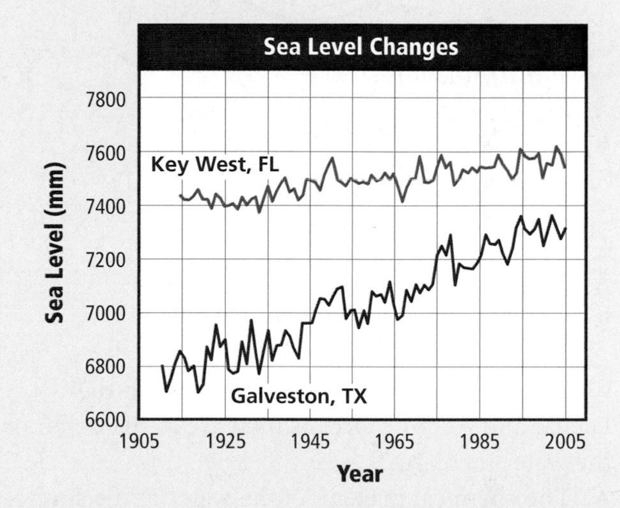

44. Estimate the sea level change for Key West, Florida, and Galveston, Texas, by roughly fitting a straight best-fit line through the data points.

45. The global mean sea level rise for the past century is estimated to be about 0.18 m. Which location has a sea level rise closer to the global mean?

46. What is one possible cause for global sea level rise? What is one reason that a local sea level change can be much larger than the global average?

Cumulative Review

47. What are the four most common foliated metamorphic rocks? **(Chapter 6)**

48. Describe the process that creates supercell thunderstorms. **(Chapter 13)**

Standardized Test Practice

Multiple Choice

1. Which ocean movement is slow-moving and occurs in deep waters?

A. surface currents **C.** density currents
B. upwelling **D.** gyres

Use the illustration below to answer Questions 2 and 3.

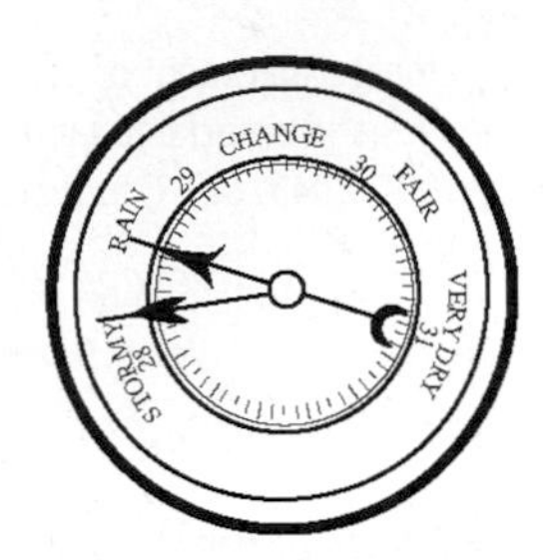

2. What instrument is shown?

A. thermometer **C.** barometer
B. hygrometer **D.** rain gauge

3. If a barometer records high pressure, what type of weather can be expected?

A. storms **C.** fair weather
B. cloudy weather **D.** hot weather

4. Why does the silica content of seawater in the Pacific and Atlantic Oceans increase as the depth of the water increases?

A. The chemical makeup of the water at the surface of the oceans dissolves more silica.
B. When shelled organisms die, their shells fall to the floor of the ocean and dissolve, releasing silica into the water.
C. Wave action at the surface of the oceans dilutes the silica.
D. Oxygen entering the surface water pushes silica down into deeper water.

5. How might volcanic activity affect changes in the climate?

A. Volcanic dust blocks incoming radiation causing lower global temperatures.
B. Volcanic dust traps radiation in the Earth's atmosphere raising global temperatures.
C. Lava running along Earth's surface increases the temperature thus increasing global temperature.
D. Lava thrown into the air increases the atmospheric temperature thus increasing global temperatures.

6. If the northern hemisphere is experiencing long hours of darkness and cold weather, what season is it in the southern hemisphere?

A. spring **C.** winter
B. summer **D.** fall

Use the illustration below to answer Questions 7–9.

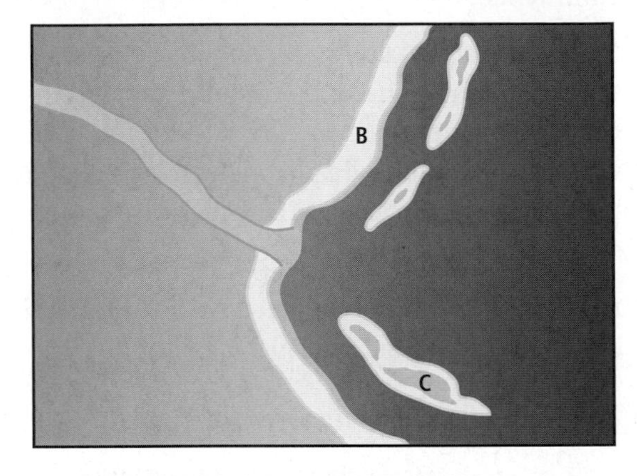

7. What shoreline feature is indicated by B?

A. a tombolo **C.** a lagoon
B. a spit **D.** a beach

8. The material in the section indicated by the B is very coarse. What conclusions can be drawn about the waves and sediment source?

A. The area is rocky and pounded by heavy waves.
B. The area is rocky and hit with light waves.
C. The area consists of fine-grained material that has been hit by heavy waves.
D. The area consists of fine-grained material that has been hit by light waves.

9. Which forms the area indicated by C?

A. large storm waves
B. gaps in the longshore bar
C. the longshore current
D. a rip current

10. Limestone that is exposed to enough heat and pressure is transformed into

A. marble
B. slate
C. quartzite
D. gneiss

Short Answer

Use the illustration below to answer Questions 11 and 12.

11. Describe the process that is occurring in the illustration.

12. How does a greenhouse mimic Earth's atmosphere?

13. How does a greenhouse benefit gardeners?

14. Describe the hypothesis of how comets might have created Earth's early oceans.

15. Discuss a location where an analog forecast could be more beneficial than a digital forecast.

16. If a sonar signal takes 3 s to be emitted and received by an oceanographer mapping the ocean floor, what is the ocean depth in that location? (Hint: sound travels 1500 m/s through water.)

17. If air is rising due to uneven heating of Earth's surface within an air mass, what can you expect to occur?

Reading for Comprehension

Underwater Hot Spots

Scientists think that underwater hot spots might host unique, previously unknown forms of life. Hydrothermal vents at ocean ridges are an essential part of the chemical balance of seawater. They support ecosystems not found anywhere at the surface and are thought to have been the sites of the early formation and evolution of life. The study of these organisms will give scientists insight to the flexibility and adaptability of life. The water that spews forth from hydrothermal vents can reach temperatures of 350°C and is rich in chemicals such as sulfur and salt. Some microorganisms have adapted to the environments on these vents creating rich underwater ecosystems that some scientists think might represent some of the earliest lifeforms on Earth.

Article obtained from: Roach, J. Hydrothermal vents found in Arctic Ocean. *National Geographic News.* January 23, 2003.

18. Which is NOT an importance of hydrothermal vents?
    A. They support unique ecosystems.
    B. They create rich underwater environments.
    C. They help underwater creatures to adapt to extremely hot temperatures.
    D. They are an essential part of the chemical balance of seawater.

19. Why is it rare to find life-forms around hydrothermal vents?
    A. They are extremely hot.
    B. They are along ocean ridges.
    C. They are not chemically balanced with the rest of the ocean.
    D. No life-form is able to survive due to the sulfur in the area.

20. Why is studying these hydrothermal vents so important to scientists?

NEED EXTRA HELP?																	
If You Missed Question . . .	1	2	3	4	5	6	7	8	9	10	11	12	13	14	15	16	17
Review Section . . .	15.3	12.3	11.2	15.2	14.3	14.3	16.1	16.1	16.1	5.2	11.1	11.1	11.1	15.1	12.4	15.1	13.1

The Dynamic Earth

Chapter 17
Plate Tectonics
BIG Idea Most geologic activity occurs at the boundaries between plates.

Chapter 18
Volcanism
BIG Idea Volcanoes develop from magma moving upward from deep within Earth.

Chapter 19
Earthquakes
BIG Idea Earthquakes are natural vibrations of the ground, some of which are caused by movement along fractures in Earth's crust.

Chapter 20
Mountain Building
BIG Idea Mountains form through dynamic processes which crumple, fold, and create faults in Earth's crust.

CAREERS IN EARTH SCIENCE

Volcanologist This **volcanologist** is monitoring volcanic activity to help forecast an eruption. Volcanologists spend much of their time in the field, collecting samples and measuring changes in the shape of a volcano.

WRITING in Earth Science
Visit glencoe.com to learn more about the work of volcanologists. Then write a short newspaper article about how volcanologists predicted a recent eruption.

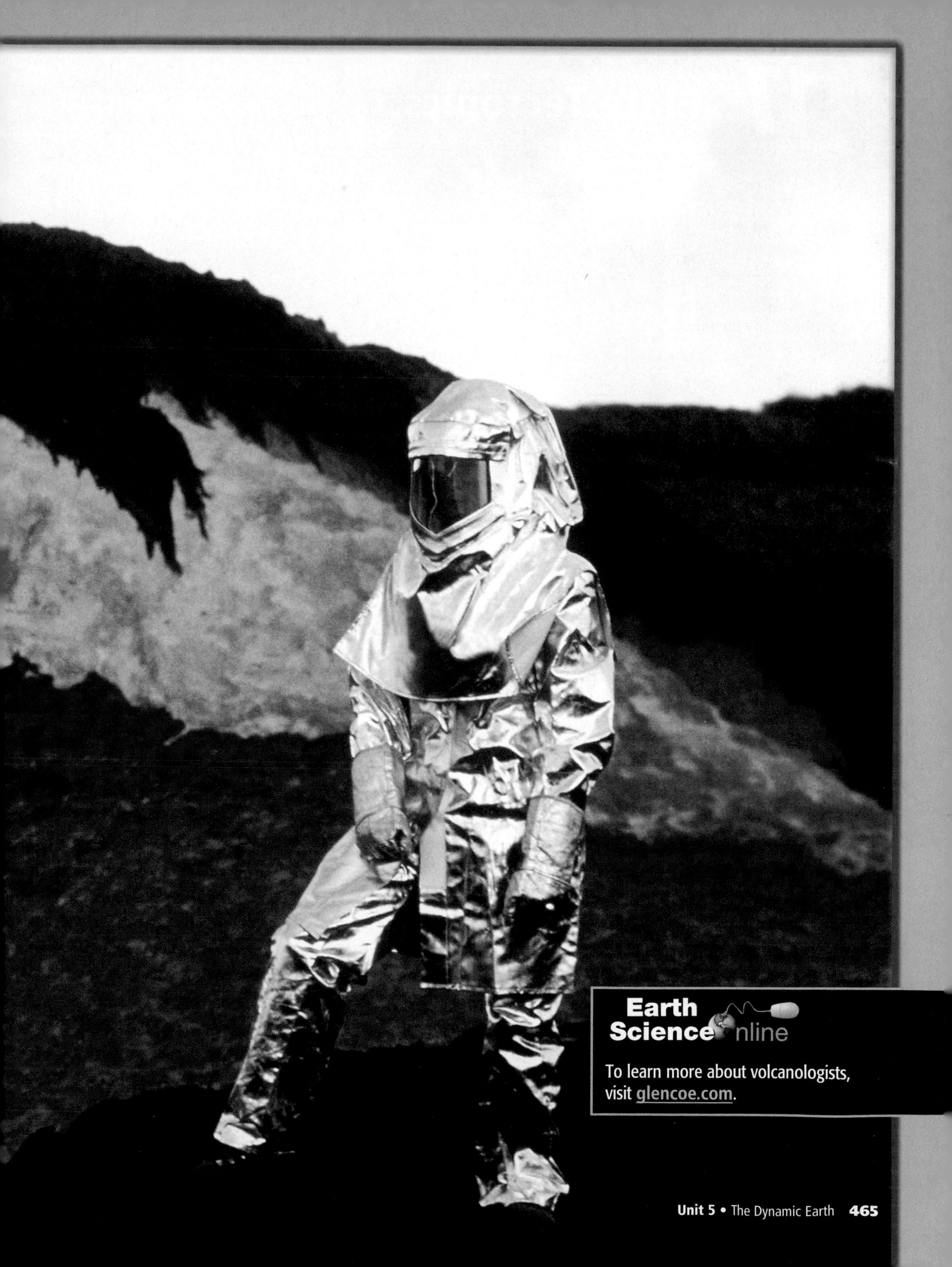

Earth Science Online

To learn more about volcanologists, visit glencoe.com.

CHAPTER 17

Plate Tectonics

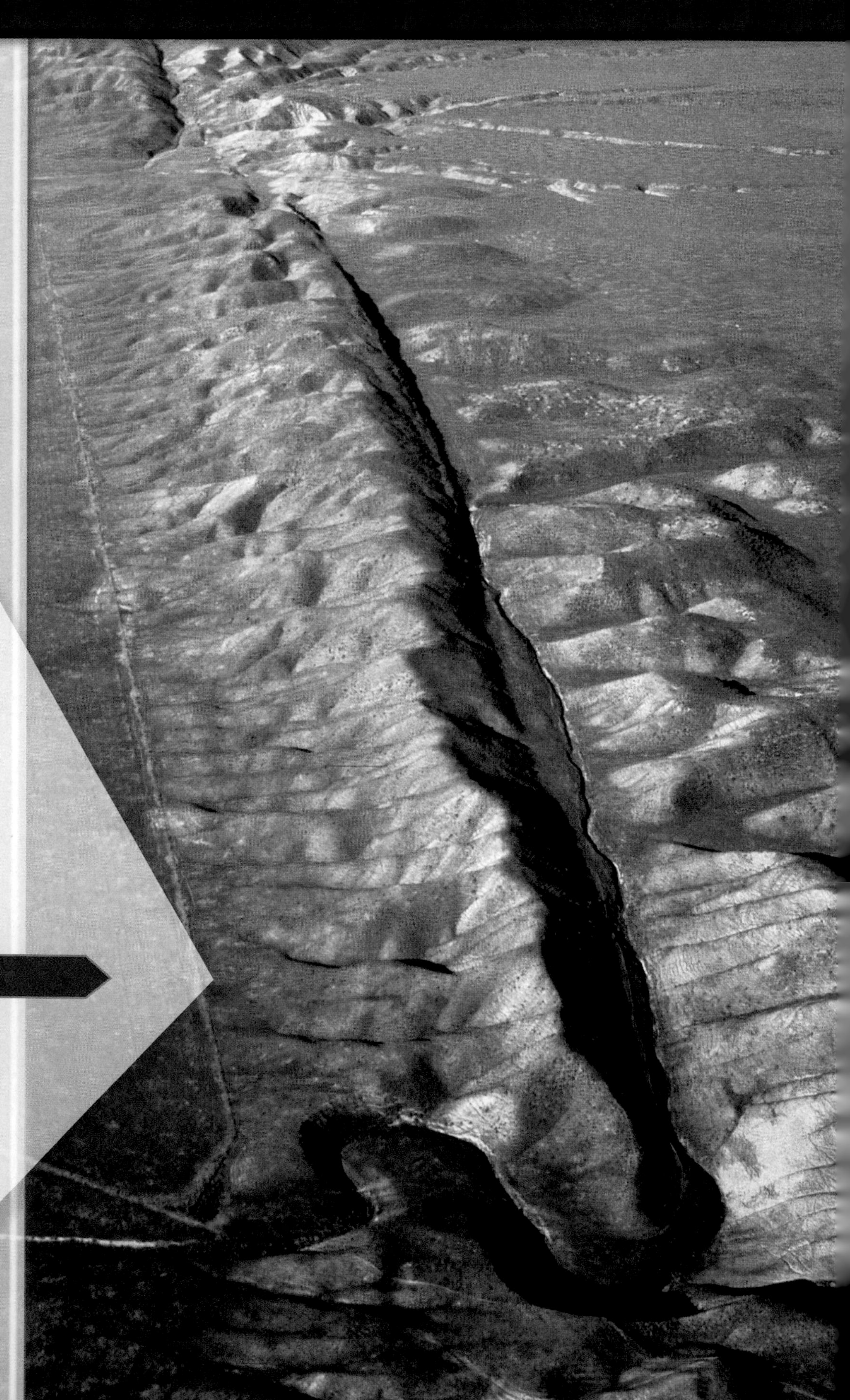

BIG Idea **Most geologic activity occurs at the boundaries between plates.**

17.1 Drifting Continents
MAIN Idea The shape and geology of the continents suggests that they were once joined together.

17.2 Seafloor Spreading
MAIN Idea Oceanic crust forms at ocean ridges and becomes part of the seafloor.

17.3 Plate Boundaries
MAIN Idea Volcanoes, mountains, and deep-sea trenches form at the boundaries between the plates.

17.4 Causes of Plate Motions
MAIN Idea Convection currents in the mantle cause plate motions.

GeoFacts

- The San Andreas Fault is a 1200-km-long gash that runs from northern California almost to Mexico.
- Each year, plate movement along the fault brings Los Angeles about 5 cm closer to San Francisco.
- In this photo, the North American Plate is on the right, the Pacific Plate is on the left.

LAUNCH Lab

Is California moving?

Southwestern California is separated from the rest of the state by a system of cracks along which movement takes place. These cracks are called faults. One of these, as you might know, is the San Andreas Fault. Movement along this fault is carrying southwestern California to the Northwest in relation to the rest of North America at a rate of about 5 cm/y.

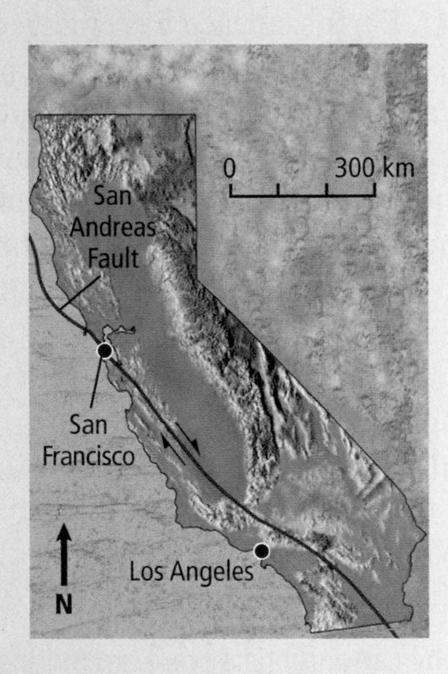

Procedure

1. Read and complete the lab safety form.
2. Use a **metric ruler** and the **map scale** to determine the actual distance between San Francisco and Los Angeles.
3. At the current rate of movement, when will these two cities be next to each other?

Analysis

1. **Infer** what might be causing the motion of these large pieces of land.
2. **Calculate** How far will southwestern California move in a 15-year period?

Plate Boundaries Make this Foldable to compare the types of plate boundaries and their features.

STEP 1 Fold up the bottom edge of a legal-sized sheet of paper about 3 cm and crease.

STEP 2 Fold the sheet into thirds.

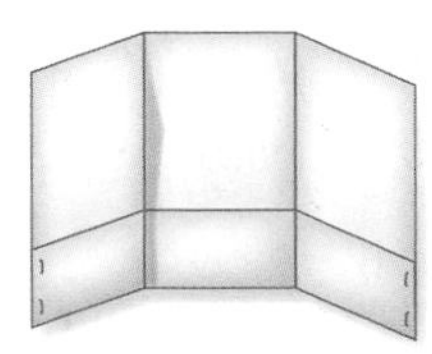

STEP 3 Glue or staple to make three pockets. Label the pockets *Divergent, Convergent,* and *Transform.*

FOLDABLES **Use this Foldable with Section 17.3.** As you read this section, summarize on index cards or quarter sheets of paper the geologic characteristics of each type of boundary and the processes associated with it.

Visit glencoe.com to

- **study entire chapters online;**
- **explore Concepts In Motion animations:**
  - **Interactive Time Lines**
  - **Interactive Figures**
  - **Interactive Tables**
- **access Web Links for more information, projects, and activities;**
- **review content with the Interactive Tutor and take Self-Check Quizzes.**

Section 17.1

Objectives

- **Identify** the lines of evidence that led Wegener to suggest that Earth's continents have moved.
- **Discuss** how evidence of ancient climates supported continental drift.
- **Explain** why continental drift was not accepted when it was first proposed.

Review Vocabulary

hypothesis: testable explanation of a situation

New Vocabulary

continental drift
Pangaea

Drifting Continents

MAIN Idea The shape and geology of the continents suggests that they were once joined together.

Real-World Reading Link When you put together a jigsaw puzzle, what features of the puzzle pieces do you use to find matching pieces? Scientists used features such as shape and position to help them piece together the way the continents were arranged millions of years ago.

Early Observations

With the exception of events such as earthquakes, volcanic eruptions, and landslides, most of Earth's surface appears to remain relatively unchanged during the course of a human lifetime. On the geologic time scale, however, Earth's surface has changed dramatically. Some of the first people to suggest that Earth's major features might have changed were early cartographers. In the late 1500s, Abraham Ortelius (or TEE lee us), a Dutch cartographer, noticed the apparent fit of continents on either side of the Atlantic Ocean. He proposed that North America and South America had been separated from Europe and Africa by earthquakes and floods. During the next 300 years, many scientists and writers noticed and commented on the matching coastlines. **Figure 17.1** shows a proposed map by a nineteenth-century cartographer.

The first time that the idea of moving continents was proposed as a scientific hypothesis was in the early 1900s. In 1912, German scientist Alfred Wegener (VAY guh nur) presented his ideas about continental movement to the scientific community.

Reading Check Infer why cartographers were among the first to suggest that the continents were once joined together.

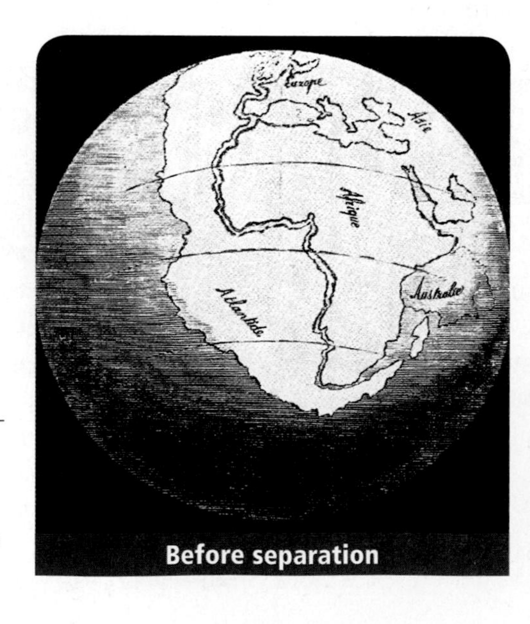

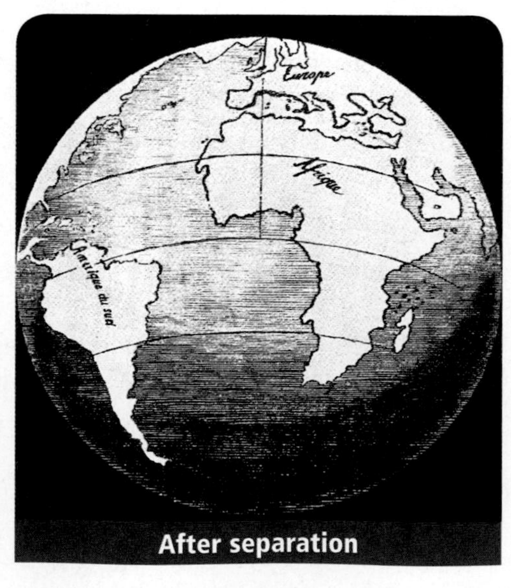

■ **Figure 17.1** Many early cartographers, such as Antonio Snider-Pelligrini, the author of these 1858 maps, noticed the apparent fit of the continents.

Continental Drift

Wegener developed an idea that he called **continental drift,** which proposed that Earth's continents had once been joined as a single landmass that broke apart and sent the continents adrift. He called this supercontinent **Pangaea** (pan JEE uh), a Greek word that means *all the earth,* and suggested that Pangaea began to break apart about 200 mya. Since that time, he reasoned, the continents have continued to slowly move to their present positions, as shown in **Figure 17.2.**

Of the many people who had suggested that continents had moved around, Wegener was the first to base his ideas on more than just the puzzlelike fit of continental coastlines on either side of the Atlantic Ocean. For Wegener, these gigantic puzzle pieces were just the beginning. He also collected and organized rock, climatic, and fossil data to support his hypothesis.

Concepts In Motion

Interactive Figure To see an animation of the breakup of Pangaea, visit glencoe.com.

Figure 17.2 Wegener hypothesized that all the continents were once joined together. He proposed that it took 200 million years of continental drift for the continents to move to their present positions.
Locate *the parts of Pangaea that became North and South America. When were they joined? When were they separated?*

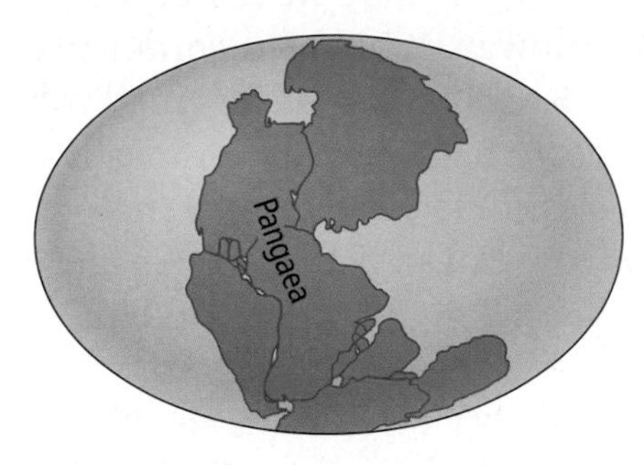

200 mya: All the continents assembled in a single landmass that Wegener named Pangaea.

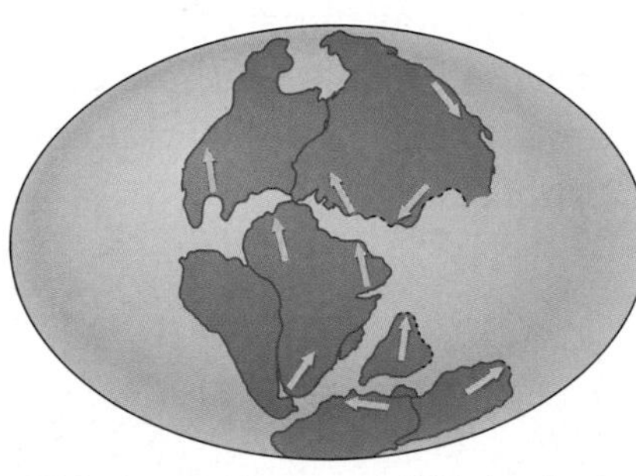

180 mya: Continental rifting breaks Pangaea into several landmasses. The North Atlantic Ocean starts to form.

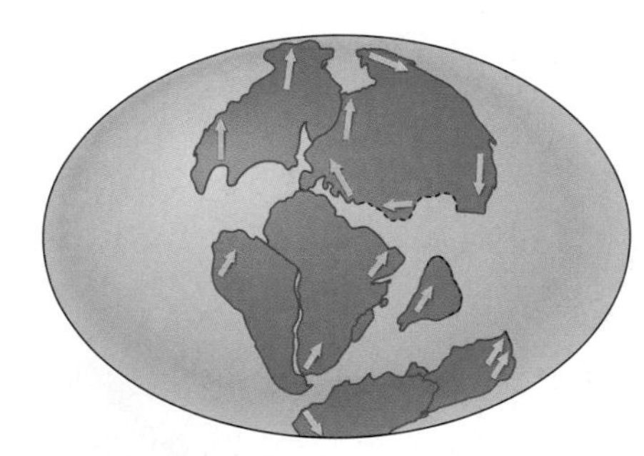

135 mya: Africa and South America begin to separate.

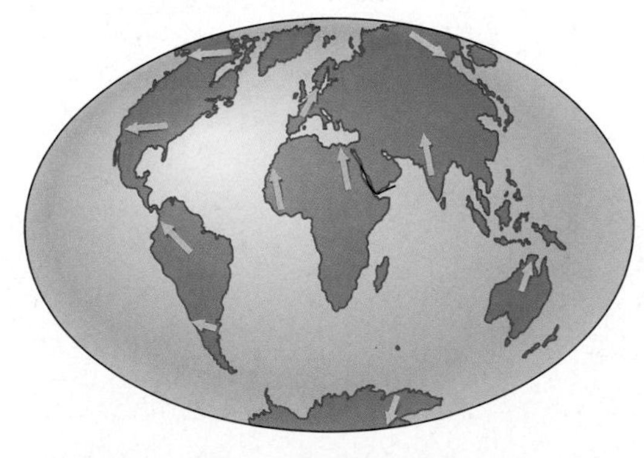

Present: India has collided with Asia to form the Himalayas and Australia has separated from Antarctica. A rift valley is forming in East Africa. Continents continue to move over Earth's surface.

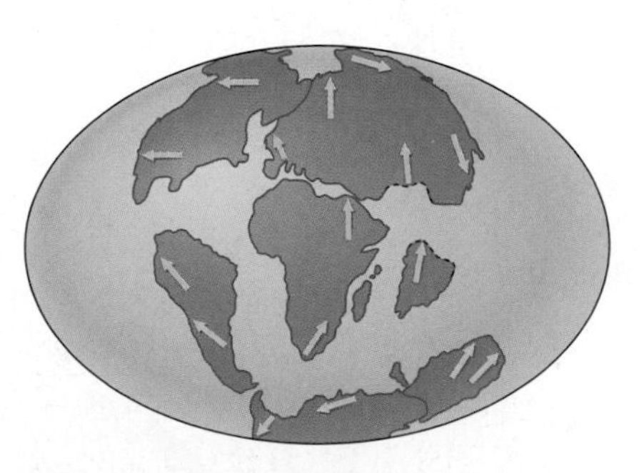

65 mya: India moves north toward Asia.

Evidence from rock formations Wegener reasoned that when Pangaea began to break apart, large geologic structures, such as mountain ranges, fractured as the continents separated. Using this reasoning, Wegener thought that there should be areas of similar rock types on opposite sides of the Atlantic Ocean. He observed that many layers of rocks in the Appalachian Mountains in the United States were identical to layers of rocks in similar mountains in Greenland and Europe. These similar groups of rocks, older than 200 million years, supported Wegener's idea that the continents had once been joined. Some of the locations where matching groups of rock have been found are indicated in **Figure 17.3.**

Evidence from fossils Wegener also gathered evidence of the existence of Pangaea from fossils. Similar fossils of several different animals and plants that once lived on or near land had been found on widely separated continents, as shown in **Figure 17.3.** Wegener reasoned that the land-dwelling animals, such as *Cynognathus* (sin ug NATH us) and *Lystrasaurus* (lihs truh SORE us) could not have swum the great distances that now exist between continents. Wegener also argued that because fossils of *Mesosaurus* (meh zoh SORE us), an aquatic reptile, had been found in only freshwater rocks, it was unlikely that this species could have crossed the oceans. The ages of these different fossils also predated Wegener's time frame for the breakup of Pangaea, and thus supported his hypothesis.

■ **Figure 17.3** Alfred Wegener used the similarity of rock layers and fossils on opposite sides of the Atlantic Ocean as evidence that Earth's continents were once joined.
Identify *groupings that suggest that there was once a single landmass.*

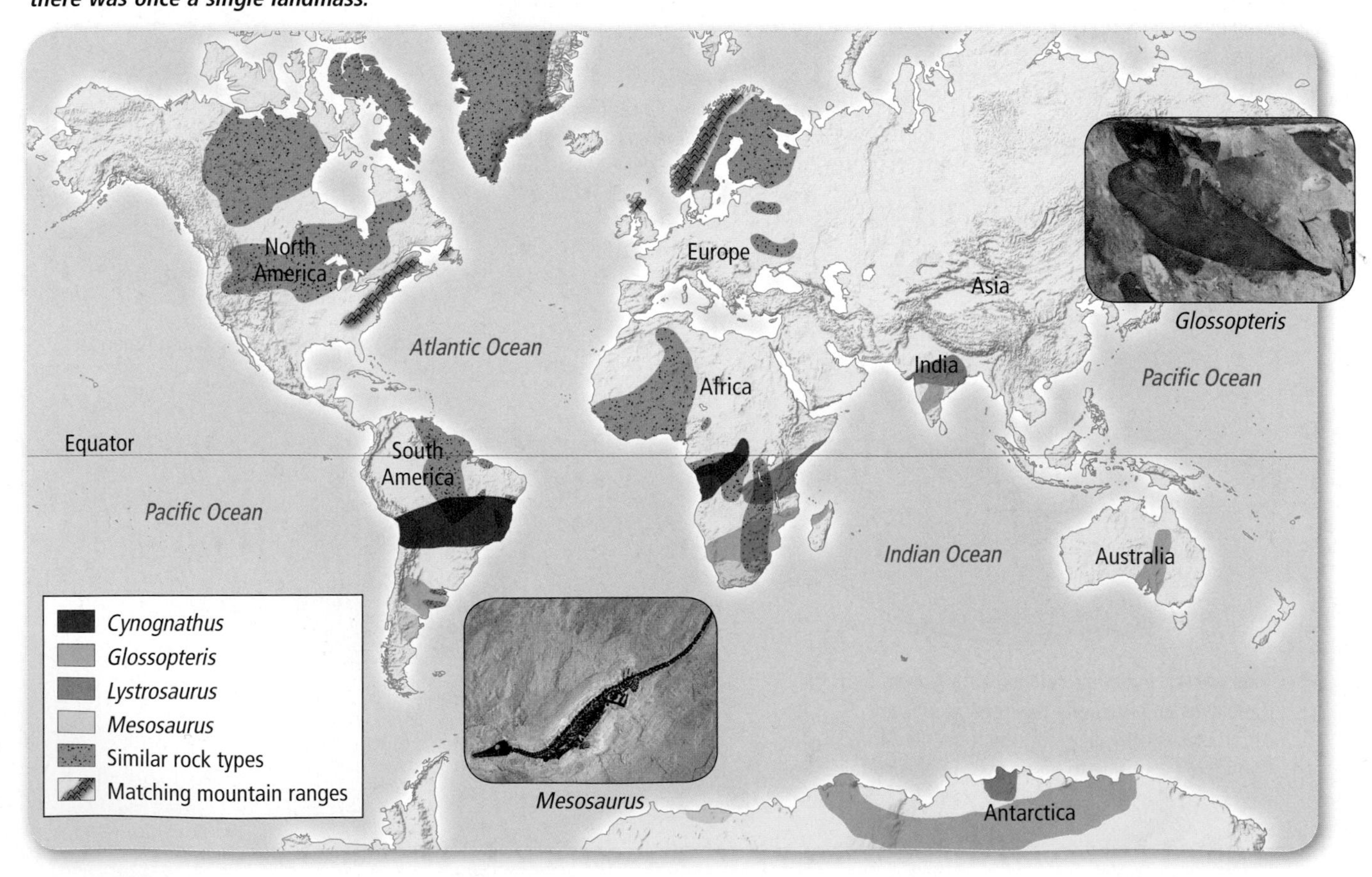

Climatic evidence Because he had a strong background in meteorology, Wegener recognized clues about ancient climates from the fossils he studied. One fossil that Wegener used to support continental drift was *Glossopteris* (glahs AHP tur us), a seed fern that resembled low shrubs, shown in **Figure 17.3.** Fossils of this plant had been found on many parts of Earth, including South America, Antarctica, and India. Wegener reasoned that the area separating these fossils was too large to have had a single climate. Wegener also argued that because *Glossopteris* grew in temperate climates, the places where these fossils had been found were once closer to the equator. This led him to conclude that the rocks containing these fossil ferns had once been joined.

Reading Check Infer how Wegener's background in meteorology helped him to support his idea of continental drift.

■ **Figure 17.4** A coal deposit in Antarctica indicates that swamp plants once thrived in this area.
Explain ***How did coal, which forms from ancient swamp material, end up in Antarctica?***

Coal deposits Recall from Chapter 6 that sedimentary rocks provide clues to past environments and climates. Wegener found evidence in these rocks that the climates of some continents had changed markedly. For example, **Figure 17.4** shows a coal deposit found in Antarctica. Coal forms from the compaction and decomposition of accumulations of ancient swamp plants. The existence of coal beds in Antarctica indicated that this frozen land once had a tropical climate. Wegener used this evidence to conclude that Antarctica must have been much closer to the equator sometime in the geologic past.

Glacial deposits Another piece of climatic evidence came from glacial deposits found in parts of Africa, India, Australia, and South America. The presence of these 290-million-year-old deposits suggested to Wegener that these areas were once covered by a thick ice cap similar to the one that covers Antarctica today. Because the traces of the ancient ice cap are found in regions where it is too warm for them to develop, Wegener proposed that they were once located near the south pole, as shown in **Figure 17.5.** Wegener suggested two possibilities to explain the deposits. Either the south pole had shifted its position, or these landmasses had once been closer to the south pole. Wegener argued that it was more likely that the landmasses had drifted apart rather than Earth changing its axis.

■ **Figure 17.5** Glacial deposits nearly 300 million years old on several continents led Wegener to propose that these landmasses might have once been joined and covered with ice. The extent of the ice is shown in white.

■ **Figure 17.6** Wegener collected further evidence for his theory on a 1930 expedition to Greenland. He died during this expedition, many years before his data became the basis for the theory of plate tectonics.

A Rejected Notion

In the early 1900s, many people in the scientific community considered the continents and ocean basins to be fixed features on Earth's surface. For the rest of his life, Wegener continued travelling to remote regions to gather evidence in support of continental drift. **Figure 17.6** shows him in Greenland on his last expedition. Although he had compiled an impressive collection of data, the continental drift hypothesis was never accepted by the scientific community.

Continental drift had two major flaws that prevented it from being widely accepted. First, it did not satisfactorily explain what force could be strong enough to push such large masses over such great distances. Wegener thought that the rotation of Earth might be responsible, but physicists were able to show that this force was not nearly enough to move continents.

Second, scientists questioned how the continents were moving. Wegener had proposed that the continents were plowing through a stationary ocean floor, but it was known that Earth's mantle below the crust was solid. So, how could continents move through something solid? These two unanswered questions—what forces could cause the movement and how continents could move through solids—were the main reasons that continental drift was rejected. It was not until the early 1960s when new technology revealed more evidence about how continents move that scientists began to reconsider Wegener's ideas. Advances in seafloor mapping and in understanding Earth's magnetic field provided the necessary evidence to show how continents move, and the source of the forces involved.

Section 17.1 Assessment

Section Summary

- The matching coastlines of continents on opposite sides of the Atlantic Ocean suggest that the continents were once joined.
- Continental drift was the idea that continents move around on Earth's surface.
- Wegener collected evidence from rocks, fossils, and ancient climates to support his theory.
- Continental drift was not accepted because there was no explanation for how the continents moved or what caused their motion.

Understand Main Ideas

1. MAIN Idea **Draw** how the continents were once adjoined as Pangaea.
2. **Explain** how ancient glacial deposits in Africa, India, Australia, and South America support the idea of continental drift.
3. **Summarize** how rocks, fossils, and climate provided evidence of continental drift.
4. **Infer** what the climate in ancient North America must have been like as a part of Pangaea.

Think Critically

5. **Interpret** Examine **Figure 17.5.** Oil deposits that are approximately 200 million years old have been discovered in Brazil. Where might geologists find oil deposits of a similar age?
6. **Evaluate** this statement: The town where I live has always been in the same place.

WRITING in Earth Science

7. Compose a letter to the editor from a scientist in the early 1900s arguing against continental drift.

Earth Science Online Self-Check Quiz glencoe.com

Objectives

- **Summarize** the evidence that led to the discovery of seafloor spreading.
- **Explain** the significance of magnetic patterns on the seafloor.
- **Explain** the process of seafloor spreading.

Review Vocabulary

basalt: a dark-gray to black fine-grained igneous rock

New Vocabulary

magnetometer
magnetic reversal
paleomagnetism
isochron
seafloor spreading

Seafloor Spreading

MAIN Idea Oceanic crust forms at ocean ridges and becomes part of the seafloor.

Real-World Reading Link Have you ever counted the rings on a tree stump to find the age of the tree? Scientists can study similar patterns on the ocean floor to determine its age.

Mapping the Ocean Floor

Until the mid-1900s, most people, including many scientists, thought that the ocean floors were essentially flat. Many people also had misconceptions that oceanic crust was unchanging and was much older than continental crust. However, advances in technology during the 1940s and 1950s showed that all of these widely accepted ideas were incorrect.

One technological advance that was used to study the ocean floor was the magnetometer. A **magnetometer** (mag nuh TAH muh tur), such as the one shown in **Figure 17.7,** is a device that can detect small changes in magnetic fields. Towed behind a ship, it can record the magnetic field generated by ocean floor rocks. You will learn more about magnetism and how it supports continental drift later in this section.

Another advancement that allowed scientists to study the ocean floor in great detail was the development of echo-sounding methods. One type of echo sounding is sonar. Recall from Chapter 15 that sonar uses sound waves to measure distance by measuring the time it takes for sound waves sent from the ship to bounce off the seafloor and return to the ship. Developments in sonar technology enabled scientists to measure water depth and map the topography of the ocean floor.

Figure 17.7 Magnetometers are devices that can detect small changes in magnetic fields. The data collected using magnetometers lowered into the ocean furthered scientists' understanding of rocks underlying the ocean floor.

■ **Figure 17.8** Sonar data revealed ocean ridges and deep-sea trenches. Earthquakes and volcanism are common along ridges and trenches.

Ocean-Floor Topography

The maps made from data collected by sonar and magnetometers surprised many scientists. They discovered that vast, underwater mountain chains called ocean ridges run along the ocean floors around Earth much like seams on a baseball. These ocean floor features, shown in **Figure 17.8,** form the longest continuous mountain range on Earth. When they were first discovered, ocean ridges generated much discussion because of their enormous length and height—they are more than 80,000 km long and up to 3 km above the ocean floor. Later, scientists discovered that earthquakes and volcanism are common along the ridges.

 **Reading Check Describe** Where are the longest continuous mountain ranges on Earth?

Maps generated with sonar data also revealed that underwater mountain chains had counterparts called deep-sea trenches, which are also shown on the map in **Figure 17.8.** Recall from Chapter 16 that a deep-sea trench is a narrow, elongated depression in the seafloor. Trenches can be thousands of kilometers long and many kilometers deep. The deepest trench, called the Marianas Trench, is in the Pacific Ocean and is more than 11 km deep. Mount Everest, the world's tallest mountain, stands at 9 km above sea level, and could fit inside the Marianas Trench with six Empire State buildings stacked on top.

These two topographic features of the ocean floor—ocean ridges and deep-sea trenches—puzzled geologists for more than a decade after their discovery. What could have formed an underwater mountain range that extended around Earth? What is the source of the volcanism associated with these mountains? What forces could depress Earth's crust enough to create trenches nearly 6 times as deep as the Grand Canyon? You will find out the answers to these questions later in this chapter.

VOCABULARY

SCIENCE USAGE V. COMMON USAGE

Depress

Science usage: to cause to sink to a lower position

Common usage: to sadden or discourage

Ocean Rocks and Sediments

In addition to making maps, scientists collected samples of deep-sea sediments and the underlying oceanic crust. Analysis of the rocks and sediments led to two important discoveries. First, the ages of the rocks that make up the seafloor varies across the ocean floor, and these variations are predictable. Rock samples taken from areas near ocean ridges were found to be younger than samples taken from areas near deep-sea trenches. The samples showed that the age of oceanic crust consistently increases with distance from a ridge, as shown in **Figure 17.9.** This trend was symmetric across the ocean ridges. Scientists also discovered from the rock samples that even the oldest parts of the seafloor are geologically young—about 180 million years old. Why are ocean-floor rocks so young compared to continental rocks, some of which are at least 3.8 billion years old? Geologists knew that oceans had existed for more than 180 million years so they wondered why there was no trace of older oceanic crust.

The second discovery involved the sediments on the ocean floor. Measurements showed that ocean-floor sediments are typically a few hundred meters thick. Large areas of continents, on the other hand, are blanketed with sedimentary rocks that are as much as 20 km thick. Scientists knew that erosion and deposition occur in Earth's oceans but did not understand why seafloor sediments were not as thick as their continental counterparts. Scientists hypothesized that the relatively thin layer of ocean sediments was related to the age of the ocean crust. Observations of ocean-floor sediments revealed that the thickness of the sediments increases with distance from an ocean ridge, as shown in **Figure 17.9.** The pattern of thickness across the ocean floor was symmetrical across the ocean ridges.

CAREERS IN EARTH SCIENCE

Marine geologist Earth scientists who study the ocean floor to understand geologic processes such as plate tectonics are marine geologists. To learn more about Earth science careers, visit glencoe.com.

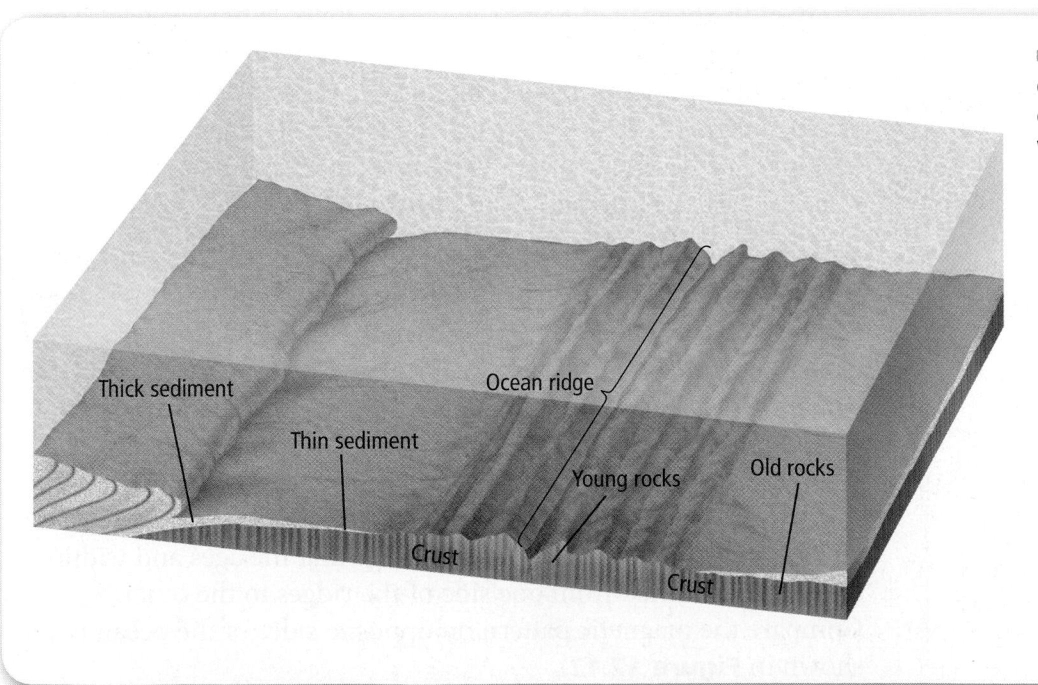

Figure 17.9 The ages of ocean crust and the thicknesses of ocean-floor sediments increase with distance from the ridge.

Figure 17.10 Earth's magnetic field is generated by the flow of molten iron in the liquid outer core. The polarity of the field changes over time from normal to reversed.

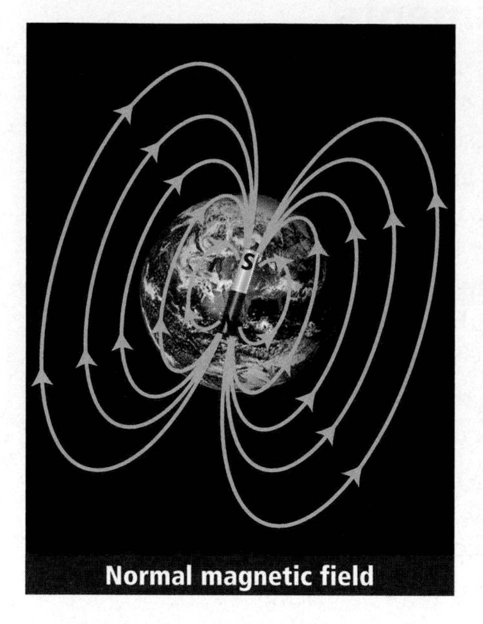

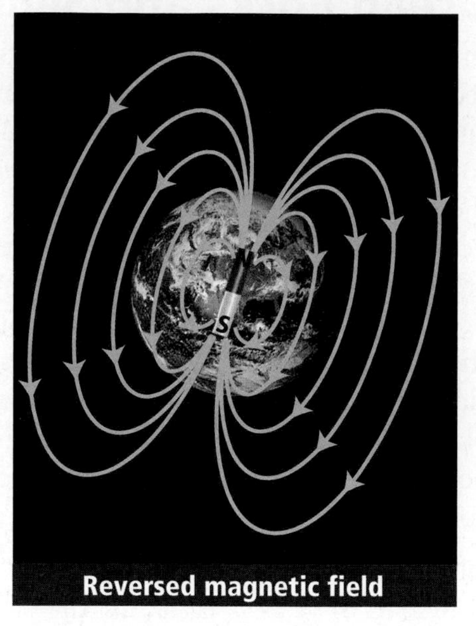

Figure 17.11 Periods of normal polarity alternate with periods of reversed polarity. Long-term changes in Earth's magnetic field, called epochs, are named as shown here. Short-term changes are called events.

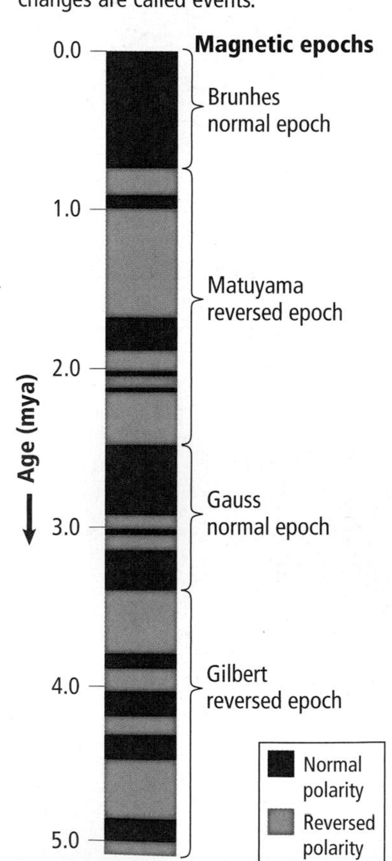

Magnetism

Earth has a magnetic field generated by the flow of molten iron in the outer core. This field is what causes a compass needle to point to the North. A **magnetic reversal** happens when the flow in the outer core changes, and Earth's magnetic field changes direction. This would cause compasses to point to the South. Magnetic reversals have occurred many times in Earth's history. As shown in **Figure 17.10,** a magnetic field that has the same orientation as Earth's present field is said to have normal polarity. A magnetic field that is opposite to the present field has reversed polarity.

Magnetic polarity time scale **Paleomagnetism** is the study of the history of Earth's magnetic field. When lava solidifies, iron-bearing minerals such as magnetite crystallize. As they crystallize, these minerals behave like tiny compasses and align with Earth's magnetic field. Data gathered from paleomagnetic studies of continental lava flows allowed scientists to construct a magnetic polarity time scale, as shown in **Figure 17.11.**

Magnetic symmetry Scientists knew that oceanic crust is mostly basaltic rock, which contains large amounts of iron-bearing minerals of volcanic origin. They hypothesized that the rocks on the ocean floor would show a record of magnetic reversals. When scientists towed magnetometers behind ships to measure the magnetic orientation of the rocks of the ocean floor, a surprising pattern emerged. The regions with normal and reverse polarity formed a series of stripes across the floor parallel to the ocean ridges. The scientists were doubly surprised to discover that the ages and widths of the stripes matched from one side of the ridges to the other. Compare the magnetic pattern on opposite sides of the ocean ridge shown in **Figure 17.12.**

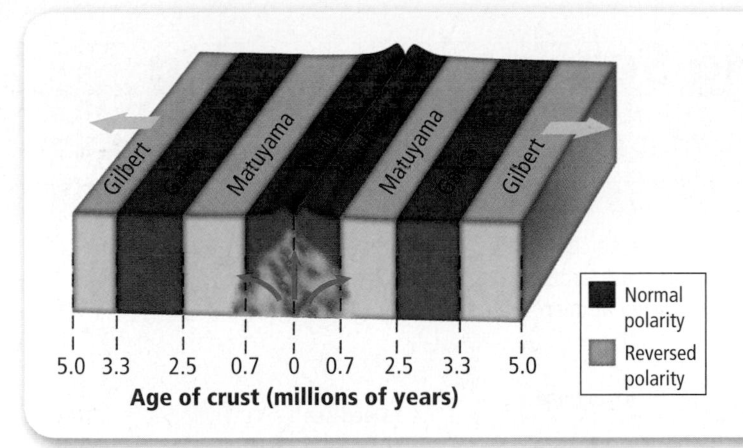

Figure 17.12 Reversals in the polarity of Earth's magnetic field are recorded in the rocks that make up the ocean floor.
Identify *the polarity of the most recently produced basalt at the ocean ridge.*

By matching the patterns on the seafloor with the known pattern of reversals on land, scientists were able to determine the age of the ocean floor from magnetic recording. This method enabled scientists to quickly create isochron (I suh krahn) maps of the ocean floor. An **isochron** is an imaginary line on a map that shows points that have the same age—that is, they formed at the same time. In the isochron map shown in **Figure 17.13,** note that relatively young ocean-floor crust is near ocean ridges, while older ocean crust is found along deep-sea trenches.

Figure 17.13 Each colored band on this isochron map of the ocean floor represents the age of that strip of the crust.
Observe *What pattern do you observe?*

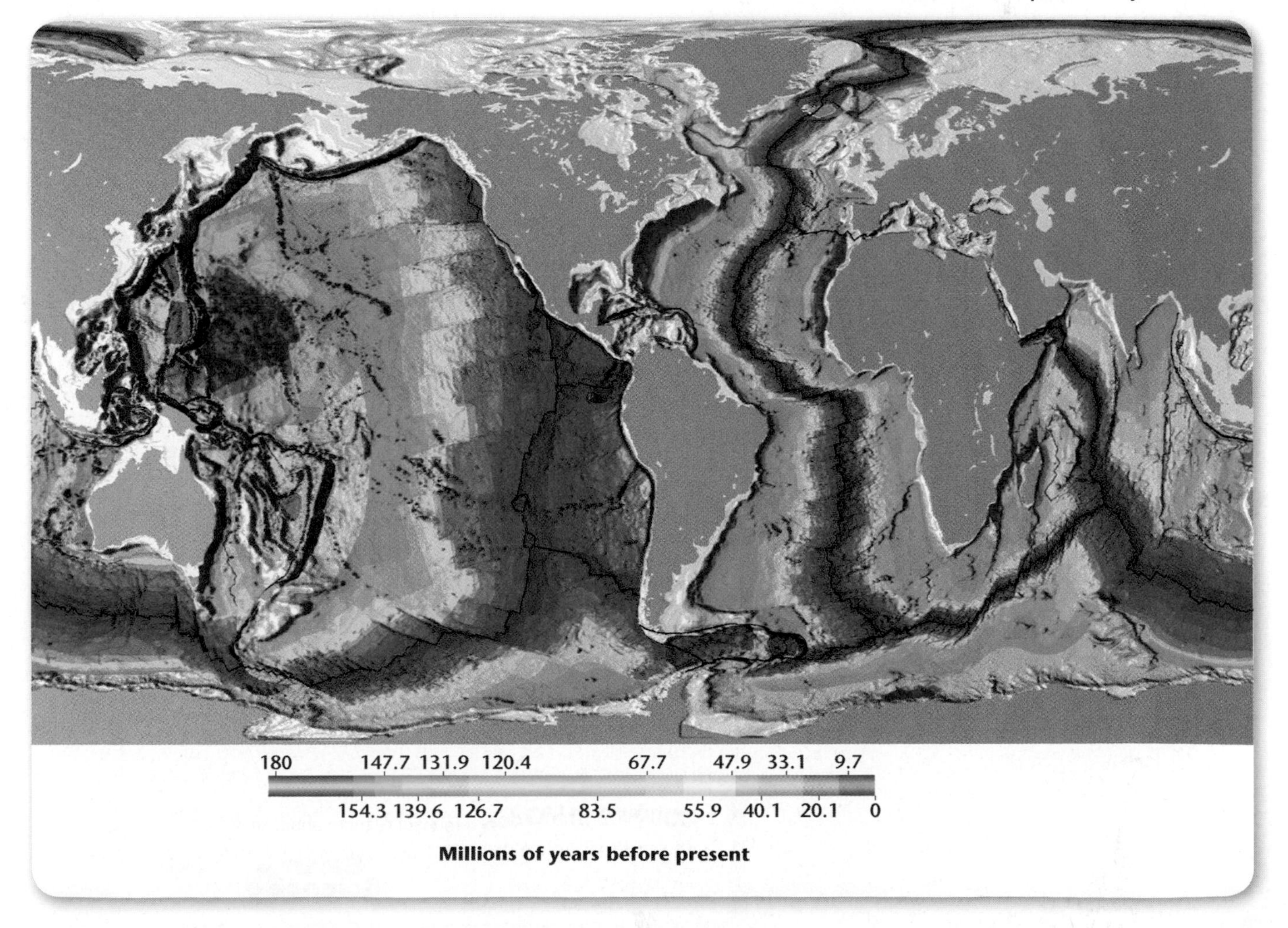

Visualizing Seafloor Spreading

Figure 17.14 Data from topographic, sedimentary, and paleomagnetic research led scientists to propose seafloor spreading. Seafloor spreading is the process by which new oceanic crust forms at ocean ridges, and slowly moves away from the spreading center until it is subducted and recycled at deep-sea trenches.

Magma intrudes into the ocean floor along a ridge and fills the gap that is created. When the molten material solidifies, it becomes new oceanic crust.

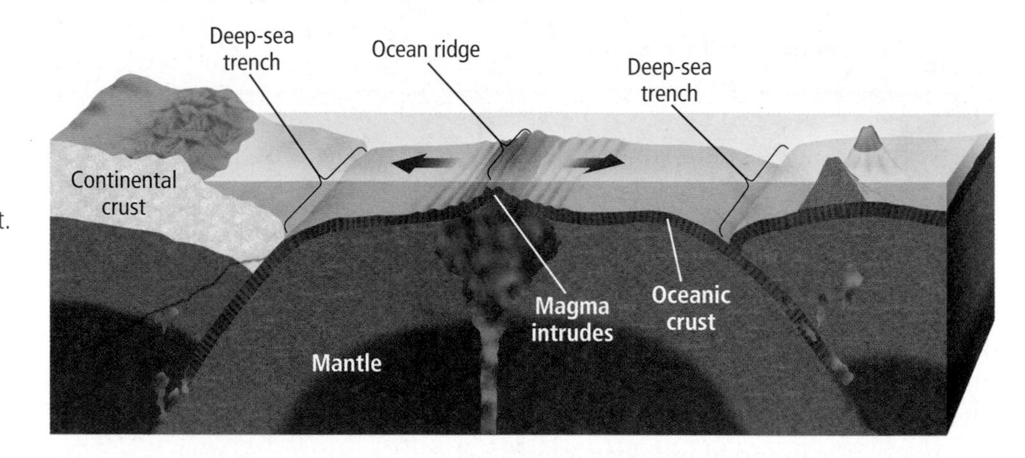

The continuous spreading and intrusion of magma result in the addition of new oceanic crust. Two halves of the oceanic crust spread apart slowly, and move apart like a conveyor belt.

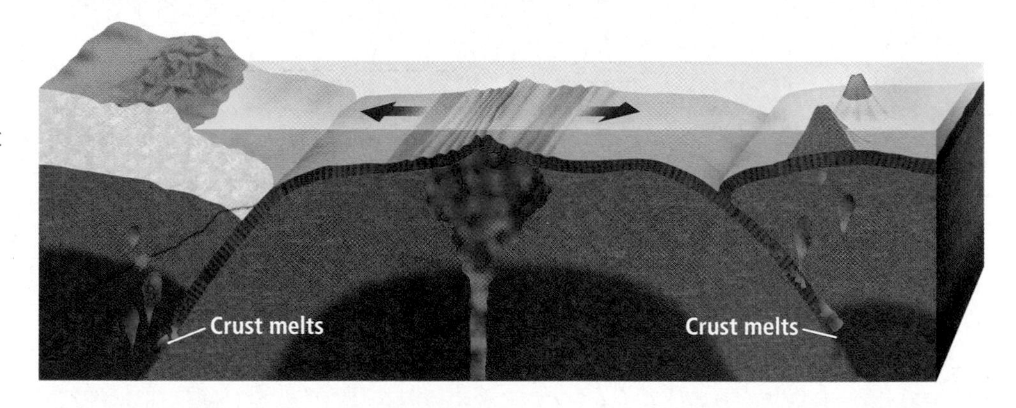

The far edges of the oceanic crust sink beneath continental crust. As it descends, water in the minerals causes the oceanic crust to melt, forming magma. The magma rises and forms part of the continental crust.

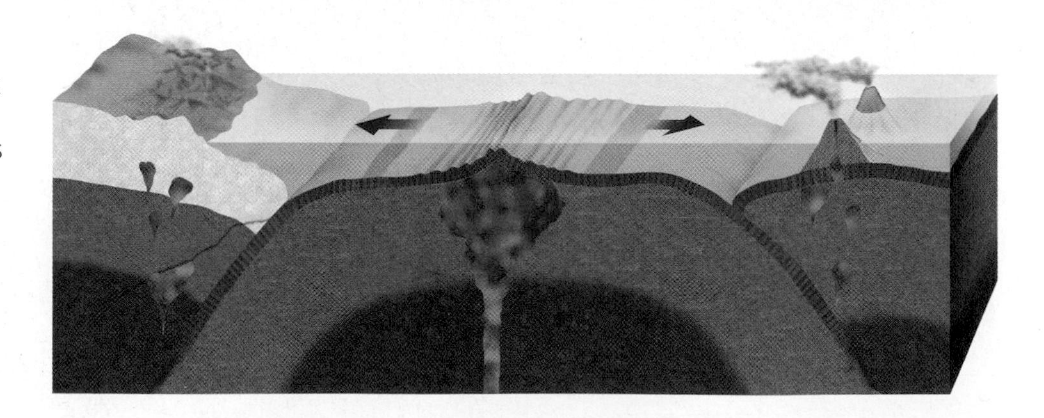

Concepts In Motion To explore more about seafloor spreading, visit glencoe.com.

Earth Science Online

Seafloor Spreading

Using all the topographic, sedimentary, and paleomagnetic data from the seafloor, seafloor spreading was proposed. **Seafloor spreading** is the theory that explains how new ocean crust is formed at ocean ridges and destroyed at deep-sea trenches. **Figure 17.14** illustrates how seafloor spreading occurs.

During seafloor spreading, magma, which is hotter and less dense than surrounding mantle material, is forced toward the surface of the crust along an ocean ridge. As the two sides of the ridge spread apart, the rising magma fills the gap that is created. When the magma solidifies, a small amount of new ocean floor is added to Earth's surface. As spreading along a ridge continues, more magma is forced upward and solidifies. This cycle of spreading and the intrusion of magma continues the formation of ocean floor, which slowly moves away from the ridge. Of course, seafloor spreading mostly happens under the sea, but in Iceland, a portion of the Mid-Atlantic Ridge rises above sea level. **Figure 17.15** shows lava erupting along the ridge.

Recall that while Wegener collected many data to support the idea that the continents are drifting across Earth's surface, he could not explain what caused the landmasses to move or how they moved. Seafloor spreading was the missing link that Wegener needed to complete his model of continental drift. Continents are not pushing through ocean crust, as Wegener proposed. In fact, continents are more like passengers that ride along while ocean crust slowly moves away from ocean ridges. Seafloor spreading led to a new understanding of how Earth's crust and rigid upper mantle move. This will be explored in the next sections.

■ **Figure 17.15** The entire island of Iceland lies on the Mid-Atlantic ocean spreading center. Because the seafloor is spreading, Iceland is growing larger. In 1783, more than 12 km^3 of lava erupted—enough to pave the entire U.S. interstate freeway system to a depth of 10 m.

Section 17.2 Assessment

Section Summary

- Studies of the seafloor provided evidence that the ocean floor is not flat and unchanging.
- Oceanic crust is geologically young.
- New oceanic crust forms as magma rises at ridges and solidifies.
- As new oceanic crust forms, the older crust moves away from the ridges.

Understand Main Ideas

1. MAIN Idea **Describe** why seafloor spreading is like a moving conveyor belt.
2. **Explain** how ocean-floor rocks and sediments provided evidence of seafloor spreading.
3. **Differentiate** between the terms *reversed polarity* and *normal polarity.*
4. **Describe** the topography of the seafloor.

Think Critically

5. **Explain** how an isochron map of the ocean floor supports the theory of seafloor spreading.
6. **Analyze** Why are magnetic bands in the eastern Pacific Ocean so far apart compared to the magnetic bands along the Mid-Atlantic Ridge?

MATH in Earth Science

7. Analyze **Figure 17.11.** What percentage of the last 5 million years has been spent in reversed polarity?

Section 17.3

Objectives

- **Describe** how Earth's tectonic plates result in many geologic features.
- **Compare and contrast** the three types of plate boundaries and the features associated with each.
- **Generalize** the processes associated with subduction zones.

Review Vocabulary

mid-ocean ridge: a major feature along the ocean floor consisting of an elevated region with a central valley

New Vocabulary

tectonic plate
divergent boundary
rift valley
convergent boundary
subduction
transform boundary

Plate Boundaries

MAIN Idea Volcanoes, mountains, and deep-sea trenches form at the boundaries between the plates.

Real-World Reading Link Imagine a pot of soup that has been allowed to cool in a refrigerator. Fats in the soup have solidified into a hard surface, but if you tilt the pot back and forth, you will see the rigid surface bending and cracking. This is similar to the relationship between different layers of Earth.

Theory of Plate Tectonics

The evidence for seafloor spreading suggested that continental and oceanic crust move as enormous slabs, which geologists describe as tectonic plates. **Tectonic plates** are huge pieces of crust and rigid upper mantle that fit together at their edges to cover Earth's surface. As illustrated in **Figure 17.16,** there are about 12 major plates and several smaller ones. These plates move very slowly—only a few centimeters each year—which is similar to the rate at which fingernails grow. Plate tectonics is the theory that describes how tectonic plates move and shape Earth's surface. They move in different directions and at different rates relative to one another and they interact with one another at their boundaries. Each type of boundary has certain geologic characteristics and processes associated with it. A divergent boundary occurs where tectonic plates move away from each other. A convergent boundary occurs where tectonic plates move toward each other. A transform boundary occurs where tectonic plates move horizontally past each other.

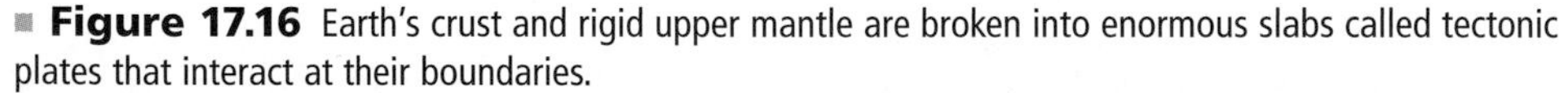

■ **Figure 17.16** Earth's crust and rigid upper mantle are broken into enormous slabs called tectonic plates that interact at their boundaries.

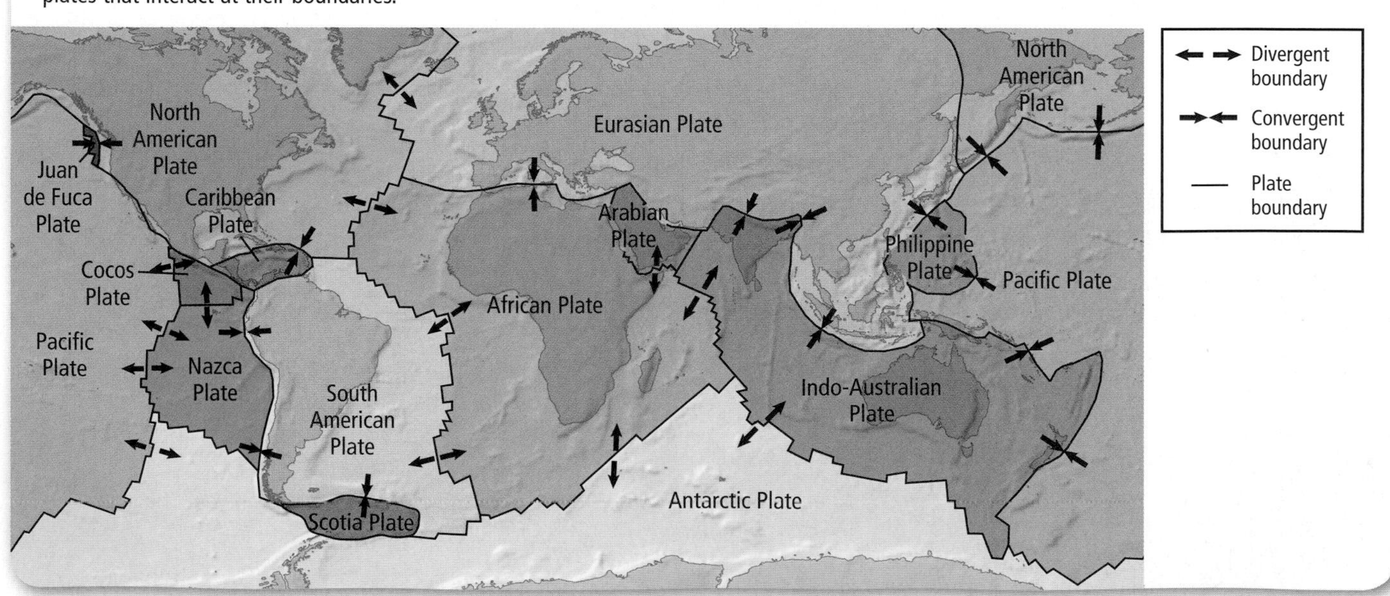

Divergent boundaries Regions where two tectonic plates are moving apart are called **divergent boundaries.** Most divergent boundaries are found along the seafloor in rift valleys. It is in this central rift that the process of seafloor spreading begins. Magma rising through the rift's faults forms a mid-ocean ridge. The mid-ocean ridge appears as a continuous mountain chain on the ocean floor. The formation of new ocean crust at most divergent boundaries accounts for the high heat flow, volcanism, and earthquakes associated with these boundaries.

Reading Check **Identify** the cause of volcanism and earthquakes associated with mid-ocean ridges.

Throughout millions of years, the process of seafloor spreading along a divergent boundary can cause an ocean basin to grow wider. Although most divergent boundaries form ridges on the ocean floor, some divergent boundaries form on continents. When continental crust begins to separate, the stretched crust forms a long, narrow depression called a **rift valley.** **Figure 17.17** shows the rift valley that is currently forming in East Africa. The rifting might eventually lead to the formation of a new ocean basin.

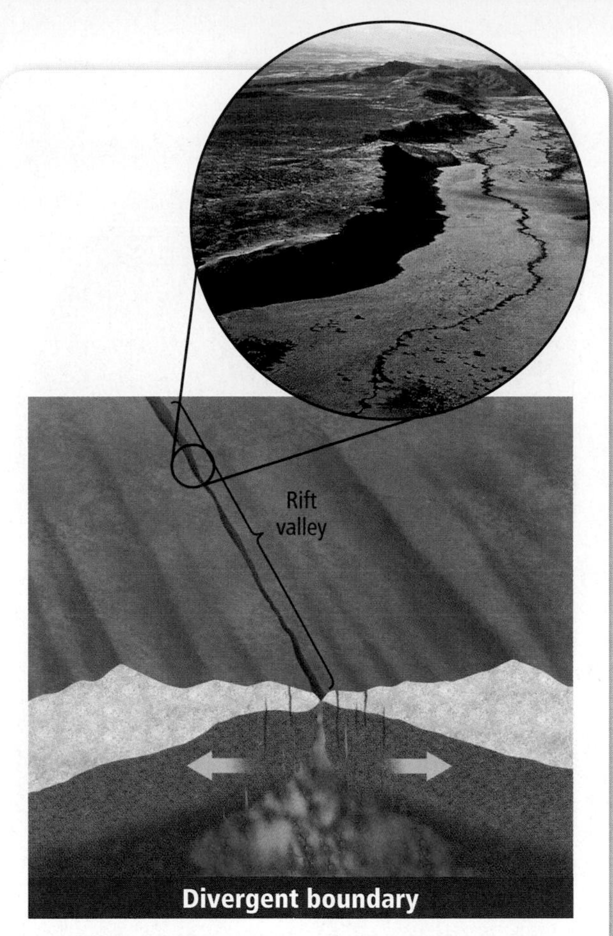

Figure 17.17 Divergent boundaries are places where plates separate. An ocean ridge is a divergent boundary on the ocean floor. In East Africa, a divergent boundary has also created a rift valley.

MiniLab

Model Ocean-Basin Formation

How did a divergent boundary form the South Atlantic Ocean? Around 150 mya, a divergent boundary split an ancient continent. Over time, new crust was added along the boundary, widening the rift between Africa and South America.

Procedure

1. Read and complete the lab safety form.
2. Use a **world map** to create **paper templates** of South America and Africa.
3. Place the two continental templates in the center of a **large piece of paper,** and fit them together along their Atlantic coastlines.
4. Carefully trace around the templates with a **pencil.** Remove the templates and label the diagram *150 mya.*
5. Use an average spreading rate of 4 cm/y and a map scale of 1 cm = 500 km to create six maps that show the development of the Atlantic Ocean at 30-million-year intervals, beginning 150 mya.

Analysis

1. **Compare** your last map with a world map. Is the actual width of the South Atlantic Ocean the same on both maps?
2. **Consider** why there might be differences between the width in your model and the actual width of the present South Atlantic Ocean.

■ **Figure 17.18** Oceanic plates are mostly basalt. Continental plates are mostly granite with a thin cover of sedimentary rock, both of which are less dense than basalt.

Convergent boundaries At **convergent boundaries,** two tectonic plates are moving toward each other. When two plates collide, the denser plate eventually descends below the other, less-dense plate in a process called **subduction.** There are three types of convergent boundaries, classified according to the type of crust involved. Recall from Chapter 1 that oceanic crust is made mostly of minerals that are high in iron and magnesium, which form dense, dark-colored basaltic rocks, such as the basalt shown in **Figure 17.18.** Continental crust is composed mostly of minerals such as feldspar and quartz, which form less-dense, lighter-colored granitic rocks. The differences in density of the crustal material affects how they converge. The three types of tectonic boundaries and their associated landforms are shown in **Table 17.1.**

Oceanic-oceanic In the oceanic-oceanic convergent boundary shown in **Table 17.1,** a subduction zone is formed when one oceanic plate, which is denser as a result of cooling, descends below another oceanic plate. The process of subduction creates an ocean trench. The subducted plate descends into the mantle, thereby recycling oceanic crust formed at the ridge. Water carried into Earth by the subducting plate lowers the melting temperature of the plate, causing it to melt at shallower depths. The molten material, called magma, is less dense, so it rises back to the surface where it often erupts and forms an arc of volcanic islands that parallel the trench. Some examples of trenches and island arcs are the Marianas Trench and Marianas Islands in the West Pacific Ocean and the Aleutian Trench and Aleutian Islands in the North Pacific Ocean. A volcanic peak in the Aleutian Island arc is shown in **Table 17.1.**

Oceanic-continental Subduction zones are also found where an oceanic plate converges with a continental plate, as shown in **Table 17.1.** Note that it is the denser oceanic plate that is subducted. Oceanic-continental convergence also produces a trench and volcanic arc. However, instead of forming an arc of volcanic islands, oceanic-continental convergence results in a chain of volcanoes along the edge of the continental plate. The result of this type of subduction is a mountain range with many volcanoes. The Peru-Chile Trench and the Andes mountain range, which are located along the western coast of South America, formed in this way.

VOCABULARY

ACADEMIC VOCABULARY

Parallel (PAIR uh lel)
extending in the same direction, everywhere equidistant, and not meeting
The commuter train runs parallel to the freeway for many kilometers.

Concepts In Motion

Interactive Table To explore more about convergent boundaries, visit glencoe.com.

Table 17.1 Summary of Convergent Boundaries

Type of Convergent Boundary	Example of Region Affected by Boundary	Example of Landform Produced
Oceanic-oceanic 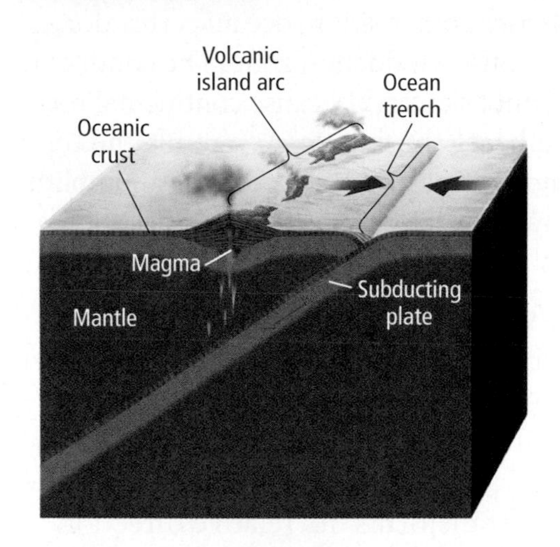	 Aleutian Islands	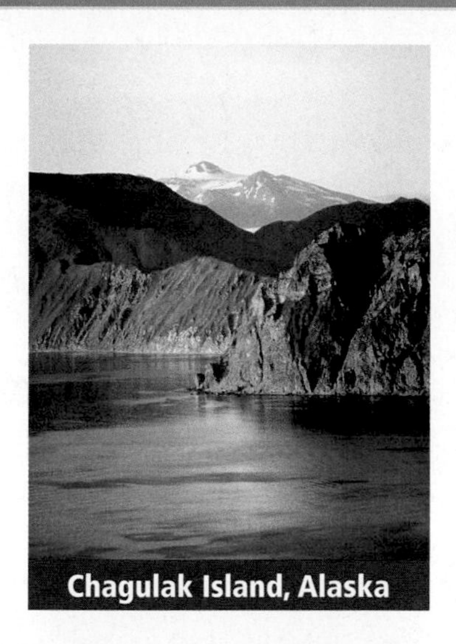 Chagulak Island, Alaska
Oceanic-continental 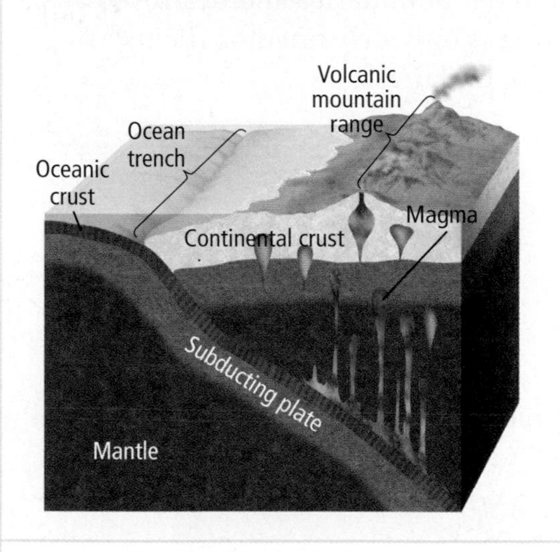	 Andes mountain range	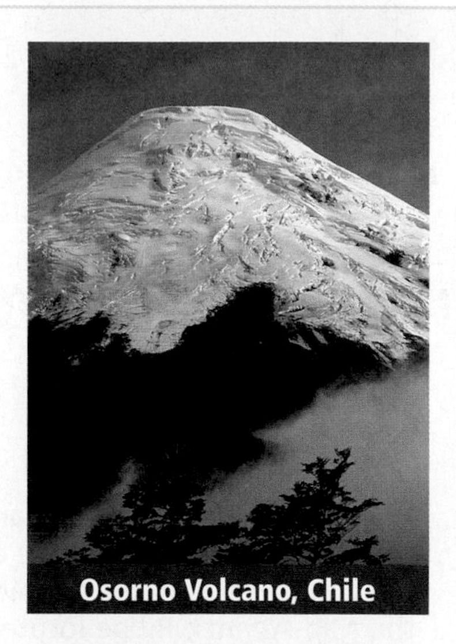 Osorno Volcano, Chile
Continental-continental 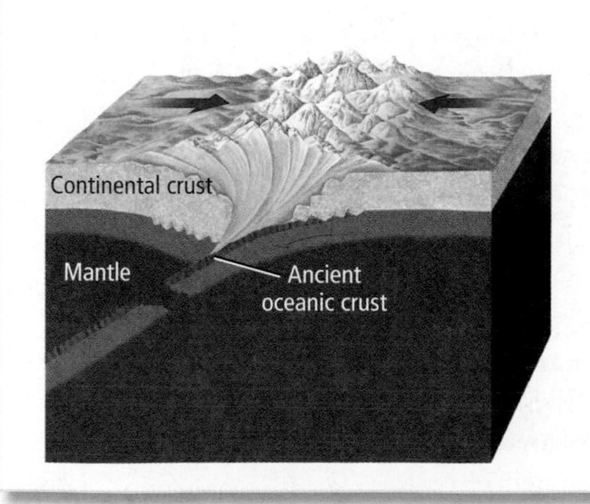	 Himalayas	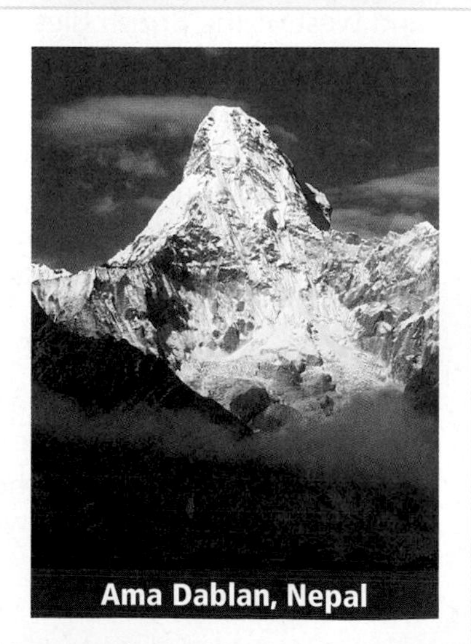 Ama Dablan, Nepal

Continental-continental The third type of convergent boundary forms when two continental plates collide. Continental-continental boundaries form long after an oceanic plate has converged with a continental plate. Recall that continents are often carried along attached to oceanic crust. Over time, an oceanic plate can be completely subducted, dragging an attached continent behind it toward the subduction zone. As a result of its denser composition, oceanic crust descends beneath the continental crust at the subduction zone. The continental crust that it pulls behind it cannot descend because continental rocks are less dense, and will not sink into the mantle. As a result, the edges of both continents collide, and become crumpled, folded, and uplifted. This forms a vast mountain range, such as the Himalayas, as shown in **Table 17.1.**

FOLDABLES Incorporate information from this section into your Foldable.

Transform boundaries A region where two plates slide horizontally past each other is a **transform boundary,** as shown in **Figure 17.19.** Transform boundaries are characterized by long faults, sometimes hundreds of kilometers in length, and by shallow earthquakes. Transform boundaries were named for the way Earth's crust changes, or transforms, its relative direction and velocity from one side of the boundary to the other. Recall that new crust is formed at divergent boundaries and destroyed at convergent boundaries. Crust is only deformed or fractured somewhat along transform boundaries.

PROBLEM-SOLVING LAB

Interpret Scientific Illustrations

How does plate motion change along a transform boundary? The figure at the right shows the Gibbs Fracture Zone, which is a segment of the Mid-Atlantic Ridge located south of Iceland and west of the British Isles. Copy this figure.

Analysis

1. **Draw** arrows on your copy to indicate the direction of seafloor movement at locations A, B, C, D, E, and F.
2. **Compare** the direction of motion for the following pairs of locations: A and D, B and E, and C and F.

Think Critically

3. **Differentiate** Which three locations are on the North American Plate?
4. **Indicate** the portion of the fracture zone that is the boundary between North America and Europe.
5. **Assess** Which two locations represent the oldest crust?

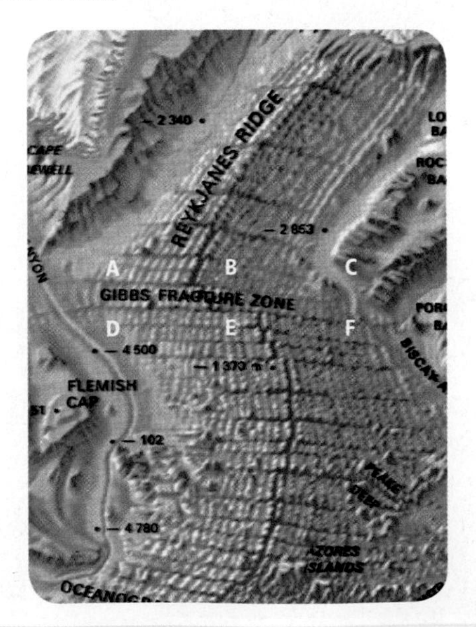

■ **Figure 17.19** Plates move horizontally past each other along a transform plate boundary. The bend in these train tracks resulted from the transform boundary running through parts of Southern California.

Most transform boundaries offset sections of ocean ridges, as you observed in the Problem-Solving Lab. Sometimes transform boundaries occur on continents. The San Andreas Fault is probably the best-known example. Recall from the Launch Lab at the beginning of this chapter that the San Andreas Fault system is part of a transform boundary that separates southwestern California from the rest of the state. Movements along this transform boundary create situations like the one shown in **Figure 17.19** and are responsible for most of the earthquakes that strike California every year.

Section 17.3 Assessment

Section Summary

- Earth's crust and rigid upper mantle are broken into large slabs of rock called tectonic plates.
- Plates move in different directions and at different rates over Earth's surface.
- At divergent plate boundaries, plates move apart. At convergent boundaries, plates come together. At transform boundaries, plates slide horizontally past each other.
- Each type of boundary is characterized by certain geologic features.

Understand Main Ideas

1. MAIN Idea **Describe** how plate tectonics results in the development of Earth's major geologic features.
2. **Summarize** the processes of convergence that formed the Himalayan mountains.
3. **List** the geologic features associated with each type of convergent boundary.
4. **Identify** the type of location where transform boundaries most commonly occur.

Think Critically

5. **Choose** three plate boundaries in **Figure 17.16,** and predict what will happen over time at each boundary.
6. **Describe** how two portions of newly formed crust move between parts of a ridge that are offset by a transform boundary.

WRITING in Earth Science

7. Write a news report on the tectonic activity that is occurring at the Aleutian Islands in Alaska.

Section 17.4

Objectives

- **Explain** the process of convection.
- **Summarize** how convection in the mantle is related to the movements of tectonic plates.
- **Compare and contrast** the processes of ridge push and slab pull.

Review Vocabulary

convection: the circulatory motion that occurs in a fluid at a nonuniform temperature owing to the variation of its density and the action of gravity

New Vocabulary

ridge push
slab pull

Causes of Plate Motions

MAIN Idea Convection currents in the mantle cause plate motions.

Real-World Reading Link You probably know a lava lamp does not contain real lava, but the materials inside a lava lamp behave much like the molten rock within Earth.

Convection

One of the main questions about the theory of plate tectonics has remained unanswered since Alfred Wegener first proposed continental drift. What force or forces cause tectonic plates to move? Many scientists now think that large-scale motion in the mantle—Earth's interior between the crust and the core—is the mechanism that drives the movement of tectonic plates.

Convection currents Recall from Chapter 11 that convection is the transfer of thermal energy by the movement of heated material from one place to another. As in a lava lamp, the cooling of matter causes it to contract slightly and increase in density. The cooled matter then sinks as a result of gravity. Warmed matter is then displaced and forced to rise. This up-and-down flow produces a pattern of motion called a convection current. Convection currents aid in the transfer of thermal energy from warmer regions of matter to cooler regions. A convection current can be observed in the series of photographs shown in **Figure 17.20.** Earth's mantle is composed of partially molten material that is heated unevenly by radioactive decay from both the mantle itself and the core beneath it. Radioactive decay heats up the molten material in the mantle and causes enormous convection currents to move material throughout the mantle.

Figure 17.20 Water cooled by the ice cube sinks to the bottom where it is warmed by the burner and rises. The process continues as the ice cube cools the water again.
Infer *what will happen to the ice cube due to convection currents.*

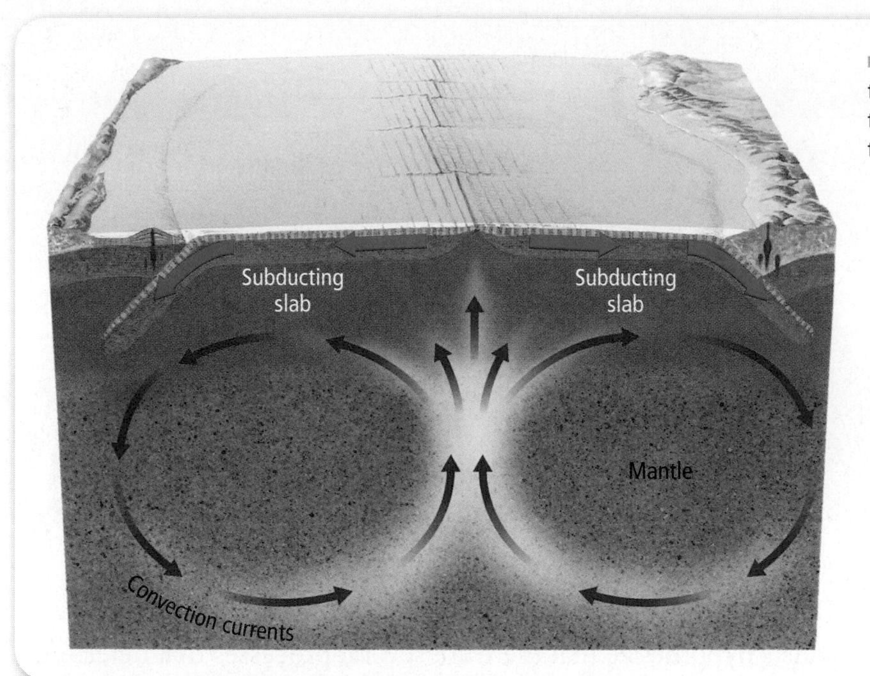

■ **Figure 17.21** Convection currents develop in the mantle, moving the crust and outermost part of the mantle, and transferring thermal energy from the Earth's interior to its exterior.

Convection in the mantle Convection currents in the mantle, illustrated in **Figure 17.21,** are thought to be the driving mechanism of plate movements. Recall that even though the mantle is a solid, much of it moves like a soft, pliable plastic. The part of the mantle that is too cold and stiff to flow lies beneath the crust and is attached to it, moving as a part of tectonic plates. In the convection currents of the mantle, cooler mantle material is denser than hot mantle material. Mantle that has cooled at the base of tectonic plates slowly sinks downward toward the center of Earth. Heated mantle material near the core is then displaced, and like the wax warmed in a lava lamp, it rises. Convection currents in the mantle are sustained by this rise and fall of material which results in a transfer of energy between Earth's hot interior and its cooler exterior. Although convection currents can be thousands of kilometers across, they flow at rates of only a few centimeters per year. Scientists think that these convection currents are set in motion by subducting slabs.

 **Reading Check Discuss** Which causes a convection current to flow: the rising of hot material, or the sinking of cold material?

Plate movement How are convergent and divergent movements of tectonic plates related to mantle convection? The rising material in the convection current spreads out as it reaches the upper mantle and causes both upward and sideways forces. These forces lift and split the lithosphere at divergent plate boundaries. As the plates separate, material rising from the mantle supplies the magma that hardens to form new ocean crust. The downward part of a convection current occurs where a sinking force pulls tectonic plates downward at convergent boundaries.

■ **Figure 17.22** Ridge push and slab pull are two of the processes that move tectonic plates over the surface of Earth.

Concepts In Motion

Interactive Figure To see an animation of ridge push and slab pull, visit glencoe.com.

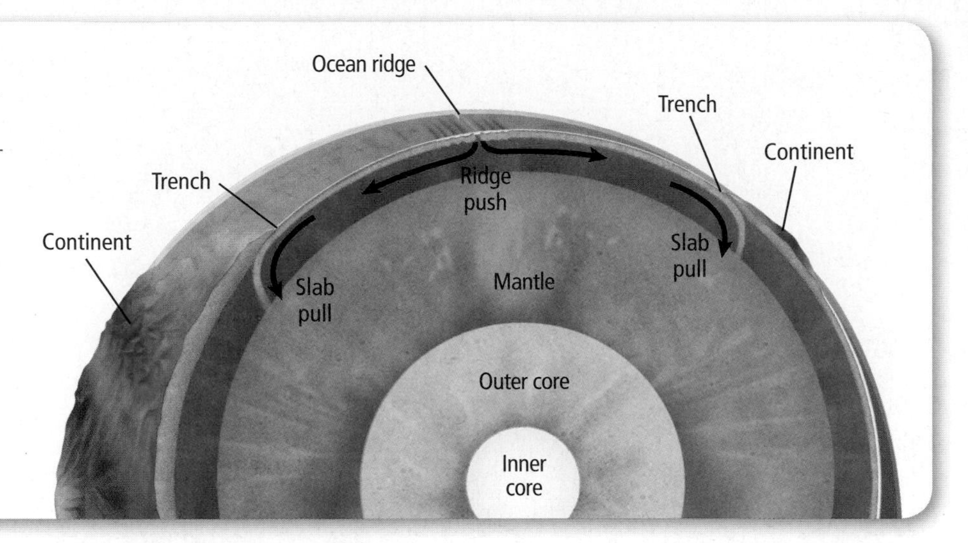

Push and Pull

Scientists hypothesize that there are several processes that determine how mantle convection affects the movement of tectonic plates. Study **Figure 17.22.** As oceanic crust cools and moves away from a divergent boundary, it becomes denser and sinks compared to the newer, less-dense oceanic crust. As the older portion of the seafloor sinks, the weight of the uplifted ridge is thought to push the oceanic plate toward the trench formed at the subduction zone in a process called **ridge push.**

A second and possibly more significant process that determines the movement of tectonic plates is called slab pull. In **slab pull,** the weight of a subducting plate pulls the trailing slab into the subduction zone much like a tablecloth slipping off the table can pull articles off with it. It is likely that combination of mechanisms such as these are involved in plate motions at subduction zones.

Section 17.4 Assessment

Section Summary

- Convection is the transfer of energy via the movement of heated matter.
- Convection currents in the mantle result in an energy transfer between Earth's hot interior and cooler exterior.
- Plate movement results from the processes called ridge push and slab pull.

Understand Main Ideas

1. MAIN Idea **Draw** a diagram comparing convection in a pot of water with convection in Earth's mantle. Relate the process of convection to plate movement.
2. **Restate** the relationships among mantle convection, ocean ridges, and subduction zones.
3. **Make** a model that illustrates the tectonic processes of ridge push and slab pull.

Think Critically

4. **Evaluate** this statement: Convection currents only move oceanic crust.
5. **Summarize** how convection is responsible for the arrangement of continents on Earth's surface.

WRITING in Earth Science

6. Write dictionary definitions for *ridge push* and *slab pull* without using those terms.

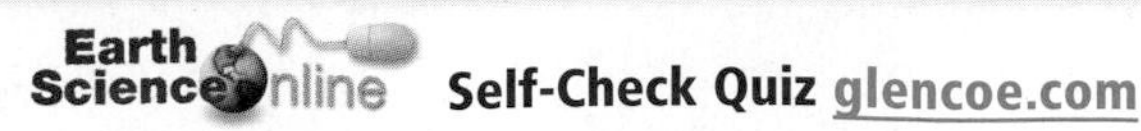

Self-Check Quiz glencoe.com

Earth Science and the Environment

Vailulu'u Seamount

The American Samoan Islands are part of an island chain in the South Pacific Ocean. Recent exploration at the edge of the island chain has revealed how tectonic processes can result in new and completely unique environments.

Discovery of a volcano In 1999, Vailulu'u (vah EEL ool oo oo), an active volcanic seamount, was discovered when oceanographers first mapped the area using remote sonar methods. The map revealed the outline of a massive volcano hollowed by a caldera. The ocean is about 5 km deep, and the ringlike ridges of the caldera come within 600 m from the ocean surface. The 1999 map showed that the caldera floor was generally flat—about 1 km below sea level. Scientists knew that the volcano was produced from a hot spot, a region of heated magma in the mantle below.

Vailulu'u revisited In 2005, a team of scientists returned to study Vailulu'u using deep-sea submersibles. Before diving, they remapped the seamount, and discovered that the floor of the caldera had changed dramatically. Sometime in the past six years, volcanic activity had developed a lava cone 300 m high, roughly the height of the Empire State Building. The cone was soon named Nafanua (nah fah NOO ah), after the Samoan goddess of war. The scientists made several trips in the submersible and discovered how tectonic activity had caused completely new ecosystems to develop.

Eel City At the top of Nafanua, they encountered 30-cm-long eels so numerous that they nicknamed the area *Eel City.* The top of the cone is too deep for sunlight to permit the growth of plants, so the scientists were puzzled about the eels' food source. Investigations revealed that the seamount had changed the local currents, depositing waves of shrimp above Nafanua.

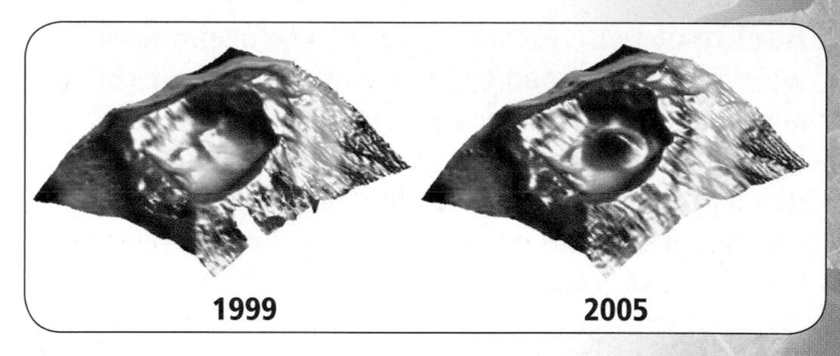

When you compare the two images, you can see the appearance of the Nafanua cone in the center of the caldera.

Moat of death Hydrothermal vents on the floor of the caldera emitted toxic chemicals, including clouds of a murky oil-like liquid containing carbon dioxide. Some of the vents released water that was a scalding 85°C. The same currents that brought shrimp to the eels were carrying fish down into the toxic environment of the caldera, which was nicknamed *the moat of death.* Yet, some life-forms were thriving. Much of the caldera floor was covered by a 1-m-thick mat of microbes, and bright red bristleworms abounded around the fish carcasses.

Birth of an island Nafanua is expected to continue growing. At its present rate, it will reach the ocean surface within a few decades and become the newest island in the Samoan chain. Earth scientists will continue to monitor the growth of Nafanua, and learn how tectonic events can help shape entire ecosystems.

WRITING in Earth Science

Investigate the biological activity and unique habitats discovered on Vailulu'u seamount. Write a newspaper article that describes the organisms and conditions on the seamount. To learn more about the Vailulu'u seamount, visit glencoe.com.

GEOLAB

MODEL PLATE BOUNDARIES AND ISOCHRONS

Background: **Isochron maps of the ocean floor were first developed using data from oceanic rocks and sediment. Isochrons are imaginary lines on a map that show the parts of Earth's surface that are the same age. When geologists first analyzed isochron maps of the ocean floor, they discovered that Earth's crust is formed along ocean ridges and recycled at the edge of oceanic crust. This discovery led to the theory known as plate tectonics. Geologists continue using maps to study the motion of tectonic plates.**

Question: *Can you determine the age of the crust and type of plate boundaries?*

Materials

paper
colored pencils
scissors
metric ruler
calculator

Safety Precautions

Procedure

1. Read and complete the lab safety form.
2. **Figure 1** shows Plate B relative to Plate A. Draw or trace the plates onto a separate sheet of paper and cut them out.

WARNING: ***Scissors can cut or puncture skin.***

3. The arrow shows the movement of the plates relative to each other. Move Plate A as shown in each part of **Figure 1.**
4. Use the symbols shown in the legend to indicate the type of plate boundary and the relative motion across the boundary for each part of **Figure 1.**
5. **Figure 2** shows two plates, A and B, separated by two ocean ridges at a transform boundary. Plates A and B are moving apart at 2 cm/y. Convert the speed 2 cm/y to km/y.
6. Trace **Figure 2** onto a separate sheet of paper. Assume the geometry of the boundaries in **Figure 2** has not changed over time. Draw isochrons on 10, 20, 30, and 40 million years.
7. Color the crust based on its age: 0–10 million years old red, 10–20 million years old yellow, 20–30 million years old green, and 30–40 million years old blue.

Analyze and Conclude

1. **Determine** the motion of a plate that would have each of the A sites that moved relative to the B plate.
2. **Apply** From your map of isochrons, what is the easiest way to identify the location of transform boundaries?
3. **Interpret** Look at **Figure 3.** From the pattern of isochrons on the ocean floor, identify the divergent plate boundaries along the Atlantic Ocean and along the Pacific Ocean.
4. **Differentiate** Which ocean is marked by wider isochrons? Based on the amount of oceanic crust produced in a given period of time, along which plate boundary is divergence happening more rapidly?
5. **Infer** The spreading center in the Pacific Ocean is not centered in the same manner as the Atlantic Ocean. Explain how this indicates the presence of convergent plate boundaries.

WRITING in Earth Science

Write a Letter Alfred Wegener never convinced the scientific community of continental drift. He died shortly before the ocean floors were mapped. Imagine you could send a message to the past. Explain to Wegener what ocean floor mapping revealed, and how plate tectonics was discovered.

Key

Use the following symbols to indicate the type of plate boundary:

- Divergent boundary
- Convergent boundary (triangles point to the plate that stays on the surface)
- Transform; arrows indicate the relative direction of motion across the boundary

Figure 1

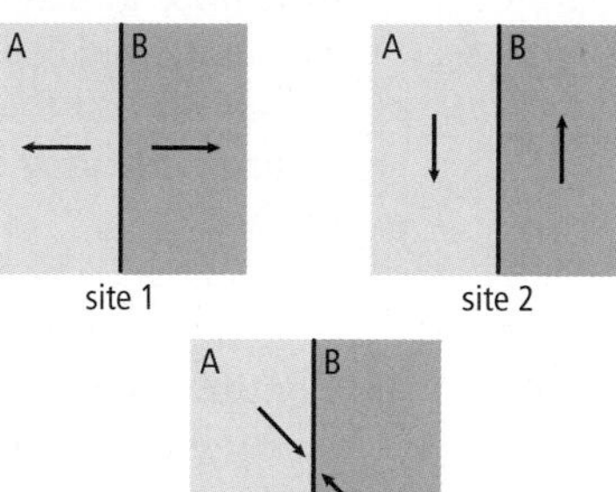

Figure 2

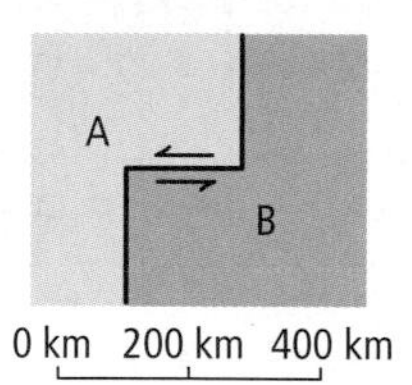

Figure 3

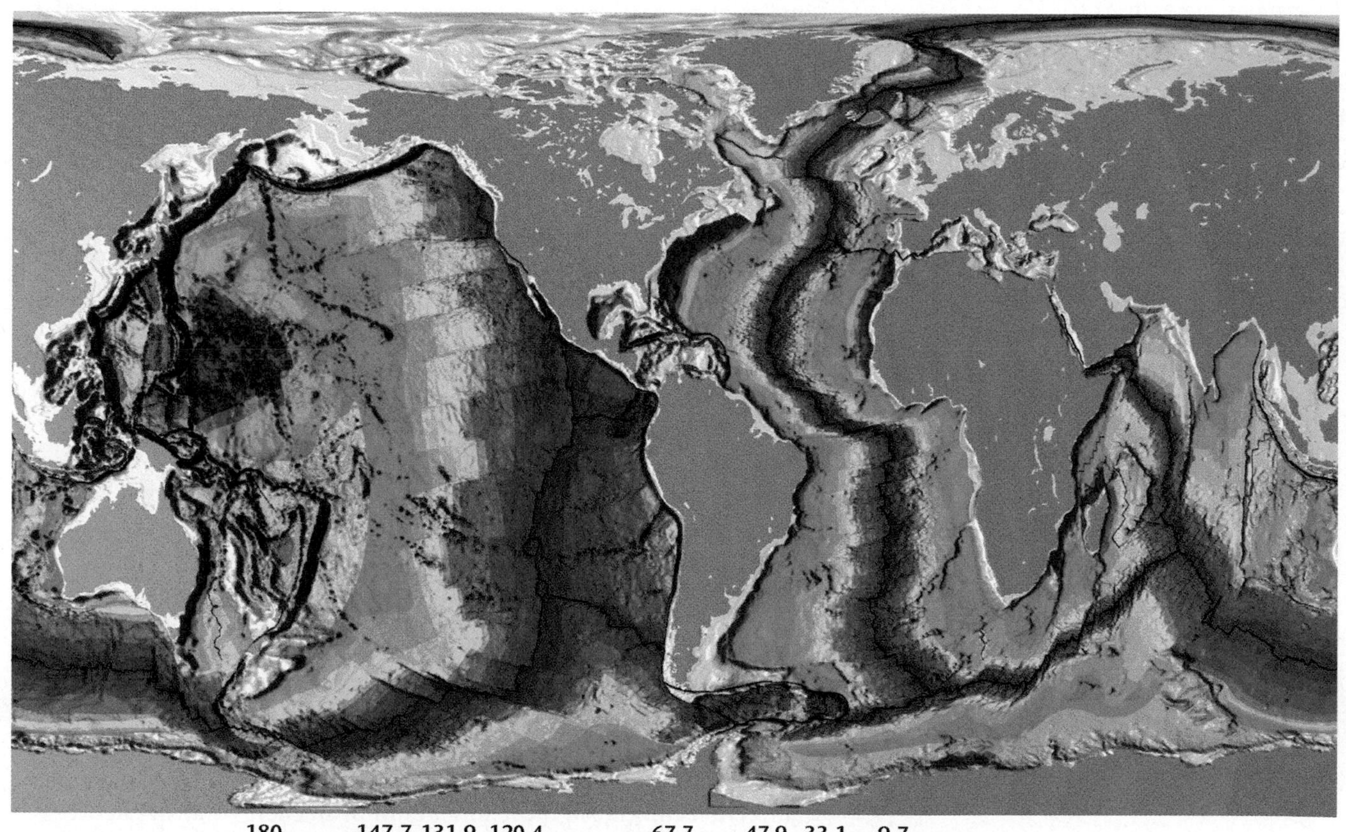

Millions of years before present

CHAPTER 17

Study Guide

Download quizzes, key terms, and flash cards from glencoe.com.

BIG Idea Most geologic activity occurs at the boundaries between plates.

Vocabulary	Key Concepts
Section 17.1 Drifting Continents	
• continental drift (p. 469) • Pangaea (p. 469)	**MAIN Idea** The shape and geology of the continents suggests that they were once joined together. • The matching coastlines of continents on opposite sides of the Atlantic Ocean suggest that the continents were once joined. • Continental drift was the idea that continents move around on Earth's surface. • Wegener collected evidence from rocks, fossils, and ancient climates to support his theory. • Continental drift was not accepted because there was no explanation for how the continents moved or what caused their motion.
Section 17.2 Seafloor Spreading	
• isochron (p. 477) • magnetic reversal (p. 476) • magnetometer (p. 473) • paleomagnetism (p. 476) • seafloor spreading (p. 479)	**MAIN Idea** Oceanic crust forms at ocean ridges and becomes part of the seafloor. • Studies of the seafloor provided evidence that the ocean floor is not flat and unchanging. • Oceanic crust is geologically young. • New oceanic crust forms as magma rises at ridges and solidifies. • As new oceanic crust forms, the older crust moves away from the ridges.
Section 17.3 Plate Boundaries	
• convergent boundary (p. 482) • divergent boundary (p. 481) • rift valley (p. 481) • subduction (p. 482) • tectonic plate (p. 480) • transform boundary (p. 484)	**MAIN Idea** Volcanoes, mountains, and deep-sea trenches form at the boundaries between the plates. • Earth's crust and rigid upper mantle are broken into large slabs of rock called tectonic plates. • Plates move in different directions and at different rates over Earth's surface. • At divergent plate boundaries, plates move apart. At convergent boundaries, plates come together. At transform boundaries, plates slide horizontally past each other. • Each type of boundary is characterized by certain geologic features.
Section 17.4 Causes of Plate Motions	
• ridge push (p. 488) • slab pull (p. 488)	**MAIN Idea** Convection currents in the mantle cause plate motions. • Convection is the transfer of energy via the movement of heated matter. • Convection currents in the mantle result in an energy transfer between Earth's hot interior and cooler exterior. • Plate movement results from the processes called ridge push and slab pull.

Vocabulary PuzzleMaker glencoe.com

CHAPTER 17 Assessment

Vocabulary Review

Replace each italicized word with the correct vocabulary term from the Study Guide.

1. *Plate tectonics* is the name given to the single continent that existed 200 mya.
2. *Continental fracture* is the idea that continents now separated by an ocean were once attached.
3. The process in which tectonic plates sink back into the mantle is called *divergence.*
4. A boundary where two plates come together is a *transform boundary.*
5. A divergent boundary within a continent forms a *trench.*

Match each of the following phrases with a vocabulary term from the Study Guide.

6. a line on a map that denotes crust that formed at the same time
7. the process that creates new ocean crust by the upwelling of magma at ocean ridges
8. the study of the history of Earth's magnetic field
9. a device that measures magnetism

Define the following vocabulary terms in complete sentences.

10. tectonic plate
11. ridge push
12. slab pull

Use what you know about the vocabulary terms on the Study Guide to describe what the terms in each pair have in common.

13. divergent boundary, transform boundary
14. subduction, convergent boundary
15. continental drift, plate tectonics
16. seafloor spreading, magnetic reversal

Understand Key Concepts

17. Which suggested to early cartographers that the continents were once joined?
    A. ocean depth
    B. position of south pole
    C. shape of continents
    D. size of Atlantic Ocean

18. What was Wegener's hypothesis called?
    A. seafloor spreading
    B. plate tectonics
    C. continental drift
    D. slab pull

Use the figure below to answer Questions 19 and 20.

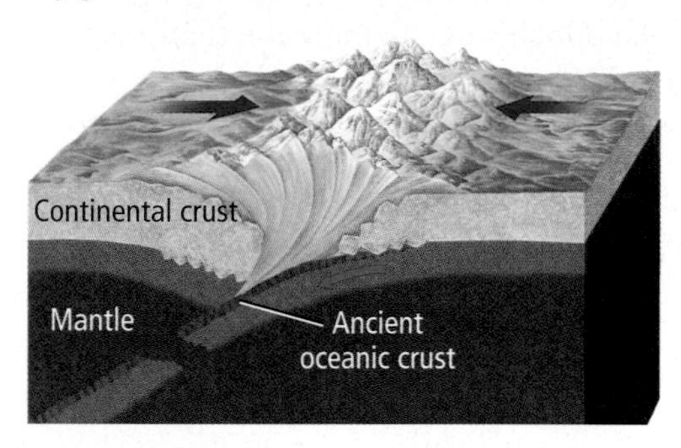

19. What type of boundary is shown?
    A. an ocean ridge
    B. a continental-continental boundary
    C. a transform boundary
    D. an oceanic-continental boundary

20. Which feature forms along this type of boundary?
    A. subduction zones
    B. oceanic trenches
    C. island arcs
    D. folded mountains

21. The weight of a subducting plate helps pull it into a subduction zone in which process?
    A. slab pull
    B. ridge push
    C. slab push
    D. ridge pull

22. Which is a convergent boundary that does not have a subduction zone?
    A. oceanic-oceanic
    B. oceanic-continental
    C. continental-continental
    D. transform

Use the figure below to answer Questions 23 and 24.

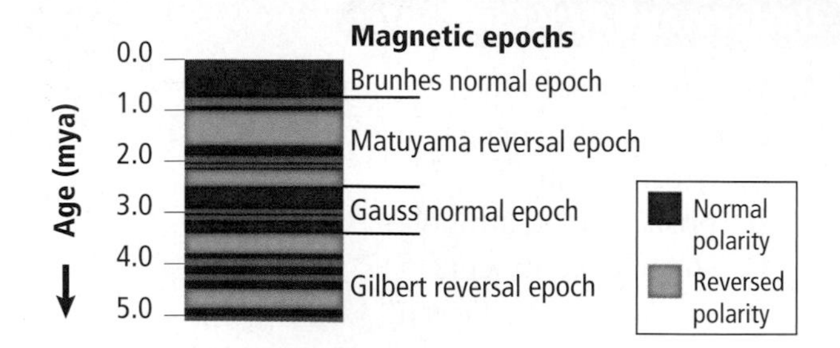

23. Approximately how long did the Gauss epoch last?
A. 5 million years
B. 3 million years
C. 1 million years
D. 100,000 years

24. Which epoch saw the most fluctuations between normal and reverse polarity?
A. Gauss
B. Matuyama
C. Gilbert
D. Brunhes

25. Generally, what is the age of oceanic crust?
A. the same age as the continental crust
B. younger than the continental crust
C. older than the continental crust
D. science has never determined its age

26. Which observation was not instrumental in formulating the hypothesis of seafloor spreading?
A. magnetization of the oceanic crust
B. depth of the ocean
C. thickness of seafloor sediments
D. identifying the location of glacial deposits

27. How fast do plates move relative to each other?
A. millimeters per day
B. centimeters per year
C. meters per year
D. centimeters per day

28. What process creates deep-sea trenches?
A. subduction
B. magnetism
C. earthquakes
D. transform boundaries

Use the photo below to answer Questions 29 and 30.

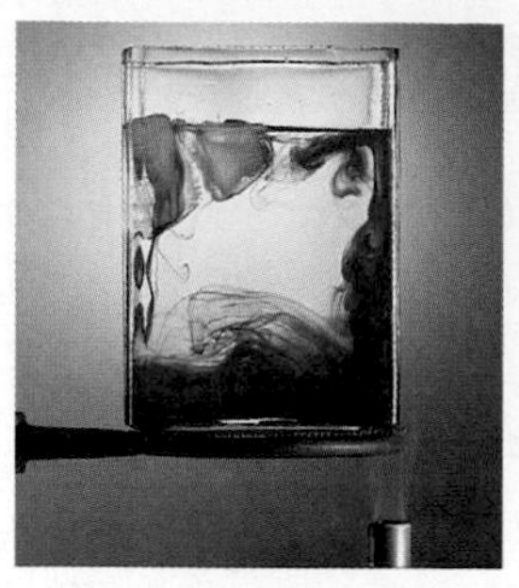

29. As shown, which direction does the icy water move?
A. up
B. down
C. remains in the same place
D. sideways

30. Which is modeled by the water movement?
A. subduction
B. continental drift
C. magnetic reversal
D. mantle convection

31. Which is not a force causing plates to move?
A. ridge push
B. slab pull
C. volcanism
D. convection

Constructed Response

32. Summarize What observations led to the proposal of continental drift?

33. CAREERS IN EARTH SCIENCE Explain why oceanographers have found that the thickness of seafloor sediments increases with increasing distance from the ocean ridge.

34. Differentiate between the magnetic field generated in Earth's core and the magnetization preserved in the oceanic crust.

35. Analyze why there are differences between continental-continental convergent boundaries and oceanic-oceanic convergent boundaries.

36. Summarize Why was the idea of moving continents more widely accepted after seafloor spreading was proposed?

Earth Science Online Chapter Test glencoe.com

Think Critically

Use the map below to answer Question 37.

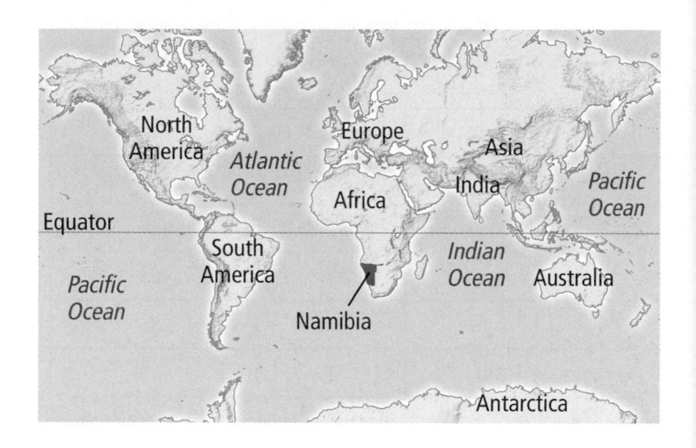

37. **Infer** If 200 million-year-old oil deposits were discovered in Namibia, where might geologists also expect to find oil deposits of a similar age? Explain.

38. **Compare and contrast** ridge push and slab pull.

39. **Summarize** How have satellite monitoring systems such as GPS made it much easier and cheaper to study the motion of tectonic plates?

40. **Consider** Do plates always stay the same shape and size? Explain.

41. **Critique** this statement: There are two kinds of tectonic plates—continental plates and oceanic plates.

Concept Mapping

42. Create a concept map using the following terms: *convergent, rift valley, divergent, transform, island arc, shallow earthquakes, mountain range,* and *plate boundary*. Refer to the *Skillbuilder Handbook* for more information.

Challenge Question

43. **Predict** Assuming that Earth's tectonic plates will continue moving in the directions shown in **Figure 17.2,** sketch a globe showing the relative positions of the continents in 60 million years.

Additional Assessment

44. *WRITING in* **Earth Science** Imagine you are on a sailboat anchored off the coast of Chile. You hear loud rumbling. Then GPS data indicates a part of the coast shifted up by about 1.5 m. Write a journal entry to describe the geologic phenomena you are seeing and experiencing.

DBQ Document–Based Questions

Data obtained from: Seismicity of the Central United States: 1990–2000. *USGS National Earthquake Information Center.*

Most earthquakes occur at plate boundaries as plates slide by each other. This map shows the location and depth of earthquakes between 1990 and 2000 in Alaska.

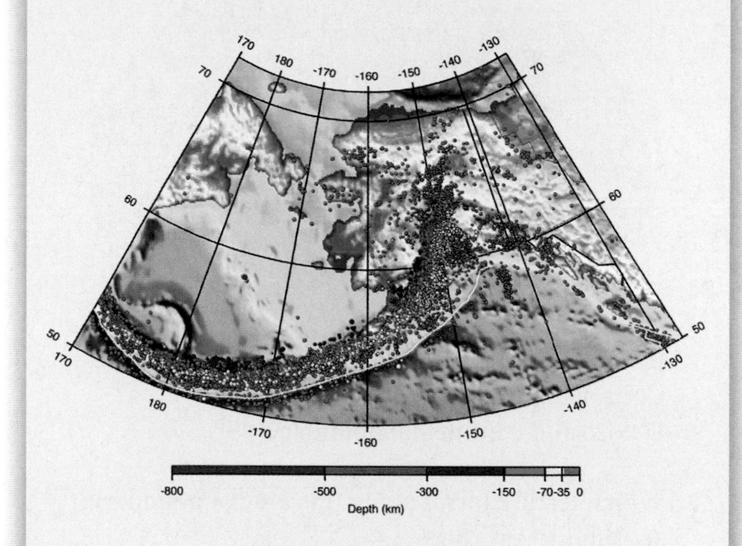

45. Identify which plate is subducting and provide evidence from the figure to support your answer.

46. Compare this map to **Figure 17.15,** which shows the location of plate boundaries. Why do parts of the plate boundaries have few or no earthquakes?

Cumulative Review

47. How do Landsat satellites collect and analyze data to map Earth's surface? **(Chapter 2)**

48. How can scientists use glaciers to study Earth's past? **(Chapter 8)**

49. Describe the major parameters used in the Köppen Classification System. **(Chapter 14)**

Standardized Test Practice

Multiple Choice

1. How does the building of jetties negatively effect coastlines?
   A. They fill in anchorage used to harbor boats with sediment.
   B. They hinder breakwater from moving sediments away from the area.
   C. They reflect energy back toward beaches, increasing erosion.
   D. They deprive beaches down the coast from the jetty of sand.

Use the diagram below to answer Questions 2 and 3.

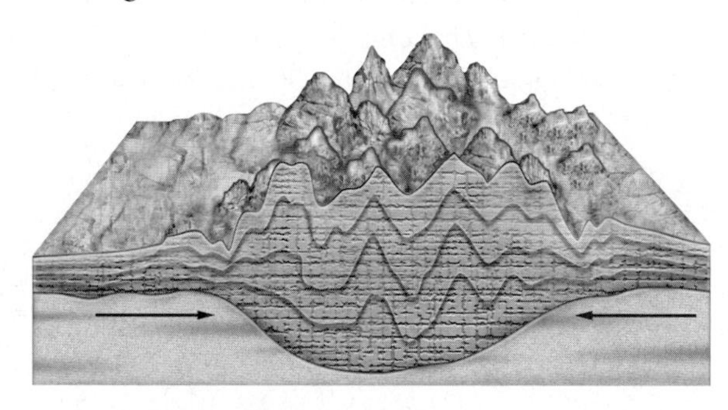

2. What type of plate boundary is shown?
   A. ocean ridge
   B. continental-continental boundary
   C. transform boundary
   D. oceanic-continental boundary

3. Which feature forms along this type of boundary?
   A. subduction zones
   B. oceanic trenches
   C. island arcs
   D. folded mountains

4. What is the best way to get out of a rip current?
   A. swim parallel to the shore
   B. swim with the rip current
   C. swim against the rip current
   D. swim under the rip current

5. The smooth parts of the ocean floor located 5 to 6 km below sea level are called the
   A. mid-ocean ridges
   B. deep-sea trenches
   C. abyssal plains
   D. continental rises

Use the table below to answer Questions 6–8.

Exercise and Heart Rates			
Subject	Resting	Fast Walk	Slow Jog
1	65	72	110
2	78	88	120
3	72	83	125
4	69	78	105
5	75	90	135
Averages	71.8	82.2	119

6. The table shows heart rates before and after a 10-min session. Which statement best summarizes the data?
   A. There is no relationship between heart rate and exercise.
   B. Exercise increased the heart rate of the participants.
   C. Heart rate increased as exercise became more strenuous.
   D. Ten minutes of exercise was not enough to increase heart rate.

7. According to the data, which subject appears to be in the best shape and why?
   A. Subject 1 because the subject had the lowest resting heart rate
   B. Subject 4 because the subject had the lowest heart rate during a slow jog
   C. Subject 5 because the subject had the fastest resting heart rate
   D. This cannot be determined from the table because not enough information is given about each of the subjects.

8. What would be the best graph to use in order to present the data found?
   A. bar graph
   B. line graph
   C. circle graph
   D. a model

Short Answer

Use the illustration below to answer Questions 9–11.

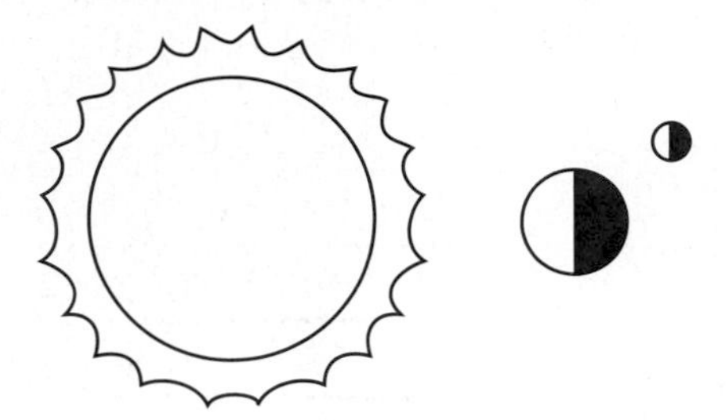

9. As the Moon moves to create a right angle along with the Sun and Earth, what occurs with the ocean's tides?

10. Describe how tides are affected when the Sun, the Moon, and Earth are aligned.

11. How do lunar tides differ from solar tides?

12. What negative impact might a major storm have on a barrier island?

13. How do atmosphere and large bodies of water affect climate in various regions?

14. How did a temperature decrease of only 5°C during the ice ages cause major changes?

Reading for Comprehension

Seafloor Maps

In 2005, the U.S. nuclear submarine *San Francisco* crashed into an uncharted underwater mountain in the South Pacific, killing one submariner and injuring dozens of others. The incident highlights a troubling nautical reality—we might know more about the geography of the Moon than that of the ocean floor. Estimates vary, but the amount of correctly mapped seafloor in the public domain is likely around 2 or 3 percent.

Although survey ships equipped with sound-based systems can accurately map the seafloor by dropping a "beam" below the ship, this method can be used to map only narrow sections at a time. Mapping all the oceans this way might take a thousand years and cost billions of U.S. dollars. However, such maps could be critical for tsunami-preparation efforts. No matter how deep the ocean, a tsunami moves along the bottom, and its path is influenced by the features of the ocean floor. Thus, understanding the location of trenches, seamounts, and other features is essential to calculations of how a tsunami will move and where and in what force it will come ashore. Other studies that could benefit from mapping include marine animal habitat and ocean mixing rates, which are essential to absorption of greenhouse gases. All are dependent on more detailed knowledge of the other 70 percent of Earth's surface.

Article obtained from: Handwerk, B. Seafloor still about 90 percent unknown, experts say. *National Geographic News.* February 17, 2003.

15. What can be inferred from this passage?

A. It is important for ships and submarines to use sonar so that they do not run into underwater mountains.

B. Mapping the seafloor is too expensive and not important enough to humans.

C. Very little is known about the seafloor, and by improving this knowledge, both humans and animals will benefit.

D. Many marine animals' lives will be disrupted if scientists continue to map the ocean floor.

16. How would knowing what is on the seafloor help an oceanographer track a tsunami?

NEED EXTRA HELP?														
If You Missed Question . . .	1	2	3	4	5	6	7	8	9	10	11	12	13	14
Review Section . . .	16.1	17.3	17.3	16.1	16.2	1.2	1.2	1.3	15.3	15.3	15.3	16.1	14.1	14.3

CHAPTER 18 Volcanism

BIG Idea **Volcanoes develop from magma moving upward from deep within Earth.**

18.1 Volcanoes
MAIN Idea The locations of volcanoes are mostly determined by plate tectonics.

18.2 Eruptions
MAIN Idea The composition of magma determines the characteristics of a volcanic eruption.

18.3 Intrusive Activity
MAIN Idea Magma that solidifies below ground forms geologic features different from those formed by magma that cools at the surface.

GeoFacts

- All the lava from Kilauea could pave a road three times around Earth.
- There are 500 active volcanoes on Earth today.
- Magma comes from the Greek word meaning *dough.*
- Many of Earth's geographic features are caused by volcanoes.

LAUNCH Lab

What makes magma rise?

Magma is molten rock that lies beneath Earth's surface. In this activity, you will model the movement of magma within Earth by making a "lava lamp."

Procedure

1. Read and complete the lab safety form.
2. Pour about 300 mL of **water** into a **600-mL beaker.**
3. Pour about 80 mL of **vegetable oil** into the beaker.
4. Sprinkle **table salt** on top of the oil while you slowly count to 5.
5. Add more salt to keep the movement going.

Analysis

1. **Identify** which component of your model represents magma.
2. **Describe** what happened to the oil before and after you added the salt.
3. **Hypothesize** what causes the "magma" to rise.

FOLDABLES™ Study Organizer

Classification of Volcanoes Make this Foldable to help you understand volcanoes.

STEP 1 Stack two sheets of notebook paper approximately 1.5 cm apart.

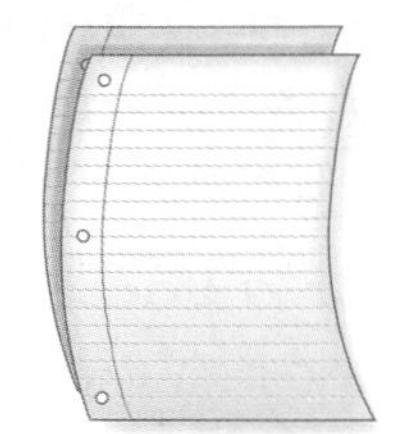

STEP 2 Fold up the bottom edges to form four tabs.

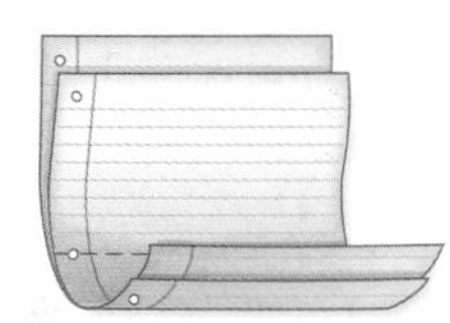

STEP 3 Staple along the folded edge. With the stapled end at the top, label the tabs as follows: *Volcano Types, Shield Volcano, Composite Volcan*o, and *Cinder Cone.*

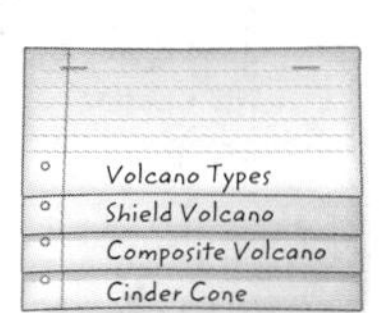

FOLDABLES **Use this Foldable with Section 18.1.** As you study the section, write about the characteristics of each kind of volcano under each tab.

Visit glencoe.com to

- **study entire chapters online;**
- **explore Concepts In Motion animations:**
  - **Interactive Time Lines**
  - **Interactive Figures**
  - **Interactive Tables**
- **access Web Links for more information, projects, and activities;**
- **review content with the Interactive Tutor and take Self-Check Quizzes.**

Section **18.1**

Objectives

- **Describe** how plate tectonics influences the formation of volcanoes.
- **Locate** major zones of volcanism.
- **Identify** the parts of a volcano.
- **Differentiate** between volcanic landforms.

Review Vocabulary

convergent: tending to move toward one point or to approach each other

New Vocabulary

volcanism
hot spot
flood basalt
fissure
conduit
vent
crater
caldera
shield volcano
cinder cone
composite volcano

Volcanoes

MAIN Idea **The locations of volcanoes are mostly determined by plate tectonics.**

Real-World Reading Link Road crews spread salt on icy winter roads because salt makes the ice melt at a lower temperature. At extremely high temperatures, rocks can melt. Often, if heated rocks are in contact with water, they melt more easily.

Zones of Volcanism

Volcanoes are fueled by magma. Recall from Chapter 5 that magma is a slushy mixture of molten rock, mineral crystals, and gases. As you observed in the Launch Lab, once magma forms, it rises toward Earth's surface because it is less dense than the surrounding mantle and crust. Magma that reaches Earth's surface is called lava. **Volcanism** describes all the processes associated with the discharge of magma, hot fluids, and gases.

As you read this, approximately 20 volcanoes are erupting. In a given year, volcanoes will erupt in about 60 different places on Earth. The distribution of volcanoes on Earth's surface is not random. A map of active volcanoes, shown in **Figure 18.1,** reveals striking patterns on Earth's surface. Most volcanoes form at plate boundaries. The majority form at convergent boundaries and divergent boundaries. Along these margins, magma rises toward Earth's surface. Only about 5 percent of magma erupts far from plate boundaries.

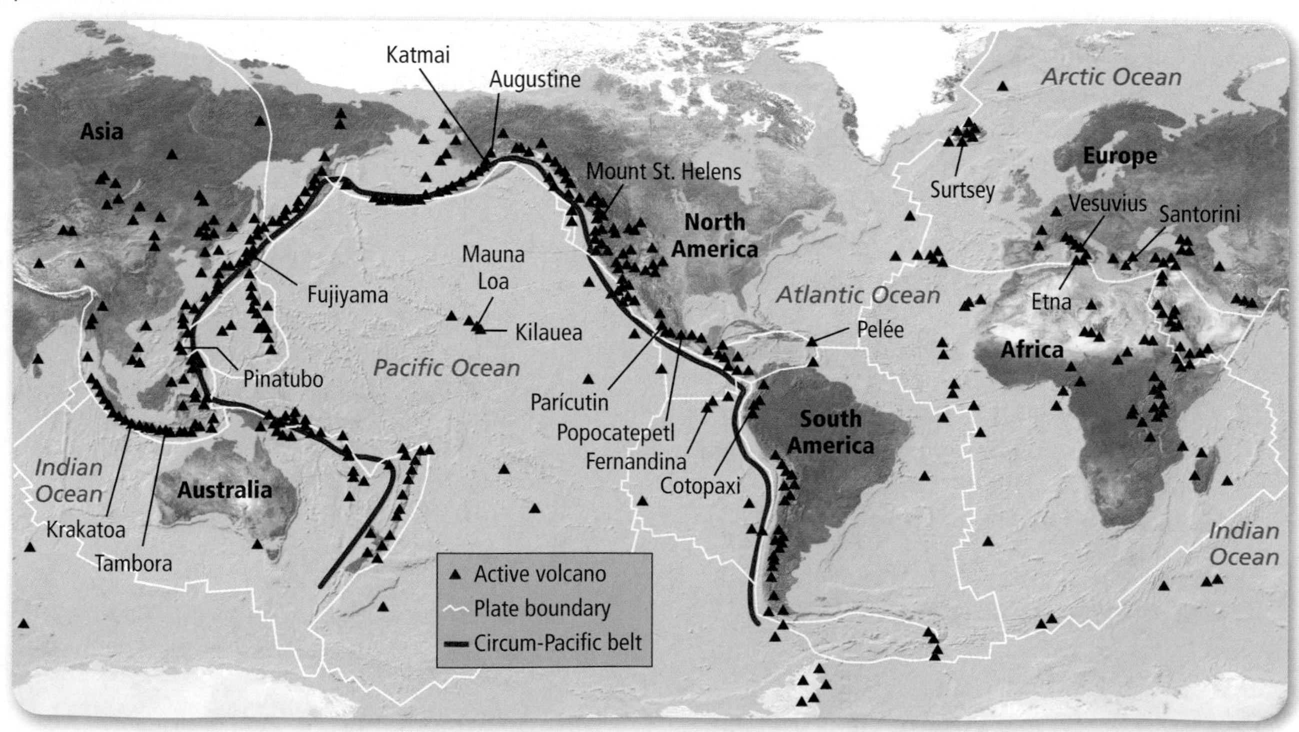

■ **Figure 18.1** Most of Earth's active volcanoes are located along plate boundaries.

Convergent volcanism Recall from Chapter 17 that tectonic plates collide at convergent boundaries, which can form subduction zones—places where slabs of oceanic crust descend into the mantle. As shown in **Figure 18.2,** an oceanic plate descends below another plate into the mantle. As the oceanic plate descends, magma forms. The magma moves upward because it is less dense than the surrounding solid material. As it rises, the magma mixes with rock, minerals, and sediment from the overlying plate. Most volcanoes located on land result from oceanic-continental subduction. These volcanoes are characterized by explosive eruptions

Reading Check **Define** What is convergent volcanism?

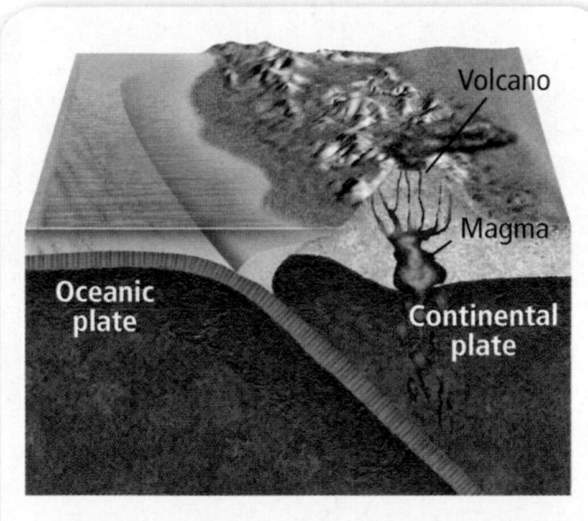

Figure 18.2 In an oceanic-continental subduction zone, the denser oceanic plate slides under the continental plate into the hot mantle. Parts of the plate melt and magma rises, eventually leading to the formation of a volcano.
***Identify** a volcano from Figure 18.1 that is associated with oceanic-continental convergence.*

Concepts In Motion
Interactive Figure To see an animation of subduction, visit glencoe.com.

Two major belts The volcanoes associated with convergent plate boundaries forms a major belt, shown in **Figure 18.1.** The Circum-Pacific Belt is also called the Pacific Ring of Fire. The name *Circum-Pacific* gives a hint about the location of the belt. *Circum* means *around* (as in circumference). The outline of the belt corresponds to the outline of the Pacific Plate. The belt stretches along the western coasts of North and South America, across the Aleutian Islands, and down the eastern coast of Asia. Volcanoes in the Cascade Range of the western United States and Mount Pinatubo in the Philippines are some of the volcanoes in the Circum-Pacific Belt. A smaller belt is called the Mediterranean Belt. It includes Mount Etna and Mount Vesuvius, two volcanoes in Italy. Its general outlines correspond to the boundaries between the Eurasian, African, and Arabian plates.

DATA ANALYSIS LAB

Based on Real Data*

Interpret the Graph

How do zones of volcanism relate to lava production? Researchers classify types of volcanic eruptions and study how much lava each type of volcano emits during an average year. The circle graphs show data from 5337 eruptions and annual lava production for each zone.

Think Critically

1. **Describe** the relationship between the type of volcanism and annual lava production.
2. **Consider** Why is it important for scientists to study this relationship?
3. **Evaluate** What could be the next step in the researchers' investigation?

Data and Observations

Number of Eruptions in Average Year

Lava Production

Convergent

Hot spot

Rift

*Data obtained from: Crisp, J. 1984. Rates of magma emplacement and volcanic output. *Journal of Volcanology and Geothermal Research* 20: 177–211.

■ **Figure 18.3** Eruptions at divergent boundaries tend to be nonexplosive. At the divergent boundary on the ocean floor, eruptions often form huge piles of lava called pillow lava.

Concepts In Motion
Interactive Figure To see an animation of divergent plate boundaries, visit glencoe.com.

Divergent volcanism Recall from Chapter 17 that at divergent plate boundaries tectonic plates move apart and new ocean floor is produced as magma rises to fill the gap. At ocean ridges, this lava takes the form of giant pillows like those in **Figure 18.3,** and is called pillow lava. Unlike the explosive volcanoes detailed in **Figure 18.4,** volcanism at divergent boundaries tends to be nonexplosive, with effusions of large amounts of lava. About two-thirds of Earth's volcanism occurs underwater along divergent boundaries at ocean ridges.

Vocabulary
Science usage v. Common usage
Plume
Science usage: an elongated column

Common usage: a large, showy feather of a bird

 **Reading Check** **Convert** the fraction of volcanism that happens underwater to a percentage.

Hot spots Some volcanoes form far from plate boundaries over hot spots. Scientists hypothesize that **hot spots** are unusually hot regions of Earth's mantle where high-temperature plumes of magma rise to the surface.

■ **Figure 18.4**
Volcanoes in Focus
Volcanoes constantly shape Earth's surface.

6000 B.C.

3000 B.C.

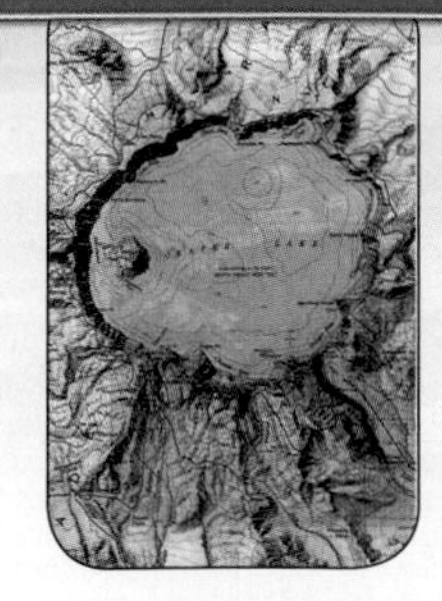

4845 B.C. Mount Mazama erupts in Oregon. The mountain collapses into a 9-km-wide depression, known today as Crater Lake (topographic map).

1630 B.C. In Greece, Santorini explodes, causing tsunamis 200 m high. Nearby, Minoan civilization on the Isle of Crete disappears.

A.D. 79 Mount Vesuvius in Italy erupts, burying two cities in ash.

Hot spot volcanoes Some of Earth's best-known volcanoes formed as a result of hot spots under the ocean. For example, the Hawaiian islands, shown in the map in **Figure 18.5,** are located over a plume of magma. As the rising magma melted through the crust, it formed volcanoes. The hot spot formed by the magma plume remained stationary while the Pacific Plate slowly moved northwest. Over time, the hot spot has left a trail of volcanic islands on the floor of the Pacific Ocean. The volcanoes on the oldest Hawaiian island, Kauai, are inactive because the island no longer sits above the stationary hot spot. Even older volcanoes to the northwest are no longer above sea level. The world's most active volcano, Kilauea, on the Big Island of Hawaii, is currently located over the hot spot. Another volcano, Loihi, is forming on the seafloor southeast of the Big Island of Hawaii and might eventually rise above the ocean surface to form a new island.

Hot spots and plate motion Chains of volcanoes that form over stationary hot spots provide information about plate motions. The rate and direction of plate motion can be calculated from the positions of these volcanoes. The map in **Figure 18.5** shows that the Hawaiian islands are at one end of the Hawaiian-Emperor volcanic chain. The oldest seamount, Meiji, is at the other end of the chain and is about 80 million years old, which indicates that this hot spot has existed for at least that many years. The bend in the chain at Daikakuji Seamount records a change in the direction of the Pacific Plate that occurred 43 mya.

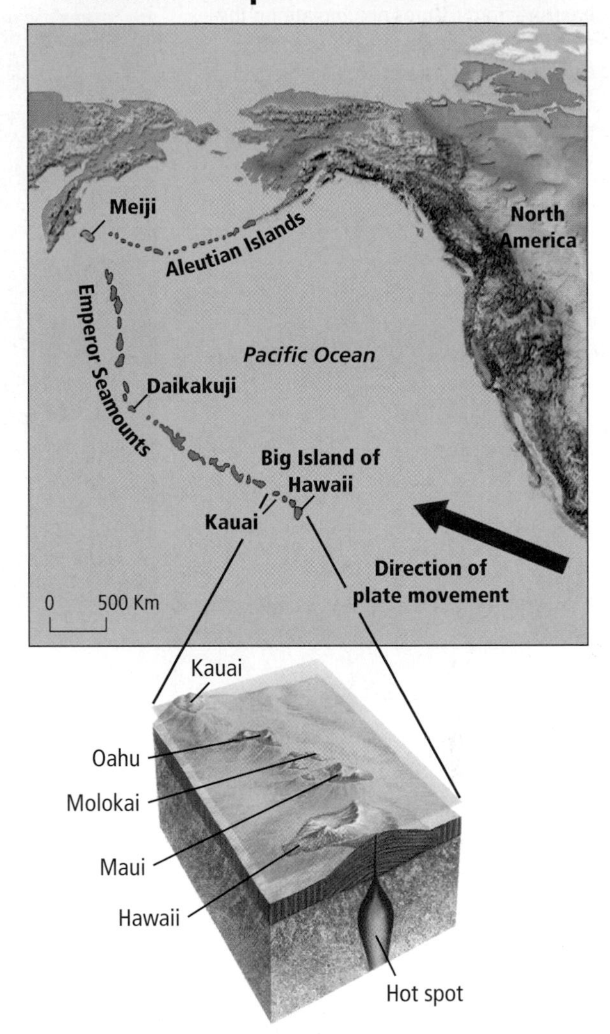

■ **Figure 18.5** The Hawaiian islands have been forming for millions of years as the Pacific Plate moves slowly over a stationary hot spot that is currently located under the Big Island of Hawaii.

1800 **1900** **2000**

1883 In Indonesia, Krakatoa erupts, destroying two-thirds of the island and generating a tsunami that kills more than 36,000 people.

1912 Katmai erupts in Alaska with ten times more force than Mount St. Helens. This eruption is one of the most powerful in recorded history.

1980 In Washington, Mount St. Helens' eruption blasts through the side of the volcano. Most of the 57 fatalities are from ash inhalation.

1991 Mount Pinatubo erupts in the Philippines, releasing 10 km^3 of ash, reducing global temperatures by 0.5°C.

Concepts In Motion

Interactive Time Line To learn more about these discoveries and others, visit glencoe.com. Earth Science Online

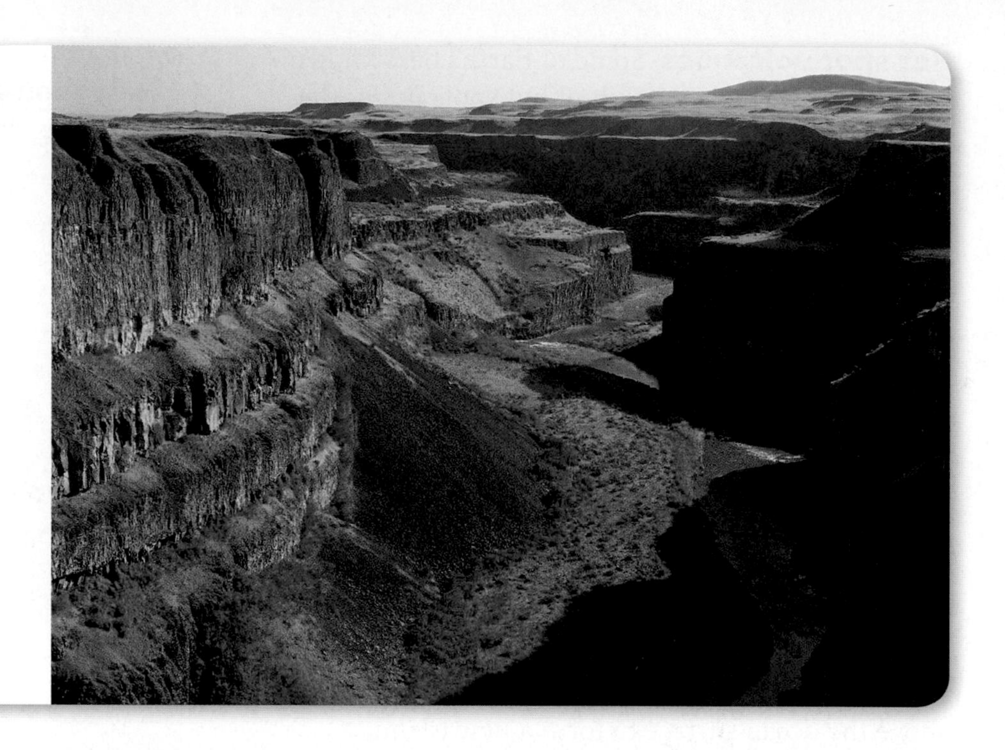

■ **Figure 18.6** Huge amounts of lava erupting from fissures accumulate on the surface, often forming layers 1 km thick. Over time, streams and other geologic forces erode the layers of basalt, leaving plateaus like this one in Palouse Canyon, Washington.

Flood basalts When hot spots occur beneath continental crust, they can lead to the formation of flood basalts. **Flood basalts** form when lava flows out of long cracks in Earth's crust. These cracks are called **fissures.** Over hundreds or even thousands of years, these fissure eruptions can form flat plains called plateaus, as shown in **Figure 18.6.** As in other eruptions, when the lava flows across Earth's surface, water vapor and other gases escape.

■ **Figure 18.7** More than 17 mya, enormous amounts of lava poured out of large fissures, producing a basaltic plateau more than 1 km thick in the northwestern part of the United States.

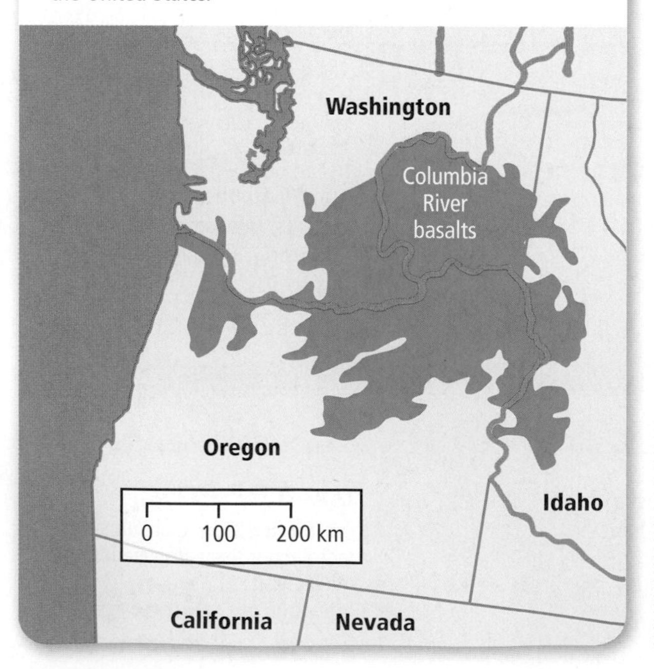

Columbia River Basalts The volume of basalt erupted by fissure eruptions can be tremendous. For example, the Columbia River basalts, located in the northwestern United States and shown on the map in **Figure 18.7,** contain 170,000 km^3 of basalt. This volume of basalt could fill Lake Superior, the largest of the Great Lakes, 15 times. However, the Columbia River Basalts are small in comparison to the Deccan Traps.

Deccan Traps About 65 mya in India, a huge flood basalt eruption created an enormous plateau called the Deccan Traps. The volume of basalt in the Deccan Traps is estimated to be about 512,000 km^3. That volume would cover the island of Manhattan with a layer 10,000 km thick, or the entire state of New York with a layer 4 km thick. Some geologists hypothesize that the eruption of the Deccan Traps caused a global change in climate that might have influenced the extinction of the dinosaurs.

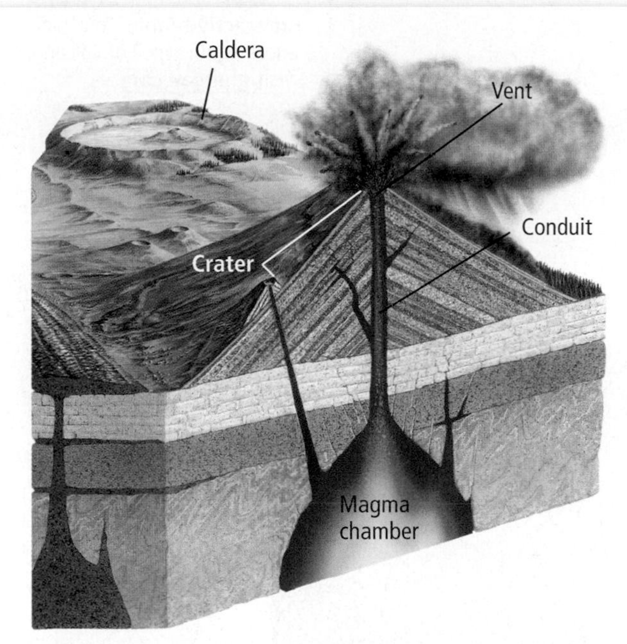

■ **Figure 18.8** Magma moves upward from deep within Earth through a conduit and erupts at Earth's surface through a vent. The area around the vent is called a crater. A caldera can form when the crust collapses into an empty magma chamber.

Concepts In Motion **Interactive Figure** To see an animation of caldera formation, visit glencoe.com.

Anatomy of a Volcano

As you read in Chapter 5, when magma reaches Earth's surface it is called lava. Lava reaches the surface by traveling through a tubelike structure called a **conduit,** and emerges through an opening called a **vent.** As lava flows through the vent and out onto the surface, it cools and solidifies around the vent. Over time, layers of solidified lava can accumulate to form a mountain known as a volcano. At the top of a volcano, around the vent, is a bowl-shaped depression called a **crater.** The crater is connected to the magma chamber by the conduit. Locate the crater, conduit, and vent of the volcano shown in **Figure 18.8.**

Volcanic craters are usually less than 1 km in diameter. Larger depressions, called **calderas,** can be up to 50 km in diameter. Calderas often form after the magma chamber beneath a volcano empties from a major eruption. The summit or the side of a volcano collapses into the emptied magma chamber, leaving an expansive, circular depression. After the surface material collapses, water sometimes fills the caldera, forming scenic lakes. The caldera known as Crater Lake in southern Oregon formed when Mount Mazama collapsed.

MiniLab

Model a Caldera

How do calderas form? Calderas are volcanic craters that form when the summit or the side of a volcano collapses into the magma chamber that once fueled the volcano.

Procedure

1. Read and complete the lab safety form.
2. Obtain a **small box,** a **10-cm length of rubber tubing,** a **clamp,** and a **balloon** from your teacher.
3. Line the box with **newspaper** and make a small hole in the box and the newspaper with **scissors.**
4. Thread the neck of the balloon through the hole, insert the rubber tubing into the neck, securing it with **tape,** inflate the balloon by blowing through the tubing, and use the clamp to close the tubing.
5. Pour six cups of **sand** over the balloon.
6. Sculpt the sand into the shape of a volcano. You might need to vary the amount of sand and type of box to reach the desired effect.
7. Remove the clamp, releasing the air from the balloon. Observe your caldera forming, and record your observations
8. Compare your caldera to your classmates'.

Analysis

1. **Sequence** the formation of the caldera.
2. **Compare** the features of a caldera with those of a crater.
3. **Infer** how the caldera will form if you vary how much you inflate the balloon.

Concepts In Motion

Interactive Table To explore more about types of volcanoes, visit glencoe.com.

Table 18.1 Types of Volcanoes

Description	Example of Volcanoes
Shield Volcanoes • Largest of the three types of volcanoes • Long, gentle slopes • Composed of layers of solidified basaltic lava • Quiet eruptions	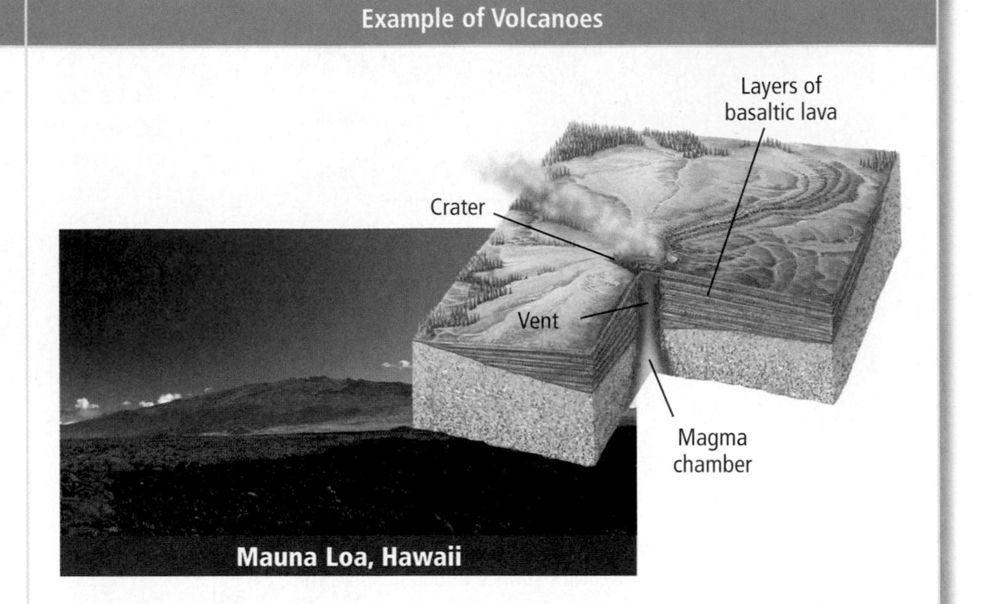 Mauna Loa, Hawaii
Cinder Cones • Smallest of the three types of volcanoes • Steep-sloped, cone-shaped • Usually composed of basaltic lava • Explosive eruptions • Usually form at edges of larger volcanoes	 Lassen Volcanic Park, California
Composite Volcanoes • Considerably larger than cinder cones • Tall, majestic mountains • Composed of layers of rock from explosive eruptions and lava flows • Cycle through periods of quiet and explosive eruptions	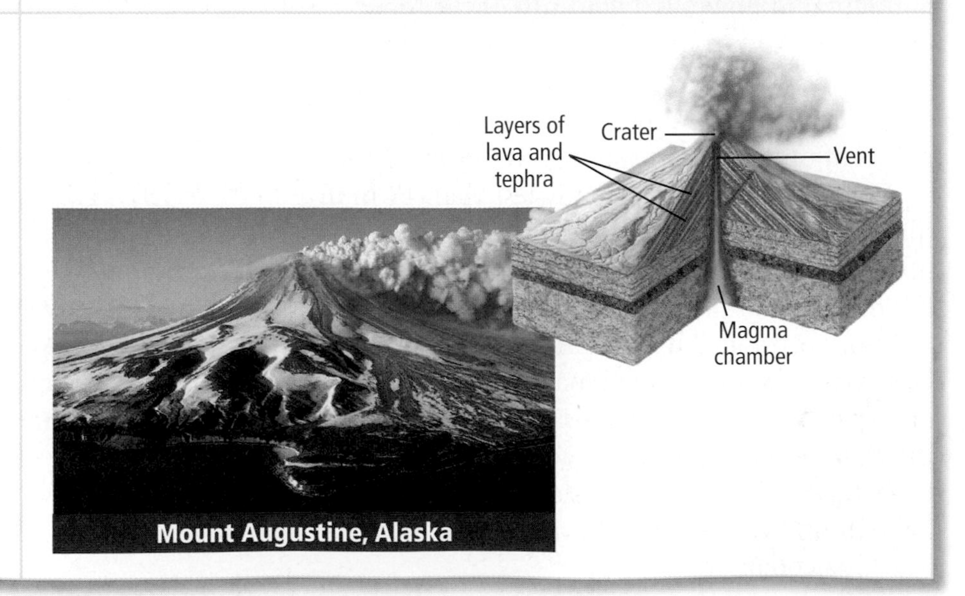 Mount Augustine, Alaska

Types of Volcanoes

The appearance of a volcano depends on two factors: the type of material that forms the volcano and the type of eruptions that occur. Based on these two criteria, three major types of volcanoes have been identified and are shown in **Table 18.1.** Each differs in size, shape, and composition.

Shield volcanoes A **shield volcano** is a mountain with broad, gently sloping sides and a nearly circular base. Shield volcanoes form when layers of lava accumulate during nonexplosive eruptions. They are the largest type of volcano. Mauna Loa, which is shown in **Table 18.1,** is a shield volcano.

Cinder cones When eruptions eject small pieces of magma into the air, **cinder cones** form as this material, called tephra, falls back to Earth and piles up around the vent. Cinder cones have steep sides and are generally small; most are less than 500 m high. The Lassen Volcanic Park cinder cone shown in **Table 18.1** is 700 m high. Cinder cones often form on or very near larger volcanoes.

Composite volcanoes **Composite volcanoes** are formed of layers of hardened chunks of lava from violent eruptions alternating with layers of lava that oozed downslope before solidifying. Composite volcanoes are generally cone-shaped with concave slopes, and are much larger than cinder cones. Because of their explosive nature, they are potentially dangerous to humans and the environment. Some examples of these are Mount Augustine in Alaska, shown in **Table 18.1,** and several in the Cascade Range of the western United States, such as Mount St. Helens.

CAREERS IN EARTH SCIENCE

Volcanologist Scientists who study eruptions, lava, magma, and the conditions under which these form are volcanologists. Some work in the field, studying active volcanoes. Many volcanologists also work in the laboratory to understand how rocks melt to form magma. To learn more about Earth science careers, visit glencoe.com.

FOLDABLES

Incorporate information from this section into your Foldable.

Section 18.1 Assessment

Section Summary

- Volcanism includes all the processes in which magma and gases rise to Earth's surface.
- Most volcanoes on land are part of two major volcanic chains: the Circum-Pacific Belt and the Mediterranean Belt.
- Parts of a volcano include a vent, magma chamber, crater, and caldera.
- Flood basalts form when lava flows from fissures to form flat plains or plateaus.
- There are three major types of volcanoes: shield, composite, and cinder cone.

Understand Main Ideas

1. MAIN Idea **Explain** how the location of volcanoes is related to the theory of plate tectonics.
2. **Identify** two volcanoes in the Mediterranean Belt.
3. **Draw** a volcano, labeling the parts.
4. **Propose** Yellowstone National Park is an area of previous volcanism. Using a map of the United States, suggest the type(s) of tectonic processes associated with this area.

Think Critically

5. **Evaluate** the following statement: Volcanoes are only found along coastlines.
6. **Decide** whether a flood basalt is or is not a volcano.

MATH in Earth Science

7. If the Pacific Plate has moved 500 km in the last 4.7 million years, calculate its average velocity in centimeters per year. Refer to the *Skillbuilder Handbook* for more information.

Section 18.2

Objectives

- **Explain** how magma type influences volcanic activity.
- **Describe** the role of pressure and dissolved gases in eruptions.
- **Recognize** classifications of material ejected by eruptions.

Review Vocabulary

basaltic: relates to a group of rocks rich in dark-colored minerals containing magnesium and iron

New Vocabulary

viscosity
tephra
pyroclastic flow

Eruptions

MAIN Idea The composition of magma determines the characteristics of a volcanic eruption.

Real-World Reading Link Have you ever shaken a can of soda and then opened it? If so, it probably sprayed your hand, clothes, and maybe even your friends. This is similar to the process that underlies explosive volcanic eruptions.

Making Magma

What makes the eruption of one volcano quiet, and the eruption of another explosively violent? The activity of a volcano depends on the composition of the magma. As shown in **Figure 18.9,** lava from an eruption can be thin and runny or thick and lumpy. In order to understand why volcanic eruptions are not all the same, you first need to understand how rocks melt to make magma.

Temperature Depending on their composition, most rocks begin to melt at temperatures between 800°C and 1200°C. Such temperatures are found in the crust and upper mantle. Recall from Chapter 5 that temperature increases with depth beneath Earth's surface. In addition to temperature, pressure and the presence of water also affect the formation of magma.

Pressure Pressure increases with depth because of the weight of overlying rocks. As pressure increases, the temperature at which a substance melts also increases. **Figure 18.10** shows two melting curves for a type of feldspar called albite. Note that at Earth's surface, albite, in the absence of water, melts at about 1100°C, but at a depth of about 12 km, its melting point is about 1150°C. At a depth of about 100 km, the melting point of dry albite increases to 1440°C. The effect of pressure explains why most of the rocks in Earth's lower crust and upper mantle do not melt.

■ **Figure 18.9** The way in which lava flows depends on the composition of the magma. Mount Etna's lava is thin and runny compared to the thick and lumpy lava that erupts at Mount St. Helens.

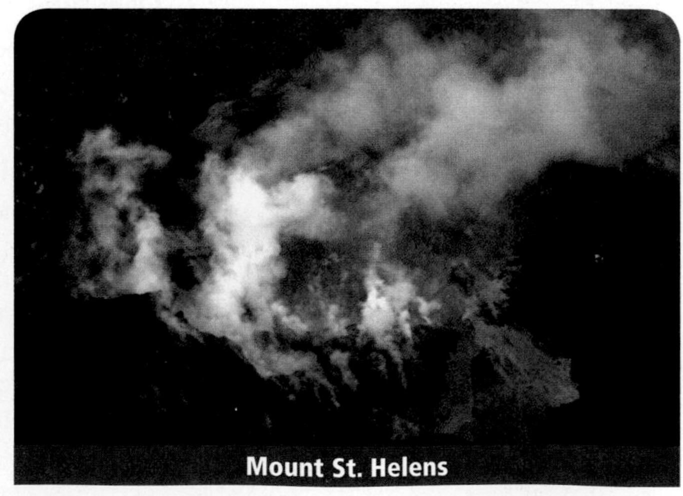

Composition of Magma

The composition of magma determines a volcano's explosivity, which is how it erupts and how its lava flows. What are the factors that determine the composition of magma? Scientists now know that the factors include magma's interaction with overlying crust, its temperature, pressure, amounts of dissolved gas, and—very significantly—the amount of silica a magma contains. Understanding the factors that determine the behavior of magma can aid scientists in predicting the explosivity of volcanic eruptions.

Dissolved gases In general, as the amount of gases in magma increases, the magma's explosivity also increases. In the same way that gas dissolved in soda gives the soda its fizz, the gases dissolved in magma give a volcano its "bang." Important gases in magma are water vapor, carbon dioxide, sulfur dioxide, and hydrogen sulfide. Water vapor is the most common dissolved gas in magma. The presence of water vapor determines where magma forms. As shown in **Figure 18.10,** minerals in the mantle, such as albite melt at high temperatures. The presence of dissolved water vapor lowers the melting temperature of minerals, causing mantle material to melt into magma. This eventually forms volcanoes and fuels their eruptions.

Viscosity The physical property that describes a material's resistance to flow is called **viscosity**. Temperature and silica content affect the viscosity of a magma. In general, cooler magma has a higher viscosity. In other words, cool magma, much like chilled honey, tends to resist flowing.

Reading Check Infer Which has a higher viscosity: syrup or water?

Magma with high silica content tends to be thick and sticky. Because it is thick, magma with high silica content tends to trap gases, which produces explosive eruptions. In general, magma with low silica content has low viscosity—it tends to be thin and runny, like warm syrup. Magma with low silica content tends to flow easily and produce quiet, nonexplosive eruptions.

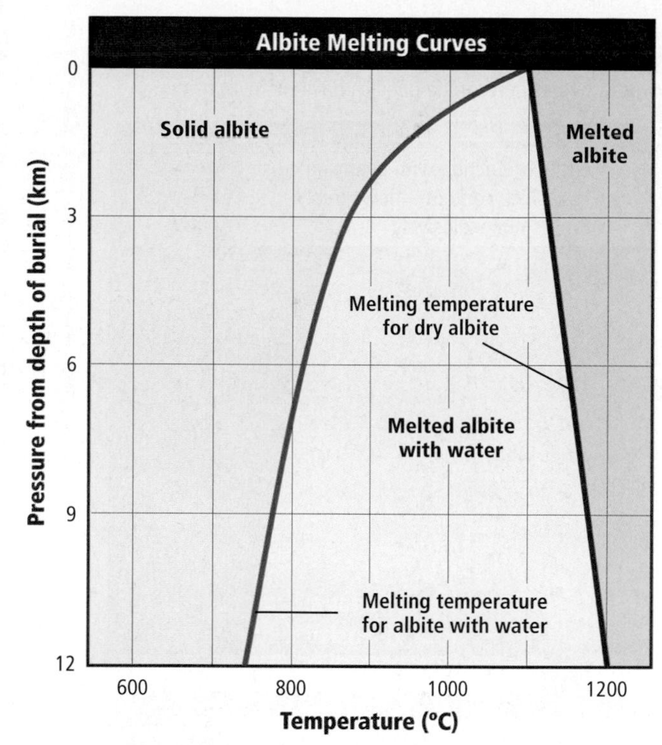

■ **Figure 18.10** Both the pressure and water content of the mineral albite affect how the mineral melts.
***Locate** the melting curve of wet albite. How does the melting point of wet albite compare to that of dry albite at a depth of 3 km? At a depth of 12 km?*

VOCABULARY
ACADEMIC VOCABULARY
Aid
to provide with what is useful or necessary in achieving an end
Glasses aid Omar in seeing clearly.

Figure 18.11 Generally, magma and lava with a low percentage of silica have low viscosity, and those with a higher percentage of silica have high viscosity.

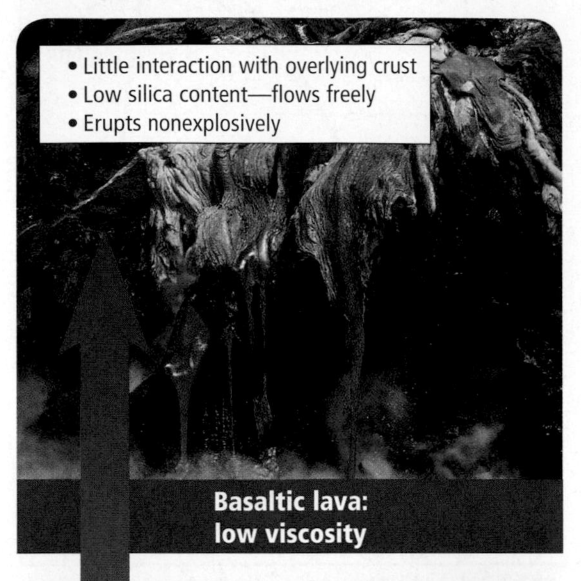

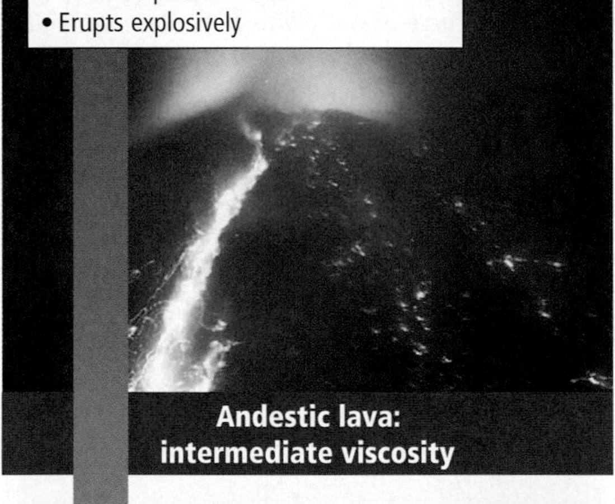

Types of Magma

The silica content of magma determines not only its explosivity and viscosity, but also which type of volcanic rock it forms as lava cools. Refer to **Figure 18.11** to summarize types of magma.

Basaltic magma When rock in the upper mantle melts, basaltic magma typically forms. Basaltic magma has the same silica content as the rock basalt—less than 50 percent silica. This magma rises from the upper mantle to Earth's surface and reacts very little with overlying continental crust or sediments. Its low silica content produces low-viscosity magma. Dissolved gases escape easily from basaltic magma. The resulting volcano is characterized by quiet eruptions. **Figure 18.12** shows how properties of magma affect the types of eruptions that occur. Volcanoes such as Kilauea and Mauna Loa actively produce basaltic magma. Surtsey, a volcano that was formed south of Iceland in 1963, is another volcano that produces basaltic magma.

Andesitic magma Andesitic (an duh SIH tihk) magma has the same silica content as the rock andesite—50 to 60 percent silica. Andesitic magma is found along oceanic-continental subduction zones. The source material for this magma can be either oceanic crust or oceanic sediments. The higher silica content results in a magma that has intermediate viscosity. Thus, the volcanoes it fuels are said to have intermediate explosivity. Colima Volcano in Mexico and Tambora in Indonesia are two examples of andesitic volcanoes. Both volcanoes have produced massive explosions that sent huge volumes of ash and debris into the atmosphere. This not only devastated the local communities, but also impacted the global environment.

Rhyolitic magma When molten material rises and mixes with the overlying continental crust rich in silica and water, it forms rhyolitic (ri uh LIH tihk) magma. Rhyolitic magma has the same composition as the rock granite—more than 60 percent silica. The high viscosity of rhyolitic magma slows down its movement. High viscosity, along with the large volume of gas trapped within this magma, makes the volcanoes fueled by rhyolitic magma very explosive. The dormant volcanoes in Yellowstone National Park in the western United States were fueled by rhyolitic magma. The most recent of these eruptions, which occurred 640,000 years ago, was so powerful that it released 1000 km^3 of volcanic material into the air.

Visualizing Eruptions

Figure 18.12 As magma rises due to plate tectonics and hot spots, it mixes with Earth's crust. This mixing causes differences in the temperature, silica content, and gas content of magma as it reaches Earth's surface. These properties of magma determine how volcanoes erupt.

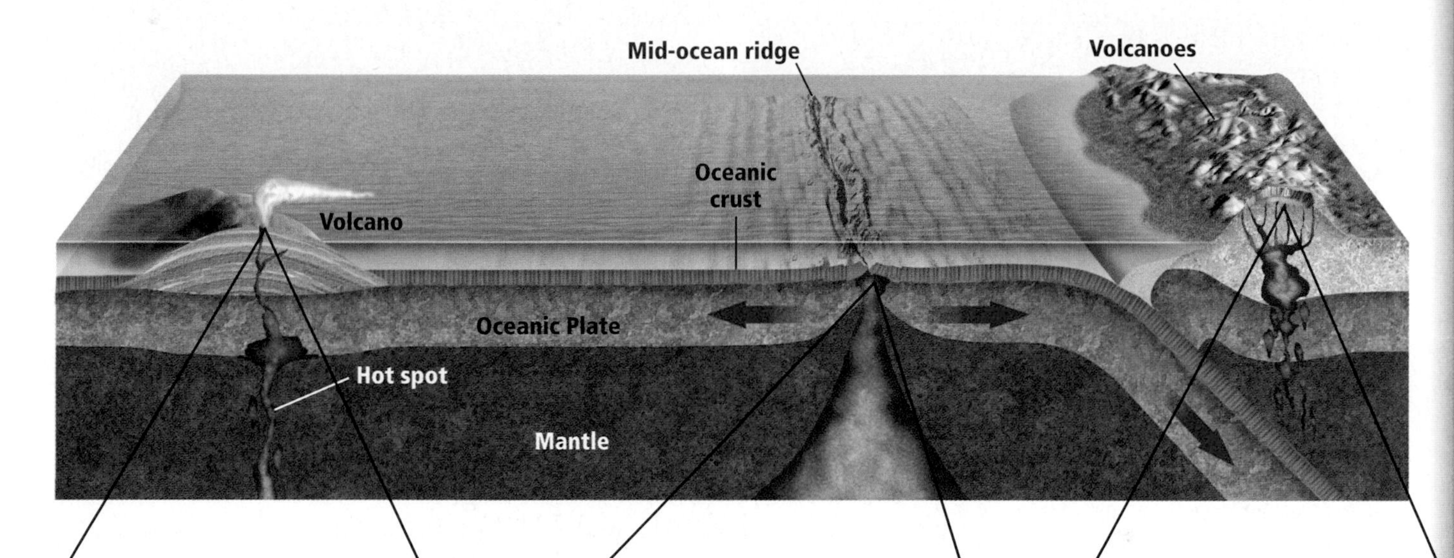

Quiet eruptions Earth's most active volcanoes are associated with hot spots under oceanic crust. Magma that upwells through oceanic crust maintains high temperature and low silica and gas contents. Lava oozes freely out of these volcanoes in eruptions that are relatively gentle.

Underwater eruptions Most pillow lava forms at diverging plate boundaries along oceanic crust. Lava oozes out of fissures in the ocean floor and forms bubble-shaped lumps as it cools.

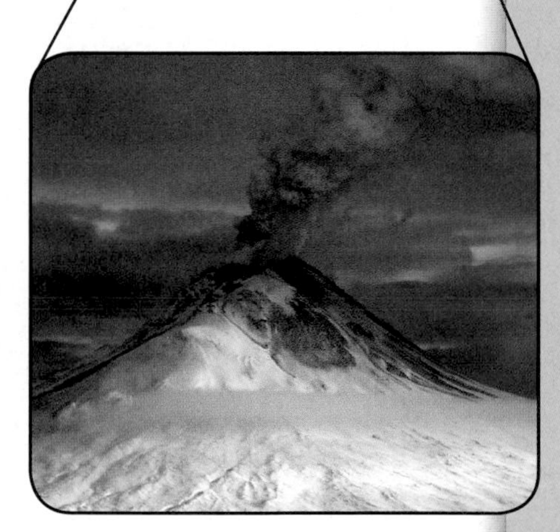

Explosive eruptions Dangerous eruptions occur where magma high in silica passes through continental crust. This magma traps gases, causing tremendous pressure to build. The release of pressure drives violent eruptions.

Concepts In Motion To explore more about plate tectonics resulting in volcanism, visit glencoe.com. Earth Science Online

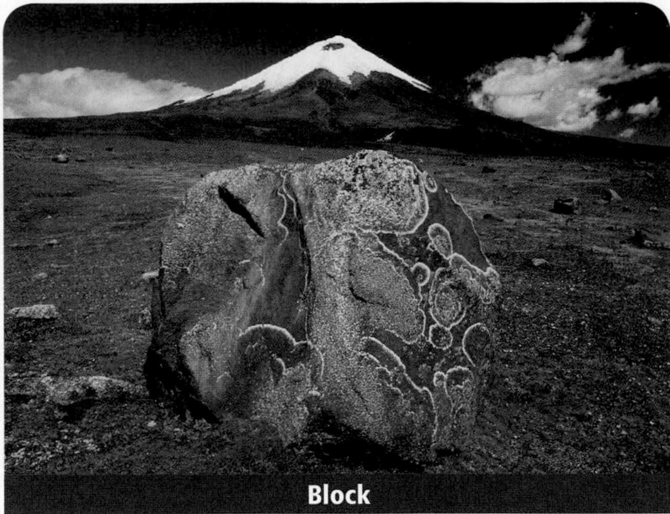

■ **Figure 18.13** Fine ash (shown actual size) is the smallest type of tephra. The 1-m-tall block shown here, ejected from Cotopaxi volcano in Ecuador, is an example of the largest category of tephra.

Compare *the two types of tephra. What do they have in common?*

Explosive Eruptions

When lava is too viscous to flow freely from the vent, pressure builds up in the lava until the volcano explodes, throwing lava and rock into the air. The erupted materials are called **tephra.** Tephra can be pieces of lava that solidified during the eruption, or pieces of the crust carried by the magma before the eruption. Tephra are classified by size. The smallest fragments, with diameters less than 2 mm, are called ash, as shown in **Figure 18.13.** The largest tephra thrown from a volcano are called blocks. The one shown in **Figure 18.13** is only about 1 m high, but some blocks can be the size of a car. Large explosive eruptions can disperse tephra over much of the planet. Ash can rise 40 km into the atmosphere during explosive eruptions and pose a threat to aircraft and can even change the weather. The 1991 eruption of Mount Pinatubo in the Philippines, shown in **Figure 18.14,** sent up a plume of ash 24 km high. Tiny sulfuric acid droplets and particles remained in the stratosphere for about two years, blocking the Sun's rays and lowering global temperatures.

■ **Figure 18.14** In 1991, the eruption of Mount Pinatubo in the Philippines sent so much ash into the stratosphere that it lowered global temperatures for two years.

Figure 18.15 A pyroclastic flow from Mount Pelée was so powerful that it destroyed the entire town of St. Pierre in only a few minutes.

Pyroclastic Flows

Some tephra cause tremendous damage and kill thousands of people. Violent volcanic eruptions can send clouds of ash and other tephra down a slope at speeds of about 80 km/h. Rapidly moving clouds of tephra mixed with hot, suffocating gases are called **pyroclastic flows.** They can have internal temperatures of more than 700°C. **Figure 18.15** shows a pyroclastic flow pouring down Mayon Volcano in Mexico in 2000. One widely known and deadly pyroclastic flow occurred in 1902 on Mount Pelée, on the island of Martinique in the Caribbean Sea. More than 29,000 people suffocated or were burned to death. What little was left of the town of St. Pierre after the eruption is shown in **Figure 18.15.**

Section 18.2 Assessment

Section Summary

- There are three major types of magma: basaltic, andesitic, and rhyolitic.
- Because of their relative silica contents, basaltic magma is the least explosive magma and rhyolitic magma is the most explosive.
- Temperature, pressure, and the presence of water are factors that affect the formation of magma.
- Rock fragments ejected during eruptions are called tephra.

Understand Main Ideas

1. MAIN Idea **Discuss** how the composition of magma determines an eruption's characteristics.
2. **Restate** how the viscosity of magma is related to its explosivity.
3. **Predict** the explosivity of a volcano having magma with high silica content and high gas content.
4. **Differentiate** between sizes of tephra.

Think Critically

5. **Compare and contrast** the tectonic processes that made Kilauea and Mount Etna.
6. **Infer** the composition of magma that fueled the A.D. 79 eruption of Mount Vesuvius that buried the town of Pompeii.

WRITING in Earth Science

7. Write a news report covering the 1902 eruption of Mount Pelée.

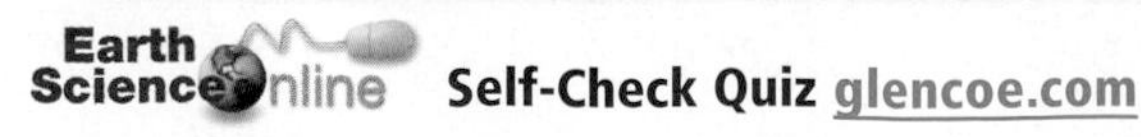

Section 18.3

Objectives

- **Compare and contrast** features formed from magma that solidifies near the surface with those that solidify deep underground
- **Classify** the different types of intrusive rock bodies.
- **Describe** how geologic processes result in intrusive rocks that appear at Earth's surface.

Review Vocabulary

igneous rock: rock formed by solidification of magma

New Vocabulary

pluton
batholith
stock
laccolith
sill
dike

Intrusive Activity

MAIN Idea Magma that solidifies below ground forms geologic features different from those formed by magma that cools at the surface.

Real-World Reading Link Have you ever been surprised when the icing on the inside of a layer cake was a different color or flavor than the icing on the outside? You might also be surprised if you could look inside Earth's layers because much volcanism cannot be seen at Earth's surface.

Plutons

Most of Earth's volcanism happens below the surface because not all magma emerges at the surface. Before it gets to the surface, rising magma can interact with the crust in several ways, as illustrated in **Figure 18.16.** Magma can force the overlying rock apart and enter the newly formed fissures. Magma can also cause blocks of rock to break off and sink into the magma, where the rocks eventually melt. Finally, magma can melt its way through the rock into which it intrudes. What happens deep in Earth as magma slowly cools? Recall from Chapter 5 that when magma cools, minerals begin to crystallize.

Over a long period of time, minerals in the magma solidify, forming intrusive igneous rock bodies. Some of these rock bodies are ribbonlike features only a few centimeters thick and several hundred meters long. Others are massive, and range in volume from about 1 km^3 to hundreds of cubic kilometers. These intrusive igneous rock bodies, called **plutons** (PLOO tahns), can be exposed at Earth's surface as a result of uplift and erosion and are classified based on their size, shape, and relationship to surrounding rocks.

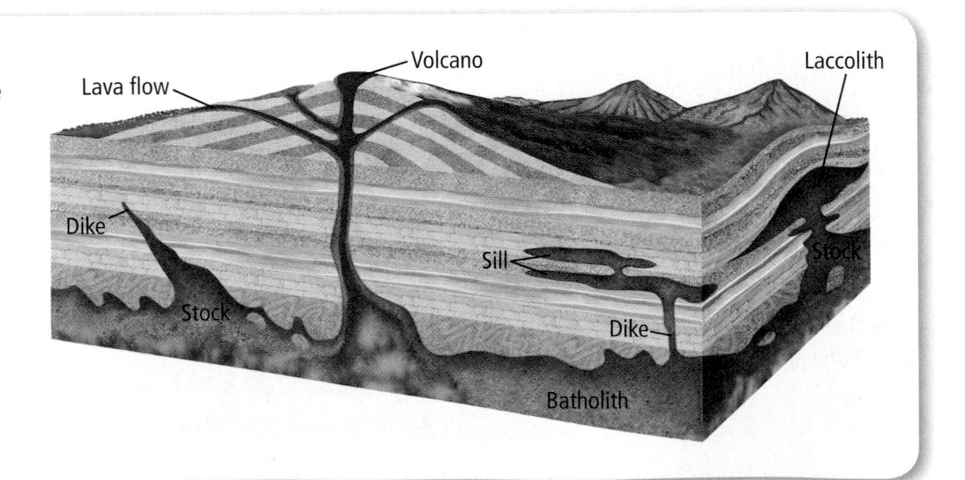

Figure 18.16 Magma moving upward solidifies and forms bodies of rock both at the surface and deep within Earth.

Batholiths and stocks The largest plutons are called batholiths. **Batholiths** (BATH uh lihths) are irregularly shaped masses of coarse-grained igneous rocks that cover at least 100 km^2 and take millions of years to form. Batholiths are common in the interior of major mountain chains.

Many batholiths in North America are composed primarily of granite—the most common rock type found in plutons. However, gabbro and diorite, the intrusive equivalents of basalt and andesite, are also found in batholiths. The largest batholith in North America is the Coast Range Batholith in British Columbia, shown in **Figure 18.17;** it is more than 1500 km long. Irregularly shaped plutons that are similar to batholiths but smaller in size are called **stocks.** Both batholiths and stocks, shown in **Figure 18.16,** cut across older rocks and generally form 5 to 30 km beneath Earth's surface.

Laccoliths Sometimes when magma intrudes into parallel rock layers close to Earth's surface, some of the rocks bow upward as a result of the intense pressure of the magma body. When the magma solidifies, a laccolith forms, as shown in **Figure 18.16.** A **laccolith** (LA kuh lihth) is a lens-shaped pluton with a round top and flat bottom. Compared to batholiths and stocks, laccoliths are relatively small; at most, they are 16 km wide. **Figure 18.17** shows a laccolith in Red and White Mountain, Colorado. Laccoliths also exist in the Black Hills of South Dakota, and the Judith Mountains of Montana, among other places.

Reading Check Contrast What is the difference between a laccolith and a batholith?

Sills A **sill** forms when magma intrudes parallel to layers of rock, as shown in **Figure 18.16.** A sill can range from only a few centimeters to hundreds of meters in thickness. **Figure 18.17** shows the Palisades Sill, which is exposed in the cliffs above the Hudson River near New York City and is about 300 m thick. The rock that was originally above the sill has eroded. What effect do you think this sill had on the sedimentary rocks into which it intruded? One effect is to lift the rock above it. Because it takes great amounts of force to lift entire layers of rock, most sills form relatively close to the surface. Another effect of sills is to metamorphose the surrounding rocks.

Figure 18.17 Batholiths, laccoliths, and sills form when magma intrudes into the crust and solidifies.

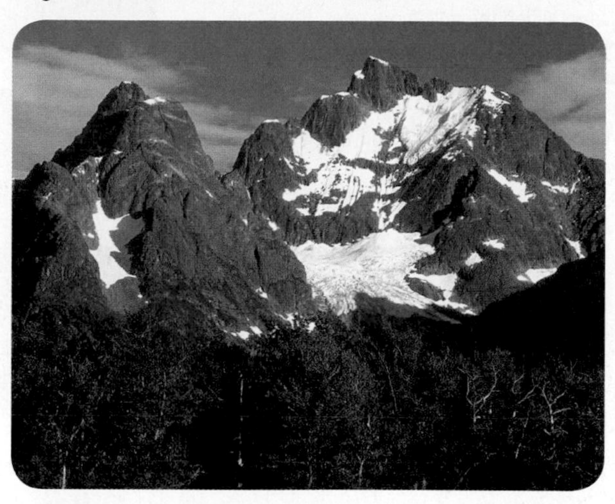

The Coast Range Batholith in British Columbia formed 5 to 30 km below Earth's surface.

Laccoliths push Earth's surface up, creating a rounded top and flat bottom.

The Palisades Sill in New York state formed more than 200 mya.

■ **Figure 18.18** Unlike sills, dikes cut across the rock into which they intrude. Sometimes dikes extend from the conduit of a volcano. When the volcano erodes, the more erosion-resistant conduit and dike are left standing. Try to imagine the volcano that once surrounded this volcanic neck in New Mexico. **Infer** ***how big the volcano must have been.***

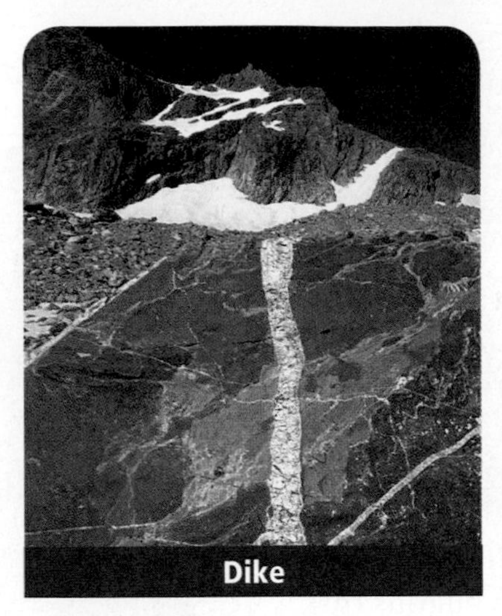

Dike

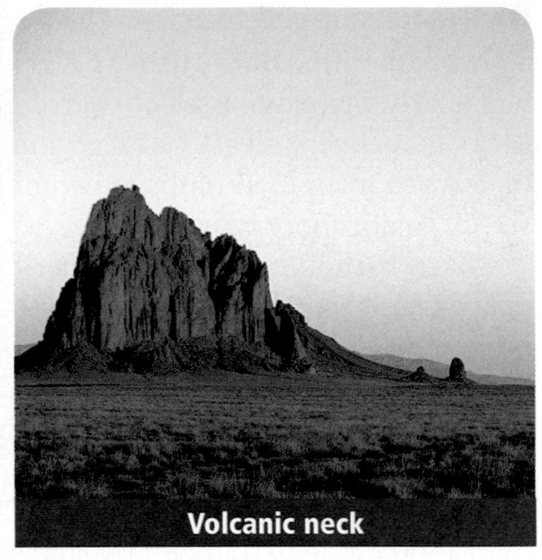

Volcanic neck

Dikes Unlike a sill, which is parallel to the rocks it intrudes, a **dike** is a pluton that cuts across preexisting rocks. Dikes often form when magma invades cracks in surrounding rock bodies. Dikes range in size from a few centimeters to several meters wide and can be tens of kilometers long. The Great Dike in Zimbabwe, Africa is an exception—it is about 8 km wide and 500 km long.

A volcanic neck occurs when the magma in a volcano conduit solidifies. Dikes are often associated with the conduit but do not always form the neck. Ship Rock in New Mexico, shown in **Figure 18.18,** has dikes extending from the neck.

Textures While the textures of sills and dikes vary, most are coarse-grained. Recall from Chapter 5 that grain size is related to the rate of cooling. The coarse-grained texture of most sills and dikes suggests that they formed deep in Earth's crust, where magma cooled slowly enough for large mineral grains to develop, as shown in **Figure 18.19.** Dikes and sills with a fine-grained texture formed closer to the surface where many crystals began growing at the same time, such as minerals of the sill in **Figure 18.19.**

■ **Figure 18.19** Plutons forming deep in Earth cool slowly, giving crystals time to grow. Larger crystals produce a coarse-grained rock. Intrusive rocks that form closer to Earth's surface cool more quickly. As a result, many crystals form rapidly at the same time, and the rock is finer-grained.

Coarse-grained dike

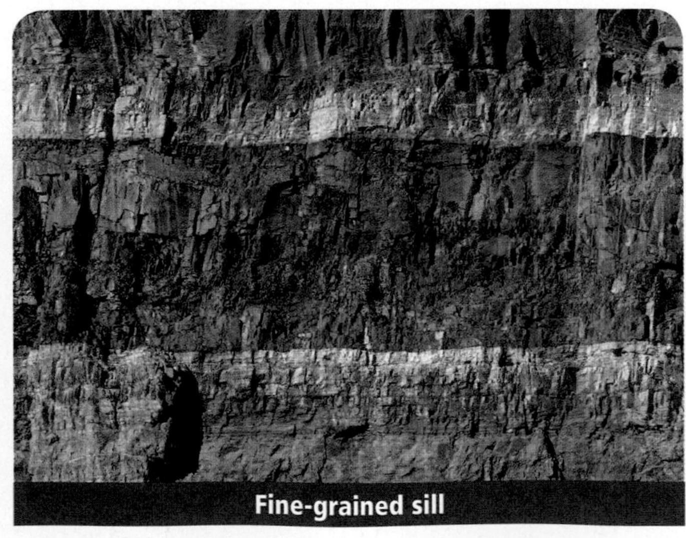

Fine-grained sill

Plutons and Tectonics

Many plutons form as the result of mountain-building processes. In fact, batholiths are found at the cores of many of Earth's mountain ranges. From where did the enormous volume of cooled magma that formed these igneous bodies come? The processes that result in batholiths are complex. Recall from Chapter 17 that many major mountain chains formed along continental-continental convergent plate boundaries. Scientists think that some of these collisions might have forced continental crust down into the upper mantle where it melted, intruded into the overlying rocks, and eventually cooled to form batholiths.

Plutons are also thought to form as a result of oceanic plate convergence. Again, recall from Chapter 17 that a subduction zone develops when an oceanic plate converges with another plate. Water from the subducted plate causes the overlying mantle to melt. Plutons often form when the melted material rises but does not erupt at the surface.

The Sierra Nevada batholith formed from at least five episodes of this type of igneous activity beneath what is now California. The famous granite cliffs found in Yosemite National Park, some of which are shown in **Figure 18.20,** are part of this vast batholith. Although they were once far below Earth's surface, uplift and erosion have brought them to their present position.

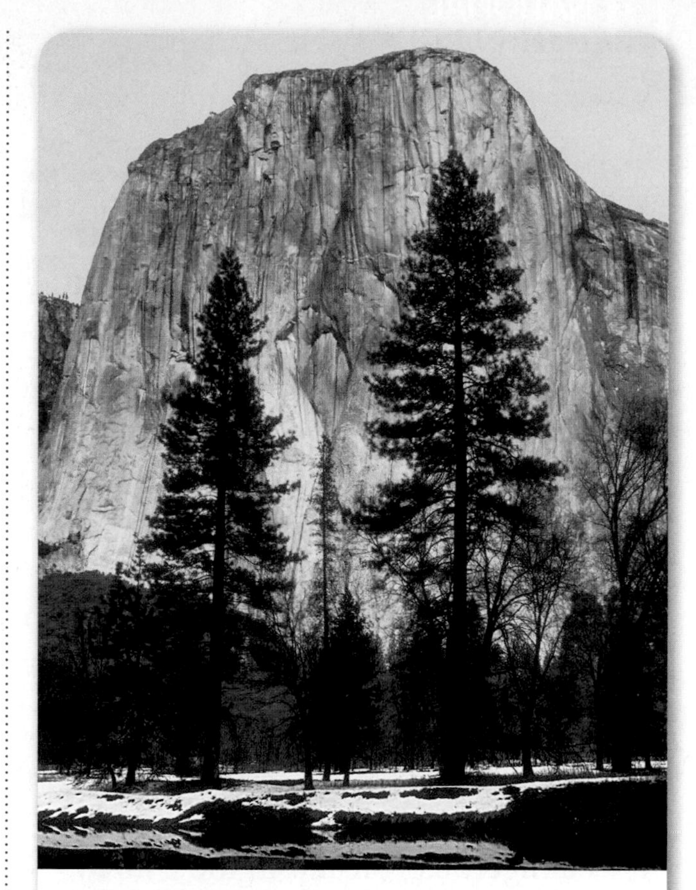

Figure 18.20 The granite cliffs that tower over Yosemite National Park in California are part of the Sierra Nevada batholith that has been exposed at Earth's surface.

Section 18.3 Assessment

Section Summary

- Intrusive igneous rocks are classified according to their size, shape, and relationship to the surrounding rocks.
- Most of Earth's volcanism happens below Earth's surface.
- Magma can intrude into rock in different ways, taking different forms when it cools.
- Batholiths form the core of many mountain ranges.

Understand Main Ideas

1. MAIN Idea **Compare and contrast** volcanic eruptions at Earth's surface with intrusive volcanic activity.
2. **Describe** the different types of plutons.
3. **Relate** the size of plutons to the locations where they are found.
4. **Identify** processes that expose plutons at Earth's surface.

Think Critically

5. **Predict** why the texture in the same sill might vary with finer grains along the margin and coarser grains toward the middle.
6. **Infer** what type of pluton might be found at the base of an extinct volcano.

WRITING in **Earth Science**

7. Write a defense or rebuttal for this statement: Of the different types of plutons, sills form at the greatest depths beneath Earth's surface.

ON SITE: HAWAIIAN VOLCANO OBSERVATORY

Volcanologists often wear helmets, climbing gear, heat-resistant clothing, gas masks, and other gear to protect themselves from dangerous conditions in and around active volcanoes. Once this volcanologist climbs down to the test site, he will put on heat-resistant gloves.

Kilauea, a shield volcano on the island of Hawaii, is one of the world's most active volcanoes and the most dangerous volcano in the United States, according to the United States Geological Survey (USGS). Scientists monitor the conditions of Kilauea at the nearby Hawaiian Volcano Observatory (HVO). The observatory also serves as a laboratory where samples gathered in and around Kilauea can be studied.

Lava collection Imagine standing next to moving lava that is 1170°C. To get a direct measurement of the temperature or to collect a sample, scientists must withstand high temperatures and watch where they step. Samples are collected with heat-resistant materials and immediately cooled in a container with water to prevent contamination from the surrounding air. To protect themselves, volcanologists wear some of the gear shown in the photo.

Seismic activity Earthquake activity beneath a volcano is an indicator of impending eruptions. One way to monitor earthquakes is to check seismic activity. Scientists place seismometers in and around the vents of volcanoes to monitor seismic activity.

Gas samples Volcanologists collect samples of gases released at vents that they will analyze for sulfur dioxide and carbon dioxide in the HVO laboratory. An increase in sulfur-dioxide or carbon-dioxide emission can indicate a potential eruption.

Ground monitoring An instrument called an electronic distance meter (EDM) helps scientists monitor the ground around volcanoes and predict an eruption. As magma rises toward Earth's surface, the ground might tilt, sink, or bulge from pressure.

Volcanologists at HVO are constantly recording data, running tests, and making advances around the world. Without their research, we might not understand volcanoes as well as we do today.

WRITING in Earth Science

Research the methods scientists use to predict time, size, and type of eruption. Visit glencoe.com for more information. Summarize your findings and share your research with your classmates.

GEOLAB

INTERNET: PREDICT THE SAFETY OF A VOLCANO

Background: Some volcanoes are explosively dangerous. Along with clouds of ash and other volcanic debris, pyroclastic flows, landslides, and mudflows are common volcanic hazards. However, an explosive volcano might not be a hazard to human life and property if it is located in a remote area or if it erupts infrequently.

Question: *What factors should be considered when evaluating a volcano?*

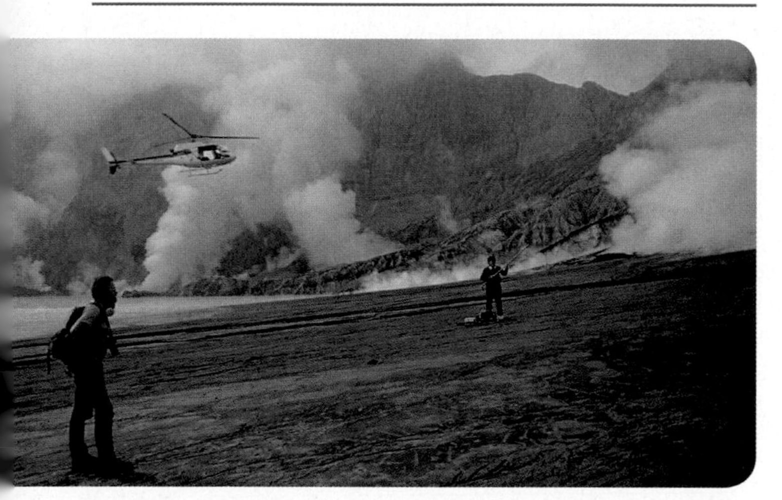

Helicopters transport researchers to remote volcanic sites. Researchers analyze data to determine hazards to humans.

Materials

Internet access to glencoe.com or volcano data provided by your teacher
current reference books with additional volcano data
markers or colored pencils

Procedure

Imagine that you work for the United States Geological Survey (USGS) and are asked to evaluate several volcanoes around the world. Your job is to determine if the volcanoes are safe for the nearby inhabitants. If the volcanoes are not safe, you must make recommendations to ensure the safety of the people around them.

1. Read and complete the lab safety form.
2. Form a team of scientists of three to four people.
3. Within your team, brainstorm some factors you might use to evaluate the volcanoes. Record your ideas. You might include factors such as eruption interval, composition of lava, approximate number of people living near the volcano, and the date of the last known eruption.
4. With your group, decide which factors you will include.
5. Use the factors you have chosen to create a data table. Make sure your teacher approves your table and your factors before you proceed.
6. Visit glencoe.com (or use the information your teacher provides) and select a country where there is a known volcano.
7. Complete your data table for your first country.
8. Repeat Steps 6 and 7 for two more countries.

Analyze and Conclude

1. **Interpret Data** Is it safe for people to live close to any of the volcanoes? Why or why not?
2. **Interpret Data** Do any of the volcanoes pose an immediate threat to the people who might live nearby? Why or why not?
3. **Conclude** Prepare to present your findings to a group of scientists from around the world. Be sure to include your predictions and recommendations, and be prepared for questions. Display your data table to help communicate your findings.

SHARE YOUR DATA

Peer Review Visit glencoe.com and post a summary of your recommendations for each of your volcanoes. Compare and contrast your data with that of other students who completed this lab.

CHAPTER 18

Study Guide

Download quizzes, key terms, and flash cards from glencoe.com.

BIG Idea Volcanoes develop from magma moving upward from deep within Earth.

Vocabulary	Key Concepts
Section 18.1 Volcanoes	
• caldera (p. 505) • cinder cone (p. 507) • composite volcano (p. 507) • conduit (p. 505) • crater (p. 505) • fissure (p. 504) • flood basalt (p. 504) • hot spot (p. 502) • shield volcano (p. 507) • vent (p. 505) • volcanism (p. 500)	**MAIN Idea** The locations of volcanoes are mostly determined by plate tectonics. • Volcanism includes all the processes in which magma and gases rise to Earth's surface. • Most volcanoes on land are part of two major volcanic chains: the Circum-Pacific Belt and the Mediterranean Belt. • Parts of a volcano include a vent, magma chamber, crater, and caldera. • Flood basalts form when lava flows from fissures to form flat plains or plateaus. • There are three major types of volcanoes: shield, composite, and cinder cone.
Section 18.2 Eruptions	
• pyroclastic flow (p. 513) • tephra (p. 512) • viscosity (p. 509)	**MAIN Idea** The composition of magma determines the characteristics of a volcanic eruption. • There are three major types of magma: basaltic, andesitic, and rhyolitic. • Because of their relative silica contents, basaltic magma is the least explosive magma and rhyolitic magma is the most explosive. • Temperature, pressure, and the presence of water are factors that affect the formation of magma. • Rock fragments ejected during eruptions are called tephra.
Section 18.3 Intrusive Activity	
• batholith (p. 515) • dike (p. 516) • laccolith (p. 515) • pluton (p. 514) • sill (p. 515) • stock (p. 515)	**MAIN Idea** Magma that solidifies below ground forms geologic features different from those formed by magma that cools at the surface. • Intrusive igneous rocks are classified according to their size, shape, and relationship to the surrounding rocks. • Most of Earth's volcanism happens below Earth's surface. • Magma can intrude into rock in different ways, taking different forms when it cools. • Batholiths form the core of many mountain ranges.

CHAPTER 18 Assessment

Vocabulary Review

Make each of the following sentences true by replacing the italicized words with terms from the Study Guide.

1. In the most explosive types of eruptions, lava accumulates to form a *shield volcano.*
2. Lava travels through a conduit to erupt through a *fissure* at the top of a volcano.
3. *Hot spots* refer to all processes associated with the discharge of magma, hot water, and steam.
4. Ash is the smallest type of *lava flow.*

Complete the sentences below using vocabulary terms from the Study Guide.

5. A(n) ________ is a bowl-shaped depression that surrounds the vent at a volcano's summit.
6. A(n) ________ forms in the depression left when an empty magma chamber collapses.
7. The type of volcano that is the smallest and has the steepest slopes is called a(n) ________.

Match each description below with the correct vocabulary term from the Study Guide.

8. any rock body that has formed at great depths underground
9. plutons having an area of more than 100 km^2; often forms the core of mountains
10. flowing cloud of tephra and lava mixed with hot, suffocating gases
11. formed when magma intrudes vertically across existing rock

Use what you know about the vocabulary terms to describe what the terms in each pair have in common.

12. laccolith, sill
13. shield volcano, flood basalt
14. fissure, conduit
15. sill, dike

Understand Key Concepts

16. Which area is surrounded by the Ring of Fire?
    A. the Atlantic Ocean
    B. the United States
    C. the Mediterranean Sea
    D. the Pacific Ocean

Use the diagram below to answer Questions 17 and 18.

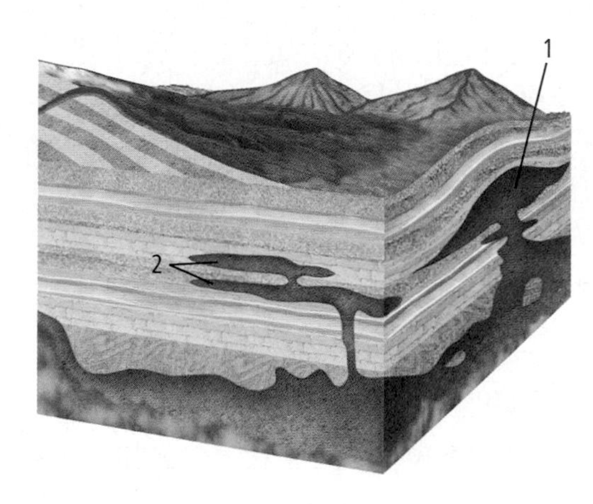

17. In the diagram, what is the structure labeled *1*?
    A. batholith
    B. laccolith
    C. dike
    D. sill

18. In the diagram, what is the structure labeled *2*?
    A. batholith
    B. laccolith
    C. dike
    D. sill

19. Which is not true?
    A. An increase in silica increases the viscosity of a magma.
    B. Andesitic magma has both an intermediate gas content and explosiveness.
    C. An increase in temperature increases a magma's viscosity.
    D. Basaltic magma has a low viscosity and retains little gas.

Use the figure below to answer Questions 20 and 21.

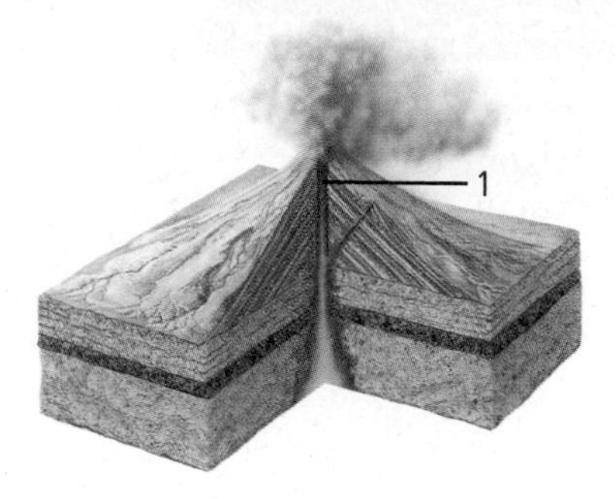

20. Which type of volcano is shown?
- **A.** shield volcano
- **B.** composite volcano
- **C.** flood basalt volcano
- **D.** cinder cone

21. What is the feature labeled *1*?
- **A.** crater
- **B.** cinder cone
- **C.** vent
- **D.** magma chamber

22. What causes the magma to rise upward in a mantle plume?
- **A.** The magma is less dense than the surrounding material.
- **B.** The magma is denser than the surrounding material.
- **C.** The magma is pulled upward by the air pressure.
- **D.** The magma is pushed upward by the surrounding rock.

23. Which type of volcanism produces the most lava annually?
- **A.** convergent
- **B.** divergent
- **C.** hot spot
- **D.** cinder cones

Constructed Response

24. Differentiate among batholiths, stocks, and laccoliths according to their relative sizes and shapes.

25. Infer A particular outcrop has a narrow ribbon of basalt that runs almost perpendicular to several layers of sandstone. What is this feature called?

26. Describe hot spots.

27. Identify one specific example of the three types of volcanoes.

28. Compare and contrast Kilauea and the Columbia River flood basalt in terms of the processes related to their development.

29. Analyze why volcanic blocks are uncommon on shield volcanoes.

Use the diagram below to answer Question 30.

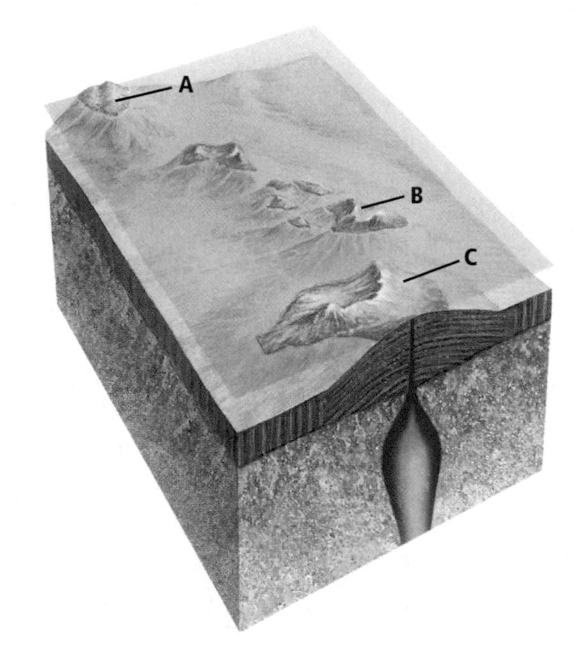

30. Distinguish which island is the oldest and in which direction the plate is moving. Explain your reasoning.

31. Decide Is the Pacific Ring of Fire an accurate name? Explain.

32. Explain the relationship between the viscosity of a magma and its temperature.

33. Explain how volcanic activity can affect global weather.

34. Draw a diagram of the three volcano types, showing their relative sizes.

35. Analyze why smaller plutons are more likely to be fine-grained, and larger plutons more likely to be coarse-grained.

Chapter Test glencoe.com

Think Critically

Use the table below to answer Questions 36 and 37.

Magma Composition and Characteristics

	Basaltic Magma	Andesitic Magma	Rhyolitic Magma
Source material	upper mantle	oceanic crusts and sediments	continental crust
Viscosity	low	intermediate	high
Gas content	1–2%	3–4%	4–6%
Silica content	about 50%	about 60%	about 70%
Location of magma	both oceanic and continental crust	continental margins associated with subduction zones	continental crust

36. Analyze and rank the types of magma in terms of explosiveness based on the data. Explain your reasoning.

37. Categorize each of the three types of volcanoes in terms of the characteristics of magma shown in the table.

38. Predict what would happen if there were no plate tectonics.

Concept Mapping

39. Create a concept map using the following terms: *pluton, vertical, batholith, cuts across, stock, parallel, laccolith, sill,* and *dike.* For more help refer to the *Skillbuilder Handbook.*

Challenge Question

40. Formulate a way to recognize the difference between an ancient lava flow and an intrusive igneous rock.

Additional Assessment

41. WRITING in **Earth Science** Imagine you are in charge of a volcano observatory. One day, GPS measurements indicate that a volcano is expanding, there have been several earthquakes, and the flux of volcanic gases has increased. Should you issue a warning of an impending eruption? Write a press release to warn people about the situation.

DBQ Document–Based Questions

Data obtained from: Takada, A. 1999. Variations in magma supply and magma partitioning: the role of tectonic settings. *Journal of Volcanic Geothermal Research* 83:93–110.

Studying the history of past eruptions yields important data for making estimations about predicting eruptions. The graph below shows the total volume of lava erupted at two Hawaiian islands over 200 years.

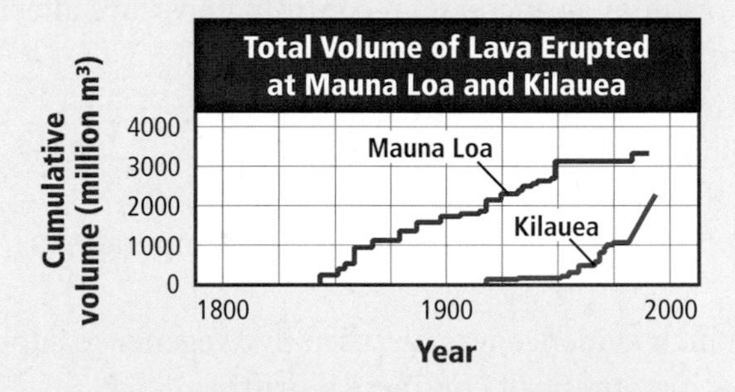

42. In what years did the two largest eruptions occur at Mauna Loa?

43. What is the average volume of lava at Mauna Loa between 1840 and 1990?

44. Can you predict when the next eruption will occur? Explain your answer.

45. Eruptions at Mauna Loa are large and last a short length of time. What feature of the graph shows this? Compare and contrast the last eruption at Kilauea with eruptions at Mauna Loa.

Cumulative Review

46. List six of the most important mineral properties used in mineral identification. **(Chapter 4)**

47. What observations support the theory of plate tectonics? **(Chapter 17)**

Standardized Test Practice

Multiple Choice

Use the figure below to answer Questions 1 and 2.

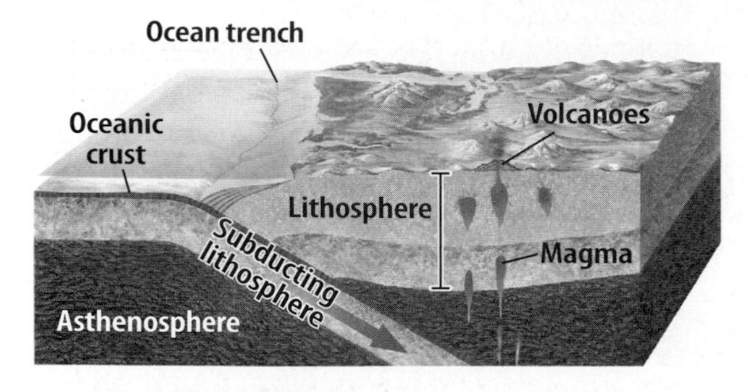

1. What process is occurring in the figure above?
   A. continental-continental divergence
   B. oceanic-continental divergence
   C. continental-continental subduction
   D. oceanic-continental subduction

2. How does an increase in confining pressure affect a rock's melting temperature?
   A. The melting temperature increases.
   B. The melting temperature decreases.
   C. The melting temperature is stabilized.
   D. It has no effect on the melting temperature.

3. Which evidence was not used by Wegener to support his hypothesis of continental drift?
   A. coal beds in America
   B. fossils of land-dwelling animals
   C. glacial deposits
   D. paleomagnetic data

4. What is the name for the constant production of new ocean floor?
   A. continental drift
   B. hot spot
   C. seafloor spreading
   D. subduction

5. The weight of a subducting plate helps pull the trailing lithosphere into a subduction zone in which process?
   A. ridge pull
   B. ridge push
   C. slab pull
   D. slab push

6. What type of model uses molded clay, soil, and chemicals to simulate a volcanic eruption?
   A. conceptual model
   B. physical model
   C. mathematical model
   D. computer model

7. Which of these processes of the water cycle is a direct effect of the Sun's energy?
   A. formation of precipitation
   B. runoff of water over soil
   C. evaporation
   D. seeping of water into soil

Use the figure below to answer Questions 8 and 9.

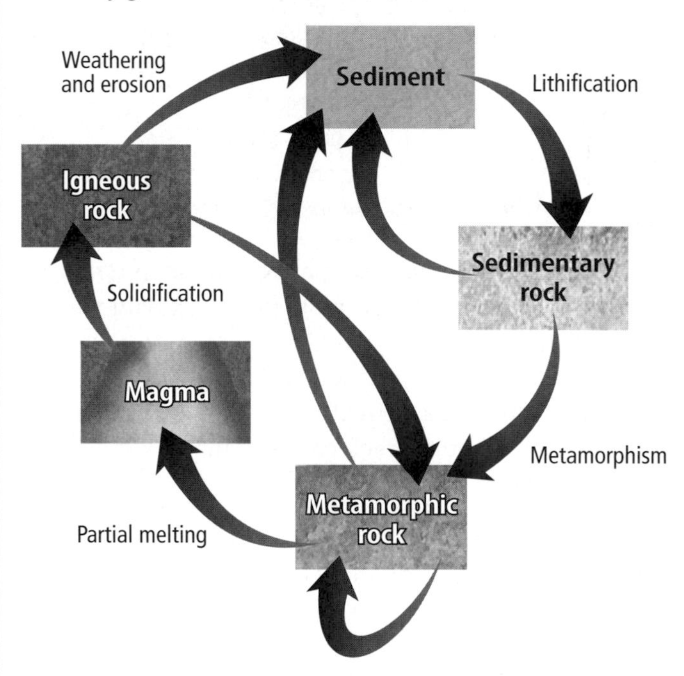

8. Which process brings rocks to Earth's surface where they can be eroded?
   A. lithification
   B. weathering
   C. solidification
   D. metamorphism

9. What rock type is produced when magma solidifies?
   A. metamorphic rock
   B. sedimentary rock
   C. igneous rock
   D. lava

Short Answer

Use the table below to answer Questions 10–12.

Notable Volcanic Eruptions			
Volcano	Date	Volume Ejected	Height of Plume
Toba	74,000 years ago	2,800 km^3	50–80 km
Vesuvius	A.D. 79	4 km^3	32 km
Tambora	1815	150 km^3	44 km
Krakatau	1883	21 km^3	36 km
Mount St. Helens	1980	1 km^3	19 km
Mount Pinatubo	1991	5 km^3	35 km

10. Order the volcanic eruptions according to the quantities of pyroclastic material produced.

11. Hypothesize why the eruption of Vesuvius in A.D. 79 was more deadly than the eruption of Mount Pinatubo in 1991, even though the eruptions were approximately the same size.

12. Calculate the difference in plume height of volcanic debris during the eruption of Tambora in 1815 compared to the plume from the 1980 eruption of Mount St. Helens.

13. Distinguish between the everyday use of the term *theory* and its true scientific meaning.

14. When a tropical rain forest is cleared, why does the soil usually become useless for growing crops after only a few years?

15. What role do glaciers play in Earth's rock cycle?

16. Write a list of numbered statements that summarizes the major steps in the water cycle.

Reading for Comprehension

Eruption of Mount Pinatubo

On June 15, 1991, Mount Pinatubo roared awake after a six-century sleep. The 1760-m volcano belched clouds of gas and ash known as pyroclastic material. Their temperature: 816°C. Streams of ash and sulfur dioxide rocketed 40 km into the stratosphere. Another blast at dawn blew away the side of the mountain. So much ash and pumice choked the air that the sky grew black by afternoon, and chunks of volcanic rock fell with a force similar to hail. That evening, earthquakes struck the already-damaged city. Pinatubo's eruption had created an underground cavern that caved in on itself.

17. What can be inferred from this text?
- **A.** Volcanoes are unpredictable and can erupt at any time.
- **B.** Volcanoes always erupt explosively.
- **C.** Volcanoes can change the surface of Earth in many ways.
- **D.** Volcanoes are always accompanied by earthquakes.

18. According to this text, which statement is false?
- **A.** Volcanoes can release gases into the stratosphere.
- **B.** The eruption of Mount Pinatubo was caused by the collapse of an underground cavern.
- **C.** The gas and ash released during the 1991 eruption was as hot as 816°C.
- **D.** Volcanic eruptions can change the shape of the mountain.

19. In the days leading up to the June 15th eruption, towns in areas surrounding Mount Pinatubo were evacuated. Based on the text above, explain why it would be necessary to evacuate these areas.

NEED EXTRA HELP?

If You Missed Question . . .	1	2	3	4	5	6	7	8	9	10	11	12	13	14	15	16
Review Section . . .	17.3	5.1	17.1	17.2	17.4	1.3	9.1	6.1	5.1	18.1	18.3	18.3	1.3	8.3	3.3	9.1

CHAPTER 19 Earthquakes

BIG Idea Earthquakes are natural vibrations of the ground, some of which are caused by movement along fractures in Earth's crust.

19.1 Forces Within Earth
MAIN Idea Faults form when the forces acting on rock exceed the rock's strength.

19.2 Seismic Waves and Earth's Interior
MAIN Idea Seismic waves can be used to make images of the internal structure of Earth.

19.3 Measuring and Locating Earthquakes
MAIN Idea Scientists measure the strength and chart the location of earthquakes using seismic waves.

19.4 Earthquakes and Society
MAIN Idea The probability of an earthquake's occurrence is determined from the history of earthquakes and knowing where and how quickly strain accumulates.

GeoFacts

- Earth experiences 500,000 earthquakes each year.
- Most earthquakes are so small that they are not felt.
- Each year, Southern California has about 10,000 earthquakes.

Start-Up Activities

LAUNCH Lab

What can cause an earthquake?

When pieces of Earth's crust suddenly move relative to one another, earthquakes occur. This movement occurs along fractures in the crust that are called faults.

Procedure

1. Read and complete the lab safety form.
2. Slide the largest surfaces of two smooth **wooden blocks** against each other. Describe the movement.
3. Cut two pieces of coarse-grained **sandpaper** so that they are about 1 cm longer than the largest surface of each block.
4. Place the sandpaper, coarse side up, against the largest surface of each block. Wrap the paper over the edges of the blocks and secure it with **thumbtacks.**
5. Slide the sandpaper-covered sides of the blocks against each other. Describe the movement.

Analysis

1. **Compare** the two movements of the wooden blocks.
2. **Apply** Which parts of Earth are represented by the blocks?
3. **Infer** which of the two scenarios shows what happens during an earthquake.

Types of Faults Make this Foldable to show the three basic types of faults.

STEP 1 Fold a sheet of paper in half. Make the back edge about 2 cm longer than the front edge.

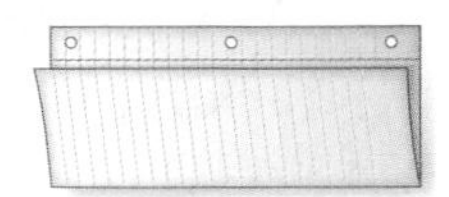

STEP 2 Fold into thirds.

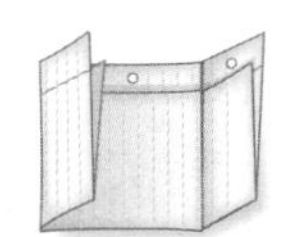

STEP 3 Unfold and cut along the folds of the top flap to make three tabs.

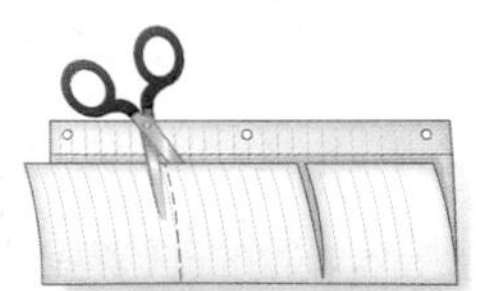

STEP 4 Label the tabs *Reverse, Normal,* and *Strike-slip*.

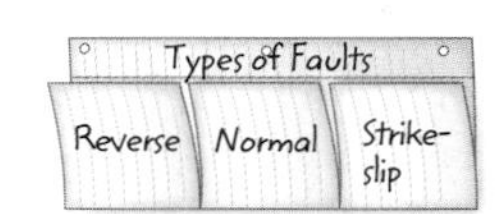

FOLDABLES **Use this Foldable with Section 19.1.** As you read this section, explain in your own words the characteristics associated with each type of fault.

Visit glencoe.com to

- **study entire chapters online;**
- **explore Concepts In Motion animations:**
  - **Interactive Time Lines**
  - **Interactive Figures**
  - **Interactive Tables**
- **access Web Links for more information, projects, and activities;**
- **review content with the Interactive Tutor and take Self-Check Quizzes.**

Section 19.1

Objectives

- **Define** stress and strain as they apply to rocks.
- **Distinguish** among the three types of movement of faults.
- **Contrast** the three types of seismic waves.

Review Vocabulary

fracture: the texture or general appearance of the freshly broken surface of a mineral

New Vocabulary

stress
strain
elastic deformation
plastic deformation
fault
seismic wave
primary wave
secondary wave
focus
epicenter

Forces Within Earth

MAIN Idea Faults form when the forces acting on rock exceed the rock's strength.

Real-World Reading Link If you bend a paperclip, it takes on a new shape. If you bend a popsicle stick, it will eventually break. The same is true of rocks; when forces are applied to rocks, they either bend or break.

Stress and Strain

Most earthquakes are the result of movement of Earth's crust produced by plate tectonics. As a whole, tectonic plates tend to move gradually. Along the boundaries between two plates, rocks in the crust often resist movement. Over time, stress builds up. **Stress** is the total force acting on crustal rocks per unit of area. When stress overcomes the strength of the rocks involved, movement occurs along fractures in the rocks. The vibrations caused by this sudden movement are felt as an earthquake. The characteristics of earthquakes are determined by the orientation and magnitude of stress applied to rocks, and by the strength of the rocks involved.

There are three kinds of stress that act on Earth's rocks: compression, tension, and shear. Compression is stress that decreases the volume of a material, tension is stress that pulls a material apart, and shear is stress that causes a material to twist. The deformation of materials in response to stress is called **strain. Figure 19.1** illustrates the strain caused by compression, tension, and shear.

Even though rocks can be twisted, squeezed, and stretched, they fracture when stress and strain reach a critical point. At these breaks, rock can move, releasing the energy built up as a result of stress. Earthquakes are the result of this movement and release of energy. For example, the 2005 earthquake in Pakistan was caused by a release of built-up compression stress. When that energy was released as an earthquake, more than 75,000 people were killed and 3 million were made homeless.

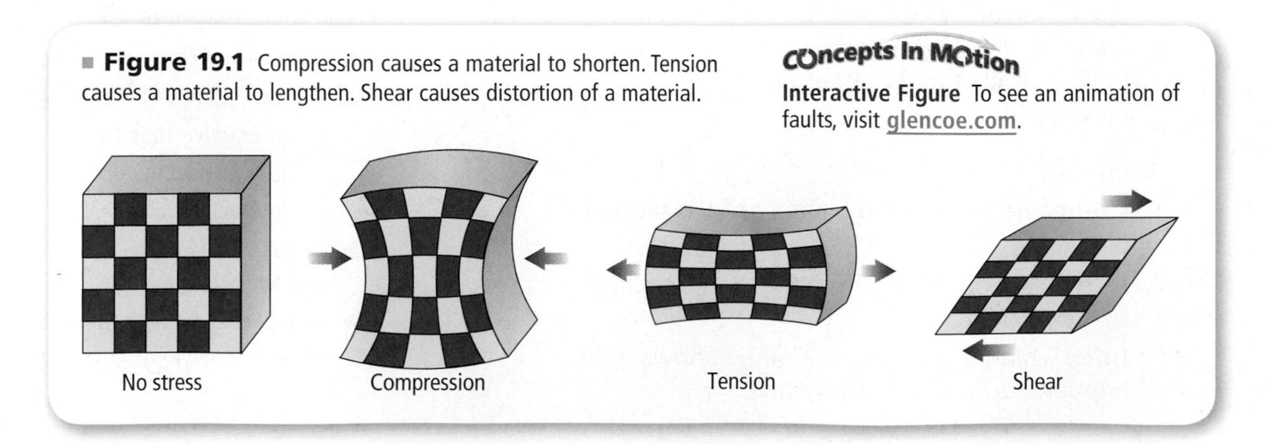

Figure 19.1 Compression causes a material to shorten. Tension causes a material to lengthen. Shear causes distortion of a material.

Concepts In Motion
Interactive Figure To see an animation of faults, visit glencoe.com.

Laboratory experiments on rock samples show a distinct relationship between stress and strain. When the stress applied to a rock is plotted against strain, a stress-strain curve, like the one shown in **Figure 19.2,** is produced. A stress-strain curve usually has two segments—a straight segment and a curved segment. Each segment represents a different type of response to stress.

Elastic deformation The first segment of a stress-strain curve shows what happens under conditions in which stress is low. Under low stress, a material shows elastic deformation. **Elastic deformation** is caused when a material is compressed, bent, or stretched. This is the same type of deformation that happens from gently pulling on the ends of a rubber band. When the stress on the rubber band is released, it returns to its original size and shape. **Figure 19.2** illustrates that elastic deformation is the result of stress and strain. If the stress is reduced to zero, as the graph shows, the deformation of the rock disappears.

Plastic deformation When stress builds up past a certain point, called the elastic limit, rocks undergo **plastic deformation,** shown by the second segment of the graph in **Figure 19.2.** Unlike elastic deformation, this type of strain produces permanent deformation, which means that the material stays deformed even when stress is reduced to zero. Even a rubber band undergoes plastic deformation when it is stretched beyond its elastic limit. At first the rubber band stretches, then it tears slightly, and finally, two pieces will snap apart. The tear in the rubber band is an example of permanent deformation. When stress increases to be greater than the strength of a rock, the rock ruptures. The point of rupture, called failure, is designated by the "X" on the graph in **Figure 19.2.**

Reading Check **Differentiate** between elastic deformation and plastic deformation.

Most materials exhibit both elastic and plastic behavior, although to different degrees. Brittle materials, such as dry wood, glass, and certain plastics, fail before much plastic deformation occurs. Other materials, such as metals, rubber, and silicon putty, can undergo a great deal of deformation before failure occurs, or they might not fail at all. Temperature and pressure also influence deformation. As pressure increases, rocks require greater stress to reach the elastic limit. At high enough temperatures, solid rock can also deform, causing it to flow in a fluidlike manner. This flow reduces stress.

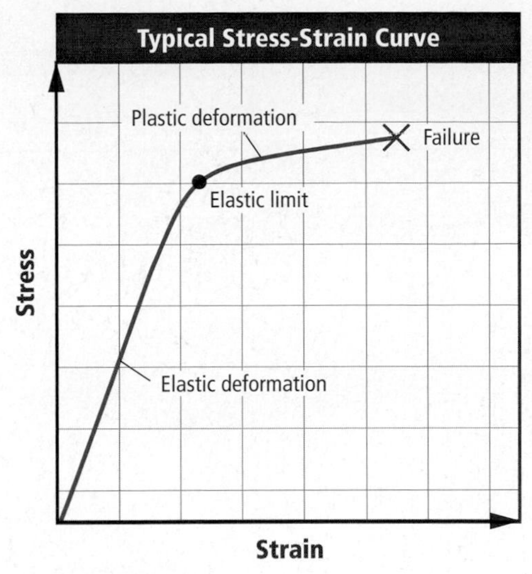

Figure 19.2 A typical stress-strain curve has two parts. Elastic deformation occurs as a result of low stress. When the stress is removed, material returns to its original shape. Plastic deformation occurs under high stress. The deformation of the material is permanent. When plastic deformation is exceeded, an earthquake occurs.
Describe ***what happens to a material at the point on the graph at which elastic deformation changes into plastic deformation.***

VOCABULARY
SCIENCE USAGE V. COMMON USAGE
Failure
Science usage: a collapsing, fracturing, or giving way under stress

Common usage: lack of satisfactory performance or effect

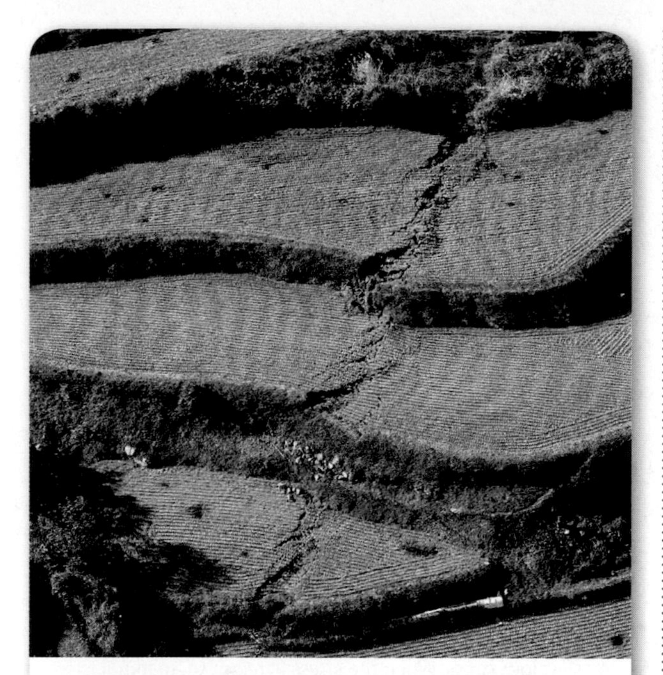

■ **Figure 19.3** A major fault passes through these rice fields on an island in Japan.
Identify *the direction of movement that occurred along this fault.*

Faults

Crustal rocks fail when stresses exceed the strength of the rocks. The resulting movement occurs along a weak region in the crustal rock called a fault. A **fault** is any fracture or system of fractures along which Earth moves. **Figure 19.3** shows a fault. The surface along which the movement takes places is called the fault plane. The orientation of the fault plane can vary from nearly horizontal to almost vertical. The movement along a fault results in earthquakes. Several historic earthquakes are described in the time line in **Figure 19.4**.

Reverse and normal faults Reverse faults form as a result of horizontal and vertical compression that squeezes rock and creates a shortening of the crust. This causes rock on one side of a reverse fault to be pushed up relative to the other side. Reverse faulting can be seen near convergent plate boundaries.

Movement along a normal fault is partly horizontal and partly vertical. The horizontal movement pulls rock apart and stretches the crust. Vertical movement occurs as the stretching causes rock on one side of the fault to move down relative to the other side. The Basin and Range province in the southwestern United States is characterized by normal faulting. The crust is being stretched apart in that area. Note in the diagrams shown in **Table 19.1** that the two areas separated by the reverse fault would be closer after the faulting than before, and that two areas at a normal fault would be farther apart after the faulting than before the faulting.

FOLDABLES
Incorporate information from this section into your Foldable.

■ **Figure 19.4**
Major Earthquakes and Advances in Research and Design

As earthquakes cause casualties and damage around the world, scientists work to find better ways to warn and protect people.

1800 — 1900 — 1950

1811–1812 Several strong earthquakes occur along the Mississippi River valley over three months, destroying the entire town of New Madrid, Missouri.

1880 Following an earthquake in Japan, scientists invent the first modern seismograph to record the intensity of earthquakes.

1906 An earthquake in San Francisco kills between 3000 and 5000 people and causes a fire that rages for three days, destroying most of the city.

1923 Approximately 140,000 people die in an earthquake and subsequent fires that destroy the homes of over a million people in Tokyo and Yokohama, Japan.

1948 An earthquake destroys Ashgabat, capital of Turkmenistan, killing nearly nine out of ten people living in the city and its surrounding areas.

Concepts In Motion **Interactive Table** To explore more about faults, visit glencoe.com.

Table 19.1 Types of Faults

Type of Fault	Type of Movement	Example
Reverse	Compression causes horizontal and vertical movement.	
Normal	Tension causes horizontal and vertical movement	
Strike-slip	Shear causes horizontal movement.	

Strike-slip faults Strike-slip faults are caused by horizontal shear. As shown in **Table 19.1,** the movement at a strike-slip fault is mainly horizontal and in opposite directions, similar to the way cars move in opposite directions on either side of a freeway. The San Andreas Fault, which runs through California, is a strike-slip fault. Horizontal motion along the San Andreas and several other related faults is responsible for many of the state's earthquakes. The result of motion along strike-slip faults can easily be seen in the many offset features that were originally continuous across the fault.

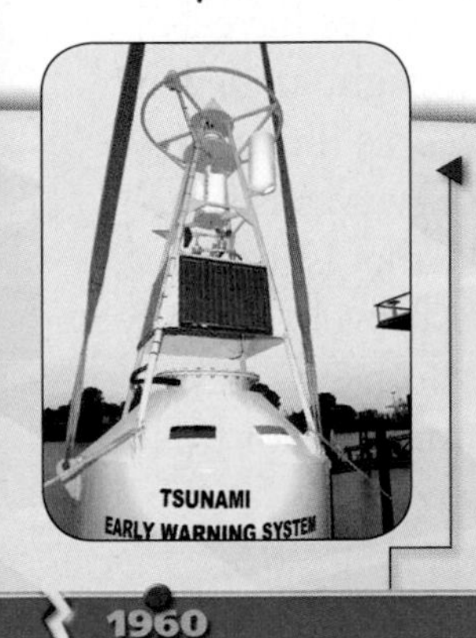

1960 | 1980 | 2000

1960 In Chile, a 9.5 earthquake generates tsunamis that hit Hawaii, Japan, New Zealand, and Samoa. This is the largest earthquake recorded.

1965 The United States, Japan, Chile, and Russia form the International Pacific Tsunami Warning System.

1972 The University of California, Berkeley creates the first modern shake table to test building designs.

1982 New Zealand constructs the first building with seismic isolation, using lead-rubber bearings to prevent the building from swaying during an earthquake.

2004 A 9.0 earthquake in the Indian Ocean triggers the most deadly tsunami in history. The tsunami travels as far as the East African Coast.

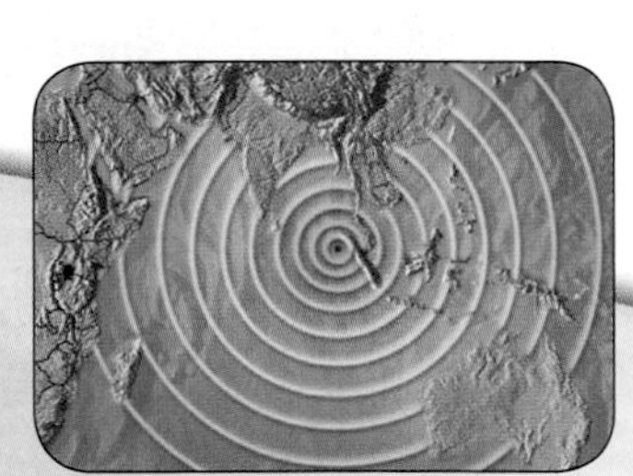

Concepts In Motion **Interactive Time Line** To learn more about these discoveries and others, visit glencoe.com. Earth Science Online

Concepts In Motion

Interactive Figure To see an animation of seismic waves, visit glencoe.com.

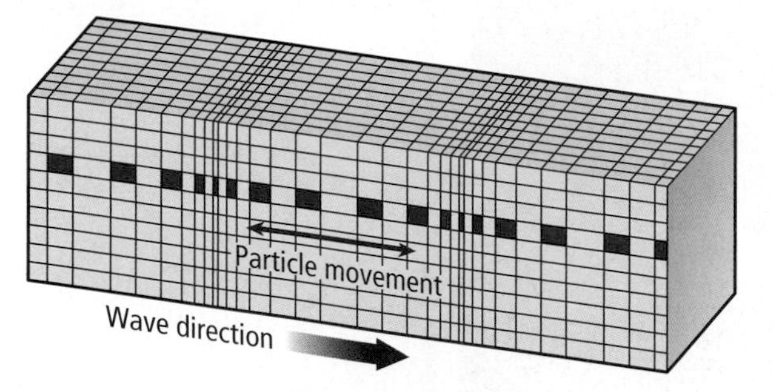

P-wave movement

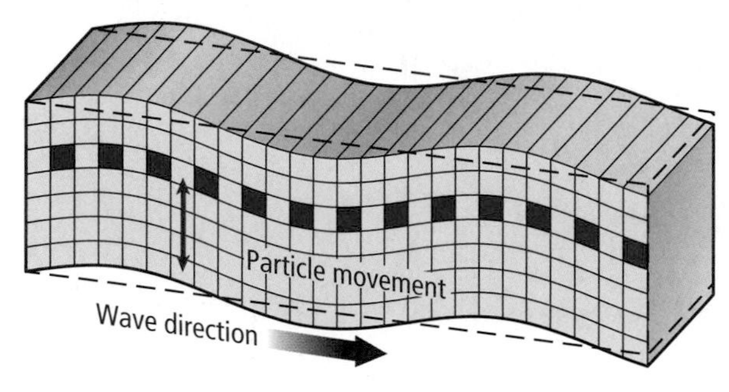

S-wave movement

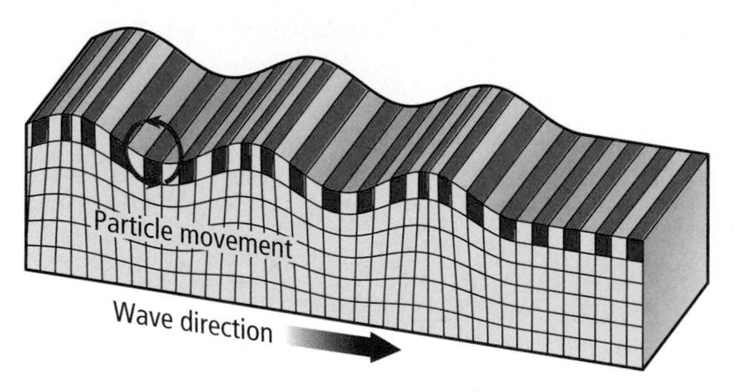

Surface wave movement

■ **Figure 19.5** Seismic waves are characterized by the types of movement they cause. Rock particles move back and forth as a P-wave passes. Rock particles move at right angles to the direction of the S-wave. A surface wave causes rock particles to move both up and down and from side to side.

Earthquake Waves

Most earthquakes are caused by movements along faults. Recall from the Launch Lab that some slippage along faults is relatively smooth. Other movements, modeled by the sandpaper-covered blocks, show that irregular surfaces in rocks can snag and lock. As stress continues to build in these rocks, they undergo elastic deformation. Beyond the elastic limit, they bend or stretch. Before that limit, an earthquake occurs when they slip or crumble.

Types of seismic waves The vibrations of the ground produced during an earthquake are called **seismic waves.** Every earthquake generates three types of seismic waves: primary waves, secondary waves, and surface waves.

Primary waves Also referred to as P-waves, **primary waves** squeeze and push rocks in the direction along which the waves are traveling, as shown in **Figure 19.5.** Note how a volume of rock, which is represented by small red squares, changes length as a P-wave passes through it. The compressional movement of P-waves is similar to the movement along a loosely coiled wire. If the coil is tugged and released quickly, the vibration passes through the length of the coil parallel to the direction of the initial tug.

Secondary waves **Secondary waves,** called S-waves, are named with respect to their arrival times. They are slower than P-waves, so they are the second set of waves to be felt. S-waves have a motion that causes rocks to move at right angles in relation to the direction of the waves, as illustrated in **Figure 19.5.** The movement of S-waves is similar to the movement of a jump rope that is jerked up and down at one end. The waves travel vertically to the other end of the jump rope. Both P-waves and S-waves pass through Earth's interior. For this reason, they are also called body waves.

Surface waves The third and slowest type of waves are surface waves, which travel only along Earth's surface. Surface waves can cause the ground to move sideways and up and down like ocean waves, as shown in **Figure 19.5.** These waves usually cause the most destruction because they cause the most movement of the ground, and take the longest time to pass.

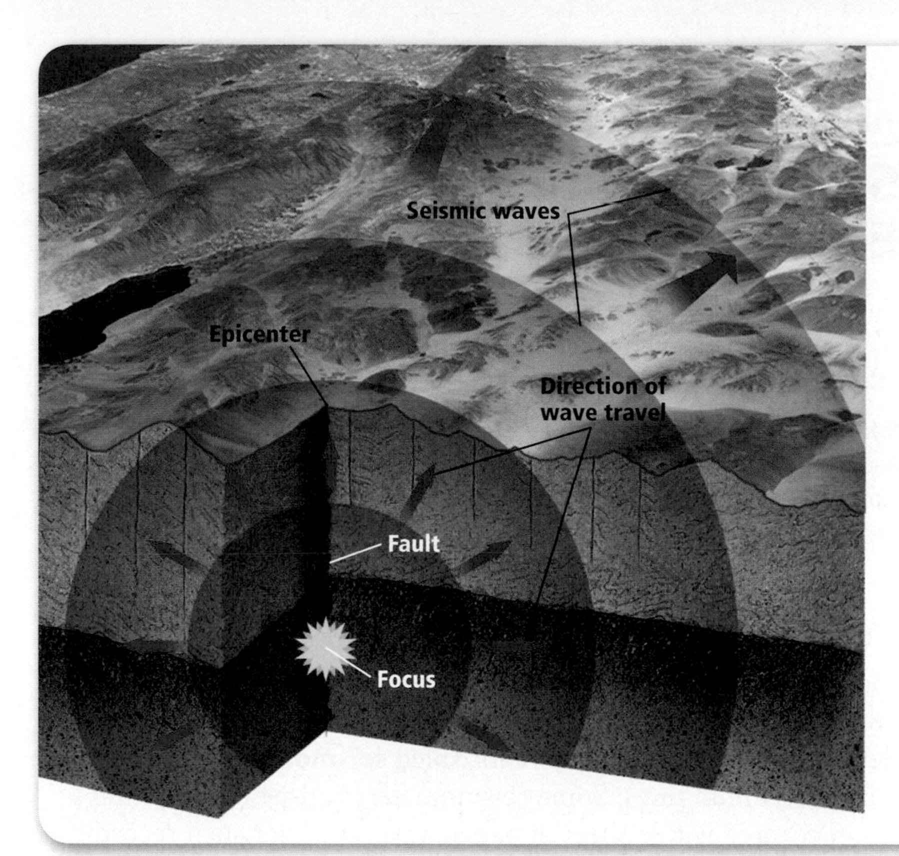

Figure 19.6 The focus of an earthquake is the point of initial fault rupture. The surface point directly above the focus is the epicenter.
Infer *the point at which surface waves will cause the most damage.*

Generation of seismic waves The first body waves generated by an earthquake spread out from the point of failure of crustal rocks. The point where the waves originate is the **focus** of the earthquake. The focus is usually several kilometers below Earth's surface. The point on Earth's surface directly above the focus is the **epicenter** (EH pih sen tur), shown in **Figure 19.6.** Surface waves originate from the epicenter and spread out.

Section 19.1 Assessment

Section Summary

- Stress is force per unit of area that acts on a material and strain is the deformation of a material in response to stress.
- Reverse, normal, and strike-slip are the major types of faults.
- The three types of seismic waves are P-waves, S-waves, and surface waves.

Understand Main Ideas

1. MAIN Idea **Describe** how the formation of a fault can result in an earthquake.
2. **Explain** why a stress-strain curve usually has two segments.
3. **Compare and contrast** the movement produced by each of the three types of faults.
4. **Draw** three diagrams to show how each type of seismic wave moves through rock. How do they differ?

Think Critically

5. **Relate** the movement produced by seismic waves to the observations a person would make of them as they traveled across Earth's surface.

WRITING in Earth Science

6. Relate the movement of seismic waves to movement of something you might see every day. Make a list and share it with your classmates.

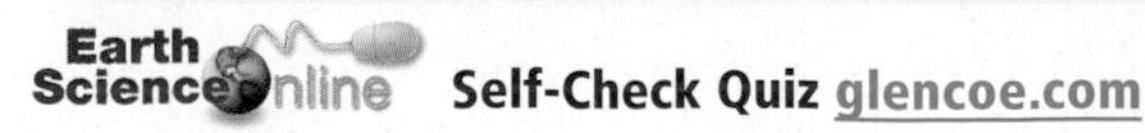
Self-Check Quiz glencoe.com

Section 19.2

Objectives

- **Describe** how a seismometer works.
- **Explain** how seismic waves have been used to determine the structure and composition of Earth's interior.

Review Vocabulary

mantle: the part of Earth's interior beneath the lithosphere and above the central core

New Vocabulary

seismometer
seismogram

Seismic Waves and Earth's Interior

MAIN Idea Seismic waves can be used to make images of the internal structure of Earth.

Real-World Reading Link When you look in a mirror, you see yourself because light waves reflect off your face to the mirror and back to your eye. Similarly, seismic waves traveling through Earth reflect off structures inside Earth, which allows these structures to be imaged.

Seismometers and Seismograms

Most of the vibrations caused by seismic waves cannot be felt at great distances from an earthquake's epicenter, but they can be detected by sensitive instruments called **seismometers** (size MAH muh turz). Some seismometers consist of a rotating drum covered with a sheet of paper, a pen or other such recording tool, and a mass, such as a pendulum. Seismometers vary in design, but all include a frame that is anchored to the ground and a mass that is suspended from a spring or wire, as shown in **Figure 19.7.** During an earthquake, the mass and the pen attached to it tend to stay at rest due to inertia, while the ground beneath shakes. The motion of the mass in relation to the frame is then registered on the paper with the recording tool, or is directly recorded onto a computer disk. The record produced by a seismometer is called a **seismogram** (SIZE muh gram). A portion of one is shown in **Figure 19.8.**

Concepts In Motion

Interactive Figure To see an animation of seismometers, visit glencoe.com.

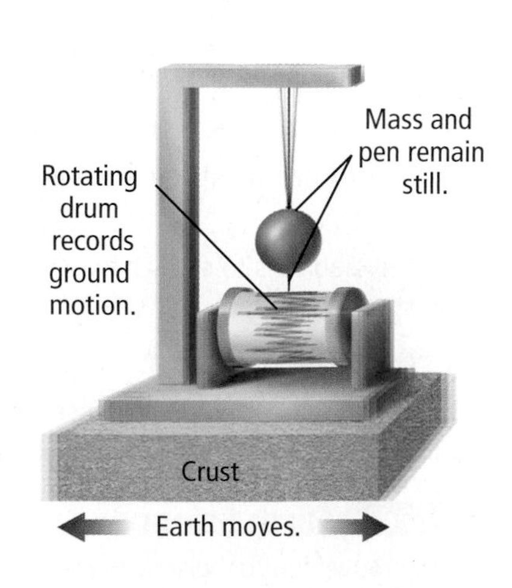

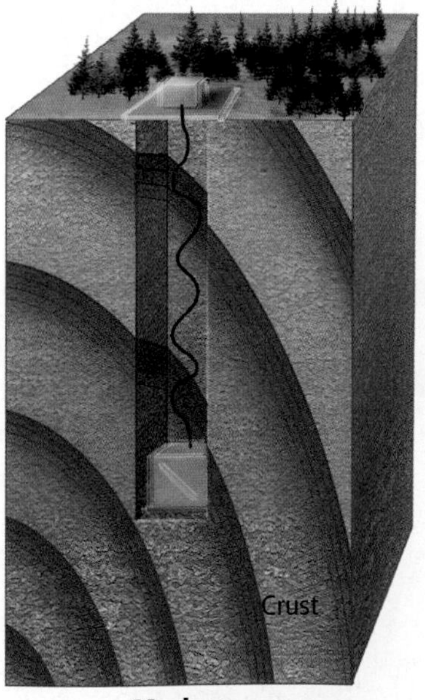

Figure 19.7 The frame of a historic seismometer is anchored to the ground. When an earthquake occurs, the frame moves but the hanging mass and attached pen do not. The mass and pen record the relative movement as the recording device moves under them. Compare this to a modern sensor and transmitter.

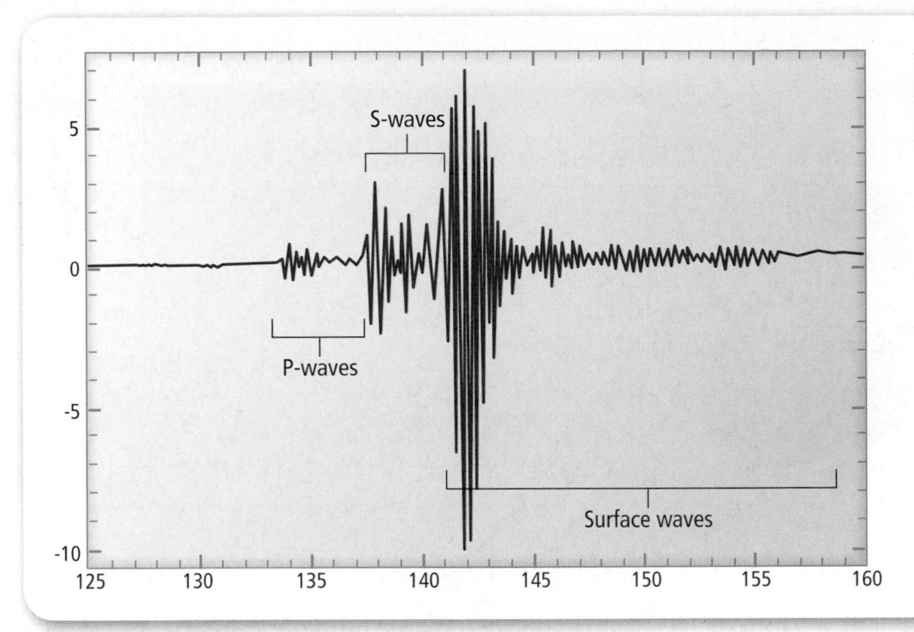

■ **Figure 19.8** Seismograms provide a record of the seismic waves that pass a certain point.

Travel-time curves Seismic waves that travel from the focus of an earthquake are recorded by seismometers housed in distant facilities. Over many years, the arrival times of seismic waves from countless earthquakes at seismic facilities around the world have been collected. Using these data, seismologists have been able to construct global travel-time curves for the arrival of P-waves and S-waves of earthquakes, as shown in **Figure 19.9.** These curves provide the average travel times of all P- and S-waves, from wherever an earthquake occurs on Earth.

 **Reading Check Summarize** how seismograms are used to construct global travel-time curves.

Distance from the epicenter Note that in **Figure 19.9,** as in **Figure 19.8,** the P-waves arrive first, then the S-waves. The surface waves arrive last. With increasing travel distance from the epicenter, the time separation between the curves for the P-waves and S-waves increases. This means that waves recorded on seismograms from more distant facilities are farther apart than waves recorded on seismograms at stations closer to the epicenter. This separation of seismic waves on seismograms can be used to determine the distance from the epicenter of an earthquake to the seismic facility that recorded the seismogram. This method of precisely locating an earthquake's epicenter will be discussed in Section 19.3.

■ **Figure 19.9** Travel-time curves show how long it takes for P-waves and S-waves to reach seismic stations located at different distances from an earthquake's epicenter.
Determine *how long it takes P-waves to travel to a seismogram 2000 km away. How long does it take for S-waves to travel the same distance?*

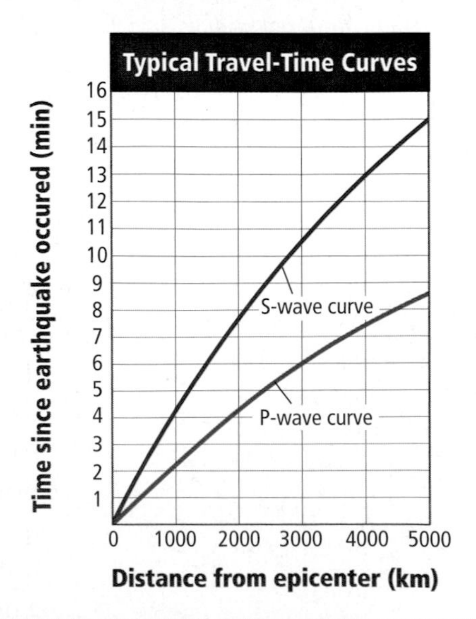

■ **Figure 19.10** Earth's layers are each composed of different materials. By examining the behavior of seismic waves moving through different kinds of rock, scientists have determined the composition of layers all the way to Earth's inner core.

Concepts In Motion

Interactive Figure To see an animation of P-waves and S-waves, visit glencoe.com.

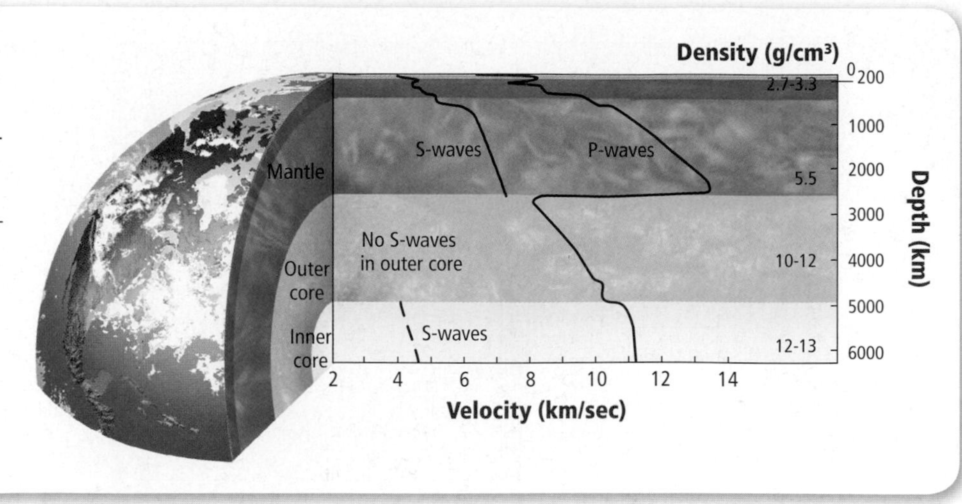

Clues to Earth's Interior

The seismic waves that shake the ground during an earthquake also travel through Earth's interior. This provides information that has enabled scientists to construct models of Earth's internal structure. Therefore, even though seismic waves can wreak havoc on the surface, they are invaluable for their contribution to scientists' understanding of Earth's interior.

Earth's internal structure Seismic waves change speed and direction at the boundaries between different materials. Note in **Figure 19.10** that as P-waves and S-waves initially travel through the mantle, they follow fairly direct paths. When P-waves strike the core, they are refracted, which means they bend. Seismic waves also reflect off of major boundaries inside Earth. By recording the travel-time curves and path of each wave, seismologists learn about differences in density and composition within Earth.

What happens to the S-waves generated by an earthquake? To answer this question, seismologists first determined that the right-angle motion of S-waves will not travel through liquid. Then, seismologists noticed that S-waves do not travel through Earth's center. This observation led to the discovery that Earth's core must be at least partly liquid. The data collected for the paths and travel times of the waves inside Earth led to the current understanding that Earth's core has an outer region that is liquid and an inner region that is solid.

Vocabulary

Academic Vocabulary

Encounter
to come upon or experience, especially unexpectedly
We had never encountered such a violent storm.

Earth's composition **Figure 19.11** shows that seismic waves change their paths as they encounter boundaries between zones of different materials. They also change their speed. By comparing the speed of seismic waves with measurements made on different rock types, scientists have determined the thickness and composition of Earth's different regions. As a result, scientists have determined that the upper mantle is peridotite, which is made mostly of the mineral olivine. The outer core is mostly liquid iron and nickel. The inner core is mostly solid iron and nickel.

Visualizing Seismic Waves

Figure 19.11 The travel times and behavior of seismic waves provide a detailed picture of Earth's internal structure. These waves also provide clues about the composition of the various parts of Earth.

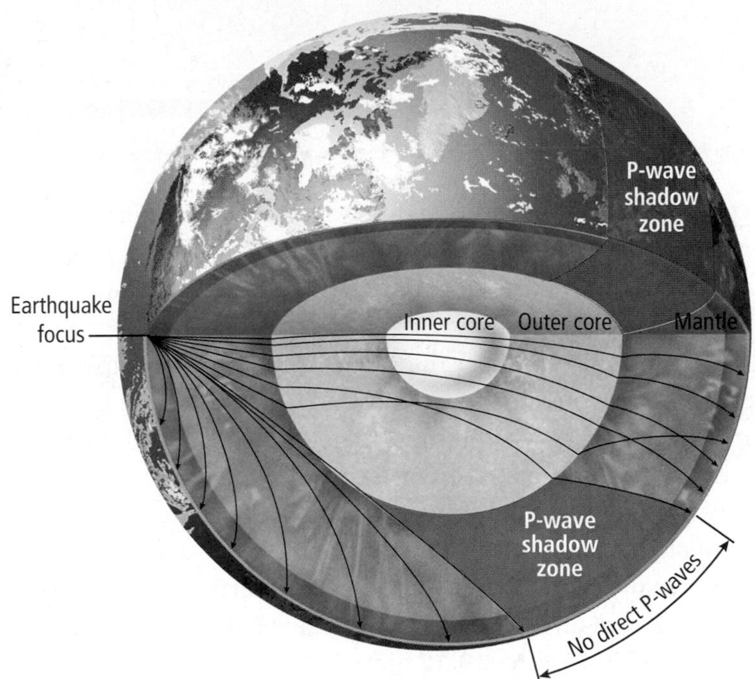

P-waves in the outer core are refracted. This generates a P-wave shadow zone on Earth's surface where no direct P-waves appear on seismograms. Other P-waves are reflected and refracted by the inner core. These can be detected by seismometers on the other side of the shadow zone.

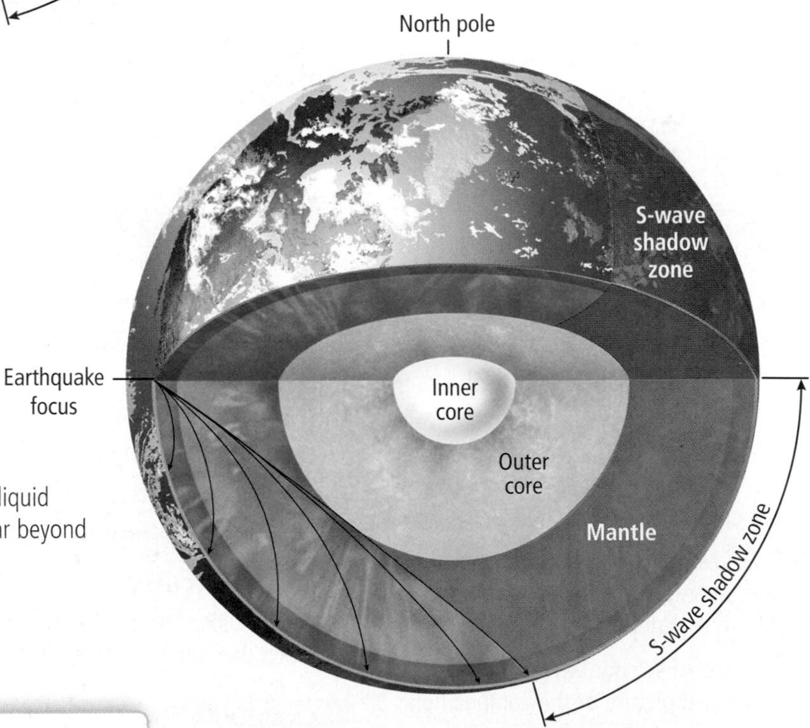

S-waves cannot travel through the liquid outer core and thus do not reappear beyond the S-wave shadow zone.

Concepts In Motion To explore more about seismic waves, visit glencoe.com.

Earth Science Online

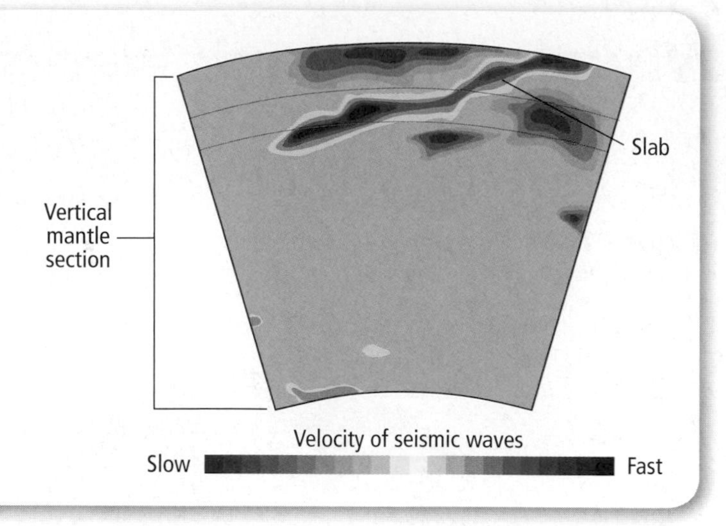

Figure 19.12 Images like this one from Japan are generated by capturing the path of seismic waves through Earth's interior. Areas of red indicate seismic waves that are traveling more slowly than average and areas of blue indicate seismic waves that are traveling faster than average. The blue area is a subducted plate.

Imaging Earth's interior Seismic wave speed and Earth's density vary with factors other than depth. Recall from Chapter 17 that cold slabs sink back into Earth at subduction zones, and recall from Chapter 18 that mantle plumes are regions where hot mantle material is rising. Because the speed of seismic waves depends on temperature and composition, it is possible to use seismic waves to create images of structures such as slabs and plumes. In general, the speed of seismic waves decreases as temperature increases. Thus, waves travel more slowly in hotter areas and more quickly in cooler regions. Using measurements made at seismometers around the world and waves recorded from many thousands of earthquakes, Earth's internal structure can be visualized, and features such as slabs can be located in images like the one in **Figure 19.12.** These images are similar to CT scans, except that the images are made using seismic waves instead of X rays.

Section 19.2 Assessment

Section Summary

- Seismometers are devices that record seismic wave activity on a seismogram.
- Travel times for P-waves and S-waves enable scientists to pinpoint the epicenters of earthquakes.
- P-waves and S-waves change speed and direction when they encounter different materials.
- Analysis of seismic waves provides a detailed picture of the composition of Earth's interior.

Understand Main Ideas

1. MAIN Idea **Explain** how P-waves and S-waves are used to determine the properties of Earth's core.
2. **Draw** a diagram of a seismometer showing how the movement of Earth is translated into a seismogram.
3. **Describe** how seismic travel-time curves are used to study earthquakes.
4. **Differentiate** between the speed of waves through hot and cold material.

Think Critically

5. **Infer** Using the seismogram in **Figure 19.8**, suggest why surface waves cause so much damage even though they are the last to arrive at a seismic station.

WRITING in **Earth Science**

6. Write a newspaper article reporting on the ways scientists have determined the composition of Earth.

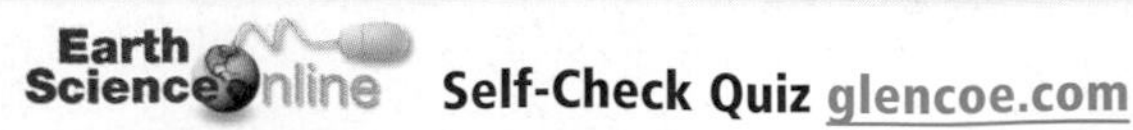

Section 19.3

Measuring and Locating Earthquakes

Objectives
- **Compare and contrast** earthquake magnitude and intensity and the scales used to measure each.
- **Explain** why data from at least three seismic stations are needed to locate an earthquake's epicenter.
- **Describe** Earth's seismic belts.

Review Vocabulary

plot: to mark or note on a map or chart

New Vocabulary

Richter scale
magnitude
amplitude
moment magnitude scale
modified Mercalli scale

MAIN Idea Scientists measure the strength and chart the location of earthquakes using seismic waves.

Real-World Reading Link When someone speaks to you from nearby, you can hear them clearly. However, the sound gets fainter as they get farther away. Similarly, the energy of seismic waves gets weaker the farther away you are from the source of an earthquake.

Earthquake Magnitude and Intensity

More than 1 million earthquakes are felt each year, but news accounts report on only the largest ones. Scientists have developed several methods for describing the size of an earthquake.

Richter scale The **Richter scale,** devised by a geologist named Charles Richter, is a numerical rating system that measures the energy of the largest seismic waves, called the **magnitude,** that are produced during an earthquake. The numbers in the Richter scale are determined by the height, called the **amplitude,** of the largest seismic wave. Each successive number represents an increase in amplitude of a factor of 10. For example, the seismic waves of a magnitude-8 earthquake on the Richter scale are ten times larger than those of a magnitude-7 earthquake. The differences in the amounts of energy released by earthquakes are even greater than the differences between the amplitudes of their waves. Each increase in magnitude corresponds to about a 32-fold increase in seismic energy. Thus, an earthquake of magnitude-8 releases about 32 times the energy of a magnitude-7 earthquake. The damage shown in **Figure 19.13** was caused by an earthquake measuring 7.6 on the Richter scale.

Figure 19.13 The damage shown here was caused by a magnitude-7.6 earthquake that struck Pakistan in December 2005.

■ **Figure 19.14** The modified Mercalli scale measures damage done by an earthquake. An earthquake strong enough to knock groceries off the store's shelves would probably be rated V using the modified Mercalli scale.

Moment magnitude scale While the Richter scale is often used to describe the magnitude of an earthquake, most earthquake scientists, called seismologists, use a scale called the moment magnitude scale. The **moment magnitude scale** is a rating scale that measures the energy released by an earthquake, taking into account the size of the fault rupture, the amount of movement along the fault, and the rocks' stiffness. Most often, when you hear about an earthquake on the news, the number given is from the moment magnitude scale.

Modified Mercalli scale Another way to describe earthquakes is with respect to the amount of damage they cause. This measure, called the intensity of an earthquake, is determined using the **modified Mercalli scale,** which rates the types of damage and other effects of an earthquake as noted by observers during and after its occurrence. This scale uses the Roman numerals I to XII to designate the degree of intensity. Specific effects or damage correspond to specific numerals; the worse the damage, the higher the numeral. A simplified version of the modified Mercalli scale is shown in **Table 19.2.** You can use the information given in this scale to rate the intensity of the earthquakes such as the one that caused the damage shown in **Figure 19.14.**

Concepts In Motion
Interactive Table To explore more about the modified Mercalli scale, visit glencoe.com.

Table 19.2 Modified Mercalli Scale

I	Not felt except under unusual conditions
II	Felt only by a few persons; suspended objects might swing.
III	Quite noticeable indoors; vibrations are like the passing of a truck.
IV	Felt indoors by many, outdoors by few; dishes and windows rattle; standing cars rock noticeably.
V	Felt by nearly everyone; some dishes and windows break and some plaster cracks.
VI	Felt by all; furniture moves; some plaster falls and some chimneys are damaged.
VII	Everybody runs outdoors; some chimneys break; damage is slight in well-built structures but considerable in weak structures.
VIII	Chimneys, smokestacks, and walls fall; heavy furniture is overturned; partial collapse of ordinary buildings occurs.
IX	Great general damage occurs; buildings shift off foundations; ground cracks; underground pipes break.
X	Most ordinary structures are destroyed; rails are bent; landslides are common.
XI	Few structures remain standing; bridges are destroyed; railroad ties are greatly bent; broad fissures form in the ground.
XII	Damage is total; objects are thrown upward into the air.

Earthquake intensity The intensity of an earthquake depends primarily on the amplitude of the surface waves generated. Like body waves, surface waves gradually decrease in size with increasing distance from the focus of an earthquake. Because of this, the intensity also decreases as the distance from a earthquake's epicenter increases. Maximum intensity values are observed in the region near the epicenter; Mercalli values decrease to I at distances far from the epicenter.

In the MiniLab, you will use the modified Mercalli scale values to make a seismic-intensity map. These maps are a visual demonstration of an earthquake's intensity. Contour lines join points that experienced the same intensity. They demonstrate how the maximum intensity is usually found near the earthquake's epicenter.

Depth of focus As you learned earlier in this section, earthquake intensity and magnitude reflect the size of the seismic waves generated by the earthquake. Another factor that determines the intensity of an earthquake is the depth of its focus. As shown in **Figure 19.15,** earthquakes can be classified as shallow, intermediate, or deep, depending on the location of the focus. Catastrophic earthquakes with high intensity values are almost always shallow-focus events.

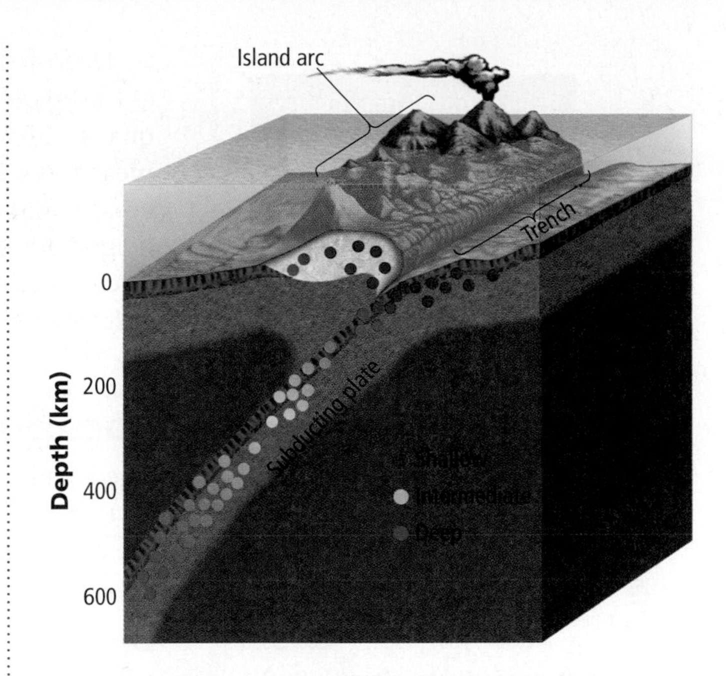

Figure 19.15 Earthquakes are classified as shallow, intermediate, or deep, depending on the location of the focus. Shallow-focus earthquakes are the most damaging.

MiniLab

Make a Map

How is a seismic-intensity map made? Seismic-intensity data plotted on contour maps give scientists a visual picture of an epicenter's location and the earthquake's intensity.

Procedure

1. Read and complete the lab safety form.
2. Trace the map onto **paper.** Mark the locations indicated by the letters on the map.
3. Plot these Mercalli intensity values on the map next to the correct letter: A, I; B, III; C, II; D, III; E, IV; F, IV; G, IV; H, V; I, V; J, V; K, VI; L, VIII; M, VII; N, VIII; O, III.
4. Draw contours on the map to connect the intensity values.

Analysis

1. **Determine** the maximum intensity value.
2. **Find** the location of the maximum intensity value.
3. **Estimate** the earthquake's epicenter.

Intensity Values of an Earthquake

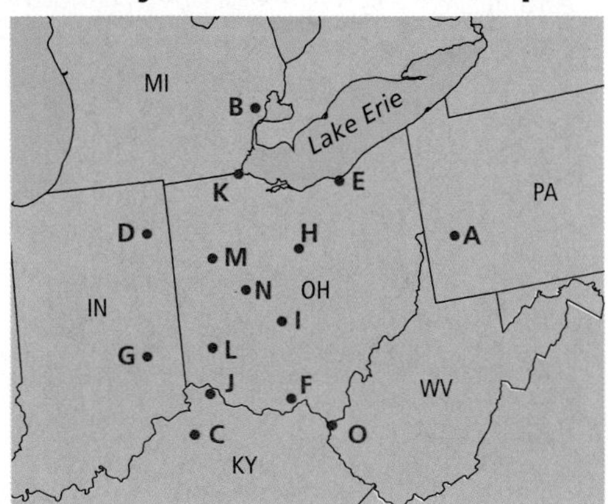

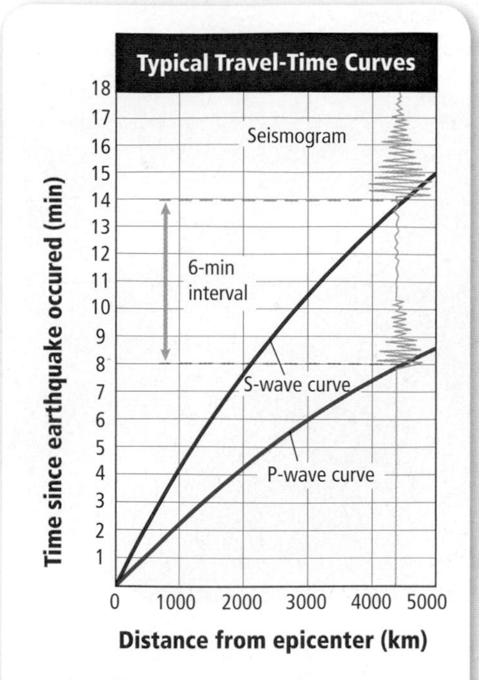

■ **Figure 19.16** This travel-time curve also shows seismographic data for an earthquake event.

■ **Figure 19.17** To locate the epicenter of an earthquake, scientists identify the seismic stations on a map, and draw a circle with the radius of distance to the epicenter from each station. The point where all the circles intersect is the epicenter.
Identify ***the epicenter of this earthquake.***

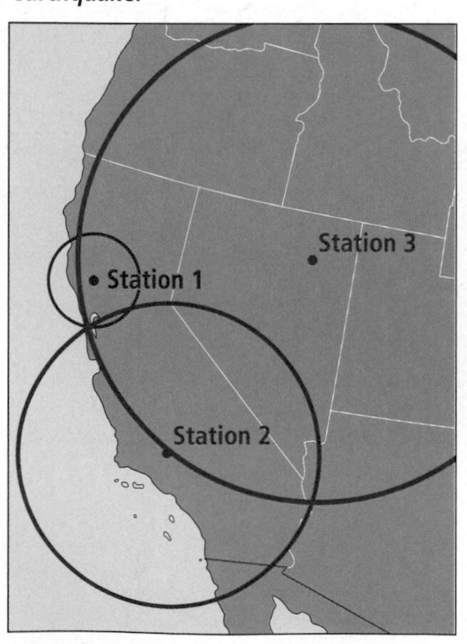

Deep-focus earthquakes generally produce smaller vibrations at the epicenter than those produced by shallow-focus earthquakes. For example, a shallow-focus, moderate earthquake that measures a magnitude-6 on the Richter scale can generate a greater maximum intensity than a deep-focus earthquake of magnitude-8. Because the modified Mercalli scale is based on intensity rather than magnitude, it is a better measure of an earthquake's effect on people.

Locating an Earthquake

The location of an earthquake's epicenter and the time of the earthquake's occurrence are usually not known at first. However, the epicenter's location, as well as the time of occurrence, can be determined using seismograms and travel-time curves.

Distance to an earthquake Just as a person riding a bike will travel faster than a person who is walking, P-waves reach a seismograph station before the S-waves. Consider the effect of the distance traveled on the time it takes for both waves to arrive. Like the bicyclist and the walker, the gap in their arrival times will be greater when the distance traveled is longer. **Figure 19.16** shows the same travel-time curve graph shown in **Figure 19.9** of Section 19.2, but this time it is joined with the seismogram from a specific earthquake. The seismometer recorded the time that elapsed between the arrival of the first P-waves and first S-waves. Seismologists determine the distance to an earthquake's epicenter by measuring the separation on any seismogram and identifying that same separation time on the travel-time graph. The separation time for the earthquake shown in **Figure 19.16** is 6 min. Based on travel times of seismic waves, the distance between the earthquake's epicenter and the seismic station that recorded the waves can only be 4500 km. This is because the known travel time over that distance is 8 min for P-waves and 14 min for S-waves. Farther from the epicenter, the gap between the travel times for both waves increases.

Reading Check **Apply** If the gap between P- and S-waves is 2 min, what can you infer about the distance from the epicenter to the seismometer?

Seismologists analyze data from many seismograms to locate the epicenter. Calculating the distance between an earthquake's epicenter and a seismic station provides enough information to determine that the epicenter was a certain distance in any direction from the seismic station. This can be represented by a circle around the seismic station with a radius equal to the distance to the epicenter. Consider the effect of adding data from a second seismic station. The two circles will overlap at two points. When data from a third seismic station is added, the rings will overlap only at one point—the epicenter, as shown in **Figure 19.17.**

Time of an earthquake The gap in the arrival times of different seismic waves on a seismogram provides information about the distance to the epicenter. Seismologists can also use the seismogram to gain information about the exact time that the earthquake occurred at the focus. The time can be determined by using a table similar to the travel-time graph shown in **Figure 19.9.** The exact arrival times of the P-waves and S-waves at a seismic station are recorded on the seismogram. Seismologists read the travel time of either wave to the epicenter from that station using graphs similar to the one shown in **Figure 19.9.** For example, consider a seimogram that registered the arrival of P-waves at exactly 10:00 A.M. If the P-waves traveled 4500 km, and took 8 min according to the appropriate travel-time curve, then it can be determined that the earthquake occurred at the focus at 9:52 A.M.

 **Reading Check** **List** the information contained in a seismogram.

Seismic Belts

Over the years, seismologists have collected and plotted the locations of numerous earthquake epicenters. The global distribution of these epicenters reveals a noteworthy pattern. Earthquake locations are not randomly distributed. The majority of the world's earthquakes occur along narrow seismic belts that separate large regions with little or no seismic activity.

DATA ANALYSIS LAB

Based on Real Data*

Interpret the Data

How can you find an earthquake's epicenter? To pinpoint the epicenter, analyze the P-wave and S-wave data recorded at seismic stations.

Analysis

1. Obtain a map of the western hemisphere from your teacher and mark the seismic stations listed in the table.
2. For each station, calculate and record the arrival time differences by subtracting the P-wave arrival time from the S-wave arrival times.
3. Use the arrival time differences and the travel-time curve **(Figure 19.9)** to find the distance between the epicenter and each seismic station. Record the distances.
4. Draw a circle around each station. Use the distance from the epicenter as the radius for each circle. Repeat for each seismic station.
5. Identify the epicenter of the earthquake.

Data and Observations

Seismic Station	P-wave Arrival Time (PST)	S-wave Arrival Time (PST)	Arrival Time Difference (min)	Distance from Epicenter (km)
Newcomb, NY	8:39:02	8:44:02		
Idaho Springs, CO	8:35:22	8:37:57		
Darwin, CA	8:35:38	8:38:17		

Think Critically

6. **Explain** why you need to find the difference in time of arrival between P- and S-waves for each seismic station.
7. **Identify** sources of error in determining an earthquake's epicenter.
8. **Explain** why data from more seismic stations would be useful for finding the epicenter.

*Data obtained from: Significant earthquakes of the world. 2006. *USGS Earthquake Center.*

Figure 19.18 Notice the pattern of global epicenter locations on the map.
Identify *Based on this map, do you live near an epicenter?*

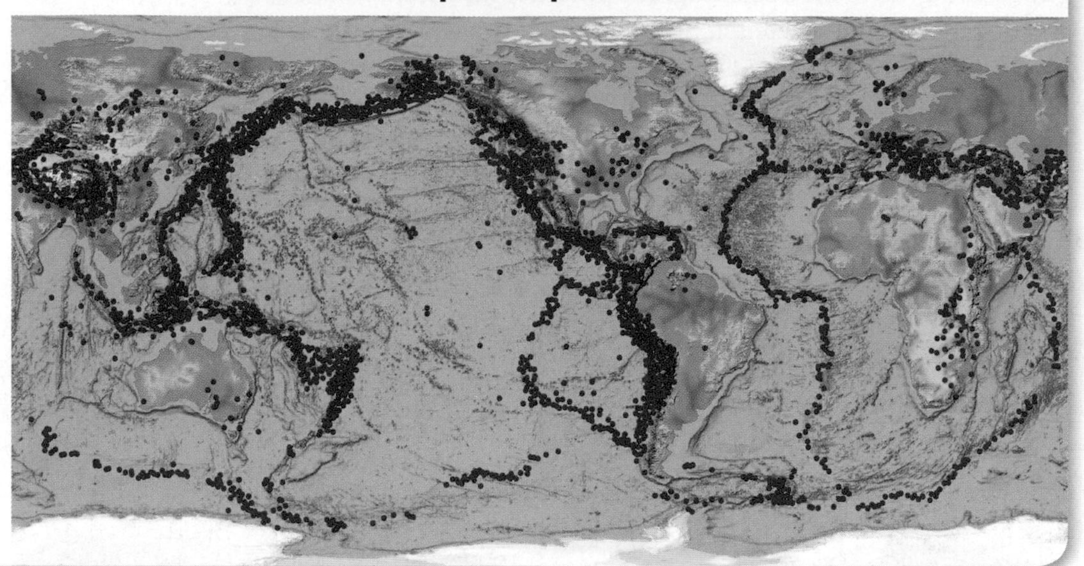

As shown in **Figure 19.18,** earthquakes occur in narrow bands. The locations of most earthquakes correspond closely with tectonic plate boundaries. In fact, almost 80 percent of all earthquakes occur on the Circum-Pacific Belt and about 15 percent on the Mediterranean-Asian Belt across southern Europe and Asia. These belts are subduction zones, where tectonic plates are colliding and one plate is forced to sink beneath another. Most of the remaining earthquakes occur in narrow bands along the crests of ocean ridges, where tectonic plates are diverging.

Section 19.3 Assessment

Section Summary

- Earthquake magnitude is a measure of the energy released during an earthquake and can be measured on the Richter scale.
- Intensity is a measure of the damage caused by an earthquake and is measured with the modified Mercalli scale.
- Data from at least three seismic stations are needed to locate an earthquake's epicenter.
- Most earthquakes occur in seismic belts, which are areas associated with plate boundaries.

Understand Main Ideas

1. MAIN Idea **Summarize** the ways that scientists can use seismic waves to measure and locate earthquakes.
2. **Compare and contrast** earthquake magnitude and intensity and the scales used to measure each.
3. **Explain** why data from at least three seismic stations makes it possible to locate an earthquake's epicenter.
4. **Describe** how the boundaries between Earth's tectonic plates compare with the location of most of the earthquakes shown in the map in **Figure 19.18.**

Think Critically

5. **Formulate** a reason why a magnitude-3 earthquake can possibly cause more damage than a magnitude-6 earthquake.

MATH in Earth Science

6. Calculate how much more energy a magnitude-9 earthquake releases compared to that of a magnitude-7 earthquake.

Earth Science Online Self-Check Quiz glencoe.com

Section 19.4

Objectives

- **Discuss** factors that affect the amount of damage caused by an earthquake.
- **Explain** some of the factors considered in earthquake-probability studies.
- **Identify** how different types of structures are affected by earthquakes.

Review Vocabulary

geology: study of materials that make up Earth and the processes that form and change these materials

New Vocabulary

soil liquefaction
tsunami
seismic gap

Earthquakes and Society

MAIN Idea The probability of an earthquake's occurrence is determined from the history of earthquakes and knowing where and how quickly strain accumulates.

Real-World Reading Link If, in your city, it rains an average of 11 days every July, how can you predict the weather in your city for July 4 ten years from now? You could estimate that there is a 11/31 chance that it will rain. In the same way, the probability of an earthquake's occurrence can be estimated from the history of earthquakes in the region.

Earthquake Hazards

Earthquakes are known to occur frequently along plate boundaries. An earthquake of magnitude-5 can be catastrophic in one region, but relatively harmless in another. There are many factors that determine the severity of damage produced by an earthquake. These factors are called earthquake hazards. Identifying earthquake hazards in an area can sometimes help to prevent some of the damage and loss of life. For example, the design of certain buildings can affect earthquake damage. As you can see in **Figure 19.19,** the most severe damage occurs to unreinforced buildings made of brittle building materials such as concrete. Wooden structures, on the other hand, are more resilient and generally sustain less damage.

Figure 19.19 Concrete buildings are often brittle and can be easily damaged in an earthquake. The building on the left shifted on its foundation after an earthquake and is held up by a single piece of wood.

■ **Figure 19.20** One type of damage caused by earthquakes is called pancaking because shaking causes a building's supporting walls to collapse and the upper floors to fall one on top of the other like a stack of pancakes.

Structural failure In many earthquake-prone areas, buildings are destroyed as the ground beneath them shakes. In some cases, the supporting walls of the ground floor fail and cause the upper floors, which initially remain intact, to fall and collapse as they hit the ground or lower floors. The resulting debris resembles a stack of pancakes; thus, the process is called pancaking. This type of structural failure, shown in **Figure 19.20,** was a tragic consequence of the earthquake in Islamabad, Pakistan, in 2005.

Reading Check **Explain** what happens when a building pancakes.

Another type of structural failure is related to the height of a building. During the 1985 Mexico City earthquake, for example, most buildings between five and 15 stories tall collapsed or were otherwise completely destroyed, as shown in **Figure 19.21.** Similar structures that were either shorter or taller, however, sustained only minor damage. The shaking caused by the earthquake had the same frequency of vibration as the natural sway of the intermediate buildings. This caused those buildings to sway the most violently during the earthquake. The ground vibrations, however, were too rapid to affect taller buildings, whose frequency of vibration was longer than those of the earthquake, and too slow to affect shorter buildings, whose frequency of vibration was shorter.

■ **Figure 19.21** Many medium-sized buildings were damaged or destroyed during the 1985 Mexico City earthquake because they vibrated with the same frequency as the seismic waves.

■ **Figure 19.22** Soil liquefaction happens when seismic vibrations cause poorly consolidated soil to liquefy and behave like quicksand. The buildings pictured here were built on this type of soil and an earthquake caused the buildings to sink into the ground.

Land and soil failure In addition to their effects on structures made by humans, earthquakes can wreak havoc on Earth's landscape. In sloping areas, earthquakes can trigger massive landslides. For example, most of the estimated 30,000 deaths caused by the magnitude-7.8 earthquake that struck in Peru in 1970 resulted from a landslide that buried several towns. In areas with sand that is nearly saturated with water, seismic vibrations can cause the ground to behave like a liquid in a phenomenon called **soil liquefaction** (lih kwuh FAK shun). It can generate landslides even in areas of low relief. It can cause trees and houses to fall over or to sink into the ground and underground pipes and tanks to rise to the surface. **Figure 19.22** shows tilted buildings that resulted when the soil under them liquefied during an earthquake.

 **Reading Check** **Summarize** how solid ground can take the properties of a liquid.

In addition to determining landslide risks, the type of ground material can also affect the severity of an earthquake in an area. Ground motion is amplified in some soft materials, such as unconsolidated sediments. It is muted in more resistant materials, such as granite. The severe damage to structures in Mexico City during the 1985 earthquake is attributed to the soft sediments on which the city is built. The thickness of the sediments caused them to resonate with the same frequency as that of the surface waves generated by the earthquake. This produced reverberations that greatly enhanced the ground motion and the resulting damage.

▪ **Figure 19.23** A tsunami is generated when an underwater fault or landslide displaces a column of water.

Concepts In Motion
Interactive Figure To see an animation of a tsunami, visit glencoe.com.

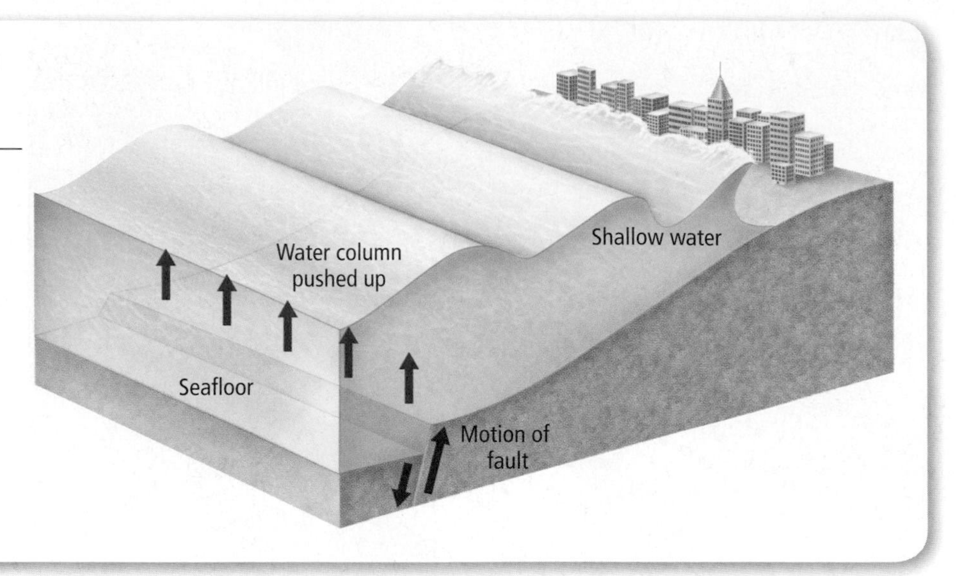

Tsunami Another type of earthquake hazard is a **tsunami** (soo NAH mee)—a large ocean wave generated by vertical motions of the seafloor during an earthquake. These motions displace the entire column of water overlying the fault, creating bulges and depressions in the water, as shown in **Figure 19.23.** The disturbance then spreads out from the epicenter in the form of extremely long waves. While these waves are in the open ocean, their height is generally less than 1 m. When the waves enter shallow water, however, they can form huge breakers with heights occasionally exceeding 30 m. These enormous wave heights, together with open-ocean speeds between 500 and 800 km/h, make tsunamis dangerous threats to coastal areas both near to and far from a earthquake's epicenter. The Indian Ocean tsunami of December 26, 2004, originated with a magnitude-9.0 earthquake in the ocean about 160 km west of Sumatra. The 30-m-tall tsunami radiated across the Indian Ocean and struck the coasts of Indonesia, Sri Lanka, India, Thailand, Somalia, and several other nations. The death toll from the tsunami exceeded 225,000, making it one of the most devastating natural disasters in modern history. The aftermath of that catastrophic event is shown in **Figure 19.24.**

VOCABULARY
SCIENCE USAGE V. COMMON USAGE
Column
Science usage: a hypothetical cylinder of water that goes from the surface to the bottom of a body of water

Common usage: a vertical arrangement of items

▪ **Figure 19.24** The destruction from the December 26, 2004, tsunami in the Indian Ocean was not isolated to the shoreline. As seen here, areas inland were devastated by the tsunami, which took at least 225,000 lives.

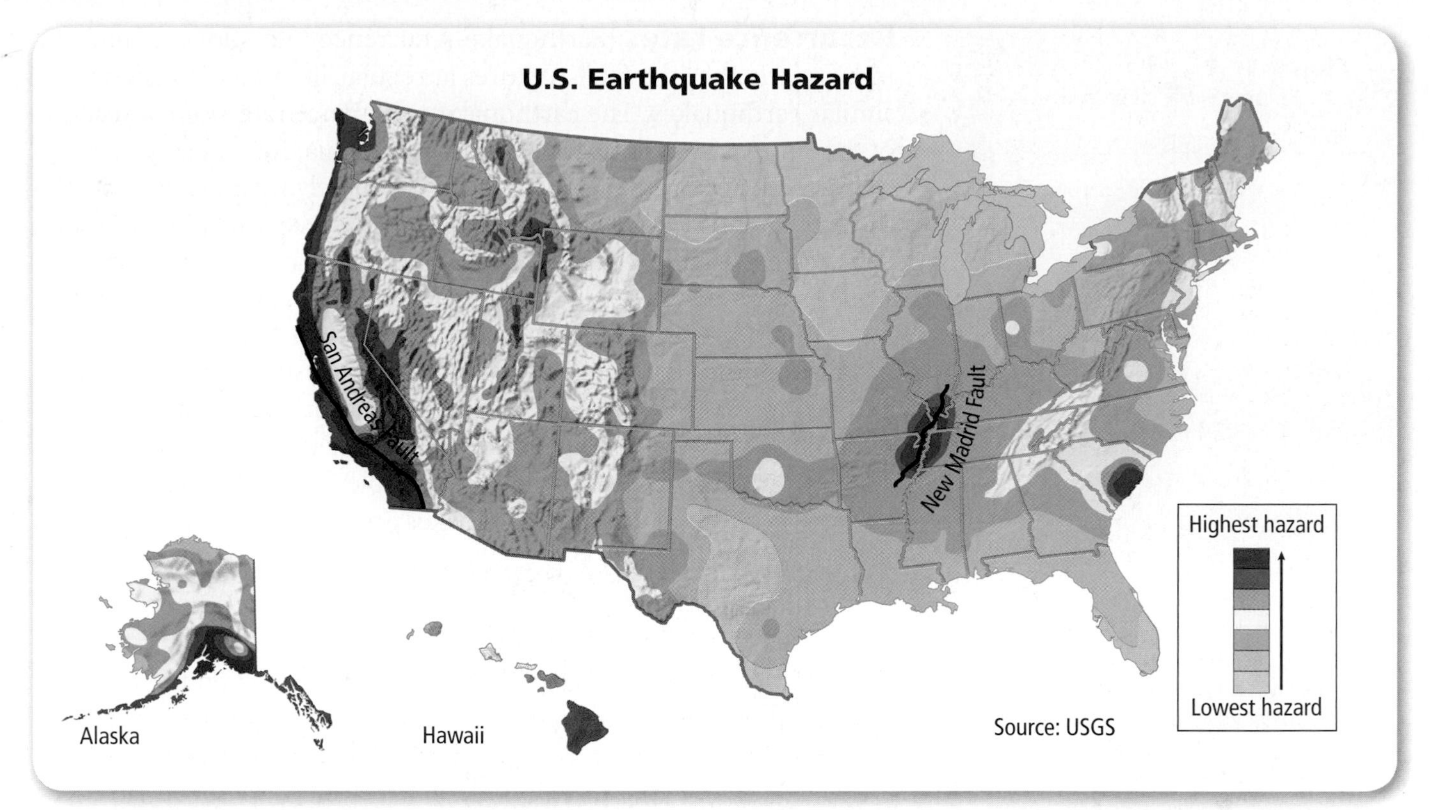

■ **Figure 19.25** Areas of high seismic risk in the United States include Alaska, Hawaii, and some of the western states.

Locate ***the areas of highest seismic risk on the map. Locate your own state. What is the seismic risk of your area?***

Earthquake Forecasting

To minimize the damage and deaths caused by earthquakes, seismologists are searching for ways to forecast these events. There is currently no completely reliable way to forecast the exact time and location of the next earthquake. Instead, earthquake forecasting is based on calculating the probability of an earthquake. The probability of an earthquake's occurrence is based on two factors: the history of earthquakes in an area and the rate at which strain builds up in the rocks.

Reading Check **Identify** the two factors seismologists use to determine the probability of an earthquake occurring in a certain area.

To read about the challenges of earthquake forecasting, go to the **National Geographic Expedition** on page 916.

Seismic risk Recall that most earthquakes occur in long, narrow bands called seismic belts. The probability of future earthquakes is much greater in these belts than elsewhere on Earth. The pattern of earthquakes in the past is usually a reliable indicator of future earthquakes in a given area. Seismometers and sedimentary rocks can be used to determine the frequency of large earthquakes. The history of an area's seismic activity can be used to generate seismic-risk maps. A seismic-risk map of the United States is shown in **Figure 19.25.** In addition to Alaska, Hawaii, and some western states, there are several regions of relatively high seismic risk in the central and eastern United States. These regions have experienced some of the most intense earthquakes in the past and probably will experience significant seismic activity in the future.

■ **Figure 19.26** This drilling rig was used to drill a hole 2.3 km deep in Parkfield, California. Once completed, the hole was rigged with instruments to record data during major and minor tremors. The goal of the project was to better understand how earthquakes work and what triggers them. This information could help scientists predict when earthquakes will occur.

Recurrence rates Earthquake-recurrence rates along a fault can indicate whether the fault ruptures at regular intervals to generate similar earthquakes. The earthquake-recurrence rate along a section of the San Andreas fault at Parkfield, California, for example, shows that a sequence of earthquakes of approximately magnitude 6 shook the area about every 22 years from 1857 until 1966. In 1987 seismologists forecasted a 90-percent probability that a major earthquake would rock the area within the next few decades. Several kinds of instruments at the drilling site, shown in **Figure 19.26,** were installed around Parkfield in an attempt to measure the earthquake as it occurred. In September, 2004, a magnitude-6 earthquake struck. Extensive data were collected before and after the 2004 earthquake. The information obtained will be invaluable for predicting and preparing for future recurrent earthquakes around the world.

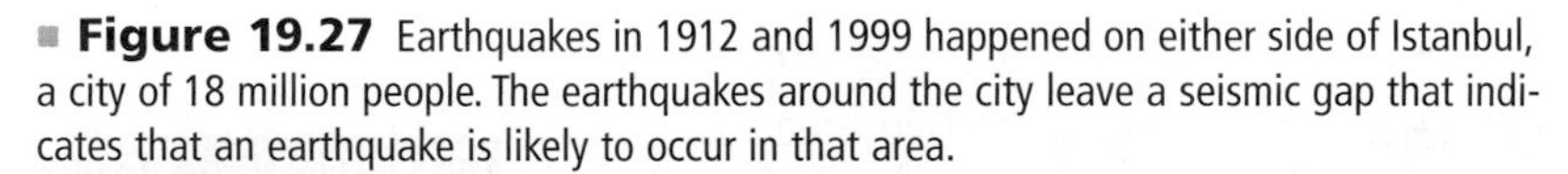

Reading Check **Infer** the significance of studying recurrence rates of earthquakes.

Seismic gaps Probability forecasts are also based on the location of seismic gaps. **Seismic gaps** are sections located along faults that are known to be active, but which have not experienced significant earthquakes for a long period of time. A seismic gap in the San Andreas Fault cuts through San Francisco. This section of the fault has not ruptured since the devastating earthquake that struck the city in 1906. Because of this inactivity, seismologists currently forecast that there is a 67-percent probability that the San Francisco area will experience a magnitude-7 or higher earthquake within the next 30 years. **Figure 19.27** shows the seismic-gap map for a fault that passes through an area of Turkey. Like the San Andreas Fault in California, there is a long history of earthquakes along the major fault shown below.

■ **Figure 19.27** Earthquakes in 1912 and 1999 happened on either side of Istanbul, a city of 18 million people. The earthquakes around the city leave a seismic gap that indicates that an earthquake is likely to occur in that area.

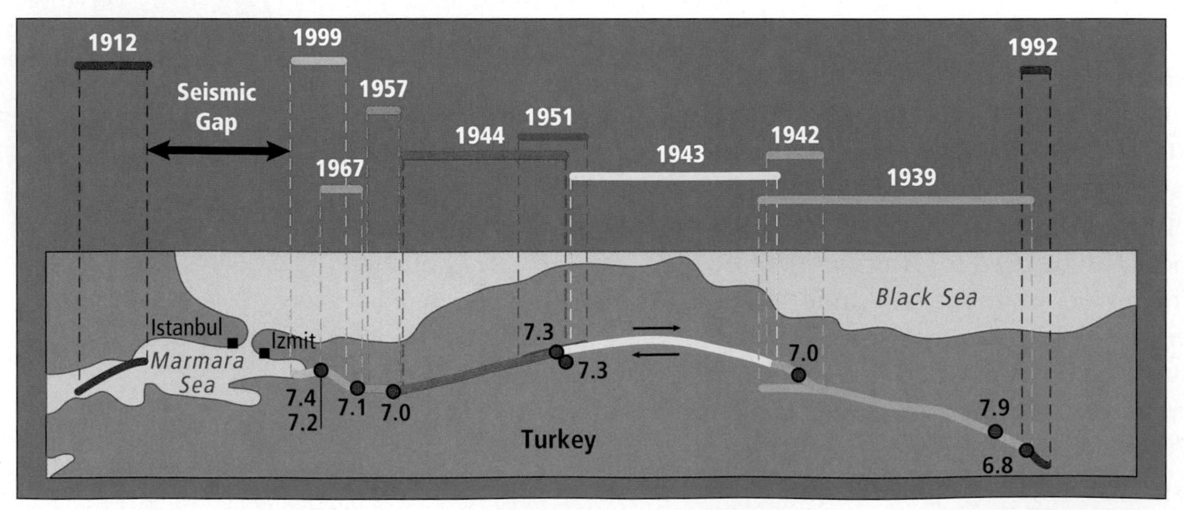

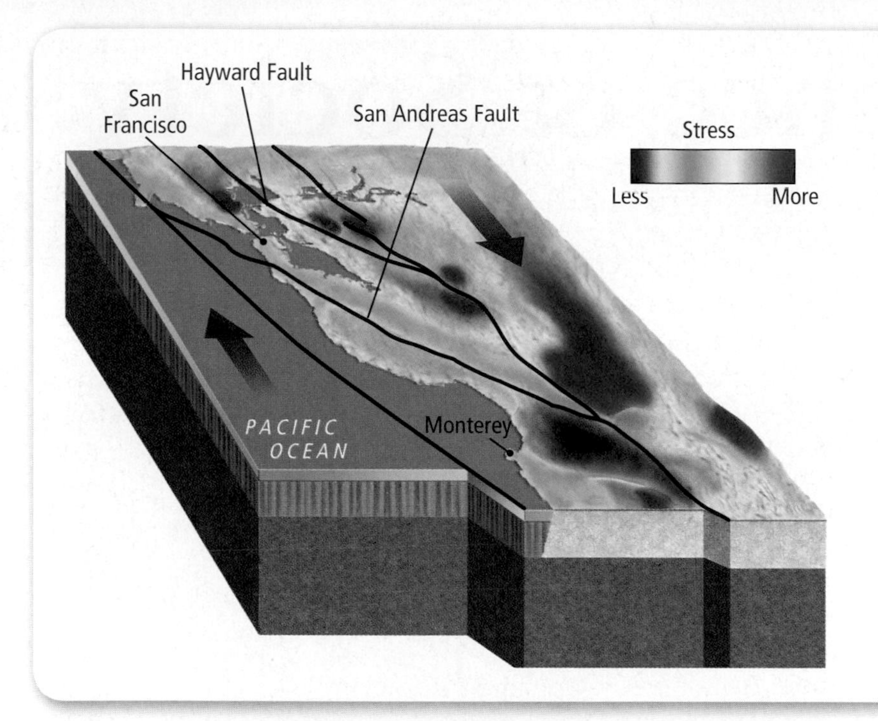

■ **Figure 19.28** Stress-accumulation maps help scientists determine the probability of an earthquake in any particular place.
Explain ***Why does stress build up in the areas indicated?***

Stress accumulation The rate at which stress builds up in rocks is another factor seismologists use to determine the earthquake probability along a section of a fault. Eventually this stress is released, generating an earthquake. Scientists use satellite-based technology such as GPS to measure the stress that accumulates along a fault. The stress accumulated in a particular part of a fault, together with the amount of stress released during the last earthquake in a particular part of the fault, can be used to develop images like **Figure 19.28.** Another factor is how much time has passed since an earthquake has struck that section of the fault.

Section 19.4 Assessment

Section Summary

- Earthquake forecasting is based on seismic history and measurements of accumulated strain.
- Earthquakes cause damage by creating vibrations that can shake Earth.
- Earthquakes can cause structural collapse, landslides, soil liquefaction, and tsunamis.
- Seismic gaps are sections along an active fault that have not experienced significant earthquakes for a long period of time.

Understand Main Ideas

1. MAIN Idea **List** some examples of how scientists determine the probability of an earthquake occurring.
2. **Summarize** the effects of the different types of hazards caused by earthquakes.
3. **Draw** before-and-after pictures of what can happen when an earthquake ruptures along a fault.
4. **Summarize** the events that lead to a tsunami.

Think Critically

5. **Assess** where an earthquake is most likely to occur: In the same place that a magnitude-7.5 earthquake occurred 20 years ago or at a location between areas that had earthquakes 20 and 60 years ago, respectively.

WRITING in Earth Science

6. Imagine you are on an international aid committee. Write a report suggesting ways to identify areas that are vulnerable to earthquakes.

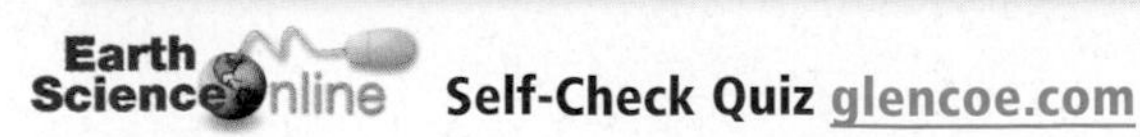

Earth Science & Society

Learning from the Past

At 5:15 on a Wednesday morning, most people were still sleeping when an earthquake struck California. The city of San Francisco was the hardest hit. It shook violently for an entire minute, toppling many buildings. In the days that followed, fire devastated entire neighborhoods.

The earthquake leveled the city Modern geologists calculate that the earthquake of April 18, 1906, had an approximate magnitude of 7.9. The total damage to San Francisco involved 490 city blocks—25,000 buildings were destroyed, 250,000 people were left homeless, and approximately 3000 were killed. Streets sank 1 m, bridges collapsed, and people were trapped under buildings.

Fires caused by broken gas lines spread through the city for three days. The efforts of the firefighters were futile, because the city's water supply had been destroyed. Contaminated drinking water put the survivors' health at risk. The food supply became limited, disease spread, and looting was rampant.

Scientists analyze the earthquake The 1906 San Francisco earthquake had monumental effect on human life and on the area. Before 1906, scientists knew very little about earthquakes and their effects. This earthquake is considered to be the beginning of modern seismology in the United States.

A theory is proposed At the time of the earthquake, the theory of plate tectonics was not yet understood, so the vast movements of land puzzled scientists. Geologists analyzed the displacement of the crust and the energy released in the movement. They proposed the elastic-rebound theory, which is still used today. They theorized that tensions had been gradually building up in Earth's crust north and south of San Francisco, along a line now known as the San Andreas Fault.

The San Francisco city hall was destroyed during the 1906 earthquake.

The tension accumulated until portions of the crust reached a limit. Like a rubber band that had been stretched too far, portions of the crust snapped. This sudden release of stored energy was the cause of the 1906 earthquake.

Preparing for the future Geologists know that tensions in the crust along the Hayward Fault, the part of the San Andreas Fault where the San Francisco earthquake is thought to have occurred, continue to build as they did before the 1906 earthquake. However, in the past century, scientists and society have worked to prepare for future earthquakes, to predict where they are likely to occur, and to design buildings that can withstand their impacts.

WRITING in Earth Science

Earthquake Expedition Create a presentation or Web site that compares and contrasts the 1906 San Francisco earthquake with the 1989 Loma Prieta earthquake. For more information on these earthquakes, visit glencoe.com.

GEOLAB

RELATE EPICENTERS AND PLATE TECTONICS

Background: The separation of P-waves and S-waves on a seismogram allows you to estimate the distance between the seismic station that recorded the data and the epicenter of that earthquake. If the distance to the epicenter, called epicentral distance, from three or more seismic stations is known, then the exact location of the earthquake's epicenter can be determined. By locating the epicenter on a map of tectonic plate boundaries, you can determine the type of plate movement that caused the earthquake.

Question: *How do seismologists locate the epicenter of an earthquake?*

Materials

U.S. map
Figure 17.16 and **Figure 19.9**
calculator
drafting compass
metric ruler

Procedure

Determine the epicenter location and the time of occurrence of an actual earthquake, using the travel times of P- and S-waves recorded at three seismic stations.

1. Read and complete the lab safety form.
2. The table gives data from three seismic stations. Use the travel-time curves in **Figure 19.9** and the P-S separation times to determine the distances from the epicenter to each seismic station. Copy the table and enter these distances in the table row *Distance from epicenter*.
3. Obtain a map of North America from your teacher. Accurately mark the three seismic station locations.
4. Use the map scale to determine the distance in centimeters represented by the *Distance from epicenter* calculated in Step 2. Enter these distances in the table row *Map distance*.
5. Use the number calculated in *Map distance* to set the compass point to a spacing that represents the distance from the first seismic station to the epicenter.

Seismic Data

Seismic station	Berkeley, CA	Boulder, CO	Knoxville, TN
P-S separation (min)	3.9	3.6	4.6
Distance from epicenter (km)			
Map distance (cm)			

6. Place the compass point on the seismic station location and draw a circle.
7. Repeat for the other two seismic stations.
8. Mark the point of intersection of the three circles. This is the epicenter of the earthquake.

Analyze and Conclude

1. **Interpret Data** Where is this epicenter located?
2. **Describe** In which major seismic belt did this earthquake occur?
3. **Interpret Data** Use **Figure 17.16** to determine which plates form the boundary associated with this earthquake.
4. **Conclude** Describe how tectonic motions caused this earthquake.

WRITING in Earth Science

Imagine You are a reporter for the newspaper based near the epicenter of this earthquake. Write an article explaining how geologic processes resulted in this earthquake. Describe whether the earthquake should have been a surprise to the residents, given its location in relation to plate boundaries.

CHAPTER 19

Study Guide

Download quizzes, key terms, and flash cards from glencoe.com.

BIG Idea **Earthquakes are natural vibrations of the ground, some of which are caused by movement along fractures in Earth's crust.**

Vocabulary	Key Concepts
Section 19.1 Forces Within Earth	
• elastic deformation (p. 529) • epicenter (p. 533) • fault (p. 530) • focus (p. 533) • plastic deformation (p. 529) • primary wave (p. 532) • secondary wave (p. 532) • seismic wave (p. 532) • strain (p. 528) • stress (p. 528)	**MAIN Idea** Faults form when the forces acting on rock exceed the rock's strength. • Stress is force per unit of area that acts on a material and strain is the deformation of a material in response to stress. • Reverse, normal, and strike-slip are the major types of faults. • The three types of seismic waves are P-waves, S-waves, and surface waves.
Section 19.2 Seismic Waves and Earth's Interior	
• seismogram (p. 534) • seismometer (p. 534)	**MAIN Idea** Seismic waves can be used to make images of the internal structure of Earth. • Seismometers are devices that record seismic wave activity on a seismogram. • Travel times for P-waves and S-waves enable scientists to pinpoint the epicenters of earthquakes. • P-waves and S-waves change speed and direction when they encounter different materials. • Analysis of seismic waves provides a detailed picture of the composition of Earth's interior.
Section 19.3 Measuring and Locating Earthquakes	
• amplitude (p. 539) • magnitude (p. 539) • modified Mercalli scale (p. 540) • moment magnitude scale (p. 540) • Richter scale (p. 539)	**MAIN Idea** Scientists measure the strength and chart the location of earthquakes using seismic waves. • Earthquake magnitude is a measure of the energy released during an earthquake and can be measured on the Richter scale. • Intensity is a measure of the damage caused by an earthquake and is measured with the modified Mercalli scale. • Data from at least three seismic stations are needed to locate an earthquake's epicenter. • Most earthquakes occur in seismic belts, which are areas associated with plate boundaries.
Section 19.4 Earthquakes and Society	
• seismic gap (p. 550) • soil liquefaction (p. 547) • tsunami (p. 548)	**MAIN Idea** The probability of an earthquake's occurrence is determined from the history of earthquakes and knowing where and how quickly strain accumulates. • Earthquake forecasting is based on seismic history and measurements of accumulated strain. • Earthquakes cause damage by creating vibrations that can shake Earth. • Earthquakes can cause structural collapse, landslides, soil liquefaction, and tsunamis. • Seismic gaps are sections along an active fault that have not experienced significant earthquakes for a long period of time.

CHAPTER 19 Assessment

Vocabulary Review

Complete the sentences below with the correct vocabulary term from the Study Guide.

1. ________ is the deformation caused by stress.

2. ________ deformation causes a material to bend and stretch.

3. The amount of energy released and the amplitude of seismic waves are measured by the scale known as the ________.

4. ________ happens when seismic vibrations cause subsurface materials to liquefy and behave like quicksand.

5. A travel-time curve shows the relationship between the travel time of a given type of wave and ________.

6. The type of seismic wave that does not pass through the outer core is called a(n) ________.

The sentences below are incorrect. Make each sentence correct by replacing the italicized word with a vocabulary term from the Study Guide.

7. A *fault plane* is a region where earthquakes are expected but none has occurred for a long time.

8. The damage caused by earthquakes is described by the *moment magnitude* scale.

9. An underwater earthquake causes the movement of a column of water, resulting in a *seismic wave.*

10. The recording made by a seismometer is called a *stress-strain curve.*

Distinguish between the vocabulary terms in each pair.

11. epicenter, focus

12. stress, strain

13. plastic deformation, elastic deformation

14. secondary wave, surface wave

15. Richter scale, moment magnitude scale

16. amplitude, magnitude

Understand Key Concepts

17. What is stress?
- **A.** speed seismic waves travel
- **B.** point at which rocks fail and generate an earthquake
- **C.** force per unit area
- **D.** measure of the deformation of rocks

Use the diagram below to answer Questions 18–20.

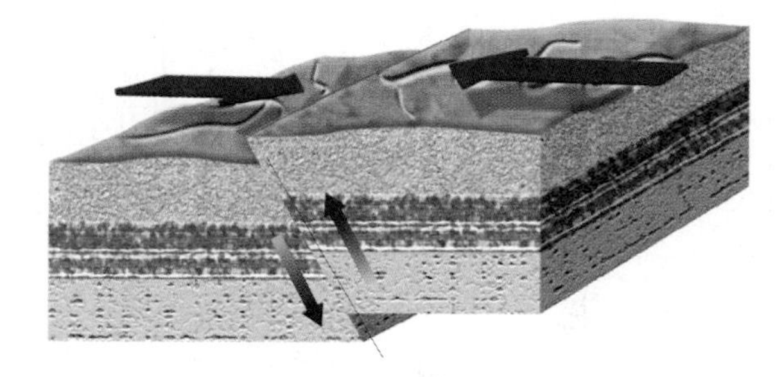

18. Which type of fault is shown?
- **A.** reverse
- **B.** normal
- **C.** shear
- **D.** strike-slip

19. Which type of force caused this fault to form?
- **A.** compression
- **B.** tension
- **C.** shear
- **D.** divergent

20. In which direction is the movement in this type of fault?
- **A.** horizontal
- **B.** horizontal and vertical
- **C.** side-to-side
- **D.** vertical

21. What happens to a rock that undergoes elastic deformation once the stress is removed?
- **A.** It returns to its original shape.
- **B.** It breaks to generate an earthquake.
- **C.** It undergoes plastic deformation.
- **D.** It does not change shape.

22. Which type of geologic material is most prone to liquefaction?
    A. granite
    B. metamorphic rock
    C. soil and loose sediment
    D. lava flows

Use the figure below to answer Questions 23–25.

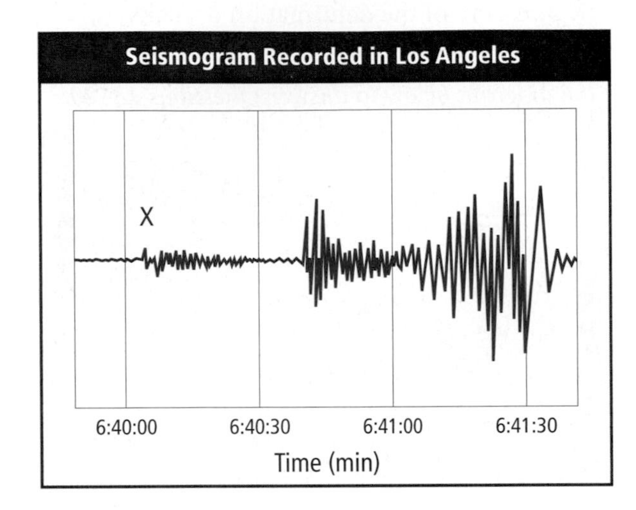

23. Which type of wave is labeled “X”?
    A. P-wave
    B. S-wave
    C. surface wave
    D. shear wave

24. At what time did the surface waves arrive at this station?
    A. 6:40:00
    B. 6:40:05
    C. 6:40:33
    D. 6:41:10

25. What can the difference in travel times between P- and S-waves be used to determine?
    A. how far away the epicenter was
    B. the type of fault
    C. the depth of the earthquake
    D. whether the core is liquid

26. Which seismic hazard is a form of structural failure?
    A. tsunami
    B. pancaking
    C. soil liquefaction
    D. seismic gap

Constructed Response

Use the figure below to answer Questions 27–29.

Some Earthquakes in Recent History

Location	Year	Richter Magnitude
Chile	1960	8.5
California	1906	7.9
Alaska	1964	8.6
Colombia	1994	6.8
Taiwan	1999	7.6

27. **Calculate** How much more energy was released by the Chilean earthquake than the Taiwan earthquake?

28. **Approximate** How much larger was the amplitude of the waves generated by the Alaskan earthquake than the Taiwan earthquake?

29. **Classify** the earthquake locations with the type of plate boundary, and suggest how the tectonic processes were probably related.

30. **Name** five states with high seismic risk.

31. **Compare and contrast** a tsunami and a surface wave.

32. **Explain** why scientists need measurements from more than two seismometers to determine the exact location of an earthquake. Make a diagram similar to **Figure 19.17** to support your answer.

33. **Describe** three different ways earthquakes can cause damage or cause harm to people.

Think Critically

34. **Summarize** the factors considered when assessing seismic risk.

35. **Evaluate** how earthquake intensity is related to the type of fault.

36. **Draw** the basic components of a seismometer.

Earth Science Online Chapter Test glencoe.com

Use the figure below to answer Questions 37 and 38.

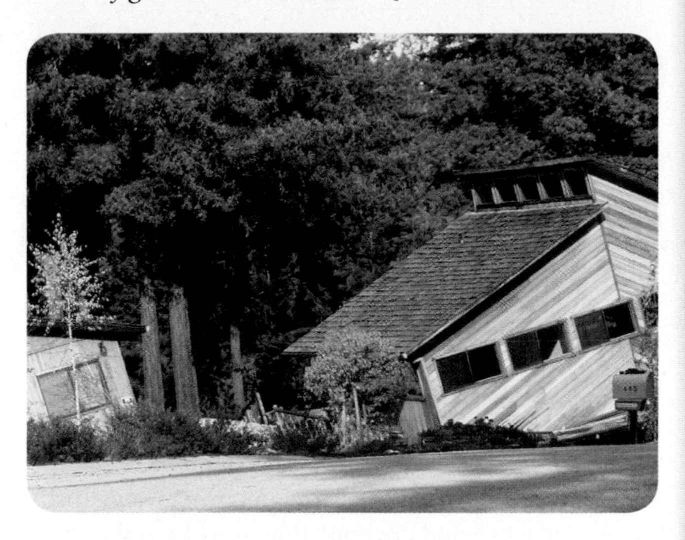

37. **Appraise** the specific type of earthquake damage shown, and propose the possible causes.

38. **Infer** the intensity of the earthquake that caused this damage, using the modified Mercalli scale.

39. **Explain** why there are three different ways to measure the size of earthquakes.

40. **Critique** this statement: If a certain area has not had an earthquake for over a hundred years, it is not likely to ever occur.

41. **Design** a house that would be structurally sound in an earthquake. Label the features, and explain how they would help prevent earthquake damage.

42. **Suggest** cost-effective ways of saving lives in an earthquake in the United States. How might your strategy be different in California and Florida?

Concept Mapping

43. Use the following terms to complete the concept map: *reverse faults, tension, types of stress, strike-slip faults, compression, shear, causes,* and *normal faults.*

Challenge Question

44. **Explain** why most earthquakes are shallow. Use the concept of plastic deformation and brittle failure and your knowledge about the temperature of Earth's interior.

Additional Assessment

45. **WRITING in Earth Science** Imagine you live along an active fault. Write a disaster plan for your school, giving guidelines on what to do before, during, and after an earthquake. Include a list of disaster kit supplies.

DBQ Document–Based Questions

Data obtained from: Fukao Y., S. Widiyantoro, and M. Obayashi. 2001. Stagnant slabs in the upper and lower mantle transition region. *Reviews of Geophysics* 39 (3): 291–323.

The figure below shows a cross section of Earth extending from the surface to the boundary between the core and the mantle. The colors show how the speed of seismic waves differs from the expected value for waves at that depth. This cross section is taken across the subduction zone off the west coast of South America. West is left, and east is right.

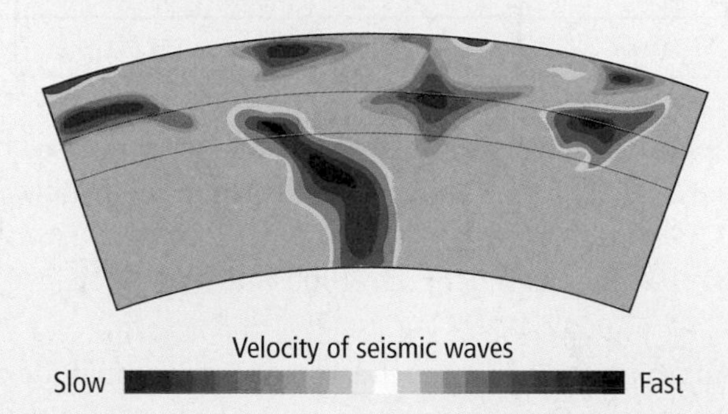

46. What properties of subsurface material could cause seismic waves to move quickly through the blue areas and more slowly through the red areas?

47. Thinking about plate tectonics, what portion of the diagram could represent a subducting plate with molten rock rising from the subduction zone to form volcanoes?

Cumulative Review

48. What is the most common extrusive igneous rock? **(Chapter 5)**

49. Describe three processes that affect the salinity of the oceans. **(Chapter 15)**

Standardized Test Practice

Multiple Choice

1. Why is only 61 percent of the northern hemisphere covered with water?
   A. Most landmasses are in the northern hemisphere.
   B. Most landmasses are in the southern hemisphere.
   C. The northern hemisphere is colder thus allowing less water to flow there.
   D. Gravity causes more water to settle in the southern hemisphere.

Use the table below to answer Questions 2 and 3.

Some Earthquakes in Recent History		
Location	**Year**	**Richter Magnitude**
Chile	1960	8.5
California	1906	7.9
Alaska	1964	8.6
Colombia	1994	6.8
Taiwan	1999	7.6

2. Approximately how much more energy was released by the earthquake in Chile than the earthquake in Taiwan?
   A. 2 times as much C. 32 times as much
   B. 10 times as much D. 1000 times as much

3. Approximately how much larger was the amplitude of the waves generated by the earthquake in Alaska than the earthquake in Taiwan?
   A. 2 times as large C. 100 times as large
   B. 10 times as large D. 1000 times as large

4. Who were the first people to propose the idea that Earth's landmasses at one time were all connected?
   A. explorers C. scientists
   B. mathematicians D. mapmakers

5. Which does NOT have any effect on Earth's tides?
   A. Earth C. the Sun
   B. the Moon D. the atmosphere

6. What is formed as turbidity currents drop sediment?
   A. continental rise C. abyssal plains
   B. continental slope D. deep-sea trenches

Use the maps below to answer Questions 7 and 8.

Ocean currents

Wind currents

7. What can be concluded by comparing the maps?
   A. Surface wind currents flow mostly to the east, and surface ocean currents flow mostly to the west.
   B. Surface wind currents flow mostly to the west, and surface ocean currents flow mostly to the east.
   C. The direction of surface ocean currents is opposite the direction of surface wind currents.
   D. The direction of surface ocean currents is related to the direction of surface wind currents.

8. The Coriolis effect is the rightward curvature of winds in the northern hemisphere and the leftward curvature of winds in the southern hemisphere. What causes the Coriolis effect?
   A. the intersection of warm and cool ocean currents
   B. the revolution of Earth around the Sun
   C. the rotation of Earth on its axis
   D. the seasonal changes in global temperature

9. Most sedimentary rocks are formed by
   A. uplifting and melting
   B. compaction and cementation
   C. eruption of volcanoes
   D. changes deep within Earth

10. Which type of volcano is potentially the most dangerous to humans and the environment?
    A. shield volcano C. cinder cone volcano
    B. composite volcano D. compact volcano

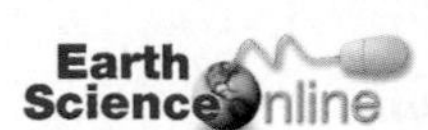

Short Answer

Use the illustration below to answer Questions 11 and 12.

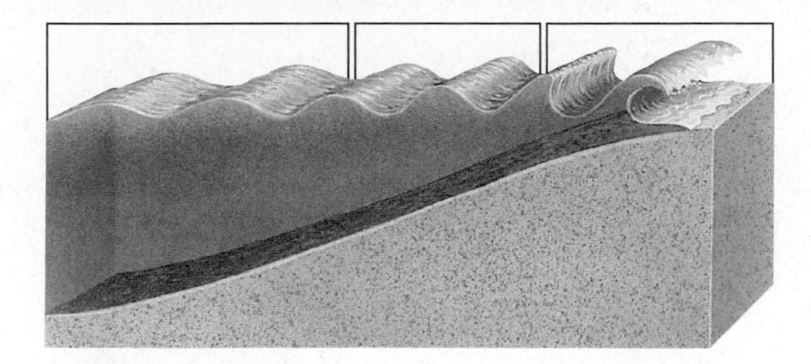

11. Describe the changes that occur as waves move closer to shore.
12. Contrast water movement and energy movement in an ocean wave.
13. What does the location of mountains on the seafloor that are not near any active volcanism suggest?
14. How does convection in the mantle cause plate motions?
15. Describe tephra and its two possible sources.
16. How are the continental shelves, covered with water after the last ice age, now benefiting humans?

Reading for Comprehension

Earthquake Detection

The belief that animals can detect incoming earthquakes has been around for centuries. In 373 B.C., historians recorded that animals, including rats, snakes, and weasels, deserted the Greek city of Helice just days before an earthquake devastated the place. Similar accounts have surfaced across the centuries since.

Catfish moving violently, chickens that stop laying eggs, and bees leaving their hive in a panic have been reported. But precisely what animals sense is a mystery. One theory is that wild and domestic creatures can feel Earth vibrate before humans can. Other ideas suggest that they detect electric changes in the air or gas released from Earth. Earthquakes are a sudden phenomenon. Seismologists have no way of knowing exactly when or where the next one will hit. An estimated 500,000 detectable earthquakes occur in the world each year. Of those, 100,000 can be felt by humans, and 100 cause damage. Researchers have long studied animals in hopes of discovering what they hear or feel before an earthquake in order to use that sense as a prediction tool. American seismologists are skeptical. Even though there have been documented cases of strange animal behavior prior to earthquakes, according to the USGS, a reproducible connection between a specific behavior and the occurrence of an earthquake has never been made.

Article obtained from: Mott, M. Can animals sense earthquakes? *National Geographic News.* November 11, 2003.

17. What can be inferred from this passage?
    A. Animals can predict earthquakes because they can feel the vibrations before humans.
    B. Animals cannot predict earthquakes.
    C. Further study and research is needed before it can be confirmed or denied that animals can predict earthquakes.
    D. Animals have been predicting earthquakes for centuries.

18. Which was NOT an animal behavior cited as proof that animals can predict earthquakes?
    A. catfish moving violently
    B. chickens laying eggs
    C. bees leaving their hives
    D. snakes deserting a city

NEED EXTRA HELP?																
If You Missed Question . . .	1	2	3	4	5	6	7	8	9	10	11	12	13	14	15	16
Review Section . . .	15.1	19.3	19.3	17.1	15.2	16.2	15.3	12.2	6.1	18.3	15.3	15.3	16.2	17.4	18.2	16.2

CHAPTER 20 Mountain Building

BIG Idea Mountains form through dynamic processes which crumple, fold, and create faults in Earth's crust.

20.1 Crust-Mantle Relationships

MAIN Idea The height of mountains is controlled primarily by the density and thickness of the crust.

20.2 Orogeny

MAIN Idea Convergence causes the crust to thicken and form mountain belts.

20.3 Other Types of Mountain Building

MAIN Idea Mountains on the ocean floor and some mountains on continents form through processes other than convergence.

GeoFacts

- The layers of a mountain record the vast geologic history of the region.
- Fossils of marine organisms have been found at the top of Mount Everest.
- The Himalayas are geologically young mountains—they are still growing.

Start-Up Activities

LAUNCH Lab

How does crust displace the mantle?

Continental and oceanic crust have different densities. Each displaces the mantle.

Procedure

1. Read and complete the lab safety form.
2. Obtain 3 **wood blocks** from your teacher. Determine the mass, volume, and density of each block. Record all of these values in a data table.
3. Half fill a **clear plastic container** with **water.** Place both of the 2-cm-thick blocks in the container.
4. Using a **ruler,** measure and record how much of each block is above the water surface.
5. Replace the 2-cm-thick blocks with the 4-cm-thick softwood block.
6. Measure and record how much of the block is above the water surface.

Analysis

1. **Describe** How do density and thickness affect the height of flotation?
2. **Infer** Which block represents oceanic crust? Continental crust?

Visit glencoe.com to

- study entire chapters online;
- explore Concepts In Motion animations:
  - Interactive Time Lines
  - Interactive Figures
  - Interactive Tables
- access Web Links for more information, projects, and activities;
- review content with the Interactive Tutor and take Self-Check Quizzes.

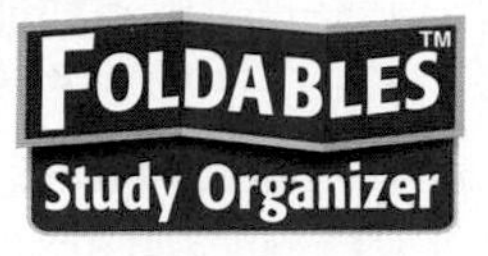

Mountain Building Processes Make this Foldable to compare the processes that form plate boundary and non-plate boundary mountains.

STEP 1 Fold a sheet of paper in half lengthwise.

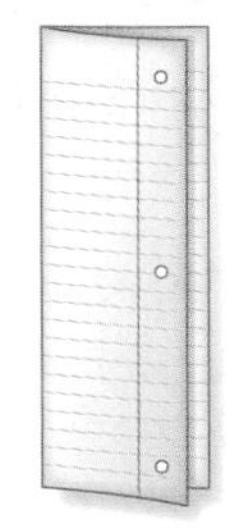

STEP 2 Fold the top down about 4 cm.

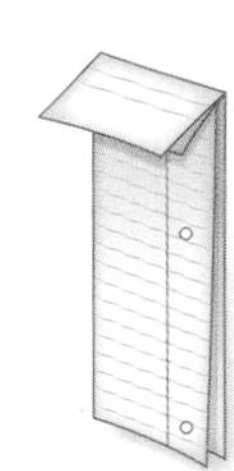

STEP 3 Unfold and draw lines along the fold lines. Label the columns *Plate Boundary Mountains* and *Non-Plate Boundary Mountains.*

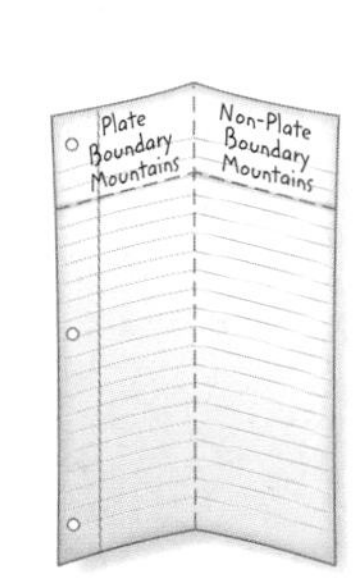

FOLDABLES **Use this Foldable with Sections 20.2 and 20.3.** As you read, record the different types of mountains and the processes that form them. Include examples and their locations.

Section 20.1

Objectives

- **Describe** the elevation distribution of Earth's surface.
- **Explain** isostasy and how it pertains to Earth's mountains.
- **Describe** how Earth's crust responds to the addition and removal of mass.

Review Vocabulary

equilibrium: a state of balance between opposing forces

New Vocabulary

topography
isostasy
root
isostatic rebound

Crust-Mantle Relationships

MAIN Idea The height of mountains is controlled primarily by the density and thickness of the crust.

Real-World Reading Link When you sit in an inflatable raft, your mass causes the raft to sink deeper into the water. When you get out of the raft, it rises. Similarly, when mountains erode, the crust rises to compensate for the mass that is removed.

Earth's Topography

When you look at a globe or a map of Earth's surface, you immediately notice the oceans and continents. From these representations of Earth, you can estimate that about 71 percent of Earth's surface is below sea level, and about 29 percent lies above sea level. What is not obvious from most maps and globes, however, is the variation in elevations of the crust, which is referred to as its **topography.** Recall from Chapter 2 that topographic maps show an area's hills and valleys. When a very large map scale is used, such as the one in **Figure 20.1,** the topography of Earth's entire crust can be shown. When Earth's topography is plotted on a graph such as **Figure 20.2,** a pattern in the distribution of elevations emerges. Note that most of Earth's elevations cluster around two main ranges of elevation. Above sea level, elevation averages around 0 to 1 km. Below sea level, elevations range between –4 and –5 km. These two ranges dominate Earth's topography and reflect the basic differences in density and thickness between continental and oceanic crust.

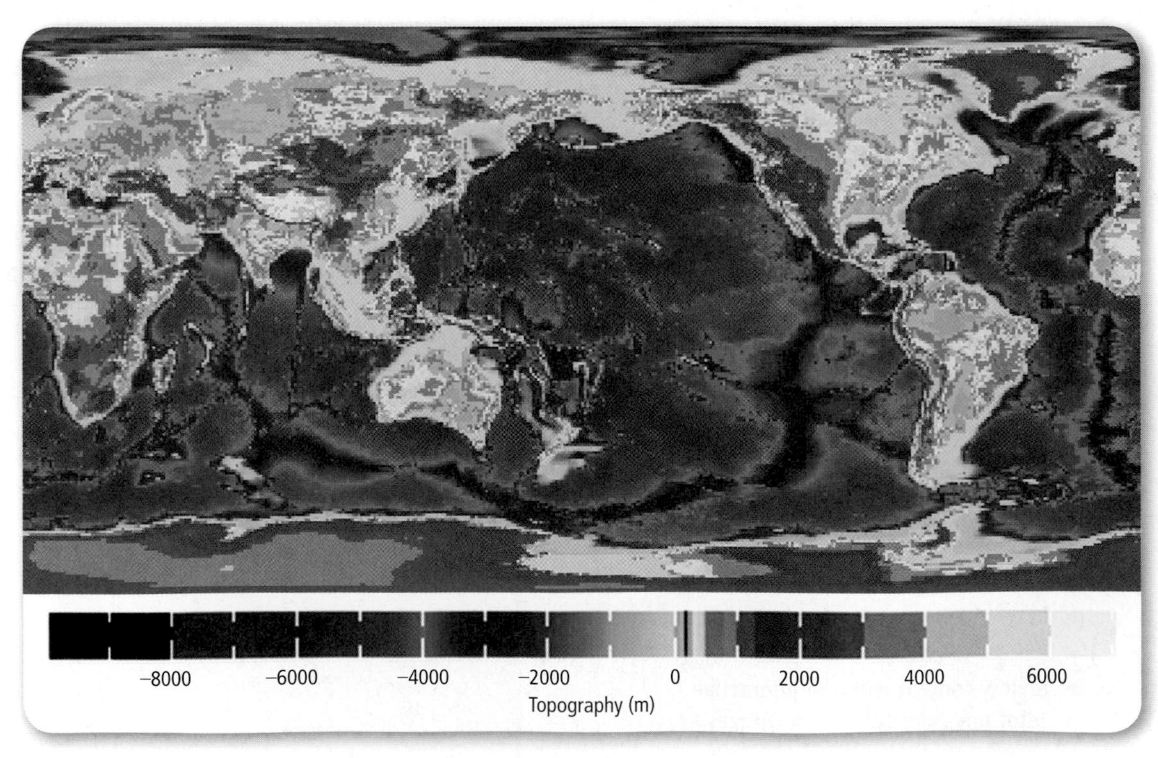

■ **Figure 20.1** Topographic maps show differences in elevation on Earth's surface.
***Interpret** the map to determine Earth's highest and lowest elevations. Where are they?*

Continental crust You observed in the Launch Lab that blocks of wood with different densities displaced different amounts of water, and thus floated at various heights above the surface of the water. You observed that blocks of higher density displaced more water than blocks of lower density. Recall from Chapter 1 that oceanic crust is composed mainly of basalt, which has an average density of about 2.9 g/cm^3. Continental crust is composed of more granitic rock, which has an average density of about 2.8 g/cm^3. The slightly higher density of oceanic crust causes it to displace more of the mantle—which has a density of about 3.3 g/cm^3—than the same thickness of continental crust.

Differences in elevation, however, are not caused by density differences alone. Also recall from the Launch Lab that when the thicker wood block was placed in the water, it displaced more water than the other two blocks. However, because of its density, it floated higher in the water than the hardwood block. Continental crust, which is thicker and less dense than oceanic crust, behaves similarly. It extends deeper into the mantle because of its thickness, and it rises higher above Earth's surface than oceanic crust because of its lower density, as shown in **Figure 20.3.**

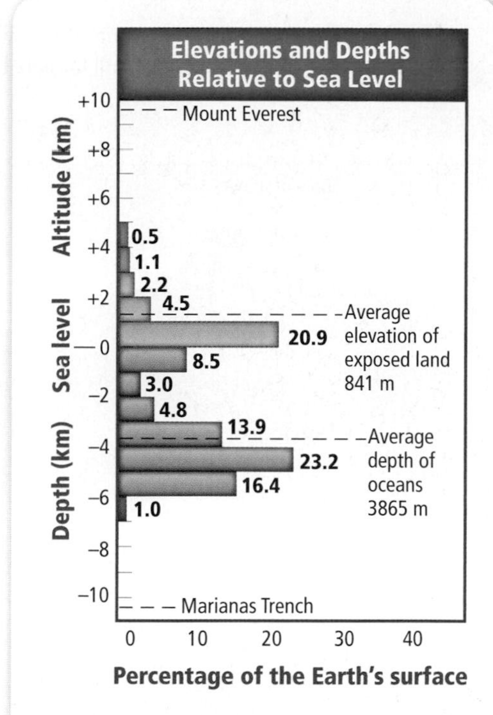

■ **Figure 20.2** About 29 percent of Earth is land and 71 percent is water.
Interpret ***At what elevation does most of Earth's surface lie? At what depth?***

Isostasy

The displacement of the mantle by Earth's continental and oceanic crust is a condition of equilibrium called **isostasy** (i SAHS tuh see). The crust and mantle are in equilibrium when the downward force of gravity on the mass of crust is balanced by the upward force of buoyancy that results from displacement of the mantle by the crust. This balance might be familiar to you if you have ever watched people get in and out of a small boat. As the people boarded the boat, it sank deeper into the water. Conversely, as the people got out of the boat, it displaced less water and floated higher in the water. A similar sinking and rising that results from the addition and removal of mass occurs within Earth's crust. Gravitational and seismic studies have detected thickened areas of continental material, called **roots,** that extend into the mantle below Earth's mountain ranges.

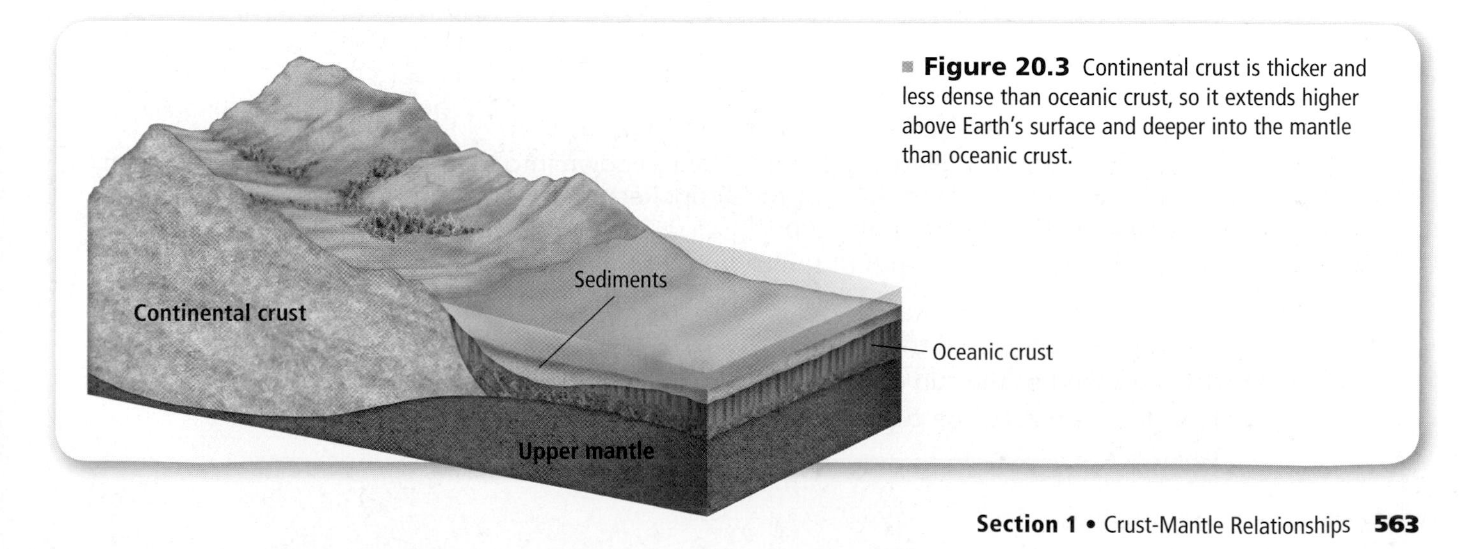

■ **Figure 20.3** Continental crust is thicker and less dense than oceanic crust, so it extends higher above Earth's surface and deeper into the mantle than oceanic crust.

■ **Figure 20.4** According to the principle of isostasy, parts of Earth's crust rise or subside until they are buoyantly supported by their roots.

Concepts In Motion

Interactive Figure To see an animation of isostasy, visit glencoe.com.

Massive roots underlie mountains.

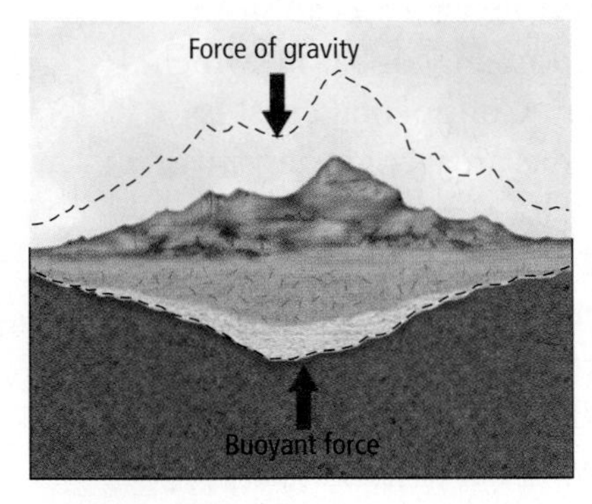

As erosion takes place, the mountain loses mass. The root rises in response to this decrease in mass.

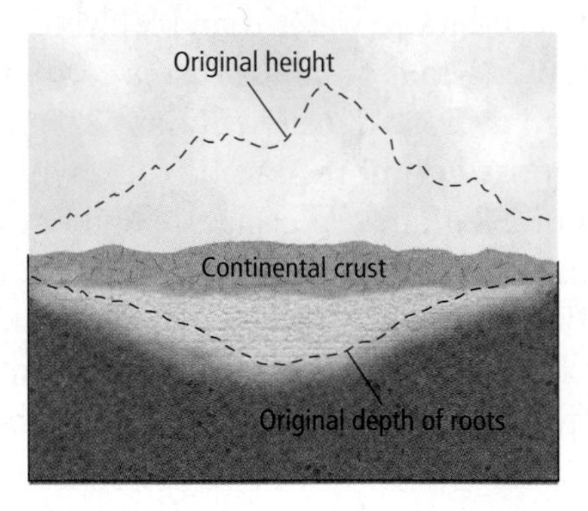

When the mountain erodes to the average continental thickness, both root and mountain are gone.

Mountain roots A mountain range requires large roots to counter the enormous mass of the range above Earth's surface. **Figure 20.4** illustrates how, according to the principle of isostasy, parts of the crust rise or subside until these parts are buoyantly supported by their roots. Continents and mountains are said to float on the mantle because they are less dense than the underlying mantle. They project into the mantle to provide the necessary buoyant support. What do you think happens when mass is removed from a mountain or mountain range? If erosion continues, the mountain will eventually disappear, exposing the roots.

MiniLab

Model Isostatic Rebound

How can isostatic rebound be measured? Isostatic rebound is the process through which the underlying material rises when the overlying mass is removed.

Procedure

1. Read and complete the lab safety form.
2. Working in groups, fill a **1000-mL beaker** with **corn syrup.**
3. Using a pencil, push a **paper or plastic cup** (open side up) down into the syrup far enough so the cup is three-fourths of the way to the bottom of the syrup. Record the depth of the bottom of the cup relative to the surface, then let go of the cup.
4. At 5-s intervals, record the new depth of the bottom of the cup.

Analysis

1. **Describe** In which direction did the cup move? Why?
2. **Explain** why the speed of the cup changes as it moves.
3. **Infer** If enough time passes, the cup stops moving. Why?

■ **Figure 20.5** Before erosion, the Appalachian Mountains were thousands of meters taller than they are now. Because of isostatic rebound, as the mountains eroded, the deep root also rose thousands of meters closer to the surface. The mountains visible today are only the roots of an ancient mountain range. They too are being eroded and will someday resemble the craton in northern Canada.

Isostasy and Erosion

The Appalachian Mountains, shown in **Figure 20.5,** in the eastern United States formed hundreds of millions of years ago when the North American continent collided with Europe and Africa. Rates of erosion on land are such that these mountains should have been completely eroded millions of years ago. Why, then, do these mountains still exist? As the mountains rose above Earth's surface, deep roots formed until isostatic equilibrium was achieved and the mountains were buoyantly supported. As peaks eroded, the mass decreased. This allowed the roots themselves to rise and erode.

A balance between erosion and the decrease in the size of the root will continue for hundreds of millions of years until the mountains disappear and the roots are exposed at the surface. This slow process of the crust's rising as the result of the removal of overlying material is called **isostatic rebound.** Erosion and rebound allows metamorphic rocks formed at great depths to rise to the top of mountain ranges such as the Appalachians.

PROBLEM-SOLVING LAB

Make and Use a Graph

Can you get a rebound? The rate of isostatic rebound changes over time. An initially rapid rate often declines to a very slow rate. The data shown indicates rebound after the North American ice sheet melted 10,000 years ago.

Analysis

1. Plot a graph with *Years before present* on the *x*-axis and *Total amount of rebound* on the *y*-axis.
2. **Describe** how the rate of isostatic rebound decreases with time by studying your graph.

Think Critically

3. **Identify** the percentage of the total rebound that occurred during the first 2000 years.
4. **Predict** how much rebound will still occur and approximately how long this will take.
5. **Compare and contrast** mountain erosion to glaciation in terms of isostatic rebound.

Data and Observations

Isostatic Rebound Data					
Years before present	8000	6000	4000	2000	0
Total amount of rebound (m)	50	75	88	94	97

■ **Figure 20.6** Mount Everest, a peak in Asia, is currently the highest mountain on Earth. A deep root supports its mass. Scientists have determined that Mount Everest has a root that is nearly 70 km thick.

Seamounts Crustal movements resulting from isostasy are not restricted to Earth's continents. For example, recall from Chapter 18 that hot spots under the ocean floor can produce individual volcanic mountains. When these mountains are underwater, they are called seamounts. On the geologic time scale, these mountains form very quickly. What do you think happens to the seafloor after these seamounts form? The seamounts are added mass. As a result of isostasy, the oceanic crust around these peaks displaces the underlying mantle until equilibrium is achieved.

You have just learned that the elevation of Earth's crust depends on the thickness of the crust as well as its density. You also learned that a mountain peak is countered by a root. Mountain roots can be many times as deep as a mountain is high. Mount Everest, shown in **Figure 20.6,** towers nearly 9 km above sea level and is the tallest peak in the Himalayas. Some parts of the Himalayas are underlain by roots that are nearly 70 km thick. As India continues to push northward into Asia, the Himalayas, including Mount Everest, continue to grow in height. Currently, the combined thickness is approximately equal to 868 football fields lined up end-to-end. Where do the immense forces required to produce such crustal thickening originate? You will read about these forces in Section 20.2.

Section 20.1 Assessment

Section Summary

- The majority of Earth's elevations are either 0 to 1 km above sea level or 4 to 5 km below sea level.
- The mass of a mountain above Earth's surface is supported by a root that projects into the mantle.
- The addition of mass to Earth's crust depresses the crust, while the removal of mass from the crust causes the crust to rebound in a process called isostatic rebound.

Understand Main Ideas

1. MAIN Idea **Relate** density and crustal thickness to mountain building.
2. **Describe** the pattern in Earth's elevations, and explain what causes the pattern in distribution.
3. **Explain** why isostatic rebound slows down over time.
4. **Infer** why the crust is thicker beneath continental mountain ranges than it is under flat-lying stretches of landscape.

Think Critically

5. **Apply** the principle of isostasy to explain how the melting of the ice sheets that once covered the Great Lakes has affected the land around the lakes.
6. **Consider** how the term *root* applies differently to mountains than it does to plants.

MATH in Earth Science

7. Suppose a mountain is being uplifted at a rate of 1 m every 1000 y. It is also being eroded at a rate of 1 cm/y. Is this mountain getting larger or smaller? Explain.

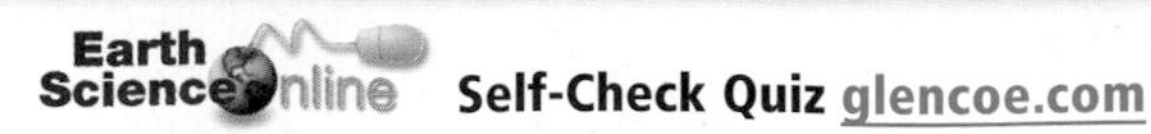
Self-Check Quiz glencoe.com

Section 20.2

Objectives

- **Identify** orogenic processes.
- **Compare and contrast** the different types of mountains that form along convergent plate boundaries.
- **Explain** how the Appalachian Mountains formed.

Review Vocabulary

island arc: a line of islands that forms over a subducting oceanic plate

New Vocabulary

orogeny
compressive force

Orogeny

MAIN Idea Convergence causes the crust to thicken and form mountain belts.

Real-World Reading Link When you push a snowplow or shovel through snow, the compression creates a thick pile of snow. Similarly, the compression caused by plate tectonics thickens the crust to form mountains.

Mountain Building at Convergent Boundaries

Orogeny (oh RAH jun nee) refers to all processes that form mountain ranges. In earlier chapters you read about many of these processes. Recall what you read in Chapter 6 about metamorphism and how rocks can be squeezed and folded. In Chapter 18 you read about rising magma and igneous intrusions, and in Chapter 19 you read about movement along faults. The result of all these processes can be broad, linear regions of deformation that you know as mountain ranges, but in geology are also known as orogenic belts. Look at **Figure 20.7** and recall from Chapter 17 what you read about the interaction of converging tectonic plates at their boundaries. Most orogenic belts are associated with convergent plate boundaries. Here, **compressive forces** squeeze the crust and cause intense deformation in the form of folding, faulting, metamorphism, and igneous intrusions. In general, the tallest and most varied orogenic belts form at convergent boundaries. However, interactions at each type of convergent boundary create different types of mountain ranges.

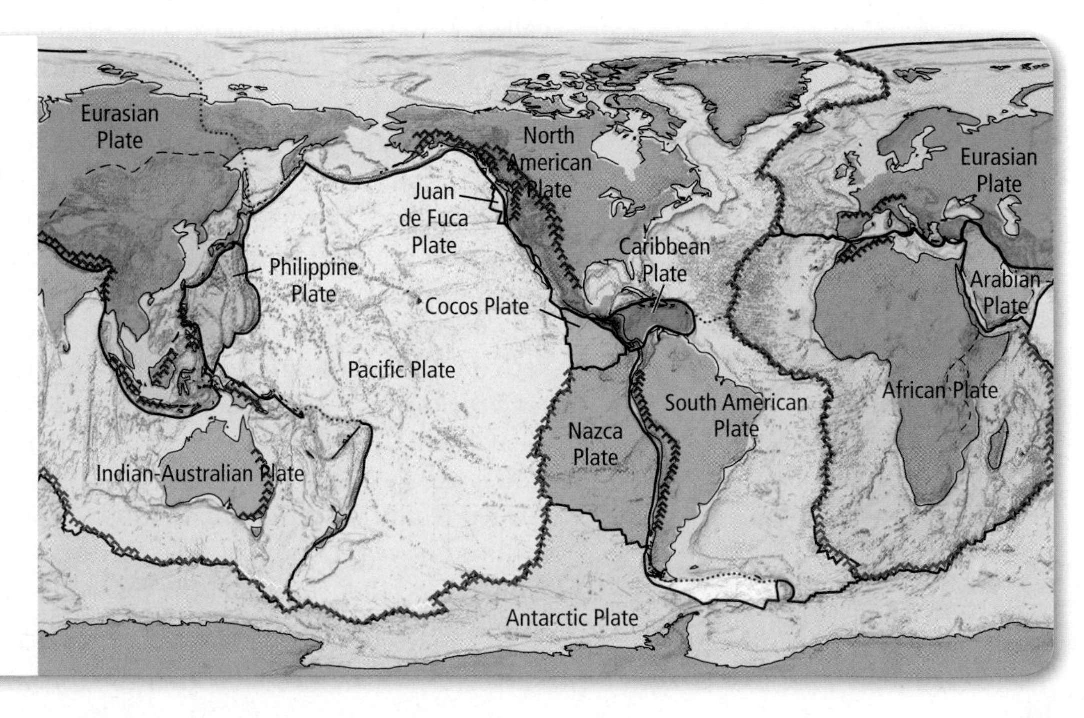

Figure 20.7 Most of Earth's mountain ranges (blue and red peaks on the map) formed along plate boundaries. **Identify** *the mountain ranges that lie along the South American Plate by comparing a world map with the one shown here.*

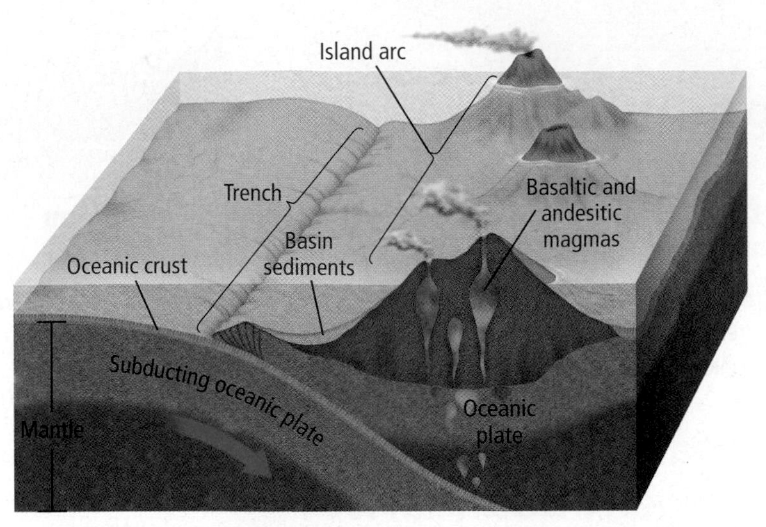

■ **Figure 20.8** Convergence between two oceanic plates results in the formation of individual volcanic peaks that make up an island arc complex. Mount Mazinga is one of several volcanic peaks that make up the island arc complex in the southern Caribbean known as the Lesser Antilles.

Lesser Antilles island arc

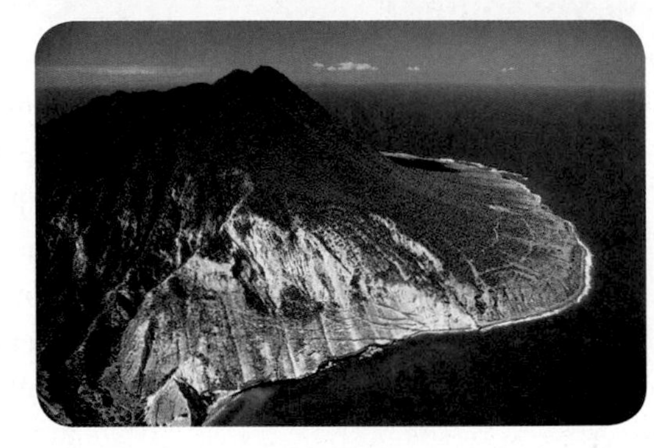

Mount Mazinga in the Lesser Antilles

Concepts In Motion
Interactive Figure To see an animation of island formation, visit glencoe.com.

Oceanic-oceanic convergence Recall from Chapter 17 that when an oceanic plate converges with another oceanic plate, one plate descends into the mantle to create a subduction zone. As parts of the subducted plate melt, magma is forced upward where it can form a series of individual volcanic peaks that together are called an island arc complex. The Aleutian Islands off the coast of Alaska and the Lesser Antilles in the Caribbean are examples of island arc complexes. The tectonic relationships and processes associated with oceanic-oceanic convergence are detailed in **Figure 20.8.**

What kinds of rocks make up island arc complexes? Often, they are a jumbled mixture of rock types. They are partly composed of the basaltic and andesitic magmas that you read about in Chapter 18. In addition to these volcanic rocks, some large island arc complexes contain sedimentary rocks. How do these sedimentary rocks eventually become part of a mountain? Recall from Chapter 17 that between an island arc and a trench is a depression, called a basin. This basin fills with sediments that have been eroded from the island arc. If subduction continues for tens of millions of years, some of these sediments can be uplifted, folded, faulted, and thrust against the existing island arc. This ultimately forms complex new masses of sedimentary and volcanic rocks. Parts of Japan formed in this way.

CAREERS IN EARTH SCIENCE

Petrologist A petrologist studies the composition and formation of rocks. Generally, petrologists specialize in a particular type of rock: igneous, sedimentary, or metamorphic. To learn more about Earth science careers, visit glencoe.com.

Oceanic-continental convergence Oceanic-continental boundaries are similar to oceanic-oceanic boundaries in that convergence along both creates subduction zones and trenches. Unlike convergence at oceanic-oceanic boundaries, convergence between oceanic and continental plates produces mountain belts that are much bigger and more complicated than island arc complexes. When an oceanic plate converges with a continental plate, the descending oceanic plate forces the edge of the continental plate upward. This uplift marks the beginning of orogeny. In addition to uplift, compressive forces can cause the continental crust to fold and thicken. As the crust thickens, higher mountains form. Deep roots develop to support these enormous masses of rocks.

Recall from Chapter 18 that volcanic mountains can form over the subducting plate. As illustrated in **Figure 20.9,** sediments eroded from such volcanic mountains can fill the low areas between the trench and the coast. These sediments, along with ocean sediments and material scraped off the descending plate, are shoved against the edge of the continent to form a jumble of highly folded, faulted, and metamorphosed rocks. The metamorphosed rocks shown in **Figure 20.9** are from Cwm Tydu, Cardigan Bay, Wales. They formed when the landmass that is now the United Kingdom collided with the North American Plate millions of years ago.

Reading Check **Compare** convergence at oceanic-continental boundaries with convergence at oceanic-oceanic boundaries.

Vocabulary
Science usage v. Common usage
Uplift
Science usage: to cause a portion of Earth's surface to rise above adjacent areas

Common usage: to improve the spiritual, social, or intellectual condition

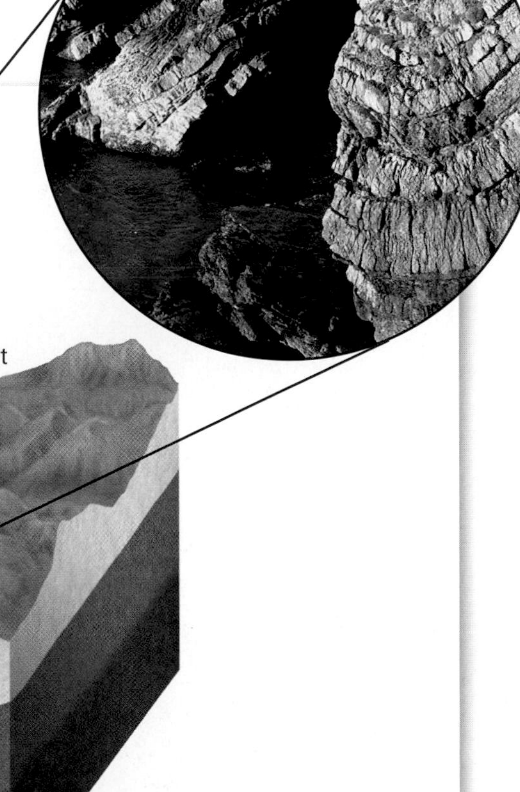

■ **Figure 20.9** At an oceanic-continental boundary, compression causes continental crust to fold and thicken. Igneous activity and metamorphism are also common along such boundaries. This sample of metamorphosed rock formed as the result of convergence of an oceanic plate with a continental plate.

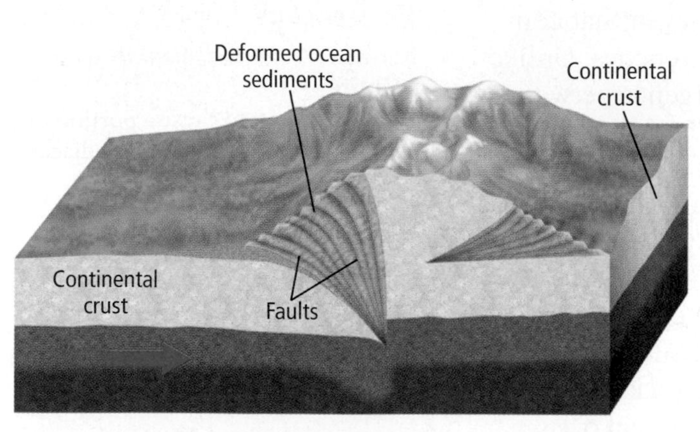

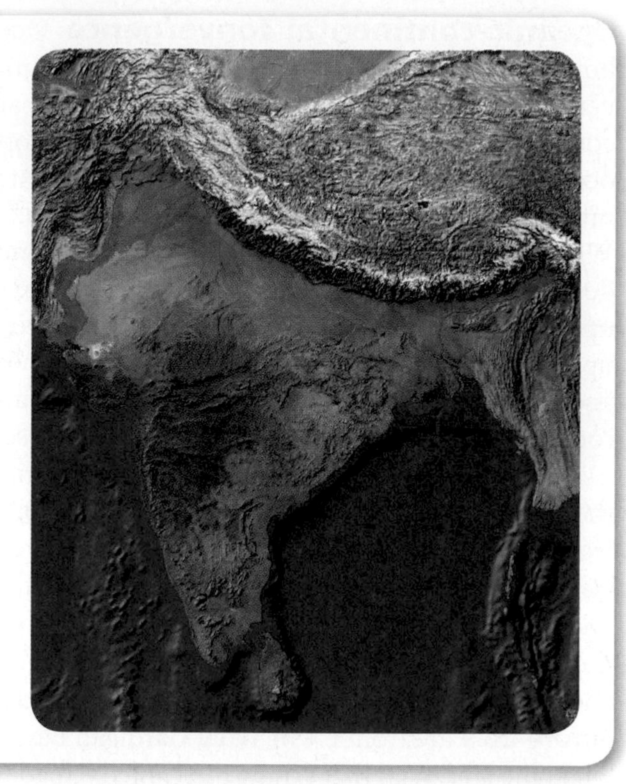

■ **Figure 20.10** Intense folding and faulting along continental-continental boundaries produce some of the highest mountain ranges on Earth. The Himalayas are the result of the convergence between the Indian and Eurasian plates.

Concepts In Motion

Interactive Figure To see an animation of convergence, visit glencoe.com.

Continental-continental convergence Earth's tallest mountain ranges, including the Himalayas, are formed at continental-continental plate boundaries. Because of its relatively low density, continental crust cannot be subducted into the mantle when two continental plates converge. Instead, the low-density continental crust becomes highly folded, faulted, and thickened as shown in **Figure 20.10.** Compressional forces break the crust into thick slabs that are thrust onto each other along low-angle faults. This process can double the thickness of the deformed crust. Deformation can also extend laterally for hundreds of kilometers into the continents involved. For example, studies of rocks in southern Tibet suggest that the original edge of Asia has been pushed approximately 2000 km eastward since the collision of Indian and Eurasian plates. The magma that forms as a result of continental-continental mountain building solidifies beneath Earth's surface to form granite batholiths.

Reading Check **Explain** why continental crust does not subduct.

Marine sedimentary rock Another common characteristic of the mountains that form when two continents collide is the presence of marine sedimentary rock near the mountains' summits. Such rock forms from the sediments deposited in the ocean basin that existed between the continents before their collision. For example, Mount Godwin Austen (also known as K2) in the western Himalayas is composed of thousands of meters of marine limestone that sits upon a granite base. The limestone represents the northern portions of the old continental margin of India that were pushed up and over the rest of the continent when India began to collide with Asia about 50 mya.

The Appalachian Mountains—A Case Study

Recall from Chapter 17 that Alfred Wegener used the matching rocks and geologic structures in the Appalachians and mountains in Greenland and northern Europe to support his hypothesis of continental drift. In addition to Wegener, many other scientists have studied the Appalachians. Based on these studies, geologists have divided the Appalachians into several distinct regions, as illustrated in **Figure 20.11.** Each region is characterized by rocks that show different degrees of deformation. For example, rocks of the Valley and Ridge Province are highly folded sedimentary rocks. In contrast, the rocks of the Piedmont Province consist of older, deformed metamorphic and igneous rocks that are overlain by relatively undeformed sedimentary layers. These regions, pictured in **Figure 20.12,** are different because they formed in different ways.

The early Appalachians The tectonic history of the Appalachians is illustrated in **Figure 20.13.** It began about 800 to 700 mya when ancestral North America separated from ancestral Africa along two divergent boundaries to form two oceans. The ancestral Atlantic Ocean was located off the western coast of ancestral Africa. A shallow, marginal sea formed along the eastern coast of ancestral North America. A continental fragment was located between the two divergent boundaries.

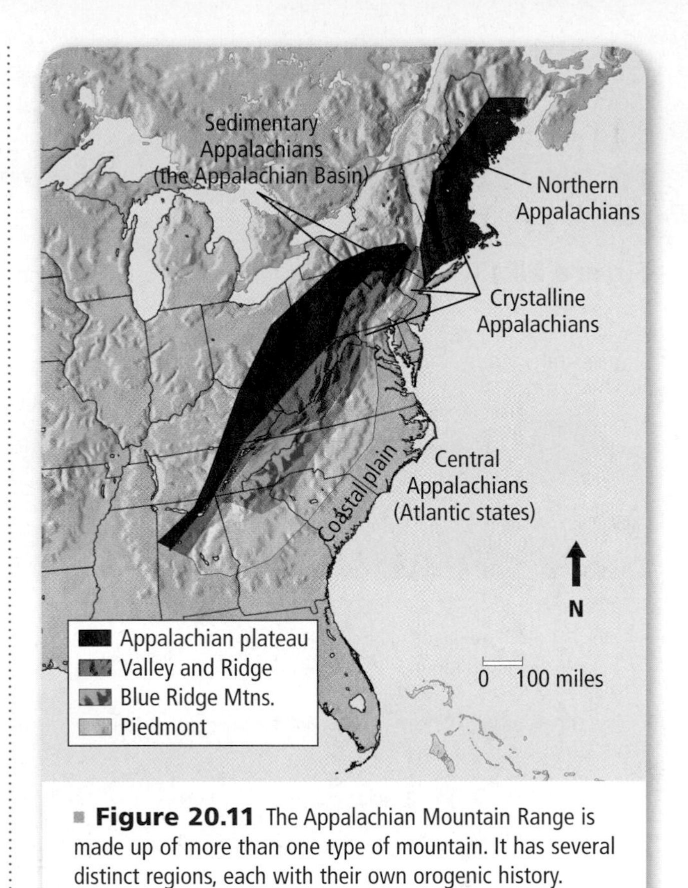

■ **Figure 20.11** The Appalachian Mountain Range is made up of more than one type of mountain. It has several distinct regions, each with their own orogenic history.

Concepts In Motion

Interactive Figure To see an animation of folding rocks, visit glencoe.com.

■ **Figure 20.12** The Valley and Ridge Province of the Appalachians has highly folded rocks. Rocks from the Piedmont Province are relatively undeformed.

Folded rock from the Valley and Ridge Province

Visualizing the Rise and Fall of the Appalachians

Figure 20.13 The Appalachians formed hundreds of millions of years ago as a result of convergence.

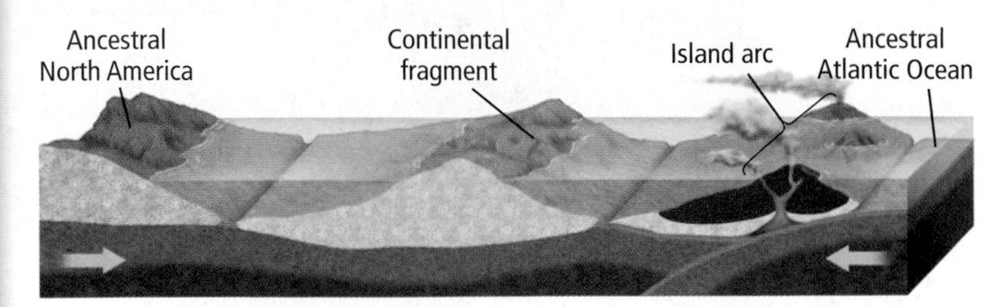

700–600 mya Convergence causes the ancestral Atlantic Ocean to begin to close. An island arc develops east of ancestral North America.

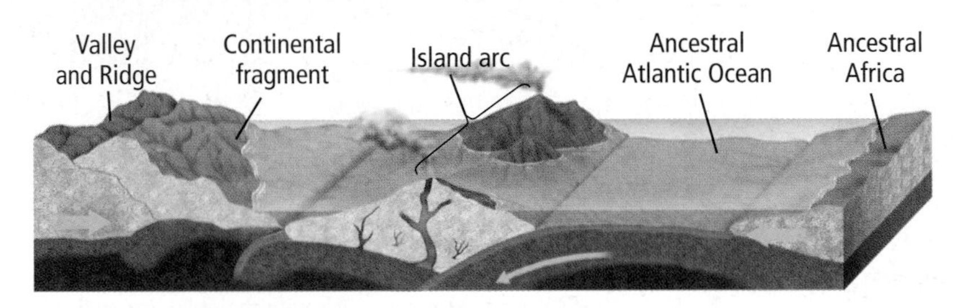

500–400 mya The continental fragment, which eventually becomes the Blue Ridge Province, becomes attached to ancestral North America.

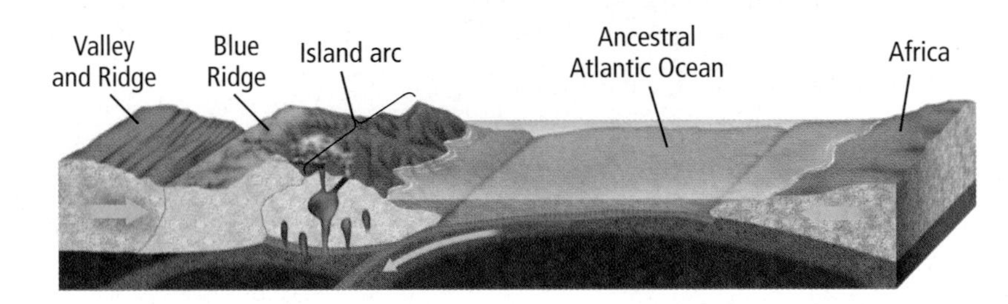

400–300 mya The island arc becomes attached to ancestral North America and the continental fragment is thrust farther onto ancestral North America. The arc becomes the Piedmont Province.

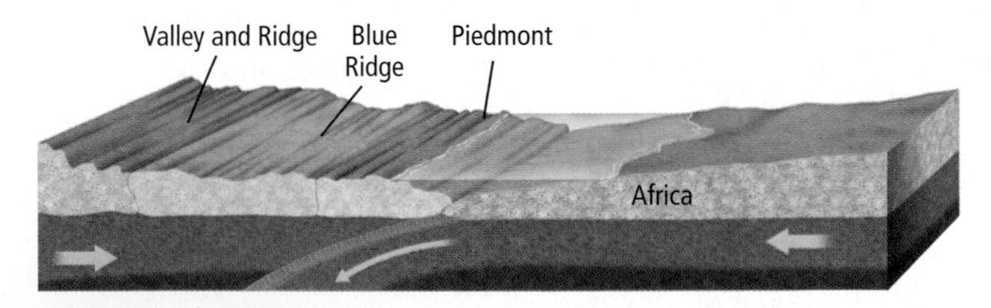

300–260 mya Pangaea forms. Ancestral Africa collides with ancestral North America to close the ancestral Atlantic Ocean. Compression forces the Blue Ridge and Piedmont rocks farther west and the folded Valley and Ridge Province forms.

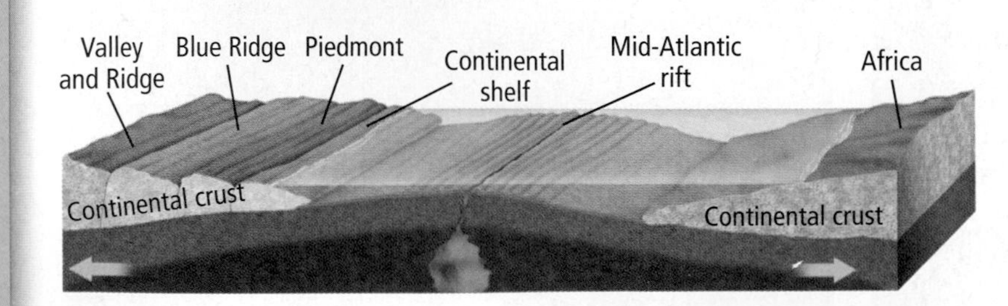

Present After the breakup of Pangaea, tension forces open the modern Atlantic Ocean and separates the continents. North America and Africa continue to move apart as the Atlantic Ocean widens.

Concepts In Motion To explore more about other mountain ranges that formed along convergent boundaries, visit glencoe.com.

Earth Science Online

About 700 to 600 mya, the directions of plate motions reversed. The ancestral Atlantic Ocean began to close as the plates converged. This convergence resulted in the formation of a volcanic island arc east of ancestral North America, as illustrated in **Figure 20.13.**

About 200 million years passed before the continental fragment became attached to ancestral North America as illustrated in **Figure 20.13.** These highly metamorphosed rocks, some of which are shown in **Figure 20.14,** were thrust over younger rocks to become the Blue Ridge Province.

The final stages of formation Between about 400 and 300 mya, the island arc became attached to North America, as illustrated in **Figure 20.13.** Evidence of this event is preserved in the Piedmont Province as a group of metamorphic and igneous rocks. These rocks were also faulted over the continent, pushing the Blue Ridge rocks farther west.

Between about 300 and 260 mya, the ancestral Atlantic Ocean closed as ancestral Africa, Europe, and South America collided with ancestral North America to form Pangaea. This collision resulted in extensive folding and faulting to form the Valley and Ridge Province, as illustrated in **Figure 20.13.** When rifting caused Pangaea to break apart about 200 mya, the modern Atlantic Ocean formed, and the continents moved to their present positions, as illustrated in **Figure 20.13.**

The Appalachian Mountains are only one example of the many mountain ranges that have formed along convergent boundaries. In Section 20.3, you will read about the orogeny that takes place along divergent plate boundaries, as well as some of the types of mountains that form far from plate margins.

Figure 20.14 Outcrops, such as this one in close-up, show the highly metamorphosed rocks found in the Blue Ridge Province.

Section 20.2 Assessment

Section Summary

- Orogeny refers to all of the processes that form mountain belts.
- Most mountain belts are associated with plate boundaries.
- Island arc complexes, highly deformed mountains, and very tall mountains form as a result of the convergence of tectonic plates.
- The Appalachian Mountains are geologically ancient; they began to form 700 to 800 mya.

Understand Main Ideas

1. MAIN Idea **Describe** how convergence relates to orogenic belts.
2. **Identify** the tectonic plate on which Mount Mazinga is situated.
3. **Explain** why you might find fossil shells at the top of a mountain.
4. **Differentiate** between the types of mountains that form at convergent plate boundaries.

Think Critically

5. **Infer** how the Aleutian Islands in Alaska formed.
6. **Evaluate** this statement: The Appalachian mountains are younger than the Himalayas.

WRITING in Earth Science

7. Write and illustrate the story of the formation of the Appalachian Mountains for a middle school student.

Section 20.3

Objectives

- **Identify** the processes associated with non-boundary mountains.
- **Describe** the mountain ranges that form along ocean ridges.
- **Compare and contrast** uplifted and fault-block mountains.

Review Vocabulary

normal fault: a crack in Earth where the rock above the fault plane has dropped down

New Vocabulary

uplifted mountain
plateau
fault-block mountain

Other Types of Mountain Building

MAIN Idea Mountains on the ocean floor and some mountains on continents form through processes other than convergence.

Real-World Reading Link When you take a cake out of the oven, the cake cools and contracts. Similarly, the ocean floor and the crust and mantle below it cool and contract as they move away from the ocean ridge.

Divergent-Boundary Mountains

When ocean ridges were first discovered, people in the scientific community were stunned simply because of their length. These underwater volcanic mountains form a continuous chain that snakes along Earth's ocean floor for over 65,000 km. In addition to their being much longer than most of their continental counterparts, these mountains formed as a result of different orogenic processes. Recall from Chapter 17 that ocean ridges are regions of broad uplift that form when new oceanic crust is created by seafloor spreading. The newly formed crust and underlying mantle at the ocean ridge are hot. When rocks are heated, they expand, which results in a decrease in density. This decrease allows the ridge to bulge upward, as illustrated in **Figure 20.15.** As the oceanic plates move away from the ridge, the newly formed crust and mantle cool and contract, and the surface of the crust subsides. As a result, the crust stands highest where the ocean crust is youngest, and the underwater mountain chains have gently sloping sides.

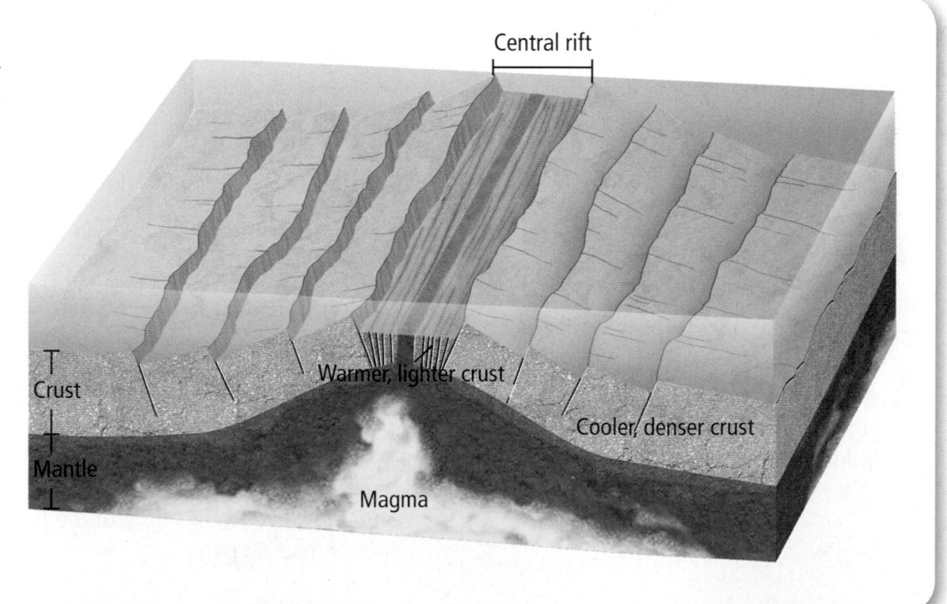

Figure 20.15 An ocean ridge is a broad, topographic high that forms as lithosphere bulges upward due to an increase in temperature along a divergent boundary.
Determine *where the ocean is deeper: near the ridge or far from it?*

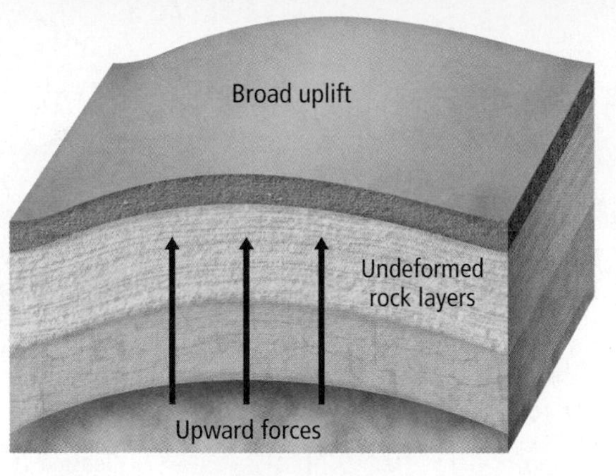

■ **Figure 20.16** The Adirondack Mountains of New York State are uplifted mountains. Uplifted mountains form when large sections of Earth's crust are forced upward without much structural deformation.

Uplifted Mountains

As illustrated in **Figure 20.16,** some mountains form when large regions of Earth have been slowly forced upward as a unit. These mountains are called **uplifted mountains.** The Adirondack Mountains in New York State, shown in **Figure 20.16,** are uplifted mountains. Generally, the rocks that make up uplifted mountains undergo less deformation than rocks associated with plate-boundary orogeny, which, as you have just read, are highly folded, faulted, and metamorphosed. The cause of large-scale regional uplift is not well understood. One popular hypothesis is that the part of the lithosphere made of mantle rocks becomes cold and dense enough that it sinks into the underlying mantle. The mantle lithosphere is replaced by hotter and less dense mantle. The lower density of the new mantle provides buoyancy which vertically lifts the overlying crust. This process has been used to explain the uplift of the Sierra Nevadas, in California, which are shown in **Figure 20.17.** When a whole region is uplifted, a relatively flat-topped area called a **plateau** can form, like the Colorado Plateau, which extends through Colorado, Utah, Arizona, and New Mexico. Erosion eventually carves these relatively undeformed, uplifted masses to form peaks, valleys, and canyons.

■ **Figure 20.17** The Sierra Nevadas are the result of regional uplift.

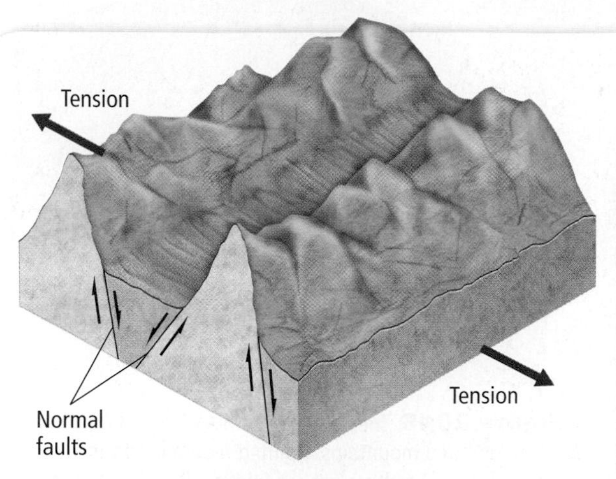

■ **Figure 20.18** Fault-block mountains are areas of Earth's crust that are higher than the surrounding landscape as the result of faulting. The Basin and Range Province consists of hundreds of mountains separated by normal faults.

Fault-Block Mountains

Another type of mountain that is not necessarily associated with plate boundaries is a fault-block mountain. Recall from Chapter 19 how movement at faults lifts land on one side of a fault and drops it on the other. **Figure 20.18** illustrates how **fault-block mountains** form between large faults when pieces of crust are tilted, uplifted, or dropped downward. The Basin and Range Province of the southwestern United States and northern Mexico, a part of which is shown in **Figure 20.18,** consists of hundreds of nearly parallel mountains separated by normal faults. The Grand Tetons in Wyoming are also fault-block mountains. You will explore the topography of this range in the GeoLab at the end of this chapter.

Section 20.3 Assessment

Section Summary

- Divergent boundaries, uplift, and faulting produce some of Earth's mountains.
- Underwater volcanic mountains at divergent boundaries form Earth's longest mountain chain.
- Regional uplift can result in the formation of uplifted mountains that are made of nearly undeformed layers of rock.
- Fault-block mountains form when large pieces of the crust are tilted, uplifted, or dropped downward between normal faults.

Understand Main Ideas

1. MAIN Idea **Explain** why all of Earth's mountains do not form at convergent plate boundaries.
2. **Identify** the kinds of rocks associated with ocean ridges.
3. **Explain** why an ocean ridge is higher than the surrounding crust.
4. **Compare** ocean ridges with fault-block mountains.
5. **Compare and contrast** the formation of uplifted and fault-block mountains.

Think Critically

6. **Formulate** criteria for identifying an uplifted mountain.

WRITING in **Earth Science**

7. Write three practice test questions for your classmates to assess their knowledge of mountain building.

Earth Science Online Self-Check Quiz glencoe.com

ON SITE: HIKING THE APPALACHIAN TRAIL

Mount Katahdin is located in Maine at one end of the Appalachian Trail.

The Appalachian Mountains sprawl from Canada to Alabama. Every year, more than 3 million people embark on a journey of the Appalachian Trail to explore at least part of this mountain range. The rich geologic history of these mountains makes this one of the most exciting and beautiful hikes America has to offer.

History The Appalachian Trail (AT) was opened in 1937 in an effort to allow people to escape the bustle of everyday life and reconnect with nature. Today, the trail covers 3499 km, from Springer Mountain in Georgia to Mount Katahdin in Maine, and can be divided into three areas—the southern, central, and northern Appalachians.

Hiking the Trail A select group of hikers each year choose what is called a thru-hike. Thru-hikers plan to hike the entire AT from start to finish. Most choose to begin in Georgia and hike north to Maine. This trip takes 6 months or longer, and requires much planning.

Southern Appalachians Springer Mountain sits 6087 km above sea level. The soil around Springer Mountain is covered with metamorphic rocks, allowing for only sparse tree growth. These uplifted mountains in the south were never impacted by glaciation as much as some of the northern summits.

Central Appalachians In the central Appalachians, hikers encounter Slide Mountain, part of the Catskill Range of New York. The Catskills are characterized by steep slopes and rounded uplands. Comprised mainly of sandstone and conglomerate, Slide Mountain has resisted some of the weathering that much of the Appalachians have endured. The peak is 6727 km above sea level, but it is difficult to enjoy the view before reaching the top due to the dense forests.

Northern Appalachians After months of hiking, imagine encountering the end of the AT, Mount Katahdin, shown in the figure. To the east, two large cirque basins, the Great Basin and North Basin, are visible. These areas were sculpted by glaciers over 25,000 years ago. Here, the high altitude combined with the northern location of the mountains, does not provide the necessary environment for thick tree growth, so the final view is breathtaking. As of 2006, only 9483 hikers have completed the thru-hike.

WRITING in Earth Science

Research a firsthand account of someone hiking the AT and the geological formations you might encounter. Visit glencoe.com for more information on the Appalachian Trail.

GEOLAB

MAPPING MAKE A MAP PROFILE

Background: A map profile, also called a topographic profile, is a side view of a geographic or geologic feature constructed from a topographic map. The Grand Tetons, a mountain range in Wyoming, formed when enormous blocks of rock were faulted along their eastern flanks, causing the blocks to tilt to the west.

Question: *How do you construct a map profile?*

Materials

metric ruler
sharp pencil
graph paper

Safety Precautions

Procedure

Contour lines are lines on a map that connect points of equal elevation. Locate the index contour lines on the map on the next page. Index contour lines are the ones in a darker color.

1. Read and complete the lab safety form.
2. On graph paper, make a grid like the one shown on the facing page.
3. Place the edge of a paper strip on the map along the profile line AA′ and mark where each major contour line intersects the strip.
4. Label each intersection point with the correct elevation.
5. Transfer the points from the paper strip to the profile grid.
6. Connect the points with a smooth line to construct a profile of the mountain range along line AA′.
7. Label the major geographic features on your profile.

Analyze and Conclude

1. **Interpret Data** Describe how the topographic profile changes with distance from Point A.
2. **Interpret Data** What is the elevation of the highest point on the map topographic profile? The lowest point?

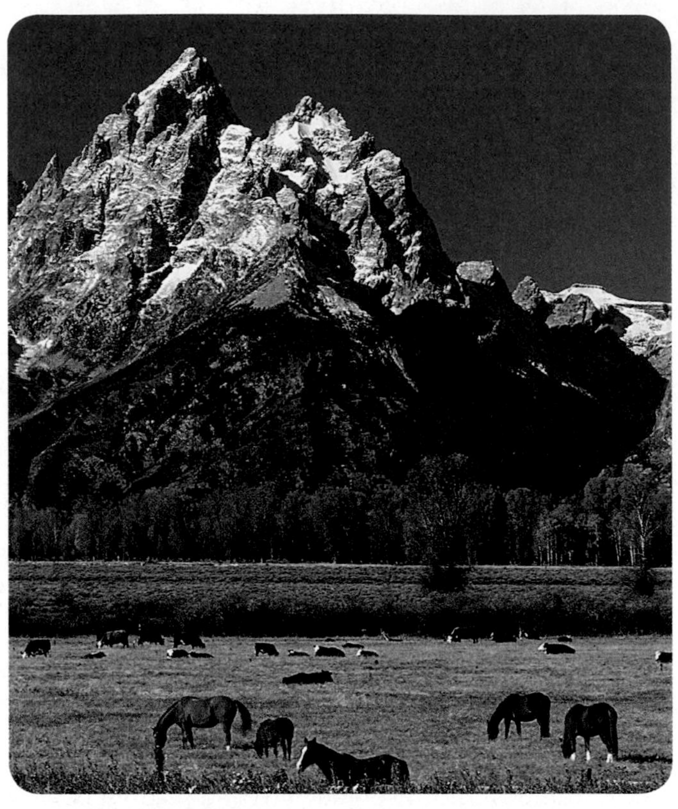

The rugged Grand Tetons stand tall above the plains in Grand Teton National Park in Wyoming.

3. **Interpret Data** What is the average elevation shown in the profile?
4. **Interpret Data** Calculate the total relief shown in the profile.
5. **Interpret Data** Is your topographic profile an accurate model of the topography along line AA′? Explain.
6. **Analyze** What determined the scale of this topographic profile?

APPLY YOUR SKILL

Apply Obtain a topographic map of your hometown and make a topographic profile. How does it compare to the profile you made in this lab?

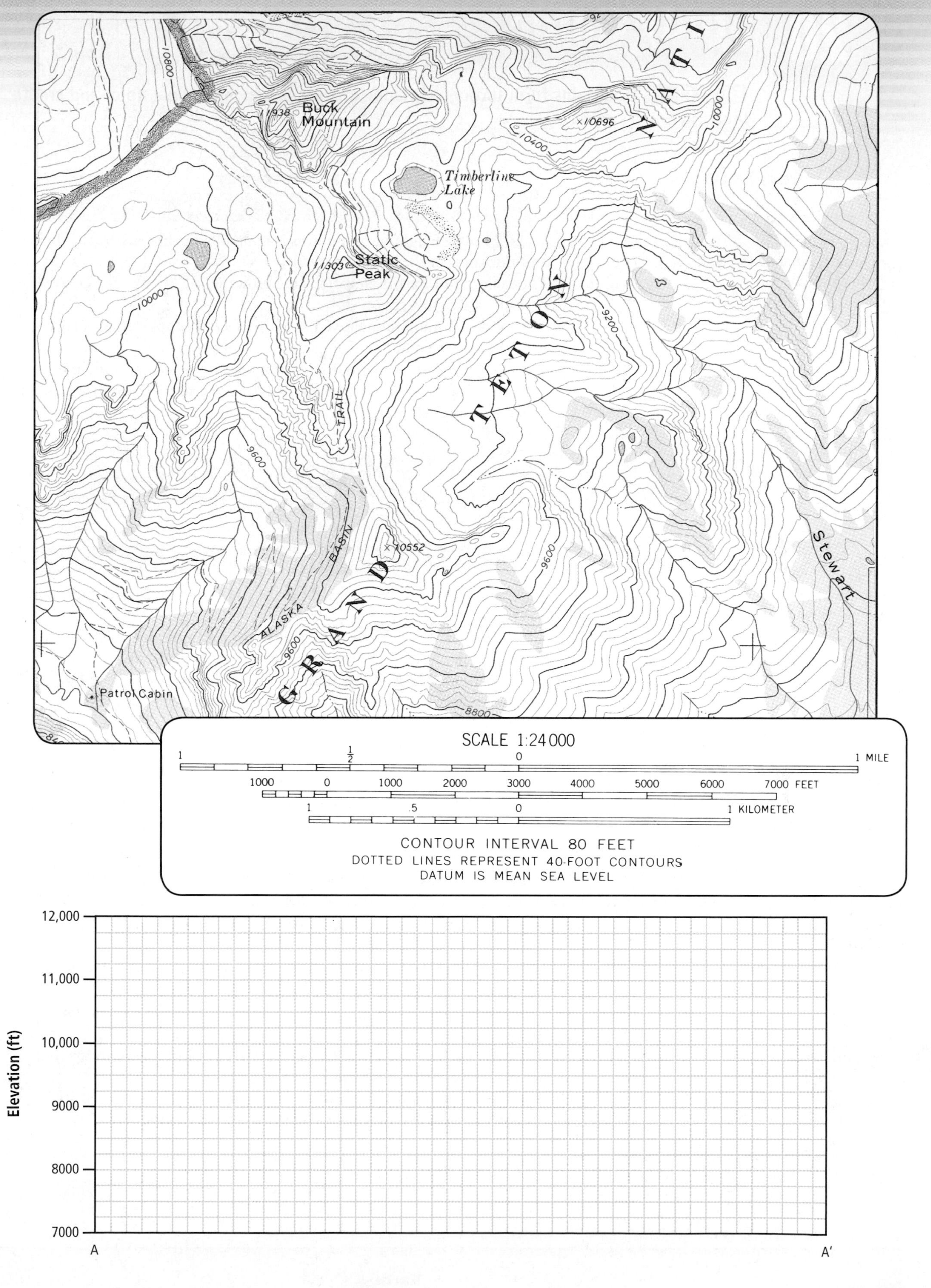

Buck Mountain
11938
×10696
10400
10000
10800
Timberline Lake
Static Peak
11303
10000
9200
TETON
TRAIL
9600
×10552
BASIN
ALASKA
GRAND
9600
9600
Stewart
Patrol Cabin
8800
SCALE 1:24 000
1 MILE
1000 0 1000 2000 3000 4000 5000 6000 7000 FEET
1 KILOMETER
CONTOUR INTERVAL 80 FEET
DOTTED LINES REPRESENT 40-FOOT CONTOURS
DATUM IS MEAN SEA LEVEL
Elevation (ft)
12,000
11,000
10,000
9000
8000
7000
A
A′

CHAPTER 20

Study Guide

Download quizzes, key terms, and flash cards from glencoe.com.

BIG Idea Mountains form through dynamic processes which crumple, fold, and create faults in Earth's crust.

Vocabulary	Key Concepts
Section 20.1 Crust-Mantle Relationships	
• isostasy (p. 563) • isostatic rebound (p. 565) • root (p. 563) • topography (p. 562)	**MAIN Idea** The height of mountains is controlled primarily by the density and thickness of the crust. • The majority of Earth's elevations are either 0 to 1 km above sea level or 4 to 5 km below sea level. • The mass of a mountain above Earth's surface is supported by a root that projects into the mantle. • The addition of mass to Earth's crust depresses the crust, while the removal of mass from the crust causes the crust to rebound in a process called isostatic rebound.
Section 20.2 Orogeny	
• compressive force (p. 567) • orogeny (p. 567)	**MAIN Idea** Convergence causes the crust to thicken and form mountain belts. • Orogeny refers to all of the processes that form mountain belts. • Most mountain belts are associated with plate boundaries. • Island arc complexes, highly deformed mountains, and very tall mountains form as a result of the convergence of tectonic plates. • The Appalachian Mountains are geologically ancient; they began to form 700 to 800 mya.
Section 20.3 Other Types of Mountain Building	
• fault-block mountain (p. 576) • plateau (p. 575) • uplifted mountain (p. 575)	**MAIN Idea** Mountains on the ocean floor and some mountains on continents form through processes other than convergence. • Divergent boundaries, uplift, and faulting produce some of Earth's mountains. • Underwater volcanic mountains at divergent boundaries form Earth's longest mountain chain. • Regional uplift can result in the formation of uplifted mountains that are made of nearly undeformed layers of rock. • Fault-block mountains form when large pieces of the crust are tilted, uplifted, or dropped downward between normal faults.

Assessment

Vocabulary Review

Match each description with the correct term from the Study Guide.

1. continental crust that extends down into the mantle
2. the process of the crust rising in response to removing mass at the surface
3. a mountain bounded by normal faults
4. a type of mountain that often shows little deformation

Fill in the blanks with the correct term from the Study Guide.

5. ________ is the process of mountain formation.
6. Folding, faulting, and metamorphism are created by ________ forces.
7. A flat-topped landform created by uplift is called a ________.

Each of the following sentences is false. Make each sentence true by replacing the italicized words with terms from the Study Guide.

8. The elevation of the crust is called *geography.*
9. *Orogenic forces* refer to the rising up of the crust when a large amount of mass is removed.

Understand Key Concepts

10. What is the approximate percentage of Earth's surface that is covered by continents?
    - **A.** 10 percent
    - **B.** 30 percent
    - **C.** 50 percent
    - **D.** 70 percent
11. What purpose do mountain roots serve?
    - **A.** help prevent mountains from eroding too quickly
    - **B.** balance the amount of crust and mantle in an area
    - **C.** serve as counterbalance to the large weight above
    - **D.** help prevent the mountain from sinking into the mantle
12. Which causes differences in elevation on Earth?
    - **A.** density and thickness of the crust
    - **B.** vertical dikes and pillow basalts
    - **C.** seamounts and hot spots
    - **D.** uplifted and faulted mountains
13. Which is not associated with orogeny at convergent boundaries?
    - **A.** island arcs
    - **B.** highly folded and faulted ranges
    - **C.** ocean ridges
    - **D.** deformed sedimentary rocks
14. During oceanic-oceanic convergence, where do island arc volcanoes form?
    - **A.** on the plate that is subducted
    - **B.** on the plate that is not subducted
    - **C.** do not form on either plate
    - **D.** form on both plates

Use the figure below to answer Question 15.

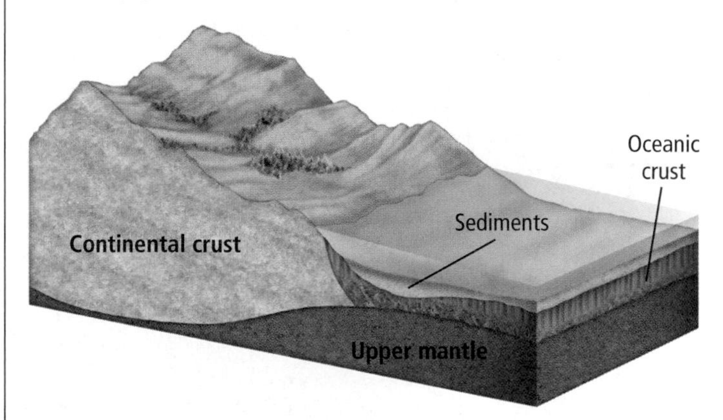

15. Organize the terms *mantle, continental crust,* and *oceanic crust* in order of increasing density.
    - **A.** mantle, oceanic crust, continental crust
    - **B.** oceanic crust, mantle, continental crust
    - **C.** continental crust, oceanic crust, mantle
    - **D.** oceanic crust, continental crust, mantle
16. At which type of plate boundary do the highest mountains form?
    - **A.** convergent; continental-continental
    - **B.** convergent; continental-oceanic
    - **C.** divergent; oceanic-oceanic
    - **D.** divergent; oceanic-continental

Use the figure below and ***Figure 20.13*** *to answer Questions 17 and 18.*

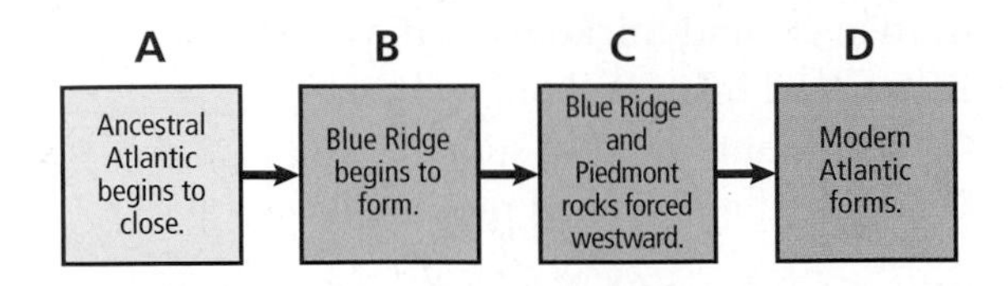

17. Which occurred between Events B and C?
- **A.** The island arc attached to North America.
- **B.** Plate motions reversed.
- **C.** Africa collided with North America.
- **D.** The island arc developed.

18. Approximately when did Event C occur?
- **A.** 800 to 700 mya
- **B.** 700 to 600 mya
- **C.** 500 to 400 mya
- **D.** 300 to 260 mya

Constructed Response

Use the figure below to answer Questions 19 and 20.

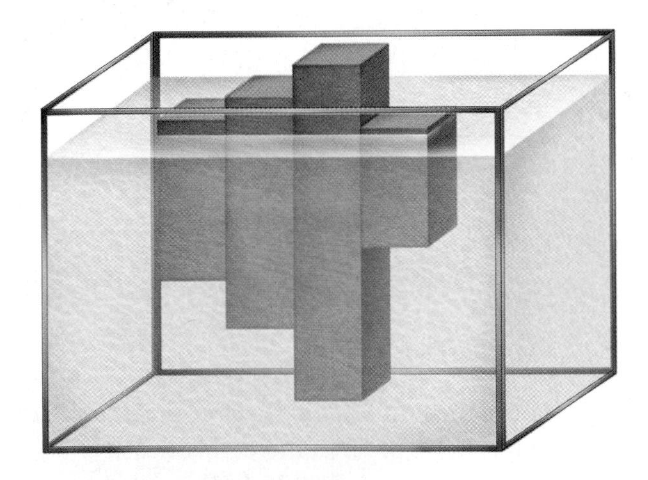

19. Restate the principle that explains what is happening in this figure.

20. Relate the largest block to a mountain described in this chapter.

21. Describe what happens to a mountain's roots as the mountain is eroded.

22. Explain why continental crust can displace more of the mantle than oceanic crust can.

23. Explain why ocean ridges rise high above the surrounding ocean floor.

24. Discuss the processes that can bring roots of mountains to the surface.

25. Describe three mechanisms of crustal thickening.

Use the figure below to answer Questions 26 and 27.

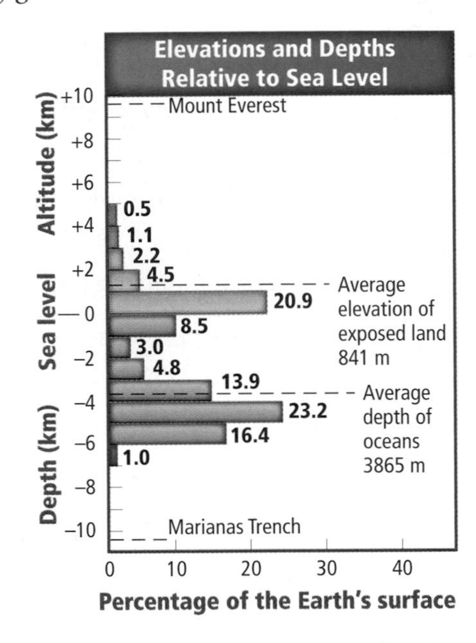

26. Calculate the difference in elevation between the average elevation of continents and the average depth of oceans.

27. Generalize State one generalization about Mount Everest and the Marianas Trench based on the data in the graph.

28. Infer where continental crust is the thickest.

29. Summarize the characteristics of an island arc.

Think Critically

30. Illustrate why the interaction between two oceanic crust plates rarely results in the formation of high mountain ranges.

31. Hypothesize The Andes Mountains run along the western coast of South America. Hypothesize about the tectonic setting in which they formed.

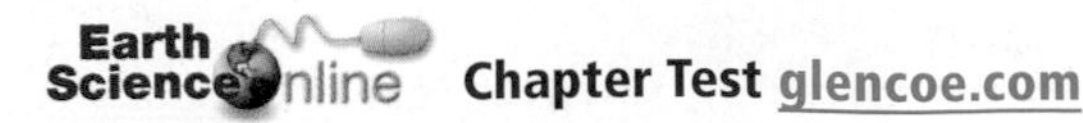

32. CAREERS IN EARTH SCIENCE Structural geologists collect data from both field observations and laboratory analysis to help interpret the structural history of an area. What can a structural geologist interpret about the history of the outcrop shown in the opening photo of this chapter?

Use the figure below to answer Question 33.

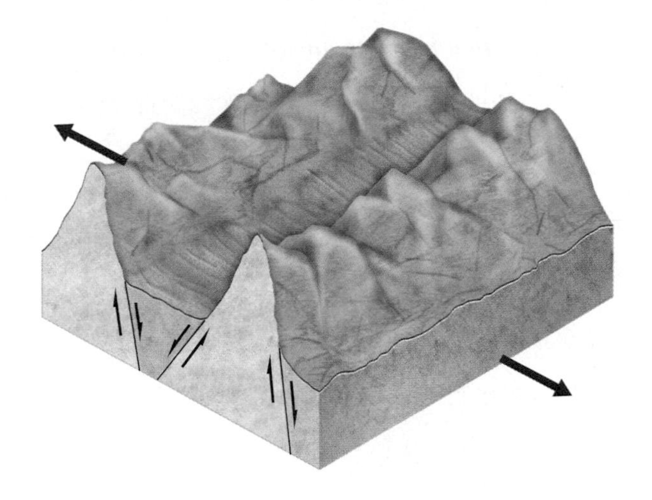

33. **Interpret** and explain what type of forces are acting on this area if fault-block mountains are forming.

34. **Decide** whether a continental crust thinner than the average thickness of 40 km would depress the mantle more or less than it does now.

35. **Evaluate** whether isostatic rebound of an area ever stops.

Concept Mapping

36. Create a concept map using the following terms: *Earth's tallest mountains, uplifted mountains, individual volcano peaks, divergent boundaries, little structural deformation, compression, fault-block mountains, ocean ridges,* and *convergent boundaries.* Refer to the *Skillbuilder Handbook* for more information.

Challenge Question

37. **Consider** whether all mountains are in a state of isostatic equilibrium. Explain your answer.

Additional Assessment

38. WRITING in Earth Science Review **Figure 20.13.** Speculate about what might happen over the next 500 million years. Draw three new pictures for 100, 300, and 500 million years from now. Write a caption for each, describing the orogenic and plate tectonic processes you think will occur.

Document–Based Questions

Data obtained from: Fischer, K. M. 2002. Waning buoyancy in the crustal roots of old mountains. *Nature* 417: 933–936.

Buoyancy is measured by the ratio of height of the continents to the depth of the root beneath the continent. In this graph, buoyancy (R) *is shown in relation to the age of the mountain belt.*

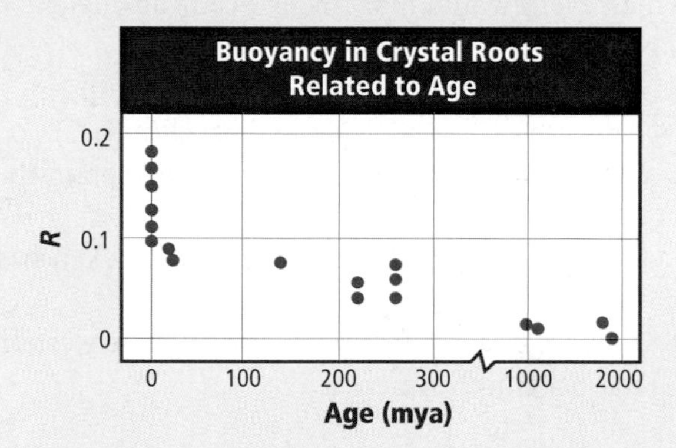

39. Describe the trend (if any) that is evident in the data.

40. Because *R* depends on the density of the crust and mantle, how might the value of *R* change if the density of continental crust were constant?

41. Provide an explanation for the change in the value of *R*. What is happening to the crustal root to cause *R* to change?

Cumulative Review

42. Explain what scientists mean when they say Earth operates as a system. **(Chapter 1)**

43. Explain why weathered rock alone is not a soil. **(Chapter 7)**

Standardized Test Practice

Multiple Choice

Use the map below to answer Questions 1 and 2.

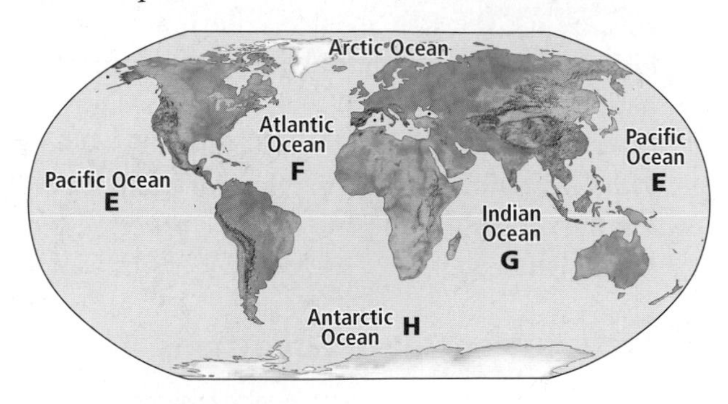

1. In which location does the El Niño southern oscillation cycle begin?
   A. E
   B. F
   C. G
   D. H

2. Which event would most likely occur during an El Nino?
   A. Much of the northwestern coast of South America experiences a cool, dry climate.
   B. Frequent and intense hurricanes develop in the Atlantic Ocean.
   C. Strong trade winds move water westward across the Pacific Ocean.
   D. A warm ocean current develops off the western coast of South America.

3. Why would a thickness of continental crust displace less mantle than the same thickness of oceanic crust?
   A. Continental crust is more dense.
   B. Continental crust is less dense.
   C. Continental crust is mainly basalt.
   D. Continental crust is closer to the mantle.

4. When do shield volcanoes form?
   A. when layer upon layer of lava accumulates during nonexplosive eruptions
   B. when layers of hardened, frothy mixtures of gas and magma formed by explosive eruptions alternate with layers of oozing lava
   C. when eruptions eject small pieces of magma into the air and the pieces fall to the ground and collect around a vent
   D. when thick lava hardens around a central vent

5. Which is NOT a method used to increase soil fertility?
   A. planting legumes
   B. adding compost to the soil
   C. planting the same crops every year
   D. using commercially produced fertilizers

Use the illustration below to answer Questions 6 and 7.

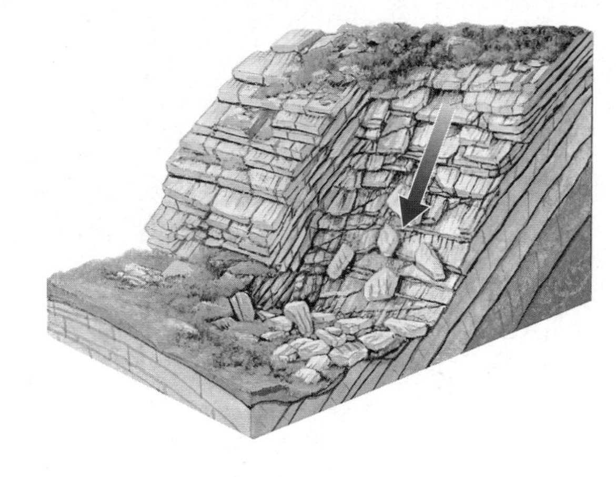

6. The rock slide shown is produced by
   A. the chemical breakdown of rocks
   B. the separation of a thin block of soil, rock, and debris from bedrock
   C. material rotating and sliding down a curved slope
   D. the melting of snow off the rocks

7. What potential damage would a rock slide have on a river?
   A. changing the chemical composition of the river
   B. the changing of the physical characteristics of the river
   C. permanently increasing water levels
   D. damming rivers and causing flooding

8. Which CANNOT form as the result of oceanic-oceanic convergence?
   A. rift zones
   B. trenches
   C. subduction zones
   D. island arc complexes

Short Answer

Use the map below to answer Questions 9–11.

9. The map above shows the paths that Atlantic hurricanes took in 2004. Why do no hurricanes form in the northeastern Atlantic?

10. What can be inferred about the impact that every hurricane formed in the Atlantic will have on North America?

11. What time of year did these hurricanes most likely occur and why?

12. Explain why, in topographical terms, people in the central United States experience very cold winters and very hot summers.

13. How do sill and dike formation differ?

14. Many parts of West Virginia have karst topography. Discuss the potential values and trade-offs of having a geologist prepare an environmental impact statement for areas where new housing developments are planned.

Reading for Comprehension

Iberian Earthquakes

The Great Portugal Quake was one of the greatest natural disasters in European history. The 8.7 magnitude earthquake that struck Portugal in 1755 killed at least 60,000 people and triggered tsunamis that wrecked seaports in Portugal, Spain, and Morocco. The earthquake's cause remained a mystery because the tectonic activity of the region was not clearly understood. The plate boundary off southern Iberia—the peninsula occupied by Spain and Portugal—is not well defined. A new study suggests that it happened as a result of subduction—the process of the oceanic lithosphere (the outer solid part of Earth) diving beneath the continental lithosphere. The study also shows continued activity in the plate system, prompting fears that another earthquake could hit the region with potentially devastating consequences, although probably not for many years to come. Recent seismic images and seafloor bathymetry instead suggest that subduction is occurring in the region, causing compressive stress to accumulate along the interface between the tectonic plates, which leads to earthquakes.

Article obtained from Lovgren, S. Great Portugal quake may have a sequel, study says. *National Geographic News.* August 30, 2004.

15. What do new images indicate was the cause of the earthquake?
A. delamination, which occurs shortly after mountain building
B. subduction in the region causing compressive stress between the tectonic plates
C. major tsunamis, which wrecked seaports
D. unknown tectonic activity in the region

16. Why is it important to research an earthquake that occurred so long ago?

NEED EXTRA HELP?														
If You Missed Question . . .	1	2	3	4	5	6	7	8	9	10	11	12	13	14
Review Section . . .	14.3	14.3	20.1	18.1	7.3	8.1	8.1	20.2	13.3	13.3	13.3	14.1	18.2	10.2

Geologic Time

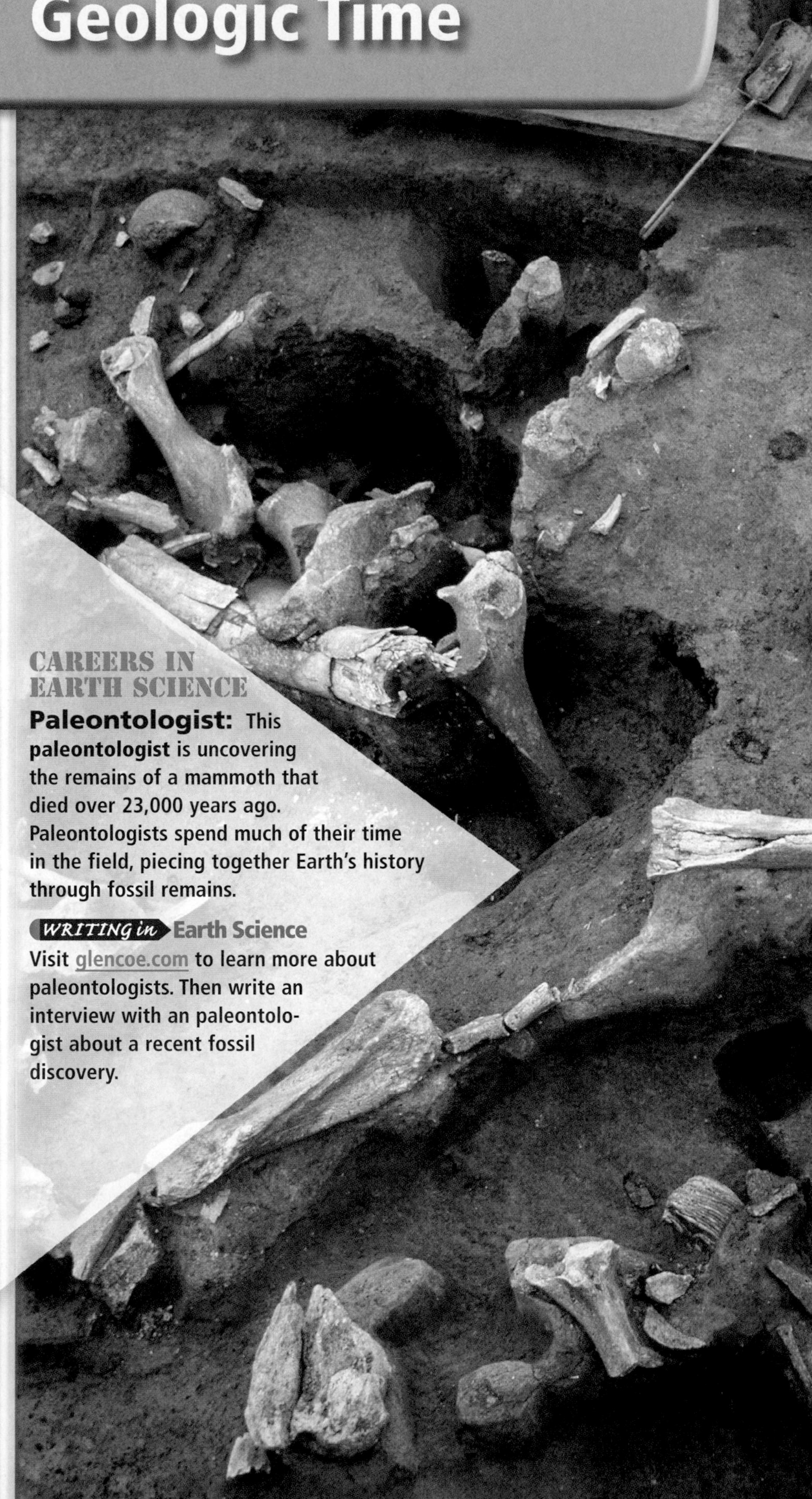

Chapter 21
Fossils and the Rock Record

BIG Idea Scientists use several methods to learn about Earth's long history.

Chapter 22
The Precambrian Earth

BIG Idea The oceans and atmosphere formed and life began during the three eons of the Precambrian, which spans nearly 90 percent of Earth's history.

Chapter 23
The Paleozoic, Mesozoic, and Cenozoic Eras

BIG Idea Complex life developed and diversified during the three eras of the Phanerozoic as the continents moved into their present positions.

CAREERS IN EARTH SCIENCE

Paleontologist: This **paleontologist** is uncovering the remains of a mammoth that died over 23,000 years ago. Paleontologists spend much of their time in the field, piecing together Earth's history through fossil remains.

WRITING in Earth Science
Visit glencoe.com to learn more about paleontologists. Then write an interview with an paleontologist about a recent fossil discovery.

Earth Science Online

To learn more about paleontologists, visit glencoe.com.

CHAPTER 21 Fossils and the Rock Record

BIG Idea Scientists use several methods to learn about Earth's long history.

21.1 The Rock Record
MAIN Idea Scientists organize geologic time to help them communicate about Earth's history.

21.2 Relative-Age Dating
MAIN Idea Scientists use geologic principles to learn the sequence in which geologic events occurred.

21.3 Absolute-Age Dating
MAIN Idea Radioactive decay and certain kinds of sediments help scientists determine the numeric age of many rocks.

21.4 Fossil Remains
MAIN Idea Fossils provide scientists with a record of the history of life on Earth.

GeoFacts

- The land that is now Badlands National Park in South Dakota was once covered by forest, then by swamp, and later by grasslands.
- Ancestors of alligators, camels, and rhinoceroses once thrived in the Badlands.
- The Badlands are considered the birthplace of vertebrate paleontology in North America.

LAUNCH Lab

How are fossils made?

Have you ever wandered through a museum and stood beneath the fossilized bones of a *Tyrannosaurus rex*? Fossilized bones provide evidence that dinosaurs and other ancient organisms existed. A fossil forms when a bone or other hard body part is quickly covered by mud, sand, or other sediments, and after long periods of time, the bones absorb minerals from Earth and become petrified.

Procedure

1. Read and complete the lab safety form.
2. Pour 500 mL of **sand** into a **plastic milk carton** with the top cut off.
3. Bury a **sponge** in the center of the sand.
4. Pour 250 mL of **hot tap water** into a **500 mL beaker.**
5. Measure 100 mL of **salt,** add the salt to the water, and use a **stirring rod** to stir the mixture vigorously.
6. Pour the water over the sand and place the container in direct sunlight for 5 to 7 days, leaving it undisturbed.
7. Dig up your fossilized sponge.

Analysis

1. **Describe** in your science journal what happened to the sponge.
2. **Explain** how this activity models the formation of a fossil.

Relative-Age v. Absolute-Age Dating Make this Foldable to compare and contrast relative-age dating to absolute-age dating of rocks.

STEP 1 Find the center of a vertical sheet of paper.

STEP 2 Fold the top and bottom to the center line to make a shutter fold.

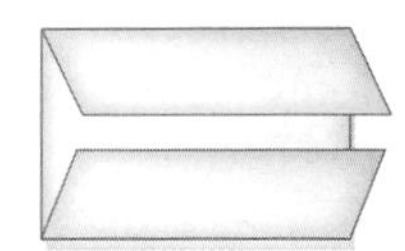

STEP 3 Label the tabs *Relative-Age Dating* and *Absolute-Age Dating.*

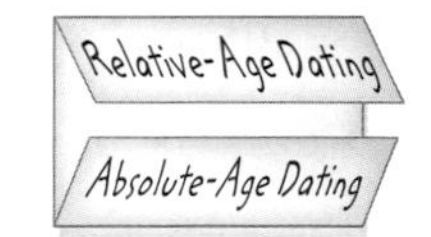

FOLDABLES **Use this Foldable with Sections 21.2 and 21.3.** As you learn about age dating of rocks, summarize that information on your Foldable. Be sure to include examples along with advantages and disadvantages of each type.

Visit glencoe.com to

- **study entire chapters online;**
- **explore Concepts In Motion animations:**
  - **Interactive Time Lines**
  - **Interactive Figures**
  - **Interactive Tables**
- **access Web Links for more information, projects, and activities;**
- **review content with the Interactive Tutor and take Self-Check Quizzes.**

Objectives

- **Explain** why scientists need a geologic time scale.
- **Distinguish** among eons, eras, periods, and epochs.
- **Characterize** the groups of plants and animals that dominated eras in Earth's history.

Review Vocabulary

fossil: the remains, trace, or imprint of a once-living plant or animal

New Vocabulary

geologic time scale
eon
Precambrian
era
period
epoch
mass extinction

The Rock Record

MAIN Idea Scientists organize geologic time to help them communicate about Earth's history.

Real-World Reading Link Imagine how difficult it would be to plan a meeting with a friend if time were not divided into units of months, weeks, days, hours, and minutes. By organizing geologic time into time units, scientists can communicate more effectively about events in Earth's history.

Organizing Time

A hike down the Grand Canyon reveals the multicolored layers of rock, called strata, that make up the canyon walls, as shown in **Figure 21.1.** Some of the layers contain fossils, which are the remains, traces, or imprints of ancient organisms. By studying rock layers and the fossils within them, geologists can reconstruct aspects of Earth's history and interpret ancient environments.

To help in the analysis of Earth's rocks, geologists have divided the history of Earth into time units. These time units are based largely on the fossils contained within the rocks. The time units are part of the **geologic time scale,** a record of Earth's history from its origin 4.6 billion years ago (bya) to the present. Since the naming of the first geologic time unit, the Jurassic (joo RA sihk), in 1795, development of the time scale has continued to the present day. Some of the units have remained unchanged for centuries, while others have been reorganized as scientists have gained new knowledge. The geologic time scale is shown in **Figure 21.2.**

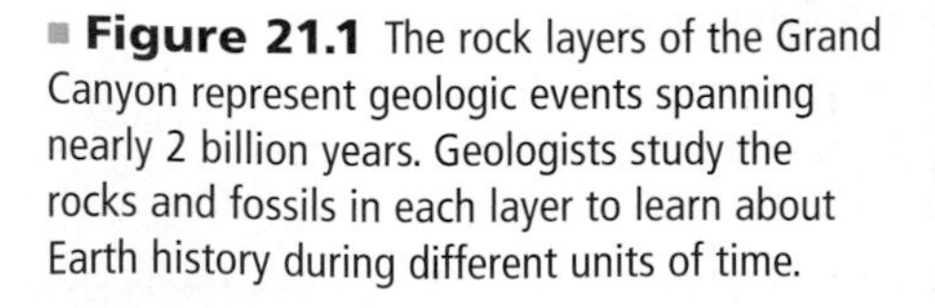

■ **Figure 21.1** The rock layers of the Grand Canyon represent geologic events spanning nearly 2 billion years. Geologists study the rocks and fossils in each layer to learn about Earth history during different units of time.

Visualizing the Geologic Time Scale

Figure 21.2 The geologic time scale begins with Earth's formation 4.6 billion years ago (bya). Geologists organize Earth's history according to groupings called eons. Each eon contains eras, which in turn contain periods. Each period in the geologic time scale contains epochs. The current geologic epoch is called the Holocene Epoch. Each unit on the scale is labeled with its range of time in millions of years ago (mya).
Identify ***the period, era, and eon representing the most modern unit of time.***

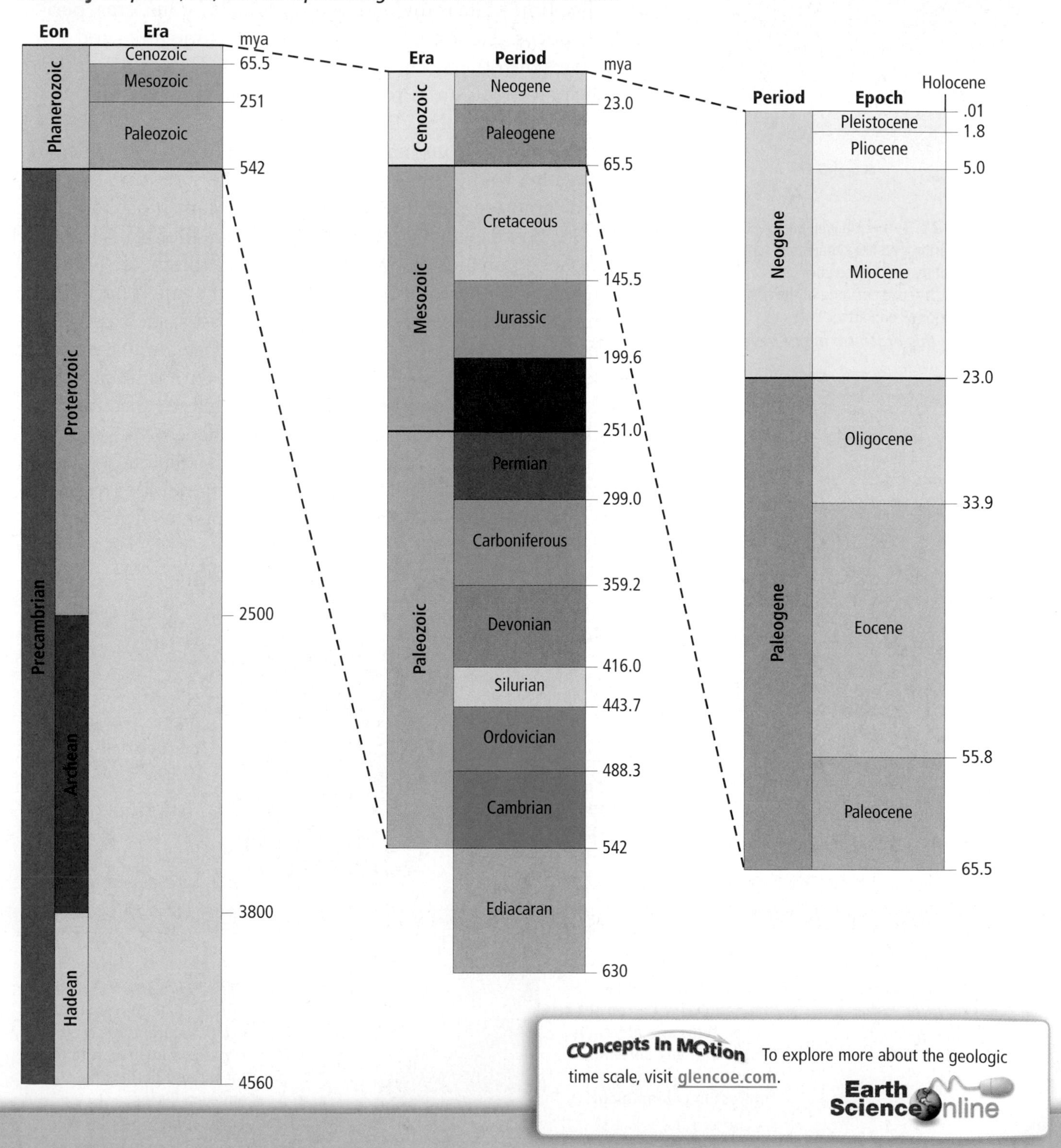

Concepts In Motion To explore more about the geologic time scale, visit glencoe.com.

Earth Science Online

■ **Figure 21.3** This is a well-preserved fossil of an arthropod-like organism, found in a sedimentary rock of the late Precambrian. It represents one of the first complex life-forms on Earth.
Infer *how this organism might have moved.*

The Geologic Time Scale

The geologic time scale enables scientists to find relationships among the geological events, environmental conditions, and fossilized life-forms that are preserved in the rock record. The oldest division of time is at the bottom of the scale, shown in **Figure 21.2.** Moving upward, each division is more recent, just as the rock layers in the rock record are generally younger toward the surface.

Reading Check **Explain** why scientists need a geologic time scale.

Eons The time scale is divided into units called eons, eras, periods, and epochs. An **eon** is the largest of these time units and encompasses the others. They consist of the Hadean (HAY dee un), the Archean (ar KEE un), Proterozoic (pro tuh ruh ZOH ihk), and Phanerozoic (fa nuh ruh ZOH ihk) Eons.

The three earliest eons make up 90 percent of geologic time, known together as the **Precambrian** (pree KAM bree un). During the Precambrian, Earth was formed and became hospitable to modern life. Fossil evidence suggests that simple life-forms began in the Archean Eon and that by the end of the Proterozoic Eon, life had evolved to the point that some organisms might have been able to move in complex ways. Most of these fossils, such as the one shown in **Figure 21.3,** were soft-bodied organisms, many of which resembled modern animals. Others had bodies with rigid parts. All life-forms until then had soft bodies without shells or skeletons.

Fossils dating from the most recent eon, the Phanerozoic, are the best-preserved, not only because they are younger, but because they represent organisms with hard parts, which are more easily preserved. The time line in **Figure 21.4** shows some important fossil and age-dating discoveries.

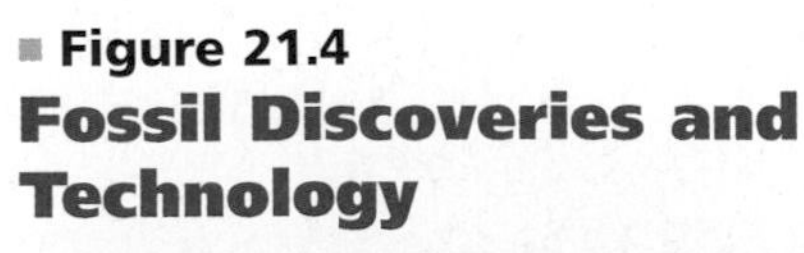

■ **Figure 21.4**
Fossil Discoveries and Technology

Fossil discoveries and dating technology have changed our understanding of life on Earth.

1796 William Smith, a canal surveyor, creates the first geologic map based on distinct fossil layers.

1820s Mary Anning, the daughter of a cabinetmaker, finds and identifies fossils of many ancient creatures, sparking great interest in paleontology.

1857 Quarry workers uncover a skeleton identified as Neanderthal, a species similar to modern humans.

1909 The discovery of the Burgess Shale fossils in the Rocky Mountains reveals the diversity of invertebrate life that thrived during the Cambrian Period.

1929 An Anasazi ruin becomes the first prehistoric site to be dated using tree-ring chronology.

Eras All eons are made up of eras, the next-largest unit of time. **Eras** are usually tens to hundreds of millions of years in duration. Like all other time units, they are defined by the different life-forms found in the rocks; the names of the eras are based on the relative ages of these life-forms. For example, in Greek, *paleo* means *old, meso* means *middle,* and *ceno* means *recent. Zoic* means *of life* in Greek; thus, *Mesozoic* means *middle life* and *Cenozoic* means *recent life.*

Periods All eras are divided into periods. **Periods** are generally tens of millions of years in duration, though some periods of the Precambrian are considerably longer. Some periods are named for the geographic region in which the rocks or fossils characterizing the age were first observed and described. Consider, for example, the Ediacaran (ee dee A kuh run) Period at the end of the Precambrian. It is named for the Ediacara Hills in Australia, shown in **Figure 21.5.** It was here that fossils typical of the period were first found, as shown in **Figure 21.4.** The Ediacaran Period was added to the geologic time scale in 2004.

Epochs Epochs (EE pahks) are even smaller divisions of geologic time. Although the time scale in **Figure 21.2** shows epochs only for periods of the Cenozoic Era, all periods of geologic time are divided into epochs. **Epochs** are generally hundreds of thousands to millions of years in duration. Rocks and sediments from the epochs of the Cenozoic Era are the most complete because there has been less time for weathering and erosion to remove evidence of this part of Earth's history. For this reason, the epochs of the Cenozoic are relatively short in duration. For example, the Holocene (HOH luh seen) Epoch, which includes modern time, began only about 11,000 years ago.

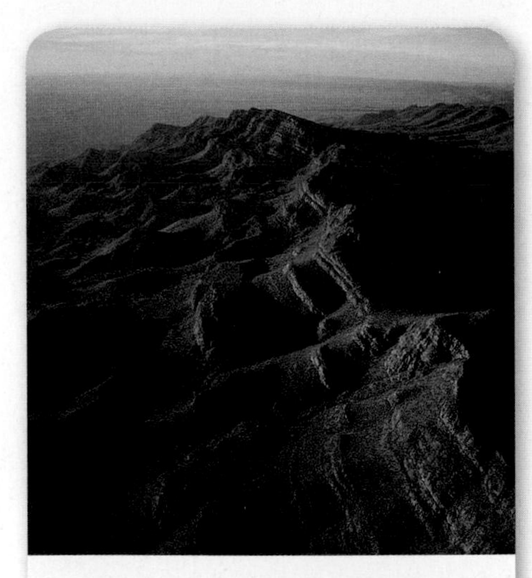

■ **Figure 21.5** The Ediacara Hills of Australia yielded the first fossils typical of the Ediacaran Period. Fossils from that time found anywhere in the world are called Ediacaran fossils.

1946 University of Chicago scientists show that the age of relatively recent organic objects and artifacts can be determined with radiocarbon dating.

1993 Fossils found in western Australia provide evidence that bacteria existed 3.5 bya.

2006 A 164-million-year-old, beaverlike fossil unearthed by Chinese researchers suggests that aquatic mammals might have thrived alongside dinosaurs.

1940 1970 2000

1987 Jenny Clack leads an expedition to Greenland that unearths fossils of animals that lived 360 mya, showing that animals developed legs prior to moving onto land.

Concepts In Motion

Interactive Time Line To learn more about these discoveries and others, visit glencoe.com. Earth Science Online

■ **Figure 21.6** Trilobites are Paleozoic fossils found all over the world. Like 90 percent of life-forms of that era, they perished during a mass extinction.

Succession of Life-Forms

During the Phanerozoic Eon, multicellular life diversified. Fossils from the Phanerozoic are abundant, while those from the Precambrian are relatively few. The word *Phanerozoic* means *visible life* in Greek. During the first era of the Phanerozoic, the Paleozoic (pay lee uh ZOH ihk), the oceans became full of many different kinds of organisms. Small, segmented animals called trilobites, shown in **Figure 21.6,** were among the first hard-shelled life-forms. Trilobites dominated the oceans in the early part of the Paleozoic Era; land plants appeared later, followed by land animals. Swamps of the Carboniferous (kar buh NIH fuh rus) Period provided the plant material that developed into the coal deposits of today. The end of the Paleozoic is marked by the largest mass extinction event in Earth's history. In a **mass extinction,** many groups of organisms disappear from the rock record at about the same time. At the end of the Paleozoic, 90 percent of all marine organisms became extinct.

The age of dinosaurs The era following the Paleozoic—the Mesozoic (mez uh ZOH ihk)—is known for the emergence of dinosaurs, but many other organisms also appeared during the Mesozoic. Large predatory reptiles ruled the oceans, and corals closely related to today's corals built huge reef systems. Water-dwelling amphibians began adapting to terrestrial environments. Insects, some as large as birds, lived. Mammals evolved and began to diversify. Flowering plants and trees emerged. The end of the Mesozoic is marked by a large extinction event. Many groups of organisms became extinct, including the non-avian dinosaurs and large marine reptiles.

The rise of mammals During the era that followed—the Cenozoic (sen uh ZOH ihk)—mammals increased both in number and diversity. Human ancestors, the first primates, emerged in the epoch called the Paleocene, and modern humans appeared in the Pleistocene (PLYS tuh seen) Epoch.

Section 21.1 Assessment

Section Summary

- Scientists organize geologic time into eons, eras, periods, and epochs.
- Scientists divide time into units based on fossils of plants and animals.
- The Precambrian makes up nearly 90 percent of geologic time.
- The geologic time scale changes as scientists learn more about Earth.

Understand Main Ideas

1. MAIN Idea **Explain** the purpose of the geologic time scale.
2. **Distinguish** among eons, eras, periods, and epochs, using specific examples.
3. **Describe** the importance of extinction events to geologists.
4. **Explain** why scientists know more about the Cenozoic than they do about other eras.

Think Critically

5. **Discuss** why scientists know so little about Precambrian Earth.

MATH in Earth Science

6. Make a bar graph that shows the relative percentage of time spanned by each era of the Phanerozoic Eon. For more help, refer to the *Skillbuilder Handbook*.

Earth Science Online Self-Check Quiz glencoe.com

Section 21.2

Objectives

- **Describe** uniformitarianism and explain its importance to geology.
- **Apply** geologic principles to interpret rock sequences and determine relative ages.
- **Compare and contrast** different types of unconformities.
- **Explain** how scientists use correlation to understand the history of a region.

Review Vocabulary

granite: a coarse-grained, intrusive igneous rock

New Vocabulary

uniformitarianism
relative-age dating
original horizontality
superposition
cross-cutting relationship
principle of inclusions
unconformity
correlation
key bed

Relative-Age Dating

MAIN Idea Scientists use geologic principles to learn the sequence in which geologic events occurred.

Real-World Reading Link If you were to put the following events into a time sequence of first to last, how would you do it? Go to school. Wake up. Put on your clothes. Eat lunch. You would probably rely on your past experiences. Scientists also use information from the past to place events into a likely time sequence.

Interpreting Geology

Recall from Section 21.1 that Earth's history stretches back billions of years. Scientists have not always thought that Earth was this old. Early ideas about Earth's age were generally placed in the context of time spans that a person could understand relative to his or her own life. This changed as people began to explore Earth and Earth processes in scientific ways. James Hutton, a Scottish geologist who lived in the late 1700s, was one of the first scientists to think of Earth as very old. He attempted to explain Earth's history in terms of geologic forces, such as erosion and sea-level changes, that operate over long stretches of time. His work helped set the stage for the development of the geologic time scale.

Uniformitarianism Hutton's work lies at the foundation of **uniformitarianism,** which states that geologic processes occurring today have been occurring since Earth formed. For example, if you stand on the shore of an ocean and watch the waves come in, you are observing a process that has not changed since the oceans were formed. The waves crashing on a shore in the Jurassic Period were much like the waves crashing on a shore today. The photo in **Figure 21.7** was taken recently on a beach in Oregon, but a beach in the Jurassic Period probably looked very similar.

Figure 21.7 An ancient Jurassic beach probably looked much like this beach in Oregon. The geologic processes that formed it are unchanged.

■ **Figure 21.8** The horizontal layers of the Grand Canyon were formed by deposition of sediment over millions of years. The principle of original horizontality states that the tilted strata at the bottom were formed horizontally.

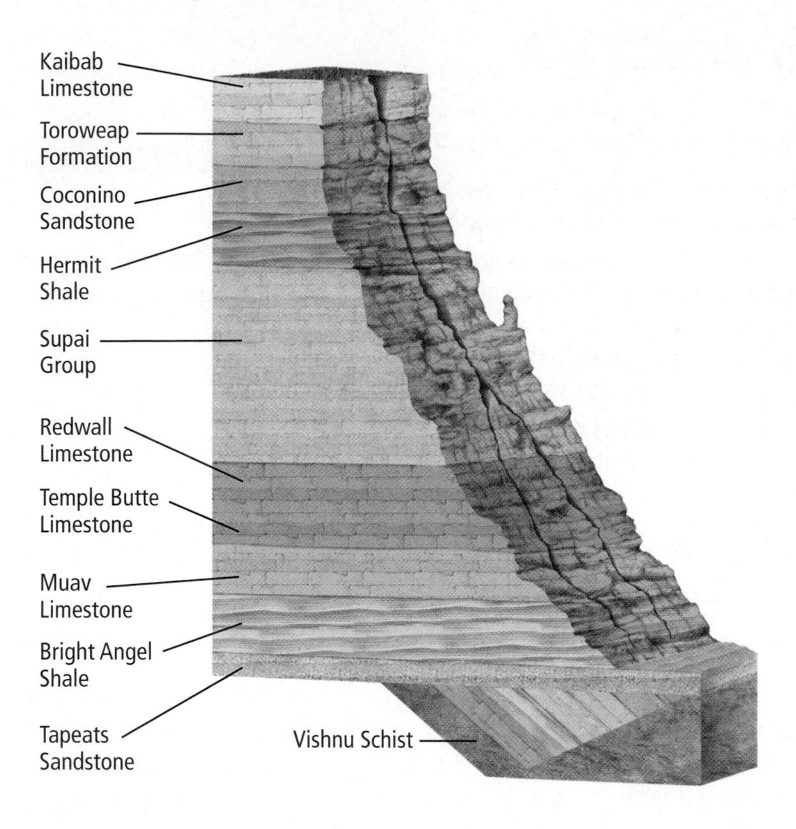

Principles for Determining Relative Age

Because of uniformitarianism, scientists can learn about the past by studying the present. One way to do this is by studying the order in which geologic events occurred using a method called **relative-age dating.** This does not allow scientists to determine exactly how many years ago an event occurred, but it gives scientists a clearer understanding about geologic events in Earth's history. Scientists use several ways to determine relative ages, called the principles of relative dating. They include original horizontality, superposition, cross-cutting relationships, and inclusions.

FOLDABLES Incorporate information from this section into your Foldable.

Original horizontality **Original horizontality** is the principle that sedimentary rocks are deposited in horizontal or nearly horizontal layers. This can be seen in the walls of the Grand Canyon, illustrated in **Figure 21.8.** Sediment is deposited in horizontal layers for the same reason that layers of sand on a beach are mostly flat; that is, gravity combined with wind and water spreads them evenly.

VOCABULARY

ACADEMIC VOCABULARY

Principle
a general hypothesis that has been tested repeatedly; sometimes also called a law
The geologic principle was illustrated in the rock layers the students observed.

Superposition Geologists cannot determine the numeric ages of most rock layers in the Grand Canyon using relative-age dating methods. However, they can assume that the oldest rocks are at the bottom and that each successive layer above is younger. Thus, they can infer that the Kaibab Limestone at the top of the canyon is much younger than the Vishnu Group, which is at the bottom. This is an application of **superposition,** the principle that in an undisturbed rock sequence, the oldest rocks are at the bottom and each consecutive layer is younger than the layer beneath it.

Cross-cutting relationships Rocks exposed in the deepest part of the Grand Canyon are mostly igneous and metamorphic. Within the metamorphic schist of the Vishnu Group in the bottom sequence are intrusions—also called dikes—of granite, as shown in **Figure 21.9.** You learned in Chapter 5 that intrusions are rocks that form when magma solidifies in existing rock. The principle of **cross-cutting relationships** states that an intrusion is younger than the rock it cuts across. Therefore, the granite intrusion in the Grand Canyon is younger than the schist because the granite cuts across the schist.

The principle of cross-cutting relationships also applies to faults. Recall from Chapter 19 that a fault is a fracture in Earth along which movement takes place. Many faults exist in earthquake-prone areas, such as California, and in ancient, mountainous regions, such as the Adirondacks of New York. A fault is younger than the strata and surrounding geologic features because the fault cuts across them.

Inclusions Relative age can also be determined where one rock layer contains pieces of rock from the layer next to it. This might occur after an exposed layer has eroded and the loose material on the surface has become incorporated into the layer deposited on top of it. The **principle of inclusions** states that the fragments, called inclusions, in a rock layer must be older than the rock layer that contains them.

As you learned in Chapter 6, once a rock has eroded, the resulting sediment might be transported and redeposited many kilometers away. In this way, a rock formed in the Triassic Period might contain inclusions from a Cambrian rock. Inclusions can also form from pieces of rock that are trapped within a lava flow.

■ **Figure 21.9** According to the principle of cross-cutting relationships, this igneous intrusion is younger than the schist it cuts across.
Infer *how the igneous intrusion was formed.*

MiniLab

Determine Relative Age

How is relative age determined? Scientists use geologic principles to determine the relative ages of rock layers.

Procedure

1. Read and complete the lab safety form.
2. Draw a diagram showing four horizontal layers of rock. Starting from the bottom, label the layers *1* through *4.*
3. Draw a vertical intrusion from Layer 1 through Layer 3.
4. Label a point at the bottom left corner of the diagram *X* and a point at the top right corner *Y.*
5. Cut the paper in a diagonal line from X to Y. Move the top-left piece 1.5 cm along the cut.

Analysis

1. **Describe** what principles you would use to determine the relative ages of the layers in your diagram.
2. **Explain** how the principle of cross-cutting relationships can help you determine the relative age of the vertical intrusion.
3. **Infer** what the XY cut represents. Is the XY cut older or younger than the surrounding layers?

Concepts In Motion

Interactive Figure To see an animation of an angular unconformity, visit glencoe.com.

■ **Figure 21.10** An unconformity is any erosional surface separating two layers of rock that have been deposited at different times. The three types of unconformities are illustrated below.

Unconformities Earth's surface is constantly changing as a result of weathering, erosion, earthquakes, volcanism, and other processes. This makes it difficult to find a sequence of rock layers in the geologic record in which a layer has not been disturbed. Sometimes, the record of a past event or time period is missing entirely. For example, if rocks from a volcanic eruption erode, the record of that eruption is lost. If an eroded area is covered at a later time by a new layer of sediment, the eroded surface represents a gap in the rock record. Buried surfaces of erosion are called **unconformities.** The rock layer immediately above an unconformity is sometimes considerably younger than the rock layer immediately below it. Scientists recognize three different types of unconformities, which are illustrated in **Figure 21.10.**

Disconformity When a horizontal layer of sedimentary rock overlies another horizontal layer of sedimentary rock, the eroded surface is called a disconformity. Disconformities can be easy to identify when the eroded surface is uneven. Where the eroded surface is smooth, disconformities are often hard to see.

Nonconformity When a layer of sedimentary rock overlies a layer of igneous or metamorphic rock, such as granite or marble, the eroded surface is easier to identify. This kind of eroded surface is called a nonconformity. Both granite and marble form deep in Earth. A nonconformity indicates a gap in the rock record during which rock layers were uplifted, eroded at Earth's surface, and new layers of rock formed on top.

Reading Check **Distinguish** between a disconformity and a nonconformity.

Angular unconformity As you learned in Chapter 20, when horizontal layers of sedimentary rock are deformed during mountain building, they are usually uplifted and tilted. During this process, the layers are exposed to weathering and erosion. If a horizontal layer of sedimentary rock is later laid down on top of the tilted, eroded layers, the resulting unconformity is called an angular unconformity. **Figure 21.10** shows how angular unconformities record the complex history of mountain formation and erosion.

Correlation The Kaibab Limestone layer rims the top of the Grand Canyon in Arizona, but it is also found more than 100 km away at the bottom of Zion National Park in Utah. How do geologists know that these layers, which are far apart from each other, formed at the same time? One method is by correlation (kor uh LAY shun). **Correlation** is the matching of unique rock outcrops or fossils exposed in one geographic region to similar outcrops exposed in other geographic regions. Through correlation of many different layers of rocks, geologists have determined that Zion, Bryce Canyon, and the Grand Canyon are all part of one layered sequence called the Grand Staircase, illustrated in **Figure 21.11.**

Key beds Distinctive rock layers are sometimes deposited over wide geographic areas as a result of a large meteorite strike, volcanic eruption, or other brief event. Because these layers are easy to recognize, they help geologists correlate rock formations in different geographic areas where the layers are exposed. A rock or sediment layer used as a marker in this way is called a **key bed.** Geologists know that the layers above a key bed are younger than the layers below it. The key-bed ash layer that marks the 1980 eruption of Mount St. Helens deposited volcanic ash over many states.

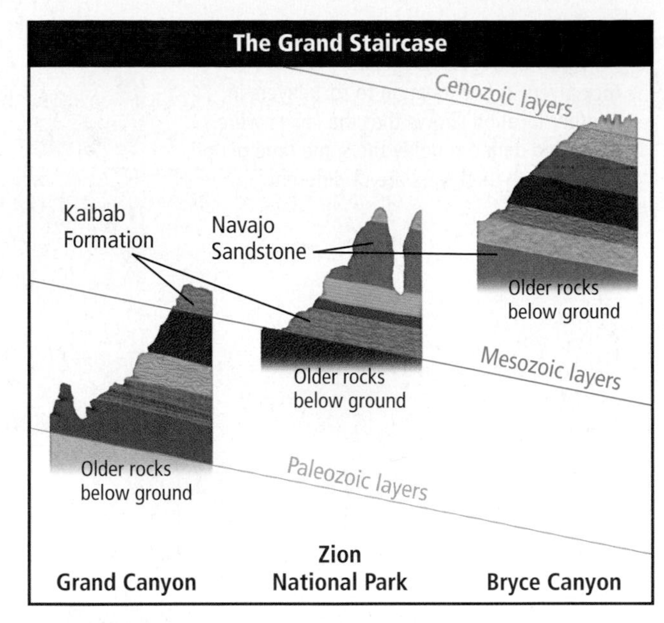

Figure 21.11 The top layers of rocks at the Grand Canyon are identical to the bottom layers at Zion National Park, and the top layers at Zion are the bottom layers at Bryce Canyon.
Infer *the makeup of the buried layer below Zion's Kaibab layer.*

PROBLEM-SOLVING LAB

Interpret the Diagram

How do you interpret the relative ages of rock layers? The diagram at right illustrates a sequence of rock layers. Geologists use the principles of relative-age dating to determine the order in which layers such as these were formed.

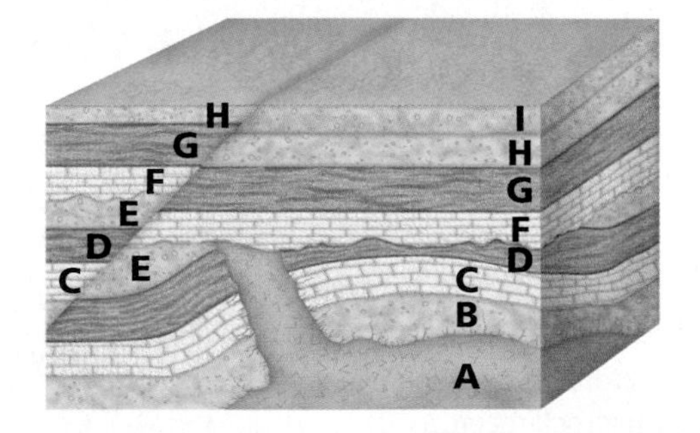

Analysis

1. **Identify** a type of unconformity between any two layers of rock. Justify your answer.
2. **Interpret** which rock layer is oldest.
3. **Infer** where inclusions might be found. Explain.
4. **Compare and contrast** the rock layers on the right and left sides of the diagram. Why do they not match?

Think Critically

5. **Apply** Which feature is younger, the dike or the folded strata? What geologic principle did you use to determine your answer?
6. **Propose** why there is no layer labeled *I* on the left side of the diagram.

■ **Figure 21.12** Correlating fossils from rock layers in one location to rock layers in another location shows that the layers were deposited during roughly the same time period, even though the layers are of different material.

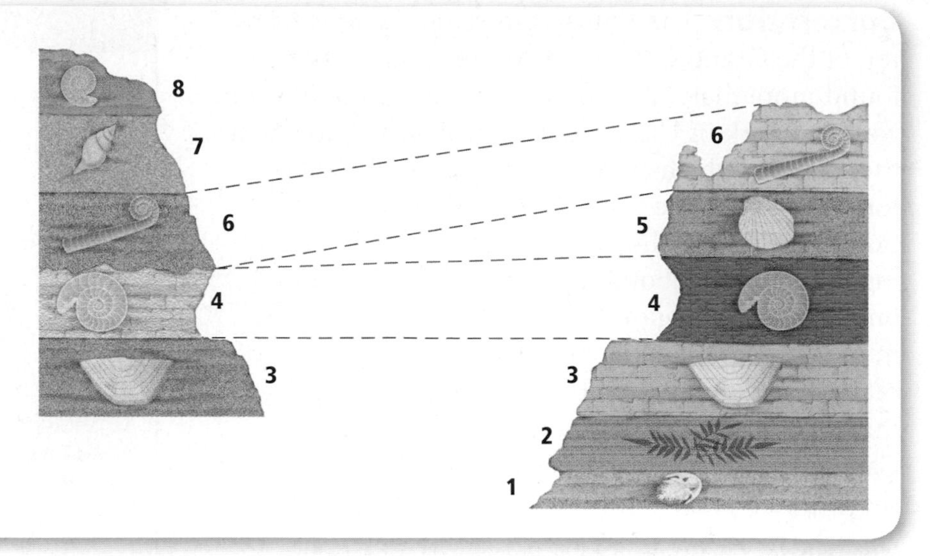

CAREERS IN EARTH SCIENCE

Petroleum Geologist Petroleum geologists use geologic principles to identify petroleum and natural gas reserves in the rock record. To learn more about Earth science careers, visit glencoe.com.

Fossil correlation Geologists also use fossils to correlate rock formations in locations that are geographically distant. As shown in **Figure 21.12,** fossils can indicate similar times of deposition even though the layers might be made of entirely different material.

The correlation of fossils and rock layers aids in the relative dating of rock sequences and helps geologists understand the history of larger geographic regions. Petroleum geologists also use correlation to help them locate reserves of oil and gas. For example, if a sandstone layer in one area contains oil, it is possible that the same layer in other areas also contains oil. It is largely through correlation that geologists have constructed the geologic time scale.

Section 21.2 Assessment

Section Summary

- The principle of uniformitarianism states that processes occurring today have been occurring since Earth formed.
- Scientists use geologic principles to determine the relative ages of rock sequences.
- An unconformity represents a gap of time in the rock record.
- Geologists use correlation to compare rock layers in different geographic areas.

Understand Main Ideas

1. MAIN Idea **Summarize** the principles that geologists use to determine relative ages of rocks.
2. **Make a diagram** to compare and contrast the three types of unconformities.
3. **Explain** how geologists use fossils to understand the geologic history of a large region.
4. **Discuss** how a coal seam might be used as a key bed.
5. **Apply** Explain how the principle of uniformitarianism would help geologists determine the source of a layer of particular igneous rock.

Think Critically

6. **Propose** how a scientist might support a hypothesis that rocks from one quarry were formed at the same time as rocks from another quarry 50 km away.

WRITING in **Earth Science**

7. Write a paragraph that explains how an event, such as a large hurricane, might result in a key bed. Use a specific example in your paragraph.

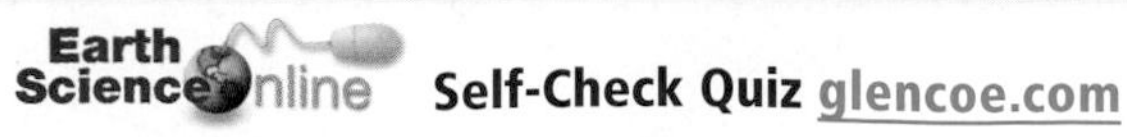

Section 21.3

Objectives

- **Compare and contrast** absolute-age dating and relative-age dating.
- **Describe** how scientists date rocks and other objects using radioactive elements.
- **Explain** how scientists can use certain non-radioactive material to date geologic events.

Review Vocabulary

isotope: one of two or more forms of an element with differing numbers of neutrons

New Vocabulary

absolute-age dating
radioactive decay
radiometric dating
half-life
radiocarbon dating
dendrochronology
varve

Absolute-Age Dating

MAIN Idea Radioactive decay and certain kinds of sediments help scientists determine the numeric age of many rocks.

Real-World Reading Link If a TV programming guide listed only the order of TV shows but not the times they aired, you would not know when to watch a program. Scientists, too, find it helpful to know exactly when events occurred.

Radioactive Isotopes

As you learned in Section 21.2, relative-age dating is a method of comparing past geologic events based on the order of strata in the rock record. In contrast, **absolute-age dating** enables scientists to determine the numerical age of rocks and other objects. In one type of absolute-age dating method, scientists measure the decay of the radioactive isotopes in igneous and metamorphic rocks in addition to some remains of organisms preserved in sediments.

Radioactive decay Radioactive isotopes emit nuclear particles at a constant rate. Recall from Chapter 3 that an element is defined by the number of protons it contains. As the number of protons changes with each emission, the original radioactive isotope, called the parent, is gradually converted to a different element, called the daughter. For example, a radioactive isotope of uranium, U-238, will decay into the daughter isotope lead-206 (Pb-206) over a specific span of time, as illustrated in **Figure 21.13.** Eventually, enough of the parent decays that traces of it are undetectable, and only the daughter product is measurable. The emission of radioactive particles and the resulting change into other isotopes over time is called **radioactive decay.** Because the rate of radioactive decay is constant regardless of pressure, temperature, or any other physical changes, scientists use it to determine the absolute age of the rock or object in which it occurs.

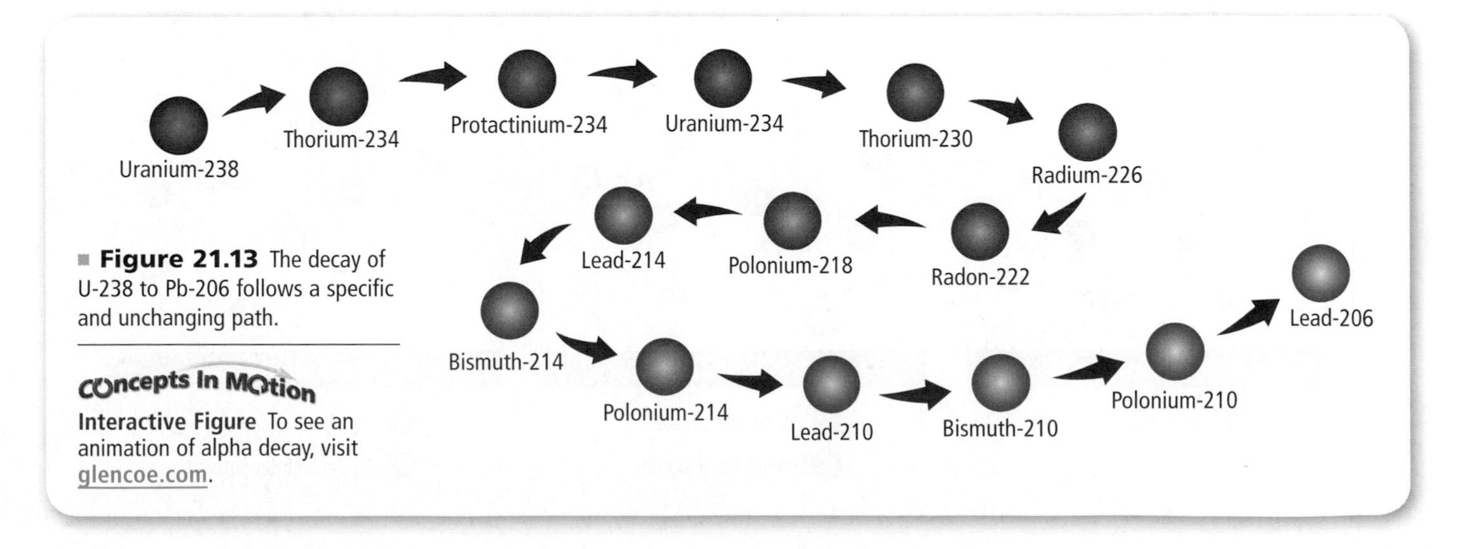

Figure 21.13 The decay of U-238 to Pb-206 follows a specific and unchanging path.

Concepts In Motion

Interactive Figure To see an animation of alpha decay, visit glencoe.com.

■ **Figure 21.14** As the number of parent atoms decreases during radioactive decay, the number of daughter atoms increases by the same amount.
Interpret ***What percentage of daughter isotope would exist in a sample containing 50 percent parent isotope?***

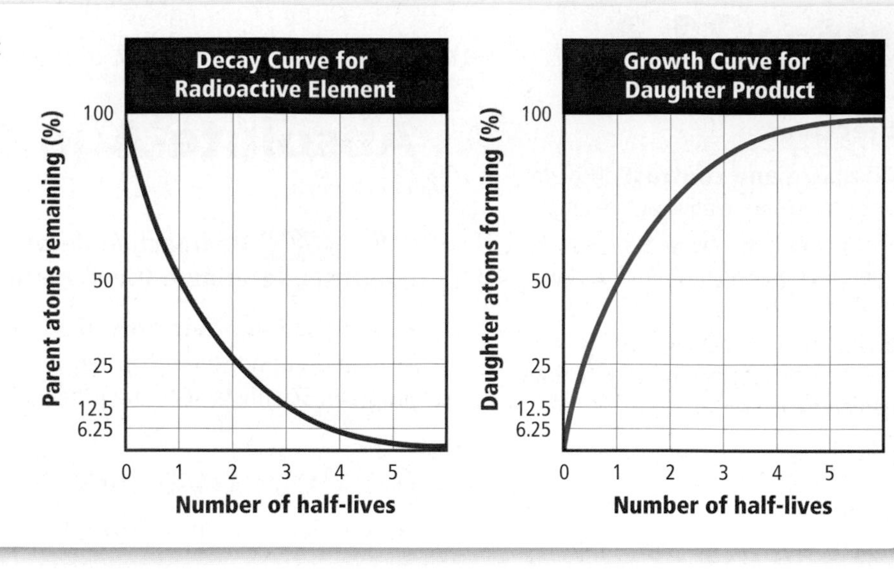

Radiometric Dating

As the number of parent atoms decreases during radioactive decay, the number of daughter atoms increases, shown in **Figure 21.14.** The ratio of parent isotope to daughter product in a mineral indicates the amount of time that has passed since the object formed. For example, by measuring this ratio in the minerals of an igneous rock, geologists pinpoint when the minerals first crystallized from magma. When scientists date an object using radioactive isotopes, they are using a method called **radiometric dating.**

Half-life Scientists measure the length of time it takes for one-half of the original isotope to decay, called its **half-life.** After one half-life, 50 percent of the parent remains, resulting in a 1:1 ratio of parent-to-daughter product. After two half-lives, one-half of the remaining 50 percent of the parent decays. The result is 25:75 percent ratio of the original parent to the daughter product—a 1:3 ratio. This process is shown in **Figure 21.15.**

Concepts In Motion
Interactive Figure To see an animation of half-lives, visit glencoe.com.

■ **Figure 21.15** After one half-life, a sample contains 50 percent parent and 50 percent daughter. After two half-lives, the sample contains 25 percent parent and 75 percent daughter.

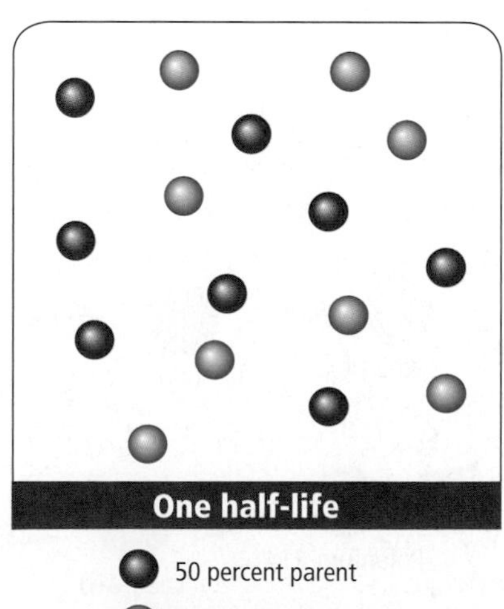

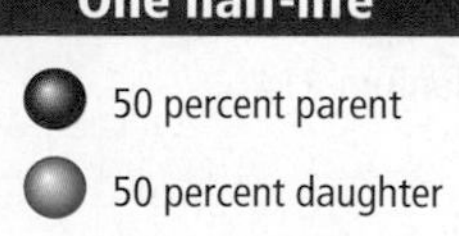

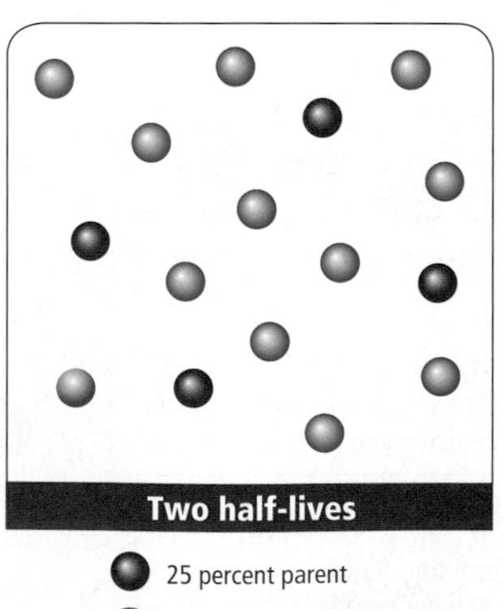

Concepts In Motion
Interactive Table To explore more about radioactive decay visit glencoe.com.

Table 21.1 Half-Lives of Selected Radioactive Isotopes

Radioactive Parent Isotope	Approximate Half-life	Daughter Product
Rubidium-87 (Rb-87)	48.6 billion years	strontium-87 (Sr-87)
Thorium-232 (Th-232)	14.0 billion years	lead-208 (Pb-208)
Potassium-40 (K-40)	1.3 billion years	argon-40 (Ar-40)
Uranium-238 (U-238)	4.5 billion years	lead-206 (Pb-206)
Uranium-235 (U-235)	0.7 billion years	lead-207 (Pb-207)
Carbon-14 (C-14)	5730 years	nitrogen-14 (N-14)

Dating rocks To date an igneous or metamorphic rock using radiometric dating, scientists examine the parent-daughter ratios of the radioactive isotopes in the minerals that comprise the rock. **Table 21.1** lists some of the radioactive isotopes they might use. The best isotope to use for dating depends on the approximate age of the rock being dated. For example, scientists might use uranium-235 (U-235), which has a half-life of 700 million years, to date a rock that is a few tens of millions of years old. Conversely, to date a rock that is hundreds of millions of years old, scientists might use U-238, which has a longer half life. If an isotope with a shorter half-life is used for an ancient rock, there might be a point when the parent-daughter ratio becomes too small to measure.

Radiometric dating is not useful for dating sedimentary rocks because, as you learned in Chapter 6, the minerals in most sedimentary rocks were formed from pre-existing rocks. **Figure 21.16** shows how geologists can learn the approximate age of sedimentary layers by dating layers of igneous rock that lie between them.

Reading Check **Explain** why radiometric dating is not useful for sedimentary rocks.

Radiocarbon dating Notice in **Table 21.1** that the half-life of carbon-14 (C-14) is much shorter than the half-lives of other isotopes. Scientists use C-14 to determine the age of organic materials, which contain abundant carbon, in a process called **radiocarbon dating.** Organic materials used in radiocarbon dating include plant and animal material such as bones, charcoal, and amber.

The tissues of all living organisms, including humans, contain small amounts of C-14. During an organism's life the C-14 decays, but is continually replenished by the process of respiration. When the organism dies, it no longer takes in C-14, so over time, the amount of C-14 decreases. Scientists can measure the amount of C-14 in organic material to determine how much time has passed since the organism's death. This method is particularly useful for dating recent geologic events for which organic remains exist.

Figure 21.16 To help them determine the age of sedimentary rocks, scientists date layers of igneous rock or volcanic ash above and below the sedimentary layers.

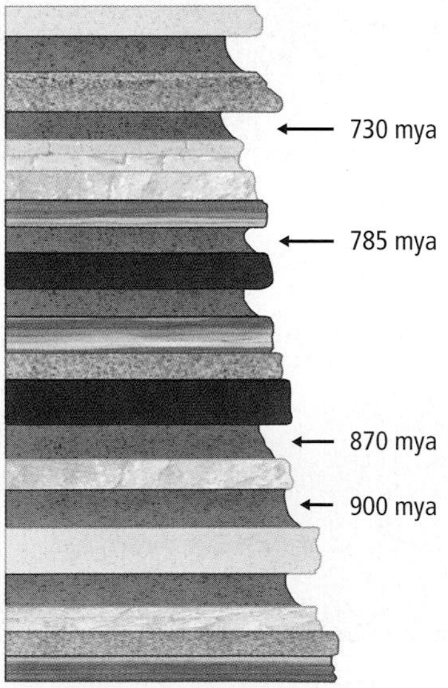

730 mya
785 mya
870 mya
900 mya

Radiometric Dating of Volcanic Ash

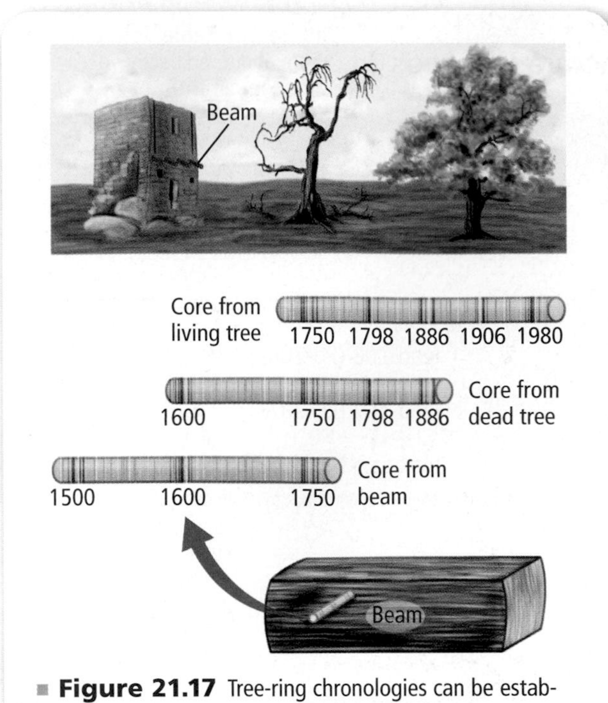

■ **Figure 21.17** Tree-ring chronologies can be established by matching tree rings from different wood samples, both living and dead. The science of using tree rings to determine absolute age is called dendrochronology. **Calculate** *the number of years represented in this tree-ring chronology.*

■ **Figure 21.18** Ice cores are stored in facilities such as the one in Denver, Colorado. Scientists use ice cores to date glacier deposits and to learn about ancient climates.

Other Ways to Determine Absolute Age

Radiometric dating is one of the most common ways for geologists to date geologic material, but other dating methods are available. Geologists can also use other materials, such as tree rings, ice cores, and lake-bottom and ocean-bottom sediments, to help determine the ages of some objects or events.

Tree rings Many trees contain a record of time in the rings of their trunks. These rings are called annual tree rings. Each annual tree ring consists of a pair of early season and late season growth rings. The width of the rings depends on certain conditions in the environment. For example, when rain is plentiful, trees grow fast and rings are wide. The harsh conditions of drought result in narrow rings. Trees from the same geographic region tend to have the same patterns of ring widths for a given time span. By matching the rings in these trees, as shown in **Figure 21.17,** scientists have established tree-ring chronologies that can span time periods up to 10,000 years.

 **Reading Check** **Describe** how tree rings can show past environmental conditions.

The science of using tree rings to determine absolute age is called **dendrochronology** and has helped geologists date relatively recent geologic events that toppled trees, such as volcanic eruptions, earthquakes, and glaciation. Dendrochronology is also useful in archaeological studies. In Mesa Verde National Park in Colorado, archaeologists used dendrochronology to determine the age of the wooden rafters in the pueblos of the Anasazi, an ancient group of Native Americans. Also, dendrochronology provides a reliable way for geologists to confirm the results from radiocarbon dating.

Ice cores Ice cores are analogous to tree rings. Like tree rings, they contain a record of past environmental conditions in annual layers of snow deposition; summer ice tends to have more bubbles and larger crystals than winter ice. Geologists use ice-core chronologies to study glacial cycles through geologic history. The National Ice Core Facility in Colorado is one of several facilities around the world that store thousands of meters of ice cores from ice sheets, such as the core shown in **Figure 21.18.** Because ice cores contain information about past environmental conditions, scientists also use them to study climate change.

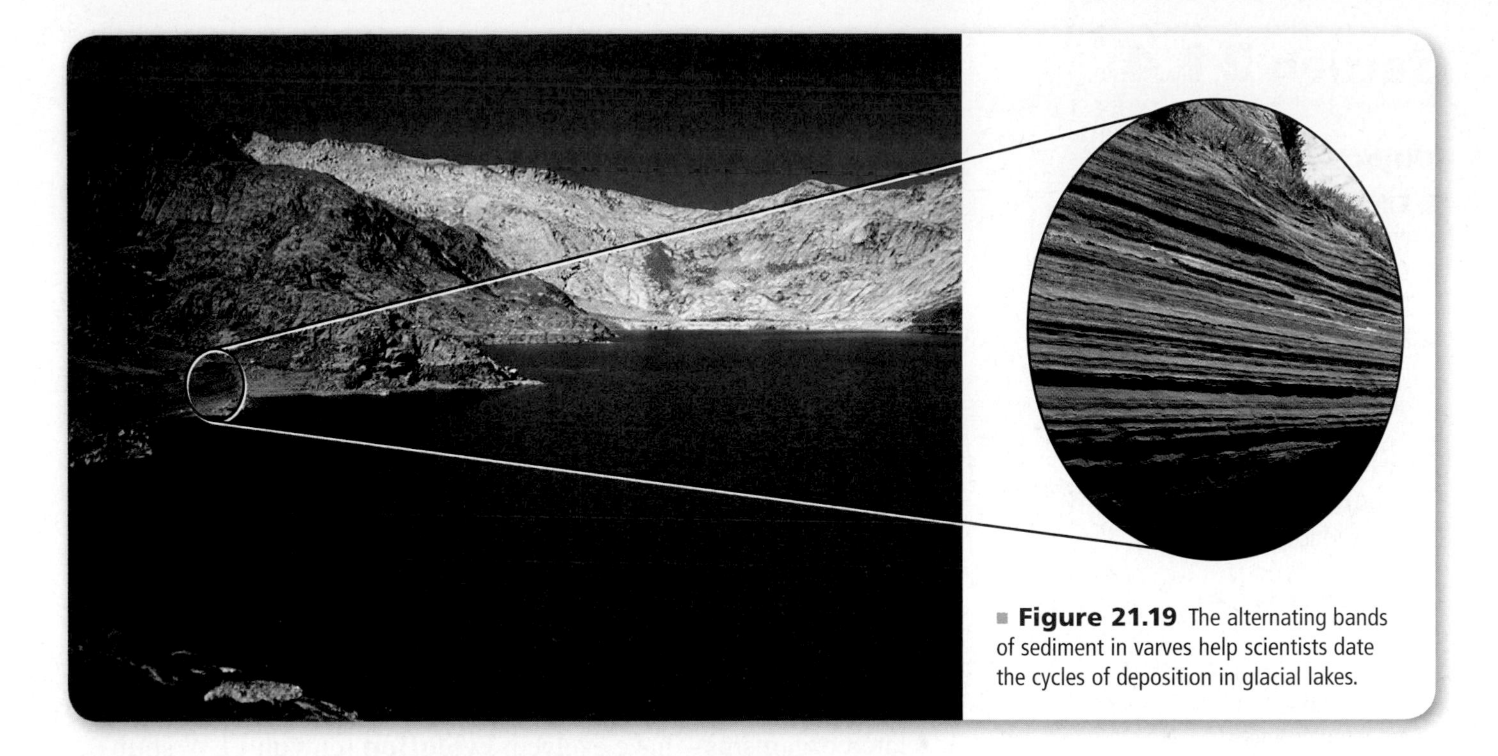

■ **Figure 21.19** The alternating bands of sediment in varves help scientists date the cycles of deposition in glacial lakes.

Varves Bands of alternating light- and dark-colored sediments of sand, clay, and silt are called **varves.** Varves represent the seasonal deposition of sediments, usually in lakes. Summer deposits are generally sand-sized particles with traces of living matter, compared to the thinner, fine-grained sediments of winter. Varves, shown in **Figure 21.19,** are typical of lake deposits near glaciers, where summer meltwaters actively carry sand into the lake, and little to no sedimentation occurs in the winter. Using varved cores, scientists can date cycles of glacial sedimentation over periods as long as 120,000 years.

Section 21.3 Assessment

Section Summary

- Techniques of absolute-age dating help identify numeric dates of geologic events.
- The decay rate of certain radioactive elements can be used as a kind of geologic clock.
- Annual tree rings, ice cores, and sediment deposits can be used to date recent geologic events.

Understand Main Ideas

1. MAIN Idea **Point out** the differences between relative-age dating and absolute-age dating.
2. **Explain** how the process of radioactive decay can provide more accurate measurements of age compared to relative-age dating.
3. **Compare and contrast** the use of U-238 and C-14 in absolute-age dating.
4. **Describe** the usefulness of varves to geologists who study glacial lake deposits.
5. **Discuss** the link between uniformitarianism and absolute-age dating.

Think Critically

6. **Infer** why scientists might choose to use two different methods to date a tree felled by an advancing glacier. What methods might the scientists use?

MATH in Earth Science

7. A rock sample contains 25 percent K-40 and 75 percent daughter product Ar-40. If K-40 has a half-life of 1.3 billion years, how old is the rock?

Section 21.4

Objectives

- **Explain** methods by which fossils are preserved.
- **Describe** how scientists use index fossils.
- **Discuss** how fossils are used to interpret Earth's past physical and environmental history.

Review Vocabulary

groundwater: water beneath Earth's surface

New Vocabulary

evolution
original preservation
altered hard part
mineral replacement
mold
cast
trace fossil
index fossil

Fossil Remains

MAIN Idea **Fossils provide scientists with a record of the history of life on Earth.**

Real-World Reading Link Think about the last time you bought souvenirs while on a vacation or at an event. You might have brought back pictures of the places you saw or the people you visited, or you might have brought back objects with inscribed names and dates. Like souvenirs, fossils are a record of the past.

The Fossil Record

Fossils are the preserved remains or traces of once-living organisms. They provide evidence of the past existence of a wide variety of life-forms, most of which are now extinct. The diverse fossil record also provides evidence that species—groups of closely related organisms—have evolved. **Evolution** (eh vuh LEW shun) is the change in species over time.

When geologists find fossils in rocks, they know that the rocks are about the same age as the fossils, and they can infer that the same fossils found elsewhere are also of the same age. Some fossils, such as the radiolarian microfossils shown in **Figure 21.20,** also provide information about past climates and environments. Radiolarians are unicellular organisms with hard shells that have populated the oceans since the Cambrian Period. When they die, their shells are deposited in large quantities in ocean sediment called radiolarian ooze.

Petroleum geologists use radiolarians and other microfossils to determine the age of rocks that might produce oil. Microfossils provide information about the ages of rocks and can indicate whether the rocks had ever been subjected to the temperatures and pressures necessary to form oil or gas.

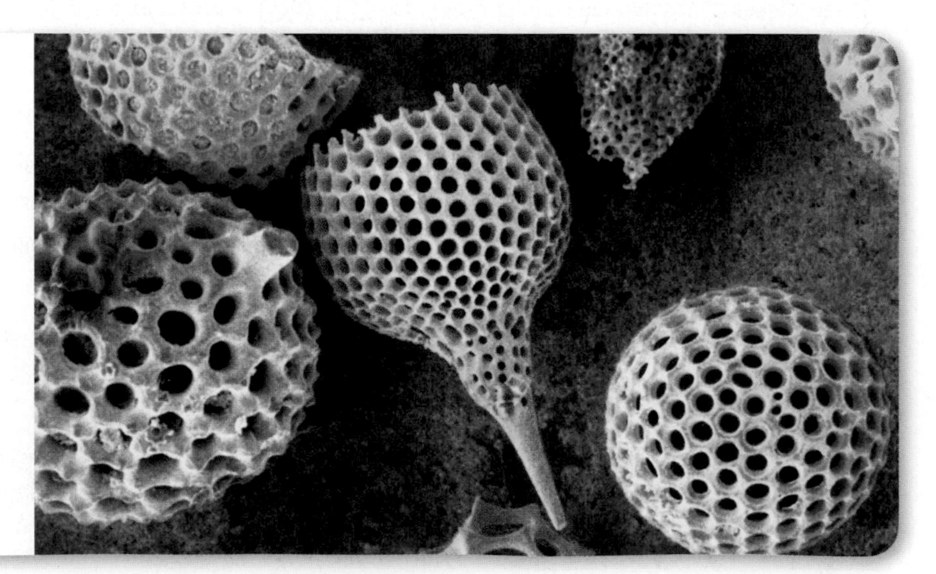

Figure 21.20 These tiny radiolarian microfossils—each no bigger than 1 mm in diameter—provide clues to geologists about ancient marine environments. This photograph is a color-enhanced SEM magnification at 80×.

Original preservation Fossils with **original preservation** are the remains of plants and animals that have been altered very little since the organisms' deaths. Such fossils are uncommon because their preservation requires extraordinary circumstances, such as either freezing, arid, or oxygen-free environments. For example, soft parts of mammoths are preserved in the sticky ooze of California's La Brea Tar Pit. Original woody parts of plants are embedded in the permafrost of 10,000-year-old Alaskan bogs. Tree sap from prehistoric trees sometimes hardens into amber that contains insects, as illustrated in **Figure 21.21.** Soft parts are also preserved when plants or animals are dried and their remains are mummified.

Original preservation fossils can be surprisingly old. For example, in 2005, a scientist from North Carolina discovered soft tissue in a 70-million-year-old dinosaur bone excavated in Montana. Scientists have since found preserved tissue in other dinosaur bones.

Reading Check **Explain** why fossils with original preservation are rare.

Figure 21.21 This insect was trapped in tree sap millions of years ago.

NATIONAL GEOGRAPHIC To read more about fossils and fossil discoveries, go to the **National Geographic Expedition** on page 922.

Altered hard parts Under most circumstances, the soft organic material of plants and animals decays quickly. However, over time, the remaining hard parts, such as shells, bones, or cell walls, can become fossils with **altered hard parts.** These fossils are the most common type of fossil, and can form from two processes.

Mineral replacement In the process of **mineral replacement,** the pore spaces of an organism's buried hard parts are filled in with minerals from groundwater. The groundwater comes in contact with the hard part and gradually replaces the hard part's original mineral material with a different mineral. A shell's calcite ($CaCO_3$), for example, might be replaced by silica (SiO_2). Mineral replacement can occur in trees that are buried by volcanic ash. Over time, minerals dissolved from the ash solidify into microscopic spaces within the wood. The result is a fossil called petrified wood, shown in **Figure 21.22.**

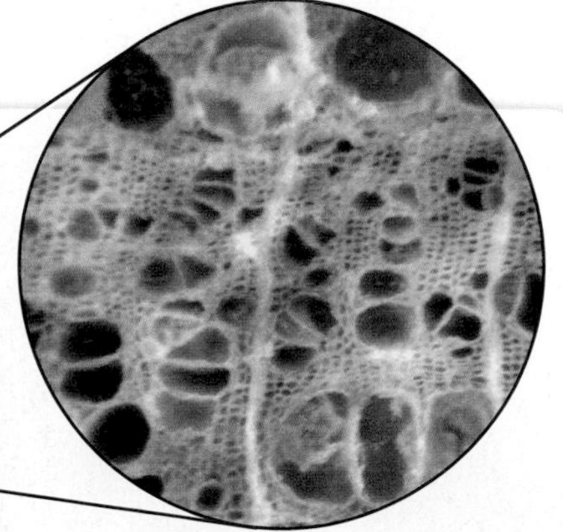

Figure 21.22 Petrified wood is an example of mineral replacement in fossils. The blowout shows that tree rings and cell walls are still evident at 100× magnification with a light microscope.
Describe *from where the minerals in the petrified wood came.*

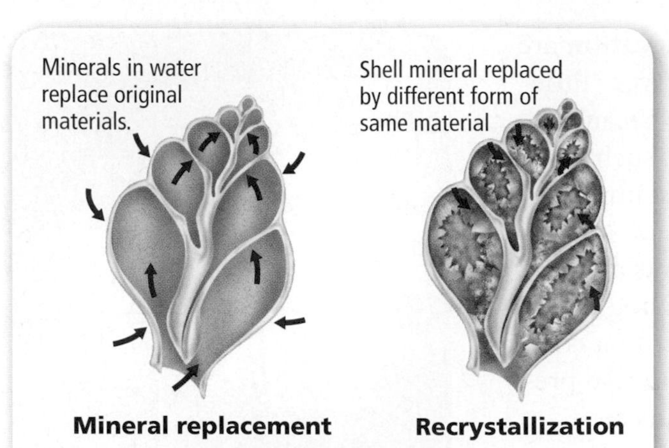

■ **Figure 21.23** During mineral replacement, the minerals in a buried hard part are replaced by other minerals in groundwater. During recrystallization, temperature and pressure change the crystal structure of the hard part's original material.
Explain ***why the internal structure of the shell changes during recrystallization.***

Recrystallization Another way in which hard parts can be altered and preserved is the process of recrystallization (ree krihs tuh luh ZAY shun). Recrystallization can occur when a buried hard part is subjected to changes in temperature and pressure over time. The process of recrystallization is similar to that of mineral replacement, although in mineral replacement the original mineral is replaced by a mineral from the water, whereas in recrystallization the original mineral is transformed into a new mineral. A snail shell, for example, is composed of aragonite ($CaCO_3$). Through recrystallization, the aragonite undergoes a change in internal structure to become calcite, the basic material of limestone or chalk. Though calcite has the same composition ($CaCO_3$) as aragonite, it has a crystal structure that is more stable than aragonite over long periods of time. **Figure 21.23** shows how mineral replacement and recrystallization differ.

Reading Check **Compare and contrast** recrystallization and mineral replacement.

Molds and casts Some fossils do not contain any original or altered material of the original organism. These fossils might instead be molds or casts. A **mold** forms when sediments cover the original hard part of an organism, such as a shell, and the hard part is later removed by erosion or weathering. A hollowed-out impression of the shell, called the mold, is left in its place. A mold might later become filled with material to create a **cast** of the shell. A mold and a cast of a distinctive animal called an ammonite are shown in **Figure 21.24.**

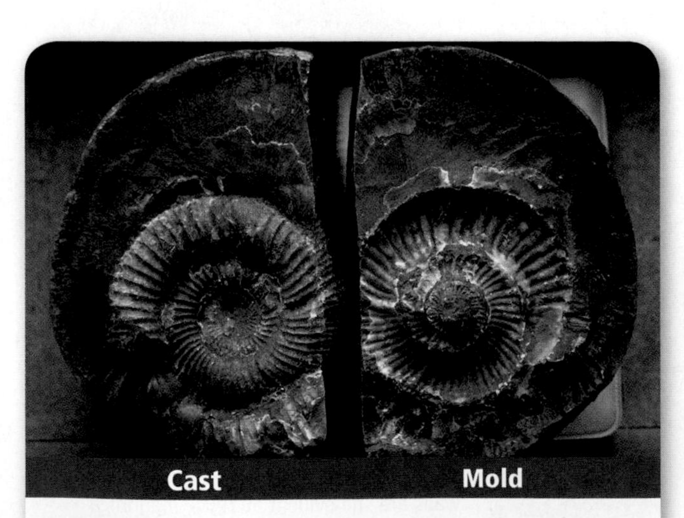

■ **Figure 21.24** A mold of this ammonite was formed when the dead animal's shell eroded. The cavity was later filled with minerals to create a cast.

Trace fossils Sometimes the only fossil evidence of an organism is indirect. Indirect fossils, called trace fossils, include traces of worm trails, footprints, and tunneling burrows. **Trace fossils** can provide information about how an organism lived, moved, and obtained food. For example, dinosaur tracks provide scientists with clues about dinosaur size and walking characteristics. Other trace fossils include gastroliths (GAS truh lihths) and coprolites (KAH pruh lites). Gastroliths are smooth, rounded rocks once present in the stomachs of dinosaurs to help them grind and digest food. Coprolites are the fossilized solid waste materials of animals. By analyzing coprolites, scientists learn about animal eating habits.

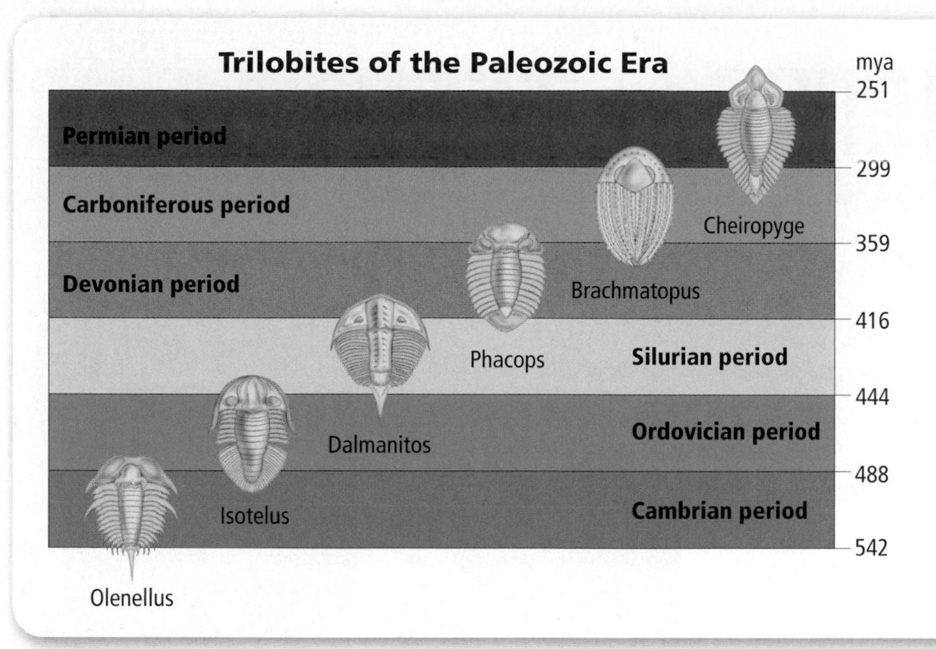

Figure 21.25 These trilobite species make excellent index fossils because each species lived for a relatively short period of time before becoming extinct.

Index Fossils

As you learned in the previous sections, fossils help scientists determine the relative ages of rock sequences through the process of correlation. Some fossils are more useful than others for relative-age dating. **Index fossils** are fossils that are easily recognized, abundant, and widely distributed geographically. They also represent species that existed for relatively short periods of geologic time. The different species of trilobites shown in **Figure 21.25** make excellent index fossils for the Paleozoic Era because each was distinct, abundant, and existed for a certain range of time. If a geologist finds one in a rock layer, he or she can immediately determine an approximate age of the layer.

Section 21.4 Assessment

Section Summary

- Fossils provide evidence that species have evolved.
- Fossils help scientists date rocks and locate reserves of oil, gas, and minerals.
- Fossils can be preserved in several different ways.
- Index fossils help scientists correlate rock layers in the geologic record.

Understand Main Ideas

1. MAIN Idea **Describe** how the fossil record helps scientists understand Earth's history.
2. **List** ways in which fossils can form, and give an example of each.
3. **Explain** how scientists might be able to determine the relative age of a layer of sediment if they find a fossilized trilobite in the layer.
4. **Compare and contrast** a mold and a cast.

Think Critically

5. **Evaluate** Why are the best index fossils widespread?

WRITING in **Earth Science**

6. Imagine that you have just visited a petrified forest. Write a letter to a friend describing the forest. Explain what the forest looks like and how it was fossilized.

EARTH SCIENCE AND TECHNOLOGY

DISCOVERING DINOSAUR TISSUES

Helicopters, explosives, and bulldozers are some of the tools paleontologists use to excavate and transport large dinosaur fossils. CT scans, microscopes, and computer modeling are among the latest technology used to analyze the soft tissues found recently in several dinosaur fossils.

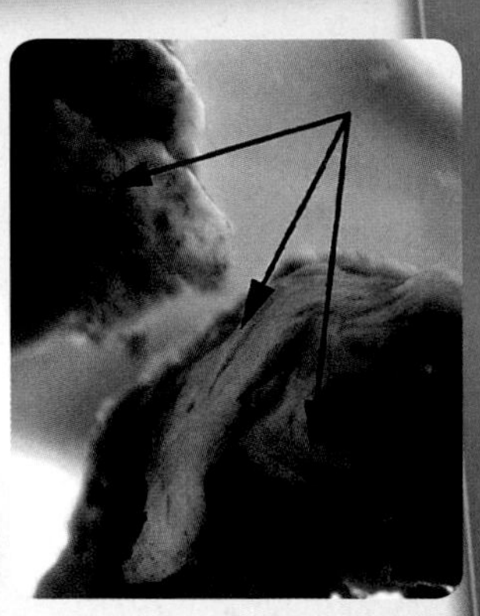
The soft tissue from the *T. rex* discovered in 2003 was almost perfectly preserved, and provides clues about how the dinosaur lived.

Soft tissue During the summer of 2000, paleontologists digging in Montana uncovered a well-preserved hadrosaur, a type of plant-eating dinosaur that lived about 77 mya. The most exciting part of the discovery came when scientists realized that the fossil contained soft tissues including skin, muscle tissue on the shoulder, and tissue from the throat—a rare find. As the fossil was uncovered, scientists found well-preserved stomach contents which revealed that the dinosaur's last meal included ferns and magnolia leaves.

Bone tissue from *Tyrannosaurus rex* In 2003, the fossil of a small *T. rex* was discovered. After excavating it, scientists realized that it was too big to transport by helicopter. As a result, they carefully broke the thighbone into two pieces. Breaking a fossil is unusual because every effort is made to keep bones intact during the transport of a specimen. However, the break led to another surprise. The bone held preserved soft tissues including the connective tissue that makes up bone, blood vessels, and possibly even blood cells.

New technology for old questions
Although other dinosaur specimens with soft tissue were discovered in the early twentieth century, the technology for preservation and analysis did not exist. These recent discoveries, coupled with modern technology, allow scientists new insights for answering old questions. Analysis of soft tissue could help scientists determine whether dinosaurs were warm-blooded or cold-blooded.

Tissue analysis can reveal more information about the diet of a species, which leads to more information about the environment at that time. For example, when the stomach contents of the hadrosaur were analyzed, scientists found over 36 types of pollen samples, including some from plants that could only survive in warm, humid conditions.

The tissues can also give clues about dinosaurs' evolutionary relationships to modern species. For example, in 2006, the proteins in the *T. rex* tissue were sequenced and produced molecular evidence of the relationship between dinosaurs and birds. The collagen found in the *T. rex* sample more closely matched the collagen in chickens than that of other organisms alive today. As more soft tissues are sequenced and analyzed, molecular evidence for more evolutionary relationships may be provided.

WRITING in Earth Science

Poster Make a poster that shows examples of the most recent dinosaur soft tissue discoveries and the types of information scientists can gather by analyzing them.
To learn more about recent dinosaur discoveries, visit glencoe.com.

GEOLAB

DESIGN YOUR OWN: INTERPRET HISTORY-SHAPING EVENTS

Background: **Volcanoes, earthquakes, mountain building, floods, and other geologic events affect the surface of Earth—and the life that inhabits it—in important ways. However, not all events affect Earth equally. Some events in Earth's history have been more critical than others in shaping Earth.**

Question: *What have been the most important events in Earth's history?*

The Sierra Nevadas that extend through California resulted from a series of Earth-shaping events.

Materials

list of Earth-shaping events found at glencoe.com or provided by your teacher
colored pencils
poster board
geologic time scale
reference books
internet access

Procedure

Imagine that NASA is planning to send a space probe to a distant galaxy. You are part of a team that has been assigned the task of listing the most important events that have shaped Earth's history. This list will be carried as part of the spaceship's payload. It will be used to help describe Earth to any possible residents of the galaxy.

1. Read and complete the lab safety form.
2. Form into groups. Each group should have three or four team members.
3. Obtain a list of Earth-shaping events from either glencoe.com or your teacher.
4. Choose two other resources where you can find at least ten more events to add to your list.
5. Brainstorm about the events that you think had the most impact on the direction that Earth's development has taken over time.
6. Discuss the best way to display your list.
7. Make sure your teacher approves your plan.
8. Put your plan into effect.

Analyze and Conclude

1. **Interpret Data** Plot your list on a copy of the geologic time scale. Compare the number of events in each era. Did more Earth-shaping events occur early in Earth's history or later on? Explain.
2. **Compare** your list with the lists of others in your class. What events do all lists share? Do these events share common features?
3. **Infer** Choose one event in the Mesozoic Era, and infer how Earth's history might have progressed had the event not occurred.
4. **Evaluate** How do extinction events influence the development of life on Earth?

SHARE YOUR DATA

Peer Review Visit glencoe.com and post a list of the ten most important events that you think shaped Earth's history. Compare your list with lists of other groups of students who have completed this lab.

CHAPTER 21

Study Guide

Download quizzes, key terms, and flash cards from glencoe.com.

BIG Idea Scientists use several methods to learn about Earth's long history.

Vocabulary	Key Concepts
Section 21.1 The Rock Record	
• eon (p. 592) • epoch (p. 593) • era (p. 593) • geologic time scale (p. 590) • mass extinction (p. 594) • period (p. 593) • Precambrian (p. 592)	**MAIN Idea** Scientists organize geologic time to help them communicate about Earth's history. • Scientists organize geologic time into eons, eras, periods, and epochs. • Scientists divide time into units based on fossils of plants and animals. • The Precambrian makes up nearly 90 percent of geologic time. • The geologic time scale changes as scientists learn more about Earth.
Section 21.2 Relative-Age Dating	
• correlation (p. 599) • cross-cutting relationship (p. 597) • key bed (p. 599) • principle of inclusions (p. 597) • original horizontality (p. 596) • relative-age dating (p. 596) • superposition (p. 596) • unconformity (p. 598) • uniformitarianism (p. 595)	**MAIN Idea** Scientists use geologic principles to learn the sequence in which geologic events occurred. • The principle of uniformitarianism states that processes occurring today have been occurring since Earth formed. • Scientists use geologic principles to determine the relative ages of rock sequences. • An unconformity represents a gap of time in the rock record. • Geologists use correlation to compare rock layers in different geographic areas.
Section 21.3 Absolute-Age Dating	
• absolute-age dating (p. 601) • dendrochronology (p. 604) • half-life (p. 602) • radioactive decay (p. 601) • radiocarbon dating (p. 603) • radiometric dating (p. 602) • varve (p. 605)	**MAIN Idea** Radioactive decay and certain kinds of sediments help scientists determine the numeric age of many rocks. • Techniques of absolute-age dating help identify numeric dates of geologic events. • The decay rate of certain radioactive elements can be used as a kind of geologic clock. • Annual tree rings, ice cores, and sediment deposits can be used to date recent geologic events.
Section 21.4 Fossil Remains	
• altered hard part (p. 607) • cast (p. 608) • evolution (p. 606) • index fossil (p. 609) • mineral replacement (p. 607) • mold (p. 608) • original preservation (p. 607) • trace fossil (p. 608)	**MAIN Idea** Fossils provide scientists with a record of the history of life on Earth. • Fossils provide evidence that species have evolved. • Fossils help scientists date rocks and locate reserves of oil, gas, and minerals. • Fossils can be preserved in several different ways. • Index fossils help scientists correlate rock layers in the geologic record.

Assessment

Vocabulary Review

Match each definition with the correct vocabulary term from the Study Guide.

1. the record of Earth's history from its origin to the present
2. a gap in the rock record caused by erosion
3. the emission of radioactive isotopes and the resulting change into other products over time
4. the largest time unit in the geologic time scale
5. the matching of unique outcrops among regions

Distinguish between the vocabulary terms in each pair.

6. period, epoch
7. altered hard part, original preservation
8. absolute-age dating, relative-age dating
9. fossil, index fossil
10. mold, cast

The sentences below are incorrect. Make each sentence correct by replacing the italicized word or phrase with a term from the Study Guide.

11. *Original horizontality* is the principle that a fault or intrusion is younger than the rock it intersects.
12. *Relative-age dating* states that processes operating today have been operating since Earth formed.
13. A *varve* is a sedimentary layer used to match rock layers across large areas.
14. *Correlation* is the change in species over time.

Understand Key Concepts

15. The end of which era is marked by the largest extinction event in Earth's history?
    A. Cenozoic
    B. Mesozoic
    C. Paleozoic
    D. Precambrian

16. How old is a mammoth's tusk if 25 percent of the original C-14 remains in the sample? The half-life of C-14 is 5730 years.
    A. 5730 years
    B. 11,460 years
    C. 17,190 years
    D. 22,920 years

Use the figure below to answer Question 17.

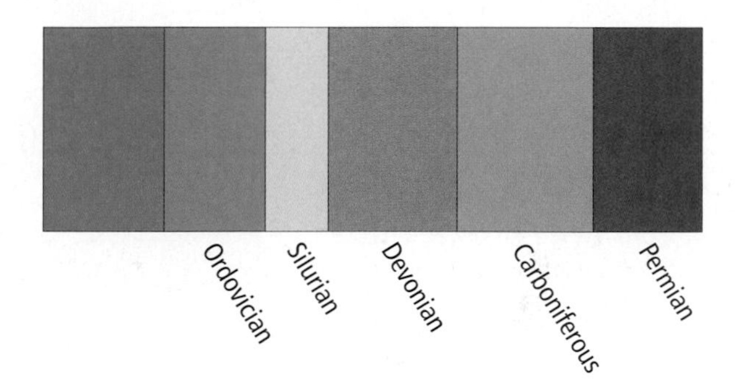

17. Which time period is missing in the diagram?
    A. Cambrian
    B. Permian
    C. Triassic
    D. Paleogene

18. Which is not a typical characteristic of an index fossil?
    A. was commonplace while alive
    B. existed for a long period of time
    C. is geographically widespread
    D. is easily recognizable

19. Which is the smallest division of geologic time?
    A. period
    B. eon
    C. era
    D. epoch

20. Which geologic principle is used when a geologist observes an outcrop of rocks and determines that the bottom layer is the oldest?
    A. uniformitarianism
    B. original horizontality
    C. superposition
    D. inclusion

21. Uranium-238 breaks down into thorium-234. Which is thorium-234 in relation to uranium-238?
A. parent
B. brother
C. son
D. daughter

Use the figure below to answer Question 22.

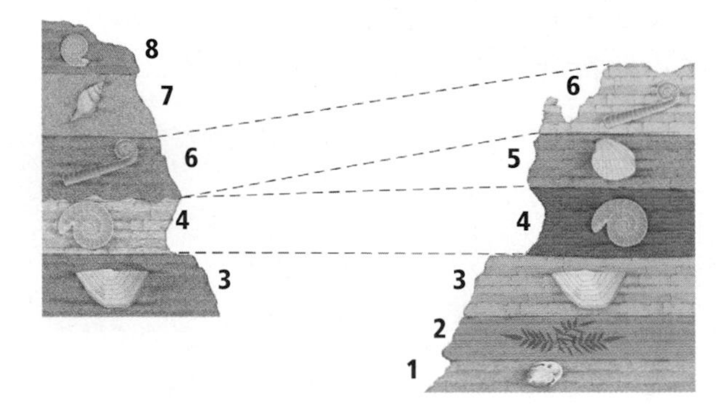

22. What does the diagram show?
A. uniformitarianism
B. inclusion
C. cross-cutting relationships
D. correlation

23. Trees that have been buried by volcanic ash are likely to be preserved in which manner?
A. original preservation
B. mummification
C. mineral replacement
D. recrystallization

24. Which are glacial lake sediments that show cycles of deposition?
A. annual rings
B. tillites
C. varves
D. unconformities

Constructed Response

25. Sequence the steps by which a mold and a cast are formed.

26. Explain why mass extinctions are important to geologists.

27. Compare and contrast absolute-age dating and relative-age dating.

28. Assess the usefulness of a universally accepted geologic time scale.

29. Explain, in your own words, why an unconformity is any gap in the rock record.

30. Argue for or against making the time units of the geologic time scale of equal duration.

31. Relate How are microscopic fossils associated with discovering oil at a particular site?

Think Critically

Use the diagram below to answer Questions 32 to 34.

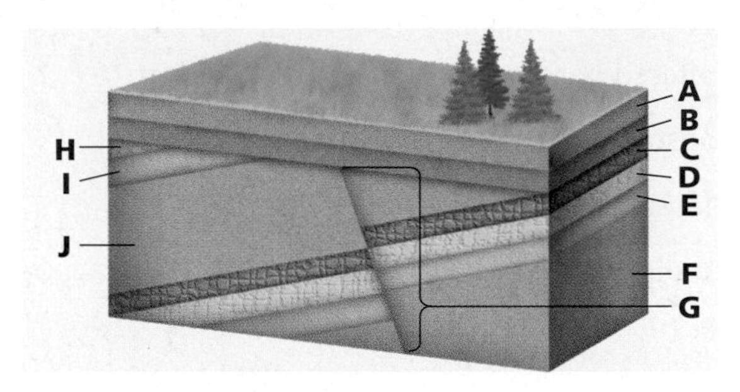

32. Identify the oldest rock layer in the diagram.

33. Find an angular unconformity in the diagram.

34. Apply List the order of geologic events in the diagram from oldest to youngest along with the geologic principles that you used.

35. Critique this statement: The principles for determining relative age are based on common sense.

36. Create One way to remember the order of words in a sequence is to create a phrase, called an acrostic, that uses the same first letter of each word in the sequence. For example, "**M**y **D**ear **A**unt **S**ally" is often used to remember the mathematical order of operations: **M**ultiply and **D**ivide before you **A**dd and **S**ubtract. Create an acrostic to help you remember the periods of the Phanerozoic Eon.

37. **Solve** The half-life of K-40 is 1.3 billion years. What is the age of an ancient igneous rock that contains a mineral with 12.5 percent K-40 and 87.5 percent Ar-40?

38. **Compare and contrast** an index fossil and a key bed.

39. **Assess** whether a clam or a spider has a better chance at becoming a fossil.

40. **Evaluate** Can radiocarbon dating be used to determine the age of a dinosaur bone? Explain.

Use the diagram below to answer Question 41.

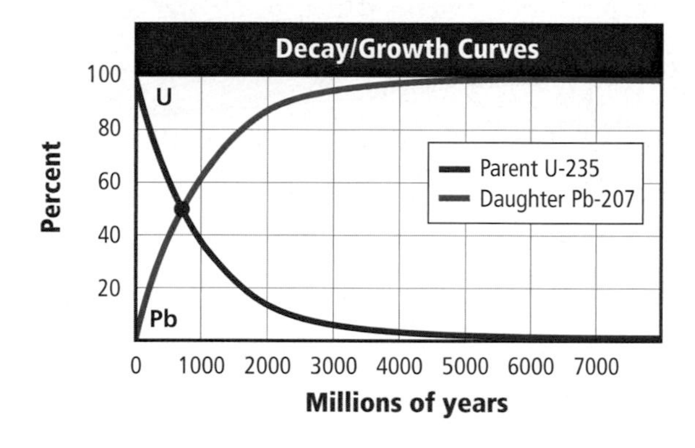

41. **Analyze** What does the red dot signify on the graph?

42. **CAREERS IN EARTH SCIENCE** A geologist discovers wood buried within sediments of a landslide that is thought to have been caused by an ancient earthquake. Explain two methods that the geologist could use to determine when the earthquake occurred.

Concept Mapping

43. Create a concept map using the following terms: *absolute-age dating, geologic time scale, relative-age dating, fossils, unconformities,* and *radiometric dating.*

Challenge Question

44. **Assess** Do you think domestic dogs might make good index fossils for future geologists? Explain.

Additional Assessment

45. **WRITING in Earth Science** Imagine that you are a bacterium that lives for only 20 minutes. Explain how your observations about the world would be different from those of a human being who lives for about 80 years. Evaluate the difference between human time and geologic time.

DBQ Document–Based Questions

Data obtained from: Bambach, R.K., et al. 2004. Origination, extinction, and mass depletion of marine diversity. *Paleobiology* 30: 522–542.

The figure below plots the diversity, measured as the number of different types, of marine animals throughout the Phanerozoic. The animals are grouped into levels of organization called genera that include closely related species. Use the data to answer the questions below.

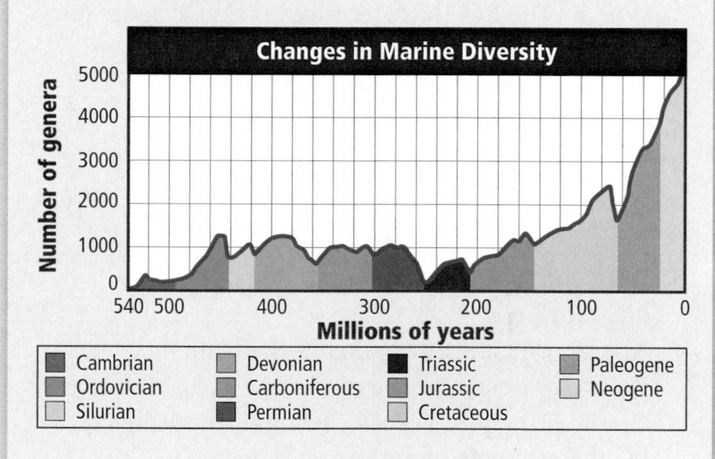

46. With what do the largest drops in diversity coincide on the geologic time scale?

47. Explain what the decreases in diversity mean.

48. Use the information in the graph to support adding one or more new eras to the Phanerozoic.

Cumulative Review

49. What subatomic particles make up the nucleus of an atom? **(Chapter 3)**

50. Why do tornadoes occur most frequently in the central United States? **(Chapter 13)**

Standardized Test Practice

Multiple Choice

1. Which type of mountains form as the result of uplift far from plate boundaries?
   A. ocean ridges
   B. fault block mountains
   C. folded mountains
   D. volcanic ranges

Use the diagram below to answer Questions 2 and 3.

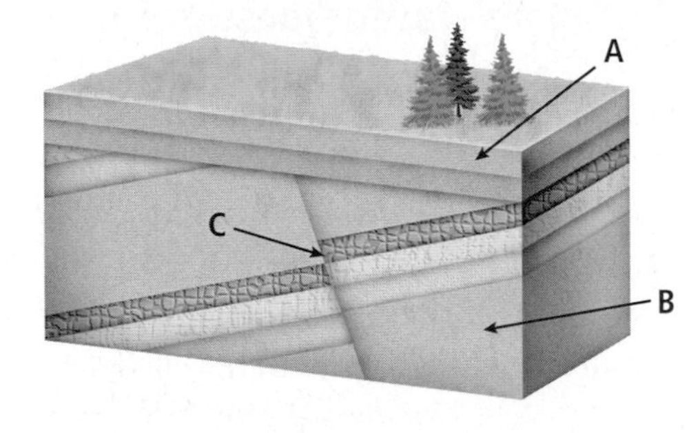

2. Which principle for determining relative age is relevant to Point A in this diagram of a rock region?
   A. the principle of original horizontality
   B. the principle of superposition
   C. the principle of cross-cutting relationships
   D. the principle of uniformitarianism

3. Which principle is relevant to Point C in this diagram?
   A. the principle of original horizontality
   B. the principle of superposition
   C. the principle of cross-cutting relationships
   D. the principle of uniformitarianism

4. What aspect of the discovery of ocean ridges was important in the scientific community?
   A. their location
   B. their volcanic activity
   C. their height
   D. their age

5. Which is not a factor affecting the formation of magma?
   A. time
   B. temperature
   C. pressure
   D. water

6. Earth's crust is broken up into a dozen or more enormous slabs called what?
   A. boundaries
   B. tectonic plates
   C. subduction zones
   D. subduction plates

Use the diagram below to answer Questions 7 and 8.

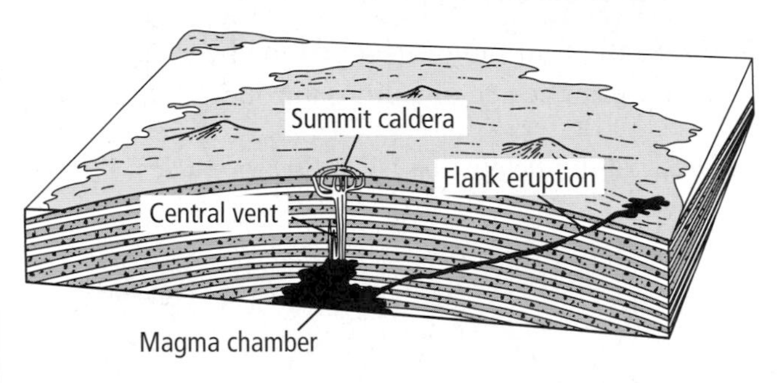

7. Which type of volcano is shown?
   A. cinder cone
   B. composite
   C. shield
   D. pyroclastic

8. What level of threat might the development of this volcano pose to humans?
   A. Low; it is built as layer upon layer and accumulates during nonexplosive eruptions.
   B. Low; it is considered to be an inactive volcano.
   C. Moderate; it forms when pieces of magma explode and build around a vent, but is rather small.
   D. High; it has a violently explosive nature.

9. What happened to the magnetic fields generated by magnetic rocks along the ocean floor?
   A. There were no reversals of the magnetic field along the ocean floor.
   B. Each side of an ocean ridge had its own magnetic pattern.
   C. Normal and reverse polarity regions formed stripes that ran perpendicular to ocean ridges.
   D. Normal and reverse polarity regions formed stripes that ran parallel to ocean ridges.

10. What does orogeny refer to?
    A. the drifting of microcontinents
    B. the building of mountain ranges
    C. the formation of volcanic islands
    D. the breaking apart of supercontinents

Short Answer

Use the map below to answer Questions 11–13.

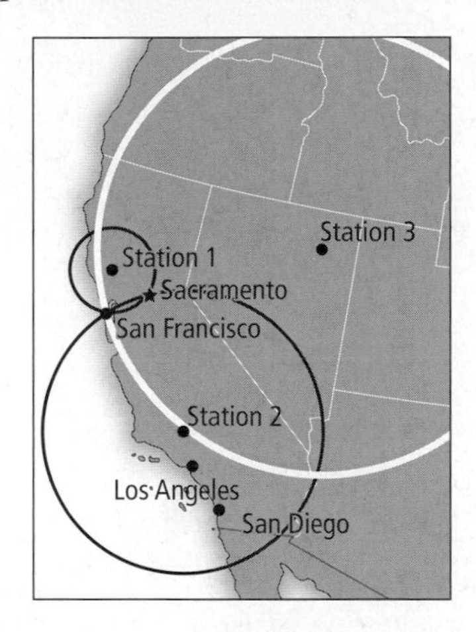

11. According to the map, where was the epicenter of the earthquake located? How can the epicenter be determined?

12. Why is it important to use three stations to locate the epicenter of an earthquake?

13. How might this earthquake affect Los Angeles?

14. The Florida peninsula gets more thunderstorms than any other part of the United States. What geographic feature of Florida causes it to get so many thunderstorms? How does this feature allow thunderstorms to form?

15. Why is the Appalachian Mountain Belt divided into several regions?

16. Describe acid precipitation in terms of the pH scale and the reason for its pH value.

Reading for Comprehension

Dating Gold

The radioactive decay of metal inside South African gold nuggets helped scientists determine the origin of the world's largest gold deposit. The placer model indicates the gold is older than surrounding rock. The hydrothermal model indicates that the hot spring fluids deposited the gold inside the rocks. It was decided to determine the age of the gold itself. If the gold is older than the rocks in which it is found, then the rocks must have built up around the gold, bolstering the placer model. If the gold is younger than the rocks, that means it must have seeped in with fluids, supporting the hydrothermal model. Two elements found inside gold, rhenium and osmium, serve as a radioactive clock. Rhenium decays into osmium over very long spans of time—it takes about 42.3 billion years for half of a sample of rhenium to transmute. By dissolving gold grains in acid and measuring the ratio of rhenium to osmium, scientists can determine the gold's age. Gold from places in the Rand is three billion years old—a quarter of a billion years older than its surrounding rock, thus supporting the placer model.

Article obtained from: Choi, C. 2002. Origin of world's largest gold deposit found? *United Press International Science News* (September): 1-2.

17. What is the half-life of rhenium?
    - A. 42.3 years
    - B. 42.3 thousand years
    - C. 42.3 million years
    - D. 42.3 billion years

18. Why was this study conducted?
    - A. to determine the origin of the gold deposit
    - B. to disprove the hydrothermal model
    - C. to support the placer model
    - D. to explain radioactive decay

NEED EXTRA HELP?

If You Missed Question . . .	1	2	3	4	5	6	7	8	9	10	11	12	13	14	15	16
Review Section . . .	20.3	21.2	21.2	21.3	18.2	17.3	18.1	18.1	17.2	20.2	19.3	19.3	19.3	13.1	20.2	7.1

CHAPTER 22

The Precambrian Earth

BIG Idea **The oceans and atmosphere formed and life began during the three eons of the Precambrian, which spans nearly 90 percent of Earth's history.**

22.1 Early Earth
MAIN Idea Several lines of evidence indicate that Earth is about 4.56 billion years old.

22.2 Formation of the Crust and Continents
MAIN Idea The molten rock of Earth's early surface formed into crust and then continents.

22.3 Formation of the Atmosphere and Oceans
MAIN Idea The formation of Earth's oceans and atmosphere provided a hospitable environment for life to begin.

22.4 Early Life on Earth
MAIN Idea Life began on Earth fewer than a billion years after Earth formed.

GeoFacts

- Stromatolites are mounded structures made by tiny organisms called cyanobacteria.
- Stromatolites dominated Precambrian oceans for billions of years.
- NASA scientists use stromatolite gas emissions as a marker to search for extraterrestrial life.

Start-Up Activities

LAUNCH Lab

How do liquids of different densities model early Earth?

Earth's core, mantle, and crust have different average densities. The core is the most dense, the crust is the least dense, and the mantle lies between. Scientists think that early in Earth's history, temperatures were hot enough for the materials that make up Earth to act like liquids.

Procedure

1. Read and complete the lab safety form.
2. Fill a **250-mL beaker** with 50 mL of **tap water.**
3. Pour 50 mL of **vegetable oil** into the beaker.
4. Pour 50 mL of **corn syrup** into the beaker.
5. Allow the mixture to sit for a few minutes.

Analysis

1. **Describe** what happened to the liquids in the beaker.
2. **Identify** which component of the experiment represents Earth's mantle, which represents Earth's crust, and which represents Earth's core.
3. **Relate** the results to the formation of layers in early Earth.

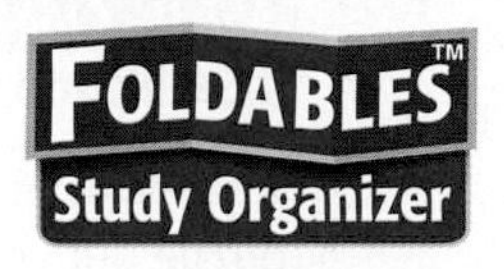

Formation of Earth's Atmosphere Make this Foldable to compare Earth's atmosphere in the early Precambrian to its atmosphere in the late Precambrian.

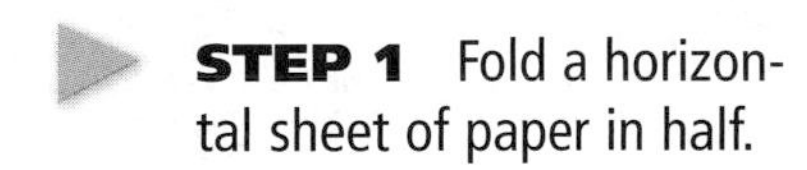

STEP 1 Fold a horizontal sheet of paper in half.

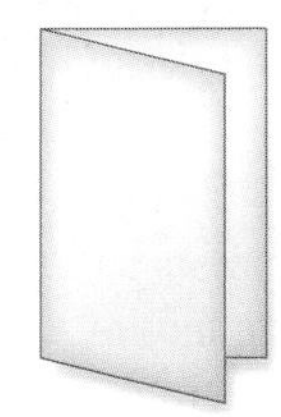

STEP 2 Unfold and fold up the bottom edge about 6 cm.

STEP 3 Staple or glue the edges and center of the bottom flap to make two pockets. Label as shown.

FOLDABLES Use this Foldable with Section 22.3. As you read this section, summarize what you learn about Earth's atmosphere on index cards or quarter-sheets of paper.

Visit glencoe.com to

- study entire chapters online;
- explore Concepts In Motion animations:
  - Interactive Time Lines
  - Interactive Figures
  - Interactive Tables
- access Web Links for more information, projects, and activities;
- review content with the Interactive Tutor and take Self-Check Quizzes.

Section 22.1

Objectives

- **Describe** the evidence that indicates Earth is 4.56 billion years old.
- **Describe** the heat sources of early Earth.

Review Vocabulary

metamorphism: changes in the mineral composition or structure of rocks caused by pressure and temperature over time

New Vocabulary

zircon
meteorite
asteroid

Early Earth

MAIN Idea Several lines of evidence indicate that Earth is about 4.56 billion years old.

Real-World Reading Link Imagine that you are putting together a jigsaw puzzle but you do not have the picture on the box. You do not know what the puzzle looks like, and you have only about 10 percent of the pieces. This is similar to the challenge that scientists face when they study the early Precambrian.

The Age of Earth

The Precambrian, which includes the Hadean, Archean, and Proterozoic Eons, is a time period that spans nearly 90 percent of Earth's history. When Earth first formed it was hot, volcanically active, and no continents existed on its surface. Rocks of Earth's earliest eon—the Hadean—do not exist, so scientists know very little about Earth's first 700 million years. The oldest existing rocks, and the earliest signs of life, are from the Archean. As illustrated in **Figure 22.1,** the earliest life-forms were simple, unicellular organisms.

Crustal rock evidence Absolute-age dating has revealed that the oldest crustal rocks are between 3.96 and 3.8 billion years in age. Evidence that Earth is older than 3.96 billion years exists in small grains of the mineral zircon ($ZrSiO_4$) found in certain metamorphosed Precambrian rocks in Australia. Because **zircon** is a stable and common mineral that can survive erosion and metamorphism, scientists often use it to age-date old rocks. Geologists theorize that the zircon in the Australian rocks is residue from crustal rocks that no longer exist. Based on radiometric dating, which shows that the zircon is at least 4.4 billion years old, Earth must also be at least this old.

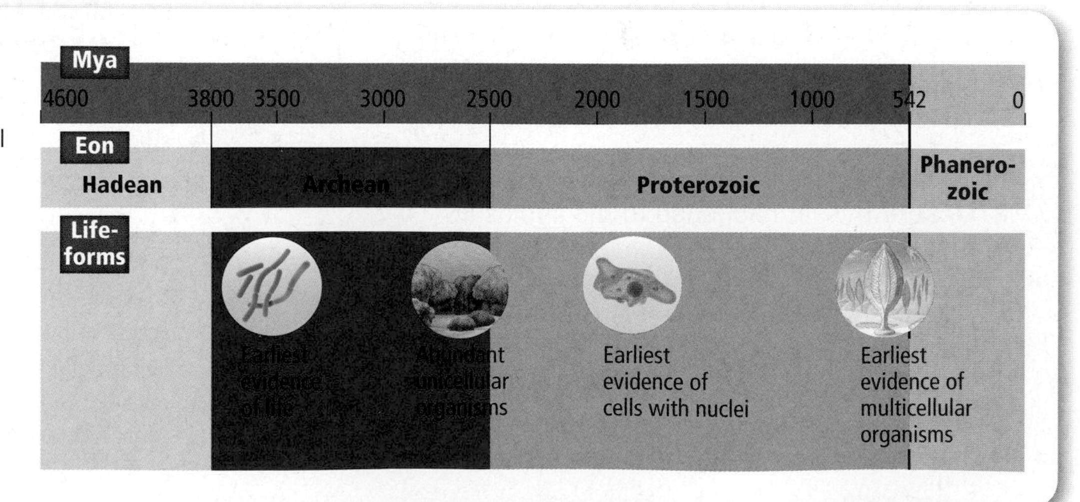

■ **Figure 22.1** The Precambrian lasted for nearly 4 billion years. Multicellular organisms did not appear until the end of the Proterozoic.

Solar system evidence Evidence from meteorites (MEE tee uh rites) and other bodies in the solar system suggests that Earth is more than 4.4 billion years old. **Meteorites** are small fragments of orbiting bodies that have fallen on Earth's surface. They have fallen to Earth throughout Earth's history, but most have been dated at between 4.7 and 4.5 billion years old. Many scientists agree that all parts of the solar system formed at the same time, so they assume that Earth and meteorites are approximately the same age.

In addition, the oldest rock samples from the Moon, collected during the *Apollo* missions in the 1970s, have been dated at 4.45 billion years old. Scientists think that the Moon formed very early in Earth's history when a massive solar system body collided with Earth. You will learn more about the Moon's formation in Chapter 28. Considering all the evidence, scientists agree that Earth is about 4.56 billion years old.

Reading Check **Explain** why scientists think that Earth is older than the oldest rocks in the crust.

Early Earth's Heat Sources

Earth was extremely hot after it formed. There were three likely sources of this heat: Earth's gravitational contraction, radioactivity, and bombardment by asteroids, meteorites, and other solar system bodies.

Gravitational contraction Scientists think that Earth formed by the gradual accumulation of small, rocky bodies in orbit around the Sun, as illustrated in **Figure 22.2.** As Earth accumulated these small bodies, it grew in size and mass. With increased mass came increased gravity. Gravity caused Earth's center to squeeze together with so much force that the pressure raised Earth's internal temperature.

Radioactivity A second source of Earth's heat was the decay of radioactive isotopes, which you learned about in Chapter 3. Scientists know that certain radioactive isotopes were more abundant in Earth's past than they are today. While some of these isotopes, such as uranium-238, are long-lasting and continue to decay today, others were short-lived and have nearly disappeared. Radioactive decay generates heat. Because there were more radioactive isotopes in early Earth, more heat was generated, making Earth hotter than it is today.

Figure 22.2 The accumulation of small orbiting bodies gradually formed Earth. As Earth grew in mass, gravity caused Earth to contract, generating heat.

Asteroid and meteorite bombardment A third source of heat in early Earth came from the impacts of meteors, asteroids (AS tuh roydz), and other objects in the solar system. **Asteroids** are metallic or silica-rich objects between 1 km and 950 km in diameter. Today, most asteroids orbit the Sun between the orbits of Mars and Jupiter. Large asteroids seldom collide with Earth. Planetary geologists estimate that only about 60 objects with diameters of 5 km or more have struck Earth during the last 600 million years. Most objects that hit Earth today are meteorites—fragments of asteroids.

However, evidence from the surfaces of the Moon and other planets suggests that for the first 500 to 700 million years of Earth's history, many more asteroids were distributed throughout the solar system than there are today and that collisions were much more frequent. The impacts of these bodies on Earth's surface generated a tremendous amount of thermal energy. Scientists think that the massive collision that likely formed the Moon generated so much heat that much of Earth melted. The debris (duh BREE) from the impacts also caused a blanketing effect, which prevented the newly generated heat from escaping to space.

CAREERS IN EARTH SCIENCE

Planetary Geologist Planetary geologists, or astrogeologists, study the planets and their places in the solar system and universe. Some planetary geologists study conditions under which extraterrestrial life might exist. To learn more about Earth science careers, visit glencoe.com.

Cooling The combined effects of gravitational contraction, radioactivity, and bombardment made Earth's beginning very hot. Eventually, Earth's surface cooled enough for an atmosphere and oceans to form. Scientists do not know exactly how long it took for this to happen, but evidence suggests that Earth cooled enough for liquid water to form within its first 200 million years. The cooling process continues even today. As much as half of Earth's internal heat remains from Earth's formation.

Section 22.1 Assessment

Section Summary

- Scientists use Earth rocks, zircon crystals, moon rocks, and meteorites to determine Earth's age.
- Likely heat sources of early Earth were gravitational contraction, radioactivity, and asteroid and meteorite bombardment.
- Cooling of Earth led to the formation of liquid water.

Understand Main Ideas

1. MAIN Idea **Summarize** the data that scientists use to determine Earth's age.
2. **Explain** why scientists think that moon rocks and meteorites are the same age as Earth.
3. **Explain** how gravitational contraction, radioactivity, and asteroid and meteorite bombardment heated early Earth.
4. **Describe** the importance of zircon as an age-dating tool.

Think Critically

5. **Evaluate** Which of Earth's early sources of heat are not major contributors to Earth's present-day internal heat?

MATH in Earth Science

6. If an average of 5000 asteroids bombarded Earth every million years during the Hadean, calculate the total number of asteroid impacts that occurred during this eon. Refer to **Figure 22.1** for information on geologic time scales.

Earth Science Online Self-Check Quiz glencoe.com

Section 22.2

Objectives

- **Summarize** the process by which Earth differentiated.
- **Explain** the origin of Earth's crust and continents.
- **Describe** how the continents grew during the Precambrian.

Review Vocabulary

magma: molten, liquid rock material found underground

New Vocabulary

differentiation
microcontinent
craton
Precambrian shield
Canadian Shield
Laurentia

Formation of the Crust and Continents

MAIN Idea The molten rock of Earth's early surface formed into crust and then continents.

Real-World Reading Link Have you ever cooked pudding? If so, you might have noticed that when the pudding cooled, a crust formed on the top. Scientists think that Earth's crust formed in a similar way.

Formation of the Crust

Because of the intense heat in early Earth, many scientists think that much of the planet consisted of hot, molten magma. As Earth cooled, the minerals and elements in this molten magma became concentrated in specific density zones.

Differentiation Scientists know that less-dense materials float on top of more-dense materials. As you observed in the Launch Lab, oil floats on water because oil is less dense than water. This same general principle operated on early molten Earth. The element with the highest density—iron—sank toward the center. In contrast, the light elements, such as silicon and oxygen, remained closer to the surface. The process by which a planet becomes internally zoned when heavy materials sink toward its center and lighter materials accumulate near its surface is called **differentiation** (dih fuh ren shee AY shun). The differentiated zones of Earth are illustrated in **Figure 22.3.**

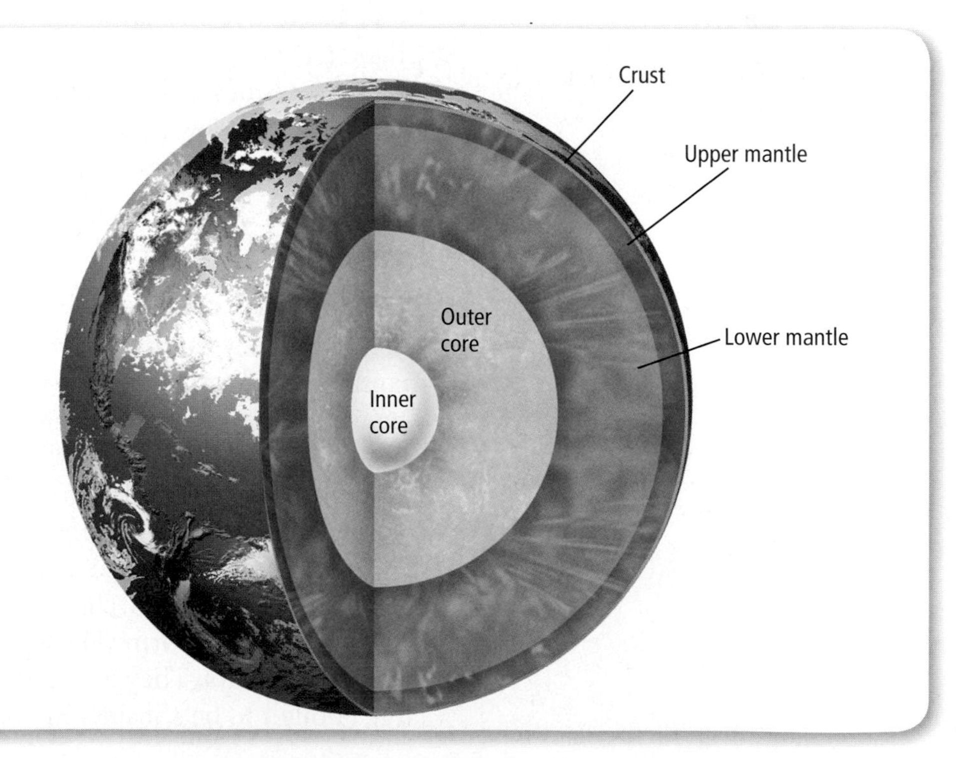

Figure 22.3 Earth differentiated into layers shortly after it formed.
Analyze *What is the densest part of Earth?*

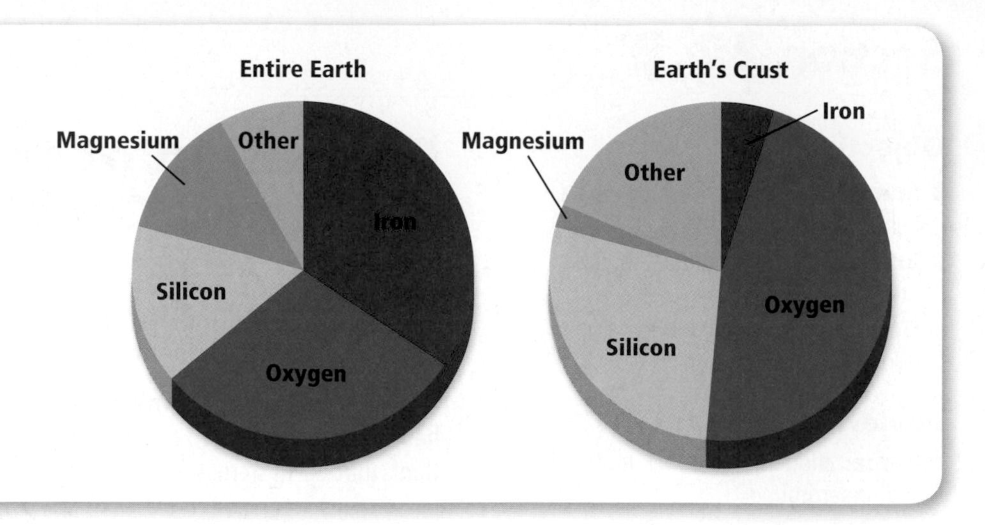

■ **Figure 22.4** Larger amounts of dense elements are found in Earth as a whole than are found in Earth's crust. **Estimate** *the percentage of iron in Earth's crust and in the entire Earth.*

Relative densities The process of differentiation explains the relative densities of parts of Earth today. **Figure 22.4** compares the proportions of elements in Earth's crust and in Earth as a whole. Notice that iron, a dense element, is much less abundant in the crust than it is in the entire Earth, while the crust has a higher proportion of less-dense elements, such as silicon and oxygen. This also explains why granite occurs on Earth's surface. Granite is composed mainly of feldspar, mica, and quartz, which, as you learned in Chapter 4, are minerals with low densities.

 **Reading Check** **Explain** why there is more iron in Earth's core than there is in the crust.

Vocabulary
Science usage v. Common usage
Differentiate
Science usage: to layer into distinct zones

Common usage: to distinguish; to mark as different

Earliest crust Some type of early crust formed as soon as Earth's upper layer began to cool. This crust was probably similar to the basaltic crust that underlies Earth's oceans today. Recall from Chapter 17 that present-day oceanic crust is recycled at subduction zones. Pieces of Earth's early crust were also recycled, though scientists do not know how the recycling occurred. Some suggest that it occurred by a process that does not occur on Earth today. Most agree that the recycling was vigorous—so vigorous that none of Earth's earliest crust exists today.

Continental crust As the early crustal pieces were returned to the mantle, they carried water. The introduction of water into the mantle was essential for the formation of the first continental crust. The water reacted with the mantle material to produce new material that was less dense than the original crustal pieces. As this material reemerged on Earth's surface, it crystallized to form small fragments of granite-containing crust. Granite makes up much of the crust that forms Earth's continents today. As volcanic activity continued during the Archean, small fragments of granite-rich crust continued to form. These crustal fragments are called **microcontinents.** They are called this because they were not large enough to be considered continents.

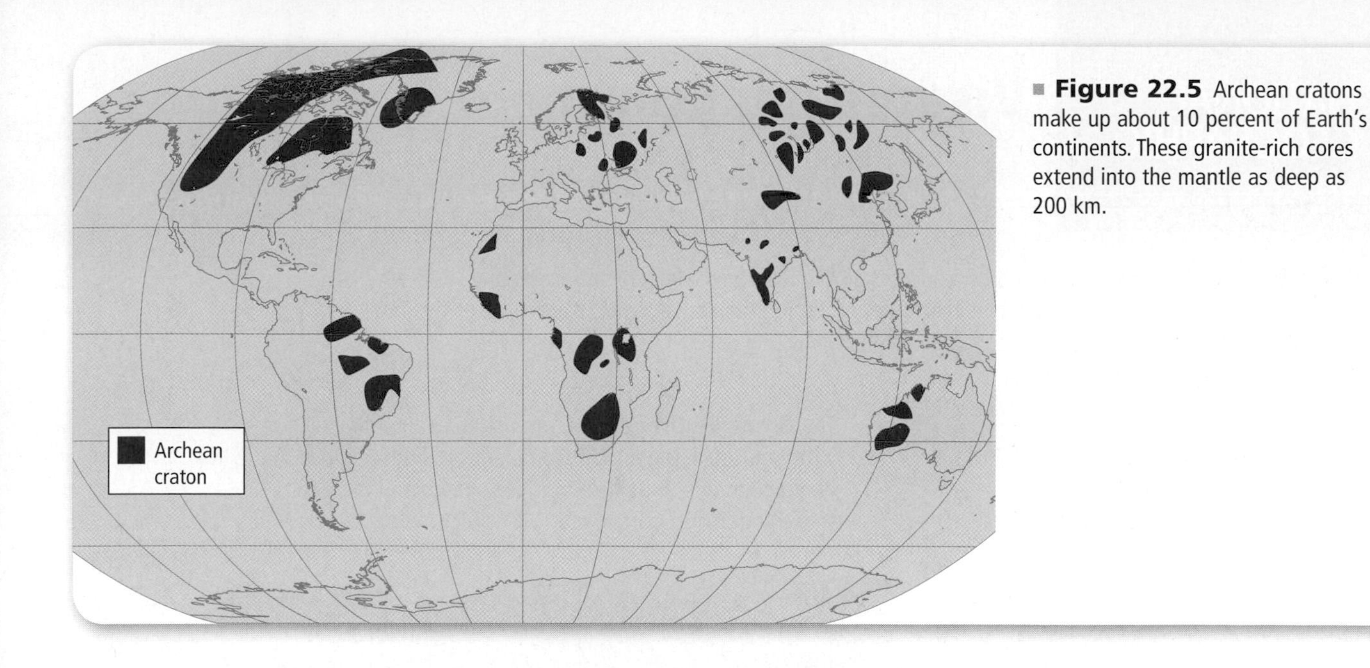

■ **Figure 22.5** Archean cratons make up about 10 percent of Earth's continents. These granite-rich cores extend into the mantle as deep as 200 km.

Cratons Most of the microcontinents that formed during the Archean and early Proterozoic still exist as the cores of today's continents. A **craton** (KRAY tahn) is the oldest and most stable part of a continent. It is attached to a part of the upper mantle that has a depth that can extend to 200 km. Cratons are made of granitic rocks, such as granite and gneiss, with alternating bands of metamorphosed basaltic rocks, which represent ancient continental collisions. As shown in **Figure 22.5,** the Archean cratons represent about 10 percent of Earth's total landmass.

Precambrian shields Most of the cratons are buried beneath sedimentary rocks. However, in some places deep erosion has exposed the rocks of the craton. This exposed area is called a **Precambrian shield.**

In North America, the Precambrian shield is called the **Canadian Shield** because much of it is exposed in Canada. The Canadian Shield also occupies a large part of Greenland, as well as the northern parts of Minnesota, Wisconsin, and Michigan. Valuable minerals such as nickel, silver, and gold are found in the rocks of the Canadian Shield. The oldest rocks in the Canadian Shield are about 3.8 billion years old. In contrast, North America's platform rocks are generally younger than about 600 million years.

Growth of the Continents

Recall from Chapter 17 that all of Earth's continents were once consolidated into a single landmass called Pangaea. Pangaea formed relatively recently in Earth's history—only about 200 mya. The plate tectonic forces that formed Pangaea have been at work at least since the end of the Archean.

Visualizing Continent Formation

Figure 22.6 North America was formed by a succession of mountain-building episodes over billions of years. This map shows mountain-building events that occurred during the Precambrian. By the end of the Precambrian, about 75 percent of North America had formed.

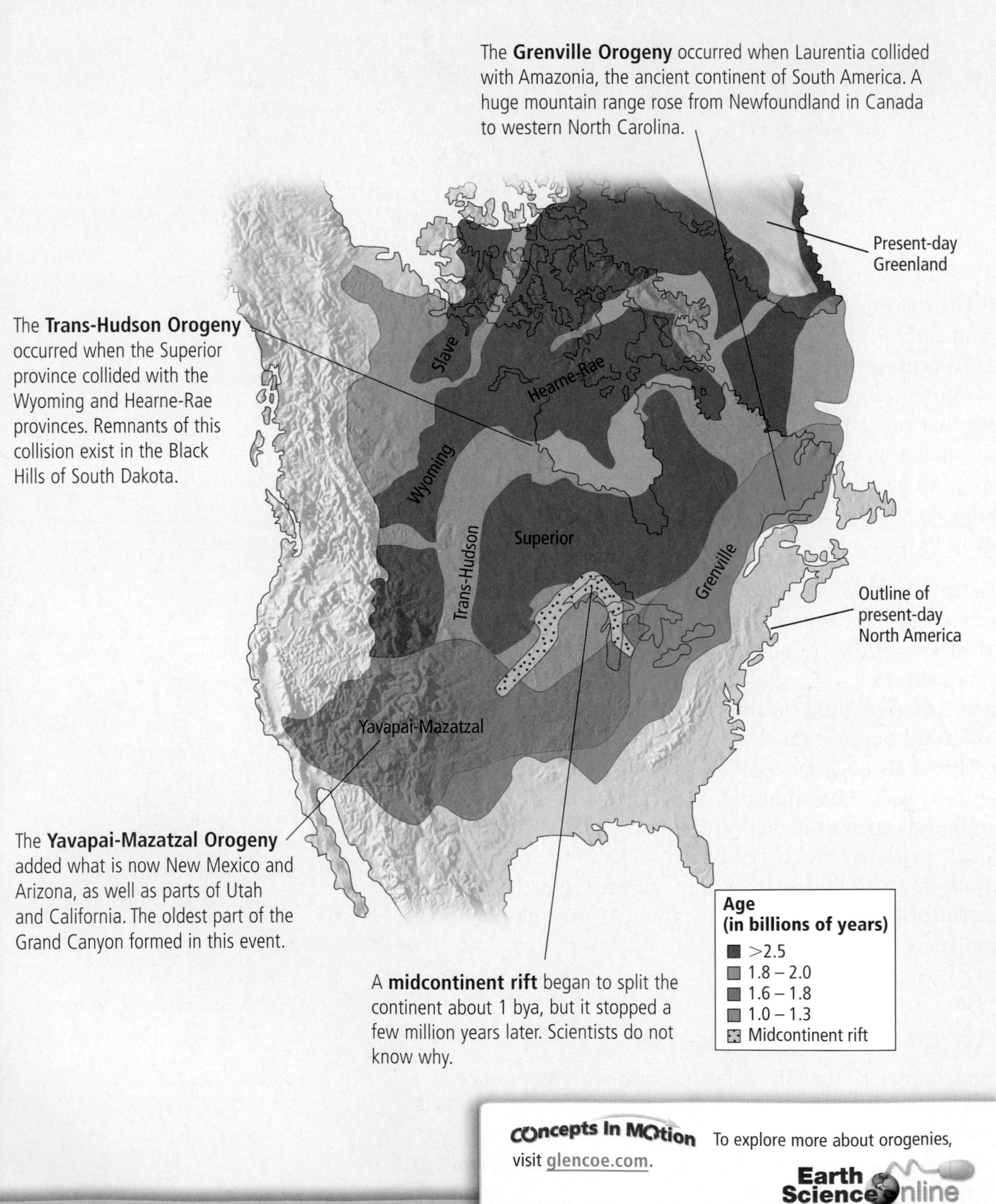

Concepts In Motion To explore more about orogenies, visit glencoe.com.

Earth Science Online

Mountain building During the Proterozoic, the microcontinents that formed during the Archean collided with each other, becoming larger but fewer in number. As they collided, they formed massive mountains. Recall from Chapter 20 that mountain-building episodes are called orogenies. The belts of rocks that are deformed by the immense energy of collisions are called orogenies. The mountain-building events that formed North America are illustrated in **Figure 22.6.**

Laurentia One of Earth's largest Proterozoic landmasses was **Laurentia** (law REN shuh). Laurentia was the ancient continent of North America. As shown in **Figure 22.7,** the growth of Laurentia involved many different mountain-building events. For example, near the end of the early Proterozoic, between 1.8 and 1.6 bya, thousands of square kilometers were added to Laurentia when Laurentia collided with a volcanic island arc. This collision is called the Yavapai-Mazatzal Orogeny.

The first supercontinent The collision of Laurentia with Amazonia occurred at the end of the Proterozoic, about 1.3 to 1 bya. This collision coincided with the formation of Earth's first supercontinent, called Rodinia (roh DIN ee ah), shown in **Figure 22.7.** Rodinia was positioned on the equator with Laurentia at its center. By the time Rodinia formed, nearly 75 percent of Earth's continental crust was in place. The remaining 25 percent was added during the three eras of the Phanerozoic eon. The breakup of this supercontinent began about 750 mya.

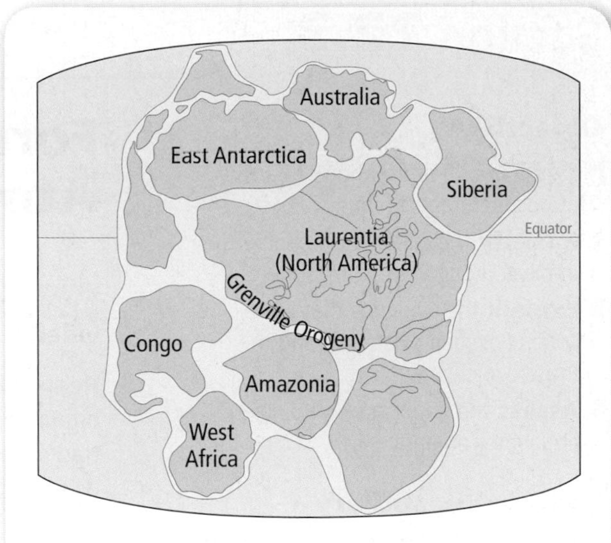

Figure 22.7 Earth's first supercontinent—Rodinia—formed when Laurentia collided with Amazonia in the Grenville Orogeny.

Section 22.2 Assessment

Section Summary

- Earth differentiated into specific density zones early in its formation.
- Plate tectonics caused microcontinents to collide and fuse throughout the Proterozoic.
- The ancient continent of Laurentia formed as a result of many mountain-building episodes.
- Earth's first supercontinent formed at the end of the Proterozoic.

Understand Main Ideas

1. MAIN Idea **Describe** how Earth's continents formed.
2. **Explain** why pieces of Earth's earliest crust do not exist today.
3. **Deduce** how a craton is like a continent's root.
4. **Discuss** how the concept of uniformitarianism helps explain why Earth formed different density zones.

Think Critically

5. **Evaluate** whether it is reasonable to call the Proterozoic the age of continent building.
6. **Infer** why little evidence of Proterozoic orogenies exists today.

WRITING in Earth Science

7. Suppose you are the North American craton. Write a short story about how Laurentia formed around you.

Section 22.3

Objectives

- **Describe** the formation of Earth's atmosphere and oceans.
- **Identify** the cause for the increase in oxygen gas in the atmosphere.
- **Explain** the evidence that atmospheric oxygen existed during the Proterozoic.
- **Assess** the importance of oxygen and water on early Earth.

Review Vocabulary

ultraviolet radiation: high-energy rays from the Sun that can damage living organisms

New Vocabulary

cyanobacteria
stromatolite
banded-iron formation
red bed

Formation of the Atmosphere and Oceans

MAIN Idea The formation of Earth's oceans and atmosphere provided a hospitable environment for life to begin.

Real-World Reading Link Have you thanked a plant lately? Plants and other organisms that produce oxygen provide nearly all the oxygen that you breathe. Had oxygen-producing organisms not existed on early Earth, it is likely that you would not be here today!

Formation of the Atmosphere

Scientists think that an atmosphere began to form on Earth during Earth's formation process. Asteroids, meteorites, and other objects that collided with Earth during this time probably contained water. The water would have vaporized on impact, forming a haze around the planet. Hydrogen and helium probably were also present, with lesser amounts of ammonia and methane. However, hydrogen and helium have small atomic masses, and many scientists think that neither gas stayed near Earth for long. Earth's gravity was, and still is, too weak to keep them from escaping to space. Some scientists also think that much of the ammonia and methane surrounding Earth might have been broken apart by the Sun's intense ultraviolet radiation, releasing more hydrogen into space.

Outgassing Once Earth was formed, its atmosphere changed with the addition of volcanic gases. Volcanic eruptions release large quantities of gases, and there was considerable volcanic activity during the Precambrian. A modern example of the volume of gases released during eruptions is shown in **Figure 22.8.**

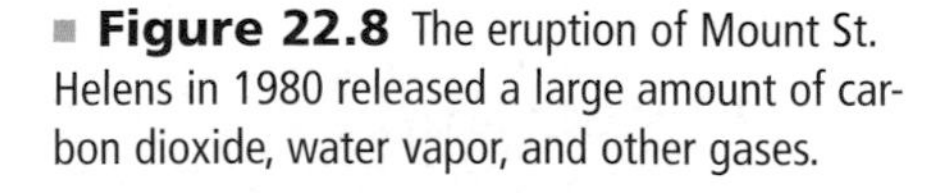

Figure 22.8 The eruption of Mount St. Helens in 1980 released a large amount of carbon dioxide, water vapor, and other gases.

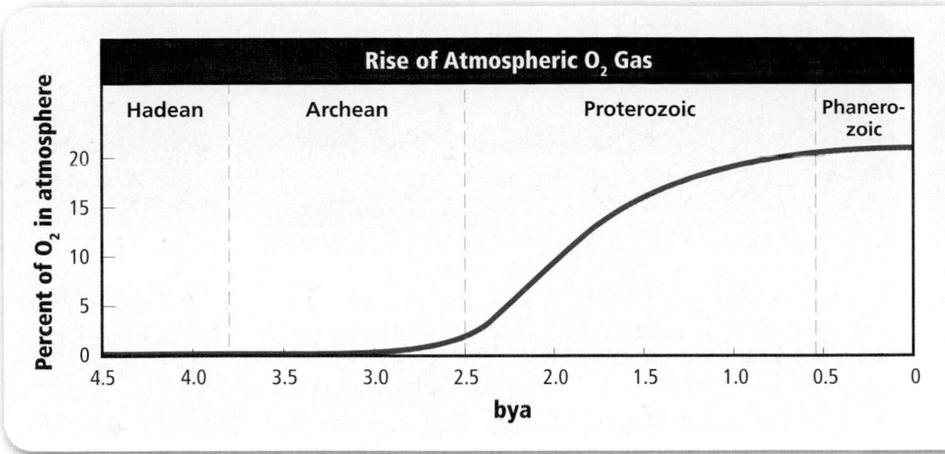

■ **Figure 22.9** There were only negligible amounts of free oxygen in Earth's atmosphere until the early Proterozoic.
Analyze ***How old was Earth when oxygen began to accumulate in its atmosphere?***

In Chapter 15, you learned that present-day volcanoes release large amounts of water vapor, carbon dioxide, and trace amounts of nitrogen and other gases in a process called outgassing. While scientists do not know the exact concentration of gases in Earth's early atmosphere, it probably contained the same gases that vent from volcanoes today.

Oxygen in the Atmosphere

One gas that volcanoes do not generally produce is oxygen. There was little oxygen in the Hadean and Archean atmospheres that was not bonded with carbon or other elements. As illustrated in **Figure 22.9,** atmospheric oxygen did not begin to accumulate until the early Proterozoic. Where did the oxygen gas come from?

FOLDABLES
Incorporate information from this section into your Foldable.

First oxygen producers The oldest known fossils that help answer this question are preserved in rocks in Australia and South Africa that are about 3.5 billion years old. These fossils appear to be traces of tiny, threadlike organisms called **cyanobacteria.** Like their present-day counterparts, ancient cyanobacteria used photosynthesis and produced the nutrients they needed to survive. In the process of photosynthesis, organisms use light energy and convert carbon dioxide and water into sugar. Oxygen gas is given off as a waste product. Today, some bacteria and protists, and most plants produce oxygen using this same process.

Reading Check **Explain** how plants produce oxygen gas.

Stromatolites Most scientists think that microscopic cyanobacteria could have produced enough oxygen to change the composition of the atmosphere that existed on Earth during the Archean. By the early Proterozoic, large, coral reef-like mounds of cyanobacteria called **stromatolites** (stroh MA tuh lites) dominated the shallow oceans that at that time covered most of Earth's continents. Stromatolites are made by billions of cyanobacteria colonies that trap and bind sediments together. The photo on the opening page of this chapter shows present-day stromatolites. These structures are similar in size and shape to Precambrian fossil stromatolites found in Glacier National Park, shown in **Figure 22.10.**

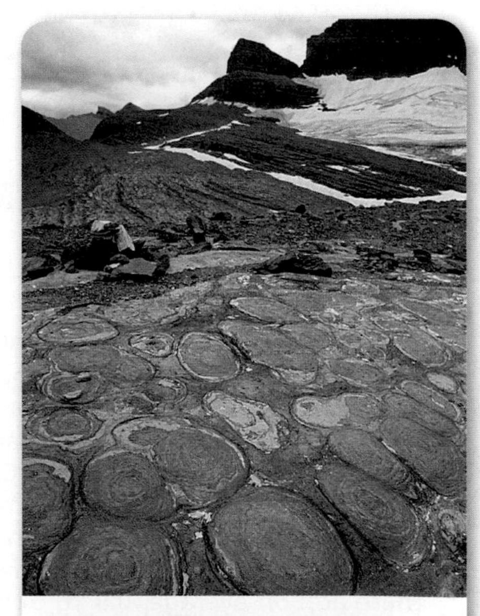

■ **Figure 22.10** These well-preserved fossil stromatolites in Glacier National Park are evidence that cyanobacteria existed during the Precambrian.

Figure 22.11 This iron mine in Brazil contains banded-iron formations that date from the Proterozoic.
Explain *how banded-iron formations are evidence of atmospheric oxygen gas.*

Evidence in rocks Scientists can verify whether there was oxygen in Earth's Archean atmosphere by looking for oxidized iron in Archean rocks. Scientists know that iron reacts with oxygen in the atmosphere to form iron oxides, more commonly called rust. Iron oxides are identified by their red color and provide evidence of oxygen in the atmosphere. The absence of iron oxides in rocks of the late Archean indicates that there was no oxygen gas in the atmosphere at that time. Had atmospheric oxygen gas been present, it would have reacted with the small grains of iron ions in the water or with the iron contained in sediments.

Banded iron By the beginning of the Proterozoic, however, cyanobacteria increased oxygen gas levels enough so that iron oxides began to form in localized areas. These locally high concentrations of iron oxides are called **banded-iron formations.** Banded-iron formations consist of alternating bands of iron oxide and chert, an iron-poor sedimentary rock. The iron oxides appear to have been deposited cyclically, perhaps in response to seasonal variations. Today, these formations are mined for iron ore. An iron mine and a banded-iron rock are shown in **Figure 22.11.**

PROBLEM-SOLVING LAB

Calculate Profits

How do you calculate mining profits? Precambrian rocks contain many important mineral deposits, such as uranium oxide, which is used in nuclear reactors. In uranium oxide deposits in southern Ontario in Canada, the ore-containing rocks cover an area 750 m long and 15,000 m wide with an average thickness of 3 m. Analysis of the deposit indicates that there are, on average, 0.9 kg of uranium oxide per metric ton of rock. Additionally, 0.3 m^3 of the uranium-bearing rock has a mass of 1 metric ton.

Analysis

1. **Solve** How many kilograms of uranium-oxide ore does this deposit contain?
2. **Compute** It will cost \$45/m^3 and 10 years to mine and extract the ore. How much will this cost?

Think Critically

3. **Assess** Assume that the current market price of uranium oxide is \$26.00/kg. Based on your answer to Question 2, can the ore be mined for a profit?

Red beds Many sedimentary rocks that date from the mid-Proterozoic, beginning about 1.8 bya, are rusty red in color. These rocks are called **red beds** because they contain so much iron oxide. The presence of red beds in mid-Proterozoic and younger rocks is strong evidence that the atmosphere by the mid-Proterozoic contained oxygen gas. 

Importance of oxygen Oxygen is important not only because most animals require it for respiration, but also because it provides protection from harmful ultraviolet radiation (UV) from the Sun. Today, only a small fraction of the Sun's UV radiation reaches Earth's surface. This is because Earth is protected by ozone in Earth's upper atmosphere.

As you learned in Chapter 11, an ozone molecule consists of three oxygen atoms bonded together. As oxygen accumulated in Earth's atmosphere, an ozone layer began to develop. Ozone filtered out much of the UV radiation, providing an environment where new life-forms could develop.

Reading Check **Describe** the importance of oxygen for the evolution of life.

Formation of the Oceans

As you learned in Chapter 15, some scientists think that the oceans reached their current size very early in Earth's history. The water that filled the oceans probably originated from the two major sources that provided water in Earth's atmosphere: volcanic outgassing, asteroids, comets, and other objects that bombarded Earth's surface. Earth's early Precambrian atmosphere was rich with water vapor from these sources. As Earth cooled, the water vapor condensed to form liquid water. Recall from Chapter 11 that condensation occurs when matter changes state from a gas to a liquid.

Rain As liquid water formed, a tremendous amount of rain fell. The rain filled the low-lying basins and eventually formed the oceans. Rainwater dissolved the soluble minerals exposed at Earth's surface and—just as they do today—rivers, runoff, and groundwater transported these minerals to the oceans. The dissolved minerals made the oceans of the Precambrian salty, just as dissolved minerals make today's oceans salty.

MiniLab

Model Red Bed Formation

Why are red beds red? Red beds contain so much iron oxide that they appear rusty red in color. Red beds that date from the mid-Proterozoic provide evidence that oxygen gas existed in the Proterozoic atmosphere.

Procedure

1. Read and complete the lab safety form.
2. Place 40 mL of **white sand** in a **150-mL beaker.**
3. Add **water** so that the total volume is 120 mL.
4. Add 15 mL of **bleach.**
   WARNING: ***Use bleach in a well-ventilated area.***
5. Place a piece of **steel wool** about the size of your thumbnail in the beaker.
6. Cover the beaker with a **petri dish,** and allow it to sit undisturbed for one day.
7. Remove the steel wool, and stir the contents of the beaker. Allow the mixture to settle for 5 min after stirring.
8. Slowly pour off the water so that the iron-oxide sediment is left behind.
9. Stir the mixture again; then spoon some of the sand onto a **watch glass,** and allow it to dry.

Analysis

1. **Describe** how the color of the sediment changed.
2. **Explain** where the iron in the experiment came from.
3. **Conclude** where, in nature, the red in rocks comes from.
4. **Assess** the function of the bleach in the experiment.

Figure 22.12 Scientists think that the channels in this canyon on Mars were carved by liquid water long ago. This image was taken from a height of 273 km by the *Mars Express Orbiter.*

Water and life The Precambrian began with an environment inhospitable to life. When it ended, much of Earth was covered with oceans that were teeming with tiny cyanobacteria and other life-forms. Life as it exists on Earth today cannot survive without liquid water.

Scientists think that Earth is not the only object in the solar system that contains or has contained water. Some scientists estimate that the asteroid Ceres contains more freshwater than Earth. Scientists also think that some surface features on Mars, such as the canyon shown in **Figure 22.12,** were carved by liquid water, and that water might still be present in Mars's interior. The moons of Saturn and Jupiter might also contain water in their interiors.

The search for life elsewhere in the solar system and universe today is centered on the search for water. Life on Earth has been found in almost every environment that contains water, from antarctic ice to hot, deep-water ocean vents. Scientists think that simple life-forms might exist in similar environments on other objects in the solar system.

Section 22.3 Assessment

Section Summary

- Earth's atmosphere and oceans began forming early in Earth's history.
- Oxygen gas began to accumulate in the Proterozoic by photosynthesizing cyanobacteria.
- Evidence for atmospheric oxygen can be found in rocks.
- The water that filled Earth's oceans most likely came from two major sources.

Understand Main Ideas

1. MAIN Idea **Explain** why an atmosphere rich in oxygen was important for the evolution of life.
2. **Explain** how scientists conclude that ancient cyanobacteria produced oxygen.
3. **Describe** the relationship between banded-iron formations and oxygen gas.
4. **Describe** where the water in Earth's oceans originated.

Think Critically

5. **Conclude** What would Earth be like if oxygen gas had not formed in the atmosphere?

MATH in Earth Science

6. If asteroids brought 1 cm of water to Earth every 50,000 years, and the average depth of Earth's oceans is 3700 m, how many years would it take to fill the ocean basins from this source?

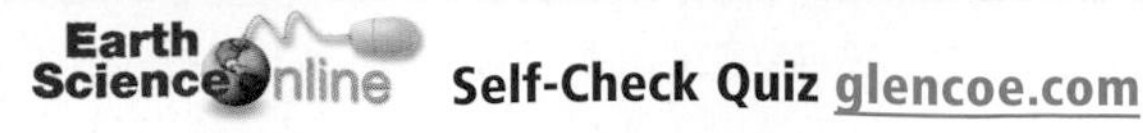

Self-Check Quiz glencoe.com

Section 22.4

Objectives

- **Describe** experimental evidence showing how life might have begun on Earth.
- **Compare and contrast** prokaryotes and eukaryotes.
- **Describe** Earth's first multicellular organisms.

Review Vocabulary

hydrothermal vent: a hole in the seafloor through which water erupts

New Vocabulary

amino acid
prokaryote
eukaryote
Ediacaran biota

Early Life on Earth

MAIN Idea Life began on Earth fewer than a billion years after Earth formed.

Real-World Reading Link If you have ever smelled ammonia, which is often used in household cleaners, you know that its pungent scent can make your nose sting. Some scientists think, however, that the presence of ammonia was necessary for life to form on Earth.

Origin of Life

You have learned that fossil evidence suggests that cyanobacteria existed on Earth as early as 3.5 bya. Though cyanobacteria are simple organisms, photosynthesis—the process by which they produce oxygen—is complex, and it is likely that cyanobacteria evolved from simpler life-forms. Most scientists think that intense asteroid and meteorite bombardment prevented life from developing on Earth until at least 3.9 bya. Where and how the first life-form developed, however, remains an active area of research.

Primordial soup During the first half of the twentieth century, scientists thought that Earth's earliest atmosphere contained hydrogen, methane, and ammonia. Some biologists suggested that such an atmosphere, with energy supplied by lightning, would give rise to an organic "primordial soup" in Earth's shallow oceans. Primordial (pry MOR dee al) means *earliest* or *original.*

In 1953, Stanley Miller and Harold Urey devised an apparatus, shown in **Figure 22.13,** to test this hypothesis. They connected an upper chamber containing hydrogen, methane, and ammonia to a lower chamber designed to catch any particles that condensed in the upper chamber. They added sparks from tungsten electrodes as a substitute for lightning. Within a week, organic molecules had formed in the lower chamber—the primordial soup!

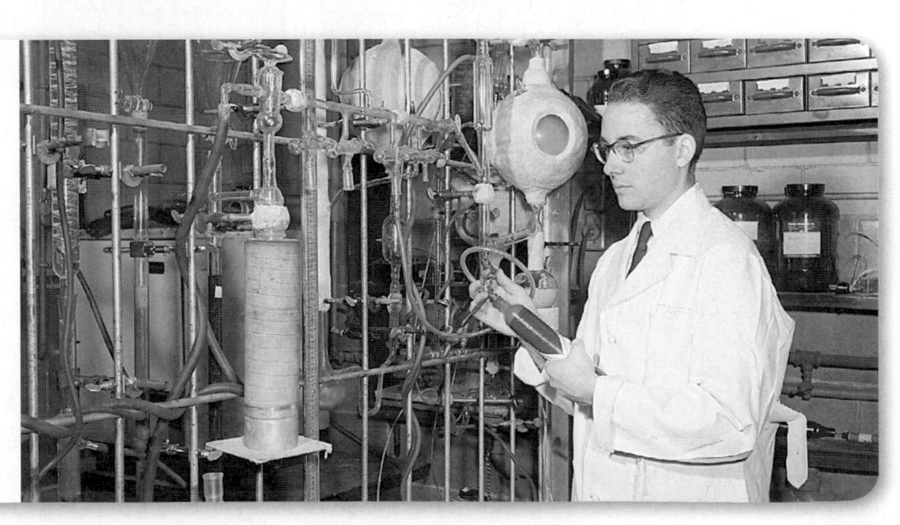

Figure 22.13 In 1953, Stanley Miller, shown here, and Harold Urey performed experiments to test whether organic molecules could form on early Earth.

Concepts In Motion

Interactive Figure To see an animation of the Miller-Urey experiment, visit glencoe.com.

Uncertainties The organic molecules that formed in Miller and Urey's experiment included **amino acids,** the building blocks of proteins. Miller and Urey were the first to show experimentally that amino acids and other molecules necessary for the origin of life could have formed in conditions thought present on early Earth. However, Earth's atmosphere contained gases like those that vent from volcanoes—carbon dioxide, water vapor, and traces of ammonia, methane, and hydrogen. When combinations of these gases are used in simulations, amino acids do not form in high quantities, leading scientists to question whether those processes were sufficient for the origin of life. Some scientists continue to explore the possibility that amino acids, and therefore life, arose in Earth's oceans under localized conditions similar to those in the Miller-Urey experiment, which is possible, given what is known about the early Earth.

VOCABULARY

ACADEMIC VOCABULARY

Simulate
to create a representation or model of something
The video game simulated the airplane's flight with impressive realism.

Other scenarios Because of uncertainties with the conditions in the Miller-Urey experiment, other scientists propose other scenarios and conduct new research into sources and conditions for the origin of life. Some of those are shown in **Table 22.1.** Some think that amino acids organized elsewhere in the universe and were transported to Earth in asteroids or comets. Their experiments show that chemical synthesis of organic molecules is possible in interstellar clouds, and amino acids have been found in meteorites. Other scientists hypothesize that amino acids originated deep in Earth or its oceans. Experiments show that conditions there are favorable for chemical synthesis, and organisms have been found at depths exceeding 3 km.

Table 22.1 How Life Might Have Begun on Earth: Three Hypotheses

Concepts In Motion
Interactive Table To explore more about the origins of life on Earth, visit glencoe.com.

	Earth's Surface	Deep Earth	Space
Hypothesis	Life originated on Earth's surface in warm, shallow oceans.	Life originated in hydrothermal vents.	Organic molecules were brought to Earth in asteroids or comets.
Requirement	Hydrogen, methane, and ammonia must be present in the atmosphere.	Life must survive at high temperatures and pressures.	Organic molecules must be present in extraterrestrial bodies.
Evidence	Simulations produce amino acids.	Simulations of deep-sea vents produce amino acids.	Some meteorites contain amino acids that survived impact.
Drawback	The composition of the early atmosphere did not have large amounts of the required gasses.	It might have been too hot for organic molecules to survive.	It is difficult to test at this time due to technical limitations.

One current area of research explores the possibility that life emerged deep in the ocean at hydrothermal vents. The energy and nutrients necessary for the origin of life are present in this environment. As shown in **Figure 22.14,** a variety of unique organisms live near hydrothermal vents.

No single theory needs to be exclusive; it is possible that all of these contributed to the origin of life. Regardless of how life arose, it is known that conditions during that time were not hospitable, and life probably had many starts and restarts on early Earth. Asteroid impacts were probably still common between 3.9 and 3.5 bya when life arose. Large impacts during this time could have vaporized many early life forms.

■ **Figure 22.14** These tubeworms tolerate extreme pressures and temperatures near hydrothermal vents 2 km below the ocean's surface.
Deduce ***why pressure is high in a hydrothermal-vent environment.***

An RNA world While experiments have shown the likelihood that amino acids existed on early Earth, scientists are still learning how the amino acids were organized into complex proteins and other molecules of life. One essential characteristic of life is the ability to reproduce. All cells require RNA and DNA to reproduce. In modern organisms, RNA carries and translates the instructions necessary for cells to function. Both RNA and DNA use proteins called enzymes to replicate.

Recent experiments have shown that RNA molecules called ribozymes can act as enzymes. They can replicate without the aid of enzymes. This suggests that RNA molecules might have been the first replicating molecules on Earth. An RNA-based world might have been intermediate between an inorganic world and today's DNA-based organic world.

Proterozoic Life

Fossil evidence indicates that unicellular organisms dominated Earth until the end of the Precambrian. These organisms are **prokaryotes** (proh KE ree ohts)—organisms that do not contain nuclei. Nuclei are separate compartments that contain DNA and RNA. Organisms whose cells contain RNA and DNA in nuclei are called **eukaryotes** (yew KE ree ohts). **Figure 22.15** illustrates how prokaryotes and eukaryotes differ in the packaging of their DNA and RNA.

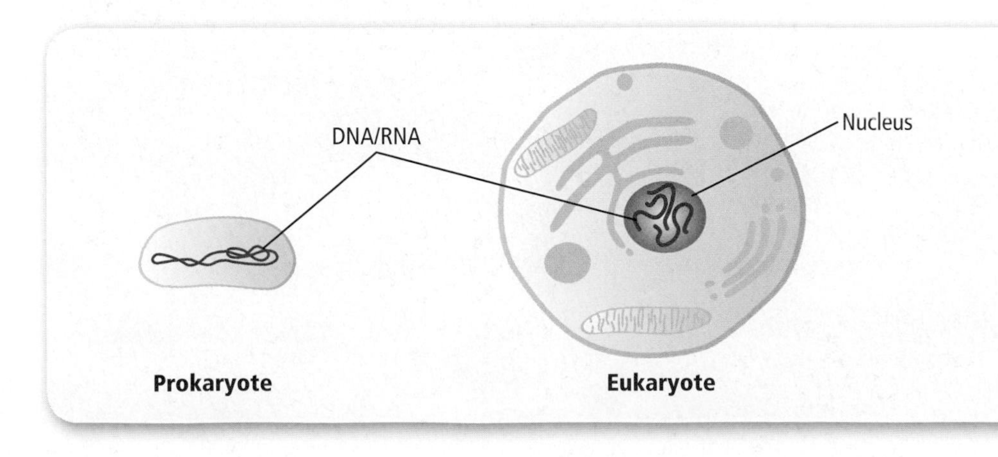

■ **Figure 22.15** Unlike prokaryotes, eukaryotes store DNA in compartments called nuclei.

■ **Figure 22.16** Sunbeams streaming through ice might have provided a refuge for some life-forms 750 mya, when ice covered Earth.

Simple eukaryotes Eukaryotes can be unicellular or multicellular, but because they contain nuclei and other internal structures, they tend to be larger than prokaryotes. This general observation is useful in determining whether a fossil represents a prokaryote or a eukaryote because it is rare for a fossil to be preserved in enough detail to determine whether its cells had nuclei. The oldest-known eukaryote fossil is unicellular. It was found in a banded-iron formation, about 2.1 billion years old, in Michigan.

Reading Check **Explain** how the relative sizes of eukaryotes and prokaryotes are useful to paleontologists.

Snowball Earth Some scientists think that a widespread glaciation event 750 mya played a critical role in the extinction of many early unicellular eukaryotes. This glaciation event was so widespread that some geologists compare Earth at that time to a giant snowball. Evidence from ancient glacial deposits around the world suggests that glacial ice might have advanced as far as the equator and that even the oceans might have been frozen. Though many organisms went extinct during this time, some life-forms survived, perhaps near hydrothermal vents or in pockets of sunlight streaming through cracks in ice, as illustrated in **Figure 22.16.**

Multicellular organisms The rock record indicates that shortly after the ice retreated toward the poles, about 630 mya, the climate warmed dramatically and the first multicellular organisms appeared in the oceans. Fossils of this time period were first discovered in 1947 in Australia's Ediacara Hills. Collectively called the **Ediacaran biota** (ee dee A kuh ruhn by OH tuh), these fossils show the impressions of large, soft-bodied eukaryotes. **Figure 22.17** shows what these organisms might have looked like.

■ **Figure 22.17** This reconstruction of an ocean in the Ediacaran Period shows how Earth's first multicellular organisms might have looked. They ranged from several centimeters to two meters in length.

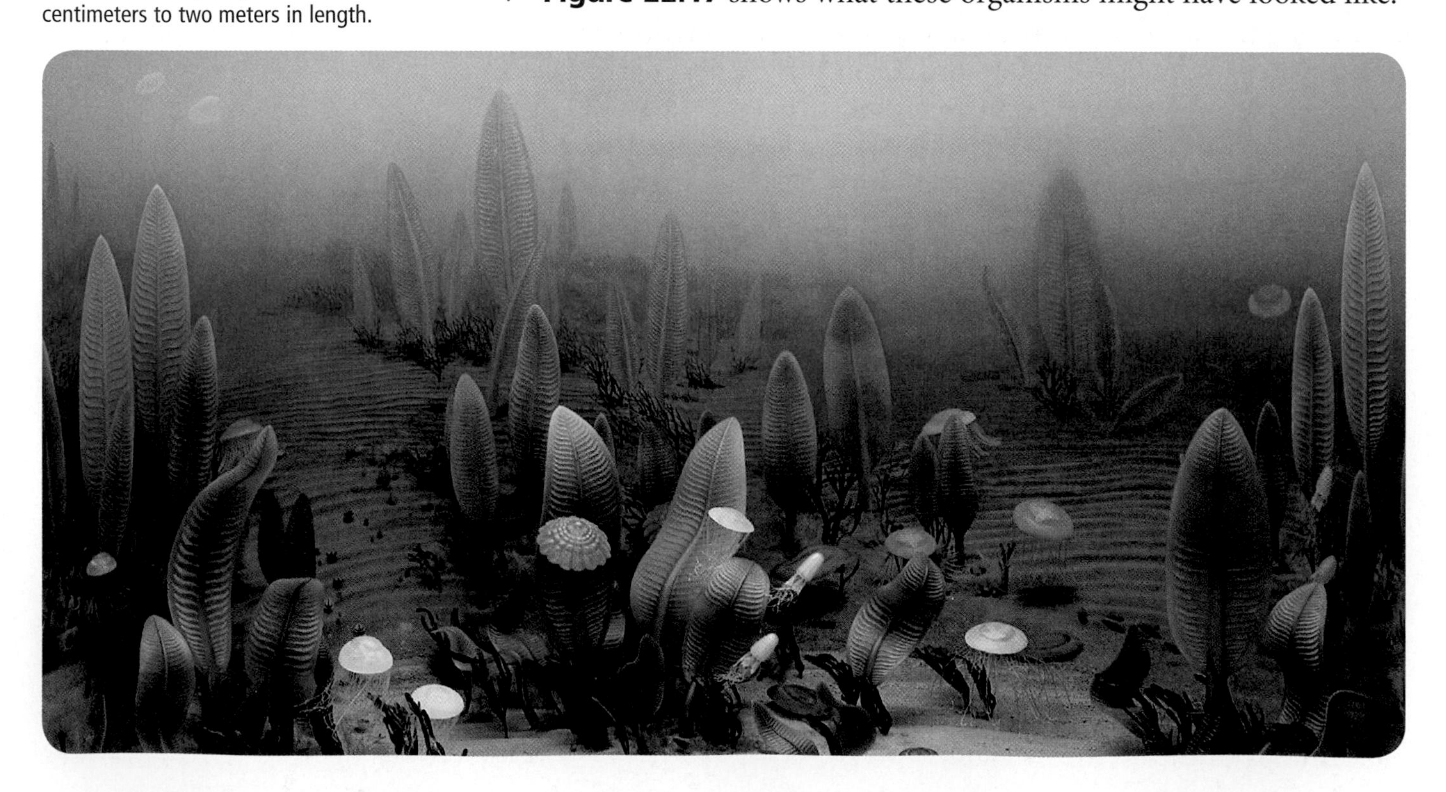

Ediacaran biota The discovery of the Ediacaran biota at first seemed to solve one of the great mysteries in geology: why there are no fossils of the ancestors of the complex and diverse animals that existed during the Cambrian period—the first period of the Paleozoic era. The Ediacaran biota seemed to provide fossil evidence of an ancestral stock of complex organisms. As shown in **Figure 22.18,** one type of Ediacaran organism appeared similar in overall body shape to sea pens. Others appeared similar to jellyfish, segmented worms, arthropods, and echinoderms—just the type of ancestral stock that geologists had been hoping to find.

However, upon closer examination, some scientists have questioned that conclusion and suggest that Ediacaran organisms are not relatives of present-day animal groups but, instead, represent unique organisms. These scientists point out that none of the Ediacaran organisms shows evidence of a mouth, anus, or gut, and there is little evidence that they could move. As a result, there is an ongoing debate in the scientific community about the precise nature of many of these fossils.

Mass extinction In recent years, geologists have found Ediacaran fossils in all parts of the world. This suggests that these organisms were widely distributed throughout the shallow oceans of the late Proterozoic. They seem to have flourished between 630 mya and 540 mya. Then, in an apparent mass extinction, most of them disappeared, and organisms more likely related to present-day organisms began to inhabit the oceans.

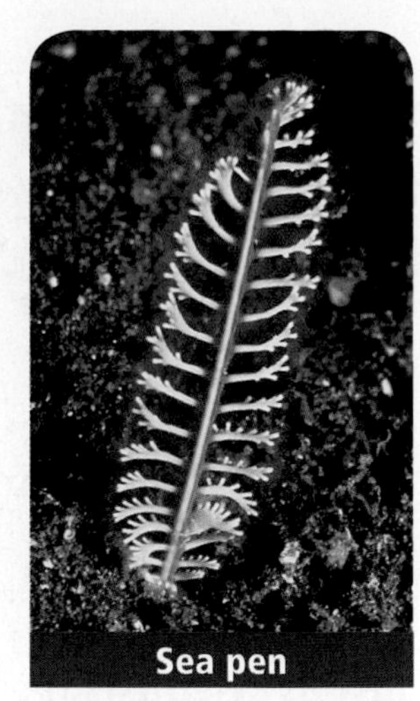

Figure 22.18 One type of Ediacaran organism resembles a present-day sea pen. Some scientists think that the two are related.

Section 22.4 Assessment

Section Summary

- Scientists think that life on Earth began between 3.9 and 3.5 bya.
- Stanley Miller and Harold Urey were the first to show experimentally that organic molecules could have formed on early Earth.
- Scientists have developed several hypotheses to explain how and where life formed.
- Eukaryotes appeared after prokaryotes.
- Earth's first multicellular organisms evolved at the end of the Precambrian.

Understand Main Ideas

1. MAIN Idea **List** three hypotheses about the origin of life, and describe the evidence for each.
2. **Explain** why scientists think that life on Earth began after 3.9 bya.
3. **Identify** the ingredients that Miller and Urey thought made up Earth's early atmosphere.
4. **Compare and contrast** eukaryotes and prokaryotes.
5. **Discuss** why some scientists think that Ediacaran organisms do not represent present-day animal groups.

Think Critically

6. **Hypothesize** one reason that the Ediacaran organisms became extinct.

WRITING in Earth Science

7. Write a newspaper article about the discovery of a new fossil outcrop that dates to the end of the Precambrian. Describe the fossil organisms found in this outcrop.

EARTH SCIENCE and TECHNOLOGY

EXPLORING MARS

On future missions to Mars, scientists hope to collect samples by drilling into the surface.

Just as scientists have questions about the history of Earth, including past climate conditions and the development of early life on Earth, they have similar questions about Mars. Scientists are using new technologies to develop devices to collect data which might help answer these questions.

Analyzing current evidence Based on evidence found by examining meteorites from Mars that have been found in various locations on Earth, scientists think that liquid water flowed on the surface of Mars at some point in the planet's history. Using high-powered microscopes, some scientists think that they have found evidence of past microscopic life on Mars in the form of bacteria in at least one of the meteorites. Photos of Mars's surface show canyons that scientists believe to have been formed when large amounts of liquid water flowed on Mars's surface in the past. The latest data collected by the Mars *Odyssey* spacecraft shows evidence of water in the form of ice under the surface of Mars.

Baseball-sized probes In an effort to collect more extensive data from Mars, scientists have developed baseball-sized probes that could be released by the thousands on the surface of Mars. The probes, equipped with cameras and sensors to collect data about the environment, would be able to move around the planet with ease compared to the rovers that have already explored a small portion of Mars. They would be able to move into regions that current rovers are unable to reach, including lava tubes, caves, and canyons. Lava tubes are areas that scientists think might contain evidence of water on Mars. The probes would be able to move down the side of the canyons, giving scientists an up-close look at the canyons walls.

Automated drills NASA scientists are currently testing a drill to send to Mars that is operated by artificial intelligence. The drill will be able to operate automatically, without control from humans, for hours at a time. Special sensors that pick up on changes in the vibration of the drill will help keep the drill from failing, such as when the drill hits rocks or becomes jammed. The device will be used to drill into Mars's surface in the search for evidence of water and life on Mars.

Sensors and soil samples Further efforts to answer the question about the past or present existence of water and life on Mars include a special sensor that has been placed on the arm of the *Phoenix* Lander, a craft scheduled to leave for Mars in 2007. The sensor will be able to analyze the frozen soil on Mars for liquid water by allowing the soil to heat up in the Sun. Scientists hope that liquid water will be able to be detected before the ice turns to vapor.

WRITING in Earth Science

Presentation Research more information about the latest technology for future exploration and data collection on Mars. Write a report that summarizes what you have learned and present your report to your class. Include handouts and illustrations as part of your presentation. To learn more about the exploration of Mars, visit glencoe.com.

GEOLAB

MAPPING: MAP CONTINENTAL GROWTH

Background: During the Precambrian, microcontinents and island arcs collided to form what would become present-day continents.

Question: *How does the distribution of the ages of rocks help geologists reconstruct the sequence of continental growth?*

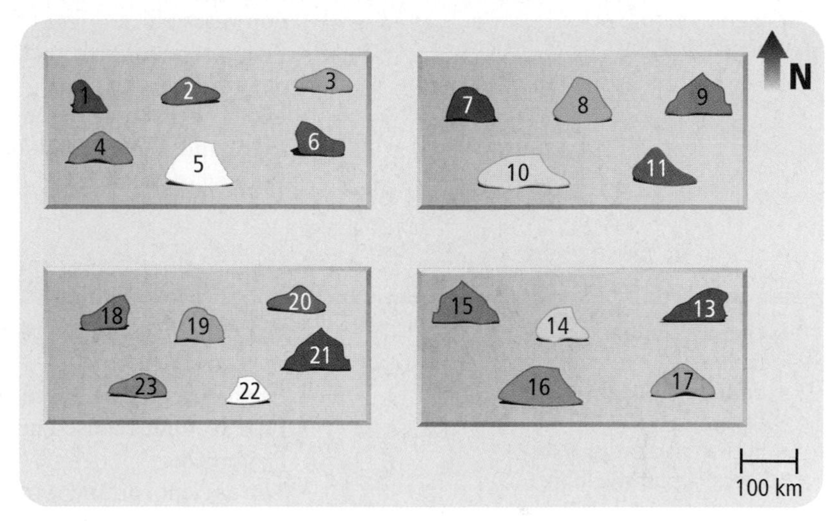

Locality Data

Materials

rock samples
paper
metric ruler
colored pencils

Safety Precautions

Procedure

Suppose you are working on a geologic survey team that is updating its geologic map of a continent. You have gathered the ages of rock samples found in various locations throughout the continent.

1. Read and complete the lab safety form.
2. Your teacher has set up locations around the classroom with a rock sample of a different age at each location. Draw a rough map of the room showing the locations and ages of all the rocks.
3. Measure and record the distance, in centimeters, between the rocks.
4. Plot your measurements on an outline map of your classroom, using a scale of 1 cm = 100 km.
5. Use a pencil to draw lines on the map, separating rocks of different ages.
6. Use colored pencils to shade in each area on the map that contains rocks of the same ages. These are your geologic age provinces.
7. Make a key for your map. Name the oldest province *Province A*, the next oldest province *Province B*, and so on for all provinces.

Analyze and Conclude

1. **Compare** your map with those of your classmates.
2. **Identify** the oldest province on your map. Where is it located in relation to the other provinces?
3. **Describe** the sequence of collision events that formed the continent represented by your map.
4. **Interpret Data** Use your map to find the likely sites of metamorphic rocks. Determine what types of metamorphism might have occurred.
5. **Interpret Data** Based on your map, where would you expect to find the highest and most rugged mountains? The most weathered mountains? Explain.

APPLY YOUR SKILL

Time Line Make a time line that shows the order of accretion of the provinces of the North American continent shown in **Figure 22.7.**

CHAPTER 22

Study Guide

Download quizzes, key terms, and flash cards from glencoe.com.

BIG Idea The oceans and atmosphere formed and life began during the three eons of the Precambrian, which spans nearly 90 percent of Earth's history.

Vocabulary	Key Concepts
Section 22.1 Early Earth	
• asteroid (p. 622) • meteorite (p. 621) • zircon (p. 620)	**MAIN Idea** Several lines of evidence indicate that Earth is about 4.56 billion years old. • Scientists use Earth rocks, zircon crystals, moon rocks, and meteorites to determine Earth's age. • Likely heat sources of early Earth were gravitational contraction, radioactivity, and asteroid and meteorite bombardment. • Cooling of Earth led to the formation of liquid water.
Section 22.2 Formation of the Crust and Continents	
• Canadian Shield (p. 625) • craton (p. 625) • differentiation (p. 623) • Laurentia (p. 627) • microcontinent (p. 624) • Precambrian shield (p. 625)	**MAIN Idea** The molten rock of Earth's early surface formed into crust and then continents. • Earth differentiated into specific density zones early in its formation. • Plate tectonics caused microcontinents to collide and fuse throughout the Proterozoic. • The ancient continent of Laurentia formed as a result of many mountain-building episodes. • Earth's first supercontinent formed at the end of the Proterozoic.
Section 22.3 Formation of the Atmosphere and Oceans	
• banded-iron formation (p. 630) • cyanobacteria (p. 629) • red bed (p. 631) • stromatolite (p. 629)	**MAIN Idea** The formation of Earth's oceans and atmosphere provided a hospitable environment for life to begin. • Earth's atmosphere and oceans began forming early in Earth's history. • Oxygen gas began to accumulate in the Proterozoic by photosynthesizing cyanobacteria. • Evidence for atmospheric oxygen can be found in rocks. • The water that filled Earth's oceans most likely came from two major sources.
Section 22.4 Early Life on Earth	
• amino acid (p. 634) • Ediacaran biota (p. 636) • eukaryote (p. 635) • prokaryote (p. 635)	**MAIN Idea** Life began on Earth fewer than a billion years after Earth formed. • Scientists think that life on Earth began between 3.9 and 3.5 bya. • Stanley Miller and Harold Urey were the first to show experimentally that organic molecules could have formed on early Earth. • Scientists have developed several hypotheses to explain how and where life formed. • Eukaryotes appeared after prokaryotes. • Earth's first multicellular organisms evolved at the end of the Precambrian.

Vocabulary PuzzleMaker glencoe.com

CHAPTER 22

Assessment

Vocabulary Review

Identify the vocabulary term from the Study Guide described by each phrase.

1. bodies that orbit the Sun between Mars and Jupiter
2. the name of the ancient continent that makes up most of North America
3. the first photosynthetic, oxygen-producing organisms on Earth
4. the process by which a planet becomes zoned with heavy materials near its center and lighter materials near its surface

Use the vocabulary term from the Study Guide to answer the following questions.

5. What are the building-blocks of protein?
6. What is the name of the Precambrian Shield in North America?
7. What are rocks called that consist of alternating bands of iron and chert?
8. What type of organism packages its DNA in nuclei?

Complete each sentence by providing the missing vocabulary term from the Study Guide.

9. The ________ was a group of organisms containing the first multicellular eukaryotes.
10. ________ is a very stable mineral often used to date Precambrian rocks.
11. A ________ is a mound made by microorganisms in shallow seas.
12. An old, stable part of a continent is called a ________.

Understand Key Concepts

13. What process contributed to the formation of Earth's early atmosphere?
    A. outgassing
    B. differentiation
    C. crystallization
    D. photosynthesis

14. Which was not a source of heat for early Earth?
    A. asteroid and meteorite bombardment
    B. hydrothermal energy
    C. gravitational contraction
    D. radioactivity

Use the figure below to answer Questions 15 and 16.

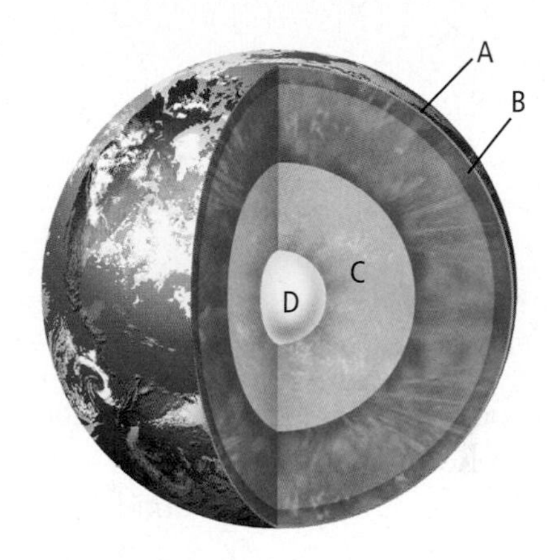

15. Which part of Earth is the most dense?
    A. A
    B. B
    C. C
    D. D

16. In which part of Earth would you find granite?
    A. A
    B. B
    C. C
    D. D

17. Why is oxygen gas important to life on Earth?
    A. It is used by plants to undergo photosynthesis.
    B. It is required by cyanobacteria and stromatolites to survive.
    C. It is a source of heat at Earth's surface.
    D. It provides protection from harmful ultraviolet radiation from the Sun.

18. Upon what age of Earth do most scientists agree?
    A. 4.56 thousand years old
    B. 45.6 million years old
    C. 4.56 billion years old
    D. 45.6 billion years old

19. A meteorite is a fragment of which object?
    A. comet
    B. asteroid
    C. planet
    D. the Moon

Use the figure below to answer Question 20.

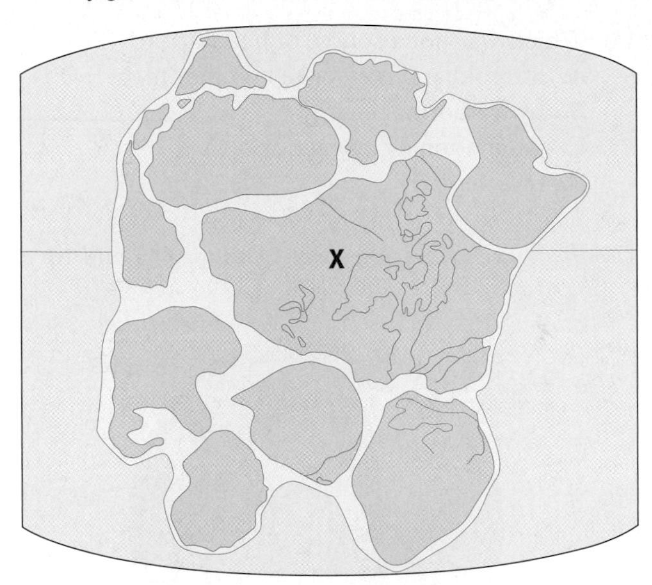

20. What is the name of the continent labeled *X* in this figure of Rodinia?
A. Baltica
B. Amazonia
C. Gondawana
D. Laurentia

21. Which is likely to give the oldest radiometric age date?
A. meteorite
B. granite
C. zircon
D. metamorphic rock

22. Which was the earliest type of life on Earth?
A. eukaryotes
B. prokaryotes
C. ribozymes
D. Ediacaran biota

23. Refer to **Figure 22.6** in the text. How old are the rocks that underlie most of the state of Arizona?
A. 1.0–1.3 billion years
B. 1.8–2.0 billion years
C. 1.6–1.8 billion years
D. > 2.5 billion years

Constructed Response

24. List the evidence that scientists use to determine Earth's age.

25. Identify sources of the gases that made up Earth's early atmosphere.

26. Explain how gravitational contraction heated early Earth.

27. Discuss how supercontinents form.

28. Explain why Earth's earliest crust no longer exists.

29. Explain why scientists think that cyanobacteria were not the first life-forms on Earth.

30. Evaluate How do red beds serve as evidence that there was oxygen gas in the atmosphere during the mid-Proterozoic?

Think Critically

31. Identify the sources of Earth's heat today.

32. Explain why there is little hydrogen or helium in Earth's atmosphere today.

33. Discuss how scientists infer that Hadean time existed.

Use the photo below to answer Question 34.

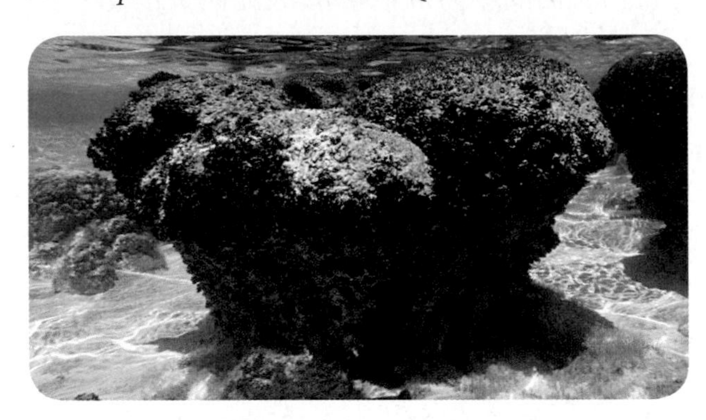

34. Discuss how the structures in the photo are related to oxygen gas in the atmosphere.

35. Assess how the concept of uniformitarianism can be used to explain your answer to Question 34.

Use the figure below to answer Question 36.

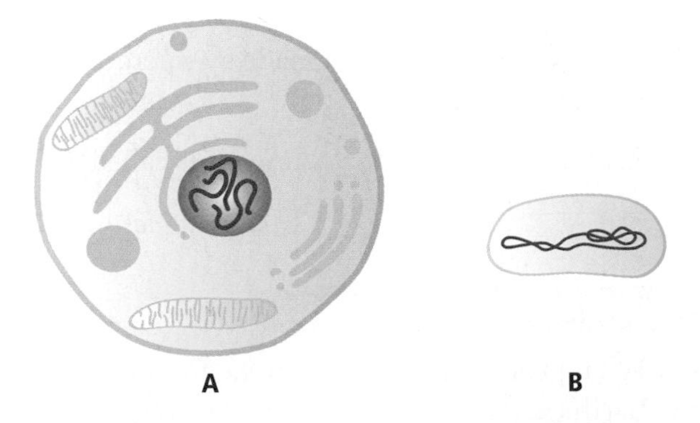

36. Identify each cell shown as a prokaryotic or a eukaryotic. Explain the differences between them.

Earth Science Online Chapter Test glencoe.com

37. **Infer** the composition of the atmosphere had there never been life on Earth.

38. **Discuss** what scientists mean when they refer to an "RNA World."

Use the photo below to answer Question 39.

39. **Explain** how the process illustrated above likely contributed to the formation of Earth's oceans.

40. **Explain** Where in North America would you look if you wanted to find evidence of Archean life?

41. **Evaluate** which is more important for the existence of life—liquid water or oxygen gas.

42. **CAREERS IN EARTH SCIENCE** Imagine that you have a rock sample from an Earthlike planet in a distant solar system. Plan an experiment that might help you determine the age of the planet.

43. **Evaluate** the significance of a Snowball Earth for the evolution of life.

Concept Mapping

44. Create a concept map showing the cause and effects of oxygen in Earth's atmosphere. Include the following key terms in the concept map: *oxygen, respiration, ozone, photosynthesis,* and *cyanobacteria.*

Challenge Question

45. **Propose** what Earth would be like if both continental crust and oceanic crust were made of the same material.

Additional Assessment

46. **WRITING in Earth Science** Suppose your spaceship has landed on a planet that you suspect has life-forms similar to cyanobacteria. Write a letter to your best friend explaining what you see outside your spaceship window.

DBQ Document–Based Questions

Data obtained from: Schopf, J.W. 1999. *Cradle of Life: The discovery of Earth's earliest fossils.* Princeton: Princeton University Press.

Earth contains rocks of varying ages, but they are not distributed evenly in time. This means that information about the geologic past is not available in the same quantity for all of Earth's past. Use the data below to answer the following questions.

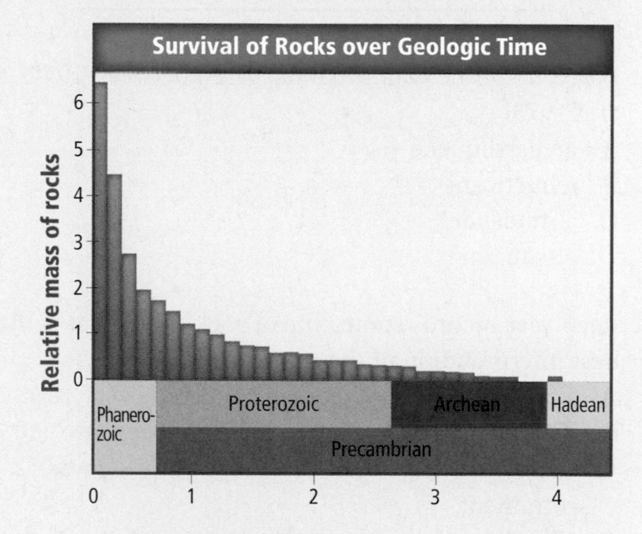

47. Which eon contains the most complete rock record?

48. What general trend is apparent about rocks through geologic time?

49. Why do you think that the trend in the data above exists?

Cumulative Review

50. Why could winter applications of salt to de-ice roads have negative effects on groundwater resources? **(Chapter 10)**

51. Explain the concept of uniformitarianism. **(Chapter 21)**

Standardized Test Practice

Multiple Choice

1. Which contains the fewest number of years?
   A. eon
   B. era
   C. period
   D. epoch

Use the table to answer Questions 2 and 3.

Fault Line Activity	
Date	**Rock Slip Measurement (mm)**
1973	5
1974	8
1975	300
1976	10

2. Based upon the data, what occurred between 1974 and 1975?
   A. an earthquake
   B. a hurricane
   C. a mudslide
   D. a tsunami

3. Each year records some type of rock slip. What is the best interpretation of the movement?
   A. Earthquakes in some form occurred every year.
   B. The rising movement should have been an indicator to scientists that an earthquake was imminent.
   C. The slippage was so slight and smooth that it was not felt.
   D. The slippage was similar to aftershocks of an earthquake.

4. What is caused by differences in air pressure?
   A. wind
   B. clouds
   C. rain
   D. thunder

5. Which is not a cause of climatic variations?
   A. latitude
   B. frontal systems
   C. topography
   D. air masses

6. In which fossil do original structures of an organism remain?
   A. mold fossil
   B. permineralized fossil
   C. cast fossil
   D. trace fossil

Use the illustrations below to answer Questions 7 and 8.

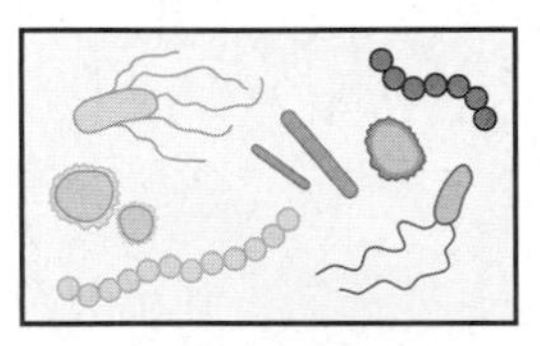

Group A

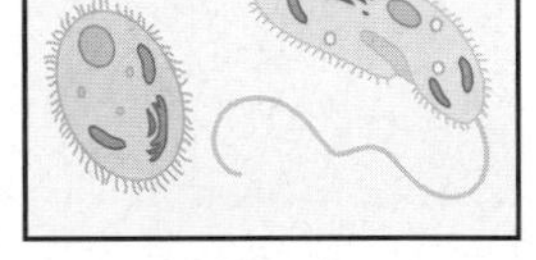

Group B

7. How do members of Group A differ from members of Group B?
   A. They are plants.
   B. They can be found in Proterozoic fossils.
   C. They contain no nuclei.
   D. They are all unicellular.

8. Where did members of Group B probably originate?
   A. glaciers
   B. hydrothermal vents
   C. Australian fauna
   D. oil deposits

9. Which type of graph would best show the number of volcanic eruptions over a period of time?
   A. line graph
   B. circle graph
   C. pictograph
   D. bar graph

10. Squeezing that causes intense deformation at plate boundaries which leads to mountain building is known as what?
   A. orogeny
   B. convergence
   C. divergence
   D. transverse motion

Short Answer

Use the illustration below to answer Questions 11 and 12.

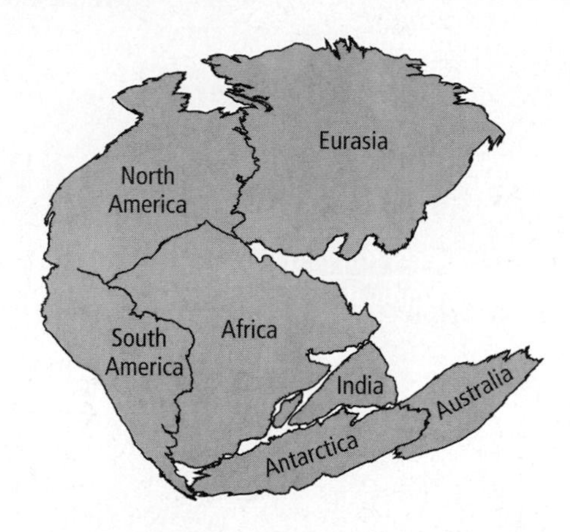

11. Who was the first scientist to propose this view of ancient Earth?

12. What features support this hypothesis?

13. How does oceanic crust differ from continental crust?

14. Why do mountains have roots?

15. Why doesn't all water become absorbed as it lies on the ground?

16. What principle for determining relative age is modeled at the Grand Canyon with its multiple layers of sedimentary rock? Explain the force that was responsible for the horizontal layering.

Reading for Comprehension

Akilia Rock Analysis

Past analysis of rocks found on Akilia, an island off the southwestern coast of Greenland, led scientists to conclude that they were at least 3.85 billion years old and contained evidence of the earliest life on the planet. "Not so," say geologists Chris Fedo and Martin Whitehouse. For billions of years, Fedo says, "[these rocks] have been squashed tens of miles underground. The rocks are so strongly deformed that understanding the original relationships among different layers is extraordinarily difficult." Some of the bands in the rock formation show irregular variations in thickness, a pattern that is not produced by sedimentation. "These bands do not have a sedimentary origin," says Fedo. "Rather," says Fedo, "we think the green bands are igneous." If the banded rocks are igneous, the Akilia rocks might have formed in the absence of life-sustaining oceans, making them much less likely to have harbored early life.

Article obtained from: Harder, B. New analysis throws age of life on earth into doubt. *National Geographic News* May 23, 2002. (Online resource accessed October 1, 2006.)

17. What can be inferred from this passage?
    A. Greenland is the oldest landmass on Earth.
    B. Scientists can better determine the age of the rock, if they know the type of rock it is.
    C. Green bands in the rock make it difficult to determine the age of the rock.
    D. These are the only scientists who think that Earth is younger than 4.56 billion years.

18. How does knowing the age of Earth help scientists in their studies?

NEED EXTRA HELP?																
If You Missed Question . . .	1	2	3	4	5	6	7	8	9	10	11	12	13	14	15	16
Review Section . . .	21.1	19.1	19.1	11.2	14.1	21.4	22.4	22.4	1.3	20.2	17.1	17.1	20.1	20.2	10.1	21.2

CHAPTER 23

The Paleozoic, Mesozoic, and Cenozoic Eras

BIG Idea **Complex life developed and diversified during the three eras of the Phanerozoic as the continents moved into their present positions.**

23.1 The Paleozoic Era
MAIN Idea Life increased in complexity during the Paleozoic while the continents collided to form Pangaea.

23.2 The Mesozoic Era
MAIN Idea Reptiles became the dominant terrestrial animals in the Mesozoic while Pangaea broke apart.

23.3 The Cenozoic Era
MAIN Idea Mammals became the dominant terrestrial animals in the Cenozoic while the continents assumed their present forms.

GeoFacts

- Relatives of New Zealand's Kauri trees first appeared in the Jurassic, nearly 200 mya.
- When injured, Kauri trees secrete resin, which hardens into amber.
- Cell structures of insects trapped in amber can be preserved for millions of years.

LAUNCH Lab

How is oil stored in rocks?

Many sedimentary rocks contain oil and water. How are these materials stored in sedimentary rocks?

Procedure

1. Read and complete the lab safety form.
2. Place an unglazed **brick** or **sandstone** sample on your table.
3. Sketch and label a magnified cross section of the brick or sandstone before you add the **oil** or **water.**
4. Using a **dropper,** slowly squeeze three to five drops per min of water or oil onto the brick or sandstone for 10 min.
5. Revise your sketch to show the view after you added the oil or water.

Analysis

1. **Infer** Observe the brick or sandstone sample. Where did the water or oil go?
2. **Compare and contrast** the appearance of the brick before and after the oil or water was added.
3. **Conclude** how rocks in nature store oil and water.

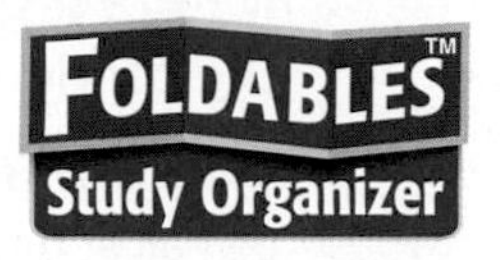

Life-Forms of the Paleozoic Make a Foldable to compare the life-forms of the early, middle, and late Paleozoic.

STEP 1 Fold a sheet of paper to the margin line.

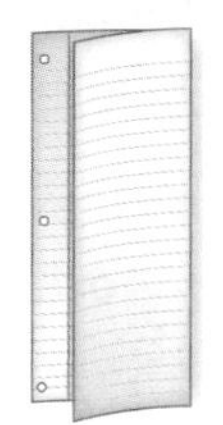

STEP 2 Fold the sheet into thirds.

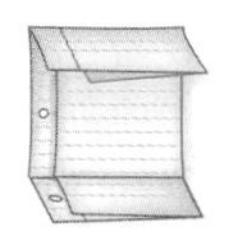

STEP 3 Cut along the fold lines of the top flap to make three tabs. Label the tabs *Early Paleozoic, Middle Paleozoic,* and *Late Paleozoic.*

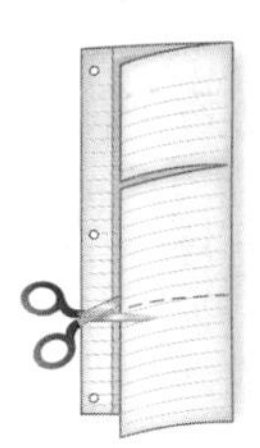

FOLDABLES **Use this Foldable with Sections 23.1, 23.2, and 23.3.** As you read, use your Foldable to describe the life-forms present, including the names of specific plants and animals.

Visit glencoe.com to

- **study entire chapters online;**
- **explore Concepts In Motion animations:**
  - **Interactive Time Lines**
  - **Interactive Figures**
  - **Interactive Tables**
- **access Web Links for more information, projects, and activities;**
- **review content with the Interactive Tutor and take Self-Check Quizzes.**

Section 23.1

Objectives

- **Define** the term passive margin.
- **Explain** how transgressions and regressions indicate sea-level changes.
- **Discuss** the tectonic forces that shaped Laurentia during the Paleozoic.
- **Summarize** the changes in Paleozoic life-forms.

Review Vocabulary

evaporite: a sediment deposit that has crystallized out of water supersaturated with dissolved minerals

New Vocabulary

paleogeography
passive margin
transgression
regression
Cambrian explosion

The Paleozoic Era

MAIN Idea Life increased in complexity during the Paleozoic while the continents collided to form Pangaea.

Real-World Reading Link Have you noticed that some things seem to happen all at once? For instance, you might notice that everyone at school is suddenly talking about a certain music group that just yesterday was unknown. In a similar way, there suddenly appeared in the Paleozoic rock record an entire collection of new, complex life-forms.

Paleozoic Paleogeography

The geologic activity of the three eras of the Phanerozoic Eon are well represented in the rock record. By studying this record, geologists can reconstruct estimates of landscapes that have long since disappeared. The ancient geographic setting of an area is called its **paleogeography** (pay lee oh jee AH gruh fee). The paleogeography of the Paleozoic Era—the first era of the Phanerozoic—is defined by the breakup of the supercontinent Rodinia. As this breakup proceeded, multicellular life evolved with increasing complexity, as illustrated in **Figure 23.1.**

Passive margins Recall from Chapter 22 that the ancient North American continent of Laurentia split off from Rodinia by the early Paleozoic. Laurentia was located near the equator and was surrounded by ocean. In addition, it was almost completely covered by a shallow, tropical sea. Throughout the Cambrian, there was no tectonic activity on Laurentia so no mountain ranges formed. The edge of a continent is called a margin. When there is no tectonic activity along a margin, it is called a **passive margin.** During the Cambrian, Laurentia was completely surrounded by passive margins—there was no tectonic activity along its edges.

■ **Figure 23.1** Life-forms became more complex during the six periods of the Paleozoic.

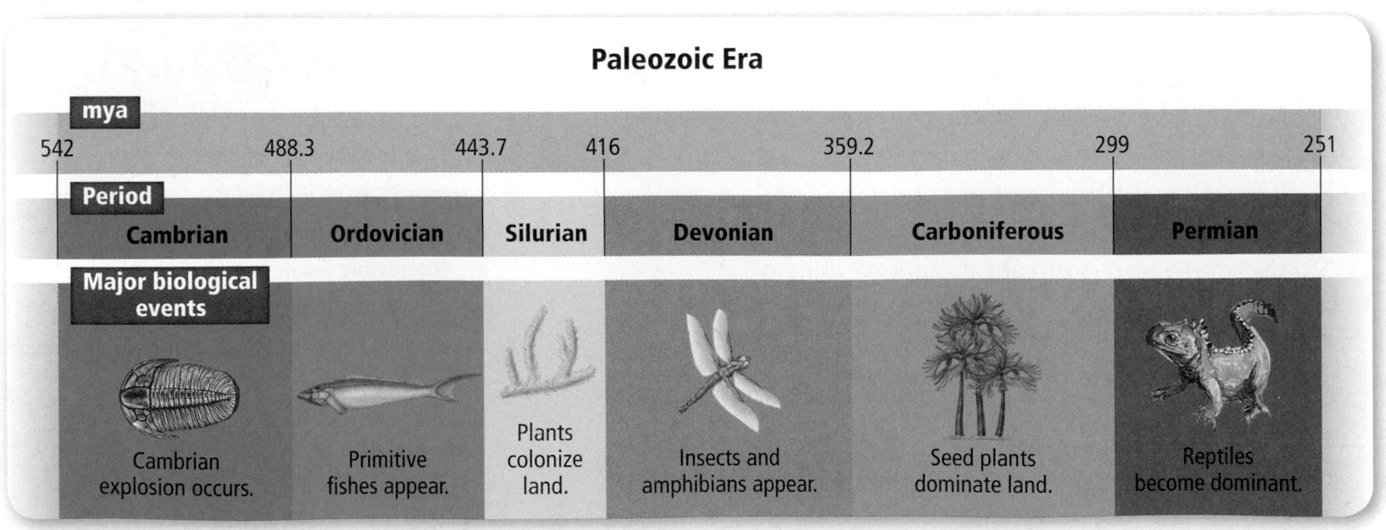

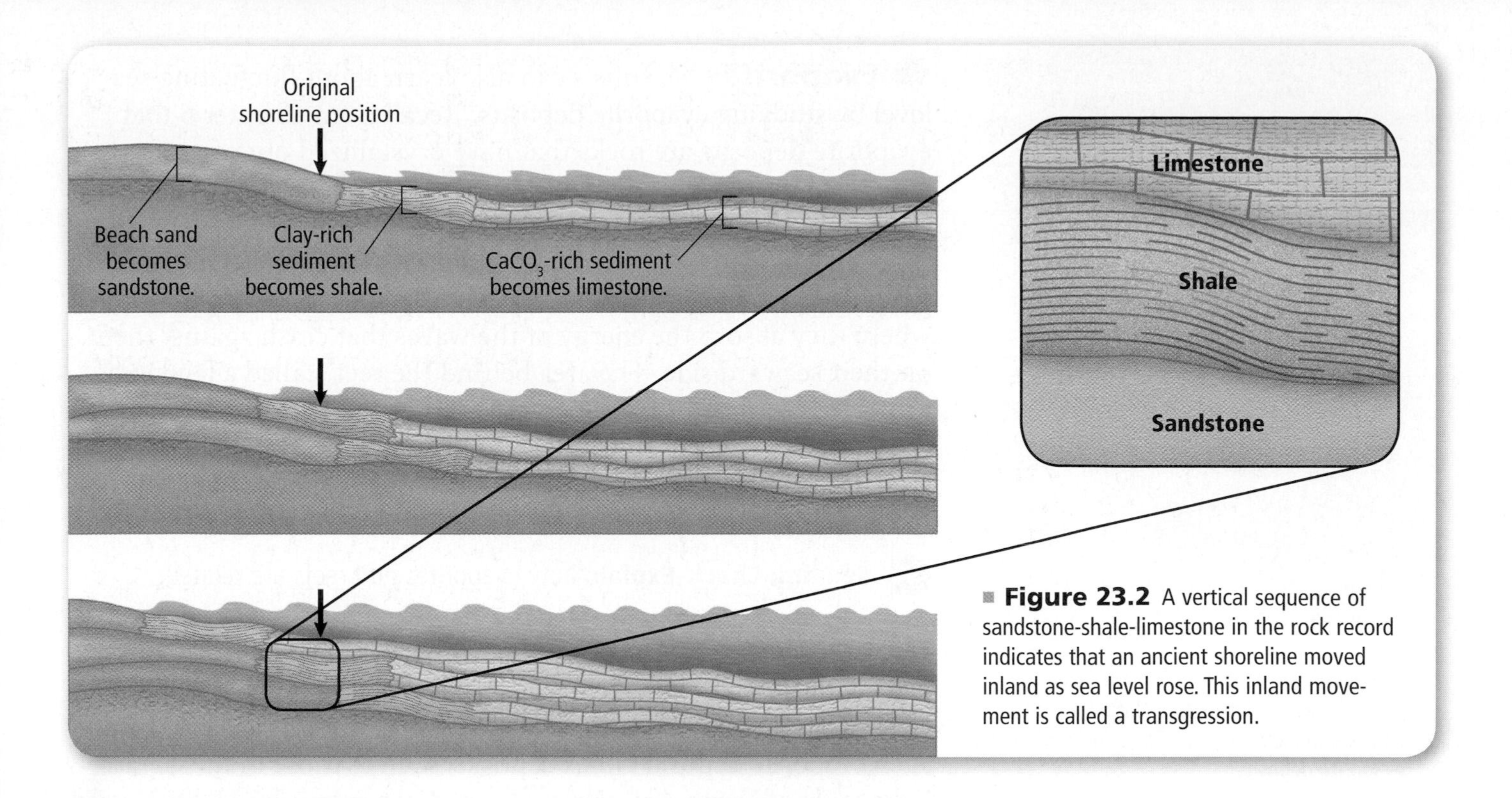

■ **Figure 23.2** A vertical sequence of sandstone-shale-limestone in the rock record indicates that an ancient shoreline moved inland as sea level rose. This inland movement is called a transgression.

Sea Level Changes in the Rock Record

Rock sequences preserved in passive margins tell paleogeographers a great deal about ancient shorelines. These sequences are useful in charting the rise and fall of sea level. To understand this, it is first necessary to understand how sediment is deposited on a shoreline.

Shoreline deposition Ocean tides wash small grains of sand and sediment ashore to make beaches. Tides also deposit offshore sediment the size of clay particles (<0.002 mm). Calcium carbonate ($CaCO_3$) sediment accumulates farther from shore as calcium muds form from sea water and organisms containing calcium carbonate die and fall to the seafloor. The sand deposited on the beaches eventually becomes sandstone, the offshore clay sediment compacts to form shale, and the calcium carbonate sediment farther offshore turns into limestone, as shown in **Figure 23.2.**

Transgression When sea levels rise or fall, the deposition of sediment shifts. As illustrated in **Figure 23.2,** a rise in sea level causes the water to move inland to an area that previously had been dry. The area where clay sediment was deposited also moves shoreward on top of the old beach. This movement is called a **transgression.** The result of the transgression is the formation of deep-water deposits overlying shallow-water deposits. This appears in the rock record as a vertical or stepwise sequence of sandstone-shale-limestone.

Regression When sea level falls, the shoreline moves seaward in a process called **regression.** This process results in shallow-water deposits overlying deep-water deposits. A stacked sequence of limestone-shale-sandstone is evidence of a regression.

VOCABULARY
SCIENCE USAGE V. COMMON USAGE

Transgression
Science usage: movement of a shoreline inland as sea level rises

Common usage: violation of a law or moral duty

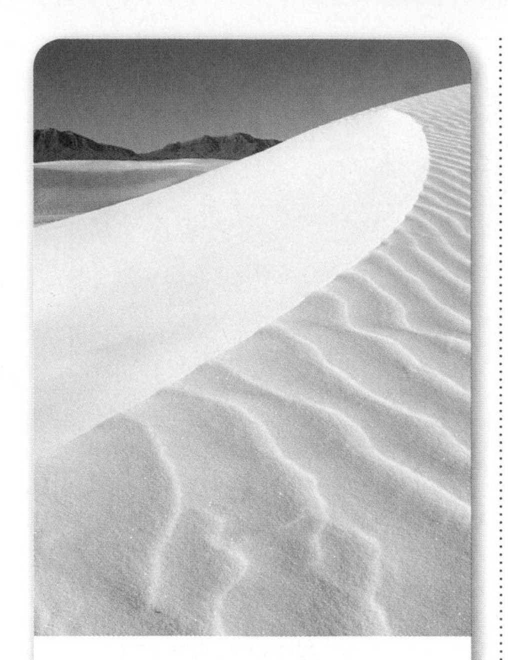

■ **Figure 23.3** The white sands of New Mexico's White Sands National Park are made of gypsum from ancient evaporite deposits.

Evaporites Scientists can also learn about fluctuating sea level by studying evaporite deposits. Recall from Chapter 6 that evaporite deposits are rocks that have crystallized out of water that is supersaturated with dissolved minerals. Some evaporite deposits can be associated with fossilized reefs.

Reefs are made of the carbonate skeletons of tropical organisms. Reefs form in long, linear mounds parallel to a continent or island, where they absorb the energy of the waves that crash against them on their seaward side. The area behind the reef, called a lagoon, is protected from the wave's energy. Water in the lagoons evaporates in the warm, tropical sunshine, and minerals such as halite and gypsum precipitate out. Over time, cycles of evaporite deposition mark changes in water level.

Reading Check **Explain** how evaporites and reefs are related.

Mineral deposits Huge amounts of gypsum and halite evaporites were deposited in Paleozoic lagoons. The white sands of White Sands National Park, shown in **Figure 23.3,** are the remains of one such evaporite deposit. Other deposits, such as those in the Great Lakes area of North America, are mined commercially. Halite is used as road salt. Gypsum is an ingredient in plaster and drywall.

Impermeability As shown in **Figure 23.4,** reef rocks tend to have large pore spaces, allowing oil and other liquids to move through them. Evaporite rocks, in contrast, are impermeable. This means that they contain very little pore space and liquid cannot move through them. When an evaporite deposit overlies a reef rock that contains oil, it seals in the oil and prevents the oil from migrating. A good example is the Permian Basin, home to the Great Permian Reef Complex in western Texas and southeastern New Mexico. The oil in this complex rarely leaks to Earth's surface because of its tight evaporite seal.

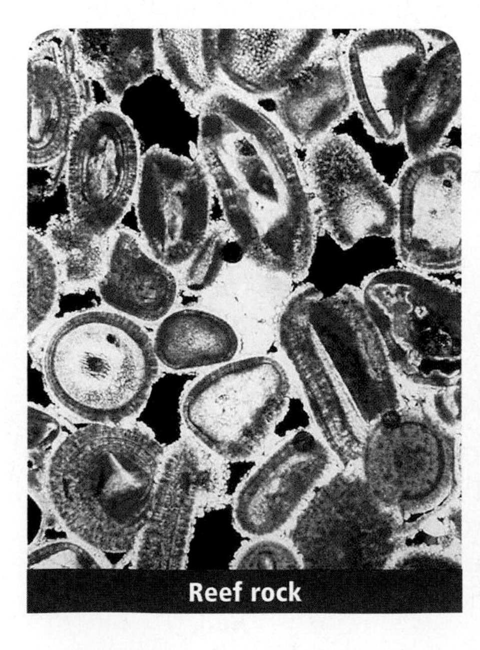

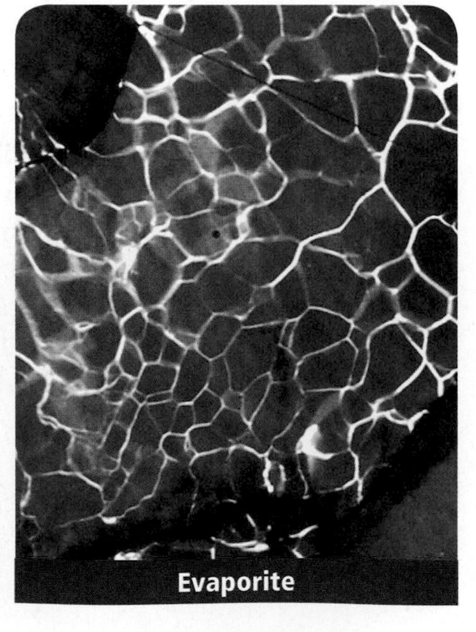

■ **Figure 23.4** Reef rocks have large pores that can contain oil or other liquids, in contrast to evaporite rock, which is impermeable to liquids.
Infer *why ancient evaporite deposits are important to petroleum geologists.*

Glaciation Scientists have determined that sea levels transgressed and regressed as many as 50 times during the late Paleozoic. Geologists have found a number of reasons for relative sea level change—climate and glaciation cycles, crustal subsidence and uplift, sedimentation rates, and plate motions. These were all factors in the transgressive and regressive cycles of the Paleozoic.

Reading Check **Explain** how glaciation affects sea level.

Mountain Building

Laurentia's margins were passive during the first period of the Paleozoic, and mountains were not forming. However, changes occurred during the Ordovician (or duh VIH shun) Period. At that time, Laurentia collided with the Taconic Island Arc, and mountains began to rise in what is now northeastern North America. This event is called the Taconic Orogeny. The Taconic Orogeny added new land and established an active volcanic zone along Laurentia's eastern margin. Remnants of this event are present in New York's Taconic Mountains.

Laurentia deformed Laurentia was further transformed in the Silurian (si LUR ee uhn) Period when Laurentia's eastern margin collided with Baltica and Avalonia. Baltica was a landmass that today is part of northern Europe and parts of Russia. Avalonia was an island ocean arc. You can see Baltica and Avalonia approaching Laurentia in **Figure 23.5.** The deformation caused by these collisions added folds, faults, and igneous intrusions to the already deformed Taconic rocks.

VOCABULARY
ACADEMIC VOCABULARY
Transform
to change in a major way
The continent was transformed by a massive orogeny.

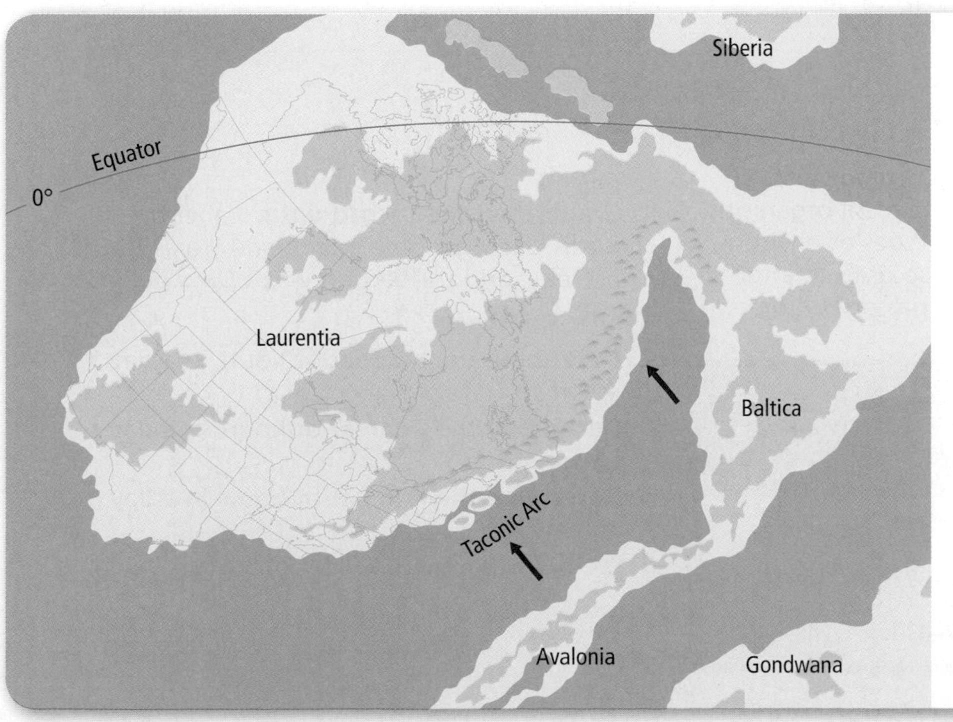

Figure 23.5 Baltica and Avalonia collided with the Taconic Island Arc during the Ordovician. This was the first of many Paleozoic tectonic events that transformed eastern Laurentia.

Ouachita Orogeny Another Laurentian mountain-building event—the Ouachita (WAH shuh taw) Orogeny—occurred during the Carboniferous Period when southeastern Laurentia began to collide with Gondwana. Recall from Chapter 17 that Gondwana was the large landmass that eventually formed the southern continents, including Africa and South America. This collision formed the Ouachita Mountains of Arkansas and Oklahoma and was so intense that it caused the crust to uplift inland as far as present-day Colorado. Vertical faults raised rocks more than 2 km, forming a mountain range that geologists call the Ancestral Rockies.

Alleghenian Orogeny As Gondwana continued to push against Laurentia, the Appalachian Mountains began to form. This event, called the Alleghenian Orogeny, was the last of the Paleozoic mountain-building events to affect eastern North America. When it was completed at the end of the Paleozoic, the Appalachians were possibly higher than the Himalayas, and one giant supercontinent—Pangaea—had formed on Earth's surface.

CAREERS IN EARTH SCIENCE

Paleoecologist Paleoecologists study the ecology and climate of ancient environments using evidence from fossils and rocks. Some paleoecologists apply this knowledge to understand future global climate change. To learn more about Earth science careers, visit glencoe.com.

Paleozoic Life

The formation of Pangaea was the major geologic story of the Paleozoic, but the Paleozoic rocks also tell another dramatic story. Multicellular animals went through extensive diversification at the beginning of this era, including the first appearance of organisms with hard parts. As you learned in Chapter 21, fossils help geologists correlate geologic landscapes and piece together geologic time. Fossils also help paleoecologists (pay lee oh ih KAH luh jists) learn about the ecology of ancient environments.

FOLDABLES

Incorporate information from this section into your Foldable.

DATA ANALYSIS LAB

Based on Real Data*

Interpret the Table

Can you find the time? Paleoecologists study the shapes and compositions of fossil organisms to interpret how and in what types of environments they lived. Fossils are also used to interpret climatic changes and the passage of time.

Time Record Data

Geologic Era	Hours Per Day	Day Per Year	Geologic Time (mya)
Cenozoic	23.5–24	365–377	0–65
Mesozoic	23.5–22.4	377–392	65–248
Paleozoic	22.4–20	392–430	248–543

Analysis

1. **Graph the time record data.** Label the *x*-axis *Geologic time (mya),* one *y*-axis *Hours per day*, and the second *y*-axis *Days per year*.

Think Critically

2. **Determine** the number of hours in a day 400 mya.
3. **Determine** the number of hours in a day 200 mya.
4. **Determine** the number of hours in a day 150 mya.
5. **Predict** when there will be 24.5 hours in a day.

*Data obtained from: Prothero, D.R., and R.H. Dott, Jr. 2004. *Evolution of the Earth.* New York: McGraw-Hill

■ **Figure 23.6** The organisms shown in this artist's reconstruction are among the Cambrian organisms that had hard parts.

Cambrian explosion Nearly every major marine group living today appeared during the first period of the Paleozoic. The geologically rapid diversification of such a large collection of organisms in the Cambrian fossil record is known as the **Cambrian explosion.** Some of the best-preserved Cambrian organisms occur in the Burgess Shale in the Canadian Rocky Mountains, and in western China. A spectacular array of fossil organisms with hard parts has been found there, including fossils of creatures like those shown in **Figure 23.6.**

Ordovician extinction At the end of the Ordovician, more than half of the marine groups that appeared in the Cambrian became extinct. Those that survived suffered large losses in their numbers. What caused this extinction? Geologists have found evidence of glacial deposits in rocks of northern Africa, which at the time was situated at the south pole. As you learned in Chapter 8, when water freezes in glaciers, sea level drops. Then, as now, most marine animals lived in the relatively shallow waters of the continental shelves. When sea level is high, the shelves are flooded and marine animals have many places to live. During regression, however, continental shelves can become too narrow to support diverse animal habitats.

Devonian extinction Following the late Ordovician extinction, marine life recovered and new species evolved, including a tremendous diversification of vertebrates, including fish and the first appearance of tetrapods on land. In the late Devonian (dih VOH nee un), another extinction event eliminated approximately 50 percent of the marine groups. Some scientists think that global cooling was again the cause and there is evidence that some continents had glaciers at this time.

MiniLab

Model Continental Shelf Area

How does shelf area change when continents collide? Colliding continents decrease the habitat areas available to marine organisms, which tend to live along the shallow shelves surrounding the continents.

Procedure

1. Read and complete the lab safety form.
2. Using 250 g of **modeling clay,** make a sphere and flatten it into a disk that is 0.5 cm thick. This represents a continent.
3. Divide another 250 g of clay into two equal spheres and flatten them as above.
4. Roll another 250 g of clay into three cylinders, each with a diameter of about 0.5 cm. Wrap the cylinders around the edges of the clay disks. These represent continental shelves.
5. Use the following formula to calculate the area of the large continent and the large continent plus the continental shelf.

   $$\text{area} = \pi r^2$$

   Subtract the continent area from the total area. This equals the area of the continental shelf.
6. Repeat Step 5 for both small models.

Analysis

1. **Assess** which has more shelf area: two small continents or one large continent. Why?
2. **Conclude** how the existence of a single large supercontinent limits the amount of habitat space for marine organisms.
3. **Explain** the relationship between reduced habitat space and extinction.

■ **Figure 23.7** This artist's reconstruction shows what a Carboniferous swamp might have looked like.
Explain *why Carboniferous swamps produced coal deposits.*

Terrestrial plants The Ordovician and Devonian extinction events appear to have affected mainly marine life. They had little effect on life-forms living on land. Simple land plants began to appear on Earth in the Ordovician. In the Carboniferous, the first plants with seeds, called seed ferns, diversified. Because seeds contain their own moisture and food sources, they enabled terrestrial plants to survive in a variety of environments.

Coal deposits Many Carboniferous plants lived in low-lying swamps, such as the one shown in **Figure 23.7.** As these plants died and sediment accumulated, they compacted to coal deposits. Swamps were also breeding grounds for insects. Fossils of the largest known insects have been found in Carboniferous sediment deposits, including dragonflies with 74-cm wingspans. Compare this to the largest known wingspan of a modern dragonfly—19 cm.

Permian changes At the end of the Permian, the largest mass extinction in the history of Earth occurred. The Permo-Triassic Extinction Event caused the extinction of nearly 95 percent of marine life-forms. Unlike the mass extinctions at the end of the Ordovician and Devonian, this extinction affect both marine and terrestrial organisms. More than 65 percent of the amphibians and almost one-third of all insects did not survive. What could have caused such a widespread catastrophe? It was probably a combination of causes. First, there was a dramatic drop in sea level from the coalescence of Pangaea closing and draining the shallow seas. A major regression would have been particularly critical for organisms inhabiting the continental shelves when there was only one continent. Other contributing factors include extreme volcanism in Siberia, low atmospheric oxygen levels, and even a meteorite impact.

Section 23.1 Assessment

Section Summary

- Scientists study sediment and evaporite deposits to learn how sea levels fluctuated in the past.
- Eastern Laurentia was transformed by many mountain-building events during the Paleozoic.
- A great diversity of multicellular life appeared during the first period of the Paleozoic.
- The largest extinction event in Earth's history occurred at the end of the Paleozoic.

Understand Main Ideas

1. MAIN Idea **Explain** how the formation of Pangaea affected the evolution of life-forms.
2. **Compare** transgression and regression.
3. **Discuss** the relationship between oil deposits and evaporites.
4. **Assess** the significance of the Cambrian explosion.

Think Critically

5. **Infer** what has happened to the Ancestral Rockies since their formation.
6. **Predict** changes in the fossil and rock record that might indicate a marine extinction event.

MATH in Earth Science

7. Is a mass extinction occurring on Earth today? If 10 million species exist today and 5.5 species become extinct every day, calculate how many years it would take for 96 percent of today's species to become extinct.

Section 23.2

Objectives

- **Discuss** how the breakup of Pangaea affected Earth's life-forms and paleogeography.
- **Explain** how the mountains of western North America formed.
- **Identify** possible causes of the extinction of the dinosaurs.

Review Vocabulary

subduction: the process by which one tectonic plate descends beneath another

New Vocabulary

phytoplankton
amniotic egg
iridium

The Mesozoic Era

MAIN Idea Reptiles became the dominant terrestrial animals during the Mesozoic while Pangaea broke apart.

Real-World Reading Link Do you like mystery novels? One of the biggest mysteries in the history of science is what caused the extinction of the dinosaurs.

Mesozoic Paleogeography

The mass extinction event that ended the Paleozoic Era ushered in new opportunities for animals and plants of the Mesozoic Era. Earth's life-forms changed drastically as new kinds of organisms, shown in **Figure 23.8,** evolved to fill empty niches. While some groups of these organisms remain on Earth today, none of the giant reptiles that dominated the land, sea, and air, and typified the period survived. The dinosaurs all became extinct at the end of the era.

Breakup of Pangaea When the Mesozoic Era began, a single global ocean and a single continent—Pangaea—defined Earth's paleogeography. During the middle Triassic Period, Pangaea began to break apart. The heat beneath Pangaea caused the continent to expand, and Pangaea's brittle lithosphere began to crack. Some of the large cracks, called rifts, gradually widened, and the landmass began spreading apart. The ocean flooded the rift valleys to form seaways, and large blocks of crust collapsed to form deep valleys. The Mesozoic climate was warm and tropical, and it remained warm enough throughout the era that glaciers did not form.

■ **Figure 23.8** Although dinosaurs are the most famous of the Mesozoic life-forms, other organisms also appeared during this era.

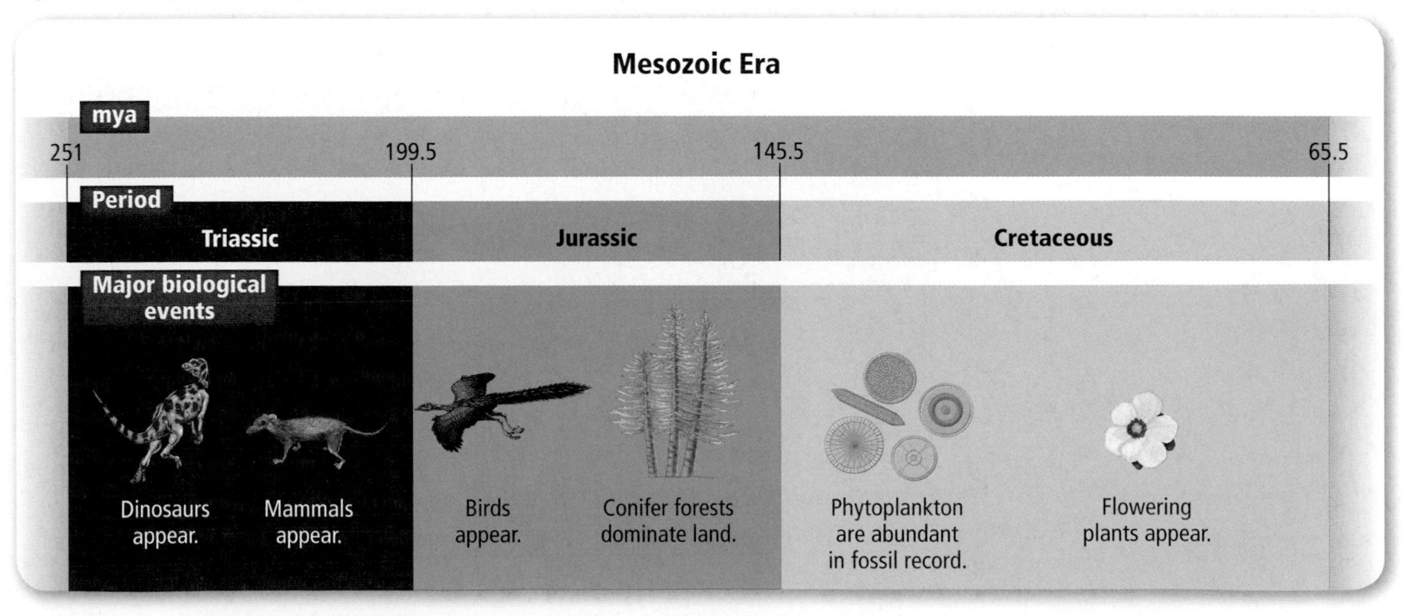

■ **Figure 23.9** The Red Sea and the Gulf of Aden are widening into a new seaway.
Identify ***the tectonic force behind the creation of this new seaway.***

Seaways As the continents continued to split apart, mid-ocean rift systems developed at the junctures, and the widening seaways became oceans. The Atlantic Ocean began forming early in the Triassic as North America rifted away from Europe and Africa. Some of the spreading areas at this juncture joined to form a long, continuous rift system called the Mid-Atlantic Ridge. As you learned in Chapter 20, this mid-ocean ridge system is still active today, erupting magma deep in the ocean as it widens. The Red Sea and Gulf of Aden, shown in **Figure 23.9,** are new seaways in East Africa that are today slowly widening by a few centimeters a year as a result of continental breakup.

Reading Check **Explain** how the Atlantic Ocean formed.

Changing sea level The formation of mid-ocean rift systems was partly responsible for a rise in sea level during the Mesozoic. The hot magma that erupted at the ridges displaced a considerable amount of seawater onto the continents. However, sea level dropped at the end of the Triassic, and desertlike conditions developed in western North America. The climate became arid and, as evidenced in ancient sand dunes, a thick blanket of sand covered some of the land. Sea level rose again during the Jurassic, and a shallow sea formed in North America's center. The ocean continued to rise during the Cretaceous (krih TAY shus), covering much of North America's interior. **Figure 23.10** shows that nearly one-third of Earth's landmasses were covered with water.

Mountain Building

You learned in Chapter 20 that the collision of continents during the Paleozoic transformed the eastern margin of Laurentia, while the continent's western margin remained passive. During the Mesozoic and early Cenozoic, the reverse was true. As the breakup of Pangaea proceeded, multiple mountain-building episodes occurred along Laurentia's western margin, while little was happening along its eastern edge.

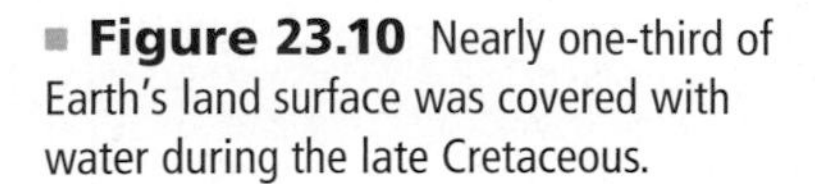

■ **Figure 23.10** Nearly one-third of Earth's land surface was covered with water during the late Cretaceous.

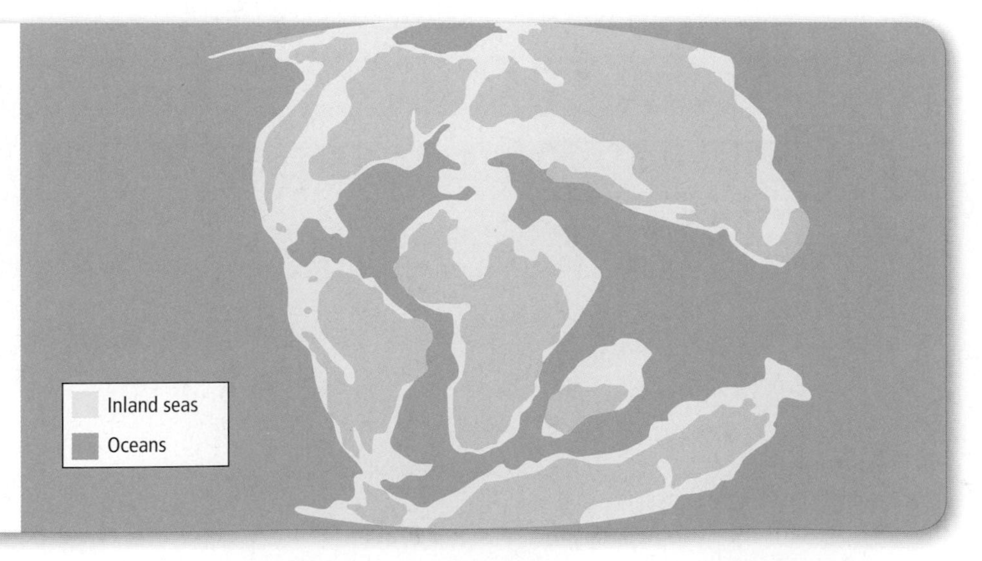

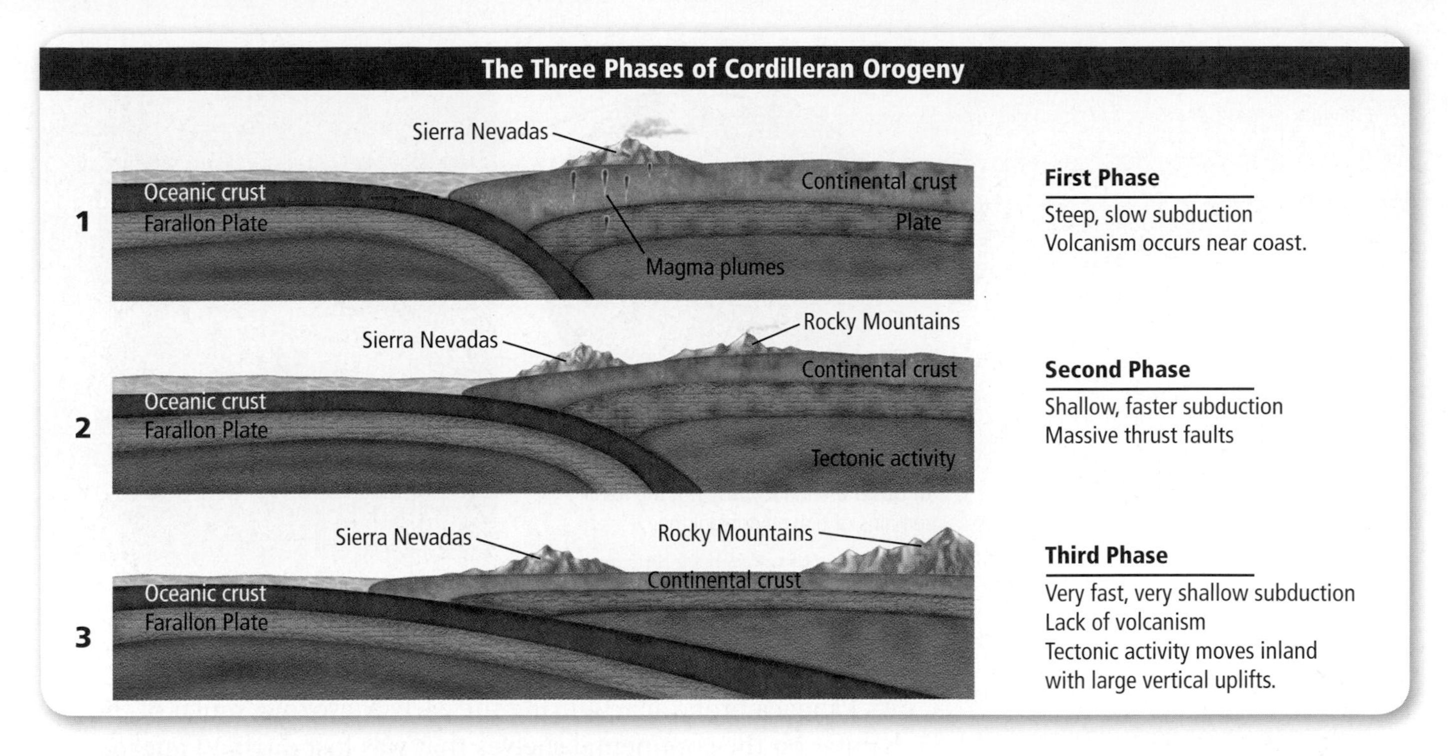

Figure 23.11 During the three phases of the Cordilleran Orogeny, mountains formed farther inland as the angle of subduction became more shallow and the speed increased, causing massive faulting and uplift.

Cordillera Much of the mountain building that occurred in western Laurentia was caused by the subduction of the oceanic Farallon Plate beneath Laurentia's western margin. As the plate descended, many structural features of the present-day Rocky Mountains, Sierra Nevadas, and other western mountain ranges were formed. Geologists call these ranges collectively the North American Cordillera (kor dee AYR uh). Cordillera means *mountain range* in Spanish. The Cordilleran Orogeny consisted of three distinct phases. As shown in **Figure 23.11,** each phase was characterized by a different rate and angle of subduction.

First phase The first phase of the Cordilleran Orogeny occurred during the late Jurassic and early Cretaceous when subduction proceeded slowly and the oceanic plate descended at a steep angle, producing magma, which rose to form the Sierra Nevadas.

Second phase The second phase of the Cordilleran Orogeny occurred during the Cretaceous when subduction increased in speed but the oceanic plate descended at a shallow angle. As a result, there was less volcanism along Laurentia's margin and more tectonic activity inland with massive thrust faults occurring in the Rocky Mountain area.

Third phase During the third phase of the Cordilleran Orogeny, which began during the late Cretaceous and continued into the Cenozoic, the angle of subduction was even more shallow than that of the second phase. The shallow angle was caused by rapid subduction. The subduction rate was so fast that some scientists suggest the oceanic plate was pushed almost horizontally beneath the North American Plate. As a result, this phase was characterized by large, vertical uplifts and a lack of volcanism. This range now extends from northern Mexico into Canada.

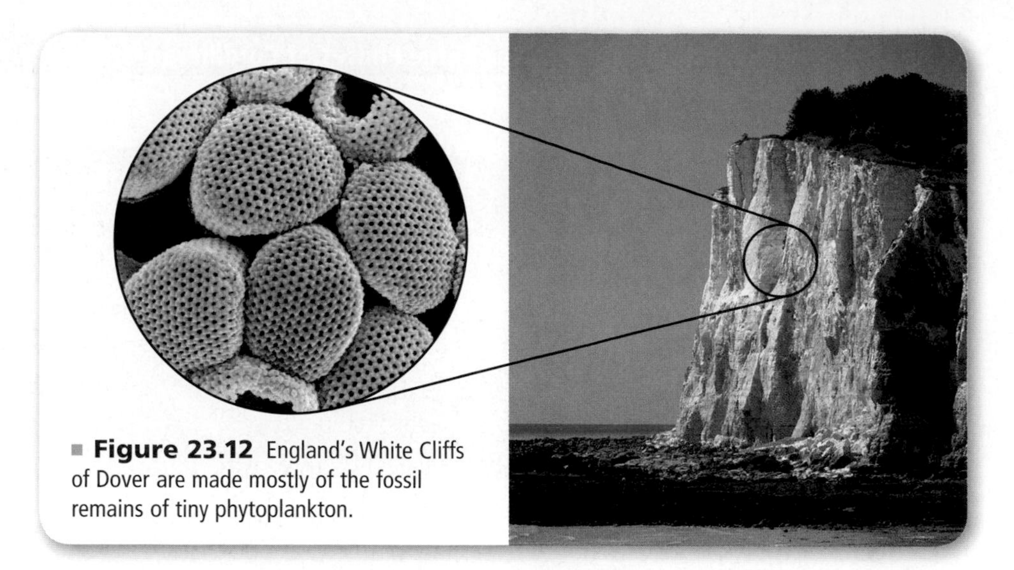

■ **Figure 23.12** England's White Cliffs of Dover are made mostly of the fossil remains of tiny phytoplankton.

Mesozoic Life

As Pangaea broke apart during the early Mesozoic, much of the habitat on the continental shelves that was lost during Pangaea's formation once again became available. New marine organisms, ranging from large predatory reptiles to tiny photosynthetic phytoplankton, evolved to fill these niches. **Phytoplankton** were, and are today, microscopic organisms at the base of the marine food chain. These organisms were abundant during the Cretaceous. The remains of their shell-like hard parts are seen in many chalk deposits around the world, including England's famous White Cliffs of Dover, shown in **Figure 23.12.**

Plant life As the cool climate that characterized the late Paleozoic came to an end during the Mesozoic, plant life changed sharply. The large, temperate swamps dried as the climate warmed. Tall cycad trees are seed plants without true flowers. These evolved during the Jurassic, along with ginkgos, pine trees, and other conifers. Flowering plants appeared during the Cretaceous.

Terrestrial animals Mammals appeared during the late Triassic, around the same time as the dinosaurs. However, the dominant Mesozoic animals were the reptiles. Unlike amphibians, whose eggs need to be laid in water to prevent drying out, reptiles can lay their eggs on dry land. These eggs, called **amniotic** (am nee AH tihk) **eggs,** contain the food and water required by developing embryos inside. Aminiotic eggs made it possible for reptiles, including dinosaurs, to roam widely.

Dinosaurs Archosaurs are a group of reptiles which includes dinosaurs and crocodilians. Archosaurs have a unique skeletal structure that allows for speed and flexibility of movement. While lizards and turtles walk with a sprawling posture, archosaurs have a hip structure that allows the legs to be held underneath the body. This enabled some archosaurs to run with an upright posture, as shown in **Figure 23.13.**

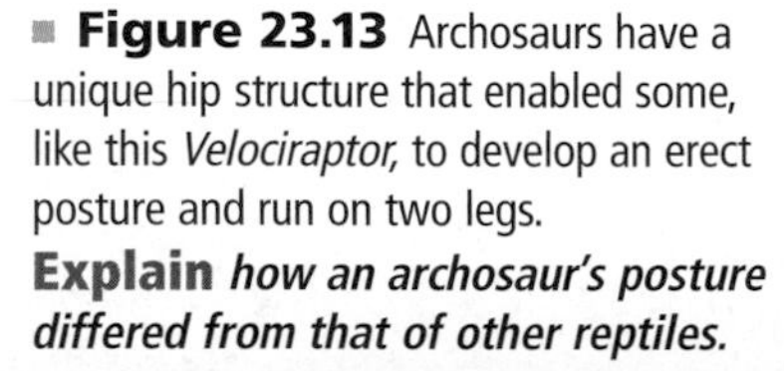

■ **Figure 23.13** Archosaurs have a unique hip structure that enabled some, like this *Velociraptor,* to develop an erect posture and run on two legs.
Explain *how an archosaur's posture differed from that of other reptiles.*

Concepts In Motion **Interactive Table** To explore more about extinctions during the Phanerozoic, visit glencoe.com.

Table 23.1 Major Extinctions in the Phanerozoic

Extinction event	End Ordovician	Late Devonian	Permo-Triassic	End Triassic	End Cretaceous
Approximate mya	439 mya	364 mya	250 mya	200 mya	65 mya
Percentage groups extinct	57 percent marine	50 percent marine	80 percent marine 70 percent land	48 percent marine	50 percent marine 56 percent land

Mass extinction At the end of the Mesozoic, an extinction event devastated terrestrial dinosaurs, most marine reptiles, plants, and many other organisms. Today, most scientists agree that the combination of massive volcanism, which stressed Earth's climate, and a large meteorite impact that occurred at the end of the Cretaceous is responsible for the extinction event. It is thought that the meteorite was at least 10 km in diameter. An impact of this size could have blown up to 25 trillion metric tons of rock into the atmosphere, causing long-lasting greenhouse warming. Evidence for this impact, which scientists think occurred in Mexico's Yucatan Peninsula, exists in a clay layer that separates Cretaceous rocks from rocks of the first period of the Cenozoic. Found worldwide, this layer contains an unusually high amount of **iridium** (ih RID ee um), a rare metal in Earth's rocks but a relatively common metal in asteroids. As shown in **Table 23.1,** this extinction event was relatively mild compared with other Phanerozoic extinction events.

Section 23.2 Assessment

Section Summary

- The breakup of Pangaea triggered a series of tectonic events that transformed western Laurentia.
- The Atlantic Ocean began to form during the Mesozoic as North America broke away from Europe.
- Dinosaurs and other new organisms evolved to fill niches left empty by the Permo-Triassic Extinction Event.
- All dinosaurs, except birds, along with many other organisms became extinct during a mass extinction event at the end of the Mesozoic.

Understand Main Ideas

1. MAIN Idea **Discuss** the significance of the Permo-Triassic Extinction Event for the animals that populated the Mesozoic.
2. **Explain** how rifts are related to the formation of oceans.
3. **Compare** the tectonic events that transformed Laurentia's western margin with the tectonic events that changed Laurentia's eastern margin.
4. **Discuss** the evidence that suggests a meteorite impact was responsible for the extinctions at the end of the Mesozoic Era.

Think Critically

5. **Deduce** what happened to the oceanic plate that subducted beneath western North America during the Mesozoic.

WRITING in Earth Science

6. Prepare a report documenting the chain of events that might have occurred once the meteorite hit Earth. Include a discussion of the effect on climate, air quality, and plant and animal life.

Earth Science Online **Self-Check Quiz** glencoe.com

Section 23.3

Objectives

- **Assess** the extent of glaciation during the Cenozoic.
- **Describe** tectonic activity in North America during the Cenozoic.
- **Explain** how climate change affected life-forms during the Cenozoic.

Review Vocabulary

San Andreas Fault: a transform fault that separates the western edge of Southern California from the rest of the state and is responsible for most of California's earthquakes

New Vocabulary

Homo sapiens
bipedal

The Cenozoic Era

MAIN Idea Mammals became the dominant terrestrial animals during the Cenozoic while the continents assumed their present forms.

Real-World Reading Link Have you ever been to a soccer game during which a player was injured? Usually another player fills in and the game goes on.

Cenozoic Paleogeography

The Cenozoic Era encompasses about 1.5 percent of Earth's total history—approximately the last 66 million years. Despite its relative shortness, scientists know more about this era than any other. Humans evolved during the Cenozoic, appearing in their present-day form during the Pleistocene Epoch. **Figure 23.14** shows that you live in the Holocene, the current epoch of the Cenozoic.

Cooling trend You learned in Section 23.2 that the Mesozoic Era was relatively warm. Earth remained warm during the earliest epoch of the Cenozoic. However, as Australia split apart from Antarctica during the Eocene (EE uh seen) Period, the worldwide climate began to cool. Scientists think that the cooling climate was caused, in part, by a change in ocean currents. When Antarctica and Australia were connected, a current of warm water flowing from the Pacific, Atlantic, and Indian Oceans moderated Antarctica's temperature. After Antarctica and Australia split apart, Antarctica was isolated over the South pole. A cold current began to flow around it, and a permanent ice cap began to grow in the Oligocene (AH luh goh seen).

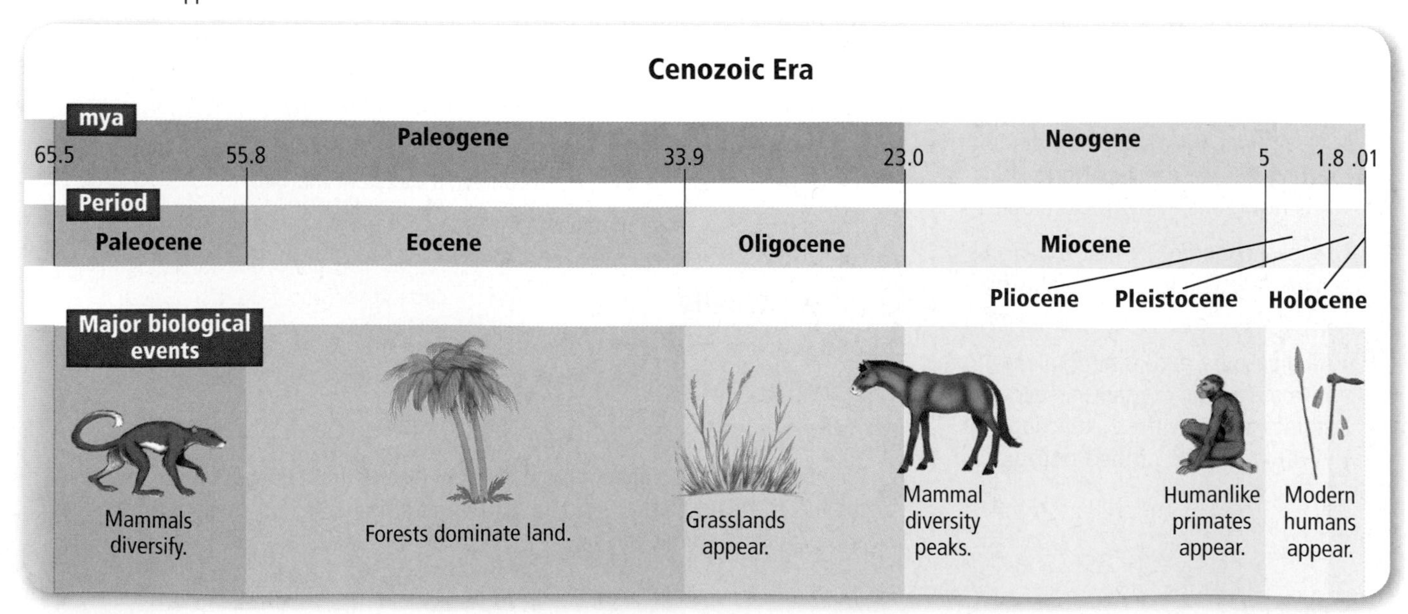

■ **Figure 23.14** Mammals diversified widely during the Cenozoic, but modern humans did not appear until the end of the era.

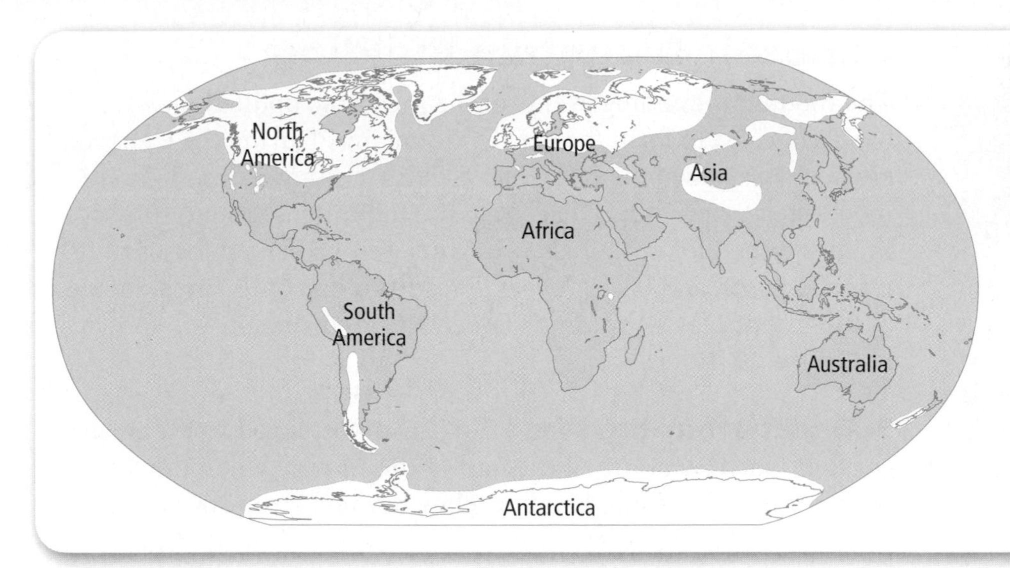

Figure 23.15 At the peak of Pleistocene glaciation, glaciers covered nearly one-third of Earth's land surfaces.
Infer *why patches of glaciation existed near the equator.*

Miocene warming In the early Miocene Period, the climate warmed again. The ice cap on Antarctica began to melt, and the ocean flooded the margins of North America. This trend reversed during the middle and late Miocene. Antarctica's ice cap stopped melting and the Arctic Ocean began to freeze, and resulted in the formation of the arctic ice cap. This set the stage for the ice ages.

Ice ages Starting in the late Pliocene (PLY uh seen) and continuing throughout the Pleistocene, ice covered much of Earth's northern hemisphere. Glaciers advanced and retreated in at least four stages over North America and the northern latitudes. During the peak of these ice ages, glaciers up to 3 km thick covered nearly one-third of Earth's land surfaces, as shown in **Figure 23.15.** In North America, the paths of the Ohio and Missouri Rivers roughly mark the southernmost point of glacier coverage. Glaciers carved out lakes and valleys, dropped huge boulders, and left behind abundant deposits of clay, sand, and gravel. In northeastern Washington State, glacial melting caused such a rush of water at the end of the last ice age that it created the largest waterfall recorded on Earth's surface. The remnants are shown in **Figure 23.16.**

Figure 23.16 This photo shows the remnants of Earth's largest waterfall in what is now Washington State. The waterfall, more than 5 km long and 120 km high, once flowed with water from glacial melting.

■ **Figure 23.17** This 38-million-year-old fossil bird was found in Wyoming's Green River Formation. The fossil is about 25 cm long.

Cenozoic Mountain Building

The mountain-building events of the Mesozoic uplifted massive blocks of crust to form the Rocky Mountains. During the Cenozoic, erosion wore down the Rockies but uplift continued. Eroded sediment filled large basins adjacent to the mountains. Today, this sediment is mined for coal. It also contains well-preserved fossils of fish, insects, plants, and birds. A fossil bird from one of the most famous of these deposits—Wyoming's Green River Formation—is shown in **Figure 23.17.**

Subduction in the West Volcanism returned to the western coast of North America at the end of the Eocene when the last remnant of the oceanic Farallon Plate began a steep subduction beneath the Pacific Northwest. As a result, the Cascade Mountains began to rise. Volcanoes in the Cascade range remain active today, as shown in **Figure 23.18.**

While subduction continued in northwestern North America, the Farallon Plate disappeared completely under what is now California. The North American Plate came into contact with another oceanic plate—the Pacific Plate—that was moving in a different direction. As a result, the San Andreas Fault formed. The San Andreas Fault is a transform boundary between the two plates. Recall from Chapter 17 that in a transform boundary, two plates slide against each other and there is no subduction. Because there is no subduction beneath central and Southern California today, there is little volcanic activity there.

Basin and Range Province The beginning of the interaction between the North American Plate and the Pacific Plate coincided with the formation of the Basin and Range Province in the southwestern United States and northern Mexico. Recall from Chapter 20 that the Basin and Range Province consists of hundreds of nearly parallel mountains. These mountains were formed when stresses in Earth's crust pulled it apart. This process, illustrated in **Figure 23.19,** continues today.

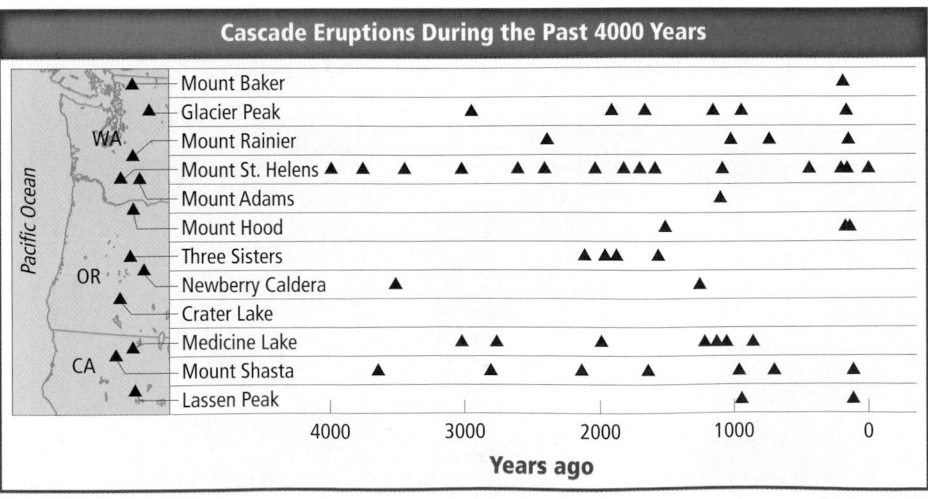

Source: USGS

■ **Figure 23.18** The Cascade Mountain Range includes active volcanoes that have erupted many times during the past 4000 years.
Conclude *which volcano is the most active.*

Visualizing the Basin and Range Province

Figure 23.19 The Basin and Range Province is a series of mountains and basins that is bordered on the west by California's Sierra Nevadas and on the east by Utah's Wasatch Mountains. During the past 25 million years, crustal stretching has increased the distance between these two points by about 80 km.

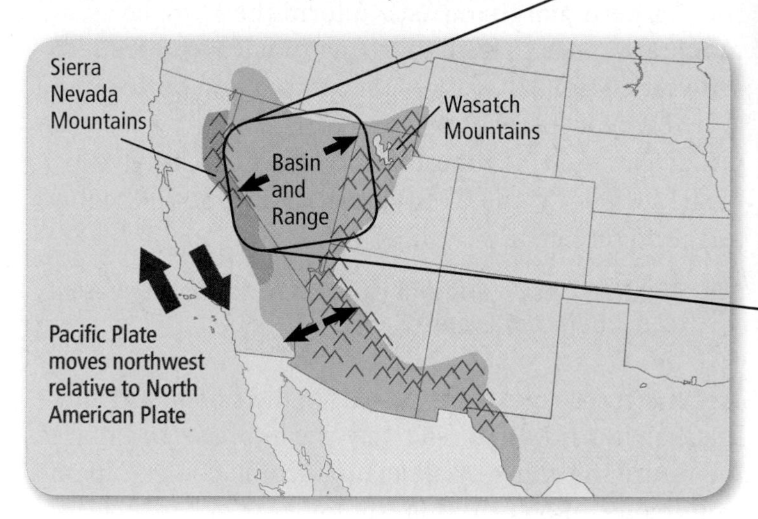

The stretching underneath the Basin and Range Province is caused, in part, by the steady movement of the Pacific Plate relative to the North American Plate. The North American Plate is being stretched to the northwest, and the Basin and Range Province is being stretched in an east-west direction.

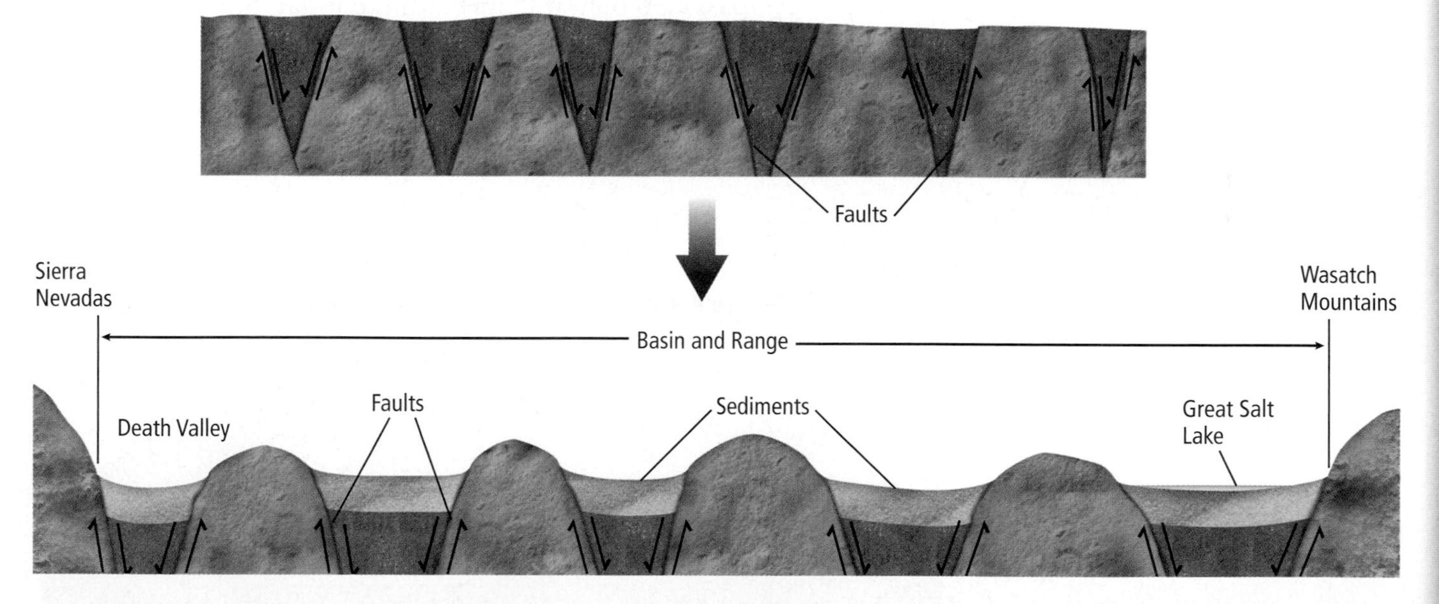

To compensate for crustal stretching, the rocks broke up into hundreds of blocks along normal fault lines. Some blocks rose to form mountains, while adjacent areas dropped to form basins. The mountains are still being pushed upward, rising as quickly as they erode, and the basins are still dropping and filling with eroded debris. The crust underneath the Basin and Range Province has stretched so much that it is one of the thinnest parts of Earth's crust today.

Concepts In Motion To explore more about the formation of the Basin and Range Province, visit glencoe.com. Earth Science Online

■ **Figure 23.20** The Himalayas appear as an abrupt junction where India crashed into Asia.

Continental collisions The final breakup of Pangaea during the early Cenozoic resulted in several separate continents. It also brought some continents together. During the Paleocene, Africa began to collide with Eurasia, creating the Alps and narrowing the ancient Tethys (TEE thus) Ocean, which once separated Eurasia and Gondwana. The remnants of this ocean now exist as four bodies of water in Europe and central Asia—the Black, Caspian, Aral, and Mediterranean Seas.

Also during the Paleogene, India began crashing into the southern margin of Asia to form the Himalayas, a mountain range that is still rising today. **Figure 23.20** shows the Himalayas as an abrupt junction where India joined Asia. The rocks on the top of Mount Everest are Ordovician marine limestone. Tectonic forces have pushed what was the Ordovician seafloor to the highest elevation on Earth.

Reading Check **Explain** why marine fossils are present on top of Mount Everest.

Tectonic forces continue Many scientists think that Earth is now in a relatively warm phase and that in the future the climate will again become cooler. No one can predict when or if this will happen. What is clear is that the tectonic forces that have shaped Earth over the past 4.56 billion years continue today. Some scientists think that in 250 million years, those forces will have largely eliminated the Atlantic Ocean and formed the continents into another supercontinent, as shown in **Figure 23.21.**

■ **Figure 23.21** The Atlantic Ocean has nearly disappeared in this hypothetical map of Earth 250 million years in the future.

Cenozoic Life

Many marine organisms, including clams, sea urchins, and sharks, survived the mass extinction at the end of the Cretaceous and populated the oceans during the Cenozoic. On land, forests dominated the early Cenozoic landscapes. As the climate cooled during the late Eocene, forests gave way to open land, and grasses appeared. By the late Oligocene, grassy savannas, like those in east Africa today, were common worldwide. The rise of grasslands led to the diversification of many new mammal groups. Because mammals are the dominant terrestrial animals, many scientists call the Cenozoic the Age of Mammals.

Ice age mammals As the ice ages began, the climate began to cool and new animals evolved in northern latitudes. Two of the most famous mammals of the late Pleistocene are the woolly mammoth and the saber-toothed cat, shown in **Figure 23.22.** By the time these animals roamed Earth, modern humans—called ***Homo sapiens***—were well established.

■ **Figure 23.22** The woolly mammoth and saber-toothed cat, shown in this artist's reconstruction, were adapted to the cool Pleistocene climate.

Humans The defining characteristic of humans is their upright, or **bipedal,** locomotion. The fossil record, while incomplete, shows that the first bipedal humanlike primates appeared about 6 mya during the late Miocene. The fossil remains of the earliest modern humans—found in Africa—are about 195,000 years old.

Migrations The migrations of early humans were undoubtedly influenced by the ice ages of the late Pleistocene. For example, scientists think that the Bering Strait, which now separates Russia and Alaska, was exposed during the late Pleistocene because much of Earth's water was frozen in glaciers. It is likely that the humans who walked across the strait were North America's first inhabitants.

Section 23.3 Assessment

Section Summary

- Ice covered nearly one-third of Earth's land surface at the peak of the Cenozoic ice ages.
- The Cascade Mountains began to rise and the San Andreas Fault formed during the Cenozoic.
- The Cenozoic is known as the Age of Mammals.
- Fossil evidence suggests that modern humans appeared during the Pleistocene.

Understand Main Ideas

1. MAIN Idea **Describe** why the Cenozoic is called the Age of Mammals.
2. **Assess** the extent of glaciation in North America.
3. **Discuss** how the Basin and Range Province and the San Andreas Fault are tectonically related.
4. **Explain** how the positions of the continents contributed to Cenozoic climate change.

Think Critically

5. **Propose** Why do you think early humans migrated?

MATH in Earth Science

6. If the glacial ice on Earth were to melt, sea level would rise about 50 m above its current level. If sea level rose at an average rate of 2 mm per year, how long would it take for all the ice on Earth to melt? Use the following relationship: distance = rate × time.

ON SITE: DIGGING FOR DINOSAURS

Junior paleontologists work to carefully remove bones from the ground at this dig in Montana.

Check this out . . . look what I found! The team of paleontologists and volunteers runs up the hill to see what I am holding. In my hand is the tip of a 150-million-year-old theropod dinosaur tooth.

Day 1 Arrive at dig site The team loads equipment, gear, and tools into the vans and drives to the dig site. The quarry where digs have been ongoing is visible and everyone is eager to begin. But first we set up camp.

The land is owned by a local rancher who, in 1985, found the first chips of fossilized bone on the site. In 2003, larger bones were found and the rancher realized the importance of this find. He contacted a paleontologist at the Judith River Dinosaur Institute to ascertain interest in digging for dinosaur fossils on the land. Digs have been conducted each summer since that initial contact with the rancher.

Day 2 Basic digging skills Most on our team are volunteers who have no experience at a dig. Other members include geologists, paleontologists, and science teachers. The primary tools used are awls, which look like ice picks, and stiff paintbrushes. Shovels, air hammers, and wheelbarrows are also used throughout the digging process.

The work of a dig is slow. Each layer of rock is picked apart carefully and the debris is brushed away. The rock is inspected to be sure no fossils are accidentally brushed away or broken off.

Each member of the team fills a scoop with debris which gets dumped into a bucket and examined. Then the process is repeated. On this day, we find fossil bones from a theropod.

Days 3–6 Digging for a *Stegosaur* A few members of the team find a few tail vertebrae and fragments of a *Stegosaur* spike. The team begins digging in the hillside and eventually unearths more tail vertebrae and limb and foot bones. Three more days of excavating and little more is found.

Day 7 Finalizing the dig The final hours of the dig are busy as everyone gathers as much data as possible. We sketch and photograph fossils, take measurements of the locations where the fossils were found, and label each fossil.

We pack the fossils and the equipment into the vans. The fossils will be catalogued and stored at the Institute. The data will be used in a research paper that will be published in a scientific journal.

WRITING in Earth Science

Make a Model Research and model the various ways that fossils are prepared both for removal from the ground and for transport. Write descriptions of your models to be displayed with the models. For more information about fossils visit glencoe.com.

GEOLAB

SOLVE DINOSAUR FOSSIL PUZZLES

Background: The discovery of a sharpened piece of stone near a prehistoric campsite can be interpreted as having once been a tool used by prehistoric humans. Shape and position of objects provides scientists with information that can be used to interpret the lifestyles of early humans. Paleontologists who study dinosaurs use the same techniques as they collect and study fossils.

Question: *By studying these fossils, what can you tell about how these dinosaurs lived and what they ate?*

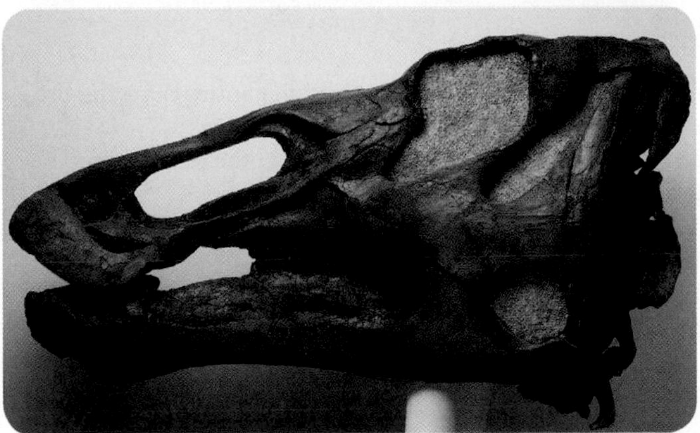

The skull on top is from an *Albertosaurus.* The bottom skull is from an *Edmontosaur.*

Materials

textbook

Internet access to glencoe.com or pictures provided by your teacher

Procedure

Imagine you are working with a team of paleontologists in a remote desert region and your team discovers the fossilized remains of two dinosaur species. Do the work of a paleontologist by studying the fossils and answering questions about how they lived.

1. Read and complete the lab safety form.
2. Compare and contrast the teeth, jaw structures, and hips of both dinosaur species.
3. Record all your observations in your science journal.

Analyze and Conclude

1. **Infer** What part of the dinosaur skeleton is most important in determining diet? Why? What is the likelihood that this part of the skeleton will be preserved?
2. **Interpret Data** Describe the diets of both dinosaur species.
3. **Interpret Data** Describe how each dinosaur species moved.
4. **Conclude** What fossil evidence from the jaws and teeth did you use to infer the diet of each dinosaur species?
5. **Conclude** What fossil evidence did you use to infer the locomotion of each dinosaur species?

APPLY YOUR SKILL

Apply Mammoths and mastodons both lived during the Pleistocene Epoch. They appear very similar in appearance yet lived slightly different lives. Examine photos of their teeth to determine where they might have lived and what they ate.

CHAPTER 23

Study Guide

BIG Idea Complex life developed and diversified during the three eras of the Phanerozoic as the continents moved into their present positions.

Vocabulary	Key Concepts
Section 23.1 The Paleozoic Era	
• Cambrian explosion (p. 653) • paleogeography (p. 648) • passive margin (p. 648) • regression (p. 649) • transgression (p. 649)	**MAIN Idea** Life increased in complexity during the Paleozoic while the continents collided to form Pangaea. • Scientists study sediment and evaporite deposits to learn how sea levels fluctuated in the past. • Eastern Laurentia was transformed by many mountain-building events during the Paleozoic. • A great diversity of multicellular life appeared during the first period of the Paleozoic. • The largest extinction event in Earth's history occurred at the end of the Paleozoic.
Section 23.2 The Mesozoic Era	
• amniotic egg (p. 658) • iridium (p. 659) • phytoplankton (p. 658)	**MAIN Idea** Reptiles became the dominant terrestrial animals during the Mesozoic while Pangaea broke apart. • The breakup of Pangaea triggered a series of tectonic events that transformed western Laurentia. • The Atlantic Ocean began to form during the Mesozoic as North America broke away from Europe. • Dinosaurs and other new organisms evolved to fill niches left empty by the Permo-Triassic Extinction Event. • All dinosaurs, except birds, along with many other organisms became extinct during a mass extinction event at the end of the Mesozoic.
Section 23.3 The Cenozoic Era	
• bipedal (p. 665) • *Homo sapiens* (p. 665)	**MAIN Idea** Mammals became the dominant terrestrial animals during the Cenozoic while the continents assumed their present forms. • Ice covered nearly one-third of Earth's land surface at the peak of the Cenozoic ice ages. • The Cascade Mountains began to rise and the San Andreas Fault formed during the Cenozoic. • The Cenozoic is known as the Age of Mammals. • Fossil evidence suggests that modern humans appeared during the Pleistocene.

CHAPTER 23

Assessment

Vocabulary Review

Match the definitions below with the correct vocabulary term from the Study Guide.

1. the ancient geographic setting of an area
2. the organisms at the base of the marine food chain
3. the increase in diversity and abundance of marine life-forms at the beginning of the Paleozoic Era
4. the movement of a shoreline seaward as sea level falls

Use a vocabulary term from the Study Guide to answer each of the following.

5. Which term is used to describe upright locomotion on two legs?
6. What are coastlines that are not experiencing tectonic activity called?

Fill in the blanks with the correct vocabulary terms from the Study Guide.

7. The movement of a shoreline inland as sea level rises is called ________.
8. ________ are primates with bipedal locomotion.
9. The ________ was a reproductive feature that allowed reptiles to migrate widely on land.

Understand Key Concepts

10. Which was the dominant terrestrial life form during the Mesozoic Era?
    A. mammals
    B. dinosaurs
    C. birds
    D. fish

11. Which term describes a shoreline that is experiencing no tectonic activity?
    A. active margin
    B. passive margin
    C. trench
    D. regression

12. During which geologic time period did the Atlantic Ocean begin to form?
    A. Triassic
    B. Cretaceous
    C. Jurassic
    D. Devonian

Use the figure below to answer Questions 13 to 15.

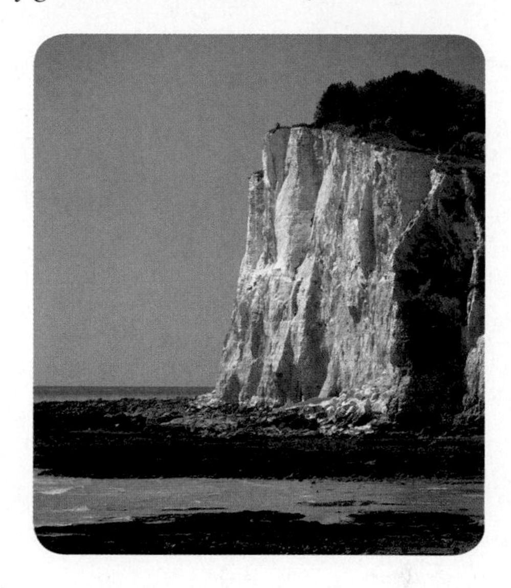

13. What formed the deposits in the photo above?
    A. asteroid residue
    B. evaporation of seawater
    C. glaciation
    D. phytoplankton

14. Where would these deposits most likely have formed?
    A. ocean floor
    B. shoreline
    C. lagoon
    D. coral reef

15. Which item could be made from this deposit?
    A. laundry detergent
    B. talcum powder
    C. chalk
    D. sponge

16. How much of Earth's land surface did glaciers cover at the height of the ice ages?
    A. 10 percent
    B. 60 percent
    C. 30 percent
    D. 90 percent

17. Which metal that is rare in Earth's rocks but relatively common in asteroids is used as evidence that there was an asteroid impact at the end of the Cretaceous?
    A. iron
    B. iridium
    C. uranium oxide
    D. zircon

18. Which supercontinent formed at the end of the Paleozoic?
    A. Rodinia
    B. Gondwana
    C. Laurasia
    D. Pangaea

Use the figure below to answer Questions 19 and 20.

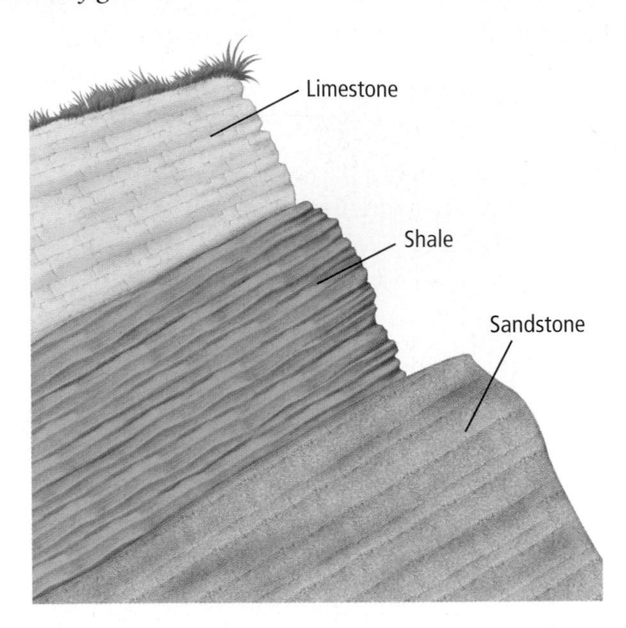

19. What does the succession of rocks in the figure above indicate?
A. a transgressive sequence where sea level rose
B. a regressive sequence where sea level fell
C. sea level fluctuated widely
D. an evaporite deposit

20. Which is a likely origin of the limestone?
A. compacted clay sediment
B. beach sand
C. remains of skeletons from phytoplankton
D. plant deposits

21. What is evidence that sea levels dropped at the end of the Triassic in North America?
A. formation of a seaway
B. presence of ancient coral reefs
C. presence of ancient sand dunes
D. presence of a passive margin along the eastern edge

Constructed Response

22. Explain how seed plants changed the landscape after they evolved during the Carboniferous.

23. Explain what skeletal feature distinguished dinosaurs from other reptiles.

24. Summarize why ancient coral reefs are good places to explore for oil.

25. Explain why there are few active volcanoes in Southern California but there is frequent earthquake activity.

26. Create a drawing that shows what happens during a regression.

27. Compare the subduction rates and angles of the three phases of the Cordilleran Orogeny.

28. Describe how sea-level change and glaciation are related.

29. Explain why volcanic ash deposits can be used as evidence of an ancient orogeny.

Think Critically

Use the figure below to answer Questions 30 and 31.

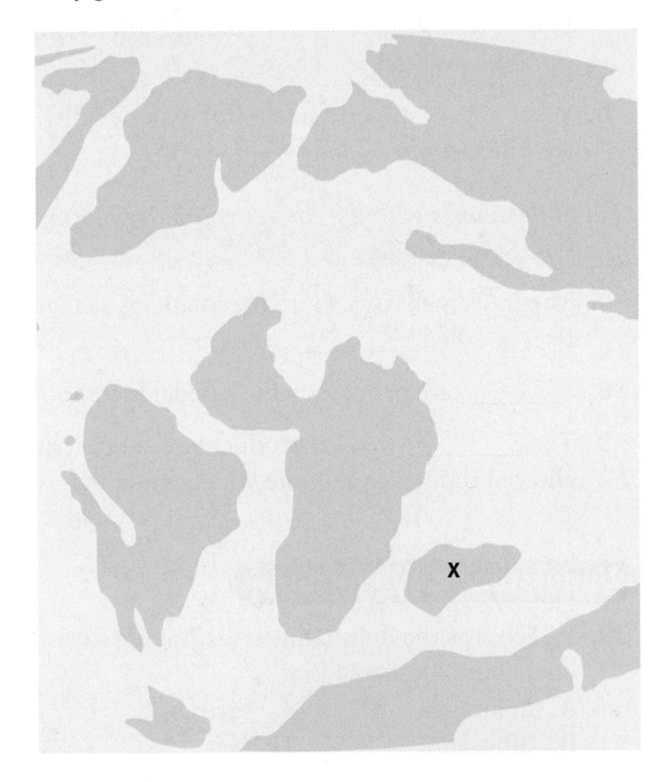

30. Explain how and where India, the continent labeled *X* in this diagram of Earth during the Mesozoic, is moving.

31. Generalize Elephants today are present naturally only in Africa and Asia. Discuss how the theory of plate tectonics might explain this.

32. Discuss how the breakup of a supercontinent might lead to the formation of a new ocean.

Earth Science Online Chapter Test glencoe.com

33. **Summarize** how the Appalachian Mountains, near the east coast of North America, are evidence that this coast was once an active margin.

Use the figure below to answer Question 34.

34. **Explain** how a major regression might stress the organisms pictured above.

35. **CAREERS IN EARTH SCIENCE** Explain why the discovery of the remains of an ancient coral reef would be exciting news to a petroleum geologist.

36. **Propose** what Earth might be like today if a meteorite had not struck it at the end of the Cretaceous.

37. **Evaluate** the relationship between extinction events and the evolution of life. Give an example.

38. **Evaluate** how Earth would be different if Antarctica were not over the south pole but were at a latitude similar to that of Australia.

39. **Evaluate** Use **Figure 23.10** to evaluate and explain sea level during the late Cretaceous.

Concept Mapping

40. Create a concept map of the three eras of the Phanerozoic with examples of the different organisms that evolved in each era.

Challenge Question

41. **Explain** how heat generated in Earth's interior can cause continents to rift apart.

Additional Assessment

42. WRITING in Earth Science Write a report that summarizes and illustrates how too much dust in the atmosphere could ultimately cause the death of a mammal population.

DBQ Document–Based Questions

Data obtained from: Evans, K.R. et al. 2005. The sedimentary record of meteorite impacts: an SEPM research conference. *The Sedimentary Record* 3:4–8.

You learned in this chapter that a meteorite impact can affect life on Earth. How common are meteorite impacts? Recent work shows that there are more impact sites preserved on Earth than you might expect.

Use the data to answer the questions below.

Impact Craters in the Continental U.S.

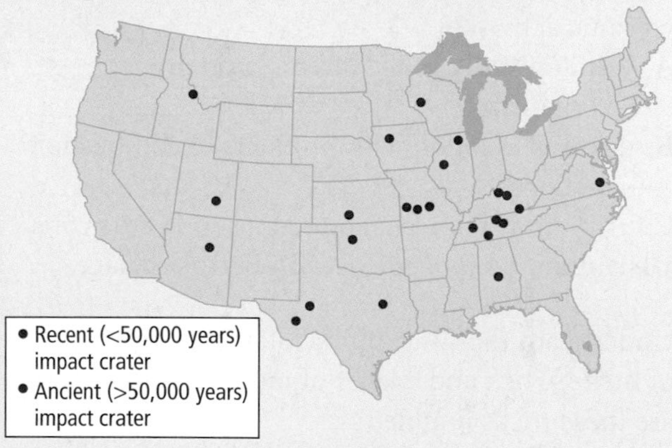

43. What percentage of impacts have occurred within the past 50,000 years?

44. Describe how the impact sites are distributed.

45. Hypothesize a reason for the distribution pattern.

Cumulative Review

46. Where are most earthquake epicenters? What is the depth of most earthquakes? Explain your answers. **(Chapter 19)**

47. Give examples of coarse-grained, medium-grained, and fine-grained clastic sedimentary rocks. **(Chapter 6)**

Standardized Test Practice

Multiple Choice

1. New ocean crust is added to Earth's tectonic plates at which type of plate boundary?
   A. convergent boundary
   B. divergent boundary
   C. deep-sea trench
   D. transformation boundary

Use the table to answer Questions 2 and 3.

Mass Extinction Theories	
Evidence for meteor impact	unusually high levels of iridium in Cretaceous-Paleogene boundary sediments; discovery of Chicxulub crater
Evidence for massive volcanic activity	volcanic eruptions during the late Cretaceous in India; unusually high levels of iridium, soot, and charcoal in Cretaceous-Paleogene boundary sediments

2. How does the presence of iridium at the Cretaceous-Paleogene boundary support the theory of massive volcanic activity?
   A. Iridium is deposited after a large fire is extinguished.
   B. Iridium is found with products of combustion reactions.
   C. Iridium is found in abundance in Earth's core.
   D. Iridium no longer exists on Earth's surface.

3. Underneath the Chicxulub crater is a large layer of melted rock and a layer of jumbled rocks. Why are these rocks jumbled?
   A. They have been broken up by water erosion.
   B. They contain a higher level of iridium.
   C. They are pieces of the surrounding rock that broke off on impact.
   D. They fell into the crater after the impact.

4. What theory of mass extinction was hypothesized for the Ordovician and Devonian eras?
   A. global cooling and glaciers
   B. global famine
   C. global disease
   D. low oxygen levels

5. Which is not a likely source of the Precambrian Earth's heat?
   A. radioactivity
   B. asteroid impact
   C. increased solar activity
   D. gravitational contraction

6. Which was not a source in the geologic record for the early presence of oxygen on Earth?
   A. red beds
   B. banded iron formations
   C. stromatolites
   D. meteorites

Use the illustration below to answer Questions 7 to 9.

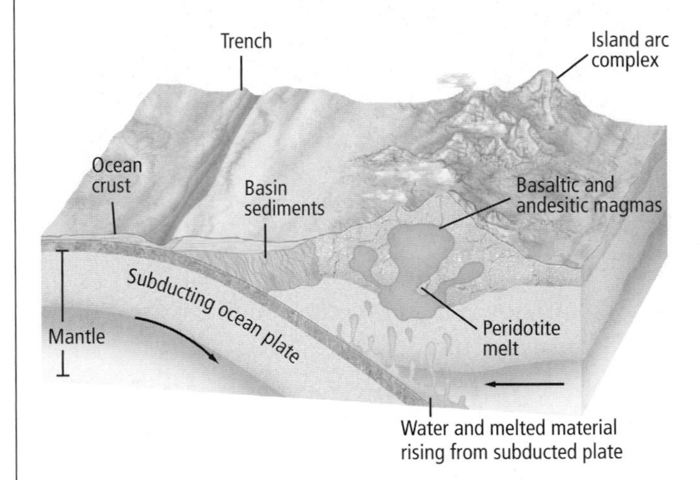

7. Which type of orogeny is shown?
   A. oceanic-oceanic convergence
   B. oceanic-continental convergence
   C. continental-continental convergence
   D. divergence

8. Which type of rock is peridotite?
   A. metamorphic
   B. igneous
   C. sedimentary
   D. volcanic

9. Where would this island arc complex most likely be located in the modern world?
   A. Antarctica
   B. Africa
   C. the Philippines
   D. North America

10. What do geologists use to help divide the history of Earth for rock study?
   A. fossils within the rocks
   B. the varying rock layers
   C. fault lines occurring in the rock layers
   D. the composition of rocks

Short Answer

Use the diagram below to answer Questions 11 to 13.

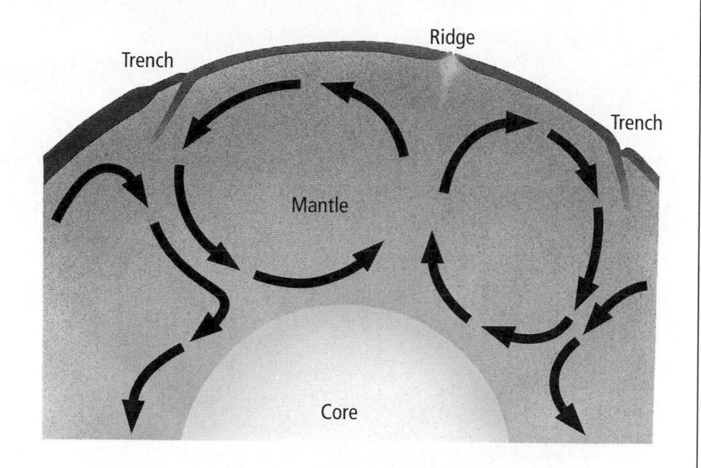

11. Describe what is being modeled in the diagram and how it affects Earth's plates.

12. How are these processes able to occur within Earth's solid mantle?

13. Why do these circulations not cause a greater amount of movement on Earth's surface?

14. Explain how James Hutton's work was linked to the principle of uniformitarianism.

15. Name two major events that occurred during Precambrian time. How much of geologic time is covered by the Precambrian?

16. Explain why hydrogen and helium, the most abundant elements in the universe, are not a significant part of Earth's atmosphere.

Reading for Comprehension

Asteriod Impact and Mass Extinction

A high-resolution map from NASA's Shuttle Radar Topography Mission has provided the most telling visible evidence to date of a 180-km-wide, 900-m-deep impact crater, which formed when Earth collided with a giant comet or asteroid 65 mya. The existence of the impact crater was first proposed in 1980. In the 1990s, satellite data and ground studies provided scientists with the evidence needed to possibly explain the ultimate demise of the dinosaurs and more than 95 percent of Earth's living species. The relatively obscure feature is all but hidden in the flat limestone plateau of Mexico's Yucatán Peninsula.

Scientists theorize three possible scenarios as to how the asteroid impact caused Earth's mass extinctions: massive quantities of dust blew into the atmosphere, blocking the Sun and stopping plant growth; sulfur released by the impact lead to global sulfuric-acid clouds that blocked the Sun and produced acid precipitation; and red-hot debris from the falling asteroid or comet triggered global wildfires.

Article obtained from: Dinosaur-killer asteroid crater imaged for first time. *National Geographic News* March 7, 2003. (Online resource accessed October 3, 2006.)

17. Which is not a possible explanation of the asteroid impact role in causing mass extinctions?
   A. Massive amounts of dust in the air blocked the Sun and killed off the plants the animals ate.
   B. The impact directly killed thousands of animals, thus limiting the reproduction rates.
   C. Sulfur produced acid rain, which killed the animals.
   D. Burning debris fell to Earth and caused global wildfires.

18. Why are the pictures of this crater taken from the NASA space shuttle so important to scientists studying Earth's history?

NEED EXTRA HELP?																
If You Missed Question . . .	1	2	3	4	5	6	7	8	9	10	11	12	13	14	15	16
Review Section . . .	17.3	23.2	23.1	23.1	22.1	22.3	20.2	19.2	20.2	21.1	17.4	17.4	17.4	21.2	21.1	22.3

UNIT 7

Resources and the Environment

Chapter 24
Earth Resources
BIG Idea People and other organisms use Earth's resources for everyday living.

Chapter 25
Energy Resources
BIG Idea People use energy resources, most of which originate from the Sun, for everyday living.

Chapter 26
Human Impact on Resources
BIG Idea The use of natural resources can impact Earth's land, air, and water.

CAREERS IN EARTH SCIENCE

Environmental Technician: These **environmental technicians** are wearing protective suits as they collect water samples for environmental testing. Environmental technicians help monitor the air, land, and water to maintain a clean environment for all living things.

WRITING in **Earth Science**
Visit glencoe.com to learn more about environmental technicians. Then write a short essay about how environmental technicians cleaned up a bay after an oil spill.

Earth Science Online

To learn more about environmental technicians, visit glencoe.com.

CHAPTER 24 Earth Resources

BIG Idea **People and other organisms use Earth's resources for everyday living.**

24.1 Natural Resources
MAIN Idea Resources are materials that organisms need; once used, some resources can be replaced, whereas others cannot.

24.2 Resources from Earth's Crust
MAIN Idea Earth's crust provides a wide variety of resources to grow food, supply building materials, and provide metals and minerals.

24.3 Air Resources
MAIN Idea The atmosphere contains gases required for life on Earth.

24.4 Water Resources
MAIN Idea Water is essential for all life, yet it is unevenly distributed on Earth's surface.

GeoFacts

- One ash tree can provide 60 baseball bats. The average major league player uses 100 bats per season.
- Safeco Field in Seattle, Washington, has 550 metric tons of clay in the infield alone.
- The retractable roof at Chase Field, in Phoenix, Arizona, was built with over 4 million kg of structural steel.

Start-Up Activities

LAUNCH Lab

What natural resources do you use in your classroom?

The materials that you use every day in your classroom, such as your paper, pencils or pens, and textbooks, all originate from multiple sources. You already know that paper comes from trees, but what about the ink? Where did other common classroom items originate?

Procedure

1. Read and complete the lab safety form.
2. Obtain a **classroom item** from your teacher.
3. Working with a partner, determine all the different components of your classroom item.
4. Next, determine where each of the components originated and classify the origin as either living or nonliving.
5. Within the living or nonliving groups, classify each as being either easily replaced or not replaceable.

Analysis

1. **Compare and contrast** your results with those of several other groups.
2. **Explain** How many items on your list were not replaceable? Why?
3. **Determine** Are any of the items on either list recyclable? Explain.
4. **Analyze** How could you make this product with more replaceable items?

FOLDABLES™ Study Organizer

Renewable v. Nonrenewable Resources Make this Foldable to compare and contrast the two main types of resources.

STEP 1 Fold up the bottom of a horizontal sheet of paper about 5 cm.

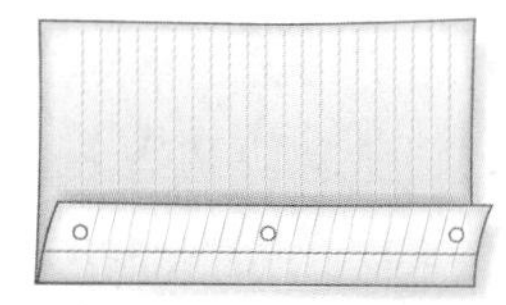

STEP 2 Fold the sheet in half.

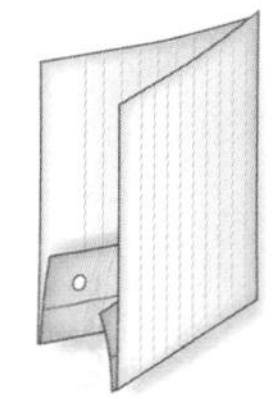

STEP 3 Open the paper and glue or staple the bottom flap to make two compartments.

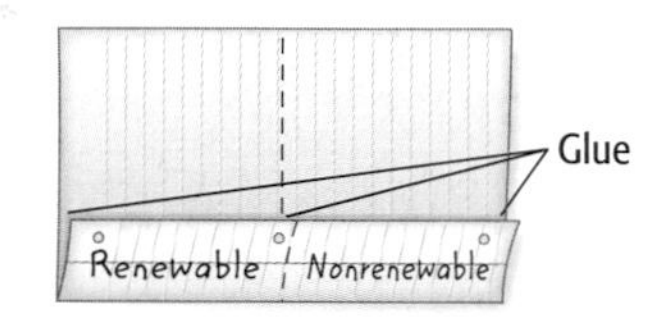

FOLDABLES **Use this Foldable with Section 24.1.** As you read about Earth's renewable and nonrenewable resources, record information on index cards or quarter-sheets of paper.

Visit glencoe.com to

- **study entire chapters online;**
- **explore Concepts In Motion animations:**
  - **Interactive Time Lines**
  - **Interactive Figures**
  - **Interactive Tables**
- **access Web Links for more information, projects, and activities;**
- **review content with the Interactive Tutor and take Self-Check Quizzes.**

Section **24.1**

Objectives

- **Distinguish** between renewable and nonrenewable resources.
- **Explain** sustainable yield.
- **Describe** how resources are unevenly distributed on Earth.

Review Vocabulary

biosphere: all of Earth's organisms and the environment in which they live

New Vocabulary

natural resource
renewable resource
sustainable yield
nonrenewable resource

Natural Resources

MAIN Idea Resources are materials that organisms need; once used, some resources can be replaced, whereas others cannot.

Real-World Reading Link Did you eat an apple or a banana for breakfast this morning? Every day, you eat food and drink water because these resources are necessary for you to live.

Resources

You and every other living thing on Earth must have certain resources to grow, develop, maintain life processes, and reproduce. The resources that Earth provides are known as **natural resources.** Natural resources include Earth's organisms, nutrients, rocks, and minerals. Natural resources might come from the soil, air, water, or deep in Earth's crust. All items that you use every day, like those shown in **Figure 24.1,** come from natural resources.

Renewable resources If you cut down a tree, you can replace that tree by planting a seedling. A tree is an example of a **renewable resource,** which is a natural resource that can be replaced by nature in a short period of time. Renewable resources include fresh air; fresh surface water in lakes, rivers, and streams; and most groundwater. When used properly, fertile soil is a renewable resource. However, if soil is exposed to wind and water erosion, the topsoil can be eroded. Renewable resources also include all living things and elements that cycle through Earth's systems, such as nitrogen, carbon, and phosphorus. Resources that exist in an inexhaustible supply, such as solar energy, are also renewable resources.

Figure 24.1 Most of the items in this photo originated as natural resources.
Identify *three resources represented in this photo.*

■ **Figure 24.2** Bamboo can be grown as a sustainable yield crop because it grows fast and needs no replanting. Bamboo can be used to produce a variety of items including flooring, cooking utensils, and clothing.

Sustainable yield of organisms Humans can use natural resources responsibly by replacing resources as they are used. The replacement of renewable resources at the same rate at which they are consumed results in a **sustainable yield.**

Organisms in the biosphere are important renewable resources. Plants and animals reproduce; therefore, as long as some mature individuals of a species survive, they can be replaced. Crops can be planted every spring and harvested every fall from the same land as long as the Sun shines, the rain falls, and the required nutrients are provided by organic matter or fertilizers. Animals that are raised for food, such as chickens and cattle, can also be replaced in short periods of time. Forests that are cut down for the production of paper products can be replanted and ready for harvest again in 10 to 20 years. Trees that are cut down for timber can be replaced after a period of up to 60 years.

Bamboo, shown in **Figure 24.2,** is one of Earth's most versatile renewable resources. Used by more than half the world's population for food, shelter, fuel, and clothing, bamboo is one of the world's fastest-growing plants. Because bamboo is a grass, it can be harvested without replanting. Bamboo grows without fertilizers or pesticides and is harvested in three to five years.

Reading Check **Identify** an example of sustainable yields.

Sunlight Some of Earth's renewable resources are not provided by Earth. The Sun provides an inexhaustible source of energy for all processes on Earth. Sunlight is considered a renewable resource because it will be available for at least the next five billion years.

VOCABULARY

ACADEMIC VOCABULARY

Mature
having completed natural growth and development
An elephant needs 13 to 20 years after birth to be mature.

FOLDABLES

Incorporate information from this section into your Foldable.

Nonrenewable resources Many homes have copper pipes that transport water to the faucets. Today, copper costs about three times more than it did five years ago. Copper is expensive because there are a limited number of copper mines, and demand continues to increase. When all the resources in the operating mines have been exhausted, no more copper will be mined unless new sources can be located. Copper is an example of a **nonrenewable resource**—a resource that exists in a fixed amount in various places in Earth's crust and can be replaced only by geological, physical, and chemical processes that take millions of years. Resources such as fossil fuels, diamonds and other gemstones, and elements such as gold, copper, and silver are therefore considered to be nonrenewable. **Figure 24.3** shows some materials you use every day and the nonrenewable resources used to make them.

CAREERS IN EARTH SCIENCE

Materials Engineer Materials engineers work with metals, stone, plastics, and combinations of materials called composites to create materials used in everyday products, including computers, television screens, golf clubs, and snowboards. To learn more about Earth science careers, visit glencoe.com.

Reading Check **Explain** why gold, fossil fuels, and gemstones are nonrenewable resources.

Distribution of Resources

You have probably noticed that natural resources are not distributed evenly on Earth. Ohio, Pennsylvania, and West Virginia have an abundance of coal. California is known for its gold deposits. Georgia has large stands of trees used for paper and lumber. Some regions of the world, such as the United States, have an abundance of different types of natural resources. Other areas might have limited types of resources, but in abundant supply. For example, Saudi Arabia and Kuwait, in the Middle East, have more petroleum reserves than other areas of the world.

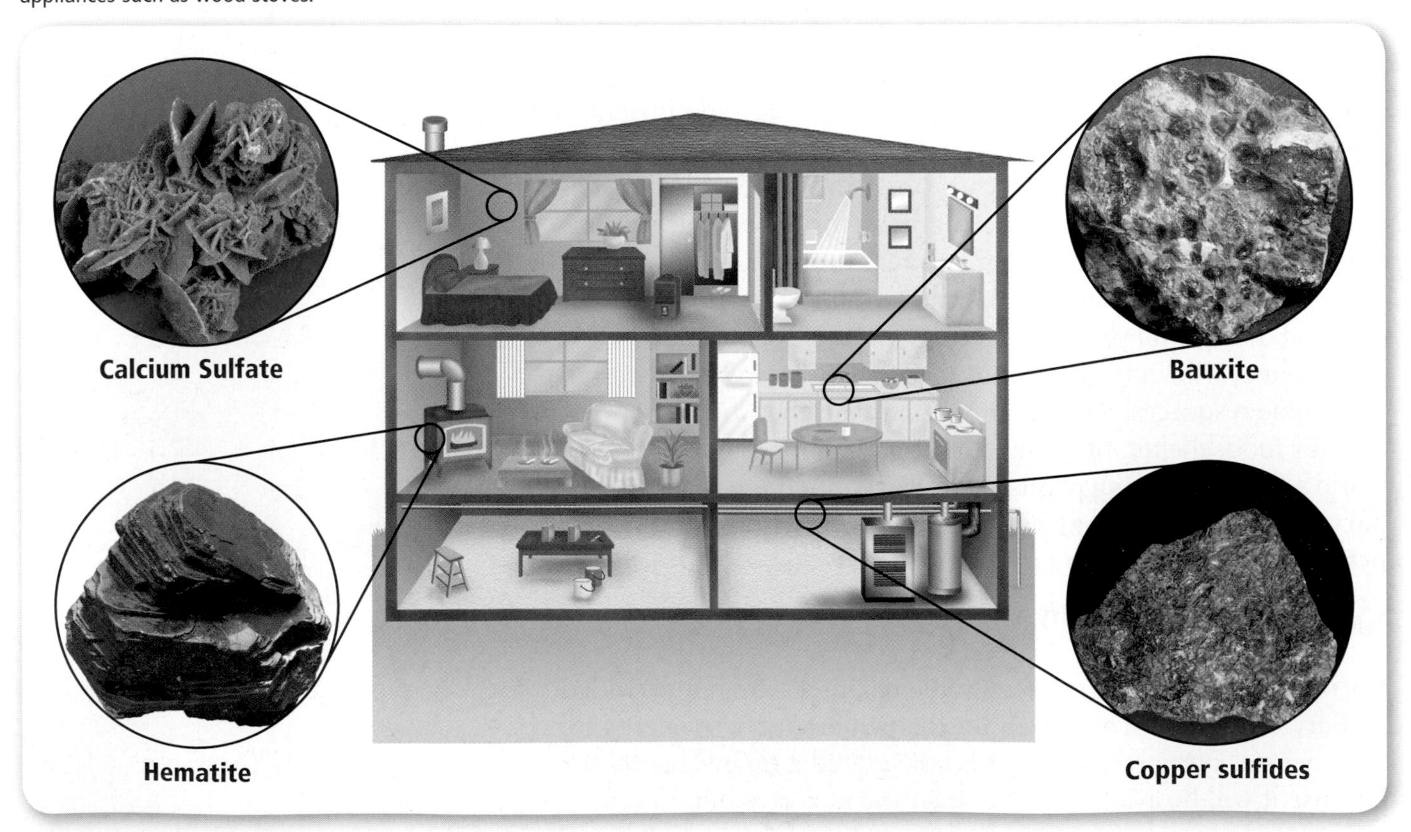

Figure 24.3 Nonrenewable resources are all around us. Aluminum from bauxite is used to make pots and pans, copper sulfides are used in copper plumbing, calcium sulfate is used to make drywall for houses and buildings, and iron from hematite is used to make appliances such as wood stoves.

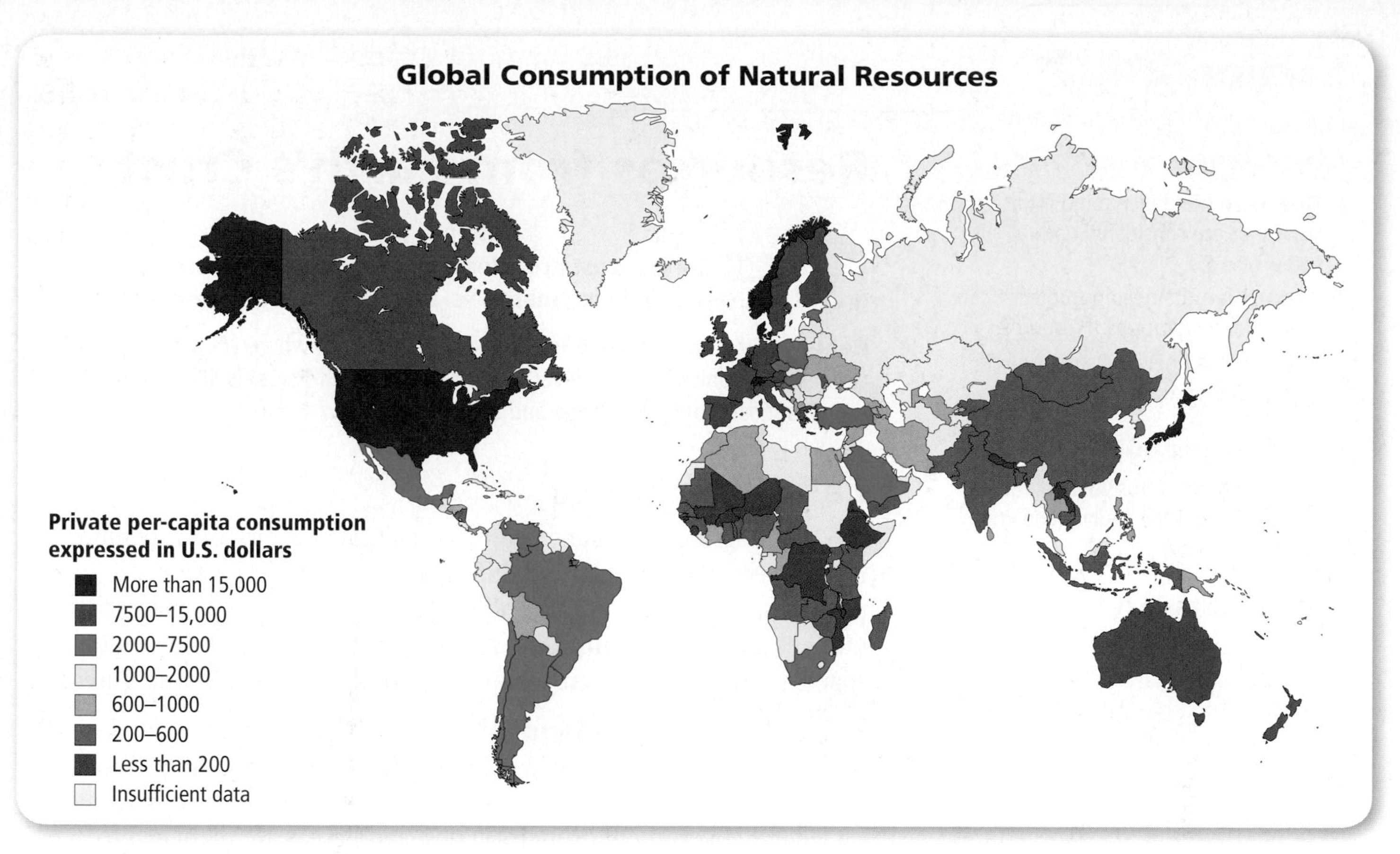

Consumption of resources Billions of people throughout the world use natural resources every day. Not only are natural resources distributed unevenly on Earth, they are likewise consumed unevenly. Although people in the United States make up only 6 percent of the world's population, they consume approximately 30 percent of Earth's mineral and energy resources each year, as shown in **Figure 24.4.** As a result, even more energy and resources are required to transport many resources from their point of origin to the places where they are being consumed.

■ **Figure 24.4** Across the globe, consumption of natural resources varies from country to country. Notice the average person in the United States consumes more than $15,000 a year in natural resources.
Determine ***How does this compare with Canada or India?***

Section 24.1 Assessment

Section Summary

- Natural resources are the resources that Earth provides, including organisms, nutrients, rocks, minerals, air, and water.
- Renewable resources can be replaced within a short period of time.
- Nonrenewable resources exist in a fixed amount and take millions of years to replace.

Understand Main Ideas

1. MAIN Idea **Explain** how organisms, including humans, use natural resources.
2. **Explain** why costs of copper and other materials continue to increase.
3. **Categorize** the following as a renewable or nonrenewable resource: trees, aluminum, cotton, gemstones, and corn. Which are produced by sustainable yield?

Think Critically

4. **Propose** why consumption of natural resources is higher in the United States. Why is it important to be aware of this?

MATH in Earth Science

5. Aluminum production from bauxite ore costs $2000 per ton, whereas aluminum recycling costs $800 per ton. What is the percent saved by recycling?

Section 24.2

Objectives

- **Describe** materials from Earth's crust that are considered natural resources.
- **Recognize** the need to protect Earth's land surface as a resource.
- **Explain** the uneven distribution of resources worldwide.

Review Vocabulary

igneous rock: intrusive or extrusive rock formed from the cooling and crystallization of magma

New Vocabulary

desertification
aggregate
bedrock
ore
tailings

Resources from Earth's Crust

MAIN Idea Earth's crust provides a wide variety of resources to grow food, supply building materials, and provide metals and minerals.

Real-World Reading Link Imagine going to a store where you can buy food, clothes, electronics, and whatever else you need. Earth's crust is like a store—it supplies most materials needed and used by humans.

Land Resources

In the springtime, many people visit garden centers and buy sand, mulch, peat moss, topsoil, and different kinds of rocks for landscaping purposes. These items are all land resources. Land provides places for humans and other organisms to live and interact. Land also provides spaces for the growth of crops, forests, grasslands, and wilderness areas.

Publicly managed land More than 828 million acres of land in the United States is managed by federal, state, and local governments. **Figure 24.5** shows that about 28 percent of land is managed by federal government agencies. These land areas are managed to support recreational uses, grazing, and mineral and energy resources. National forests are managed for sustainable yield and provide recreational spaces. Wilderness areas are places that are maintained in their natural state and protected from development.

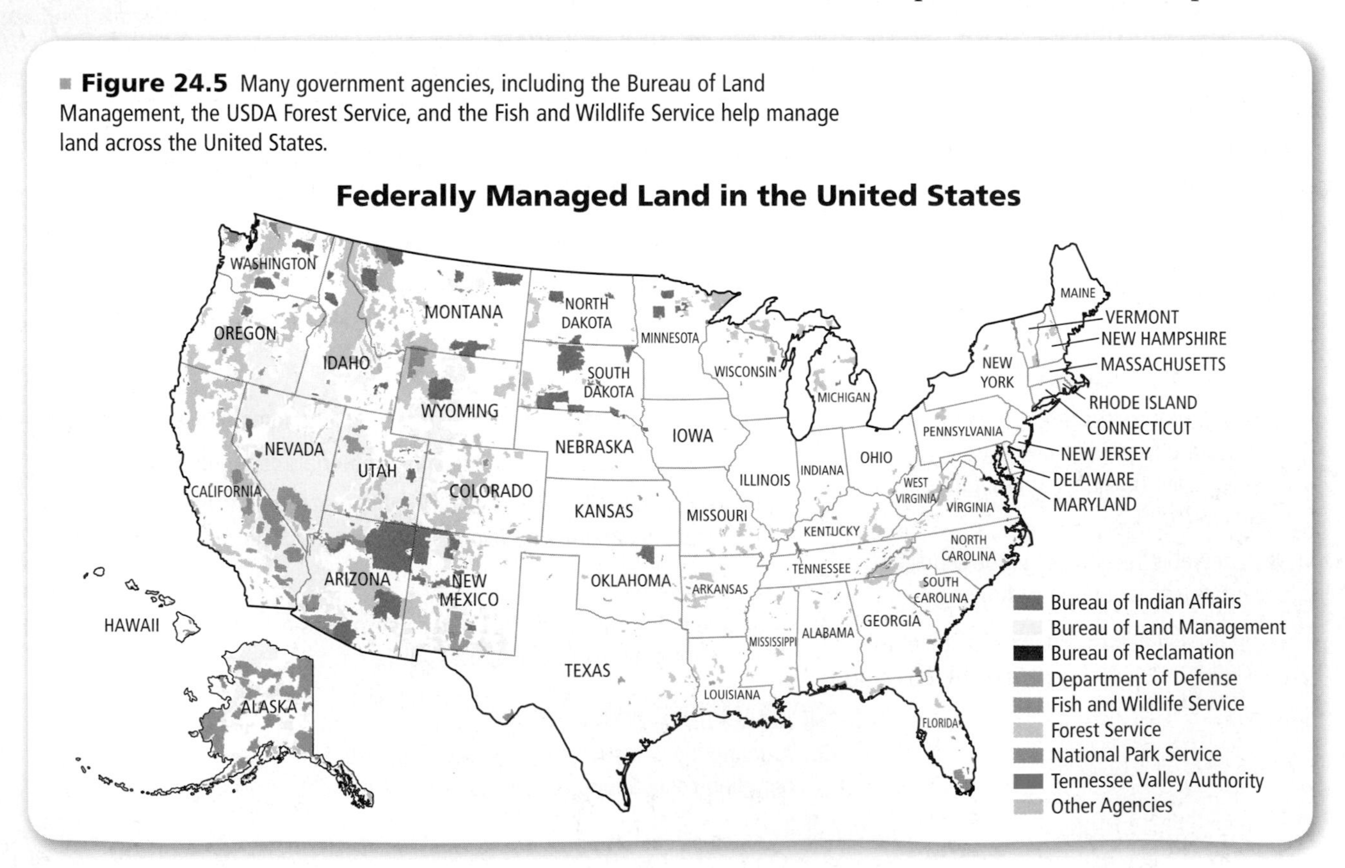

Figure 24.5 Many government agencies, including the Bureau of Land Management, the USDA Forest Service, and the Fish and Wildlife Service help manage land across the United States.

National parks The national park system in the United States preserves scenic and unique natural landscapes, preserves and interprets the country's historic and cultural heritage, protects wildlife habitats and wilderness areas, and provides areas for various types of recreation. About 49 percent of the land in the national park system is designated as wilderness.

National wildlife refuges National wildlife refuges provide protection of habitats and breeding areas for wildlife, and some provide protection for endangered species. Other uses of the land in wildlife refuges, such as fishing, trapping, farming, and logging, are permitted as long as they are compatible with the purpose of the refuge.

Soil You learned in Chapter 7 how soil forms. In some parts of Earth's crust, it can take up to 1000 years to form just a few centimeters of topsoil, yet it can be lost in a matter of minutes as a result of erosion by wind or water. Plowing and leaving the ground without plant cover can increase topsoil loss.

The loss of topsoil makes soil less fertile and less able to hold water, which results in loss of crops. Today, topsoil is eroding more quickly than it forms on about one-third of Earth's croplands. Each decade, Earth loses about 7 percent of its topsoil, yet the eroded croplands must feed an ever-increasing human population.

In arid and semiarid areas of the world, the loss of topsoil leads to **desertification,** which is the process whereby productive land becomes desert. Desertification can occur when too many grazing animals are kept on arid lands, or when trees and shrubs are cut down for use as fuel in areas with few energy resources.

Desertification is a growing problem in Africa, as shown in **Figure 24.6.** It is also a growing problem in the Middle East, in the western half of the United States, and in Australia. Desertification can be prevented by reducing overgrazing and by planting trees and shrubs to anchor soil and retain water.

Reading Check Describe activities that can lead to erosion of topsoil.

VOCABULARY

ACADEMIC VOCABULARY

Compatible
capable of existing or performing in harmonious, agreeable, or congenial combination with another or others
My sister and her roommate are not compatible; they argue all the time.

Figure 24.6 Desertification is a growing concern in many areas. Clearcutting and over-farming have led some parts of Africa to be considered in great risk of desertification.

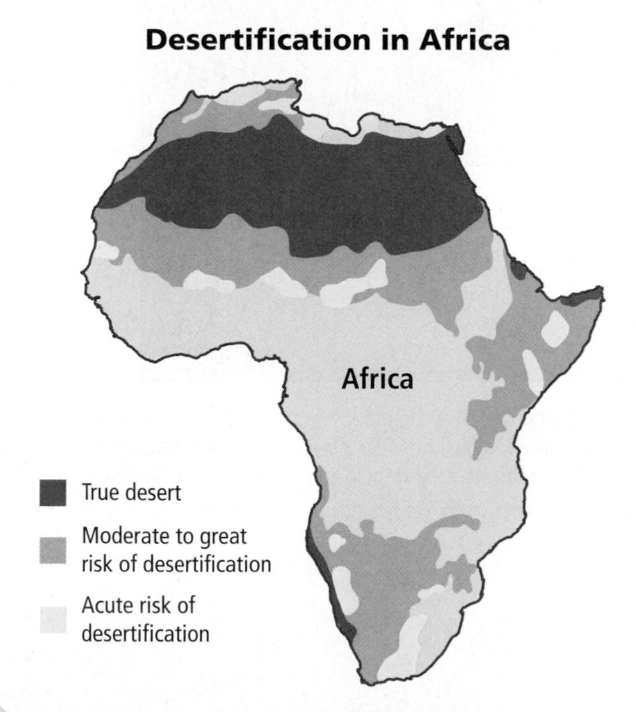

■ **Figure 24.7** Different layers of Earth's surface have value as resources. Topsoil provides nutrients for crop production, aggregate can be used to help construct roads and sidewalks.

Aggregates

Have you ever observed the construction of a highway? You might have seen workers place layers of materials on the ground before they began to build the highway surface. In some instances, the materials used for this first layer come from **aggregate,** which is sand and gravel and crushed stone that can naturally accumulate on or near Earth's surface.

You learned in Unit 3 how Earth processes transport materials. Some aggregates are transported by water and are found on floodplains in river valleys and in alluvial fans in mountainous areas. Other aggregates were deposited by glacial activity in moraines, eskers, kames, and outwash plains. Aggregates used in construction are often mixed with cement, lime, or other materials to form concrete, mortar, or asphalt.

 **Reading Check** **Define** aggregate.

Bedrock

In Chapter 7, you learned that underneath topsoil is a layer of soil consisting of inorganic matter, including broken-down rock, sand, silt, clay, and gravel, as shown in **Figure 24.7.** This deeper soil layer lies on a base of unweathered parent rock called bedrock. **Bedrock** is solid rock, and it can consist of limestone, granite, marble, or other rocks that can be mined in quarries. Slabs of bedrock are often cut from quarry faces. Large pieces of bedrock are used in the construction of buildings, monuments, flooring, and fireplaces. Bedrock is crushed for use as stone aggregate.

Ores

An **ore** is a natural resource that can be mined for a profit; that is, it can be mined as long as its value on the market is greater than the cost of its extraction. For example, the mineral hematite is an iron ore because it contains 70 percent iron by weight. Other minerals such as limonite also contain iron, but they are not considered ores because the percentage of iron contained in them is too low to make extraction profitable. Ores can be classified by the manner in which they formed. Some ores are associated with igneous rocks, and other ores are formed from processes that occur at Earth's surface.

Settling of crystals Iron, chromium, and platinum are examples of metals that are extracted from ores associated with igneous rocks. Chromium and platinum come from ores that form when minerals crystallize and settle to the bottom of a cooling body of magma. Chromite ore deposits are often found near the bases of igneous intrusions. One of the largest deposits of chromite is found in the Bushveldt Complex in South Africa.

Hydrothermal fluids The most important sources of metallic ore deposits are hydrothermal fluids. Hot water and other fluids might be part of the magma that is injected into surrounding rock during the last stages of magma crystallization. Because atoms of metals such as copper and gold do not fit into the crystals of minerals during the cooling process, they become concentrated in the remaining magma. Eventually, a solution rich in metals and silica moves into the surrounding rocks to create ore deposits known as hydrothermal veins, shown in **Figure 24.8.** Hydrothermal veins commonly form along faults and joints.

Chemical precipitation Ores of manganese and iron most commonly originate in layers formed through chemical precipitation. Iron ores in sedimentary rocks are often found in bands made up of alternating layers of iron-bearing minerals and chert shown in **Figure 24.8.** The origin of these ores, called banded iron formations, is not fully understood. Scientists think that banded iron formations resulted from an increase in atmospheric oxygen during the Precambrian.

Placer deposits Some sediments, such as grains of gold and silver, are more dense than other sediments. When stream velocity decreases, as, for example, when a stream flows around a bend, heavy sediments are sometimes dropped by the water and deposited in bars of sand and gravel. Sand and gravel bars that contain heavier sediments, such as gold nuggets, gold dust, diamonds, platinum, gemstones, and rounded pebbles of tin and titanium oxides, are known as placer deposits. Some of the gold found during the Gold Rush in California during the late 1840s was located in placer deposits.

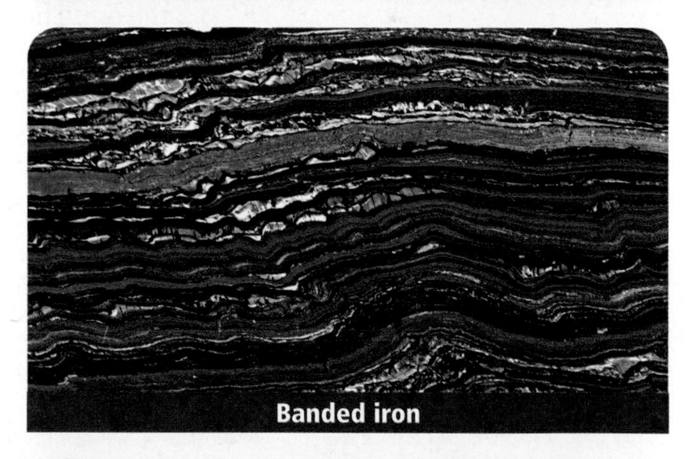

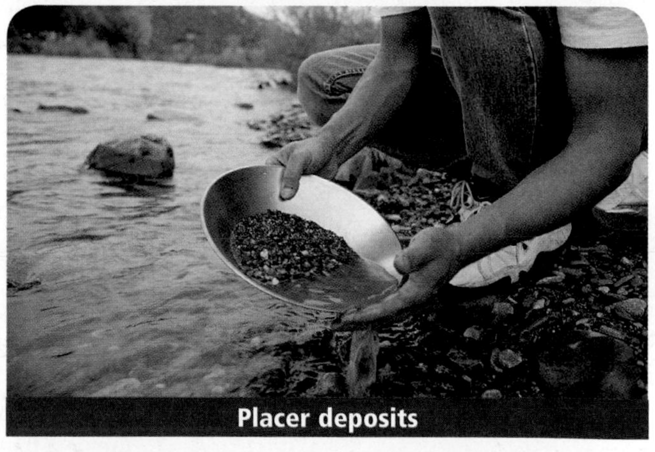

Figure 24.8 The chromite bands in the Bushveldt Complex are up to 0.5 m thick. Ores are also found in hydrothermal veins, banded formations, and placer deposits.

■ **Figure 24.9** Waste rock, such as this tailings pile in New Mexico, is discarded after minerals are extracted.

Effects of Mining

Although many of the resources that you have learned about in this section can be extracted with little impact on the surrounding environment, the extraction of others can have lasting impacts. Mines that are used to remove materials from the ground surface destroy the original ground contours. Open-pit mines can leave behind waste rock, shown in **Figure 24.9,** that can weather over time. The extraction of mineral ores often involves grinding parent rock to separate the ore. The material left after the ore is extracted, called **tailings,** might release harmful chemicals into groundwater or surface water.

Mining sometimes exposes other materials, such as mercury and arsenic, that can form acids as they weather and pollute groundwater. In addition to causing environmental problems, mining itself is a dangerous activity. In fact, the National Safety Council has identified mining as the most dangerous occupation in the United States: it has one of the highest yearly death rates of all occupations.

Section 24.2 Assessment

Section Summary

- Loss of topsoil can lead to desertification.
- Aggregates, composed of sand, gravel, and crushed stone, can be found in glacial deposits.
- An ore is a resource that can be mined at a profit. Ores can be associated with igneous rocks or formed by processes on Earth's surface.

Understand Main Ideas

1. MAIN Idea **Describe** three natural resources derived from Earth's crust.
2. **Explain** why topsoil loss is considered a worldwide problem.
3. **Identify** three reasons it is important to protect Earth's land resources.
4. **Explain** the relationship between ore and tailings.
5. **Determine** where placer materials might have originated.

Think Critically

6. **Predict** what would happen if a land resource, such as aluminum, was depleted.

WRITING in Earth Science

7. Create a three-fold pamphlet explaining the purposes and use of national parks and National Wildlife Refuge lands.

Earth Science Online Self-Check Quiz glencoe.com

Section 24.3

Objectives

- **Recognize** that the atmosphere is a resource.
- **Illustrate** carbon and nitrogen cycles.
- **Describe** natural sources of air pollution.

Review Vocabulary

photosynthesis: a process used by certain organisms to make food using energy from the Sun and carbon dioxide from the air

New Vocabulary

nitrogen-fixing bacteria
pollutant

Air Resources

MAIN Idea **The atmosphere contains gases required for life on Earth.**

Real-World Reading Link Fish and other aquatic organisms have gills, which are specialized structures used to extract dissolved oxygen from the water. Humans, however, need to breathe air to get the oxygen their cells need. Scuba divers carry tanks with compressed air when they swim under water.

Origin of Oxygen

Most organisms on Earth require oxygen or carbon dioxide to maintain their life processes, but oxygen has not always been a part of Earth's atmosphere. As you recall from Chapter 22, scientists think that 4.6 to 4.5 bya Earth's atmosphere was similar to the mixture of gases released by erupting volcanoes. These gases included carbon dioxide, nitrogen, and water vapor. As Earth's crust cooled and became more solid, rains washed most of the carbon dioxide out of the atmosphere and into the oceans. Early life-forms in the seas used carbon dioxide during photosynthesis and released oxygen. Over time, oxygen in the atmosphere built up to levels that allowed the evolution of more complex organisms that required oxygen for life processes, as shown in **Figure 24.10.**

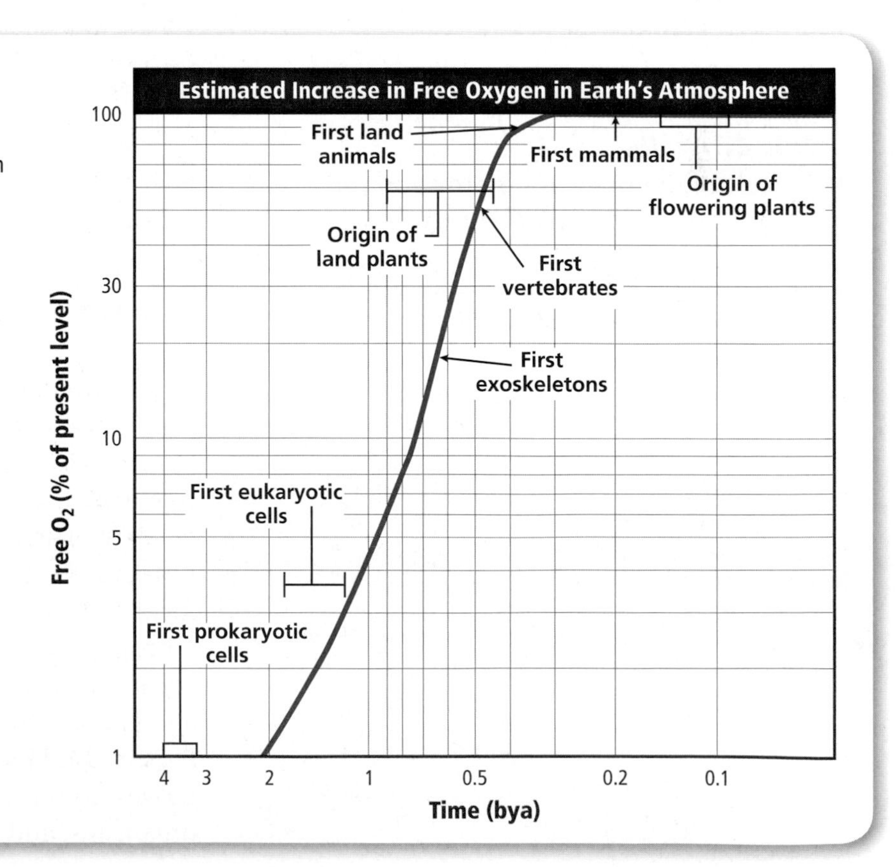

Figure 24.10 Scientists think that prokaryotes first appeared about 4 bya. It was not until 2 bya that eukaryotes appeared on Earth. Notice the difference in oxygen (O_2) gas levels between when prokaryotes and eukaryotes appeared.
Determine *When did land plants first appear?*

DATA ANALYSIS LAB

Based on Real Data*

Interpret Graphs

What is the rate of deforestation in the Amazon? Many experts are concerned about the loss of the forest cover in tropical rain forests worldwide. In the Amazon River Basin, scientists estimate that one hectare (ha, about 2.47 acres) of forest is cut down each hour.

Data and Observations

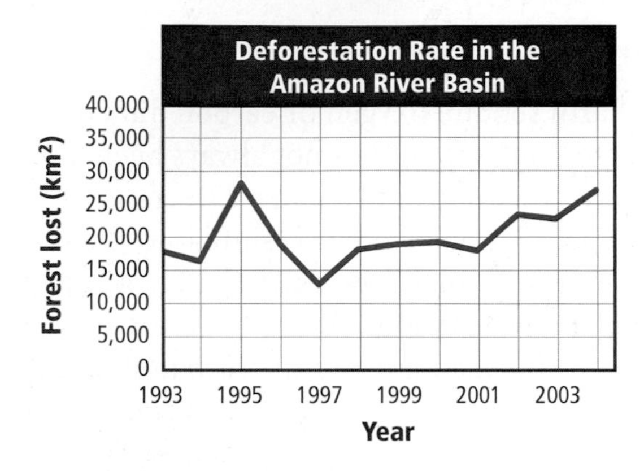

Analysis

1. How many square kilometers of the Amazon River Basin have been deforested since 2002?
2. According to the graph, what year was the peak in deforestation of the Amazon River Basin?

Think Critically

3. **Calculate** the rates of deforestation for the periods 1993 to 1998 and 1999 to 2004.
4. **Compare** the rates of deforestation for the periods from 1993 to 1998 and 1999 to 2004.
5. **Predict** what will happen to the Amazon Rain Forest over the next 30 years.
6. **Explain** how loss of rain forest could affect the carbon cycle.

*Data obtained from: Estimated Annual Deforestation Rate After 1988. *The National Institute for Space Sciences.*

Cycles of Matter

The law of conservation of mass states that the amount of matter on Earth never changes. Earth's elements cycle among organisms and the nonliving environment. In Chapter 11, you learned about how water cycles on Earth. Earth's atmosphere plays a significant role in other cycles, such as the nitrogen and carbon cycles.

Earth's cycles are in delicate balance. When fossil fuels burn, the carbon that was stored in them for millions of years is released into Earth's atmosphere. Clearing forests results in fewer trees to take in carbon and release oxygen.

Carbon cycle Life on Earth would not exist without carbon because carbon is the key element in the sugars, starches, proteins, and other compounds that make up living things. The carbon cycle is illustrated in **Figure 24.11.** During photosynthesis, green plants and algae convert carbon dioxide and water into carbohydrates and release oxygen back into the air. These carbohydrates are used as a source of energy for all organisms in a food web. Other organisms release carbon dioxide back into the air during respiration.

Carbon is also stored when organic matter is buried underground and, over millions of years, is converted to peat, coal, oil, or natural gas deposits. Carbon dioxide gas is released into the atmosphere when the fossil fuel is burned for energy.

Reading Check **Explain** how photosynthesis and respiration cycle carbon between living things and Earth's atmosphere.

Nitrogen cycle Nitrogen is an element that organisms need to produce proteins. Nitrogen makes up 78 percent of the atmosphere, but plants and animals cannot use nitrogen directly from the atmosphere. Some species of bacteria, called **nitrogen-fixing bacteria,** live in water or soil, or grow on the roots of some plants and can capture nitrogen gas. The nitrogen-fixing bacteria convert the nitrogen into a form that can be used by plants to build proteins. Nitrogen continues through the food chain as one organism eats another. As organisms excrete waste and later die, the nitrogen returns to the soil and air. **Figure 24.11** shows the nitrogen cycle. Nitrogen moves from the atmosphere to the soil, to living organisms, and then back to the atmosphere.

Visualizing Carbon and Nitrogen Cycles

Figure 24.11 All life-forms depend on carbon and nitrogen in many different ways, as shown.

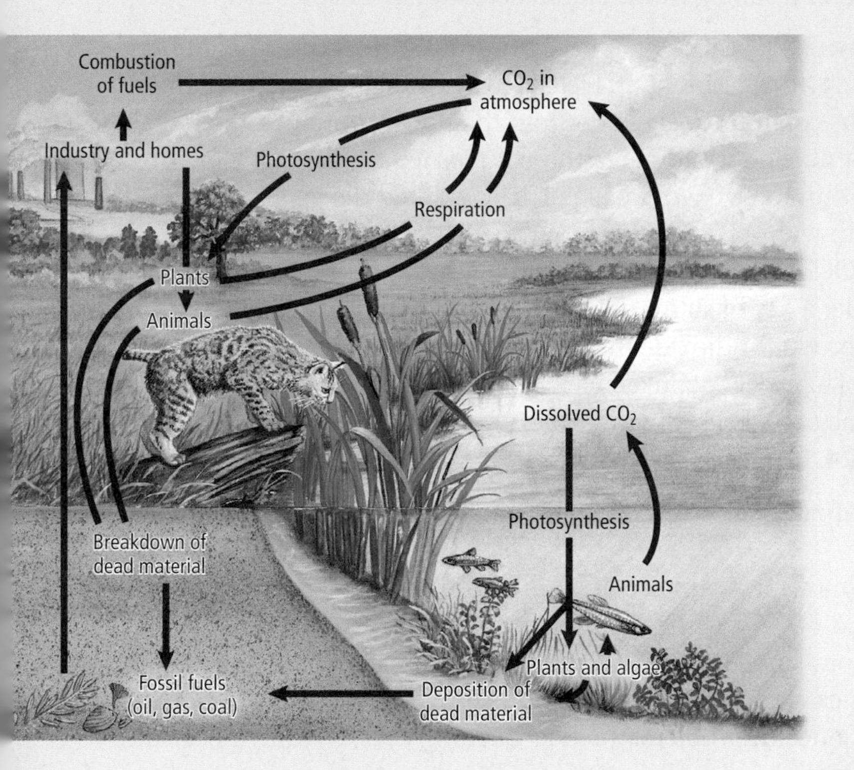

Humans have influenced the carbon cycle through the combustion of fuels. When fuels such as coal or oil are burned, one by-product of this combustion is carbon dioxide. Once released, the carbon dioxide enters the atmosphere and continues in the carbon cycle.

Nitrogen-fixing bacteria are an integral part of the nitrogen cycle. When animals produce waste, or when plants or animals die and begin to decompose, one by-product of this process is nitrogen. Nitrogen-fixing bacteria can break down the nitrogen, making it accessible for use by other plants and animals.

Atmospheric nitrogen
Land animals
Plants
Animal wastes
Aquatic animals
Nitrogen-fixing bacteria (plant roots)
Nitrogen-fixing bacteria
Decomposers
Excretion
Nitrogen-fixing bacteria
Nitrogen-fixing soil bacteria
Loss to deep sediment
Soil nitrates
Denitrifying bacteria

Concepts In Motion To explore more about the carbon and nitrogen cycles, visit glencoe.com.

Earth Science Online

■ **Figure 24.12** Vog, shown here over Kilauea, is formed when sulfur dioxide and other particulates emitted from a volcano mix with oxygen and moisture in the presence of sunlight.

Natural Air Pollution Sources

A **pollutant** is a substance that enters Earth's geochemical cycles and can harm the well-being of living things or adversely affect their activities. Air pollution can come from natural or human sources and can affect air outside or inside buildings. Natural sources of air pollution include volcanoes, fires, and radon. You will learn more about pollution sources resulting from human activities in Chapter 26.

Volcanoes Volcanoes can be significant sources of air pollution. On May 18, 1980, Mount St. Helens in Washington State shot an enormous column of ash 24 km into the sky. It continued to eject ash for about nine hours. Some of the ash reached the eastern United States within three days. Small particles entered the jet stream and circled Earth within two weeks. Mount St. Helens started erupting again in early October, 2004, and has been pumping out between 45,000 and 270,000 kg a day of sulfur dioxide. Italy's Mount Etna produces 100 times more sulfur dioxide than Mount St. Helens and is located in the middle of a heavily populated area. This sulfur helps to create acid rain and a type of bluish smog that volcanologists call vog, shown in **Figure 24.12,** which can cover large areas of land.

Reading Check **Describe** how volcanoes contribute to air pollution.

Fires Smoke is a mixture of gases and fine particles produced when wood and other organic matter burn. The most significant health threat from smoke comes from fine particles. These microscopic particles can get into your eyes and respiratory system, where they can cause health problems such as burning eyes, a runny nose, and illnesses such as chronic bronchitis. People with chronic lung disease can be at risk of serious injury from smoke.

Forest fires can release thousands of tons of carbon monoxide, a gas that interferes with oxygen transport in your blood. Gases from forest fires can also contribute to particulate and smog pollution hundreds of kilometers from the burning forest. In 2004, a large fire in Alaska and Canada, shown in **Figure 24.13,** added about 30 billion kg of carbon monoxide to the atmosphere—about as much as was released during human activities in the United States that year.

■ **Figure 24.13** Forest fires can release dangerous gases into the atmosphere. People with respiratory problems can be at risk of injury from high levels of smoke and gas.

Radon The gas known as radon-222 (Rn-222) is colorless, odorless, tasteless, and naturally occurring. Rn-222 is produced by the radioactive decay of Uranium-238 (U-238). Small amounts of U-238 are found in most soils and rocks, and in underground deposits, mainly in the northern third of the United States. Usually, radon gas from such deposits seeps upward through the soil and is released into the atmosphere, where it is diluted to harmless levels. However, when buildings are constructed with hollow concrete blocks, or when they have cracks in their foundations, radon gas can enter and build up to high levels indoors, as shown in **Figure 24.14.** Once indoors, radon gas decays into other radioactive particles that can be inhaled.

Radon is responsible for about 21,000 lung cancer deaths every year. About 2900 of these deaths occur among people who have never smoked. Because it is impossible to see or smell a buildup of radon gas in a building, the EPA suggests that people test the radon levels in their homes and offices.

Reading Check **Explain** why radon is so dangerous.

Figure 24.14 There are many ways radon can enter a home or building. Once inside, radon is colorless and odorless, making it difficult to detect. For this reason, many homes are equipped with radon detectors that have an alarm if levels exceed safety. Although radon often enters through cracks in the foundation, or through drains or other openings in the basement, they can also enter through other pathways such as showerheads.

Figure 24.15 When acid rain falls on a forest, the pH of the soil changes. As a result, the growth of the trees can be slowed. They can also become susceptible to disease, which causes large stands of trees to be damaged.
Predict *What will happen to this forest if acid rain continues to fall on it?*

Transport and Dilution

As air in the lower atmosphere moves across Earth's surface, it collects both naturally occurring and human-made pollutants. These pollutants are often transported, diluted, transformed, or removed from the atmosphere.

Some pollutants are carried downwind from their origin. Transport depends on wind direction and speed, topographical features, and the altitude of the pollutants. For example, hills, valleys, and buildings interrupt the flow of winds and thus influence the transport of pollutants. Many of the pollutants in the acid precipitation that falls in the mountain ranges of North Carolina, shown in **Figure 24.15,** were transported from coal-burning power plants in the midwestern states. If air movement in the troposphere is turbulent, some pollutants are diluted and spread out, which reduces the damage they cause.

Some air pollutants undergo physical changes. For example, dry particles might clump together and become heavy enough to fall back to Earth's surface. These and other air pollutants are removed from the atmosphere in the form of snow, mist, fog, and rain.

Section 24.3 Assessment

Section Summary

- Earth's early atmosphere had no oxygen; it was supplied over time by photosynthetic organisms.
- Oxygen, carbon, and nitrogen cycle from living organisms to the nonliving environment.
- Volcanoes, fires, and radon are natural sources of air pollution.

Understand Main Ideas

1. MAIN Idea **Explain** why the atmosphere is considered a natural resource.
2. **Compare and contrast** the carbon and nitrogen cycles.
3. **Describe** how coal-burning power plants in the Midwest can cause acid precipitation in New York.

Think Critically

4. **Predict** what might happen if there were no nitrogen-fixing bacteria on Earth.
5. **Apply** How might increasing the energy efficiency of a home lead to increased radon levels indoors?

MATH in Earth Science

6. About 21,000 people die from lung cancer related to radon each year. Of these, 2900 have never smoked. What percentage of people who die from radon-related lung cancer have never smoked?

Earth Science Online Self-Check Quiz glencoe.com

Section 24.4

Objectives

- **Explain** why the properties of water are important for life on Earth.
- **Analyze** how water is distributed and used on Earth.
- **Identify** ways in which humans can reduce the need for freshwater resources.

Review Vocabulary

aquifer: rock that holds enough water and transmits it rapidly enough to be useful as a water source

New Vocabulary

hydrogen bond
desalination

Water Resources

MAIN Idea Water is essential for all life, yet it is unevenly distributed on Earth's surface.

Real-World Reading Link What did you eat for dinner last night? How much water did it take to prepare the meal? Water is not only used to prepare, cook, and clean up, but it is also needed to grow the food that you eat.

Properties of Water

About 71 percent of Earth's surface is covered by water. The world's oceans help regulate climate, provide habitats for marine organisms, dilute and degrade many pollutants, and even have a role in shaping Earth's surface. Freshwater is an important resource for agriculture, transportation, recreation, and numerous other human activities. In addition, the organisms that live on Earth are made up mostly of water. Most animals are about 50 to 65 percent water by mass, and even trees can be composed of up to 60 percent water.

Liquid water What properties of water allow it to be so versatile? Water has a high boiling point, 100°C, and a low freezing point, 0°C. As a result, water remains liquid in most of the environments on Earth. Water can exist as a liquid over a wide range of temperatures because of the hydrogen bonds between water molecules. **Hydrogen bonds** form when the positive ends of some water molecules are attracted to the negative ends of other water molecules. Hydrogen bonds, shown in **Figure 24.16,** also cause water's surface to contract and allow water to adhere to and coat a solid. These properties enable water to rise from the roots of a plant through its stem to its leaves.

Thermal energy storage capacity Liquid water can store a large amount of thermal energy without a significant increase in temperature. This property protects aquatic organisms from rapid temperature changes, and it also contributes to water's ability to regulate Earth's climate. Because of this same property, water is used as a coolant for automobile engines, power plants, and other thermal energy-generating processes. Have you ever perspired heavily while participating in an outdoor activity on a hot day? Evaporation of perspiration from your skin helps you cool off because large quantities of thermal energy are released as the water in the perspiration changes into water vapor.

Water as a solvent Liquid water can dissolve a variety of compounds. This enables water to carry nutrients into, and waste products out of, the tissues of living things. The diffusion of water across cell membranes enables cells to regulate their internal pressure.

Figure 24.16 The attractions between the slightly positive and slightly negative ends of water molecules are called hydrogen bonds.

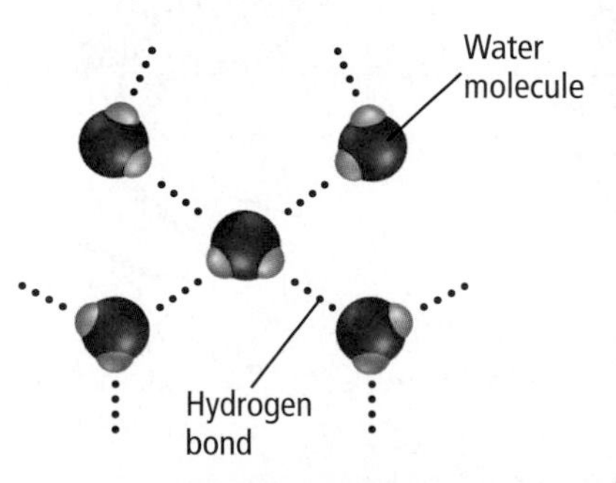

Figure 24.17 In a rock formation where weathering has previously occurred, water can enter cracks in the formation. When the water freezes, it expands, causing the cracks to widen.

Solid water Unlike most liquids, water expands when it freezes. Because ice has a lower density than liquid water, it floats on top of water. As a result, bodies of water freeze from the top down. If water did not have this property, ponds and streams would freeze solid, and aquatic organisms would die each winter. **Figure 24.17** shows that expansion of water as it freezes can also fracture rocks. Thus, ice formation in cracks in Earth's surface becomes part of the weathering process.

Location of Freshwater Resources

Freshwater resources are not distributed evenly across Earth's landmasses. The eastern United States receives ample precipitation, and most freshwater in these states is used for cooling, energy production, and manufacturing. By contrast, southwestern states often have little precipitation. In the southwestern United States, the largest use of freshwater is for agricultural uses such as irrigation. Water tables in these areas might drop as people continue to use the groundwater faster than it can be recharged.

Water distribution is a continuing problem worldwide, even though most continents have plenty of water. Since the 1970s, scarcity of water has resulted in the deaths of more than 24,000 people each year. In areas where water is scarce, women and children often walk long distances each day to collect water for domestic uses. Millions of people also try to survive on land that is prone to drought. About 25 countries, primarily in Africa, experience chronic water shortages. **Figure 24.18** shows projected water stress levels across the globe for the year 2025. These stress levels are predicted in large part by projected population growth, as well as other factors.

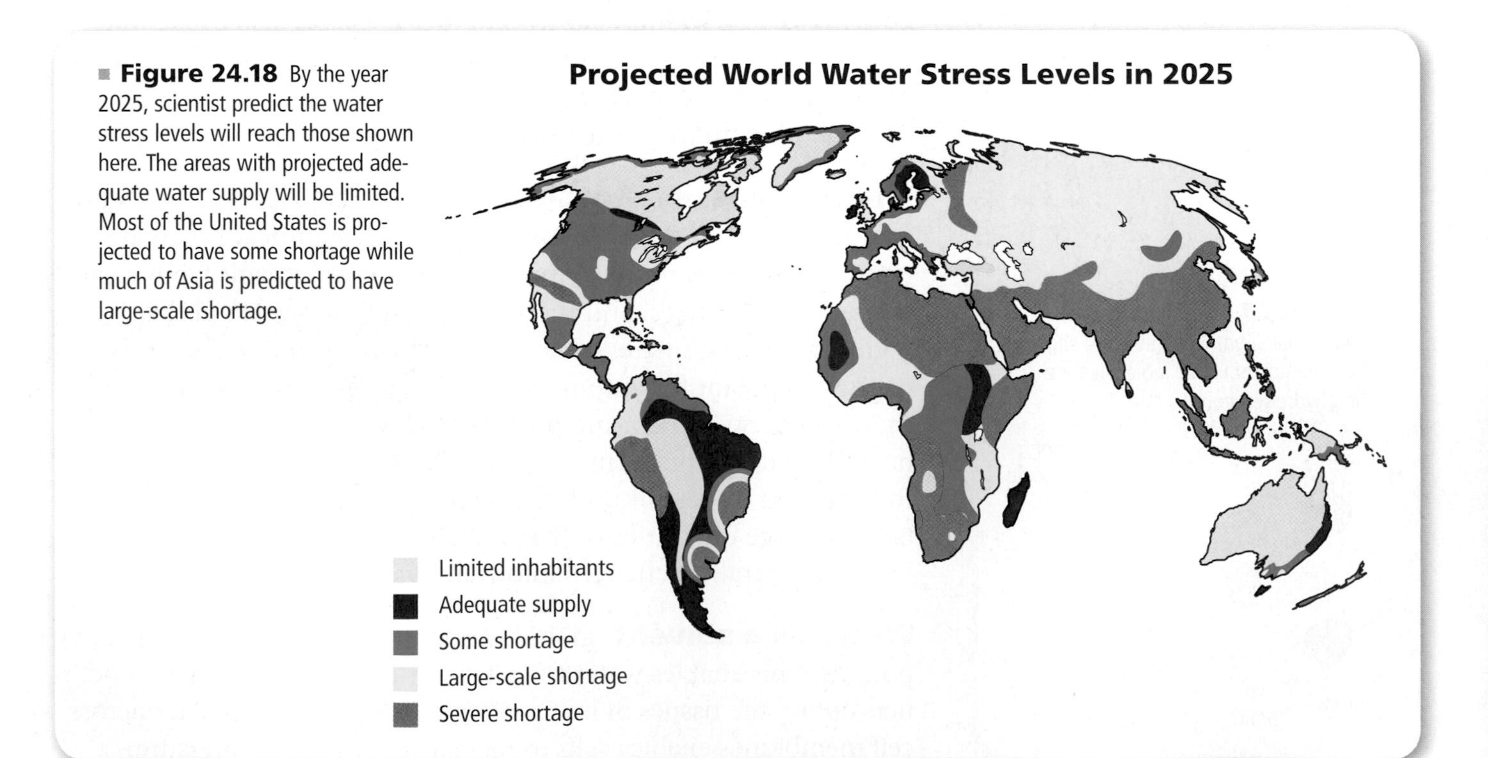

Figure 24.18 By the year 2025, scientist predict the water stress levels will reach those shown here. The areas with projected adequate water supply will be limited. Most of the United States is projected to have some shortage while much of Asia is predicted to have large-scale shortage.

Use of Freshwater Resources

Recall from Chapter 10 that the upper surface of groundwater is called the water table, and that the water-saturated rock through which groundwater flows is called an aquifer. Aquifers are refilled naturally as rain percolates through soil and rock.

In the United States, about 23 percent of all freshwater used is groundwater pumped from aquifers. Water moves through aquifers at a rate of only about 1 m/y. If the withdrawal rate of an aquifer exceeds its natural recharge rate, the water table around the withdrawal point is lowered, called drawdown. If too many wells are drilled into the same aquifer in a limited area, the drawdown can lower the water table, and, as a result, wells might run dry.

Worldwide consumption Uses of freshwater vary worldwide, but about 70 percent of the water withdrawn each year is used to irrigate 18 percent of the world's croplands. About 23 percent of freshwater is used for cooling purposes in power plants, for oil and gas production, and in industrial processing. Domestic and municipal uses account for only 7 percent of the freshwater withdrawal.

Managing Freshwater Resources

Most countries manage their supplies of freshwater by building dams, transporting surface water, or tapping groundwater. The dam shown in **Figure 24.19** was built to hold back the floodwaters of the Yangtze River in China. Called the Three Gorges Project, the structural construction of this dam was completed in 2006, and it is expected to provide freshwater and supply power to 150 million people by 2009. However, when full, water held by the dam will displace about one million people who live nearby.

MiniLab

Determine the Hardness of Water

How easily are soap suds produced? Water contains different minerals depending on its source. When water has a high mineral content, it is referred to as "hard."

Procedure

1. Read and complete the lab safety form.
2. Obtain six **clean baby food jars with lids.** Label them *A* through *F*.
3. Measure 20 mL of one **water sample.** Pour the water into the jar marked *A*.
4. Repeat Step 3 four more times, using a different water sample for jars B through E.
5. Measure 20 mL of **distilled water.** Pour this water into jar F.
6. Make a data table in your science journal. In the first column, write the letters *A* through *F*.
7. Place one drop of **liquid soap** in sample jars A through E. Do not place any soap in jar F. Tighten the lids. Shake each jar vigorously for five seconds.
8. Using the following rating scale, record in your data table the amount of suds in each jar: 1—no suds, 2—few suds, 3—moderate amount of suds, 4—lots of suds.

Analysis

1. **Order** the water samples in order from hardest to softest.
2. **Explain** What is the difference between hard and soft water?
3. **Determine** What are some disadvantages of hard water?
4. **Analyze** What was the purpose of sample F?

Figure 24.19 Dams are often built to contain freshwater resources in rivers. While this provides a readily available source of freshwater for human use, there are many other factors involved that make the damming of rivers controversial, including the flooding of farmland and displacement of people.

Dams and reservoirs Building dams is one of the primary ways that countries manage their freshwater resources. Large dams are built across river valleys, and the reservoirs behind dams capture the river's flow as well as rain and melting snow. Because the runoff is captured, flooding downstream is controlled. The water held in these reservoirs can be released as necessary to provide water for irrigation; municipal uses, such as in homes and businesses; or to produce hydroelectric power. Reservoirs also provide opportunities for recreational activities, such as fishing and boating. Dams and reservoirs currently control between 25 and 50 percent of the total runoff on every continent.

Reading Check **Explain** several advantages of building dams.

Transporting surface water If you were to visit Europe or the Middle East, you would likely see many ancient aqueducts. The Romans built aqueducts 2000 years ago to bring water from other locations to their cities. Today, many countries use aqueducts, tunnels, and underground pipes to move water from areas where it is plentiful to areas that need freshwater.

The State Water Project in California, illustrated in **Figure 24.20,** is one example of the benefits, as well as the costs, of transporting surface water. In California, about 75 percent of the precipitation occurs north of the city of Sacramento, yet 75 percent of the state's population lives south of that city. The California Water Project uses a system of dams, pumps, and aqueducts to transport water from northern California to southern California. Eighty-two percent of this water is used for agriculture. The residents of Los Angeles and San Diego are withdrawing groundwater faster than it is being replenished. As a result, there is a demand for even more water to be diverted to the south. Conflicts over the transport of surface water could increase as human populations increase.

Figure 24.20 A system of dams, pumps, and aqueducts moves water in California from the North, where there is more rainfall, to the South, where the climate is more arid.

■ **Figure 24.21** Desalination can be accomplished using several different methods. One method, called distillation, removes salt by boiling the water. Another process involves pumping the water through a filtration system to remove the salt. In some places water is desalinated in plants like this one.

Concepts In Motion

Interactive Figure To see an animation of distillation, visit glencoe.com.

Desalination With all the water available in the oceans, some countries have explored the possibility of removing salt from seawater to provide freshwater in a process called **desalination.** Several methods are available to desalinate seawater. One way is through distillation—water is first heated until it evaporates, and then it is condensed and collected. This evaporation process leaves the salts behind. Most countries that use desalination to produce freshwater use solar energy to evaporate seawater. Although the evaporation of seawater by solar energy is a slow process, it is an inexpensive way to provide needed freshwater. Some desalination plants, shown in **Figure 24.21,** use fuel to distill seawater, but because this process is expensive, it is used primarily to provide drinking water.

Section 24.4 Assessment

Section Summary

- Water has unique properties that allow life to exist on Earth.
- Water is not evenly distributed on Earth's surface.
- Water management methods distribute freshwater resources more evenly through the use of dams, aqueducts, and wells.

Understand Main Ideas

1. MAIN Idea **Describe** how the distribution of freshwater resources affects humans.
2. **Explain** why the thermal energy storage capacity of water is important to life on Earth.
3. **Explain** why water in a pond freezes from the top down.

Think Critically

4. **Propose** Do you think the process of desalination is a good option for areas like the southwestern United States where there is a high demand for freshwater? Explain your reasoning.
5. **Analyze** What are two things you could do to reduce your daily water usage?

WRITING in Earth Science

6. Imagine there is a large river near your hometown. For years, residents have used the river to fish, canoe, and swim. Recently a group has proposed damming the river to provide a clean, renewable energy source. Write two newspaper editorials—one in support of the construction of a dam and one against it.

Earth Science & Society

The Price of Water

When you go to the water fountain to get a drink, do you ever wonder where the water comes from? Depending on where you live, your water could come from groundwater or surface water, from a well or a water treatment plant.

The source of our water Water might seem like an abundant resource—after all, nearly 75 percent of our planet is covered with it. However, less than 1 percent of all the water on Earth is suitable for everyday uses such as drinking, cooking, and irrigation. Because water is a limited resource, its source is becoming a very important issue.

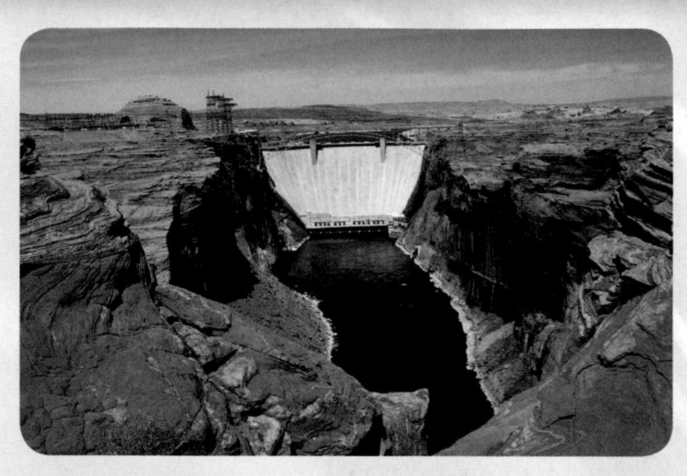

The Glen Canyon Dam on the Colorado River is one of a series of dams that controls the river's flow.

A green desert The hot, dry climate of the southwestern United States is probably the last place you would expect to see green lawns and palm trees lining the streets. Most of the area is classified as arid due to the low amounts of yearly rainfall. Yet, as many cities in this area continue to grow in population, the demand for water continues to increase.

Many cities in the Southwest draw from the same groundwater source. Often, more water is withdrawn than can be replaced by the yearly rainfall, causing the water supply to run low. Some larger cities are attempting to fix this problem by using water from rivers, streams, and lakes for residential use.

Drinking it dry Over 80 years ago, residents of some western states recognized the need for water from the Colorado River. In 1922, The Colorado River Compact was established to regulate who could use the water and how much they were allowed to use.

Today, 25 million people use water from the Colorado River. As the demand for water upstream increases, less water is available for use downstream. By the time the Colorado River reaches the U.S./Mexican border, it is a small trickle. This reduced flow has caused tension between Mexico and the United States. Residents of northern Mexico argue that they have as much right to the water of the Colorado River as those upriver.

Environmental implications By harnessing the river for public use, some of the natural ecosystems that depend on the river have been impacted. Some areas of the river have been dammed, as shown in the figure, or diverted, jeopardizing native fish species.

As the river flows south and the flow of water decreases, valuable nutrients and sediments are no longer carried to the Colorado River Delta. Plant and animal species that once thrived in this area can no longer survive.

WRITING in Earth Science

Research To learn more about sustainable water use, visit glencoe.com. Does your city have a sustainable level of water use? Write an essay explaining if your city's water usage is sustainable.

GEOLAB

DESIGN YOUR OWN: MONITOR DAILY WATER USAGE

Background: The average American uses between 300 and 380 L of water per day. Think about all the ways you use water each day, from brushing your teeth to washing your clothes.

Question: *How much water do you use each day?*

Materials

water usage table
calculator

Procedure

1. Read and complete the lab safety form.
2. Obtain a water usage table from your teacher.
3. Complete the column labeled *estimations.* Your estimations should be how many liters of water you might use in one day for each of the activities.
4. For the next five days, record your water usage and complete the table.

Analyze and Conclude

1. **Calculate** the number of liters of water you used each day to flush the toilet.
2. **Calculate** the number of liters you used each day to shower.
3. **Calculate** the total daily average number of liters of water you used.
4. **Analyze** For what purposes did you use the most water? Was this the same for all of your classmates?
5. **Predict** how this water usage might change during different seasons.
6. **Recommend** two ways you could reduce the total amount of water you use each day.

TRY AT HOME

Revise Utilizing the two recommendations you made in Question 6, record your daily water usage for another five days. Were you able to reduce your total water usage? Why or why not? For more information on water conservation visit glencoe.com.

Water Usage Activity	Liters Per Use	Estimations	Day 1	Day 2	Day 3	Day 4	Day 5	Total
Flushing the toilet	23 L/flush							
Showering	26.5 L/min							
Bathing	26.5 L/min							
Dishwasher	57 L/load							
Washing machine	227 L/load							
Bathroom sink	7.5 L/min							
Kitchen sink	11 L/min							
							Total liters used	

CHAPTER 24

Study Guide

Download quizzes, key terms, and flash cards from glencoe.com.

BIG Idea People and other organisms use Earth's resources for everyday living.

Vocabulary	Key Concepts
Section 24.1 Natural Resources	
• natural resource (p. 678) • nonrenewable resource (p. 680) • renewable resource (p. 678) • sustainable yield (p. 679)	**MAIN Idea** Resources are materials that organisms need; once used, some resources can be replaced, whereas others cannot. • Natural resources are the resources that Earth provides, including organisms, nutrients, rocks, minerals, air, and water. • Renewable resources can be replaced within a short period of time. • Nonrenewable resources exist in a fixed amount and take millions of years to replace.
Section 24.2 Resources from Earth's Crust	
• aggregate (p. 684) • bedrock (p. 684) • desertification (p. 683) • ore (p. 684) • tailings (p. 686)	**MAIN Idea** Earth's crust provides a wide variety of resources to grow food, supply building materials, and provide metals and minerals. • Loss of topsoil can lead to desertification. • Aggregates, composed of sand, gravel, and crushed stone, can be found in glacial deposits. • An ore is a resource that can be mined at a profit. Ores can be associated with igneous rocks or formed by processes on Earth's surface.
Section 24.3 Air Resources	
• nitrogen-fixing bacteria (p. 688) • pollutant (p. 690)	**MAIN Idea** The atmosphere contains gases required for life on Earth. • Earth's early atmosphere had no oxygen; it was supplied over time by photosynthetic organisms. • Oxygen, carbon, and nitrogen cycle from living organisms to the nonliving environment. • Volcanoes, fires, and radon are natural sources of air pollution.
Section 24.4 Water Resources	
• desalination (p. 697) • hydrogen bond (p. 693)	**MAIN Idea** Water is essential for all life, yet it is unevenly distributed on Earth's surface. • Water has unique properties that allow life to exist on Earth. • Water is not evenly distributed on Earth's surface. • Water management methods distribute freshwater resources more evenly through the use of dams, aqueducts, and wells.

Earth Science Online Vocabulary PuzzleMaker glencoe.com

Assessment

Vocabulary Review

Complete each sentence with the correct vocabulary term from the Study Guide.

1. Coal and oil are ________ resources because it is not possible to replace them in a short period of time.
2. Bamboo is an example of a(n) ________ because it is possible to use it indefinitely without a reduction in the supply.
3. A mixture of sand, gravel, and crushed stone is called a(n) ________.

Replace the underlined phrase with the correct vocabulary term from the Study Guide.

4. <u>Ore</u> is solid rock found underneath the loose soil and rocks in Earth's crust.
5. <u>Soil</u> is the residue of rock material left behind after the ore is removed.
6. The removal of salt from seawater is called <u>nitrification</u>.
7. The overuse of land resources might result in fertile land undergoing the process of <u>soil formation</u>.

Define each vocabulary term in a complete sentence.

8. pollutant
9. sustainable yield
10. ore

Identify the vocabulary term from the Study Guide that best fits each definition below.

11. the resources Earth provides
12. bacteria that live in soil or water and capture nitrogen gas
13. when the positive ends of some water molecules are attracted to the negative ends of other water molecules

Understand Key Concepts

14. Which resource can be replaced at a sustainable rate?
    A. iron
    B. wheat
    C. gold
    D. diamonds

15. Why are nitrogen-fixing bacteria important?
    A. They are prey for larger animals.
    B. They are part of the carbon cycle.
    C. Plants and animals cannot use nitrogen directly from the atmosphere.
    D. They are part of photosynthesis.

Use the figure below to answer Questions 16 and 17.

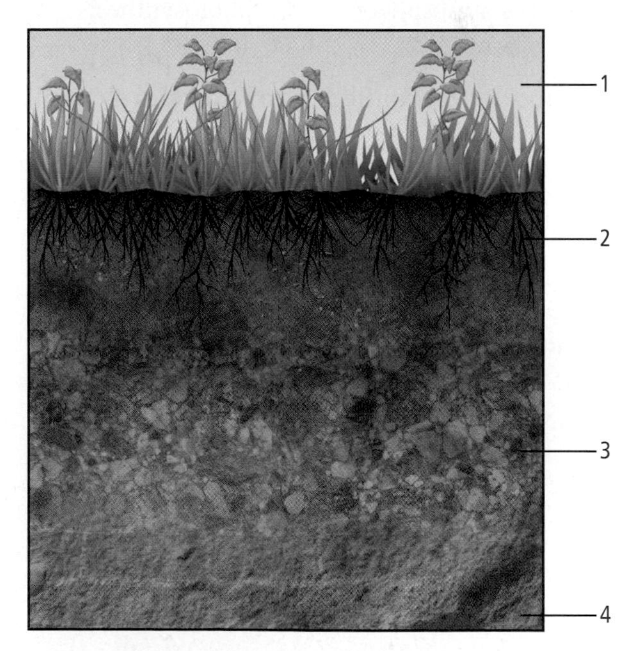

16. Which labeled area represents where aggregates are found?
    A. 1
    B. 2
    C. 3 and 4
    D. 4

17. Which layer is labeled *2*?
    A. topsoil
    B. bedrock
    C. aggregate
    D. ore

18. Which occurs when the velocity of water carrying sediments is reduced?
   - **A.** Bedrock dissolves.
   - **B.** Fine sand is pushed up and moved.
   - **C.** Heavy sediments are deposited.
   - **D.** New minerals form.

19. Which resource is found in an unlimited supply?
   - **A.** sunlight
   - **B.** gemstones
   - **C.** lumber
   - **D.** fish

20. What condition could result from overgrazing of cattle?
   - **A.** soil formation
   - **B.** chemical precipitation
   - **C.** aggregate buildup
   - **D.** desertification

21. Which is not part of the nitrogen cycle?
   - **A.** the atmosphere
   - **B.** plants
   - **C.** photosynthesis
   - **D.** soil

22. Why is water considered to be a polar molecule?
   - **A.** A water molecule has a pole.
   - **B.** Each water molecule has a positive and negative pole.
   - **C.** Water molecules form in the polar region.
   - **D.** Water molecules are attracted to magnets.

Use the figure below to answer Question 23.

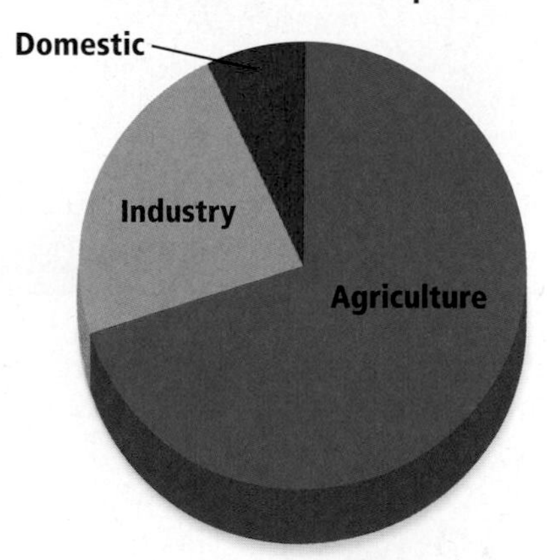

23. Industry is responsible for approximately what percent of the world's water consumption?
   - **A.** 7 percent
   - **B.** 23 percent
   - **C.** 70 percent
   - **D.** 100 percent

Constructed Response

24. **Explain** why beef and chicken purchased in the grocery store are considered renewable resources.

Use the figure below to answer Questions 25 and 26.

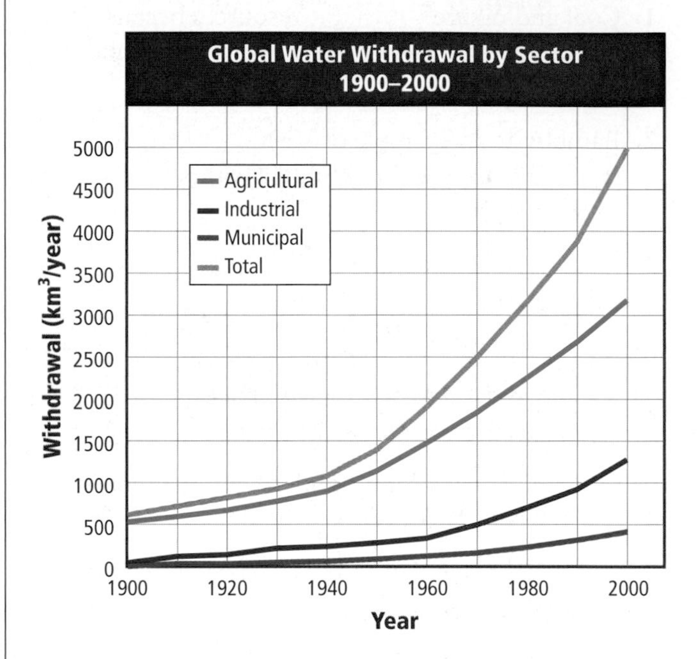

25. **Explain** why the current worldwide rate of freshwater withdrawal has changed since 1900.

26. **Determine** What type of water withdrawal has increased the most since 1900?

27. **Explain** why the loss of the forest cover in the Amazon River Basin is a world-wide concern.

28. **Explain** what issues would have to be considered if construction of a dam were proposed.

29. **Identify** several regions of the world that have a shortage of freshwater.

30. **Identify** In what states of matter can water naturally be found on Earth?

31. **Differentiate** between Earth's atmospheric composition billions of years ago and its composition today.

32. **Explain** how a substance could be both a pollutant and a requirement for life on Earth.

Think Critically

33. Explain If Earth processes recycle water resources, why is water pollution a problem?

34. Consider how the study of a landfill could provide insight into how efficiently our natural resources are being used.

Use the figure below to answer Question 35.

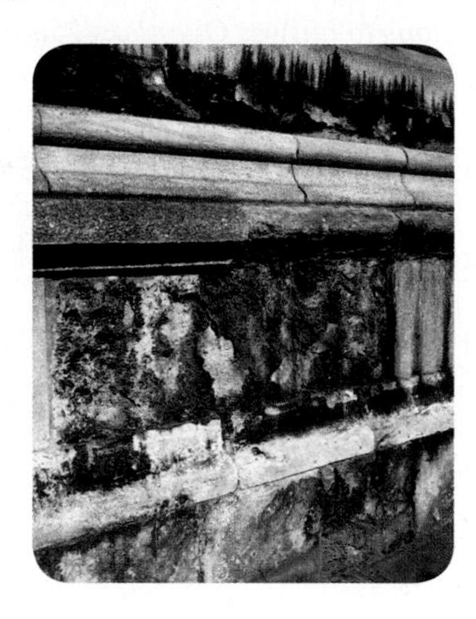

35. Infer Based on the conditions discussed in this chapter, what caused damage to this statue? How?

36. Explain how early miners applied the principle of density to finding valuable deposits of natural resources.

37. Predict what would happen to carbon in the atmosphere if photosynthesis decreased.

38. CAREERS IN EARTH SCIENCE Research and describe one job in your community or a nearby city that is closely related to providing or protecting the local water resource.

Concept Mapping

39. Make a concept map using the section titles and vocabulary words from the sections. For more help, refer to the *Skillbuilder Handbook.*

Challenge Question

40. Determine the source of water supply for your school. What procedure would you follow to answer this question?

Additional Assessment

41. *WRITING in* **Earth Science** Research your local parks and preserves. Is there an area near you that has been proposed for development? Are there endangered species in your area? Write a letter to your local congressional representative detailing what action you think should be taken.

DBQ Document–Based Questions

Data obtained from: Weibe, K., and N. Gollehon, eds. 2006. Agriculture resources and environmental indicators. *USDA* (July):134-143.

The 17 western states account for 77 percent of all irrigated land in the U.S. The average annual amount of water applied ranges from 150 acre-feet to over 2500 acre-feet. In an effort to conserve water, the USDA has suggested methods of water conservation.

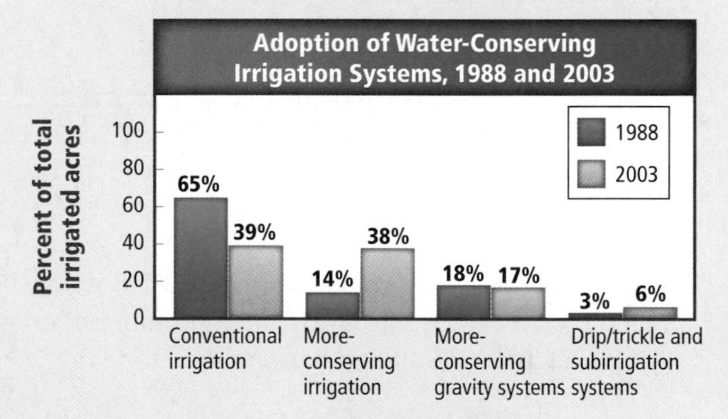

42. In 2003, which two methods of irrigation were most commonly used?

43. What percentage of acres were watered by the conventional irrigation system in 1988? In 2003?

44. What percentage acres were watered by the drip/trickle method in 2003?

Cumulative Review

45. Why is volcanic activity associated with convergent plate boundaries? **(Chapter 20)**

46. Explain the source of the heat that causes the geysers and hot springs at Yellowstone National Park. **(Chapter 23)**

Standardized Test Practice

Multiple Choice

1. What is it called when the sea level rises and shorelines move inland?
   A. regression
   B. passive margin
   C. transgression
   D. Laurentia

Use the table to answer Questions 2 and 3.

Fossil Identification Key	
1	a. Spiral shape; go to Step 2
	b. No spiral shape: go to Step 3
2	a. Less than 6 cm across: gastropod
	b. More than 6 cm across: cephalopod
3	a. Circular: crinoid columnal
	b. Branching: bryozoan

2. Mia has a fossil that is about 7 cm across and has a spiral shape. What kind of fossil did Mia find?
   A. gastropod
   B. cephalopod
   C. crinoid columnal
   D. byrozoan

3. If Mia found this fossil and chiseled it from a sedimentary rock, what type of fossil most likely is it?
   A. cast
   B. amber
   C. index fossil
   D. trace fossil

4. Which statement best explains why scientists do not rely on fossil evidence to study the Precambrian?
   A. Precambrian life-forms have not had time to fossilize.
   B. During the early Precambrian, there were no life-forms on Earth.
   C. A global event destroyed all life-forms at some point during the Precambrian.
   D. Life-forms on Earth during the Precambrian were too soft-bodied and left very few fossil imprints.

5. How does volcanic activity during early Earth explain the formation of the oceans?
   A. Volcanic eruptions caused major depressions in Earth's surface to collect water.
   B. Volcanic gas contains water vapor that cooled and condensed into liquid water, filling ocean basins.
   C. Volcanic gases created clouds which produced rain that filled ocean basins.
   D. Volcanic material blocked the Sun's rays, killing plant life that helped absorb water, and the runoff formed oceans.

6. Which is the correct succession of life-forms during the Phanerozoic Eon?
   A. ocean organisms, land plants, land animals
   B. land plants, land animals, oceanic organisms
   C. land plants, oceanic organisms, land animals
   D. land animals, land plants, oceanic organisms

Use the illustrations to answer Questions 7 and 8.

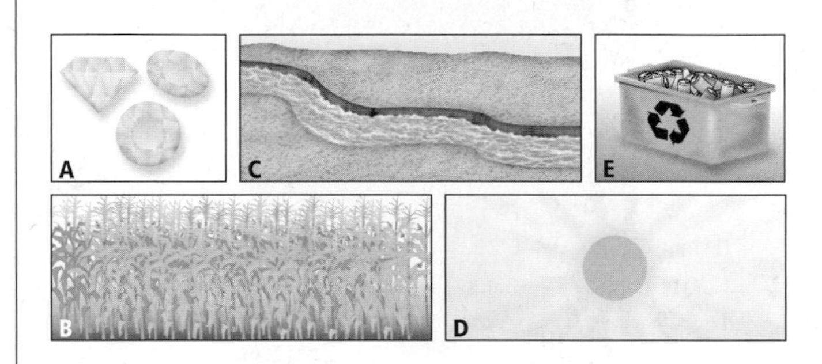

7. Which shows a nonrenewable resource?
   A. A
   B. B
   C. C
   D. D

8. Which resource is replaced through natural processes more quickly than it is used?
   A. B
   B. C
   C. D
   D. E

9. Why is radioactive decay useful in the absolute-age dating of rocks?
   A. It will only break down the fossils within the rock and not the rock itself.
   B. It will only break down the rock and not the fossils contained in the rock.
   C. It is constant regardless of environment, pressure, temperature, or any other physical changes.
   D. It fluctuates depending on environment, pressure, temperature, or any other physical changes.

10. What was formed in North America when Gondwana and Laurasia collided?
    A. Himalaya Mountains
    B. Appalachian Mountains
    C. Andes Mountains
    D. Great Permian Reef

Short Answer

Use the illustration below to answer Questions 11 and 12.

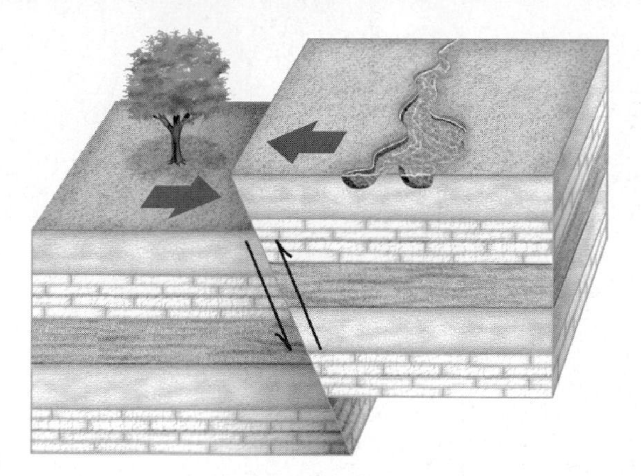

11. What type of fault is shown, and how is it formed?

12. Describe how rock surfaces along this fault lead to an earthquake.

13. Why are uplifted mountains unique?

14. Discuss how sources of heat on early Earth made conditions inhospitable to life.

15. What is the purpose of the geologic time scale?

16. Describe the formation of the Rocky Mountains during the Mesozoic Era.

Reading for Comprehension

Native Landscapes

Landscaping with native plants improves the environment. Native plants are hardy because they have adapted to the local conditions. Once established, native plants do not need pesticides, fertilizers, or watering.

A native landscape does not need to be mowed like a conventional lawn. This reduces the demand for nonrenewable resources and improves the water and air quality. The periodic burning required for maintenance of a prairie landscape mimics the natural prairie cycle and is much better for the environment. Landscaping with native wildflowers and grasses helps return the area to a healthy ecosystem. Diverse varieties of animals are attracted to native plants, enhancing biodiversity in the area.

Article obtained from: Green landscaping: greenacres. Green Landscaping with Native Plants. *United States Environmental Protection Agency*. October 2006.

17. Why is periodic burning good for a prairie landscape?
    - **A.** It gets rid of any unwanted weeds.
    - **B.** It mimics the natural prairie cycle.
    - **C.** It gets rid of any possible pests on the plants.
    - **D.** It provides a chance to create a new setting.

18. What can be inferred from this passage?
    - **A.** Landscaping with native plants is the best option for planting in an area.
    - **B.** Only native plants will survive in their given environment.
    - **C.** Native landscaping works only in prairie settings.
    - **D.** Planting native landscapes can be costly and time consuming, but it is very important.

19. Why would the Environmental Protection Agency be interested in sharing this information?
    - **A.** to reduce the number of nonnative plants sold
    - **B.** to help conserve nonrenewable resources and protect the environment from harsh chemicals
    - **C.** to provide avid gardeners with new approaches to creating their gardens
    - **D.** to identify inexpensive ways of gardening for novice gardeners

NEED EXTRA HELP?																
If You Missed Question . . .	1	2	3	4	5	6	7	8	9	10	11	12	13	14	15	16
Review Section . . .	23.1	21.4	21.4	22.4	22.3	21.1	24.1	24.1	21.3	23.2	19.1	19.1	20.3	22.1	21.1	23.3

CHAPTER 25

Energy Resources

BIG Idea People use energy resources, most of which originate from the Sun, for everyday living.

25.1 Conventional Energy Resources
MAIN Idea Biomass and fossil fuels store energy from the Sun.

25.2 Alternative Energy Resources
MAIN Idea Many resources other than fossil fuels can be developed to meet the energy needs of people on Earth.

25.3 Conservation of Energy Resources
MAIN Idea Using energy efficiently reduces the consumption of nonrenewable resources.

GeoFacts

- The solar cube at the Discovery Science Center in Santa Ana, California, is 10 stories high and provides a percentage of energy used to run the center.
- Enough sunlight falls on Earth's surface each minute to meet the world's energy demands for an entire year.
- Silicon from one metric ton of sand, used in solar cells, produces as much electricity as burning 500,000 metric tons of coal.

Start-Up Activities

LAUNCH Lab

Can you identify sources of energy?

Energy cannot be created or destroyed, but it can change form and be transferred. Thus, the same energy can be used repeatedly.

Procedure

WARNING: ***Allow the beaker to cool before moving it at the end of the activity.***

1. Read and complete the lab safety form.
2. Add 200 mL of **water** to a **250-mL glass beaker.**
3. Place the beaker on a **hot plate.**
4. Turn the hot plate on high. Observe what happens to the water as it heats up and begins to boil.

Analysis

1. **Describe** what happened to the energy as it was used to heat and boil the water.
2. **Infer** where the energy went when the water began to boil.
3. **Determine** Where did the energy to boil the water come from? Trace the electricity from your school to its source.

Alternative Energy Resources Make the following Foldable to explain some important alternatives to traditional energy resources.

STEP 1 Collect four sheets of paper and layer them 2 cm apart vertically. Keep the left and right edges even.

STEP 2 Fold up the bottom edges of the sheets to form seven equal tabs. Crease the fold to hold the tabs in place.

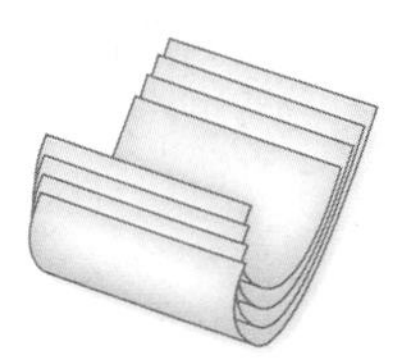

STEP 3 Staple along the fold. Label the tabs *Solar Energy, Water Energy, Geothermal, Wind, Nuclear, Biomass,* and *Other.*

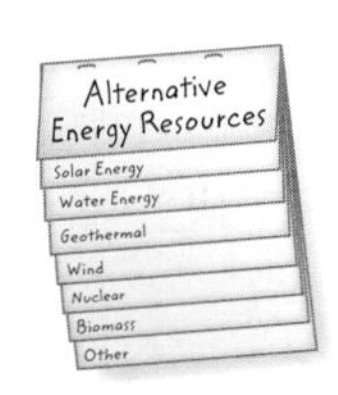

FOLDABLES **Use this Foldable with Section 25.2.** As you read this section, describe the types of resources available and explain how they differ from traditional resources.

Visit glencoe.com to

- **study entire chapters online;**
- **explore Concepts In Motion animations:**
  - **Interactive Time Lines**
  - **Interactive Figures**
  - **Interactive Tables**
- **access Web Links for more information, projects, and activities;**
- **review content with the Interactive Tutor and take Self-Check Quizzes.**

Section 25.1

Objectives

- **Explain** why the Sun is the source of most energy on Earth.
- **Identify** materials that are used as fuels.
- **Illustrate** how coal forms.

Review Vocabulary

fault: fracture in Earth's crust along which movement occurs

New Vocabulary

fuel
biomass fuel
hydrocarbon
peat
fossil fuel

Conventional Energy Resources

MAIN Idea Biomass and fossil fuels store energy from the Sun.

Real-World Reading Link What kinds of activities do you engage in each morning? In the kitchen, you might toast bread or use a microwave oven to heat up your breakfast. You might ride a bus to school or drive a car. All of these activities require energy, and the food you eat, such as toast, provides your body with the energy it needs to function.

Earth's Main Energy Source

The energy that humans and all other organisms use comes mostly from the Sun. How is solar energy used by organisms? Plants are producers—they capture the Sun's light energy in the process of photosynthesis. The light energy is converted into a form that can be used for maintenance, growth, and reproduction by the plant. When other organisms called consumers eat producers, they use that stored energy for their own life processes. For example, when a rabbit eats grass, it consumes the energy stored by the plant. The rabbit stores energy as well, and this energy can be transferred to other organisms when the rabbit is eaten, when the rabbit produces waste, or when it dies and decomposes back into the ground. **Figure 25.1** shows how trapped light energy can be transferred from plants to humans.

Humans use energy to keep them warm in cold climates, to cook food, to pump water, and to provide light. There are many different fuel sources available to humans to provide this energy. Most of these fuels also store energy that originated from the Sun.

Figure 25.1 Humans need energy to live. When you eat a bowl of cereal, you use energy derived from the Sun. The wheat plant harnessed the Sun's light energy through photosynthesis. Some of this energy was stored in the seed of the wheat which humans can consume to get energy they need to survive.

Biomass Fuels

Fuels are materials that are consumed to produce energy. The total amount of living matter in an ecosystem is its biomass. Therefore, fuels derived from living things are called **biomass fuels.** Biomass fuels, shown in **Figure 25.2,** are renewable resources.

One type of fuel available for human use is derived directly from plant material. Plant materials burn readily because of the presence of **hydrocarbons**—molecules with hydrogen and carbon bonds only. Hydrocarbons are the result of the combination of carbon dioxide and water during photosynthesis. When plant materials burn, carbon dioxide is released as a waste product.

Wood Humans have been using wood for fuel for thousands of years. Billions of people, mostly in developing countries of the world, use wood as their primary source of fuel for heating and cooking. Unfortunately, the need to use wood as a fuel has resulted in deforestation of many areas of the world. As forests near villages are cut down for fuel, people travel farther to gather the wood they need. In some parts of the world, this demand for wood has led to the complete removal of forests, which can result in erosion and the loss of topsoil.

Field crops Another biomass fuel commonly used in developing countries is field crops. The simplest way to use field crops, such as corn, hay, and straw, as fuel is to burn them. Crop residues left after harvest, including the stalks, hulls, pits, and shells from corn, grains, and nuts, are other sources of energy.

Fecal material Feces are the solid wastes of animals. In many cases, dried feces contain undigested pieces of grass that help the material to burn. Feces from cows often meet the energy needs of people in developing countries with limited forest resources. Some people collect animal fecal matter for fuel and dry it on the outside walls of their stables or compounds as shown in **Figure 25.2.**

Reading Check **Explain** how field crops, fecal material, and wood are all examples of biomass fuels.

VOCABULARY

SCIENCE USAGE V. COMMON USAGE

Consume

Science usage: to use up completely

Common usage: to eat

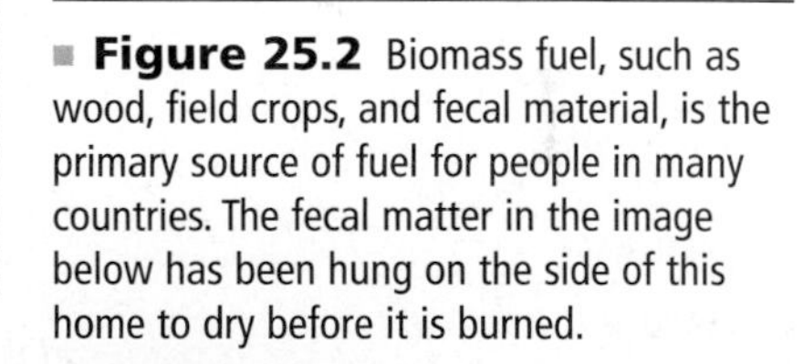

Figure 25.2 Biomass fuel, such as wood, field crops, and fecal material, is the primary source of fuel for people in many countries. The fecal matter in the image below has been hung on the side of this home to dry before it is burned.

■ **Figure 25.3** Peat has been harvested for fuel for centuries from bogs like this one in Ireland.

Peat Bogs are poorly drained areas with spongy, wet ground that is composed mainly of dead and decaying plant matter. When plants in a bog die, they fall into the water. Bog water is acidic and has low levels of oxygen; these conditions slow down or stop the growth of the bacteria that decompose dead organic matter, including plants. As a result, dead and partially decayed plant material accumulates on the bottom of the bog. Over time, as the plant material is compressed by the weight of water and by other sediments that accumulate, it becomes a light, spongy material called **peat,** shown in **Figure 25.3.** Most of the peat used as fuel today is thousands of years old.

Peat has been used as a low-cost fuel for centuries because it can be cut easily out of a bog, dried in sunlight, and then burned in a stove or furnace to produce heat. Highly decomposed peat burns with greater fuel efficiency than wood. Today, peat is used to heat many homes in Ireland, England, parts of northern Europe, and the United States.

Fossil Fuels

Energy sources that formed over geologic time as a result of the compression and incomplete decomposition of plants and other organic matter are called **fossil fuels.** Although coal, oil, and natural gas originally formed from once-living things, these energy sources are considered nonrenewable. Recall from Chapter 24 that nonrenewable resources are used at a rate faster than they can be replaced. Fossil fuels are nonrenewable resources because their formation occurs over thousands or even millions of years, but we are using them at a much faster rate.

Fossil fuels mainly consist of hydrocarbons and can be transported wherever energy is needed and used on demand. This is why most industrialized countries, including the United States, depend primarily on coal, natural gas, and petroleum to fuel electric power plants and vehicles. Although fossil fuels are diverse in their appearance and composition, all of them originated from organic matter trapped in sedimentary rock.

VOCABULARY

ACADEMIC VOCABULARY

Diverse
made up of distinct characteristics, qualities, or elements
The United States has diverse weather—the Northwest is cool and wet, while the Southwest is hot and dry.

Coal Coal is the most abundant of all the fossil fuels. Recall from Chapter 6 that coal forms from peat over millions of years. As compression continues, the hydrogen and oxygen in peat are lost and only carbon remains. The greater the carbon concentrations in coal, the hotter it burns. Most coal reserves in the United States are bituminous coal, therefore, many of the electricity-generating plants in the United States burn this type of coal. Study **Figure 25.4** to learn how the different types of coal form.

Visualizing Coal

Figure 25.4 Coal forms from the compression of organic material over time.

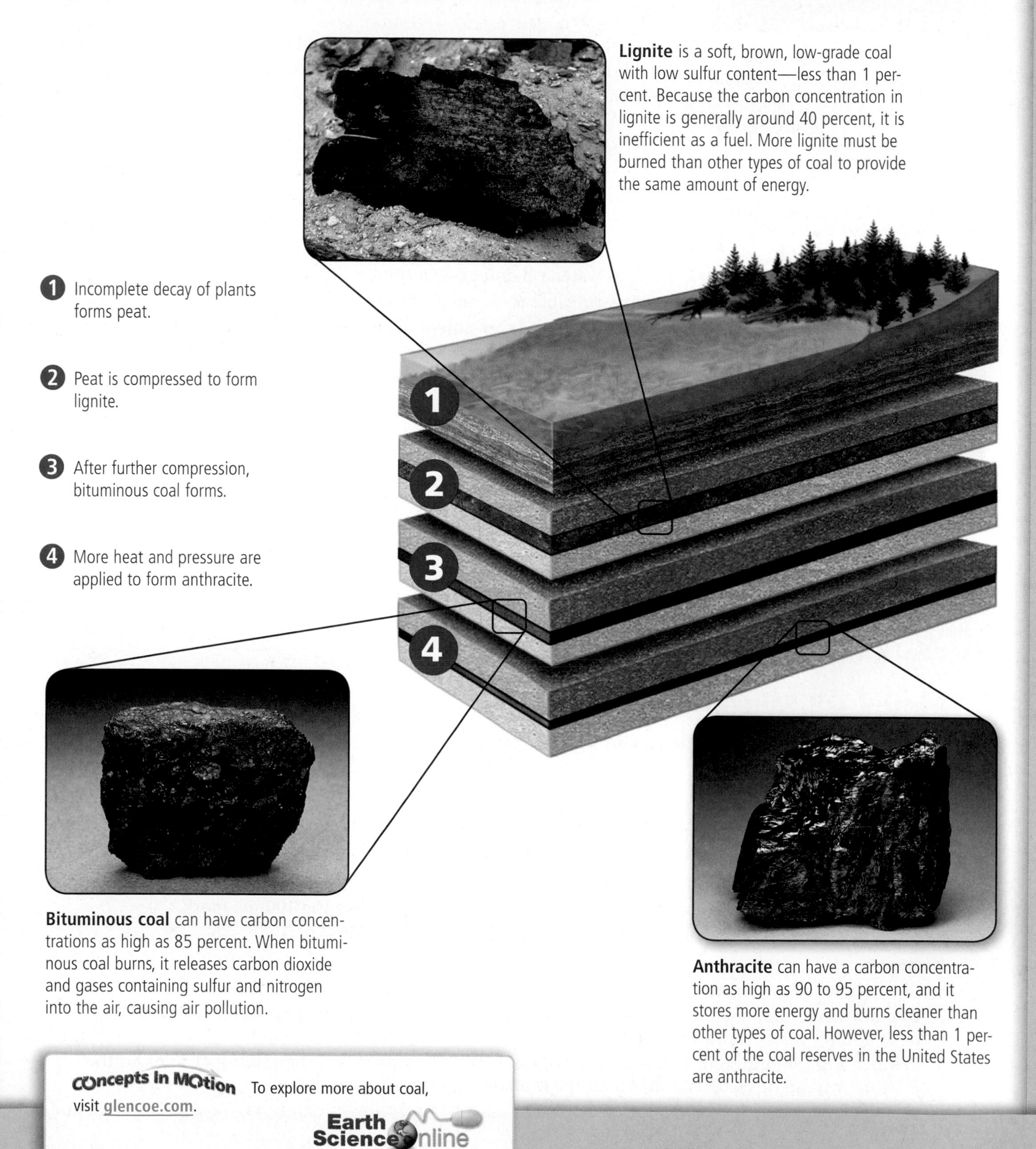

Lignite is a soft, brown, low-grade coal with low sulfur content—less than 1 percent. Because the carbon concentration in lignite is generally around 40 percent, it is inefficient as a fuel. More lignite must be burned than other types of coal to provide the same amount of energy.

1. Incomplete decay of plants forms peat.
2. Peat is compressed to form lignite.
3. After further compression, bituminous coal forms.
4. More heat and pressure are applied to form anthracite.

Bituminous coal can have carbon concentrations as high as 85 percent. When bituminous coal burns, it releases carbon dioxide and gases containing sulfur and nitrogen into the air, causing air pollution.

Anthracite can have a carbon concentration as high as 90 to 95 percent, and it stores more energy and burns cleaner than other types of coal. However, less than 1 percent of the coal reserves in the United States are anthracite.

Concepts In Motion To explore more about coal, visit glencoe.com.

Earth Science Online

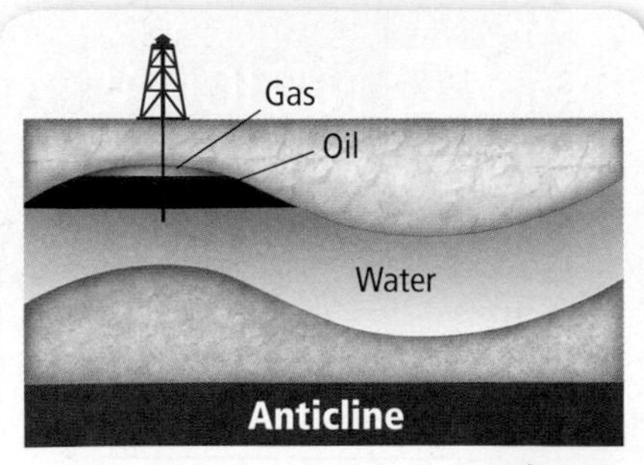

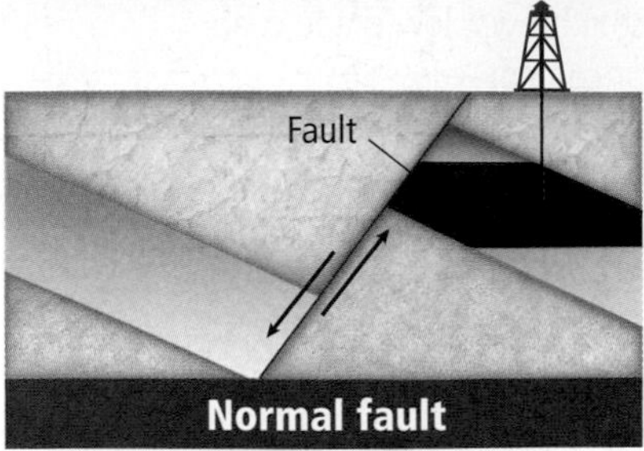

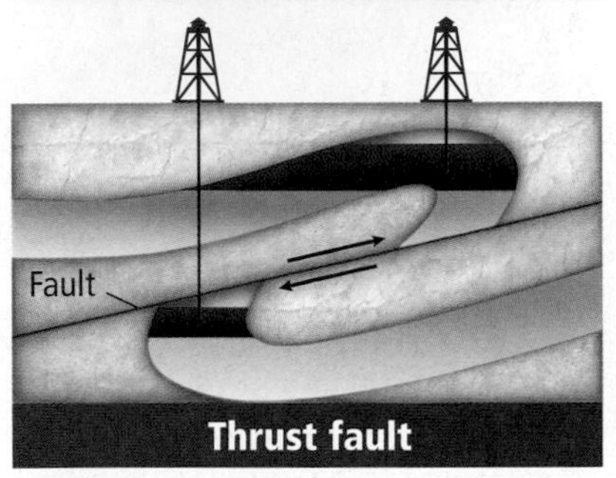

■ **Figure 25.5** These diagrams show typical structural traps for oil and gas deposits.

Petroleum and natural gas Most petroleum deposits formed from microscopic organisms in oceans. Dead and decaying organisms were buried beneath layers of clay and mud. Many layers of clay and mud increased the pressure and temperature, forming liquid oil, also called crude oil. Crude oil that is collected on Earth's surface or pumped out of the ground is refined into a wide variety of petroleum products, such as gasoline, diesel fuel, and kerosene.

Natural gas forms along with oil and is found beneath layers of solid rock. The rock prevents the gas from escaping to Earth's surface.

Migration Rock containing pores or spaces that liquid can move through is called permeable rock. Crude oil and natural gas migrate sideways and upward from their place of formation. As they migrate, they accumulate in permeable sedimentary rocks such as limestone and sandstone. Because petroleum is less dense than water, oil and gas continue to rise until they reach a barrier of impermeable rock, such as slate or shale, that prevents their continued upward movement. This barrier effectively seals the reservoir and creates a trap for the petroleum. Geologic formations such as faults and anticlines—folds of rock—can trap petroleum deposits, as shown in **Figure 25.5.**

Reading Check **Describe** how oil migrates upward through sedimentary rock.

MiniLab

Model Oil Migration

How does oil move through layers of porous rocks?

Procedure

1. Read and complete the lab safety form.
2. Pour 20 mL of **cooking oil** into a **100-mL graduated cylinder.**
3. Pour **sand** into the graduated cylinder until the sand-oil mixture reaches the 40-mL mark.
4. Add a layer of **colored aquarium gravel** above the sand until the gravel reaches the 70-mL mark.
5. Pour **tap water** into the graduated cylinder until the water reaches the 100-mL mark.
6. Observe the graduated cylinder for 5 min. Record your observations.

Analysis

1. **Identify** what the cooking oil, sand, and aquarium gravel represent.
2. **Explain** what happened when you added water to the mixture in the graduated cylinder. Why does adding water cause this change?
3. **Predict** what might occur in the graduated cylinder if you added a carbonated soft drink to the mixture instead of water. What would the bubbles represent?

Oil shale Some petroleum resources are trapped in different types of rocks. For example, oil shale is a fine-grained rock that contains a solid, waxy mixture of hydrocarbon compounds called kerogen. Oil shale can be mined, then crushed and heated until the kerogen vaporizes. The kerogen vapor can then be condensed to form a heavy, slow-flowing, dark-brown oil known as shale oil. Shale oil is processed to remove nitrogen, sulfur, and other impurities before it can be sent through the pipelines to a refinery.

The largest deposits of oil shale in the world are found in the Green River Formation, shown in **Figure 25.6.** This geologic formation contains an estimated 800 billion barrels of recoverable oil, which is three times greater than the proven oil reserves of Saudi Arabia. People in the United States use about 20 million barrels of oil per day. If oil shale could be used to meet a quarter of that demand, the estimated 800 billion barrels of recoverable oil from the Green River Formation would last for more than 400 years.

Historically, the cost of oil derived from oil shale has been significantly higher than pumped oil. Recently, prices for crude oil have again risen to levels that might make oil-shale-based oil production commercially viable, and both governments and industries are interested in pursuing the development of oil shale.

Figure 25.6 Oil shale is found primarily in sedimentary rocks. One of the most abundant sources of oil shale known is the Green River Formation in Utah, shown on the map as the dark green regions.

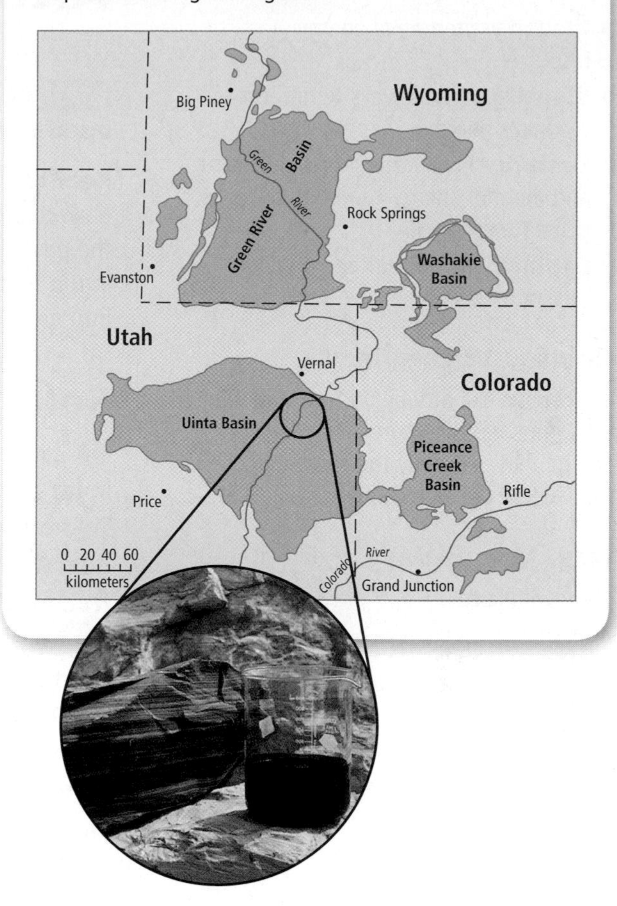

Section 25.1 Assessment

Section Summary

- The Sun is the source of most energy on Earth.
- Humans have used materials derived from living things, such as wood, as renewable fuels for thousands of years.
- Fossil fuels formed from organisms that lived millions of years ago.

Understand Main Ideas

1. MAIN Idea **Explain** how energy stored in coal was obtained from the Sun.
2. **List** four types of biomass fuels.
3. **Illustrate** how coal forms.
4. **Discuss** how two uses of energy in your home can be traced back to the Sun.

Think Critically

5. **Evaluate** this statement: Anthracite is usually found deeper in Earth's crust than lignite.
6. **Debate** whether scientists should research the prospect of obtaining oil from the Green River Formation.

WRITING in Earth Science

7. Research different ways coal can be mined and write a report on the positive and negative effects of mining.

Section 25.2

Objectives

- **Identify** alternative energy resources.
- **Identify** various ways to harness the Sun's energy.
- **Describe** how water, wind, nuclear, and thermal energy can be used to generate electricity.
- **Explain** why nuclear energy might be controversial.

Review Vocabulary

electron: subatomic partical that has little mass, but has a negative electric charge that is exactly the same magnitude as the positive charge of a proton

New Vocabulary

photovoltaic cell
hydroelectric power
geothermal energy
nuclear fission

Alternative Energy Resources

MAIN Idea Many resources other than fossil fuels can be developed to meet the energy needs of people on Earth.

Real-World Reading Link Have you ever walked barefoot across dark-colored pavement on a hot day? The thermal energy from the Sun caused the pavement to heat up and might have burned your feet. Scientists are working to find the most efficient ways to convert this thermal energy from the Sun into electricity for human use.

Solar Energy

Have you ever used a calculator with a solar collector? These solar-powered calculators use the Sun's energy to provide power. As you learned in Section 25.1, the Sun is the source of most of the energy on Earth. The main advantages of solar energy are that it is free and it doesn't cause pollution.

As you also learned in Section 25.1, many of the fuels used today are renewable resources, including wood. Most people, however, rely heavily on nonrenewable fossil fuels for their energy needs. Nonrenewable fossil fuels (oil, coal, and natural gas) are used to generate approximately 85 percent of the total energy consumed for electricity, heat, and transportation in the United States.

However, the supply of fossil fuels on Earth is limited. **Figure 25.7** shows that at the present rate of consumption, scientists estimate that oil and natural gas reserves might last only another 50 years. Although coal will last longer, burning coal releases harmful gases into the atmosphere, as you will learn in Chapter 26. Scientists, private companies, and government agencies are all studying renewable resources, such as solar energy, as alternatives to traditional energy resources, including fossil fuels.

You have learned that plants transfer the energy provided by the Sun to other organisms through food webs. Solar energy can also be used directly to meet human energy needs through passive and active solar heating.

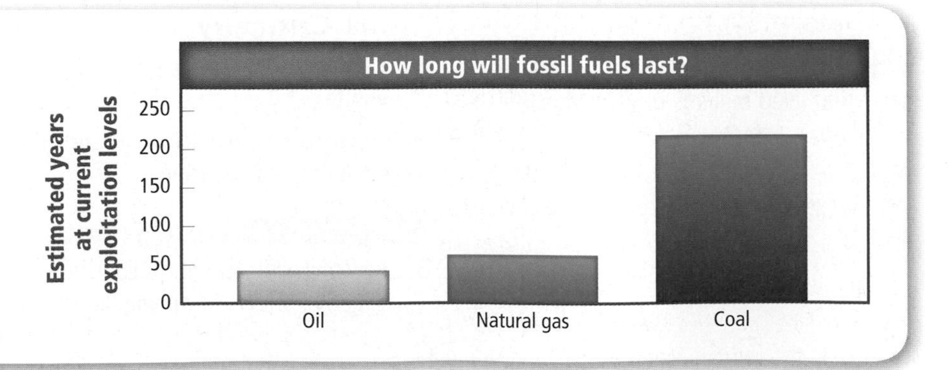

■ **Figure 25.7** At current consumption rates, available oil reserves might last only 50 years.

Passive solar heating If you have ever sat in a car that has been in the sunlight, you know that the Sun can heat up the inside of a car just by shining through the windows and on the surface of the car. In the same way, the Sun's energy can be captured in homes. Thermal energy from the Sun enters through windows, as shown in **Figure 25.8.** Floors and walls made of concrete, adobe, brick, stone, or tile have heat-storing capacities and can help to hold the thermal energy inside the home. These materials collect solar energy during the daytime and slowly release it during the evening as the surroundings cool.

In some warm climates, these materials alone can provide enough energy to keep a house warm. Solar energy that is trapped in materials and slowly released is called passive solar heating. Passive solar designs can provide up to 70 percent of the energy needed to heat a house. Although a passive solar house can be slightly more expensive to build than a traditional home, the cost of operating such a house is 30 to 40 percent lower.

 **Reading Check** **Explain** the process of heating a home using passive solar heating.

Active solar heating Even in areas that do not receive consistent sunlight, the Sun's energy can still be used for heating. Active solar-heating systems include collectors such as solar panels that absorb solar energy, and fans or pumps that distribute that energy throughout the house.

If kept away from trees, solar panels mounted on the roof can have unobstructed exposure to the Sun. Energy collected by these solar panels can be used to heat a house directly, or it can be stored for later use in insulated tanks that contain rocks, water, or a heat-absorbing chemical. Solar panels, shown in **Figure 25.8,** mounted on a roof can heat water up to 65°C, which is hot enough to wash dishes and clothing.

Passive and active solar heating rely on direct sunlight. Using direct sunlight is relatively easy, but energy is also needed during hours of darkness, or in areas that are often overcast. Solar energy is difficult to store for later use. An economical and practical method of storing large amounts of solar energy for long periods of time has not yet been developed.

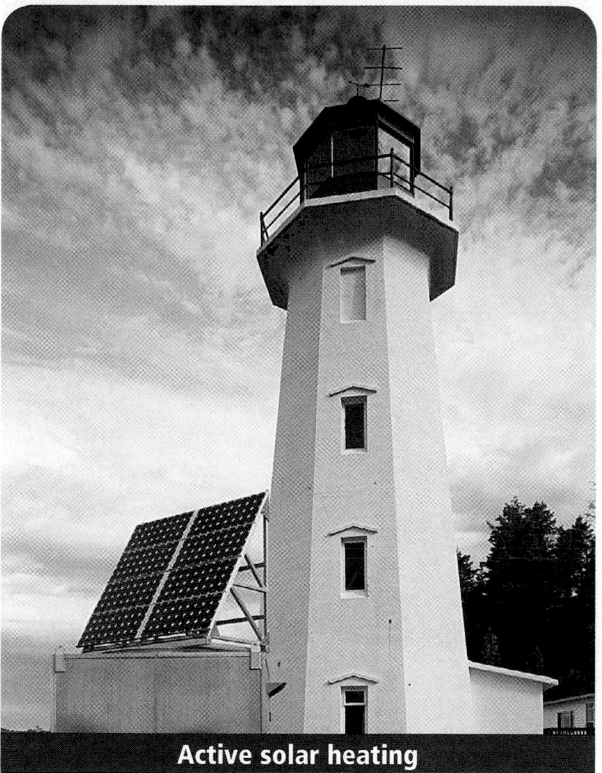

Figure 25.8 Solar heating is considered a good alternative to conventional energy resources because it is clean and readily available in some areas. However, sunlight is available during limited hours each day and it is difficult to store for later use. More research needs to be done to make solar power a reasonable alternative for more people.

FOLDABLES
Incorporate information from this section into your Foldable.

Photovoltaic cells Solar energy can be converted into electric energy by using a **photovoltaic cell,** a structure that is made of two layers of two types of silicon. The cell absorbs energy from the sunlight that strikes it. The electricity produced by photovoltaic cells can be stored in batteries. Photovoltaic cells are reliable, quiet, and typically last more than 30 years. Large-scale groups of panels can be set up in deserts and in other land areas that are not useful for other human purposes.

One example of this is a solar power tower. The solar power tower generates electricity by harnessing the solar heating of the desert surface. A glass canopy surrounds the tower and acts as a greenhouse to heat the earth beneath it. The heat creates a self-contained wind field, driving a network of 32 turbines, which generate electricity. Other advances in technology, such as those shown in **Figure 25.9,** might make renewable energy sources more accessible for future generations.

Energy from Water

Hydroelectric power is generated by converting the energy of free-falling water to electricity. When a dam is built across a large river to create a reservoir, the water stored in the reservoir can flow through pipes at controlled rates and cause turbines to spin to produce electricity. Hydroelectric power can also be generated from free-flowing water, such as the Niagara River. Today, hydroelectric power provides about 20 percent of the world's electricity and 6 percent of its total energy. Approximately 10 percent of the electricity used in the United States is generated by water, while Canada obtains more than 70 percent of its electricity from this source. Many of the hydroelectric power resources of North America and Europe have been developed, but sites have not yet been developed in Africa, South America, and Asia.

■ Figure 25.9
Development of Alternative Energy Sources

Countries develop new sources of energy to meet their growing needs.

1933–1935 The United States builds the Hoover Dam and the Grand Coulee Dam to produce hydroelectric power, the country's main energy source, second to coal, until 1984.

1800 | 1900 | 1925 | 1950

1800 Holland boasts 9000 windmills that are used chiefly for land drainage and grinding grain.

1919 Ethanol, used as fuel for early automobiles, is banned during Prohibition in the United States. Gasoline becomes the primary source of motor fuel.

1952 Coal, which had replaced wood in much of Europe due to deforestation, causes a smog that kills 4000 Londoners. England enacts new antipollution laws.

Energy from the oceans Ocean water is another potential source of energy. The energy of motion in waves, which is created primarily by wind, can be used to generate electricity. Barriers built across estuaries or inlets can capture the energy associated with the ebb and flow of tides for use in tidal power plants.

Geothermal Energy

Geothermal energy doesn't come from the Sun. Instead, it originates from Earth's internal heat. Steam produced when water is heated by hot magma beneath Earth's surface can be used to turn turbines and generate electricity. A geothermal power plant is shown in **Figure 25.10.** Energy produced by naturally occurring heat, steam, and hot water is called **geothermal energy.** While some geothermal energy escapes from Earth in small amounts that are barely noticeable, large amounts of geothermal energy are released at other surface locations. In these areas, which usually coincide with plate boundaries, geothermal energy can be used to produce electricity.

Figure 25.10 Geothermal energy plants produce clean energy by harnessing the naturally occurring heat often found at plate boundaries.
Analyze *Is geothermal energy a renewable resource? Explain.*

Concepts In Motion
Interactive Figure To see an animation of geothermal power, visit glencoe.com.

Wind Energy

Windmills in the Netherlands have been capturing wind power for human use for more than 2000 years. The windmills used today are more accurately called wind turbines because they convert the energy of the wind into electrical energy. Wind turbines currently provide 3 percent of the electricity used in Denmark. Experts suggest that wind power could supply more than 10 percent of the world's electricity by the year 2050.

1957 The first large-scale commercial nuclear power plant in the country begins operating in Shippingport, Pennsylvania.

1969 Iceland builds its first geothermal power plant. Today, geothermal energy heats 87 percent of the country's homes and supplies 17 percent of its energy needs.

1995 The United States' program uses landfill gas to make electricity, reducing certain greenhouse gas emissions.

1997 The first hybrid car to run on a gasoline engine and an electrical motor is mass-produced and released in Japan.

2005 Ninety percent of all homes in Israel use solar panels to heat water. Other countries have adopted this technology in recent decades.

1960 — 1980 — 2000

Concepts In Motion
Interactive Time Line To learn more about these discoveries and others, visit glencoe.com. Earth Science Online

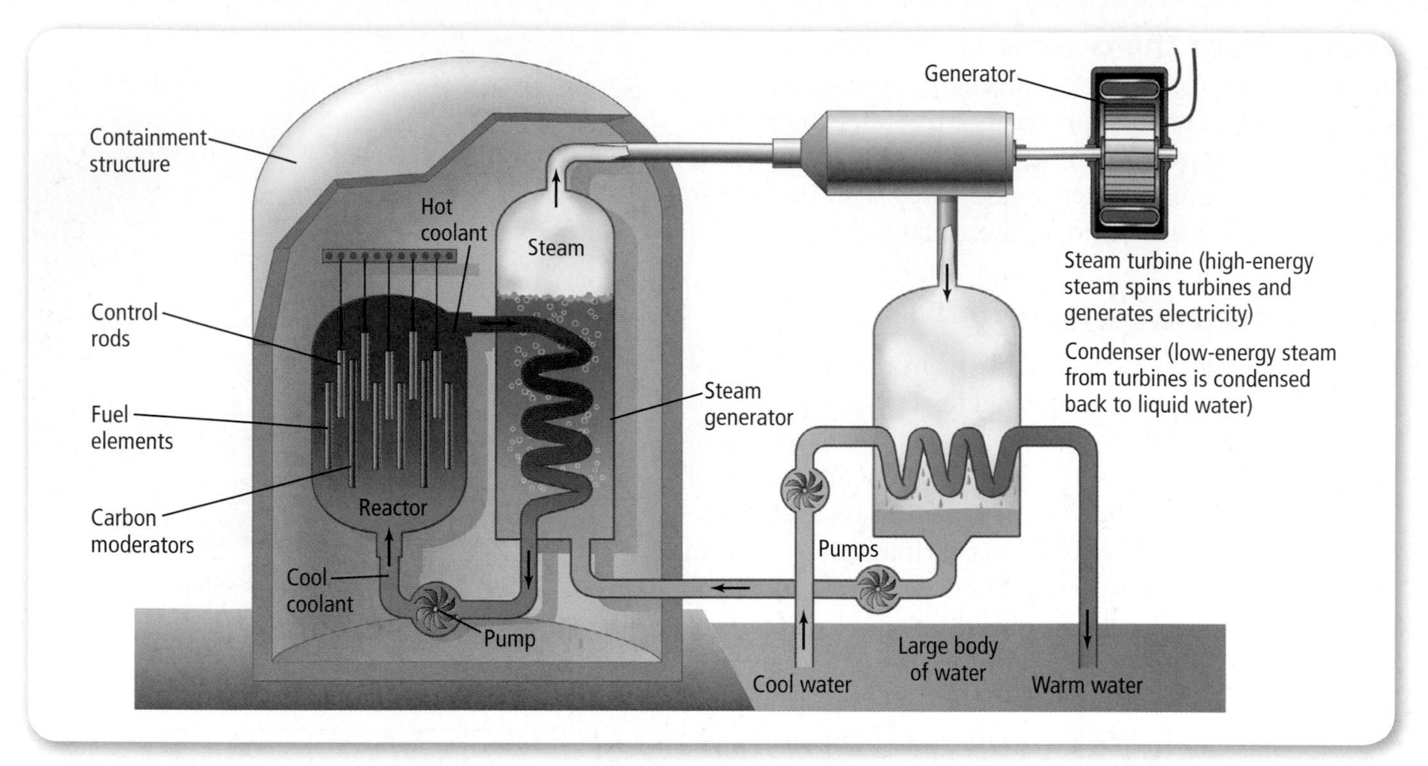

■ **Figure 25.11** Nuclear reactors rely on fission to generate heat. Heated water is converted to steam which turns a turbine to generate electricity.
Identify *how many separate systems are in this reactor.*

Concepts In Motion
Interactive Figure To see an animation of a nuclear fission reactor, visit glencoe.com.

Nuclear Energy

As you learned in Chapter 3, atoms lose particles in the process of radioactive decay. One process by which atomic particles are emitted is called nuclear fission. **Nuclear fission** is the process in which a heavy nucleus (mass number greater than 200) divides to form smaller nuclei and one or two neutrons. This process releases a large amount of energy. Radioactive elements consist of atoms that have a natural tendency to undergo nuclear fission. Uranium is one such radioactive element that is commonly used in the production of nuclear energy. Nuclear energy is one other energy source that does not come directly from the Sun.

In the late 1950s, power companies in the United States began developing nuclear power plants similar to the one shown in **Figure 25.11.** Scientists suggested that nuclear power could produce electricity at a much lower cost than coal and other types of fossil fuels. Another advantage is that nuclear power plants do not produce carbon dioxide or any other greenhouse gases. After 50 years of development, however, 445 nuclear reactors are currently producing only 17 percent of the world's electricity. Construction of new nuclear power plants in Europe has come to a halt, and new nuclear plants have not been built in the United States since 1978.

What happened to using nuclear energy as a new source of power? High operating costs, poor reactor designs, and public concerns about radioactive wastes contributed to the decline of nuclear power. In addition, nuclear accidents, such as those at Three Mile Island in Pennsylvania, in 1979, and at Chernobyl, Ukraine, in 1986, alerted people to the hazards of nuclear power plants. Because of its hazards, nuclear power has not been developed further in the United States as an alternative energy source.

Biofuels

You learned in Section 25.1 that biomass fuels include wood, dried field crops, and fecal materials from animals. Biomass is a renewable energy resource as long as the organisms that provide the biomass are replaced. Scientists are developing ways to produce fuels similar to gasoline from crops such as corn and soybeans. These fuels are called biofuels.

Ethanol Ethanol is a liquid produced by fermenting crops such as barley, wheat, and corn, which is shown in **Figure 25.12.** Ethanol can be blended with gasoline to reduce consumption of fossil fuels. Ethanol fuels burn more cleanly than pure gasoline. Most cars today can use fuels with up to 10 percent ethanol. Some vehicles, called flexible fuel vehicles, can run on mixtures containing 85 percent ethanol.

Figure 25.12 Biofuels, like biomass fuels, are derived from renewable resources. Crops like corn can be processed to create ethanol, a cleaner burning fuel than gasoline.

Biodiesel Biodiesel can be manufactured from vegetable oils, animal fats, or recycled restaurant greases. Biodiesel is safe, biodegradable, and reduces air pollution. Blends of 20 percent biodiesel with 80 percent petroleum diesel (B20) can generally be used in unmodified diesel engines; however, it is currently more expensive than regular diesel.

Section 25.2 Assessment

Section Summary

- Alternative energy resources can supplement dwindling fossil fuel reserves.
- Solar energy is unlimited, but technological advances are needed to find solutions to collect and store it.
- Nuclear energy is produced when atoms of radioactive elements emit particles in the process known as nuclear fission.
- Biofuels can help reduce consumption of fossil fuels.

Understand Main Ideas

1. MAIN Idea **Identify** one alternative energy resource that is associated with each of Earth's systems: the atmosphere, hydrosphere, biosphere, and geosphere.
2. **Compare** passive solar energy and active solar energy.
3. **Infer** which alternative energy source would have the least impact on the environment if the required technology could be developed to harness and use it. Explain.

Think Critically

4. **Analyze** In theory, solar energy could supply all of the world's energy needs. Why isn't it used to do so?
5. **Evaluate** the advantages and disadvantages of nuclear energy.

WRITING in **Earth Science**

6. Write a newspaper article that describes how alternative energy resources can be used where you live.

Earth Science Online **Self-Check Quiz** glencoe.com

Section 25.3

Objectives

- **Identify** ways to conserve energy resources.
- **Discuss** how increasing energy efficiency can help preserve fossil fuels.
- **Describe** ways to use energy more efficiently.

Review Vocabulary

renewable resource: a resource that is replaced through natural processes at a rate equal to or greater than the rate at which it is used

New Vocabulary

energy efficiency
cogeneration
sustainable energy

Conservation of Energy Resources

MAIN Idea **Using energy efficiently reduces the consumption of nonrenewable resources.**

Real-World Reading Link Think of runners on a cross-country team or a swimmer in a 400-m event. They don't sprint to start, instead they pace themselves so they have enough energy to finish the race. Energy resources can be used in this way, too.

Global Use of Energy Resources

As you learned in Chapter 24, fossil fuels are nonrenewable and are in limited supply. Yet people on Earth consume these resources at increasing rates. **Figure 25.13** shows global consumption of natural resources, both renewable and nonrenewable. However, consumption is not equal in all parts of the world. Developing countries, for example, obtain 41 percent of their energy from a renewable resource, compared to industrialized countries where renewable resources account for only about 10 percent of the energy used.

Using renewable energy resources that are locally available conserves the fuel that would be used to transport and process resources at a different location. Using a variety of energy resources rather than a single, nonrenewable energy resource, such as fossil fuels, can also help conserve resources. For example, a community that has hydroelectric energy resources might also use solar energy to generate electricity during months when water levels are low.

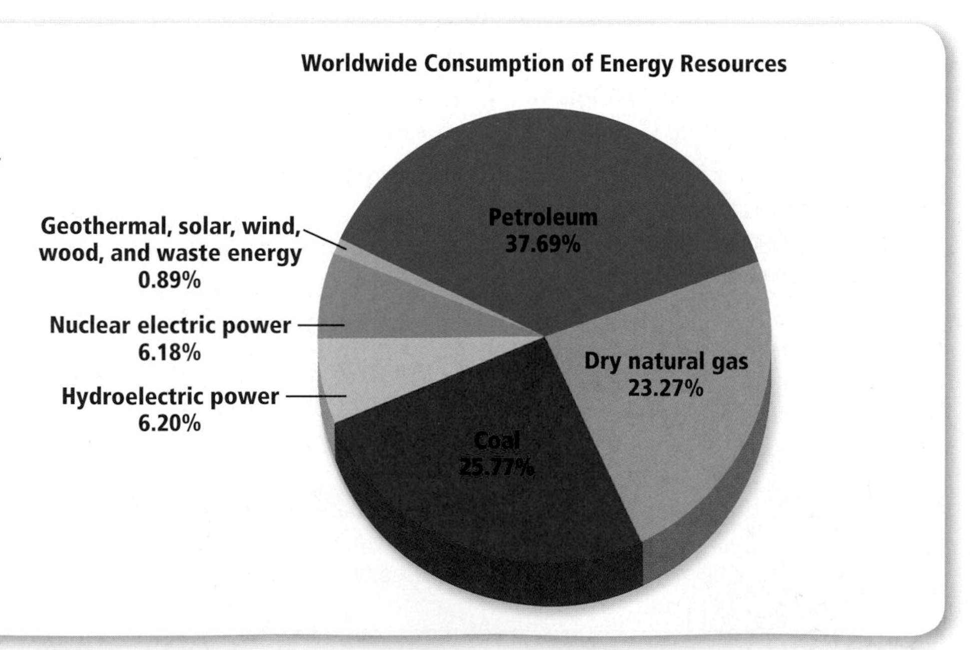

Figure 25.13 Petroleum is the most widely used energy resource worldwide, followed closely by coal and natural gas.
Explain *Why do you think nonrenewable resources account for almost 87 percent of the world energy consumption?*

Energy Efficiency

Energy is the ability to do work. The amount of work produced compared to the amount of energy used is called **energy efficiency.** Energy resources do not produce 100 percent of the potential work that is stored in the energy source.

When a car uses gasoline, some of the energy stored in the gasoline is converted to mechanical energy that moves the car, while some of the energy is used to power accessories, like the car's air conditioner. Most of the energy in the gasoline is lost as heat. Decreasing heat loss is one way that more of the stored energy can be converted to do work. To find ways to use resources more efficiently, scientists study exactly how energy resources are used and where improvements are needed. Using resources more efficiently is a type of conservation. For example, adding insulation to a house reduces heat loss, so less energy is needed to heat the air inside.

 **Reading Check** **Explain** energy efficiency.

Improving efficiency in industry Most of the electricity in the United States is generated by burning fossil fuels to heat water, forming steam. Recall that increasing the temperature of a gas also increases pressure. It is the steam pressure that spins the turbines that drive the generators to create electricity. Unfortunately, this is an inefficient process. Approximately one-third of the energy potential within the original fuel source can be converted into steam pressure.

Improving efficiency in transportation Transportation is necessary to move people, food, and other goods from one place to another. Although most transportation currently relies on oil, conservation practices can help reduce dependency on oil resources used for transportation. **Table 25.1** lists some of the advantages of public transportation, which is one way people can improve energy efficiency in transportation.

CAREERS IN EARTH SCIENCE

Environmental Consultant An environmental consultant interprets environmental data, conducts field surveys, and conducts environmental impact assessments. From this information, they make suggestions to businesses of how to limit their environmental impacts and meet governmental regulations. To learn more about Earth science careers, visit glencoe.com.

Concepts In Motion **Interactive Table** To explore more about public transportation, visit glencoe.com.

Table 25.1 Advantages of Public Transportation
Using public transportation to get to work can save a person between \$300 and \$3000 in fuel costs per year.
Using public transportation saves more than 3 billion liters of gasoline every year—equal to all energy used by U.S. manufacturers of computers and electronic equipment.
If Americans used public transportation for roughly 10 percent of daily travel needs, the United States would reduce its dependence on imported oil from the Persian Gulf by more than 40 percent.
During the past ten years, U.S. public transportation use has grown by 25.1 percent—a faster rate than highway travel (22.5 percent).

Commuting efficiently People who live in metropolitan areas can improve energy efficiency by using public transportation. Major U.S. cities, such as New York, use subways or elevated trains to move people. In Europe, mass transportation includes long-distance rail systems, as well as electric trams and trolleys. When it is necessary to drive private automobiles, carpooling can reduce the number of vehicles on the highways. Some metropolitan areas encourage carpooling by providing express lanes for cars with multiple passengers.

VOCABULARY

ACADEMIC VOCABULARY

Efficient

productive without waste

The automobile was more efficient when the proper tune-ups had been done.

Automobiles The use of fuel-efficient vehicles is another way to reduce the amount of petroleum resources consumed. Automobile manufacturers can build vehicles that achieve high rates of fuel efficiency without sacrificing performance. The future of this industry is promising as hybrid, fuel cell, and electric technologies begin to reach the consumer market. Also, less energy is needed to move something that weighs less. Smaller cars use less gasoline. Another way to conserve gasoline is to drive slower than 100 km/h (62 mph) on the freeway and use alternate forms of transportation.

Getting more for less Increased demand for fuels requires a greater supply and results in higher costs. Electricity is costly to produce, and it is not usually used efficiently in homes or industry. In the United States, approximately 43 percent of the energy used to fuel motor vehicles and to heat homes and businesses is lost as thermal energy. If energy were used more efficiently, less energy would be needed, thus decreasing the total cost of energy.

DATA ANALYSIS LAB

Based on Real Data*

Make and Use Graphs

What proportion of energy resource types are used to heat homes? Natural gas, electricity, heating oil, propane, and kerosene are used to heat American homes. The table shows percentages used to heat different types of homes.

Think Critically

1. **Compare** the sources of energy used by plotting the data on a graph. Be sure to use different colors for the different types of energy. Place the percentages on the *y*-axis and the source on the *x*-axis.
2. **Infer** why single-family homes use natural gas more than other types of dwellings.
3. **Infer** why heating oil, propane, and kerosene are not widely used as energy sources for homes.

Data and Observations

Energy Sources for American Homes (%)

Energy Source	Single-Family Dwellings	Multi-Family Dwellings	Mobile Homes
Natural gas	60	48	32
Electricity	23	42	43
Heating oil	8	7	3
Propane	5	0	15
Kerosene	1	0	4
Other	3	3	3

*Data obtained from: The National Energy Education Development Project. 2004. *Secondary Energy Infobook.*

Harnessing waste thermal energy Generating electricity produces waste thermal energy that can be recovered. The simultaneous production of two usable forms of energy is called **cogeneration.** Cogeneration captures the excess thermal energy (steam) for domestic or industrial heating. It can also be used in a large air-conditioner unit. It turns a turbine connected to a compressor that chills water sent to an air handler unit in a different building. Excess thermal energy can also be used to generate electricity that operates electrical devices within the power plant, such as sulfur-removing scrubbers on smokestacks. While industries use one-third of all energy produced in the United States, cogeneration has allowed some industries to increase production while reducing energy use. Cogeneration has enabled central Florida to operate the nation's cleanest coal-powered electric facility. The power station shown in **Figure 25.14** utilizes cogeneration for an oil refinery and chemical plant.

■ **Figure 25.14** This cogeneration power station helps reduce energy use at an oil refinery and chemical plant in Hampshire, UK.

Sustainable Energy

Energy resources on Earth are interrelated, and they affect one another. **Sustainable energy** involves the global management of Earth's natural resources to meet current and future energy needs. A good management plan incorporates both conservation and energy efficiency. New technology that extends the supply of fossil fuels is a vital part of such a plan. Global cooperation can help maintain the necessary balance between protection of the environment and economic growth. The achievement of these goals will depend on the commitment made by all so that future generations have access to the energy resources required to maintain a high quality of life on Earth.

Section 25.3 Assessment

Section Summary

- Energy resources will last longer if conservation and energy-efficiency measures are developed and used.
- Energy efficiency results in the use of fewer resources to provide more usable energy.
- Cogeneration, in which two usable forms of energy are produced at the same time from the same process, can help save resources.
- Sustainable energy can help meet current and future energy needs.

Understand Main Ideas

1. MAIN Idea **Summarize** why the conservation and efficient use of energy resources is important.
2. **List** three ways in which you could conserve electric energy in your home.
3. **Compare** energy consumption between developing and industrialized countries.
4. **Analyze** Why is it important to conserve resources instead of seeking new sources of fossil fuels for energy?

Think Critically

5. **Illustrate** how cogeneration can save energy resources.

MATH in Earth Science

6. If the global consumption of coal were reduced by 25 percent, what would the percentage consumption of coal be? Refer to **Figure 25.13** for more information.

Earth Science and the Environment

Bacteria Power!

Bacteria are all around us—some are helpful while others cause disease. Without bacteria, life would be very different. Humans have bacteria that live in the stomach and intestines to help digest food. Other bacteria cause illnesses such as strep throat and tuberculosis.

Pollution-eating bacteria Through research, scientists have discovered bacteria that can eat pollution, and other bacteria that can produce energy that can be harnessed for human use. Bacteria in the genus *Desulfitobacterium* have long been studied for their unique appetites. They eat pollution, such as toxic waste, and change it into less toxic or even nontoxic products. Recently, scientists worked with *Desulfitobacterium* successfully to find a species of bacteria that could break down freshwater pollution.

Microbial power plants Not only are *Desulfitobacterium* able to consume toxic waste, they are also able to produce energy at a constant rate. While scientists have known of the bacteria's ability to break down different toxins and produce energy as a by-product, this was the first time it was discovered that bacteria could do both at once. The energy that the bacteria produced could be harnessed to run small electrical devices.

Desulfitobacterium are able to survive extreme heat, radiation, and other environments that would easily wipe out other bacterial populations. Imagine that a fuel cell containing *Desulfitobacterium* is placed in an area where it will not be used for many years, and where it is exposed to harsh environments. If *Disulfitobacterium* was used as the power source for the fuel cell, it could exist in a stage similar to hibernation until it was needed or until conditions improved.

In the future, *Desulfitobacterium* might be used to power a wastewater treatment plant, such as this one, while helping to reclaim the wastewater being processed.

Diverse diets The metabolic capabilities of *Desulfitobacterium* bacteria are unique. The bacteria have a diverse diet, so they can use many different sources, including wastewater, chemical pollutants, and pesticides, to produce electricity.

While this biotechnology is still in the early stages of discovery and development, there are many exciting opportunities to be explored. It is possible that a bacterial colony could be used to reclaim wastewater while producing electricity to power the water treatment plant at the same time.

WRITING in Earth Science

Brochure You are marketing a fuel cell that uses these bacteria. Create a brochure explaining the potential uses of these fuel cells and why this biotechnology is important in today's world.

GEOLAB

DESIGN YOUR OWN: DESIGN AN ENERGY-EFFICIENT BUILDING

Background: **Buildings can be designed to conserve heat. Some considerations involved in the design of a building that conserves heat include the materials that will be used in construction, the materials that will store heat, and the overall layout of the building. By using a more energy-efficient design and more energy-efficient materials, consumers can decrease their monthly gas or electric bills and conserve natural resources.**

Question: *How can a building be designed to conserve heat?*

Possible Materials

glass or clear plastic squares
sturdy cardboard boxes
scissors
tape
glue
thermometers
paint
paper
aluminum foil
polystyrene
stone
mirrors
fabric
light source

Safety Precautions

Procedure

1. Read and complete the lab safety form.
2. Working in groups of three to four, brainstorm a list of design features that might contribute to the heat efficiency of a building and consider how you might incorporate some of these features into your building.
3. Design your building.
4. Make a list of heat-conserving issues that you addressed.
5. Decide which materials you will use to build your house. Collect those materials.
6. Construct the building and a control building for comparison.
7. Devise a way to test the heat-holding ability of each building.
8. Perform the test on each building. To test the buildings' heat efficiency, it may be necessary to heat the buildings and determine how long heat is conserved within each one. **WARNING:** ***Make sure the heat source is far enough away from the building materials so that they do not burn or melt.***
9. Record your data in a table. Then, make a graph of your data.
10. Make modifications to the design to improve the building's efficiency.

Analyze and Conclude

1. **Conclude** Was the building you designed more energy-efficient than the control building?
2. **Analyze** What problems did you encounter, and how did you solve them?
3. **Analyze** How did your observations affect decisions that you might make if you were to repeat this lab? Why do you think your design worked or did not work?
4. **Predict** Would your design work in a home in your community? In a community with a different climate? Why or why not?
5. **Compare and contrast** the building you designed and the control building.
6. **Compare and contrast** your design and the designs of your classmates.
7. **Determine** how your design could be improved.
8. **Predict** how using different energy sources might affect your results.

TRY AT HOME

Apply How could you incorporate some of your design elements into your own home? Discuss your lab with an adult at home and make suggestions to conserve heat. Visit glencoe.com for more information on heat-efficient designs.

CHAPTER 25

Study Guide

Download quizzes, key terms, and flash cards from glencoe.com.

BIG Idea People use energy resources, most of which originate from the Sun, for everyday living.

Vocabulary	Key Concepts
Section 25.1 Conventional Energy Resources	
• biomass fuel (p. 709) • fossil fuel (p. 710) • fuel (p. 709) • hydrocarbon (p. 709) • peat (p. 710)	**MAIN Idea** Biomass and fossil fuels store energy from the Sun. • The Sun is the source of most energy on Earth. • Humans have used materials derived from living things, such as wood, as renewable fuels for thousands of years. • Fossil fuels formed from organisms that lived millions of years ago.

Lignite

Bituminous

Anthracite

Vocabulary	Key Concepts
Section 25.2 Alternative Energy Resources	
• geothermal energy (p. 717) • hydroelectric power (p. 716) • nuclear fission (p. 718) • photovoltaic cell (p. 716)	**MAIN Idea** Many resources other than fossil fuels can be developed to meet the energy needs of people on Earth. • Alternative energy resources can supplement dwindling fossil fuel reserves. • Solar energy is unlimited, but technological advances are needed to find solutions to collect and store it. • Nuclear energy is produced when atoms of radioactive elements emit particles in the process known as nuclear fission. • Biofuels can help reduce consumption of fossil fuels.
Section 25.3 Conservation of Energy Resources	
• cogeneration (p. 723) • energy efficiency (p. 721) • sustainable energy (p. 723)	**MAIN Idea** Using energy efficiently reduces the consumption of nonrenewable resources. • Energy resources will last longer if conservation and energy-efficiency measures are developed and used. • Energy efficiency results in the use of fewer resources to provide more usable energy. • Cogeneration, in which two usable forms of energy are produced at the same time from the same process, can help save resources. • Sustainable energy can help meet current and future energy needs.

Assessment

Vocabulary Review

Write a sentence defining each of the following vocabulary terms.

1. fuel
2. peat
3. fossil fuel
4. energy efficiency
5. geothermal energy
6. cogeneration

Fill in the blanks with an appropriate vocabulary term from the Study Guide.

7. ________ is a form of energy generated by the conversion of free-falling water to electricity.
8. Solar energy is converted into electric energy through the use of ________.
9. Molecules with hydrogen and carbon bonds are called ________.

Replace the underlined words with the correct vocabulary term from the Study Guide.

10. <u>The process in which a heavy nucleus divides to form smaller nuclei</u> results in a release of a large amount of energy.
11. Global management of Earth's natural resources to meet human needs will allow people to have <u>all the energy they need to live</u>.
12. <u>Fuels formed from organic matter</u> are burned in developing countries as a source of heat.

Understand Key Concepts

13. Which is the primary source of energy on Earth?
   A. oil
   B. coal
   C. the Sun
   D. wood

14. When a consumer eats a producer, from where are they gaining energy?
   A. Earth
   B. the plant
   C. the Sun
   D. the ground

Use the figure below to answer Questions 15 and 16.

15. Which best describes the type of resource illustrated in the figure?
   A. biomass
   B. biofuel
   C. solar heating
   D. fossil fuel

16. What type of resource is shown?
   A. fossil fuel
   B. renewable resource
   C. nonrenewable resource
   D. cogeneration

17. Which is not derived from living things?
   A. petroleum
   B. coal
   C. peat
   D. nuclear power

18. Which form of energy commonly coincides with tectonic plate boundaries?
   A. fossil fuels
   B. geothermal energy
   C. wind energy
   D. biomass fuels

Use the diagram below to answer Questions 19 and 20.

19. Which process happens in Layer 1?
    - **A.** Vegetation accumulates and forms peat.
    - **B.** Bituminous coal forms from lignite.
    - **C.** Lignite forms from accumulated vegetation.
    - **D.** Anthracite forms from bituminous coal.

20. Which is formed in Layer 4?
    - **A.** anthracite
    - **B.** bituminous coal
    - **C.** lignite
    - **D.** peat

21. Which is one reason nuclear power plants are not widespread?
    - **A.** Nuclear power is not energy efficient.
    - **B.** Nuclear reactors emit greenhouse gases.
    - **C.** Nuclear reactions occur only on the Sun.
    - **D.** Negative public perception of nuclear power.

Constructed Response

22. **Describe** three ways fossil fuels are used for energy.
23. **Draw and label** a diagram to explain passive solar heating.
24. **Identify** one form of energy not derived from the Sun.
25. **Describe** the formation of lignite.
26. **Explain** how organisms living on Earth in this era could become fossil fuels.
27. **Analyze** why a substance such as water is good to use in passive heating situations.
28. **List** three ways to conserve oil.

Think Critically

29. **Explain** why lignite, which has a carbon concentration of 40 percent, burns less efficiently than anthracite, which has a carbon concentration of 90 to 95 percent.
30. **Distinguish** What characteristics of water allow it to be used to produce energy as well as store energy?
31. **Explain** why biomass fuels are more widely used than oil for fuel in developing countries.

Use the graph below to answer Questions 32 and 33.

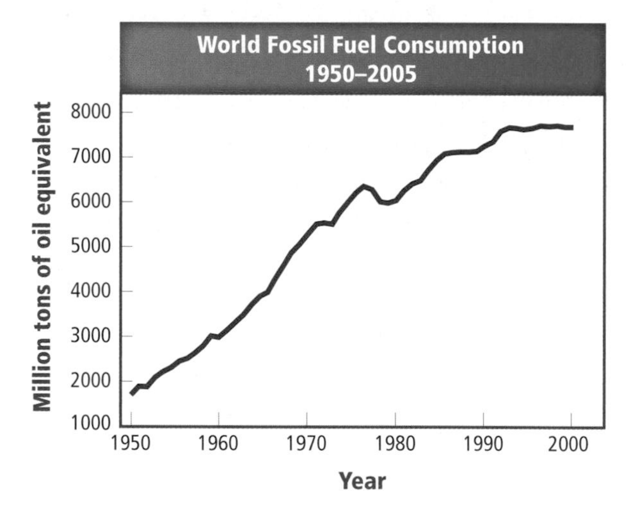

32. **Calculate** how many more tons of fossil fuels were used in 2000 compared to 1960.
33. **Predict** Do you think the trend shown on the graph would be the same for developing and industrialized countries if they were shown separately? Why?
34. **Explain** why not all organic resources are considered renewable. Give an example of a renewable and nonrenewable organic resource.
35. **Predict** What might be some negative consequences of a nation being dependent on foreign energy resources?

Earth Science Online Chapter Test glencoe.com

36. **Compare and Contrast** How might the fuel-use by people living in the northeastern United States differ from fuel-use by people who live in the southern and southwestern United States?

37. **Imagine** that you and your friends took a trip to a deserted island that had no plants larger than small shrubs. Describe how you would seek a fuel source from the island.

38. **Evaluate** the potential for using more solar energy in your community. Which type of solar energy collection would work best? Is solar energy an effective energy source for your community? Why or why not?

39. **Analyze** why biomass fuels are not widely used in the United States.

40. **Compare and contrast** nuclear energy with energy that comes from petroleum.

41. **Imagine** you are eating a cheeseburger. Explain all the ways you are gaining energy derived from the Sun.

42. **Predict** what might happen to gas prices, assuming oil continues to be used at the current rate and an alternative fuel source is not discovered. Explain.

43. **Explain** why a wood-burning stove is not an efficient way to heat a home.

Concept Mapping

44. Make a concept map to organize information about alternative energy resources using the following terms: *geothermal energy, hydroelectric power, solar energy, wind power, tidal power,* and *biomass fuels*. For more help, refer to the *Skillbuilder Handbook*.

Challenge Question

45. **Apply** If a standard home costs $150,000 to build and costs $2300 per year to heat, and the same home, built with materials designed to use passive solar heat, costs $180,000 to build, but $400 per year to heat, how long will it take to make up the price difference between the two houses?

Additional Assessment

46. WRITING in **Earth Science** Write a letter to the editor for a local newspaper to convince others to recycle. Include specific examples and how those actions will assist in extending the limited supply of a particular natural resource.

DBQ Document–Based Questions

Data obtained from: Annual Energy Review 2005. July 2006. *Energy Information Administration* (EIA-0384).

Energy Consumption (quadrillion btu)				
Year	Fossil Fuels	Nuclear	Renewable	Total
2000	84.96	7.86	6.17	98.99
2001	83.18	8.03	5.35	96.56
2002	83.99	8.14	5.93	98.06
2003	84.39	7.96	6.14	98.49
2004	86.23	8.22	6.22	100.67
2005	85.96	8.13	6.06	100.15

47. Compare and contrast the consumption of renewable energy resources with the consumption of other energy resources.

48. In 2001, what percentage of the total energy consumed in the United States was fossil fuels? Based on the data, has that percentage changed significantly in the first part of this decade?

49. What percentage of the total energy consumption for 2005 was comprised of fossil fuels?

Cumulative Review

50. Name the molecule that is necessary for life that was absent from Earth's early atmosphere. **(Chapter 22)**

51. Bedrock is found everywhere in Earth's crust. Explain whether or not an abundance of bedrock would diminish the concern over availability as a resource. **(Chapter 24)**

Standardized Test Practice

Multiple Choice

1. Which is the most expensive and least used method of providing water to areas in the United States?
   A. tapping groundwater C. desalination
   B. aqueducts D. dams

Use the illustration to answer Questions 2 and 3.

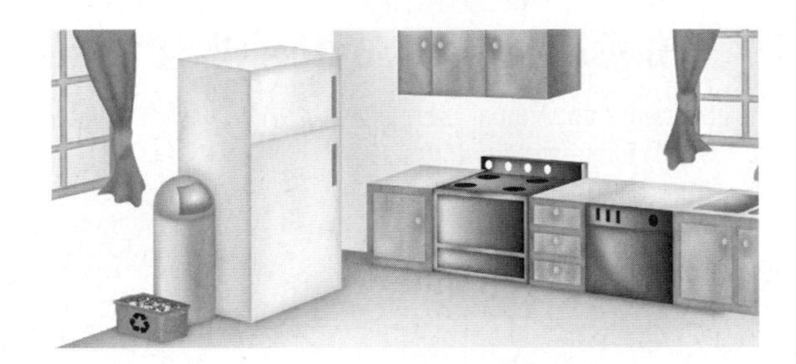

2. How could this kitchen be made more energy efficient?
   A. by maintaining older appliances instead of replacing them with newer ones
   B. by replacing the kitchen cabinets
   C. by washing the dishes in the dishwasher instead of the sink
   D. by replacing the old windows with newer ones

3. If this kitchen were located in a home in Arizona, which alternative energy source could be used?
   A. peat C. ethanol
   B. solar energy D. hydroelectric power

4. Which relationship between geologic structures and plate boundaries is most accurate?
   A. Explosive volcanoes most often occur near convergent boundaries.
   B. Folded mountains commonly develop at divergent boundaries.
   C. Rift valleys are usually produced at convergent boundaries.
   D. Volcanic arcs are usually found along transform boundaries.

5. Besides being a requirement for respiration, why else is oxygen important in the atmosphere?
   A. It provides protection from ultraviolet rays emitted by the Sun.
   B. It regulates climate and weather patterns on Earth.
   C. It is the major component of wind to cool Earth.
   D. It allows rays from the Sun to filter in and warm Earth.

6. What was the goal of Stanley Miller's research?
   A. to refute the belief that life could have existed on early Earth
   B. to explain the formation of oxygen on early Earth
   C. to test the primordial soup hypothesis
   D. to do an analysis of the atmosphere present on early Earth

Use the illustration below to answer Questions 7 and 8.

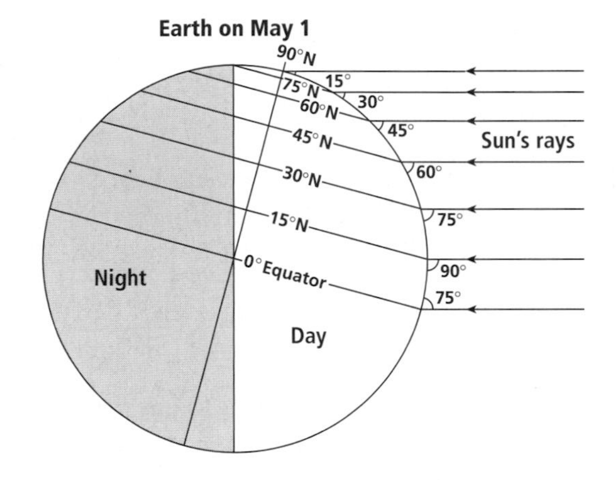

7. Which change can be expected to occur at 45° N over the next 30 days?
   A. The duration of solar radiation will decrease and the temperature will decrease.
   B. The duration of solar radiation will decrease and the temperature will increase.
   C. The duration of solar radiation will increase and the temperature will decrease.
   D. The duration of solar radiation will increase and the temperature will increase.

8. Where would the risk of sunburn be highest?
   A. the equator C. 45° N
   B. 15° N D. 75° N

9. Besides the formation of Pangaea, what other major event occurred during the Paleozoic?
   A. the first major volcanic eruption
   B. the first appearance of life
   C. the appearance of complex life
   D. the mass extinction of all life

Short Answer

Use the illustration below to answer Questions 10 to 12.

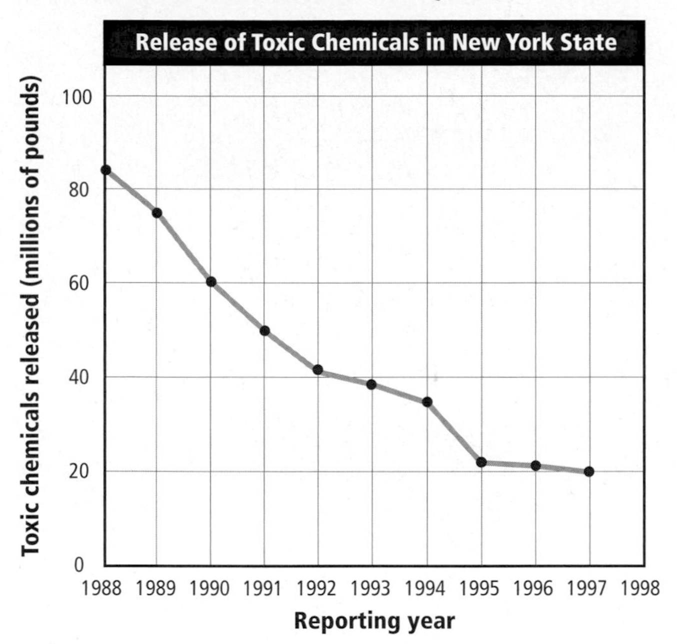

10. About how much change has occurred in the amount of toxic chemical released from 1988 to 1997?

11. During what four-year period was the greatest drop in toxic chemicals released? What is one possible explanation for this major drop?

12. State one possible explanation for why the amount of toxic chemicals released remained relatively constant between 1995 and 1997.

13. How were coral reefs formed?

14. How does the process of relative-age dating differ from the process of absolute-age dating?

15. What do scientists hypothesize as the cause of the cooling trend during the Cenozoic Era?

Reading for Comprehension

Vegetable Oil Fuels

Chemists and advocates for alternative energy technologies are training their sights on the grease used to cook french fries. Unlike petroleum-based products, vegetable oils are biodegradable, non-toxic, and are derived from a renewable resource. One problem, however, is the high development cost of vegetable-derived motor oils relative to petroleum-based products. Advocates for the use of vegetable oils say they are easier on the environment because they are much more biodegradable than conventional, petroleum-based oils. When spilled or disposed of on the ground, vegetable oil will decompose by upward of 98 percent. Petroleum based products only decompose 20 to 40 percent. Additionally, vegetable oils are a renewable resource.

Article obtained from: Roach, J. Vegetable oil—the new fuel? *National Geographic News.* April 22, 2003. (Online resource accessed October 7, 2006.)

16. What can be inferred from this passage?
- **A.** Petroleum-based oils are better than vegetable oils as energy sources.
- **B.** Vegetable oils will not be able to be used in car engines.
- **C.** Although vegetable oils are better for the environment, it will be some time before they replace the use of petroleum-based oils.
- **D.** Even though vegetable oils are better for the environment than petroleum-based oils, it is still better not to use them for energy sources.

17. When spilled or disposed of, what percent of vegetable oils will decompose?
- **A.** 3 percent
- **B.** 75 percent
- **C.** 98 percent
- **D.** 100 percent

NEED EXTRA HELP?															
If You Missed Question . . .	1	2	3	4	5	6	7	8	9	10	11	12	13	14	15
Review Section . . .	24.4	25.3	25.2	18.1	22.3	22.4	12.1	12.1	23.1	24.3	24.3	24.3	23.1	21.3	23.3

CHAPTER 26 Human Impact on Resources

BIG Idea The use of natural resources can impact Earth's land, air, and water.

26.1 Populations and the Use of Natural Resources
MAIN Idea More demands are placed on natural resources as the human population increases.

26.2 Human Impact on Land Resources
MAIN Idea Extraction of minerals, farming, and waste disposal can have negative environmental impacts.

26.3 Human Impact on Air Resources
MAIN Idea Manufacturing processes and the burning of fossil fuels can pollute Earth's atmosphere.

26.4 Human Impact on Water Resources
MAIN Idea Pollution controls and conservation protect water resources.

GeoFacts

- Climbers have left more than 50 tons of trash on the top of Mount Everest.
- In 1993, laws were passed requiring climbers to bring down the nonbiodegradable items they take up.
- Many groups remove trash, including used oxygen tanks, batteries, and other nonbiodegradable items.

Start-Up Activities

LAUNCH Lab

What resources are used in classroom items?

As you learned in Chapter 24, natural resources include air, water, land, and living organisms. Use of natural resources can have global impacts.

Procedure

1. Read and complete the lab safety form.
2. Working in groups of two or three, make a pile of 15 **items** from your classroom.
3. Make a data table for your items. Record as much of the following information as you can.
   - What resources were used to make the item?
   - Are the resources renewable or nonrenewable?
   - Where was the item made?

Analysis

1. **Observe** How many different resources are represented by the items in your collection?
2. **Calculate** What percent of your 15 items were renewable and what percent were nonrenewable resources?

Visit glencoe.com to

- study entire chapters online;
- explore Concepts In Motion animations:
  - Interactive Time Lines
  - Interactive Figures
  - Interactive Tables
- access Web Links for more information, projects, and activities;
- review content with the Interactive Tutor and take Self-Check Quizzes.

Sources of Water Pollution Make this Foldable to compare the two main types of water-pollution sources.

STEP 1 Fold a sheet of paper in half from top to bottom.

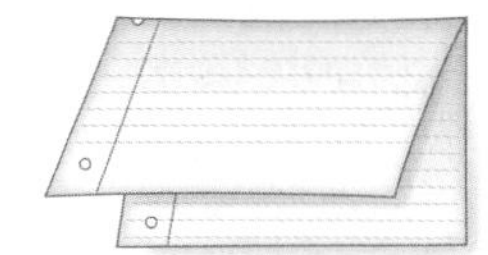

STEP 2 Fold in half again, as shown.

STEP 3 Unfold once and cut along the fold line of the top flap to make two tabs.

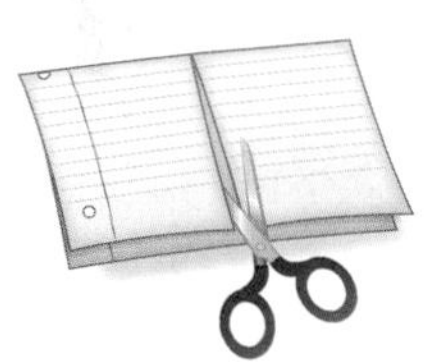

STEP 4 Label the tabs *Point Sources* and *Nonpoint Sources.*

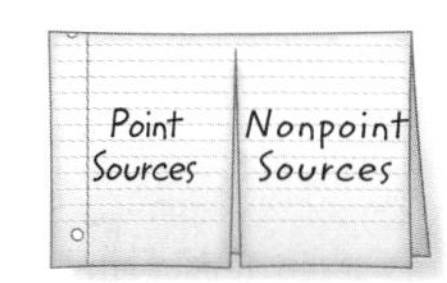

FOLDABLES **Use this Foldable with Section 26.4.** As you read this section, explain the two main types of pollution and give some examples of each.

Section 26.1

Objectives

- **Summarize** the typical pattern of population growth of organisms.
- **Describe** what happens to populations when they reach carrying capacity.
- **Identify** environmental factors that affect population growth.

Review Vocabulary

population: individual organisms of a single species that share the same geographic location at the same time

New Vocabulary

exponential growth
carrying capacity
density-independent factor
density-dependent factor

Populations and the Use of Natural Resources

MAIN Idea More demands are placed on natural resources as the human population increases.

Real-World Reading Link How many pets do you have? If you only have one pet, it might not take much time or money to care for it. What if you had eight pets? The amount of time and money you would need to properly care for your pets might put a strain on your money and activities. Similarly, as world population grows, it puts a strain on available natural resources.

Resources and Organisms

Like all organisms, humans need natural resources to grow, reproduce, and maintain life. Among the resources that organisms require are air, food, water, and shelter. To meet their basic needs, most organisms are adapted to their immediate environment. They live in balance with the natural resources provided within their environment. For example, songbirds live in grassy meadows, forage for grass seeds to eat, weave nests out of dried grasses and twigs, and drink water from ponds or streams nearby.

Other organisms, however, alter their environment to better meet their needs. For example, beavers build dams, like the one in **Figure 26.1,** across streams to create ponds where none previously existed. Such alteration of the environment has both positive and negative impacts: it kills some trees and displaces both aquatic and terrestrial organisms, but at the same time, it creates a new wetland environment for other organisms. Of all organisms, however, humans have an unequaled capacity to modify their environments. This capacity allows humans to live in every terrestrial environment on Earth. As a result, humans also have the greatest impact on Earth's natural resources.

Figure 26.1 Beavers can alter their environments to suit their needs. Notice how, by damming the stream, the beavers have changed the water level.
Infer *How might this affect the other organisms living in this environment?*

Population Growth

Population growth is defined as an increase in the size of a population over time. A graph of a growing population resembles a J-shaped curve at first. Whether the population is one of dandelions in a lawn, squirrels in a city park, or herring gulls on an island, the initial increase in population is small because the number of adults capable of reproducing is low.

As the number of reproducing adults increases, however, the rate of population growth increases rapidly. As shown in **Figure 26.2,** the population then experiences **exponential growth,** which is a pattern of growth in which a population grows faster as it increases in size.

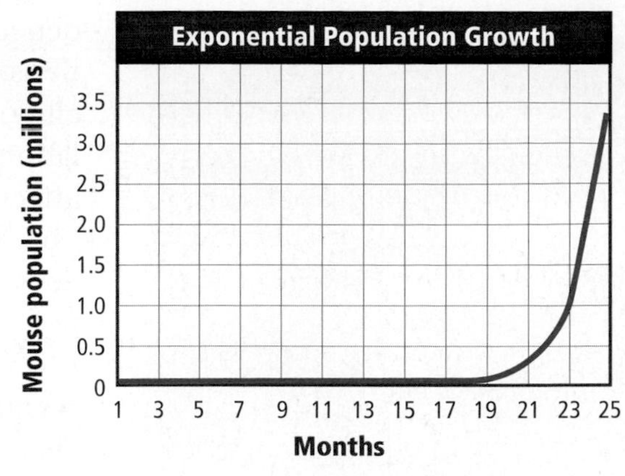

Figure 26.2 If two mice were allowed to reproduce in perfect conditions and all their offspring survived, the population would grow slowly at first, but then would accelerate quickly.

 **Reading Check** **Explain** exponential growth. Why is this an important concept to understand in relation to how organisms affect their environment?

Limits to population growth If the population graphed in **Figure 26.2** were studied for an extended period of time, what do you think would happen to the size of the population? Would it continue to grow exponentially? Many of Earth's natural resources are in limited supply, and therefore, most populations cannot continue to grow forever.

Eventually, one or more limiting factors, such as the availability of food, water, or shelter will cause a population to stop increasing. This leveling-off of population size results in an S-shaped curve, similar to the one in **Figure 26.3.**

Carrying capacity The number of organisms that any given environment can support is its **carrying capacity.** When population size has not yet reached the carrying capacity of a particular environment, the population will continue to grow for several reasons. First, there will be more births than deaths because of adequate resources. Second, because of the availability of resources, more individuals might move to the area than die or leave.

If the population size temporarily exceeds the carrying capacity, the number of deaths will increase, or the number of births will decrease until the population size returns to the carrying capacity. A population at the carrying capacity for its environment is in equilibrium. The population will continue to fluctuate around the carrying capacity as long as natural resources remain available.

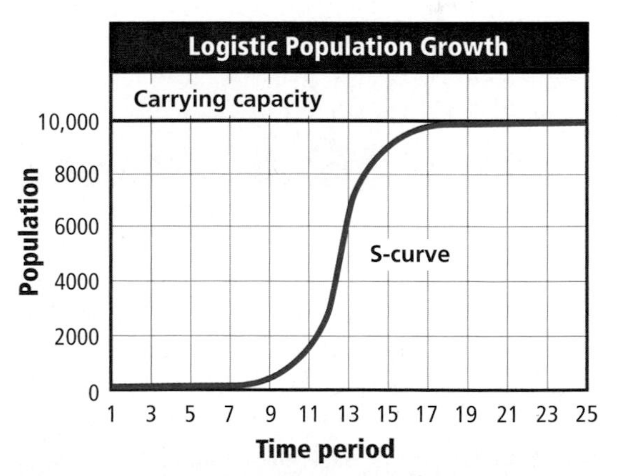

Figure 26.3 A population grows exponentially until it reaches its carrying capacity. Carrying capacity is limited by the resources available to the population. **Predict** *what would happen if the population exceeded carrying capacity.*

Concepts In Motion

Interactive Figure To see an animation of carrying capacity, visit glencoe.com.

■ **Figure 26.4** A forest fire is one example of a density-independent factor of population growth. Fires can affect trees, birds, mammals, and other populations. Fires, like the one that occurred here, can also encourage new growth.

Environmental limits Environmental factors that do not depend on population size, such as storms and fires, are **density-independent factors.** Density-independent factors affect all populations that they come in contact with, regardless of population size, as **Figure 26.4** shows. Environmental factors that affect population growth, such as disease, predators, and competition for food, are called **density-dependent factors.** Density-dependent factors are often biotic factors that increasingly affect a population as the population's size increases.

One example of a density-dependent factor can be seen in populations of white-tailed deer in the United States. White-tailed deer are found in most of the continental United States. In some areas, deer populations have grown in recent years due largely to a decrease of natural predators. Although the population is increasing, the amount of food available to the deer does not change. For this reason, every year many deer starve to death. The density-dependent factor in this example is the lack of food due to over-population.

Human Population Growth

During your lifetime, you might have seen an increase in the number of cars, houses, and roads around you. The human population on Earth is growing. The growth curve is still in the J-shaped stage. The human population is expected to continue to grow for at least another 50 years at its current rate. The human population has not yet reached carrying capacity, but the current rate of growth cannot continue forever. As the population increases, demand for natural resources will also continue to increase steadily. Use of natural resources has already had global environmental implications.

Section 26.1 Assessment

Section Summary

- All organisms use resources to maintain their existence. The use of these resources has an impact on the environment.
- As populations increase, the demand for resources increases. Because resources are limited, populations will stop growing when they reach carrying capacity.
- Populations grow exponentially at early stages. Earth is currently experiencing a human population explosion.

Understand Main Ideas

1. MAIN Idea **Explain** how an increasing human population places more demands on Earth's natural resources.
2. **Identify** three limiting factors that keep populations from growing indefinitely.
3. **Compare** density-dependent and density-independent factors that limit population growth.

Think Critically

4. **Predict** how a small population of bacteria placed in a petri dish with limited nutrients will change over time. Draw a graph to represent the population growth.

MATH in Earth Science

5. If a city has 300,000 residents and an average birth rate of 1.5 children per person, how many people will there be in the next generation?

Earth Science Online Self-Check Quiz glencoe.com

Section 26.2

Objectives

- **Describe** the environmental impact of mineral extraction.
- **Discuss** the environmental problems created by agriculture and forestry, and list possible solutions.
- **Explain** how urban development affects soil and water.

Review Vocabulary

erosion: movement of weathered materials from one location to another by agents such as water, wind, glaciers, and gravity

New Vocabulary

reclamation
deforestation
pesticide
bioremediation

Human Impact on Land Resources

MAIN Idea Extraction of minerals, farming, and waste disposal can have negative environmental impacts.

Real-World Reading Link Do you spend much time talking on the telephone, listening to a digital music player, or using a computer? Perhaps you use a microwave oven to heat after-school snacks. Many of the materials in these items are derived from land resources.

Mining for Resources

How much land per year do you think is necessary to provide the raw materials that you use? Each year, a typical person in the United States consumes resources equal to the renewable yield from approximately 5 ha of forest and farmland. Many of these raw materials come from under the surface of Earth. To access these resources for human use, they must be extracted through one of many mining techniques.

Mining techniques can have a significant impact on Earth's surface. Modern societies require huge amounts of land resources, including iron, aluminum, copper, sand, gravel, and limestone. Unfortunately, the extraction of these resources often disturbs large areas of Earth's surface, as shown in **Figure 26.5.** Groundwater can become polluted, natural habitats can be disturbed or destroyed, and air quality can suffer. Finding a balance between the need for mineral resources and controlling the environmental change caused by extraction can be difficult, but scientists, in conjunction with mining companies, have created ways to reduce the impact of mining on the environment.

Figure 26.5 Mines, such as the one shown here, can have negative environmental impacts such as topsoil erosion.
Determine *Where would the eroded topsoil go? What other environmental impacts might a mine like this one have?*

■ **Figure 26.6** White-tailed deer and other native wildlife species thrive on this reclaimed strip mine land in eastern Wyoming.

Restoring the land In the United States, the Surface Mining Control and Reclamation Act of 1977 requires mining companies to restore the land to its original contours and to replant vegetation in a process called **reclamation.** However, vegetation cannot grow without topsoil. Mining companies can scrape the topsoil off of the land surface prior to mining and stockpile it for reclamation after materials have been removed. **Figure 26.6** shows a strip-mined area that has been reclaimed. Although reclamation repairs much of the damage that surface mining causes, it can be extremely difficult to restore land to its original contours and vegetation.

Reading Check **Explain** why it is important to have legislation requiring mining companies to restore land to its original contours.

■ **Figure 26.7** Runoff from a nearby mine pollutes this river. The presence of metals, including iron, causes the orange color in the runoff.

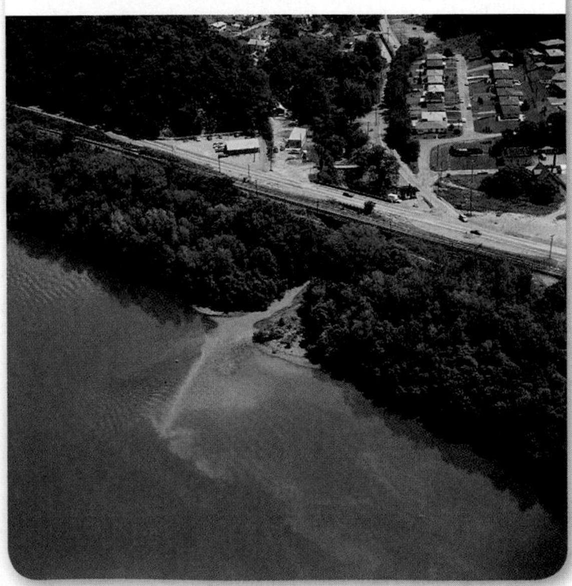

Underground mining Underground mining, also called subsurface mining, is used where mineral resources lie deep under the ground. Underground mining is less disruptive to the land surface than surface mining, but it still has impacts on the environment. For example, although the underground mines cannot be seen, the mountains of waste rock dug from under the ground are stockpiled on the surface. The water in **Figure 26.7** is orange because precipitation seeps through mine waste piles and causes a decrease in pH, dissolving many harmful metals in the waste. When the runoff from the piles reaches the stream, which has a higher pH, the higher pH causes the metals to come out of solution and discolor the water. Although many mining companies build large holding ponds to contain polluted water until it can be treated, these ponds sometimes leak.

Forestry

Clearing forested land is another way in which topsoil is lost. Worldwide, thousands of hectares of forests are cut down annually for firewood, charcoal, paper, and lumber. In parts of the world, the clearing of forested land results in **deforestation,** which is the removal of trees from a forested area without adequate replanting. Deforestation often involves clear-cutting, the complete removal of all the trees in an area. Clear-cutting can cause erosion of topsoil.

Fortunately, the negative environmental impacts of deforestation can be minimized through the practices of selective logging and the retention of buffer zones of trees along streambeds. In selective logging, workers remove only designated trees. This practice reduces the amount of ground left bare and thus helps prevent erosion.

VOCABULARY

ACADEMIC VOCABULARY

Adequate
sufficient to satisfy a requirement or meet a need
She filled her gas tank to ensure she had an adequate supply of gas for the trip.

Urban Development

As the human population continues to increase, more people live in cities and towns. Agricultural land located near cities is being converted to suburban housing. As people populate areas that were once agricultural or rural, stores and industry follow. Seventy percent of the population in North America lives in urban and suburban areas, and an estimated 5 billion people worldwide will be living in cities and towns by the year 2025.

The development of urban areas has many environmental impacts. When towns and cities expand into rural areas, natural habitats are lost to roads, houses, and other buildings. Development leaves less land for agricultural use, which puts pressure on the remaining farmland for increased production. Other problems are created when concrete and asphalt cover large areas. Because there are fewer opportunities for rainwater to soak into the ground, groundwater supplies are not recharged, and flooding increases during heavy rains.

Solid waste Each person in the United States generates an average of 1.5 kg of solid waste per day. Where does it all go? Much of it is buried in landfills. **Figure 26.8** shows the percentages and material types that were disposed in landfills in the United States in 2003.

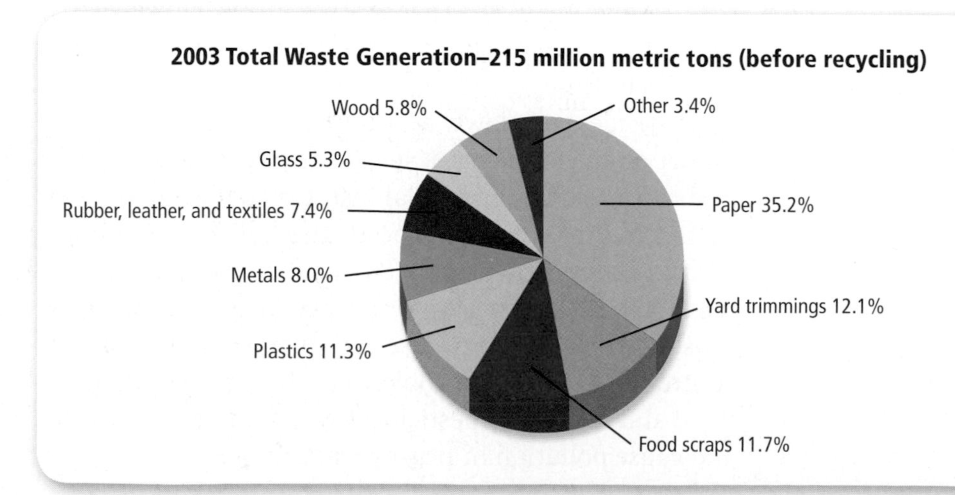

Figure 26.8 This circle graph shows the total solid waste generated in the United States in 2003.
Determine ***What material composed the highest percentage of solid waste in the United States in 2003? Why do you think this was the case? Could this material be recycled?***

MiniLab

Model Nutrient Loss

How does soil lose nutrients when subjected to farming, strip-mining, or development? If an area of soil has no plant cover for an extended period of time, the nutrients in the soil can be washed away by rainfall.

Procedure

1. Read and complete the lab safety form.
2. Place a **coffee filter** inside a **funnel.**
3. Place the funnel so that it is resting inside a **100-mL beaker.**
4. Pour a **sand mixture** into the coffee filter.
5. Measure 50 mL of **water** and pour it into the sand mixture.
6. Record your observations.
7. Carefully remove the funnel with the sand mixture and discard the water as instructed by your teacher.
8. Replace the funnel with the sand mixture in the 100-mL beaker.
9. Repeat Steps 5 through 8 four times.
10. Discard the sand mixture and funnel as instructed by your teacher.

Analysis

1. **Observe** What did you notice about the water as you repeated the investigation?
2. **Analyze** Why were you instructed to discard the water in between investigations?
3. **Predict** If the material in the sand represents nutrients needed for plant growth, and the sand represents soil, what would happen if you tried to grow crops in the soil at the end of your investigation?
4. **Infer** In what other ways might nutrient loss impact the environment?
5. **Apply** How could you slow the process of nutrient loss?

Improper disposal of wastes can result in the contamination of land and water resources. Heavy metals, such as lead and mercury, and poisonous chemicals, such as arsenic, are by-products of many industrial processes and can pollute the soil and groundwater. Some of this type of contamination has been caused by industries that operated before the dangers of improper waste disposal were known.

Agriculture

Vegetation, including agricultural crops, needs the nutrients from topsoil to grow. It can take thousands of years for topsoil to form, and thus, once it is lost, it is hard to replace. Whenever fields are plowed and the plants whose roots hold the soil in place are removed, topsoil can be eroded by wind and water and nutrients can be lost. The addition of fertilizers can help replace some of the nutrients, but there are other substances in topsoil that fertilizers cannot provide.

Topsoil contains trace minerals as well as organisms such as earthworms and nitrogen-fixing bacteria. Earthworms burrow into soil, providing oxygen and space for plant roots to grow, and nitrogen-fixing bacteria take nitrogen out of the air and make it available to plants. Topsoil also has an abundance of organic matter, including fecal material from organisms that live in the soil as well as decaying organisms. Organic matter helps hold moisture, reduces erosion, and releases nutrients back into the soil. Soil erosion can be reduced and fertility can be increased by using a variety of farming practices as shown in **Figure 26.9.**

Effects of pesticides Chemicals applied to farm fields to control weeds, insects, and fungi are called **pesticides.** Pesticides have played an important role in boosting food production worldwide by eliminating or controlling organisms that destroy crops. However, some pesticides remain in the environment for long periods of time.

Pesticides can slowly accumulate in organisms higher on the food chain, such as fishes and birds. Some pesticides also kill beneficial insect predators along with destructive insects. When pesticides kill decomposers, such as worms, the overall fertility of topsoil deteriorates. Insects can develop resistance to an insecticide, causing some farmers to use ever-increasing amounts to control pests. Further problems can be created when wind and rain carry pesticides away from farm fields and cause pollution in nearby waterways.

Visualizing Agricultural Practices

Figure 26.9 Using agricultural conservation practices can help protect precious nutrients in the soil as well as help reduce topsoil loss. Contour farming, crop rotation, and no-till farming are shown below.

Contour farming is often done on hillsides or other areas prone to erosion. Farmers plant crops with the contour of the earth, slowing the flow of runoff and helping to prevent erosion.

No-till farming Farmers leave the unused portion of the crops on the field instead of plowing them under each year. In this image, the crop in the previous year was wheat. After the seeds were harvested, the stalks were left on the field to prevent erosion and maintain topsoil.

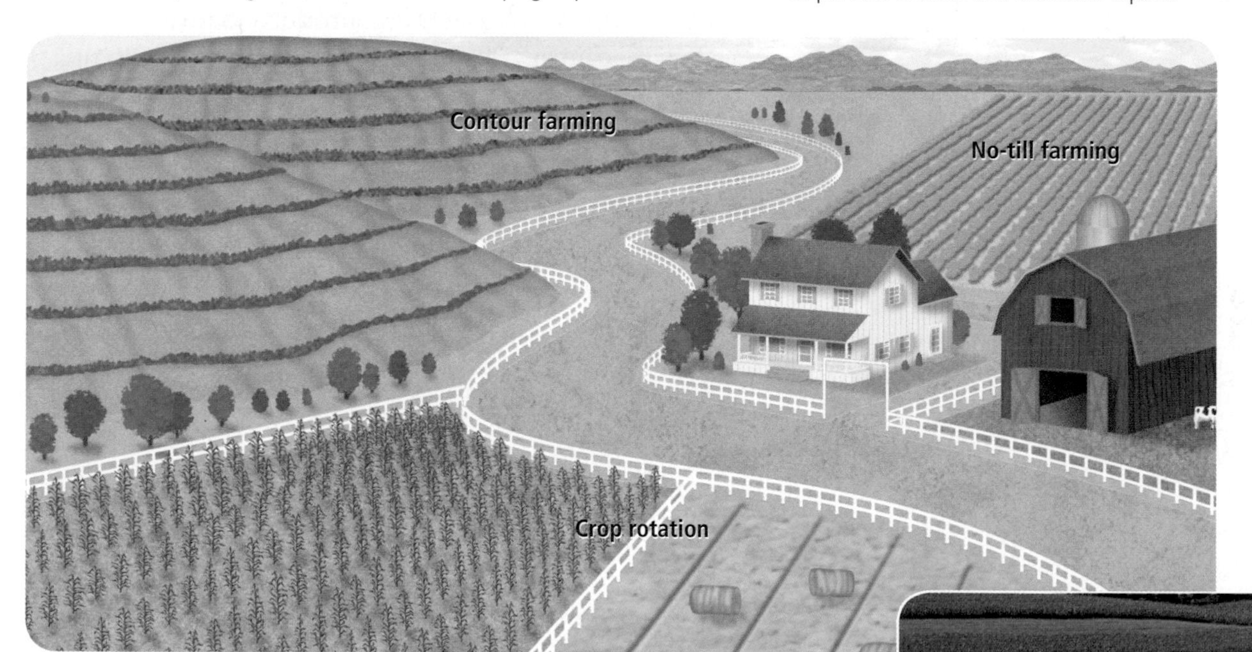

Crop rotation involves planting different crops in succession. For instance, a farmer might plant alfalfa in a field one year, followed by corn the next year, and winter wheat the following year. Crop rotation helps to maintain the soil's nutrient balance and also helps to reduce the number of crop-specific pests.

Concepts In Motion To explore more about agricultural practices, visit glencoe.com.

Earth Science Online

■ **Figure 26.10** Barriers such as this one are often used on construction sites to prevent erosion and reduce loss of topsoil.

Conservation

People are becoming aware of the need to protect the environment, and communities are making increased efforts to do so as urban development continues. For example, developers are often required to place barriers, like those shown in **Figure 26.10,** around construction sites to catch sediment from increased erosion. In the United States, wetlands are now recognized as valuable ecosystems and are protected from development.

Waste disposal remains a problem because of the immense volume of trash. Modern landfills are carefully designed to minimize leakage of toxic liquids. Impermeable clay or plastic layers are placed beneath a landfill, and trash is compacted by machines and buried under a layer of dirt to reduce volume and eliminate windblown trash. Vents in landfills release methane and other gases that are generated as the garbage decomposes.

Several methods are available for cleaning up industrial waste sites. Contaminated soil can be removed and disposed of at hazardous waste landfills. Soil can also be incinerated to destroy the toxic chemicals. The drawbacks to this method are that it can be expensive to treat large volumes of soil, and it can produce toxic ash.

Bioremediation is the use of organisms to clean up or break down toxic wastes. These organisms actually eat pollutants for food, neutralizing their negative impacts on the environment. Bioremediation is useful for contamination caused by spilled gasoline and oil.

Section 26.2 Assessment

Section Summary

- Humans require large amounts of land resources.
- The extraction of resources can disrupt Earth's surface.
- Growing populations increase the demand for food and result in increased urban development.
- Agriculture, poor forestry practices, and urban development can cause habitat loss, increased erosion, and water pollution.
- Human impact on land resources can be minimized through the use of modern techniques.

Understand Main Ideas

1. MAIN Idea **Describe** how extracting resources, growing food, and urban development contribute to land and water pollution.
2. **Propose** ways that land can be restored after it is strip-mined for coal.
3. **Predict** How many items will you throw away during lunch today? How much will you throw away in one week? One month? How does this relate to the impact of urban development on the environment?

Think Critically

4. **Suggest** methods of development that will reduce soil erosion and damage to streams.

WRITING in Earth Science

5. Write an article for your school paper suggesting ways everyone can reduce the amount of waste that they produce.

Earth Science Online Self-Check Quiz glencoe.com

Section 26.3

Objectives

- **Relate** the greenhouse effect and global warming.
- **Sequence** the reactions that occur as CFCs cause ozone depletion.
- **Identify** the causes and effects of acid precipitation.

Review Vocabulary

greenhouse effect: heating of Earth's surface by certain atmospheric gases, which helps keep Earth warm enough to sustain life

New Vocabulary

photochemical smog
ozone hole
acid precipitation

Human Impact on Air Resources

MAIN Idea Manufacturing processes and the burning of fossil fuels can pollute Earth's atmosphere.

Real-World Reading Link If you've ever enjoyed a campfire, you might have wondered what happens to the wood as it burns. When fuel is burned for energy or when products are manufactured, particles and gases such as carbon dioxide are released to Earth's atmosphere.

Global Impacts of Air Pollution

It has become evident that human activities can affect Earth on a global scale. The global atmospheric effects of air pollution include global warming, ozone depletion, and acid precipitation.

Global warming Recall from Chapter 14 that the greenhouse effect is a natural phenomenon in which Earth's atmosphere traps thermal energy in the troposphere to warm Earth. A phenomenon related to the greenhouse effect is global warming, which is the increase in Earth's average surface temperature.

Over time, Earth has experienced periods of global warming and cooling. Some scientists think that the current period of global warming that Earth is experiencing is related to increased levels of carbon dioxide in the atmosphere. Some of this increase is thought to have been caused by humans.

Human activities, especially the burning of fossil fuels, contribute to increased levels of carbon dioxide. Fossil fuels contain carbon, and when they are burned, the carbon combines with oxygen to form carbon dioxide. Since the beginning of the industrial revolution, around 1850, humans have been burning fossil fuels at an ever-increasing rate. **Figure 26.11** shows how atmospheric carbon dioxide has increased since 1960.

Figure 26.11 The graph shows the increased levels of carbon dioxide in the atmosphere since 1960, based on data gathered in Mauna Loa, in Hawaii.
Explain *What are some possible contributing factors to this rise in carbon dioxide levels? What impacts might this have on the environment?*

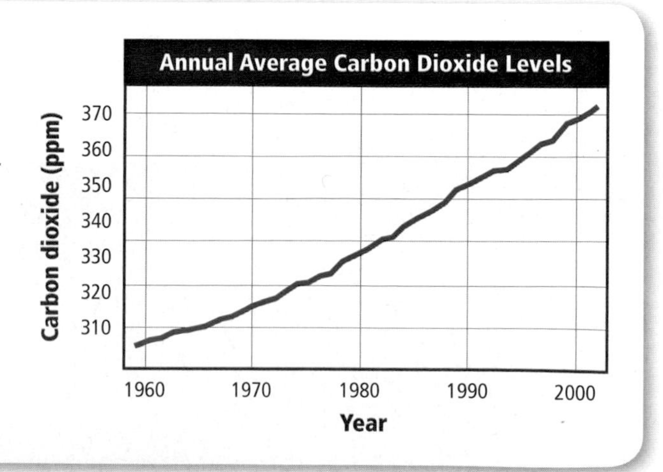

Studies indicate that Earth's mean surface temperature has risen about 0.5°C in the last century. Some scientists predict that if concentrations of carbon dioxide and other greenhouse gases continue to increase, average global temperatures could rise between 1 and 3.5°C in the next 100 years. Other scientists, however, assert that humans have not kept weather records long enough to tell to what extent the present rate of global warming is an artificial or a natural phenomenon. They argue that the increase in Earth's temperature could be part of a natural pattern of climatic change.

Photochemical smog On sunny days, you might notice a yellow-brown haze near densely populated areas. This haze is a type of air pollution, called **photochemical smog** that forms mainly from automobile exhaust in the presence of sunlight. **Figure 26.12** shows how air pollutants from car exhaust form ground-level ozone. Recall from Chapter 11 that in the upper atmosphere, solar radiation converts oxygen gas into ozone. Ozone in the upper atmosphere is beneficial because it absorbs and filters out harmful ultraviolet (UV) radiation. However, ground-level ozone can irritate the eyes, noses, throats, and lungs of humans and other animals. It also has harmful effects on plants. When smog occurs in a city, the air becomes harmful to breathe, especially for those who already have some difficulty breathing.

Air pollution also occurs in the form of particulate matter. The solid particles of materials such as ash, dust, and pollen range in size from microscopic bits to large grains. When humans breathe in particulates, they can lodge in lung tissues and cause breathing difficulties and lung disease.

VOCABULARY

ACADEMIC VOCABULARY

Particulate
of or relating to minute separate particles
People with asthma might be more sensitive when there is a high level of particulate matter in the air.

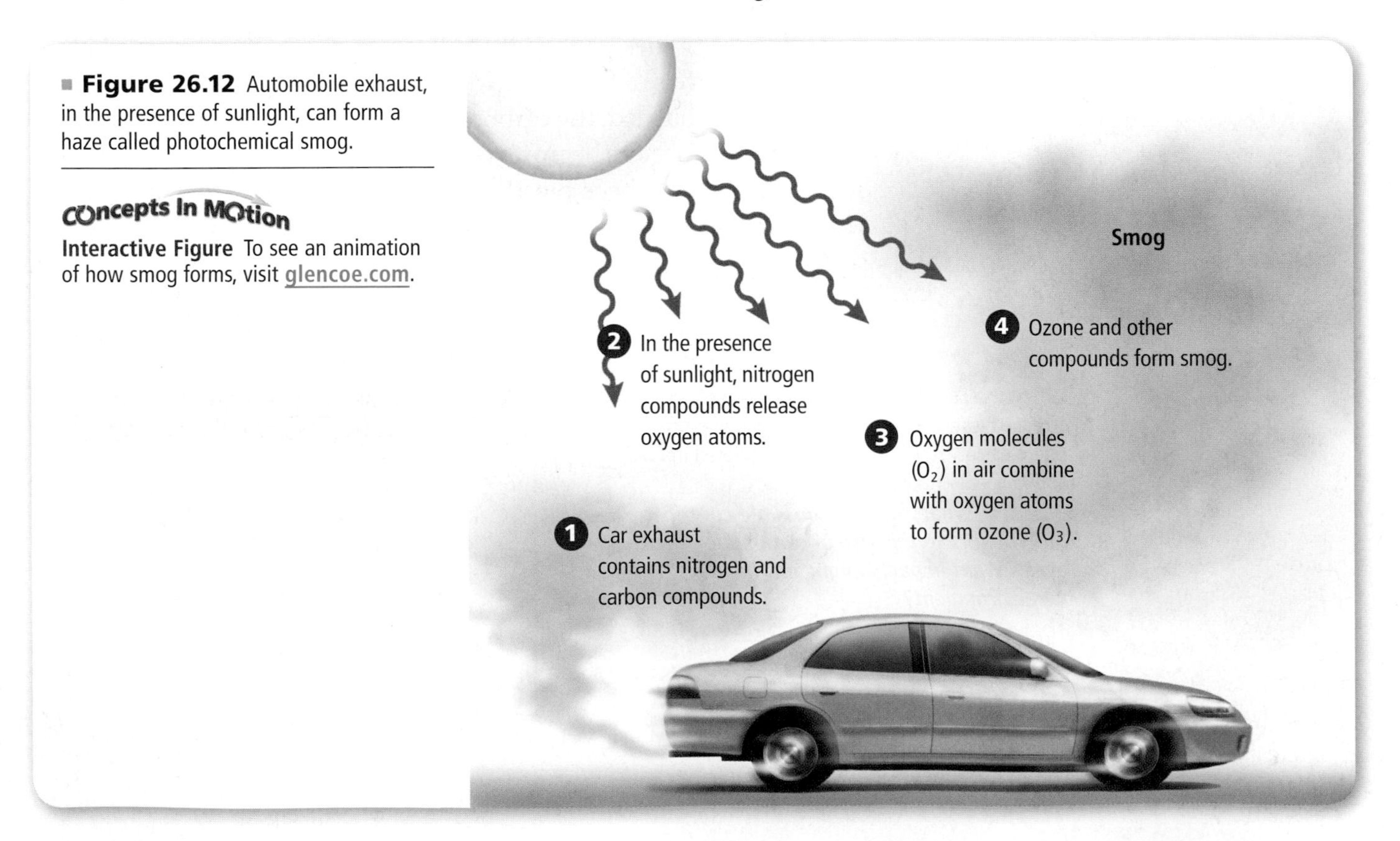

Figure 26.12 Automobile exhaust, in the presence of sunlight, can form a haze called photochemical smog.

Concepts In Motion
Interactive Figure To see an animation of how smog forms, visit glencoe.com.

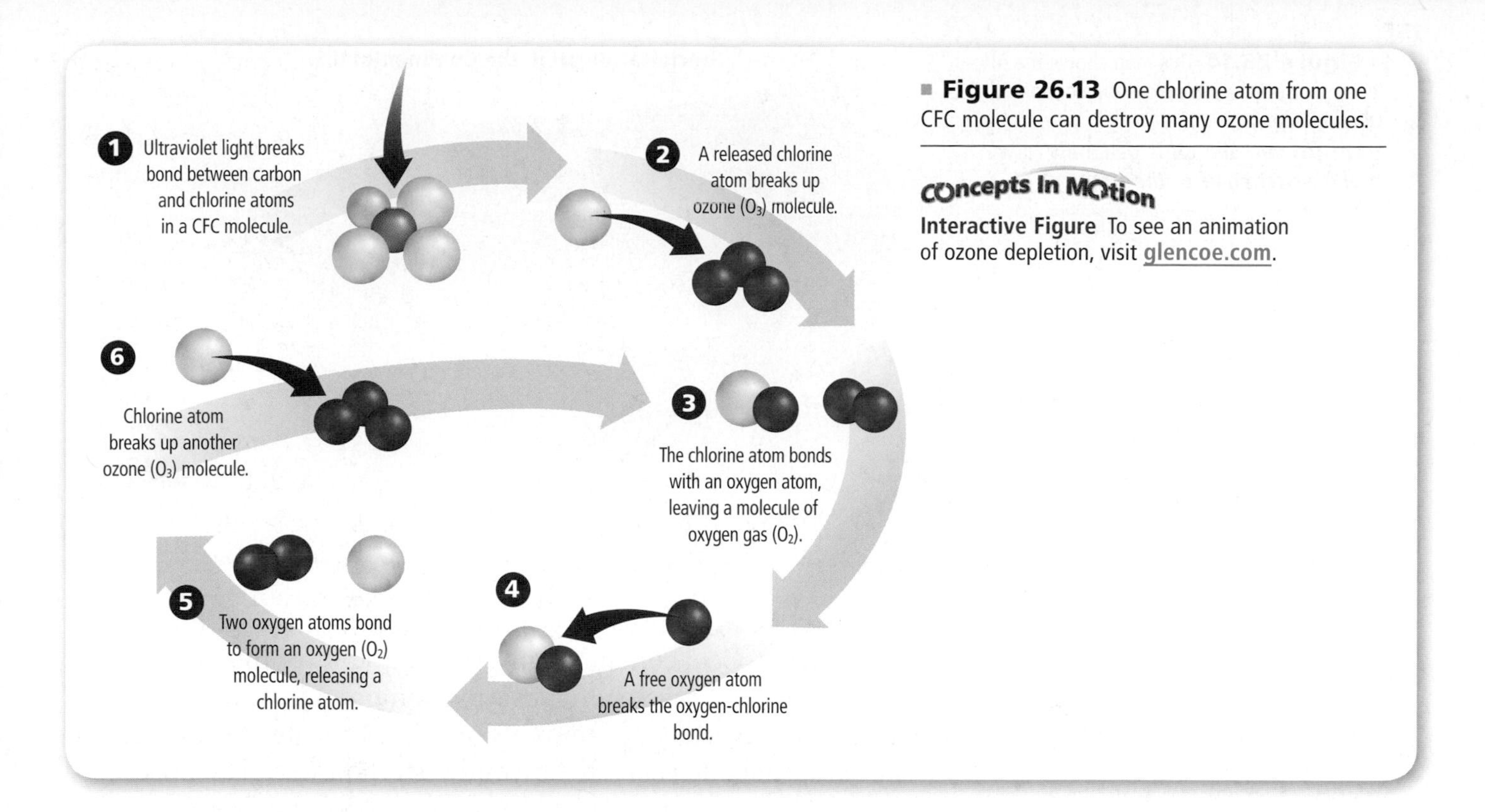

■ **Figure 26.13** One chlorine atom from one CFC molecule can destroy many ozone molecules.

Concepts In Motion

Interactive Figure To see an animation of ozone depletion, visit glencoe.com.

Ozone depletion Recall from Chapter 11 that the ozone layer in the stratosphere serves as a protective shield as it absorbs and filters out harmful UV radiation. UV radiation has been linked to eye damage, skin cancer, and reduced crop yields.

In the early 1970s, scientists suggested that chlorofluorocarbons (CFCs) could destroy ozone in the upper atmosphere. All of the CFCs present in the atmosphere are a result of human activity. CFCs are released from cleaning agents, old refrigerators that are not disposed of properly, and propellants in aerosol cans.

Although CFCs are stable and harmless near Earth's surface, they destroy ozone molecules, as shown in **Figure 26.13,** when they migrate into the upper atmosphere. Since the mid-1980s, atmospheric studies have detected a thinning of the ozone layer, including an extremely thin area over Antarctica. This hole was publicized in the news media as an **ozone hole,** which is a seasonal decrease in ozone over Earth's polar regions.

Acid precipitation Another major air pollution problem is acid precipitation, which is defined as precipitation with a pH of less than 5.0. Recall from Chapter 3 that pH is a measure of the acidity of a substance on a scale of 0 to 14, with 7 being neutral.

Natural precipitation has a pH of about 5.0 to 5.6, which is slightly acidic. **Acid precipitation** forms when sulfur dioxide and nitrogen oxides combine with atmospheric moisture to create sulfuric acid and nitric acid. Acid precipitation includes acidic rain, snow, fog, mist, and gas. Although volcanoes and marshes add sulfur gases to the atmosphere, 90 percent of the sulfur emissions in eastern North America are of human origin.

Figure 26.14 This map shows the pH levels of precipitation across the continental United States.

Explain *why the pH is generally lower in the eastern half of the country.*

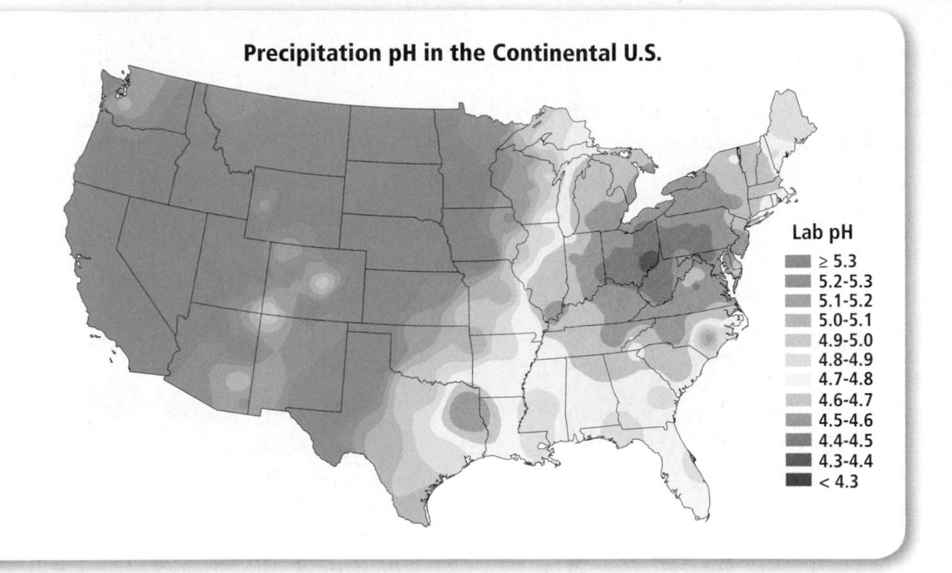

Burning coal One cause of acid precipitation is coal-burning power plants. Coal contains significant amounts of the mineral pyrite (FeS_2) and other sulfur-bearing compounds. When sulfur-rich coal is burned, large amounts of sulfur dioxide are released. The sulfur dioxide generated by midwestern power plants rises high into the air and is carried by winds toward the East Coast, where they mix with precipitation and fall to the ground. The distribution of acid precipitation is shown in **Figure 26.14.**

Effects of acid precipitation When acid rain enters surface waters, it damages aquatic ecosystems and vegetation and affects plants and soil. Trees affected by acid precipitation might not be killed outright, but might become more susceptible to damage and disease. Acid precipitation also depletes the soil of nutrients and damages buildings and statues by accelerating weathering.

DATA ANALYSIS LAB

Based on Real Data*

Interpret the Data

Are you breathing cleaner air? This table lists changes in emissions in the United State since the Clean Air Act of 1972.

Think Critically

1. **Graph** the data from the table. Put years on the *x*-axis and the pollutant emissions per year on the *y*-axis. Use different colors for each pollutant.
2. **Infer** why emissions of lead have declined so drastically since 1970.
3. **Evaluate** Could you estimate the reductions for 2005 by looking at the graph? Explain why or why not.

Pollutant (millions of tons)	1970	1980	1990	2000	2004
Particulate matter <10 microns	12.2	6.2	3.2	2.3	2.5
Sulfur dioxide	31.2	25.9	23.1	16.3	15.2
Nitrogen oxides	26.9	27.1	25.2	22.3	18.8
Volatile organic compounds	33.7	30.1	23.1	16.9	15.0
Carbon monoxide	197.3	177.8	143.6	102.4	87.2
Lead	0.221	0.074	0.005	0.003	0.003

*Data obtained from: Air Emission Trends—Continued Progress Through 2004. *U.S. Environmental Protection Agency.*

Reducing Air Pollution

Air pollution is difficult to control because it travels through the air to neighboring regions. Solving air pollution problems requires the cooperation of both state and national governments. In the last decade, the governments of many nations have met in an attempt to reduce global air pollution, especially that which is caused by carbon dioxide and CFCs. Since 1970, the United States Congress passed clean air laws which set specific reduction goals and enforcement policies for many types of air pollution. Because of this and other regulations, there have been significant reductions in air pollutants in the United States since 1970.

Controlling the source Many coal-burning power plants have installed devices to reduce emissions of particulate matter and sulfur dioxide, such as the one shown in **Figure 26.15.** In North America and western Europe, the use of low-sulfur coal and natural gas has helped to reduce such emissions. However, scientists agree that the most effective way to reduce air pollution is to remove older, highly polluting vehicles from roadways. It is estimated that just 10 percent of the motor vehicles in operation produce 50 to 60 percent of the air pollution generated by gasoline-powered engines. Switching to newer cars with more efficient engines could significantly reduce air pollution throughout the world.

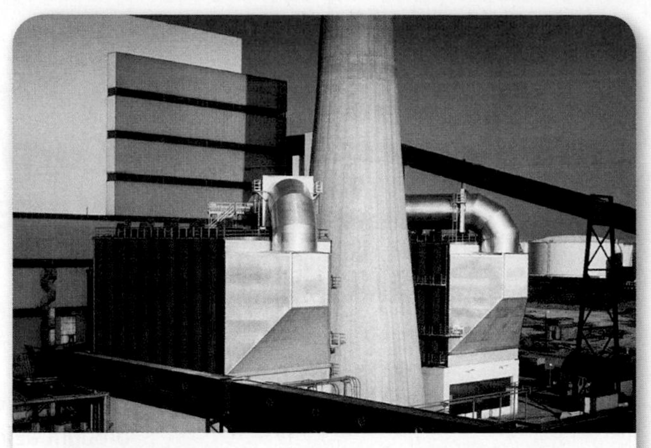

Figure 26.15 Scrubbers are often required to clean out the smoke stacks on coal plants. Scrubbers help remove gases and particulate matter before they enter the air.

Section 26.3 Assessment

Section Summary

- Many human activities create air pollution. Air pollution can cause human health problems.
- CFCs are a major cause of ozone depletion.
- Clean air laws and cleaner automobiles have resulted in a decrease in air pollution emissions since 1970.

Understand Main Ideas

1. MAIN Idea **Name** two forms of pollutants found in air. What are some of the natural and human sources of these pollutants?
2. **Relate** global warming and the greenhouse effect.
3. **Describe** how CFCs cause ozone depletion.
4. **List** some of the causes and effects of acid precipitation on ecosystems.

Think Critically

5. **Predict** The atmosphere of Venus is 90 percent carbon dioxide. Based on this information, what could you infer about the average surface temperature of Venus? Explain your answer.

MATH in Earth Science

6. If carbon monoxide emissions were reduced from 102 to 87 million metric tons in one year, what would be the percent decrease?

Section 26.4

Objectives

- **Identify** ways to conserve water.
- **Summarize** the types and sources of water pollution.
- **Describe** some methods of controlling water pollution.

Review Vocabulary

runoff: water that flows downslope on Earth's surface and can enter a stream, river, or lake

New Vocabulary

point source
nonpoint source

Human Impact on Water Resources

MAIN Idea Pollution controls and conservation protect water resources.

Real-World Reading Link Imagine camping at a remote location. You have brought along the supply of water. What would happen if extra friends joined the group or half the water supply were spilled?

Use of Water Resources

Humans depend on water in many ways. Most people use freshwater in their homes for bathing, drinking, cooking, and washing. The irrigation of crops also requires water. Because water supplies are not distributed evenly on Earth, some areas have less water than is needed.

Water conservation Is there a leaky faucet in your home? In the United States alone, 20 to 35 percent of the water taken from public water supplies is lost through leaky toilets, bathtubs, and faucets.

When there is not enough water to go around, people have two choices: decrease demand or develop new supplies. When new supplies are not readily available or are too expensive to develop, water conservation can help. Because large amounts of water are used for crops, efficient irrigation practices can greatly reduce water usage. Monitoring soil moisture to irrigate only when the soil is dry, using equipment that places water near plant roots to reduce evaporation as shown in **Figure 26.16,** and raising water prices have all been effective in minimizing the amount of irrigation water. Industries can also conserve water by recycling cooling water and wastewater, or by using conservation practices.

■ **Figure 26.16** Farmers develop methods of water conservation such as the drip irrigation system shown above. In a drip irrigation system, the water is released slowly so more is absorbed into the soil and less is lost to runoff and evaporation.

Water Pollution

Pollution is another area in which humans have an impact on water supplies. Some supplies of water have been polluted by human activities and are no longer usable. Water-pollution sources are grouped into two main types. **Figure 26.17** shows that **point sources** originate from a single point of origin, such as a sewage-treatment plant or an industrial site, while **nonpoint sources** generate pollution from widespread areas.

Most water used for domestic purposes, including showering, laundry, cooking, and using the bathroom, is treated at a sewage treatment facility. Treated sewage is then released through a point source to a receiving stream. In the past, treated sewage still contained contaminants. Fortunately, methods to treat sewage have greatly improved. Point sources also include wastes that enter streams from illegal dumping, accidental spills, and industries that use water in manufacturing processes and discharge waste into streams and rivers.

Precipitation can absorb air pollutants and deposit them far from their source. Runoff can wash pesticides and fertilizers into streams as it flows over farms or lawns. It can also wash oil, gasoline, and other chemicals from roads and parking lots. Each of these is an example of nonpoint-source pollution.

Reading Check Compare point-source and nonpoint-source pollution.

Point source

Nonpoint source

■ **Figure 26.17** Point-source pollution comes from a single source, while nonpoint-source pollution is generated from a widespread area.

Pollution of groundwater Leaking chemical-storage barrels, underground gasoline-storage tanks, landfills, road salts, nitrates from fertilizers, sewage from septic systems, and other pollutants can seep into the ground and pollute underground water supplies. Polluted groundwater might find its way into the drinking-water supplies of people who rely on wells. Once groundwater is contaminated, the pollutants can be difficult to remove.

Pollution in the oceans Although human activities have the greatest impact on freshwater supplies, pollution of ocean waters is also a concern. Nearly 50 percent of the U.S. population lives near coastlines. Pollutants from such cities often end up in estuaries and other nearshore regions. Pollution of nearshore zones can affect organisms because many depend on estuaries for breeding and raising young.

Another common ocean pollutant is mercury. Mercury released into the air and water from burning coal and manufacturing is ingested by fish. The fish are then eaten by larger predators and the mercury is passed along the food chain. Mercury has been detected in bears that do not live near polluted waters because they have eaten salmon that migrate from the oceans.

CAREERS IN EARTH SCIENCE

Hydrologist An Earth scientist who studies the distribution, circulation, and physical properties of underground and surface waters is a hydrologist. They sometimes work in offices, helping companies comply with environmental regulations. To learn more about Earth science careers, visit glencoe.com.

To read about how frogs indicate environmental health, go to the **National Geographic Expedition** on page 928.

Reducing Water Pollution

In recent decades, many steps have been taken to prevent and reduce water pollution as people have found that it is much cheaper and more efficient to prevent pollution than it is to clean it up later. Two major laws have been passed in the United States to combat water pollution: the Safe Drinking Water Act and the Clean Water Act.

The Safe Drinking Water Act In 1974, the Safe Drinking Water Act was passed. This act was designed to ensure that everyone in the United States has access to safe drinking water. Progress is being made, but many water supplies still do not consistently meet the standards. In 1998, 20 percent of public water supplies were in violation of the act at least once in a one-year period. The goal of the Safe Drinking Water Act is to reduce this number to less than 5 percent by the year 2008.

The Clean Water Act The primary federal law that protects U.S. waters is the Clean Water Act of 1972. The act was amended in 1977, 1981, and again in 1987. The two main goals of the Clean Water Act are to eliminate discharge of pollutants into rivers, streams, lakes, and wetlands, and to restore water quality to levels that allow for recreational uses of waters, including fishing and swimming.

Is the Clean Water Act working? Since 1972, the number of people served by sewage-treatment plants has increased from 85 million to 190 million. During that same time period, the annual rate of wetland losses has decreased from 146,000 ha/y to about 32,000 ha/y. Two-thirds of the nation's waters are now safe for swimming and fishing, compared to only one-third in 1972. Continued monitoring and improvement are still necessary. In 2000, the EPA reported that 39 percent of the nation's rivers that were tested were polluted and 45 percent of lakes tested were polluted.

Section 26.4 Assessment

Section Summary

- Humans use water to irrigate crops, for industry, cooking, bathing, and drinking.
- Conserving water can stretch limited supplies.
- Water can be polluted from point sources and nonpoint sources. Groundwater and oceans can also become polluted.
- The United States has passed laws to limit water pollution.

Understand Main Ideas

1. MAIN Idea **Identify** ways surface waters can be polluted.
2. **Determine** how residents of a city might reduce water consumption.
3. **Analyze** What are some of the positive impacts of the Clean Water Act?
4. **Predict** What are some ways to minimize the need for irrigation?

Think Critically

5. **Infer** which type of pollution is easier to eliminate: point sources or nonpoint sources? Give an example of each type and explain how it might be controlled.

WRITING in Earth Science

6. Write your own Clean Water Act. What regulations would you place on businesses or homes? How quickly would you expect change?

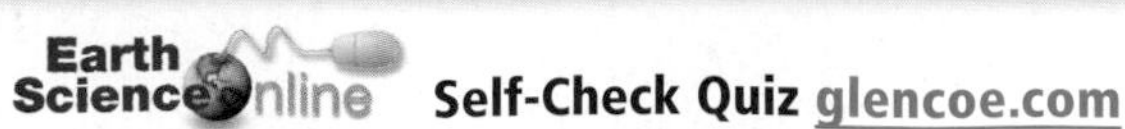

EARTH SCIENCE AND TECHNOLOGY

MEASURING AND MODELING GLOBAL WARMING

You have seen movies and TV shows that talk about the catastrophic future that awaits Earth as a result of global warming. But what is causing global warming?

Global warming At any time, Earth's surface temperature varies greatly from place to place. One way to specify Earth's temperature is to calculate average temperatures over Earth's surface over a specific time period. Temperature data show that over the past 100 years, the average annual global temperature has increased by about 0.6°C. This increase in average global temperature is called global warming.

The data The temperature data used to calculate average global temperatures come from thermometer measurements on land and at sea. Other types of data are used to reconstruct historical climate records. These data are obtained from ice cores, sediment cores from the bottoms of lakes and oceans, tree rings, and corals.

Possible causes Earth's surface temperature is maintained through a balancing act between the rate at which Earth absorbs energy and the rate at which Earth radiates energy. Increasing the concentration of greenhouse gases could increase Earth's temperature by increasing the amount of energy that Earth absorbs. Data show that since the middle of the nineteenth century, the concentration of atmospheric carbon dioxide has risen by about 100 ppm (about 35 percent) around the same period as humans began burning fossil fuels. Carbon dioxide and average global temperature have increased over approximately the same time period.

However, other factors such as solar variations and volcanic eruptions can also cause global temperatures to change. Is global warming due to human-caused increases in greenhouse gases, natural causes, or both?

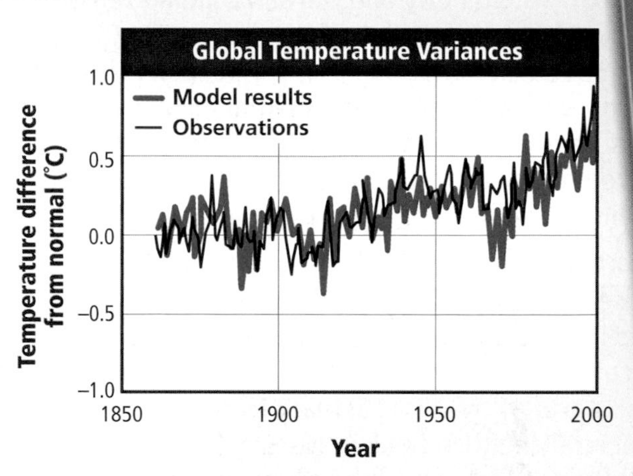

Calculated (black line) and measured (blue line) global average temperatures agree best when both natural effects and greenhouse gas increases are included in the computer climate model calculations.

Computer climate models The effects of possible causes on average global temperatures can be estimated using computer climate models. These programs calculate how changes in greenhouse gases, solar radiation, and other factors change surface temperatures over Earth's surface.

What gives the best agreement? Computer climate models calculate the average global annual temperature that would result from various changes. The best agreement with the temperature data comes when greenhouse gas changes and natural causes together are used to calculate average global temperatures. This indicates that natural changes and greenhouse gas increases together probably have caused the temperature changes currently observed.

WRITING in Earth Science

Debate Visit glencoe.com to learn more about global warming data and how the data are used. Prepare for a class debate about international policies that are based on global warming data. For information about conducting a debate, see the *Skillbuilder Handbook*.

GEOLAB

MAPPING: PINPOINT A SOURCE OF POLLUTION

Background: Iris City and the surrounding region are shown in the map on the facing page. Iris City is a medium-sized city of about 100,000 people. It is experiencing many types of environmental impacts. Iris City obtains its drinking water from Opal Lake. Studies of the lake have detected increased levels of nitrogen, phosphorus, hydrocarbons, sewage, and silt.

The northwest end of Opal Lake is experiencing increased development while the remainder of the watershed is a combination of forest and logging clear-cuts. Last spring, blooms of cyanobacteria choked parts of the Vista Estuary Nature Preserve. Commercial shellfish beds in Iris Bay have been closed because of sewage contamination.

A natural-gas power plant has been proposed for location A, near the Vista Cutoff, an abandoned channel of the Vista River. The plant would provide jobs as well as generate electricity. The company plans to divert 25 percent of the Vista River through the Vista Cutoff.

The Lucky Mine was abandoned 60 years ago. A mining company has applied for permits to reopen the mine. An estimated 1 million grams of gold can be recovered using modern techniques.

You will work with a small group of students to make recommendations to the residents of Iris City. Included in your recommendations should be: possible pollution sources for Opal Lake, possible causes of the cyanobacteria bloom, recommendations for the development of the natural-gas power plant, and the opening of Lucky Mine.

Question: *How can the residents of Iris City best manage their water supply?*

Materials
metric ruler
science notebook

Procedure
1. Read and complete the lab safety form.
2. Working in small groups, brainstorm possible sources of pollution in Opal Lake and Iris Bay.
3. Discuss what steps the residents of Iris City might take to protect their drinking water.
4. Research common causes of cyanobacteria blooms. Discuss what might be causing the bloom in the Vista Estuary Nature Preserve.
5. Discuss the positive and negative aspects of diverting water from the Vista River through the Vista Cutoff. What are other possible impacts (both positive and negative) in the development of the natural-gas power plant?
6. Discuss the possibility of reopening Lucky Mine. If it is reopened, brainstorm ways the mining company might minimize negative environmental impacts.
7. Prepare to present your recommendations to the class.

Analyze and Conclude
1. **Identify** What did your group list as possible sources of pollution for Opal Lake and for Iris Bay? Are these point sources or nonpoint sources of pollution?
2. **Experiment** How would you determine if you were correct about the source of the cyanobacteria bloom in Vista Estuary? What tests might you run to confirm your predictions?
3. **Research** What information did you discover in your research on cyanobacteria that applied to Iris City? List all the possible sources your group identified.
4. **Research** You have identified the possible sources of pollution for Opal Lake. Reports indicate increased levels in nitrogen, phosphorus, hydrocarbons, sewage, and silt. Research how this might affect the health of the residents that obtain their water from the lake.

WRITING in Earth Science
Write a report for Iris City, detailing all your group's findings and recommendations. For more information on water quality, visit glencoe.com.

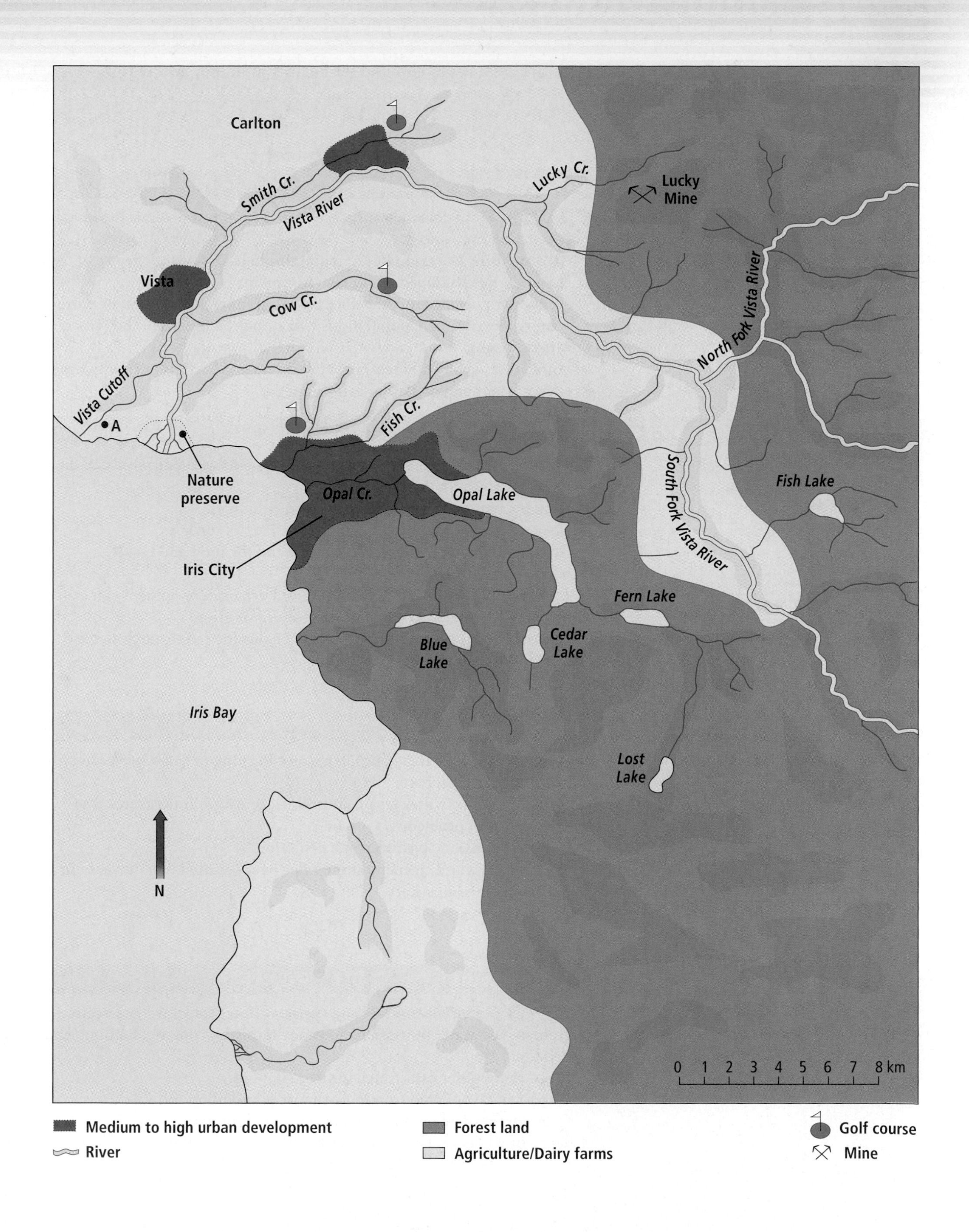

Carlton
Smith Cr.
Vista River
Lucky Cr.
Lucky Mine
Vista
Cow Cr.
North Fork Vista River
Vista Cutoff
A
Fish Cr.
Nature preserve
Opal Cr.
Opal Lake
Fish Lake
South Fork Vista River
Iris City
Fern Lake
Blue Lake
Cedar Lake
Iris Bay
Lost Lake
N
0 1 2 3 4 5 6 7 8 km
Medium to high urban development
River
Forest land
Agriculture/Dairy farms
Golf course
Mine

CHAPTER 26

Study Guide

BIG Idea The use of natural resources can impact Earth's land, air, and water.

Vocabulary	Key Concepts
Section 26.1 Populations and the Use of Natural Resources	
• carrying capacity (p. 735) • density-dependent factor (p. 736) • density-independent factor (p. 736) • exponential growth (p. 735)	**MAIN Idea** More demands are placed on natural resources as the human population increases. • All organisms use resources to maintain their existence. The use of these resources has an impact on the environment. • As populations increase, the demand for resources increases. Because resources are limited, populations will stop growing when they reach carrying capacity. • Populations grow exponentially at early stages. Earth is currently experiencing a human population explosion.
Section 26.2 Human Impact on Land Resources	
• bioremediation (p. 742) • deforestation (p. 739) • pesticide (p. 740) • reclamation (p. 738)	**MAIN Idea** Extraction of minerals, farming, and waste disposal can have negative environmental impacts. • Humans require large amounts of land resources. • The extraction of resources can disrupt Earth's surface. • Growing populations increase the demand for food and result in increased urban development. • Agriculture, poor forestry practices, and urban development can cause habitat loss, increased erosion, and water pollution. • Human impact on land resources can be minimized through the use of modern techniques.
Section 26.3 Human Impact on Air Resources	
• acid precipitation (p. 745) • ozone hole (p. 745) • photochemical smog (p. 744)	**MAIN Idea** Manufacturing processes and burning of fossil fuels can pollute Earth's atmosphere. • Many human activities create air pollution. Air pollution can cause human health problems. • CFCs are a major cause of ozone depletion. • Clean air laws and cleaner automobiles have resulted in a decrease in air pollution emissions since 1970.
Section 26.4 Human Impact on Water Resources	
• nonpoint source (p. 749) • point source (p. 749)	**MAIN Idea** Pollution controls and conservation protect water resources. • Humans use water to irrigate crops, for industry, cooking, bathing, and drinking. • Conserving water can stretch limited supplies. • Water can be polluted from point sources and nonpoint sources. Groundwater and oceans can also become polluted. • The United States has passed laws to limit water pollution.

CHAPTER 26 Assessment

Vocabulary Review

Complete the sentences below using the vocabulary terms from the Study Guide.

1. A type of air pollution that forms from car exhaust in the presence of sunlight is called ________.

2. The ________ is the total number of organisms that an environment can support.

3. The use of organisms to clean up toxic waste is called ________.

4. The removal of trees from a forested area without adequate replanting is called________.

5. Restoring land that had been previously mined to its original contours is called ______.

Match each description below with the correct vocabulary term from the Study Guide.

6. a pattern of growth in which a population grows faster as it increases in size

7. chemicals applied to plants to kill insects, fungi, and weeds

8. water pollution that comes from widely spread areas

9. a factor, such as lack of food, that affects a population

Each of the following sentences is false. Make each sentence true by replacing the italicized word with terms from the Study Guide.

10. A seasonal change that appears over Earth's polar regions is called *acid precipitation.*

11. Any environmental factor that does not depend on the number of members in a population, such as storms, droughts, floods, fires, and pollution, is a *density-dependent factor.*

12. The *ozone hole* forms when sulfur dioxide and nitrogen oxides combine with atmospheric moisture to create sulfuric acid and nitric acid.

Understand Key Concepts

13. Which diagram represents mouse population growth?

A.

C.

B.

D.

14. Which environmental impact might result from deforestation?
A. erosion of topsoil
B. increased photochemical smog
C. a decrease in the size of the ozone hole
D. bioremediation of toxic waste

15. Which is a problem associated with the expansion of highly populated areas?
A. the loss of natural resources
B. the expense of building
C. the concentration of resources
D. the availability of cultural resources

16. Which federal law protects the U.S. water supply?
A. the Clean Water Act
B. the Clean Ocean Act
C. the Clean Air Act
D. the Clean Hydrosphere Act

17. Which is an example of point-source pollution?
A. acid rain
B. runoff from a parking lot
C. sewage outfall
D. topsoil from a farm field

18. What is the major source of photochemical smog in the United States?
A. point sources
B. car exhaust
C. power plants
D. acid precipitation

Use the figure below to answer Question 19.

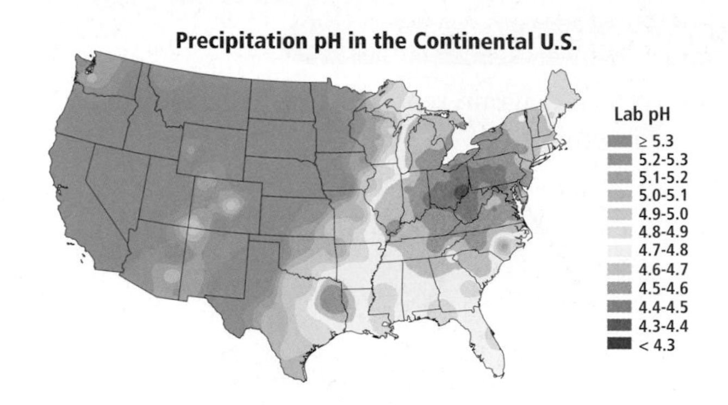

19. In which part of the United States is precipitation the most acidic?
- **A.** southwest
- **B.** northwest
- **C.** southeast
- **D.** northeast

20. Acid precipitation comes mainly from which source?
- **A.** coal-fired power plants
- **B.** particulate pollution
- **C.** CFCs
- **D.** carbon dioxide

21. What was the Safe Drinking Water Act designed to do?
- **A.** clean United States' drinking water
- **B.** ensure access to safe drinking water
- **C.** increase the amount of drinking water available in the United States
- **D.** decrease the cost of drinking water in the United States

22. Which is most likely to cause groundwater pollution?
- **A.** leaking underground storage tank
- **B.** acid precipitation
- **C.** deforestation
- **D.** particulate matter

23. Which is a common ocean pollutant?
- **A.** carbon
- **B.** ozone
- **C.** neon
- **D.** mercury

Constructed Response

Use the figure below to answer Question 24.

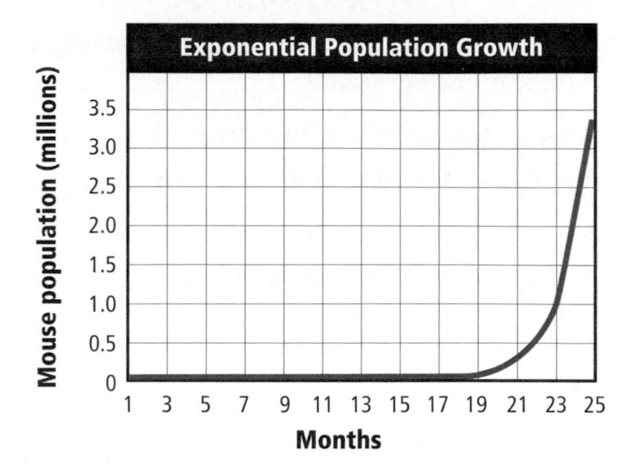

24. Determine What would the graph look like if you extended it 10 more years?

25. Describe two effects acid rain might have on an ecosystem.

26. Illustrate Draw a line graph that represents a warming rate increase of 3°C every 100 years. What is assumed regarding current Earth conditions and the availability of natural resources over this period of time?

27. Predict what conditions might exist on Earth if global warming continues to occur.

28. Analyze What are some ways to reduce photochemical smog?

29. Compare the possible causes of ozone depletion and global warming.

30. Analyze Why might ground-level ozone be worse on a sunny weekday than on a sunny weekend?

31. Interpret If mining for resources can have negative environmental impacts, why do humans still continue this practice?

32. Analyze Imagine you have a farm where you grow crops, including corn and alfalfa. Most of your fields are located on moderate-to-steep slopes. What farming methods could you use to reduce erosion? Explain.

Think Critically

33. **Propose** a plan for a model city that addresses land use and pollution issues.

34. **Distinguish** between point-source and nonpoint-source pollution. Why is it important to understand how each impacts the environment?

35. **Propose** an alternative solution for disposal of our increasing amount of solid waste.

36. **Interpret** How can you prevent nonpoint-source pollution if there isn't one single place the pollution is generated?

37. **Analyze** How have humans adapted to their environment? What lifestyle changes and adaptations do people make based on the environment in which they live?

38. **Propose** two ways that you could conserve water in your everyday life.

39. **Analyze** A population of birds inhabits an island. They feed on seeds and berries and live in nests in trees. One season, there is no crop of berries and the number of birds on the island decreases. The next season, there is a hurricane and many of the birds' nests are knocked out of the trees. Which of these factors is density-dependent and which is density-independent? Why?

Concept Mapping

40. Use these terms to make a concept map to organize the major ideas in Section 26.2: *erosion, topsoil loss, water pollution, waste rock,* and *mineral extraction.* Refer to the *Skillbuilder Handbook* for more information.

Challenge Question

41. **Apply** As undeveloped land is developed into cities and towns, many wild animals lose their habitat. In addition, animals can be cut off from their natural hunting and breeding ranges. For animals that normally have large ranging habitats, brainstorm ways humans can protect them from extinction and still develop new land.

Additional Assessment

42. *WRITING in* **Earth Science** Write a radio announcement as part of an Earth Day celebration designed to encourage the public to be mindful of the potential negative effects human activity can have on Earth. Try to incorporate a catchy jingle or phrase to linger with the listener.

DBQ Document–Based Questions

Data obtained from: Municipal Solid Waste Generation, Recycling, and Disposal in the United States: Facts and figures 2003. Environmental Protection Agency, Tables 13 and 19.

This bar graph is based on data from the Environmental Protection Agency showing commonly recycled materials and the percentage of the products that were recycled in 2003.

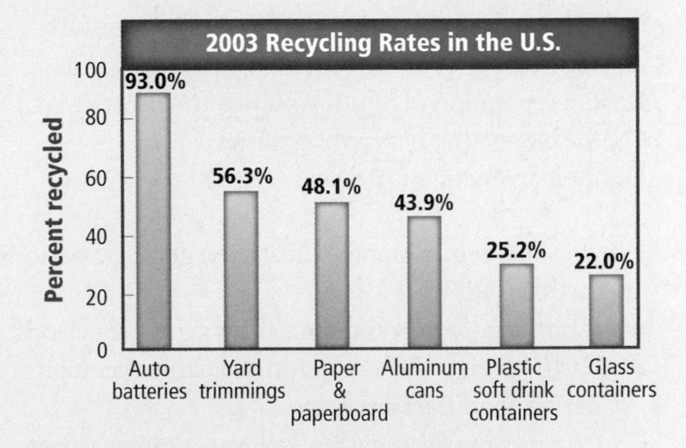

43. Which material has the highest recycle rate?

44. Which product has the lowest recycle rate?

45. Why do you think a higher percentage of aluminum cans are recycled than glass containers?

Cumulative Review

46. During what processes is latent heat stored and released? **(Chapter 11)**

47. What is the relationship between plate tectonics and the mantle underneath these plates? **(Chapter 17)**

48. How can energy from water be harnessed for human use? **(Chapter 25)**

Standardized Test Practice

Multiple Choice

1. Which type of coal burns the most efficiently?
   A. peat
   B. lignite
   C. bituminous coal
   D. anthracite

Use the illustrations below to answer Questions 2 and 3.

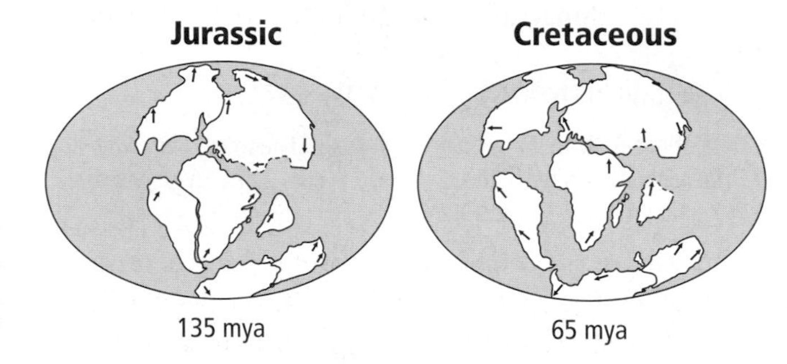

2. What was the most drastic move that occurred to land from the Jurassic Period to the Cretaceous Period?
   A. the nearing of South America and Antarctica
   B. the separation of South America from Africa
   C. the elongating of North America
   D. the narrowing of Africa

3. Which statement about the illustrated geologic periods on Earth is TRUE?
   A. It rains more now than in earlier geologic periods.
   B. Earth's oceans were saltier in the earlier geologic period than they are today.
   C. Changes in ocean currents played a major role in climate change.
   D. Earth's rivers were larger in the earlier periods than they are today.

4. Potassium-40 (K-40) decays into argon-40 (Ar-40) with a half-life of 1.3 billion years. A scientist studies a mineral sample and finds that the ratio of K-40 to Ar-40 is 1:3. How old is the rock?
   A. 0.6 billion years old
   B. 1.3 billion years old
   C. 2.6 billion years old
   D. 3.9 billion years old

5. Which is not a good way to conserve transportation energy?
   A. Drive at a lower speed.
   B. Make frequent stops.
   C. Work from home.
   D. Use a hybrid or electric car.

6. If fire destroys a wooded area, what limit to population growth has the area experienced?
   A. equilibrium
   B. a density-dependent factor
   C. a density-independent factor
   D. carrying capacity

Use the table below to answer Questions 7 and 8.

Amount of Land Created by Mt. Kilauea Eruptions	
Volcanic Activity Episode	**Net Amount of New Land Area (km2)**
1–48b	17.2
48	34.3
49	3.7
50	0.11
51–52	0.01
53	2.3
54	0.24
55	59.5

7. Which statement best describes the eruption of Mount Kilauea?
   A. The same amount of lava flows from Mount Kilauea during every period of activity.
   B. Mount Kilauea erupts on a regular basis and in a similar way each time.
   C. The eruptions of Mount Kilauea are alternately explosive and quiet.
   D. The amount of new land area produced by Mount Kilauea during a period of activity varies.

8. How might scientists be able to use this data?
   A. to predict the occurrence of another earthquake
   B. to monitor land growth due to eruptions
   C. to predict the intensity of the earthquake based on the land growth
   D. to predict the amount of new land that will form with the next eruption

9. Which is not considered a source of natural air pollution?
   A. volcanoes
   B. fires
   C. radon
   D. smog

Short Answer

Use the chart below to answer Questions 10 and 11.

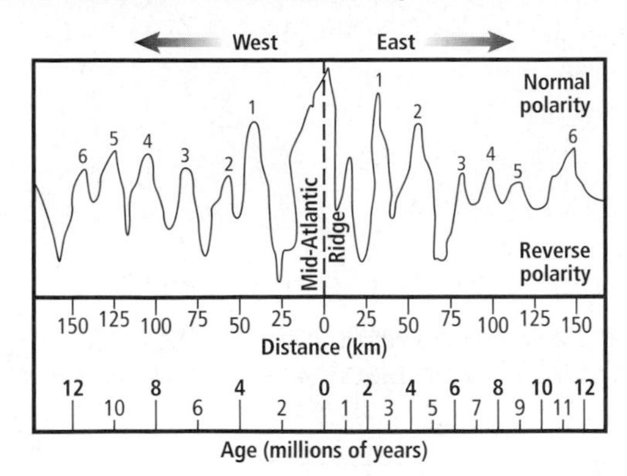

10. A group of scientists used a magnetometer and other equipment to obtain this magnetic field profile of part of the ocean floor. What information does the profile give?

11. What can the scientists conclude about how the ocean floor is formed near the Mid-Atlantic Ridge?

12. Explain why living things are considered to be renewable resources.

13. Why is photosynthesis important to humans?

14. Describe the process of differentiation in the formation of Earth's landmasses.

Reading for Comprehension

Acid Deposition

Acid deposition penetrates deeply into the fabric of an ecosystem, changing the chemistry of the soil and streams and narrowing the space where certain plants and animals can survive. Because there are so many changes, it takes many years for ecosystems to recover from acid deposition, even after emissions are reduced and the rain pH is restored to normal. However, there are some things that people can do to restore lakes and streams more quickly. Limestone or lime (a naturally occurring basic compound) can be added to acidic lakes to "cancel out" the acidity. Liming tends to be expensive, has to be done repeatedly to keep the water from returning to its acidic condition, and is considered a short-term remedy in only specific areas, rather than an effort to reduce or prevent pollution. Furthermore, it does not solve the broader problems of changes in soil chemistry and forest health in the watershed, and it does nothing to address visibility reductions, materials damage, and risk to human health.

Article obtained from: Acid rain. Reducing acid rain. *Environmental Protection Agency.* October 4, 2996. (Online resource accessed October 17, 2006.)

15. Which is not a limitation to using liming to improve lake acidity?
A. It does not improve soil chemistry.
B. It does not improve forest health.
C. It does not reduce risk to human health.
D. It does not act as a short-term remedy.

16. Why is adding limestone to an acidic lake beneficial?
A. It cancels out the acidity of the lake.
B. It restores naturally occurring bases in the water.
C. It can be done only once to fix the problem.
D. It allows for the return of wildlife.

17. Aside from liming's drawbacks, why is it still a good option for improving water condition?

NEED EXTRA HELP?														
If You Missed Question . . .	1	2	3	4	5	6	7	8	9	10	11	12	13	14
Review Section . . .	25.1	23.3	23.3	21.3	25.3	26.1	18.2	18.2	24.3	17.2	17.2	24.1	24.3	22.2

UNIT 8 Beyond Earth

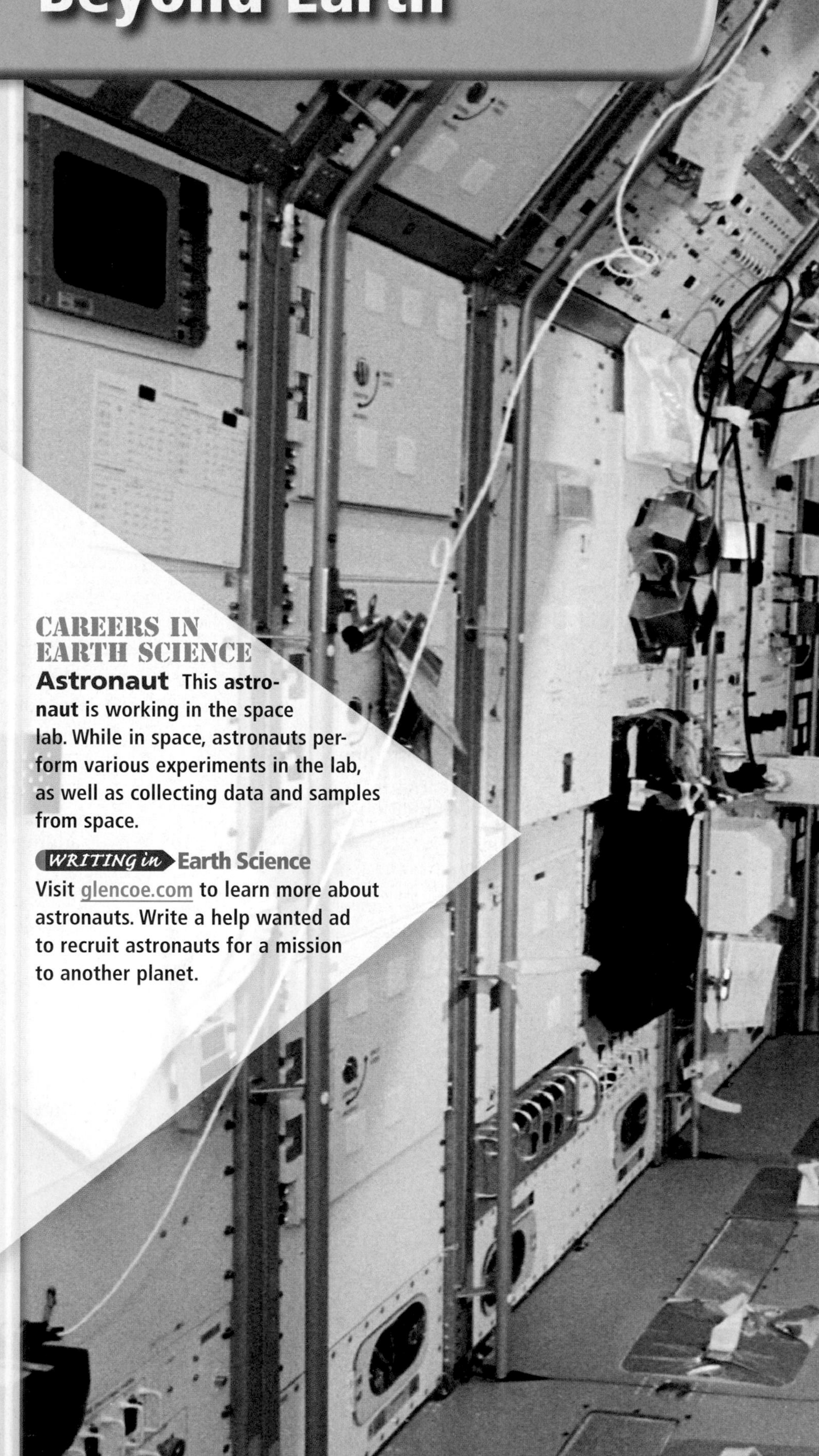

Chapter 27

The Sun-Earth-Moon System

BIG Idea The Sun, Earth, and the Moon form a dynamic system that influences all life on Earth.

Chapter 28

Our Solar System

BIG Idea Using the laws of motion and gravitation, astronomers can understand the orbits and the properties of the planets and other objects in the solar system.

Chapter 29

Stars

BIG Idea The life cycle of every star is determined by its mass, luminosity, magnitude, temperature, and composition.

Chapter 30

Galaxies and the Universe

BIG Idea Observations of galaxy expansion, cosmic background radiation, and the Big Bang theory describe an expanding universe that is 13.7 billion years old.

CAREERS IN EARTH SCIENCE

Astronaut This astronaut is working in the space lab. While in space, astronauts perform various experiments in the lab, as well as collecting data and samples from space.

WRITING in Earth Science

Visit glencoe.com to learn more about astronauts. Write a help wanted ad to recruit astronauts for a mission to another planet.

Earth Science Online

To learn more about astronauts, visit glencoe.com.

CHAPTER 27 The Sun-Earth-Moon System

BIG Idea **The Sun, Earth, and the Moon form a dynamic system that influences all life on Earth.**

27.1 Tools of Astronomy
MAIN Idea Radiation emitted or reflected by distant objects allows scientists to study the universe.

27.2 The Moon
MAIN Idea The Moon, Earth's nearest neighbor in space, is unique among the moons in our solar system.

27.3 The Sun-Earth-Moon System
MAIN Idea Motions of the Sun-Earth-Moon system define Earth's day, month, and year.

GeoFacts

- The volume of the Sun equals 1.3 million Earths.
- Earth is 5×10^6 km closer to the Sun in January than it is in July.
- Finding water on the Moon might make permanent lunar bases possible.

LAUNCH Lab

How can the Sun-Earth-Moon system be modeled?

The Sun is about 109 times larger in diameter than Earth, and Earth is about 3.7 times larger in diameter than the Moon. The distance between Earth and the Moon is 30 times Earth's diameter. The Sun is 390 times farther from Earth than is the Moon.

Procedure

1. Read and complete the lab safety form.
2. Calculate the diameters of Earth and the Sun using a scale in which the Moon's diameter is equal to 1 cm.
3. Using this scale, calculate the distances between Earth and the Moon and Earth and the Sun.
4. Cut out **paper** circles to represent your scaled Earth and Moon, and place them at the scaled distance apart.

Analysis

1. **Compare** the diameters of your cutout Earth and Moon to the distance between them.
2. **Infer** why your model does not have a scaled Sun placed at the scaled Sun distance.

Visit glencoe.com to

- study entire chapters online;
- explore Concepts In Motion animations:
  - Interactive Time Lines
  - Interactive Figures
  - Interactive Tables
- access Web Links for more information, projects, and activities;
- review content with the Interactive Tutor and take Self-Check Quizzes.

Phases of the Moon Make the following Foldable to help you learn about the major phases of the Moon.

STEP 1 Fold four sheets of paper in half from top to bottom.

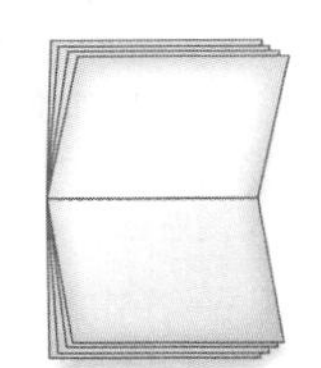

STEP 2 On two sheets of paper, make 3-cm cuts along the fold toward the center on each side.

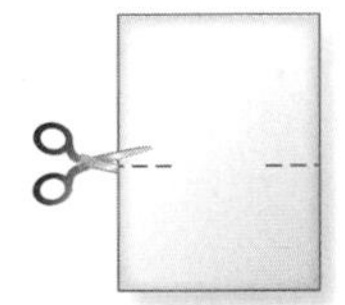

STEP 3 Cut a slit approximately 16 cm long along the fold line in the remaining two sheets of paper.

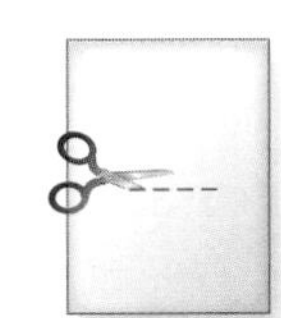

STEP 4 Slip the first two sheets through the slit in the second two sheets to make a 16-page booklet.

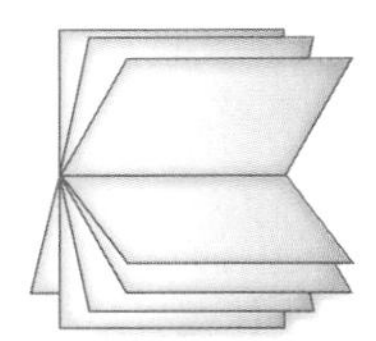

FOLDABLES Use this Foldable with Section 27.3. Draw each major phase of the Moon in order on the bottom pages of your Foldable. Indicate the positions of the Sun, the Moon, and Earth. Include a sketch of how the Moon appears from Earth during that phase. As you turn the pages of your completed book, you will see how the Moon appears to change shape and position. Take notes on the top pages.

Section 27.1

Objectives

- **Define** electromagnetic radiation.
- **Explain** how telescopes work.
- **Describe** how space exploration helps scientists learn about the universe.

Review Vocabulary

refraction: occurs when a light ray changes direction as it passes from one material into another

New Vocabulary

electromagnetic spectrum
refracting telescope
reflecting telescope
interferometry

Tools of Astronomy

MAIN Idea Radiation emitted or reflected by distant objects allows scientists to study the universe.

Real-World Reading Link Have you ever used a magnifying lens to read fine print? If so, you have used a tool that gathers and focuses light. Scientists use telescopes to gather and focus light from distant objects.

Radiation

The radiation from distant bodies throughout the universe that scientists study is called electromagnetic radiation. Electromagnetic radiation consists of electric and magnetic disturbances traveling through space as waves. Electromagnetic radiation includes visible light, infrared and ultraviolet radiation, radio waves, microwaves, X rays, and gamma rays.

You might be familiar with some forms of electromagnetic radiation. For example, overexposure to ultraviolet waves can cause sunburn, microwaves heat your food, and X rays help doctors diagnose and treat patients. All types of electromagnetic radiation, arranged according to wavelength and frequency, form the **electromagnetic spectrum,** shown in **Figure 27.1.**

Wavelength and frequency Electromagnetic radiation is classified by wavelength—the distance between peaks on a wave. Notice in **Figure 27.1** that red light has a longer wavelength than blue light, and radio waves have a much longer wavelength than gamma rays. Electromagnetic radiation is also classified according to frequency, the number of waves or oscillations that pass a given point per second. The visible light portion of the spectrum has frequencies ranging from red to violet, or 4.3×10^{14} to 7.5×10^{14} Hertz (Hz)—a unit equal to one cycle per second.

■ **Figure 27.1** The electromagnetic spectrum identifies the different radiation frequencies and wavelengths.

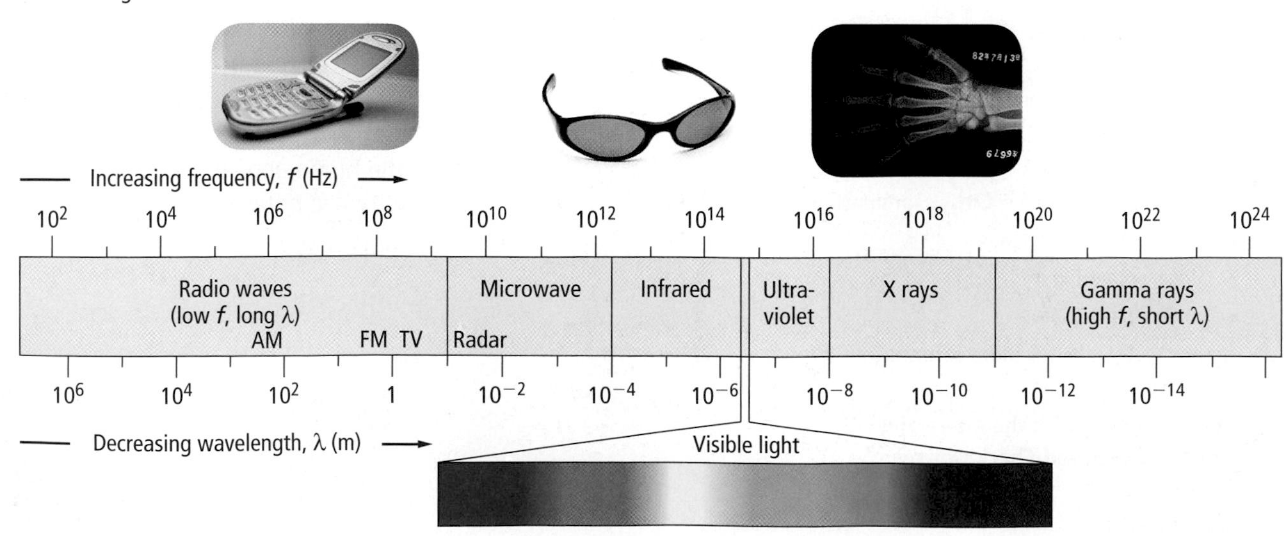

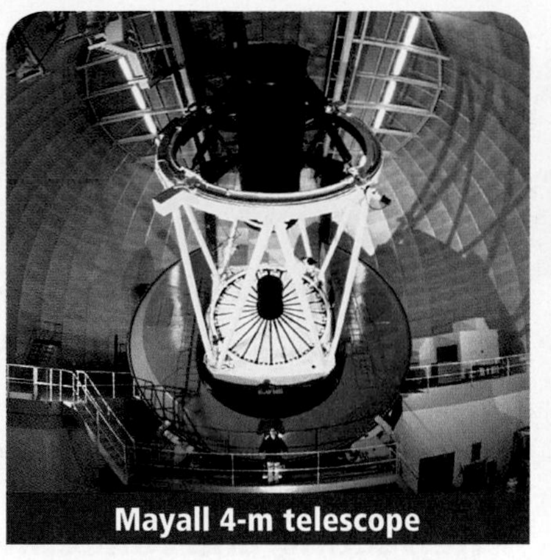

Mayall Observatory

Figure 27.2 This photo of the Pinwheel galaxy was taken by the Mayall 4-m telescope, shown with its observatory.

Frequency is related to wavelength by the mathematical relationship $c = \lambda f$, where c is the speed of light (3.0×10^8 m/s), λ is the wavelength, and f is the frequency. Note that all types of electromagnetic radiation travel at the speed of light in a vacuum. Astronomers choose their tools based on the type of radiation they wish to study. For example, to see stars forming in interstellar clouds, they use special telescopes that are sensitive to infrared wavelengths, and to view remnants of supernovas, they often use telescopes that are sensitive to UV, X-ray, and radio wavelengths.

Telescopes

Objects in space emit radiation in all portions of the electromagnetic spectrum. Telescopes, such as the one shown in **Figure 27.2,** give us the ability to observe wavelengths beyond what the human eye can detect. In addition, a telescope collects more electromagnetic radiation from distant objects and focuses it so that an image of the object can be recorded. The pupil of a typical human eye has a diameter of up to 7 mm when it is adapted to darkness; a telescope's opening, which is called its aperture, might be as large as 10 m in diameter. Larger apertures can collect more electromagnetic radiation, making dim objects in the sky appear much brighter.

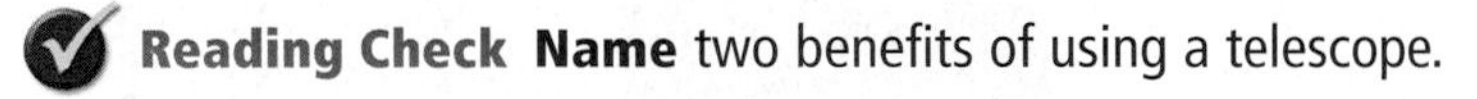
Reading Check **Name** two benefits of using a telescope.

Another way that telescopes surpass the human eye in collecting electromagnetic radiation is with the aid of cameras, or other imaging devices, to create time exposures. The human eye responds to visible light within one-tenth of a second, so objects too dim to be perceived in that time cannot be seen. Telescopes can collect light over periods of minutes or hours. In this way telescopes can detect objects that are too faint for the human eye to see. Also, astronomers can add specialized equipment. A photometer, for example, measures the intensity of visible light and a spectrophotometer displays the different wavelengths of radiation.

CAREERS IN EARTH SCIENCE

Space Engineer Space engineers design and monitor probes used to explore space. Engineers often design probes to collect information and samples from objects in the solar system. They also study the data collected. To learn more about Earth science careers, visit glencoe.com.

To read about new telescopes scientists are using to study space, go to the **National Geographic Expedition** on page 934.

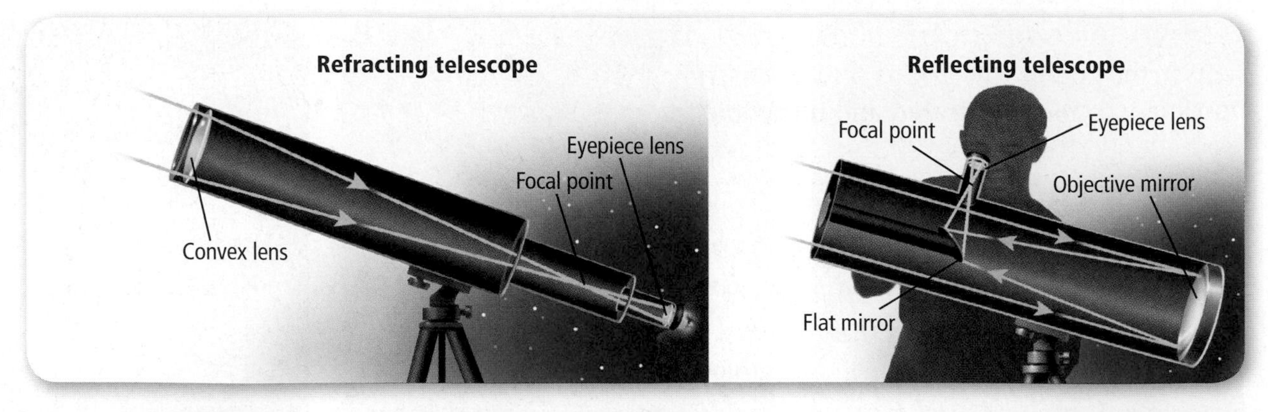

■ **Figure 27.3** Refracting telescopes use a lens to collect light. Reflecting telescopes use a mirror to collect light.

Refracting and reflecting telescopes Two different types of telescopes are used to focus visible light. The first telescopes, invented around 1600, used lenses to bring visible light to a focus and are called **refracting telescopes,** or refractors. The largest lens on such telescopes is called the objective lens. In 1668, a new telescope that used mirrors to focus light was designed. Telescopes that bring visible light to a focus with mirrors are called **reflecting telescopes,** or reflectors. **Figure 27.3** illustrates how simple refracting and reflecting telescopes work. Telescope technology has changed over time, as shown in **Figure 27.4.**

Although both refracting and reflecting telescopes are still in use today, most astronomers use reflectors because mirrors can be made larger than lenses and can collect more light.

Reading Check **Compare** refracting and reflecting telescopes.

Most telescopes used for scientific study are located in observatories far from city lights, usually at high elevations where there is less atmosphere overhead to blur images. Some of the best observatory sites in the world are located high atop mountains in the southwestern United States, along the peaks of the Andes mountain range in Chile, and on the summit of Mauna Kea, a volcano on the island of Hawaii.

■ **Figure 27.4**

Development of Astronomy

Humanity's curiosity about the night sky was limited to Earth-bound explorations until the first probe was sent into space in 1957.

38,000 B.C. — 5000 B.C. — 500 B.C. — A.D. 0 — 1000

38,000 B.C. Cro-Magnon people sketch moon phases on tools made out of bones.

4236 B.C. After lunar and solar calendars predict agricultural seasons, Egyptians adopt a 365-day calendar based on the movement of the star Sirius.

410 B.C. The first prophecies based on the positions of the five visible planets, the Moon, and the Sun were written for individuals in Mesopotamia.

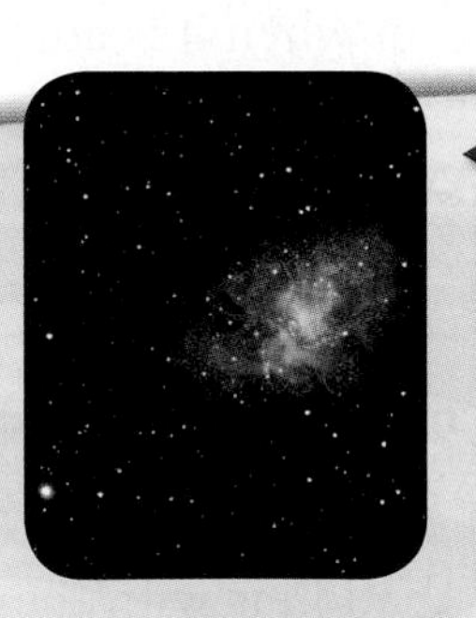

A.D. 900s Arab astronomers greatly improve the accuracy of the Greek astrolabe—a tool for celestial navigation that determines time and location.

1054 Chinese astronomers document the explosion of the supernova that creates the Crab nebula, believing it foretells the arrival of a wealthy visitor to the emperor.

Telescopes using non-visible wavelengths For all telescopes, the goal is to bring as much electromagnetic radiation as possible into focus. Infrared and ultraviolet radiation can be focused by mirrors in a way similar to that used for visible light. X rays cannot be focused by normal mirrors, and thus special designs must be used. Gamma rays cannot be focused, so telescopes designed to detect this type of radiation can determine only the direction from which the rays come.

A radio telescope collects the longer wavelengths of radio waves with a large dish antenna, which resembles a satellite TV dish. The dish plays the same role as the primary mirror in a reflecting telescope by reflecting radio waves to a point above the dish. There, a receiver converts the radio waves into electric signals that can be stored in a computer for analysis.

The data are converted into visual images by a computer. The resolution of the images produced can be improved using a process called **interferometry,** which is a technique that uses the images from several telescopes to produce a single image. By combining the images from several telescopes, astronomers can create a highly detailed image that has the same resolution of one large telescope with a dish diameter as large as the distance between the two telescopes. One example of this is the moveable telescopes shown in **Figure 27.5.** Both radio and optical telescopes can be linked this way.

Figure 27.5 The Very Large Array is situated near Socorro, New Mexico. The dish antennae of this radio telescope are mounted on tracks so they can be moved to improve resolution.

Space-Based Astronomy

Astronomers often send instruments into space to collect information because Earth's atmosphere interferes with most radiation. It blurs visual images and absorbs infrared and ultraviolet radiation, X rays, and gamma rays. Space-based telescopes allow astronomers to study radiation that would be blurred by our atmosphere. American, European, Soviet, Russian, and Japanese space programs have launched many space-based observatories to collect data.

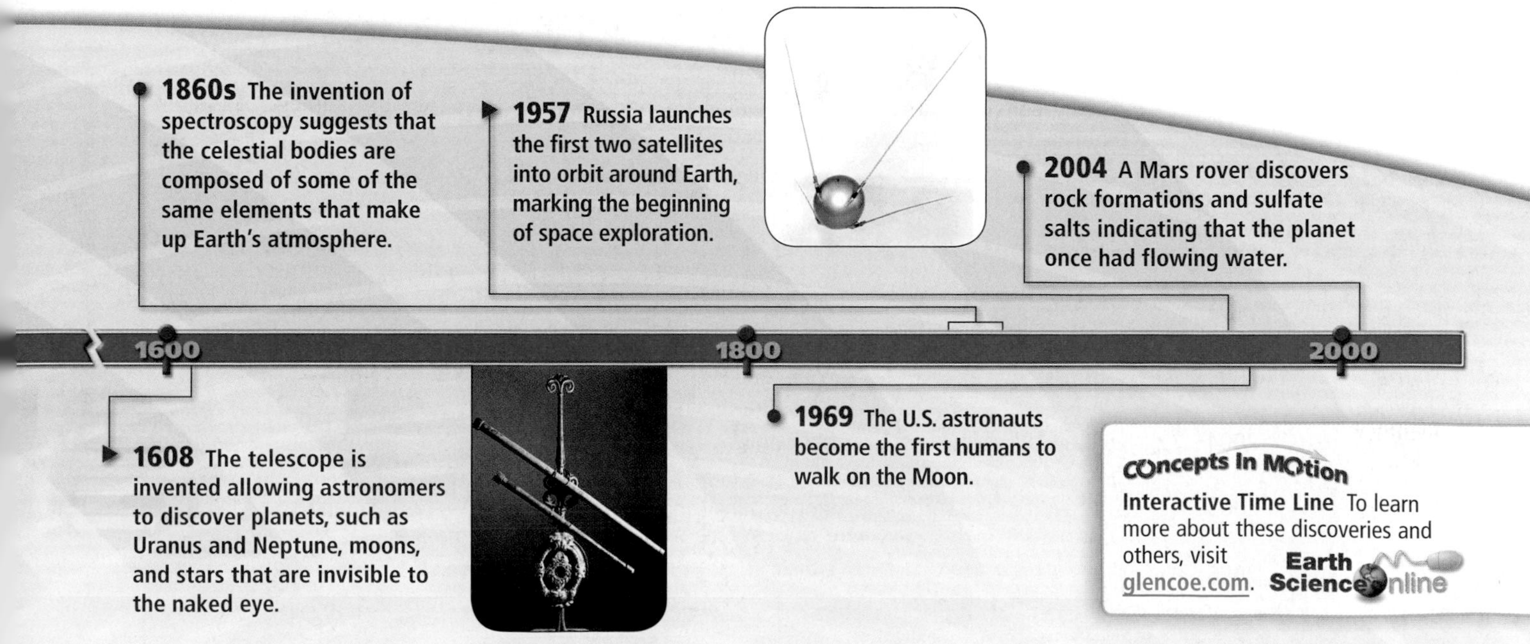

Concepts In Motion

Interactive Time Line To learn more about these discoveries and others, visit glencoe.com. Earth Science Online

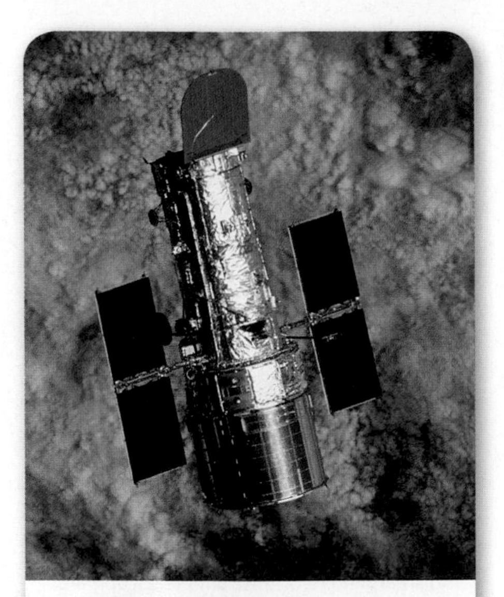

■ **Figure 27.6** The *Hubble Space Telescope* has been used to observe a comet crashing into Jupiter as well as to detect the farthest known galaxy.

Hubble Space Telescope Orbiting Earth every 97 minutes, one of the best-known space-based observatories—the *Hubble Space Telescope (HST)*—shown in **Figure 27.6,** was launched in 1990. *HST* was designed to obtain sharp visible-light images without atmospheric interference, and also to make observations in infrared and ultraviolet wavelengths. The *James Webb Space Telescope* is planned for 2013. It will only observe in the infrared range and is not a replacement for *HST*. Some other space-based telescopes that observe in wavelengths that are blocked by Earth's atmosphere are shown in **Table 27.1.**

Spacecraft In addition to making observations from above Earth's atmosphere, spacecraft can be sent directly to the bodies being observed. Robotic probes are spacecraft that can make close-up observations and sometimes land to collect information directly. Probes are practical only for objects within our solar system, because other stars are too far away. In 2005, the *Cassini* spacecraft arrived at Saturn, where it went into orbit for a detailed look at its moons and rings; the *Mars Reconnaissance Orbiter* reached Mars in 2006 and began orbiting the red planet to use its high-resolution cameras to search for places where life might have evolved; and *New Horizons* was launched in 2006, on its way to Pluto and the region beyond. *New Horizons* is armed with visible, infrared, and ultraviolet cameras, as well as equipment to measure magnetic fields.

Table 27.1 Orbiting Telescopes

Name	Launch	Wavelengths	Studies	Host
Chandra	1999	X ray	wide ranging	NASA
Newton	1999	X ray	wide ranging, black holes	ESA
MAP	2001	microwave	early universe	NASA
Integral	2002	X ray, gamma ray	wide ranging, neutron stars	ESA, Russia, NASA
CHIPSat	2003	X ray	interstellar plasma	NASA
Galex	2003	UV	survey	JPL, NASA
MOST	2003	visible	observe stars	Canada
Spitzer	2003	IR	wide range	NASA
Swift	2004	X ray, UV, visible	back holes	NASA
Suzaku	2005	X ray	star-forming regions	Japan
Akari	2006	IR	survey	Japan

Human spaceflight Before humans can safely explore space, scientists must learn about the effects of space, such as weightlessness and radiation. The most recent human studies have been accomplished with the space shuttle program, which began in 1981. Shuttles are used to place and service satellites, such as the *HST* and the *Chandra X-ray Telescope.* The space shuttle provides an environment for scientists to study the effects of weightlessness on humans, plants, the growth of crystals, and other phenomena. However, because shuttle missions last a maximum of just 17 days, long-term effects must be studied in space stations. A multicountry space station called the *International Space Station,* shown in **Figure 27.7,** is the ideal environment for studying the effects of space on humans. Crews have lived aboard the *International Space Station* since 2000. The crew members conduct many different experiments in this weightless environment.

■ **Figure 27.7** This view of the *International Space Station* was taken from the Space Shuttle *Discovery.* The Caspian Sea is visible in the background.
Review ***What types of studies can be carried out in the space station?***

Spinoff technology Space-exploration programs not only benefit astronomers and space exploration, but they also benefit society. Many technologies that were originally developed for use in space programs are now used by people throughout the world. Did you know that the technology for the space shuttle's fuel pumps led to the development of pumps used in artificial hearts? Or that the Apollo program that put humans on the Moon led to the development of cordless tools? In fact, more than 1400 different NASA technologies have been passed on to commercial industries for common use; these are called spinoffs.

Section 27.1 Assessment

Section Summary

- Telescopes collect and focus electromagnetic radiation emitted or reflected from distant objects.
- Electromagnetic radiation is classified by wavelength and frequency.
- The two main types of optical telescopes are refractors and reflectors.
- Space-based astronomy includes the study of orbiting telescopes, satellites, and probes.
- Technology originally developed to explore space is now used by people on Earth.

Understand Main Ideas

1. MAIN Idea **Explain** how electromagnetic radiation helps scientists study the universe.
2. **Distinguish** between refracting and reflecting telescopes and how they work.
3. **Report** on how interferometry affects the images that are produced by telescopes.
4. **Examine** the reasons why astronomers send telescopes and probes into space.

Think Critically

5. **Assess** the benefits of technology spinoffs to society.
6. **Consider** the advantages and disadvantages of using robotic probes to study distant objects in space.

MATH in Earth Science

7. Calculate the wavelength of radiation with a frequency of 10^{12} Hz. [*Hint: Use the equation* $c = \lambda f$.]

Section 27.2

Objectives

- **Describe** the history of lunar exploration.
- **Recognize** lunar properties and structures.
- **Identify** features of the Moon.
- **Explain** the theory of how the Moon formed.

Review Vocabulary

lava: magma that flows onto the surface from the interior of an astronomical body

New Vocabulary

albedo
highland
mare
impact crater
ejecta
ray
rille
regolith

The Moon

MAIN Idea The Moon, Earth's nearest neighbor in space, is unique among the moons in our solar system.

Real-World Reading Link How many songs, poems, and stories do you know that mention the Moon? The Moon is a familiar object in the night sky and much has been written about it.

Exploring the Moon

Astronomers have learned much about the Moon from observations with telescopes. However, most knowledge of the Moon comes from explorations by space probes, such as *Lunar Prospector* and *Clementine,* and from landings by astronauts. The first step toward reaching the Moon was in 1957, when the Soviet Union launched the first artificial satellite, *Sputnik I.* Four years later, Soviet cosmonaut Yuri A. Gagarin became the first human in space.

That same year, the United States launched the first American, Alan B. Shepard, Jr., into space during Project Mercury. This was followed by Project Gemini that launched two-person crews. Finally, on July 20, 1969, the Apollo program landed Neil Armstrong and Edwin "Buzz" Aldrin on the Moon during the Apollo 11 mission. Astronauts of the Apollo program explored several areas of the Moon, often using special vehicles, such as the *Lunar Roving Vehicle* shown in **Figure 27.8** After a gap of many years, scientists hope to return to the Moon before 2029. In the planning stages are a new spacecraft and lander that can carry more astronauts. Also, astronauts hope to remain longer on the Moon and eventually establish a permanent base there.

Reading Check **Identify** the source of most information about the Moon.

Figure 27.8 Apollo 15 astronauts used the *Lunar Roving Vehicle (LRV)* to explore the Moon's surface.
Explain *how the* LRV *might have resulted in improved mission performance.*

The Lunar Surface

Although the Moon is the brightest object in our night sky, the lunar surface is dark. The **albedo** of the Moon, the percentage of incoming sunlight that its surface reflects, is very small—only about 7 percent. In contrast, Earth has an average albedo of nearly 31 percent. Sunlight that is absorbed by the surface of the Moon produces extreme differences in temperature. Because the Moon has no atmosphere to absorb heat, sunlight can heat the Moon's surface to 400 K (127°C), while the temperature of its unlit surface can drop to a chilly 100 K (−173°C).

The "man in the Moon" pattern seen from Earth is produced by the Moon's surface features. Lunar **highlands** are heavily cratered regions of the Moon that are light in color and mountainous. Other regions called **maria** (MAH ree uh) (singular, mare [MAH ray]) are dark, smooth plains, which average 3 km lower in elevation. Maria have few craters.

Reading Check **Explain** what lunar features produce the "man in the Moon."

Lunar craters The craters on the Moon, called **impact craters,** formed when objects from space crashed into the lunar surface. The material blasted out during these impacts fell back to the Moon's surface as **ejecta.** Some craters have long trails of ejecta, called **rays,** that radiate outward from the impact site, as shown in **Figure 27.9.** Rays are visible as light-colored streaks. Although the maria are mostly smooth, they do have a few scattered craters and rilles. **Rilles** are meandering, valleylike structures that might be collapsed lava tubes. In addition, there are mountain ranges near some of the maria.

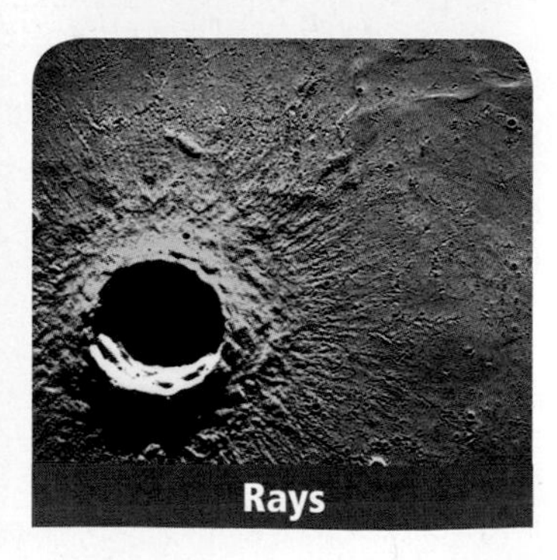

■ **Figure 27.9** You can see some of the details for the maria and highlands in the view of the full moon. Craters, ejecta, rilles, and rays are visible in close-up views of the Moon's surface.

Concepts In Motion
Interactive Table To explore more about the Moon and Earth, visit glencoe.com.

Table 27.2 The Moon and Earth

	The Moon	Earth
Mass (kg)	7.349×10^{22}	5.974×10^{24}
Radius (km)	1737.4	6378.1
Volume (km^3)	2.197×10^{10}	1.083×10^{12}
Density (kg/m^3)	3340	5515

Lunar properties Earth's moon is unique among all the moons in the solar system. First, it is one of the largest moons compared to the radius and mass of the planet it orbits, as shown in **Table 27.2.** Also, it is a solid, rocky body, in contrast with the icy compositions of other moons of the solar system. Finally, the Moon's orbit is farther from Earth relative to the distance of most moons from the planets they orbit. **Figure 27.10** shows a photo of Earth and the Moon taken from space.

Composition The Moon is made up of minerals similar to those of Earth—mostly silicates. Recall from Chapter 4 that silicates are compounds containing silicon and oxygen that make up 96 percent of the minerals in Earth's crust. The highlands, which cover most of the lunar surface, are predominately lunar breccias (BRE chee uhs), which are rocks formed by the fusion of smaller pieces of rock during impacts. Unlike sedimentary breccias on Earth, most of the lunar breccias are composed of plagioclase feldspar, a silicate containing high quantities of calcium and aluminum but low quantities of iron. The maria are predominately basalt, but unlike basalt on Earth, they contain no water.

Reading Check **Describe** the compositions of the lunar highlands and maria.

History of the Moon

The entire lunar surface is old—radiometric dating of rocks from the highlands indicates an age between 3.8 and 4.6 billion years—about the same age as Earth. Based on the ages of the highlands and the frequency of the impact craters that cover them, scientists theorize that the Moon was heavily bombarded during its first 800 million years. This caused the breaking and heating of surface rocks and resulted in a layer of loose, ground-up rock called **regolith** on the surface. The regolith averages several meters in thickness, but it varies greatly depending on location.

Figure 27.10 This photograph of the view of the Moon and Earth was taken by *Mariner 10* on its way to Venus.

Layered structure Scientists infer from seismic data that the Moon, like Earth, has a layered structure, which consists of the crust, upper mantle, lower mantle, and core, as illustrated in **Figure 27.11.** The crust varies in thickness and is thickest on the far side. The far side of the Moon is the side that is always facing away from Earth. The Moon's upper mantle is solid, its lower mantle is thought to be partially molten, and its core is solid iron.

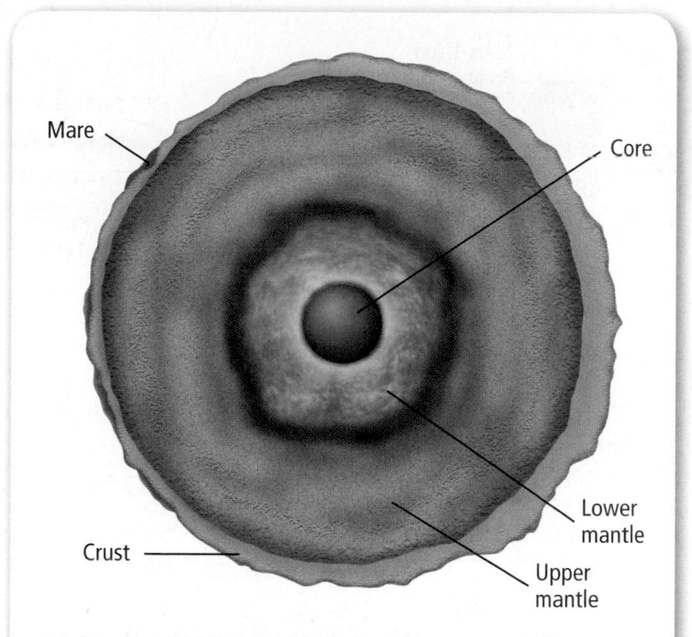

■ **Figure 27.11** Scientists deduce the structure of the Moon's interior from seismic data obtained from seismometers left on the Moon's surface.

Formation of maria After the period of intense bombardment that formed the highlands, lava welled up from the Moon's interior and filled in the large impact basins. This lava fill created the dark, smooth plains of the maria. Scientists estimate the maria formed between 3.1 and 3.8 bya, making them younger than the highlands. Flowing lava in the maria scarred the surface with rilles. Rilles are much like lava tubes found on Earth, through which lava flows in underground streams. The maria have remained relatively free of craters because fewer impacts have occurred on the Moon since they formed.

Often lava did not fill the basins completely and left the rims of the basins above the lava. This left behind the mountain ranges that now surround many maria. As shown in **Figure 27.12,** there are virtually no maria on the far side of the Moon, which is covered almost completely with highlands. Scientists hypothesize that this is because the crust is thicker on the far side, which made it difficult for lava to reach the lunar surface. You will determine the relative ages of the Moon's surface features in this chapter's GeoLab.

Tectonics Seismometers measure strength and frequency of moonquakes. Seismic data show that on average, the Moon experiences an annual moonquake that would be strong enough to cause dishes to fall out of a cupboard if it happened on Earth. Despite these moonquakes, scientists think that the Moon is not tectonically active. The Moon has no active volcanoes and no significant magnetic field. Scientists know from the locations and shapes of mountains on the Moon that they were not formed tectonically, as mountain ranges on Earth are formed. Lunar mountains are actually higher elevations that surround ancient impact basins filled with lava.

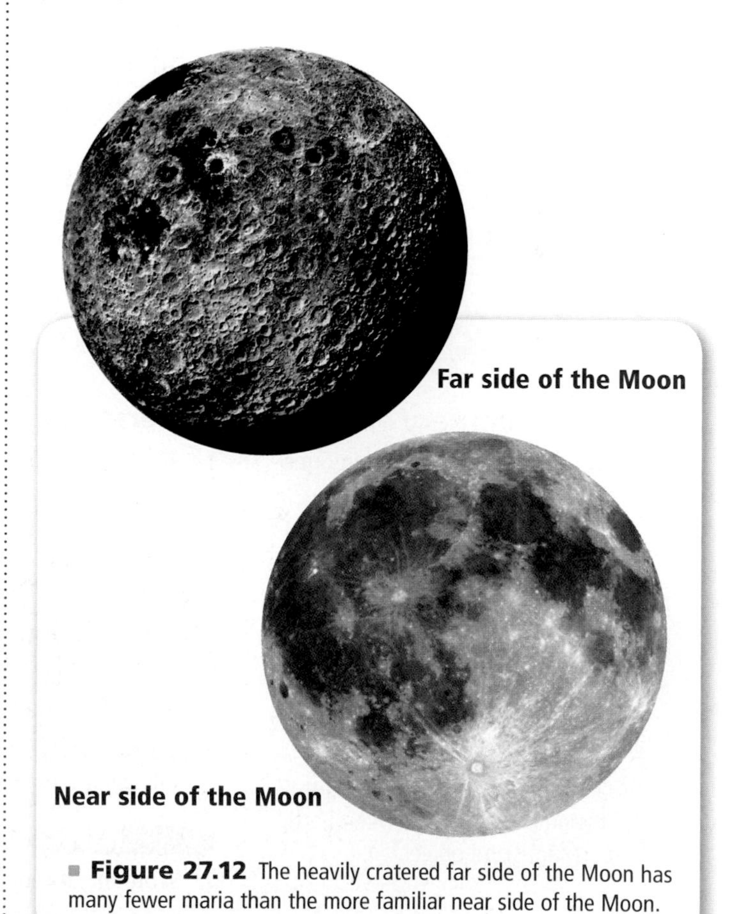

■ **Figure 27.12** The heavily cratered far side of the Moon has many fewer maria than the more familiar near side of the Moon.

■ **Figure 27.13** The impact theory of the Moon's formation states that material ejected from Earth and from the striking object eventually merged to form the Moon.

Concepts In Motion

Interactive Figure To see an animation of the Moon impact theory, visit glencoe.com.

Formation

Several theories have been proposed to explain the Moon's unique properties. The theory that is accepted by most astronomers today was developed using computer simulations. This theory is known as the impact theory.

According to the impact theory, the Moon formed as the result of a collision between Earth and a Mars-sized object about 4.5 bya when the solar system was forming. This computer model suggests that the object struck Earth with a glancing blow. The impact caused materials from the incoming body and Earth's outer layers to be ejected into space, where over time they merged to form the Moon, as illustrated by **Figure 27.13.** According to the model, the Moon is made up of a small amount of iron at the core, and mostly silicate material that came from Earth's mantle and crust. This explains why the Moon's crust is so similar to Earth's crust in chemical composition. This theory has been accepted as similarities have been found between bulk samples of rock taken from Earth and from the Moon.

Section 27.2 Assessment

Section Summary

- Astronomers have gathered information about the Moon using telescopes, space probes, and astronaut exploration.
- Like Earth's crust, the Moon's crust is composed mostly of silicates.
- Surface features on the Moon include highlands, maria, ejecta, rays, and rilles. It is heavily cratered.
- The Moon probably formed about 4.5 bya in a collision between Earth and a Mars-sized object.

Understand Main Ideas

1. MAIN Idea **Compare and contrast** the Moon and the moons of other planets.
2. **Classify** the following according to age: maria, highlands, and rilles.
3. **Explain** how scientists determined that the Moon has no tectonics.
4. **Distinguish** the steps involved in the impact theory of lunar formation.

Think Critically

5. **Infer** how the surface of the Moon would look if the crust on the far side were the same thickness as the crust on the near side.
6. **Summarize** the major ideas in this section using an outline format. Include the following terms: *highlands, crust, lava, maria, craters, tectonics,* and *impact theory.*

WRITING in Earth Science

7. Write the introductory paragraph to an article entitled *History of the Moon.*

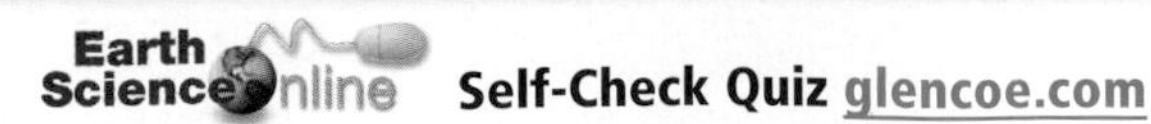

Section 27.3

The Sun-Earth-Moon System

Objectives

- **Identify** the relative positions and motions of the Sun, Earth, and Moon.
- **Describe** the phases of the Moon.
- **Distinguish** between solstices and equinoxes.
- **Explain** eclipses of the Sun and Moon.

Review Vocabulary

revolution: the time it takes for a planetary body to make one orbit around another, larger body

New Vocabulary

ecliptic plane
solstice
equinox
synchronous rotation
solar eclipse
perigee
apogee
lunar eclipse

MAIN Idea Motions of the Sun-Earth-Moon system define Earth's day, month, and year.

Real-World Reading Link Have you ever tried to guess the time by judging the Sun's position? If so, you were observing an effect of the motions of the Sun-Earth-Moon system.

Daily Motions

From the vantage point of Earth, the most obvious pattern of motion in the sky is the daily rising and setting of the Sun, the Moon, stars, and everything else that is visible in the night sky. The Sun rises in the east and sets in the west, as do the Moon, planets, and stars. These daily motions result from Earth's rotation. The Sun, the Moon, planets, and stars do not orbit around Earth every day. It only appears that way because we observe the sky from a planet that rotates. But how do we know that Earth rotates?

Earth's rotation There are two relatively simple ways to demonstrate that Earth is rotating. One is to use a Foucault pendulum, like the one shown in **Figure 27.14.** A Foucault pendulum swings in a constant direction. But as Earth turns under it, the pendulum seems to shift its orientation. The second way is to observe the way that air on Earth is diverted from a north-south direction to an east-west direction by the Coriolis effect.

Day length The time period from one noon to the next is called a solar day. Our timekeeping system is based on the solar day. But the length of a day as we observe it is roughly four minutes longer than the time it takes Earth to rotate once on its axis. As Earth rotates, it also moves in its orbit and has to turn a little farther each day to align again with the Sun.

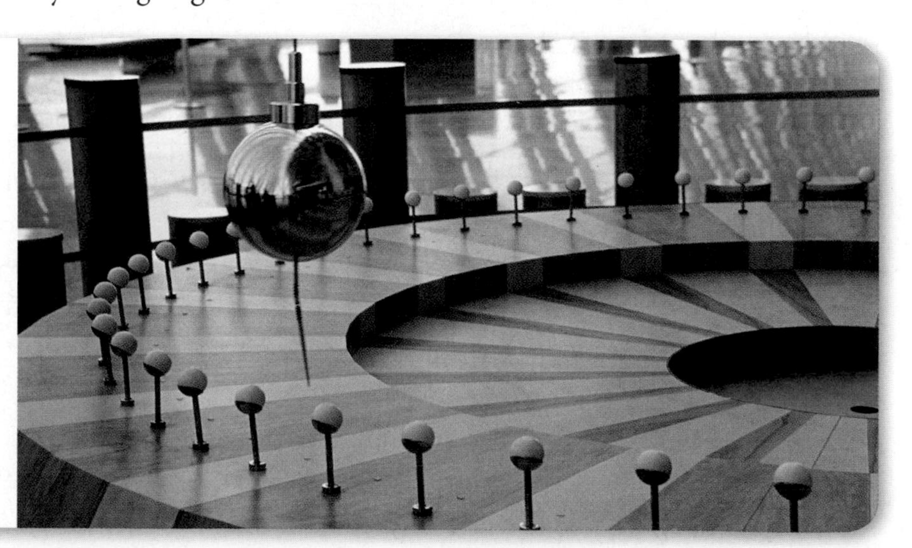

■ **Figure 27.14** This Foucault pendulum is surrounded by pegs. As Earth rotates under it, the pendulum knocks over the pegs, showing the progress of the rotation.

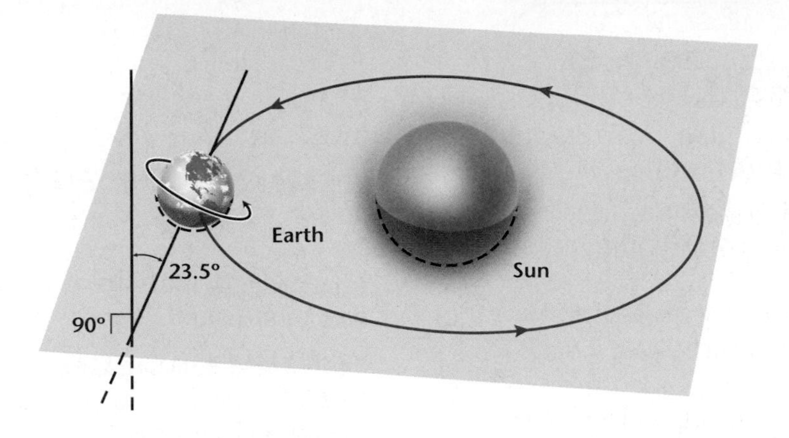

■ **Figure 27.15** Earth's nearly circular orbit around the Sun lies on the ecliptic plane. When looking toward the horizon and the plane of the ecliptic, different stars are visible during the year.
Predict *Do the positions of stars vary when you look overhead?*

Annual Motions

Earth orbits the Sun in a slightly elliptical orbit, as shown in **Figure 27.15.** The plane of Earth's orbit is called the **ecliptic plane.** As Earth rotates, the Sun, planets, and constellations appear to move across the sky in a path known as the ecliptic. As Earth moves in its orbit, different constellations are visible.

VOCABULARY
ACADEMIC VOCABULARY
Cycle
recurring sequence of events or phenomena
The cycle of seasons repeats every year.

The effects of Earth's tilt Earth's axis is tilted relative to the ecliptic at approximately 23.5°. As Earth orbits the Sun, the orientation of Earth's axis remains fixed in space so that, at a given time, the northern hemisphere of Earth is tilted toward the Sun, while at another point, six months later, the northern hemisphere is tipped away from the Sun. A cycle of the seasons is a result of this tilt and Earth's orbital motion around the Sun. Another effect is the changing angle of the Sun above the horizon from summer to winter. More hours of daylight cause the summer months to be warmer.

MiniLab

Predict the Sun's Summer Solstice Position

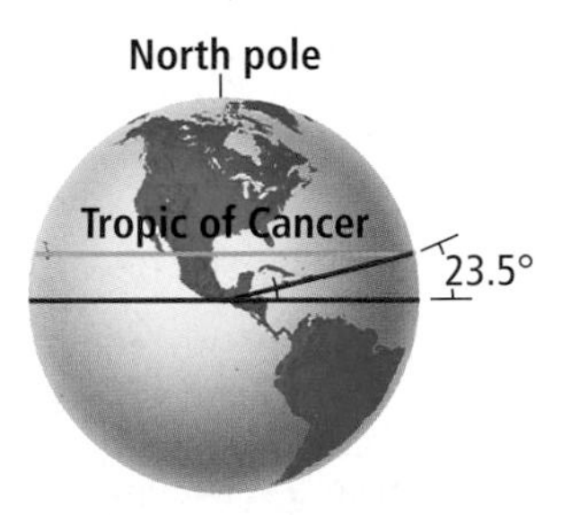

How can the Sun's position during the summer solstice be determined at specific latitudes? At summer solstice for the northern hemisphere, the Sun is directly overhead at the Tropic of Cancer.

Procedure

1. Read and complete the lab safety form.
2. Draw a straight line to represent the equator and mark the center of the line with a dot.
3. Use a **protractor** to measure the angle of latitude of the Tropic of Cancer from the equator line. Draw a line at that angle from the line's center dot.
4. Find your home latitude and measure that angle of latitude on your diagram. Draw a line from the line center for this location.
5. Measure the angle between the line for the Tropic of Cancer and the line for your location. Subtract that angle from 90°. This gives you the angle above the horizon for the maximum height of the Sun on the solstice at your location.

Analysis

1. **Describe** how the position of the Sun varies with latitude on Earth.
2. **Consider** the angle that would illustrate the winter solstice for the northern hemisphere.

Solstices Earth's orbit around the Sun and the tilt of Earth's axis are illustrated in **Figure 27.16.** Positions 1 and 3 correspond to the solstices. At a **solstice,** the Sun is overhead at its farthest distance either north or south of the equator. The lines of latitude that correspond to these positions on Earth have been identified as the Tropic of Cancer and the Tropic of Capricorn. The area between these latitudes is commonly known as the tropics. Position 1 corresponds to the summer solstice in the northern hemisphere when the Sun is directly overhead at the Tropic of Cancer, 23.5° north latitude. At this time, around June 21 each year, the number of daylight hours reaches its maximum, and the Sun is in the sky continuously within the region of the Arctic Circle. On this day, the number of daylight hours in the southern hemisphere is at its minimum, and the Sun does not appear in the region within the Antarctic Circle.

 **Reading Check** **Identify** where the Sun is directly overhead at the summer solstice in the northern hemisphere.

As Earth moves past Position 2, the Sun's altitude decreases in the northern hemisphere until Earth reaches Position 3, known as winter solstice for the northern hemisphere. Here the Sun is directly overhead at the Tropic of Capricorn, 23.5° south latitude. This happens around December 21. On this day, the number of daylight hours in the northern hemisphere is at its minimum and the Sun does not appear in the region within the Arctic Circle. Then, as Earth continues around its orbit past Position 4, the Sun's altitude increases again until it returns to Position 1. Notice that the summer and winter solstices are reversed for those living in the southern hemisphere-June 21 is the winter solstice and December 21 is the summer solstice.

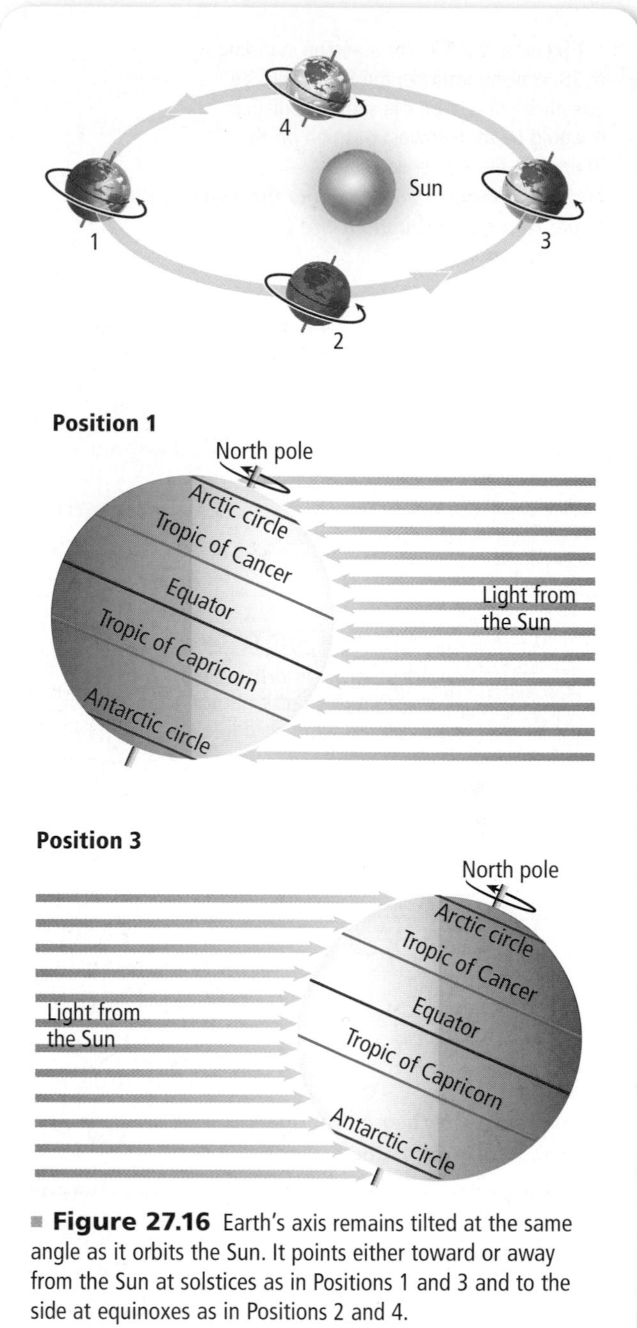

Figure 27.16 Earth's axis remains tilted at the same angle as it orbits the Sun. It points either toward or away from the Sun at solstices as in Positions 1 and 3 and to the side at equinoxes as in Positions 2 and 4.
Identify *the correct term for each position for each hemisphere.*

Equinoxes Positions 2 and 4, where Earth is midway between solstices, represent the equinoxes, a term meaning *equal nights.* At an **equinox,** Earth's axis is perpendicular to the Sun's rays and at noon the Sun is directly overhead at the equator. Those living in the northern hemisphere refer to Position 2 as the autumnal equinox, and Position 4 as the vernal equinox. Those in the southern hemisphere do the reverse—Positions 2 and 4 are the vernal and autumnal equinoxes, respectively.

■ **Figure 27.17** For a person standing at 23.5° north latitude, the Sun would be directly overhead on the summer solstice. It would be at its lowest position on the horizon at the winter solstice.
***Draw** a diagram showing how the Sun's angle changes throughout the year at your latitude.*

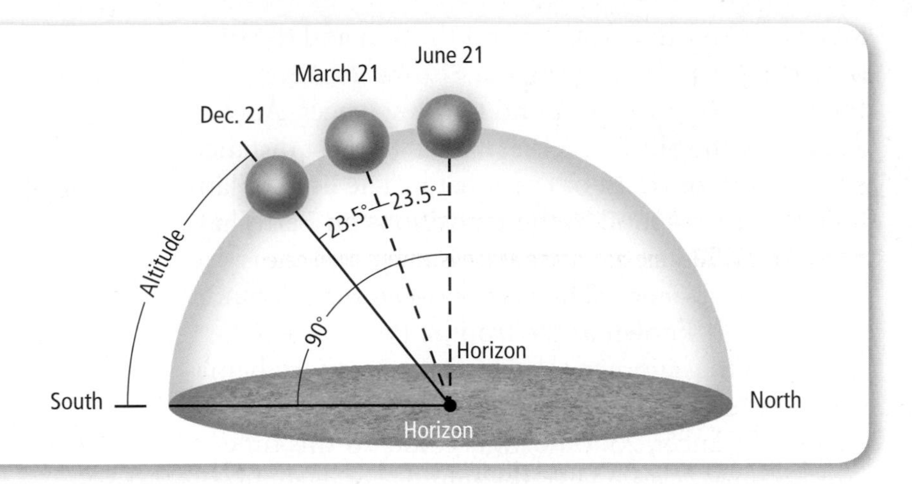

Changes in altitude The Sun's maximum height at midday, called its zenith, varies throughout the year depending on the viewer's location. For example, on the summer solstice, a person located at 23.5° north latitude sees the Sun's zenith directly overhead. At the equinox, it appears lower, and at the winter solstice, it is at its lowest position, shown in **Figure 27.17.** Then it starts moving higher again to complete the cycle.

FOLDABLES
Incorporate information from this section into your Foldable.

Phases of the Moon

Just as the Sun appears to change its position in the sky throughout the year, the Moon also changes position relative to the ecliptic plane as it orbits Earth. The Moon's cycle is more complex, as you will learn later in this section. More striking are the changing views of the illuminated side of the Moon as it orbits Earth. The sequential changes in the appearance of the Moon are called lunar phases, and are shown in **Figure 27.18.**

 **Reading Check** **Explain** what is meant by the term *lunar phases.*

As you have read, the light given off by the Moon is a reflection of the Sun's light. In fact, one half of the Moon is illuminated at all times. How much of this lighted half is visible from Earth varies as the Moon revolves around Earth. When the Moon is between Earth and the Sun, for instance, the side that is illuminated is not visible from Earth. This phase is called a new moon.

VOCABULARY
SCIENCE USAGE V. COMMON USAGE
Altitude
Science usage: angular elevation of a celestial body above the horizon

Common usage: vertical elevation of a body above a surface

Waxing and waning Starting at the new moon, as the Moon moves in its orbit around Earth, more of the sunlit side of the Moon becomes visible. This increase in the visible sunlit surface of the Moon is called the waxing phase. The waxing phases are called waxing crescent, first quarter, and waxing gibbous. Then, as the Moon moves to the far side of the Earth from the Sun, the entire sunlit side of the Moon faces Earth. This is known as a full moon.

After the full moon, the portion of the sunlit side that is visible begins to decrease. This is called the waning phase. The waning phases are named similarly to the waxing phases, that is, waning gibbous and waning crescent. When exactly half of the sunlit portion is visible, it is called the third quarter.

Visualizing the Phases of the Moon

Figure 27.18 One-half of the Moon is always illuminated by the Sun's light, but the entire lighted half is visible from Earth only at full moon. The rest of the time you see portions of the lighted half. These portions are called lunar phases.

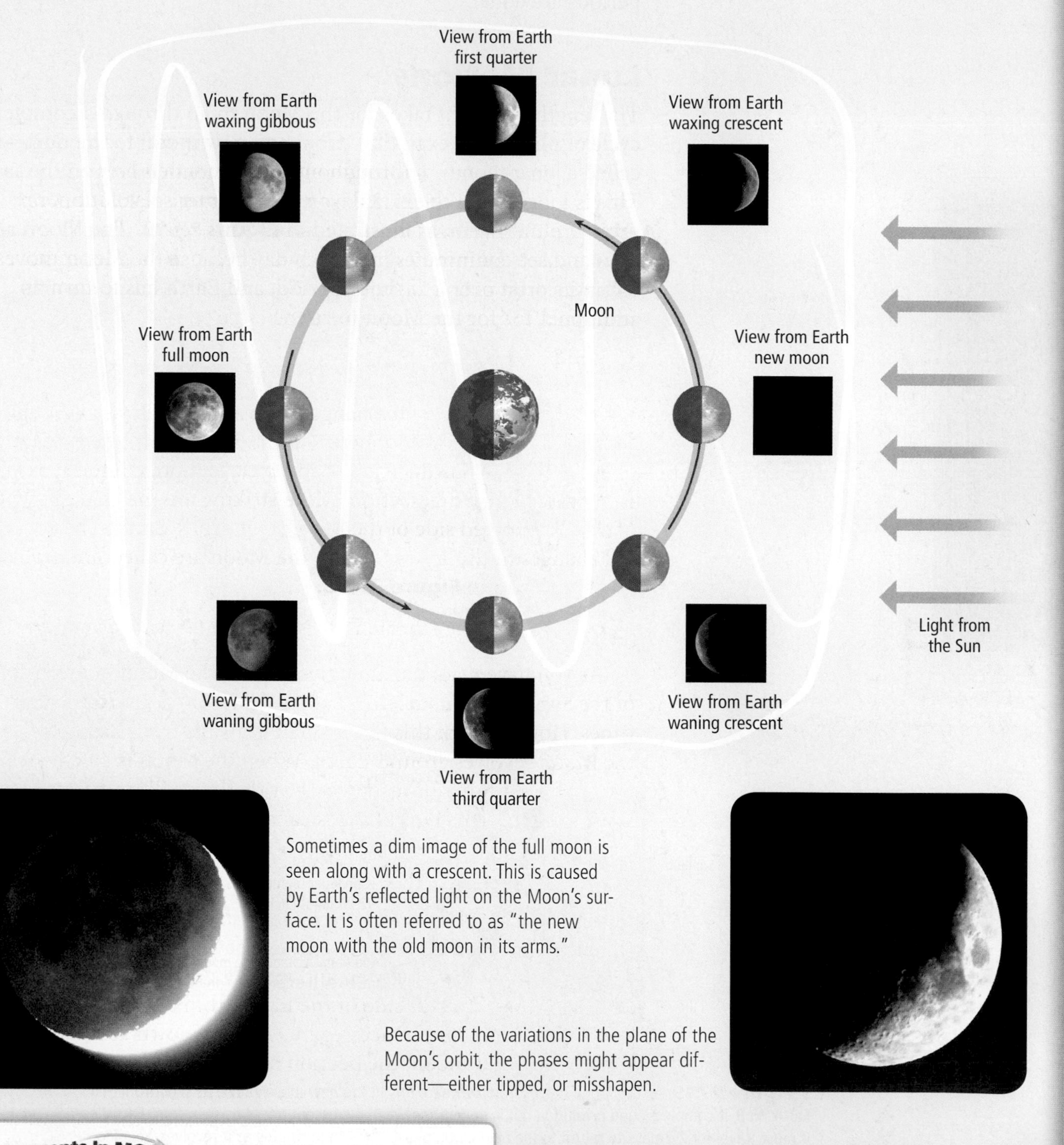

Sometimes a dim image of the full moon is seen along with a crescent. This is caused by Earth's reflected light on the Moon's surface. It is often referred to as "the new moon with the old moon in its arms."

Because of the variations in the plane of the Moon's orbit, the phases might appear different—either tipped, or misshapen.

Concepts In Motion To explore more about lunar phases, visit glencoe.com. Earth Science Online

Synchronous rotation You might have noticed that the surface features of the Moon always look the same. As the Moon orbits Earth, the same side faces Earth at all times. This is because the Moon rotates with a period equal to its orbital period, in other words, the Moon spins exactly once each time it goes around Earth. This is no coincidence. Scientists theorize that Earth's gravity slowed the Moon's original spin until the Moon reached **synchronous rotation,** the state at which its orbital and rotational periods are equal.

Lunar Motions

The length of time it takes for the Moon to go through a complete cycle of phases, for example—from one new moon to the next—is called a lunar month. The length of a lunar month is about 29.5 days. This is longer than the 27.3 days it takes for one revolution, or orbit, around Earth, as illustrated in **Figure 27.19.** The Moon also rises and sets 50 minutes later each day because the Moon moves 13° in its orbit over a 24-hour period, and Earth has to turn an additional 13° for the Moon to rise.

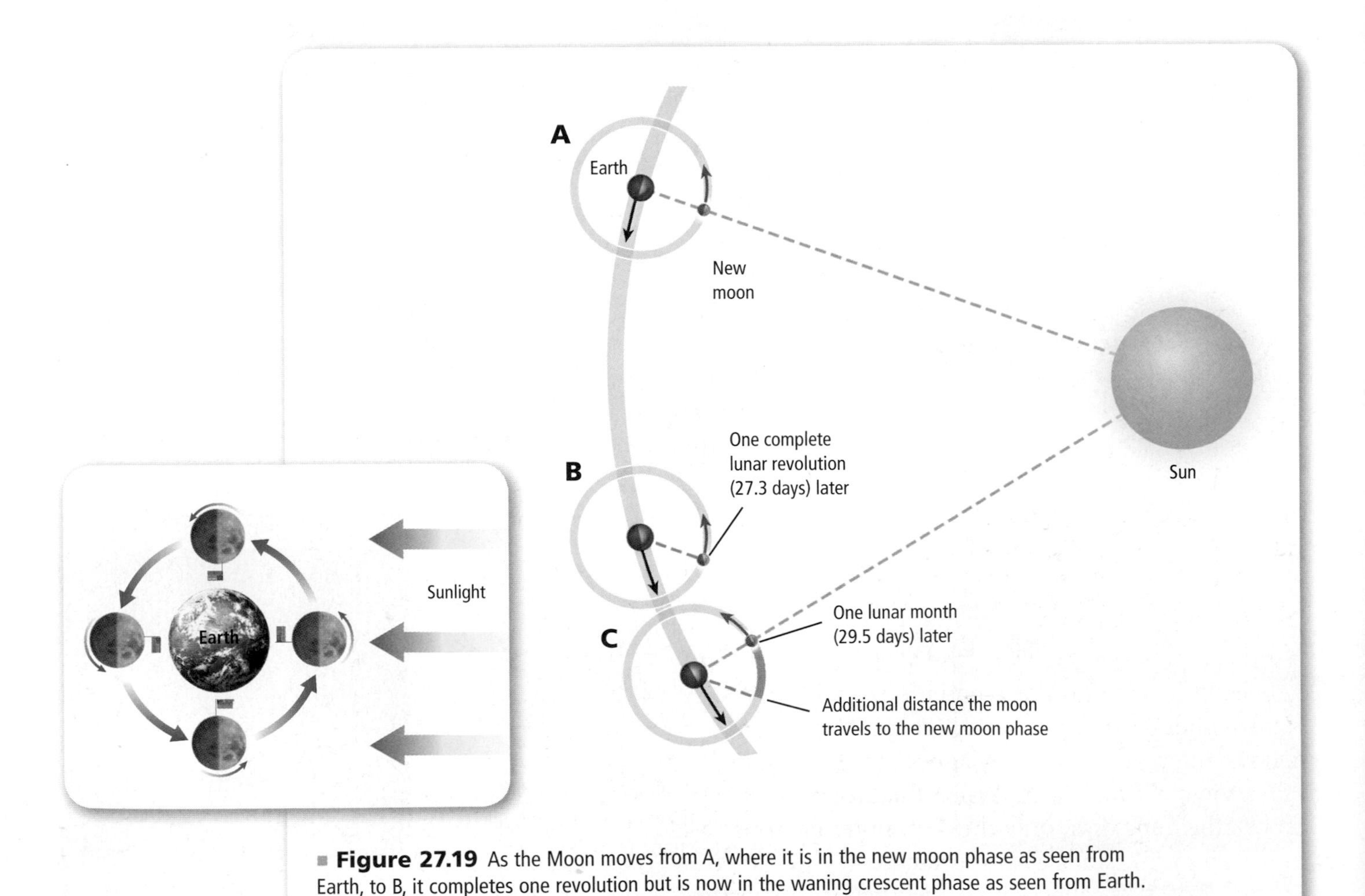

■ **Figure 27.19** As the Moon moves from A, where it is in the new moon phase as seen from Earth, to B, it completes one revolution but is now in the waning crescent phase as seen from Earth. It must travel for 2.2 days to return to the new moon phase. The Moon rotates as it revolves, keeping the same side facing Earth, as shown in the inset.

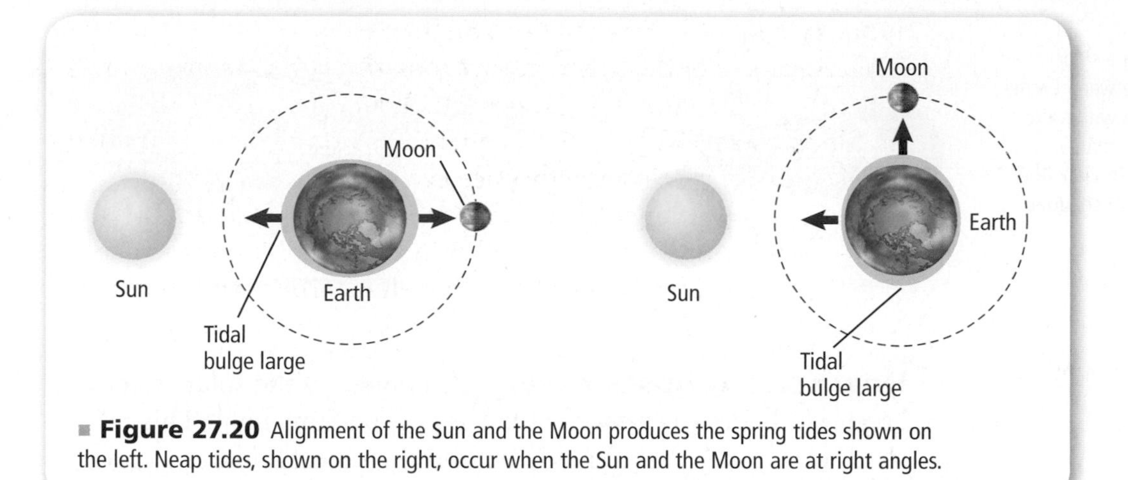

Figure 27.20 Alignment of the Sun and the Moon produces the spring tides shown on the left. Neap tides, shown on the right, occur when the Sun and the Moon are at right angles.

Tides One effect the Moon has on Earth is causing ocean tides. The Moon's gravity pulls on Earth along an imaginary line connecting Earth and the Moon, and this creates bulges of ocean water on both the near and far sides of Earth. Earth's rotation also contributes to the formation of tides, as you learned in Chapter 15. As Earth rotates, these bulges remain aligned with the Moon, so that a person at a shoreline on Earth's surface would observe that the ocean level rises and falls every 12 hours.

Spring and neap tides The Sun's gravitational pull also affects tides, but the Sun's influence is half that of the Moon's because the Sun is farther away. However, when the Sun and the Moon are aligned along the same direction, their effects are combined, and tides are higher than normal. These tides, called spring tides, are especially high when the Moon is nearest Earth and Earth is nearest the Sun in their slightly elliptical orbits. When the Moon is at a right angle to the Sun-Earth line, the result is lower-than-normal tides, called neap tides. The Sun and the Moon alignments during spring and neap tides are shown in **Figure 27.20.**

Solar Eclipses

A **solar eclipse** occurs when the Moon passes directly between the Sun and Earth and blocks the Sun from view. Although the Sun is much larger than the Moon, it is far enough away that they appear to be the same size when viewed from Earth. When the Moon perfectly blocks the Sun's disk, only the dim, outer gaseous layers of the Sun are visible. This spectacular sight, shown in **Figure 27.21,** is called a total solar eclipse. A partial solar eclipse is seen when the Moon blocks only a portion of the Sun's disk.

Figure 27.21 The stages of a total solar eclipse are seen in this multiple-exposure photograph. **Explain** *why the Moon seems to cross the Sun at an angle rather than directly right to left.*

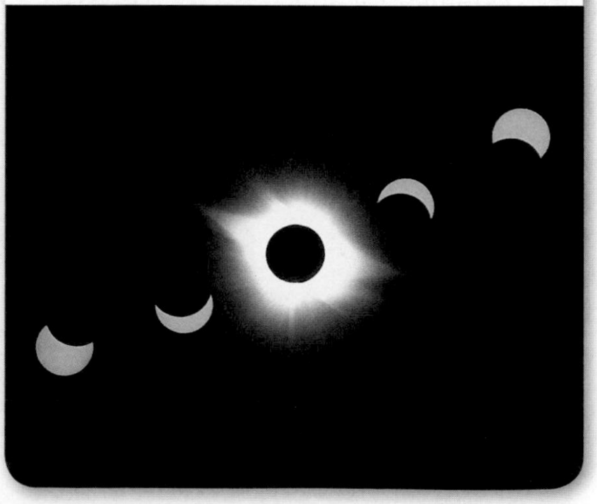

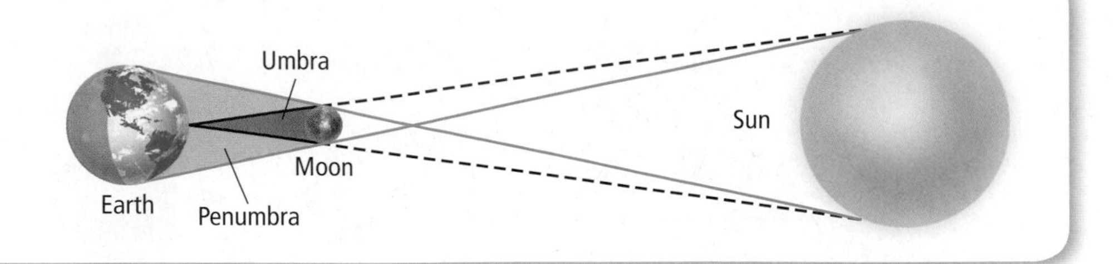

Figure 27.22 During a solar eclipse, the Moon passes between Earth and the Sun. Those on Earth within the darkest part of the Moon's shadow (umbra) see a total eclipse. Those within the lighter part, or penumbral shadow, see only a partial eclipse.

Concepts In Motion

Interactive Figure To see an animation of an eclipse, visit glencoe.com.

How solar eclipses occur Each object in the solar system creates a shadow as it blocks the path of the Sun's light. This shadow is totally dark directly behind the object and has a cone shape. During a solar eclipse, the Moon casts a shadow on Earth as it passes between the Sun and Earth. This shadow consists of two regions, as illustrated in **Figure 27.22.** The inner, cone-shaped portion, which blocks the direct sunlight, is called the umbra. People who witness an eclipse from within the umbra shadow see a total solar eclipse. That means they see the Moon completely cover the face of the Sun. The outer portion of this shadow, where some of the Sun's light still reaches, is called the penumbra. People in the region of the penumbra shadow see a partial solar eclipse, where a part of the Sun's disk is blocked by the Moon. Typically, the umbral shadow is never wider than 270 km, so a total solar eclipse is visible from a very small portion of Earth, whereas a partial solar eclipse is visible from a much larger portion.

PROBLEM-SOLVING LAB

Interpret Scientific Illustrations

How can you predict how a solar eclipse will look to an observer at various positions?

The diagram below shows the Moon eclipsing the Sun. The Sun will appear differently to observers located at Points A through E.

Analysis

1. **Observe** the points in relation to the position of the Moon's umbra and penumbra.

Think Critically

2. **Draw** how the solar eclipse would appear to an observer at each labeled point.
3. **Design** a data table to display your drawings.
4. **Classify** the type of solar eclipse represented in each of your drawings.

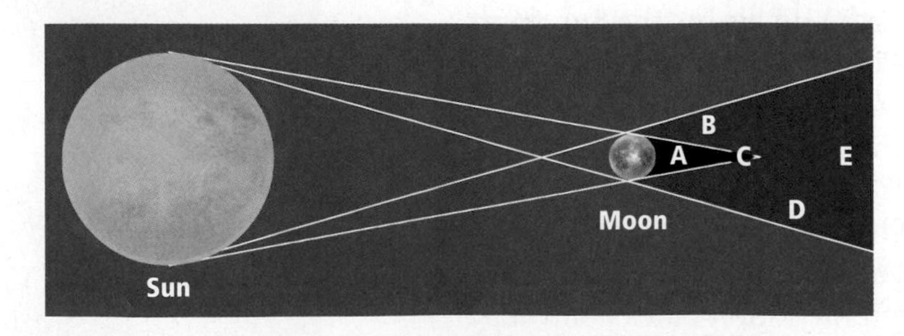

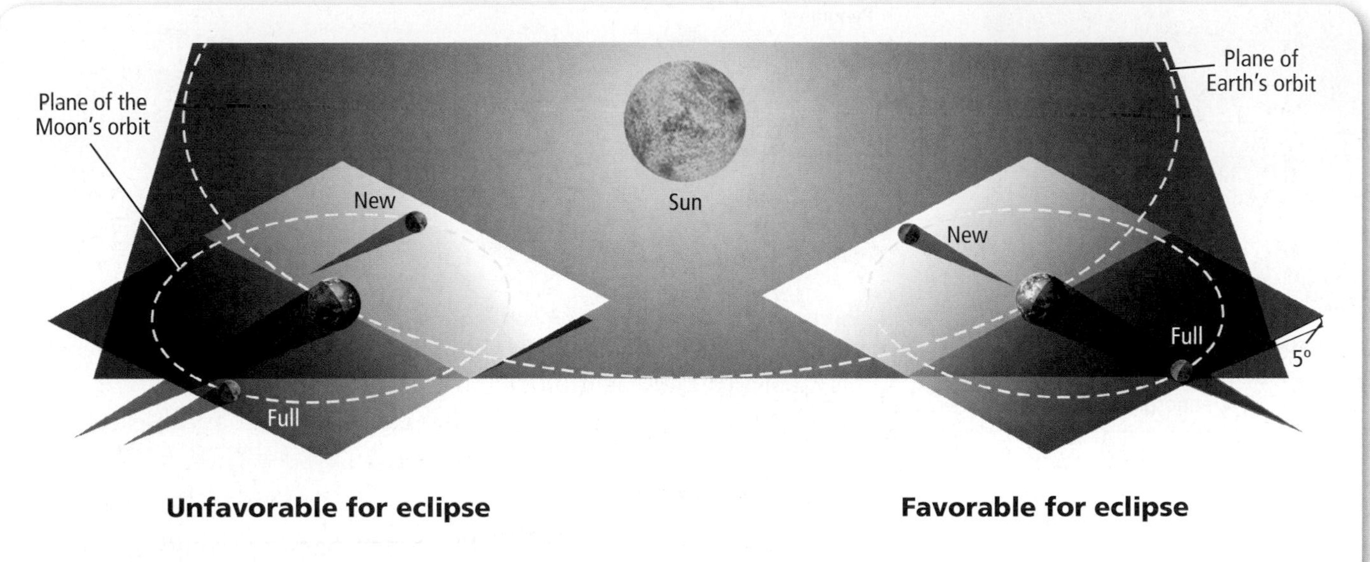

■ **Figure 27.23** Eclipses can take place only when Earth, the Moon, and the Sun are perfectly aligned. This can happen only when the Moon's orbital plane and ecliptic plane intersect along the Sun-Earth line, as shown in diagram on the right. In the left diagram, this does not happen, and the Moon's shadow misses Earth.

Effects of tilted orbits You might wonder why a solar eclipse does not occur every month when the Moon passes between the Sun and Earth during the new moon phase. This does not happen because the Moon's orbit is tilted 5° relative to the ecliptic plane. Normally, the Moon passes above or below the Sun as seen from Earth, so no solar eclipse takes place. Only when the Moon crosses the ecliptic plane is it possible for the proper alignment for a solar eclipse to occur, but even that does not guarantee a solar eclipse. The plane of the Moon's orbit also rotates slowly around Earth, and a solar eclipse occurs only when the intersection of the Moon and the ecliptic plane is in a line with the Sun and Earth, as **Figure 27.23** illustrates.

 **Reading Check** **Determine** why a total solar eclipse does not occur every month.

Annular eclipses Not only does the Moon move above and below the plane of Earth and the Sun, but the Moon's distance from Earth increases and decreases as the Moon moves in its elliptical orbit around Earth. The closest point in the Moon's orbit to Earth is called **perigee,** and the farthest point is called **apogee.** When the Moon is near apogee, it appears smaller from Earth, and thus will not completely block the disk of the Sun during an eclipse. This is called an annular eclipse because, as **Figure 27.24** shows, a ring of the Sun, called the annulus, appears around the dark Moon. Earth's orbit also has a perigee and apogee. When Earth is nearest the Sun and the Moon is at apogee from Earth, the Moon would not block the Sun entirely. The opposite is true for Earth at apogee to the Sun and the Moon at perigee to Earth.

■ **Figure 27.24** An annular eclipse takes place when the Moon is too far away for its umbral shadow to reach Earth. A ring, or annulus, is left uncovered.
Predict ***Would annular eclipses occur if the Moon's orbit were a perfect circle?***

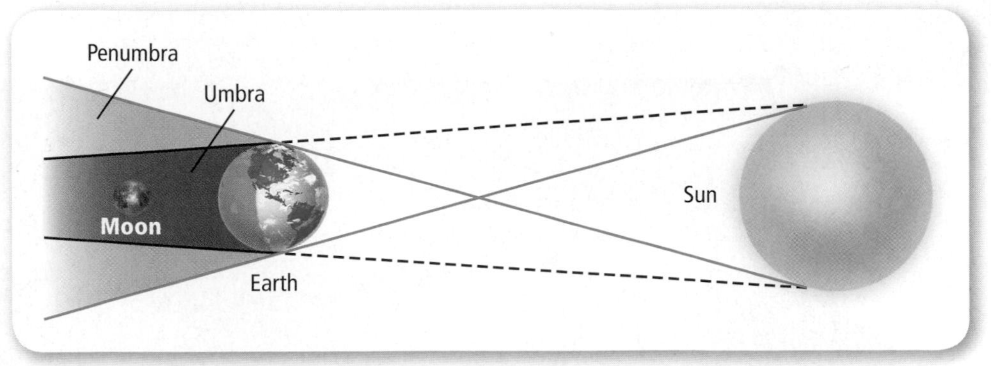

■ **Figure 27.25** When the Moon is completely within Earth's umbra, a total lunar eclipse takes place, as shown in the diagram. The darkened Moon often has a reddish color, as shown in the photo, because Earth's atmosphere bends and scatters the Sun's light.

Lunar Eclipses

A **lunar eclipse** occurs when the Moon passes through Earth's shadow. As illustrated in **Figure 27.25,** this can happen only at the time of a full moon when the Moon is on the opposite side of Earth from the Sun. The shadow of Earth has umbral and penumbral portions, just as the Moon's shadow does. A total lunar eclipse occurs when the entire Moon is within Earth's umbra. This lasts for approximately two hours. During a total lunar eclipse, the Moon is faintly visible, as shown in **Figure 27.25,** because sunlight that has passed near Earth has been filtered and refracted by Earth's atmosphere. This light can give the eclipsed Moon a reddish color as Earth's atmosphere bends the red light into the umbra, much like a lens. Like solar eclipses, lunar eclipses do not occur every full moon because the Moon in its orbit usually passes above or below the Sun as seen from Earth.

Section 27.3 Assessment

Section Summary

- Earth's rotation defines one day, and Earth's revolution around the Sun defines one year.
- Seasons are caused by the tilt of Earth's spin axis relative to the ecliptic plane.
- The gravitational attraction of both the Sun and the Moon causes tides.
- The Moon's phases result from our view of its lighted side as it orbits Earth.
- Solar and lunar eclipses occur when the Sun's light is blocked.

Understand Main Ideas

1. MAIN Idea **State** one proof that Earth rotates, one proof Earth rotates in 24 hours, and make one observation that proves it revolves around the Sun in one year.
2. **Compare** solar and lunar eclipses, including the positions of the Sun, Earth, and Moon.
3. **Diagram** the waxing and waning phases of the Moon.
4. **Analyze** why the Moon has a greater effect on Earth's tides than the Sun, even though the Sun is more massive.

Think Critically

5. **Relate** what you have learned about lunar phases to how Earth would appear to an observer on the Moon. Diagram the positions of the Sun, Earth, and the Moon and draw how Earth would appear in several positions to explain your answer.

MATH in Earth Science

6. Consider what would happen if Earth's axis were tilted 45°. At what latitudes would the Sun be directly overhead on the solstices and the equinoxes?

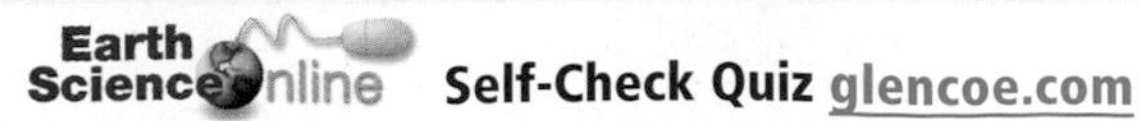

ON SITE: LIVING IN SPACE

This astronaut is working in microgravity. Notice the footholds and handholds the astronauts use to stay in place.

An orbiting space shuttle and everything aboard it, including the astronauts, are falling continuously around Earth. The result is apparent weightlessness—they experience microgravity conditions. Performing everyday tasks, such as sleeping and exercising, is challenging in microgravity. What would it be like to float in space?

Disorientation Some astronauts experience space sickness during the first few days in microgravity. This happens because the brain is confused by the mismatched visual, sensory, and pressure messages that it receives. To help their bodies prepare for microgravity, astronauts train in special airplanes where they experience short periods of free-fall weightlessness. Once sensory systems have adjusted to microgravity, space sickness subsides.

Sleeping How would you sleep without gravity to keep you in bed? Astronauts on the orbiting space shuttle and the *International Space Station* spend about eight hours a day sleeping. Although astronauts can sleep in any orientation they prefer, they must be anchored to something—a wall, a seat, or a bed. This prevents them from floating and bumping into other things while they sleep, which might harm them as well as other astronauts and equipment.

Exercising In orbit, exercising is particularly important to the overall health of astronauts. On Earth, muscles work against the force of gravity to move, maintain balance, and support our bodies. In microgravity, muscles are underused and begin to atrophy, meaning they lose tone and mass.

Supporting the weight of the body on Earth is one of the functions of bones. Scientists know that gravity is important in the process of bone maintenance and formation. In microgravity, bone formation is disrupted and bones lose important minerals. Without proper amounts of these minerals, bones become weaker and the risk of fracture increases. Astronauts exercise each day while in space while strapped to exercise equipment.

WRITING in Earth Science

Interview Suppose you are a newspaper reporter and you will interview an astronaut who has returned from space. Write at least five interview questions about how microgravity affects the human body and the completion of everyday tasks. Include questions about the astronaut's personal experiences. To learn more about space travel, visit glencoe.com.

GEOLAB

MAPPING: DETERMINE RELATIVE AGES OF LUNAR FEATURES

Background: Recall from Chapter 21 that an intrusion or a fault can cut across an older geologic feature. This principle of crosscutting relationships is also used to determine the relative ages of surface features on the Moon. By observing which features cut across others, you can infer which features are older and which are younger.

Question: *How can you use images of the Moon to interpret relative ages of lunar features?*

Materials

paper
metric ruler

Procedure

1. Read and complete the lab safety form.
2. Review the information about the history of the Moon and the lunar surface starting on page 772.
3. Observe Photo 1 and identify the older of the craters in the crater Pairs A-D and C-B using the principle of crosscutting relationships.
4. Observe Photo 2. Identify and list the features in order of their relative ages.
5. Observe Photo 3. Identify the mare, rille, and craters. Then list the features in order of their relative ages.
6. Observe Photo 4. Identify the features using your knowledge of crosscutting relationships and lunar history. Then list the features in order of their relative ages.

Analyze and Conclude

1. **Summarize** the problems you had in identifying and choosing the ages of the features.
2. **Select** Based on information from all the photos, what features are usually the oldest? The youngest?
3. **Explain** whether scientists could use this process to determine the exact age difference between two overlapping craters. Why or why not?
4. **Identify** the relative-age dating that scientists use to analyze craters on Earth.
5. **Evaluate** If the small crater in Photo 2, labeled *A*, is 44 km across, what is the scale for that photo? At that scale, what is the size of the large crater labeled *F*?
6. **Judge** Which would be older, a crater that had rays crossing it, or the crater that caused the rays? Explain.
7. **Estimate** If the crater labeled *A* in Photo 1 is 17 km across, how long is the chain of craters in the photo?
8. **Infer** What might have caused the chain of craters in Photo 1?

WRITING in Earth Science

Guidebook Using what you learned in this lab, prepare a guidebook that contains instructions for identifying and determining relative ages of lunar features. For more information on lunar features, visit glencoe.com.

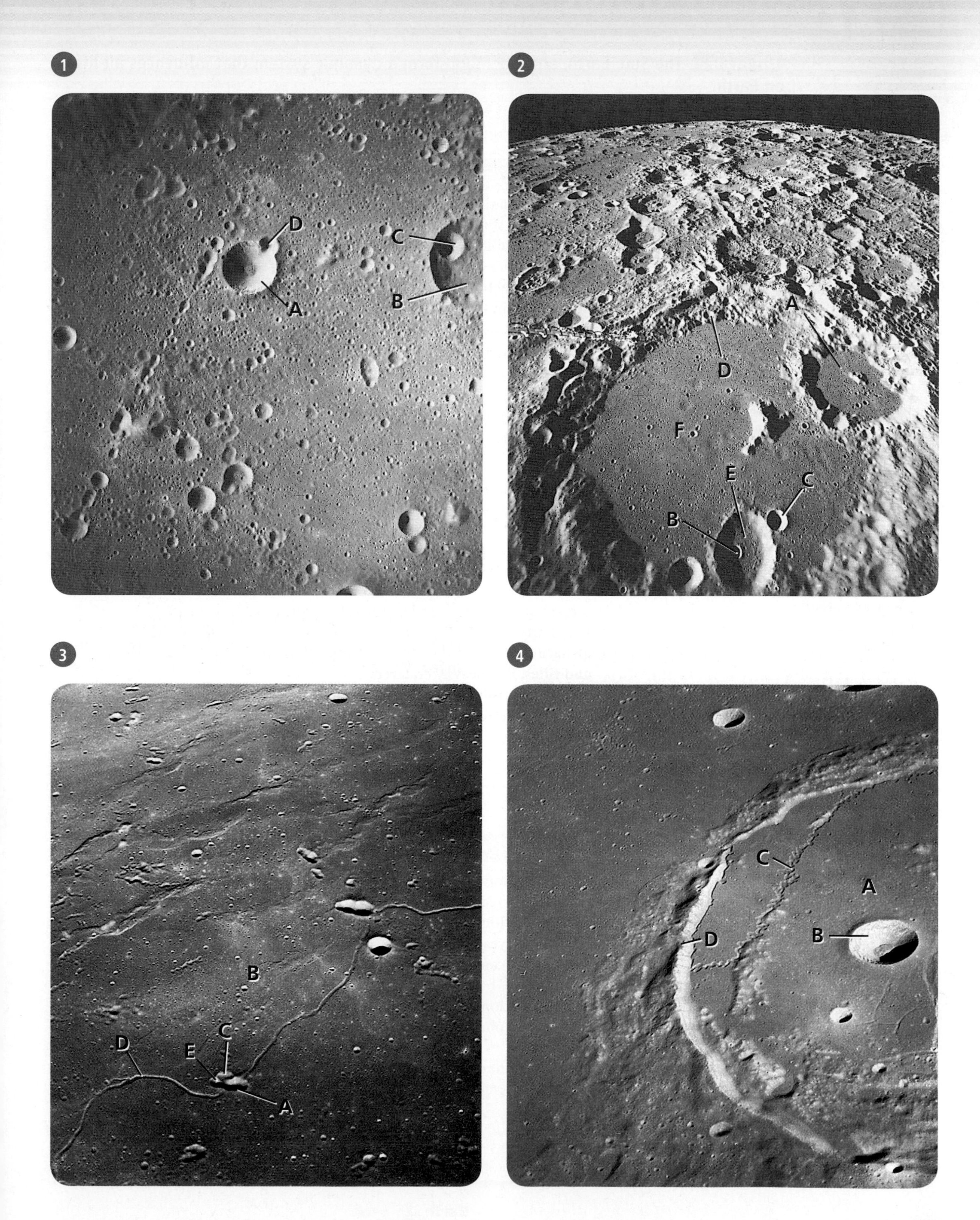
1
D
C
A
B
2
A
D
F
E
C
B
3
B
D
C
E
A
4
C
A
B
D

Study Guide

BIG Idea The Sun, Earth, and the Moon form a dynamic system that influences all life on Earth.

Vocabulary	Key Concepts
Section 27.1 Tools of Astronomy	
• electromagnetic spectrum (p. 764) • interferometry (p. 767) • reflecting telescope (p. 766) • refracting telescope (p. 766)	**MAIN Idea** Radiation emitted or reflected by distant objects allows scientists to study the universe. • Telescopes collect and focus electromagnetic radiation emitted or reflected from distant objects. • Electromagnetic radiation is classified by wavelength and frequency. • The two main types of optical telescopes are refractors and reflectors. • Space-based astronomy includes the study of orbiting telescopes, satellites, and probes. • Technology originally developed to explore space is now used by people on Earth.
Section 27.2 The Moon	
• albedo (p. 771) • ejecta (p. 771) • highland (p. 771) • impact crater (p. 771) • mare (p. 771) • ray (p. 771) • regolith (p. 772) • rille (p. 771)	**MAIN Idea** The Moon, Earth's nearest neighbor in space, is unique among the moons in our solar system. • Astronomers have gathered information about the Moon using telescopes, space probes, and astronaut exploration. • Like Earth's crust, the Moon's crust is composed mostly of silicates. • Surface features on the Moon include highlands, maria, ejecta, rays, and rilles. It is heavily cratered. • The Moon probably formed about 4.5 bya in a collision between Earth and a Mars-size object.
Section 27.3 The Sun-Earth-Moon System	
• apogee (p. 783) • ecliptic plane (p. 776) • equinox (p. 777) • lunar eclipse (p. 784) • perigee (p. 783) • solar eclipse (p. 781) • solstice (p. 777) • synchronous rotation (p. 780)	**MAIN Idea** Motions of the Sun-Earth-Moon system define Earth's day, month, and year. • Earth's rotation defines one day, and Earth's revolution around the Sun defines one year. • Seasons are caused by the tilt of Earth's spin axis relative to the ecliptic plane. • The gravitational attraction of both the Sun and the Moon causes tides. • The Moon's phases result from our view of its lighted side as it orbits Earth. • Solar and lunar eclipses occur when the Sun's light is blocked.

Assessment

Vocabulary Review

Fill in the blanks with the correct vocabulary term from the Study Guide.

1. Linking telescopes to improve the detail in the images obtained is called ________.
2. A telescope that uses curved lenses to focus visible light is called a(n) ________.
3. A(n) ________ can take place only when the Moon is in the new moon phase.

Each of the following sentences is false. Make each sentence true by replacing the italicized words with vocabulary terms from the Study Guide.

4. The Moon's *perigee* is the amount of sunlight that its surface reflects.
5. The far side of the Moon has many more *maria* than the near side.
6. *Interferometry* explains why the same side of the Moon is always visible from Earth.

Match each description below with the correct vocabulary term from the Study Guide.

7. a device that uses a mirror to collect light from distant objects
8. the point in the Moon's orbit when it is farthest from Earth
9. loose, ground-up rock, such as the layer covering much of the surface of the Moon

Understand Key Concepts

10. Which is the highest point in the sky that the Sun reaches on a given day?
    A. ecliptic
    B. solstice
    C. tropic
    D. zenith

Use the diagram below to answer Questions 11 and 12.

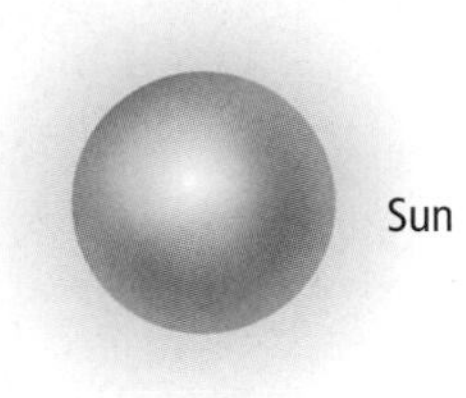

11. In the diagram above, which season is it in the northern hemisphere?
    A. autumn
    B. spring
    C. summer
    D. winter

12. When Earth is in the position shown in the diagram, at which place on Earth is the Sun most likely to be directly overhead at midday?
    A. Arctic Circle
    B. equator
    C. Tropic of Cancer
    D. Tropic of Capricorn

13. Which type of electromagnetic radiation has a longer wavelength than visible light?
    A. gamma ray
    B. X ray
    C. radio wave
    D. ultraviolet ray

14. Which geographic features on the Moon are most likely to be the oldest?
    A. craters
    B. highlands
    C. maria
    D. regolith

15. What is the mineral composition of most moon rocks?
    A. basalts containing water
    B. feldspar with high iron content
    C. sedimentary breccias
    D. silicates

Use the diagram below to answer Questions 16 and 17.

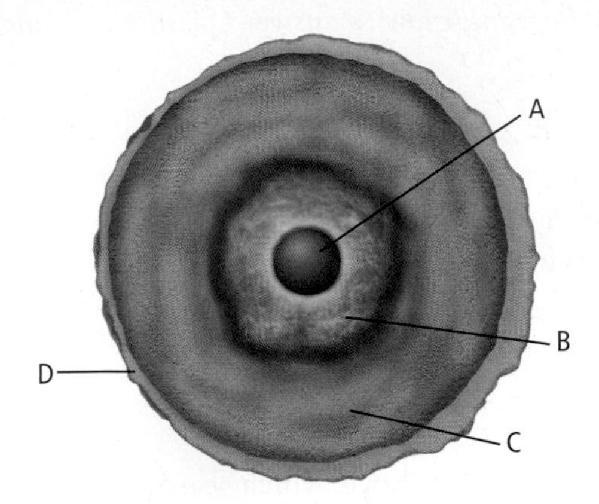

16. Which area of the Moon is physically molten?
    A. core
    B. lower mantle
    C. upper mantle
    D. crust

17. Which area of the Moon is probably solid iron?
    A. core
    B. lower mantle
    C. upper mantle
    D. crust

Constructed Response

18. **Describe** the advantages of placing telescopes in space.

Use the illustration below to answer Question 19.

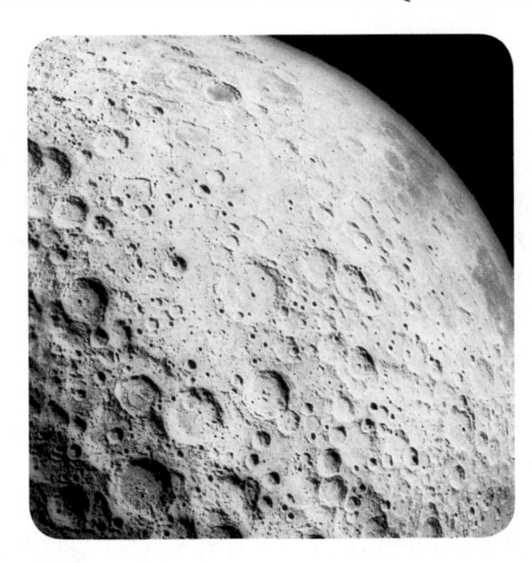

19. **Identify** What part of the lunar surface is most likely shown in this photograph?

20. **Distinguish** between rays and rilles, including where they are found and how they are formed on the Moon.

21. **Summarize** the ways in which Earth's Moon is unusual among all the moons in the solar system.

22. **Assess** the advantages of human missions compared with using robotic spacecraft to explore space.

Think Critically

Use the illustration below to answer Question 23.

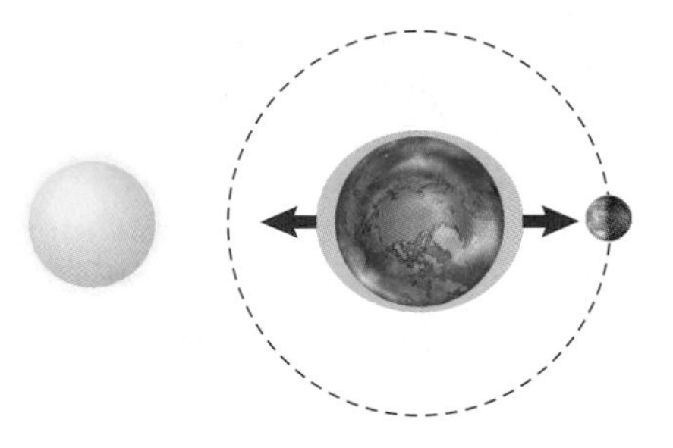

23. **Draw** a diagram similar to the one above to illustrate ocean tides that are not neap or spring.

24. **Infer** Why are lunar breccias not sedimentary like most breccias found on Earth?

25. **Contrast** the geological history of maria with that of the highlands.

26. **Consider** What would seasons be like if Earth were not tilted on its axis?

27. **Draw** a diagram showing the altitude of the Sun at summer solstice viewed from a position of 40° north latitude.

28. **Infer** Would ocean tides exist if Earth had no moon? If so, describe what they would be like.

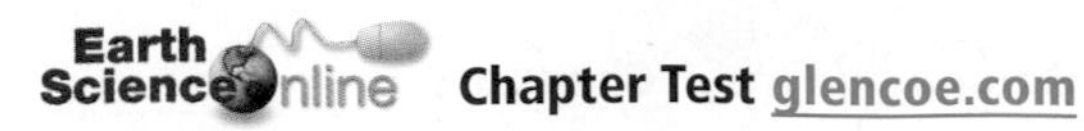

Use the illustration below to answer Questions 29 and 30.

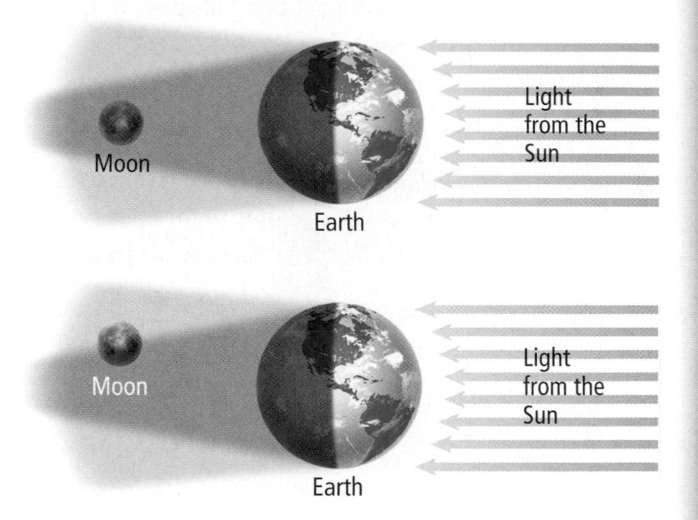

29. **List** the types of shadows as well as the types of eclipses that will be seen by an observer on the unlit side of Earth in each scenario.

30. **Infer** the view of the Sun from the Moon in each scenario.

31. **Appraise** Based on what you know about how maria formed, where would you expect to find the highest concentration of iron?

32. **Compare and contrast** the Moon's interior structure in **Figure 27.11** with Earth's interior structure in **Figure 1.3**.

Concept Mapping

33. Create a concept map using the following terms: *the Moon, albedo, Earth, phases, impact theory, highlands, maria, rilles, craters, rays, breccia,* and *regolith.* Refer to the *Skillbuilder Handbook* for more information.

Challenge Question

34. **Describe** the interrelationship between the Sun, Earth, and the Moon regarding tides and eclipses.

Additional Assessment

35. WRITING in Earth Science Imagine that you are the science officer on a scouting mission from another planet. You just observed the impact that formed Earth's Moon. Write a report describing the event.

Document–Based Questions

Data obtained from: Lang, T. et al. 2004. Cortical and trabecular bone mineral loss from the spine and hip in long-duration spaceflight. *Journal of Bone and Mineral Research* 19 (6).

Bone loss in the lower extremities and spine is a serious problem for astronauts who spend long periods in microgravity. The data below shows the percent loss of bone mineral per month from 13 crew members of the International Space Station.

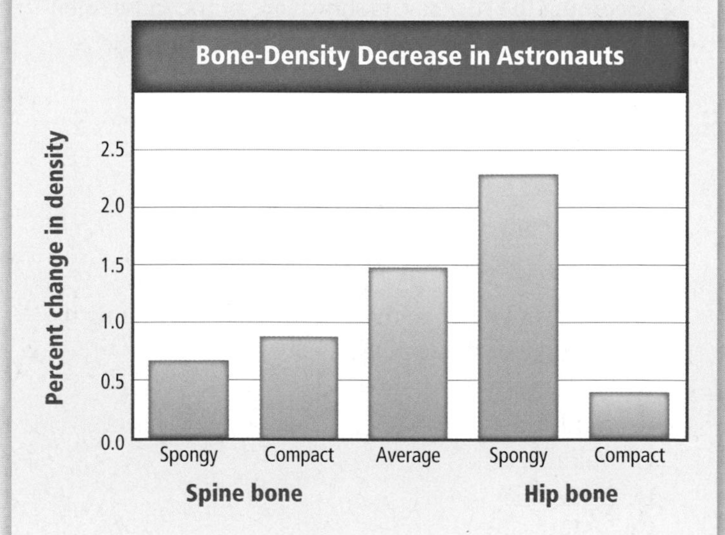

36. Evaluate which body area showed the highest overall rate of bone loss.

37. Compare bone loss of the two types of bone in the hip. Which has the highest rate of loss? By how much?

Cumulative Review

38. What is the source of CFCs and how do CFCs cause ozone depletion? **(Chapter 26)**

39. What are the most common minerals in granite? In basalt? **(Chapter 5)**

Standardized Test Practice

Multiple Choice

1. Which is not considered a renewable resource?
   A. brick
   B. stone
   C. copper
   D. wood

Use the geologic cross section below to answer Questions 2 and 3.

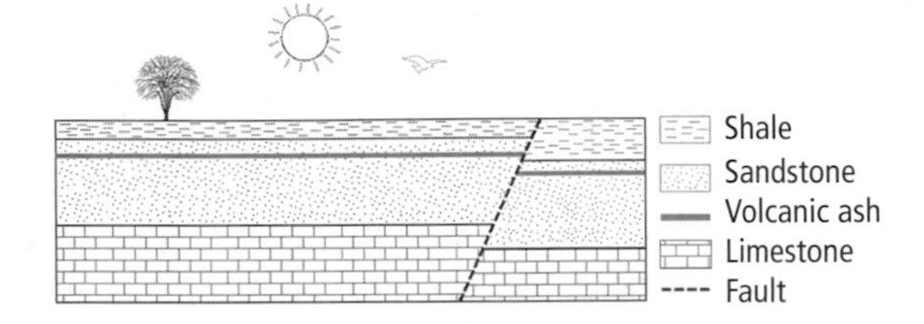

2. Assuming the rock layers shown are in the same orientation that they were deposited, which layer is the oldest?
   A. shale
   B. sandstone
   C. volcanic ash
   D. limestone

3. Which layer would be most helpful in determining the absolute age of these rocks?
   A. shale
   B. sandstone
   C. volcanic ash
   D. limestone

4. Which fossil fuel was originally known as rock oil?
   A. petroleum
   B. natural gas
   C. coal
   D. oil shale

5. In which process does the weight of a subducting plate help pull the trailing lithosphere into a subduction zone?
   A. slab pull
   B. ridge pull
   C. slab push
   D. ridge push

6. What is debris from an impact that falls back to the surface of the Moon called?
   A. rilles
   B. maria
   C. ejecta
   D. albedo

Use the illustrations below to answer Questions 7 to 9.

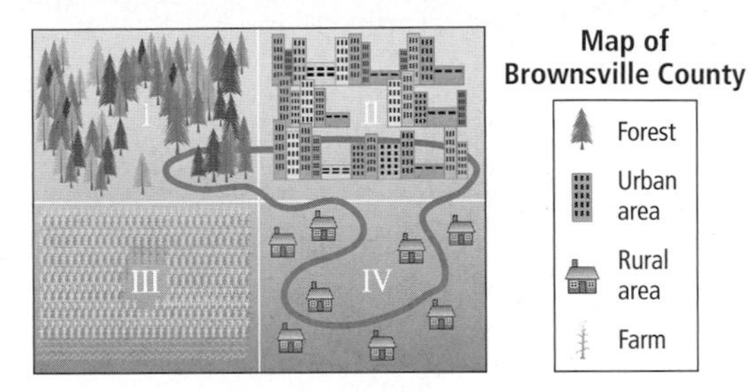

7. Which area of Brownsville is most likely to have problems with flooding during heavy rains?
   A. I
   B. II
   C. III
   D. IV

8. If Brownsville County decided to clear area I in order to expand area III, Brownsville might develop problems with topsoil erosion and pesticide pollution. What might be one way to minimize harmful effects?
   A. deforestation
   B. clear-cutting
   C. monoculture
   D. selective logging

9. What will happen if the size of Brownsville's human population reaches the carrying capacity for its environment?
   A. There will be more births than deaths.
   B. The death rate will increase and the birth rate will increase.
   C. The population will reach equilibrium.
   D. The death rate will increase and the birth rate will decrease.

10. The Marianas Islands in the Pacific Ocean were formed by volcanic action. Which is a TRUE statement?
    A. There are glaciers near the Marianas Islands.
    B. Tectonic plates collide near the Marianas Islands.
    C. The Marianas Islands are larger than most islands.
    D. The Marianas Islands are uninhabited.

Earth Science Online Standardized Test Practice glencoe.com

Short Answer

Use the diagram below to answer Questions 11 to 13.

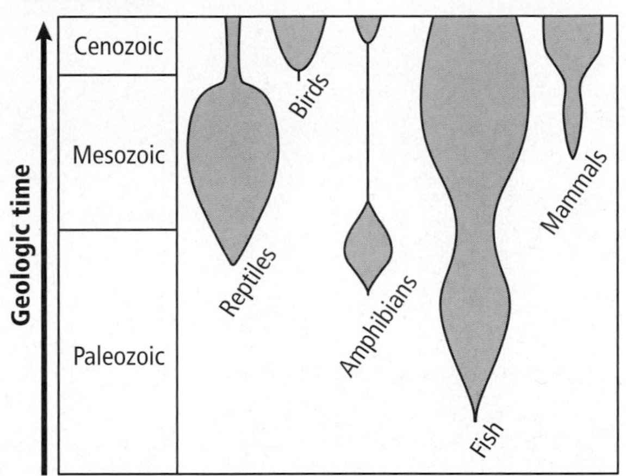

11. If a wider bar represents more species of that type of organism, explain the change in diversity of amphibians from their introduction to present time.

12. What can be inferred about the conditions on Earth for living things from the beginning of the Cenozoic Era to present time?

13. How might the idea that oceans developed before land be supported by looking at this diagram?

14. How does passive solar heating differ from active solar heating?

15. Why is improving the energy efficiency of automobiles important?

16. What two major flaws did scientists of Wegener's day cite as reasons to reject his hypothesis of continental drift?

Reading for Comprehension

Space Observatories

Why put observatories in space? Most telescopes are on the ground where you can deploy a heavier telescope and fix it more easily. The trouble is that earthbound telescopes must look through the Earth's atmosphere which blocks out a broad range of the electromagnetic spectrum, allowing a narrow band of visible light to reach the surface. Telescopes that explore the universe using light beyond the visible spectrum, such as those onboard the *CHANDRA X-Ray Observatory* need to be carried above the absorbing atmosphere. The Earth's atmosphere also blurs the light it lets through. The blurring is caused by varying density and continual motion of air. By orbiting above Earth's atmosphere, the *Hubble Space Telescope* gets clearer images.

Article obtained from: Astronomy picture of the day. *Hubble* Floats Free. *NASA*. November 24, 2002. (Online resource accessed October 17, 2006.).

17. What is a benefit of earthbound telescopes?
- **A.** They can be larger and are more easily fixed.
- **B.** They are able to capture the entire electromagnetic spectrum.
- **C.** They can use larger mirrors.
- **D.** They can capture the visible light reaching Earth's surface.

18. What can be inferred from this passage?
- **A.** Earthbound telescopes have no benefits for scientific study.
- **B.** Using telescopes outside the Earth's atmosphere produces the clearest pictures.
- **C.** The *Hubble Space Telescope* needs to have larger mirrors to take better pictures.
- **D.** It is impossible to fix telescopes orbiting outside Earth's atmosphere.

NEED EXTRA HELP?																
If You Missed Question . . .	1	2	3	4	5	6	7	8	9	10	11	12	13	14	15	16
Review Section . . .	24.1	21.2	21.3	25.1	17.4	27.2	26.2	26.2	26.1	17.3	23.1	23.3	23.1	25.2	25.3	17.1

CHAPTER 28 Our Solar System

BIG Idea **Using the laws of motion and gravitation, astronomers can understand the orbits and the properties of the planets and other objects in the solar system.**

28.1 Formation of the Solar System
MAIN Idea The solar system formed from the collapse of an interstellar cloud.

28.2 The Inner Planets
MAIN Idea Mercury, Venus, Earth, and Mars have high densities and rocky surfaces.

28.3 The Outer Planets
MAIN Idea Jupiter, Saturn, Uranus, and Neptune have large masses, low densities, and many moons and rings.

28.4 Other Solar System Objects
MAIN Idea Rocks, dust, and ice compose the remaining 2 percent of the solar system.

GeoFacts

- It is likely that Jupiter was the first planet in the solar system to form.
- It rains sulfuric acid on Venus.
- Mercury's days are two-thirds the length of its years.

LAUNCH Lab

What can be learned from space missions?

Most of the planets in our solar system have been explored by uncrewed space probes. You can learn about these missions and their discoveries by using a variety of resources. Both the agencies that sponsor missions and the scientists involved usually provide extensive information about the design, operation, and scientific goals of the missions.

Procedure

1. Read and complete the lab safety form.
2. Go to glencoe.com and find information on missions to four different planets.
3. Draw a table listing some of the key aspects of each mission. Include the type of mission (flyby, lander, or orbiter), the scientific goals, the launch date, and the date of arrival at the planet.

Analysis

1. **Summarize** in a table what scientists learned from each mission or what they hope to learn.
2. **Determine** which missions are still in progress, which ones have gone beyond their mission life, and which ones have been completed.
3. **Suggest** other missions that could be conducted in the future.

The Planets Make the following Foldable that features the planets of our solar system.

STEP 1 Fold a sheet of paper in half.

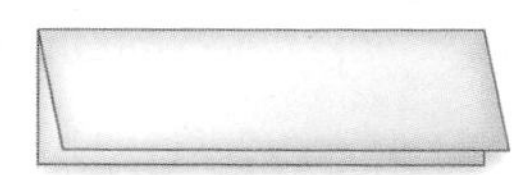

STEP 2 Fold in half and then in half again to form eight sections.

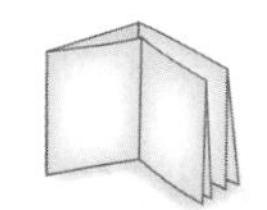

STEP 3 Cut along the long fold line, stopping before you reach the last two sections.

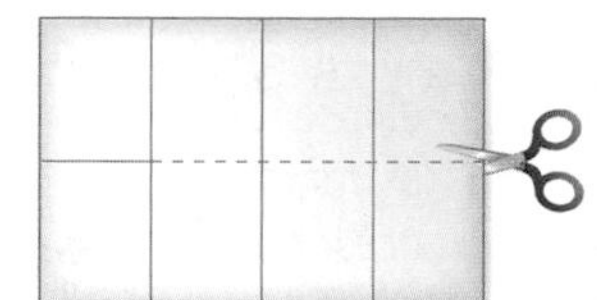

STEP 4 Refold the paper into an accordion book. You might want to glue the double pages together.

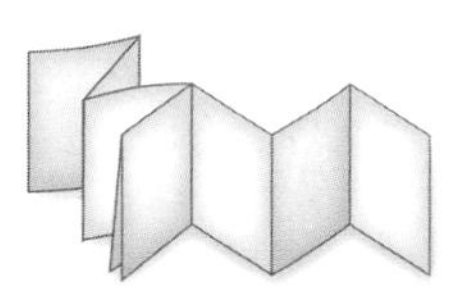

FOLDABLES **Use this Foldable with Sections 28.1, 28.2, and 28.3.** As you read these sections, summarize the main characteristics of the planets.

Visit glencoe.com to

- **study entire chapters online;**
- **explore Concepts In Motion animations:**
  - **Interactive Time Lines**
  - **Interactive Figures**
  - **Interactive Tables**
- **access Web Links for more information, projects, and activities;**
- **review content with the Interactive Tutor and take Self-Check Quizzes.**

Section 28.1

Objectives

- **Explain** how the solar system formed.
- **Describe** early concepts of the structure of the solar system.
- **Describe** how our current knowledge of the solar system developed.
- **Relate** gravity to the motions of the objects in the solar system.

Review Vocabulary

focus: one of two fixed points used to define an ellipse

New Vocabulary

planetesimal
retrograde motion
ellipse
astronomical unit
eccentricity

Formation of the Solar System

MAIN Idea The solar system formed from the collapse of an interstellar cloud.

Real-World Reading Link If you have ever made a snowman by rolling a snowball over the ground, you have demonstrated how planets formed from tiny grains of matter.

Formation Theory

Theories of the origin of the solar system rely on direct observations and data from probes. Scientific theories must explain observed facts, such as the shape of the solar system, differences among the planets, and the nature of the oldest planetary surfaces—asteroids, meteorites, and comets.

A Collapsing Interstellar Cloud

Stars and planets form from interstellar clouds, which exist in space between the stars. These clouds consist mostly of hydrogen and helium gas with small amounts of other elements and dust. Dust makes interstellar clouds look dark because it blocks the light from stars within or behind the clouds. Often, starlight reflects off of the dust and partially illuminates the clouds. Also, stars can heat clouds, making them glow on their own. This is why interstellar clouds often appear as blotches of light and dark, as shown in **Figure 28.1.** This interstellar dust can be thought of as a kind of smog that contains elements formed in older stars, which expelled their matter long ago.

At first, the density of interstellar gas is low—much lower than the best vacuums created in laboratories. However, gravity slowly draws matter together until it is concentrated enough to form a star and possibly planets. Astronomers think that the solar system began this way. They have also observed planets around other stars, and hope that studying such planet systems will provide clues to how our solar system formed.

Figure 28.1 Stars form in collapsing interstellar clouds, such as in the Eagle nebula, pictured here.

■ **Figure 28.2** The interstellar cloud that formed our solar system collapsed into a rotating disk of dust and gas. When concentrated matter in the center acquired enough mass, the Sun formed in the center and the remaining matter gradually condensed, forming the planets.

Collapse accelerates At first, the collapse of an interstellar cloud is slow, but it gradually accelerates and the cloud becomes much denser at its center. If rotating, the cloud spins faster as it contracts, for the same reason that ice skaters spin faster as they pull their arms close to their bodies—centripetal force. As the collapsing cloud spins, the rotation slows the collapse in the equatorial plane, and the cloud becomes flattened. Eventually, the cloud becomes a rotating disk with a dense concentration of matter at the center, as shown in **Figure 28.2.**

 **Reading Check** **Explain** why the rotating disk spins faster as it contracts.

Matter condenses Astronomers think our solar system began in this manner. The Sun formed when the dense concentration of gas and dust at the center of a rotating disk reached a temperature and pressure high enough to fuse hydrogen into helium. The rotating disk surrounding the young Sun became our solar system. Within this disk, the temperature varied greatly with location; the area closest to the dense center was still warm, while the outer edge of the disk was cold. This temperature gradient resulted in different elements and compounds condensing, depending on their distance from the Sun. This also affected the distribution of elements in the forming planets. The inner planets are richer in the higher melting point elements and the outer planets are composed mostly of the more volatile elements. That is why the outer planets and their moons consist mostly of gases and ices. Eventually, the condensation of materials into liquid and solid forms slowed.

VOCABULARY
ACADEMIC VOCABULARY
Collapse
to fall down, give way, or cave in
The hot-air balloon collapsed when the fabric was torn.

To read more about ways that astronomers are studying the formation of the solar system, go to the **National Geographic Expedition** on page 934.

Concepts In Motion
Interactive Table To explore more about the planets, visit glencoe.com.

Table 28.1 Physical Data of the Planets

Planet	Diameter (km)	Relative Mass (Earth = 1)	Average Density (kg/m³)	Atmosphere	Distance from the Sun (AU)	Moons
Mercury	4,880	0.06	5430	none	0.39	0
Venus	12,104	0.821	5240	CO_2, N_2	0.72	0
Earth	12,742	1.00	5520	N_2, O_2, H_2O	1.00	1
Mars	6,778	0.21	3930	CO_2, N_2, Ar	1.52	2
Jupiter	139,822	317.8	1330	H_2, He	5.2	63
Saturn	116,464	95.2	700	H_2, He	9.58	47
Uranus	50,724	14.5	1300	H_2, He, CH_4	19.2	27
Neptune	49,248	17.1	1760	H_2, He, CH_4	30.04	13

Planetesimals

FOLDABLES
Incorporate information from this section into your Foldable.

Next, the tiny grains of condensed material started to accumulate and merge, forming larger particles. These particles grew as grains collided and stuck together and as gas particles collected on their surfaces. Eventually, colliding particles in the early solar system merged to form **planetesimals**—objects hundreds of kilometers in diameter. Growth continued as planetesimals collided and merged. Sometimes, collisions destroyed planetesimals, but the overall result was a smaller number of larger bodies—the planets. Some of their properties are given in **Table 28.1.**

Gas giants form The first large planet to develop was Jupiter. Jupiter increased in size through the merging of icy planetesimals that contained mostly lighter elements. It grew larger as its gravity attracted additional gas, dust, and planetesimals. Saturn and the other gas giants formed similarly, but they could not become as large because Jupiter had collected so much of the available material. As each gas giant attracted material from its surroundings, a disk formed in its equatorial plane, much like the disk of the early solar system. In this disk, matter clumped together to form rings and satellites.

CAREERS IN EARTH SCIENCE

Planetologist A planetologist applies the theories and methods of sciences, such as physics, chemistry, and geology, as well as mathematics, to study the origin, composition, and distribution of matter in planetary systems. To learn more about Earth science careers, visit glencoe.com.

Terrestrial planets form Planets also formed by the merging of planetesimals in the inner part of the main disk, near the young Sun. These were composed primarily of elements that resist vaporization, so the inner planets are rocky and dense, in contrast to the gaseous outer planets. Also, scientists think that the Sun's gravitational force swept up much of the gas in the area of the inner planets and prevented them from acquiring much of this material from their surroundings. Thus, the inner planets did not develop satellites.

Debris Material that remained after the formation of the planets and satellites is called debris. Eventually, the amount of interplanetary debris diminished as it crashed into planets or was diverted out of the solar system. Some debris that was not ejected from the solar system became icy objects known as comets. Other debris formed rocky planetesimals known as asteroids. Most asteroids are found in the area between Jupiter and Mars known as the asteroid belt, shown in **Figure 28.3.** They remain there because Jupiter's gravitational force prevented them from merging to form a planet.

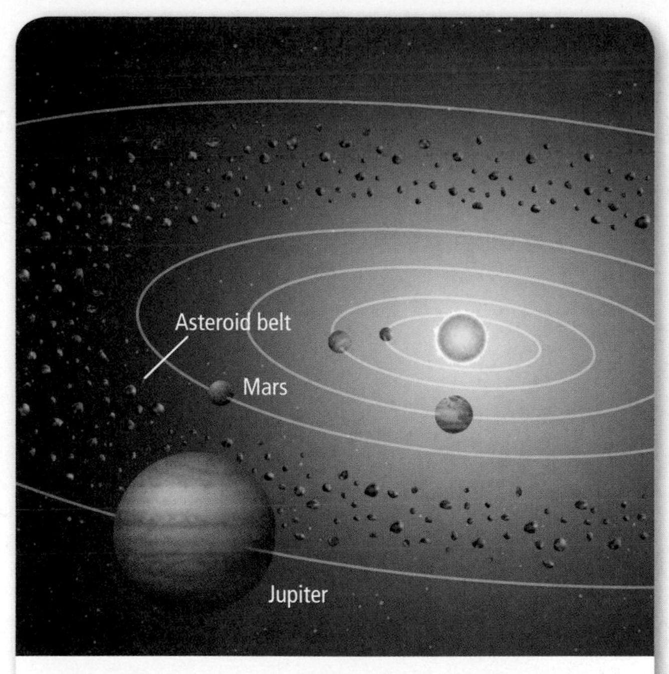

■ **Figure 28.3** Thousands of asteroids have been detected in the asteroid belt, which lies between Mars and Jupiter.

Modeling the Solar System

Ancient astronomers assumed that the Sun, planets, and stars orbited a stationary Earth in an Earth-centered model of the solar system. They thought this explained the most obvious daily motion of the stars and planets rising in the east and setting in the west. But as you learned in Chapter 27, this does not happen because these bodies orbit Earth, but rather that Earth spins on its axis.

This geocentric (jee oh SEN trihk), or Earth-centered, model could not readily explain some other aspects of planetary motion. For example, the planets might appear farther to the east one evening, against the background of the stars, than they had the previous night. Sometimes a planet seems to reverse direction and move back to the west. The apparent backward movement of a planet is called **retrograde motion.** The retrograde motion of Mars is shown in the time-lapse image and diagram in **Figure 28.4.** The search for a simple explanation of retrograde motion motivated early astronomers to keep searching for a better explanation for the design of the solar system.

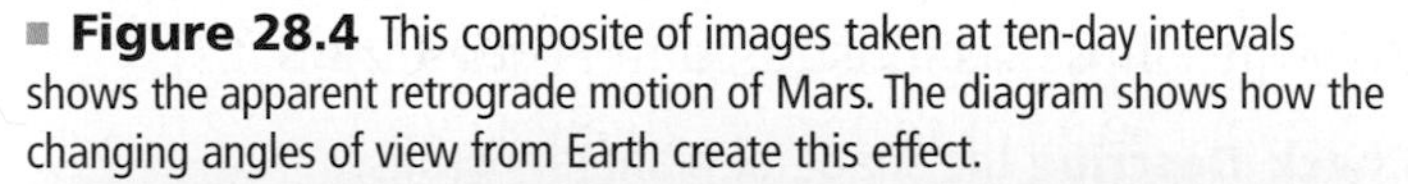

■ **Figure 28.4** This composite of images taken at ten-day intervals shows the apparent retrograde motion of Mars. The diagram shows how the changing angles of view from Earth create this effect.

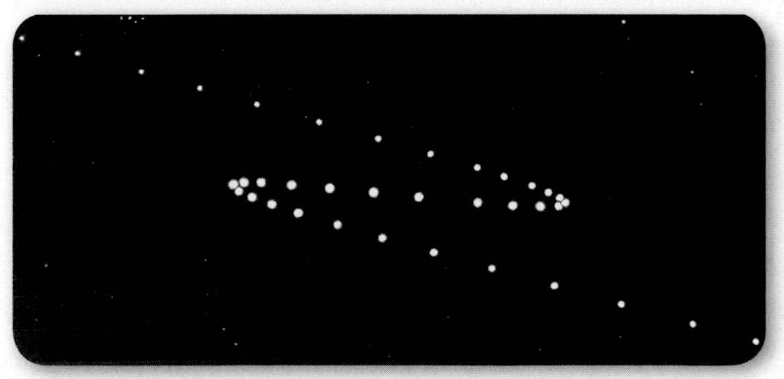

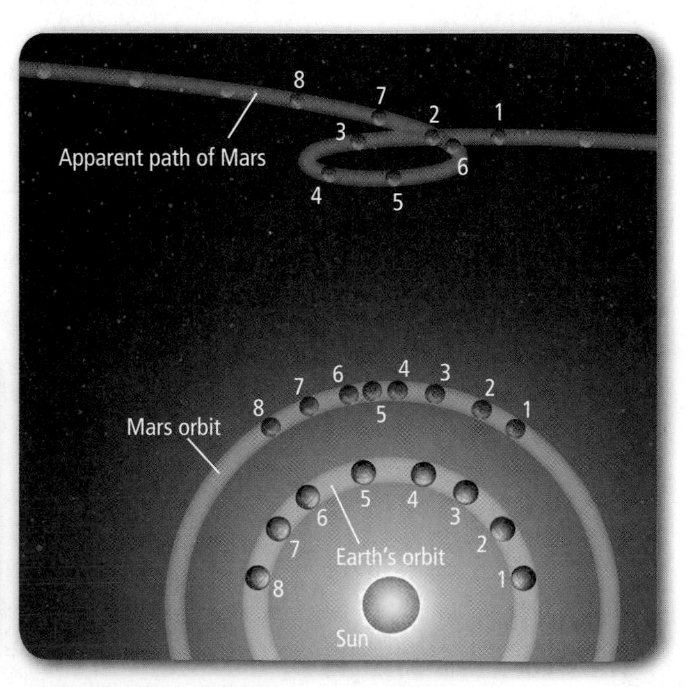

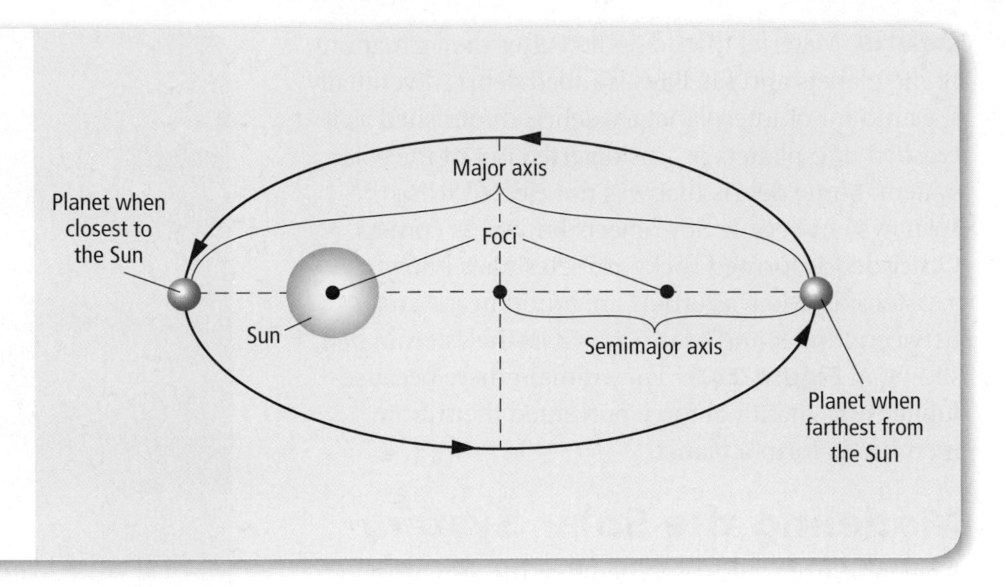

■ **Figure 28.5** This diagram shows the geometry of an ellipse using an exaggerated planetary orbit. The Sun lies at one of the two foci. The minor axis of the ellipse is its shorter diameter. The major axis of the ellipse is its longer diameter, which equals the distance between a planet's closest and farthest points from the Sun. Half of the semimajor axis represents the average distance of the planet to the Sun.

Heliocentric model In 1543, Polish scientist Nicolaus Copernicus suggested that the Sun was the center of the solar system. In this Sun-centered, or heliocentric (hee lee oh SEN trihk) model, Earth and all the other planets orbit the Sun. In a heliocentric model, the increased gravity of proximity to the Sun causes the inner planets to move faster in their orbits than do the outer planets. It also provided a simple explanation for retrograde motion.

Kepler's first law Within a century, the ideas of Copernicus were confirmed by other astronomers, who found evidence that supported the heliocentric model. For example, Tycho Brahe (TIE coh BRAH), a Danish astronomer, designed and built very accurate equipment for observing the stars. From 1576–1601, before the telescope was used in astronomy, he made accurate observations of the planets' positions. Using Brahe's data, German astronomer Johannes Kepler demonstrated that each planet orbits the Sun in a shape called an ellipse, rather than a circle. This is known as Kepler's first law of planetary motion. An **ellipse** is an oval shape that is centered on two points instead of a single point, as in a circle. The two points are called the foci (singular, focus). The major axis is the line that runs through both foci at the maximum diameter of the ellipse, as illustrated in **Figure 28.5.**

VOCABULARY
SCIENCE USAGE V. COMMON USAGE
Law
Science usage: a general relation proved or assumed to hold between mathematical expressions

Common usage: a rule of conduct prescribed as binding and enforced by a controlling authority

Reading Check Describe the shape of planetary orbits.

Each planet has its own elliptical orbit, but the Sun is always at one focus. For each planet, the average distance between the Sun and the planet is its semimajor axis, which equals half the length of the major axis of its orbit, as shown in **Figure 28.5.** Earth's semimajor axis is of special importance because it is a unit used to measure distances within the solar system. Earth's average distance from the Sun is 1.496×10^8 km, or 1 **astronomical unit** (AU). Distance in space is often measured in Au. For example, Mars is 1.52 Au from the Sun.

Eccentricity A planet in an elliptical orbit does not orbit at a constant distance from the Sun. The shape of a planet's elliptical orbit is defined by **eccentricity,** which is the ratio of the distance between the foci to the length of the major axis. You will investigate this ratio in the MiniLab. The orbits of most planets are not very eccentric; in fact, some are almost perfect circles.

The eccentricity of a planet can change slightly. Earth's eccentricity today is about 0.02, but the gravitational attraction of other planets can stretch the eccentricity to 0.05, or cause it to fall to 0.01.

Kepler's second and third laws In addition to discovering the shapes of planetary orbits, Kepler showed that planets move faster when they are closer to the Sun. He demonstrated this by proving that an imaginary line between the Sun and a planet sweeps out equal amounts of area in equal amounts of time, as shown in **Figure 28.6.** This is known as Kepler's second law.

The length of time it takes for a planet or other body to travel a complete orbit around the Sun is called its orbital period. In Kepler's third law of planetary motion, he determined the mathematical relationship between the size of a planet's ellipse and its orbital period. This relationship is written as follows:

$$P^2 = a^3$$

P is time measured in Earth years, and a is length of the semimajor axis measured in astronomical units.

MiniLab

Explore Eccentricity

How is eccentricity of an ellipse calculated? Eccentricity is the ratio of the distance between the foci to the length of the major axis. Eccentricity ranges from 0 to 1; the larger the eccentricity, the more extreme the ellipse.

Procedure

WARNING: *Use caution when handling sharp objects.*

1. Read and complete the lab safety form.
2. Tie a piece of **string** to form a circle that will fit on a piece of **cardboard.**
3. Place a sheet of **paper** on the cardboard.
4. Stick two **pins** in the paper a few centimeters apart and on a line that passes through the center point of the paper.
5. Loop the string over the pins, and keeping the string taut, use a pencil to trace an ellipse around the pins.
6. Use a **ruler** to measure the major axis and the distance between the pins. Calculate the eccentricity.

Analysis

1. **Identify** what the two pins represent.
2. **Explain** how the eccentricity changes as the distance between the pins changes.
3. **Predict** the kind of figure formed and the eccentricity if the two pins were at the same location.

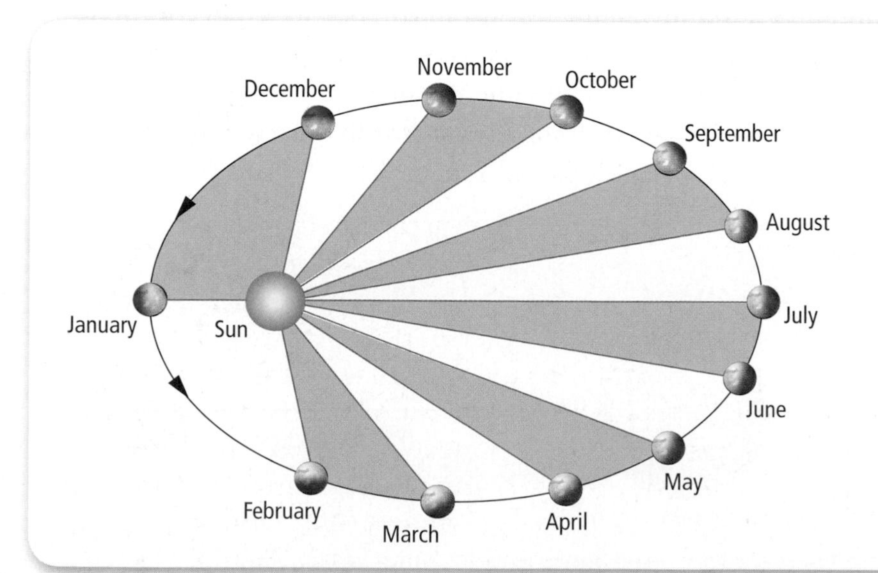

Figure 28.6 Kepler's second law states that planets move faster when close to the Sun and slower when farther away. This means that a planet sweeps out equal areas in equal amounts of time. (Note: *not drawn to scale*)

■ **Figure 28.7** Galileo would probably be astounded to see Jupiter's moons in the composite image above. Still, his view of Jupiter and its moons proved a milestone in support of heliocentric theory.

Galileo While Kepler was developing his ideas, Italian scientist Galileo Galilei became the first person to use a telescope to observe the sky. Galileo made many discoveries that supported Copernicus's ideas. The most famous of these was his discovery that four moons orbit the planet Jupiter, proving that not all celestial bodies orbit Earth, and demonstrating that Earth was not necessarily the center of the solar system. Galileo's view of Jupiter's moons, similar to the chapter opener photo, is compared with our present-day view of them, shown in **Figure 28.7.** The underlying explanation for the heliocentric model remained unknown until 1684, when English scientist Isaac Newton published his law of universal gravitation.

Gravity

Newton first developed an understanding of gravity by observing falling objects. He described falling as downward acceleration produced by gravity, an attractive force between two objects. He determined that both the masses of and the distance between two bodies determined the force between them. This relationship is expressed in his law of universal gravitation, illustrated in **Figure 28.8,** and that is stated mathematically as follows:

$$F = \frac{Gm_1m_2}{r^2}$$

F is the force measured in newtons, G is the universal gravitation constant (6.6726×10^{-11} m^3/ kg·s^2), m_1 and m_2 are the masses of the bodies in kilograms, and r is the distance between the two bodies in meters.

Gravity and orbits Newton realized that this attractive force could explain why planets move according to Kepler's laws. He observed the Moon's motion and realized that its direction changes because of the gravitational attraction of Earth. In a sense, the Moon is constantly falling toward Earth. If it were not for this attraction, the Moon would continue to move in a straight line and would not orbit Earth. The same is true of the planets and their moons, stars, and all orbiting bodies throughout the universe.

Concepts In Motion

Interactive Figure To see an animation of gravitational attraction, visit glencoe.com.

■ **Figure 28.8** The gravitational attraction between these two objects is 3.3×10^{-10} N.
Predict *the effect of doubling the masses of both objects, and check your prediction using Newton's equation.*

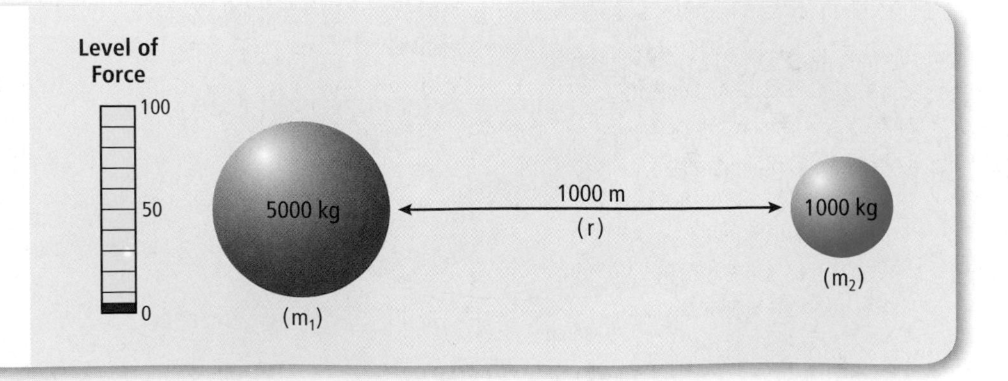

Center of mass Newton also determined that each planet orbits a point between it and the Sun called the center of mass. For any planet and the Sun, the center of mass is just above or within the surface of the Sun, because the Sun is much more massive than any planet. **Figure 28.9** shows how this is similar to the balance point on a seesaw.

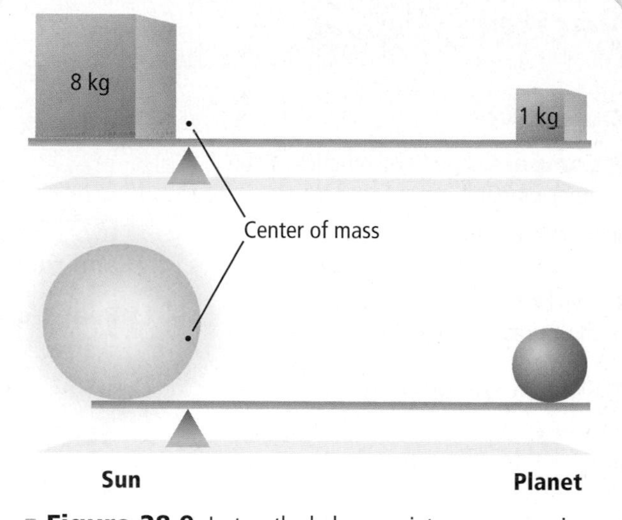

Figure 28.9 Just as the balance point on a seesaw is closer to the heavier box, the center of mass between two orbiting bodies is closer to the more massive body.

Present-Day Viewpoints

Astronomers traditionally divided the planets into two groups: the four smaller, rocky, inner planets, Mercury, Venus, Earth, and Mars; and the four outer gas planets, Jupiter, Saturn, Uranus, and Neptune. It was not clear how to classify Pluto, because it is different from the gas giants in composition and orbit. Pluto also did not fit the present-day theory of how the solar system developed. Then in the early 2000s, astronomers discovered a vast number of small, icy bodies inhabiting the outer reaches of the solar system, thousands of AU beyond the orbit of Neptune. At least one of these is larger than Pluto.

These discoveries have led many astronomers to rethink traditional views of the solar system. Some already define it in terms of three zones: Zone 1, Mercury, Venus, Earth, Mars; Zone 2, Jupiter, Saturn, Uranus, Neptune; and Zone 3, everything else, including Pluto. In science, views change as new data becomes available and new theories are proposed. Astronomy today is a rapidly changing field.

Section 28.1 Assessment

Section Summary

- A collapsed interstellar cloud formed the Sun and planets from a rotating disk.
- The inner planets formed closer to the Sun than the outer planets, leaving debris to produce asteroids and comets.
- Copernicus created the heliocentric model and Kepler defined its shape and mechanics.
- Newton explained the forces governing the solar system bodies and provided proof for Kepler's laws.
- Present-day astronomers divide the solar system into three zones.

Understand Main Ideas

1. MAIN Idea **Describe** the formation of the solar system.
2. **Explain** why retrograde motion is an apparent motion.
3. **Describe** how the gravitational force between two bodies is related to their masses and the distance between them.
4. **Compare** the shape of two ellipses having eccentricities of 0.05 and 0.75.

Think Critically

5. **Infer** Based on what you have learned about Kepler's third law, which planet moves faster in its orbit: Jupiter or Neptune? Explain.

MATH in Earth Science

6. Use Newton's law of universal gravitation to calculate the force of gravity between two students standing 12 m apart. Their masses are 65 kg and 50 kg.

Section 28.2

Objectives

- **Compare** the characteristics of the inner planets.
- **Survey** some of the space probes used to explore the solar system.
- **Explain** the differences among the terrestrial planets.

Review Vocabulary

albedo: the amount of sunlight that reflects from the surface

New Vocabulary

terrestrial planet
scarp

The Inner Planets

MAIN Idea Mercury, Venus, Earth, and Mars have high densities and rocky surfaces.

Real-World Reading Link Just as in a family in which brothers and sisters share a strong resemblance, the inner planets share many characteristics.

Terrestrial Planets

The four inner planets are called **terrestrial planets** because they are similar in density to Earth and have solid, rocky surfaces. Their average densities, obtained by dividing the mass of a planet by its volume, range from about 3.5 to just over 5.5 g/cm^3. Average density is an important indicator of internal conditions, and densities in this range indicate that the interiors of these planets are compressed.

Mercury

Mercury is the planet closest to the Sun, and for this reason it is difficult to see from Earth. During the day it is lost in the Sun's light and it is more easily seen at sunset and sunrise. Mercury is about one-third the size of Earth and has a smaller mass. Mercury has no moons. Radio observations in the 1960s revealed that Mercury has a slow spin of 1407.6 hours. In one orbit around the Sun, Mercury rotates one and one-half times, as shown in **Figure 28.10.** As Mercury spins, the side facing the Sun at the beginning of the orbit faces away from the Sun at the end of the orbit. This means that two complete Mercury years equal three complete Mercury days.

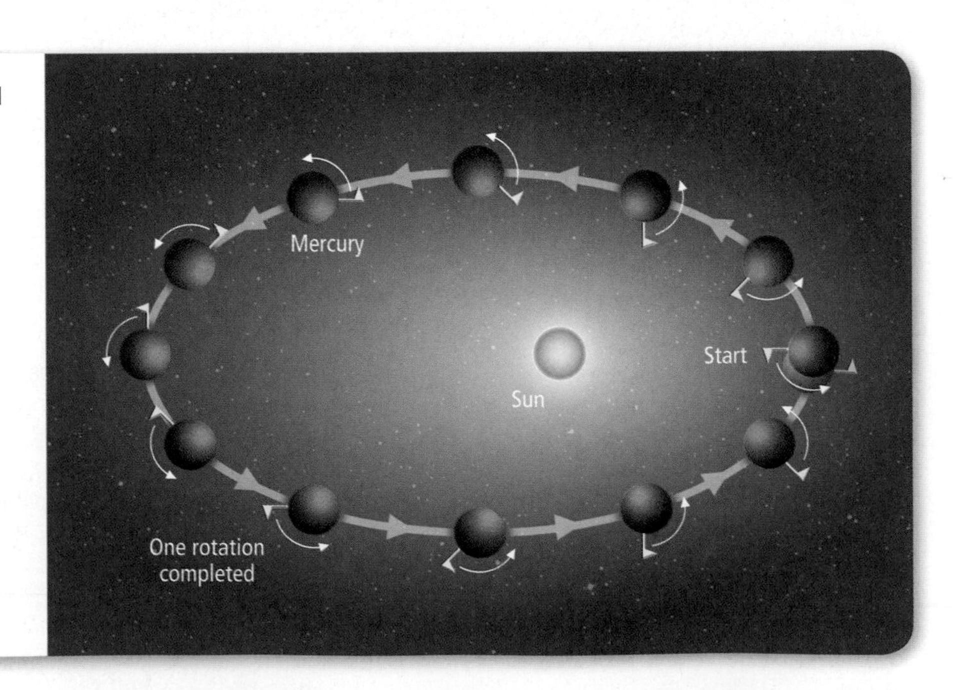

Figure 28.10 Because of Mercury's odd rotation, its day lasts for two-thirds of its year. **Compare** *Mercury's orbital motion with that of Earth's Moon.*

■ **Figure 28.11** This mosaic of Mercury's heavily cratered surface was made by *Mariner 10.* Craters range in size from 100 to 1300 km in diameter.

Atmosphere Unlike Earth and the other planets, Mercury's atmosphere is constantly being replenished by the solar wind. What little atmosphere does exist is composed primarily of oxygen and sodium atoms deposited by the Sun. The daytime surface temperature on Mercury is 700 K (427°C), while temperatures at night fall to 100 K (−173°C). This is the largest day-night temperature difference among the planets.

Surface Most knowledge about Mercury is based on the radio observations from Earth, and images from U.S. space probe *Mariner 10,* which passed close to Mercury three times in 1974 and 1975. Images from *Mariner 10* show that Mercury's surface, like that of the Moon, is covered with craters and plains, as shown in **Figure 28.11.** The plains on Mercury's surface are smooth and relatively crater free. Scientists think that the plains formed from lava flows that covered cratered terrain, much like the maria formed on the Moon. The surface gravity of Mercury is much greater than that of the Moon, resulting in smaller crater diameters and shorter lengths of ejecta.

Mercury has a planetwide system of cliffs called **scarps,** such as the one shown in **Figure 28.12.** Though similar to those on Earth, Mercury's scarps are much higher. Scientists hypothesize that the scarps developed as Mercury's crust shrank and fractured early in the planet's geologic history. Scientists will learn more about the surface of Mercury with the arrival of the Japanese-European *Messenger* mission in 2011.

Reading Check **Compare** the surfaces of the Moon and Mercury.

Interior Without seismic data, scientists have no way to analyze the interior of Mercury. However, its high density suggests that Mercury has a large nickel-iron core. Mercury's small magnetic field indicates that some of its core is molten.

■ **Figure 28.12** Discovery, the largest scarp on Mercury, is 550 km long and 1.5 km high.

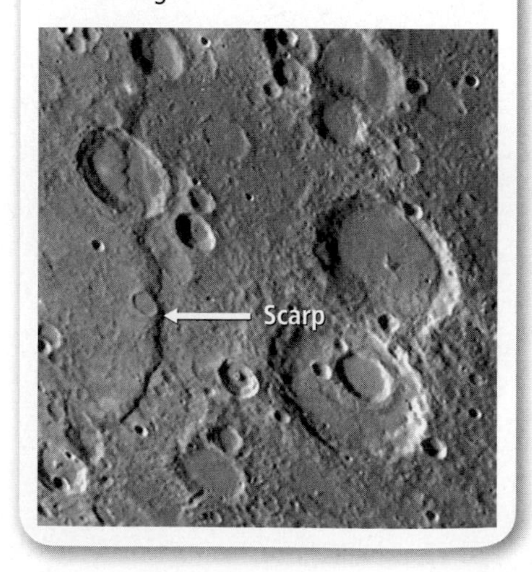

■ **Figure 28.13** The structure of Mercury's interior, which contains a proportionally larger core than Earth, suggests that Mercury was once much larger.

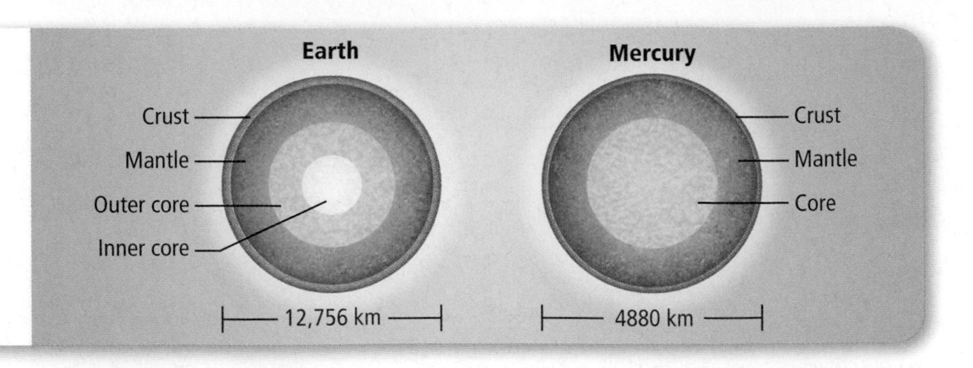

Early Mercury Mercury's small size, high density, and probable molten interior resemble what Earth might be like if its crust and mantle were removed, as shown in **Figure 28.13.** These observations suggest that Mercury was originally much larger, with a mantle and crust similar to Earth's, and that the outer layers might have been lost in a collision with another celestial body early in its history.

Venus

Venus and Mercury are the only two planets closer to the Sun than Earth. Like Mercury, Venus has no moons. Venus is the brightest planet in the sky because it is close to Earth and because its albedo is 0.75—the highest of any planet. Venus is the first bright "star" to be seen after sunset in the western sky, or the last "star" to be seen before sunrise in the morning, depending on which side of the Sun it is on. For these reasons it is often called either the evening or morning star.

Thick clouds around Venus prevent astronomers from observing the surface directly. However, astronomers learned much about Venus from spacecraft launched by the United States and the Soviet Union. Some probes landed on the surface of the planet, and others flew by. Then, the 1978 *Pioneer-Venus* and 1989 *Magellan* missions of the United States used radar to map 98 percent of the surface of Venus. A view of the surface was obtained using a type of radar imaging and combining images from *Magellan* spacecraft with those produced by the radio telescope in Arecibo, Puerto Rico. This view, shown in **Figure 28.14,** uses false colors to outline the major landmasses. In 2006, a European space probe, called *Venus Express,* went into orbit around Venus. Its mission was to gather atmospheric data for about one and one-half years.

■ **Figure 28.14** Radar imaging revealed the surface of Venus. Highlands are shown in red, and valleys are shown in blue. Large highland regions are like continents on Earth.
Infer ***What do green areas represent?***

Retrograde rotation Radar measurements show that Venus rotates slowly—a day on Venus is equivalent to 243 Earth days. Also, Venus rotates clockwise, unlike most planets that spin counterclockwise. This backward spin, called retrograde rotation, means that an observer on Venus would see the Sun rise in the west and set in the east. Astronomers theorize that this retrograde rotation might be the result of a collision between Venus and another body early in the solar system's history.

Atmosphere Venus is the planet most similar to Earth in physical properties, such as diameter, mass, and density, but its surface conditions and atmosphere are vastly different from those on Earth. The atmospheric pressure on Venus is 92 atmospheres (atm), compared to 1 atm at sea level on Earth. If you were on Venus, the pressure of the atmosphere would make you feel like you were under 915 m of water.

The atmosphere of Venus is composed primarily of carbon dioxide and nitrogen, somewhat similar to Earth's atmosphere. Venus also has clouds, as shown in **Figure 28.15,** an image taken of the night side of Venus by *Venus Express*. Instead of being composed of water vapor and ice, as on Earth, clouds on Venus consist of sulfuric acid.

■ **Figure 28.15** Clouds swirl around Venus in this image taken using ultraviolet wavelengths.

Greenhouse effect Venus also experiences a greenhouse effect similar to Earth's, but Venus's is more efficient. As you learned in Chapter 14, greenhouse gases in Earth's atmosphere trap infrared radiation and keep Earth much warmer than it would be if it had no atmosphere. The concentration of carbon dioxide is so high in Venus's atmosphere that it keeps the surface extremely hot—hot enough to melt lead. In fact, Venus is the hottest planet, with an average surface temperature of about 737 K (464°C), compared with Earth's average surface temperature of 288 K (15°C). It is so hot on the surface of Venus that no liquid water can exist.

PROBLEM-SOLVING LAB

Apply Kepler's Third Law

How well do the orbits of the planets conform to Kepler's third law? For the six planets closest to the Sun, Kepler observed that $P^2 = a^3$, where P is the orbital period in years and a is the semimajor axis in AU.

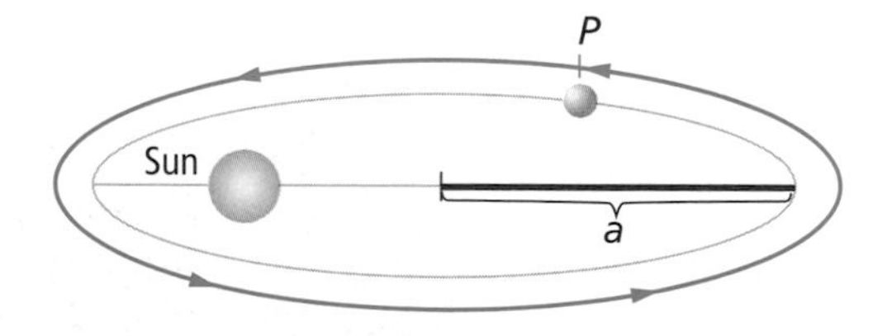

Analysis

1. Use this typical planet orbit diagram and the data from the *Reference Handbook* to confirm the relationship between P^2 and a^3 for each of the planets.

Think Critically

2. **Prepare** a table showing your results and how much they deviate from predicted values.
3. **Determine** which planets conform most closely to Kepler's law and which do not seem to follow it.
4. **Consider** Would Kepler have formulated this law if he had been able to study Uranus and Neptune? Explain.
5. **Predict** the orbital period of an asteroid orbiting the Sun at 2.5 AU.
6. **Solve** Find the semimajor axis of Halley's comet, which has an orbital period of 76 years.

Surface The *Magellan* orbiter used radar reflection measurements to map the surface of Venus. This revealed that Venus has a surface smoothed by volcanic lava flows and with few impact craters. The most recent volcanic activity took place about 500 mya. Unlike Earth, there is little evidence of current tectonic activity on Venus, and there is no well-defined system of crustal plates.

Interior Because the size and density of Venus are similar to Earth's, it is probable that the internal structure is similar also. Astronomers theorize that Venus has a liquid metal core that extends halfway to the surface. Despite this core, Venus has no measurable magnetic field, probably because of its slow rotation.

Earth

Earth, shown in **Figure 28.16,** has many unique properties when compared with other planets. Its distance from the Sun and its nearly circular orbit allow water to exist on its surface in all three states—solid, liquid, and gas. Liquid water is required for life, and Earth's abundance of water has been important for the development and existence of life on Earth. In addition, Earth's mild greenhouse effect and moderately dense atmosphere of nitrogen and oxygen provide conditions suitable for life.

Earth is the most dense and the most tectonically active of the terrestrial planets. It is the only planet where plate tectonics occurs. Unlike Venus and Mercury, Earth has a moon, probably acquired by an impact, as you learned in Chapter 27.

Mars

Mars is often referred to as the red planet because of its reddish surface color, as shown in **Figure 28.16.** Mars is smaller and less dense than Earth and has two irregularly shaped moons—Phobos and Deimos. Mars has been the target of a lot of recent exploration —*Mars Odyssey* and *Global Surveyor* in 2001, *Exploration Rovers*, *Reconnaissance Orbiter*, and *Mars Express* in 2003.

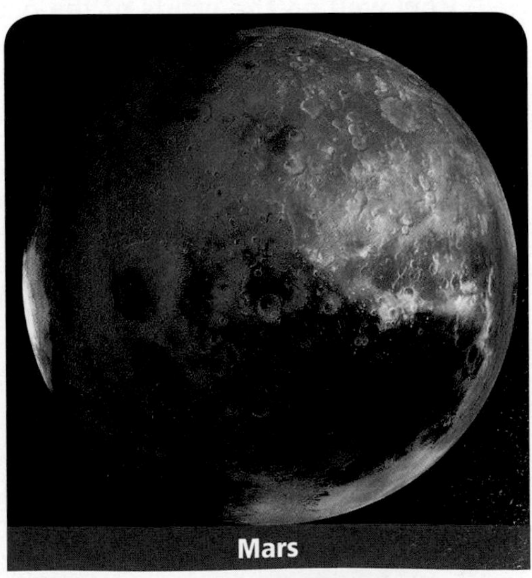

■ **Figure 28.16** Earth's blue seas and white clouds contrast sharply with the reddish, barren Mars.

Figure 28.17 Orbital probes and landers have provided photographic details of the Martian features and surface, such as Olympus Mons and Gusev crater.

Atmosphere Both Mars and Venus have atmospheres of similar composition. The density and pressure of the atmosphere on Mars are much lower; therefore Mars does not have a strong greenhouse effect like Venus does. Although the atmosphere is thin, it is turbulent—there is constant wind, and dust storms can last for weeks at a time.

Surface The southern and northern hemispheres of Mars vary greatly, as shown in **Figure 28.17.** The southern hemisphere is a heavily cratered, highland region resembling the highlands of the Moon. The northern hemisphere has sparsely cratered plains. Scientists theorize that great lava flows covered the once-cratered terrain of the northern hemisphere. Four gigantic shield volcanoes are located near the equator, near a region called the Tharsis Plateau. The largest volcano on Mars is Olympus Mons. The base of Olympus Mons is larger than the state of Colorado, and the volcano rises 3 times higher than Mount Everest in the Himalayas.

Tectonics An enormous canyon, Valles Marineris, shown in **Figure 28.18,** lies on the Martian equator, splitting the Tharsis Plateau. This canyon is 4000 km long—almost 10 times the length of the Grand Canyon on Earth and more than 3 times its depth. It probably formed as a fracture during a period of tectonic activity 3 bya, when the Tharsis Plateau was uplifted. The gigantic volcanoes were caused during the same period by upwelling of magma at a hot spot, much like the Hawaiian Island chain was formed. However, with no plate movement on Mars, magma accumulated in one area.

Erosional features Other Martian surface features include dried river and lake beds, outflow channels, and runoff channels. These erosional features suggest that liquid water once existed on the surface of Mars. Astronomers think that the atmosphere was once much warmer, thicker, and richer in carbon dioxide, allowing liquid water to flow on Mars. Although there is a relatively small amount of ice at the poles, astronomers continue to search for water at other locations on the Martian surface.

Figure 28.18 Valles Marineris is a 4000-km-long canyon on Mars.

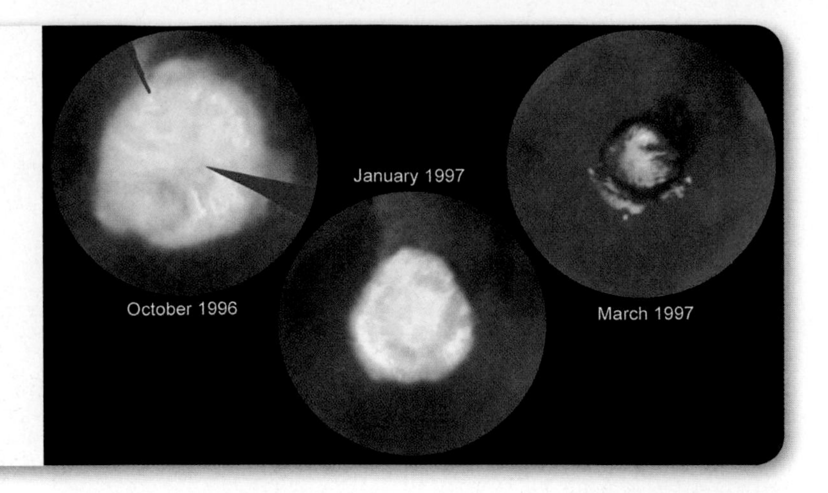

■ **Figure 28.19** These images of Mars's northern ice cap were taken three months apart by the *Hubble Space Telescope* in 1997.
Interpret ***What do these images indicate about the orientation of Mars's axis?***

Ice caps Ice caps cover both poles on Mars. The caps grow and shrink with the seasons. Martian seasons are caused by a combination of a tilted axis and a slightly eccentric orbit. Both caps are made of carbon dioxide ice, sometimes called dry ice. Water ice lies beneath the carbon dioxide ice in the northern cap, shown in **Figure 28.19,** and is exposed during the northern hemisphere's summer when the carbon dioxide ice evaporates. There might also be water ice beneath the southern cap, but the carbon dioxide ice does not completely evaporate to expose it.

Interior The internal structure of Mars remains unknown. Astronomers hypothesize that there is a core of iron, nickel, and possibly sulfur that extends somewhere between 1200 km and 2400 km from the center of the planet. Because Mars has no magnetic field, astronomers think that the core is probably solid. Above the solid core is a mantle. There is no evidence of current tectonic activity or tectonic plates on the surface of the crust.

Section 28.2 Assessment

Section Summary

- Mercury is heavily cratered and has high cliffs. It has a hot surface and no real atmosphere.
- Venus has clouds containing sulfuric acid and an atmosphere of carbon dioxide that produces a strong greenhouse effect.
- Earth is the only planet that has all three forms of water on its surface.
- Mars has a thin atmosphere. Surface features include four volcanoes and channels that suggest that liquid water once existed on the surface.

Understand Main Ideas

1. MAIN Idea **Identify** the reason that the inner planets are called terrestrial planets.
2. **Summarize** the characteristics of each of the terrestrial planets.
3. **Compare** the average surface temperatures of Earth and Venus, and describe what causes them.
4. **Describe** the evidence that indicates there was once tectonic activity on Mercury, Venus, and Mars.

Think Critically

5. **Consider** what the inner planets would be like if impacts had not shaped their formation and evolution.

MATH in Earth Science

6. Using the *Reference Handbook,* create a graph showing the distance from the Sun for each terrestrial planet on the *x*-axis and their orbital periods in Earth days on the *y*-axis. For more help, refer to the *Skillbuilder Handbook*.

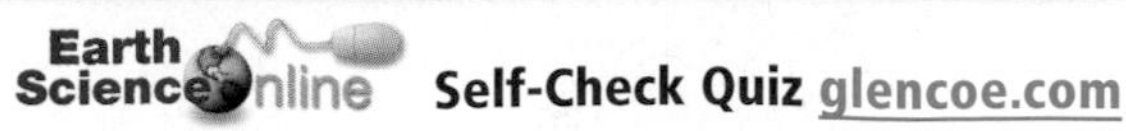

Section 28.3

Objectives

- **Compare and contrast** the gas giant planets.
- **Identify** the major moons.
- **Explain** the formation of moons and rings.
- **Compare** the composition of the gas giant planets to the composition of the Sun.

Review Vocabulary

asteroid: metallic or silicate-rich objects that orbit the Sun in a belt between Mars and Jupiter

New Vocabulary

gas giant planet
liquid metallic hydrogen
belt
zone

The Outer Planets

MAIN Idea Jupiter, Saturn, Uranus, and Neptune have large masses, low densities, and many moons and rings.

Real-World Reading Link Just as the inner planets resemble a family that shares many physical characteristics, the outer planets also show strong family resemblances.

The Gas Giant Planets

Jupiter, Saturn, Uranus, and Neptune are known as the gas giants. The **gas giant planets** are all very large, ranging from 15 to more than 300 times the mass of Earth, and from about 4 to more than 10 times Earth's diameter. Their interiors are either gases or liquids, and they might have small, solid cores. They are made primarily of lightweight elements such as hydrogen, helium, carbon, nitrogen, and oxygen, and they are very cold at their surfaces. The gas giants have many satellites as well as ring systems.

Jupiter

Jupiter is the largest planet, with a diameter one-tenth that of the Sun and 11 times larger than Earth's. Jupiter's mass makes up 70 percent of all planetary matter in the solar system. Jupiter appears bright because its albedo is 0.52. Telescopic views of Jupiter show a banded appearance, as a result of flow patterns in its atmosphere. Nestled among Jupiter's cloud bands is the Great Red Spot, an atmospheric storm that has raged for more than 300 years. This is shown in **Figure 28.20.**

Rings The *Galileo* spacecraft observed Jupiter and its moons during a 5-year mission in the 1990s. It revealed two faint rings around the planet in addition to a 6400-km-wide ring around Jupiter that had been discovered by *Voyager 1.* A portion of Jupiter's faint ring system is also shown in **Figure 28.20.**

Figure 28.20 Jupiter's cloud bands contain the Great Red Spot. The planet is circled by three faint rings that are probably composed of dust particles.

Jupiter's cloud bands

Jupiter's rings

Atmosphere and interior Jupiter has a density of 1326 kg/m^3, which is low for its size, because it is composed mostly of hydrogen and helium in gaseous or liquid form. Below the liquid hydrogen is a layer of **liquid metallic hydrogen,** a form of hydrogen that has properties of both a liquid and a metal, which can exist only under conditions of very high pressure. Electric currents exist within the layer of liquid metallic hydrogen and generate Jupiter's magnetic field. Models suggest that Jupiter might have an Earth-sized solid core containing heavier elements.

Rotation Jupiter rotates very rapidly for its size; it spins once on its axis in a little less than 10 hours, giving it the shortest day in the solar system. This rapid rotation distorts the shape of the planet so that the diameter through its equatorial plane is 7 percent larger than the diameter through its poles. Jupiter's rapid rotation causes its clouds to flow rapidly as well, in bands of alternating dark and light colors called belts and zones. **Belts** are low, warm, dark-colored clouds that sink, and **zones** are high, cool, light-colored clouds that rise. These are similar to cloud patterns in Earth's atmosphere caused by Earth's rotation.

Moons Jupiter has more than 60 moons, most of which are extremely small. Jupiter's four largest moons, Io, Europa, Ganymede, and Callisto, are called Galilean satellites after their discoverer. Three of them are bigger than Earth's Moon, and all four are composed of ice and rock. The ice content is lower in Io and Europa, which are shown in **Figure 28.21,** because they have been squeezed and heated by Jupiter's gravitational force more than the outer Galilean moons. In fact, Io is almost completely molten inside and undergoes constant volcanic eruptions. Gravitational heating has melted Europa's ice in the past, and astronomers hypothesize that it still has a subsurface ocean of liquid water. Cracks and water channels mark Europa's icy surface.

■ **Figure 28.21** Jupiter's gravity heats Europa and Io, causing some visible effects: volcanic eruptions on Io and melting and refreezing of Europa's icy surface causing it to be crisscrossed by cracks and water channels.

Reading Check **Explain** why scientists think that Europa has an ocean of liquid water beneath its surface.

Jupiter's smaller moons were discovered by a series of space probes beginning with *Pioneer 10* and *Pioneer 11* in the 1970s followed by *Voyager 1* and *Voyager 2* that also detected Jupiter's rings. Most of the information on Jupiter and its moons came from the *Galileo* space probe that arrived at Jupiter in 1995. Jupiter's four small, inner moons are thought to be the source of Jupiter's rings. Scientists think that the rings are produced as meteoroids strike these moons and release fine dust into Jupiter's orbit.

Gravity assist A technique first used to help propel *Mariner 10* to Venus and Mars was to use the Sun's gravity to boost the speed of the satellite. Today it is common for satellites to use a planet's gravity to help propel them deeper into space. Jupiter is the most massive planet, and so any satellite passing deeper into space than Jupiter uses its gravity to give it an assist. Recent flybys on their way to Saturn and Pluto by the *Cassini* and *New Horizons* missions used that assist.

Saturn

Saturn, shown in **Figure 28.22,** is the second-largest planet in the solar system. Five space probes have visited Saturn, including *Pioneer 10*, *Pioneer 11*, and *Voyagers 1* and *2*. In 2004, the United States' *Cassini* mission arrived at Saturn and began to orbit the planet.

Atmosphere and interior Saturn is slightly smaller than Jupiter and its average density is lower than that of water. Like Jupiter, Saturn rotates rapidly for its size and has a layered cloud system. Saturn's atmosphere is mostly hydrogen and helium with ammonia ice near the cloud tops. The internal structure of Saturn is probably similar to Jupiter's—fluid throughout, except for a small, solid core. Saturn's magnetic field is 1000 times stronger than Earth's and is aligned with its rotational axis. This is highly unusual among the planets.

Rings Saturn's most striking feature is its rings, which are shown in **Figure 28.22.** Saturn's rings are much broader and brighter than those of the other gas giant planets. They are composed of pieces of ice that range from microscopic particles to house-sized chunks. There are seven major rings, and each ring is made up of narrower rings, called ringlets. The rings contain many open gaps.

These ringlets and gaps are caused by the gravitational effects of Saturn's many moons. The rings are thin—less than 200 m thick—because rotational forces keep the orbits of all the particles confined to Saturn's equatorial plane. The ring particles have not combined to form a large satellite because Saturn's gravity prevents particles located close to the planet from sticking together. This is why the major moons of the gas giant planets are always beyond the rings.

Origin of the rings Until recently, astronomers thought that the ring particles were left over from the formation of Saturn and its moons. Now, many astronomers think it is more likely that the ring particles are debris left over from collisions of asteroids and other objects, or from moons broken apart by Saturn's gravity.

Moons Saturn has more than 55 satellites, including the giant Titan, which is larger than the planet Mercury. Titan is unique among planetary satellites because it has a dense atmosphere made of nitrogen and methane. Methane can exist as a gas, a liquid, and a solid on Titan's surface. In 2005, *Cassini* released the *Huygens* (HOY gens) probe into Titan's atmosphere. *Cassini* detected plumes of ice and water vapor ejected from Saturn's moon Enceladus, suggesting geologic activity.

■ **Figure 28.22** Saturn's rings are made of chunks of rock and ice that can be as small as dust particles or as large as a house. A close-up view reveals ringlets and gaps.
Explain *why the ring particles orbit Saturn in the same plane.*

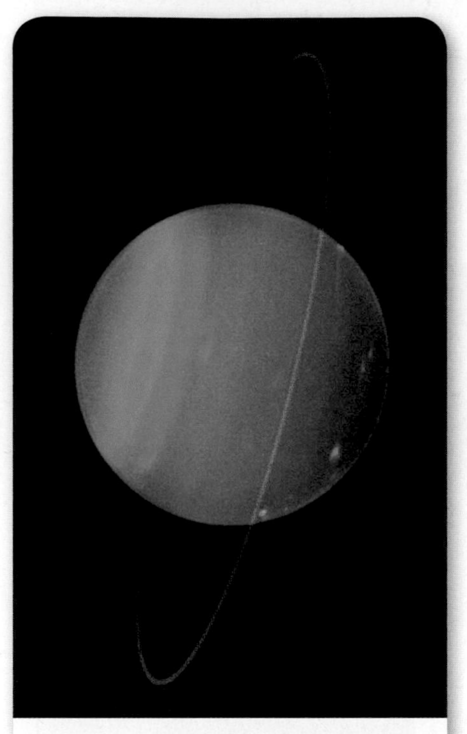

■ **Figure 28.23** The blue color of Uranus is caused by methane in its atmosphere, which reflects blue light.

Uranus

Uranus was discovered accidentally in 1781, when a bluish object was observed moving relative to the stars. In 1986, *Voyager 2* flew by Uranus and provided detailed information about the planet, including the existence of new moons and rings. Uranus's average temperature is 58 K (–215°C).

Atmosphere Uranus is 4 times larger and 15 times more massive than Earth. It has a blue, velvety appearance, shown in **Figure 28.23,** which is caused by methane gas in Uranus's atmosphere. Most of Uranus's atmosphere is composed of helium and hydrogen, which are colorless. There are few clouds, and they differ little in brightness and color from the surrounding atmosphere contributing to Uranus's featureless appearance. The internal structure of Uranus is similar to that of Jupiter and Saturn; it is completely fluid except for a small, solid core. Uranus also has a strong magnetic field.

Moons and rings Uranus has at least 27 moons and a faint ring system. Many of Uranus's rings are dark—almost black and almost invisible. They were discovered only when the brightness of a star behind the rings dimmed as Uranus moved in its orbit and the rings blocked the starlight.

Rotation The rotational axis of Uranus is tipped so far that its north pole almost lies in its orbital plane, as shown in **Figure 28.24.** Astronomers hypothesize that Uranus was knocked sideways by a massive collision with a passing object, such as a large asteroid, early in the solar system's history. Each pole on Uranus spends 42 Earth years in darkness and 42 Earth years in sunlight due to this tilt.

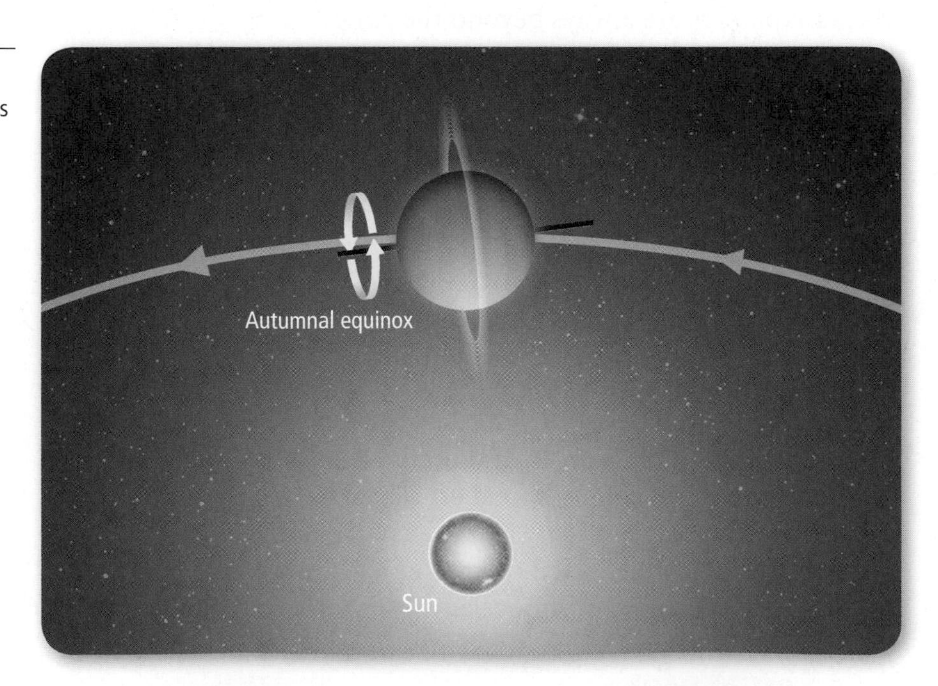

■ **Figure 28.24** The axis or rotation of Uranus is tipped 98 degrees. This view shows its position at an equinox.
Draw ***a diagram showing its position at the other equinox and solstices.***

Neptune

The existence of Neptune was predicted before it was discovered, based on small deviations in the motion of Uranus and the application of Newton's universal law of gravitation. In 1846, Neptune was discovered where astronomers had predicted it to be. Few details can be observed on Neptune with an Earth-based telescope, but *Voyager 2* flew past Neptune in 1989 and took the image of its cloud-streaked atmosphere, shown in **Figure 28.25.** Neptune is the last of the gas giant planets and orbits the Sun almost 4.5 billion km away.

Atmosphere Neptune is slightly smaller and denser than Uranus, but its radius is about 4 times as large as Earth's. Other similarities between Neptune and Uranus include their bluish color caused by methane in the atmosphere, their atmospheric compositions, temperatures, magnetic fields, interiors, and particle belts or rings. Unlike Uranus, however, Neptune has distinctive clouds and atmospheric belts and zones similar to those of Jupiter and Saturn. In fact, Neptune once had a persistent storm, the Great Dark Spot, similar to Jupiter's Great Red Spot, but the storm disappeared in 1994.

Moons and rings Neptune has 13 moons, the largest of which is Triton. Triton has a retrograde orbit, which means that it orbits backward, unlike other large satellites in the solar system. Triton, as shown in **Figure 28.25,** has a thin atmosphere and nitrogen geysers. The geysers are caused by nitrogen gas below Triton's south polar ice, which expands and erupts when heated by the Sun. Neptune's six rings are composed of microscopic dust particles, which do not reflect light well. Therefore, Neptune's rings are not as visible from Earth as Saturn's rings.

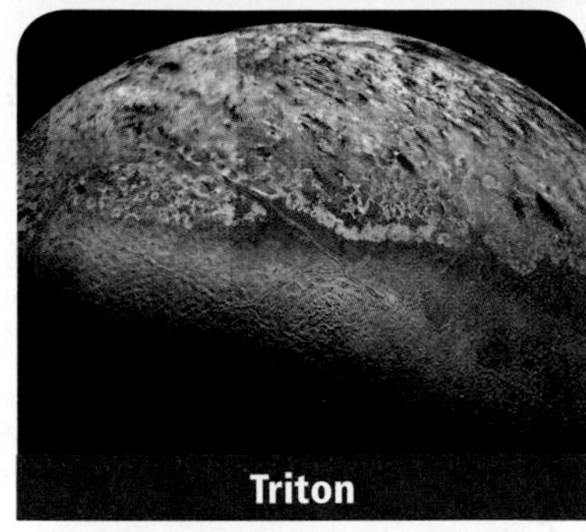

Figure 28.25 *Voyager 2* took the image of Neptune above showing its cloud streaks, as well as this close-up view of Neptune's largest moon, Triton. Dark streaks indicate the sites of nitrogen geysers on Triton.

Section 28.3 Assessment

Section Summary

- The gas giant planets are composed mostly of hydrogen and helium.
- The gas giant planets have ring systems and many moons.
- Some moons of Jupiter and Saturn have water and experience volcanic activity.
- All four gas giant planets have been visited by space probes.

Understand Main Ideas

1. MAIN Idea **Create** a table that lists the gas giant planets and their characteristics.
2. **Compare** the composition of the gas giant planets to the Sun.
3. **Compare** Earth's Moon with the moons of the gas giant planets.

Think Critically

4. **Evaluate** Where do you think are the most likely sites on which to find extraterrestrial life? Explain.

WRITING in Earth Science

5. Research and describe one of the *Voyager* missions to interstellar space.

Earth Science Online Self-Check Quiz glencoe.com

Section 28.4

Objectives

- **Distinguish** between planets and dwarf planets.
- **Identify** the oldest members of the solar system.
- **Describe** meteoroids, meteors, and meteorites.
- **Determine** the structure and behavior of comets.

Review Vocabulary

smog: air polluted with hydrocarbons and nitrogen oxides

New Vocabulary

dwarf planet
meteoroid
meteor
meteorite
Kuiper belt
comet
meteor shower

Other Solar System Objects

MAIN Idea Rocks, dust, and ice compose the remaining 2 percent of the solar system.

Real-World Reading Link The radio might have been your favorite source of music until digital music players became available. Similarly, improvements in technology lead to a change in Pluto's rank as a planet when astronomers discovered many more objects that had similar characteristics to Pluto.

Dwarf Planets

In the early 2000s, astronomers began to detect large objects in the region of the planet Pluto, about 40 AU from the Sun, called the Kuiper belt. Then in 2003, one object, now known as Eris, was discovered that appeared to be the same size, or larger, than Pluto. At this time, the scientific community began to take a closer look at the planetary status of Pluto and other solar system objects.

Ceres In 1801, Giuseppe Piazzi discovered a large object in orbit between Mars and Jupiter. Scientists had predicted that there was a planet somewhere in that region, and it seemed that this discovery was it. However, Ceres, shown in **Figure 28.26,** was extremely small for a planet. In the following century, hundreds—now hundreds of thousands—of other objects were discovered in the same region. Therefore, Ceres was no longer thought of as a planet, but as the largest of the asteroids in what would be called the asteroid belt.

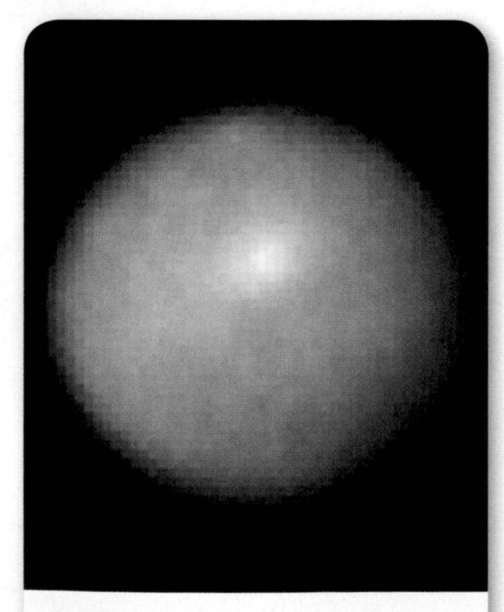

Figure 28.26 Imaged from the *Hubble Space Telescope,* the newly described dwarf planet, Ceres, is the largest body in the asteroid belt.

Pluto Since its discovery by Clyde Tombaugh in 1930, Pluto has been an unusual planet. It is not a terrestrial or gas planet; it is made of rock and ice. It does not have a circular orbit; its orbit is long, elliptical, and overlaps the orbit of Neptune. And it is smaller than Earth's Moon. It is one of many similar objects that exist outside of the orbit of Neptune. It has three moons, two of which orbit at widely odd angles from the plane of the ecliptic.

How many others? With the discovery of objects close to and larger than Pluto's size, the International Astronomical Union (IAU) faced a dilemma. Should Eris be named the tenth planet? Or should there be a change in the way these new objects are classified? For now, the answer is change. Pluto, Eris, and Ceres have been placed into a new classification of objects in space called dwarf planets. The IAU has defined a **dwarf planet** as an object that, due to its own gravity, is spherical in shape, orbits the Sun, is not a satellite, and has not cleared the area of its orbit of smaller debris. Currently the IAU has limited this classification to Pluto, Eris, and Ceres, but there are at least 12 other objects whose classifications are undecided, some of which are shown in **Figure 28.27.**

Visualizing the Kuiper Belt

Figure 28.27 Recent findings of objects beyond Pluto, in a vast disk called the Kuiper belt, have forced scientists to rethink what features define a planet.
(Note: *Buffy (XR190) is a nickname used by its discoverer. EL61 is an official number assigned to an unnamed body.*)

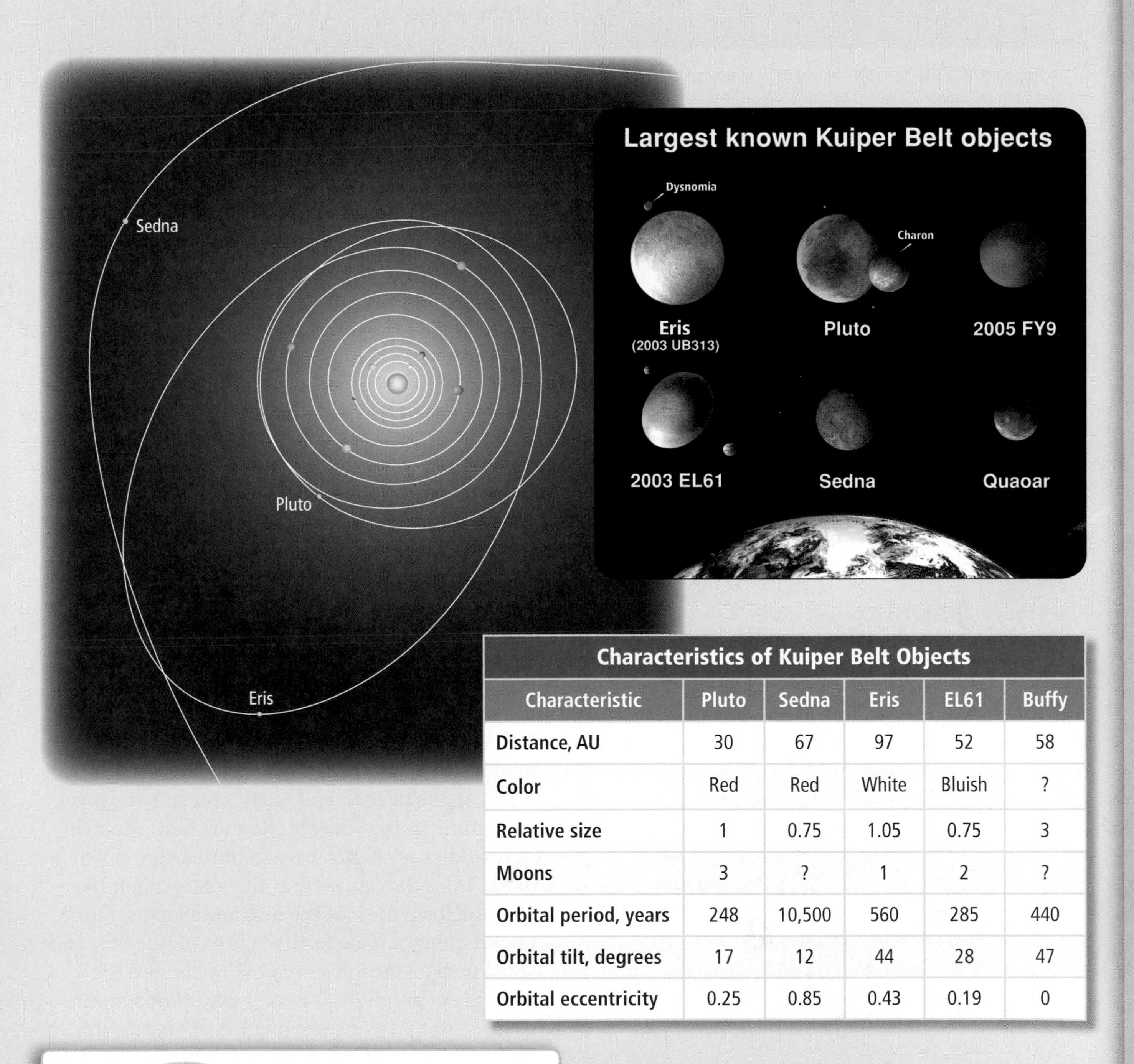

Characteristics of Kuiper Belt Objects					
Characteristic	**Pluto**	**Sedna**	**Eris**	**EL61**	**Buffy**
Distance, AU	30	67	97	52	58
Color	Red	Red	White	Bluish	?
Relative size	1	0.75	1.05	0.75	3
Moons	3	?	1	2	?
Orbital period, years	248	10,500	560	285	440
Orbital tilt, degrees	17	12	44	28	47
Orbital eccentricity	0.25	0.85	0.43	0.19	0

Concepts In Motion To explore more about the Kuiper belt objects, visit glencoe.com.

Earth Science Online

■ **Figure 28.28** Asteroid Ida and its tiny moon, Dactyl, are shown in this image gathered by the *Galileo* spacecraft.

Small Solar System Bodies

Once the IAU defined planets and dwarf planets, they had to identify what was left. In the early 1800s, a name was given to the rocky planetesimals between Mars and Jupiter—the asteroid belt. Objects beyond the orbit of Neptune have been called trans-Neptunian objects (TNOs), Kuiper belt objects (KBOs), comets, and members of the Oort cloud. But what would the collective name for these objects be? The IAU calls them small solar system bodies.

Asteroids There are thousands of asteroids orbiting the Sun between Mars and Jupiter. They are rocky bodies that vary in diameter and have pitted, irregular surfaces. Some asteroids have satellites of their own, such as the asteroid Ida, shown in **Figure 28.28.** Astronomers estimate that the total mass of all the known asteroids in the solar system is equivalent to only about 0.08 percent of Earth's mass.

Reading Check **Describe** the asteroid belt.

As asteroids orbit, they occasionally collide and break into fragments. When an asteroid fragment, or any other interplanetary material, enters Earth's atmosphere it is called a **meteoroid.** As a meteoroid passes through the atmosphere, it is heated by friction and burns, producing a streak of light called a **meteor.** If the meteoroid does not burn up completely and part of it strikes the ground, the part that hits the ground is called a **meteorite.** When large meteorites strike Earth, they produce impact craters. Any craters visible on Earth must be young, otherwise they would have been erased by erosion.

Kuiper belt Like the rocky asteroid belt, another group of small solar system bodies that are mostly made of rock and ice lies outside the orbit of Neptune in the **Kuiper** (KI pur) **belt.** Most of these bodies probably formed in this region—30 to 50 AU from the Sun—from the material left over from the formation of the Sun and planets. Some, however, might have formed closer to the Sun and were knocked into this area by Jupiter and the other gas giant planets. Eris, Pluto, Pluto's moon Charon, and an ever-growing list of objects are being detected within this band; however, none of them has been identified as a comet. Comets come from the farthest limits of the solar system, the Oort cloud, shown in **Figure 28.29.**

■ **Figure 28.29** The Kuiper belt appears as the outermost limit of the planetary disk. The Oort cloud surrounds the Sun, echoing its solar sphere.

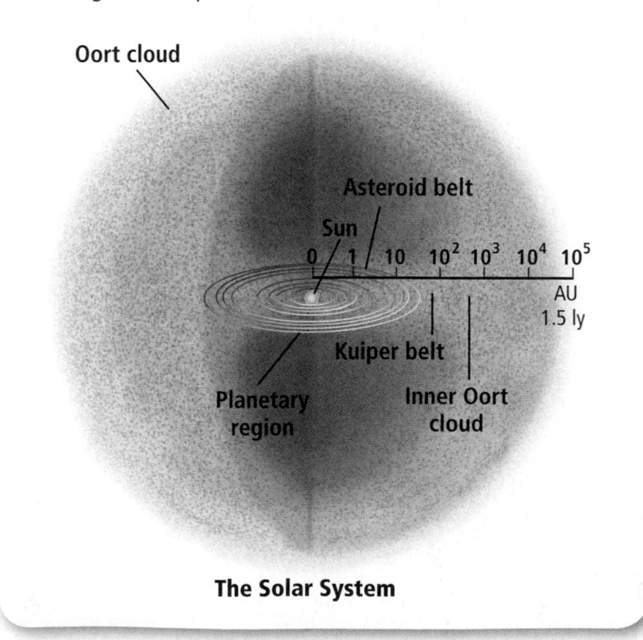

Comets

Comets are small, icy bodies that have highly eccentric orbits around the Sun. Ranging from 1 to 10 km in diameter, most comets orbit in a continuous distribution that extends from the Kuiper belt to 100,000 AU from the Sun. The outermost region is known as the Oort cloud and expands into a sphere surrounding the Sun. Occasionally, a comet is disturbed by the gravity of another object and is thrown into the inner solar system.

Comet structure When a comet comes within 3 AU of the Sun, it begins to evaporate. It forms a head and one or more tails. The head is surrounded by an envelope of glowing gas, and it has a small solid core. The tails form as gas and dust are pushed away from the comet by particles and radiation from the Sun. This is why comets' tails always point away from the Sun, as shown in **Figure 28.30.**

Periodic comets Comets that repeatedly return to the inner solar system are known as periodic comets. One example is Halley's comet, which has a 76-year period—it appeared last in 1985, and is expected to appear again in 2061. Each time a periodic comet comes near the Sun, it loses some of its matter, leaving behind a trail of particles. When Earth crosses the trail of a comet, particles left in the trail burn in Earth's upper atmosphere producing bright streaks of light called a **meteor shower.** In fact, most meteors are caused by dust particles from comets.

Figure 28.30 A comet's tail always points away from the Sun and is driven by a stream of particles and radiation. The comet Hale-Bopp was imaged when its orbit brought it close to the Sun in 1997.

Section 28.4 Assessment

Section Summary

- Dwarf planets, asteroids, and comets formed from the debris of the solar system formation.
- Meteoroids are planetesimals that enter Earth's atmosphere.
- Mostly rock and ice, the Kuiper belt objects are currently being detected and analyzed.
- Periodic comets are in regular, permanent orbit around the Sun, while others might pass this way only once.
- The outermost regions of the solar system house the comets in the Oort cloud.

Understand Main Ideas

1. MAIN Idea **Identify** the kinds of small solar system bodies and their compositions.
2. **Compare** planets and dwarf planets.
3. **Distinguish** among meteors, meteoroids, and meteorites.
4. **Explain** why a comet's tail always points away from the Sun.
5. **Compare and contrast** the asteroid belt and the Kuiper belt.

Think Critically

6. **Infer** why comets have highly eccentric orbits.

WRITING in Earth Science

7. Suppose you are traveling from the outer reaches of the solar system toward the Sun. Write a scientifically accurate description of the things you see.

EARTH SCIENCE AND TECHNOLOGY

WATER IN THE SOLAR SYSTEM

In recent years, data collected by spacecraft have shown evidence of water in places in our solar system other than Earth. Scientists think there might be water, either in a liquid or solid state, on Earth's Moon, under the poles of Mercury and Mars, on several of Jupiter's moons, and on at least one of Saturn's moons. Further investigation and data collection is planned by NASA to confirm these findings.

Earth's Moon Several spacecraft have collected evidence that leads scientists to believe there is subsurface ice at the Moon's poles. In 1994 and 1998, spacecraft possibly detected ice and water. Scientists hope to find definitive evidence of water under the surface of the Moon with the *Lunar Reconnaissance Orbiter (LRO)*, set to launch in 2008. A probe will take samples to test for water.

Mercury's poles Because Mercury's axis is not tilted, the interior temperatures of large craters at the poles do not ever rise above –212°C. Radar images lead scientists to think that ice exists in these craters. In August 2004, NASA launched the *Messenger* spacecraft that will reach Mercury in 2011. *Messenger* is equipped with a spectrometer that will be used to detect hydrogen, which is part of water, at Mercury's poles.

Mars's north pole Using a spectrometer, the *Odyssey* spacecraft recorded high levels of hydrogen just beneath the surface at Mars's north pole in 2002. Scientists think that water exists there in the form of ice, and that the soil might be comparable to the permafrost found at high latitudes on Earth. The *Phoenix Lander*, scheduled to launch in 2007 and reach Mars in 2008, is equipped with a robotic arm that will drill into the surface of Mars at the pole. A specialized sensor will be able to detect if there is any water in the soil.

This colorized image shows the plume of liquid water ejected from Enceladus's surface in geyserlike eruptions.

Jupiter's moons Ganymede, Europa, and Callisto all have icy surfaces. However, based on readings of the magnetic fields and high-resolution photos taken by the spacecraft *Galileo* of all three moons, scientists hypothesize that they each have a subsurface ocean. NASA has proposed a new mission to Jupiter called *Jupiter Icy Moons Orbiter* (JIMO) that would launch in 2015 and orbit the three moons. The main goals of the mission would be to learn more about the history of the moons, map their surfaces, and confirm the existence of subsurface oceans.

Saturn's moon—Enceladus The spacecraft *Cassini* has recorded geyserlike eruptions of liquid water coming from the surface of Enceladus, even though the moon's average temperature is –201°C. The figure above shows a colorized image of such an eruption.

WRITING in Earth Science

Poster Research more information about where in the solar system water might exist. Make a poster that shows the bodies of the solar system and if water might be found on them. Include captions that explain what type of exploration is planned. To learn more about the water in the solar system, visit glencoe.com.

GEOLAB

DESIGN YOUR OWN: MODEL THE SOLAR SYSTEM

Background: Models are useful for understanding the scale of the solar system.

Question: *How can you choose a scale that will easily demonstrate relative sizes of objects and distances between them in the solar system?*

Materials

calculator
tape measure
meterstick
marker
masking tape
common round objects in a variety of sizes

Safety Precautions

Procedure

1. Read and complete the lab safety form.
2. Develop a plan to make a model showing the relative sizes of objects in the solar system and the distances between them.
3. Make sure your teacher approves your plan before you begin.
4. Design a data table for the information needed to complete your model. Include the original data and the scale data.
5. Select a scale for your model using SI units. Remember your model should have the same scale throughout.
6. Calculate the relative sizes and distances of the objects you plan to model.
7. Select the materials and quantities of each, and build your model according to the scale you selected.

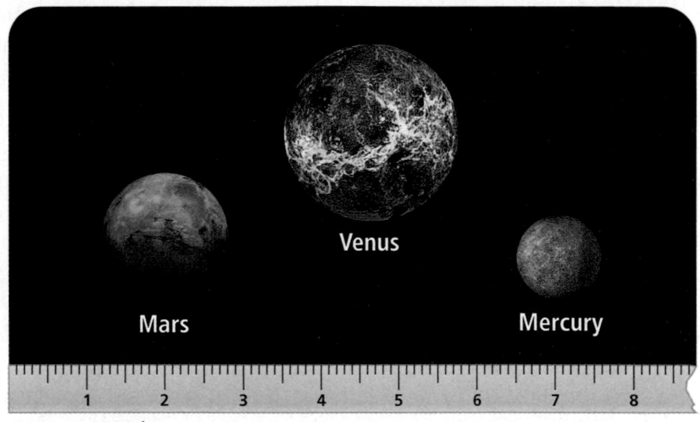

1 cm = 4000 km

Remember that the scale used has to include the largest and smallest objects and should be easy to produce.

Analyze and Conclude

1. **Think Critically** Why did the scale you chose work for your model?
2. **Explain** why you chose this scale.
3. **Observe and Infer** What possible problems could result from using a larger or smaller scale?
4. **Compare and Contrast** Compare your model with those of your classmates. Describe the advantages or disadvantages of your scale.

APPLY YOUR SKILL

Project Proxima Centauri, the closest star to the Sun, is about 4.01×10^{13} km from the Sun. Based on your scale, how far would Proxima Centauri be from the Sun in your model? If you modified your scale to better fit Proxima Centauri, how would this change the distance between Pluto and the Sun?

CHAPTER 28

Study Guide

Download quizzes, key terms, and flash cards from glencoe.com.

BIG Idea Using the laws of motion and gravitation, astronomers can understand the orbits and the properties of the planets and other objects in the solar system.

Vocabulary	Key Concepts
Section 28.1 Formation of the Solar System	
• astronomical unit (p. 800) • eccentricity (p. 801) • ellipse (p. 800) • planetesimal (p. 798) • retrograde motion (p. 799)	**MAIN Idea** The solar system formed from the collapse of an interstellar cloud. • A collapsed interstellar cloud formed the Sun and planets from a rotating disk. • The inner planets formed closer to the Sun than the outer planets, leaving debris to produce asteroids and comets. • Copernicus created the heliocentric model and Kepler defined its shape and mechanics. • Newton explained the forces governing the solar system bodies and provided proof for Kepler's laws. • Present-day astronomers divide the solar system into three zones.
Section 28.2 The Inner Planets	
• scarp (p. 805) • terrestrial planet (p. 804)	**MAIN Idea** Mercury, Venus, Earth, and Mars have high densities and rocky surfaces. • Mercury is heavily cratered and has high cliffs. It has a hot surface and no real atmosphere. • Venus has clouds containing sulfuric acid and an atmosphere of carbon dioxide that produces a strong greenhouse effect. • Earth is the only planet that has all three forms of water on its surface. • Mars has a thin atmosphere. Surface features include four volcanoes and channels that suggest that liquid water once existed on the surface.
Section 28.3 The Outer Planets	
• belt (p. 812) • gas giant planet (p. 811) • liquid metallic hydrogen (p. 812) • zone (p. 812)	**MAIN Idea** Jupiter, Saturn, Uranus, and Neptune have large masses, low densities, and many moons and rings. • The gas giant planets are composed mostly of hydrogen and helium. • The gas giant planets have ring systems and many moons. • Some moons of Jupiter and Saturn have water and experience volcanic activity. • All four gas giant planets have been visited by space probes.
Section 28.4 Other Solar System Objects	
• comet (p. 819) • dwarf planet (p. 816) • Kuiper belt (p. 818) • meteor (p. 818) • meteorite (p. 818) • meteoroid (p. 818) • meteor shower (p. 819)	**MAIN Idea** Rocks, dust, and ice compose the remaining 2 percent of the solar system. • Dwarf planets, asteroids, and comets formed from the debris of the sola system formation. • Meteoroids are planetesimals that enter Earth's atmosphere. • Mostly rock and ice, the Kuiper belt objects are currently being detected and analyzed. • Periodic comets are in regular, permanent orbit around the Sun, while others might pass this way only once. • The outermost regions of the solar system house the comets in the Oort cloud.

CHAPTER 28

Assessment

Vocabulary Review

Each of the following sentences is false. Make each sentence true by replacing the italicized words with terms from the Study Guide.

1. Rapid shrinkage of Mercury's crust produced features on its surface called *rilles.*

2. The pattern of light and dark bands on Jupiter's surface are called belts and *flows.*

3. A *meteor* is a rocky object that strikes Earth's surface.

4. A *meteorite* formed as particles of dust and gas stuck together in the early solar system.

5. The apparent backward movement of Mars as Earth passes it in its orbit is *synchronous rotation.*

6. A *light-year* is a unit of measurement used to measure distances within the solar system.

Match each phrase below with the correct term from the Study Guide.

7. a small icy object having a highly eccentric orbit around the Sun

8. Mercury, Venus, Earth, and Mars

9. multiple streaks of light caused by dust particles burning in Earth's atmosphere

10. a measure of orbital shape

11. a new solar system body classification

Understand Key Concepts

12. Who first proposed the heliocentric model of the solar system?
A. Copernicus
B. Galileo
C. Kepler
D. Newton

Use the diagram below to answer Question 13.

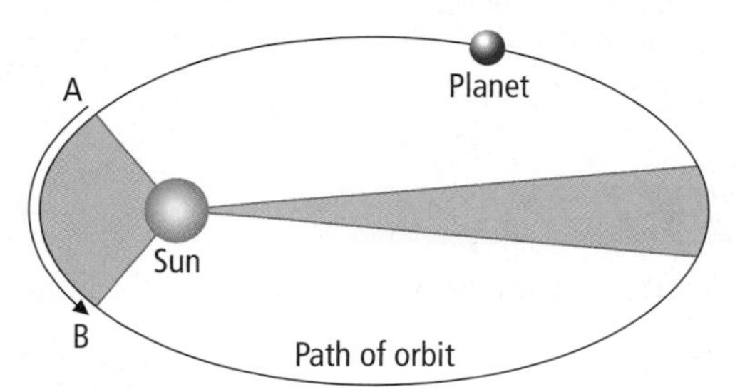

13. Which law of planetary motion does this diagram demonstrate?
A. Kepler's first law
B. Kepler's second law
C. Kepler's third law
D. Newton's law of universal gravitation

14. Which best describes a planet's retrograde motion?
A. apparent motion
B. orbital motion
C. real motion
D. rotational motion

15. Which scientist determined each planet orbits a point between it and the Sun, called the center of mass?
A. Copernicus
B. Galileo
C. Kepler
D. Newton

Use the diagram below to answer Question 16.

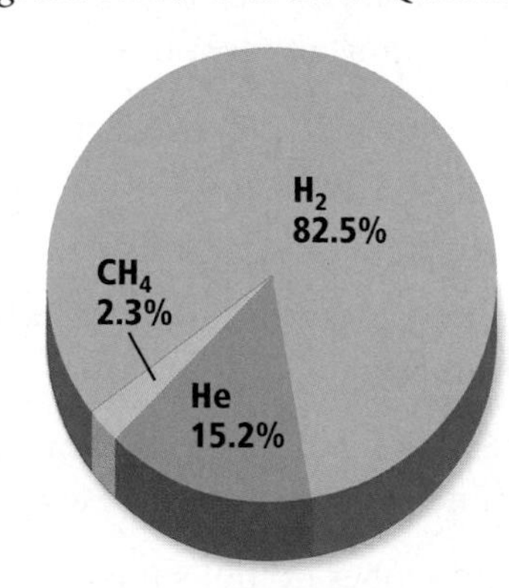

16. The atmospheric composition of which planet is shown above?
A. Jupiter
B. Mars
C. Neptune
D. Venus

Earth Science Online Chapter Test glencoe.com

17. Where do most meteorites originate?
    A. asteroid belt
    B. Kuiper belt
    C. Oort cloud
    D. Saturn's rings

Constructed Response

Use the photo below to answer Questions 18 and 19.

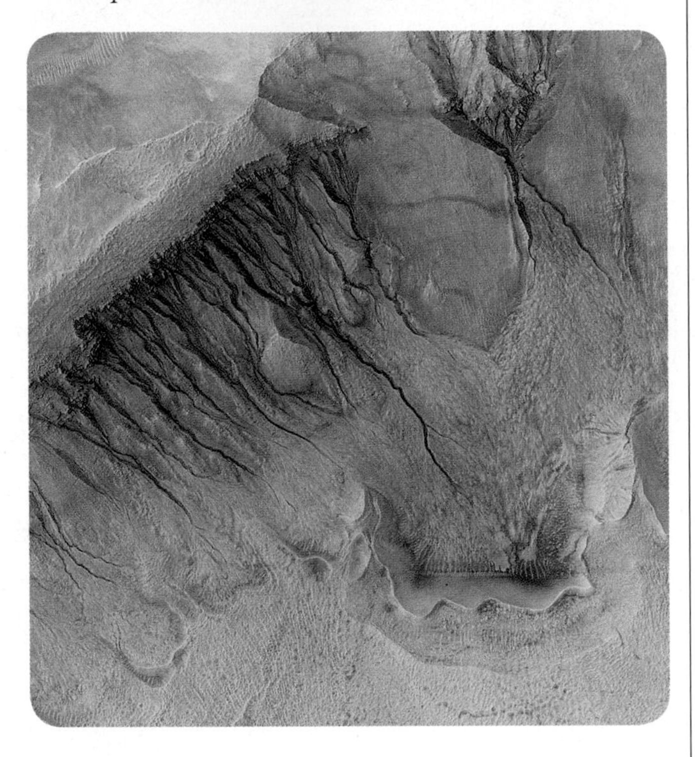

18. **Identify** these features shown on the surface of Mars and explain what most likely caused them.
19. **Infer** Based on what you have learned about Mars, state whether new features like these could be made now. Explain.
20. **Compare** Pluto and Eris and determine their common features.
21. **Compare** Sedna and EL61 to the dwarf planets and determine which features are common to each.
22. **Explain** why probes do not survive on the surface of Venus.
23. **Compare** the pivot point on a seesaw and a center of mass between two orbiting bodies.
24. **Calculate** Find the shape of an ellipse having an eccentricity of 0.9.

Use the diagram below to answer Questions 25 and 26.

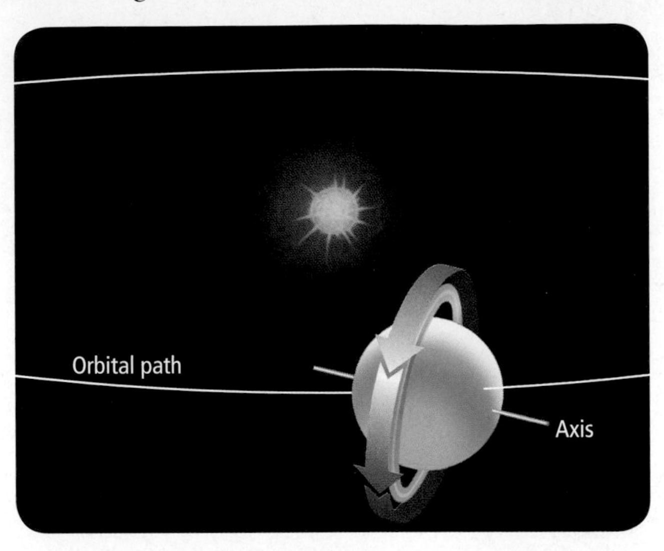

25. **Identify** the planet shown here and explain why scientists think its rotational axis is like this.
26. **Infer** how the seasons would be affected if Earth had an axis tilt similar to Uranus.

Think Critically

27. **Explain** The atmospheres of Mars and Venus contain similar percentages of CO_2, but Venus has a much higher surface temperature because of the greenhouse effect. Why doesn't this happen on Mars?
28. **CAREERS IN EARTH SCIENCE** Most astronomers do not spend long hours peering through telescopes. They operate telescopes remotely using computers and spend most of their time analyzing data. What subjects would astronomers find most useful in addition to astronomy?
29. **Discuss** the theory of formation of the rings of Saturn and the other gas giant planets.
30. **Infer** the role gravity plays in the formation of the rings of the gas giant planets.
31. **Infer** what might happen to Halley's comet as it continues to lose mass with each orbit of the Sun.
32. **Explain** why scientists think Jupiter's moon Europa might have liquid water beneath its surface.

Chapter Test glencoe.com

Use the table below to answer Questions 33 to 35.

Planet	Radius (km)	Orbital Eccentricity	Semimajor Axis (AU)
Mercury	2439.7	0.2056	0.39
Venus	6051.8	0.0067	0.72
Earth	6378.1	0.0167	1.00
Mars	3397	0.0935	1.52
Jupiter	71,492	0.0489	5.20
Saturn	60,298	0.0565	9.54
Uranus	25,559	0.047	19.19
Neptune	24,766	0.009	30.07

33. Interpret Which of the planets has an orbit that most closely resembles a perfect circle?

34. Compare Which two planets have the most similar radii?

35. Evaluate Which two planets' orbits are separated by the greatest distance?

36. Discuss the relationship between asteroids and planetesimals.

37. Explain Why were Ceres and Pluto identified as the first dwarf planets?

38. Compare and contrast the asteroid belt and the Kuiper belt.

Concept Mapping

39. Create a concept map using the following terms: *interstellar cloud, gas, dust, disk, particles, planetesimals, terrestrial planets, gas giant planets, satellites, debris, asteroids, meteoroids,* and *comets.*

Challenge Question

40. Consider Pluto's orbit sometimes brings it within the orbit of Neptune. Why is it unlikely that the two will collide? Explain.

Additional Assessment

41. *WRITING in* **Earth Science** Write a paragraph to explain to a friend how science develops over time. Discuss the relationship between Kepler's laws and Newton's law of universal gravitation.

DBQ Document–Based Questions

Data obtained from: *Physics World.* 2001. (January): 25.

Astronomers have detected planets around more than 200 stars. Although the planets themselves are too small to see directly, astronomers can detect them by measuring the Doppler shift in the star's light as it orbits its common center of mass with the unseen planet. The diagram below shows how this works.

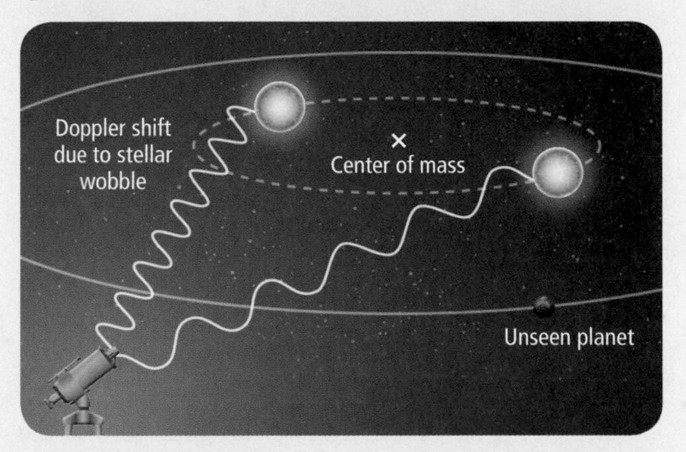

42. Based on the diagram, what is the rotational direction of the star? Explain.

43. Based on what you know about the center of mass, which planet in our solar system would be most likely to be detectable from other star systems using this method?

Cumulative Review

44. Name an example of a felsic, igneous rock. **(Chapter 5)**

45. Describe the relationship between ejecta and rays on the Moon's surface. **(Chapter 27)**

Standardized Test Practice

Multiple Choice

1. When foxes reach the brink of extinction in an area, what happens to the population of rabbits in the area?
   A. The rabbit population also becomes extinct.
   B. The rabbit population increases indefinitely.
   C. The rabbit population increases beyond the carrying capacity of the area, then decreases.
   D. The rabbit population decreases beyond the carrying capacity of the area, and then quickly increases.

Use the diagram below to answer Questions 2 and 3.

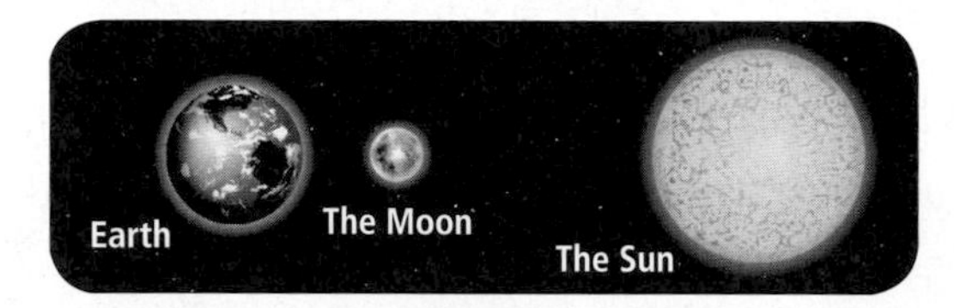

2. What results on Earth when the Sun and the Moon are aligned along the same direction?
   A. spring tides
   B. neap tides
   C. the autumnal equinox
   D. the summer solstice

3. If the Moon in this diagram were passing directly between the Sun and Earth, blocking the view of the Sun, what would you experience on Earth?
   A. a lunar eclipse
   B. a solar eclipse
   C. umbra
   D. penumbra

4. Earth's main energy source is
   A. fossil fuels
   B. hydrocarbons
   C. the Sun
   D. wind

5. Which describes life during the early Proterozoic Era?
   A. simple, unicellular life forms
   B. complex, unicellular life forms
   C. simple, multicellular life forms
   D. complex, multicellular life forms

6. Which is not considered a biomass fuel?
   A. peat
   B. coal
   C. fecal material
   D. wood

Use the illustration below to answer Questions 7 and 8.

7. Which type of fossil preservation is shown?
   A. trace fossil
   B. original remains
   C. carbon film
   D. altered hard parts

8. By studying the fossils, which is not something scientists can learn about the organism that left these prints?
   A. movement
   B. size
   C. habitat
   D. walking characteristics

9. When minerals in rocks fill a space left by a decayed organism, what type of fossil is formed?
   A. trace fossil
   B. cast fossil
   C. petrified fossil
   D. amber-preserved fossil

10. How are Mercury and the Moon similar?
    A. Both are covered with craters and plains.
    B. Both have the same night-to-day temperature difference.
    C. They have the same strength of surface gravity.
    D. Both have an extensive nickel-iron core.

Short Answer

Use the table below to answer Questions 11 to 13.

Apparent Temperature Index					
	Relative Humidity (%)				
Air Temperature (F°)		80	85	90	95
	85	97	99	102	105
	80	86	87	88	89
	75	78	78	79	79
	70	71	71	71	71

11. If the air temperature is 24°C and the relative humidity is 85%, what would the apparent temperature feel like?

12. What can be inferred about the effect relative humidity has on apparent temperature as the air temperature increases?

13. In the fall, when temperatures are moderate, how should a person plan for temperature with relative humidity factored in?

14. Although a hybrid car still requires fuel to run, why is it considered a better use of energy resources?

15. What are some steps mining companies are taking to be less destructive to the environment?

Reading for Comprehension

Tau Gruis, The Newest Planet

An international team of researchers has discovered the 100th "extrasolar" planet. This newest planet orbits the star Tau Gruis, 100 light-years from Earth, in the southern hemisphere's constellation Grus (the crane). In order to actually detect a planet, a planet must be seen going around its orbit at least once. Although scientists have been watching Tau Gruis since 1998, this is the first time that they have been able to confirm the presence of its large planet. This is an indication that there is a considerable distance between the star and the planet. Soon after the first extrasolar planets were found, beginning in 1995, most planets were found in orbit close to their host stars. Planets closer to their suns orbit at a much faster rate, and therefore take much less time to detect. Starting out, planets close in to their parent stars were found. But as the planet search program has matured, more planets farther out and in nearly circular orbits are being found. This means that scientists are getting closer to detecting more systems that are similar to our own solar system.

Article obtained from: Brendle, A. Hundredth planet outside solar system discovered. *National Geographic News.* September 17, 2002.

16. What can be inferred from this passage?
A. Our solar system is unique.
B. Detecting planets is virtually impossible.
C. As technology improves, more planets will be found.
D. Large planets are harder to find than small planets.

17. What must happen in order for an object to be considered a planet?
A. The object must go around its orbit at least once.
B. It must orbit its parent star at a particular speed.
C. It must be a particular size.
D. It must be within 100 light-years of Earth.

NEED EXTRA HELP?															
If You Missed Question . . .	1	2	3	4	5	6	7	8	9	10	11	12	13	14	15
Review Section . . .	26.1	27.3	27.3	25.1	22.4	25.1	21.4	21.4	21.4	28.2	11.2	11.2	11.2	25.3	26.2

CHAPTER 29 Stars

BIG Idea The life cycle of every star is determined by its mass, luminosity, magnitude, temperature, and composition.

29.1 The Sun
MAIN Idea The Sun contains most of the mass of the solar system and has many features typical of other stars.

29.2 Measuring the Stars
MAIN Idea Stellar classification is based on measurement of light spectra, temperature, and composition.

29.3 Stellar Evolution
MAIN Idea The Sun and other stars follow similar life cycles, leaving the galaxy enriched with heavy elements.

GeoFacts

- The last gasp of a dying star, the Butterfly nebula erupts as a pair of jet exhausts.
- A runaway thermonuclear reaction results in a star exploding into a supernova, throwing matter away from the collapsed core.
- When a massive star collapses, it can become a pulsar, a rapidly rotating object that has a magnetic field a trillion times that of Earth.

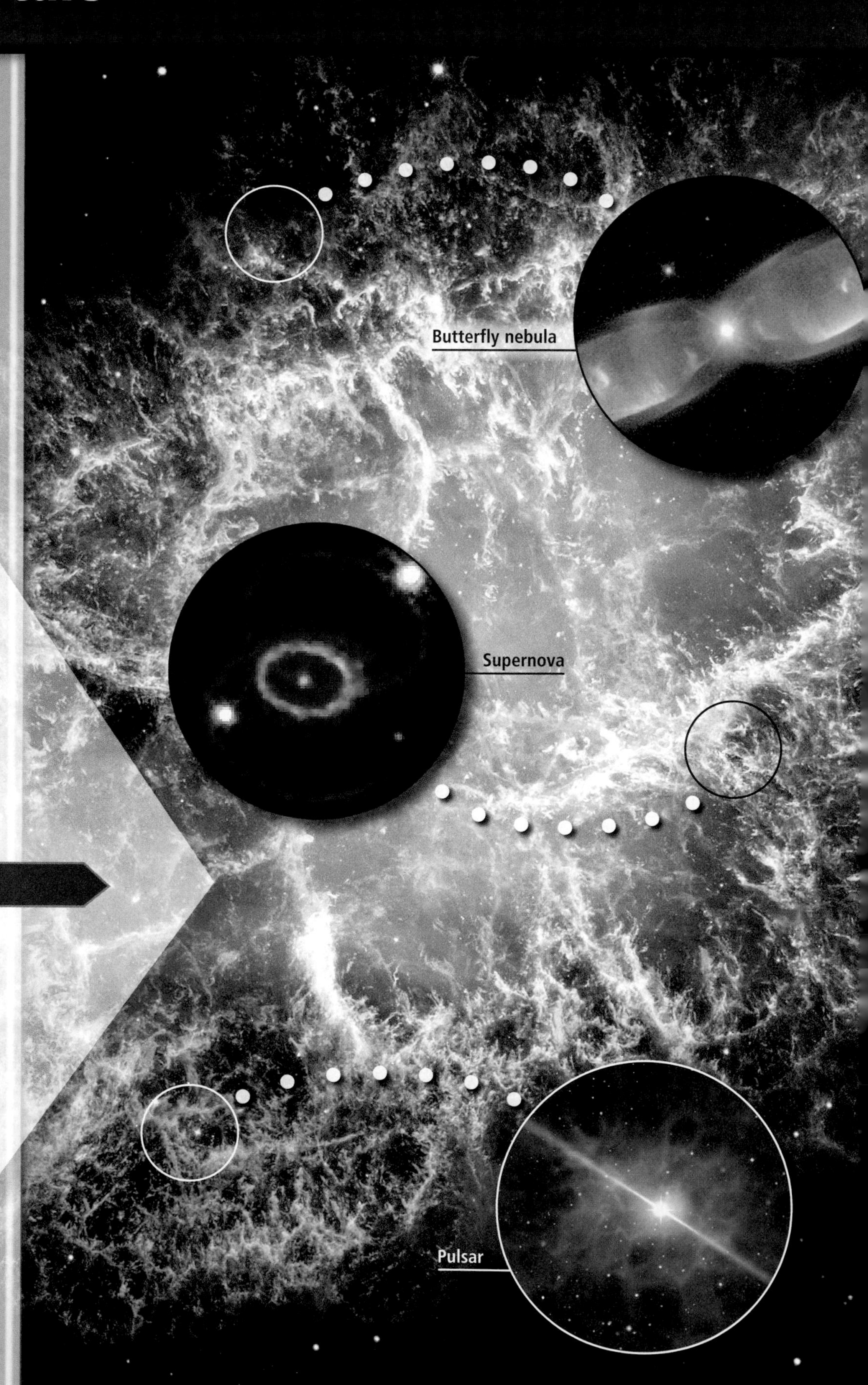

LAUNCH Lab

How can you observe sunspots?

Although the Sun is an average star, it undergoes many complex processes. Sunspots are dark spots that are visible on the surface of the Sun. They can be observed moving across the face of the Sun as it rotates.

Procedure

WARNING: ***Do not look directly at the Sun. Do not look through the telescope at the Sun. You could damage your eyes.***

1. Read and complete the lab safety form.
2. Observe the Sun through the **telescope** that your teacher has set up. Note that the telescope is pointed directly at the Sun, but the eyepiece is casting the image of the Sun on a **clipboard**.
3. Move the clipboard back and forth until you have the largest image of the Sun on the **paper.** Trace the outline of the Sun on your paper.
4. Trace sunspots that appear as dark areas on the Sun's image. Repeat this step at the same time each day for a week.
5. Measure the movement of sunspots.

Analysis

1. **Calculate** Use your data to determine the Sun's period of rotation.
2. **Determine** What is the estimated rate of motion of the largest sunspot?

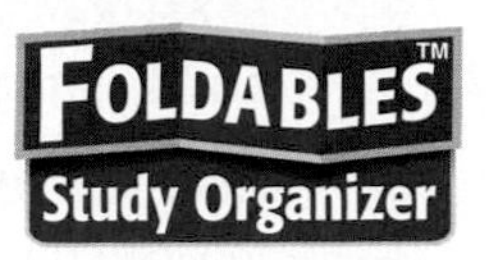

Stars Make the following Foldable that features the key vocabulary terms associated with stars.

STEP 1 Fold a sheet of paper in half lengthwise.

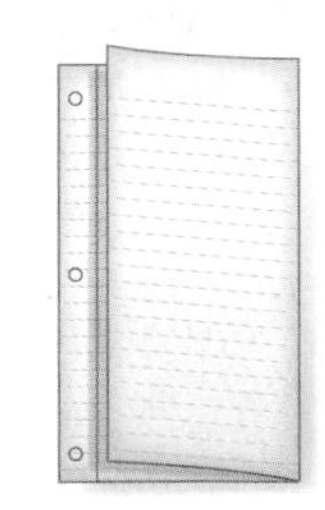

STEP 2 Cut along every third or fourth line of the top flap to form nine tabs.

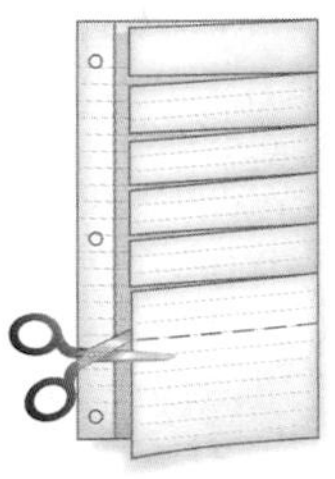

STEP 3 Label the tabs as you read.

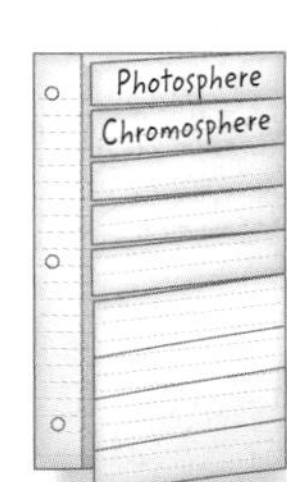

FOLDABLES **Use this Foldable with Section 29.1.** As you read this section, record key vocabulary terms and their definitions.

Visit glencoe.com to

- **study entire chapters online;**
- **explore Concepts In Motion animations:**
  - **Interactive Time Lines**
  - **Interactive Figures**
  - **Interactive Tables**
- **access Web Links for more information, projects, and activities;**
- **review content with the Interactive Tutor and take Self-Check Quizzes.**

Section 29.1

Objectives

- **Describe** the layers and features of the Sun.
- **Explain** the process of energy production in the Sun.
- **Define** the three types of spectra.

Review Vocabulary

magnetic field: the portion of space near a magnetic or current-carrying body where magnetic forces can be detected

New Vocabulary

photosphere
chromosphere
corona
solar wind
sunspot
solar flare
prominence
fusion
fission

The Sun

MAIN Idea The Sun contains most of the mass of the solar system and has many features typical of other stars.

Real-World Reading Link Have you ever had a sunburn from being outside too long on a sunny day? The Sun is more than 150 million km from Earth, but the Sun's rays are so powerful that humans still wear sunscreen for protection.

Properties of the Sun

The Sun is the largest object in the solar system, in both diameter and mass. It would take 109 Earths, or almost 10 Jupiters, lined up edge to edge, to fit across the Sun. The Sun is about 330,000 times as massive as Earth and 1048 times the mass of Jupiter. In fact, the Sun contains more than 99 percent of all the mass in the solar system. It should not be surprising, then, that the Sun's mass controls the motions of the planets and other objects.

The Sun's average density is similar to the densities of the gas giant planets, represented by Jupiter in **Table 29.1.** Astronomers deduce densities at specific points inside the Sun, as well as other information, by using computer models that explain the observations they make. These models show that the density in the center of the Sun is about 1.50×10^5 kg/m^3, which is about 13 times the density of lead. A pair of dice as dense as the Sun's center would have a mass of about 1 kg.

Unlike lead, which is a solid, the Sun's interior is gaseous throughout because of its high temperature—about 1×10^7 K in the center. At this temperature, all of the gases are completely ionized, meaning the interior is composed only of atomic nuclei and electrons. This state of matter is known as plasma. Though partially ionized, the outer layers of the Sun are not hot enough to be plasma. The Sun produces the equivalent of 4 trillion trillion 100-W lightbulbs of light each second. The small amount that reaches Earth is equal to 1.35 kilowatt/m^2.

Table 29.1 Relative Properties of the Sun

Concepts In Motion **Interactive Table** To explore more about the Sun, visit glencoe.com.

	Sun	Earth	Jupiter
Diameter (km)	1.4×10^6	1.3×10^4	1.4×10^5
Mass (kg)	2.0×10^{30}	6.0×10^{24}	1.9×10^{27}
Density (kg/m^3)	1.4×10^3	5.5×10^3	1.3×10^3

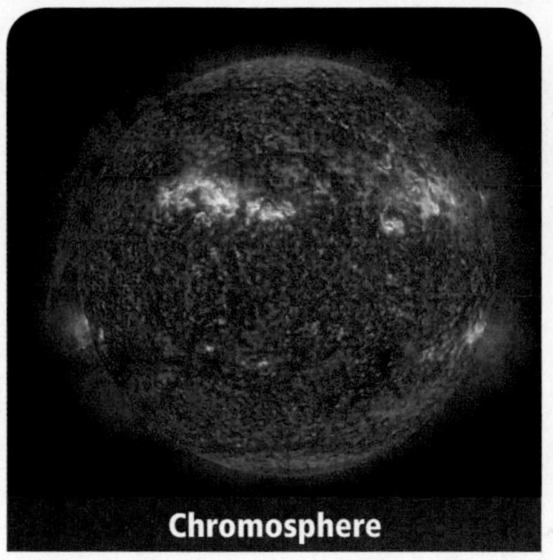

■ **Figure 29.1** Sunspots appear dark on the photosphere, the visible surface of the Sun. The chromosphere of the Sun appears red with prominences and flares suspended in the thin layer. The white-hot areas are almost 6000 K while the darker, red areas are closer to 3000 K. **Deduce** ***why the images look so different.***

The Sun's Atmosphere

You might ask how the Sun could have an atmosphere when it is already gaseous. The outer regions are organized into layers, like a planetary atmosphere separated into different levels, and each layer emits energy at wavelengths resulting from its temperature.

FOLDABLES Incorporate information from this section into your Foldable.

Photosphere The **photosphere,** shown in **Figure 29.1,** is the visible surface of the Sun. It is approximately 400 km thick and has an average temperature of 5800 K. It is also the innermost layer of the Sun's atmosphere. You might wonder how it is the visible surface of the Sun if it is the innermost layer. This is because most of the visible light emitted by the Sun comes from this layer. The two outermost layers are transparent at most wavelengths of visible light. Additionally, the outermost two layers are dim in the wavelengths they emit.

 **Reading Check** **Explain** why the innermost layer of the Sun's atmosphere is visible.

Chromosphere Outside the photosphere is the **chromosphere,** which is approximately 2500 km thick and has a temperature of nearly 30,000 K. Usually, the chromosphere is visible only during a solar eclipse when the photosphere is blocked. However, astronomers can use special filters to observe the chromosphere when the Sun is not eclipsed. The chromosphere appears red, as shown in **Figure 29.1,** because its strongest emissions are in a single band in the red wavelength.

Corona The outermost layer of the Sun's atmosphere, called the **corona,** extends several million kilometers from the outside edge of the chromosphere and has a temperature range of 1 million to 2 million K. The density of the gas in the corona is very low, which explains why the corona is so dim that it can be seen only when the photosphere is blocked by either special instruments, as in a coronagraph, or by the Moon during an eclipse, as shown in **Figure 29.2.** The temperature is so high in these outer layers of the solar atmosphere that the radiation emitted most is of ultraviolet wavelengths for the chromosphere, and X rays for the corona.

■ **Figure 29.2** The Sun's hottest and outermost layer, the corona, is only seen when the disk of the Sun is blocked as by this solar eclipse.

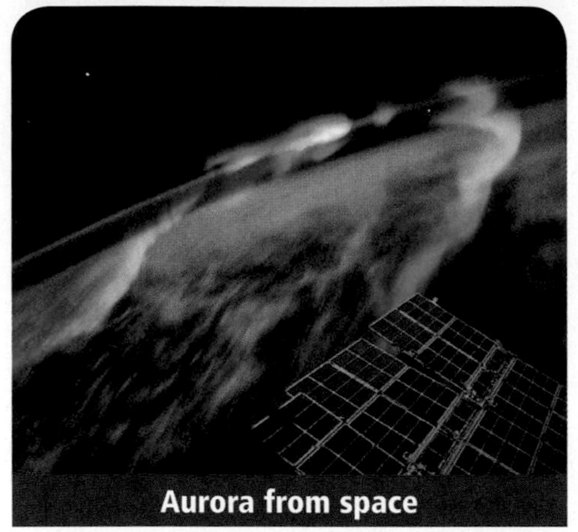

■ **Figure 29.3** The aurora is the result of particles from the Sun colliding with gases in Earth's atmosphere. It is best viewed from regions around the poles of Earth.
Infer ***When can you see the aurora?***

Solar wind The corona of the Sun does not have an abrupt edge. Instead, gas flows outward from the corona at high speeds and forms the **solar wind.** As this wind of charged particles, called ions, flows outward through the entire solar system, it bathes each planet in a flood of particles. At 1 AU—Earth's distance from the Sun—the solar wind flows at a speed of about 400 km/s. The charged particles are deflected by Earth's magnetic field and are trapped in two huge rings, called the Van Allen belts. The high-energy particles in these belts collide with gases in Earth's atmosphere and cause the gases to give off light. This light, called the aurora, can be seen from Earth or from space, as shown in **Figure 29.3.** The aurora are generally seen from Earth in the polar regions.

Solar Activity

While the solar wind and layers of the Sun's atmosphere are permanent features, other features on the Sun change over time in a process called solar activity. Some of the Sun's activity includes fountains and loops of glowing gas. Some of this gas has structure—a certain order in both time and place. This structure is driven by magnetic fields.

The Sun's magnetic field and sunspots The Sun's magnetic field disturbs the solar atmosphere periodically and causes new features to appear. The most obvious features are **sunspots,** shown in **Figure 29.4,** which are dark spots on the surface of the photosphere. Sunspots are bright, but they appear darker than the surrounding areas on the Sun because they are cooler. They are located in regions where the Sun's intense magnetic fields penetrate the photosphere. Magnetic fields create pressure that counteracts the pressure from the hot, surrounding gas. This stabilizes the sunspots despite their lower temperature. Sunspots occur in pairs with opposite magnetic polarities—with a north and a south pole similar to a magnet.

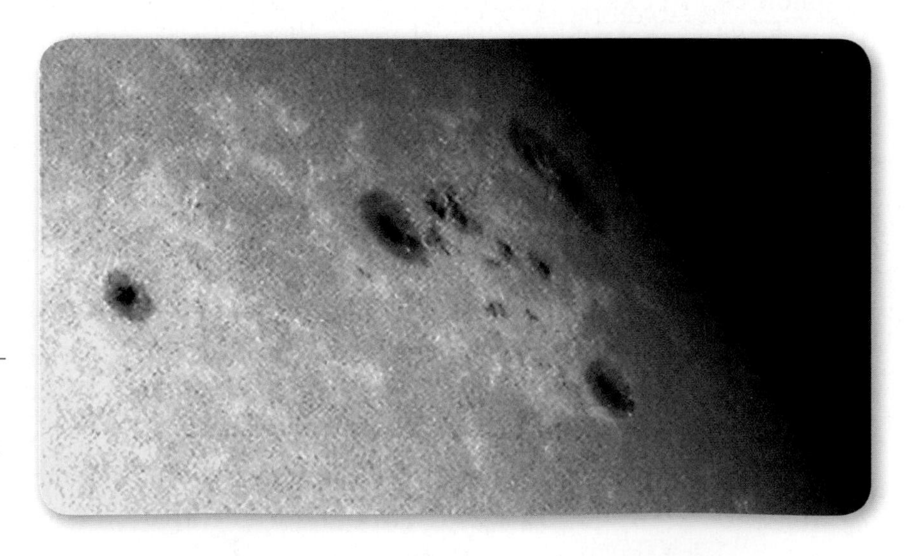

■ **Figure 29.4** Sunspots are dark spots on the surface of the photosphere. Each sunspot is accompanied by a bright, granular structure. The light and dark areas are associated with the Sun's magnetic field. Sunspots typically last about two months.

Solar activity cycle Astronomers have observed that the number of sunspots changes regularly, reaching a maximum number every 11.2 years. At this point, the Sun's magnetic field reverses, so that the north magnetic pole becomes the south magnetic pole and vice versa. Because sunspots are caused by magnetic fields, the polarities of sunspot pairs reverse when the Sun's magnetic poles reverse. Therefore, when the polarity of the Sun's magnetic field is taken into account, the length of the cycle doubles to 22.4 years. Thus, the solar activity cycle starts with minimum spots and progresses to maximum spots. The magnetic field then reverses in polarity, and the spots start again at a minimum number and progress to a maximum number. The magnetic field then switches back to the original polarity and completes the solar activity cycle.

Reading Check **Determine** how often the Sun's magnetic poles reverse themselves.

Other solar features Coronal holes, only detectable in X-ray photography and shown in **Figure 29.5,** are often located over sunspot groups. Coronal holes are areas of low density in the gas of the corona and are the main regions from which the particles that comprise the solar wind escape.

Highly active solar flares are also associated with sunspots, as shown in **Figure 29.5.** **Solar flares** are violent eruptions of particles and radiation from the surface of the Sun. Often, the released particles escape the surface of the Sun in the solar wind and Earth gets bombarded with the particles a few days later. The largest recorded solar flare, which occurred in April 2001, hurled particles from the Sun's surface at 7.2 million km/h.

Another active feature, sometimes associated with flares, is a **prominence,** which is an arc of gas that is ejected from the chromosphere, or is gas that condenses in the inner corona and rains back to the surface. **Figure 29.5** shows an image of a prominence. Prominences can reach temperatures greater than 50,000 K and can last from a few hours to a few months. Like flares, prominences are also associated with sunspots and the magnetic field, and occurrences of both vary with the solar-activity cycle.

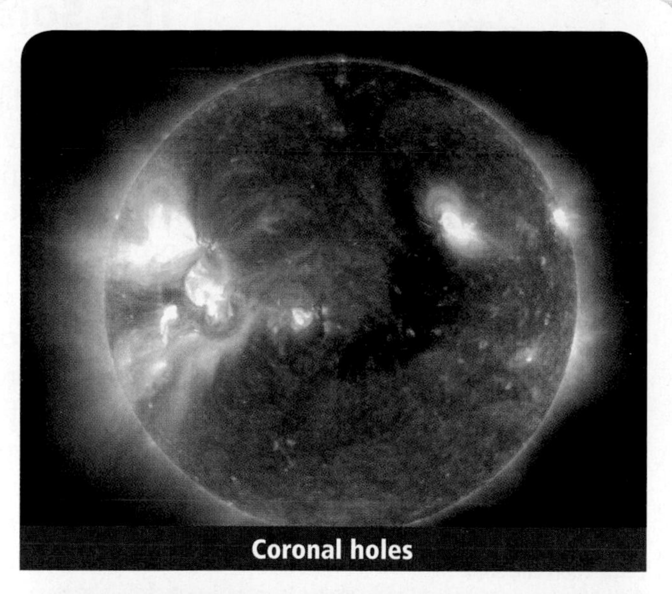

Figure 29.5 Features of the Sun's surface include coronal holes into the surface and solar flares and prominences that erupt from the surface.

■ **Figure 29.6** Energy in the Sun is transferred mostly by radiation from the core outward to about 86 percent of its radius. The outer layers transfer energy in convection currents.

The Solar Interior

You might be wondering where all the energy that causes solar activity and light comes from. Fusion occurs in the core of the Sun, where the pressure and temperature are extremely high. **Fusion** is the combination of lightweight, atomic nuclei into heavier nuclei, such as hydrogen fusing into helium. This is the opposite of the process of **fission,** which is the splitting of heavy atomic nuclei into smaller, lighter nuclei, like uranium into lead.

Energy production in the Sun In the core of the Sun, helium is a product of the process in which hydrogen nuclei fuse. The mass of the helium nucleus is less than the combined mass of the four hydrogen nuclei, which means that mass is lost during the process. Albert Einstein's special theory of relativity shows that mass and energy are equivalent, and that matter can be converted into energy and vice versa. This relationship can be expressed as $E = mc^2$, where E is energy measured in joules, m is the quantity of mass that is converted to energy measured in kilograms, and c is the speed of light measured in m/s. This theory explains that the mass lost in the fusion of hydrogen to helium is converted to energy, which powers the Sun. At the Sun's rate of hydrogen fusing, it is about halfway through its lifetime, with approximately 5 billion years left. Even so, the Sun has used only about 3 percent of its hydrogen.

Energy transport If the energy of the Sun is produced in the core, how does it get to the surface before it travels to Earth? The answer lies in the two zones in the solar interior illustrated in **Figure 29.6.** In the inner portion of the Sun, extending to about 86 percent of its radius, energy is transferred by radiation. This is the radiation zone. Above that, in the convection zone, energy is transferred by gaseous convection currents. As energy moves outward, the temperature is reduced from a central value of about 1×10^7 K to its photospheric value of about 5800 K. Leaving the Sun's outermost layer, energy moves in a variety of wavelengths in all directions. A tiny fraction of that immense amount of solar energy eventually reaches Earth.

■ **Figure 29.7** Energy excites the elements of a substance so that it emits different wavelengths of light.
Infer *what the colors of a spectrum represent.*

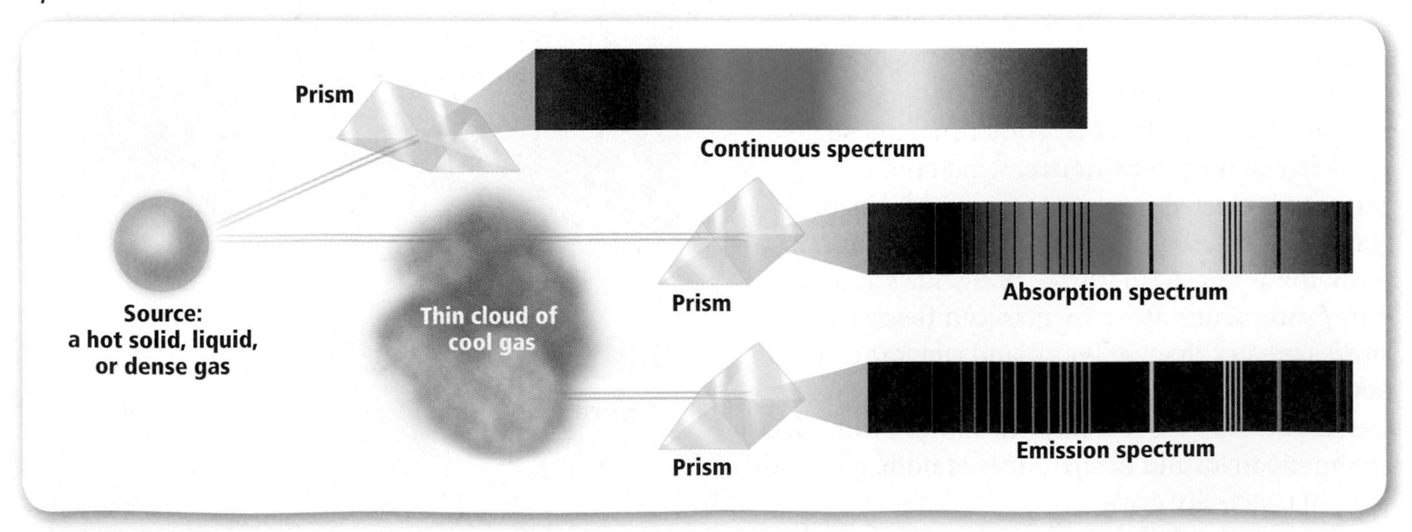

Solar energy on Earth The quantity of energy that arrives on Earth every day from the Sun is enormous. Above Earth's atmosphere, 1354 J of energy is received in 1 m^2/s (1354 W/m^2). In other words, 13 100-W lightbulbs could be operated with the solar energy that strikes a 1-m^2 area. However, not all of this energy reaches the ground because some is absorbed and scattered by the atmosphere, as you learned in Chapter 11.

Spectra

You are probably familiar with the rainbow that appears when white light is shined through a prism. This rainbow is a spectrum (plural, spectra), which is visible light arranged according to wavelengths. There are three types of spectra: continuous, emission, and absorption, as shown in **Figure 29.7.**

A spectrum that has no breaks in it, such as the one produced when light from an ordinary bulb is shined though a prism, is called a continuous spectrum. A continuous spectrum can also be produced by a glowing solid or liquid, or by a highly compressed, glowing gas. The spectrum from a noncompressed gas contains bright lines at certain wavelengths. This is called an emission spectrum, and the lines are called emission lines. The wavelengths of the visible lines depend on the element being observed because each element has its own characteristic emission spectrum.

Reading Check **Describe** continuous and emission spectra.

A spectrum produced from the Sun's light shows a series of dark bands. These dark spectral lines are caused by different chemical elements that absorb light at specific wavelengths. This is called an absorption spectrum, and the lines are called absorption lines. Absorption is caused by a cooler gas in front of a source that emits a continuous spectrum. The pattern of the dark absorption lines of an element is exactly the same as the bright emission lines for that same element. Thus, by comparing laboratory spectra of different gases with the dark lines in the solar spectrum, it is possible to identify the elements that make up the Sun's outer layers. You will experiment with identifying spectral lines in the GeoLab at the end of this chapter.

DATA ANALYSIS LAB

Based on Real Data*

Interpret Data

Can you identify elements in a star? Astronomers study the composition of stars by observing their absorption spectra. Each element in a star's outer layer produces a set of lines in the star's absorption spectrum. From the pattern of lines, astronomers can determine what elements are in a star.

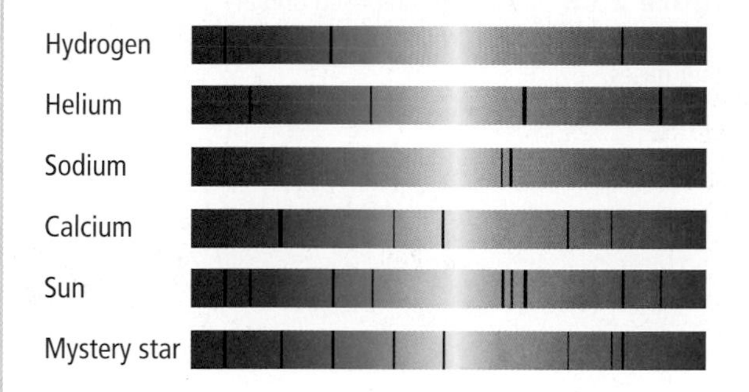

Analysis

1. Study the spectra of the four elements.
2. Examine the spectra for the Sun and the mystery star.
3. To identify the elements of the Sun and the mystery star, use a ruler to help you line up the spectral lines with the known elements.

Think Critically

4. **Identify** the elements that are present in the part of the absorption spectrum shown for the Sun.
5. **Identify** the elements that are present in the absorption spectrum for the mystery star.
6. **Determine** which elements are common to both stars.

*James B. Kaler. Professor Emeritus of Astronomy. University of Illinois. 1998.

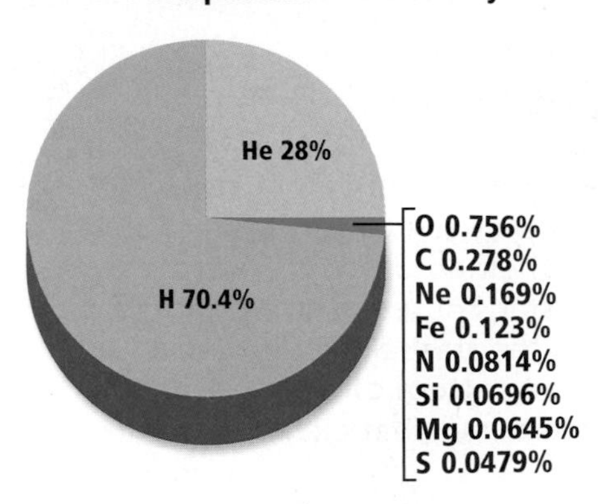

■ **Figure 29.8** The Sun is composed primarily of hydrogen and helium with small amounts of other gases.

Solar Composition

Although scientists cannot take samples from the Sun directly, they have learned a great deal about the Sun from its spectra. Using the lines of the absorption spectra like fingerprints, astronomers have identified the elements that compose the Sun. Sixty or more elements have been identified as solar components. The Sun consists of hydrogen (H), at about 70.4 percent by mass, helium, (He) 28 percent, and a small amount of other elements, as illustrated in **Figure 29.8.** This composition is similar to that of the gas giant planets. It suggests that the Sun and the gas giants represent the composition of the interstellar cloud from which the solar system formed. While the terrestrial planets have lost most of the lightweight gases, as you learned in Chapter 28, their heavier element composition probably came from a contribution to the interstellar cloud of by-products from long-extinct stars.

The Sun's composition represents that of the galaxy as a whole. Most stars have proportions of the elements similar to the Sun. Hydrogen and helium are the predominant gases in stars and in the rest of the universe. Even dying stars still have hydrogen and helium in their outer layers, because their internal temperatures might only fuse about 10 percent of their total hydrogen into helium. All other elements are in small proportions compared to hydrogen and helium. The larger the star's mass at its inception, the more heavy elements it will produce in its lifetime. But, as you will read in this chapter, there are different results when a star dies. As stars die, they return as much as 50 percent of their mass back into interstellar space, to be recycled into new generations of stars and planets.

Section 29.1 Assessment

Section Summary

- Most of the mass in the solar system is found in the Sun.
- The Sun's average density is approximately equal to that of the gas giant planets.
- The Sun has a layered atmosphere.
- The Sun's magnetic field causes sunspots and other solar activity.
- The fusion of hydrogen into helium provides the Sun's energy and composition.
- The different temperatures of the Sun's outer layers produce different spectra.

Understand Main Ideas

1. MAIN Idea **Identify** which features of the Sun are typical of stars.
2. **Describe** the outer layers of gas above the Sun's visible surface.
3. **Classify** the different types of spectra by how they are created.
4. **Describe** the process of fusion in the Sun.
5. **Compare** the composition of the Sun in **Figure 29.8** to the gas giant planets' compositions in Chapter 28.

Think Critically

6. **Infer** how the Sun would affect Earth if Earth did not have a magnetic field.
7. **Relate** the solar activity cycle with solar flares and prominences.

WRITING in **Earth Science**

8. Create a trifold brochure relating the layers and characteristics of the Sun.

Section 29.2

Measuring the Stars

Objectives

- **Determine** how distances between stars are measured.
- **Distinguish** between brightness and luminosity.
- **Identify** the properties used to classify stars.

Review Vocabulary

wavelength: the distance from one point on a wave to the next corresponding point

New Vocabulary

constellation
binary star
parsec
parallax
apparent magnitude
absolute magnitude
luminosity
Hertzsprung-Russell diagram
main sequence

MAIN Idea **Stellar classification is based on measurement of light spectra, temperature, and composition.**

Real-World Reading Link As you ride in a car on the highway at night and as a car approaches you, its lights seem to get larger and brighter. Distant stars might be just as large and just as bright as nearer ones, but the distance causes them to appear small and dim.

Groups of Stars

Long ago, many civilizations looked at the brightest stars and named groups of them after animals, mythological characters, or everyday objects. These groups of stars are called **constellations.** Today, astronomers group stars by the 88 constellations named by ancient peoples. Some constellations are visible throughout the year, depending on the observer's location. In the northern hemisphere, you can see constellations that appear to rotate around the north pole. These constellations are called circumpolar constellations. Ursa Major, also known as the Big Dipper, is a circumpolar constellation for the northern hemisphere.

Unlike circumpolar constellations, the other constellations can be seen only at certain times of the year because of Earth's changing position in its orbit around the Sun, as illustrated in **Figure 29.9.** For example, the constellation Orion can be seen in the northern hemisphere's winter, and the constellation Hercules can be seen in the northern hemisphere's summer. For this reason, constellations are classified as summer, fall, winter, and spring constellations. The most familiar constellations are the ones that are part of the zodiac. These twelve constellations lie in the ecliptic plane along the same path where the planets are seen. Different constellations can be seen in the northern and southern hemispheres, but the zodiac can be seen in both. Ancient people used the constellations to know when to prepare for planting, harvest, and ritual celebrations.

Figure 29.9 Different constellations are visible in the sky due to Earth's movement around the Sun.

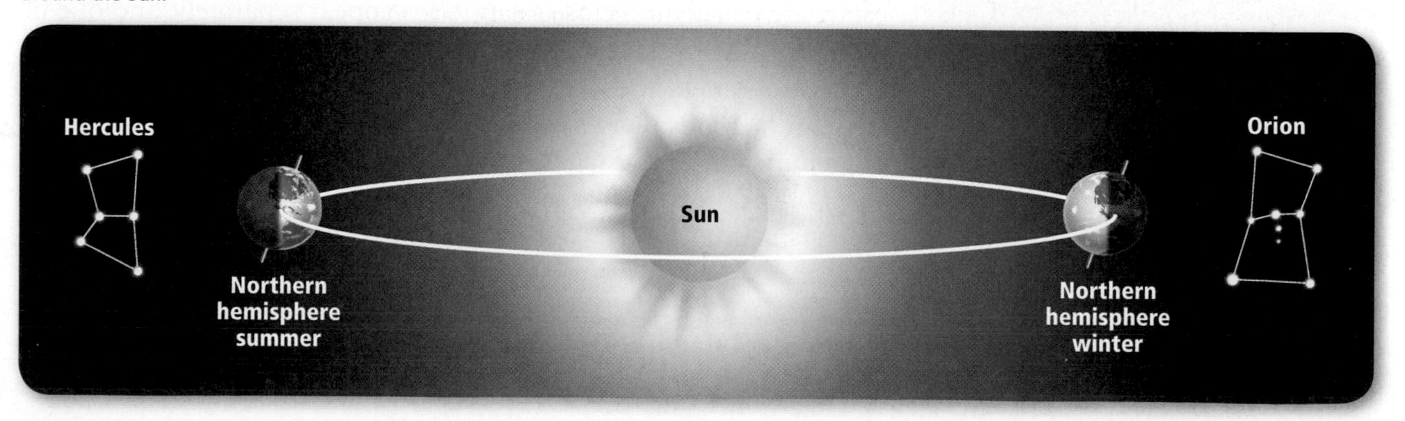

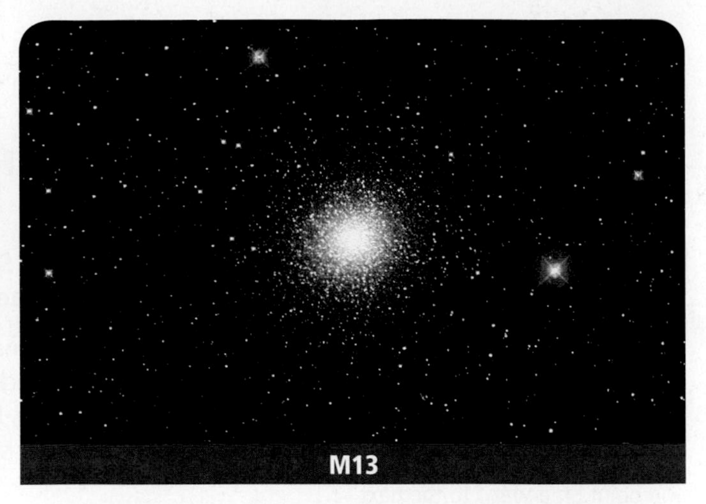

■ **Figure 29.10** Star clusters are groups of stars that are gravitationally bound to one another. The Pleiades is an open cluster group and M13 is a globular cluster.

Star clusters Although the stars in constellations appear to be close to each other, few are gravitationally bound to one other. The reason that they appear to be close together is that human eyes cannot distinguish how far or near stars are. Two stars could appear to be located next to each other in the sky, but one might be 1 trillion km from Earth, and the other might be 2 trillion km from Earth. However, by measuring distances to stars and observing how their gravities interact with each other, scientists can determine which stars are gravitationally bound to each other. A group of stars that are gravitationally bound to each other is called a cluster. The Pleiades (PLEE uh deez) in the constellation Taurus, shown in **Figure 29.10,** is an open cluster because the stars are not densely packed. In contrast, a globular cluster is a group of stars that are densely packed into a spherical shape, such as M13 in the constellation Hercules, also shown in **Figure 29.10.** Different kinds of clusters are explained in **Figure 29.12.**

Reading Check **Distinguish** between open and globular clusters.

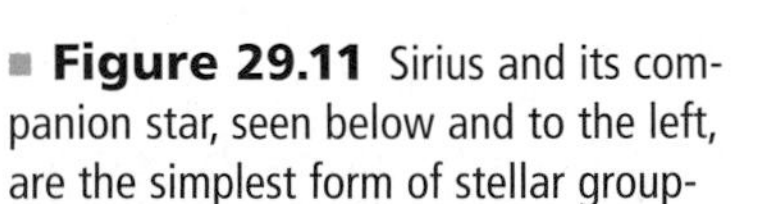

■ **Figure 29.11** Sirius and its companion star, seen below and to the left, are the simplest form of stellar grouping, known as a binary.

Binaries When only two stars are gravitationally bound together and orbit a common center of mass, they are called **binary stars.** More than half of the stars in the sky are either binary stars or members of multiple-star systems. The bright star Sirius is half of a binary system, shown in **Figure 29.11.** Most binary stars appear to be single stars to the human eye, even with a telescope. The two stars are usually too close together to appear separately, and one of the two is often much brighter than the other.

Astronomers are able to identify binary stars through the use of several methods. For example, even if only one star is visible, accurate measurements can show that its position shifts back and forth as it orbits the center of mass between it and the unseen companion star. Also, the orbital plane of a binary system can sometimes be seen edgeways from Earth. In such cases, the two stars alternately block each other and cause the total brightness of the two-star system to dip each time one star eclipses the other. This type of binary star is called an eclipsing binary.

Visualizing Star Groupings

Figure 29.12 When you look into the night sky, the stars seem to be randomly spaced from horizon to horizon. Upon closer inspection, you begin to see groups of stars that seem to cluster in one area. Star clusters are gravitationally bound groups of stars, which means that their gravities interact to hold the stars in a group.

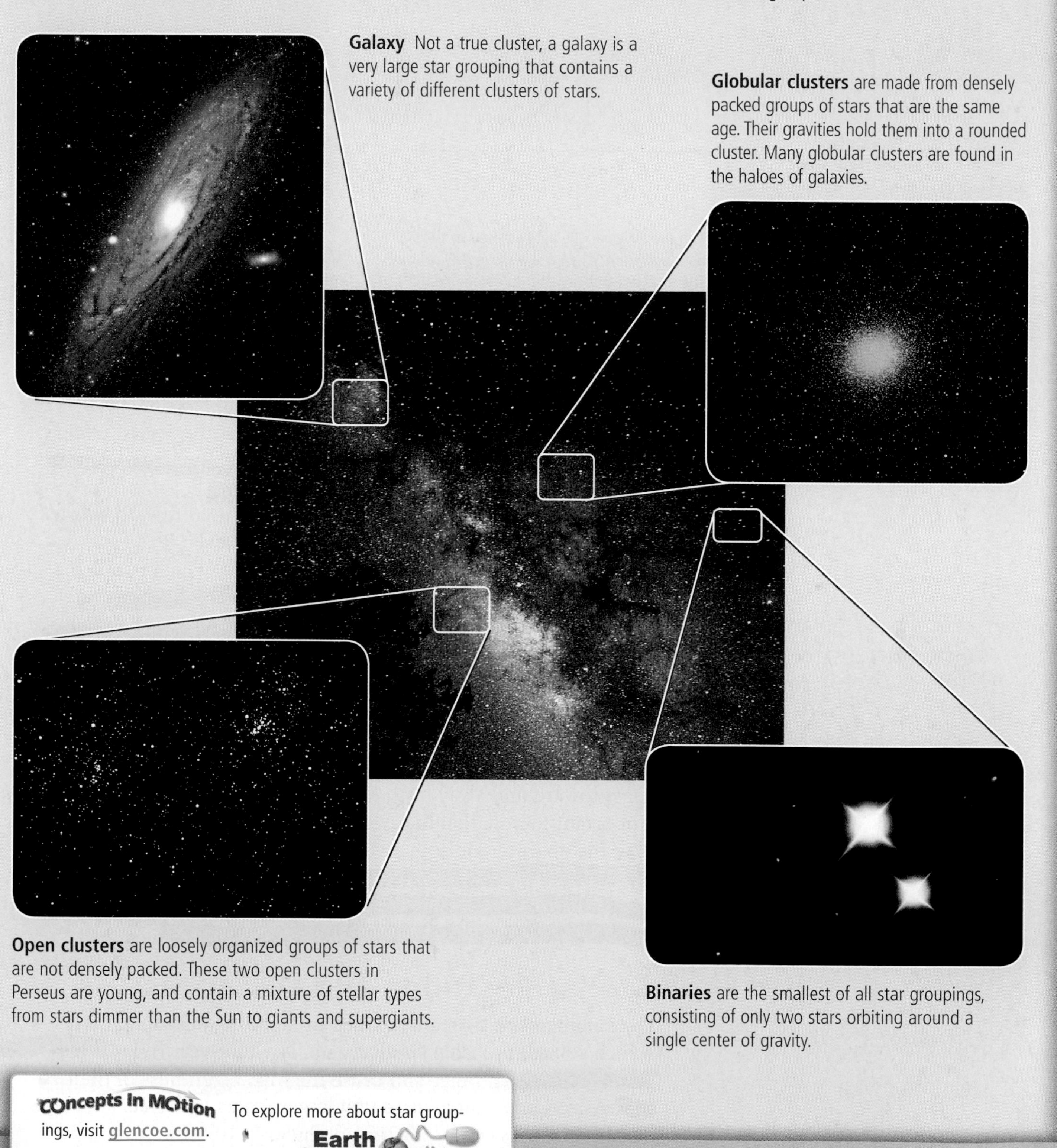

Galaxy Not a true cluster, a galaxy is a very large star grouping that contains a variety of different clusters of stars.

Globular clusters are made from densely packed groups of stars that are the same age. Their gravities hold them into a rounded cluster. Many globular clusters are found in the haloes of galaxies.

Open clusters are loosely organized groups of stars that are not densely packed. These two open clusters in Perseus are young, and contain a mixture of stellar types from stars dimmer than the Sun to giants and supergiants.

Binaries are the smallest of all star groupings, consisting of only two stars orbiting around a single center of gravity.

Concepts In Motion To explore more about star groupings, visit glencoe.com.

Earth Science Online

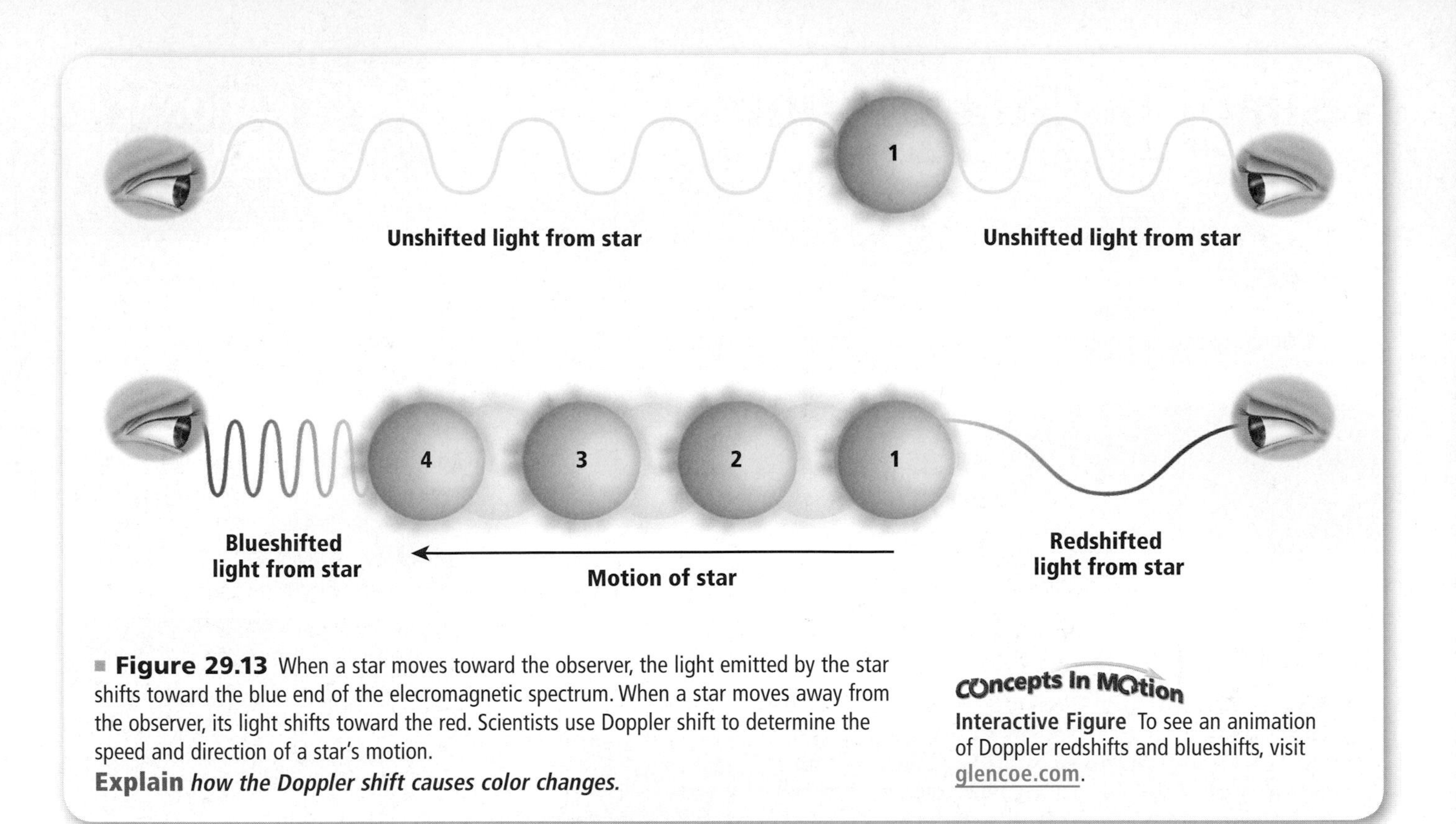

■ **Figure 29.13** When a star moves toward the observer, the light emitted by the star shifts toward the blue end of the elecromagnetic spectrum. When a star moves away from the observer, its light shifts toward the red. Scientists use Doppler shift to determine the speed and direction of a star's motion.
Explain *how the Doppler shift causes color changes.*

Concepts In Motion
Interactive Figure To see an animation of Doppler redshifts and blueshifts, visit glencoe.com.

Doppler shifts The most common way to tell that a star is one of a binary pair is to find subtle wavelength shifts, called Doppler shifts. As the star moves back and forth along the line of sight, as shown in **Figure 29.13**, its spectral lines shift. If a star is moving toward the observer, the spectral lines are shifted toward shorter wavelengths, which is called a blueshift. However, if the star is moving away, the wavelengths become longer, which is called a redshift. The higher the speed, the larger the shift, thus careful measurements of spectral line wavelengths can be used to determine the speed of a star's motion. Because there is no Doppler shift for motion that is at a right angle to the line of sight, astronomers can learn only about the portion of a star's motion that is directed toward or away from Earth. The Doppler shift in spectral lines can be used to detect binary stars as they move about their center of mass toward and away from Earth with each revolution. It is also important to note that there is no way to distinguish whether the star, the observer, or both are moving. A star undergoing periodic Doppler shifts can only be interpreted as one of a binary. Stars identified in this way are called spectroscopic binaries. Binaries can reveal much about the individual properties of stars.

Stellar Positions and Distances

Astronomers use two units of measure for long distances. One, which you are probably familiar with, is a light-year (ly). A light-year is the distance that light travels in one year, equal to 9.461×10^{12} km. Astronomers often use a unit larger than a light-year—a parsec. A **parsec** (pc) is equal to 3.26 ly, or 3.086×10^{13} km.

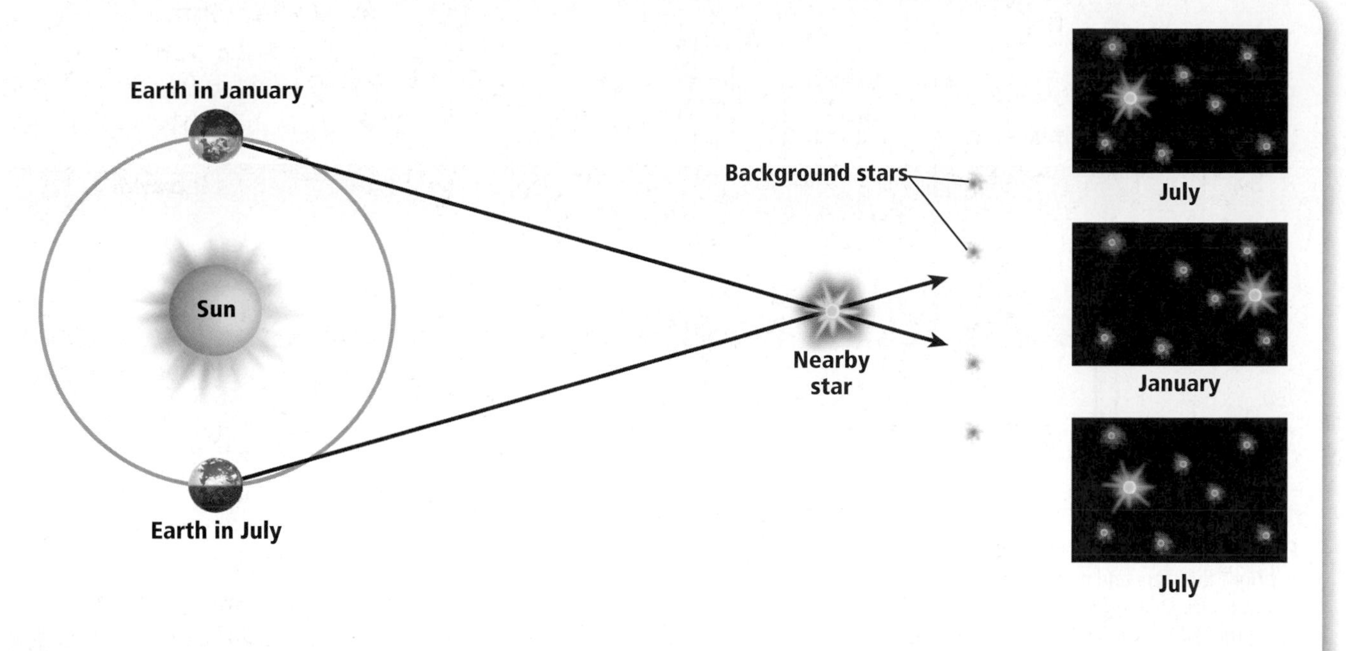

■ **Figure 29.14** As Earth orbits the Sun, nearby stars appear to change position in the sky compared to faraway stars. Earth reaches its maximum change in position at six months, so the angle measured to the star from these two positions is also at the maximum. This shift in observation position is called parallax and can be used to estimate the distance to the star being observed.
Predict *the position of the star in September.*

Concepts In Motion
Interactive Figure To see an animation of parallax, visit glencoe.com.

Parallax Precise position measurements are important for determining distances to stars. When determining the distance of stars from Earth, astronomers must account for the fact that nearby stars shift in position as observed from Earth. This apparent shift in position caused by the motion of the observer is called **parallax.** In this case, the motion of the observer is the change in position of Earth as it orbits the Sun. As Earth moves from one side of its orbit to the opposite side, a nearby star appears to be shifting back and forth, as illustrated in **Figure 29.14.** The closer the star, the larger the shift. The distance to a star can be estimated from its parallax shift by measuring the angle of the change. Using the parallax technique, astronomers could find accurate distances to stars up to only 100 pc, or approximately 300 ly, until recently. With advancements in technology, such as the *Hipparcos* satellite, astronomers can find accurate distances up to 500 pc by using parallax.

 **Reading Check** **Identify** the motion of the observer in the diagram.

VOCABULARY
ACADEMIC VOCABULARY
Precise
exactly or sharply defined or stated
The builder's accurate measurements ensured that all of the boards were cut to the same, precise length.

Basic Properties of Stars

The basic properties of a star are mass, diameter, and luminosity, which are all related to each other. Temperature is another property and is estimated by finding the spectral type of a star. Temperature controls the nuclear reaction rate and governs the luminosity, or apparent magnitude. The absolute magnitude compared to the apparent magnitude is used to find the distance to a star.

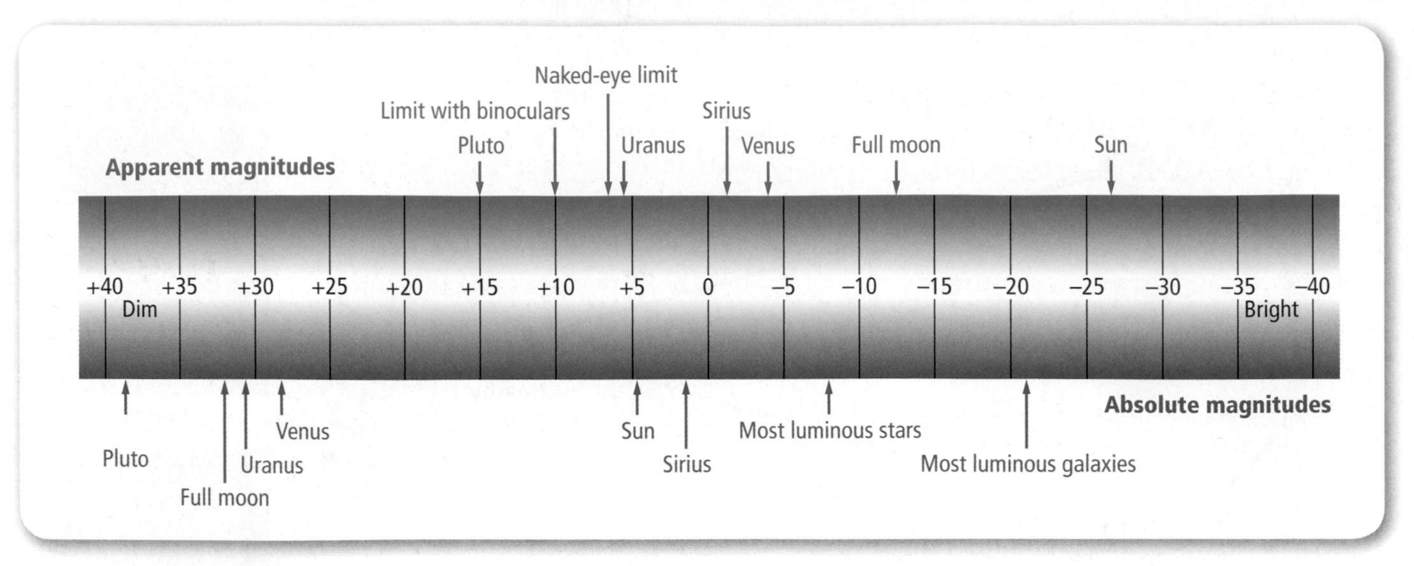

■ **Figure 29.15** Apparent magnitude is how bright the stars and planets appear in the sky from Earth. Absolute magnitude takes into account the distance to that star or planet and makes adjustments for distance.

Magnitude One of the most basic observable properties of a star is how bright it appears, or the **apparent magnitude.** The ancient Greeks established a classification system based on the brightness of stars. The brightest stars were given a ranking of +1, the next brightest +2, and so on. Today's astronomers still use this system, but they have refined it. In this system, a difference of 5 magnitudes corresponds to a factor of 100 in brightness. Thus, a magnitude +1 star is 100 times brighter than a magnitude +6 star.

Absolute magnitude Apparent magnitude does not indicate the actual brightness of a star because it does not account for distance. A faint star can appear to be very bright because it is relatively close to Earth, while a bright star can appear to be faint because it is far away. To account for these phenomena, astronomers have developed another classification system for brightness. **Absolute magnitude** is how bright a star would appear if it were placed at a distance of 10 pc. The classification of stars by absolute magnitude allows comparisons that are based on how bright the stars would appear at equal distances from an observer. The disadvantage of absolute magnitude is that it can be calculated only when the actual distance to a star is known. The apparent and absolute magnitudes for several objects are shown in **Figure 29.15.**

VOCABULARY
SCIENCE USAGE V. COMMON USAGE
Magnitude
Science usage: a number representing the apparent brightness of a celestial body

Common usage: the importance, quality, or caliber of something

Luminosity Apparent magnitudes do not give an actual measure of energy output. To measure the energy output from the surface of a star per second, called its power or **luminosity,** an astronomer must know both the star's apparent magnitude and how far away it is. The brightness observed depends on both a star's luminosity and distance from Earth, and because brightness diminishes with the square of the distance, a correction must be made for distance. Luminosity is measured in units of energy emitted per second, or watts. The Sun's luminosity is about 3.85×10^{26} W. This is equivalent to 3.85×10^{24} 100-W lightbulbs. The values for other stars vary widely, from about 0.0001 to more than 1 million times the Sun's luminosity. No other stellar property varies as much.

Classification of Stars

You have learned that the Sun has dark absorption lines at specific wavelengths in its spectrum. Other stars also have dark absorption lines in their spectra and are classified according to their patterns of absorption lines. Spectral lines provide information about a star's temperature and composition.

Temperature Stars are assigned spectral types in the following order: O, B, A, F, G, K, and M. Each class is subdivided into more specific divisions with numbers from 0 to 9. For example, a star can be classified as being a type A4 or A5.

The classes were originally based only on the pattern of spectral lines, but astronomers later discovered that the classes also correspond to stellar temperatures, with the O stars being the hottest and the M stars being the coolest. Thus, by examination of a star's spectra, it is possible to estimate its temperature.

The Sun is a type G2 star, which corresponds to a surface temperature of about 5800 K. Surface temperatures range from about 50,000 K for the hottest O stars to as low as 2000 K for the coolest M stars. **Figure 29.16** shows how spectra from some different star classes appear.

Temperature is also related to luminosity and absolute magnitude. Hotter stars put out more light than stars with lower temperatures. In most normal stars, the temperature corresponds to the luminosity. Since the temperature is not affected by its distance, by measuring the temperature and luminosity, distance is known.

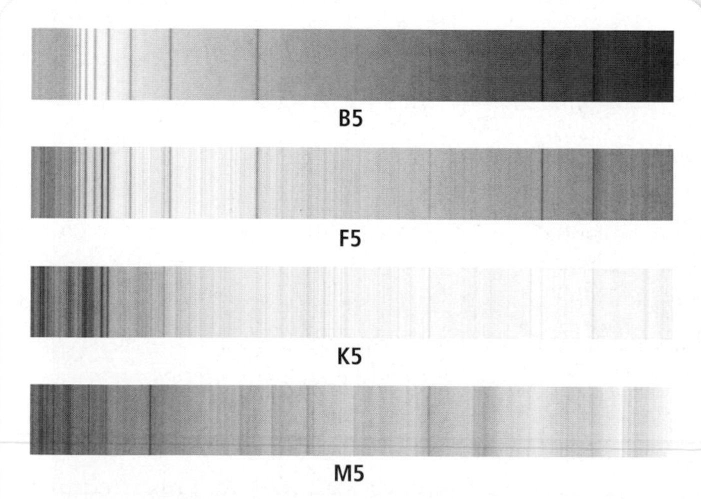

■ **Figure 29.16** These are typical absorption spectra of a class B5 star, class F5 star, class K5 star, and a class M5 star. The black stripes are absorption lines telling us each star's element composition.

MiniLab

Model Parallax

How does parallax angle change with distance? If a star is observed at six-month intervals in its orbit, it will appear to have moved because Earth is 300 million km away from the location of the first observation. The angle to the star is different and the apparent change in position of the star is called parallax.

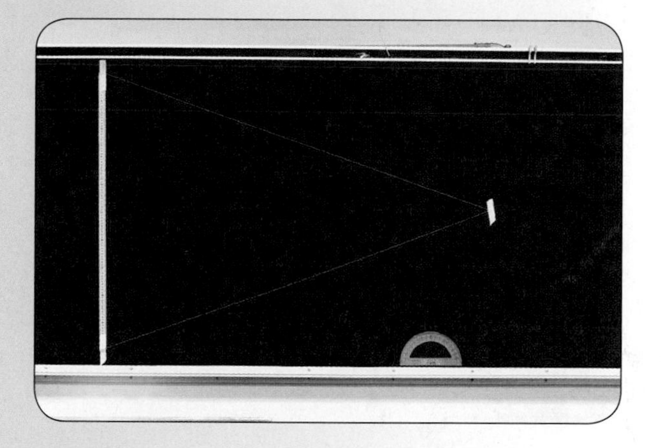

Procedure

1. Read and complete the lab safety form.
2. Place a **meterstick** at a fixed position and attach a **4-m piece of string** to each end.
3. Stand away from the meterstick and hold the two strings together to form a triangle. Be sure to hold the strings taut. Measure your distance from the meterstick. Record your measurement.
4. Measure the angle between the two pieces of string with a **protractor.** Record your measurement of the angle.
5. Repeat Steps 3 and 4 for different distances from the meterstick by shortening or lengthening the string.
6. Make a graph of the angles versus their distance from the meterstick.

Analysis

1. **Interpret** what the length of the meterstick represents. What does the angle represent?
2. **Analyze** what the graph shows. How does parallax angle depend on distance?
3. **Explain** how the angles that you measured are similar to actual stellar parallax angles.

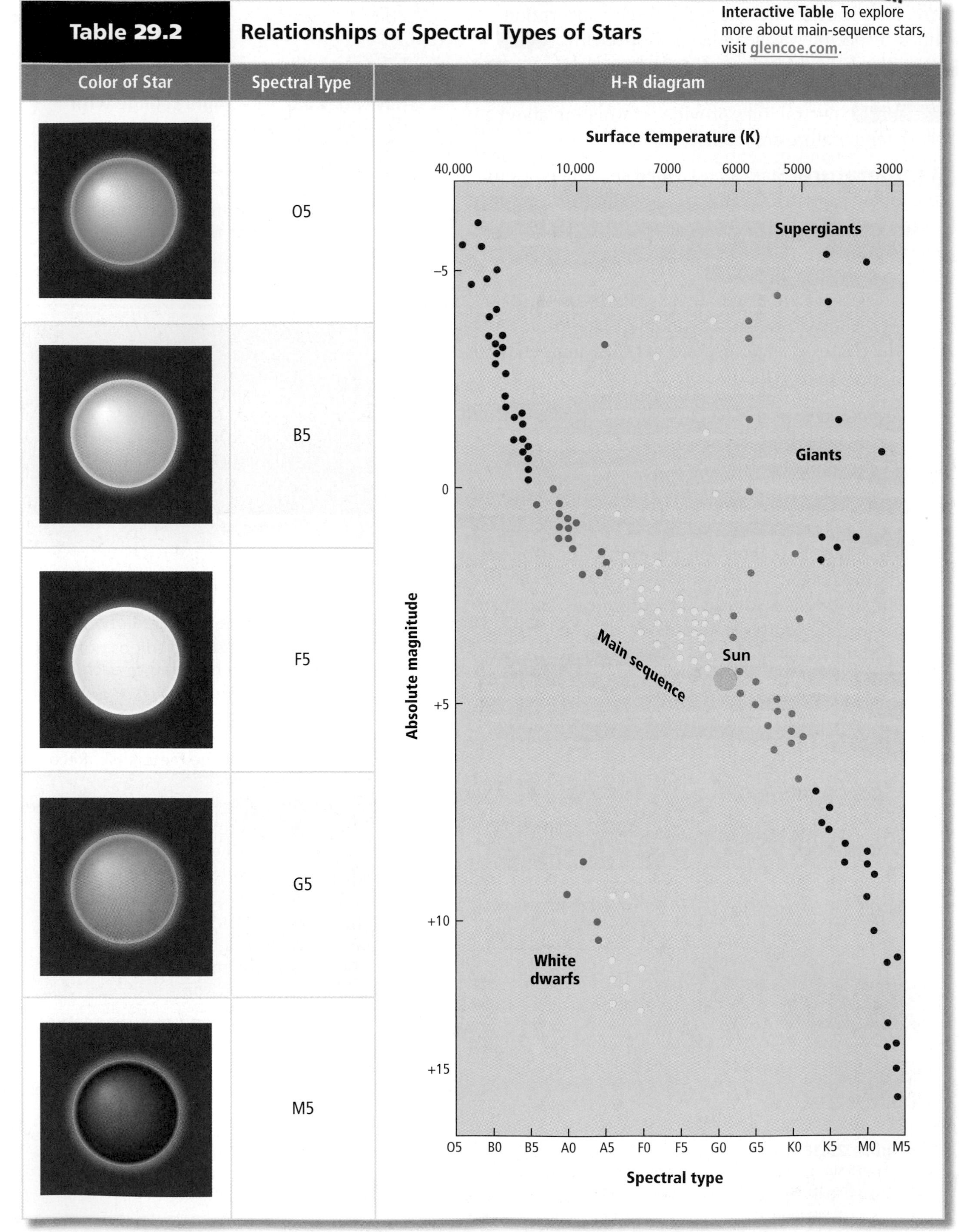

Concepts In Motion
Interactive Table To explore more about main-sequence stars, visit glencoe.com.
Table 29.2
Relationships of Spectral Types of Stars
Color of Star
Spectral Type
H-R diagram
O5
B5
F5
G5
M5
Surface temperature (K)
40,000
10,000
7000
6000
5000
3000
Supergiants
Giants
Main sequence
Sun
White dwarfs
Absolute magnitude
–5
0
+5
+10
+15
O5
B0
B5
A0
A5
F0
F5
G0
G5
K0
K5
M0
M5
Spectral type

Composition All stars, including the Sun, have nearly identical compositions, despite the differences in their spectra. The differences in the appearance of their spectra are almost entirely a result of temperature differences, shown in **Table 29.2.** Hotter stars have fairly simple visible spectra, while cooler stars have spectra with more lines. The coolest stars have bands in their spectra due to molecules such as titanium oxide in their atmospheres. Typically, about 73 percent of a star's mass is hydrogen (H), about 25 percent is helium (He), and the remaining 2 percent is composed of all the other elements. While there are some variations in the composition of stars, particularly in the final 2 percent, all stars have this general composition.

H-R diagrams The properties of mass, luminosity, temperature, and diameter are closely related. Each class of star has a specific mass, luminosity, magnitude, temperature, and diameter. These relationships can be demonstrated on a graph called the **Hertzsprung-Russell diagram** (H-R diagram) on which absolute magnitude is plotted on the vertical axis and temperature or spectral type is plotted on the horizontal axis, as shown in **Table 29.2.** Spectroscopists first plotted this graph in the early twentieth century. An H-R diagram with luminosity plotted on the vertical axis looks similar to the one in **Table 29.2** and is used to calculate the evolution of stars.

Most stars occupy the region in the diagram called the **main sequence,** which runs diagonally from the upper-left corner, where hot, luminous stars are represented, to the lower-right corner, where cool, dim stars are represented. **Table 29.3** shows some properties of main-sequence stars.

CAREERS IN EARTH SCIENCE

Spectroscopist The main job of an astronomer is to select the stars and objects to observe, but there are other scientists who are affiliated with an observatory. Scientists who make and analyze the spectra from stars are called spectroscopists. To learn more about Earth science careers, visit glencoe.com.

Table 29.3 Properties of Main-Sequence Stars

Concepts In Motion

Interactive Table To explore more about main-sequence star properties, visit glencoe.com.

Spectral Type	Mass*	Surface Temperature (K)	Luminosity*	Radius*
O5	40.0	40,000	5×10^5	18.0
B5	6.5	15,500	800	3.8
A5	2.1	8500	20	1.7
F5	1.3	6580	2.5	1.2
G5	0.9	5520	0.8	0.9
K5	0.7	4130	0.2	0.7
M5	0.2	2800	0.008	0.3

* These properties are relative to the Sun.

Main sequence About 90 percent of stars, including the Sun, fall along a broad strip of the H-R diagram called the main sequence. While stars are in the main sequence, they are fusing hydrogen in their cores. The interrelatedness of the properties of these stars indicates that all these stars have similar internal structures and functions. As stars evolve off the main sequence, they begin to fuse helium in their cores and burn hydrogen around the core edges.

The Sun lies near the center of the main sequence, being of average temperature and luminosity. A star's mass determines almost all its other properties, including its main-sequence lifetime. The more massive a star is, the higher its central temperature and the more rapidly it burns its hydrogen fuel. This is due primarily to the ratio of radiation pressure to gravitational pressure. Higher pressures cause the fuels to burn faster. As a consequence, the star runs out of hydrogen faster than a lower-mass star.

Red giants and white dwarfs The stars plotted at the upper right of the H-R diagram in **Table 29.2** are cool, yet luminous. Because cool surfaces emit much less radiation per square meter than hot surfaces do, these cool stars must have large surface areas to be so bright. For this reason, these larger, cool, luminous stars are called red giants. Red giants are so large—more than 100 times the size of the Sun in some cases—that Earth would be swallowed up if the Sun were to become a red giant! Conversely, the dim, hot stars plotted in the lower-left corner of the H-R diagram must be small, or they would be more luminous. These small, dim, hot stars are called white dwarfs. A white dwarf is about the size of Earth but has a mass about as large as the Sun's. You will learn how all the different stars are formed in Section 29.3.

Section 29.2 Assessment

Section Summary

- Stars exist in clusters held together by their gravity.
- The simplest cluster is a binary.
- Parallax is used to measure distances to stars.
- The brightness of stars is related to their temperature.
- Stars are classified by their spectra.
- The H-R diagram relates the basic properties of stars: class, temperature, and luminosity.

Understand Main Ideas

1. MAIN Idea **Relate** the stellar temperature to the classification of a star.
2. **Explain** the difference between apparent and absolute magnitudes.
3. **Explain** how parallax is used to measure the distance to stars.
4. **Compare and contrast** luminosity and magnitude.
5. **Contrast** the apparent magnitude and the absolute magnitude of a star.
6. **Compare** a light-year and a parsec.

Think Critically

7. **Design** a model to explain parallax.
8. **Explain** the relationship between radius and mass using **Table 29.3.**

MATH in Earth Science

9. Compare Orion's brightest stars, Rigel (B class) and Betelgeuse (M class), by mass, temperature, luminosity, and radius, using **Table 29.3** as a reference.

Earth Science Online Self-Check Quiz glencoe.com

Section 29.3

Objectives

- **Determine** the effect of mass on a star's evolution.
- **Identify** the features of massive and regular star life cycles.
- **Explain** how the universe is affected by the life cycles of stars.

Review Vocabulary

evolution: a radical change in composition over a star's lifetime

New Vocabulary

nebula
protostar
neutron star
pulsar
supernova
black hole

Stellar Evolution

MAIN Idea The Sun and other stars follow similar life cycles, leaving the galaxy enriched with heavy elements.

Real-World Reading Link A campfire glows brightly as long as it has fuel to burn. When the fuel is depleted, the light becomes dimmer, and the fire extinguishes. Unlike a campfire, stars shine because of nuclear reactions in their interior. Stars also die out when their nuclear fuel is gone.

Basic Structure of Stars

Mass governs a star's temperature, luminosity, and diameter. In fact, astronomers have discovered that the mass and the composition of a star determine nearly all its other properties.

Mass effects The more massive a star is, the greater the gravity pressing inward, and the hotter and more dense the star must be inside to balance its own gravity. The temperature inside a star governs the rate of nuclear reactions, which in turn determines the star's energy output—its luminosity. The balance between gravity squeezing inward and outward pressure is maintained by heat due to nuclear reactions and compression. This balance is called hydrostatic equilibrium and it must hold for any stable star, as illustrated in **Figure 29.17,** otherwise the star would expand or contract. This balance is governed by the mass of a star.

Pressure from the heat of nuclear reactions and compression

Gravity

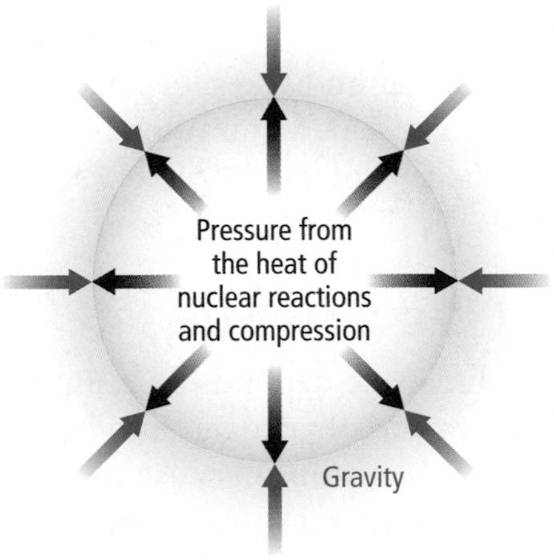

Figure 29.17 When the pressure from radiation and fusion is balanced by gravity, a star is stable and will not expand or contract.

Fusion Inside a star, conditions vary in much the same way that they do inside the Sun. The density and temperature increase toward the center, where energy is generated by nuclear fusion. Stars on the main sequence produce energy by fusing hydrogen into helium, as the Sun does. Stars that are not on the main sequence either fuse elements other than hydrogen in their cores or do not undergo fusion at all.

Stellar Evolution

A star changes as it ages because its internal composition changes as nuclear-fusion reactions in the star's core convert one element into another. With a change in the core composition, the star's density increases, its temperature rises, and its luminosity increases. As long as the star is stable and converting hydrogen to helium, it is considered a main-sequence star. Eventually, when the nuclear fuel runs out, the star's internal structure and mechanism for producing pressure must change to counteract gravity. The changes a star undergoes during its evolution begin with its formation.

■ **Figure 29.18** Temperatures will continue to build as gravity pulls the infalling matter to the center of the rotating disk. The center region is a protostar until fusion initiates and a star ignites.
Infer ***what happens to the remaining material in the disk.***

Concepts In Motion
Interactive Figure To see an animation of star formation, visit glencoe.com.

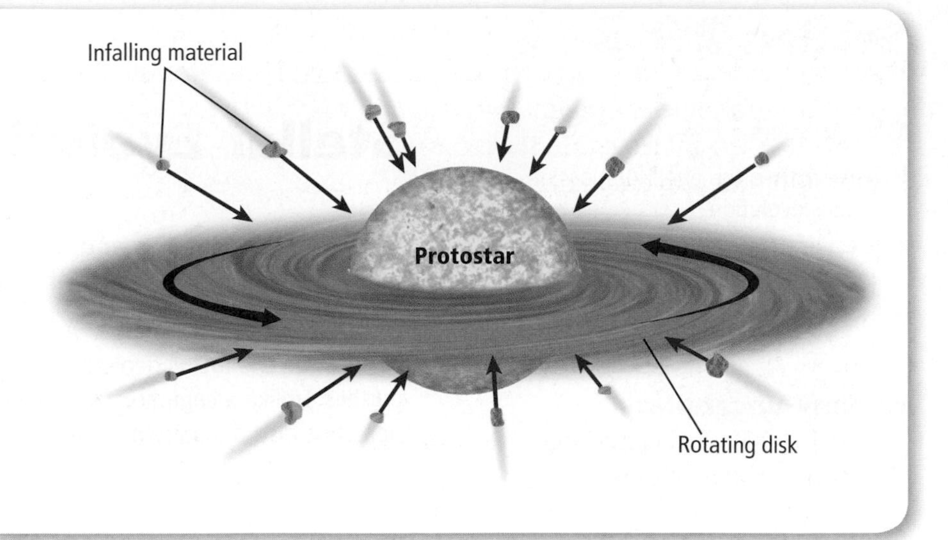

Star formation All stars form in much the same manner as the Sun did. The formation of a star begins with a cloud of interstellar gas and dust, called a **nebula** (plural, nebulae), which collapses on itself as a result of its own gravity. As the cloud contracts, its rotation forces it into a disk shape with a hot, condensed object at the center, called a **protostar,** as illustrated in **Figure 29.18.** Friction from gravity continues to increase the temperature of the protostar, until the condensed object reaches the ignition temperature for nuclear reactions and becomes a new star. A protostar is brightest at infrared wavelengths.

Reading Check **Infer** what causes the disk shape to form.

Fusion begins When the temperature inside a protostar becomes hot enough, nuclear fusion reactions begin. The first reaction to ignite is always the conversion of hydrogen to helium. Once this reaction begins, the star becomes stable because it then has sufficient internal heat to produce the pressure needed to balance gravity. The object is then truly a star and takes its place on the main sequence according to its mass. A new star often illuminates the gas and dust surrounding it, as shown in **Figure 29.19.**

■ **Figure 29.19** Using the Spitzer telescope's infrared wavelengths, protostars are imaged inside the Elephant Trunk nebula.

Life Cycles of Stars Like the Sun

What happens next during a star's life cycle depends on its mass. For example, as a star like the Sun converts hydrogen into helium in its core, it gradually becomes more luminous because the core density and temperature rise slowly and increase the reaction rate. It takes about 10 billion years for a star with the mass of the Sun to convert all of the hydrogen in its core into helium. Thus, such a star has a main-sequence lifetime of 10 billion years. From here, the next step in the life cycle of a small mass star is to become a red giant.

Red giant Only about the innermost 10 percent of a star's mass can undergo nuclear reactions because temperatures outside of this core never become hot enough for reactions to occur. Thus, when the hydrogen in its core is gone, a star has a helium center and outer layers made of hydrogen-dominated gas. Some hydrogen continues to react in a thin layer at the outer edge of the helium core, as illustrated in **Figure 29.20.** The energy produced in this layer forces the outer layers of the star to expand and cool. The star then becomes a red giant because its luminosity increases while its surface temperature decreases due to the expansion.

While the star is a red giant, it loses gas from its outer layers. The star is so large that its surface gravity is low, and thus the outer layers can be released by small expansions and contractions, or pulsations, of the star due to instability. Meanwhile, the core of the star becomes hot enough, at 100 million K, for helium to react and form carbon. The star contracts back to a more normal size, where it again becomes stable for awhile. The helium-reaction phase lasts only about one-tenth as long as the earlier hydrogen-burning phase. Afterward, when the helium in the core is depleted, the star is left with a core made of carbon.

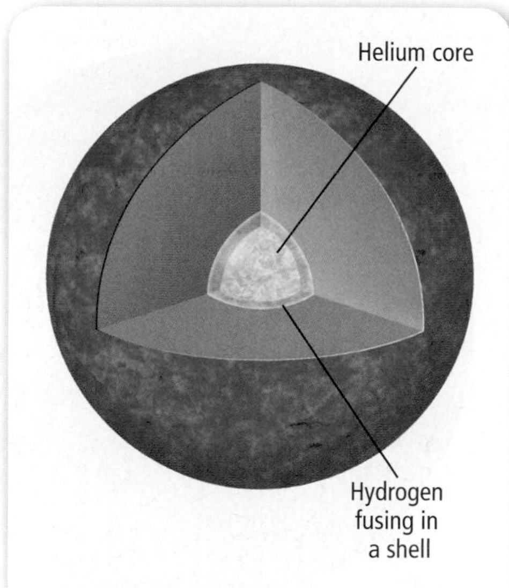

Figure 29.20 In a red giant's central region, helium is converted to carbon. In the spherical shell just outside, hydrogen continues to be converted to helium. The low temperature of the outer atmosphere due to expansion and cooling causes the red color.

Concepts In Motion
Interactive Figure To see an animation of the helium core, visit glencoe.com.

The final stages A star with the same mass as the Sun never becomes hot enough for carbon to fuse, so its energy production ends. The outer layers expand again and are expelled by pulsations that develop in the outer layers. This shell of gas is called a planetary nebula. In the center of a planetary nebula, shown in **Figure 29.21,** the core of the star becomes exposed as a small, hot object about the size of Earth. The star is then a white dwarf made of carbon.

Internal pressure in white dwarfs A white dwarf is stable despite its lack of nuclear reactions because it is supported by the resistance of electrons being squeezed together, and does not require a source of heat to be maintained. This pressure counteracts gravity and can support the core as long as the mass of the remaining core is less than about 1.4 times the mass of the Sun. The main-sequence lifetime of such a star is much longer, however, because low-mass stars are dim and do not deplete their nuclear fuel rapidly. The electron pressure does not require ongoing reactions, so it can last indefinitely. The white dwarf gradually cools, eventually losing its luminosity and becoming an undetectable black dwarf.

Figure 29.21 The star at the center of the Eskimo nebula, now a white dwarf, was the source of the remnant gases surrounding it.

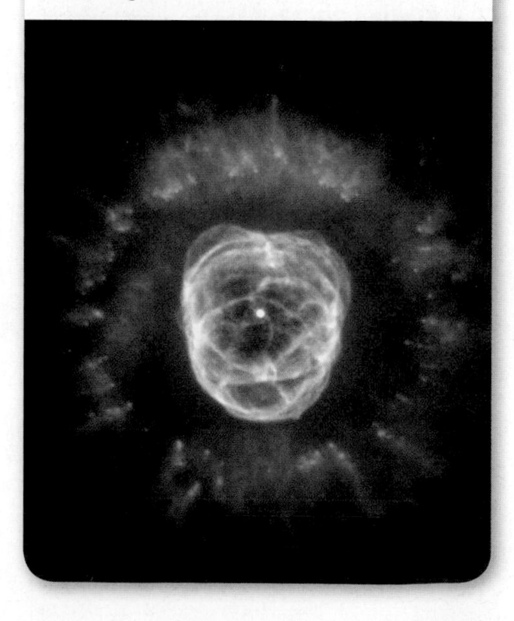

Life Cycles of Massive Stars

For stars more massive than the Sun, evolution is different. A more massive star begins its life in the same way, with hydrogen being converted to helium, but it is much higher on the main sequence. The star's lifetime in this phase is short because the star is very luminous and uses up its fuel quickly.

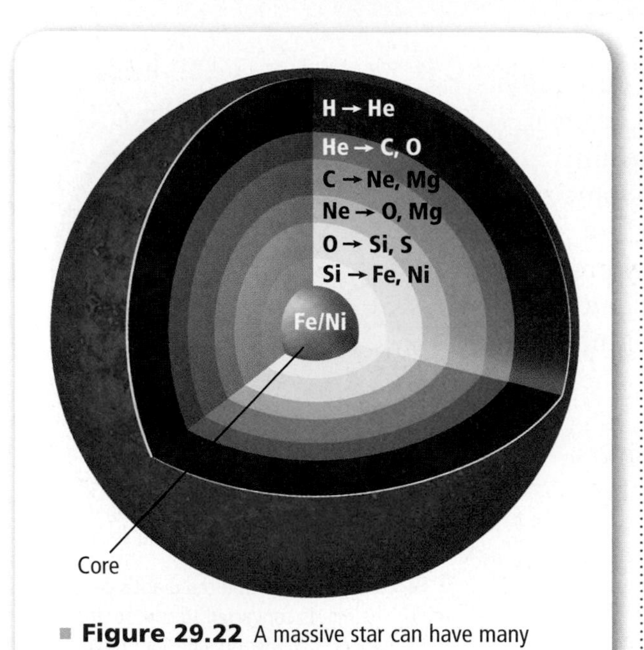

■ **Figure 29.22** A massive star can have many shells fusing different elements. These stars are the source of heavier elements in the universe.

Supergiant A massive star undergoes many more reaction phases and thus produces a rich stew of many elements in its interior. The star becomes a red giant several times as it expands following the end of each reaction stage. As more shells are formed by the fusion of different elements, illustrated in **Figure 29.22,** the star expands to a larger size and becomes a supergiant, such as Betelgeuse in the Orion constellation.

Supernova formation A star that begins with a mass between about 8 and 20 times the Sun's mass will end up with a core that is too massive to be supported by electron pressure. Such a star comes to a violent end. Once reactions in the core of the star have created iron, no further energy-producing reactions can occur, and the core of the star violently collapses in on itself, as illustrated in **Figure 29.23.** Protons and electrons in the core merge to form neutrons. Like electrons, a neutron's resistance to being squeezed close together creates a pressure that halts the collapse of the core, and the core becomes a collapsed star remnant—a **neutron star.** A neutron star has a mass of 1.5 to 3 times the Sun's mass but a radius of only about 10 km. Its density is extremely high—about 100 trillion times the density of water—and is comparable to that of an atomic nucleus.

Pulsar Some neutron stars are unique in that they have a pulsating pattern of light. The magnetic fields of these stars focus the light they emit into cones. Then as these stars rotate on their axes, the light from each spinning neutron star is observed as a series of pulses of light, as each of the cones sweeps out a path in Earth's direction. This pulsating star is known as a **pulsar.**

■ **Figure 29.23** When the outer layers of a star collapse into the neutron core, the central mass of neutrons creates a pressure that causes this mass to explode outward as a supernova, leaving a neutron star.
Compare *the diameter of a supergiant with that of a neutron star.*

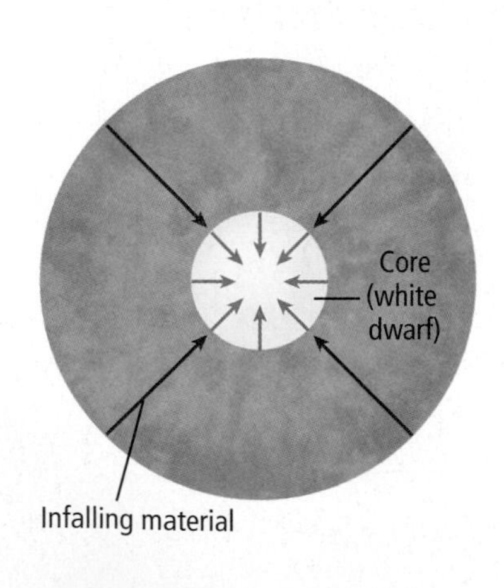

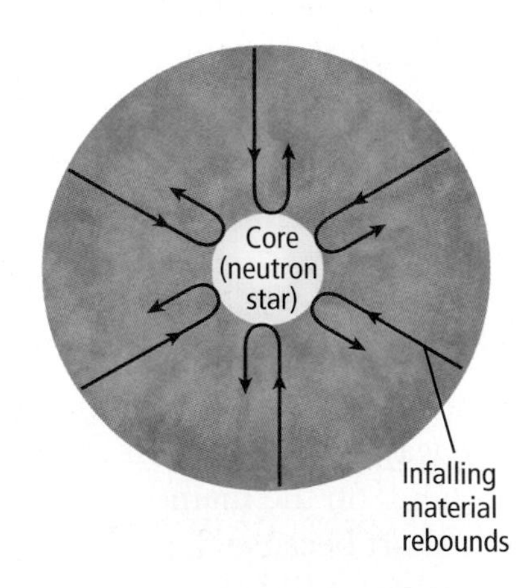

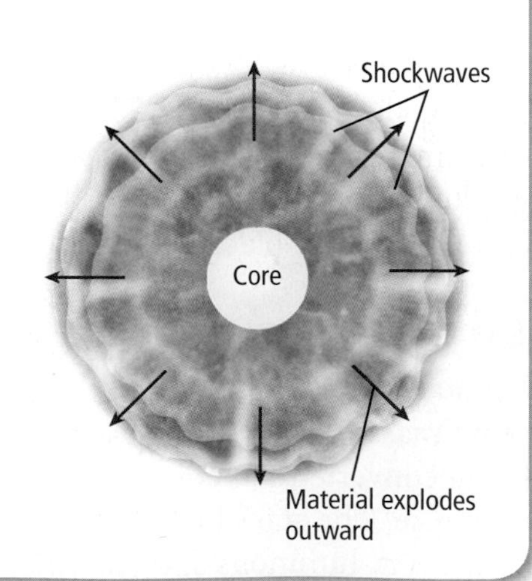

Supernova A neutron star forms quickly while the outer layers of the star are still falling inward. This infalling gas rebounds when it strikes the hard surface of the neutron star and explodes outward. The entire outer portion of the star is blown off in a massive explosion called a **supernova** (plural, supernovae). This explosion creates elements that are heavier than iron and enriches the universe. **Figure 29.24** shows photos of before and during a supernova explosion. A distant supernova explosion might be brighter than the galaxy in which it is found.

Black holes Some stars are too massive to form neutron stars. The pressure from the resistance of neutrons being squeezed together cannot support the core of a star if the star's mass is greater than about three times the mass of the Sun. A star that begins with more than 20 times the Sun's mass will end up above this mass limit, and it cannot form a neutron star. The resistance of neutrons to being squeezed is not great enough to stop the collapse, and the core of the star continues to collapse, compacting matter into a smaller volume. The small, extremely dense object that remains is called a **black hole** because its gravity is so immense that nothing, not even light, can escape it. Astronomers cannot observe what goes on inside a black hole, but they can observe the X-ray-emitting gas that spirals into it.

■ **Figure 29.24** The region of sky in the Large Magellanic Cloud seemed ordinary before one of its stars underwent a supernova explosion.

Section 29.3 Assessment

Section Summary

- The mass of a star determines its internal structure and its other properties.
- Gravity and pressure balance each other in a star.
- If the temperature in the core of a star becomes high enough, elements heavier than hydrogen can fuse together.
- A supernova occurs when the outer layers of the star bounce off the neutron star core, and explode outward.

Understand Main Ideas

1. MAIN Idea **Explain** how mass determines a star's evolution.
2. **Infer** how hydrostatic equilibrium in a star is determined by mass.
3. **Determine** how the lifetimes of stars depend on their masses.
4. **Determine** why only the most massive stars are important contributors in enriching the galaxy with heavy elements.

Think Critically

5. **Explain** how the universe would be different if massive stars did not explode at the end of their lives.
6. **Distinguish** whether there is a balance between pressure and gravity in main-sequence stars, white dwarfs, neutron stars, and black holes.

WRITING in Earth Science

7. Write a description of an observation of a supernova in another galaxy.

EARTH SCIENCE AND TECHNOLOGY

SPACE WEATHER AND EARTH SYSTEMS

Powerful hurricanes and tornadoes can cause millions of dollars worth of damage to homes and other structures. These types of strong storms can be responsible for loss of human life and the disruption of major electrical and communication systems in an area. There are also weather conditions in space. What effects do solar storms have on Earth?

Space weather Solar flares and coronal mass ejections create powerful solar storms that release billions of high-energy particles into space that travel at speeds of up to 2000 km/s. Some of these particles slam into Earth's magnetosphere—over which particles from space normally flow—much like water flows around a large rock in the middle of a river. Earth's magnetosphere normally deflects particles from the Sun, but during intense solar storms, highly charged particles cause disruptions in many of Earth's communication and electrical systems.

Monitoring space weather Two U.S. government agencies, NASA and NOAA, monitor and provide daily updates on space weather, including predictions about solar flare and solar storm occurrences. Power companies, the Federal Aviation Administration, and the U.S. Department of Defense use the data to help minimize the damage to sensitive equipment caused by solar storms.

Communication Communication satellites, locating systems, and military signals rely on radio waves that are bounced off Earth's ionosphere. The ionosphere is a layer of highly charged particles which is especially vulnerable to highly energized particles from the Sun.

A widespread coronal mass ejection blasts more than a billion tons of matter into space at millions of kilometers per hour. Fortunately one this large is rare.

These high-energy particles can interfere with radio signals and disrupt transmissions.

Satellites Solar storms can cause satellites to fall out of orbit due to temperature and density changes in Earth's upper atmosphere. They must be moved to higher orbits in response to this phenomenon. Communication satellites can also be knocked out by electric particle buildup as well.

Electricity Power companies routinely receive information about possible solar storms in order to avoid service disruption to customers. Solar storms can knock out power by inducing currents in electrical lines. In 1989, in Quebec, Canada, a solar storm caused a nine-hour blackout that affected 6 million people and cost the power company over 10 million dollars in repairs.

WRITING in Earth Science

Pamphlet Research more information about space weather and create a pamphlet that answers frequently asked questions about it. Include information about the causes and why it is important to monitor space weather. To learn more about space weather, visit glencoe.com.

GEOLAB

IDENTIFY STELLAR SPECTRAL LINES

Background: An astronomer studying a star or other type of celestial object often starts by identifying the lines in the object's spectrum. The identity of the spectral lines gives information about the chemical composition of the distant object, along with data on its temperature and other properties.

Question: *How can you identify stellar spectral lines based on two previously identified lines?*

Materials

ruler

Procedure

1. Read and complete the lab safety form.
2. Find the difference between the two labeled spectral line values on Star 1.
3. Accurately measure the distance between the two labeled spectral lines.
4. Set up a conversion scale by dividing the spectral difference by the measured distance.
   For example: 1 mm = 12 nm
5. Measure the distance from one of the labeled spectral lines to each of the unlabeled spectral lines.
6. Convert these distances to nm. Add or subtract your value to the original spectral line value. If the labeled line is to the right of the line measured, then subtract. Otherwise, add. This is the value of the wavelength.

Possible Elements and Wavelengths	
Element/Ion	**Wavelength (nm)**
H	383.5, 388.9, 397.0, 410.2, 434.1, 486.1, 656.3
He	402.6, 447.1, 492.2, 587.6, 686.7
He^+	420.0, 454.1, 468.6, 541.2, 656.0
Na	475.2, 498.3, 589.0, 589.6
Ca^+	393.4, 480.0, 530.7

7. Compare your wavelength measurements to the table of wavelengths emitted by elements, and identify the elements in the spectrum.
8. Repeat this procedure for Star 2.

Analyze and Conclude

1. **Identify** Can you see any clues in the star's spectrum about which elements are most common in the stars? Explain.
2. **Explain** Do both stars contain the same lines for all the elements in the table?
3. **Evaluate** How do the thicker absorption lines of some elements in a star's spectrum affect the accuracy of your measurements? Is there a way to improve your measurements? Explain.

INQUIRY EXTENSION

Design Your Own Obtain spectra from various sources, such as sunlight, fluorescent, and incandescent light. Compare their emission lines to those from this lab. What elements are common to each?

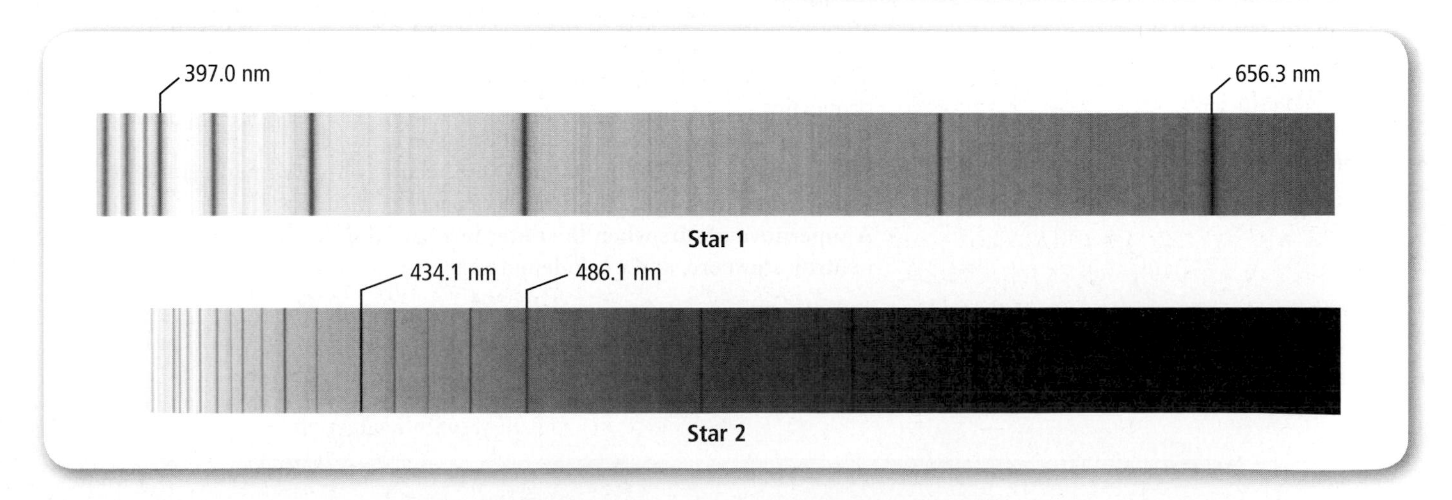

CHAPTER 29

Study Guide

Download quizzes, key terms, and flash cards from glencoe.com.

BIG Idea The life cycle of every star is determined by its mass, luminosity, magnitude, temperature, and composition.

Vocabulary	Key Concepts
Section 29.1 The Sun	
• chromosphere (p. 831) • corona (p. 831) • fission (p. 834) • fusion (p. 834) • photosphere (p. 831) • prominence (p. 833) • solar flare (p. 833) • solar wind (p. 832) • sunspot (p. 832)	**MAIN Idea** The Sun contains most of the mass of the solar system and has many features typical of other stars. • Most of the mass in the solar system is found in the Sun. • The Sun's average density is approximately equal to that of the gas giant planets. • The Sun has a layered atmosphere. • The Sun's magnetic field causes sunspots and other solar activity. • The fusion of hydrogen into helium provides the Sun's energy and composition. • The different temperatures of the Sun's outer layers produce different spectra.
Section 29.2 Measuring the Stars	
• absolute magnitude (p. 842) • apparent magnitude (p. 842) • binary star (p. 838) • constellation (p. 837) • Hertzsprung-Russell diagram (p. 845) • luminosity (p. 842) • main sequence (p. 845) • parallax (p. 841) • parsec (p. 840)	**MAIN Idea** Stellar classification is based on measurement of light spectra, temperature, and composition. • Stars exist in clusters held together by their gravity. • The simplest cluster is a binary. • Parallax is used to measure distances to stars. • The brightness of stars is related to their temperature. • Stars are classified by their spectra. • The H-R diagram relates the basic properties of stars: class, temperature, and luminosity.
Section 29.3 Stellar Evolution	
• black hole (p. 851) • nebula (p. 848) • neutron star (p. 850) • protostar (p. 848) • pulsar (p. 850) • supernova (p. 851)	**MAIN Idea** The Sun and other stars follow similar life cycles, leaving the galaxy enriched with heavy elements. • The mass of a star determines its internal structure and its other properties. • Gravity and pressure balance each other in a star. • If the temperature in the core of a star becomes high enough, elements heavier than hydrogen can fuse together. • A supernova occurs when the outer layers of the star bounce off the neutron star core, and explode outward.

Assessment

Vocabulary Review

Match the definitions below to the correct vocabulary term on the Study Guide.

1. the outermost layer of the Sun's atmosphere, having a temperature of about 1 million K
2. combining of lightweight nuclei such as hydrogen into heavier nuclei
3. dark spots where the surface is cooler on the photosphere of the Sun
4. the apparent shift in position of an object that results from the motion of the observer
5. the outward flow of charged particles from the Sun's corona flowing throughout the solar system
6. two stars that are gravitationally bound and orbit a common center of mass
7. the power or energy output from the surface of a star in units per second
8. an explosion that blows away the outer portion of a star

Distinguish between the following pairs of terms.

9. eclipsing binary, spectroscopic binary
10. giant stars, main-sequence stars
11. apparent magnitude, absolute magnitude
12. black hole, neutron star
13. fission, fusion

Define these terms in your own words.

14. constellation
15. prominence
16. main sequence
17. nebula
18 supernova
19. black hole
20. protostar

Understand Key Concepts

Use the diagram below to answer Question 21.

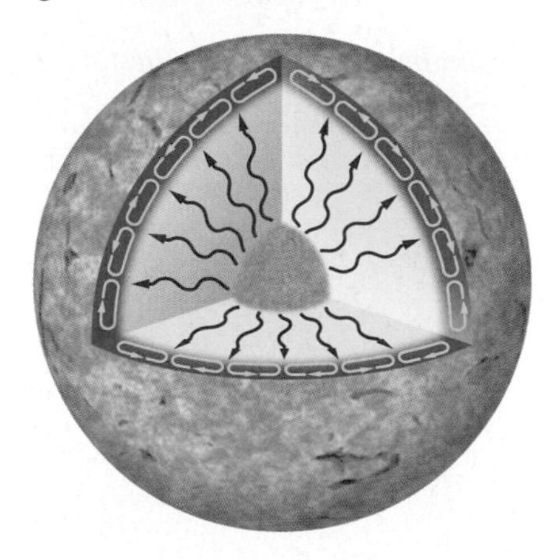

21. Starting at the center, which is the correct order of solar layers?
   A. radiation zone, core, convection currents
   B. core, convection currents, radiation zone
   C. core, radiation zone, convection currents
   D. convection currents, mantle, radiation zone

22. Why do sunspots appear dark?
   A. They are cooler than their surroundings.
   B. They are holes in the interior of the Sun.
   C. They do not have strong magnetic fields.
   D. They are hotter than their surroundings.

23. Why is the Sun's composition similar to that of the gas giant planets?
   A. They all formed at the same time.
   B. They both lost heavy elements.
   C. They all formed from the same interstellar cloud.
   D. They both gained heavy elements.

24. How is the Sun's magnetic behavior associated with its activity cycle?
   A. The magnetic field turns off when the activity cycle turns on.
   B. The activity cycle is coordinated with the peak number of sunspots.
   C. The activity cycle is independent of the number of solar flares.
   D. Solar flares are not coordinated with magnetic storms on Earth.

25. Which is NOT true about binary stars?
  A. They usually appear as one star.
  B. They move about a common center of mass.
  C. They are the most common stars in a galaxy.
  D. They are always of equal brightness.

Use the diagram below to answer Question 26.

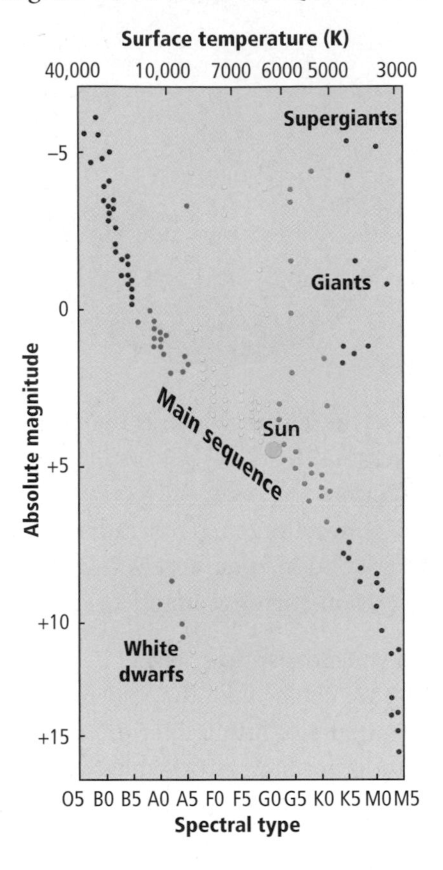

26. Which is true about the spectral classification system of stars?
  A. An A star is cooler than an M star but hotter than an F star.
  B. An O star is cooler than a B star yet hotter than an F star.
  C. A K star is hotter than both a G star and an M star.
  D. A G star is cooler than a B star and hotter than a K star.

27. Which two key stellar properties determine all the other stellar properties?
  A. radius and diameter
  B. mass and radius
  C. composition and mass
  D. diameter and composition

28. Which are in the proper time order?
  A. main-sequence star, red giant, white dwarf, planetary nebula
  B. planetary nebula, red giant, white dwarf, main-sequence star
  C. main-sequence star, white dwarf, planetary nebula, red giant
  D. planetary nebula, main-sequence star, white dwarf, red giant

Constructed Response

29. CAREERS IN EARTH SCIENCE Deduce what astronomers can tell about how stars of different masses evolve, by observing stars in clusters.

30. Detail how, if Earth's orbit were twice the diameter it is now, that would affect stellar parallax and our ability to measure distances.

31. Explain why we say the solar cycle lasts 22 years and not 11.

Use the image below to answer Questions 32 and 33.

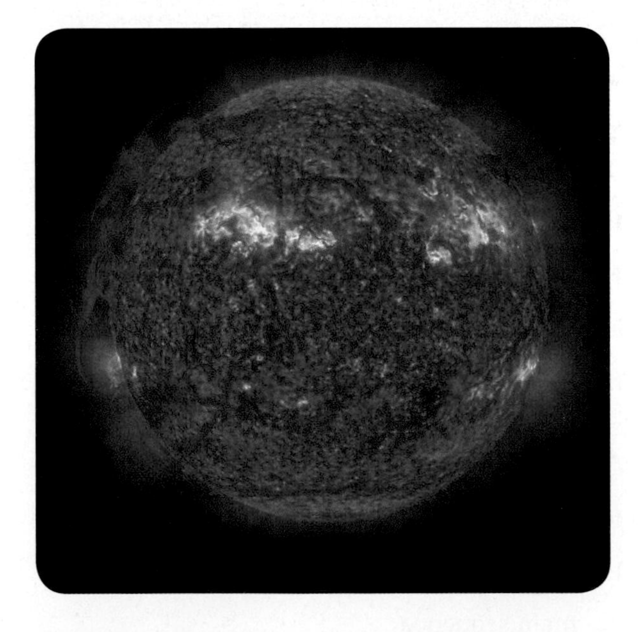

32. Identify the visible layers of the Sun in this photo.

33. Identify the light and dark areas of the Sun's surface in the photo.

34. Explain the relationship between the solar prominences and the Sun's magnetic field.

Think Critically

35. **Deduce** why it is hotter at the center of the Sun than on the surface.

36. **Predict** the layering and composition of stars other than the Sun.

37. **Explain** how the density of the Sun is so great and yet is still in the gaseous state.

Use the diagram below to answer Questions 38 and 39.

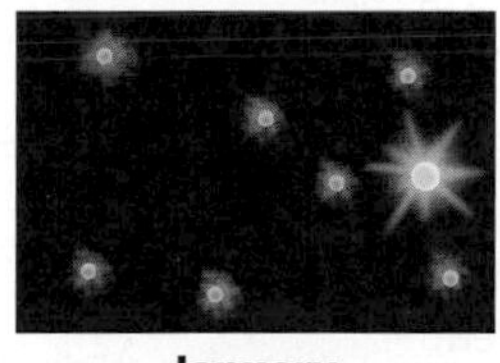
January

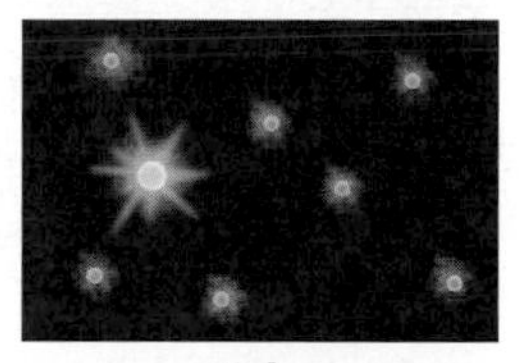
July

38. **Draw** the relative positions of Earth, the Sun, and the star in March and November, based upon the observation in the diagram.

39. **Infer** how parallax helps scientists determine magnitude and luminosity.

40. **Infer** why the parsec has become the standard unit for expressing distance to the stars rather than the AU or light-year.

41. **Compare** a B5 star to the Sun using the H-R diagram.

42. **Compare** a supernova, a neutron star, and a pulsar.

43. **Explain** the difference between a planetary nebula and a supernova.

Concept Mapping

44. Make a concept map linking the terms *fusion, luminosity, protostar*, and one other vocabulary term.

Challenge Question

45. **Organize** a procedure for discovering whether a star is binary.

Additional Assessment

46. WRITING in Earth Science The person who developed the modern system of spectral classification was Annie J. Cannon. Research her work and write about her role in forging new pathways for women in science.

Document–Based Questions

Data obtained from: Massey, P., et al. 2002. Orbits of four very massive binaries in the R136 cluster. *The Astrophysical Journal* 565:982–993.

Binary stars rotate around one another. The radial velocity is the rate of the stars in a binary pair moving toward and away from an observer. Subtract the lowest velocity from the highest velocity for each star, and divide by two to find the average velocity.

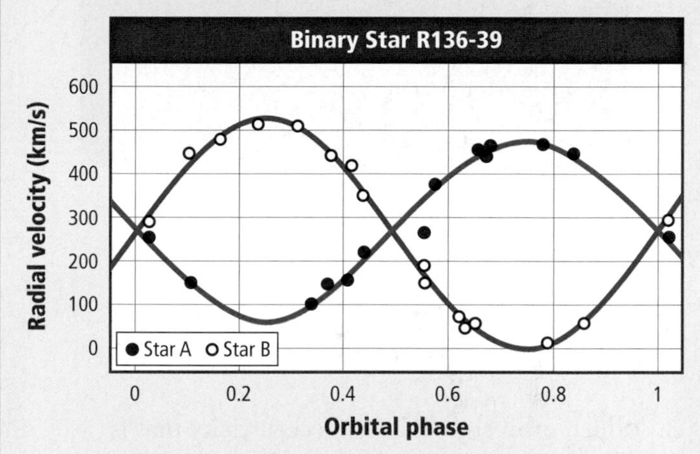

47. If the star with the larger mass has a lower average velocity, which star has the greater mass?

48. When the paths of the stars cross, there is an eclipse for the observer. At what points in the orbital phase are there eclipses?

Cumulative Review

49. Which of the mineral groups is most abundant in Earth's crust? **(Chapter 4)**

50. Briefly describe how air masses form. **(Chapter 12)**

51. What structures are formed by magmas that intrude the crust but do not erupt at the surface? **(Chapter 18)**

52. What makes an interstellar cloud collapse to start the star-formation process? **(Chapter 28)**

Standardized Test Practice

Multiple Choice

1. In December, the south pole is tilted closer to the Sun than at any other time of the year, and the north pole is tilted its farthest from the Sun. What is the northern hemisphere experiencing at that time?
   A. the winter solstice
   B. the summer solstice
   C. the vernal equinox
   D. the autumnal equinox

Use the diagram below to answer Questions 2 and 3.

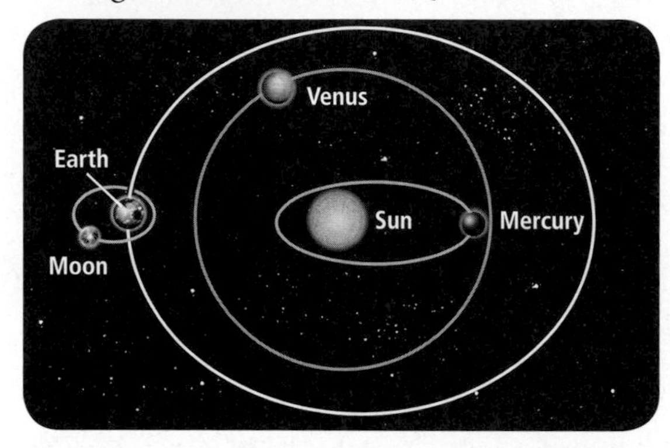

2. Which planet is moving fastest in its orbit?
   A. Mercury
   B. Venus
   C. Earth
   D. the Sun

3. Which orbit shown has an eccentricity that is closest to 0?
   A. Mercury
   B. Venus
   C. Earth
   D. the Moon

4. A bed of sedimentary rock is formed by sediments that were deposited at a rate of 1 cm/year. If the bed is 350 m thick, how long did it take for the whole bed to be deposited?
   A. 350 years
   B. 3500 years
   C. 35,000 years
   D. 350,000 years

5. Which gas giant planet is the largest?
   A. Jupiter
   B. Saturn
   C. Uranus
   D. Neptune

6. Which energy source does not come from the Sun?
   A. wind
   B. water
   C. geothermal
   D. ocean

Use the graph below to answer Questions 7–9.

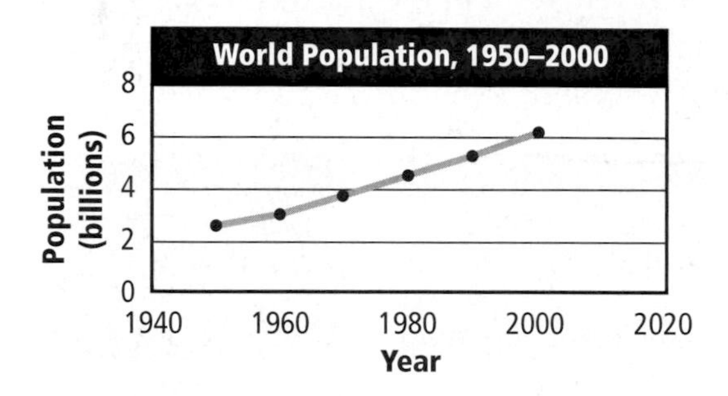

7. Which can you conclude from the graph?
   A. In 80 years, it will not be possible to feed the population.
   B. World population increases at a rate of 1 billion people every 10 years.
   C. There were approximately 2.5 billion people in the world in 1940.
   D. At the present rate of growth, the population will exceed 7 billion before 2020.

8. Based on this graph, what can be assumed about the carrying capacity of the world?
   A. The world is in a state of equilibrium.
   B. The world has not reached its carrying capacity.
   C. The world has reached its carrying capacity.
   D. The world has exceeded its carrying capacity.

9. On the graph, what is the year considered?
   A. the constant
   B. the dependent variable
   C. the independent variable
   D. the variable

10. What causes sunspots on the Sun?
   A. intense magnetic fields poking through the photosphere
   B. charged particles flowing into the solar system
   C. spots on the surface of the photosphere, which are hotter than the surrounding areas
   D. areas of low density in gas of the Sun's corona

Short Answer

Use the illustration below to answer Questions 11–13.

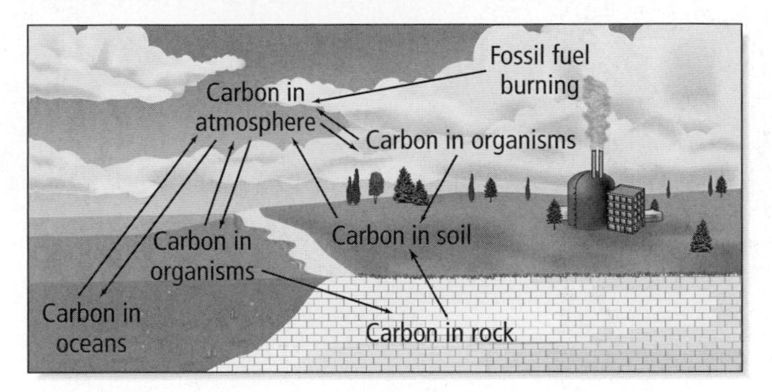

11. Describe the process shown above.
12. Why is burning fossil fuels an important part of this process?
13. Why are there two arrows between carbon in the atmosphere and carbon in organisms?
14. Describe how Earth's atmosphere would be different if there were no life on Earth.
15. Why would a minor temperature increase caused by global warming pose a threat to Earth?
16. Why is an express lane for cars with multiple passengers a good form of energy conservation?

Reading for Comprehension

The Sun's Impact on Climate

Sunspots alter the amount of energy Earth gets from the Sun, but not enough to impact global climate change, a new study suggests. The Sun's role in global warming has long been a matter of debate and is likely to remain a contentious topic.

Scientists have pondered the link between the Sun and Earth's climate since the time of Galileo. There has been an intuitive perception that the Sun's variable degree of brightness—the coming and going of sunspots for instance—might have an impact on climate. Most climate models already incorporate the effects of the Sun's waxing and waning power on Earth's weather. The number of spots cycles over time, reaching a peak every 11 years, but sunspot-driven changes to the Sun's power are too small to account for the climatic changes observed in historical data. The difference in brightness between the high point of a sunspot cycle and its low point is less than 0.1 percent of the Sun's total output.

Article obtained from: Handwerk, B. Don't blame Sun for global warming, study says. *National Geographic News.* September 13, 2006.

17. What can be inferred from this passage?
    A. Sunspots on the Sun do not affect global climate change.
    B. Sunspots greatly alter the amount of energy Earth gets from the Sun.
    C. It has long been thought that sunspots change Earth's climate.
    D. The amount of energy output from a sunspot changes drastically during its cycle.
18. Approximately how much does a sunspot cycle change the energy output of the Sun?
    A. 11 percent
    B. 1.0 percent
    C. 0.1 percent
    D. 0.01 percent
19. While a sunspot does change the amount of energy Earth gets from the Sun, why does it not impact climate?

NEED EXTRA HELP?																
If You Missed Question . . .	1	2	3	4	5	6	7	8	9	10	11	12	13	14	15	16
Review Section . . .	27.3	28.1	28.1	6.1	28.3	25.2	26.1	26.1	1.2	29.1	24.3	24.3	24.3	22.3	25.3	25.3

CHAPTER 30 Galaxies and the Universe

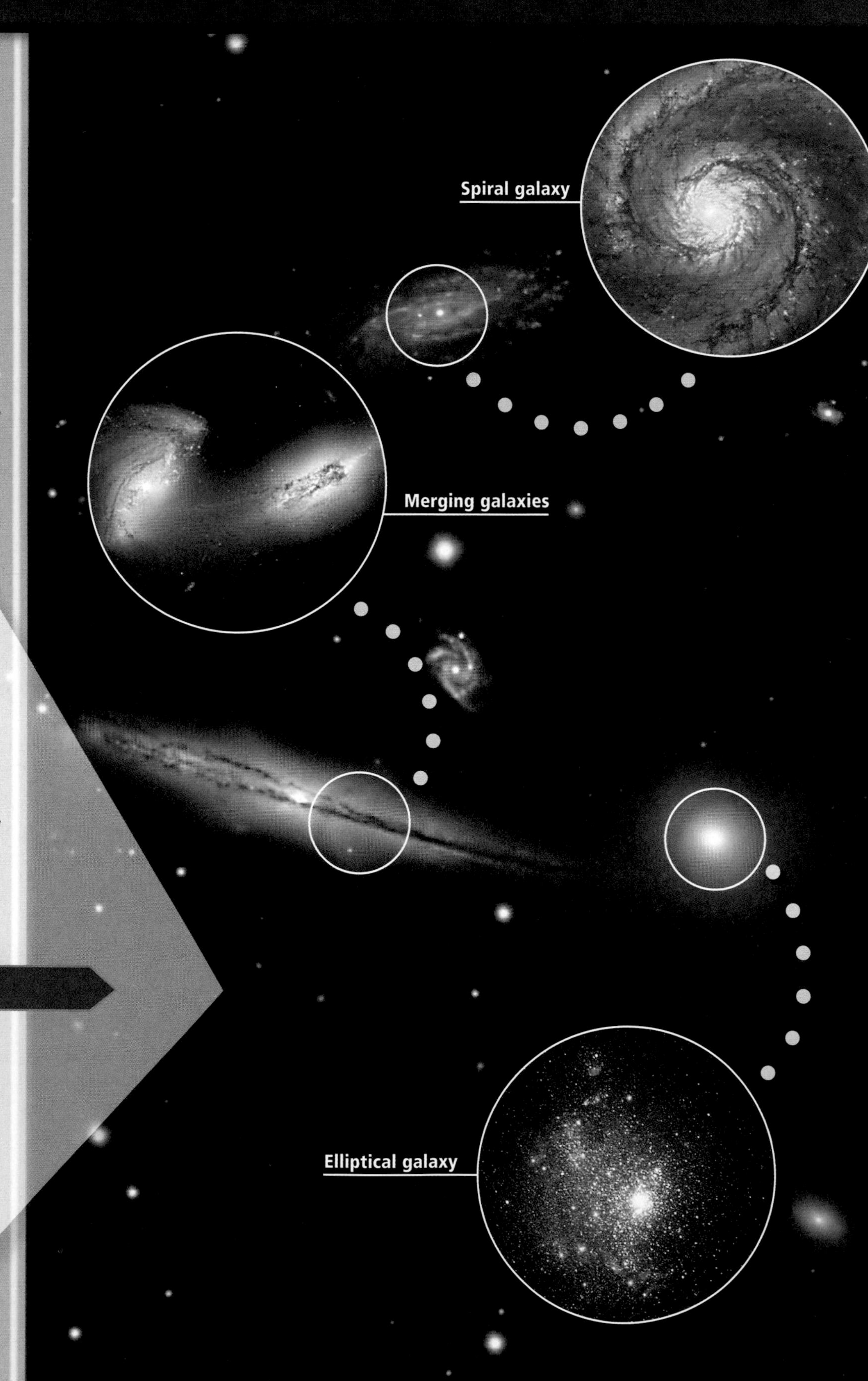

BIG Idea **Observations of galaxy expansion, cosmic background radiation, and the Big Bang theory describe an expanding universe that is about 14 billion years old.**

30.1 **The Milky Way Galaxy**
MAIN Idea Stars with varying light output allowed astronomers to map the Milky Way, which has a halo, spiral arms, and a massive galactic black hole at its center.

30.2 **Other Galaxies in the Universe**
MAIN Idea Finding galaxies with different shapes reveals the past, present, and future of the universe.

30.3 **Cosmology**
MAIN Idea The Big Bang theory was formulated by comparing evidence and models to describe the beginning of the universe.

GeoFacts

- Distance measurements in the universe are usually expressed in light-years, measured by how far light travels in one year. The nearest galaxy like ours is nearly 2 million light-years away.
- Each galaxy contains billions of stars, and there are billions of galaxies.
- Galaxies come in a variety of shapes, and most are either spiral or elliptical.

LAUNCH Lab

How big is the Milky Way?

Our solar system seems large when compared to the size of Earth. However, the Milky Way dwarfs the size of our solar system.

Procedure

1. Read and complete the lab safety form.
2. The Milky Way has a diameter of approximately 8.25×10^9 AU. What is the diameter of the Milky Way in light-years? (206,265 AU = 3.26 ly)
3. Given that the Kuiper belt has a diameter of 50 AU, what is the diameter of the Kuiper belt in ly?
4. If you were to apply the scale 1 mm = 1 ly, how large would the Milky Way be?
5. The Sun is located 28,000 ly from the center of the Milky Way. Based on the scale that you used in Question 4, what would be the distance, in millimeters, from the center of the Milky Way to the Sun?
6. If you included the Kuiper belt in your model, how many millimeters across would its orbit be?

Analysis

1. **Observe** In your science journal, describe what your model of the Milky Way would look like if you actually built it.
2. **Explain** why it would be a problem to show the size of our solar system in comparison to the Milky Way.
3. **Explain** how you would change your model to include the size of Earth.

Types of Galaxies Make this Foldable to show how galaxies are classified.

STEP 1 Fold the top of a horizontal sheet of paper down about 2 cm.

STEP 2 Fold the sheet into thirds. Unfold and draw lines along all fold lines.

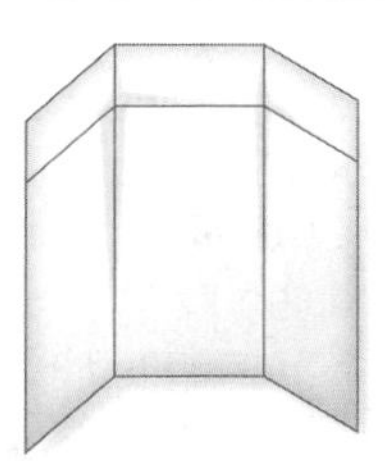

STEP 3 Label the columns *Spiral, Elliptical,* and *Irregular*.

Spiral	Elliptical	Irregular

FOLDABLES **Use this Foldable with Section 30.2.** As you read this section, record information about each type of galaxy, including sketches when appropriate.

Visit glencoe.com to

- study entire chapters online;
- explore Concepts In Motion animations:
  - Interactive Time Lines
  - Interactive Figures
  - Interactive Tables
- access Web Links for more information, projects, and activities;
- review content with the Interactive Tutor and take Self-Check Quizzes.

Section 30.1

Objectives

- **Determine** the size and shape of our galaxy.
- **Distinguish** the different kinds of variable stars.
- **Identify** the different kinds of stars in a galaxy and their locations.

Review Vocabulary

galaxy: any of the very large groups of stars and associated matter found throughout the universe

New Vocabulary

variable star
RR Lyrae variable
Cepheid variable
halo
Population I star
Population II star
spiral density wave

The Milky Way Galaxy

MAIN Idea Stars with varying light output allowed astronomers to map the Milky Way, which has a halo, spiral arms, and a massive black hole at its center.

Real-World Reading Link From inside your home, you have only a few ways to find out what is going on outside. You can look out a window or door, use a telephone or a computer, or bring in news and entertainment on a radio or TV. Similarly, scientists also have a few ways to learn about the stars in the galaxy around us.

Discovering the Milky Way

When looking at the Milky Way galaxy, it is difficult to see its size and shape because not only is the observer too close, but he or she is also inside the galaxy. Observing the band of stars stretching across the sky, you are looking at the edge of a disk from the inside of the disk. However, it is difficult to tell how big the galaxy is, where its center is, or what Earth's location is within this vast expanse of stars. Though astronomers have answers to these questions, they are still refining their measurements.

Variable stars In the 1920s, astronomers focused their attention on mapping out the locations of globular clusters of stars. These huge, spherical star clusters are located above or below the plane of the galactic disk. Astronomers estimated the distances to the clusters by identifying variable stars in them. **Variable stars** are located in the giant branch of the Hertzsprung-Russell diagram, discussed in Chapter 29, and pulsate in brightness because of the expansion and contraction of their outer layers. Variable stars are brightest at their largest diameters and and dimmest at their smallest diameters. **Figure 30.1** shows the dim and bright extremes of a variable star.

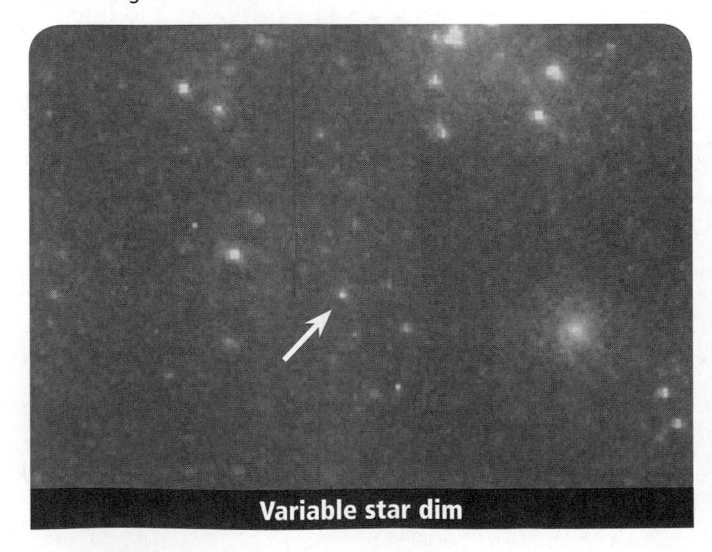

Figure 30.1 The diameters of variable stars change over a period of 1 to 100 days, causing them to brighten and dim.

Types of variables For certain types of variable stars, there is a relationship between a star's luminosity and its pulsation period, which is the time between its brightest pulses. The longer the period of pulsation takes, the greater the luminosity of the star. **RR Lyrae variables** are stars that have periods of pulsation between 1.5 hours and 1 day, and on average, they have the same luminosity. **Cepheid variables,** however, have pulsation periods between 1 and 100 days, and the luminosity as much as doubles from dimmest to brightest. By measuring the star's period of pulsation, astronomers can determine the star's absolute luminosity. This, in turn, allows them to compare the star's luminosity (energy) to its apparent magnitude (brightness) and calculate how far away the star must be to appear this dim or bright.

The galactic center After reasoning there were globular clusters orbiting the center of the Milky Way, astronomers then used RR Lyrae variables to determine the distances to them. They discovered that these clusters are located far from our solar system, and that their distribution in space is centered on a distant point 28,000 light-years (ly) away. The galactic center is a region of high star density, shown in **Figure 30.2,** much of which is obscured by interstellar gas and dust. The direction of the galactic center is toward the constellation Sagittarius. The other view of the Milky Way that is shown is along the disk into space.

Reading Check **Describe** how astronomers located the galactic center of the Milky Way.

The Shape of the Milky Way

Only by mapping the galaxy with radio waves have astronomers been able to determine its shape. This is because radio waves are long enough that they can penetrate the interstellar gas and dust without being scattered or absorbed. By measuring radio waves as well as infrared radiation, astronomers have discovered that the galactic center is surrounded by a nuclear bulge, which sticks out of the galactic disk much like the yolk in a fried egg. Around the nuclear bulge and disk is the **halo,** a spherical region where globular clusters are located, as illustrated in **Figure 30.2.**

Figure 30.2 The top two images are views of the Milky Way—one toward the outer galaxy and one close to the center. The third figure is an artist's concept of what the Milky Way galaxy looks like from space.

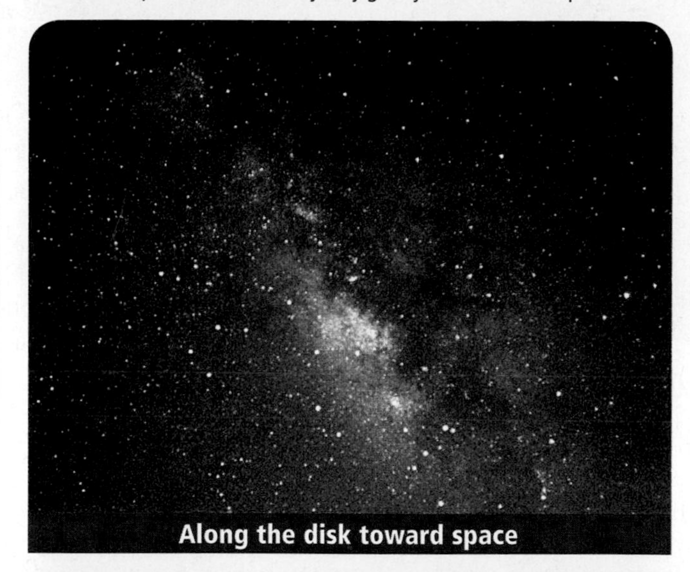

Along the disk toward space

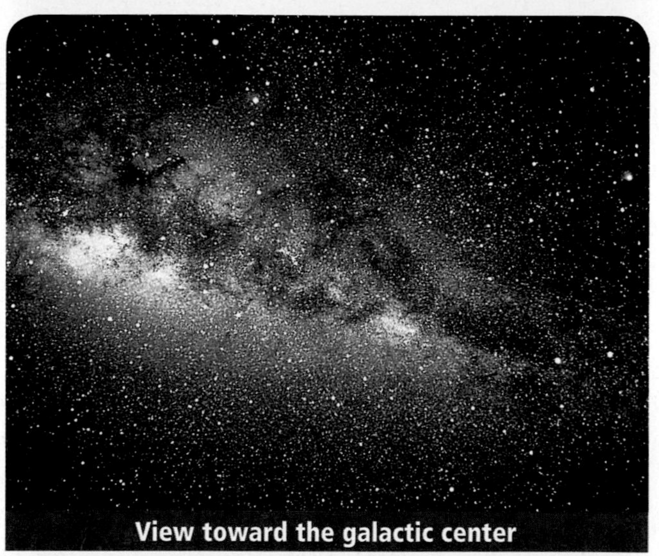

View toward the galactic center

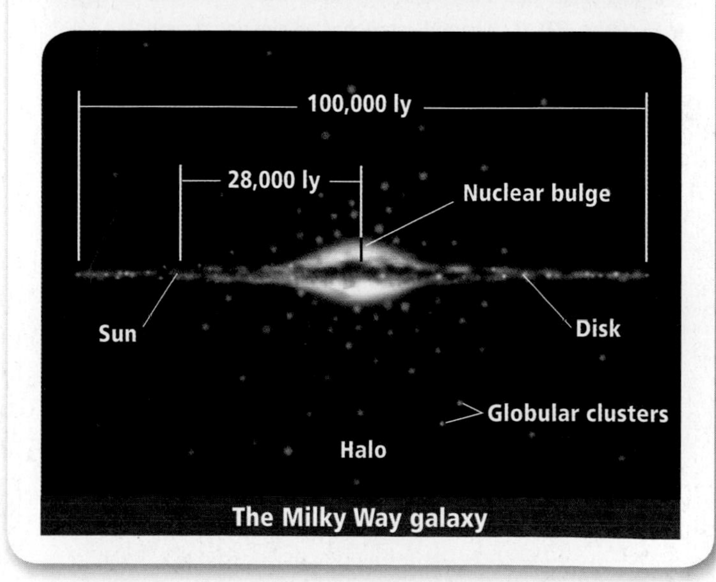

The Milky Way galaxy

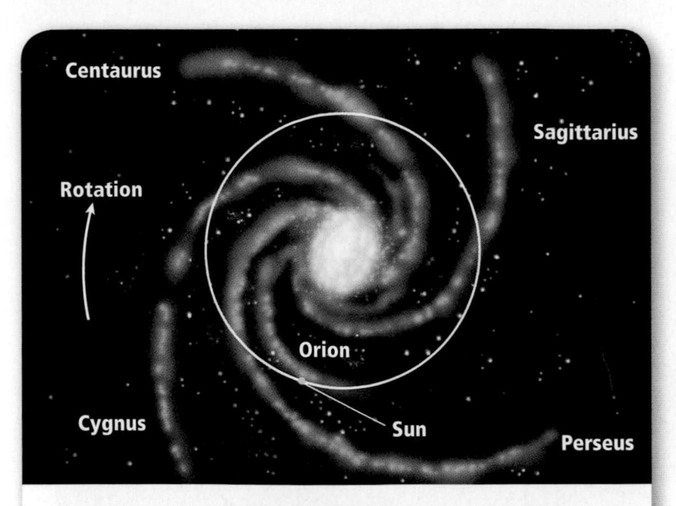

■ **Figure 30.3** The Sun is located on the minor Orion spiral arm and follows an orbital path around the nuclear center as shown. (Note: *Drawing is not to scale.*)
Infer ***how the arms were named.***

Spiral arms Knowing that the Milky Way galaxy has a disklike shape with a central bulge, astronomers speculated that it might also have spiral arms, as do many other galaxies. This was difficult to prove. Because of the distance, astronomers have no way to get outside of the galaxy and look down on the disk. Astronomers decided to use hydrogen atoms to look for the spiral arms.

To locate the spiral arms, hydrogen emission spectra are helpful for three reasons. First, hydrogen is the most abundant element in space; second, the interstellar gas, composed mostly of hydrogen, is concentrated in the spiral arms; and third, the 21-cm wavelength of hydrogen emission can penetrate the interstellar gas and dust and be detected all the way across the galactic disk.

Using the hydrogen emission as a guide, astronomers have identified four major spiral arms and numerous minor arms in the Milky Way. Using these data, scientists discovered that the Sun is located in the minor Orion arm at a distance of about 28,000 ly from the galactic center. The Sun's orbital speed is about 220 km/s, and thus its orbital period is about 240 million years. In its 5-billion-year life, the Sun has orbited the galaxy approximately 20 times. **Figure 30.3** shows the orbit that the Sun follows in a spinning galaxy.

Reading Check **Explain** how astronomers used the Milky Way's hydrogen emission spectrum to locate the arms.

Nuclear bulge or bar? Many spiral galaxies have a barlike shape rather than having a round disk to which the arms are attached. Recent radio observation of interstellar gas indicates that the Milky Way has a slightly elongated shape. Astronomers theorize that the gas density in the halo determines whether a bar will form. **Figure 30.4** shows a barred galaxy.

Using a variety of wavelengths, astronomers are discovering what the center of the Milky Way looks like. The nuclear bulge of a galaxy is typically made up of older, red stars. The bar in a galaxy center, however, is associated with younger stars and a disk that forms from neutral hydrogen gas. Star formation does continue to occur in the bulge, and most stars are about 1000 AU apart compared to 207,000 AU separation in the locale of the Sun. Recent measurements of 30 million stars in the Milky Way indicate a bar about 27,000 ly in length.

■ **Figure 30.4** A barred galaxy has an elongated central bulge.

Mass of the Milky Way

The mass located within the circle of the Sun's orbit through the galaxy, outlined in **Figure 30.3,** is about 100 billion times the mass of the Sun. Using this figure, astronomers have concluded that the galaxy contains about 100 billion stars within its disk.

Mass of the halo Evidence of the movement of outer disk stars and gas suggests that as much as 90 percent of the galaxy's mass is contained in the halo. Some of this unseen matter is probably in the form of dim stellar remnants such as white dwarfs, neutron stars, or black holes, but the nature of the remainder of this mass is unknown. As you will read in Section 30.2, the nature of unseen matter extends to other galaxies and to the universe as a whole. **Figure 30.5** shows the halo of the Sombrero galaxy.

■ **Figure 30.5** The galaxy halo is populated by older, dimmer stars, while the central bulge is populated by newer, brighter stars, as shown in this view of the Sombrero galaxy.

A galactic black hole Weighing in at a few million to a few billion times the mass of the Sun, supermassive black holes occupy the centers of most galaxies. When the center of the galaxy is observed at infrared and radio wavelengths, several dense star clusters and supernova remnants stand out. Among them is a complex source called Sagittarius A (Sgr A), with sub-source called Sgr* (Sagittarius star), which appears to be an actual point around which the whole galaxy rotates.

Careful studies of the motions of the stars that orbit close to Sagittarius A* (pronounced A star) indicate that this region has about 2.6 million times the mass of the Sun but is smaller than our solar system. Data gathered by the *Chandra X-Ray Observatory* reveal intense X-ray emissions. Astronomers think that Sagittarius A* is a supermassive black hole that glows brightly because of the hot gas surrounding it and spiraling into it. This black hole probably formed early in the history of the galaxy, at the time when the galaxy's disk was forming. Gas clouds and stars within the disk probably collided and merged to form a single, massive object that collapsed to form a black hole. **Figure 30.6** illustrates how a supermassive black hole develops. This kind of black hole should not be confused with the much smaller, stellar black hole, which is usually made from the collapsing core of a massive star.

■ **Figure 30.6** The formation of a supermassive black hole begins with the collapse of a dense gas cloud. The accumulation of mass releases photons of many wavelengths, and perhaps even a jet of matter, as shown here.

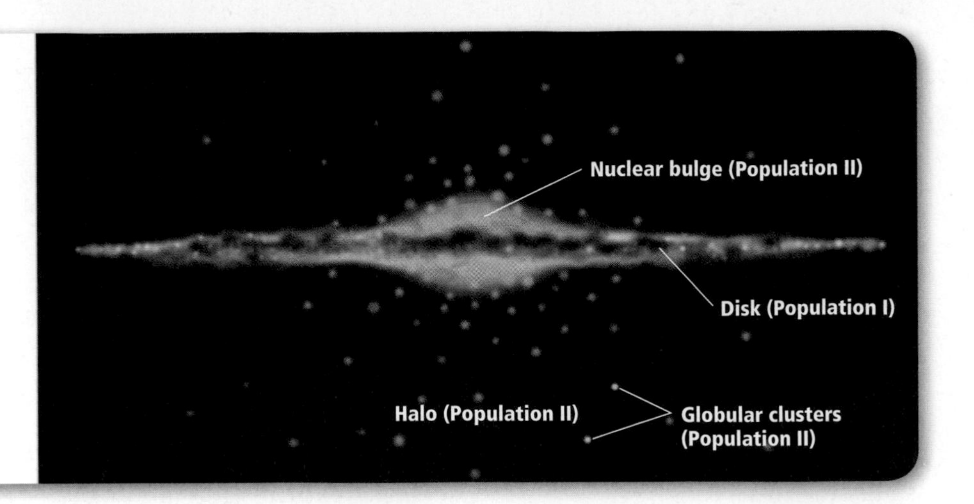

■ **Figure 30.7** Globular clusters and the nuclear bulge contain old stars poor in heavy elements. The disk contains young stars that have a higher heavy element content. (Note: *Drawing is not to scale.*)

Stellar populations in the Milky Way Even though the basic compositions of all stars are the same, there are several distinct differences in detail. The differences among stars include differences in location, motion, and age, leading to the notion of stellar populations. The population of a star provides information about its galactic history. In fact, the galaxy could be divided into two components: the round part made up of the halo and bulge noted in **Figure 30.7,** where the stars are old and contain only traces of heavy elements; and the disk, especially the spiral arms. To astronomers, heavy elements are any elements with a mass larger than helium.

Astronomers divide stars in these two regions into two classes. **Population I stars** are in the disk and arms and have small amounts of heavy elements. **Population II stars** are found in the halo and bulge and contain even smaller traces of heavy elements. Refer to **Table 30.1** for more details.

Population I Most of the young stars in the galaxy are located in the spiral arms of the disk, where the interstellar gas and dust are concentrated. Most star formation takes place in the arms. Population I stars tend to follow circular orbits with low (flat) eccentricity, and their orbits lie close to the plane of the disk. Finally, Population I stars have normal compositions, meaning that approximately 2 percent of their mass is made up of elements heavier than helium. The Sun is a Population I star.

Table 30.1 Population I and II Stars of the Milky Way

Concepts In Motion
Interactive Table To explore more about Population I and II stars, visit glencoe.com.

	Location in Galaxy	Percent of H & He	Percent Heavy Elements	Age (years)	Type of Star	Type of Galaxy	Example
Population I stars	disk arms and open clusters	98	2.0	<10 billion	young sequence stars	spiral and irregular	Sun, most giants, and supergiants
Population II stars	bulge and halo	99.9	0.1	>10 billion	old main-sequence stars (type K and M)	elliptical and spiral halos and bulges	HD 92531 and most white dwarfs

Population II There are few stars and little interstellar material currently forming in the halo or the nuclear bulge of the galaxy, and this is one of the distinguishing features of Population II stars. Age is another. The halo of the Milky Way contains the oldest known objects in the galaxy—globular clusters. These clusters are estimated to be 12 to 14 billion years old. Stars in the globular clusters have extremely small amounts of elements that are heavier than hydrogen and helium. All stars contain small amounts of these heavy elements, but in globular clusters, the amounts are mere traces. Stars like the Sun are composed of about 98 percent hydrogen and helium, whereas in globular cluster stars, this composition can be as high as 99.9 percent. This indicates their extreme age. The nuclear bulge of the galaxy also contains stars with compositions like those of globular cluster stars. **Table 30.1** points out some other comparisons of Population I and II stars.

Formation and Evolution of the Milky Way

The fact that the halo and nuclear bulge are made exclusively of old stars suggests that these parts of the galaxy formed first, before the disk that contains only younger stars. Astronomers therefore hypothesize that the galaxy began as a spherical cloud in space. The first stars formed while this cloud was round. This explains why the halo, which contains the oldest stars, is spherical. The nuclear bulge, which is also round, represents the inner portion of the original cloud. The cloud eventually collapsed under the force of its own gravity, and rotation forced it into a disklike shape. Stars that formed after this time have orbits lying in the plane of the disk. They also contain greater quantities of heavy elements because they formed from gas that had been enriched by previous generations of massive stars. In **Figure 30.8,** the nuclear bulge makes up the hat of the Sombrero galaxy.

Figure 30.8 Easily seen through small telescopes, the Sombrero galaxy gets its name from the bright glow of the nuclear bulge and the dust and gas lanes along the outer edge of its disk.
Predict *which type of stars would be found in the nuclear bulge.*

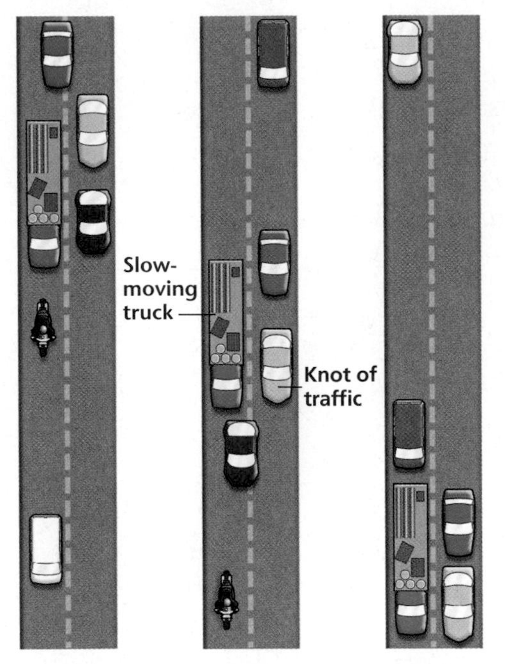

Figure 30.9 A slow truck on a highway causing a build up of cars around it illustrates one theory as to how spiral density waves maintain spiral arms in a galaxy.

Spiral Arms

Most of the main features of the galaxy are understood by astronomers, except for the way in which the spiral arms are retained. The Milky Way is subject to gravitational tugs by neighboring galaxies and is periodically disturbed by supernova explosions from within, both of which can create or affect spiral arms. There are several hypotheses about why galaxies keep this spiral shape.

One hypothesis is that a kind of wave called a spiral density wave is responsible. A **spiral density wave** has spiral regions of alternating density, which rotate as a rigid pattern. As the wave moves through gas and dust, it causes a temporary buildup of material, like a slow truck on the highway causes a buildup of cars, shown in **Figure 30.9.**

A second hypothesis is that the spiral arms are not permanent structures but instead are continually forming as a result of disturbances such as supernova explosions. The Milky Way has a broken spiral-arm pattern, which most astronomers think fits this second model best. However, some galaxies have a prominent two-armed pattern, that was more likely created by density waves.

A third possibility is considered for faraway galaxies. It suggests that the arms are only visible because they contain hot, blue stars that stand out more brightly than dimmer, redder stars. When viewed in UV wavelengths, the arms stand out, but when viewed in infrared wavelengths, they seem to disappear.

Section 30.1 Assessment

Section Summary

- The discovery of variable stars aided in determining the shape of the Milky Way.
- RR Lyrae and Cepheid are two types of variable stars used to measure distances.
- The nuclear bulge and halo of the Milky Way is a globular cluster of old stars.
- The spiral arms of the Milky Way are made of younger stars and gaseous nebulae.
- Population I stars are found in the spiral arms, while Population II stars are in the central bulge and halo.

Understand Main Ideas

1. MAIN Idea **Explain** How did astronomers determine where Earth is located within the Milky Way?
2. **Determine** What do measurements of the mass of the Milky Way indicate?
3. **Analyze** How are Population I stars and Population II stars different?
4. **Summarize** How can variable stars be used to determine the distance to globular clusters?

Think Critically

5. **Explain** If our solar system were slightly above the disk of the Milky Way, why would astronomers still have difficulty determining the shape of the galaxy?
6. **Hypothesize** What would happen to the stellar orbits near the center of the Milky Way galaxy if there were no black hole?

WRITING in Earth Science

7. Write a description of riding a spaceship from above the Milky Way galaxy into its center. Point out all of the galaxy's parts and star types.

Section 30.2

Objectives

- **Describe** how astronomers classify galaxies.
- **Identify** how galaxies are organized into clusters and superclusters.
- **Describe** the expansion of the universe.

Review Vocabulary

elliptical: relating to or shaped like an ellipse or oval

New Vocabulary

dark matter
supercluster
Hubble constant
radio galaxy
active galactic nucleus
quasar

Other Galaxies in the Universe

MAIN Idea **Finding galaxies with different shapes reveals the past, present, and future of the universe.**

Real-World Reading Link Have you ever read an old newspaper to find out what life was like in the past? Astronomers observe distant, older galaxies to get an idea of what the universe was like long ago.

Discovering Other Galaxies

Long before they knew what galaxies were, astronomers observed many objects scattered throughout the sky. Some astronomers hypothesized that these objects were nebulae or star clusters within the Milky Way. Others hypothesized that they were distant galaxies that were as large as the Milky Way.

The question of what these objects were was answered by Edwin Hubble in 1924, when he discovered Cepheid variable stars in the Great Nebula in the Andromeda constellation. Using these stars to measure the distance to the nebula, Hubble showed that they were too far away to be located in our own galaxy. The Andromeda nebula then became known as the Andromeda galaxy, shown in **Figure 30.10.**

FOLDABLES Incorporate information from this section into your Foldable.

Properties of galaxies Masses of galaxies range from the dwarf ellipticals, which have masses of approximately 1 million times the mass of the Sun; to large spirals, such as the Milky Way, with masses of around 100 billion times the mass of the Sun; to the largest galaxies, called giant ellipticals, which have masses as high as 1 trillion times that of the Sun. Measurements of the masses of many galaxies indicate that they have extensive halos containing more mass than is visible, just as the Milky Way does. **Figure 30.10** shows a large spiral and several smaller galaxies.

Figure 30.10 Andromeda is a spiral galaxy like the Milky Way. The bright elliptical object and the sphere-shaped object near the center are small galaxies orbiting the Andromeda galaxy.

Luminosities of galaxies also vary over a wide range, from the dwarf spheroidals—not much larger or more brilliant than a globular cluster—to supergiant elliptical galaxies, more than 100 times more luminous than the Milky Way. All galaxies show evidence that an unknown substance called dark matter dominates their masses. **Dark matter** is thought to be made up of a form of subatomic particle that interacts only weakly with other matter.

Classification of galaxies Hubble went on to study galaxies and categorize them according to their shapes.

Disklike galaxies Hubble classified the disklike galaxies with spiral arms as spiral galaxies. These were subdivided into normal spirals and barred spirals. As shown in **Figures 30.11** and **30.13,** barred spirals have an elongated central region—a bar—from which the spiral arms extend, while normal spirals do not have bars. A normal spiral is denoted by the letter *S,* and a barred spiral is denoted by *SB.* Normal and barred spirals are further subdivided by how tightly the spiral arms are wound and how large and bright the nucleus is. The letter *a* represents tightly wound arms and a large, bright nucleus. The letter *c* represents loosely wound arms and a small, dim nucleus. Thus, a normal spiral with class *a* arms and nucleus is denoted *Sa,* while a barred spiral with class *a* arms and nucleus is denoted *SBa.* Galaxies with flat disks that do not have spiral arms are denoted as *S0.*

Elliptical galaxies In addition to spiral galaxies, there are galaxies that are not flattened into disks and do not have spiral arms, as shown in **Figure 30.13.** Called elliptical galaxies, they are divided into subclasses based on the apparent ratio of their major and minor axes. Round ellipticals are classified as *E0,* while elongated ellipticals are classified as *E7.* Others are denoted by the letter *E* followed by a numeral *1* through *6.* The classification of both spiral and elliptical galaxies can be summarized by Hubble's tuning-fork diagram, which is illustrated in **Figure 30.12.**

Barred Spiral Galaxy

Arm
Nucleus
Bar
Bar
Arm

Spiral Galaxy

Arm
Nucleus
Arm

■ **Figure 30.11** Measurements have indicated that the Milky Way's central region might be a bar, not a spiral.

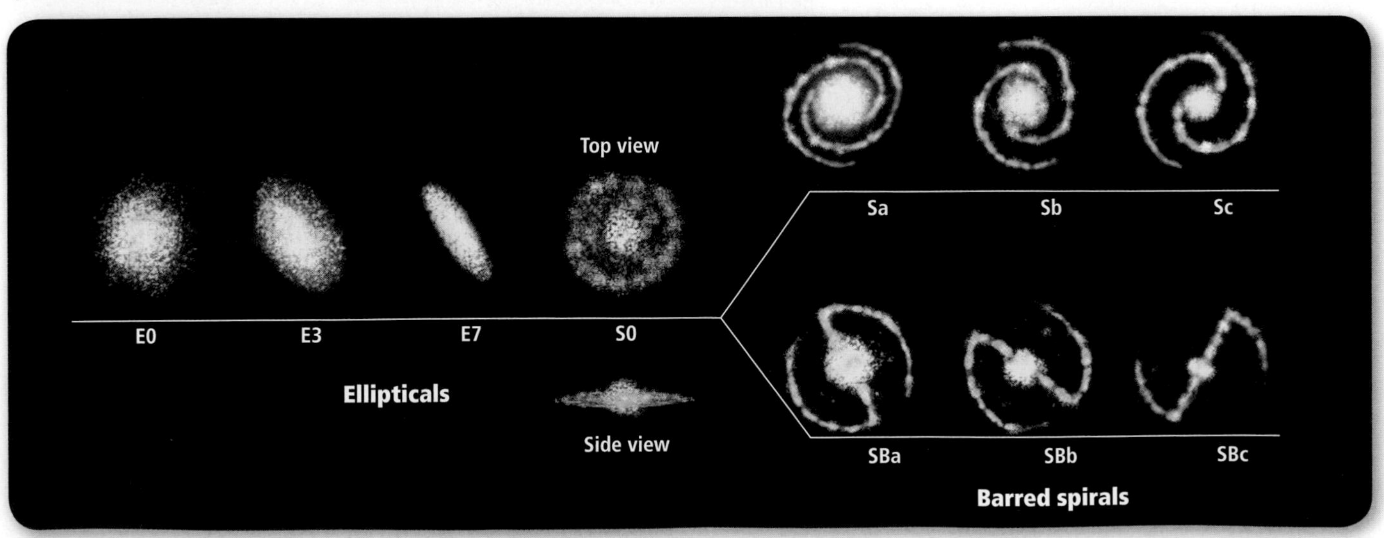

■ **Figure 30.12** The Hubble tuning-fork diagram summarizes Hubble classification for normal galaxies.
Explain *How is an S0 galaxy related to both spirals and ellipticals?*

Visualizing the Local Group

Figure 30.13 All of the stars easily visible in the night sky belong to a single galaxy, the Milky Way. Just as stars compose galaxies, galaxies are gravitationally drawn into galactic groups, or clusters. The 30 galaxies closest to Earth are members of the Local Group of galaxies.

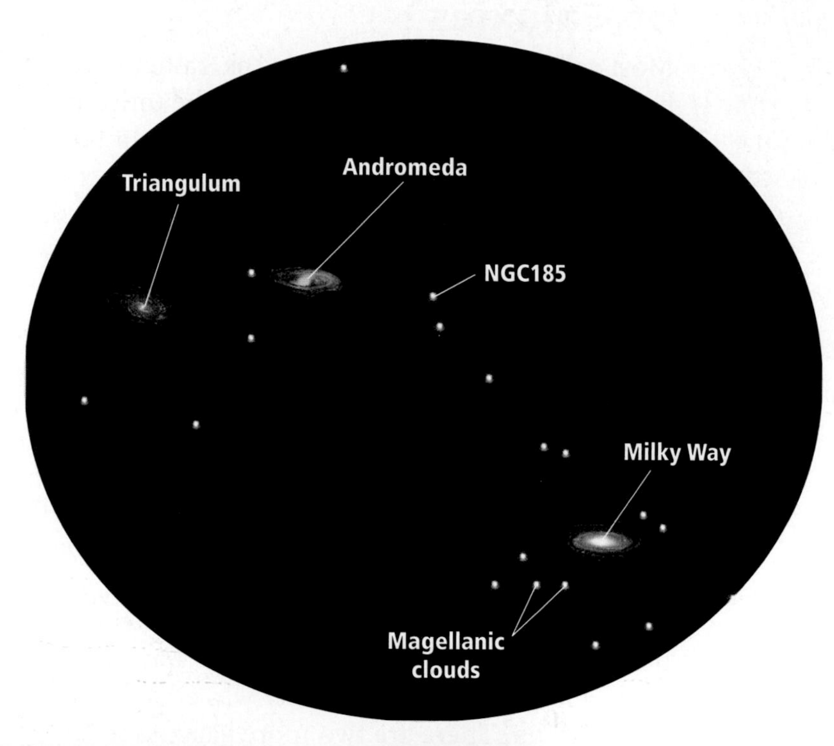

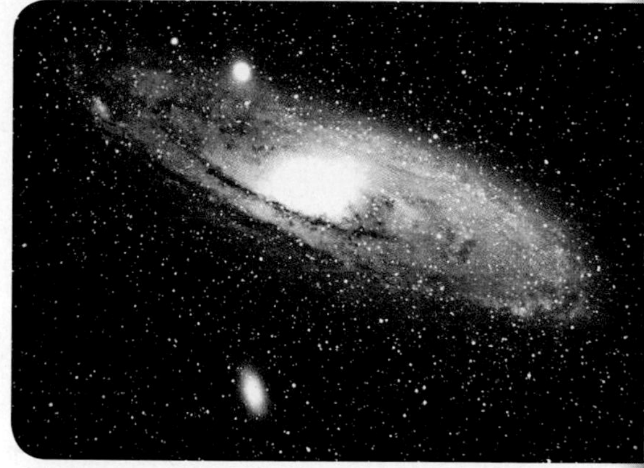

▲ **Spiral galaxies** The two largest galaxies in the Local Group, Andromeda and the Milky Way, are large, flat disks of interstellar gas and dust with arms of stars extending from the disk.

▲ **Barred spiral galaxies** Sometimes the flat disk that forms the center of a spiral galaxy is elongated into a bar shape. Recent evidence suggests that the Milky Way galaxy has a bar.

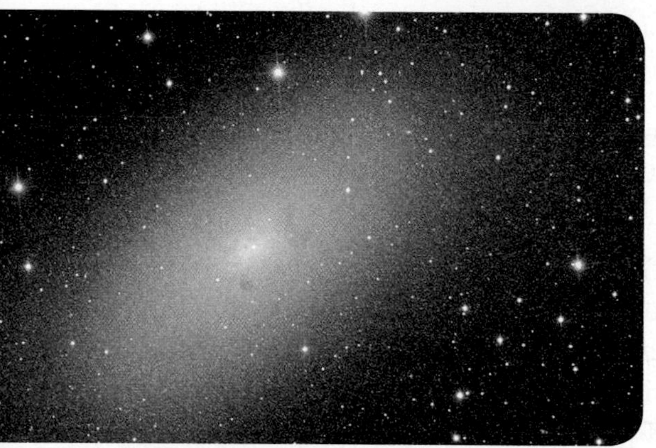

▲ **Elliptical galaxies** like NGC 185 are nearly spherical in shape and consist of a tightly packed group of relatively old stars. Nearly half of the Local Group are ellipticals.

Irregular galaxies Some galaxies are neither spiral or elliptical. Their shape seems to follow no set pattern, so astronomers have given them the classification of irregular. ▶

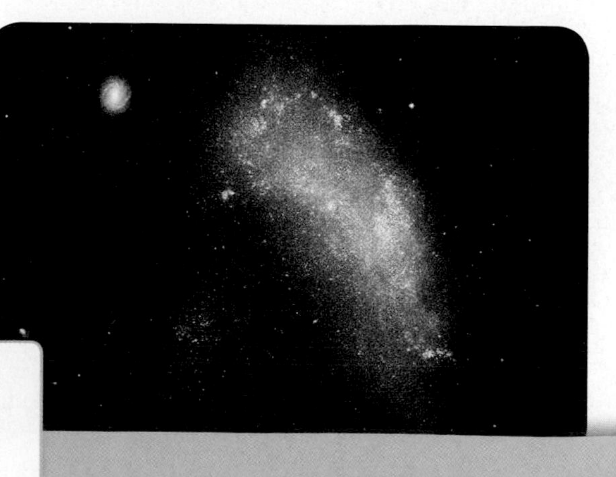

Concepts In Motion To explore more about the Local Group and galaxy types, visit glencoe.com. Earth Science Online

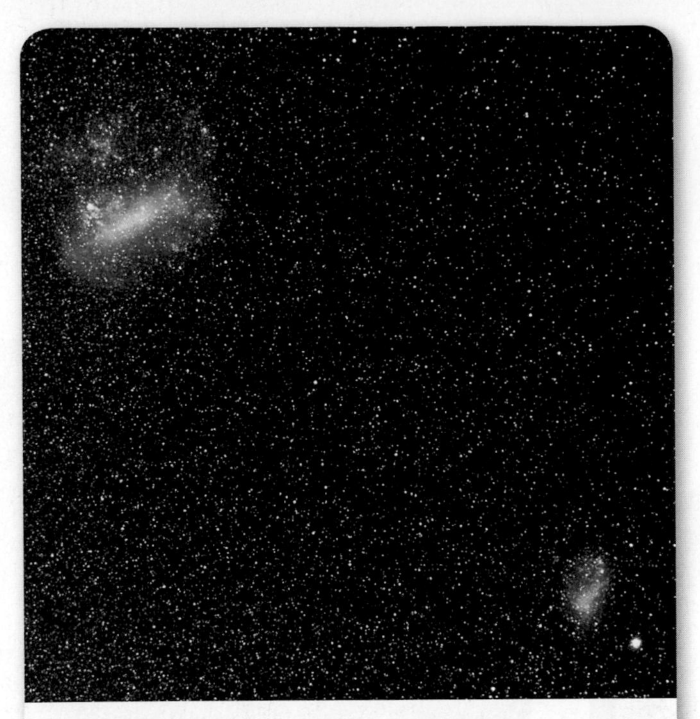

■ **Figure 30.14** The Large and Small Magellanic Clouds are small galaxies that orbit the Milky Way.

Irregular galaxies Some galaxies do not have distinct shapes. These irregular galaxies are denoted by *Irr.* The Large and Small Magellanic Clouds, shown in **Figure 30.14,** the nearest neighbors of the Milky Way, are irregular galaxies.

Groups and Clusters of Galaxies

Most galaxies are located in groups, rather than being spread uniformly throughout the universe. **Figure 30.13** shows some of the features of the Local Group of galaxies.

Local Group The Milky Way belongs to a small cluster of galaxies called the Local Group. The diameter of the Local Group is roughly 2 million ly. There are about 40 known members, of which the Milky Way and Andromeda galaxies are the largest. Most of the members are dwarf ellipticals that are companions to the larger galaxies. The closest galaxies to the Milky Way are the Large and Small Magellanic Clouds and the small Sagittarius galaxy which is merging with the Milky Way. There are several dim galaxies that recently have been found behind the dust and gas of the Milky Way. In 2006, a newly discovered galaxy was added to the Local Group. There are two more galaxies that could be added in the future and six other galaxies that are near the Local Group.

 **Reading Check** **Identify** the kinds of galaxies in the Local Group.

■ **Figure 30.15** The nearby Virgo cluster of approximately 2000 galaxies has a gravity so strong it is pulling the Milky Way toward it.

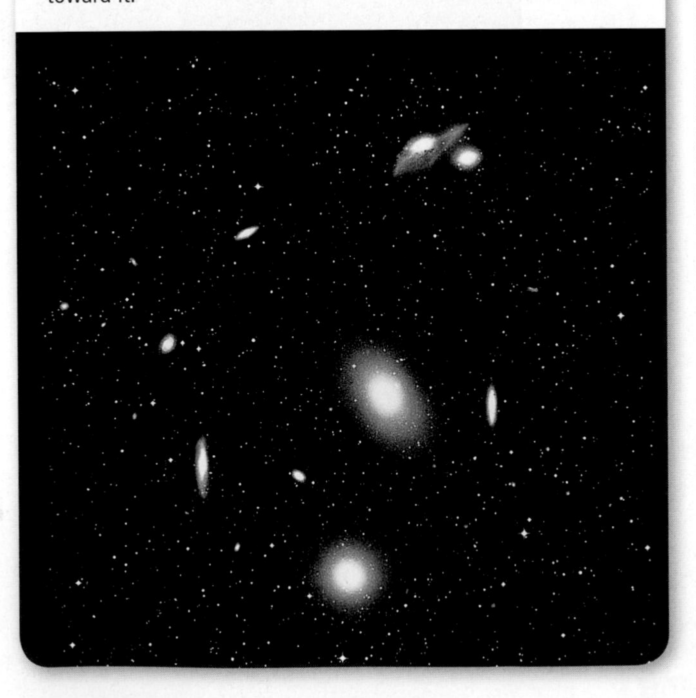

Large clusters Galaxy clusters larger than the Local Group might have hundreds or thousands of members and diameters in the range of about 5 to 30 million ly. The Virgo cluster is shown in **Figure 30.15.** Most of the galaxies in the inner region of a large cluster are ellipticals, while there is a more even mix of ellipticals and spirals in the outer portions.

In regions where galaxies are as close together as they are in large clusters, gravitational interactions among galaxies have many important effects. Galaxies often collide and form strangely shaped galaxies, as shown in **Figure 30.16,** or they form galaxies with more than one nucleus, such as the Andromeda galaxy.

Masses of clusters For clusters of galaxies, the mass determined by analyzing the motion of member galaxies is always much larger than the sum of the total masses of each the galaxies, as determined by their total luminosity. This suggests that most of the mass in a cluster of galaxies is invisible, which provides astronomers with strong evidence that the universe contains a great amount of dark matter.

Superclusters Clusters of galaxies are organized into even larger groups called **superclusters.** These gigantic formations, hundreds of millions of light-years in size, can be observed only when astronomers map out the locations of many galaxies ranging over huge distances. These superclusters appear in sheetlike and threadlike shapes, giving the appearance of a gigantic bubble bath with galaxies located on the surfaces of the bubbles, and the inner air pockets void of galaxies.

■ **Figure 30.16** This galactic merger that began 40 mya will be complete in a few billion years.

The Expanding Universe

In 1929, Edwin Hubble made another dramatic discovery. It was known at the time that most galaxies have redshifts in their spectra, indicating that all galaxies are moving away from Earth. Hubble measured the redshift and distances of many galaxies and found that the redshift of a galaxy depends on its distance from Earth. The farther away a galaxy is, the faster it is moving away. In other words, the universe is expanding.

MiniLab

Model Expansion

What does a uniform expansion look like? The discovery of redshifts of distant galaxies indicated that the universe is rapidly expanding.

Procedure

1. Read and complete the lab safety form.
2. Use a **felt-tipped marking pen** to make four dots in a row, each separated by 1 cm, on the surface of an uninflated **balloon.** Label the dots *1, 2, 3,* and *4.*
3. Partially inflate the balloon. *Do not tie the neck.* With a piece of **string** and a **meterstick,** measure the distance from Dot 1 to each of the other dots. Record your measurements.
4. Inflate the balloon more, and again measure the distance from Dot 1 to each of the other dots. Record your measurements.
5. Repeat Step 4 with the balloon fully inflated.

Analysis

1. **Identify** whether the dots are still separated from each other by equal distances after you fully inflated the balloon.
2. **Determine** how far each dot moved away from Dot 1 following each change in inflation.
3. **Infer** what the result would be if you had measured the distances from Dot 4 instead of Dot 1. From Dot 2?
4. **Explain** how this activity illustrates uniform expansion of the universe.

PROBLEM-SOLVING LAB

Make and Use Graphs

How was the Hubble constant derived? Plotting the distances and speeds for a number of galaxies created the expansion constant for Hubble's Law.

Analysis

1. Use the data to construct a graph. Plot the distance on the *x*-axis and the speed on the *y*-axis.
2. Use a ruler to draw a straight line through the center of the band of points on the graph, so that approximately as many points lie above the line as lie below it. Make sure your line starts at the origin.
3. Measure the slope by choosing a point on the line and dividing the speed at that point by the distance.

Galaxy Data

Distance (Mpc)	Speed (km/s)	Distance (Mpc)	Speed (km/s)
3.0	210	26.5	2087
8.3	450	33.7	2813
10.9	972	36.8	2697
16.2	1383	38.7	3177
17.0	1202	43.9	3835
20.4	1685	45.1	3470
21.9	1594	47.6	3784

Think Critically

4. **State** What does the slope represent?
5. **Gauge** How accurate do you think your value of *H* is? Explain.
6. **Consider** How would an astronomer improve this measurement of *H*?

Implications of redshift You might infer that Earth is at the center of the universe, but this is not the case. An observer located in any galaxy, at any place in the universe, will observe the same thing in a medium that is uniformly expanding—all points are moving away from all other points, and no point is at the center. At greater distances the expansion increases the rate of motion.

A second inference is that the universe is changing with time. If it is expanding now, it must have been smaller and denser in the past. In fact, there must have been a time when all contents of the universe were compressed together. The Big Bang theory has been proposed to explain this expansion.

Hubble's law Hubble determined that the universe is expanding by making a graph comparing a galaxy's distance to the speed at which it is moving. The result is a straight line, which can be expressed as a simple equation, $v = Hd$, where v is the velocity at which a galaxy is moving away measured in kilometers per second; d is the distance to the galaxy measured in megaparsecs (Mpc), where 1 Mpc = 3,260,000 ly; and H is a number called the **Hubble constant,** measured in kilometers per second per megaparsec. H represents the slope of the line.

Measuring *H* Determining the value of *H* requires finding distances and speeds for many galaxies and constructing a graph to find the slope. This is a difficult task because it is hard to measure accurate distances to the most remote galaxies. Hubble could obtain only a crude value for *H*. Obtaining an accurate value for *H* was one of the key goals of astronomers who designed the *Hubble Space Telescope (HST)*. It took nearly ten years after the launch of the *HST* to gather enough data to pinpoint the value of *H*. Currently, the best measurements indicate a value of approximately 70 km/s/Mpc.

New way to measure distance Once the value of *H* is known, it can be used to find distances to faraway galaxies. By measuring the speed at which a galaxy is moving, astronomers use the graph to determine the corresponding distance of the galaxy. This method works for the most remote galaxies that can be observed and allows astronomers to measure distances to the edge of the observable universe.

The only galaxies that do not seem to be moving apart are those within a cluster. The internal gravity of the galactic cluster keeps them from separating.

Active Galaxies

Galaxies that emit large amounts of energy from their cores are called active galaxies. The core of an active galaxy where highly energetic objects or activities are located is called the **active galactic nucleus** (AGN). An AGN emits as much or more energy than the rest of the galaxy. The output of this energy often varies over time, sometimes as little as a few days. About 10 percent of all known galaxies are active, including radio galaxies and quasars.

■ **Figure 30.17** In addition to radio lobes, M87 has a jet that emits visible light.

Radio galaxies Radio-telescope surveys of the sky have revealed a number of galaxies that are extremely luminous. These galaxies, called **radio galaxies,** are often giant elliptical galaxies that emit as much or more energy in radio wavelengths than they do in wavelengths of visible light. Radio galaxies have many unusual properties. The radio emission usually comes from two huge lobes of very hot gas located on opposite sides of the visible galaxy. These lobes are linked to the galaxy by jets of hot gas. The type of emission that comes from these regions indicates that the gas is ionized, and that electrons in the gas jets are traveling near the speed of light. Many radio galaxies have jets that can be observed only at radio wavelengths. One of the brightest of the radio galaxies, a giant elliptical called M87, shown in **Figure 30.17,** also has a jet of gas that emits visible light extending from the galactic center out toward one of the radio-emitting lobes.

Reading Check **Describe** the unusual properties of a radio galaxy.

Quasars In the 1960s, astronomers discovered objects that looked like ordinary stars, but some emitted strong radio waves. Most stars do not. Also, whereas most stars have spectra with absorption lines, these new objects had mostly emission lines in their spectra. These starlike objects with emission lines in their spectra were called **quasars.** Quasars are very luminous, very distant active galaxies. Many quasars vary in brightness over a period of a few days. Two quasars are shown in **Figure 30.18.**

The emission lines of quasars are those of common elements, such as hydrogen, shifted far toward longer wavelengths. Once astronomers had identified the large spectral-line shifts of quasars, they wondered whether they could have redshifts caused by the expansion of the universe.

■ **Figure 30.18** Quasars are old and distant celestial objects that emit several thousand times more energy than does our entire galaxy.
Recall ***What other objects emit jets of matter?***

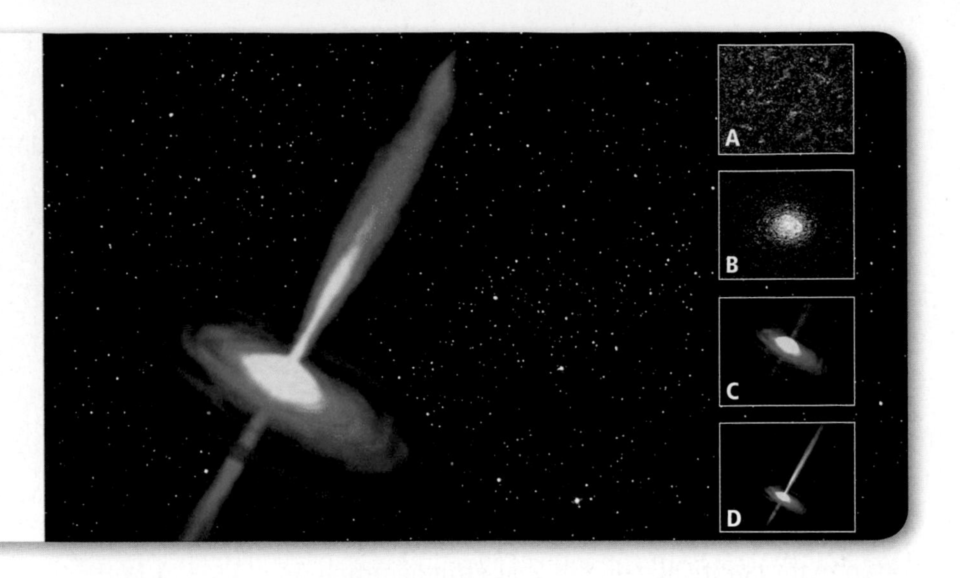

■ **Figure 30.19** An interstellar gas cloud (A) collapses gravitationally (B) on its way to forming a galaxy. The nucleus (C) forms a black hole as the gas there is compressed. Magnetic fields of the rapidly rotating disk surrounding the black hole form two highly energetic jets (D) that are perpendicular to the disk's equatorial plane.

Quasar redshift The redshift of quasars was much larger than any that had been observed in galaxies up to that time, which would mean that the quasars were much farther away than any known galaxy. At first, some astronomers doubted that quasars were far away, but in the decades since quasars were discovered, more evidence supports the hypothesis that quasars are distant. One piece of supporting evidence indicates that those quasars associated with clusters of galaxies have the same redshift, verifying that they are the same distance away. Another more important discovery is that most quasars are nuclei of very dim galaxies, shown in **Figure 30.19.** The quasars appear to be extra-bright AGNs—so much brighter than their surrounding galaxies that astronomers could not initially see those galaxies.

Reading Check **Explain** how astronomers determined distances to quasars.

CAREERS IN EARTH SCIENCE

Computer Programmer Many astronomers use equipment that does not observe light. A computer programmer writes programs astronomers can use to observe spectra, calculate, and decipher the data collected by telescopes. To learn more about Earth science careers, visit glencoe.com.

Looking back in time Because quasars are distant, it takes their light a long time to reach Earth. Therefore, observing a quasar is seeing it as it was a long time ago. For example, it takes light from the Sun approximately 8 minutes to reach Earth. When you observe the Sun, you are seeing it as it was 8 minutes earlier. When you observe the Andromeda galaxy, you see the way it looked 2 million years earlier. The most remote quasars are several billion light-years away, which indicates the stage you see is from billions of years ago. If quasars are extra-bright AGNs, then the many distant ones are nuclei of galaxies as they existed when the universe was young. This suggests that many galaxies went through a quasar stage when they were young. Consequently, today's AGNs might be former quasars that are not as energetic as they were long ago.

Looking far back into time, the early universe had many quasars. Current theory suggests that they existed around supermassive black holes that pulled gas into the center, where in a violent swirl, friction heated the gas to extreme temperatures resulting in the bright light energy that was first detected.

Source of power The AGNs and quasars emit far more energy than ordinary galaxies, but they are as small as solar systems. This suggests that all of these objects are supermassive black holes. Recall that the black hole thought to exist in the core of our own galaxy has a mass of about 1 million Suns. The black holes in the cores of AGNs and quasars are much more massive—up to hundreds of millions of times the mass of the Sun. The beams of charged particles that stream out of the cores of radio galaxies and form jets are probably created by magnetic forces. As material falls into a black hole, the magnetic forces push the charged particles out into jets. There is evidence that similar beams or jets occur in other types of AGNs and in quasars. In fact, radio-lobed quasars have jets that are essentially related to radio galaxies.

Figure 30.20 shows a supermassive black hole. In modeling a supermassive black hole of this magnitude, the mass of nearly 3 billion Suns would be needed to pull the stars in this galaxy into the center. A plasma jet, ejected from the nucleus, extends nearly 5000 ly into space.

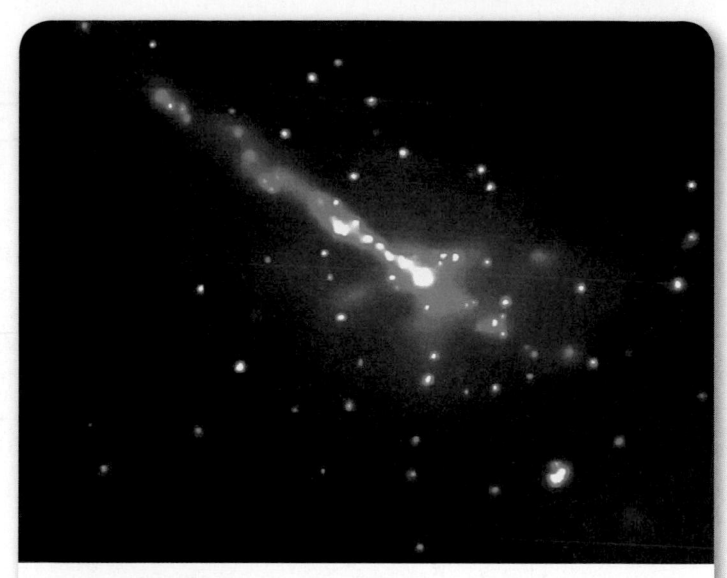

■ **Figure 30.20** A jet of energetic X-ray particles is emitted from the AGN, which probably hides a supermassive black hole. The other white areas are probably X-ray-emitting neutron stars or black hole binaries.

Section 30.2 Assessment

Section Summary

- Galaxies can be elliptical, disk-shaped, or irregular.
- Galaxies range in mass from 1 million Suns to more than a trillion Suns.
- Many galaxies seem to be organized in groups called clusters.
- Quasars are the nuclei of faraway galaxies that are dim and seen as they were long ago, due to their great distances.
- Hubble's law helped astronomers discover that the universe is expanding.

Understand Main Ideas

1. MAIN Idea **Explain** how astronomers discovered that there are other galaxies beyond the Milky Way.
2. **Summarize** why astronomers theorize that most of the matter in galaxies and clusters of galaxies is dark matter.
3. **Explain** why it is difficult for astronomers to accurately measure a value for the Hubble constant, *H*. Once a value is determined, describe how it is used.
4. **Explain** the differences among normal spiral, barred spiral, elliptical, and irregular galaxies.

Think Critically

5. **Deduce** how the nighttime sky would look from Earth if the Milky Way were an elliptical galaxy.
6. **Infer** how black holes result in similarities between AGNs and quasars.

MATH in Earth Science

7. Convert the distance across the Milky Way to Mpc if the diameter of the Milky Way is 100,000 ly. What is the distance in Mpc across a supercluster of galaxies whose diameter is 200 million ly? (1 Mpc = 3,260,000 ly)

Section 30.3

Objectives

- **Distinguish** the different models of the universe.
- **Compare and contrast** how expansion is relative to each of the models.
- **Explain** the importance of the Hubble constant.

Review Vocabulary

radiation: the process of emitting radiant energy in the form of waves or particles

New Vocabulary

cosmology
Big Bang theory
cosmic background radiation

Cosmology

MAIN Idea The Big Bang theory was formulated by comparing evidence and models to describe the beginning of the universe.

Real-World Reading Link Manipulating a magnet and iron filings can help you model Earth's magnetic field. Cosmologists use particle accelerators to help create models of the early universe.

Big Bang Model

The study of the universe—its nature, origin, and evolution—is called **cosmology.** The mathematical basis for cosmology is general relativity, from which equations were derived that describe both the energy and matter content of the universe. These equations, combined with observations of density and acceleration, led to the most accurate model so far—the Big Bang model. The fact that the universe is expanding implies that it had a beginning. The theory that the universe began as a point and has been expanding since is called the **Big Bang theory.** Although the name might seem to imply explosion into space, the theory describes an expansion of space itself while gravity holds matter in check. Review the effects of expansion by checking results from the MiniLab in Section 30.2.

Outward expansion Similar to a star's internal fusion pressure opposing the effort of a gravitational force to collapse the star, the universe has two opposing forces. In the Big Bang model, the momentum of the outward expansion of the universe is opposed by the inward force of gravity acting on the matter of the universe to slow that expansion, as illustrated in **Figure 30.21.** What ultimately will happen depends on which of these two forces is stronger.

When the rate of expansion of the universe is known, it is possible to calculate the time since the expansion started and determine the age of the universe. When the distance to a galaxy and the rate at which it is moving away from Earth are known, it is simple to calculate how long ago that galaxy and the Milky Way were together. In astronomical terms, if the value of *H,* the expansion (Hubble) constant, is known, then the age of the universe can be determined. Corrections are needed to allow for the fact that the expansion has not been constant—it has slowed since the beginning and is now accelerating.

Based on the best value for *H* that has been calculated from *Hubble Space Telescope* data and the data on the cosmic background radiation, the age of the universe can be pinpointed to 13.7 billion years. This fits with what astronomers know about the age of the Milky Way galaxy, which is estimated to be between 12 and 14 billion years old, based on the ages of the oldest star clusters.

■ **Figure 30.21** The universe is either open, flat, or closed, depending on whether gravity or the momentum of expansion dominates.

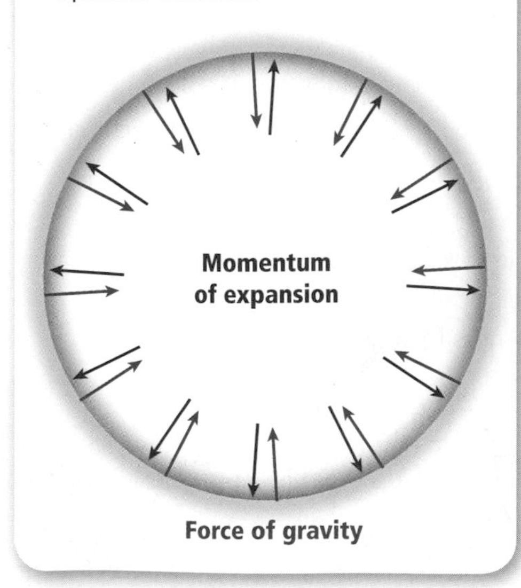

Possible outcomes Based on the Big Bang theory, there are three possible outcomes for the universe, as shown in **Figure 30.22.** The average density of the universe is an observable quantity with vast implications to the outcome.

Open universe An open universe is one in which the expansion will never stop. This would happen if the density of the universe is insufficient for gravity to ever halt the expansion.

Closed universe A closed universe will result if the expansion stops and turns into a contraction. That would mean the density is high enough that eventually the gravity caused by the mass will halt the expansion of the universe and pull all of the mass back to the original point of origin.

Flat universe A flat universe results if the expansion slows to a halt in an infinite amount of time, but never contracts. This means that while the universe would continue to expand, its expansion would be so slow that it would seem to stop.

Critical density All three outcomes are based on the premise that the rate of expansion has slowed since the beginning of the universe, but the density of the universe is what is unknown. At the critical density, there is a balance, so that the expansion will come to a halt in an infinite amount of time. The critical density, about 6×10^{-27} kg/m^3, means that, on average, there are only two hydrogen atoms for every cubic meter of space. When astronomers attempt to count the galaxies in certain regions of space and divide by the volume, they get an even smaller value. So they would conclude that the universe is open, except that the dark matter has not been included. But even the best estimates of dark matter density are not enough to conclude that the universe is a closed system.

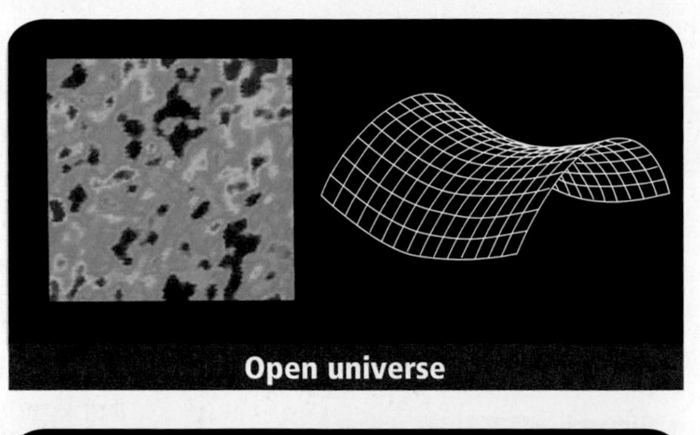

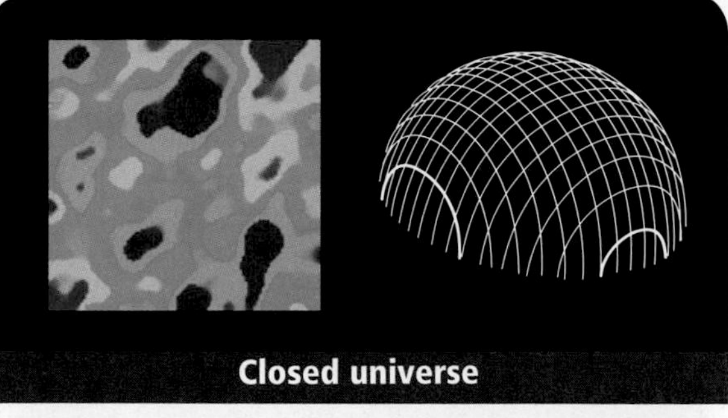

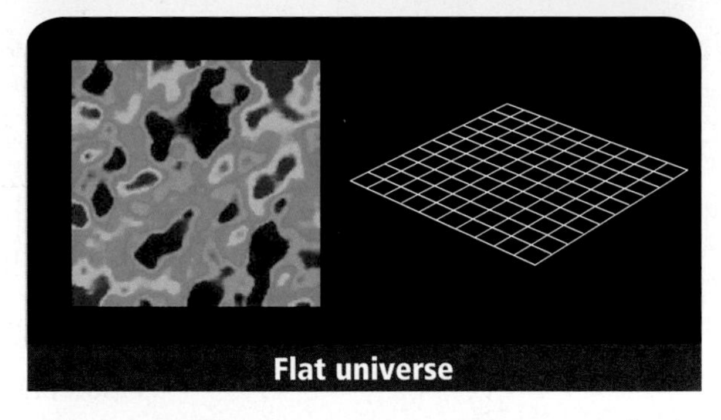

■ **Figure 30.22** There are three possible outcomes for the future of the universe. It could continue to expand forever and be open, it could snap back at the end and be a closed system, or it could be flat and just die out like a glowing ember. The green squares show the estimated cosmic background radiation necessary for each result. See **Figure 30.24.**

Cosmic Background Radiation

Scientists hypothesize that if the universe began in a highly compressed state before the Big Bang, it would have been extremely hot. Then as the universe expanded, the temperature cooled. After about 750,000 years, the universe was filled with electromagnetic radiation in the form of short-wavelength radiation. With continued expansion, the wavelengths became longer. Today this radiation is in the form of microwaves.

■ **Figure 30.23** The cosmic background radiation was discovered by accident with this radio antenna at Bell Labs in Holmdel, New Jersey.

Discovery In 1965, scientists discovered a persistent background noise in their radio antenna, shown in **Figure 30.23.** This noise was caused by weak radiation, called the **cosmic background radiation,** that appeared to come from all directions in space and corresponded to an emitting object having a temperature of about 2.735 K (–270°C). This was very close to the temperature predicted by the Big Bang theory, and the radiation was interpreted to be from the beginning of the Big Bang.

Mapping the radiation Since the discovery of the cosmic background radiation, extensive observations have confirmed that it matches the properties of the predicted leftover radiation from the early, hot phase in the expansion of the universe. Earth's atmosphere blocks much of the radiation, so it is best observed from high-altitude balloons or satellites. An orbiting observatory called the *Wilkinson Microwave Anisotropy Probe (WMAP),* launched by NASA in 2001, mapped the radiation in greater detail, as shown in **Figure 30.24.** The peak of the radiation it measured has a wavelength of approximately 1 mm; thus, it is microwave radiation in the radio portion of the electromagnetic spectrum.

Reading Check **Identify** what discovery helped solidify the Big Bang theory.

VOCABULARY

SCIENCE USAGE V. COMMON USAGE

Cosmic

Science usage: of or relating to the universe in contrast to Earth alone

Common usage: characterized by greatness of thought or intensity

Acceleration of the expansion The data produced by *WMAP* have provided enough detail to refine cosmological models. In particular, astronomers have found small wiggles in the radiation representing the first major structures in the universe. This helped to pinpoint the time at which the first galaxies and clusters of galaxies formed and also the age of the universe. According to every standard model, the expansion of the universe is slowing down due to gravity. However, the debate about the future of the universe based on this model came to a halt with the surprising discovery that the expansion of the universe is now accelerating. Astronomers have labeled this acceleration dark energy. Although they do not know its cause, they can determine the rate of acceleration and estimate the amount of dark energy.

■ **Figure 30.24** Temperature differences of one-millionth of a degree can be noted in the map of cosmic background radiation.
Write ***one-millionth of a degree in scientific notation.***

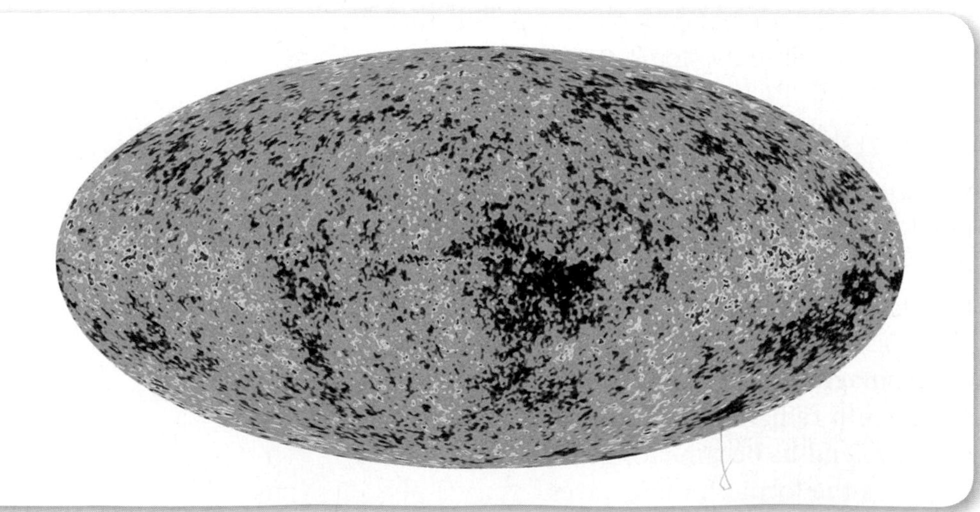

Contents of the Universe

All the evidence is now pointing in the same direction, and astronomers can say with a high degree of precision of what the universe is composed. Their best clue comes from the radiation left in space from the universe's beginning. The ripples left during the time of cooling of the universe's beginning radiation set the density at that point of time and dictated how matter and energy would separate. This in turn laid the groundwork for future galaxies. **Figure 30.25** gives one view into the universe.

Dark matter and energy Cosmologists estimate that the universe is composed of dark matter (21 percent), dark energy (75 percent), and luminous matter. If you compare the universe to Earth, dark energy is like the water covering the surface of Earth. That would be like saying that 70 percent of Earth is covered with something that is not identified.

What is unknown today is the nature of the dark matter and dark energy. Dark matter is thought to consist of subatomic particles, but of the known particles, none display the right properties to explain or fully define dark matter. And although scientists recognize the effects of dark energy, they still do not know what it is.

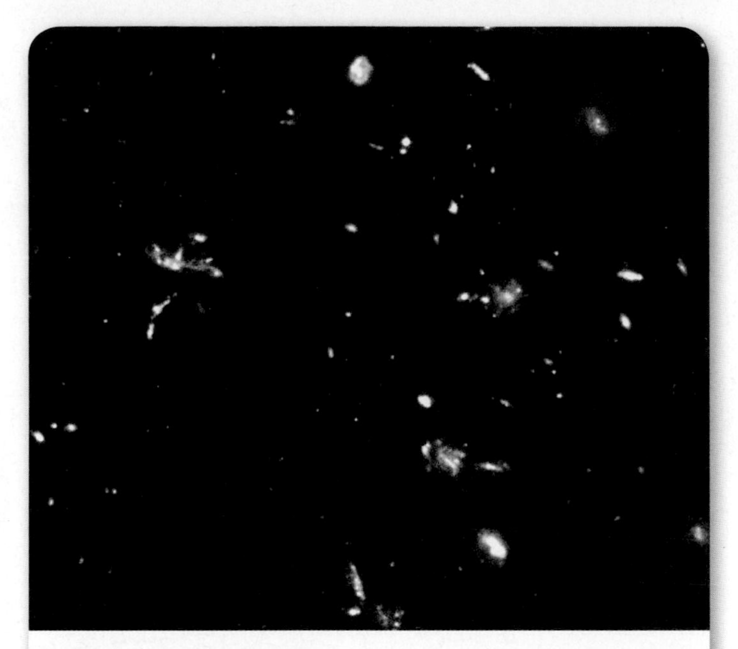

■ **Figure 30.25** In this view of deep space, galaxies appear as glowing flecks. Astronomers estimate that only 4 percent of the universe is composed of luminous matter.

Section 30.3 Assessment

Section Summary

- The study of the universe's origin, nature, and evolution is cosmology.
- The Big Bang model of the universe came from observations of density and acceleration.
- The critical density and the amount of dark energy of the universe will determine whether the universe is open or closed.
- Cosmic background radiation gives support to the Big Bang theory of the universe.
- Mapping the cosmic background radiation has indicated the existence of dark matter and dark energy.

Understand Main Ideas

1. MAIN Idea **Compare and Contrast** What are the differences among the three possible outcomes of the universe?
2. **Describe** how the age of the universe can be calculated using the Big Bang model.
3. **Explain** why dark matter is important in determining the density of matter in the universe.
4. **Explain** why the cosmic background radiation was an important discovery.

Think Critically

5. **Determine** What does dark matter have to do with the critical density of the universe?
6. **Analyze** All of the models tell us that the universe should be slowing down, but instead it is speeding up. How does this affect our model of the universe?

WRITING in Earth Science

7. Write one paragraph summarizing the evidence for the Big Bang model of the universe.

BLACK HOLES ARE GREEN?

Black holes seem to come straight from the pages of a science fiction book. They are incredibly dense cosmic bodies from which nothing—not even light—can escape. The gravitational pull attracts whatever ventures close enough.

Finding black holes Black holes are extremely difficult to see because they do not emit light, and those that are produced by a collapsed massive star can be very small (only 2 to 3 times the mass of the Sun). Astronomers know where black holes might be located due to the effects of the matter falling into them.

Supermassive black holes In the centers of some galaxies exists a different kind of black hole—a supermassive one. These black holes are huge; they can consist of more mass than a million, even a billion, Suns.

Scientists think that supermassive black holes are created when large volumes of interstellar gases collapse in on themselves. Once matter passes into a spherical boundary surrounding the black hole, called the event horizon, it is pulled into the black hole, never to escape.

Energy Before the matter gets pulled into the event horizon, however, it gathers energy through friction and from the magnetic field of the black hole. That energy is released in the form of diffuse light or focused jets.

The jets release about 1000 times more energy than the diffuse light, either in the form of radio waves or energetic X rays. The jets race outward from the black holes almost at the speed of light, creating empty bubbles in their wake. These bubbles can span thousands of light-years. Scientists used these bubbles to discover the fuel efficiency of the supermassive black holes.

In this image taken by the *Chandra X-Ray Telescope,* X rays shine from heated material falling into a black hole.

Black holes are "green" Recent research into supermassive black holes has uncovered an interesting fact: They are the most fuel-efficient engines in the entire universe. In fact, a physicist at Stanford University reported that "if you could develop a car that was as energy efficient as a supermassive black hole, it would get about one billion miles per gallon of gas!"

Astronomers think that the energy released from supermassive black holes actually prevents star formation. The heat that they produce prevents gases from cooling and potentially forming billions of new stars, effectively limiting the size of each galaxy.

WRITING in Earth Science

Summary Visit glencoe.com to learn more about black holes. Summarize what you learn in a newspaper article about black holes that is interesting and scientifically accurate.

GEOLAB

INTERNET: CLASSIFY GALAXIES

Background: Edwin Hubble developed rules for classifying galaxies according to their telescopic image shapes. Modern astronomers are also interested in the classification of galaxies. Information used for classification can indicate whether a certain type of galaxy is more likely to form than another and helps astronomers unravel the mystery of galaxy formation in the universe. Using the Internet and sharing data with your peers, you can learn how galaxies are classified.

Question: *How can different galaxies be classified?*

Materials

internet access to glencoe.com or galaxy images provided by your teacher

Visit a local library or observatory to gather images of galaxies and information about them.

Procedure

1. Read and complete the lab safety form.
2. Find a resource with multiple images of galaxies and, if possible, names or catalog numbers for the galaxies. Visit glencoe.com for links to sites that have galaxy images.
3. Choose one of the following types of galaxies to start your classification: spiral, elliptical, or irregular galaxies.
4. Sketch or gather images and information, such as catalog numbers and names of galaxies.
5. Sort the images by basic types: spiral, elliptical, or irregular galaxies.
6. Complete the data table. Add any additional information you think is important.

Galaxy Data			
Galaxy Name	Image or Sketch of Galaxy	Classification	Notes
NGC 3486		Sc	

Analyze and Conclude

1. **Differentiate** Which galaxy classes were the most difficult to find?
2. **Identify** How many of each galaxy class did you find?
3. **Calculate** the percentages of the total number of galaxies of each type. Do you think this reflects the actual percentage of each type in the universe? Explain.
4. **Discuss** Were there any galaxies that didn't fit the classification scheme? If so, why?
5. **List** What problems did you have with galaxies seen edge-on?
6. **Illustrate** Reconstruct the tuning fork diagram with images that you find.

INQUIRY EXTENSION

Share Your Data With your classmates, calculate the percentage of each type of galaxy. Based on the results, decide if your results are typical or atypical. Determine how your class might find actual percentages of galaxies by type.

CHAPTER 30 Study Guide

Download quizzes, key terms, and flash cards from glencoe.com.

BIG Idea Observations of galaxy expansion, cosmic background radiation, and the Big Bang theory describe an expanding universe that is about 14 billion years old.

Vocabulary	Key Concepts
Section 30.1 The Milky Way Galaxy	
• Cepheid variable (p. 863) • halo (p. 863) • Population I star (p. 866) • Population II star (p. 866) • RR Lyrae variable (p. 863) • spiral density wave (p. 868) • variable star (p. 862)	**MAIN Idea** Stars with varying light output allowed astronomers to map the Milky Way, which has a halo, spiral arms, and a massive galactic black hole at its center. • The discovery of variable stars aided in determining the shape of the Milky Way. • RR Lyrae and Cepheid are two types of variable stars used to measure distances. • The nuclear bulge and halo of the Milky Way is a globular cluster of old stars. • The spiral arms of the Milky Way are made of younger stars and gaseous nebulae. • Population I stars are found in the spiral arms, while Population II stars are in the central bulge and halo.
Section 30.2 Other Galaxies in the Universe	
• active galactic nucleus (p. 875) • dark matter (p. 870) • Hubble constant (p. 874) • quasar (p. 875) • radio galaxy (p. 875) • supercluster (p. 873)	**MAIN Idea** Finding galaxies with different shapes reveals the past, present, and future of the universe. • Galaxies can be elliptical, disk-shaped, or irregular. • Galaxies range in mass from 1 million Suns to more than a trillion Suns. • Many galaxies seem to be organized in groups called clusters. • Quasars are the nuclei of faraway galaxies that are dim and seen as they were long ago, due to their great distances. • Hubble's law helped astronomers discover that the universe is expanding.
Section 30.3 Cosmology	
• Big Bang theory (p. 878) • cosmic background radiation (p. 880) • cosmology (p. 878)	**MAIN Idea** The Big Bang theory was formulated by comparing evidence and models to describe the beginning of the universe. • The study of the universe's origin, nature, and evolution is cosmology. • The Big Bang model of the universe came from observations of density and acceleration. • The critical density and the amount of dark energy of the universe will determine whether the universe is open or closed. • Cosmic background radiation gives support to the Big Bang theory of the universe. • Mapping the cosmic background radiation has indicated the existence of dark matter and dark energy.

CHAPTER 30 Assessment

Vocabulary Review

The sentences below are false. Correct each sentence by replacing the italicized words with the correct vocabulary term from the Study Guide.

1. Surrounding the central bulge, this spherical region of a galaxy is known as the *Hubble constant.*
2. Radio emissions coming from two huge lobes of very hot gas located on opposite sides of the visible galaxy are evidence for *cosmology.*
3. *Population I* is the weak radiation that appears to come from all directions in space and corresponds to an object having a temperature of about 2.735 K.
4. These gigantic *quasars* are hundreds of millions of light-years in size and can be observed only when astronomers map out the locations of many galaxies ranging over large distances.
5. This *Cepheid variable* is the invisible substance that makes up to 21 percent of the universe.
6. One theory of how galaxy arms are maintained involves the *RR Lyrae variables.*
7. *Radio galaxy* is the study of the origin and history of the universe.
8. A *supercluster* is a star whose magnitude changes are produced by expansion and shrinking of its outer layers.

Distinguish between the terms in each of the following pairs.

9. RR Lyrae variable, Cepheid variable
10. quasar, radio galaxy
11. dark matter, cosmic background radiation
12. halo, active galactic nucleus
13. cosmology, Hubble constant

In the set of terms below, select the term that does not belong and explain why it does not belong.

14. RR Lyrae, Cepheid, Population II, quasar

Understand Key Concepts

15. Which are the oldest objects in the Milky Way?
 A. globular clusters
 B. spiral arms
 C. Cepheid variables
 D. Population I stars

Use the diagram below to answer Question 16.

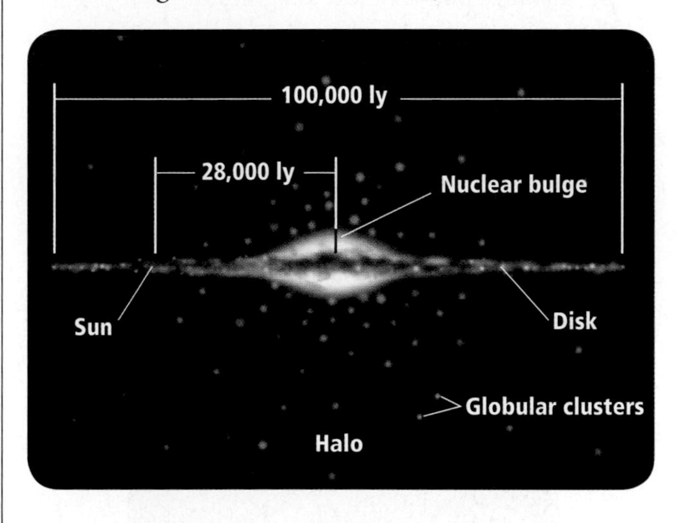

16. Where in the Milky Way are new stars being formed?
 A. in the nuclear bulge
 B. in globular clusters
 C. in the spiral arms of the disk
 D. in the halo

17. Where does the energy emitted by AGNs and quasars most likely originate?
 A. material falling into a supermassive black hole
 B. a neutron star
 C. a supernova explosion
 D. a pulsar

18. What is the origin of the cosmic background radiation?
 A. It is emitted by stars.
 B. It is a remnant of the Big Bang.
 C. It is emitted by radio galaxies.
 D. It is dark energy.

19. In the Big Bang model, which describes a universe that will stop expanding and begin to contract?
 A. open
 B. flat
 C. closed
 D. elliptical

20. Which does the existence of cosmic background radiation support?
A. critical density
B. Hubble constant
C. the inflationary model
D. the Big Bang theory

21. Which two measurements are required to determine the Hubble constant?
A. distance and speed
B. distance and absolute magnitude
C. apparent magnitude and speed
D. apparent and absolute magnitudes

22. Without doing any calculations, what can astronomers determine from a variable star's period of pulsation?
A. distance
B. apparent magnitude
C. luminosity
D. age

Use the diagram below to answer Questions 23 and 24.

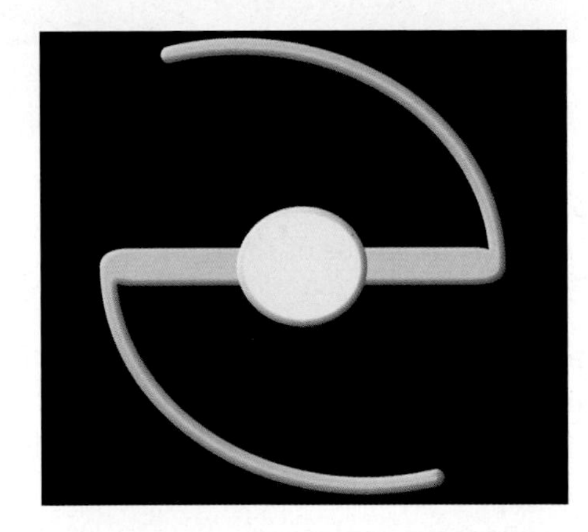

23. Which kind of galaxy is illustrated above?
A. spiral
B. barred spiral
C. elliptical
D. irregular

24. Which designation would the tuning fork diagram assign this galaxy?
A. S0
B. SB
C. Sa
D. E3

Use the diagram to answer Question 25.

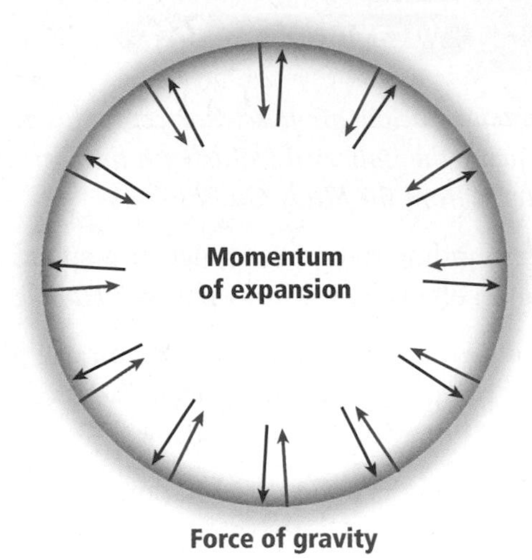

25. Which would cause the universe to collapse in on itself to make a closed universe?
A. force of gravity
B. critical density
C. momentum of outward expansion
D. Hubble constant

Constructed Response

26. **Interpret** the relationship between mass and density and the expansion of the universe.

27. **Discuss** Why are pulsating variable stars useful for finding distances to globular clusters?

28. **Explain** How do astronomers observe the spiral structure of the Milky Way?

29. **CAREERS IN EARTH SCIENCE** Why do astronomers think that there is a great amount of mass in the halo of the Milky Way?

30. **Explain** Why are the stars in globular clusters classified as Population II stars?

31. **Relate** the classification of a galaxy to its shape.

32. **Compare** active galactic nuclei with quasars.

33. **Explain** What do redshifts and Hubble's law tell us about the motion of galaxies?

34. **Discuss** how astronomers determined that dark matter exists.

Earth Science Online Chapter Test glencoe.com

Think Critically

35. Infer How would a star that forms in the Milky Way a few billion years in the future compare with the Sun?

Use the graph below to answer Question 36.

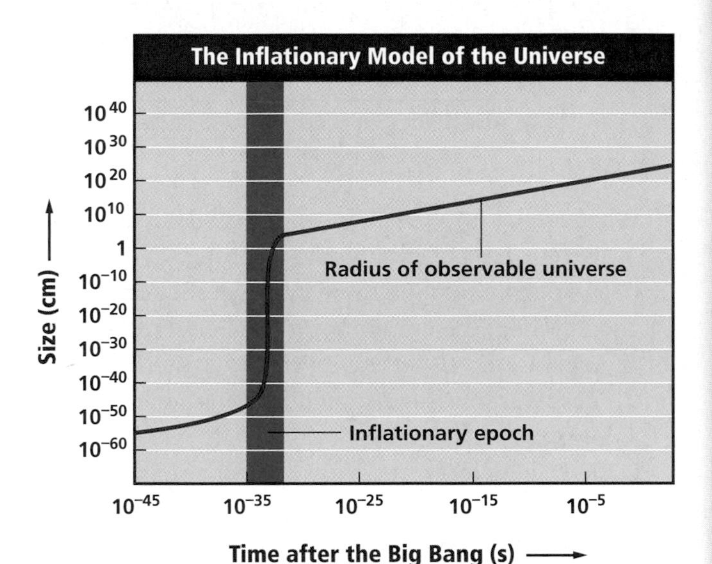

36. Explain what happened to the universe during the 10^{-35} s portion of the graph.

37. Compare the importance of variable stars and cosmic background radiation to the determination of the shape of the universe.

38. Identify the cause-and-effect relationship between Population I and Population II stars.

Concept Mapping

39. Use the following terms to construct a concept map to organize the major ideas in this chapter: *cosmic background radiation, quasars, Hubble's law, black holes, galaxy clusters,* and *Big Bang theory.*

Challenge Question

40. Infer the difficulty of determining the outcome of the universe resulting from the presence of dark matter.

Additional Assessment

41. WRITING in Earth Science Write an essay explaining the necessity for continuing space-based satellite telescope use and development.

DBQ Document–Based Questions

Data obtained from: Silk, J. 1998. The SETI module: cosmology. Syracuse University.

The graph below shows the changes in the strength of the major forces in the universe from the Big Bang until the present.

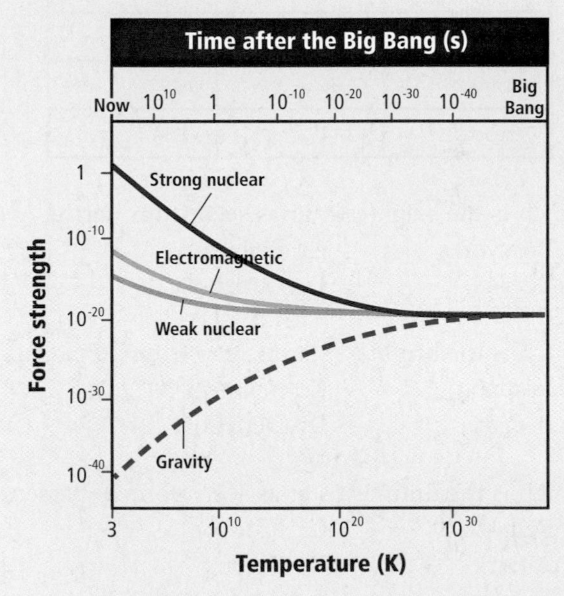

42. At what time and temperature did gravity and the strong-electro-weak forces separate?

43. What has happened to the force of gravity since the Big Bang? To the other forces?

44. One-billionth of a second (10^{-9}) after the Big Bang initiation, atoms began to form. At what temperature did this occur?

Cumulative Review

45. Which two basic stellar properties are displayed on an H-R diagram? **(Chapter 29)**

46. Why is it unwise to try to forecast weather by simply extrapolating current conditions beyond a few hours? **(Chapter 12)**

Standardized Test Practice

Multiple Choice

1. What is a piece of interplanetary material that burns up in Earth's atmosphere called?
   A. a meteorite
   B. an asteroid
   C. a meteor
   D. a meteoroid

Use the table below to answer Questions 2 to 4.

Stellar Magnitudes		
Star	Apparent Magnitude	Absolute Magnitude
Procyon	+0.38	+2.66
Altair	+0.77	+2.22
Becrux	+1.25	-3.92
Bellatrix	+1.64	-1.29
Denebola	+2.14	+1.54

2. Which is the brightest star as seen from Earth?
   A. Procyon
   B. Becrux
   C. Bellatrix
   D. Denebola

3. Which is the brightest star as seen from 10 parsecs?
   A. Procyon
   B. Becrux
   C. Bellatrix
   D. Denebola

4. Which is the dimmest star as seen from 10 parsecs?
   A. Bellatrix
   B. Altair
   C. Procyon
   D. Becrux

5. What two measurements are required to determine the Hubble constant?
   A. distance and speed
   B. distance and absolute magnitude
   C. apparent magnitude and speed
   D. apparent and absolute magnitude

6. What does Kepler's first law state?
   A. Each planet revolves around the Sun in a circular path.
   B. Each planet revolves around the Sun in an elliptical path.
   C. Planets closer to the Sun move faster than planets farther away.
   D. Planets closer to the Sun move slower than planets farther away.

Use the graph below to answer Questions 7 to 9.

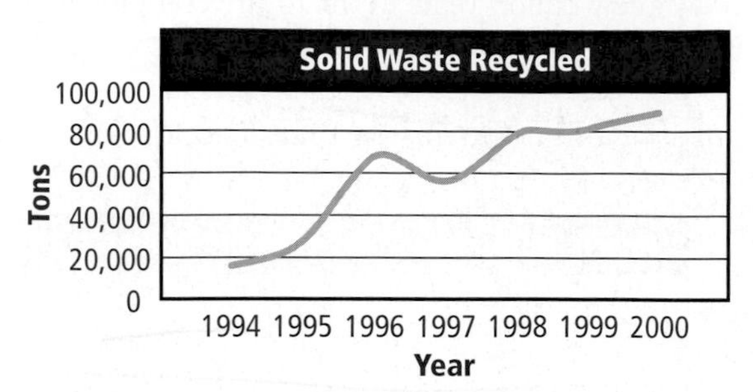

7. What can be implied about the graph above?
   A. Before 1994, recycling did not exist.
   B. As people became more aware of the benefits of recycling, the amount of waste being recycled increased.
   C. Less waste was consumed in 1994, so less waste was recycled.
   D. Recycling interests began to decrease in 1998.

8. What could not account for the sharp increase in recycling between 1995 and 1996?
   A. implementation of recycling laws
   B. increased public awareness
   C. more convenience for recycling
   D. less production of materials that need recycling

9. Which of the following years had the greatest increase in amount of material recycled?
   A. 2000–2001
   B. 1998–1999
   C. 1996–1997
   D. 1995–1996

10. Carbon-14 has a radioactive decay half-life of 5730 years. Which item would carbon-14 be most useful for dating?
    A. a rock from the Moon
    B. a Native American fire pit
    C. a jawbone from a triceratops
    D. a granite rock from the Canadian Shield

11. In which region of the Milky Way galaxy is 90 percent of its mass located?
    A. spiral arms
    B. halo
    C. nuclear bulge
    D. disk

Short Answer

Use the illustration below to answer Questions 12 to 14.

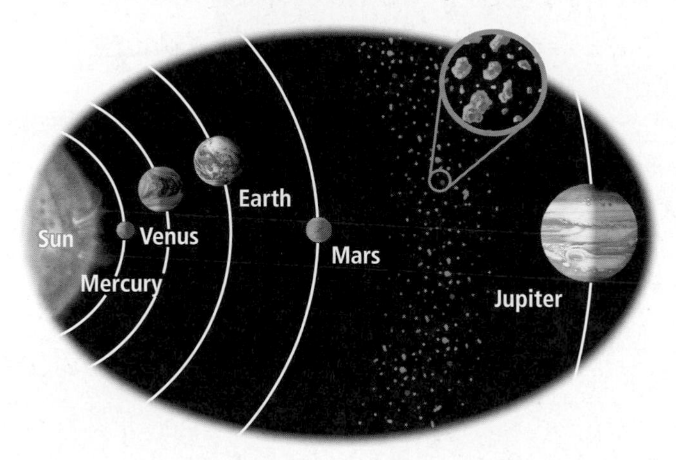

12. What do the lines through the planets represent?
13. Name and describe the material located between Mars and Jupiter.
14. Explain why this material did not form into a planet.
15. Compare and contrast refracting and reflecting telescopes. Which one is used more widely today? Why?
16. Describe the geocentric model of the solar system.
17. Why is Earth's Moon unique among all moons in the solar system?

Reading for Comprehension

First Stars in the Universe

NASA researchers say they have detected what might be the faint infrared glow of the first stars in the universe. Known as population III stars, the distant bodies are thought to have formed just 200 million years after the big bang.

The original stars formed from gas and dust in the void of space and are thought to have been many times more massive than today's stars. The ancient stars remain invisible to telescopes and have never before been detected. Using NASA's orbiting *Spitzer Space Telescope* the stars have been identified indirectly by measuring the enduring energy that they once radiated into the void of space. As the universe expands, starlight is stretched into longer, redder wavelengths. Most emissions from the first stars in the universe would appear today as infrared light. The universe is filled with background radiation known as the cosmic infrared background (CIB). This includes radiation from all stars—young and old, near and far. If these earliest stars were massive and formed in the standard cosmological mode, they should have left a signature in the fluctuations of the CIB.

Article obtained from: Handwerk, B. First stars in universe detected? *National Geographic News.* November 2, 2005.

18. What can be inferred from this passage?
    A. These stars are still present in space.
    B. Telescopes are not a good way to view stars.
    C. These first stars formed at the same time as the Big Bang.
    D. The first stars no longer exist, but we are just now seeing their radiation.

19. What are scientists seeing that confirms the existence of these stars?
    A. their visible light finally reaching Earth
    B. the faint infrared glow from their emissions
    C. the gas and dust particles of the stars
    D. radiation from the existing stars

20. Infer why basic telescopes are not able to find these stars but the *Spitzer Space Telescope* can.

NEED EXTRA HELP?																	
If You Missed Question . . .	1	2	3	4	5	6	7	8	9	10	11	12	13	14	15	16	17
Review Section . . .	28.1	29.2	29.2	29.2	30.2	28.1	26.1	26.1	26.1	21.3	30.1	28.1	28.4	28.4	27.1	28.1	27.2

eXpeditions!

Swim the Okavango...

Explore the African Landscape...

Dig for Dinosaurs...

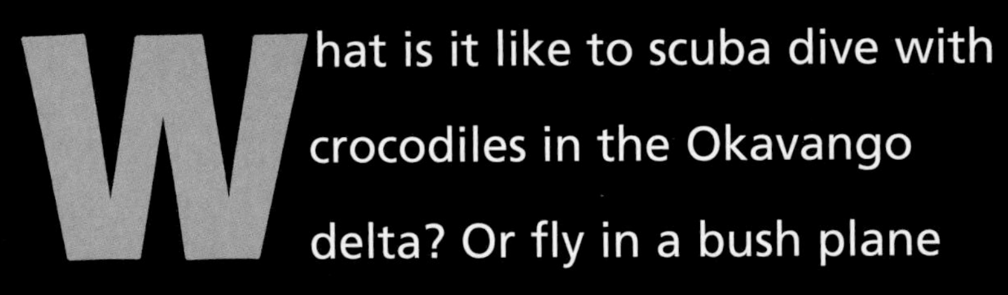

What is it like to scuba dive with crocodiles in the Okavango delta? Or fly in a bush plane over the African continent? Or dig for dinosaurs in China? The **National Geographic Expeditions** allow you to share in the excitement and adventures of explorers, scientists, and environmentalists as they venture into the unknown. Each Expedition takes you on a journey that enriches your learning about our dynamic planet.

Table of Contents

For more information on these Expeditions, visit glencoe.com. You can also link to original National Geographic articles that cover these topics and more.

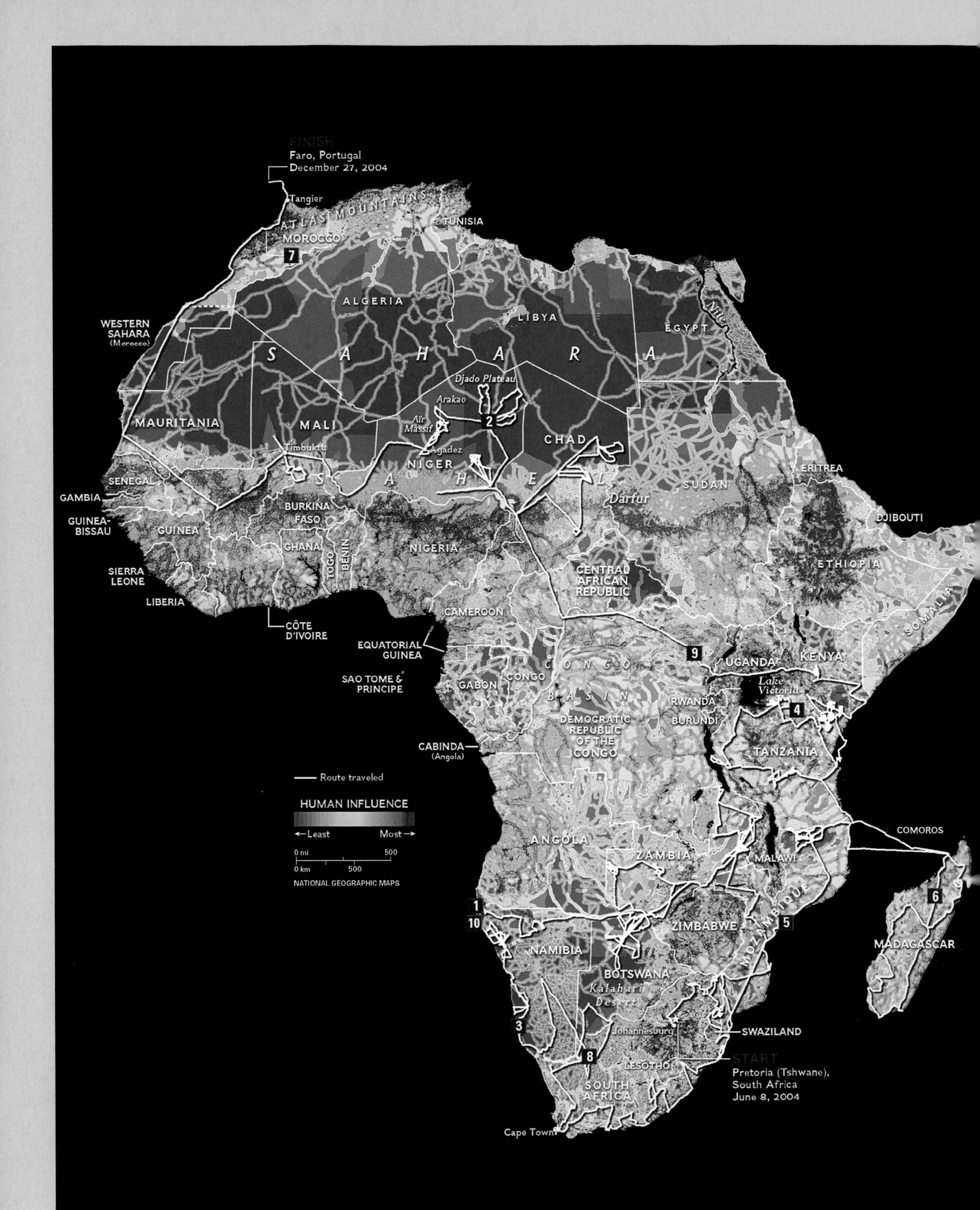
FINISH
Faro, Portugal
December 27, 2004
Tangier
ATLAS MOUNTAINS
TUNISIA
MOROCCO
7
ALGERIA
LIBYA
EGYPT
Nile
WESTERN SAHARA
(Morocco)
S A H A R A
Djado Plateau
Arakao
2
Aïr Massif
MAURITANIA
MALI
CHAD
Timbuktu
Agadez
NIGER
S A H E L
ERITREA
SUDAN
Darfur
SENEGAL
GAMBIA
BURKINA FASO
DJIBOUTI
GUINEA-BISSAU
GUINEA
GHANA
TOGO
BENIN
NIGERIA
ETHIOPIA
SIERRA LEONE
CENTRAL AFRICAN REPUBLIC
LIBERIA
CAMEROON
CÔTE D'IVOIRE
SOMALIA
EQUATORIAL GUINEA
9
UGANDA
KENYA
C O N G O
SAO TOME & PRINCIPE
GABON
CONGO
Lake Victoria
B A S I N
RWANDA
4
BURUNDI
DEMOCRATIC REPUBLIC OF THE CONGO
CABINDA
(Angola)
TANZANIA
Route traveled
HUMAN INFLUENCE
Least
Most
0 mi
500
0 km
500
NATIONAL GEOGRAPHIC MAPS
COMOROS
ANGOLA
ZAMBIA
MALAWI
1
6
10
ZIMBABWE
MOZAMBIQUE
5
NAMIBIA
MADAGASCAR
BOTSWANA
Kalahari Desert
3
Johannesburg
SWAZILAND
8
LESOTHO
START
Pretoria (Tshwane), South Africa
June 8, 2004
SOUTH AFRICA
Cape Town

Africa by Air

Tracing the Human Footprint

ABOVE: Biologist J. Michael Fay catches some sleep after eight hours in the air. On board this and another Cessna, Fay, pilots, photographers, and others risk their lives to map humanity's impact on tropical forests, savannas, and deserts. Along the way on this "Megaflyover" they struggle against malaria, sandstorms, and brushfires and try to keep the planes, computers, digital cameras, and GPS systems running.

LEFT: The map of Africa shows the zig-zag route taken by Michael Fay during a joint project—the Human Footprint—of the Wildlife Conservation Society (WCS) and the National Geographic Society.

From a low-flying plane is how J. Michael Fay sees the land on a mild December morning, as an heirloom Cessna 182 carrying him and three others approaches the Aïr Massif, a vast range of highlands standing up from the Sahara. The Cessna is painted scarlet and specially equipped for collecting data. The plane looks like a toy, or an enameled piñata, but it bears serious purposes, not candy. With a young Austrian pilot named Mario Scherer at the controls, and Fay in the right seat amid a rat's nest of custom-rigged digital hardware and cables, it caresses the topography, circling here, dipping a wing there, rising nervily through high notches to put peaks close at eye level on each side. Mounted in its right door is a high-resolution digital camera that automatically, every 20 seconds, takes a vertical shot of the ground. The photos, each tagged with Global Positioning Satellite (GPS) data registering exact time, latitude, longitude, and altitude, are uploaded into a computer on Fay's lap, through which he can add notes. A similar computer, scrolling out a map along the plane's flight line, rests under his left elbow. Fay's attention flicks constantly, tirelessly, between the computer screens and the terrain passing below.

ABOVE: The Namib Desert is just one of the 104 terrestrial ecoregions identified in Africa. Among Earth's driest places—and perhaps the oldest desert—it boasts orange sand dunes made sharp and steep by the blowing wind.

BELOW: The Ituri Forest in the Congo is a vastly different type of ecoregion from the Namib Desert in terms of the amount of rainfall, types of vegetation, and animal life.

The Realities of Land Use

Today is our tenth day of survey flying in Niger, and the 187th day since Fay and his chief pilot, Peter Ragg, departed from an airfield in South Africa. Fay's aerial enterprise is closely linked with an ambitious initiative of the Wildlife Conservation Society—the Human Footprint project. That project involves a program of multidimensional mapping to show gradients of wildness and human impacts around the world.

Fay himself, a restless individualist with a surprisingly good nose for politics, wants nothing less than to change the way the world perceives and uses ecosystems and natural resources—starting with perceptions in Washington, D.C. The ultimate goal of his Africa Megaflyover, he says, is to convince "the powers that be, in particular the U.S. Congress," that integrating natural resource management into American foreign policy is "a very, very smart thing to do. And a good investment."

Wherever humans live at high population densities, making unsustainable demands on natural systems, he notes, you eventually see ecological breakdown. Unmet needs and tensions lead toward conflict. A pilot himself, he recognized the value of low-altitude flying to illuminate the realities of land use.

A bush plane shows you patterns you'll never perceive from the ground. It allows flexibly targeted coverage ("Let's circle that spot again") and the capture of fine details you can't get from a satellite. Africa, the continent he knows and loves best after 25 years of working there, was the logical place.

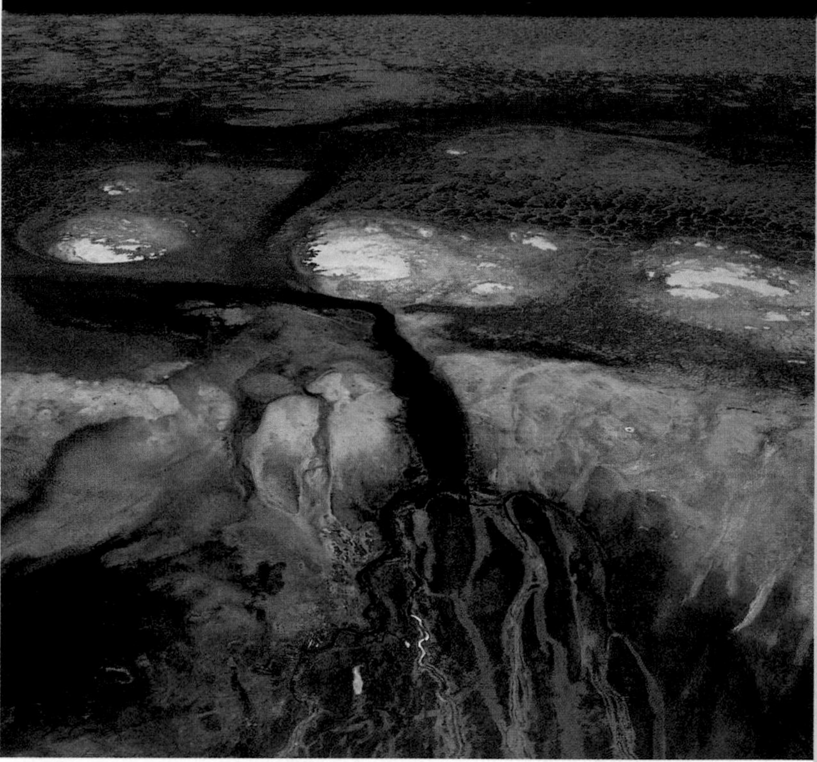

ABOVE: Lake Natron straddles Tanzania and Kenya. Along Kenya's section of the lake, microorganisms living in the water's salty crust create a palate of pink. Pigments in the organisms cause the feathers of the flamingos that depend on the lake for food to turn pink.

What Is Africa?

Of course Africa isn't really a place; it's a million places. Nowadays it encompasses 47 countries (not counting Madagascar and other islands) with a total population of 900 million humans. It can also be parsed into 104 terrestrial ecoregions, each unique in its physical and climatic features. Each one harbors a distinct plant and animal community. Ecoregions in many cases transcend national boundaries. Within or near all these ecoregions live people whose most elemental struggles and aspirations transcend ecological boundaries as well as national ones. Africans want better and fuller employment. They want food security and education for their children. They want good governance, free of oppression and corruption. They want fair, sensible arrangements for the management of wild landscapes and natural resources—arrangements chosen and controlled by Africans. They want peace.

Along with the human struggles come human impacts. Although some areas of Africa are less heavily inhabited than they might be, others are overburdened, eroded and blighted by the presence and demands of too many people. Because the African landmass is so large, climate change may affect its interior regions by bringing considerably higher temperatures and worse droughts and floods. This contributes to increased desertification and new patterns of disease. Poaching wildlife, both for subsistence and commercially, is an old problem but still serious. Timber harvesting, even when done selectively, often brings workers who empty a forest of its fauna for bush meat.

None of these concerns is unique to Africa. But Africa particularly deserves special attention. Africa's glories and successes deserve special attention, too. African peoples produce magnificent art, graceful cultures, terrific music, great works of the mind, and astonishing acts of political and moral courage.

Documenting Ecological Dimensions

Fay's intent is to document the ecological dimensions of that variousness. His conceptual starting point was the World Wildlife Fund map of 104 African ecoregions and the Human Footprint project, conceived by Eric W. Sanderson and a team of colleagues at the WCS and Columbia University. Sanderson's group used nine different geographic data sets (measuring factors such as road density, railways, population density, nighttime lighting) to represent the weight of human influence all over the planet, including Africa.

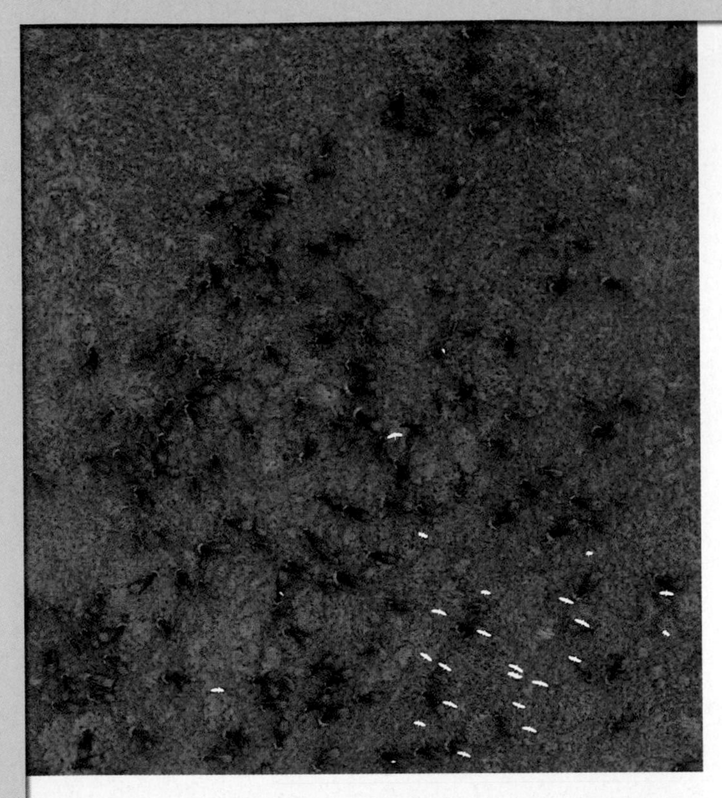

ABOVE: A herd of buffalo is seen wallowing in the muddy swamps of the Zambezi Delta in Mozambique. Protection efforts there seem to have buffalo populations increasing.

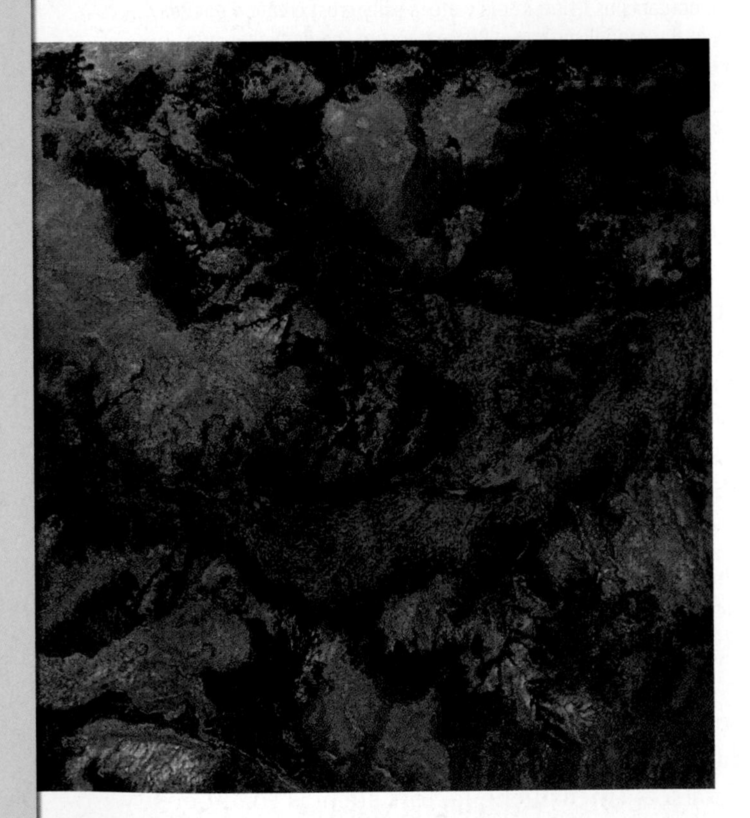

ABOVE: View of an area near the Mahajamba River on Madagascar reveals a rocky terrain. Madagascar is considered a hot spot for conservation because its unique flora and fauna are found nowhere else and there are tremendous population and resource pressures on the land.

Fay wanted to cover as many of the 104 regions as time, budget, and politics would allow. Then he would present an enormous body of data—between what is possible to what is actually happening—to decision-makers and say: Here's some information that might be relevant to your resources-and-security planning.

Fay recruited Ragg, an experienced bush pilot (and, in an earlier life, a successful optometrist in Austria), who offered his flying skills and the use of his two vintage airplanes, one for primary data gathering, one for support. Ragg in turn enlisted his fellow Austrian, Mario Scherer, who had found African bush flying a lively change from his recent work as a war-crimes investigator in Kosovo. Fay drummed up support from various sources—the Human Footprint lab at WCS, the WILD Foundation, the Bateleurs (an Africa-based organization of bush pilots volunteering for conservation), and, as chief financial sponsor, the National Geographic Society.

Beginning the Megaflyover

The first takeoff was on June 8, 2004, from Swartkop Air Force Base near Pretoria. Soon after,—OK, it was five minutes—Fay's network of digital gizmos suffered an outage. The camera quit, the computers went to battery power, and he sniffed a hint of electrical fire. Oh well, he thought, better a data-system meltdown than full-on engine failure within sight of the runway. He re-rigged.

Hopping his way across southern Africa and then northward on a chain of one-day flights, Fay arranged collaborations wherever possible. He assisted local conservationists, field scientists, or national agencies with their aerial-survey needs as well as adding data to his own comprehensive trove. Wherever he went, Fay tried to complement the aerial data-gathering with contacts, conversations, and observations on the ground.

Many computer crashes, camera shutdowns, and other minor problems have followed that first glitch above Swartkop. Most were easily repaired. There have also been a few dire aviation scares, caused by high winds, drastic loss of oil pressure, and other forms of mischance.

By the time I [David Quammen] met them in Niger, Fay and his pilots had flown 600 hours, crisscrossing 16 countries, usually at about 150 meters (500 feet) above the ground. One of the Cessnas had gotten a new engine. Both planes needed maintenance.

Theme of Absence

From the air over Niger we enjoyed some notable sights. A pair of addaxes skittered like sand crabs along a linear dune. Seven Barbary sheep galloped up sausage-like towers of dark sandstone along the Djado Plateau. Camels stood stuporous and serene in the middle of nowhere. Near one village we gawked down at a cluster of saltmaking pits. Each pit, a nice disk, variously sized, shone azure or turquoise or coppery green from the mineral solutions of their individual sumps—all together a necklace of bright-colored jewels.

ABOVE: An old abandoned town north of Dirkou, Niger, is built of blocks of salt. The houses without roofs reveal room after room, some only a few feet square. Newer towns are built of mud brick and have satellite dishes and telephones.

LEFT: Casablanca, Morocco, is an enormous place, with miles and miles of buildings. Water availability is evident from the green vegetation.

Mostly what we observed and recorded, though, were variations on a theme of absence. Some days we flew a 650-kilometer (400-mile) loop without glimpsing a single animal, and dozens of miles without spotting so much as a plant. Even absence is a form of data. Niger is a country desolated by recent human-caused losses. The addax is nearly extinct here, for instance, and the Barbary sheep, and the desert cheetah. Their disappearance from remote habitat areas may correlate with the presence of four-by-four tracks, indicating unimpeded access by poachers. Such tracks show clearly from 150 meters (500 feet) up.

Another sight came into view: a large green oval. It was a pond, evidently spring-fed from beneath the sands. Water? Fay peered down for a moment, having noticed something, tapped a note into his computer: "no animal tracks." It hadn't struck me, but of course: A water hole out here should attract gazelles and other mammals from many miles around—attract them, that is, if any exist. He tapped again: "4x4." Meaning, tire marks. An absence of animal sign, a presence of human sign. Cause and effect? Anyway, data. The challenge for Fay is that he must deliver meaning from the mountainous pile of facts and photographs he has collected.

eXpeditions activity

Consider the kind of variation there must be to have 104 distinct ecoregions in Africa. Verbally illustrate an ecoregion by listing and describing some of its factors.

Where the Elements Reign

State of Rock

RIGHT: Blasted by wind, broken by water, the Colorado Plateau spreads across 336,700 square kilometers (130,000 square miles) of Arizona, New Mexico, Utah, and Colorado. This arid expanse, best seen by air, has been called useless by some, a landscape that conspires against human settlement. For others it's nature's grandest work in progress.

LEFT: Hoodoos are columns of rock in fantastic shapes that are found in western North America. These appear in Utah's Bryce Canyon National Park.

Bizarre. Is that the right word for the Colorado Plateau, this thirsty sprawl of gaudy-hued stone festooned with such names as Hell Roaring Canyon, Scorpion Gulch, and Horsethief Point?

Edward Abbey began his classic *Desert Solitaire* with the simple "This is the most beautiful place on earth." Fiery rock can do that to a person. Others trying to understand the attraction of the plateau country apply adjectives like "amazing" and "awesome." In truth, a single adjective may not suffice. All the same, as I [Mike Edwards] fly over the plateau on a May morning, looking down on whalebacks of slickrock, on crashing waves of rock, on minarets and pyramids of rock hewn by water and wind—how could any word fit better than "bizarre"?

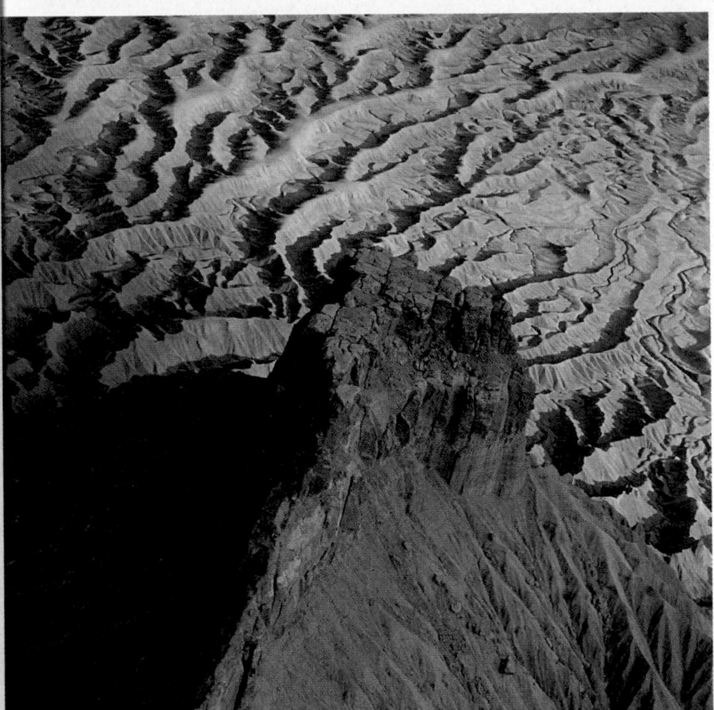

TOP LEFT: Dawn casts a Martian glow over the Buttes of the Cross in Glen Canyon National Recreation Area. Miners once scoured this backcountry for uranium. Today the mines are silent.

TOP RIGHT: A water-carved fist jabs across a canyon in Utah's Capitol Reef National Park.

ABOVE: Factory Butte looms over unproductive curves and folds of shale, salty sediments deposited by an ancient sea.

WATER'S TATTOO

Desert this is, but water's tattoo is everywhere. Spidery little arroyos coalesce into bigger arroyos that plunge into the still deeper groove of a river, maybe into the thousand-foot-deep canyon of the Escalante, a scalpel-cut in red rock, so narrow that the stream and its fringe of willows and tamarisks are invisible unless you're dead-on overhead.

Most of the collected runoff, if it hasn't vaporized or died in a mudflat, swells the Colorado River. By the time the river courses into Arizona and roars into the plateau country's most dazzling feature, the Grand Canyon, it is plowing a furrow more than a mile deep. Pretty impressive digging, this, considering that the precipitation in parts of the plateau averages only six inches a year.

Water was also present at the creation, in far greater abundance. Tens of millions of years ago, seas, swamps, and rivers deposited dozens of layers of rock: limestones, mudstones, shales, many reddened by traces of iron. In those eons the plateau country was flat and much lower than its heights today, which are typically 1,524 meters (5,000 feet) above sea level. Winds also contributed raw material, the makings of sandstone layers hundreds of feet thick. The whole shebang was thrust upward by forces within the Earth.

Colliding tectonic plates tilted and bent layers like cardboard. Everywhere then as now, water attacked the soft stones, carving canyons. That's Plateau Geology 101, slightly abbreviated.

Powell's Hundred-Day Journey

Winging 762 meters (2,500 feet) over this rockscape, I feel a tenuous kinship with John Wesley Powell. On his scary hundred-day journey down the Green and Colorado Rivers in 1869, that indefatigable adventurer-scientist surmounted cliff tops to reconnoiter the uncharted territory he had penetrated. Of course, I'm riding shotgun in a Cessna while Powell had to pull himself up one-armed from a riverine chasm to a lookout; he lost his right arm at Shiloh in the Civil War. But we gaped at the same sights. "The landscape everywhere . . . ," he wrote, "is of rock—cliffs of rock, tables of rock, plateaus of rock, crags of rock—ten thousand strangely carved forms."

Clearly Powell was awed by the fantastical convolutions of this land. He became the chief expositor of the "plateau province," as he called it, documenting not only the geology but also the ways and lineages of its Indians, meanwhile campaigning for sensible husbanding of the water.

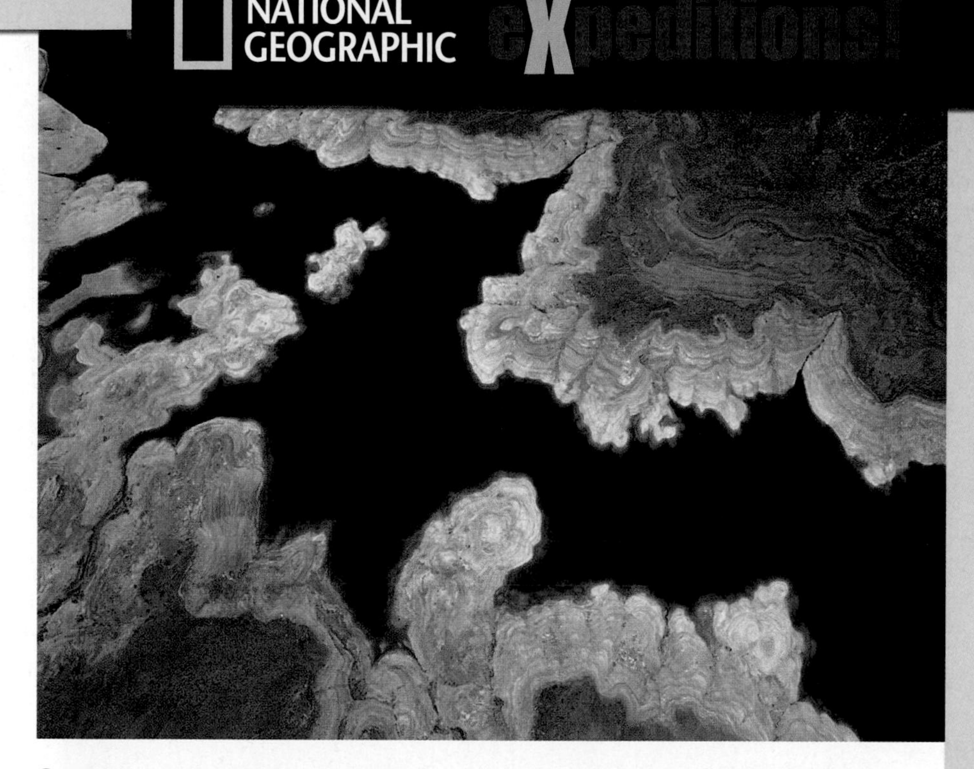

ABOVE: A bathtub ring of bleached rock—a sign of severe drought—lines Lake Powell, the country's second largest human-made lake and canteen for much of the Southwest. Long years of drought and demand from distant cities have depleted the lake's reserves. The years September 1999–September 2004 are the driest five years in a century.

RIGHT MIDDLE: Swelling beneath Arizona's Coyote Buttes, sandstone waves evoke white water.

RIGHT BOTTOM: This view of Earth's outer layers shows fractures in sandstone that form car-size blisters.

ABOVE: Photographer Adriel Heisey found a paradise of form as he soared over bands of shale and sandstone. "Everywhere I turned," he says, "there were geometric patterns that defied my ability to comprehend."

BELOW: Sunlight pours over Jacobs Chair, a Utah butte named for a cattleman who drowned fording a storm-swollen creek nearby. In the late 1800s Mormon pioneers left a trail of names as they hacked a wagon route across southern Utah during the grueling Hole-in-the-Rock expedition.

Protecting The Landscape

Powell hadn't glimpsed some of the craziest rock shapes—the pinnacle-like hoodoos of Bryce Canyon or the vaulting spans of Arches, canonized by Edward Abbey. Bryce and Arches are two of the roughly 30 parks and other national preserves that make the heart of the province one of the nation's most protected regions. It isn't perfect protection; environmentalists decry its insufficiencies while locals, ingrained with the Westerner's mistrust of bureaucracy, grumble about overkill. In the largest unit, the 1.7-million-acre national monument Grand Staircase–Escalante, created only in 1996, off-road vehicles plow tracks that won't disappear for decades. On the other hand, many mining claims have been relinquished or bought out by the U.S. government, and no new claims are permitted. And while allowing multiple uses, including ranching (yes, there's a little grass here and there), the Bureau of Land Management is charged with superintending the monument to protect its attributes.

Protecting Earth's History

Gaudy vistas are only one of those attributes. The plateau is a time machine nonpareil, holding who knows what secrets. When the rocks at the bottom of the Grand Canyon are counted in, the swath of Earth's history exposed by water's relentless gouging of the plateau is reckoned by geologists to reach back 1.7 billion years, more than a third of Earth's existence.

ABOVE: Standing tall against the cold, Castle Rock (on the right), and other buttes bear a dusting of snow in Monument Valley. With each season the elements shape the plateau anew, shattering canyon walls and cliff faces, chiseling subtle changes into the enduring symbols of the West.

One afternoon, puffing along behind Alan Titus, a BLM paleontologist, I dropped back a mere 75 million years or so. Titus assured me we were tramping through a swamp where ferns and magnolias had flourished, although it looked awfully like a forest of stunted piñons and junipers. "The whistles and shrieks you hear are not birds," Titus said, coaxing my imagination into play. "They're dinosaurs." Soon he had me seeing huge crocodiles and snakes sloshing lazily in warm pools. And then, guiding me to a row of dark, roundish objects half-buried in the soil, Titus said: "You're looking at the remains of an animal that's been extinct for 65 million years or more"—the fossilized vertebrae of a duck-billed dinosaur that lived in the Cretaceous period. Titus says this specimen was 7.6 m (25 feet) long.

"I'm waiting for that big tyrannosaur," he said. "I want to find one before I retire."

On another afternoon I climbed to a high cliff's edge. The setting sun infused the rocky layers vaulting away to the horizon with a crimson incandescence—the kind of glow, surely, that compelled Edward Abbey to pronounce these rocks beautiful. A hot, dry wind came up, gusting stronger and stronger, and as it assaulted the cliff faces it whined and screamed. Sounded just like dinosaur shrieks.

eXpeditions activity

Research the status of and plans for conservation on the Colorado Plateau. Discuss whether changes in weather have affected the status and plans and how the plans may affect people who depend on the Colorado River for their main source of water.

Africa's Miracle Delta

Okavango

RIGHT: The miracle happens in slow motion, for this part of southern Africa is so flat (a gradient of about a hundredth degree) that the floodwaters take three months to reach the delta and four more to traverse its 240-kilometer (150-mile) length. Yet by the time its force is spent, the flood has increased the Okavango's wetland area by about two times, creating an oasis up to half the size of Lake Erie. It is one of the largest inland deltas on Earth. From space the delta looks like the footprint of a bird. Water flows into the system through the leg, called the Panhandle, a strip of land 96 kilometers (70 miles) long and 14 kilometers (9 miles) wide along which the Okavango River meanders in lazy loops. Forward-pointing toes channel water through the delta.

LEFT: Suspended in an ethereal realm of water lilies, and light, a river Bushman, pole in hand, peers into the emerald forest of Botswana's Okavango River. As if by magic it ebbs and flows with seasonal floods before vanishing in the Kalahari Desert. The result: an oasis for wild things above and below the surface.

The Miracle Is This: Under cloudless skies at the driest time of Botswana's year, when rain is both a fading memory and a distant promise, a flood comes to the Okavango Delta. Generated by rainfall 1600 kilometers (1000 miles) and two countries away in the highlands of Angola, the flood wave snakes down the Okavango River and spreads across the delta, swelling its lagoons and channels and spilling outward to inundate its floodplains. In a land withered by drought, this gift of water is like unction, and all nature responds to it.

"My surface gives you life. Below is death."

The delta's deepest, most diverse underwater habitats lie in the Panhandle. The flood peaks here in April, raising the level of the Okavango River by six feet. In May the level has started to drop. Sediment borne on the flood wave has settled, and the water in Ncamasere channel, an offshoot of the main river midway down the Panhandle, becomes clean and clear. And deadly. The waters of the delta are full of crocodiles. The Bayei people, one of several Okavango tribes, say as much in a poem they teach their children: "I am the river. My surface gives you life. Below is death." For photographer David Doubilet and me [Kennedy Warne], going below the surface was an essential part of our work.

We wanted to see the delta as few had dared to see it before—a croc's-eye view. People in passing boats, noticing our wet suits and scuba gear, didn't hesitate to give their opinion on croc-watching: you're out of your minds. Perhaps we were, but it was winter, and we reasoned that because crocodiles are reptiles, their metabolism would be sluggish.

The larger crocodiles spent much of the day basking on the riverbanks in well-used haul-outs, usually with chutes down which they slid into the water if disturbed. In the cool of the night the warmth-loving crocs came to life for the hunt, floating at the water's edge. Their eyes gleamed blood-red in our spotlight as we motored up the channel. Although Nile crocodiles are one of only a handful of predators that actively hunt humans, I figured that if I initiated an encounter, thus denying the animal its advantage of surprise, I would retain the upper hand.

Crocodiles are the delta's most feared aquatic predator, but locals say that hippopotamuses cause more deaths and injuries. Accidental meetings in narrow channels are often the trigger for an attack. Hippos can bite a canoe in half with one snap of their jaws, and their teeth can puncture an aluminum boat as if it were a can.

BELOW: In the crocodile's lair, photographer Jennifer Hayes explores caverns formed by floating mats of papyrus in the deep waters of the Ncamasere channel in the Panhandle. Croc tracks were everywhere.

The two-ton vegetarians aren't slowpokes, either. Guy Lobjoit, an Okavango fishing guide, told me he once had a hippo keep up with him while he was doing nearly 32 km/h (20 mph) in his runabout.

Okavango's Bounty

People have been living with the dangers and the bounty of the delta for at least 100,000 years. The seasonal floodplain, the webbing between the delta's toes, is a rich part of the Okavango larder. Here the floodwater forms a lake six inches to a foot deep, dotted with countless islands. The water brings a flush of plant growth, which in turn attracts wildlife into these fertile, sun-warmed shallows. The local people make good use of the molapo, as the floodplain is called. During the flood they fish, and in the dry season they graze cattle. All year round they harvest fruits, cut thatching grass and reeds, and hunt game on these productive lands.

At Guma, near the top of the delta, a Bayei man known simply as Madala, Old One, and a young fishing guide called Fish took me into the molapo during the flood season to show me something of their way of life. We journeyed by mokoro, or dugout canoe, the mode of transport in the delta. The mokoro that Fish poled was made from kiaat, a teak-like timber, with metal patches covering cracks he called its wounds. Madala's canoe was fiberglass. He explained that the new synthetic canoes are more stable than the traditional wooden ones. More sustainable too, as trees suitable for mokoro-making are a limited resource in the delta.

Poling is a hypnotically beautiful way to travel. Each thrust of the wooden pole moved the mokoro through beds of reed and sedge that rustled against the hull. The foghorn snort of a hippo warned us to avoid its channel.

As we poked along, Fish would point to various plants and describe their properties. The root of the star apple makes an excellent toothbrush; the bark of the rain tree can be ground up and thrown into the water to paralyze fish; chewed sickle bush leaves are good for treating snakebite. Madala cut a tall papyrus stem and pounded the fleshy white base against his palm to soften it before handing it to me to eat. It was sweet, fibrous, and refreshing, reminiscent of fresh coconut.

We made camp under the boughs of a sycamore fig. While Madala set his net in a lagoon thick with water lilies, Fish waded into the floodplain to spear small fish with a porcupine quill. I climbed a baobab tree to collect its maraca-shaped fruits containing a white pulp that substitutes well for cream of tartar. Madala mixed it with water to make a tangy sauce. That night we rolled balls of cornmeal porridge with our fingers and dipped them in a casserole of freshly caught bream, water lily fruit, and heart of palm. Other than the presence of a few tourists—and a carton of long-life milk for our tea—I suspected that little in this scene had changed since the first European explorers visited the Okavango over 150 years ago.

ABOVE: Following the water, hundreds of African buffalo graze in the lower delta during the flood. Herds here reach into the thousands, their numbers regulated more by grass than predators.

Water . . . Life is coming

One thing that has changed—and continues to change—is the path the water takes through the delta. In 1849, much of the flow was down the western channel system and into Lake Ngami. In the 1880s the water flow, responding to a range of subtle landscape cues, began to favor the eastern channels. The sluggish western channel became choked with vegetation, and Lake Ngami dried up. The Batawana people, Botswana's dominant tribe, followed the water, shifting their main settlement to a lush site on the delta's southern edge. They called the place Maun, "place of reeds."

Today Maun is a town of 45,000, with barely a reed to be found. Water flow seems to be moving westward once more, and floods, which follow a natural cycle of higher and lower volumes, have diminished in size. The result is this commercial gateway to the delta has a water shortage. It has become a place of dust. Not surprisingly, when the annual flood does reach Maun (though there is no guarantee that it will), the whole town celebrates. On a breathless July day—the sky the eggshell blue of the Botswana flag, the air full of the smell of wild sage—I watched as the flood crept down the broad, dry bed of the river that runs through town. Children dug furiously with sticks in the sand to encourage the trickle to run faster. Some leaped back and forth across the steadily widening stream, laughing for joy. Others just let it run over their bare feet, looking at it as if it was the first time they had seen water. "The water is coming," I heard a father explain to his daughter. "The fish are coming. The water lilies are coming. Life is coming."

On a bank of the river, behind a twig fence that didn't look as if it could keep out a goat, let alone a cow or a hippo, a man who told me his name was Flay Million Dube walked around his vegetable plot. He told me, "I'm not working today because I'm so happy." He had just been down to the river to wash his face and hands in the new water, he said. Tomorrow he would put fresh, cool mud around his beds of spinach, broccoli, and kings onion.

The dry season

By October the time of sadness has come. The flood has vanished, ten billion tons of water sucked up into the atmosphere whence it came. People cast thirsty glances at the sky, where glowering thunderclouds build in the afternoons, but the summer rains are still two months away. The floodplains dry out, and water levels in the channels and lagoons drop to their lowest levels. As the delta shrinks, life retreats.

○ **LEFT:** One of the many elephants that came from the south, lumbering toward the widening ribbon of water, trunk cocked in an S, snuffing the sweet elixir. Standing at the water's edge, the thirsty animal sucked up trunkfuls and gushed it into its mouth, spilling barely a drop.

Maun broils in temperatures of 100 plus. Hot winds sandblast the town, and the sky becomes white with dust. The Thamalakane River, where I had witnessed the arrival of the new water three months earlier, was again bone-dry. Flay Million Dube's garden was bare soil, not a plant to be seen. No children played in the riverbed. Only a few dust devils whirled in the heat haze.

Not since the 1960s has the Thamalakane flowed all year round, delivering water to the delta's outlet, the once mighty Boteti River. Fifty miles southeast of Maun, at a camp called Meno A Kwena, all that remains of the Boteti is groundwater, the legacy of floods past. Larger animals can dig for it, but with each successive year of low flood volumes the water table drops a little farther out of reach.

David Dugmore, who runs the camp, has made it a personal mission to provide water for at least some of the thirsty animals—pumping groundwater to fill a small water hole. But his is only one small relief station in a vast arid landscape. Maintaining the supply line is also a problem, he told me, pointing to lion tooth marks in the pipe that runs from pump to pool. "The lions are so desperate for water they bite into the pipe, working their way along until they reach the water hole."

An hour's drive down sandy tracks brought us to a group of hippopotamuses stranded in a pond. There was no water for miles upriver or down, so the hippos were marooned. There was little grazing to be had, and it was with relief that we saw a wildlife ranger drive up and unload half a dozen hay bales, which he cut open and spread beside the pool. The hippos trotted out of the water and began to munch. Were it not for their daily handout, they would starve.

Is climate change casting its long shadow over the miracle delta? Apparently not, according to researchers, who have detected an 18-year oscillation in rainfall in the region and an 80-year cycle of high and low flood volumes. We're reaching the end of the 40-year low part of the cycle, they say, and should see larger floods in the future, peaking in mid-century. Rainfall should also increase over the next few years.

River and rain contribute in roughly equal measure to the delta's water budget. The summer rains have the function of recharging the groundwater aquifer. If the rains are good, little floodwater is needed to bring the water table to the surface, and the bulk of the inflowing water then spills into the seasonal floodplains, creating a large flooded area. If the rains are poor, much of the floodwater soaks into the ground, filling the gap left by lack of rain, and the area of inundation is reduced.

Influences on the Okavango

Terence McCarthy, a professor in the School of Geosciences at the University of the Witwatersrand, in Johannesburg, speaks of the delta as a living organism with a circulatory system. McCarthy and his colleagues, who have been studying the delta since 1985, have discovered that one of the largest contributions to the life of the delta is made by one of its smallest inhabitants: termites. Their colonies are giant construction companies that have transformed the Okavango Delta from a piece of flat real estate into a mosaic of an estimated 150,000 islands. It stems from the termites' need for air-conditioning. Some species build above-ground air vents to control the temperature in their networks of galleries and tunnels. These turrets, sometimes ten feet high, and their surrounding earthworks are above flood level, providing dry, fertile sites on which trees can become established.

"IT WAS STRANGE TO THINK THAT THE WATER FLOWING BENEATH ME WAS BRINGING LIFE TO A DISTANT DELTA."

Trees can be thought of as kidneys of the delta, cleansing the system by removing its salts. They do this by sucking water out of the ground and pumping it into the atmosphere by transpiration. In the process, soluble salts are deposited around the tree roots—a "toxic waste storage system," McCarthy calls it. Without the delta's millions of tree pumps, the 400,000 tons of salts carried in yearly by the Okavango River would be poisoning the delta. By concentrating salts in the soil and groundwater beneath them, trees not only keep the water in the delta fresh but also expand the size of their island platforms.

Most channels in the delta have a life expectancy of about a hundred years. During that time sandy sediment gradually raises the height of the channel bed, slowing the current and allowing the fringing stands of papyrus to spread into the channel. Clumps of papyrus eventually break off and jam the channel until it becomes completely blocked. At this point the hippos come to the aid of the delta's circulatory system, breaking through papyrus jams and forming new channel connections. It is only because the delta is so flat that water follows such randomly created corridors—the paths the hippos have trod.

Biological influences are part of a system as intricate and responsive as any on Earth. Yet the delta is not immune to human disturbance. The chief threats lie upstream, in the two countries with which Botswana shares the inflowing water. Angola and Namibia both experienced long, brutal wars in the latter part of the 20th century and now look to rivers to help build their economies. Two aspects of development, the increased use of agricultural fertilizers and the production of hydroelectricity, could have disastrous downstream effects on the delta.

Papyrus can thrive in nutrient-poor conditions. Enrichment of the delta through fertilizer runoff from irrigated farmland upstream could cause rampant growth of papyrus and lead to wholesale channel blockage. "If the Panhandle becomes blocked," said Map Ives, of a large tourism company, "it's good night Okavango Delta." Damming the rivers that supply the delta would be equally catastrophic. Scientists such as Terence McCarthy point out that dams deprive rivers of sediment that is vital to the functioning of the delta. More than 200,000 tons of it is deposited in the delta's upper reaches each year, raising the channel beds and starting the process of channel switching by which the Okavango renews itself. Without an annual injection of sand, channels would be scoured out instead of built up, becoming ever deeper and swifter. Channel switching would cease; whole sections of the delta would be lost.

I was near the grazing country of the Bié Plateau. It was November, and the summer rains were starting. It was strange to think that the water flowing beneath me was bringing life to a distant delta. But it was: In a few weeks the flood would start to rise in the Panhandle. Relief would come to the Okavango's parched plains. The miracle would begin again.

---

eXpeditions activity

Research the use of fertilizers and the dams built north of the Panhandle since 2004. How have they affected the Okavango delta? What other measures might be taken to provide water to the area in a way that will not harm the ecosystem?

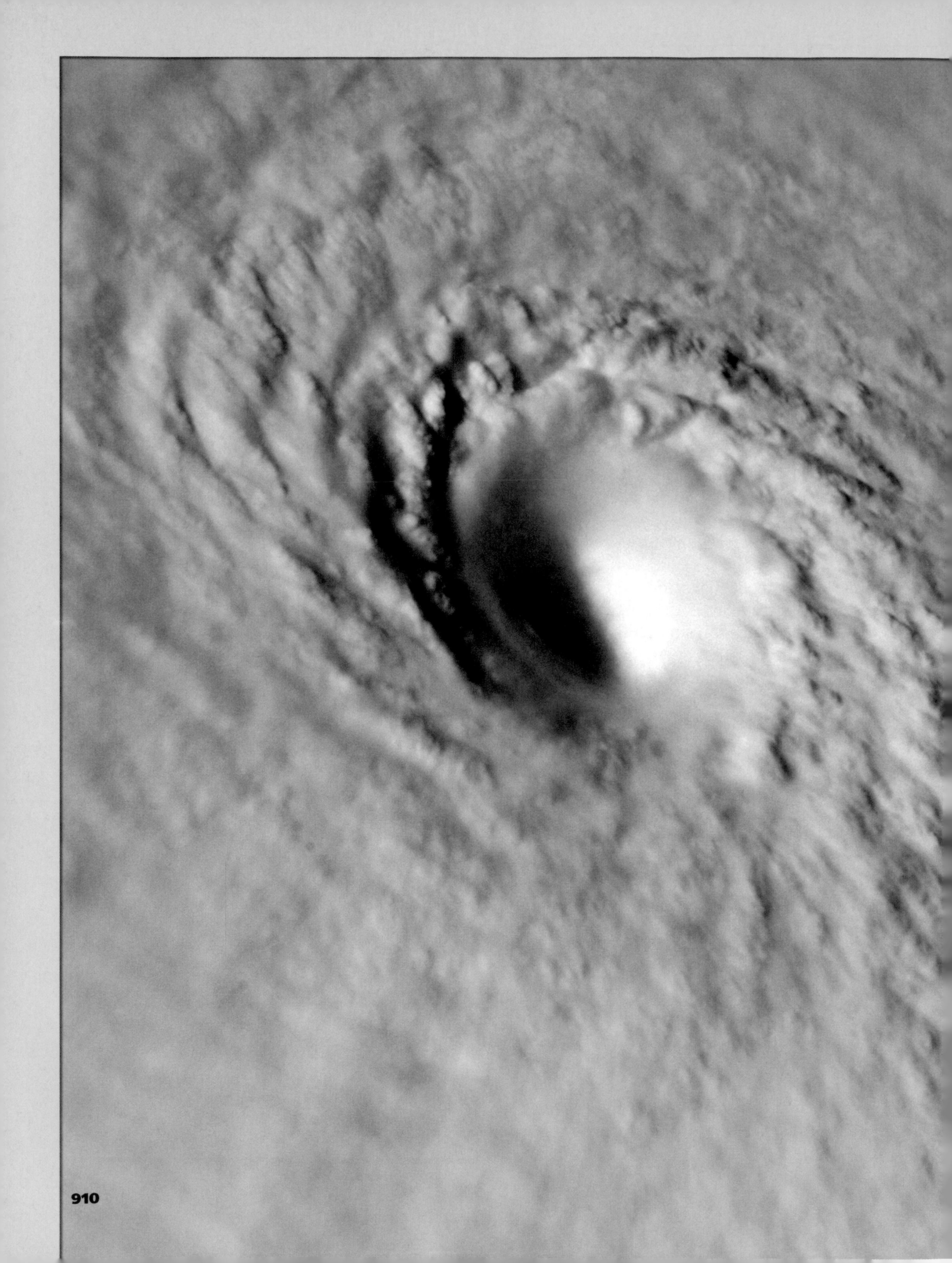

No End in Sight

Super Storms

ABOVE: Year 2005: Never before had a hurricane caused as much economic damage as Katrina. Never before had the Atlantic seen 27 named tropical storms—so many that the list of storm names had to be extended with Greek letters. Seven made landfall in the United States. Never had 15 hurricanes been spotted in one season, including four Category 5 storms. Each image shows an area 1,191 km (740 miles) wide.

LEFT: As Hurricane Wilma spun toward the Yucatán Peninsula on October 19, 2005 (the day this image was shot from the *International Space Station*), a hurricane hunter plane recorded an atmospheric pressure of 882 millibars in the eye of the storm, driving winds of 240 km/h (175 mph).

When the fiercest hurricane ever recorded in the Atlantic is bearing down on you, a salvaged armchair under a wood-and-tin awning might seem a poor choice of shelter. But that's where Don E. ("I'd rather keep my last name out of it") was parked when Wilma hit South Florida at 6:30 A.M., October 24, 2005. For Don and a buddy, it was the start of the workday at Jimbo's Place, a bait shop down by the water on Miami's Virginia Key. "Once we got out here, it was kind of too late to do anything but ride it out," Don says.

Jimbo's looks like nothing so much as an abandoned shack. But whether through good luck or unexpectedly sound construction, it survived Wilma's fury. The winds had ebbed from 280 km/h (175 mph) at sea to 193 km/h (120 mph) by the time the storm hit, but Wilma still left almost all of South Florida without power.

Season of Record Breakers

Wilma was a record breaker in a season of unsettling records. Katrina, at the end of August, killed more than a thousand people and left much of New Orleans and the neighboring coast in ruins. The damage exceeded a hundred billion dollars—the costliest natural disaster in U.S. history—and the toll in fractured lives is incalculable. Rita, in September, rivaled Wilma in intensity and ravaged the Gulf Coast through western Louisiana and East Texas.

Days after Wilma, one visitor to Jimbo's was already worrying about what future hurricane seasons might bring. Sharan Majumdar, 34, is a hurricane researcher at the University of Miami's Rosenstiel School of Marine and Atmospheric Science, just across the highway from Jimbo's. He is one of a cadre of scientists trying to understand nature's most powerful storms and more reliably predict their surges, ebbs, and lurching paths from birth to landfall.

Majumdar says he can't really blame his fellow patrons at Jimbo's for deciding to stay put during Wilma. Forecasts can get hurricane tracks wrong by hundreds of miles and wind speeds by tens of miles per hour. As a result, Majumdar says, "people often return after an evacuation to find nothing really happened." The solution, he says, is to improve forecasting through better science. "That's the only way to get people to trust the warnings."

"IT WOULD BE REALLY NICE TO SAY WHAT YOU NEED TO MAKE A HURRICANE...AND WE REALLY CAN'T DO THAT YET."

Why Hurricanes Form

Like all weather, hurricanes are fueled by heat—the heat of sun-drenched tropical seas, which powers the storms by sending warm, moist air rushing toward the frigid upper atmosphere like smoke up a chimney.

As surrounding air is sucked in at the base of the storm, Earth's rotation gives it a twist, creating a whorl of rain bands. These whiptails of thunderstorm activity are strongest where they converge in a ring of rising, spinning air, the eyewall, which encloses the cloud-free eye.

Hurricanes (called typhoons in the western Pacific and tropical cyclones in the Indian Ocean) can propel themselves to an altitude of 15,240 meters (50,000 feet) or more, where the rising air finally vents itself in spiraling exhaust jets of cirrus clouds. The largest ever, the 1979 Pacific typhoon Tip, sent gale-force winds across more than 2,174 kilometers (1,350 miles). Even an average hurricane packs some 1.5 trillion watts of power in its winds, or half the world's electrical generating capacity.

Starting this great weather engine requires surface waters of 80 degrees or more, moist air, and little wind shear—a difference in wind speed at the surface and aloft that can tear apart a developing hurricane. But those ingredients often produce nothing more than a tropical disturbance—an unremarkable cluster of thunderstorms.

"Disturbances look very similar day to day," says David Nolan of the Rosenstiel School, "and then all of a sudden you get a big burst of convection, then within six hours it becomes a depression, then it becomes a hurricane, then it's flooding my apartment." Katrina soaked Nolan's 14th-floor Miami Beach home as the storm crossed Florida on its fateful course to New Orleans and the Gulf Coast. "It would be really nice to say what you need to make a hurricane," he adds. "And we really can't do that yet."

One thing was clear in 2005: Conditions were ideal for making hurricanes. Yet 2005 was just a continuation of the upward trend that began in 1995. Because of a tropical climate shift that brought warmer waters and reduced wind shear, the Atlantic has spawned unusual numbers of hurricanes for nine of the past eleven seasons. "We're 11 years into the cycle of high activity and landfall," National Oceanic & Atmospheric Administration (NOAA) meteorologist Gerry Bell says, "but I can't tell you if it will last another ten years, or thirty."

Hurricane Tracking Technology

Weather satellites make it easy for meteorologists to keep tabs on hurricanes. But ordinary satellite images show only the cloud tops. Space-borne infrared sensors can reveal more detail, charting the size and shape of the warm eye, and satellite radar and microwave sensors can map the rain. Hurricane hunter aircraft actually fly right into Atlantic hurricanes. But they only probe conditions at altitudes of several thousand feet, above the worst turbulence, Jack Beven of the National Hurricane Center (NHC) in Miami says—"not at the surface, where they really matter to people."

In 2005, scientists flew a robotic aircraft straight into the maelstrom when tropical storm Ophelia was off the mid-Atlantic coast. The craft, called Aerosonde, swooped and circled for ten hours, monitoring winds and the flow of heat and moisture from the ocean into the storm.

A clearer picture

Below are forecasts of wind speed for Hurricane Katrina for the morning of August 28, 2005, when it intensified to Category 5. The current model runs at a 12-km (7.5-mile) resolution; HWRF will run at a 9-km (5.6-mile) resolution. Researchers are experimenting with even higher resolution models (bottom). The finer grid captures critical features that the older models overlook.

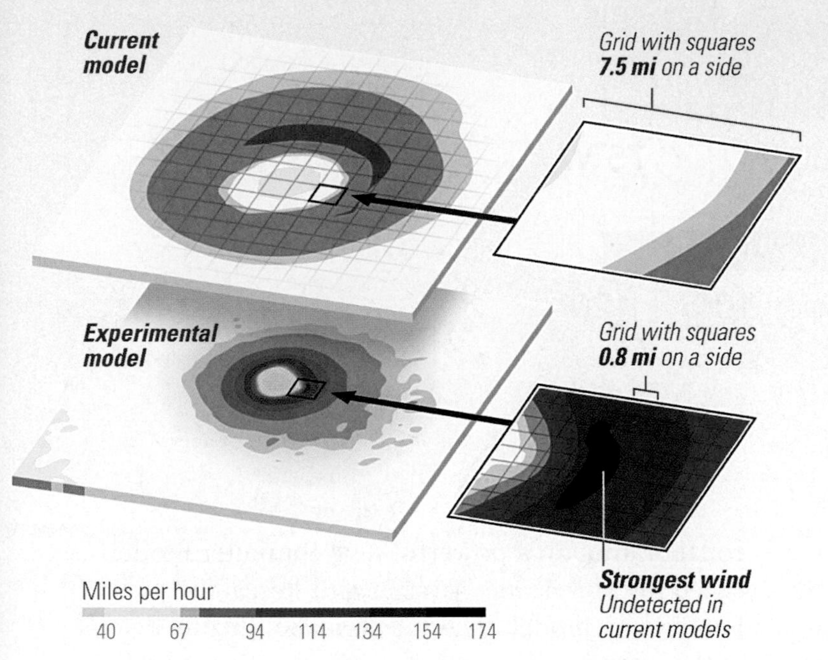

A fuller picture

Accurate forecasts of a storm's track and intensity require the best possible picture of the ocean and atmosphere. Unlike current models, HWRF will rely on real-time wave, ocean, and coastal data to improve forecasts.

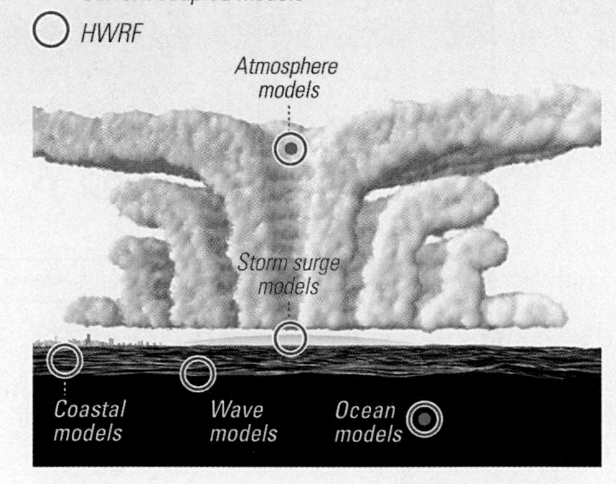

ABOVE: Meteorologists work at improving computer models to better forecast hurricane conditions. In 2007, a high-resolution NOAA Hurricane Weather Research and Forecasting (HWRF) model became operational.
ABOVE LEFT: Compare the model used in 2005 on the morning of August 28 when Hurricane Katrina intensified to Category 5 with an experimental model with an even higher resolution than the 2007 HWRF model.

That was a test, but forecasters routinely probe the heart of storms with shorter-lived devices called dropsondes. Released from high-flying aircraft into hurricanes and the surrounding winds, these instrument-packed tubes descend by parachute. "They take about 15 minutes from 12,192 meters (40,000 feet) to splash," Majumdar says. Along the way, they measure temperature, pressure, humidity, and wind every half second, transmitting it all to the airplane before they hit the water.

By cranking dropsonde data into computer models that can simulate a storm and how it is likely to evolve, researchers have sharpened their forecasts of storm tracks. Three-day forecasts of Atlantic storm positions were off by an average of 708 kilometers (440 miles) in the 1970s; by 2005 the average error was 278 kilometers (173 miles). But one-day forecasts were still wide of the mark by an average of 113 kilometers (70 miles)—more than enough to keep coastal dwellers second-guessing the experts.

Shifts in Storm Intensity

Storm intensity is proving even harder to forecast. Three-day wind-speed forecasts, off by an average of 37 kilometers per hour (23 miles per hour) in the early 1990s, had improved only marginally by 2005.

Hurricanes regularly surprise observers with their mood shifts. In a matter of hours, a Category 5 storm (winds over 249 km/h) can fade to a Category 3 (179–209 km/h), or a mere tropical storm can explode into a killer. "Intensity changes are the things that really hurt people," says NOAA's Bell.

The state of the ocean below a storm explains some intensity shifts. In 1995, tropical storm Opal was inching toward Category 1 status—an entry-level hurricane—as it made its way through the western Gulf of Mexico. Then, in just 14 hours, it surged to Category 4. Satellite readings of the warm sea surface showed nothing unusual.

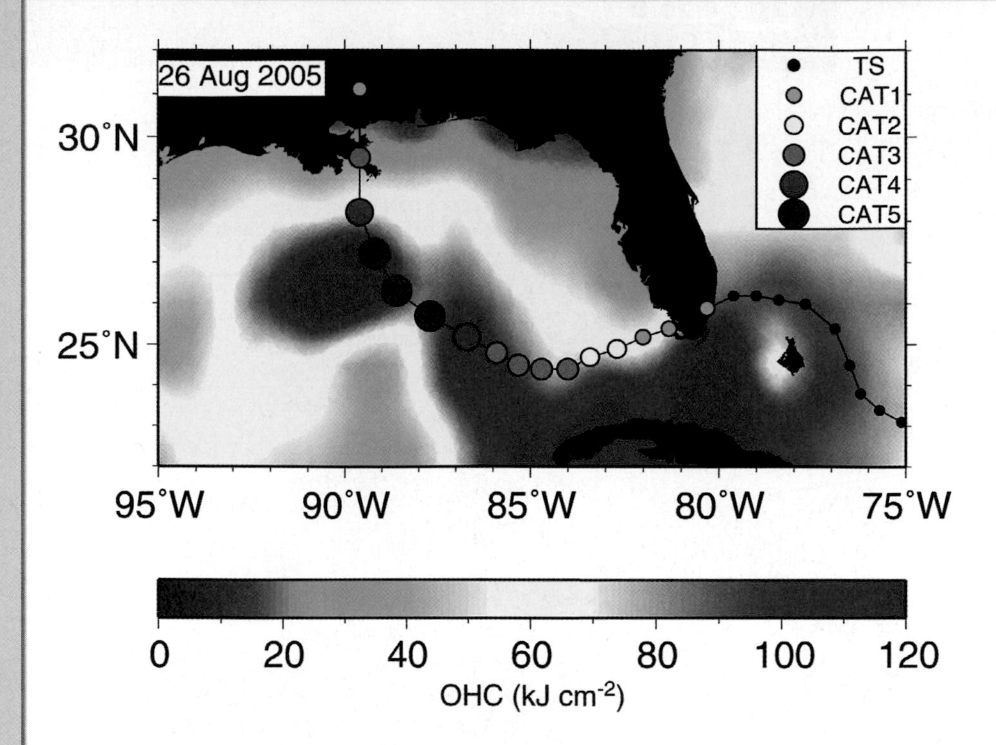

LEFT: Hurricanes Katrina and Rita strengthened dramatically when they crossed the Loop Current in the Gulf of Mexico. Ocean probes showed that the Loop Current's warmth extended to a depth of 91 meters (300 feet), increasing the supply of heat to the storms. The image shows the Loop Current three days before Katrina's landfall. The storm intensified as it traveled over warmer waters (dark red).

But Nick Shay of the Rosenstiel School and his colleagues discovered that the warm layer wasn't limited to the top few yards of the ocean, as it usually is in the Gulf. Cold water at greater depths acts as a brake on hurricane intensity when the winds churn it to the surface. But Opal had strayed across a pool of warm water extending hundreds of feet down. No matter how hard the wind blew, it stirred up more hurricane fuel, causing the storm to intensify.

The tropical ocean is littered with these deep warm pockets, and their importance was underscored last year by both Katrina and Rita, which shot up to Category 5 when they passed over a deep band of warm Gulf water called the Loop Current. Satellites can detect subsurface warmth by looking for subtle bulges in the sea surface, Shay says. "It's not really rocket science, but here's something that works and improves intensity forecasts by 5 to 15 percent."

Waves, on the other hand, can blunt a storm. Whipped up by a hurricane, they can reach heights of more than a hundred feet, exerting a drag on the winds that created them. "Heat adds fuel, but waves slow the winds down—they're fighting each other," says Shuyi Chen of the Rosenstiel School, who is collaborating on a powerful new computer model, called the Hurricane Weather and Research Forecasting model, that will simulate the fine details of the interplay between atmosphere, waves, and ocean. "You can get a forecast one to two categories wrong if you don't get the waves right."

Forecasters also need to understand a hurricane's internal workings. Katrina, for example, had grown into a certifiable monster by the morning of Sunday, August 28. Sucking energy from the Loop Current, the storm had screamed from the low end of Category 3 to a peak of 282 km/h (175 mph), well into Category 5, in just 12 hours. And then, swiftly and remarkably, the storm took a breather. In satellite images late Sunday, hours before landfall, a huge bite appeared in the southern side of the eyewall.

Scientists probing the storm with aircraft and radar in a project called RAINEX worked out what had happened. Katrina's ferocious rain bands had converged toward the heart of the storm, cutting off the eyewall's moisture supply. The old eyewall broke up and a new one formed farther out—an inertial brake that slowed the storm just as a skater's arms slow her spin when she thrusts them outward.

Storm Surges

From a washed-out stretch along the Mississippi coast, almost four months after Katrina, the view inland took your breath away. The once lush coastline was still a litter of debris. Water was the primary agent of destruction here. Most hurricane casualties come not from wind but from rain, waves, and, as the scene here made harshly evident, surge—the vast mound of seawater that is pushed in front of the storm, rising 8.5 meters (28 feet) or more in the case of Katrina.

Coastal waters are shallow, easily plowed up by inrushing winds. Bays and estuaries can funnel and intensify surge, while barrier islands and wetlands can buffer it. Coastal development weakens those defenses. Channels crisscross the marshlands, dredged for boat traffic. They let salt water into the back marshes, killing vegetation that holds them together. Add all the dikes and levees that hem in the Mississippi, cutting off the sediment that once replenished the marshes, and the result is staggering: More than 20 percent of Louisiana's coastal wetlands reverted to open water from the 1950s through 2000, 70 square kilometers (27 square miles) every year.

Global Warming Debate

Just over the horizon of scientific certainty lies the possibility that things might get worse. Kerry Emanuel, a meteorologist at the Massachusetts Institute of Technology, concluded that during the past three decades, the storms have grown almost twice as destructive. Emanuel's results, published weeks before Katrina, were soon joined by another study, led by Peter Webster of the Georgia Institute of Technology. Webster concluded that the strongest storms—Categories 4 and 5—have become nearly twice as common over 35 years.

But the claim that hurricanes are growing stronger as a result has set off a tempest of its own. William Gray of Colorado State University, a pioneer hurricane forecaster, has called it "plain wrong." He and the NHC's Christopher Landsea say Emanuel and Webster's statistics are fuzzy and that data on past storms can't be trusted. Until weather satellites became common in the 1970s, many tropical storms at sea went unrecorded, and since then changes in sensing technology have made it difficult to compare hurricane strengths.

ABOVE: The remains of an oceanfront house now stand in the ocean off Dauphin Island, Alabama. Ferocious and more frequent storms have made low-lying barrier islands more vulnerable.

Emanuel agrees that the data aren't perfect; "The only way to get a better answer would be to have a longer record of reliable data," which would make any trends stand out. To improve the record, Landsea has been analyzing hurricanes back to the mid-1800s, trying to gauge their intensity from accounts of storm surge and wind damage.

While the debates go on, hurricanes will continue to strike increasingly populous coasts. That, says Landsea, is reason enough to worry. "The changes in society are as important, if not more important than global warming, or even natural cycles," he says. "When you double some vulnerable populations every 20 to 30 years, that's what's going to cause disasters. We've got a huge problem even if hurricanes don't change at all."

eXpeditions activity

For each of the following factors—ocean temperature, wind shear, waves, land barriers, and moisture in the air—describe how lesser or greater measures affect the intensity of hurricanes.

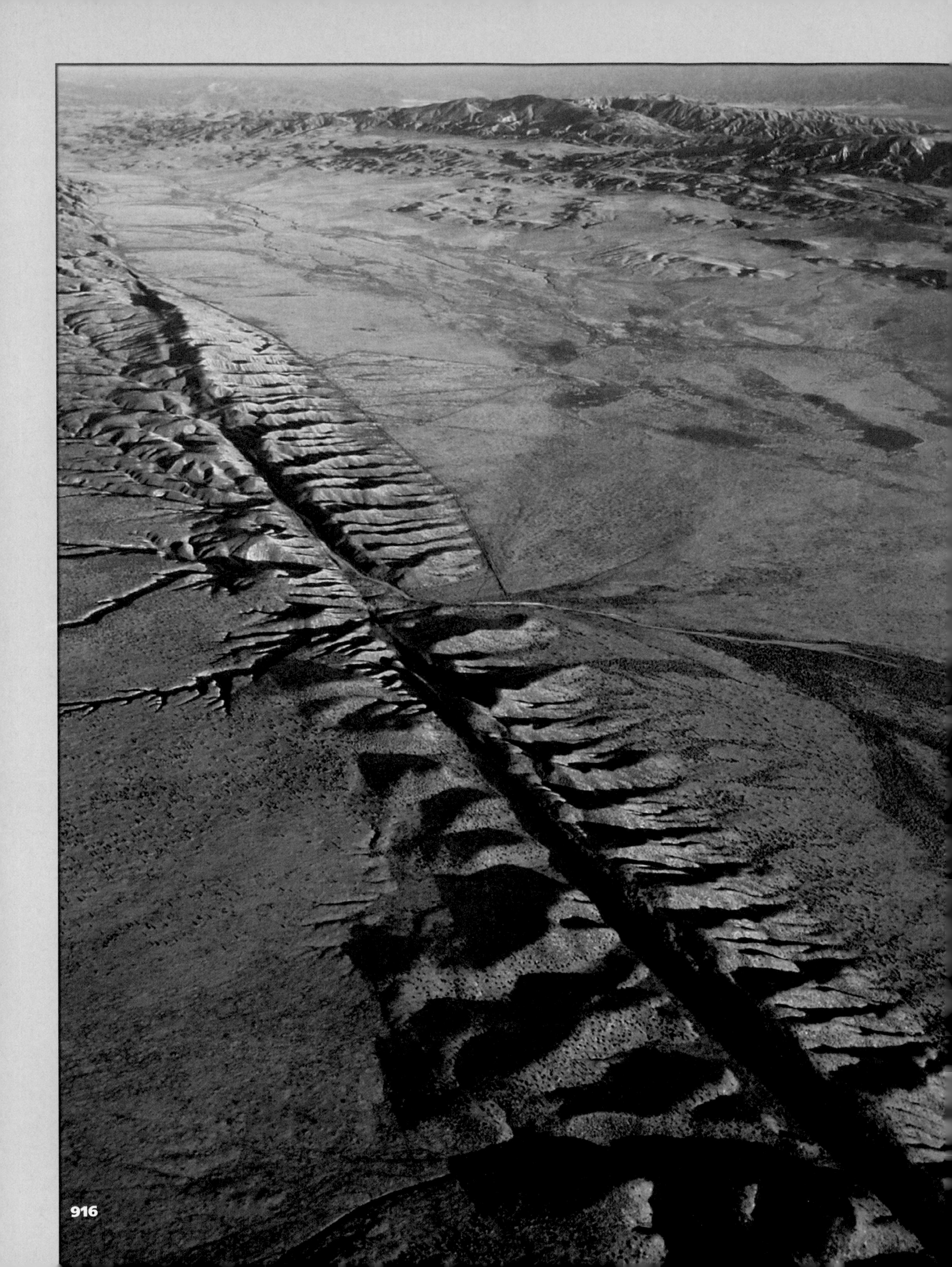

Where on Earth will it strike?

The Next Big One

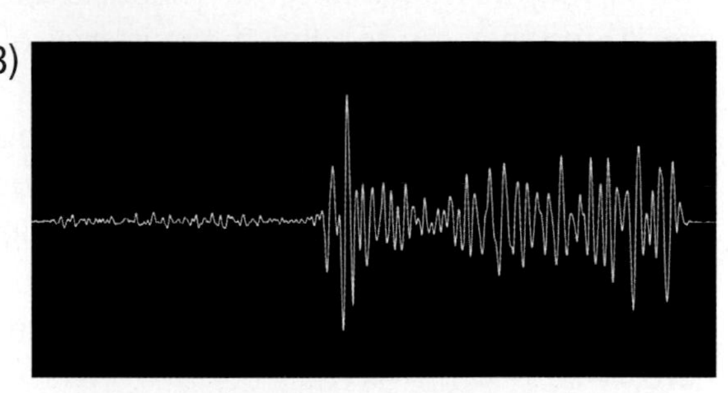

RIGHT: (A) Fires consume San Francisco after the 1906 earthquake.
(B) A seismograph of the event, occurring at 5:12 A.M., April 18, 1906, at a magnitude of 7.8, was recorded in Germany, 9,012 kilometers (5,600 miles) away.

LEFT: Part of the San Andreas Fault can be seen from the air above Carrizo Plain in south-central California. Thousands of years of earthquakes have left a zippered scar.

It's been a hundred years plus since the last big one in California, the 1906 San Francisco earthquake, which helped give birth to modern earthquake science. A century plus later, we have a highly successful theory, called plate tectonics, that explains why 1906-type earthquakes happen—along with why continents drift, mountains rise, and volcanoes line the Pacific Rim. Plate tectonics may be one of the signature triumphs of the human mind. And yet scientists still can't say when an earthquake will happen. They can't even come close.

Other Questions

Some of the simplest questions about earthquakes remain hard to answer. Why do they start? What makes them stop? Does a fault tend to slip a little—telegraphing its malign intent—before it breaks catastrophically? Why do some small quakes grow into bigger quakes, while others stay small?

And there's the broader question: Are there clear patterns, rules, and regularities in earthquakes, or are they inherently random and chaotic? Maybe, as Berkeley seismologist Robert Nadeau says, "A lot of the randomness is just lack of knowledge." But any look at a seismic map shows that faults don't follow neat and orderly lines across the landscape. There are places, such as Southern California, where they look like a shattered windshield. All that cracked, unstable crust seethes with stress. When one fault lurches, it can dump stress on other faults. UCLA seismologist David Jackson, a leader of the chaos camp, says the field of earthquake science is "waking up to complexity."

Yet at the moment, earthquake prediction remains a matter of myth, in which birds and snakes and fish and bunny rabbits somehow sniff out the coming calamity. What scientists can do right now is make good maps of fault zones and figure out which ones are probably due for a rupture. And they can make forecasts.

A forecast might say that, over a certain number of years, there's a certain likelihood of a certain magnitude earthquake in a given spot. And that you should bolt your house to its foundation and lash the water heater to the wall.

Turning forecasts into predictions—"a magnitude 7 earthquake is expected here three days from now"—may be impossible, but scientists are doing everything they can to solve the mysteries of earthquakes.

They break rocks in laboratories, studying how stone behaves under stress. They hike through ghost forests where dead trees tell of long-ago tsunamis. They make maps of precarious, balanced rocks to see where the ground has shaken in the past, and how hard.

They dig trenches across faults, searching for the active trace. They have wired up fault zones with so many sensors it's as though the Earth is a patient in intensive care. Surely, we tell ourselves—trying hard to be persuasive—there must be some way to impose order and decorum on all that slippery ground.

BELOW: When a fault like the San Andreas ruptures, adjacent plates in the Earth's crust suddenly lurch past each other. Two kinds of vibrations radiate from the hypocenter, the spot deep in the Earth where the rupture originates: P waves, which compress and stretch the rock as they go, and S waves, which shake the rock from side to side. Seismic stations around the world record these vibrations, allowing geologists to locate the hypocenter and the surface location directly above it, the epicenter. At the Earth's surface, P and S waves produce the heaving and rolling surface waves of an earthquake, which are usually most severe along the rupturing fault.

Studying the San Andreas

We've been trying ever since the Earth humbled San Francisco. In April 1906 the city was the commercial and financial powerhouse of the West with some 400,000 citizens. All that changed at 5:12 A.M., April 18. Something happened deep under the seafloor just off the Golden Gate, out near the shipping channel. Along an ancient crack in the Earth, two slabs of rock began moving in opposite directions. An earthquake will unzip a fault at two miles per second. This one broke north and south. In some places the slip was just 1.8–2.1 m (6–7 feet), but elsewhere the ground lurched fully 4.9 m in a snap. The fault broke for 435 kilometers (296 miles), from Shelter Cove, way up in the redwood country of northern California, all the way south to the old mission town of San Juan Bautista.

It wasn't the worst earthquake in history, but it was sensational. It inspired a kind of war on earthquakes, using the weapons of science. Until the San Francisco earthquake, geologists weren't sure how earthquakes and faults were connected. Many believed that faults were the by-products of earthquakes, not their source.

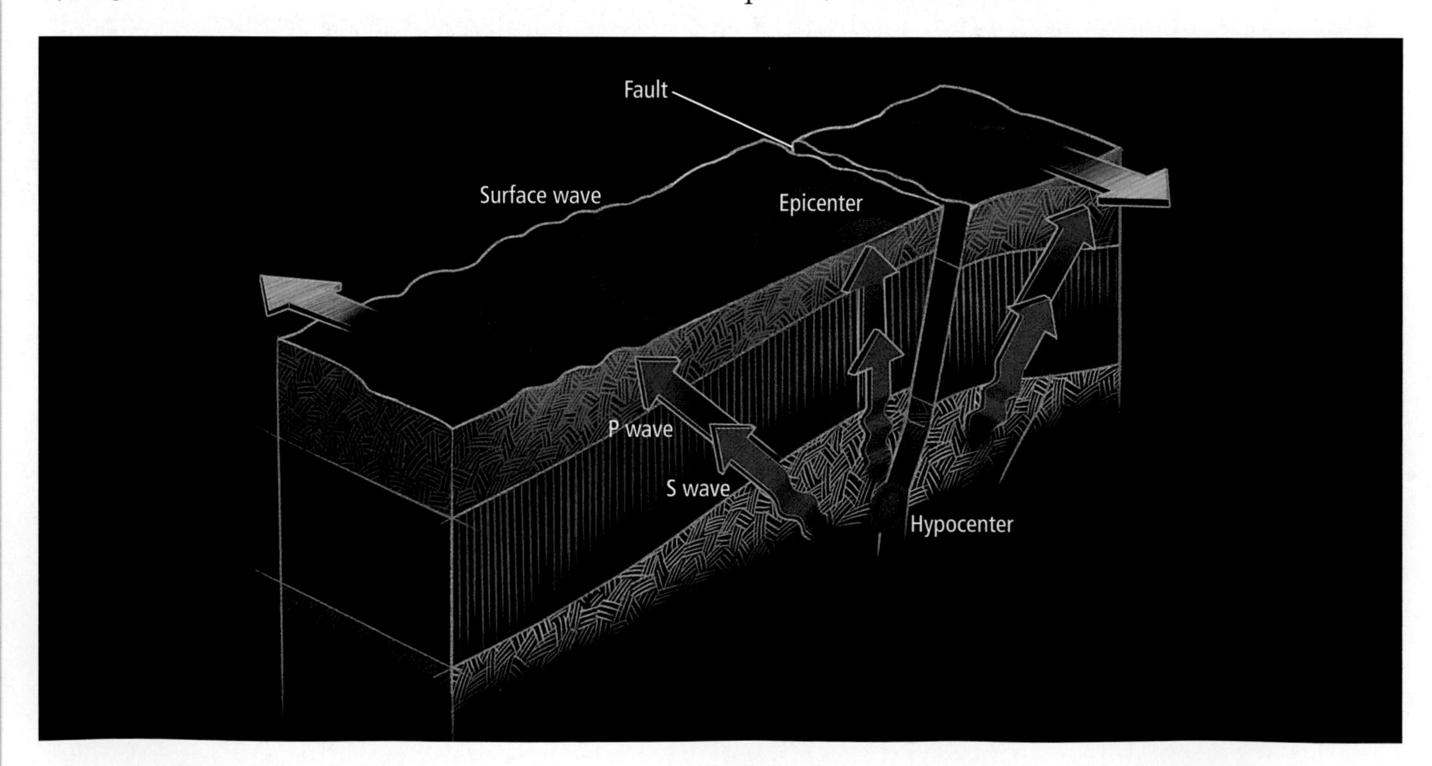

The great Berkeley geologist Andrew Lawson had discovered the San Andreas Fault more than a decade earlier, naming it after the San Andreas Valley. But he thought it was just a trivial thing not much more than a dozen miles in length, responsible for the narrow valley that holds San Andreas Lake and Crystal Springs Reservoir on the San Francisco Peninsula.

But earthquakes are teachable moments. When the fires died down and San Francisco started to rebuild, Lawson and a team of colleagues set out to solve the mystery of the Great Earthquake. They literally walked the "mole tracks" where the fault rupture had churned across barnyards and meadows. Then they continued south for 966 kilometers (600 miles), reading the landscape, discovering the unbroken sections of the fault. This fault just kept going and going, all the way down past Los Angeles.

In the course of the investigation, a scientist named Harry Fielding Reid figured out why earthquakes happen. Reid studied all the reports of ground motion, of roads and fence lines offset by the fault, and came up with the key concept of "elastic rebound." The surface of the Earth isn't perfectly stiff. It bends. Land at some distance from a locked fault will slowly stretch in opposite directions, but the fault itself will remain locked, under increasing strain. Finally the fault breaks, and the land springs back violently, releasing accumulated strain.

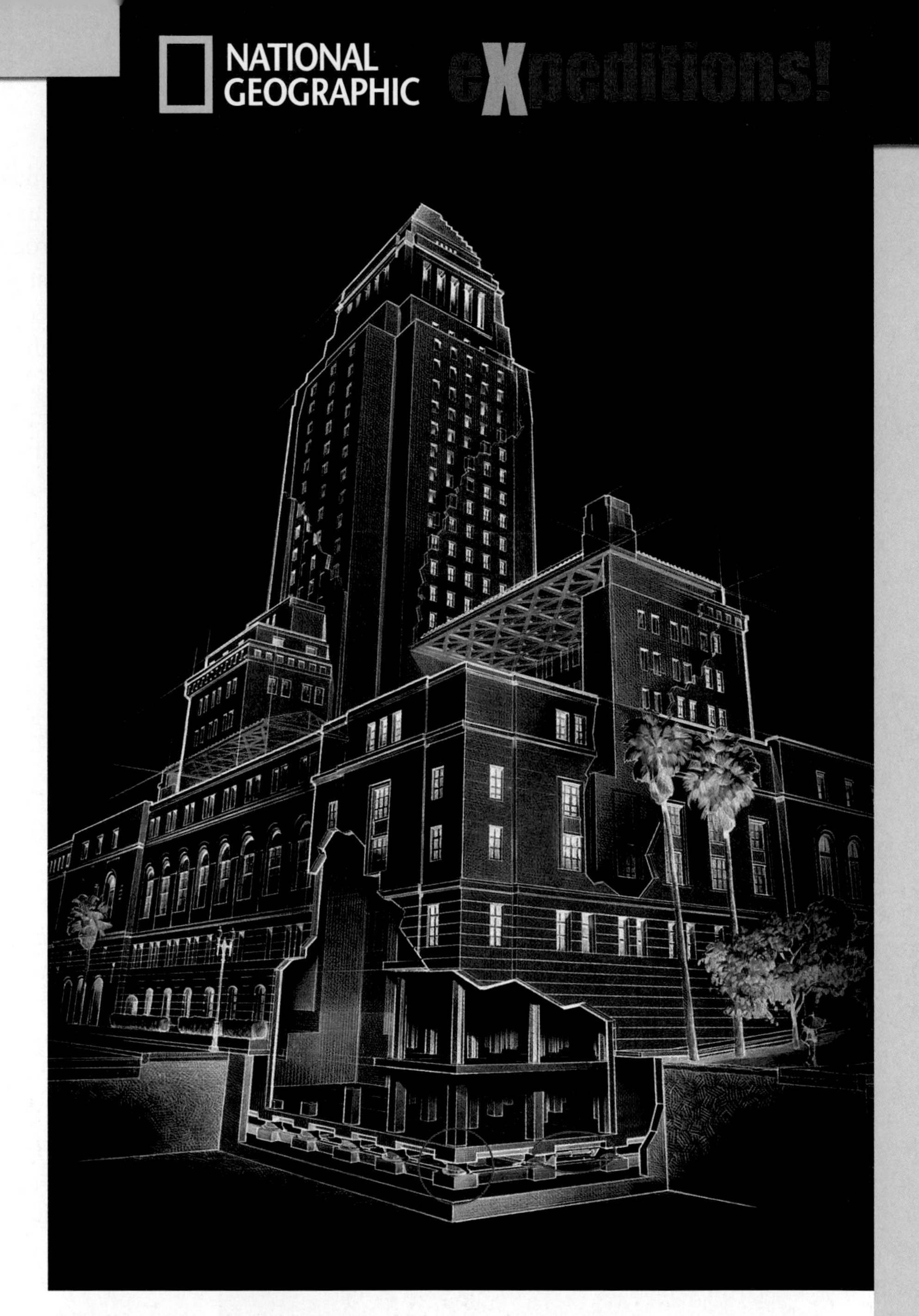

ABOVE: The Los Angeles City Hall suffered severe damage during the 1994 Northridge earthquake as cracks opened throughout the building and exterior tiles crashed to the ground. A three-year restoration and safety upgrade included reinforcing walls in the tower with steel bars and connecting the north and south wings to the main structure with steel bracing. The most dramatic changes were made at the base: Engineers dug a moat around the building to separate it from the heaving ground and installed a suspension system on the foundation to reduce shaking in future quakes.

BELOW: The Yasaka Pagoda in Kyoto, Japan, has survived more than five centuries of earthquakes. After a jolt, the building sways as each story rocks independently around a central, anchoring pillar. With a new appreciation of this ancient design, engineers are now adapting it for modern use. In other ways, the Japanese are preparing for earthquakes. Under Tokyo's streets, construction crews prepare quake resistant ducts for the city's water, electricity, telephone, and sewage lines. By minimizing damage to the main utility lines, services should be able to be restored quickly after an earthquake.

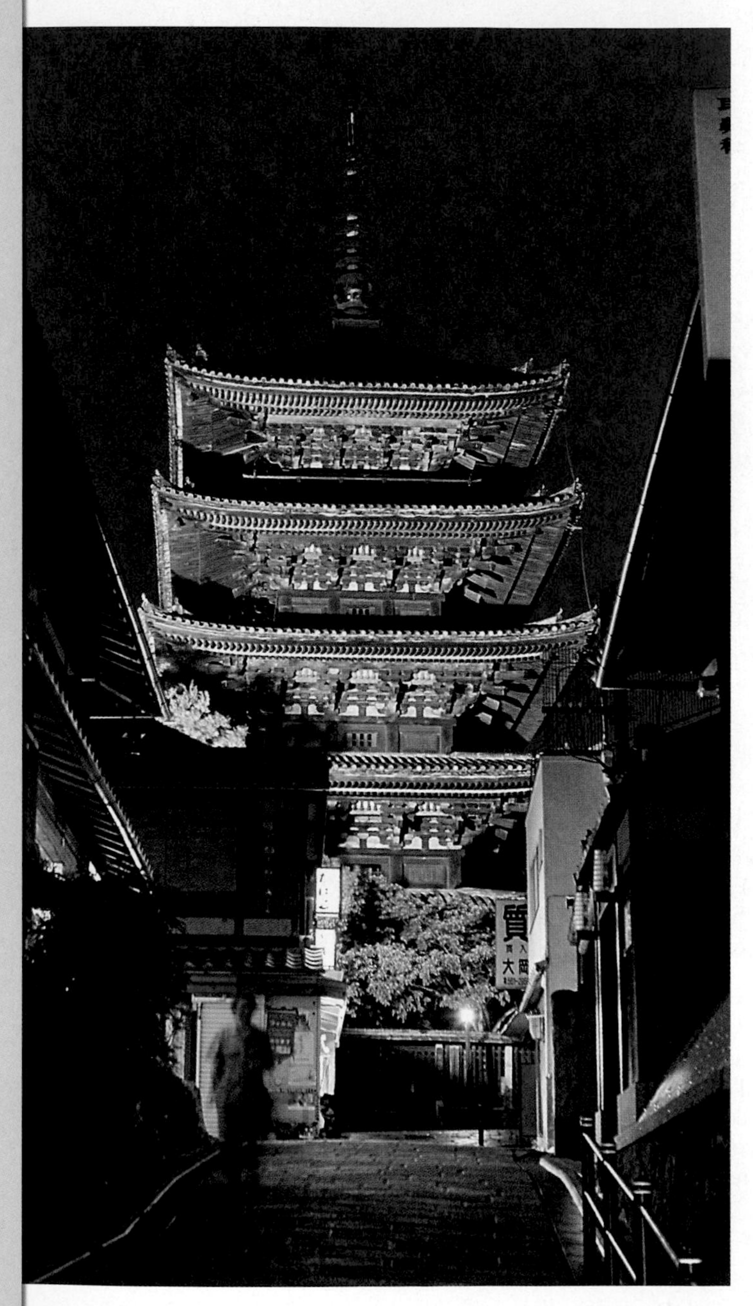

An earthquake, says Bill Ellsworth of the U.S. Geological Survey in Menlo Park, California, is "a relaxation process" —from the standpoint of the planet at least. Lawson, Reid, and their colleagues had no way of understanding the ultimate source of the forces behind earthquakes.

But by the late 1960s, scientists had come to realize that the Earth is divided into about 15 plates of crust, constantly shifting as new rock forms at mid-ocean ridges and old crust dives into the Earth's interior at subduction zones in the deep sea.

Suddenly the Himalayas were revealed as a crash site, with India slamming into Asia. And the San Andreas was not just a long strike-slip fault: It was a plate boundary, where the North American and Pacific plates grind slowly past each other at a rate—precisely measured by GPS—of two inches a year.

But except for a section called the "creeping zone" in central California, the San Andreas is locked. Around San Francisco, the fault hasn't budged since 1906. North of Los Angeles, a long stretch of the fault has been stuck since 1857. Near Palm Springs, there's been no action on the fault since about 1680. At some point the San Andreas will have another relaxation event. When that happens, despite all the measurements and all the scientific conferences, nearly everyone will be caught by surprise.

Carol Prentice, a geologist with the U.S. Geological Survey in Menlo Park, would love to know when, exactly, the San Andreas had a major quake prior to 1906. You sometimes read that the San Andreas breaks every 150 years or 200 years or 250 years, but that is not hard data. That's an informed guess.

On the Point Reyes Peninsula, a knuckle of land north of San Francisco, Tina Niemi is digging for an answer. In the compacted sediment and peat of a trench dug across the fault trace, the University of Missouri geologist can discern a faint fracture, a line that slants across the trench wall from upper left to lower right. The line isn't perfectly straight; it jogs and splays. Along with other clues, these kinks suggest that something has jolted the soil here as many as 12 times over the past 3,000 years.

Niemi doesn't see any simple pattern to the quakes—not in time, not in magnitude. "Our data support more of a model for irregular occurrence," she says.

Nearby faults add another level of uncertainty. High in the Santa Cruz Mountains near Palo Alto you can stand on the San Andreas not far from the epicenter of the 1989 magnitude 6.9 Loma Prieta earthquake. That quake was strong enough to destroy freeways and bridges and kill scores of people, but it never ruptured the surface. To this day, no one is sure how much of the quake to blame on the San Andreas and how much on other, unknown faults.

A Closer Look

"With faults, you don't have the luxury of tinkering under the hood to see what's what," writes USGS seismologist Susan Hough in her book *Earthshaking Science*. But some scientists want to sneak a look. Their idea: Drill the San Andreas. Find the biggest oil drilling rig in California and ram huge steel pipes into the depths of the fault and send a bunch of gadgets down there to sample the rock and record its twitching.

The project is under way near Parkfield, a village in a dusty central California valley. Parkfield's claim to fame is earthquakes. At the Parkfield Cafe there's a sign that says, "If you feel a shake or a quake get under your table and eat your steak." The quakes aren't actually very strong here. They tend to be magnitude 6.

There has been a string of them. After the M6 in 1966, scientists realized that these quakes had occurred fairly regularly, roughly every 22 years, and so in the early 1980s the notion arose that there ought to be another Parkfield quake around 1988.

Scientists wired the fault every which way, hoping to detect signs of building strain, moving water, or some other quake precursor. But year after year, the quake refused to show. It became something of an embarrassment for everyone who argued that earthquakes follow patterns.

Finally, on September 28, 2004, an M6 struck near Parkfield, although its epicenter was miles farther south than expected. A camera had been set up to catch the fault rupturing from north to south, but it broke from south to north. "We missed Parkfield by over ten years—and that was an earthquake in a barrel," said UCLA's David Jackson, he of the chaos camp.

Most disappointing to scientists was the lack of any precursors. They pored over the data and could find no evidence of anything unusual on the fault prior to the September 28 rupture. Maybe there was a very tiny change in crustal strain a day before the quake—but even that wasn't certain. The unsettling notion arose that the jig was up, that these things are just flat-out unpredictable, random, weird.

But science marches on—and digs deeper. At Parkfield there are still seismometers and GPS stations everywhere, and now there's even that 56-meter (185-foot) oil-drilling rig. By late summer 2005 it had punctured the fault and reached its terminal depth of two miles.

"In a sense we're testing the predictability of earthquakes," says Mark Zoback of Stanford University, part of the drilling team. Of the chaotic versus linear debate, he says, "we're the guys who are trying to find out which side is right." His rig is the next best thing to sending a person down into the fault directly, although even the rig can't get instruments down to the six-mile depths where many large earthquakes start.

eXpeditions activity

Research earthquakes greater than an M6 that have occurred in the last year. Place these on a world map. Identify those that have been forecast and those that have been surprises. Were any of them predicted? Discuss your findings with other students and develop a consensus of whether the idea of chaos in earthquake occurrence or a linear pattern seems to exist.

Jewels in the Ash

A thick rain of volcanic ash sends a rat-sized mammal (*Gobiconodon zofiae*) and three dinosaurs (*Dilong paradoxus*) fleeing for their lives. Based on the features of a 125-million-year-old fossil preserved by such ashfalls, these tyrannosaurs exhibit a downy covering of protofeathers, the first found among their family. The evolutionary precursors of true feathers, protofeathers were hairlike and probably developed for insulation. Frightful teeth of *Dilong paradoxus* resemble those of *Tyrannosaurus rex,* its more recent and much larger cousin. One of the oldest tyrannosaurs yet found, *D. paradoxus* was unearthed in China's Liaoning Province. Preserved in clay, silt, and ash, a stunning array of dinosaurs, birds, mammals, fish, and insects make the fossil bed among the most prolific in the world.

LEFT: Guarding long-buried treasure, a monitor at a dig site looks out for thieves who illegally remove fossils for sale to collectors. He sleeps in the one-room shack on the rim, keeping an ear cocked for voices in the dark.

Liaoning Opens a Window on the Mesozoic

In Liaoning Province, located in the rolling farm country of northeastern China, peasant farmers make only a few hundred dollars a year. They know they can make many times that amount by selling just one prized fossil on the black market. Even if recovered, illegally removed fossils have diminished scientific value, says Xu Xing of the Institute of Vertebrate Paleontology and Paleoanthropology (IVPP) in Beijing: "If it isn't collected right, a fossil loses its context—the layer it was found in and its relationship to other fossils."

Xu, also a postdoctoral fellow at the American Museum of Natural History in New York, waxes enthusiastic about what has become one of Earth's most celebrated fossil beds. Discoveries there are casting light on life during the Mesozoic, specifically 130 million to 110 million years ago—a time distinguished by the diversification of dinosaurs, mammals, birds, and flowering plants.

"Liaoning opens a window on the late Mesozoic that is more complete and more in-depth than anywhere else on Earth," Xu asserts. The reason is the diversity and great abundance of terrestrial plants and animals and their fossilization. At most sites only bone can be found. At Liaoning the fine particles of ash and mud that covered animals preserved soft body parts and prevented decomposition by sealing off oxygen.

Some scientists call Liaoning a Mesozoic Pompeii, evoking the ancient Roman city where humans were entombed by the eruption of Mount Vesuvius. But in its own way Liaoning is even more remarkable. Repeated volcanic eruptions created a layer cake of fossil beds spanning millions of years. So far, more than 60 species of plants, nearly 90 species of vertebrates, and about 300 species of invertebrates have been identified. Paleontologists marveled at dinosaur fossils with stomach contents identifiable as the bones of lizards and mammals, and at bird fossils containing plant seeds.

Predator-Prey Relationships

Liaoning is situated within a vast region whose primeval flora and fauna are referred to as the Jehol biota. The area was characterized by a warm climate and numerous lakes. These conditions provided a fruitful environment for plants and animals to differentiate and flourish. So many individual fossils have been found that scientists are able to study population dynamics, succession within communities of interacting species, and even predator-prey relationships.

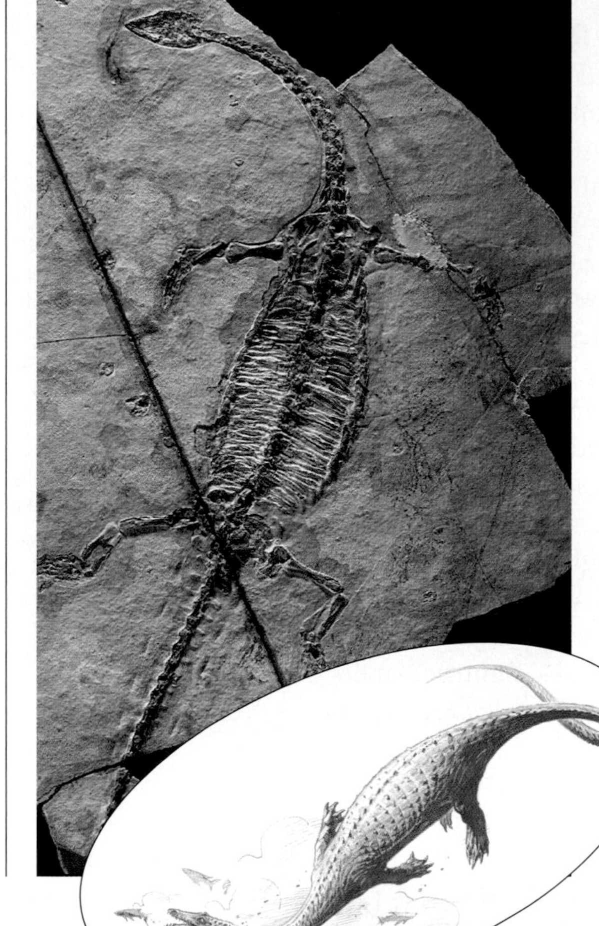

LEFT: A small fish lies next to the head of a fossilized specimen of *Hyphalosaurus lingyuanensis,* as if predation had been cut off by the sudden death of both predator and prey. That the animal, just under four feet long, was indeed a fish-eater is suggested by its small head, needlelike teeth, and pointed snout. *Hyphalosaurus,* which inhabited freshwater, bears a resemblance to *Plesiosaurus,* a marine creature, in an example of convergent evolution—when distantly related creatures evolve similar traits in similar environments.

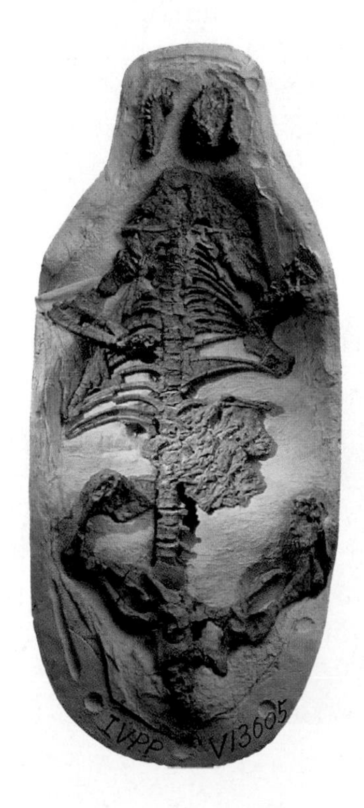

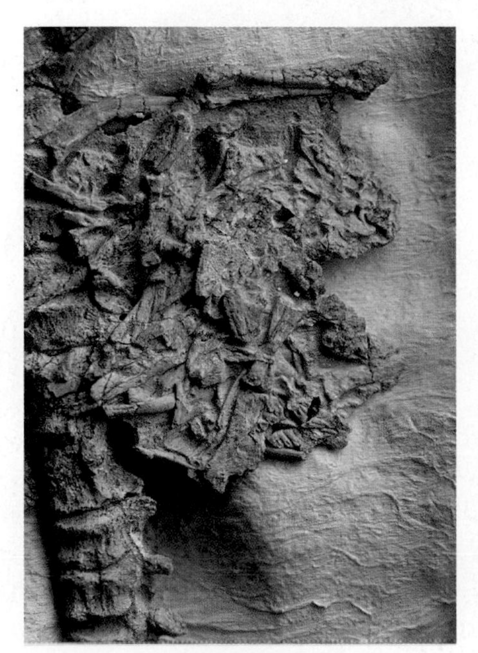

LEFT: Scientists who have long known that dinosaurs preyed on mammals were stunned by the discovery of a fossil that turned the tables. Found by villagers, the cat-sized mammal's skeleton contains the remains of its last meal: a young beaked dinosaur called *Psittacosaurus.* Uncertainties remain about the behavior of *Repenomamus robustus,* says Hu Yaoming of the IVPP. "Did it catch the dinosaur after a pursuit? If so, that suggests it was active in daytime when the dinosaur would also be active. But if it was a scavenger, it could have eaten it day or night."

With large pointy teeth and powerful jaws designed for catching and ripping prey, *R. robustus* shows that Mesozoic mammals could compete with the smaller dinosaurs for territory—and food. Its stomach contents, shown in the inset at the right, included a hind limb of the young beaked dinosaur, along with a forelimb and several teeth. These bones are well articulated—still joined to other bones in their natural positions—suggesting that *R. robustus,* rather than chewing, swallowed food in chunks.

Bird-Dinosaur Relationships

"The site preserved not just bones but often whole skeletons," says paleontologist Hans-Dieter Sues of the Smithsonian Institution, "and some birds were preserved so well you can distinguish between male and female. Liaoning is unique."

During the 1990s Liaoning jumped from the pages of scholarly journals onto front pages everywhere through a series of spectacular discoveries of archaic birds and—more intriguingly—dinosaurs with feathers. These fossils bolstered the once controversial but now widely endorsed theory that modern-day birds descended from dinosaurs. They also provide much new evidence in the ongoing debate about how flight originated.

Fossils are being uncovered faster than paleontologists can describe the specimens and spread the new knowledge through scientific papers. And Liaoning promises to provide fresh discoveries for many years to come.

BELOW: In death a pterosaur rests with a wing bone in its mouth, perhaps from the natural collapse of the wing, perhaps, as some scientists speculate, from a struggle before volcanic gases snuffed out its life. The pointed beak and the sharp and slender front teeth suggest that it preyed on fish.

Long before bird and bats took wing, the skies were ruled by pterosaurs, reptiles that were Earth's first flying vertebrates. They arose 230 million years ago during the later part of the Triassic Period and thrived for 165 million years until going extinct at the end of the Cretaceous Period.

The wingspan of *Hoopterus gracilis* was nearly four and a half feet, easily exceeding the ten-inch span of *Pterodactylus elegans,* perhaps the smallest pterosaur, but falling far short of *Quetzalcoatlus northropi,* whose wings stretched at least 11 meters (36 feet) tip to tip. The pterosaurs' light, hollow bones aided takeoff. But their fragility also made fossilization difficult—best achievable in the soft ooze of the seafloor or lake beds like those at Liaoning. The discovery of *H. gracilis* and other pterosaurs at Liaoning, some preserved with body coverings of fuzz, extends the known range of pterosaurs.

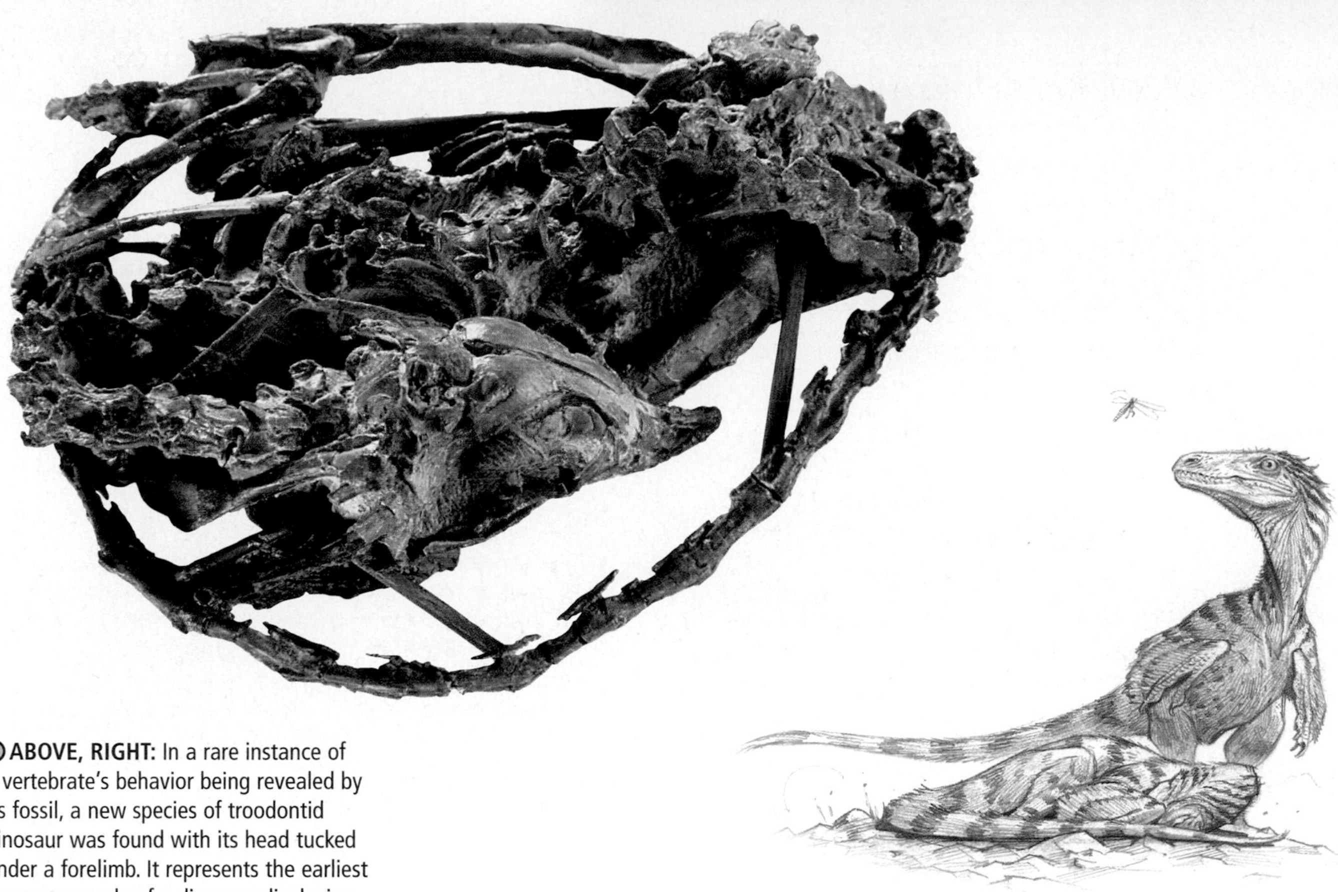

ABOVE, RIGHT: In a rare instance of a vertebrate's behavior being revealed by its fossil, a new species of troodontid dinosaur was found with its head tucked under a forelimb. It represents the earliest known example of a dinosaur displaying the sleeping posture exhibited by modern-day birds. The "tuck-in" pose would have preserved body heat, suggesting that, like birds, at least some dinosaurs were warm-blooded. Delighted paleontologists named the pigeon-sized creature *Mei long,* meaning "soundly sleeping dragon." Though it is not known how the dinosaur died, it probably was killed instantly by a thick deposit of volcanic ash or by volcanic gas followed by a covering of ash and mud. *Mei long* adds to the ever increasing evidence of bird-dinosaur kinship.

eXpeditions activity

Reread the article and make a chart of recent discoveries made at the Liaoning fossil site, naming the discoveries, evidence, and new knowledge, as in the example given. Compare your chart with that of a partner. Discuss any differences and how your charts could be changed to be more accurate.

Discovery	Evidence	New Knowledge
new species of troodontid dinosaur	skeleton with head tucked under forelimb	earliest known example of posture of modern-day bird, suggesting some dinosaurs were warm blooded

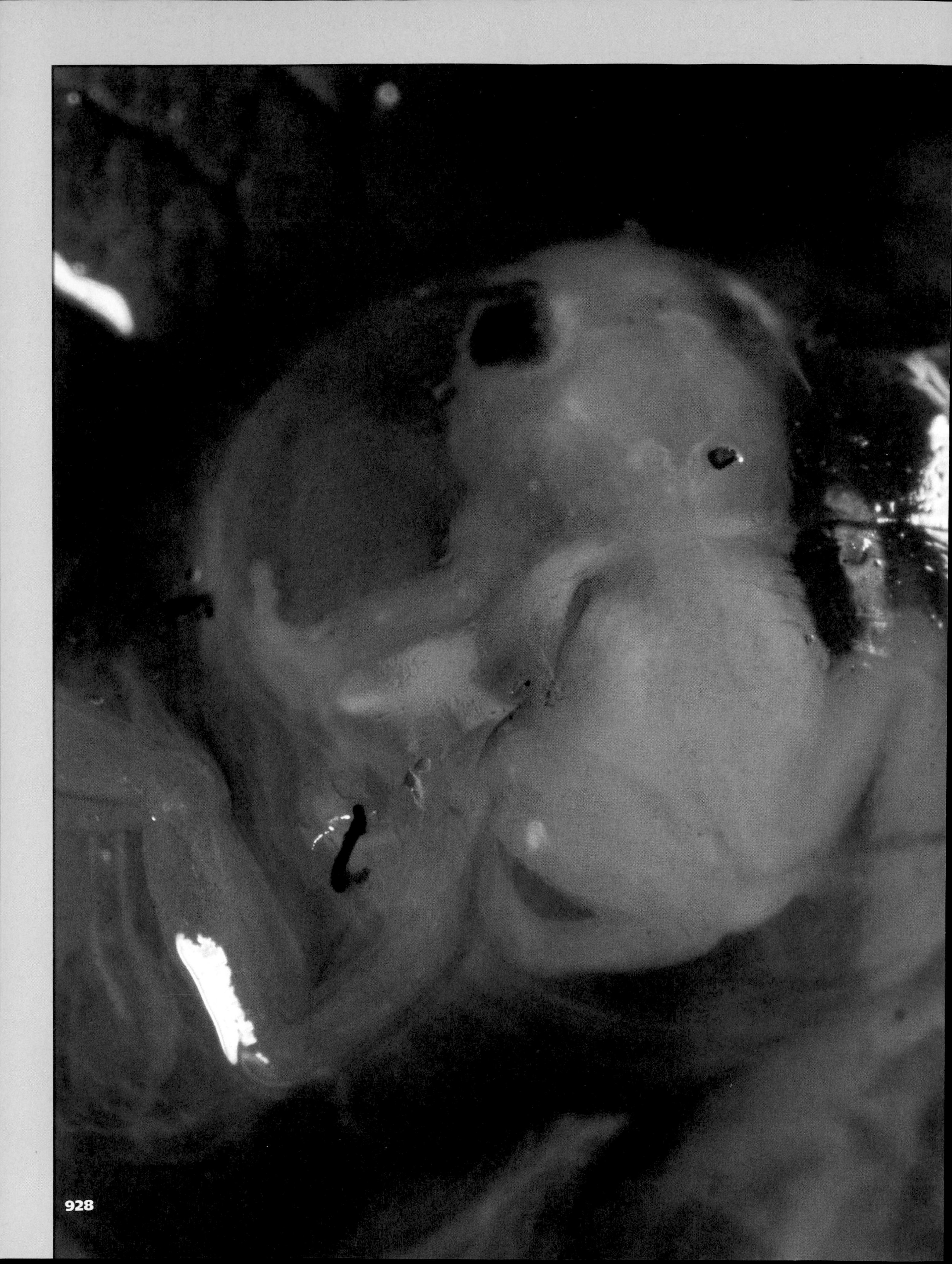

The Fragile World of Frogs

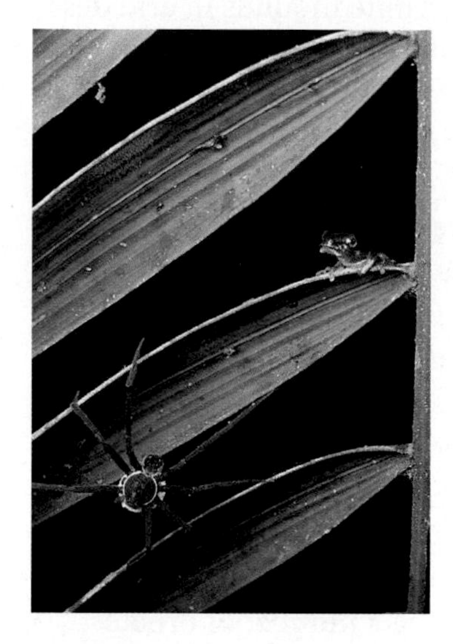

ABOVE: This young red-eyed tree frog *(Agalychnis callidryas)* has newly emerged from a pond. As a tadpole, this frog survived dragonflies, water beetles, fish, and shrimps. As an adult, it must evade bats, birds, snakes—and humanity's hunger for land.

LEFT: Life and strife begin together as a tiny froglet *(Oreophryne sp.)* hatches fully formed, bypassing the tadpole stage. Miraculously diverse—and in serious decline—frogs signal that something's amiss in the natural world.

Stephen Richards, an Australian herpetologist, has spent a good deal of his life traversing the woods of Papua New Guinea (PNG) and Queensland, Australia, searching out frogs—and has dozens of new species to show for his efforts. "If you want to catch frogs, you're going to be out at night." But not just any night. Rain is generally a necessity too. To a frog the steady spatter is a signal that it's time to find a mate. Male frogs gather near streams and ponds to call to the females. The males' songs have another, coincidental use: They help scientists like Richards find them. Richards said, "If they're not calling, they can be very difficult to find." Finding frogs has always been a basic herpetological skill. But in the past two decades that skill has taken on added importance. Increasingly it has become a key tool for monitoring the health of frog populations and species and for helping to identify the causes of some species' precipitous decline. Here, in PNG, the frogs seem healthy, their populations stable. "But what do we really know about their populations?" asked Richards. "We're just now beginning to identify the species that live here. That's the situation in many places. We could be losing frogs that we've never seen."

Variations in Frog Habitats

Like a man on a mission, Richards waded into a stream, clambered up a waterfall, and disappeared into the woods. Only moments later he reappeared near the top of the fall. "Look at this!" he shouted over the rushing water. "[This female had] started her journey to the stream, where she'll lay her eggs," he said. Her chosen male will fertilize the eggs as they emerge from her cloaca. "But look where she plans to put her eggs—here in this rushing current. It seems the most inhospitable place for a frog to lay eggs or for tadpoles to live."

The torrent-stream frog's tadpoles thrive in this tumultuous environment because they have evolved enlarged mouths with suction lips that enable them to stick to the surface of a rock while grazing on the algae growing there. Where the going was tough for an animal like a frog, they'd been extremely successful, so much so that there were numerous torrent-stream species here. One of the joys of frog-watching is discovering the many variations frogs have evolved for the basics of life.

Frogs were always turning up in places—dry deserts, cold mountains, plunging waterfalls—where they shouldn't be. Their ancestors, the first amphibians, arose some 350 million years ago. Amphibians were the first animals with backbones to walk on land. Some were large, crocodile-size creatures, but over time they evolved into the three orders of smaller animals known today: the Anura (frogs and toads), Caudata (salamanders and newts), and Gymnophiona (caecilians, wormlike creatures that live in leaf litter or streams).

Anurans, the frogs and toads, are the most successful living amphibians, with some 4500 species now known, a count that's been steadily growing because in the past few years scientists have ferreted out more than 50 new species annually. Like most amphibians, frogs spend at least part of their lives in water. It might seem that that would limit where they can live, but frogs are found from Arctic tundra to the driest of deserts and from sea-level mangrove swamps to the 18,000-foot-high Tibetan Plateau.

Newly laid green eggs of the red-eyed tree frog, adjoin a clutch whose embryos will soon hatch. If attacked by snakes or wasps, the tadpoles can pop out early, escaping to water below.

To survive in such extreme environments, frogs have evolved an impressive range of adaptations. The North American wood frog (*Rana sylvatica*), for instance, can survive freezing temperatures for as long as seven months, relying on a natural antifreeze in its blood to protect its organs. Some species in the dry forests of South America secrete a waxy coating to protect themselves from drying out, while the water-holding frogs of Australia store water in their bladder and under their skin for use during droughts. Another Australian frog, shaped like a fat turtle, spends most of its life burrowed beneath termite mounds in arid deserts, where it feeds on nothing but termites.

Reproductive Behaviors

Equally intriguing are frogs' reproductive behaviors. Many frogs and toads have a multi-stage life, that passage from tadpole to adult called metamorphosis. There are at least a dozen ways frogs can make that journey. The standard method begins with fertilized eggs in a pond; the tadpoles hatch, feed on algae, and change into frogs. Many other species have evolved methods that bypass the pond.

In Queensland's Paluma Range National Park, Adam Felton, a colleague of Richards's at James Cook University, has been watching a number of ornate nursery frog (*Cophixalus ornatus*) males. "*Cophixalus* is a microhylid," Felton explains, "and all Australian microhylids have what we call direct development." In other words the microhylid froglets develop inside the egg. Felton shines his headlamp on a patch of earth. "See this little hollow? That's the male's nest, and those little clear balls are the eggs he's guarding—or should be, if he was here." Dad should be hunkering down on top of those eggs. "That protects them, keeps them moist, and may even prevent fungi and bacteria from growing on them." In South America, male Darwin's frogs slurp up the fertilized eggs and hold them in their vocal sacs until the froglets emerge.

In other species females provide the parental care. The female poison dart frog (*Dendrobates pumilio*) not only attends her eggs but also transports the hatched tadpoles to pools in tree holes or within the inner leaves of bromeliads. She visits her offspring to lay unfertilized eggs for them to eat. The female Surinam toad (*Pipa pipa*), an aquatic species, converts her entire back to a nursery. She and the male swim end over end, transferring the fertilized eggs to pouches in her back. Her skin grows over the eggs, sealing them in until the froglets hatch.

Even tadpoles left on their own sometimes have protective devices. In Panama, Stan Rand, a herpetologist with the Smithsonian Tropical Research Institute, guides me to a pond where the tree frog *Agalychnis callidryas* has laid masses of gelatinous eggs on palm fronds. These are a favorite food of an arboreal snake, but when the snake bumps the eggs, the tadpoles hatch prematurely.

"It's a 50-50 chance for them," says Rand. "They'll still face predators in the pond, but if they don't hatch, they'll be eaten by the snake."

As wonderful as all these adaptations are, none is—or was—as remarkable as that of Australia's gastric-brooding frogs, *Rheobatrachus*. Researchers invariably mention the two species in this genus as the most astonishing example of what frogs can do. The females of this two-inch-long stream dweller swallowed their fertilized eggs or tadpoles, shut down their digestive systems, and hatched their young in their stomachs. About a month later the mother opened her mouth and regurgitated her tiny froglets.

ABOVE: No one has seen a gastric brooder in the wild since 1981, and none are in captivity. They are apparently extinct.

Die-offs in Australia

Keith McDonald, the chief ranger with Queensland Parks and Wildlife Service, had helped monitor the two known populations of gastric brooders shortly after they were first discovered about 25 years before. "I'd been watching this population; I went back three months later," McDonald recalls. He searched for frogs; none were to be found.

Since the 1970s, more than a dozen Queensland frog species, especially the stream-dwelling types, have experienced sudden, massive die-offs. In some cases, such as in these remote Queensland mountains, certain frog populations vanished in a few short months. Something in the environment was adversely affecting frogs, but no one was certain what it was or how many factors were to blame.

Now, after intensive study and monitoring programs, researchers have some strong clues. Some of the best leads for the mass deaths have come from the Queensland rain forests. Richards and others realized something was wrong with the Queensland frogs shortly after he started a study project in Paluma Range National Park in 1989. That year four species of frogs lived in the clear waters of Birthday Creek. Less than two years later two of the frog species—a lace-lid tree frog and a torrent frog similar to those in PNG—had vanished.

Elsewhere in Queensland, McDonald had noticed that the population of another species, a day frog, which had lived alongside the gastric brooders, was beginning to dwindle. "We'd never paid much attention to the day frogs because we were all fascinated by the gastric brooders," says McDonald. "And the day frogs were about as common as fleas on a dog's back. . . . Then *phhht!* They were gone too." In 1993 McDonald and other researchers happened to find some of the day frogs as they were dying, one of the first times that frogs had been spotted in the throes of a massive die-off. They collected some and sent their remains to Rick Speare, an infectious disease specialist at James Cook University.

Die-offs in the Americas

While the Australian frogs were disappearing, frogs in the rain forests of Costa Rica's mountains also began to dwindle. Most alarmingly the splendid Costa Rican golden toad (*Bufo periglenes*) vanished. Not one has been seen in the wild since 1989. Other frog species disappeared in that habitat too, but not until 1996 did a scientist chance upon a dying population. That year Karen Lips, a herpetologist at Southern Illinois University, found dead and dying frogs in the high rain forests of Panama, and like McDonald, she dispatched them to a disease expert. These specimens, together with those from Australia and others from the National Zoo in Washington, D.C., led to the discovery of a previously unknown frog killer: a waterborne organism called a chytrid (KI trid) fungus. Most chytrids are simply decomposers of plant materials. Some are known to live as parasites on plants and invertebrates, but this species is the first discovered to kill vertebrates. Scientists do not know how the chytrid kills frogs. Speare suspects that the fungus may release a toxin as it eats the keratin (a protein) in the frog's skin.

"We're looking at a newly introduced disease, an emerging pathogen," Speare said. The fungus has been found on 44 Australian frog species and appears to have caused the extinction of four, including the gastric brooders. And several U.S. frog species are now known to have died from the disease. The chytrid is suspected as the main cause of the sudden declines in frog populations in mountainous regions of Central and South America. But researchers still do not know where the chytrid originated or how it arrived in the Americas or Australia or how they can stop its spread in the wild. The discovery of the lethal chytrid is all the more troubling because many of the frogs at highest risk of succumbing to the disease live in places that people have set aside for their protection.

In the dense rain forest of Panama's Omar Torrijos National Park, Karen Lips hikes up a rocky stream, counting frogs and toads with her graduate assistant, Jeanne Robertson. Lips thinks the chytrid killed Costa Rica's toads and other montane frogs. She found frogs dying from the disease at another Panamanian site and worries that it may have reached this park.

Our first day of surveying, however, dispels her fear. Only a few minutes down the trail, Robertson calls out "Frog!" simultaneously bending down to grab her prize. She reveals a three-inch-long golden frog. There's nothing subtle about this hue, and Lips explains that it serves as a warning to potential predators: The frog's skin is laden with lethal poisons. We find numerous frogs, including several tiny golden frog babies.

Lips's survey will also begin fleshing out the baseline list of frogs found in this park. "That's the thing about frogs," she said. "There's still so much basic work that needs to be done. Dozens of new species are found every year, and we know so little about any of them."

BELOW: Lips and Robertson catch, weigh, and measure the frogs, then gently scissor off one back toe—which does not cripple the frog. Lips takes the toe clips because they may help us learn more about chytrids. In particular, researchers hope to discover why some species are susceptible to it and others are not.

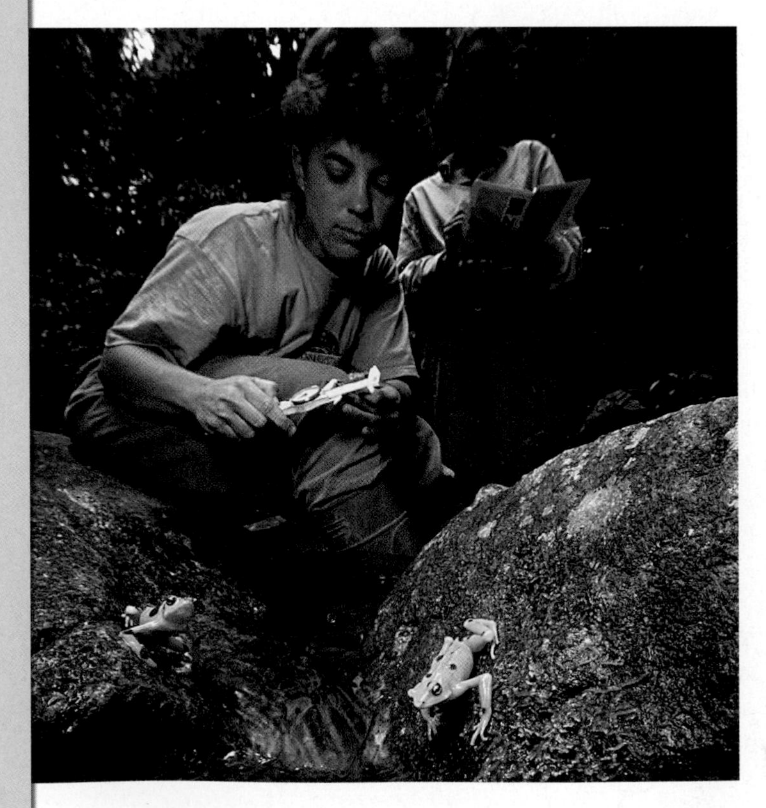

Protecting Frogs

The Chiricahua leopard frog (*Rana chiricahuensis*), once common throughout the Southwest, has nearly vanished from several desert canyons—but has found protection on a rancher's land. Matt Magoffin discovered a healthy population of them

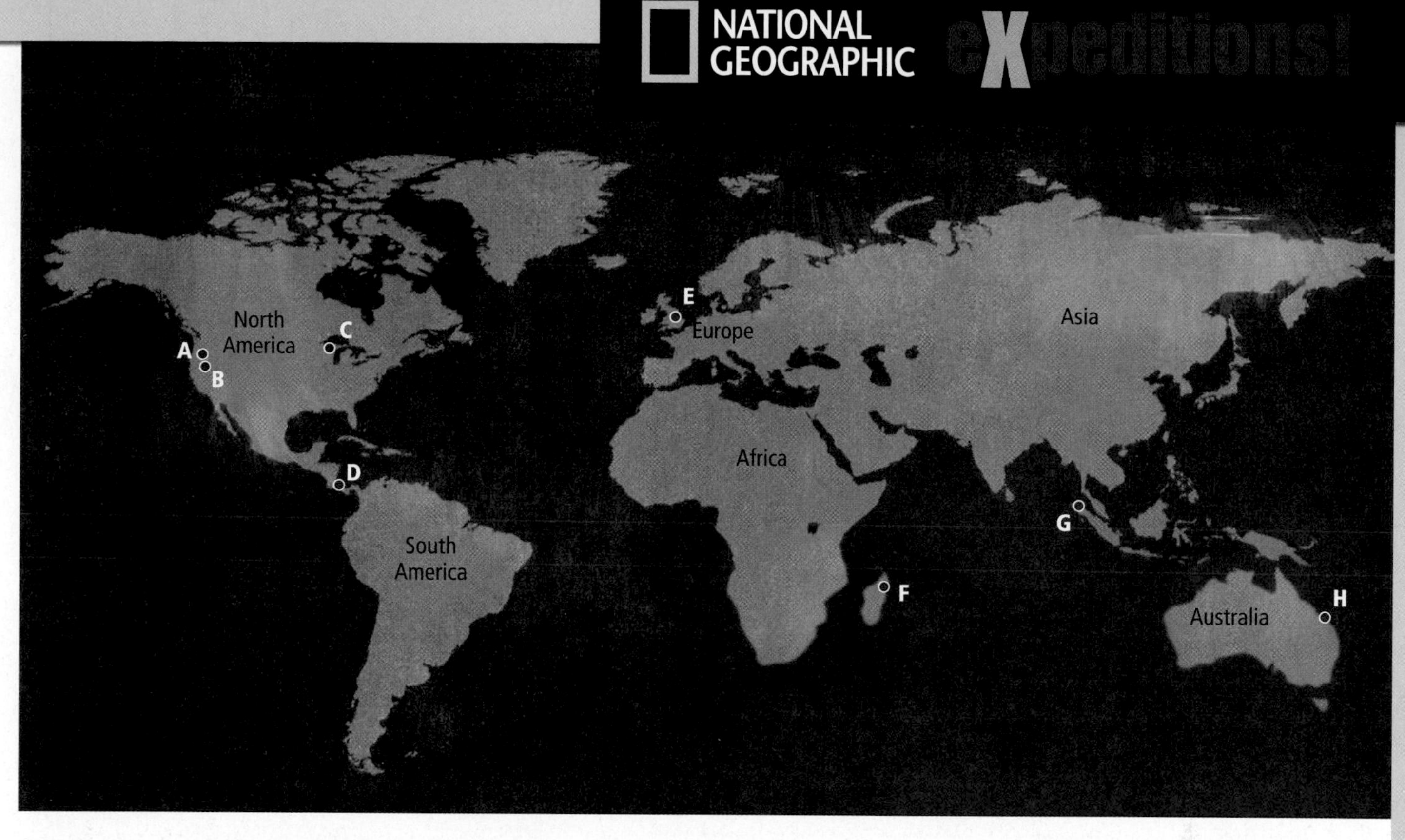

ABOVE: Examples of environmental changes that contribute to a decline in frog populations:
(A) ULTRAVIOLET RADIATION: Ultraviolet light can alter DNA cells and suppress immune responses. Biologists think UV radiation contributed to massive die-offs of frogs' eggs in Oregon.
(B) INTERSPECIES COMPETITION: Non-native fish and voracious bull-frogs stocked for food and sport devour native frogs and their prey. Such actions have nearly wiped out the mountain yellow-legged frog in California.
(C) PARASITES, PESTICIDES: Gruesome deformities have appeared in Canada and the northern U.S. Trematode parasites and pesticides and their by-products are suspected.
(D) CLIMATE CHANGE: Populations of the golden toad, once found in Costa Rica, collapsed when dry weather pushed moisture-giving clouds beyond the habitat. Biologists blame global warming.
(E) POLLUTION: The highly permeable eggs and skin of frogs easily admit toxic substances. Acid rain has caused frog declines in Britain, Canada, Scandinavia, and Eastern Europe. Heavy metal, fertilizer compounds, and agricultural chemicals also take a toll.
(F) HABITAT DESTRUCTION: As humans cut trees, drain marshes, pave meadows, and dam rivers, frogs die. The world over, habitat loss and alteration is by far the greatest cause of death for frogs—creatures of limited range that can't easily relocate.
(G) OVERHARVESTING: Millions of frogs die each year to supply restaurants with frog legs. Most come from Indonesia. India and Bangladesh banned exports after frog declines led to a rise in mosquitoes, malaria, and pesticide use.
(H): DISEASE: *Saprolegnia,* chytrid, iridovirus—such diseases are affecting frogs worldwide.

in a large cattle tank. During a bad drought Magoffin trucked a thousand gallons of water every week to the tank to keep the frogs going. Now conservationists are using the Magoffin frogs to restock other areas.

Frogs—Sentinel Species

Frogs have thrived from the tundra to the tropics for 190 million years. Now they are vanishing—rapidly and perhaps irretrievably. Why? Habitat loss is only the most obvious villain. Additional causes remain unclear, particularly in Asia, Africa, and other regions where research has been minimal. Scientists warn that climate change, pollution, and other factors may be acting together to deform and kill frogs. Frogs are regarded as "sentinel species" in ecosystems because of their intimate contact with air, water, and earth. "We share the planet," says Mike Lannoo of the Declining Amphibian Populations Task Force. "If something out there is affecting frogs, there's a chance it's affecting us too."

eXpeditions activity

Research the current status of three of the frogs discussed in the article. Make a list of changes in environmental conditions since this article was written that contribute to an increase in the frogs' populations or add to their decline.

NASA's Spitzer Space Telescope

Night Vision

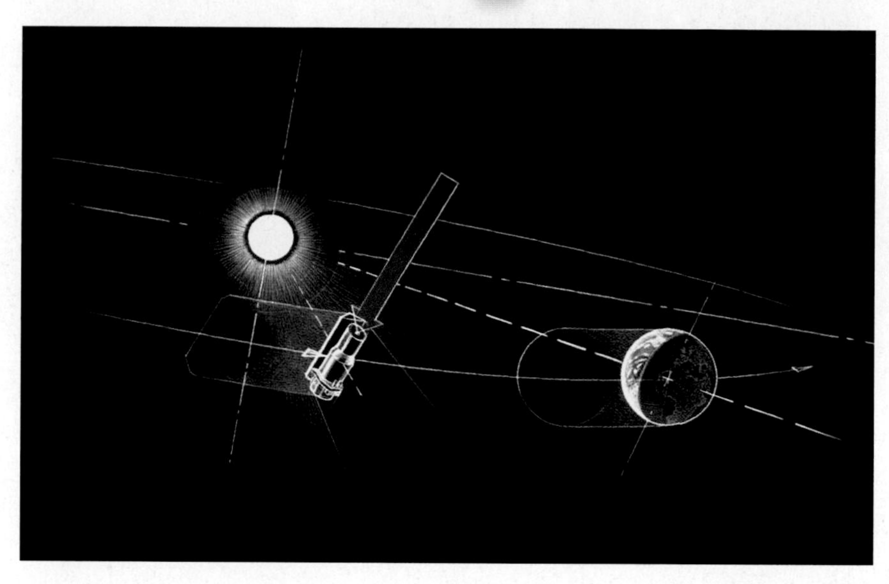

ABOVE: Most space telescopes orbit the Earth, but *Spitzer* orbits the Sun, trailing Earth by 41.8 million kilometers (26 million miles)—a gap that increases by 17.7 million kilometers (11 million miles) every year. The distance keeps the telescope far from the disruptive heat of our planet and reduces the Earth and Moon to mere dots, giving *Spitzer* an uninterrupted view of large sections of the sky.

LEFT: New stars blaze in a cloud of dust and gas 50 light-years across and 7000 light-years away. Ultraviolet radiation from a nearby massive star sculpted the cloud into pillars and canyons; then gravity squeezed the denser clumps until stars burst to life, like those atop the pillars. The *Spitzer Space Telescope* made this image at infrared wavelengths invisible to human eyes.

There's a lot hiding in the universe's dark corners. Interstellar dust clouds and inky stretches of deep space can appear dull to ordinary telescopes. But to a car-sized telescope 41.8 million kilometers (26 million miles) from Earth, they are alive with light—infrared light, or heat rays. Since its launch in August 2003, says Robert Kennicutt, an astronomer at the University of Arizona, NASA's *Spitzer Space Telescope* "has opened up half the universe to us."

Nursery of Stars

In the process, it has exposed cosmic birthplaces. Stars take shape in clouds of gas and dust, and planets emerge in disks of debris around new stars. Early galaxies are also swathed in dust. Little visible light gets out, but these objects still emit heat—and infrared radiation. "If you only look in visible light at these objects, you don't even see the tip of the iceberg—you see the tip of the tip of the iceberg," says Charles Lawrence of the Jet Propulsion Laboratory in Pasadena, California. "We look in the infrared because that's where the photons are."

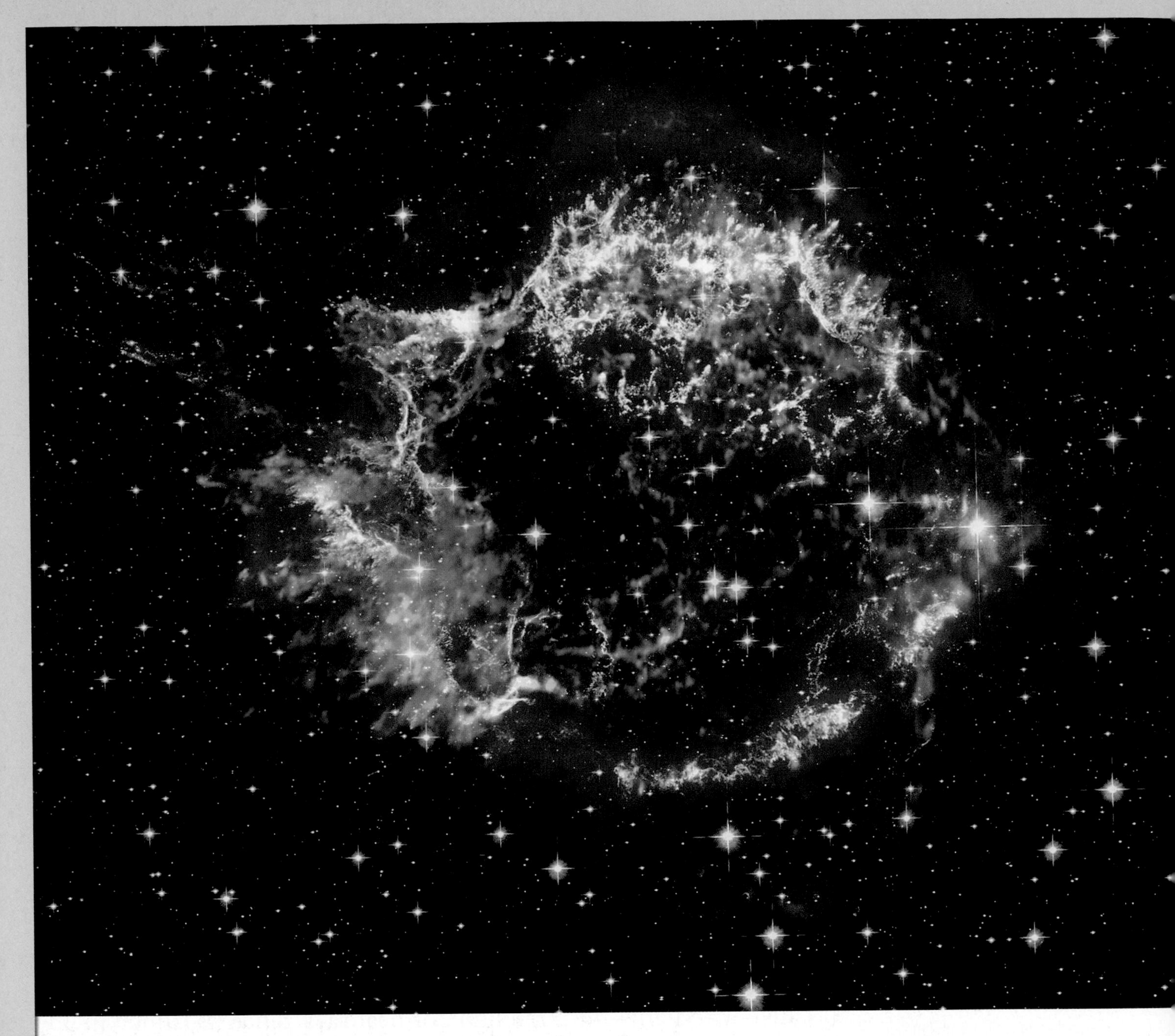

○ **ABOVE:** A star exploded 325 years ago, leaving a debris cloud 15 light-years wide called Cassiopeia A. This composite image combines data from three space telescopes. The *Chandra X-Ray Observatory* mapped hot gas (blue and green), rich in iron and silicon from the exploded star. The *Hubble Space Telescope* captured wisps of cooler gas (yellow). *Spitzer* data (red) revealed a shell of dust from interstellar space heated by the blast's shock wave.

Catching those photons, or light particles, meant going into space, because Earth's atmosphere blocks most infrared. Lyman Spitzer, the American astrophysicist for whom the telescope is named, pointed out the advantages of space telescopes back in 1946. Since then, instruments such as the legendary *Hubble Space Telescope* have proved him right. But the *Spitzer* telescope's infrared vision is the keenest ever, thanks to a mirror nearly three feet across, sensitive detectors cooled almost to absolute zero, and an orbit far from Earth's distracting heat.

RIGHT: Infrared light concentrated by *Spitzer*'s mirrors falls on detectors chilled close to absolute zero to detect the faint heat of distant objects. Ingenious passive cooling technologies—including a solar panel that doubles as a sun shield—help save the liquid helium coolant. Three instruments capture and analyze different infrared frequencies: the Infrared Array Camera (IRAC), Infrared Spectrograph (IRS), and Multiband Imaging Photometer (MIPS).

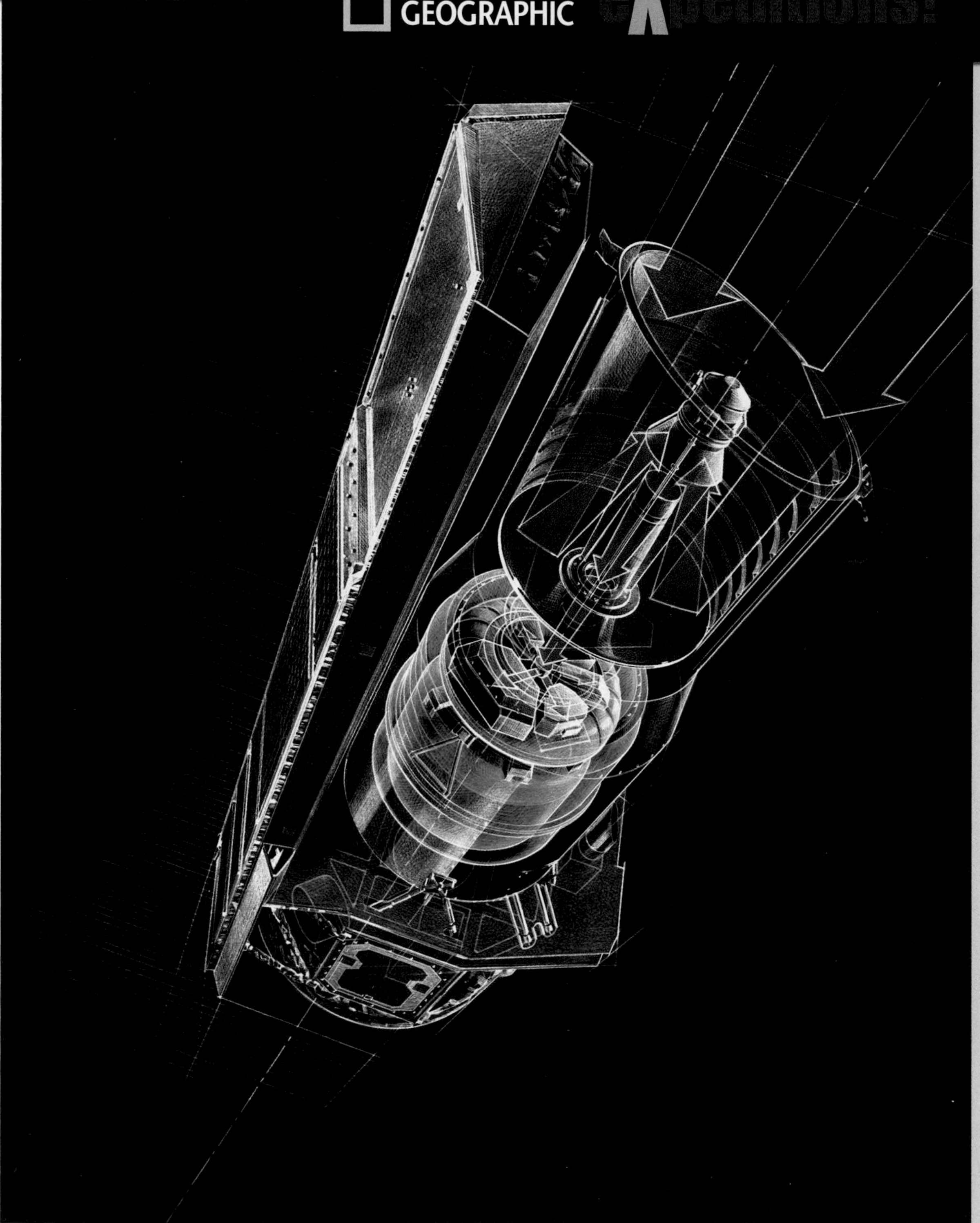

How and Where Planets Form

Already the telescope has gleaned clues about how and where planets form, and even spotted two of them by picking up their infrared glow. It is helping astronomers understand how light and radiation from existing stars can trigger the collapse of gas clouds to form new stars. And in the far reaches of space, *Spitzer* is finding young galaxies glowing in the infrared. "We've made major progress in searching for galaxies at the beginning of the universe," says Giovanni Fazio of the Harvard-Smithsonian Center for Astrophysics. "I'm like a child let loose in a toy store."

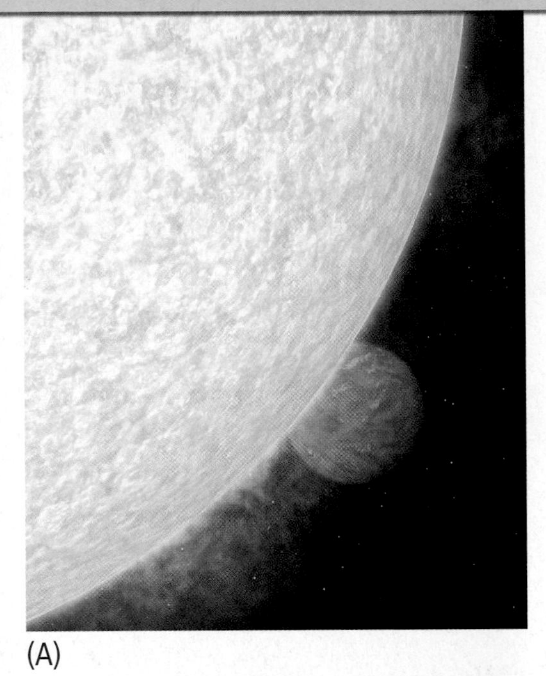
(A)

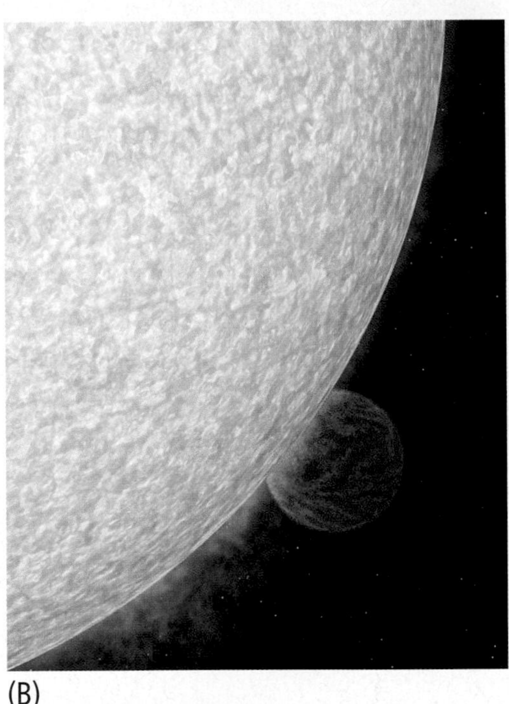
(B)

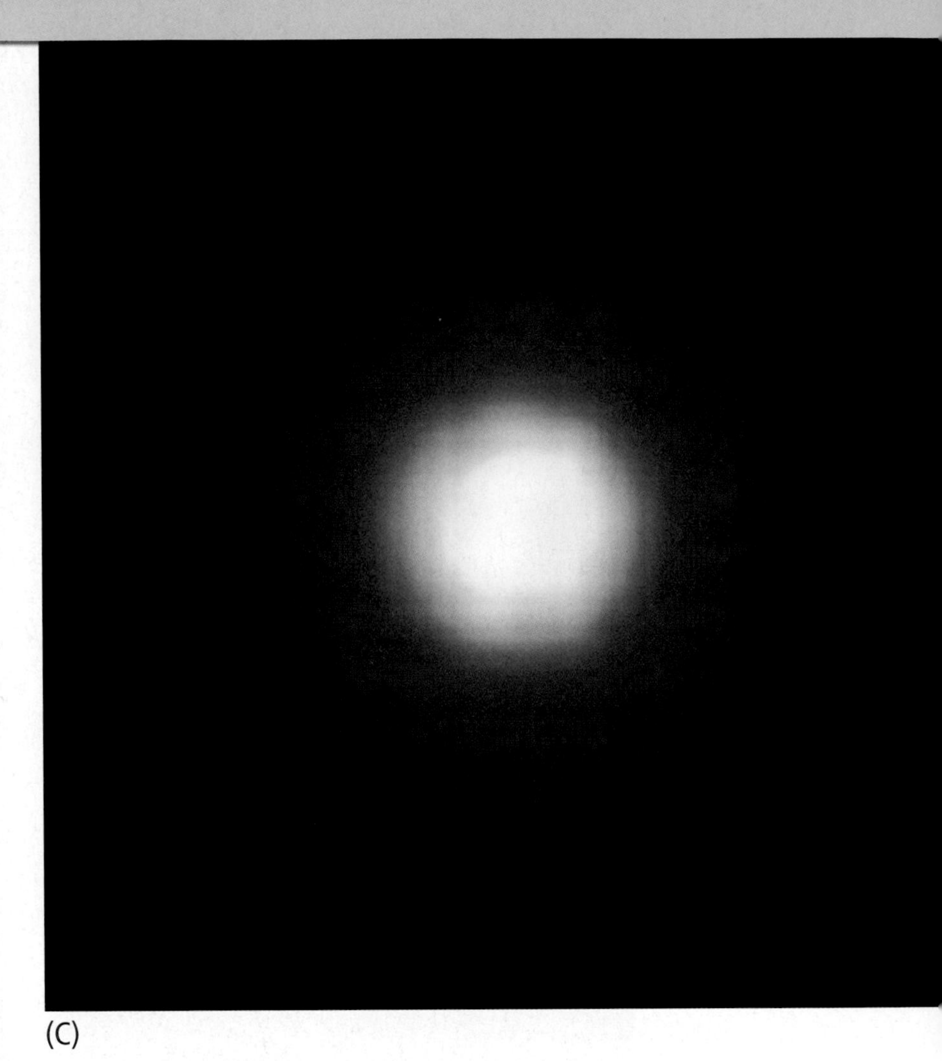
(C)

A: A planet in visible light is shown according to an artist's concept.

B: *Spitzer* shows an infrared image of a planet.

C: The disk of debris surrounding Vega is at least 20 times the size of our solar system.

Now You See It . . .

Astronomers have detected more than 150 planets around other stars without actually seeing their light. But in late 2004, *Spitzer* captured infrared light from two Jupiter-size planets. Both lie so close to their stars that they orbit in three days and are heated to 727°C (1340°F) or more.

In visible light, each planet is lost in the star's glare. But in the infrared, each emits its own light. Key to detecting this infrared glow was the fact that the planets disappear behind their stars on each orbit. As astronomers monitored each system with *Spitzer*, they saw the light dim as the planet vanished, then brighten as it emerged, adding its light to the star's. "We could use the same trick to study light from even smaller planets," says David Charbonneau of the Harvard-Smithsonian Center for Astrophysics, who led a team that detected light from one planet.

Unseen planets probably lurk in a disk of debris imaged by *Spitzer* around the star Vega. The disks are made of fine dust, perhaps kicked up by debris from planets shattered by giant collisions.

ABOVE: Dust girdling the Sombrero galaxy is a circle of shadow in a visible-light image.

BELOW: The circle shines when infrared data from *Spitzer* are added. The glow—the result of stars heating the dust—reveals clumpy regions where new stars are forming. The bright spot at the galaxy's center results from a different heat source: a titanic black hole sucking in matter.

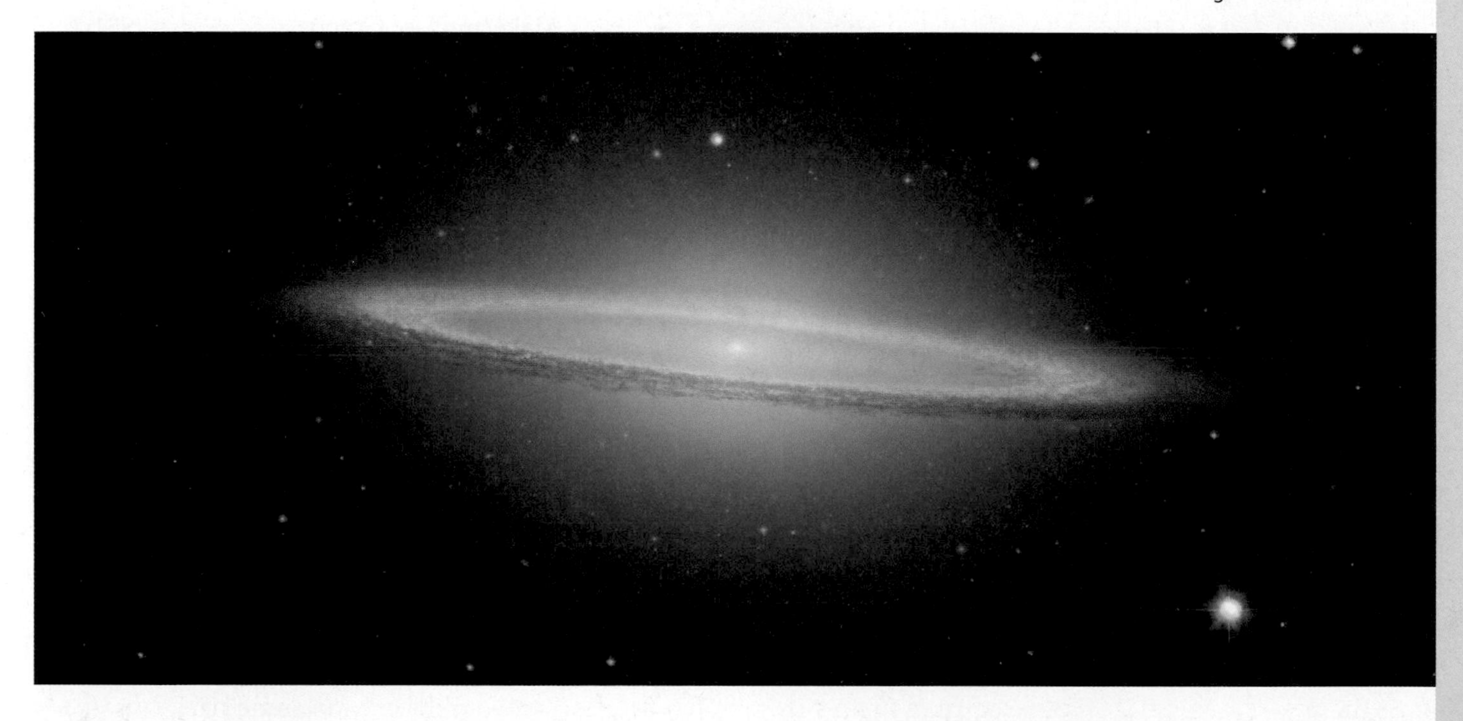

Golden Age of Astrophysics

The fun should continue until *Spitzer* runs out of the liquid helium that helps cool it, in about 2008. Early in the next decade, NASA plans to launch the *James Webb Space Telescope*, a much larger infrared observatory. "This is the golden age of astrophysics," says Lawrence. "A thousand years from now we'll look back and say that."

eXpeditions activity

Research the *James Webb Space Telescope*. Then prepare a presentation about what its location will be in space, special features that distinguish it from previous telescopes, and what astrophysicists expect from the telescope.

For students and parents/guardians

The *Skillbuilder Handbook* and the *Reference Handbook* are designed to help you as students achieve success as you embark on the adventure of learning Earth science. These reference pages will also enable your parents or guardians to help you in this exciting journey. There are many ways of learning new information. Completing the exercises will help you learn key science skills, such as interpreting what you read and organizing information in a clear, easy-to-understand way.

Table of Contents

Make Comparisons

Why Learn this Skill?

Suppose you want to buy a portable MP3 music player, and you must choose among three different models. You would probably compare the characteristics of the three models, such as price, amount of memory, sound quality, and size to determine which model is best for you.

In the study of Earth science, you often compare the structures and functions of one type of rock or planet with another. You will also compare scientific discoveries or events from one time period with those from a different time period. This helps you gain an understanding of how the past has affected the present.

Learn the Skill

When making comparisons, you examine two or more groups, situations, events, or theories. You must first decide what items will be compared and determine which characteristics you will use to compare them. Then identify any similarities and differences.

For example, comparisons can be made between the two minerals shown on this page. The physical properties of halite can be compared to the physical properties of quartz.

Practice the Skill

Create a table with the title *Mineral Comparison*. Make two columns. Label the first column *Halite*, and the second column *Quartz*. List all of your observations of these two minerals in the appropriate column of your table. Similarities you might point out are that both minerals are solids that occur as crystals, and both are inorganic compounds. Differences might include that halite has a cubic crystal structure, whereas quartz has a hexagonal crystal structure.

When you have finished the table, answer these questions.

1. What items are being compared? How are they being compared?
2. What properties do the minerals have in common?
3. What properties are unique to each mineral?

Apply the Skill

Make Comparisons Read two editorial articles in a science journal or magazine that express different viewpoints on the same issue. Identify the similarities and differences between the two points of view.

Halite

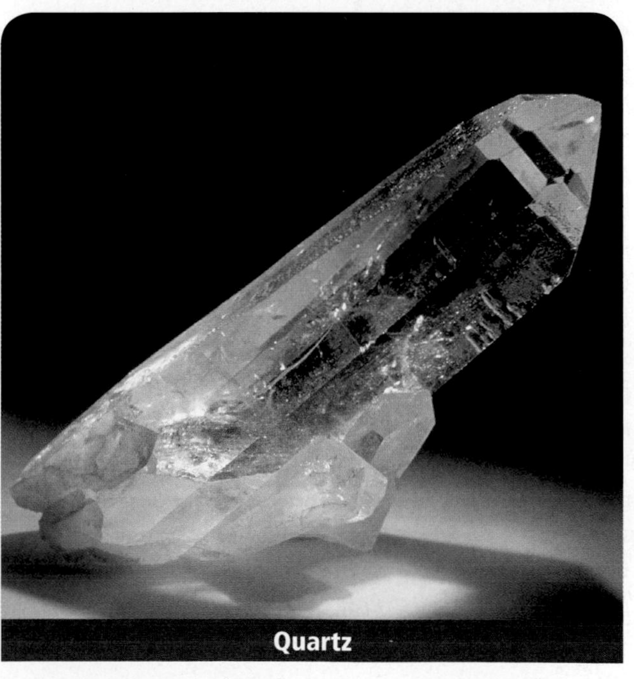

Quartz

Analyze Information

Why Learn this Skill?

Analyzing, or looking at separate parts of something to understand the entire piece, is a way to think critically about written work. The ability to analyze information is important when determining which ideas are more useful than others.

Learn the Skill

To analyze information, use the following steps:

- Identify the topic being discussed.
- Examine how the information is organized—identify the main points.
- Summarize the information in your own words, and then make a statement based on your understanding of the topic and what you already know.

Practice the Skill

Read the following excerpt from *National Geographic*. Use the steps listed above to analyze the information and answer the questions that follow.

His name alone makes Fabien Cousteau, grandson of the late Jacques, a big fish in the world of underwater exploration. Now he's taking that big-fish status to extremes. The Paris-born, New York-based explorer had become a virtual shark, thanks to his new shark-shaped submarine. He uses the sub to dive incognito among the oceans' top predators, great white sharks.

Created at a cost of more than $100,000, the 4.3-meter-long contraption is designed to look and move as much like the real thing as possible. It carries a single passenger, who fits inside lying down, propped up on elbows to navigate and observe. "This is akin to being the first human being in the space capsule in outer space," Cousteau said. "It's pretty similar. You have no idea what's going to happen; it's a prototype."

Cousteau used the submarine to make a documentary intended to demystify the notion that great white sharks are ruthless, mindless killers. Great whites have been around for more than 400 million years. Anything that has survived that long isn't "stupid," he said.

Cousteau calls the sub Troy, *in reference to the mythical Trojan horse statue, in which Greek soldiers were spirited into the fortress kingdom of Troy. Propelled by a wagging tail and covered in a flexible, skinlike material, the sub—created by Cousteau and a team of scientists and engineers—swims silently. The steel-ribbed, womb-like interior is filled with water, requiring Cousteau to wear a wet suit and use scuba gear to breathe.*

Fabien Cousteau enters his shark-shaped submarine.

Importantly, Troy *allows Cousteau to be a shark, not shark bait. At the heart of the project is a desire to observe what great white sharks do when people aren't around to watch. Prior to this, most shark observations have come from humans sitting in cages and enticing the predators with bait—conditions that spawn unnatural behaviors, Cousteau said. "Now all of the sudden we can see what they do as white sharks rather than as trained circus animals," he said.*

While Cousteau is reluctant to guess what the sharks thought when Troy *invaded their space, the explorer said they seemed to act naturally. Some even puffed their gills and gaped toward* Troy*—actions thought to be communication signals. And though a few sharks made aggressive gestures, none of the predators attacked the shark-shaped sub.*

1. What topic is being discussed?
2. What are the main points of the article?
3. Summarize the information in this article, and then provide your analysis based on this information and your own knowledge.

Apply the Skill

Analyze Information Find a short, informative article on a new scientific discovery or new application of science technology, such as hybrid-car technology. Analyze the information and make a statement of your own.

Synthesize Information

Why Learn this Skill?

The skill of synthesizing involves combining and analyzing information gathered from separate sources or at different times to make logical connections. Being able to synthesize information can be a useful skill for you as a student when you need to gather data from several sources for a report or a presentation.

Learn the Skill

Follow these steps to synthesize information:

- Select important and relevant information.
- Analyze the information and build connections.
- Reinforce or modify the connections as you acquire new information.

Suppose you need to write a research paper on global levels of atmospheric carbon dioxide (CO_2) levels. You need to synthesize what you learn to inform others. You can begin by detailing the ideas and information from sources you already have about global levels of atmospheric carbon dioxide. A table such as **Table SH.1** could help you categorize the facts from these sources.

Table SH.1 Global Levels of Atmospheric CO_2

Year	Global Atmospheric CO_2 Concentration (ppm)	Year	Global Atmospheric CO_2 Concentration (ppm)
1745	279	1935	307
1791	280	1949	311
1816	284	1958	312
1843	287	1965	318
1854	288	1974	330
1874	290	1984	344
1894	297	1995	361
1909	299	1998	367
1921	302	2005	385

Then you might select an additional article about greenhouse gases, such as the one below.

According to the National Academy of Scientists, Earth's surface temperature has risen about one degree Fahrenheit in the past 100 years. This increase in temperature can be correlated to an increase in the concentration of carbon dioxide and other greenhouse gases in the atmosphere. How might this increase in temperature affect Earth's climate?

Carbon dioxide is one of the greenhouse gases that helps keep temperatures on Earth warm enough to support life. However, a buildup of carbon dioxide and other greenhouse gases such as methane and nitrous oxide can lead to global warming, an increase in Earth's average surface temperature. Since the industrial revolution in the 1800s, atmospheric concentrations of carbon dioxide have increased by almost 30 percent, methane concentrations have more than doubled, and nitrous oxide concentrations have increased approximately 15 percent. Scientists attribute these increases to the burning of fossil fuels for automobiles, industry, and electricity, as well as deforestation, increased agriculture, landfills, and mining.

Practice the Skill

Use the table and the passage on this page to answer these questions.

1. What information is presented in the table?
2. What is the main idea of the passage? What information does the passage add to your knowledge about the topic?
3. By synthesizing the two sources and using your own knowledge, what conclusions can you draw about global warming?

Apply the Skill

Synthesize Information Find two sources of information on the same topic and write a short report. In your report, answer these questions: What kinds of sources did you use? What are the main ideas of each source? How does each source add to your understanding of the topic? Do the sources support or contradict each other?

Take Notes and Outline

Why Learn this Skill?

One of the best ways to remember something is to write it down. Taking notes—writing down information in a brief and orderly format—not only helps you remember, but also makes studying easier.

Learn the Skill

There are several styles of note-taking, but the goal of every style is to explain information and put it in a logical order. As you read, identify and summarize the main ideas and details that support them and write them in your notes. Paraphrase—that is, state in your own words—the information rather than copying it directly from the text. Use note cards or develop a personal "shorthand"—using symbols to represent words—to represent the information in a compact manner.

You might also find it helpful to create an outline when taking notes. When outlining material, first read the material to identify the main ideas. In textbooks, look at the section headings for clues to main topics. Then identify the subheadings. Place supporting details under the appropriate headings. The basic pattern for outlines is shown below:

MAIN TOPIC
- I. FIRST IDEA OR ITEM
  - A. FIRST DETAIL
    - 1. SUBDETAIL
    - 2. SUBDETAIL
  - B. SECOND DETAIL
- II. SECOND IDEA OR ITEM
  - A. FIRST DETAIL
  - B. SECOND DETAIL
    - 1. SUBDETAIL
    - 2. SUBDETAIL
- III. THIRD IDEA OR ITEM

Practice the Skill

Read the following excerpt from *National Geographic*. Use the steps you just read about to take notes and create an outline. Then answer the questions that follow.

Dinosaur fans still have a lot to look forward to. According to a new estimate of dinosaur diversity, the 21st century will bring an avalanche of new discoveries. "We only know about 29 percent of all dinosaurs out there to be found," said study co-author Peter Dodson, a paleobiologist and anatomy professor at the University of Pennsylvania in Philadelphia.

Dodson and statistics professor Steve Wang of Swarthmore College, in Swarthmore, Pennsylvania, made a statistical analysis of an exhaustive database of all known dinosaur genera (the taxonomic group one notch above species). They then used this data to estimate the total number of genera preserved in the fossil record.

The pair predicts that scientists will eventually discover 1,844 dinosaur genera in total—at least 1,300 more than the 527 recognized today from remains other than isolated teeth. What's more, the duo believes that 75 percent of these dinos will be discovered within the next 60 to 100 years and 90 percent within 100 to 140 years, based on an analysis of historical discovery patterns.

The tally applies only to specimens preserved as fossils. Many other types of dinosaurs likely roamed the Earth during the dinosaurs' 160-million-year reign, but remains from these species will never be known to science, the researchers say.

1. What is the main topic?
2. What are the first, second, and third ideas?
3. Name two details for each of the ideas.
4. Name two subdetails for each of the details.

Apply the Skill

Take Notes and Outline Scan a science journal for a short article about a new laboratory technique. Take notes by using shorthand or by creating an outline. Summarize the article using only your notes.

Understand Cause and Effect

Why Learn this Skill?

In order to understand an event, you should look for how that event or chain of events came about. When scientists are unsure of the cause for an event, they often design experiments. Although there might be an explanation, an experiment should be performed to be certain the cause created the event you observed. This process examines the causes and effects of events.

Learn the Skill

Calderas can form when the summit or side of a volcano collapses into the magma chamber that once fueled the volcano. An empty magma chamber can *cause* the volcano to collapse. The caldera that forms is the *effect,* or result. The figure below shows how one event—the **cause**—led to another—the **effect.**

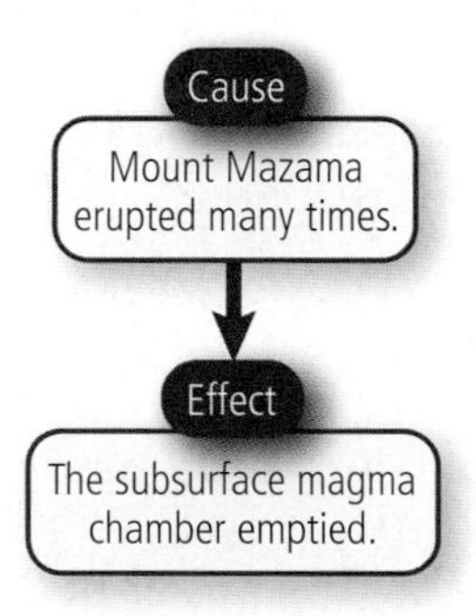

You can often identify cause-and-effect relationships in sentences from clue words such as the following.

because	**produced**
due to	**as a result**
so that	**that is why**
therefore	**for this reason**
thus	**consequently**
led to	**in order to**

Read the sample sentences below.

"The volcano collapsed into the partially empty magma chamber. As a result, a depression was formed where the volcano once stood."

In the example above, the cause is the collapse of the volcano. The cause-and-effect clue words "as a result" tell you that the depression is the effect of the collapsing volcano.

In a chain of events, an effect often becomes the cause of other events. The next chart shows the complete chain of events that occur when a caldera forms.

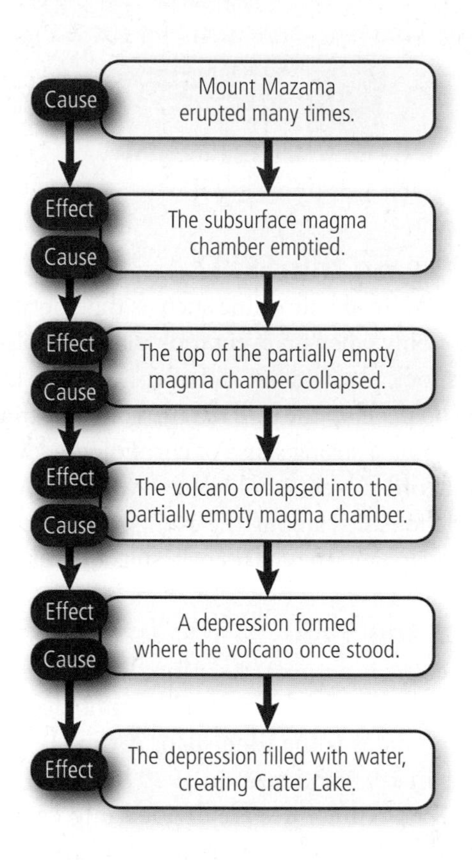

Practice the Skill

Make a chart like the one above showing which events listed below are causes and which are effects.

1. As water vapor rises, it cools and changes back to a liquid.
2. Droplets inside clouds join to form bigger drops.
3. Water evaporates from oceans, lakes, and rivers.
4. Water vapor rises into the atmosphere.
5. Water droplets become heavy and fall as rain or snow.

Apply the Skill

Understand Cause and Effect Read an account of a recent scientific event or discovery in a science journal. Determine at least one cause and one effect of that event. Show the chain of events in a chart.

Skillbuilder Handbook

Read a Time Line

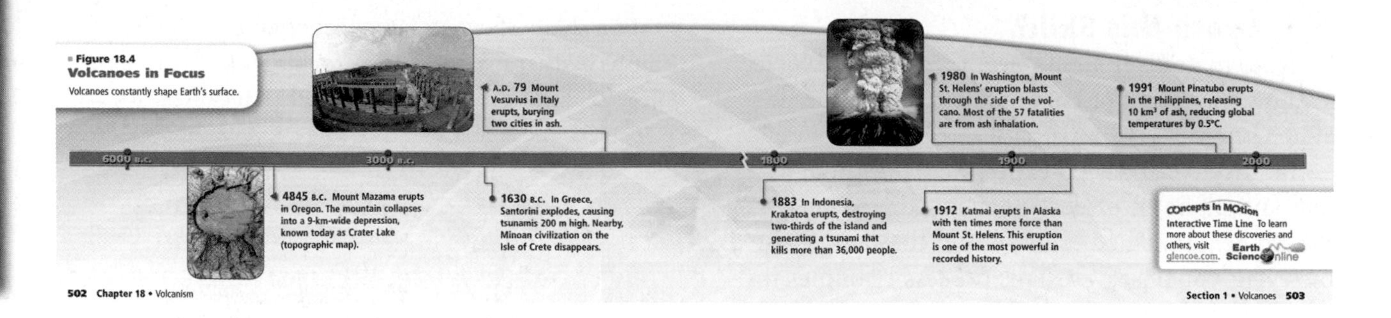

Why Learn this Skill?

When you read a time line such as the one above, you see not only when an event took place, but also what events took place before and after it. A time line can help you develop the skill of chronological thinking. Developing a strong sense of chronology—when and in what order events took place—will help you examine relationships among the events. It will also help you understand the causes or results of events.

Learn the Skill

A time line is a linear chart that list events that occurred on specific dates. The number of years between dates is the time span. A time line that begins in 1910 and ends in 1920 has a ten-year time span. Some time lines are divided into centuries. The twentieth century includes the 1900s, the nineteenth century includes the 1800s, and so on.

Time lines are usually divided into smaller parts, or time intervals. On the two time lines below, the first time line has a 300-year time span divided into 100-year time intervals. The second time line has a six-year time span divided into two-year time intervals.

Practice the Skill

Study the time line above and then answer these questions.

1. What time span and intervals appear on this time line?
2. How much more powerful was Katmai's eruption than Mount St. Helens' eruption?
3. How many years after Santorini erupted did Vesuvius erupt?
4. How many years apart were Krakatoa's eruption and Mt. Pinatubo's eruption?

Apply the Skill

Read a Time Line Sometimes a time line shows events that occurred during the same period but are related to two different subjects. The time line above shows events related to volcanoes between 6000 B.C. and A.D. 2000. Copy the time line and events onto a piece of paper. Then use a different color to add in events related to earthquakes during this same time span. Refer to Chapter 19 for help.

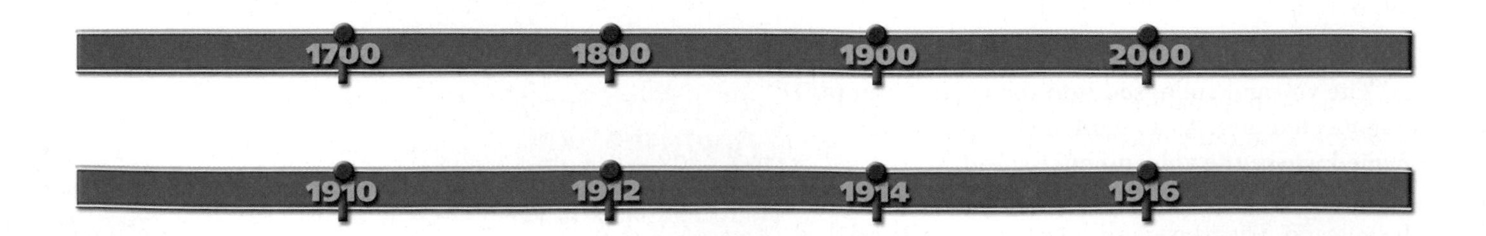

Analyze Media Sources

Why Learn this Skill?

To stay informed, people use a variety of media sources, including print media, broadcast media, and electronic media. The Internet has become an especially valuable research tool. It is convenient to use, and the information it contains is plentiful. Whichever media source you use to gather information, it is important to analyze the source to determine its accuracy and reliability.

Learn the Skill

There are a number of issues to consider when analyzing a media source. The most important one is to check the accuracy of the source and content. The author and publishers or sponsors should be credible and clearly indicated. To analyze print media or broadcast media, ask yourself the following questions.

- Is the information current?
- Are the sources revealed?
- Is more than one source used?
- Is the information biased?
- Does the information represent both sides of an issue?
- Is the information reported firsthand or secondhand?

For electronic media, ask yourself these questions in addition to the ones above.

- Is the author credible and clearly identified?
- Are the facts on the Web site documented?
- Are the links within the Web site appropriate and current?
- Does the Web site contain links to other useful resources?

Practice the Skill

To practice analyzing print media, choose two articles on global warming, one from a newspaper and the other from a newsmagazine. Then answer these questions.

1. What points are the authors of the articles trying to make? Were they successful? Can the facts be verified?
2. Did either article reflect a bias toward one viewpoint or another? List any unsupported statements.
3. Was the information reported firsthand or secondhand? Do the articles seem to represent both sides fairly?
4. How many sources can you identify in the articles? List them.

To practice analyzing electronic media, visit glencoe.com and select Web links. Choose one link from the list, read the information on that Web site, and then answer these questions.

1. Who is the author or sponsor of the Web site?
2. What links does the Web site contain? How are they appropriate to the topic?
3. What sources were used for the information on the Web site?

Apply the Skill

Analyze Media Sources Think of a national issue on which public opinion is divided. Read newspaper features, editorials, and Web sites, and monitor television reports about the issue. Which news sources more fairly represents the issue? Which news sources have the most reliable information? Can you identify any biases? Can you verify the credibility of the news source?

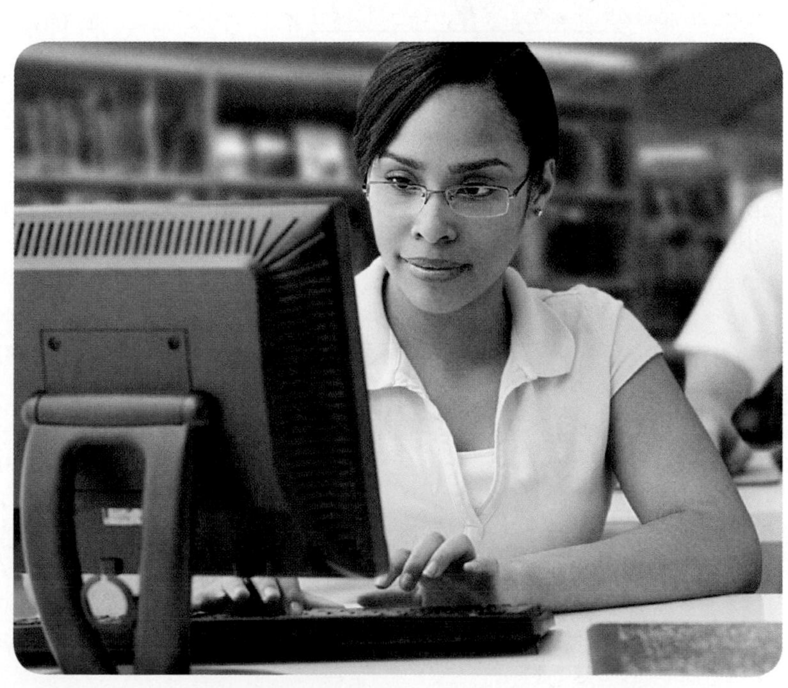

Use Graphic Organizers

Why Learn this Skill?

While you read this textbook, you will be looking for important ideas or concepts. One way to arrange these ideas is to create a graphic organizer. In addition to Foldables™, you will find various other graphic organizers throughout your book. Some organizers show a sequence, or flow, of events. Other organizers emphasize the relationship among concepts. Developing your own organizers while you read will help you better understand and remember what you read.

Learn the Skill

An **events chain concept map** is used to describe a sequence of events, such as a stage of a process or procedure. When making an events-chain map, first identify the event that starts the sequence and add events in chronological order until you reach an outcome.

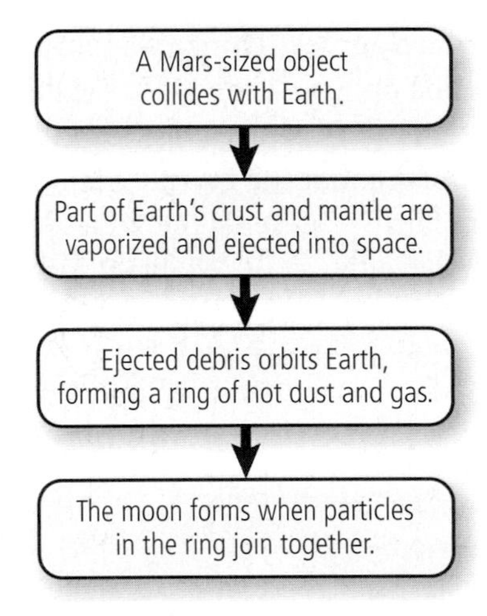

In a **cycle concept map,** the series of events do not produce a final outcome. The event that appears to be the final event relates back to the initiating event. Therefore, the cycle repeats itself.

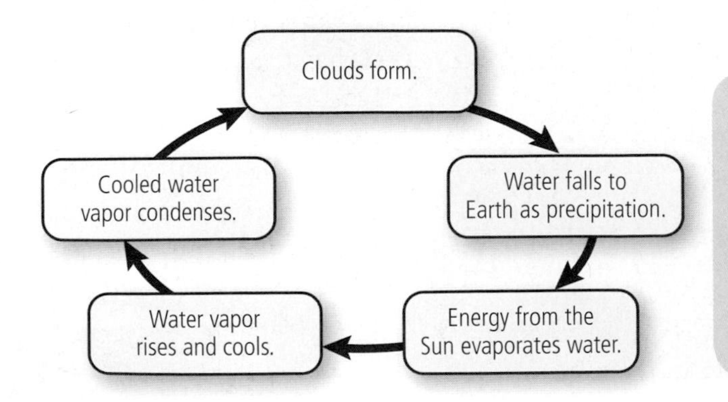

A **network tree concept map** shows the relationship among concepts, which are written in order from general to specific. The words written on the lines between the circles, called linking words, describe the relationships among the concepts. The concepts and the linking words can form sentences.

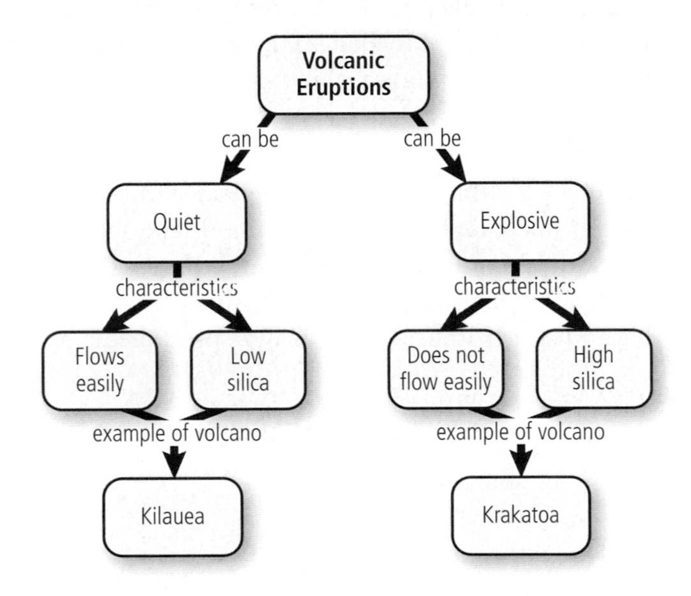

Practice the Skill

1. Create an events chain concept map of the events in sedimentary rock formation. Refer to Chapter 6 for help.
2. Create a cycle concept map of the nitrogen cycle. Make sure that the cycle shows the event that appears to be the final event relating back to the starting event. Refer to Chapter 24 for help.
3. Create a network tree concept map with these words: *Cenozoic, trilobites, eras, Paleozoic, mammals, dinosaurs, first land plants, Gondwana, Mesozoic, early Pangaea, late Pangaea.* Add linking words to describe the relationships between the concepts. Refer to Chapters 21, 22, and 23 for help.

Apply the Skill

Use Graphic Organizers Create an events chain concept map of the scientific method. Create a cycle concept map of the water cycle. Create a network tree concept map of pollution that includes air and water, sources of each pollution type, and examples of each type of pollution.

Debate Skills

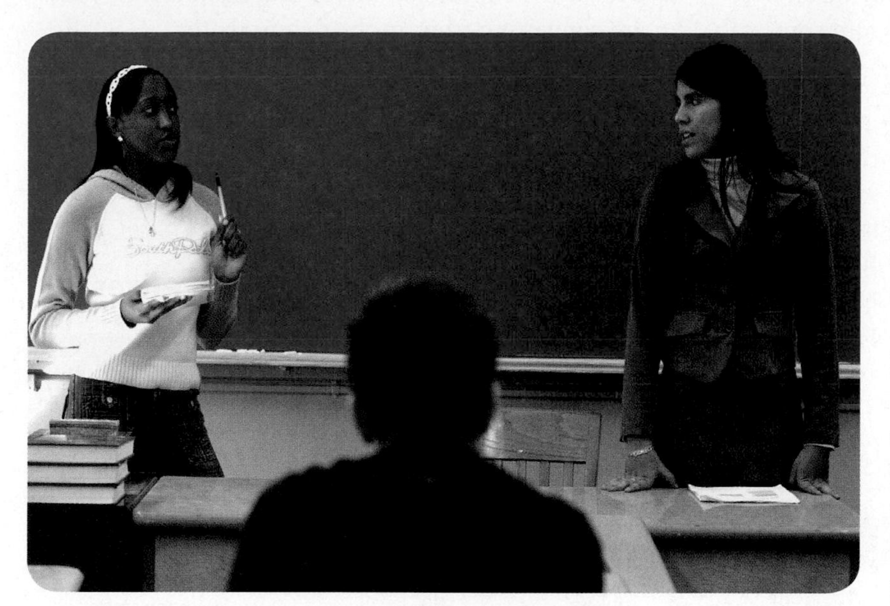

New research always is leading to new scientific theories. There are often opposing points of view on how this research is conducted, how it is interpreted, and how it is communicated. *The Earth Science and Society* features in your book offer a chance to debate a current controversial topic. Here is an overview on how to conduct a debate.

Choose a Position and Research

First, choose an Earth science issue that has at least two opposing viewpoints. The issue can come from current events, your textbook, or your teacher. These topics could include global warming or fossil fuel use. Topics are stated as affirmative declarations such as "Global warming is not detrimental to the environment."

One speaker will argue the positive position—the viewpoint that supports the statement—and another speaker will argue the negative position—the viewpoint that disputes the statement. Either individually or with a group, choose your position for the debate. The viewpoint that you choose does not have to reflect your personal belief. The purpose of debate is to create a strong argument supported by scientific evidence.

After choosing your position, conduct research to support your viewpoint. Use the Internet, find articles in your library, or use your textbook to gather evidence to support your argument.

A strong argument contains scientific evidence, expert opinions, and your own analysis of the issue. Research the opposing position also. Becoming aware of what points the other side might argue will help you to strengthen the evidence for your position.

Hold the Debate

You will have a specific amount of time, determined by your teacher, in which to present your argument. Organize your speech to fit within the time limit: explain the viewpoint that you will be arguing, present an analysis of your evidence, and conclude by summing up your most important points. Try to vary the elements of your argument. Your speech should not be a list of facts, a reading of a newspaper article, or a statement of your personal opinion, but an organized analysis of your evidence presented in your own manner of speaking. It is also important to remember that you must never make personal attacks against your opponent. Argue the issue. You will be evaluated on your overall presentation, organ-ization and development of ideas, and strength of support for your argument.

Additional Roles There are other roles that you can play in a debate. You can act as the timekeeper. The timekeeper times the length of the debaters' speeches and gives quiet signals to the speaker when time is almost up (usually a hand signal).

You can also act as a judge. There are important elements to look for when judging a speech: an introduction that tells the audience what position the speaker will be arguing, strong evidence that supports the speaker's position, and organization. It is helpful to take notes during the debate to summarize the main points of each side's argument. Then, decide which debater presented the strongest argument for his or her position. You can have a class discussion about the strengths and weaknesses of the debate and other viewpoints on this issue that could be argued.

Experimental data is often expressed using numbers and units. The following sections provide an overview of the common system of units and some calculations involving units.

Measure in SI

The International System of Measurements, abbreviated SI, is accepted as the standard for measurement throughout most of the world. The SI system contains seven base units. All other units of measurement can be derived from these base units.

Table SH.2 SI Base Units

Measurement	Unit	Symbol
Length	meter	m
Mass	kilogram	kg
Time	second	s
Electric current	ampere	A
Temperature	kelvin	K
Amount of substance	mole	mol
Intensity of light	candela	cd

Some units are derived by combining base units. For example, units for volume are derived from units of length. A liter (L) is a cubic decimeter (dm^3, or dm × dm × dm). Units of density (g/L) are derived from units of mass (g) and units of volume (L).

When units are multiplied by factors of ten, new units are created. For example, if a base unit is multiplied by 1000, the new unit has the prefix *kilo-*. One thousand meters is equal to one kilometer. Prefixes for some units are shown in **Table SH.3.**

To convert a given unit to a unit with a different factor of ten, multiply the unit by a conversion factor. A conversion factor is a ratio equal to one. The equivalents in **Table SH.3** can be used to make such a ratio. For example, 1 km = 1000 m. Two conversion factors can be made from this equivalent.

$$\frac{1000 \text{ m}}{1 \text{ km}} = 1 \quad \text{and} \quad \frac{1 \text{ km}}{1000 \text{ m}} = 1$$

To convert one unit to another factor of ten, choose the conversion factor that has the unit you are converting from in the denominator.

$$1 \cancel{\text{km}} \times \frac{1000 \text{ m}}{1 \cancel{\text{km}}} = 1000 \text{ m}$$

A unit can be multiplied by several conversion factors to obtain the desired unit.

Table SH.3 Common SI Prefixes

Prefix	Symbol	Equivalents
mega-	m	1×10^{6} base units
kilo-	k	1×10^{3} base units
hecto-	h	1×10^{2} base units
deka-	da	1×10^{1} base units
deci-	d	1×10^{-1} base units
centi-	c	1×10^{-2} base units
milli-	m	1×10^{-3} base units
micro-	μ	1×10^{-6} base units
nano-	n	1×10^{-9} base units
pico-	p	1×10^{-12} base units

Practice Problem 1 How would you convert 1000 micrometers to kilometers?

Convert Temperature

The following formulas can be used to convert between Fahrenheit and Celsius temperatures. Notice that each equation can be obtained by algebraically rearranging the other. Therefore, you only need to remember one of the equations.

Conversion of Fahrenheit to Celsius

$$°C = \frac{(°F) - 32}{1.8}$$

Conversion of Celsius to Fahrenheit

$$°F = 1.8(°C) + 32$$

Make and Use Tables

Tables help visually organize data so that it can be interpreted more easily. Tables are composed of several components—a title describing the contents of the table, columns and rows that separate and organize information, and headings that describe the information in each column or row.

Table SH.4 Glacier Movement Rates

Depth (m)	Distance (m)	Average Speed (m/day)
0	13.1	0.198
20	13.1	0.198
60	12.8	0.194
100	12.2	0.185
140	11.2	0.170
180	9.6	0.145

Looking at this table, you should not only be able to pick out specific information, but you should also notice trends.

Practice Problem 2 If scientists drilled another 40 m into the glacier, what would the speed of the glacier's movement be at that depth?

Make and Use Graphs

Scientists often organize data in graphs. The types of graphs typically used in science are the line graph, the bar graph, and the circle graph.

Line Graphs A line graph is used to show the relationship between two variables. The independent variable is plotted on the horizontal axis, called the *x*-axis. The dependent variable is plotted on the vertical axis, called the *y*-axis. The dependent variable *(y)* changes as a result of a change in the independent variable *(x)*.

Suppose your class wanted to collect data about humidity. You could make a graph of the amount of water vapor that air can hold at various temperatures. **Table SH.5** shows the data.

Table SH.5 Amount of Water Vapor in Air at Various Temperatures

Air Temperature (˚C)	Air (g/m^3)
10	10
20	18
30	31
40	50
50	80

To make a graph of the amount of water vapor in air, start by determining the dependent and independent variables. The average amount of water vapor found per cubic meter of air is the dependent variable and is plotted on the *y*-axis. The independent variable, air temperature, is plotted on the *x*-axis.

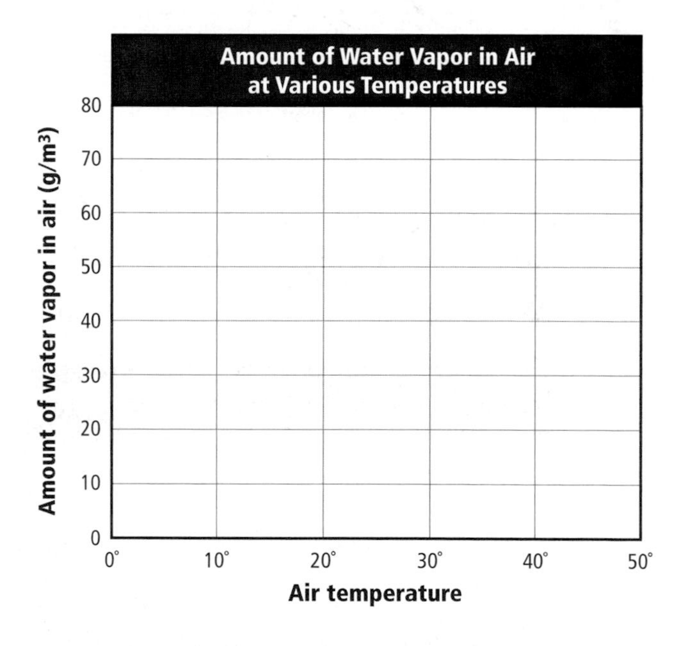

Plain or graph paper can be used to construct graphs. Draw a grid on your paper or a box around the squares that you intend to use on your graph paper. Give your graph a title and label each axis with a title and units. In this example, label the *x*-axis *Air temperature*. Because the lowest temperature was 10 and the highest was 50, you know that you will have to start numbers on the *y*-axis at least at 0 and number to at least 50. You decide to start numbering at 0 and number by equally spaced intervals of ten.

Label the y-axis of your graph *Amount of water vapor in air (g/m³)*. Begin plotting points by locating 0°C on the x-axis and 5 g/m³ on the y-axis. Where an imaginary vertical line from the x-axis and an imaginary horizontal line from the y-axis meet, place the first data point. Place other data points using the same process. After all the points are plotted, draw a "best fit" straight line through all the points.

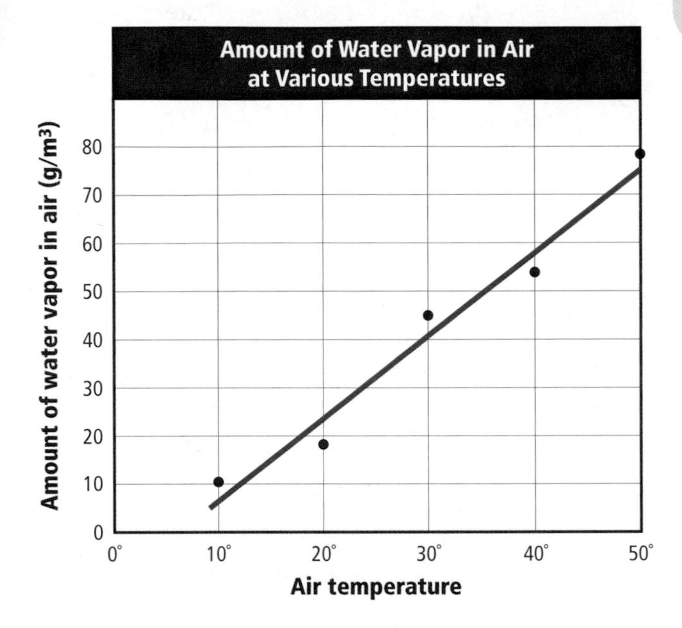

Practice Problem 3 According to the graph, does the amount of water vapor in air increase or decrease with air temperature?

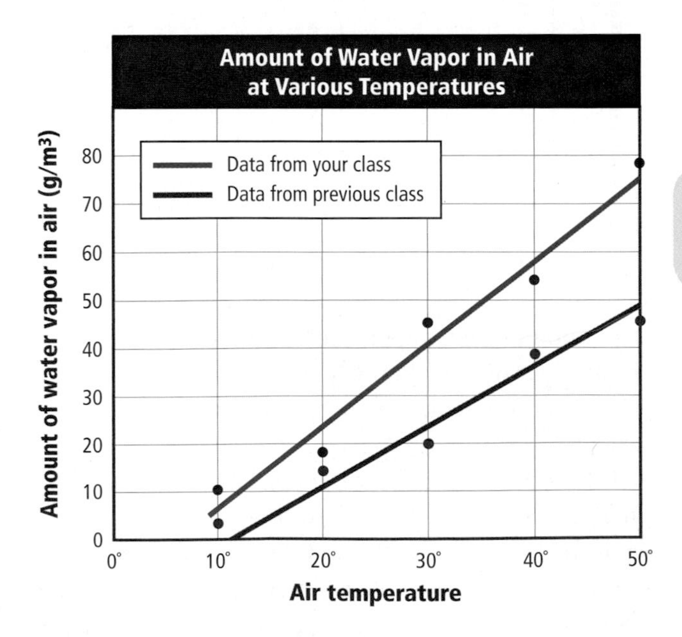

What if you wanted to compare the data about humidity collected by your class with similar data collected a year ago by a different class? The data from the other class can be plotted on the same graph to make the comparison. Include a key with different lines indicating different sets of data.

Practice Problem 4 How did the data from your class compare to the data from the previous class?

Bar Graphs A bar graph displays a comparison of different categories of data by representing each category with a bar. The length of the bar is related to the category's frequency. To make a bar graph, set up the x-axis and y-axis as you did for the line graph. Plot the data by drawing thick bars from the x-axis up to the y-axis point.

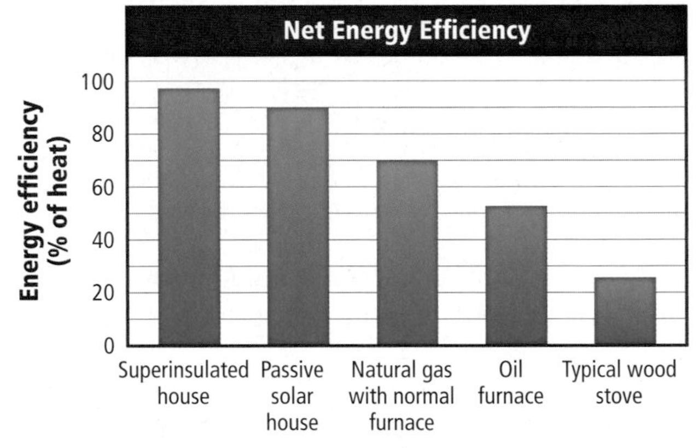

Look at the graph above. The independent variable is the energy efficiency. The dependent variable is the heating method.

Practice Problem 5 Which type of heating method has the second greatest efficiency? Is this more than twice as efficient as the lowest efficiency? Explain.

Bar graphs can also be used to display multiple sets of data in different categories at the same time. A bar graph that displays two sets of data is called a double-bar graph. Double-bar graphs have a legend to denote which bars represent each set of data. The graph below is an example of a double-bar graph.

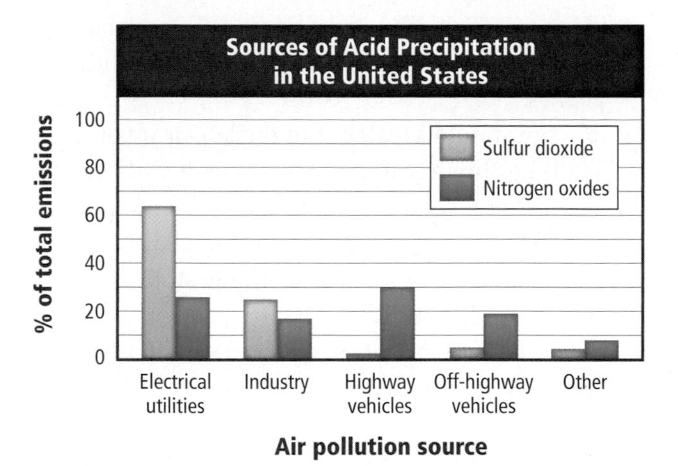

Circle Graphs A circle graph consists of a circle divided into sections that represent parts of a whole. When all the sections are placed together, they equal 100 percent of the whole.

Suppose you want to make a circle graph to show the percentage of solid wastes generated by various industries in the United States each year. The total amount of solid waste generated each year is estimated at ten billion metric tons. The whole circle graph will therefore represent this amount of solid waste. You find that 7.5 billion metric tons of waste is generated by mining and oil and gas production. The total amount of solid waste generated each year by mining and oil and gas production makes up one section of the circle graph, as follows.

$$\text{Segment of circle for total waste} = \frac{\text{waste from mining and oil and gas production}}{\text{total waste}}$$

$$= \frac{7.5}{10}$$

$$= 0.75 \times 360^\circ$$

$$= 270^\circ$$

To draw your circle graph, you will need a compass and a protractor. First, use the compass to draw a circle.

Then, draw a straight line from the center to the edge of the circle. Place your protractor on this line, and mark the point on the circle where 270° angle will intersect the circle. Draw a straight line from the center of the circle to the intersection point. This is the section for the waste generated from mining and oil and gas production.

Now, try to perform the same operation for the other data to find the number of degrees of the circle that each represents, and draw them in as well: agriculture, 1.3 billion metric tons; industry, 0.95 billion metric tons; municipal, 0.15 billion metric tons; and sewage sludge, 0.1 billion metric tons.

Complete your graph by labeling the sections of the graph and giving the graph a title. Your completed graph should look similar to the one below.

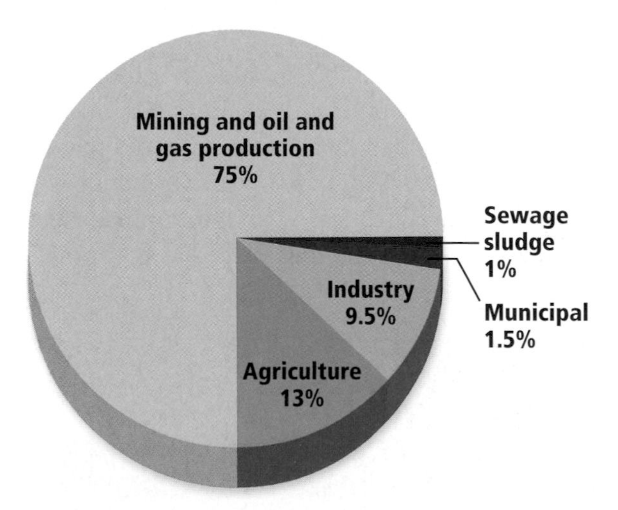

Practice Problem 6 There are 25 varieties of flowering plants growing around the high school. Construct a circle graph showing the percentage of each flower's color. Two varieties have yellow blooms, five varieties have blue-purple blooms, eight varieties have white blooms, and ten varieties have red blooms.

Safety in the Laboratory

The Earth science laboratory is a safe place to work if you are careful to observe the following important safety rules. You are responsible for your own safety and for the safety of others. The safety rules given here will protect you and others from harm in the laboratory. While carrying out procedures in any of the activities or GeoLabs, take note of the safety symbols and warning statements.

Safety Rules

1. Always read and complete the lab safety form and obtain your teacher's permission before beginning an investigation.
2. Study the procedure outline in the text. If you have questions, ask your teacher. Make sure that you understand all safety symbols shown on the page.
3. Use the safety equipment provided for you. Safety goggles and an apron should be worn during all investigations that involve the use of chemicals.
4. When heating test tubes, always slant them away from yourself and others.
5. Never eat or drink in the lab, and never use lab glassware as food or drink containers. Never inhale chemicals. Do not taste any substances or draw any material into a tube or pipet with your mouth.
6. If you spill any chemical, wash it off immediately with water. Report the spill immediately to your teacher.
7. Know the location and proper use of the fire extinguisher, eye wash, safety shower, fire blanket, fire alarm, and first aid kit. First aid procedures in the science laboratory are listed in **Table RH.1.**
8. Keep materials away from flames. Tie back hair and loose clothing when you are working with flames.
9. If a fire should break out in the lab, or if your clothing should catch fire, smother it with the fire blanket or a coat, get under a safety shower, or use the fire department's recommendation for putting out a fire on your clothing: stop, drop, and roll. NEVER RUN.
10. Report any accident or injury, no matter how small, to your teacher.

Clean-Up Procedures

1. Turn off the water and gas. Disconnect electrical devices.
2. Return all materials to their proper places.
3. Dispose of chemicals and other materials as directed by your teacher. Place broken glass and solid substances in the proper containers. Never discard materials in the sink.
4. Clean your work area.
5. Wash your hands thoroughly after working in the laboratory.

Table RH.1 First Aid in the Science Laboratory

Injury	Safe Response
Burns	Apply cold water. Call your teacher immediately.
Cuts and bruises	Stop any bleeding by applying direct pressure. Cover cuts with a clean dressing. Apply cold compresses to bruises. Call your teacher immediately.
Fainting	Leave the person lying down. Loosen any tight clothing and keep crowds away. Call your teacher immediately.
Foreign matter in eye	Flush with plenty of water. Use an eyewash bottle or fountain.
Poisoning	Note the suspected poisoning agent and call your teacher immediately.
Any spills on skin	Flush with large amounts of water or use safety shower. Call your teacher immediately.

Safety Symbols

Safety symbols in the following table are used in the lab activities to indicate possible hazards. Learn the meaning of each symbol. **It is recommended that you wear safety goggles and apron at all times in the lab. This might be required in your school district.**

SAFETY SYMBOLS	HAZARD	EXAMPLES	PRECAUTION	REMEDY
DISPOSAL	Special disposal procedures need to be followed.	certain chemicals, living organisms	Do not dispose of these materials in the sink or trash can.	Dispose of wastes as directed by your teacher.
BIOLOGICAL	Organisms or other biological materials that might be harmful to humans	bacteria, fungi, blood, unpreserved tissues, plant materials	Avoid skin contact with these materials. Wear mask or gloves.	Notify your teacher if you suspect contact with material. Wash hands thoroughly.
EXTREME TEMPERATURE	Objects that can burn skin by being too cold or too hot	boiling liquids, hot plates, dry ice, liquid nitrogen	Use proper protection when handling.	Go to your teacher for first aid.
SHARP OBJECT	Use of tools or glassware that can easily puncture or slice skin	razor blades, pins, scalpels, pointed tools, dissecting probes, broken glass	Practice common-sense behavior and follow guidelines for use of the tool.	Go to your teacher for first aid.
FUME	Possible danger to respiratory tract from fumes	ammonia, acetone, nail polish remover, heated sulfur, moth balls	Make sure there is good ventilation. Never smell fumes directly. Wear a mask.	Leave foul area and notify your teacher immediately.
ELECTRICAL	Possible danger from electrical shock or burn	improper grounding, liquid spills, short circuits, exposed wires	Double-check setup with teacher. Check condition of wires and apparatus. Use GFI-protected outlets.	Do not attempt to fix electrical problems. Notify your teacher immediately.
IRRITANT	Substances that can irritate the skin or mucous membranes of the respiratory tract	pollen, moth balls, steel wool, fiberglass, potassium permanganate	Wear dust mask and gloves. Practice extra care when handling these materials.	Go to your teacher for first aid.
CHEMICAL	Chemicals that can react with and destroy tissue and other materials	bleaches such as hydrogen peroxide; acids such as sulfuric acid, hydrochloric acid; bases such as ammonia, sodium hydroxide	Wear goggles, gloves, and an apron.	Immediately flush the affected area with water and notify your teacher.
TOXIC	Substance may be poisonous if touched, inhaled, or swallowed.	mercury, many metal compounds, iodine, poinsettia plant parts	Follow your teacher's instructions.	Always wash hands thoroughly after use. Go to your teacher for first aid.
FLAMMABLE	Open flame may ignite flammable chemicals, loose clothing, or hair.	alcohol, kerosene, potassium permanganate, hair, clothing	Avoid open flames and heat when using flammable chemicals.	Notify your teacher immediately. Use fire safety equipment if applicable.
OPEN FLAME	Open flame in use, may cause fire.	hair, clothing, paper, synthetic materials	Tie back hair and loose clothing. Follow teacher's instructions on lighting and extinguishing flames.	Always wash hands thoroughly after use. Go to your teacher for first aid.

Eye Safety
Proper eye protection must be worn at all times by anyone performing or observing science activities.

Clothing Protection
This symbol appears when substances could stain or burn clothing.

Animal Safety
This symbol appears when safety of animals and students must be ensured.

Radioactivity
This symbol appears when radioactive materials are used.

Handwashing
After the lab, wash hands with soap and water before removing goggles.

Physiographic Map of Earth

Reference Handbook

Topographic Map Symbols

Roads and Railroads

Primary highway, hard surface

Secondary highway, hard surface

Light-duty road, hard or improved surface

Unimproved road

Railroad: single track and multiple track

Railroads in juxtaposition

Buildings and Structures

Buildings

School, church, and cemetery — cem

Barn and warehouse

Wells, not water (with labels) — o oil o gas

Tanks: oil, water, etc. (labeled if water) — water

Open-pit mine, quarry, or prospect

Tunnel

Benchmark — BM Δ 293

Bridge

Campsite

Habitats

Marsh (swamp)

Wooded marsh

Woods or brushwood

Vineyard

Submerged marsh

Mangrove

Coral reef, rocks

Orchard

Urban area

Perennial streams

Elevated aqueduct

Water well and spring

Small rapids

Large rapids

Intermittent lake

Intermittent stream

Glacier

Large falls

Dry lake bed

Surface Elevations

Spot elevation — ×7369

Water elevation — 670

Index contour — 100

Intermediate contour

Depression contour

Boundaries

National

State

County, parish, municipal

Civil township, precinct, town, barrio

Incorporated city, village, town, hamlet

Reservation, national or state

Small park, cemetery, airport, etc.

Land grant

Township or range line, United States land survey

Township or range line, approximate location

Weather Map Symbols

Sample Plotted Report at Each Station

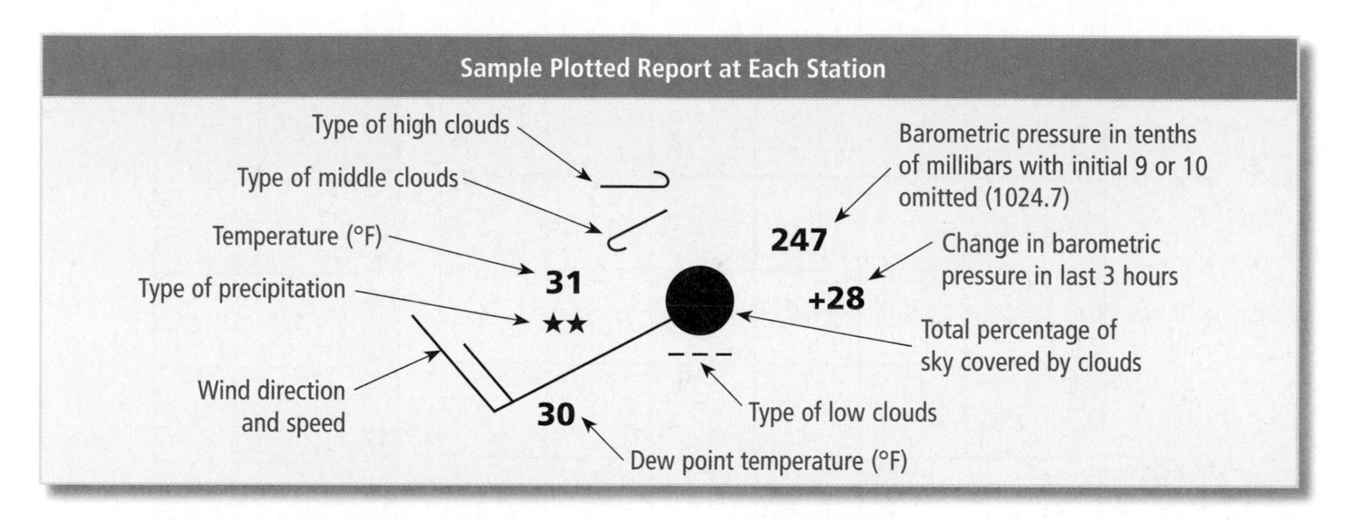

Symbols Used in Plotting Report

Precipitation	Wind Direction and Speed	Sky Coverage	Fronts and Pressure Sysyems	
Fog	0 calm	No cover	(H) or High	Center of high- or
Snow	1–2 knots	1/10 or less	(L) or Low	low-pressure system
Rain	3–7 knots	2/10 to 3/10		Cold front
Thunderstorm	8–12 knots	4/10		Warm front
Drizzle	13–17 knots	1/2		Occluded front
Showers	18–22 knots	6/10		Stationary front
	23–27 knots	7/10		
	48–52 knots	Overcast with openings		
	1 knot = 1.852 km/h	Completely overcast		

Clouds

Some Types of High Clouds	Some Types of Middle Clouds	Some Types of Low Clouds
Scattered cirrus	Thin altostratus layer	Cumulus of fair weather
Dense cirrus in patches	Thick altostratus layer	Stratocumulus
Veil of cirrus covering entire sky	Thin altostratus in patches	Fractocumulus of bad weather
Cirrus not covering entire sky	Thin altostratus in bands	Stratus of fair weather

PERIODIC TABLE OF THE ELEMENTS

Element — Hydrogen; Atomic number — 1; Symbol — H; Atomic mass — 1.008; State of matter

State of matter: Gas, Liquid, Solid, Synthetic

Metal, Metalloid, Nonmetal, Recently observed

Period	1	2	3	4	5	6	7	8	9	10	11	12	13	14	15	16	17	18
1	Hydrogen 1 **H** 1.008																	Helium 2 **He** 4.003
2	Lithium 3 **Li** 6.941	Beryllium 4 **Be** 9.012											Boron 5 **B** 10.811	Carbon 6 **C** 12.011	Nitrogen 7 **N** 14.007	Oxygen 8 **O** 15.999	Fluorine 9 **F** 18.998	Neon 10 **Ne** 20.180
3	Sodium 11 **Na** 22.990	Magnesium 12 **Mg** 24.305											Aluminum 13 **Al** 26.982	Silicon 14 **Si** 28.086	Phosphorus 15 **P** 30.974	Sulfur 16 **S** 32.066	Chlorine 17 **Cl** 35.453	Argon 18 **Ar** 39.948
4	Potassium 19 **K** 39.098	Calcium 20 **Ca** 40.078	Scandium 21 **Sc** 44.956	Titanium 22 **Ti** 47.867	Vanadium 23 **V** 50.942	Chromium 24 **Cr** 51.996	Manganese 25 **Mn** 54.938	Iron 26 **Fe** 55.847	Cobalt 27 **Co** 58.933	Nickel 28 **Ni** 58.693	Copper 29 **Cu** 63.546	Zinc 30 **Zn** 65.39	Gallium 31 **Ga** 69.723	Germanium 32 **Ge** 72.61	Arsenic 33 **As** 74.922	Selenium 34 **Se** 78.96	Bromine 35 **Br** 79.904	Krypton 36 **Kr** 83.80
5	Rubidium 37 **Rb** 85.468	Strontium 38 **Sr** 87.62	Yttrium 39 **Y** 88.906	Zirconium 40 **Zr** 91.224	Niobium 41 **Nb** 92.906	Molybdenum 42 **Mo** 95.94	Technetium 43 **Tc** (98)	Ruthenium 44 **Ru** 101.07	Rhodium 45 **Rh** 102.906	Palladium 46 **Pd** 106.42	Silver 47 **Ag** 107.868	Cadmium 48 **Cd** 112.411	Indium 49 **In** 114.82	Tin 50 **Sn** 118.710	Antimony 51 **Sb** 121.757	Tellurium 52 **Te** 127.60	Iodine 53 **I** 126.904	Xenon 54 **Xe** 131.290
6	Cesium 55 **Cs** 132.905	Barium 56 **Ba** 137.327	Lanthanum 57 **La** 138.905	Hafnium 72 **Hf** 178.49	Tantalum 73 **Ta** 180.948	Tungsten 74 **W** 183.84	Rhenium 75 **Re** 186.207	Osmium 76 **Os** 190.23	Iridium 77 **Ir** 192.217	Platinum 78 **Pt** 195.08	Gold 79 **Au** 196.967	Mercury 80 **Hg** 200.59	Thallium 81 **Tl** 204.383	Lead 82 **Pb** 207.2	Bismuth 83 **Bi** 208.980	Polonium 84 **Po** 208.982	Astatine 85 **At** 209.987	Radon 86 **Rn** 222.018
7	Francium 87 **Fr** (223)	Radium 88 **Ra** (226)	Actinium 89 **Ac** (227)	Rutherfordium 104 **Rf** (261)	Dubnium 105 **Db** (262)	Seaborgium 106 **Sg** (266)	Bohrium 107 **Bh** (264)	Hassium 108 **Hs** (277)	Meitnerium 109 **Mt** (268)	Darmstadtium 110 **Ds** (281)	Roentgenium 111 **Rg** (272)	Ununbium * 112 **Uub** (285)	Ununtrium * 113 **Uut** (284)	Ununquadium * 114 **Uuq** (289)	Ununpentium * 115 **Uup** (288)	Ununhexium * 116 **Uuh** (291)		Ununoctium * 118 **Uuo** (294)

The number in parentheses is the mass number of the longest lived isotope for that element.

* The names and symbols for elements 112, 113, 114, 115, 116, and 118 are temporary. Final names will be selected when the elements' discoveries are verified.

Series														
Lanthanide series	Cerium 58 **Ce** 140.115	Praseodymium 59 **Pr** 140.908	Neodymium 60 **Nd** 144.242	Promethium 61 **Pm** (145)	Samarium 62 **Sm** 150.36	Europium 63 **Eu** 151.965	Gadolinium 64 **Gd** 157.25	Terbium 65 **Tb** 158.925	Dysprosium 66 **Dy** 162.50	Holmium 67 **Ho** 164.930	Erbium 68 **Er** 167.259	Thulium 69 **Tm** 168.934	Ytterbium 70 **Yb** 173.04	Lutetium 71 **Lu** 174.967
Actinide series	Thorium 90 **Th** 232.038	Protactinium 91 **Pa** 231.036	Uranium 92 **U** 238.029	Neptunium 93 **Np** (237)	Plutonium 94 **Pu** (244)	Americium 95 **Am** (243)	Curium 96 **Cm** (247)	Berkelium 97 **Bk** (247)	Californium 98 **Cf** (251)	Einsteinium 99 **Es** (252)	Fermium 100 **Fm** (257)	Mendelevium 101 **Md** (258)	Nobelium 102 **No** (259)	Lawrencium 103 **Lr** (262)

Table RH.2	Relative Humidity %									
Dry-Bulb Temperature	Dry-Bulb Temperature Minus Wet-Bulb Temperature (°C)									
	1	2	3	4	5	6	7	8	9	10
0°C	81	64	46	29	13					
1°C	83	66	49	33	18					
2°C	84	68	52	37	22	7				
3°C	84	69	55	40	25	12				
4°C	85	71	57	43	29	16				
5°C	85	72	58	45	32	20				
6°C	86	73	60	48	35	24	11			
7°C	86	74	61	49	38	26	15			
8°C	87	75	63	51	40	29	19	8		
9°C	87	76	65	53	42	32	21	12		
10°C	88	77	66	55	44	34	24	15	6	
11°C	89	78	67	56	46	36	27	18	9	
12°C	89	78	68	58	48	39	29	21	12	
13°C	89	79	69	59	50	41	32	22	15	7
14°C	90	79	70	60	51	42	34	26	18	10
15°C	90	80	71	61	53	44	36	27	20	13
16°C	90	81	71	63	54	46	38	30	23	15
17°C	90	81	72	64	55	47	40	32	25	18
18°C	91	82	73	65	57	49	41	34	27	20
19°C	91	82	74	65	58	50	43	36	29	22
20°C	91	83	74	66	59	51	44	37	31	24
21°C	91	83	75	67	60	53	46	39	32	26
22°C	92	83	76	68	61	54	47	40	34	28
23°C	92	84	76	69	62	55	48	42	36	30
24°C	92	84	77	69	62	56	49	43	37	31
25°C	92	84	77	70	63	57	50	44	39	33
26°C	92	85	78	71	64	58	51	46	40	34
27°C	92	85	78	71	65	58	52	47	41	36
28°C	93	85	78	72	65	59	53	48	42	37
29°C	93	86	79	72	66	60	54	49	43	38
30°C	93	86	79	73	67	61	55	50	44	39
31°C	93	86	80	73	67	62	56	50	45	40
32°C	93	86	80	74	68	62	57	51	46	41

Table RH.3	Minerals with Metallic Luster						
Mineral (Formula)	Color	Streak	Hardness	Specific Gravity	Crystal System	Breakage Pattern	Uses and Other Properties
Bornite (Cu_5FeS_4)	bronze, tarnishes to dark blue purple	gray-black	3	4.9–5.4	tetragonal	uneven fracture	source of copper called "peacock ore" because of the purple shine when it tarnishes
Chalcopyrite ($CuFeS_2$)	brassy to golden yellow	greenish black	3.5–4	4.2	tetragonal	uneven fracture	main ore of copper
Chromite ($FeCr_2O_4$)	black or brown	brown to black	5.5	4.6	cubic	irregular fracture	ore of chromium, stainless steel, metallurgical bricks
Copper (Cu)	copper red	copper red	3	8.5–9	cubic	hackly	coins, pipes, gutters, wire, cooking utensils, jewelry, decorative plaques; malleable and ductile
Galena (PbS)	gray	gray to black	2.5	7.5	cubic	cubic cleavage perfect	source of lead, used in pipes, shields for X rays, fishing equipment sinkers
Gold (Au)	pale to golden yellow	yellow	2.5–3	19.3	cubic	hackly	jewelry, money, gold leaf, fillings for teeth, medicines; does not tarnish
Graphite (C)	black to gray	black to gray	1–2	2.3	hexagonal	basal cleavage (scales)	pencil lead, lubricants for locks, rods to control some small nuclear reactions, battery poles
Hematite (specular) (Fe_2O_3)	black or reddish brown	red or reddish brown	6	5.3	hexagonal	irregular fracture	source of iron; roasted in a blast furnace, converted to "pig" iron, made into steel
Magnetite (Fe_3O_4)	black	black	6	5.2	cubic	conchoidal fracture	source of iron, naturally magnetic, called lodestone
Pyrite (FeS_2)	light, brassy yellow	greenish black	6.5	5.0	cubic	uneven fracture	source of iron, "fool's gold," alters to limonite
Pyrrhotite ($Fe_{1-X}S$)* *contains one less atom of Fe than S	bronze	gray-black	4	4.6	hexagonal	uneven fracture	an ore of iron and sulfur; may be magnetic
Silver (Ag)	silvery white, tarnishes to black	light gray to silver	2.5	10–12	cubic	hackly	coins, fillings for teeth, jewelry, silverplate, wires; malleable and ductile

Table RH.4 Minerals with Nonmetallic Luster

Mineral (Formula)	Color	Streak	Hardness	Specific Gravity	Crystal System	Breakage Pattern	Uses and Other Properties
Augite ((Ca, Na)(Mg, Fe, Al)(Al, Si)$_2$O$_6$)	black	colorless	6	3.3	monoclinic	2-directional cleavage	square or 8-sided cross section
Corundum (Al_2O_3)	colorless, blue, brown, green, white, pink, red	colorless	9	4.0	hexagonal	fracture	gemstones: ruby is red, sapphire is blue; industrial abrasive
Feldspar (orthoclase) ($KAlSi_3O_8$)	colorless, white to gray, green, yellow	colorless	6	2.5	monoclinic	two cleavage planes meet at 90° angle	insoluble in acids; used in the manufacture of porcelain
Feldspar (plagioclase) ($NaAlSi_3O_8$) ($CaAl_2Si_3O_8$)	gray, green, white	colorless	6	2.5	triclinic	two cleavage planes meet at 86° angle	used in ceramics; striations present on some faces
Fluorite (CaF_2)	colorless, white, blue, green, red, yellow, purple	colorless	4	3–3.2	cubic	cleavage	used in the manufacture of optical equipment; glows under ultraviolet light
Garnet (Mg, Fe, Ca, Mn)$_3$, (Al, Fe, Cr)$_2$, (SiO$_4$)$_3$	deep yellow-red, green, black	colorless	7.5	3.5	cubic	conchoidal fracture	used in jewelry; also used as an abrasive
Hornblende Ca$_2$Na (Mg, Fe^2)$_4$, (Al, Fe$_3$, Ti)$_3$, Si$_8$O$_{22}$ (O, OH)$_2$	green to black	gray to white	5–6	3.4	monoclinic	cleavage in two directions	will transmit light on thin edges; 6-sided cross section
Limonite (hydrous iron oxides)	yellow, brown, black	yellow, brown	5.5	2.7–4.3	N/A	conchoidal fracture	source of iron; weathers easily, coloring matter of soils
Olivine ((Mg, Fe)$_2$ SiO$_4$)	Olive green	colorless	6.5	3.5	orthorhombic	Conchoidal fracture	Gemstones, refractory sand
Quartz (SiO_2)	Colorless, various colors	colorless	7	2.6	hexagonal	Conchoidal fracture	Used in glass manufacture, electronic equipment, radios, computers, watches, gemstones
Topaz (Al$_2$SiO$_4$ (F, OH)$_2$)	Colorless, white, pink, yellow, pale blue	colorless	8	3.5	orthorhombic	Basal cleavage	Valuable gemstone

Table RH.5	Rocks	
Rock Type	**Rock Name**	**Characteristics**
Igneous (intrusive)	granite	large mineral grains of quartz, feldspar, hornblende, and mica; usually light in color
	diorite	large mineral grains of feldspar, hornblende, and mica; less quartz than granite; intermediate in color
	gabbro	large mineral grains of feldspar, hornblende, augite, olivine, and mica; no quartz; dark in color
Igneous (extrusive)	rhyolite	small or no visible grains of quartz, feldspar, hornblende, and mica; light in color
	andesite	small or no visible grains of quartz, feldspar, hornblende, and mica; less quartz than rhyolite; intermediate in color
	basalt	small or no visible grains of feldspar, hornblende, augite, olivine, and mica; no quartz; dark in color; vessicles may be present
	obsidian	glassy texture; no visible grains; volcanic glass; fracture is conchoidal; color is usually black, but may be red-brown or black with white flecks
	pumice	frothy texture; floats; usually light in color
Sedimentary (clastic)	conglomerate	coarse-grained; gravel- or pebble-sized grains
	sandstone	sand-sized grains 1/16 to 2 mm in size; varies in color
	siltstone	grains smaller than sand but larger than clay
	shale	smallest grains; usually dark in color
Sedimentary (chemical or biochemical)	limestone	major mineral is calcite; usually forms in oceans, lakes, rivers, and caves; often contains fossils; effervesces in dilute HCl
	coal	occurs in swampy, low-lying areas; compacted layers of organic material, mainly plant remains
Sedimentary (chemical)	rock salt	commonly forms by the evaporation of seawater
Metamorphic	gneiss	well-developed banding because of alternating layers of different minerals, usually of different colors; common parent rock is granite
	schist	well-developed parallel arrangement of flat, sheetlike minerals, mainly micas; common parent rocks are shale and phyllite
	phyllite	shiny or silky appearance; may look wrinkled; common parent rocks are shale and slate
	slate	harder, denser, and shinier than shale; common parent rock is shale
Metamorphic (nonfoliated)	marble	interlocking calcite or dolomite crystals; common parent rock is limestone
	soapstone	composed mainly of the mineral talc; soft with a greasy feel
	quartzite	hard and well-cemented with interlocking quartz crystals; common parent rock is sandstone

Solar System Charts

The Planets

	Mercury	Venus	Earth	Mars	Jupiter	Saturn	Uranus	Neptune
Mass (kg)	3.302×10^{23}	4.8685×10^{24}	5.9736×10^{24}	6.4185×10^{23}	1.8986×10^{27}	5.6846×10^{26}	8.6832×10^{25}	1.0243×10^{26}
Equatorial radius (km)	2439.7	6051.8	6378.1	3397	71,492	60,268	25,559	24,764
Mean density (kg/m^3)	5427	5243	5515	3933	1326	700	1300	1760
Albedo	0.056	0.750	0.306	0.250	0.343	0.342	0.300	0.290
Semimajor axis (km)	5.791×10^{7}	1.0821×10^{8}	1.4960×10^{8}	2.2792×10^{8}	7.7857×10^{8}	1.43353×10^{9}	2.87246×10^{9}	4.49506×10^{9}
Orbital period (Earth days)	87.969	224.701	365.256	686.980	4330.6	10,755.7	30,687.2	60,189
Orbital inclination (degrees)	7.00	3.39	0.00	1.850	1.304	2.485	0.772	1.769
Orbital eccentricity	0.2056	0.0067	0.0167	0.0935	0.0484	0.0541	0.0472	0.0859
Rotational period (hours)	1407.6	5832.5[R]	23.9345	24.6229	9.9250	10.656	17.24[R]	16.11
Axial tilt (degrees)	0.01	177.36	23.45	25.19	3.13	26.73	97.86	29.58
Average surface temperature (K)	440	737	288	210	163	133	78	73
Number of known moons*	0	0	1	2	63	56	27	13

*Number as of 2007.
[R] indicates retrograde rotation.

The Moon	
Mass (kg)	7.349×10^{22}
Equatorial radius (km)	1737.4
Mean density (kg/m3)	3340
Albedo	0.067
Semimajor axis (km)	3.844×10^{5}
Orbital period (Earth days)	27.3217
Lunar period (Earth days)	29.53
Orbital inclination (degrees)	5.145
Orbital eccentricity	0.0549
Rotational period (hours)	655.728

The Sun	
Mass (kg)	1.99×10^{30}
Equatorial radius (km)	6.96×10^{5}
Mean density (kg/m^3)	1409
Absolute magnitude	14.83
Luminosity (W)	384.6
Spectral type	G2
Rotational period (hours)	609.12
Average temperature (K)	5778

Glossary/Glosario

A multilingual science glossary at glencoe.com includes Arabic, Bengali, Chinese, English, Haitian Creole, Hmong, Korean, Portuguese, Russian, Tagalog, Urdu, and Vietnamese.

Pronunciation Key

Use the following key to help you sound out words in the glossary.

a	**back** (BAK)	**ew**	**food** (FEWD)
ay	**day** (DAY)	**yoo**	**pure** (PYOOR)
ah	**father** (FAH thur)	**yew**	**few** (FYEW)
ow	**flower** (FLOW ur)	**uh**	**comma** (CAHM uh)
ar	**car** (CAR)	**u** (+con)	**rub** (RUB)
e	**less** (LES)	**sh**	**shelf** (SHELF)
ee	**leaf** (LEEF)	**ch**	**nature** (NAY chur)
ih	**trip** (TRIHP)	**g**	**gift** (GIHFT)
i (i+con+e)	**idea, life** (i DEE uh, LIFE)	**j**	**gem** (JEM)
oh	**go** (GOH)	**ing**	**sing** (SING)
aw	**soft** (SAWFT)	**zh**	**vision** (VIHZH un)
or	**orbit** (OR but)	**k**	**cake** (KAYK)
oy	**coin** (COYN)	**s**	**seed, cent** (SEED, SENT)
oo	**foot** (FOOT)	**z**	**zone, raise** (ZOHN, RAYZ)

Como usar el glosario en espanol:
1. Busca el termino en ingles que desees encontrar.
2. El termino en espanol, junto con la definicion, se encuentran en la columna de la derecha.

Glossary/Glosario

A

English

abrasion: (p. 203) process of erosion in which windblown or waterborne particles, such as sand, scrape against rock surfaces or other materials and wear them away.

absolute-age dating: (p. 601) method that enables scientists to determine tha actual age of certain rocks and other objects.

absolute magnitude: (p. 842) brightness an object would have if it were placed at a distance of 10 pc; classification system for stellar brightness that can be calculated only when the actual distance to a star is known.

abyssal plain: (p. 451) smooth, flat part of the seafloor covered with muddy sediments and sedimentary rocks that extends seaward from the continental margin.

acid: (p. 71) solution containing a substance that produces hydrogen ions: (H^+) in water.

acid precipitation: (p. 745) any precipitation with a pH of less than 5.0 that forms when sulfur dioxide and nitrogen oxides combine with moisture in the atmosphere to produce sulfuric acid and nitric acid.

Español

abrasión: (pág. 203) proceso erosivo en que las partículas por el viento o el agua, como la arena, chocan y raspan superficies rocosas u otros materiales y los desgastan.

datación absoluta: (pág. 601) permite a los científicos determinar la antigüedad real de ciertas rocas y objetos.

magnitud absoluta: (pág. 842) brillo que tendría un objeto si estuviera a una distancia de 10 pc; sistema de clasificación del brillo estelar que se puede calcular sólo cuando se conoce la distancia verdadera hasta la estrella.

llanura abisal: (pág. 451) parte plana y lisa del fondo del mar cubierta con sedimentos fangosos y rocas sedimentarias y que se extiende desde el margen continental hacia el mar.

ácido: (pág. 71) solución que contiene una sustancia que produce iones hidrógeno (H^+) en agua.

precipitación ácida: (pág. 745) toda precipitación con un pH menor que 5.0 que se forma cuando se combinan el dióxido de azufre y óxidos de nitrógeno con la humedad en la atmósfera para producir ácido sulfúrico o ácido nítrico.

active galactic nucleus (AGN): (p. 875) a galaxy's core in which highly energetic objects or activities are located.

aggregate: (p. 684) mixture of sand, gravel, and crushed stone that accumulates naturally; found in floodplains, alluvial fans, or glacial deposits.

air mass: (p. 316) large volume of air that has the characteristics of the area over which it forms.

air-mass thunderstorm: (p. 346) type of thunderstorm in which air rises because of unequal heating of Earth's surface within a single air mass and is most common during the afternoon and evening.

albedo: (p. 771) percentage of sunlight that is reflected by the surface of a planet or a satellite, such as the Moon.

altered hard part: (p. 607) fossil whose organic material has been removed and whose hard parts have been changed by recrystallization or mineral replacement.

amino acid: (p. 634) a building block of proteins.

Amniotic (am nee AH tihk) egg: (p. 658) egg with a shell, providing a complete environment for a developing embryo.

amplitude: (p. 539) the size of the seismic waves; an increase of 1 in the scale represents an increase in amplitude of a factor of 10.

analog forecast: (p. 331) weather forecast that compares current weather patterns to patterns that occurred in the past.

anemometer (a nuh MAH muh tur): (p. 325) weather instrument used to measure wind speed.

apogee: (p. 783) farthest point in the Moon's elliptical orbit to Earth.

apparent magnitude: (p. 842) classification system based on how bright a star appears to be; does not take distance into account so cannot indicate how bright a star actually is.

aquiclude: (p. 255) impermeable layer that is a barrier to groundwater; such as silt, clay, and shale.

aquifer: (p. 255) permeable underground layer through which groundwater flows relatively easily.

núcleo galáctico activo (NGA): (pág. 875) centro de la galaxia donde se ubican cuerpos o suceden eventos con gran cantidad de energía.

agregado: (pág. 684) mezcla natural de arena, grava y piedra triturada que se acumula naturalmente; se encuentra en llanuras aluviales, abanicos aluviales o depósitos glaciales.

masa de aire: (pág. 316) gran volumen de aire que tiene las características del área sobre la que se forma.

tormenta eléctrica de masa de aire: (pág. 346) tipo de tormenta en que el aire asciende debido al calentamiento desigual de la superficie terrestre bajo una misma masa de aire; es más común durante la tarde y la noche.

albedo: (pág. 771) porcentaje de luz solar que refleja la superficie de un planeta o un satélite, como por ejemplo, la Luna.

partes duras alteradas: (pág. 607) fósiles cuya materia orgánica ha desaparecido y cuyas partes duras han sido transformadas por recristalización o sustitución de minerales.

aminoácido: (pág. 634) unidad básica de las proteínas.

huevo amniótico: (pág. 658) huevo con cascarón; provee un ambiente completo para el embrión en desarrollo.

amplitud: (pág. 539) la magnitud de las ondas sísmicas; un aumento de 1 unidad en esta escala representa un aumento en amplitud de un factor de 10.

pronóstico análogo: (pág. 331) pronóstico del tiempo que compara los patrones actuales del clima con patrones ocurridos en el pasado.

anemómetro: (pág. 325) instrumento meteorológico que se utiliza para medir la velocidad de viento.

apogeo: (pág. 783) punto de la órbita elíptica de la Luna en que ésta se encuentra más alejada de la Tierra.

magnitud aparente: (pág. 842) sistema de clasificación basado el brillo aparente de una estrella; no toma en cuenta la distancia y por lo tanto no indica el brillo real de la estrella.

acuiclusos: (pág. 255) capas impermeables que sirven de barrera a las aguas subterráneas, como por ejemplo limo, arcilla o esquisto.

acuífero: (pág. 255) capa subterránea permeable por la cual el agua subterránea fluye de manera relativamente fácil.

artesian well: (p. 264) fountain of water that spurts above the land surface when a well taps a deep, confined aquifer containing water under pressure.

pozo artesiano: (pág. 264) fuente de agua que brota hacia la superficie terrestre, cuando un pozo conecta con un acuífero profundo y confinado que contiene agua bajo presión.

asteroid (AS tuh royd): (p. 622) metallic or silica-rich object, 1 km to 950 km in diameter, that bombarded early Earth, generating heat energy; **(p.795)** rocky remnant of the early solar system found mostly between the orbits of Mars and Jupiter in the asteroid belt.

asteroide: (pág. 622) cuerpo metálico o rico en sílice que mide de 1 a 950 km de diámetro y que bombardeó la Tierra primitiva generando energía calórica; **(pág. 795)** restos rocosos del sistema solar primitivo que se hallan principalmente entre las órbitas de Marte y Júpiter, en el cinturón de asteroides.

astronomical unit (AU): (p. 800) the average distance between the Sun and Earth, 1.496×10^8 km.

unidad astronómica (UA): (pág. 800) la distancia promedio entre el Sol y la Tierra, equivale a 1.496×10^8 km.

astronomy: (p. 6) study of objects beyond Earth's atmosphere.

astronomía: (pág. 6) el estudio de los cuerpos que se encuentran más allá de la atmósfera de la Tierra.

atmosphere: (p. 8) blanket of gases surrounding Earth that contains about 78 percent nitrogen, 21 percent oxygen, and 1 percent other gases such as argon, carbon dioxide, and water vapor.

atmósfera: (pág. 8) manto de gases que rodea la Tierra; está compuesta aproximadamente por 78 por ciento de nitrógeno, 21 por ciento de oxígeno y 1 por ciento de otros gases como el argón, el dióxido de carbono y el vapor del agua.

atomic number: (p. 62) number of protons contained in an atom's nucleus.

número atómico: (pág. 62) número de protones que contiene el núcleo de un átomo.

avalanche: (p. 198) landslide that occurs in a mountainous area when snow falls on an icy crust, becomes heavy, slips off, and slides swiftly down a mountainside.

avalancha: (pág. 198) deslizamiento que ocurre en un área montañosa cuando la nieve cae sobre una capa helada, aumenta de peso, se desprende y se resbala rápidamente montaña abajo.

B

banded-iron formations: (p. 630) alternating bands of iron oxide and chert; an iron-poor sedimentary rock.

formaciones de hierro en bandas: (pág. 630) bandas alternadas de óxido ferroso y pedernal; roca sedimentaria deficiente en hierro.

barometer: (p. 324) instrument used to measure air pressure.

barómetro: (pág. 324) instrumento que se usa para medir la presión atmosférica.

barrier island: (p. 442) long ridges of sand or other sediment deposited or shaped by the longshore current, that are separated from the mainland and can be up to tens of kilometers long.

barrera litoral: (pág. 442) grandes lomas de arena u otro sedimento que son depositadas, o que adquieren su forma, por la acción de las corrientes litorales; están separadas del continente y pueden llegar a medir decenas de kilómetros de largo.

basaltic rock: (p. 118) rock that is dark colored, has lower silica contents, and is rich in iron and magnesium; contains mostly plagioclase and pyroxene.

roca basáltica: (pág. 118) roca oscura con bajo contenido en sílice pero rica en hierro y magnesio; contiene principalmente plagioclasa y piroxenos.

base: (p. 72) substance that produces hydroxide ions (OH^-) in water.

base: (pág. 72) sustancia que produce iones hidróxido (OH^-) en agua.

base level: (p. 233) the elevation at which a stream enters another stream or body of water.

nivel base: (pág. 233) elevación a la cual una corriente entra a otra corriente o masa de agua.

batholith: (p. 515) coarse-grained, irregularly shaped, igneous rock mass that covers at least 100 km^2, generally forms 10–30 km below Earth's surface, and is common in the interior of major mountain chains.

batolito: (pág. 515) masa rocosa ígnea de grano grueso y de forma irregular que cubre por lo menos 100 km^2; generalmente se forma de 10 a 30 km bajo la superficie terrestre y es común en el interior de las principales cadenas montañosas.

beach: (p. 438) area in which loose sediment is deposited and moved about by waves along the shore.

playa: (pág. 438) área en que sedimentos sueltos son depositados y transportados por las olas a lo largo de la costa.

bedding: (p. 137) horizontal layering in sedimentary rock that can range from a few millimeters to several meters thick.

estratificación: (pág. 137) capas horizontales de roca sedimentaria que pueden medir de un milímetro a varios metros de grosor.

bed load: (p. 228) describes sediments that are too heavy or large to be kept in suspension or solution and are pushed or rolled along the bottom of a streambed.

carga de fondo: (pág. 228) término que describe los sedimentos que no se mantienen en suspensión, o en solución, porque son demasiado pesados o grandes y son empujados o arrastrados sobre el fondo del cauce de una corriente.

bedrock: (p. 684) unweathered, solid parent rock that can consist of limestone, marble, granite, or other quarried rock.

roca firme: (pág. 684) roca madre sólida no meteorizada que puede consistir en piedra caliza, mármol, granito o alguna otra piedra de cantera.

belt: (p. 812) low, warm, dark-colored cloud that sinks and flows rapidly in the Jovian atmosphere.

cinturón: (pág. 812) nube baja, tibia y oscura que desciende y fluye rápidamente en la atmósfera joviana.

Big Bang theory: (p. 878) theory that proposes that the universe began as a single point and has been expanding ever since.

teoría de la Gran Explosión: (pág. 878) propone que el universo empezó en un solo punto y se ha estado expan-diendo desde entonces.

binary star: (p. 838) describes two stars that are bound together by gravity and orbit a common center of mass.

estrella binaria: (pág. 838) describe dos estrellas unidas por la gravedad que giran alrededor de un centro común de masa.

biomass fuels (p. 709) fuels derived from living things; renewable resources.

biocombustible: (pág. 709) combustibles derivados de los seres vivos; recursos renovables.

bioremediation: (p. 742) use of organisms to clean up toxic waste.

biorremediación: (pág. 742) uso de organismos para limpiar desechos tóxicos.

biosphere: (p. 9) all of Earth's organisms and the environments in which they live.

biosfera: (pág. 9) incluye a todos los organismos de la Tierra y los ambientes en que éstos viven.

bipedal: (p. 665) walking upright on two legs.

bipedalismo: (pág. 665) que camina erguido sobre dos piernas.

black hole: (p. 851) small, extremely dense remnant of a star whose gravity is so immense that not even light can escape its gravity field.

agujero negro: (pág. 851) restos de una estrella muy densos y pequeños cuya gravedad es tan grande que ni la luz puede escapar de su campo de gravedad.

Bowen's reaction series: (p. 114) sequential, predictable, dual-branched pattern in which minerals crystallize from cooling magma.

serie de reacción de Bowen: (pág. 114) patrón de dos ramas, predecible y secuencial que siguen los minerales al cristalizarse a partir de magma que se enfría.

breaker: (p. 422) collapsing wave that forms when a wave reaches shallow water and becomes so steep that the crest topples forward.

rompiente: (pág. 422) ola que se colapsa; se forma cuando una ola alcanza aguas poco profundas y se vuelve tan empinada que la cresta de la ola se cae hacia adelante.

C

caldera: (p. 505) large crater, up to 50 km in diameter, that can form when the summit or side of a volcano collapses into the magma chamber during or after an eruption.

caldera: (pág. 505) cráter grande, de hasta 50 km de diámetro, que se forma cuando la cumbre o la ladera de un volcán se desploman en la cámara de magma durante o después de una erupción.

Cambrian explosion: (p. 653) sudden appearance of a diverse collection of organisms in the Cambrian fossil record.

explosión del Cámbrico: (pág. 653) aparición repentina de un conjunto diverso de organismos en el registro fósil del Cámbrico.

Canadian shield: (p. 625) name given to the Precambrian shield in North America because much of it is exposed in Canada.

escudo canadiense: (pág. 625) nombre que recibe el escudo Precámbrico en Norteamérica porque la mayor parte está expuesto en Canadá.

carrying capacity: (p. 735) number of organisms that a specific environment can support.

capacidad de carga: (pág. 735) número de organismos que un ambiente específico puede sustentar.

cartography: (p. 30) science of mapmaking.

cartografía: (pág. 30) ciencia de la elaboración de mapas.

cast: (p. 608) fossil formed when an earlier fossil of a plant or animal leaves a cavity that becomes filled with minerals or sediment.

molde: (pág. 608) fósil que se forma cuando un fósil precedente de una planta o un animal forma una cavidad que se rellena con minerales o sedimentos.

cave: (p. 260) a natural underground opening connected to Earth's surface, usually formed when groundwater dissolves limestone.

caverna: (pág. 260) cavidad subterránea abierta a la superficie terrestre, generalmente se forma cuando el agua subterránea disuelve la piedra caliza.

cementation: (p. 137) process of sedimentary rock formation that occurs when dissolved minerals precipitate out of groundwater and either a new mineral grows between the sediment grains or the same mineral grows between and over the grains.

cementación: (pág. 137) proceso de formación de roca sedimentaria que ocurre cuando los minerales disueltos del agua subterránea se precipitan y se forma un nuevo mineral entre los granos de sedimento o se acumula el mismo mineral entre y sobre los granos.

chemical bond: (p. 67) force that holds the atoms of elements together in a compound.

enlace químico: (pág. 67) fuerza que mantiene unidos los átomos de los elementos en un compuesto.

chemical reaction: (p. 70) change of one or more substances into other substances.

reacción química: (pág. 70) sucede cuando una o más sustancias se convierten en otras sustancias.

chemical weathering: (p. 166) process by which rocks and minerals undergo changes in their composition due to chemical reactions with agents such as acids, water, oxygen, and carbon dioxide.

meteorización química: (pág. 166) proceso mediante el cual las rocas y los minerales experimentan cambios en su composición, debido a reacciones químicas con agentes como ácidos, agua, oxígeno o dióxido de carbono.

chromosphere: (p. 831) layer of the Sun's atmosphere above the photosphere and below the corona that is about 2500 km thick and has a temperature around 30,000 K at its top.

cromosfera: (pág. 831) capa de la atmósfera del Sol situada encima de la fotosfera y debajo de la corona; mide aproximadamente 2500 km de ancho y tiene una temperatura cercana a 30,000 K en su parte superior.

cinder cone: (p. 507) steep-sided, generally small volcano that is built by the accumulation of tephra around the vent.

cono de carbonilla: (pág. 507) volcán generalmente pequeño, de laderas muy inclinadas, que se forma debido a la acumulación de tefrita alrededor de la chimenea.

cirque: (p. 209) deep depression scooped out by a valley glacier.

circo: (pág. 209) depresión profunda formada por un glaciar de valle.

cirrus (SIHR us): (p. 301) high clouds made up of ice crystals that form at heights of 6000 m; often have a wispy, indistinct appearance.

cirro: (pag. 301) nubes altas formadas por cristales de hielo que se forman a alturas de 6000 m; con frecuencia parecen espigas borrosas.

clastic: (p. 141) rock and mineral fragments produced by weathering and erosion and classified according to particle size and shape.

clástico: (pág. 141) describe los fragmentos de roca y de mineral producidos por la meteorización y la erosión; se clasifican según su tamaño y forma de partícula.

clastic sedimentary rock: **(p. 141)** most common type of sedimentary rock, formed from the abundant deposits of loose sediments that accumulate on Earth's surface; classified according to the size of their particles.

cleavage: **(p. 92)** the manner in which a mineral breaks along planes where atomic bonding is weak.

climate: **(p. 314)** the long-term average of variation in weather for a particular area.

climatology: **(p. 376)** study of Earth's climate in order to understand and predict climatic change, based on past and present variations in temperature, precipitation, wind, and other weather variables.

coalescence (ko uh LEH sunts): **(p. 302)** process that occurs when cloud droplets collide and form larger droplets, which eventually become too heavy to remain aloft and can fall to Earth as precipitation.

cogeneration: **(p. 723)** production of two usable forms of energy at the same time from the same process, which can conserve resources and generate income.

cold wave: **(p. 364)** extended period of below-average temperatures caused by large, high-pressure systems of continental polar or arctic origin.

comet: **(p. 819)** small, eccentrically orbiting body made of rock and ice which have one or more tails that point away from the Sun.

composite volcano: **(p. 507)** generally cone-shaped with concave slopes; built by violent eruptions of volcanic fragments and lava that accumulate in alternating layers.

compound: **(p. 66)** substance composed of atoms of two or more different elements that are chemically combined.

compressive force: **(p. 567)** squeezing force that can cause the intense deformation—folding, faulting metamorphism, and igneous intrusions—associated with mountain building.

condensation: **(p. 75)** process by which a cooling gas changes into a liquid and releases thermal energy.

condensation nucleus: **(p. 297)** small particle in the atmosphere around which cloud droplets can form.

roca sedimentaria clástica: **(pág. 141)** el tipo más común de roca sedimentaria; se forma a partir de los abundantes depósitos de sedimentos sueltos que se acumulan sobre la superficie de la Tierra; se clasifican según el tamaño de sus partículas.

crucero: **(pág. 92)** la forma en la cuál un mineral se rompe a lo largo de los planos donde los enlaces atómicos son débiles.

clima: **(pág. 314)** promedio durante un largo periodo de las variaciones en las condiciones del tiempo de un área determinada.

climatología: **(pág. 376)** estudio del clima de la Tierra para entender y pronosticar los cambios climáticos; se basa en variaciones pasadas y presentes de temperatura, precipitación, viento y otras variables del tiempo.

coalescencia: **(pág. 302)** proceso que ocurre cuando las gotas de nube chocan entre sí, formando gotas cada vez más grandes; estas gotas puede llegar a ser demasiado pesadas para seguir suspendidas en el aire y entonces caen a la Tierra como precipitación.

cogeneración: **(pág. 723)** producción simultánea de dos formas útiles de energía a partir del mismo proceso; puede ayudar a conservar recursos y obtener ganancias.

onda fría: **(pág. 364)** período prolongado de temperaturas más bajas que el promedio, causado por grandes sistemas de alta presión de origen polar continental o ártico.

cometa: **(pág. 819)** cuerpo pequeño de órbita excéntrica compuesto por roca y hielo y que contiene una o más colas que apuntan hacia el lado opuesto al Sol.

volcán compuesto: **(pág. 507)** volcán que en general tiene forma cónica y laderas cóncavas; se forma por erupciones violentas de fragmentos y lava volcánicos que se acumulan creando capas alternadas.

compuesto: **(pág. 66)** sustancia compuesta por átomos de dos o más elementos diferentes unidos químicamente.

fuerzas de compresión: **(pág. 567)** fuerzas de aplastamiento que pueden causar intensas deformaciones como plegamientos, fallas, metamorfismo e intrusiones ígneas; asociadas con la formación de montañas.

condensación: **(pág. 75)** proceso por el cual un gas enfriador se transforma en un líquido y libera energía térmica.

núcleos de condensación: **(pág. 297)** partículas pequeñas de la atmósfera alrededor de las cuales se pueden formar las gotas de nubes.

conduction: **(p. 288)** the transfer of thermal energy between objects in contact by the collisions between the particles in the objects.

conducción: **(pág. 288)** transferencia de energía entre cuerpos en contacto debida a la colisión entre las partículas de los cuerpos.

conduit: **(p. 505)** a tubelike structure that allows lava to reach the surface.

conducto: **(pág. 505)** estructura tubular que permite que la lava llegue a la superficie.

conic projection: **(p. 35)** map that is highly accurate for small areas, made by projecting points and lines from a globe onto a cone.

proyección cónica: **(pág. 35)** mapa de gran exactitud para áreas pequeñas que se elabora mediante la proyección de puntos y líneas de un globo a un cono.

constellation: **(p. 837)** group of stars that forms a pattern in the sky that resembles an animal, mythological character, or everyday object.

constelación: **(pág. 837)** grupo de estrellas que forman en el firmamento un patrón que semeja un animal, un personaje mitológico o un objeto cotidiano.

contact metamorphism: **(p. 149)** local effect that occurs when molten rock meets solid rock.

metamorfismo de contacto: **(pág. 149)** efecto local que ocurre cuando la roca fundida se encuentra con roca sólida.

continental drift: **(p. 469)** Wegener's hypothesis that Earth's continents were joined as a single landmass, called Pangaea, that broke apart about 200 mya and slowly moved to their present positions.

deriva continental: **(pág. 469)** hipótesis de Wegener que propone que los continentes de la Tierra estaban unidos en una sola masa terrestre, llamada Pangaea, la cual se separó hace aproximadamente 200 millones de años y que los fragmentos resultantes se movieron lentamente a sus ubicaciones actuales.

continental glacier: **(p. 208)** glacier that forms over a broad, continent-sized area of land and usually spreads out from its center.

glaciar continental: **(pág. 208)** glaciar que se forma sobre una amplia área del tamaño de un continente y que generalmente se extiende a partir de su centro.

continental margin: **(p. 447)** area where edges of continents meet the ocean; represents the shallowest part of the ocean that consists of the continental shelf, the continental slope, and the continental rise.

margen continental: **(pág. 447)** área donde los límites de los continentes se unen con el océano; representa la parte menos profunda del océano y consiste en la plataforma continental, el talud continental y el pie del talud continental.

continental rise: **(p. 449)** gently sloping accumulation of sediments deposited by a turbidity current at the foot of a continental margin.

pie del talud continental: **(pág. 449)** acumulación de sedimentos, con pendiente leve, depositados por una corriente de turbidez al pie de un margen continental.

continental shelf: **(p. 447)** shallowest part of a continental margin, with an average depth of 130 m and an average width of 60 km, that extends into the ocean from the shore and provides a nutrient-rich home to large numbers of fish.

plataforma continental: **(pág. 447)** parte más superficial del margen continental, tiene una profundidad promedio de 130 m y una anchura promedio de 60 km, se extiende hacia el océano desde la costa y proporciona un lugar rico en nutrientes a un gran número de peces.

continental slope: **(p. 448)** sloping oceanic region found beyond the continental shelf that generally marks the edge of the continental crust and may be cut by sub-marine canyons.

talud continental: **(pág. 448)** región oceánica inclinada que se encuentra más allá de la plataforma continental; generalmente marca el límite de la corteza continental y puede estar seccionada por cañones submarinos.

contour interval: **(p. 36)** difference in elevation between two side-by-side contour lines on a topographic map.

intervalo entre curvas de nivel: **(pág. 36)** diferencia en la elevación entre dos curvas de nivel contiguas en un mapa topográfico.

contour line: **(p. 36)** line on a topographic map that connects points of equal elevation.

curva de nivel: **(pág. 36)** curva en un mapa topográfico que conecta puntos de igual elevación.

control: (p. 12) standard for comparison in an experiment.

convection: (p. 288) the transfer of thermal energy by the movement of heated material from one place to another.

convergent boundary: (p. 482) place where two tectonic plates are moving toward each other; is associated with trenches, islands arcs, and folded mountains.

Coriolis effect: (p. 318) effect of a rotating body that influences the motion of any object or fluid; on Earth, air moving north or south from the equator appears to move right or left, respectively; the combination of the Coriolis effect and Earth's heat imbalance creates the trade winds, polar easterlies, and prevailing westerlies.

corona: (p. 831) top layer of the Sun's atmosphere that extends from the top of the chromosphere and ranges in temperature from 1 million to 2 million K.

correlation: (p. 599) matching of rock outcrops of one geographic region to another.

cosmic background radiation: (p. 880) weak radiation that is left over from the early, hot stages of the Big Bang expansion of the universe.

cosmology: (p. 878) study of the universe, including its current nature, origin, and evolution, based on observation and the use of theoretical models.

covalent bond: (p. 67) attraction of two atoms for a shared pair of electrons that holds the atoms together.

crater: (p. 505) bowl-shaped depression that forms around the central vent at the summit of a volcano.

craton (KRAY tahn): (p. 625) continental core formed from Archean or Proterozoic microcontinents; deepest (as far as 200 km into the mantle) and most stable part of a continent.

creep: (p. 195) slow, steady downhill movement of loose weathered Earth materials, especially soils, causing objects on a slope to tilt.

crest: (p. 421) highest point of a wave.

control: (pág. 12) estándar de comparación en un experimento.

convección: (pág. 288) transferencia de energía térmica debido al movimiento de material caliente de un lado a otro.

límite convergente: (pág. 482) lugar donde dos placas tectónicas se mueven aproximándose cada vez más entre sí; está asociado con fosas abisales, arcos insulares y montañas plegadas.

efecto de Coriolis: (pág. 318) efecto producido por un cuerpo en rotación que influye en el movimiento de todo cuerpo objeto o fluido; en la Tierra, las corrientes aire que se mueven desde el norte o desde el sur parecen desplazarse hacia la derecha o hacia la izquierda, respectivamente; la combinación del efecto de Coriolis y el desequilibrio térmico de la Tierra originan los vientos alisios, los vientos polares del este y los vientos dominantes del oeste.

corona: (pág. 831) capa superior de la atmósfera del Sol que se extiende desde la parte superior de la cromosfera y tiene un rango de temperatura de 1 a 2 millones K.

correlación: (pág. 599) correspondencia entre los afloramientos rocosos de una región geográfica y otra.

radiación cósmica de fondo: (pág. 880) radiación residual débil proveniente de las calientes etapas iniciales de la expansión del universo causada por la Gran Explosión.

cosmología: (pág. 878) estudio del universo; abarca su naturaleza actual, su origen y evolución y se basa en la observación y el uso de modelos teóricos.

enlace covalente: (pág. 67) atracción de dos átomos hacia un par compartido de electrones que mantienen a los átomos unidos.

cráter: (pág. 505) depresión en forma de tazón que generalmente se forma alrededor de la abertura central en la cumbre de un volcán.

cratón: (pág. 625) zona central de un continente formada a partir de microcontinentes del arcaico o del Proterozoico; son la parte más profunda (penetran hasta 200 km hacia el manto) y estable de un continente.

deslizamiento: (pág. 195) movimiento cuesta abajo constante y lento de materia meteorizada suelta de la Tierra, especialmente los suelos, lo que ocasiona que se inclinen los objetos en una ladera.

cresta: (pág. 421) punto más alto de una onda.

cross-bedding: (p. 138) depositional feature of sedimentary rock that forms as inclined layers of sediment are carried forward across a horizontal surface.

estratificación cruzada: (pág. 138) característica de la depo-sitación de roca sedimentaria que se forma a medida que capas inclinadas de sedimento son arrastradas hacia delante, a lo largo de una superficie horizontal.

cross-cutting relationships: (p. 597) the principle that an intrusion is younger than the rock it cuts across.

relaciones de corte transversal: (pág. 597) principio que establece que una intrusión es menos antigua que la roca que atraviesa.

crystal: (p. 87) solid in which atoms are arranged in repeating patterns.

cristal: (pág. 87) sólido cuyos átomos están ordenados en patrones repetitivos.

crystalline structure: (p. 73) regular geometric pattern of particles in most solids, giving a solid a definite shape and volume.

estructura cristalina: (pág. 73) patrón geométrico y regular que tienen las partículas en la mayoría de los sólidos; dan al sólido una forma y volumen definidos.

cumulus (KYEW myuh lus): (p. 301) puffy, lumpy-looking clouds that usually occur below 2000 m.

cúmulo: (pág. 301) nubes esponjosas con aspecto de madejas de algodón que generalmente se hallan a alturas menores de 2000 m.

cyanobacteria: (p. 629) microscopic, photosynthetic prokaryotes that formed stromatolites and changed early Earth's atmosphere by generating oxygen.

cianobacterias: (pág. 629) organismos procariotas fotosintéticos microscópicos que formaron estromatolitos y modificaron la atmósfera primitiva de la Tierra al producir oxígeno.

D

dark matter: (p. 870) invisible material thought to be made up of a form of subatomic particle that interacts only weakly with other matter.

materia oscura: (pág. 870) sustancia invisible formada por algún tipo de partícula subatómica que interactúa débilmente con otros tipos de material.

deep-sea trench: (p. 451) elongated, sometimes arc-shaped depression in the seafloor that can extend for thousands of kilometers, is the deepest part of the ocean basin, and is found primarily in the Pacific Ocean.

fosa abisal: (pág. 451) depresión alargada y en algunas ocasiones con forma de arco, que se puede extender miles de kilómetros; es la parte más profunda de la cuenca oceánica y se halla principalmente en el océano Pacífico.

deflation: (p. 202) lowering of land surface caused by wind erosion of loose surface particles, often leaving coarse sediments behind.

deflación: (pág. 202) depresión de la superficie terrestre causada por la erosión eólica de partículas superficiales sueltas; a menudo sólo contiene sedimentos gruesos.

deforestation: (p. 739) removal of trees from a forested area without adequate replanting, often using clear-cutting, which can result in loss of topsoil and water pollution.

deforestación: (pág. 739) eliminación de árboles de un área forestal, sin realizar una adecuada reforestación; a menudo es resultado de una corta a hecho, lo que puede ocasionar la pérdida del mantillo y la contaminación de las aguas.

delta: (p. 236) triangular deposit, usually made up of silt and clay particles, that forms where a stream enters a large body of water.

delta: (pág. 236) depósito triangular compuesto generalmente por partículas de limo y arcilla, que se forma en el sitio donde una corriente de agua entra a una gran masa de agua.

dendrochronology: (p. 604) science of using tree rings to determine absolute age; helped to date relatively recent geologic events and environmental changes.

dendrocronología: (pág. 604) ciencia que usa los anillos de crecimiento anual de los árboles para determinar la edad absoluta; permite datar eventos geológicos y cambios ambientales relativamente recientes.

density current: (p. 427) movement of ocean water that occurs in depths too great to be affected by surface winds and is generated by differences in water temperature and salinity.

corriente de densidad: (pág. 427) movimiento de las aguas oceánicas que ocurre a grandes profundidades, no se ve afectado por los vientos superficiales y es generado por las diferencias en temperatura y salinidad del agua.

density-dependent factor: (p. 736) environmental factor, such as disease, predators, or lack of food, that increasingly affects a population as the population's size increases.

factor dependiente de la densidad: (pág. 736) factor ambiental como las enfermedades, los depredadores o la falta de alimento, que afecta con creciente intensidad a una población a medida que aumenta el tamaño de su población.

density-independent factor: (p. 736) environmental factor that does not depend on population size, such as storms, flood, fires, or pollution.

factor independiente de la densidad: (pág. 736) factor ambiental, como las tempestades, las inundaciones, los incendios o la contaminación, que no son afectados por el tamaño de la población.

dependent variable: (p. 12) factor in an experiment that can change if the independent variable is changed.

variable dependiente: (pág. 12) factor de un experimento que puede cambiar al variar la variable independiente.

deposition: (p. 171) occurs when eroded materials are dropped in another location.

depositación: (pág. 171) ocurre cuando los materiales erosionados son depositados en otro sitio.

desalination: (p. 697) process that removes salt from seawater in order to provide freshwater.

desalinización: (pág. 697) proceso de eliminación de la sal del agua marina para obtener agua dulce.

desertification: (p. 683) process by which productive land becomes desert; in arid areas can occur through the loss of topsoil.

desertificación: (pág. 683) proceso mediante el cual las tierras productivas se convierten en desierto; en áreas áridas puede ocurrir debido a la pérdida del mantillo del suelo.

dew point: (p. 295) temperature to which air is cooled at a constant pressure to reach saturation, at which point condensation can occur.

punto de rocío: (pág. 295) temperatura a la cual el aire que se enfría a una presión constante alcanza la saturación, punto en el cual ocurre la condensación.

differentiation (dih fuh ren shee AY shun): (p. 623) process in which a planet becomes internally zoned, with the heavy materials sinking toward the center and the lighter materials accumulating near its surface.

diferenciación: (pág. 623) proceso en que un planeta se divide internamente en zonas, los materiales pesados se hunden hacia el centro, mientras que los materiales más ligeros se acumulan cerca de su superficie.

digital forecast: (p. 331) weather forecast that uses numerical data to predict how atmospheric variables change over time.

pronóstico digital: (pág. 331) pronóstico del tiempo que se basa en datos numéricos para predecir el cambio de las variables atmosféricas con el tiempo.

dike: (p. 516) pluton that cuts across preexisting rocks and often forms when magma invades cracks in surrounding rock bodies.

dique: (pág. 516) plutón que atraviesa las rocas preexistentes; suele formarse cuando el magma invade las grietas de los cuerpos rocosos circundantes.

discharge: (p. 229) measure of a volume of stream water that flows over a specific location in a particular amount of time.

descarga: (pág. 229) medida del volumen de agua corriente que fluye sobre una ubicación dada en cierto lapso de tiempo.

divergent boundary: (p. 481) place where two of Earth's tectonic plates are moving apart; is associated with volcanism, earthquakes, and high heat flow, and is found primarily on the seafloor.

límite divergente: (pág. 481) lugar donde dos placas tectónicas terrestres se alejan entre sí; se asocia con actividad volcánica, terremotos, un alto flujo de calor y se hallan principalmente en el fondo marino.

divide: (p. 227) elevated land that divides one watershed from another.

divisoria: (pág. 227) terreno elevado que separa una cuenca hidrográfica de otra.

Doppler effect: (p. 327) change in the wave frequency that occurs due to the relative motion of the wave as it moves toward or away from an observer.

efecto Doppler: (pág. 327) cambio en la frecuencia de onda que ocurre debido al movimiento relativo de la onda a medida que se acerca o se aleja de un observador.

downburst: (p. 351) violent downdrafts that are concentrated in a local area.

reventón: (pág. 351) violentos chorros de viento descendientes que se concentran en un área local.

drawdown: (p. 263) difference between the water level in a pumped well and the original water-table level.

tasa de agotamiento: (pág. 263) diferencia entre el nivel de agua en un pozo artesanal en uso y el nivel original del manto freático.

drought: (p. 362) extended period of well-below-average rainfall, usually caused by shifts in global wind patterns, allowing high-pressure systems to remain for weeks or months over continental areas.

sequía: (pág. 362) período prolongado con precipitación muy por debajo del promedio, generalmente es causado por cambios en los patrones globales de vientos, lo que permite que los sistemas de alta presión permanezcan sobre áreas continentales durante semanas o meses.

drumlin: (p. 210) elongated landform that results when a glacier moves over an older moraine.

drumlin: (pág. 210) formación alargada de tierra que se forma cuando un glaciar se mueve sobre una morrena más antigua.

dune: (p. 204) pile of windblown sand that develops over time, whose shape depends on sand availability, wind velocity and direction, and amount of vegetation present.

duna: (pág. 204) pila de arena formada a lo largo del tiempo por el arrastre de partículas por el viento, cuya forma depende de la disponibilidad de arena, la velocidad y dirección del viento y la cantidad de vegetación presente.

dwarf planet: (p. 816) an object that, due to its own gravity, is spherical in shape, orbits the Sun, is not a satellite, and has not cleared the area of its orbit of smaller debris.

planeta menor: (pág. 816) cuerpo que debido a su propia gravedad tiene forma esférica, tiene una órbita alrededor del Sol, no es un satélite y no ha eliminado restos más pequeños del área de su órbita.

eccentricity: (p. 801) ratio of the distance between the foci to the length of the major axis; defines the shape of a planet's elliptical orbit.

excentricidad: (pág. 801) razón de la distancia entre los focos y la longitud del eje mayor; define la forma de la órbita elíptica de un planeta.

E

ecliptic plane: (p. 776) plane of Earth's orbit around the Sun.

plano de la eclíptica: (pág. 776) plano de la órbita de la Tierra alrededor del Sol.

Ediacaran biota (ee dee A kuh ruhn • by OH tuh): (p. 636) fossils of various multicellular organisms from about 630 mya.

biota Ediacarana: (pág. 636) fósiles de diversos organismos multicelulares de hace cerca de 630 millones de años.

ejecta: (p. 771) material that falls back to the lunar surface after being blasted out by the impact of a space object.

eyecta: (pág. 771) material que cae de regreso a la superficie lunar luego de ser expulsado por el impacto de un cuerpo espacial.

elastic deformation: (p. 529) causes materials to bend and stretch; proportional to stress, so if the stress is reduced or returns to zero the strain or deformation is reduced or disappears.

deformación elástica: (pág. 529) ocasiona que los materiales se doblen y se estiren; es proporcional al grado de tensión, por lo que si la tensión se reduce o desaparece, la deformación también se reduce o desaparece.

El Niño: (p. 388) a band of anomalously warm ocean temperatures that occasionally develops off the western coast of South America and can cause short-term climatic changes felt worldwide.

El Niño: (pág. 388) una banda de agua oceánica que tiene temperaturas anómalamente cálidas que en ocasiones se desarrolla frente a la costa occidental de Sudamérica; puede causar cambios climáticos a corto plazo que afectan a todo el mundo.

electromagnetic spectrum: (p. 764) all types of electromagnetic radiation arranged according to wavelength and frequency.

electron: (p. 61) tiny atomic particle with little mass and a negative electric charge; an atom's electrons are equal in number to its protons and are located in a cloudlike region surrounding the nucleus.

element: (p. 60) natural or artificial substance that cannot be broken down into simpler substances by physical or chemical means.

ellipse (p. 800) an oval that is centered on two points called foci; the shape of planets' orbits.

energy efficiency: (p. 721) a type of conservation in which the amount of work produced is compared to the amount of energy used.

environmental science: (p. 7) study of the interactions of humans with environment.

eon: (p. 592) longest time unit in the geologic time scale.

epicenter (EH pih sen tur): (p. 533) point on Earth's surface directly above the focus of an earthquake.

epoch: (p. 593) time unit in the geological time scale, smaller than a period, measured in hundreds of thousands to millions of years.

equator: (p. 30) imaginary line that lies at 0° latitude and circles Earth midway between the north and south poles, dividing Earth into the northern hemisphere and the southern hemisphere.

equinox: (p. 777) time of year during which Earth's axis is at a 90° angle to the Sun; both hemispheres receive exactly 12 hours of sunlight and the Sun is directly overhead at the equator.

era: (p. 593) second-longest time unit in the geologic time scale, measured in tens to hundreds of millions of years, and defined by differences in lifeforms that are preserved in rocks.

erosion: (p. 171) removal and transport of weathered materials from one location to another by agents such as water, wind, glaciers, and gravity.

esker: (p. 210) long, winding ridge of layered sediments deposited by streams that flow beneath a melting glacier.

estuary: (p. 414) coastal area of lowest salinity often occurs where the lower end of a freshwater river or stream enters the ocean.

espectro electromagnético: (pág. 764) clasificación de todos los tipos de radiación electromagnética de acuerdo con su frecuencia y longitud de onda.

electrón: (pág. 61) partícula atómica diminuta con masa pequeña y carga eléctrica negativa; los electrones están ubicados en una región con forma de nube que rodea al núcleo del átomo y su número es igual al número de protones del átomo.

elemento: (pág. 60) sustancia natural o artificial que no puede separarse en sustancias más simples por medios físicos o químicos.

elipse: (pág. 800) óvalo centrado en dos puntos llamados focos; la forma de las órbitas de los planetas.

eficiencia energética: (pág. 721) tipo de conservación en el cual la cantidad de trabajo producido se compara con la cantidad de energía utilizada.

ciencias ambientales: (pág. 7) estudio de las interacciones del hombre con su entorno.

eon: (pág. 592) unidad más larga de tiempo en la escala de tiempo geológico.

epicentro: (pág. 533) punto en la superficie terrestre ubicado directamente encima del foco de un sismo.

época: (pág. 593) unidad de tiempo en la escala de tiempo geológico, es más pequeña que un período y se mide en millones a centenares de millares de años.

ecuador: (pág. 30) línea imaginaria que yace en la latitud 0° y que circunda la Tierra entre los polos norte y sur, dividiendo a la Tierra en dos hemisferios iguales: norte y sur.

equinoccio: (pág. 777) epoca del año durante la cual el eje de la Tierra forma un ángulo de 90° con el Sol, ambos hemisferios reciben exactamente 12 horas de luz solar y el Sol se halla exactamente sobre el ecuador.

era: (pág. 593) segunda unidad más grande de tiempo en la escala del tiempo geológico; se mide en decenas a centenas de millones de años y se define según las diferencias en las formas de vida preservadas en las rocas.

erosión: (pág. 171) eliminación y transporte de materiales meteorizados de un lugar a otro por agentes como el agua, el viento, los glaciares y la gravedad.

ésker: (pág. 210) formación larga y sinuosa de sedimentos estratificados, depositados por corrientes que fluyen debajo de un glaciar que se derrite.

estuario: (pág. 414) área costera de agua salobre que se forma en el sitio donde la desembocadura de un río o corriente de agua dulce entra al océano; provee una fuente excelente de alimento y refugio para organismos marinos comercialmente importantes.

eukaryote (yew KE ree oht): (p. 635) organism composed of one or more cells each of which usually contains a nucleus; larger and more complex than a prokaryote.

eucariota: (pág. 635) organismo compuesto por unas o más células nucleadas; generalmente es más grande y más complejo que un procariota.

eutrophication: (p. 239) process by which lakes become rich in nutrients from the surrounding watershed, resulting in a change in the kinds of organisms in the lake.

eutroficación: (pág. 239) proceso de aumento de la cantidad de nutrientes que contiene un lago, alimentado por los nutrientes provenientes de las cuenca circundante, lo que causa un cambio en los tipos de organismos que habitan el lago.

evaporation: (p. 74) vaporization—change of state from a liquid to a gas, involving thermal energy.

evaporación: (pág. 74) vaporización: cambio de estado de un líquido a gas que implica energía térmica.

evaporite: (p. 143) the layers of chemical sedimentary rocks that form when concentrations of dissolved minerals in a body of water reach saturation due to the evaporation of water; crystal grains precipitate out of solution and settle to the bottom.

evaporita: (pág. 143) capas de roca química sedimentaria que se forman cuando la concentración de minerales disueltos en una masa de agua alcanza el punto de saturación debido a la evaporación del agua; los cristales se precipitan de la solución y se asientan en el fondo.

evolution (eh vuh LEW shun): (p. 606) the change in species over time.

evolución: (pág. 606) cambios de las especies a lo largo del tiempo.

exfoliation: (p. 165) mechanical weathering process in which outer rock layers are stripped away, often resulting in dome-shaped formations.

exfoliación: (pág. 165) proceso de meteorización mecánica que causa la eliminación de los estratos rocosos exte-riores, a menudo produce formaciones en forma de domo.

exosphere: (p. 286) outermost layer of Earth's atmosphere that is located above the thermosphere with no clear boundary at the top; transitional region between Earth's atmosphere and outer space.

exosfera: (pág. 286) capa más externa de la atmósfera terrestre, está localizada por encima de la termosfera y no tiene un límite definido en su parte más alejada; región de transición entre la atmósfera de la Tierra y el espacio exterior.

exponential growth: (p. 735) pattern of growth in which a population of organisms grows faster as it increases in size, resulting in a population explosion.

crecimiento exponencial: (pág. 735) patrón de crecimiento en que una población de organismos crece cada vez más rápido a medida que aumenta de tamaño, causando una explosión demográfica.

extrusive rock: (p. 118) fine-grained igneous rock that is formed when molten rock cools quickly and solidifies on Earth's surface.

roca extrusiva: (pág. 118) roca ígnea de grano fino que se forma cuando la roca fundida se enfría rápidamente y se solidifica en la superficie terrestre.

eye: (p. 356) calm center of a tropical cyclone that develops when the winds around its center reach at least 120 km/h.

ojo: (pág. 356) centro de calma de un ciclón tropical que se desarrolla cuando los vientos a su alrededor alcanzan por lo menos 120 km/h.

eyewall: (p. 356) band where the strongest winds in a hurricane are usually concentrated, surrounding the eye.

pared del ojo de huracán: (pág. 356) banda que rodea el ojo de un huracán donde generalmente se concentran los vientos más fuertes.

F

fault: (p. 530) fracture or system of fractures in Earth's crust that occurs when stress is applied too quickly or stress is too great; can form as a result of horizontal compression (reverse fault), horizontal shear (strike-slip fault), or horizontal tension (normal fault).

falla: (pág. 530) fractura o sistema de fracturas en la corteza terrestre que ocurren en sitios donde se aplica tensión rápidamente o donde la tensión es demasiado grande; se puede formar como resultado de una compresión horizontal (falla invertida), un cizallamiento horizontal (falla de transformación) o una tensión horizontal (falla normal).

fault-block mountain: **(p. 574)** mountain that forms when large pieces of crust are tilted, uplifted, or dropped downward between large normal faults.

fission: **(p. 834)** process in which heavy atomic nuclei split into smaller, lighter atomic nuclei.

fissure: **(p. 504)** are long cracks in Earth.

flood: **(p. 230)** potentially devastating natural occurrence in which water spills over the sides of a stream's banks onto adjacent land areas.

flood basalt: **(p. 504)** huge amounts of lava that erupt from fissures.

floodplain: **(p. 230)** broad, flat, fertile area extending out from a stream's bank that is covered with water during floods.

focus: **(p. 533)** point of the initial fault rupture where an earthquake originates that usually lies at least several kilometers beneath Earth's surface.

foliated: **(p. 146)** metamorphic rock, such as schist or gneiss, whose minerals are squeezed under high pressure and arranged in wavy layers and bands.

fossil fuel: **(p. 710)** nonrenewable energy resource formed over geologic time from the compression and partial decomposition of organisms that lived millions ofyears ago.

fractional crystallization: **(p. 115)** process in which different minerals crystallize from magma at different temperatures, removing elements from magma.

fracture: **(p. 93)** when a mineral breaks into pieces with arclike, rough, or jagged edges.

front: **(p. 322)** boundary between two air masses of differing densities; can be cold, warm, stationary, or occluded and can stretch over large areas of Earth's surface.

frontal thunderstorm: **(p. 346)** type of thunderstorm usually produced by an advancing cold front, which can result in a line of thunderstorms hundreds of kilometers long, or, more rarely, an advancing warm front, which can result in a relatively mild thunderstorm.

frost wedging: **(p. 164)** mechanical weathering process that occurs when water repeatedly freezes and thaws in the cracks of rocks, often resulting in rocks splitting.

montañas de bloque de falla: **(pág. 574)** montañas que se forman cuando trozos grandes de corteza se inclinan, se elevan o se hunden entre fallas normales grandes.

fisión: **(pág. 834)** proceso mediante el cual los núcleos atómicos pesados se dividen en núcleos más livianos y pequeños.

fisura: **(pág. 504)** grandes grietas en la Tierra.

inundación: **(pág. 230)** acontecimiento natural potencialmente devastador en que el agua se desborda de las riberas de una corriente y cubre los terrenos adyacentes.

basalto de meseta: **(pág. 504)** grandes cantidades de lava que salen por las fisuras.

llanura aluvial: **(pág. 230)** área fértil, plana y ancha que se extiende desde las riberas de una corriente y queda cubierta por agua durante las inundaciones.

foco: **(pág. 533)** punto inicial de ruptura de la falla donde se origina un terremoto; generalmente se halla varios kilómetros debajo de la superficie terrestre.

foliada: **(pág. 146)** roca metamórfica, como el esquisto o el gneis, cuyos minerales son comprimidos bajo presiones altas, formando ordenadas capas y bandas onduladas.

combustible fósil: **(pág. 710)** recurso energético no renovable que se forma a lo largo del tiempo geológico, a partir de la compresión y descomposición parcial de organismos que vivieron hace millones de años.

cristalización fraccionaria: **(pág. 115)** proceso en el cual diferentes minerales se cristalizan a diferentes temperaturas a partir del magma, eliminando elementos del magma.

fractura: **(pág. 93)** sucede cuando un mineral se rompe en pedazos con bordes ásperos, arqueados o serrados.

frente: **(pág. 322)** límite entre dos masas de aire con diferentes densidades; puede ser frío, cálido, estacionario u ocluido y puede extenderse sobre grandes áreas de la superficie de la Tierra.

tormenta frontal: **(pág. 346)** tipo de tormenta que es producida generalmente por el avance de un frente frío, pudiendo producir una línea de tormentas de cientos de kilómetros de largo, o en menor frecuencia por el avance de un frente cálido, produciendo tormentas relativamente ligeras.

erosión periglaciar: **(pág. 164)** proceso mecánico de meteorización que ocurre cuando el agua se congela y se descongela, en repetidas ocasiones, en las grietas de las rocas, ocasionando el rompimiento de las mismas.

fuel: (p. 709) material, such as wood, peat, or coal, burned to produce energy.

combustible: (pág. 709) materiales como la leña, la turba o el carbón, que se queman para producir energía.

Fujita tornado intensity scale: (p. 353) classifies tornados according to their wind speed, duration, and path of destruction on a scale ranging from F0 to F5.

escala Fujita de intensidad de tornados: (pág. 353) clasifica los tornados según la velocidad de sus vientos, su duración y el daño que causan a su paso, en una escala que va de F0 a F5.

fusion: (p. 834) The combining of lightweight nuclei into heavier nuclei; occurs in the core of the Sun where temperatures and pressure are extremely high.

fusión: (pág. 834) combinación de núcleos livianos para formar núcleos más pesados: sucede en el núcleo del Sol donde las temperaturas y la presión son extremadamente altas.

G

gas giant planet: (p. 811) large, gaseous planet that is very cold at its surface; has ring systems, many moons, and lacks solid surfaces—Jupiter, Saturn, Uranus, and Neptune.

gigantes gaseosos: (pág. 811) planetas grandes y gaseosos con superficies muy frías; tienen sistemas de anillos, muchas lunas y carecen de superficie sólida: Júpiter, Saturno, Urano y Neptuno.

gem: (p. 101) rare, precious, highly prized mineral that can be cut, polished, and used for jewelry.

gema: (pág. 101) mineral sumamente valioso, precioso y escaso que se puede cortar, pulir y utilizar en joyería.

Geographic Information System (GIS): (p. 44) a mapping system that uses worldwide databases from remote sensing to create layers of information that can be superimposed upon each other to form a comprehensive map.

Sistema de Información Geográfica (SIG): (pág. 44) sistema para la elaboración de mapas que usa bases de datos mundiales obtenidos por sensores remotos, para crear capas de información que se pueden superponer para elaborar mapas que combinen dicha información.

geologic map: (p. 38) a map that shows the distribution, arrangement, and types of rocks below the soil, and other geologic features.

mapa geológico: (pág. 38) mapa que muestra la distribución, el orden y los tipos de roca del subsuelo, así como otras características geológicas.

geologic time scale: (p. 590) record of Earth's history from its origin 4.6 bya to the present.

escala del tiempo geológico: (pág. 590) registro de la historia de la Tierra desde su origen, hace 4.6 billones de años, hasta el presente.

geology: (p. 7) study of materials that make up Earth and the processes that form and change these materials, and the history of the planet and its life-forms since its origin.

geología: (pág. 7) estudio de los materiales que conforman la Tierra y de los procesos de formación y cambio de estos materiales, así como la historia del planeta y sus formas de vida desde su origen.

geosphere: (p. 8) the part of Earth from its surface to its center.

geosfera: (pág. 8) región que abarca desde la superficie hasta el centro de la Tierra.

geothermal energy: (p. 717) energy produced by Earth's naturally occurring heat, steam, and hot water.

energía geotérmica: (pág. 717) energía producida naturalmente en la Tierra por el calor, el vapor y el agua caliente.

geyser: (p. 258) explosive hot spring that erupts regularly.

géiser: (pág. 258) manantial termal explosivo que hace erupción regularmente.

glacier: (p. 207) large, moving mass of ice that forms near Earth's poles and in mountainous regions at high elevations.

glaciar: (pág. 207) enormes masas móviles de hielo que se forman cerca de los polos de la Tierra o en grandes elevaciones en regiones montañosas.

glass: (p. 73) solid that consists of densely packed atoms with a random arrangement and lacks crystals or has crystals that are not visible.

vidrio: (pág. 73) sólido formado por átomos densamente comprimidos en un ordenamiento aleatorio; carece de cristales o sus cristales no son visibles.

Global Positioning System (GPS): (p. 44) satellite-based navigation system that permits a user to pinpoint his or her exact location on Earth.

global warming: (p. 393) rise in global temperatures, which might be due to increases in atmospheric CO_2 from deforestation and burning of fossil fuels

gnomonic (noh MAHN ihk) projection: (p. 35) map useful in plotting long-distance trips by boat or plane, made by projecting points and lines from a globe onto a piece of paper that touches the globe at a single point.

graded bedding: (p. 138) type of bedding in which particle sizes become progressively heavier and coarser toward the bottom layers.

granitic rock: (p. 118) light-colored rock that has high silica content, and contains quartz and potassium feldspar.

greenhouse effect: (p. 393) natural heating of Earth's surface by certain atmospheric gases, which helps keep Earth warm enough to sustain life.

gully erosion: (p. 172) erosion that occurs when a rill channel widens and deepens.

guyot: (p. 452) large, extinct, basaltic volcanoes with flat, submerged tops.

Sistema de posicionamiento global (SPG): (pág. 44) sistema de navegación por satélite que permite al usuario localizar su ubicación exacta sobre la Tierra.

calentamiento global: (pág. 393) aumento en las temperaturas globales, que es probablemente producto del aumento en el CO_2 atmosférico, causado por la deforestación y la quema de combustibles fósiles

proyección gnomónica: (pág. 35) mapa útil para trazar viajes de distancias largas por barco o por avión; se elabora proyectando los puntos y las líneas de un globo sobre una hoja de papel que toca el globo en un solo punto.

estratificación graduada: (pág. 138) característica de la depositación de rocas sedimentarias en la cual las partículas son progresivamente más pesadas y gruesas hacia las capas inferiores de la estratificación.

roca granítica: (pág. 118) roca de color claro que tiene un alto contenido de sílice y contiene cuarzo y potasio feldespato.

efecto invernadero: (pág. 393) calentamiento natural de la superficie terrestre por ciertos gases atmosféricos; ayuda a mantener en la Tierra una temperatura lo suficientemente cálida para mantener la vida.

erosión en barrancos: (pág. 172) erosión que ocurre cuando el cauce de un arroyuelo se ensancha y profundiza.

guyot: (pág. 452) grandes volcanes basálticos extintos cuya cima es plana y está sumergida.

H

half-life: (p. 602) period of time it takes for a radioactive isotope, such as carbon-14, to decay to one-half of its original amount.

halo: (p. 863) spherical region where globular clusters are located; surrounds the Milky Way's nuclear bulge and disk.

hardness: (p. 91) measure of how easily a mineral can be scratched, which is determined by the arrangement of a mineral's atoms.

heat island: (p. 385) urban area where climate is warmer than in the surrounding countryside due to factors such as numerous concrete buildings and large expanses of asphalt.

heat wave: (p. 362) extended period of above-average temperatures caused by large, high-pressure systems that warm by compression and block cooler air masses.

vida media: (pág. 602) período de tiempo que demora un isótopo radiactivo, como el carbono 14, en desintegrarse a la mitad de su cantidad radiactiva original.

halo: (pág. 863) región esférica donde se ubican los cúmulos globulares; rodea el disco y el núcleo central de la Vía Láctea.

dureza: (pág. 91) medida de la facilidad con la que un mineral es rayado; está determinada por el ordenamiento de los átomos del mineral.

isla de calor: (pág. 385) área urbana donde el clima es más caliente que en el área rural circundante, debido a factores como los numerosos edificios de concreto y las grandes extensiones de asfalto.

ola de calor: (pág. 362) período extenso de temperaturas más altas que el promedio; es causado por grandes sistemas de alta presión que se calientan por compresión y bloquean las masas de aire más frías.

Hertzsprung-Russell diagram (H-R diagram): (p. 845) graph that relates stellar characteristics—class, mass, temperature, magnitude, diameter, and luminosity.

highland: (p. 771) light-colored, mountainous, heavily cratered area of the Moon, composed mostly of lunar breccias.

***Homo sapiens:* (p. 665)** species to which modern humans belong.

hot spot: (p. 502) unusually hot area in Earth's mantle where high-temperature plumes of mantle material rise toward the surface.

hot spring: (p. 258) thermal spring with temperatures higher than that of the human body.

Hubble constant: (p. 874) value (*H*) used to calculate the rate at which the universe is expanding; measured in kilometers per second per megaparsec.

humidity: (p. 294) amount of water vapor in the atmosphere at a given location on Earth's surface.

hydrocarbon: (p. 709) molecules with hydrogen and carbon bonds only; the result of the combination of carbon dioxide and water during photosynthesis.

hydroelectric power: (p. 716) power generated by converting the energy of free-falling water to electricity.

hydrogen bond: (p. 693) forms when the positive ends of some water molecules are attracted to the negative ends of other water molecules; cause water's surface to contract and allow water to adhere to and coat a solid.

hydrosphere: (p. 8) all the water in Earth's oceans, lakes, seas, rivers, and glaciers plus all the water in the atmosphere.

hydrothermal metamorphism: (p. 149) occurs when very hot water reacts with rock, altering its mineralogy and chemistry.

hygrometer (hi GRAH muh tur): (p. 325) weather instrument used to measure humidity.

hypothesis: (p. 10) a testable explanation of a situation.

diagrama de Hertzsprung-Russell (diagrama H-R): (pág. 845) gráfica que relaciona características estelares: incluyendo la clase, la masa, la temperatura, la magnitud, el diámetro y la luminosidad.

tierras altas: (pág. 771) áreas de la Luna de color claro, con muchos cráteres y montañas, compuestas en su mayor parte de brechas lunares.

***Homo sapiens:* (pág. 665)** especie a la cual pertenecen los seres humanos modernos.

punto caliente: (pág. 502) área muy caliente del manto de la Tierra donde plumas de material del manto a gran temperatura ascienden a la superficie.

fuente caliente: (pág. 258) manantial termal con temperaturas más altas que las del cuerpo humano.

constante de Hubble: (pág. 874) valor (*H*) que sirve para calcular la velocidad de expansión del universo; se mide en kilómetros por segundo por megaparsec.

humedad: (pág. 294) cantidad de vapor de agua en el aire en un sitio determinado de la Tierra.

hidrocarburo: (pág. 709) molécula que sólo contiene enlaces entre átomos de hidrógeno y de carbono; es producto de la unión del dióxido de carbono y el agua durante la fotosíntesis.

energía hidroeléctrica: (pág. 716) se genera al convertir la energía de una caída de agua en electricidad.

enlace de hidrógeno: (pág. 693) se forma cuando el extremo positivo de algunas moléculas de agua son atraídas por el extremo negativo de otras moléculas de agua; ocasiona que la superficie del agua se contraiga y permite al agua adherirse y recubrir un sólido.

hidrosfera: (pág. 8) toda el agua en los océanos, los lagos, los mares, los ríos y los glaciares de la Tierra, además de toda el agua en la atmósfera.

metamorfismo hidrotérmico: (pág. 149) ocurre cuando agua muy caliente reacciona con la roca, alterando su mineralogía y su química.

higrómetro: (pág. 325) instrumento meteorológico que se usa para medir la humedad.

hipótesis: (pág. 10) explicación de una situación que se puede poner a prueba.

I

ice age: (p. 387) period of extensive glacial coverage, producing long-term climatic changes, where average global temperatures decreased by 5°C.

igneous rock: (p. 112) intrusive or extrusive rock formed from the cooling and crystallization of magma or lava.

glaciación: (pág. 387) período de formación de una amplia cobertura glacial que produce cambios climáticos de largo plazo en que las temperaturas globales promedio desminuyen 5°C.

roca ígnea: (pág. 112) roca intrusiva o extrusiva formada a partir del enfriamiento y cristalización del magma o lava.

impact crater: (p. 771) crater formed when space material impacted on Moon's surface.

cráter de impacto: (pág. 771) cráter que se forma cuando material proveniente del espacio impacta la superficie de la Luna.

inclusion: (p. 597) the principle that fragments, called inclusions, in a rock layer must be older than the rock layer that contains them.

inclusión: (pág. 597) principio que establece que los fragmentos, llamados inclusiones, contenidos por un estrato rocoso deben ser más antiguos que la roca que los contiene.

independent variable: (p. 12) factor that is manipulated by the experimenter in an experiment.

variable independiente: (pág. 12) factor que es manipulado por el investigador en un experimento.

index fossils: (p. 609) remains of plants or animals that were abundant, widely distributed, and existed briefly that can be used by geologists to correlate or date rock layers.

fósiles guía: (pág. 609) restos de plantas o animales que fueron abundantes, tuvieron una amplia distribución y existieron poco tiempo, que sirven a los geólogos para correlacionar o para datar estratos rocosos.

infiltration: (p. 253) Process by which precipitation that has fallen on land surfaces enters the ground and becomes groundwater.

infiltración: (pág. 253) proceso mediante el cual la precipi-tación que cae sobre la superficie terrestre entra al suelo y se convierte en agua subterránea.

interferometry: (p. 767) process that links separate telescopes so they act as one telescope, producing more detailed images as the distance between them increases.

interferometría: (pág. 767) proceso que combina telescopios separados para que funcionen como un solo telescopio, produciendo imágenes más detalladas al aumentar la distancia entre ellos.

International Date Line: (p. 33) the 180° meridian, which serves as the transition line for calendar days.

línea internacional de cambio de fecha: (pág. 33) el meridiano 180°; sirve como la línea de transición para los días del calendario.

intrusive rock: (p. 118) coarse-grained igneous rock that is formed when molten rock cools slowly and solidifies inside Earth's crust.

roca intrusiva: (pág. 118) roca ígnea de grano grueso que se forma cuando la roca fundida se enfría lentamente y se solidifica en el interior de la corteza terrestre.

ion: (p. 64) an atom that gains or loses an electron.

ion: (pág. 64) átomo que gana o pierde un electrón.

ionic bond: (p. 68) attractive force between two ions with opposite charge.

enlace iónico: (pág. 68) fuerza de atracción entre dos iones con cargas opuestas.

iridium (ih RID ee um): (p. 659) metal that is rare in rocks at Earth's surface but is relatively common in asteroids.

iridio: (pág. 659) metal escaso en las rocas de la superficie terrestre, pero relativamente común en los meteoritos y los asteroides.

isobar: (p. 329) line on a weather map connecting areas of equal pressure

isobara: (pág. 329) línea de un mapa meteorológico que conecta áreas con igual presión.

isochron (I suh krahn): (p. 477) imaginary line on a map that shows points of the same age; formed at the same time.

isocrona: (pág. 477) línea imaginaria en un mapa que conecta puntos con la misma antigüedad; que se formaron al mismo tiempo.

isostasy (I SAHS tuh see): (p. 563) condition of equilibrium that describes the displacement of Earth's mantle by Earth's continental and oceanic crust.

isostasia: (pág. 563) condición de equilibrio que describe el desplazamiento del manto terrestre por las cortezas continental y oceánica de la Tierra.

isostatic rebound: (p. 565) slow process of Earth's crust rising as the result of the removal of overlaying material.

rebote isostático: (pág. 565) proceso lento de elevación de la corteza terrestre producto de la eliminación del material sobreyacente.

isotherm: (p. 329) line on a weather map connecting areas of equal temperature.

isoterma: (pág. 329) línea en un mapa meteorológico que conecta áreas con la misma temperatura.

isotope: (p. 62) an atom of an element that has a different mass number than the element but the same chemical properties.

isótopo: (pág. 62) átomo de un elemento que tiene un distinto número de masa que el elemento, pero las mismas propiedades químicas.

J

jet stream: (**p. 321**) narrow wind band that occurs above large temperature contrasts and can flow as fast as 185 km/h.

corriente de chorro: (**pág. 321**) banda de vientos estrecha situada por encima de áreas con grandes contrastes de temperatura y que puede alcanzar una rapidez de 185 km/h.

K

kame: (**p. 210**) a conical mound of layered sediment deposited by streams that flow beneath a melting glacier.

kame: (**pág. 210**) montículo cónico de sedimento estratificado que es depositado por corrientes que fluyen bajo un glaciar que se derrite.

karst topography: (**p. 261**) irregular topography with sinkholes, sinks, and sinking streams caused by groundwater dissolution of limestone.

topografía cárstica: (**pág. 261**) topografía irregular con sumideros, hundimientos y corrientes que desaparecen, causada por la disolución de la piedra caliza por el agua subterránea.

kettle: (**p. 212**) a lake formed when runoff and precipitation filled a kettle hole, which is a depression that formed when an ice block from a continental glacier became covered with sediment and melted.

marmita: (**pág. 212**) lago que se forma cuando la escorrentía y la precipitación llenan el hueco de una marmita, que es la depresión que se forma cuando un bloque de hielo de un glaciar continental queda cubierto con sedimento y se derrite.

key bed: (**p. 599**) a rock or sediment layer that serves as a time marker in the rock record and results from volcanic ash or meteorite-impact debris that spread out and covered large areas of Earth.

estrato guía: (**pág. 599**) capa de sedimento que sirve como marcador de tiempo del registro geológico; está formado por cenizas volcánicas o por los restos del impacto de un meteorito que se esparcen y cubren grandes áreas de la Tierra.

kimberlite: (**p. 123**) rare, ultramafic rock that can contain diamonds and other minerals formed only under very high pressures.

kimberlita: (**pág. 123**) roca ultramáfica poco común que puede contener diamantes y otros minerales que sólo se forman bajo presiones muy altas.

Köppen classification system: (**p. 383**) classification system for climates, divided into five types, based on the mean monthly values of temperature and precipitation and types of vegetation.

sistema de clasificación de Köppen: (**pág. 383**) sistema de clasificación de los climas; los clasifica en cinco tipos básicos en base a los valores mensuales promedio de temperatura y precipitación y a los tipos de vegetación.

Kuiper (KI pur) belt: (**p. 818**) small solar system bodies that are mostly rock and ice, lies outside the orbit of Neptune, 30 to 50 AU from the Sun, most probably formed in this region.

cinturón de Kuiper: (**pág. 818**) pequeños cuerpos del sistema solar formados principalmente por roca y hielo, yacen más allá de la órbita de Neptuno, entre 30 a 50 UA del Sol, y es muy probable que se hayan formado en esta región.

L

laccolith (LA kuh lihth): (**p. 515**) relatively small, mushroom-shaped pluton that forms when magma intrudes into parallel rock layers close to Earth's surface.

lacolito: (**pág. 515**) plutón relativamente pequeño con forma de champiñón que se forma cuando se introduce el magma entre estratos rocosos paralelos, cerca de la superficie terrestre.

lake: (**p. 238**) natural or human-made body of water that can form when a depression on land fills with water.

lago: (**pág. 238**) masa de agua, natural o hecha por el hombre, que se forma cuando una depresión terrestre se llena de agua.

Landsat satellite: (**p. 41**) information-gathering satellite that uses visible light and infrared radiation to map Earth's surface.

satélite Landsat: (**pág. 41**) satélite que recoge información, usando luz visible y radiación infrarroja para mapear la superficie terrestre.

landslide: (p. 197) rapid downslope movement of a mass of loose soil, rock, or debris that has separated from the bedrock; can be triggered by an earthquake.

derrumbe: (pág. 197) rápido desplazamiento cuesta abajo de una masa de tierra, rocas o escombros sueltos que se han separado del lecho rocoso; puede ser causado por un terremoto.

latent heat: (p. 295) stored energy in water vapor that is not released to warm the atmosphere until condensation takes place.

calor latente: (pág. 295) energía almacenada en el vapor de agua que no es liberada para calentar la atmósfera, hasta que ocurre la condensación.

latitude: (p. 30) distance in degrees north and south of the equator.

latitud: (pág. 30) distancia en grados hacia el norte o el sur del ecuador.

Laurentia (law REN shuh): (p. 627) ancient continent formed during the Proterozoic that is the core of modern-day North America.

Laurencia: (pág. 627) antiguo continente que se formó durante el Proterozoico y que en la actualidad corresponde al centro de Norteamérica.

lava: (p. 112) magma that flows out onto Earth's surface.

lava: (pág. 112) magma que fluye por la superficie terrestre.

Le Système International d'Unités (SI): (p. 13) replacement for the metric system; based on a decimal system using the number 10 as the base unit; includes the meter: (m), second: (s), and kilogram: (kg).

Le Système Internacional d'Unités/Sistema Internacional de Unidades (SI): (pág. 13) sustituto del sistema métrico; se basa en el sistema decimal por lo que usa el número 10 como unidad base: incluye el metro: (m), el segundo: (s) y el kilogramo: (kg).

liquid metallic hydrogen: (p. 812) form of hydrogen with both liquid and metallic properties that exists as a layer in the Jovian atmosphere.

hidrógeno metálico líquido: (pág. 812) forma de hidrógeno con propiedades de líquido y de metal que forma una capa en la atmósfera joviana.

lithification: (p. 136) the physical and chemical processes that transform sediments into sedimentary rocks.

litificación: (pág. 136) procesos físicos y químicos que transforman los sedimentos en roca sedimentaria.

loess (LESS): (p. 206) thick, windblown, fertile deposit of silt that contains high levels of nutrients and minerals.

loes: (pág. 206) amplio depósito fértil de limo que es arrastrado por el viento y contiene niveles altos de nutrientes y minerales.

longitude: (p. 31) distance in degrees east and west of the prime meridian.

longitud: (pág. 31) distancia en grados hacia el este o el oeste del primer meridiano.

longshore bar: (p. 440) submerged sandbar located in the surf zone of most beaches.

barra litoral: (pág. 440) barra de arena sumergida ubicada en la zona de oleaje de la mayoría de las playas.

longshore current: (p. 441) current that flows parallel to the shore, moves large amounts of sediments, and is formed when incoming breakers spill over a longshore bar.

corriente litoral: (pág. 441) corriente que fluye paralela a la costa, transporta grandes cantidades de sedimentos y se forma cuando las olas rompen a lo largo de una larga barra litoral.

luminosity: (p. 842) energy output from the surface of a star per second; measured in watts.

luminosidad: (pág. 842) energía que irradia la superficie de una estrella por segundo; se mide en vatios.

lunar eclipse: (p. 784) when Earth passes between the Sun and the Moon, and Earth's shadow falls on the Moon; occurs only during a full moon,

eclipse lunar: (pág. 784) sucede cuando la Tierra pasa entre el Sol y la Luna y la sombra de la Tierra cae sobre la Luna; ocurre sólo durante la luna llena.

luster: (p. 90) the way that a mineral reflects light from its surface; two types—metallic and nonmetallic.

lustre: (pág. 90) manera en que la superficie de un mineral refleja la luz; existen dos tipos: metálico o no metálico.

M

magnetic reversal: (p. 476) when Earth's magnetic field changes polarity between normal and reversed.

inversión magnética: (pág. 476) sucede cuando el campo magnético de la Tierra cambia polaridad entre normal e invertida.

magnetometer (mag nuh TAH muh tur): (p. 473) device used to map the ocean floor that detects small changes in magnetic fields.

magnitude: (p. 539) measure of the energy released during an earthquake, which can be described using the Richter scale.

main sequence: (p. 845) in an H-R diagram, the broad, diagonal band that includes about 90 percent of all stars and runs from hot, luminous stars in the upper-left corner to cool, dim stars in the lower-right corner.

map legend: (p. 39) key that explains what the symbols on a map represent.

map scale: (p. 39) ratio between the distances shown on a map and the actual distances on Earth's surface.

maria (MAH ree uh): (p. 771) dark-colored, smooth plains on the Moon surface.

mass extinction: (p. 594) occurs when an unusually large number of organisms disappear from the rock record at about the same time.

mass movement: (p. 194) downslope movement of Earth materials due to gravity that can occur suddenly or very slowly, depending on the weight of the material, its resistance to sliding, and whether a trigger, such as an earthquake, is involved.

mass number: (p. 62) combined number of protons and neutrons in the nucleus of an atom.

matter: (p. 60) anything that has volume and mass.

Maunder minimum: (p. 390) period of very low sunspot activity that occurred between 1645 and 1716 and closely corresponded with a cold climatic episode known as the "Little Ice Age."

meander: (p. 234) curve or bend in a stream formed when a stream's slope decreases, water builds up in the stream channel, and moving water erodes away the sides of the streambed.

mechanical weathering: (p. 164) process that breaks down rocks and minerals into smaller pieces but does not involve any change in their composition.

Mercator projection: (p. 34) map with parallel lines of latitude and longitude that shows true direction and the correct shapes of landmasses but distorts areas near the poles.

magnetómetro: (pág. 473) aparato que sirve para mapear el fondo marino; detecta cambios pequeños en los campos magnéticos.

magnitud: (pág. 539) medida de la energía liberada durante un sismo; se puede describir usando la escala de Richter.

secuencia principal: (pág. 845) la ancha banda diagonal de un diagrama H-R que contiene cerca del 90 por ciento de todas las estrellas; contiene desde estrellas calientes y luminosas en la esquina superior izquierda, hasta estrellas frías de brillo débil en la esquina inferior derecha.

leyenda del mapa: (pág. 39) clave que explica los símbolos en un mapa.

escala del mapa: (pág. 39) razón entre las distancias que se muestran en un mapa y las distancias reales en la superficie terrestre.

mar: (pág. 771) planicie lunar lisa y de color oscuro.

extinción masiva: (pág. 594) ocurre cuando un número insólitamente grande de organismos desaparece del registro geológico aproximadamente al mismo tiempo.

movimiento de masa: (pág. 194) movimiento cuesta abajo de materiales terrestres debido a la gravedad; puede ocurrir de manera repentina o muy lentamente: dependiendo del peso del material, la resistencia del material a deslizarse y de si ha ocurrido algún evento que lo desencadene, como un sismo.

número de masa: (pág. 62) número combinado de protones y neutrones en el núcleo de un átomo.

materia: (pág. 60) todo aquello que tiene volumen y masa.

mínimo de Maunder: (pág. 390) período de muy baja actividad de manchas solares, ocurrido entre 1645 y 1716, que se correspondió con un episodio climático frío llamado "La Pequeña Glaciación."

meandro: (pág. 234) curva o desviación en una corriente; se forma cuando disminuye la pendiente de la corriente, por lo que el agua se acumula en el cauce y el movimiento del agua erosiona los costados del cauce.

meteorización mecánica: (pág. 164) proceso de rompimiento de rocas y minerales en trozos más pequeños que no afecta la composición del material.

proyección de Mercator: (pág. 34) mapa con líneas de latitud y longitud paralelas que muestra la dirección real y las formas correctas de las masas terrestres, aunque las áreas cercanas a los polos aparecen distorsionadas.

mesosphere: **(p. 284)** layer of Earth's atmosphere above the stratopause.

metallic bond: **(p. 68)** positive ions of metal held together by the negative electrons between them; allows metals to conduct electricity.

meteor: **(pp. 818, 621)** streak of light produced when a meteoroid falls toward Earth and burns up in Earth's atmosphere.

meteorite **(MEE tee uh rite):** **(p. 818)** a small fragment of an orbiting body that has fallen to Earth, generating heat; does not completely burn up in Earth's atmosphere and strikes Earth's surface, sometimes causing an impact crater.

meteoroid: **(p. 818)** piece of interplanetary material that falls toward Earth and enters its atmosphere.

meteorology: **(p. 6)** the study of the atmosphere, which is the air surrounding Earth.

meteor shower: **(p. 819)** occurs when Earth intersects a cometary orbit and comet particles burn up as they enter Earth's upper atmosphere.

microclimate: **(p. 385)** localized climate that differs from the surrounding regional climate.

microcontinent: **(p. 624)** a small fragment of granite-rich crust formed during the Archean.

mid-ocean ridge: **(p. 451)** chain of underwater mountains that run throughout the ocean basins, have a total length over 65,000 km, and contain active and extinct volcanoes.

mineral: **(p. 86)** naturally occurring, inorganic solid with a specific chemical composition and a definite crystalline structure.

mineral replacement: **(p. 607)** the process where pore spaces of an organism's buried parts are filled in with minerals from groundwater.

modified Mercalli scale: **(p. 540)** measures earthquake intensity on a scale from I to XII; the higher the number, the greater the damage the earthquake has caused.

mold: **(p. 608)** fossil that can form when a shelled organism decays in sedimentary rock and is removed by erosion or weathering, leaving a hollowed-out impression.

molecule: **(p. 67)** combination of two or more atoms joined by covalent bonds.

mesosfera: **(pág. 284)** capa de la atmósfera terrestre ubicada encima de la estratopausa.

enlace metálico: **(pág. 68)** iones metálicos positivos que se mantienen unidos debido la carga negativa de los electrones que se encuentran entre ellos; permite a los metales conducir electricidad.

estrella fugaz: **(pág. 818, 621)** rayo luminoso que se produce cuando un meteoroide cae a la Tierra y se quema en la atmósfera terrestre.

meteorito: **(pág. 818)** fragmento pequeño de un cuerpo en órbita que cae a la Tierra generando calor; como no se quema completamente en la atmósfera, choca con la superficie terrestre y produce un cráter de impacto.

meteoroide: **(pág. 818)** trozo de material interplanetario que cae a la Tierra y entra a la atmósfera terrestre.

meteorología: **(pág. 6)** estudio de la atmósfera, la capa de aire que rodea la Tierra.

lluvia de estrellas: **(pág. 819)** ocurre cuando la Tierra interseca la órbita de un cometa y las partículas del cometa se queman al entrar a las capas superiores de la atmósfera terrestre.

microclima: **(pág. 385)** clima localizado que difiere del clima regional circundante.

microcontinentes: **(pág. 624)** trozos pequeños de corteza rica en granito que se formaron durante el Arcaico.

dorsales medioocéanicas: **(pág. 451)** cadenas montañosas submarinas que se extienden a través de las cuencas oceánicas, tienen una longitud total de más de 65,000 km y contienen innumerables volcanes activos y extintos.

mineral: **(pág. 86)** sólido inorgánico natural con una composición química específica y una estructura cristalina definida.

sustitución de minerales: **(pág. 607)** proceso en que los poros de las partes enterradas de un organismo se llenan con los minerales provenientes de aguas subterráneas.

escala de Mercalli modificada: **(pág. 540)** mide la intensidad de un sismo en una escala de I a XII; a medida que aumenta el número, mayor es el daño causado.

molde: **(pág. 608)** fósil que se forma cuando un organismo con concha se descompone en roca sedimentaria y es removido por erosión o meteorización, quedando una impresión hueca.

molécula: **(pág. 67)** combinación de dos o más átomos unidos por enlaces covalentes.

moment magnitude scale: (p. 540) scale used to measure earthquake magnitude—taking into account the size of the fault rupture, the rocks' stiffness, and amount of movement along the fault—using values that can be estimated from the size of several types of seismic waves.

escala de magnitud momentánea: (pág. 540) escala que sirve para medir la intensidad de un sismo (tomando en cuenta el tamaño de la ruptura de la falla, la rigidez de la roca y la cantidad del movimiento a lo largo de la falla) usando valores estimados a partir de la magnitud de varios tipos de ondas sísmicas.

moraine: (p. 210) ridge or layer of mixed debris deposited by a melting glacier.

morrena: (pág. 210) loma o estrato de detritos mezclados que deposita un glaciar al derretirse.

mountain thunderstorm: (p. 346) occurs when an air mass rises from orographic lifting, which involves air moving up the side of a mountain.

tormenta orográfica: (pág. 346) sucede cuando una masa de aire sube por ascenso orográfico, lo que implica el ascenso por la ladera de una montaña.

mudflow: (p. 196) rapidly flowing, often destructive mixture of mud and water that may be triggered by an earthquake, intense rainstorm, or volcanic eruption.

flujo o corriente de lodo: (pág. 196) mezcla de lodo y agua que fluye rápidamente y que a menudo es destructiva; puede ser causada por un terremoto, una lluvia intensa o una erupción volcánica.

N

natural resource: (p. 678) resources provided by Earth, including air, water, land, all living organisms, nutrients, rocks, and minerals.

recursos naturales: (pág. 678) recursos que provee la Tierra: incluyendo el aire, el agua, la tierra, todos los organismos vivos, los nutrientes, las rocas y los minerales.

neap tide: (p. 424) tide that occurs during first- or third-quarter Moon, when the Sun, the Moon, and Earth form a right angle; this causes solar tides to diminish lunar tides, causing high tides to be lower than normal and low tides to be higher than normal.

marea muerta: (pág. 424) durante el primero o el tercer cuartos lunares, el Sol, la Luna y la Tierra se encuentran en ángulo recto, causando que las mareas solares reduzcan la intensidad de las mareas lunares, lo que provoca que la marea alta sea menor que lo normal y la marea baja sea mayor que lo normal.

nebula: (p. 848) large cloud of interstellar gas and dust that collapses on itself, due to its own gravity, and forms a hot, condensed object that will become a new star.

nebulosa: (pág. 848) extensa nube de gas y polvo interestelares que se colapsa en sí misma debido a su propia gravedad, formando un cuerpo condensado caliente que se convertirá en una estrella nueva.

neutron: (p. 60) tiny atomic particle that is electrically neutral and has about the same mass as a proton.

neutrón: (pág. 60) partícula atómica diminuta, eléctricamente neutra; tiene una masa similar a la de un protón.

neutron star: (p. 850) collapsed, dense core of a star that forms quickly while its outer layers are falling inward, has a radius of about 10 km, a mass 1.5 to 3 times that of the Sun, and contains mostly neutrons.

estrella de neutrones: (pág. 850) núcleo denso y colapsado de una estrella que se forma rápidamente, al mismo tiempo que sus capas exteriores se contraen; tiene un radio aproximado de 10 km, una masa de 1.5 a 3 veces la del Sol y contiene principalmente neutrones.

nitrogen-fixing bacteria: (p. 688) bacteria found in water or soil; can grow on the roots of some plants, capture nitrogen gas, and change into a form that plants use to build proteins.

bacteria fijadora de nitrógeno: (pág. 688) bacteria que habita el suelo o el agua; puede crecer en las raíces de algunas plantas, capturar el gas nitrógeno y convertirlo a una forma que las plantas pueden usar para fabricar proteínas.

nonfoliated: (p. 147) metamorphic rocks like quartzite and marble, composed mainly of minerals that form with blocky crystal shapes.

no foliada: (pág. 147) roca metamórfica, como la cuarcita y el mármol, compuesta principalmente de minerales que forman bloques cristalinos.

nonpoint source: **(p. 749)** water-pollution source that generates pollution from widely spread areas, such as runoff from roads.

nonrenewable resource: **(p. 680)** resource that exists in Earth's crust in a fixed amount and can be replaced only by geologic, physical, or chemical processes that take hundreds of millions of years.

normal: **(p. 377)** standard value for a location, including rainfall, wind speed, and temperatures, based on meteorological records compiled for at least 30 years.

nuclear fission: **(p. 718)** the process in which a heavy nucleus divides to form smaller nuclei and one or two neutrons and produces a large amount of energy.

nucleus (NEW klee us): **(p. 60)** positively charged center of an atom, made up of protons and neutrons and surrounded by electrons in energy levels.

fuente no puntual: **(pág. 749)** fuente de contaminación del agua que genera contaminación a partir de áreas muy extensas, como la escorrentía de los caminos.

recurso no renovable: **(pág. 680)** recurso que existe en la corteza terrestre en una cantidad fija y que sólo puede ser regenerado por procesos geológicos, físicos o químicos que demoran centenas de millones de años.

normales: **(pág. 377)** valores estándar para un sitio: incluyen la lluvia, la velocidad del viento y las temperaturas; se basan en los registros meteorológicos recopilados durante por lo menos 30 años.

fisión nuclear: **(pág. 718)** proceso de división de un núcleo pesado en núcleos más pequeños y uno o dos neutrones, produciendo una gran cantidad de energía.

núcleo: **(pág. 60)** centro del átomo, tiene carga positiva, está compuesto por protones y neutrones y está rodeado por electrones localizados en niveles de energía.

O

oceanography: **(p. 7)** study of Earth's oceans including the creatures that inhabit its waters, its physical and chemical properties, and the effects of human activities.

ore: **(pp. 100, 684)** mineral that contains a valuable substance that can be mined at a profit.

original horizontality: **(p. 596)** the principle that sedimentary rocks are deposited in horizontal or nearly horizontal layers.

original preservation: **(p. 607)** describes a fossil with soft and hard parts that have undergone very little change since the organism's death.

orogeny (oh RAH juh nee): **(p. 567)** cycle of processes that form all mountain ranges, resulting in broad, linear regions of deformation that you know as mountain ranges but in geology are known as orogenic belts.

orographic lifting: **(p. 299)** cloud formation that occurs when warm, moist air is forced to rise up the side of a mountain.

outwash plain: **(p. 210)** area at the leading edge of a glacier, where outwash is deposited by meltwater streams.

oxidation: **(p. 166)** chemical reaction of oxygen with other substances.

ozone hole: **(p. 745)** a seasonal decrease on ozone over Earth's polar regions.

oceanografía: **(pág. 7)** estudio de los océanos de la Tierra: incluyendo sus propiedades físicas y químicas, los seres que los habitan y los efectos de las actividades humanas sobre ellos.

mena: **(pág. 100, 684)** mineral que contiene una sustancia valiosa que se puede extraer con fines de lucro.

horizontalidad original: **(pág. 596)** principio que establece que las rocas sedimentarias se depositan formando estratos horizontales o casi horizontales.

preservación de material original: **(pág. 607)** describe un fósil cuyas partes blandas y duras han sufrido muy pocos cambios desde la muerte del organismo.

orogenia: **(pág. 567)** ciclo de procesos que forman todas las cadenas montañosas, dando como resultado grandes regiones lineares de deformación llamadas cadenas montañosas, pero que en geología se conocen como cinturones orogénicos.

ascenso orográfico: **(pág. 299)** formación de nubes que se produce cuando el aire húmedo caliente es forzado a ascender por la ladera de una montaña.

llanura aluvial: **(pág. 210)** área en el borde frontal de un glaciar donde las corrientes del agua que se derrite depositan los derrubios.

oxidación: **(pág. 166)** reacción química del oxígeno con alguna otra sustancias.

agujero de ozono: **(pág. 745)** disminución estacional del ozono sobre las regiones polares de la Tierra.

P

paleogeography (pay lee oh jee AH gruh fee): (p. 648) the ancient geographic setting of an area.

paleogeografía: (pág. 648) características geográficas antiguas de un área.

paleomagnetism: (p. 476) study of Earth's magnetic record using data gathered from iron-bearing minerals in rocks that have recorded the orientation of Earth's magnetic field at the time of their formation.

paleomagnetismo: (pág. 476) estudio del registro magnético de la Tierra; utiliza la información recogida a partir de minerales ferrosos en las rocas porque este tipo de minerales registran la orientación del campo magnético de la Tierra en el momento en que se forman.

Pangaea (pan JEE uh): (p. 469) ancient landmass made up of all the continents that began to break apart about 200 mya.

Pangaea: (pág. 469) antigua masa terrestre compuesta por todos los continentes, los cuales se empezaron a separar hace cerca de 200 millones de años.

parallax: (p. 841) apparent positional shift of an object caused by the motion of the observer.

paralaje: (pág. 841) cambio aparente de la posición de un cuerpo causado por el movimiento del observador.

parsec (pc): (p. 840) the distance equal to 3.26 ly and 3.086×10^{13} km.

parsec: (pág. 840) distancia de 3.26 ly y 3.086×10^{13} km.

partial melting: (p. 114) process in which different minerals melt into magma at different temperatures, changing its composition.

fundición parcial: (pág. 114) proceso en el cual diferentes minerales se funden en el magma a diferentes tempe-raturas, cambiando su composición.

passive margin: (p. 648) edge of a continent along which there is no tectonic activity.

margen pasivo: (pág. 648) límite de un continente a lo largo del cual no ocurre actividad tectónica.

peat: (p. 710) light, spongy, organic fossil fuel derived from moss and other bog plants.

turba: (pág. 710) combustible fósil liviano, esponjoso y orgánico derivado del musgo y otras plantas de ciénegas.

pegmatite: (p. 122) vein deposits of extremely large-grained minerals that can contain rare ores such as lithium and beryllium.

pegmatita: (pág. 122) vetas de minerales de grano extremadamente grueso que pueden contener minerales raros como el litio y el berilio.

perigee: (p. 783) closest point in the Moon's elliptical orbit to Earth.

perigeo: (pág. 783) punto más cercano a la Tierra en la órbita elíptica de la Luna.

period: (p. 593) third-longest time unit in the geologic time scale, measured in tens of millions of years.

período: (pág. 593) tercera unidad de tiempo más grande en la escala del tiempo geológico; se mide en decenas de millones de años.

permeability: (p. 255) ability of a material to let water pass through, is high in material with large, well-connected pores and low in material with few pores or small pores.

permeabilidad: (pág. 255) capacidad de un material de permitir el paso del agua; es grande en materiales con poros grandes y bien conectados y baja en materiales con pocos poros o con poros pequeños.

pesticide: (p. 741) chemical applied to plants to kill insects and weeds.

pesticida: (pág. 741) sustancia química que se aplica a las plantas para eliminar insectos y malas hierbas.

photochemical smog: (p. 744) a type of air pollution, a yellow-brown haze formed mainly from automobile exhaust in the presence of sunlight.

smog fotoquímico: (pág. 744) tipo de contaminación del aire; niebla color amarillo marrón que se forma debido principalmente a las emisiones de los autos en presencia de la luz solar.

photosphere: (p. 831) lowest layer of the Sun's atmosphere that is also its visible surface, has an average temperature of 5800 K, and is about 400 km thick.

fotosfera: (pág. 831) capa más baja de la atmósfera solar; corresponde a su superficie visible, tiene una temperatura promedio de 5800 K y mide aproximadamente 400 km de ancho.

photovoltaic cell: **(p. 716)** thin, transparent wafer that converts sunlight into electrical energy and is made up of two layers of two types of silicon.

phytoplankton: **(p. 658)** microscopic organisms that are the basis of marine food chains; abundant during the Cretaceous and the remains of their shell-like hard parts are found in chalk deposits worldwide.

planetesimal: **(p. 798)** space object built of solid particles that can form planets through collisions and mergers.

plasma: **(p. 74)** hot, highly ionized, electrically conducting gas.

plastic deformation: **(p. 529)** premanent deformation caused by strain when stress exceeds a certain value.

plateau: **(p. 573)** a relatively flat-topped area.

pluton (PLOO tahn): **(p. 514)** intrusive igneous rock body, including batholiths, stocks, sills, and dikes, formed through mountain-building processes and oceanic-oceanic collisions; can be exposed at Earth's surface due to uplift and erosion.

point source: **(p. 749)** water-pollution source that generates pollution from a single point of origin, such as an industrial site.

polar easterlies: **(p. 320)** global wind systems that lie between latitudes 60° N and 60° S and the poles and is characterized by cold air.

polar zones: **(p. 378)** areas of Earth where solar radiation strikes at a low angle, resulting in temperatures that are nearly always cold; extend from 66.5° north and south of the equator to the poles.

pollutant: **(p. 690)** substance that enters Earth's geochemical cycles and can harm the health of living things or adversely affect their activities.

population I stars: **(p. 866)** stars in the disk and arms that have small amounts of heavy elements.

population II stars: **(p. 866)** stars in the halo and bulge that contain traces of heavy elements.

porosity: **(p. 142)** percentage of open spaces between grains in a material.

celdas fotovoltaicas: **(pág. 716)** láminas delgadas y transpa-rentes que convierten la luz solar en energía eléctrica; están compuestas de dos capas con dos tipos de silicio.

fitoplancton: **(pág. 658)** organismos microscópicos que son la base de las cadenas alimenticias marinas; fueron muy abundantes durante el Cretáceo y los restos de sus caparazones se encuentran en depósitos de carbonato de calcio por todo el mundo.

planetesimal: **(pág. 798)** cuerpo espacial formado por partículas sólidas y los cuales pueden formar planetas mediante choques y fusiones.

plasma: **(pág. 74)** gas caliente, altamente ionizado y conductor de electricidad.

deformación dúctil: **(pág. 529)** cuando la presión excede cierto valor; la tensión producida causa una deformación permanente.

altiplanicie: **(pág. 573)** área relativamente plana en la parte más alta.

plutones: **(pág. 514)** cuerpos rocosos ígneos intrusivos: incluye batolitos, macizos magmáticos, intrusiones y diques formados durante los procesos orogénicos y durante la colisión de placas oceánicas; pueden quedar expuestos a la superficie terrestre debido a levantamientos y erosión.

fuente puntual: **(pág. 749)** fuente de contaminación de agua que genera contaminación a partir de un solo punto de origen, por ejemplo, una zona industrial.

vientos polares del este: **(pág. 320)** sistemas globales del viento que se encuentran entre los polos y las latitudes 60°N y 60°S; se caracterizan por tener aire frío.

zonas polares: **(pág. 378)** áreas de la Tierra donde la radiación solar llega con un ángulo bajo, ocasionando que las temperaturas casi siempre sean frías; se extienden desde los 66.5° hasta los polos, en ambos hemisferios.

contaminante: **(pág. 690)** sustancia que entra a los ciclos geoquímicos de la Tierra y puede causar daños a la salud de los seres vivos o afectar adversamente sus actividades.

estrellas de la población I: **(pág. 866)** aquellas ubicadas en el disco y los brazos y que contienen pequeñas cantidades de elementos pesados.

estrellas de la población II: **(pág. 866)** aquellas ubicadas en el halo y en el núcleo y que contienen trazas de elementos pesados.

porosidad: **(pág. 142)** porcentaje de espacios abiertos entre los granos de una roca.

porphyritic (por fuh RIH tihk) texture: (p. 120) rock texture characterized by large, well-formed crystals surrounded by finer-grained crystals of the same or different mineral.

textura porfírica: (pág. 120) textura rocosa caracterizada por cristales grandes bien formados, rodeados por cristales de grano más fino del mismo mineral o de uno diferente.

Precambrian (pree KAM bree un): (p. 592) unit of geologic time consisting of the first three eons during which Earth formed and became hospitable to life.

Precámbrico: (pág. 592) unidad del tiempo geológico que consiste en los primeros tres eones; periodo durante el cual la Tierra se formó y adquirió condiciones aptas para la vida.

Precambrian shield: (p. 625) the top of a craton exposed at Earth's surface

escudo Precámbrico: (pág. 625) parte alta de un cratón que está expuesta en la superficie de la Tierra.

precipitation: (p. 302) all solid and liquid forms of water—including rain, snow, sleet, and hail—that fall from clouds.

precipitación: (pág. 302) toda forma líquida o sólida de agua: lluvia, nieve, aguanieve o granizo, que cae de las nubes.

prevailing westerlies: (p. 320) global wind system that lies between 30° and 60° north and south latitudes, where surface air moves toward the poles in an easterly direction.

vientos dominantes del oeste: (pág. 320) sistema de vientos globales ubicado entre los 30° y los 60° de latitud, en ambos hemisferios, donde el aire superficial se desplaza hacia los polos en dirección este.

primary wave: (p. 532) seismic wave that squeezes and pushes rocks in the same direction that the wave travels, known as a P-wave.

onda primaria: (pág. 532) onda sísmica que comprime y empuja las rocas en la misma dirección en que viaja la onda; se conocen como ondas P.

prime meridian: (p. 31) imaginary line representing 0° longitude, running from the north pole, through Greenwich, England, to the south pole.

primer meridiano: (pág. 31) línea imaginaria que representa la longitud 0°; va desde el polo norte hasta el polo sur, pasando por Greenwich, Inglaterra.

prokaryote (proh KE ree oht): (p. 635) unicellular organism that lacks a nucleus.

procariota: (pág. 635) organismo unicelular que carece de núcleo.

prominence: (p. 833) arc of gas ejected from the chromosphere, or gas that condenses in the Sun's inner corona and rains back to the surface, that can reach temperatures over 50,000 K and is associated with sunspots.

protuberancia solar: (pág. 833) arco de gas expulsado de la cromosfera o gas que se condensa en la corona interna del Sol y que se precipita de nuevo sobre su superficie; puede alcanzar temperaturas mayores a los 50,000 K y está asociada a la presencia de manchas solares.

proton: (p. 60) tiny atomic particle that has mass and a positive electric charge.

protón: (pág. 60) partícula atómica diminuta que tiene masa y una carga eléctrica positiva.

protostar: (p. 848) hot, condensed object at the center of a nebula that will become a new star when nuclear fusion reactions begin.

protoestrella: (pág. 848) cuerpo condensado, caliente, ubicado en el centro de una nebulosa, que se convertirá en una estrella nueva cuando inicien las reacciones de fusión nuclear.

pulsar: (p. 850) a spinning neutron star that exhibits a pulsing pattern.

pulsar: (pág. 850) estrella de neutrones giratoria que exhibe un patrón de pulsaciones.

pyroclastic flow: (p. 513) swift-moving, potentially deadly clouds of gas, ash, and other volcanic material produced by a violent eruption.

flujo piroclástico: (pág. 513) nubes de gas, cenizas y otros materiales volcánicos, potencialmente mortales, que se desplazan rápidamente y que son producidas por una erupción violenta.

Q

quasar: (p. 875) starlike, very bright, extremely distant object with emission lines in its spectra.

cuásares: (pág. 875) cuerpos semejantes a estrellas, muy brillantes y extremadamente lejanos, con líneas de emisión en sus espectros.

R

radiation: (p. 287) the transfer of thermal energy electromagnetic waves; the transfer of thermal energy from the Sun to Earth by radiation.

radioactive decay: (p. 601) emission of radioactive particles and its resulting change into other isotopes over time.

radiocarbon dating: (p. 603) determines the age of relatively young organic objects; objects that are alive or were once alive.

radio galaxy: (p. 875) very bright, often giant, elliptical galaxy that emits as much or more energy in the form of radio wavelengths as it does wavelengths of visible light.

radiometric dating: (p. 602) process used to determine the absolute age of a rock or fossil by determining the ratio of parent nuclei to daughter nuclei within a given sample.

radiosonde (RAY dee oh sahnd)**:** (p. 326) balloon-borne weather instrument whose sensors measure air pressure, humidity, temperature, wind speed, and wind direction of the upper atmosphere.

ray: (p. 771) long trail of ejecta that radiates outward from a Moon crater.

recharge: (p. 263) process by which water from precipitation and runoff is added to the zone of saturation.

reclamation: (p. 738) process in which a mining company restores land used during mining operations to its original contours and replants vegetation.

red bed: (p. 631) a sedimentary rock deposit that contains oxidized iron; provides evidence that free oxygen existed in the atmosphere during the Proterozoic.

reflecting telescope: (p. 766) telescope that uses mirrors to focus visible light.

refracting telescope: (p. 766) telescope that uses lenses to focus visible light.

regional metamorphism: (p. 149) process that affects large areas of Earth's crust, producing belts classified as low, medium, or high grade, depending on pressure on the rocks, temperature, and depth below the surface.

regolith: (p. 772) layer of loose, ground-up rock on the lunar surface.

radiación: (pág. 287) transferencia de energía mediante ondas electromagnéticas; la transferencia de energía térmica del Sol a la Tierra por radiación.

desintegración radiactiva: (pág. 601) emisión de partículas atómicas que a lo largo del tiempo produce nuevos isótopos.

datación radiocarbónica: (pág. 603) permite determinar la edad de cuerpos orgánicos relativamente recientes, cuerpos que están vivos o que alguna vez estuvieron vivos.

radiogalaxia: (pág. 875) galaxia elíptica muy brillante, a menudo gigantesca, cuya emisión de energía en forma de ondas de radio es similar a la que emite como ondas de luz visible.

datación radiométrica: (pág. 602) proceso que permite establecer la edad absoluta de una roca o un fósil, al determinar la razón entre los núcleos originales y los núcleos derivados de una muestra dada.

radiosonda: (pág. 326) instrumento meteorológico que se monta en un globo y cuyos sensores miden la presión atmosférica, la humedad, la temperatura, así como la velocidad y dirección del viento en la atmósfera superior.

rayo: (pág. 771) largo rastro de eyecta que irradia de un cráter lunar.

recarga: (pág. 263) proceso mediante el cual el agua de la precipitación y de la escorrentía entra a la zona de saturación.

recuperación: (pág. 738) proceso en que una compañía minera restaura los terrenos usados en las actividades mineras a sus contornos originales y reforesta con nueva vegetación.

lecho rojo: (pág. 631) depósito de roca sedimentaria que contiene hierro oxidado; es evidencia de que había oxígeno libre en la atmósfera durante el Proterozoico.

telescopio reflector: (pág. 766) telescopio que usa espejos para enfocar la luz visible.

telescopio refractor: (pág. 766) telescopio que usa lentes para enfocar la luz visible.

metamorfismo regional: (pág. 149) proceso que afecta grandes áreas de la corteza terrestre; produce cinturones de bajo, medio o alto grado, dependiendo de la presión sobre las rocas, la temperatura y la profundidad bajo la superficie.

regolito: (pág. 772) estrato de roca suelta y molida en la superficie lunar.

regression: (p. 649) occurs when sea level falls, causing the shoreline to move seaward, and results in shallower-water deposits overlying deeper-water deposits.

rejuvenation: (p. 237) process during which a stream resumes downcutting toward its base level, increasing its rate of flow.

relative-age dating: (p. 596) establishing the order of past geologic events.

relative humidity: (p. 294) ratio of water vapor contained in a specific volume of air compared with how much water vapor that amount of air actually can hold; expressed as a percentage.

remote sensing: (p. 41) process of gathering data about Earth from instruments far above the planet's surface.

renewable resource: (p. 678) natural resource, such as fresh air and most groundwater, that can be replaced by nature in a short period of time.

residual soil: (p. 177) soil that develops from parent material which is similar to local bedrock.

retrograde motion: (p. 799) a planet's apparent backward movement in the sky.

return stroke: (p. 348) a branch channel of positively charged ions that rushes upward from the ground to meet the stepped leader.

Richter scale: (p. 539) numerical rating system used to measure the amount of energy released during an earthquake.

ridge push: (p. 488) tectonic process associated with convection currents in Earth's mantle that occurs when the weight of an elevated ridge pushes an oceanic plate toward a subduction zone.

rift valley: (p. 481) long, narrow depression that forms when continental crust begins to separate at a divergent boundary.

rill erosion: (p. 172) erosion in which water running down the side of a slope carves a small stream channel.

rille: (p. 771) valleylike structure that meanders across some regions of the Moon's maria.

rock cycle: (p. 151) continuous, dynamic set of processes by which rocks are changed into other types of rock.

root: (p. 563) thickened areas of continental material, detected by gravitational and seismic studies.

regresión: (pág. 649) ocurre cuando baja el nivel del mar, provocando que la costa avance hacia el mar, ocasiona que depósitos de agua más superficiales cubran depósitos de agua más profundos.

rejuvenecimiento: (pág. 237) proceso en que una corriente reanuda la erosión hacia su nivel base, aumentando su tasa de flujo.

datación relativa: (pág. 596) ordenamiento por antigüedad de eventos geológicos pasados.

humedad relativa: (pág. 294) razón del vapor de agua que contiene un volumen específico de aire, en comparación con la cantidad de vapor de agua que ese volumen de aire podría contener, expresado como porcentaje.

percepción remota: (pág. 41) proceso de recopilación de datos sobre la Tierra con instrumentos alejados de la superficie del planeta.

recurso renovable: (pág. 678) recurso natural, como el aire y la mayoría de las aguas subterráneas, que la naturaleza puede reemplazar en un período corto de tiempo.

suelo residual: (pág. 177) suelo que se desarrolla a partir del material original y es similar a la roca madre local.

movimiento retrógrado: (pág. 799) movimiento aparentemente en retroceso de un planeta en el cielo.

descarga de retorno: (pág. 348) un canal con iones de carga positiva que asciende desde el suelo para encontrarse con la descarga líder o guía escalonada.

escala de Richter: (pág. 539) escala numérica que se emplea para medir la cantidad de energía liberada durante un sismo.

empuje de la dorsal: (pág. 488) proceso tectónico asociado con las corrientes de convección en el manto de la Tierra, que ocurre cuando el peso de una cordillera elevada empuja una placa oceánica hacia una zona de subducción.

valle del rift: (pág. 481) depresión larga y estrecha que se forma cuando la corteza continental se empieza a separar en un límite divergente.

erosión por surcos: (pág. 172) erosión en la cual el agua que corre cuesta abajo forma un canal pequeño.

surco: (pág. 771) formación tipo valle que serpentea a través de algunas regiones de los mares lunares.

ciclo de las rocas: (pág. 151) conjunto de procesos continuos y dinámicos a través de los cuales las rocas se transforman en otros tipos de roca.

raíz: (pág. 563) gruesas áreas de material continental que son detectadas en estudios sísmicos o gravitacionales.

RR Lyrae variable: (p. 863) stars with pulsation periods ranging from 1.5 hours to 1 day, generally having the same luminosity, regardless of pulsation period length.

runoff: (p. 225) water that flows downslope on Earth's surface and may enter a stream, river, or lake; its rate is influenced by the angle of the slope, vegetation, rate of precipitation, and soil composition.

estrellas variables tipo RR Lyrae: (pág. 863) estrellas con períodos de pulsación que duran de 1.5 horas a 1 día; en general tienen la misma luminosidad, independientemente de la duración de la pulsación.

escorrentía: (pág. 225) agua que corre cuesta abajo sobre la superficie terrestre y que puede incorporarse a una corriente, río o lago; su tasa de flujo está influida por el ángulo de la pendiente, la vegetación, la tasa de precipi-tación y la composición del suelo.

S

Saffir-Simpson hurricane scale: (p. 358) classifies hurricanes according to wind speed, potential for property damage, and potential for flooding in terms of the effect on the height of sea level on a scale ranging from Category 1 to Category 5.

salinity: (p. 413) measure of the amount of salts dissolved in seawater, which is 35 ppt, or 3.5% on average.

saturation: (p. 294) the point at which water molecules leaving the water's surface equals the rate of water molecules returning to the surface.

scarp: (p. 805) cliff on Mercury; similar to those on Earth but much higher.

scientific law: (p. 19) a principle that describes the behavior of a natural phenomenon.

scientific methods: (p. 10) a series of problem-solving procedures that help scientists conduct experiments.

scientific model: (p. 18) an idea, a system, or a mathematical expression that represents the idea being explained.

scientific notation: (p. 16) a method used by scientists to express a number as a value between 1 and 10 multiplied by a power of 10.

scientific theory: (p. 19) an explanation based on many observations during repeated experiments; valid only if consistent with observations, can be used to make testable predictions, and is the simplest explanation; can be changed or modified with the discovery of new data.

sea-breeze thunderstorm: (p. 346) local air-mass thunderstorm that commonly occurs along a coastal area because land and water store and release thermal energy differently.

escala de huracanes Saffir-Simpson: (pág. 358) clasifica los huracanes según la velocidad de sus vientos, el daño potencial que pueden causar a la propiedad y el potencial que tienen de causar inundaciones debido a su efecto sobre el nivel del mar, en una escala que va desde la Categoría 1 hasta la Categoría 5.

salinidad: (pág. 413) medida de la cantidad de sales disueltas en el agua de mar; en promedio es de 35 ppt ó 3.5%.

saturación: (pág. 294) sucede en el punto en el cual la tasa de salida de moléculas de agua en la superficie es igual a la tasa de retorno de las moléculas a la superficie.

escarpes: (pág. 805) fracturas en la superficie de Mercurio, similares a las de la Tierra, pero con mayor pro-fundidad.

ley científica: (pág. 19) principio que describe el comportamiento de un fenómeno natural.

métodos científicos: (pág. 10) serie de procedimientos para resolver problemas que ayudan a los científicos a realizar experimentos.

modelo científico: (pág. 18) idea, sistema o expresión matemática que representa la idea que se quiere explicar.

notación científica: (pág. 16) método que usan los científicos para expresar un número como un valor entre 1 y 10 multiplicado por una potencia de 10.

teoría científica: (pág. 19) explicación basada en muchas observaciones realizadas durante experimentos repetidos; sólo es válida si es consistente con las observaciones, permite hacer predicciones comprobables y es la explicación más sencilla; puede ser modificada debido al descubrimiento de nuevos hechos.

tormenta eléctrica de brisa marina: (pág. 346) tormenta local de masa de aire que ocurre comúnmente a lo largo de un área costera; ocurren porque la tierra y el agua almacenan y liberan energía térmica de manera distinta.

seafloor spreading: **(p. 479)** the hypothesis that new ocean crust is formed at mid-ocean ridges and destroyed at deep-sea trenches; occurs in a continuous cycle of magma intrusion and spreading.

expansión del suelo marino: **(pág. 479)** hipótesis que propone que la nueva corteza oceánica se forma en las dorsales medioceánicas y se destruye en las fosas submarinas profundas; ocurre según un ciclo continuo de intrusión y expansión del magma.

sea level: **(p. 410)** level of the oceans' surfaces, which has risen at a rate of about 3 mm per year.

nivel del mar: **(pág. 410)** nivel de la superficie del océano; actualmente sube a una velocidad de 3 mm por año.

seamount: **(p. 452)** basaltic, submerged volcano on the seafloor that is more than 1 km high.

montaña submarina: **(pág. 452)** volcán basáltico sumergido en el fondo marino que mide más de 1 km de altura.

season: **(p. 388)** short-term periods with specific weather conditions caused by regular variations in temperature, hours of daylight, and weather patterns that are due to the tilt of Earth's axis as it revolves around the Sun, causing different areas of Earth to receive different amounts of solar radiation.

estación: **(pág. 388)** períodos de corto plazo con específicas de tiempo causados por variaciones regulares en temperatura, horas de luz solar y patrones meteorológicos, provocadas por la inclinación del eje de la Tierra cuando gira alrededor del Sol, lo que ocasiona que las distintas áreas de la Tierra reciban diferentes cantidades de radiación solar.

secondary wave: **(p. 532)** seismic wave that causes rock particles to move at right angles to the direction of the wave, known as an S-wave.

onda secundaria: **(pág. 532)** onda sísmica que ocasiona que las partículas de las rocas se muevan en ángulo recto con respecto a la dirección de la onda.

sediment: **(p. 134)** small pieces of rock that are moved and deposited by water, wind, glaciers, and gravity.

sedimentos: **(pág. 134)** partículas pequeñas de roca que el agua, el viento, los glaciares y la gravedad mueven y depositan.

seismic gap: **(p. 550)** place along an active fault that has not experienced an earthquake for a long time.

vacío sísmico: **(pág. 550)** lugar a lo largo de una falla activa que no ha sufrido un terremoto durante mucho tiempo.

seismic wave: **(p. 532)** the vibrations of the ground during an earthquake.

onda sísmica: **(pág. 532)** vibraciones del terreno durante un sismo.

seismogram (SIZE muh gram): **(p. 534)** record produced by a seismometer that can provide individual tracking of each type of seismic wave.

sismograma: **(pág. 534)** registro producido por un sismógrafo que proporciona un registro individual de cada tipo de onda sísmica.

seismometer (size MAH muh tur): **(p. 534)** instrument used to measure horizontal or vertical motion during an earthquake.

sismógrafo: **(pág. 534)** instrumento que sirve para medir los movimientos horizontales y verticales durante un sismo.

shield volcano: **(p. 507)** broad volcano with gently sloping sides built by nonexplosive eruptions of basaltic lava that accumulates in layers.

volcán de escudo: **(pág. 507)** volcán ancho, de laderas con inclinación suave, formado por erupciones no explosivas de lava basáltica que se acumula en estratos.

side-scan sonar: **(p. 407)** technique that directs sound waves at an angle to the seafloor or deep-lake floor, allowing underwater topographic features to be mapped.

sonar de escaneo lateral: **(pág. 407)** técnica que dirige las ondas sonoras en ángulo hacia el fondo del mar o de un lago profundo, lo que permite trazar el relieve topográfico submarino.

silicate: **(p. 96)** mineral that contains silicon (Si), oxygen (O), and usually one or more other elements.

silicato: **(pág. 96)** mineral que contiene silicio (Si), oxígeno (O) y generalmente uno o más elementos adicionales.

sill: **(p. 515)** pluton that forms when magma intrudes parallel rock layers.

intrusión: **(pág. 515)** plutón que se forma cuando el magma penetra estratos rocosos paralelos.

sinkhole: **(p. 261)** depression in Earth's surface formed when a cave collapses or bedrock is dissolved by acidic rain or moist soil.

slab pull: **(p. 488)** tectonic process associated with convection currents in Earth's mantle that occurs as the weight of the subducting plate pulls the trailing lithosphere into a subduction zone.

slump: **(p. 198)** mass movement that occurs when Earth materials in a landslide rotate and slide along a curved surface, leaving a crescent-shaped scar on a slope.

soil: **(p. 176)** loose covering of weathered rock and decayed organic matter overlying Earth's bedrock that is characterized by texture, fertility, and color and whose composition is determined by its parent rock and environmental conditions.

soil horizon: **(p. 178)** distinct layer within a soil profile.

soil liquefaction (lih kwuh FAK shun): **(p. 547)** process associated with seismic vibrations that occur in areas of sand that is nearly saturated; resulting in the ground behaving like a liquid.

soil profile: **(p. 178)** vertical sequence of soil layers, containing A-horizon B-horizon C-horizon.

solar eclipse: **(p. 781)** when the Moon passes between Earth and the Sun and the Moon casts a shadow on Earth, blocking Earth's view of the Sun; can be partial or total.

solar flare: **(p. 833)** violent eruption of radiation and particles from the Sun's surface that is associated with sunspots.

solar wind: **(p. 832)** wind of charged particles (ions) that flows throughout the solar system and begins as gas flowing outward from the Sun's corona at high speeds.

solstice: **(p. 777)** period when the Sun is overhead at its farthest distance either north or south of the equator.

solution: **(p. 71)** homogeneous mixture whose components cannot be distinguished and can be classified as liquid, gaseous, solid, or a combination; **(p. 228)** the method of transport for materials that are dissolved in a stream's water.

sonar: **(p. 43)** use of sound waves to detect and measure objects underwater.

source region: **(p. 316)** area over which an air mass forms.

sumidero: **(pág. 261)** depresión en la superficie terrestre que se forma cuando una caverna se colapsa o cuando el lecho rocoso es disuelto por lluvia ácida o suelo húmedo.

tracción de placa: **(pág. 488)** proceso tectónico asociado con las corrientes de convección del manto de la Tierra, que ocurre cuando el peso de la placa subductora jala la litosfera hacia una zona de subducción.

deslizamiento rotacional: **(pág. 198)** movimiento en masa que ocurre cuando los materiales terrestres de un derrumbe giran y se deslizan a lo largo de una superficie curva, dejando una cicatriz con forma de medialuna en la pendiente.

suelo: **(pág. 176)** cubierta suelta de roca meteorizada y materia orgánica en descomposición que cubre el lecho rocoso terrestre; se caracteriza por su textura, fertilidad y color y su composición está determinada por la roca madre y las condiciones ambientales.

horizonte del suelo: **(pág. 178)** capa distintiva dentro de un perfil del suelo.

licuefacción del suelo: **(pág. 547)** proceso asociado con las vibraciones sísmicas que ocurren en las áreas arenosas casi saturadas; el resultado es que el suelo actúa como un líquido.

perfil del suelo: **(pág. 178)** sucesión vertical de capas del suelo, comprende los horizontes A (mantillo), B (subsuelo) y C (material original meteorizado).

eclipse solar: **(pág. 781)** sucede cuando la Luna pasa entre la Tierra y el Sol y la Luna proyecta su sombra sobre la Tierra, bloqueando la luz del Sol; puede ser parcial o total.

erupción solar: **(pág. 833)** violenta erupción de radiación y partículas desde la superficie del Sol que está asociada con las manchas solares.

viento solar: **(pág. 832)** viento de partículas cargadas (iones) que fluye a través del sistema solar y comienza como un gas que es despedido a gran velocidad por la corona del Sol.

solsticio: **(pág. 777)** sucede cuando el Sol se halla en el horizonte a su mayor distancia al norte o al sur del ecuador.

solución: **(pág. 71)** mezcla homogénea cuyos componentes no se pueden distinguir; puede clasificarse como líquida, gaseosa, sólida o una combinación de éstas; **(pág. 228)** el método de transporte de materiales que están disueltos en las aguas de una corriente.

sonar: **(pág. 43)** uso de ondas sonoras para detectar y medir objetos submarinos.

región fuente: **(pág. 316)** área sobre la cual se forma una masa de aire.

specific gravity: (p. 95) ratio of the mass of a substance to the mass of an equal volume of H_2O at 4°C.

spiral density wave: (p. 868) spiral regions of alternating density which rotates as a rigid pattern.

spring: (p. 256) natural discharge of groundwater at Earth's surface where an aquifer and an aquiclude come in contact.

spring tide: (p. 424) during full or new moon, the Sun, the Moon, and Earth are all aligned; this causes solar tides to enhance lunar tides, causing high tides to be higher than normal and low tides to be lower than normal.

stalactite: (p. 261) cone-shaped or cylindrical dripstone deposit of calcium carbonate that hangs like an icicle from a cave's ceiling.

stalagmite: (p. 261) mound-shaped dripstone deposit of calcium carbonate that forms on a cave's floor beneath a stalactite.

station model: (p. 329) record of weather data for a specific place at a specific time, using meteorological symbols.

stepped leader: (p. 348) The channel of partially charged air; the breakdown in charges in between positive and negative regions.

stock: (p. 515) irregularly shaped pluton that is similar to a batholith but smaller, generally forms 5–30 km beneath Earth's surface, and cuts across older rocks.

storm surge: (p. 359) occurs when powerful, hurricane-force winds drive a mound of ocean water toward shore, where it washes over the land, often causing enormous damage.

strain: (p. 528) deformation of materials in response to stress.

stratosphere: (p. 284) layer of Earth's atmosphere that is located above the tropopause and is made up primarily of concentrated ozone.

stratus (STRAY tus): (p. 301) a layered sheetlike cloud that covers much or all of the sky in a given area.

streak: (p. 93) color a mineral leaves when it is rubbed across an unglazed porcelain plate or when it is broken up and powdered.

stream bank: (p. 232) ground bordering each side of a stream that keeps the moving water confined.

gravedad específica: (pág. 95) razón de la masa de una sustancia con relación a la masa de un volumen igual de H_2O a 4°C.

ondas de densidad espirales: (pág. 868) regiones en espiral con densidad variable que giran siguiendo un patrón rígido.

manantial: (pág. 256) descarga natural de agua subterránea en la superficie terrestre, en el punto donde un acuífero y un acuicluso entran el contacto.

marea viva: (pág. 424) durante la luna nueva o la luna llena, el Sol, la Luna y la Tierra se encuentran alineados; esto ocasiona que la marea solar aumente el efecto de la marea lunar y provoca que la marea alta sea más alta que lo normal y que la marea baja sea más baja que lo normal.

estalactita: (pág. 261) depósito rocoso de carbonato de calcio, de forma cónica o cilíndrica, que se forma por goteo y que cuelga como un carámbano del techo de una caverna.

estalagmita: (pág. 261) depósito de carbonato de calcio, con forma de montículo, que se forma por goteo en el piso de una caverna, debajo de una estalactita.

código meteorológico: (pág. 329) registro de los datos del tiempo para un lugar específico en un tiempo dado, usando símbolos meteorológicos.

guía escalonada: (pág. 348) el canal con aire parcialmente cargado; la separación de cargas que forma regiones positivas y negativas.

macizo magmático: (pág. 515) plutón de forma irregular, similar a un batolito pero más pequeño; generalmente se forma de 5 a 30 km bajo la superficie terrestre y atraviesa rocas más antiguas.

marejada ciclónica: (pág. 359) ocurre cuando poderosos vientos huracanados arrojan una gran masa de agua del océano hacia la costa, desparramándose por el terreno y causando a menudo un daño enorme.

tensión: (pág. 528) deformación de los materiales en res-puesta a un estrés.

estratosfera: (pág. 284) capa de la atmósfera terrestre ubicada por encima de la tropopausa; está compuesta principalmente de ozono concentrado.

estrato: (pág. 301) nube con forma de capas delgadas que cubre la mayoría o todo el cielo en cierta área.

veta: (pág. 93) color que deja un mineral cuando es frotado contra un plato de porcelana sin barnizar o cuando se rompe y se pulveriza.

margen de una corriente de agua: (pág. 232) terreno que limita ambos lados de una corriente, manteniendo confinada la corriente de agua en movimiento.

stream channel: (p. 232) narrow pathway carved into sediment or rock by the movement of surface water.

stress: (p. 528) forces per unit area that act on a material—compression, tension, and shear.

stromatolite (stroh MA tuh lite): (p. 629) large mat or mound composed of billions of photosynthesizing cyanobacteria that dominated shallow oceans during the Proterozoic.

subduction: (p. 482) process by which one tectonic plate slips beneath another tectonic plate.

sublimation: (p. 75) process by which a solid slowly changes to a gas without first entering a liquid state.

sunspot: (p. 832) dark spot on the surface of the photosphere that typically lasts two months, occurs in pairs, and has a penumbra and an umbra.

supercell: (p. 350) extremely powerful, self-sustaining thunderstorm characterized by intense, rotating updrafts.

supercluster: (p. 873) gigantic threadlike or sheetlike cluster of galaxies that is hundreds of millions of light-years in size.

supernova: (p. 851) massive explosion that occurs when the outer layers of a star are blown off.

superposition: (p. 596) the principle that, in an undisturbed rock sequence, the oldest rocks are on the bottom and each consecutive layer is younger than the layer beneath it.

surface current: (p. 425) wind-driven movement of ocean water that primarily affects the upper few hundred meters of the ocean.

suspension: (p. 228) the method of transport for all particles small enough to be held up by the turbulence of a stream's moving water.

sustainable energy: (p. 723) involves global management of Earth's natural resources to ensure that current and future energy needs will be met without harming the environment.

sustainable yield: (p. 679) replacement of renewable resources at the same rate at which they are consumed.

synchronous rotation: (p. 780) the state at which the Moon's orbital and rotational periods are equal.

cauce fluvial: (pág. 232) estrecha vía labrada en el sedimento, o en la roca, por el movimiento del agua en la superficie.

estrés: (pág. 528) fuerza por unidad de área que actúa sobre un material: puede ser por compresión, tensión o cizallamiento.

estromatolitos: (pág. 629) montículos grandes compuestos de billones de cianobacterias fotosintéticas que dominaron los océanos superficiales durante el Proterozoico.

subducción: (pág. 482) proceso en que una placa tectónica se desliza por debajo de otra.

sublimación: (pág. 75) proceso en que un sólido se convierte lentamente en gas, sin convertirse primero al estado líquido.

mancha solar: (pág. 832) mancha oscura en la superficie de la fotosfera que normalmente dura dos meses, ocurren en pares y tienen una penumbra y una umbra.

supercelda: (pág. 350) tormenta autosostenible extremadamente poderosa, caracterizada por tener intensas cor-rientes ascendentes giratorias.

supercúmulo: (pág. 873) cúmulo gigantesco de galaxias con forma de filamento o lámina que mide centenares de millones de años luz.

supernova: (pág. 851) enorme explosión que ocurre cuando estallan las capas exteriores de una estrella.

superposición: (pág. 596) principio que establece que en una sucesión rocosa no perturbada, los estratos rocosos más antiguos se encuentran en el fondo y que cada capa sucesiva es más reciente que la capa subyacente.

corriente superficial: (pág. 425) movimiento de las aguas del océano producido por el viento, que afecta principalmente los primeros cientos de metros superiores de las aguas del océano.

suspensión: (pág. 228) método de transporte de todas las partículas que son suficientemente pequeñas como para ser mantenidas en el agua por la turbulencia de la corriente del agua en movimiento.

energía sostenible: (pág. 723) implica la administración global de los recursos naturales de la Tierra para asegurar que se satisfagan las necesidades energéticas actuales y futuras, sin causar daños al ambiente.

rendimiento sostenible: (pág. 679) regeneración de los recursos renovables a la misma velocidad con que se consumen.

rotación sincronizada: (pág. 780) estado en que los periodos de la órbita y de rotación de la Luna son iguales.

T

tailings: **(p. 686)** material left after mineral ore has been extracted from parent rock; can release harmful chemicals into groundwater or surface water.

escombreras: **(pág. 686)** material que queda después de que se ha extraído la mena de la roca madre; puede liberar sustancias químicas tóxicas hacia las aguas subterráneas y superficiales.

tectonic plate: **(p. 480)** huge pieces of Earth's crust that cover its surface and fit together at their edges.

placa tectónica: **(pág. 480)** enormes fragmentos de corteza que cubren la superficie terrestre; sus límites se corresponden entre sí.

temperate zone: **(p. 378)** area of Earth that extends between 23.5° and 66.5° north and south of the equator and has moderate temperatures.

zonas templadas: **(pág. 378)** áreas de la Tierra que se extienden entre los 23.5° y los 66.5°, al norte y al sur del ecuador; experimentan temperaturas moderadas.

temperature inversion: **(p. 292)** increase in temperature with height in an atmospheric layer, which inverts the temperature-altitude relationship and can worsen air-pollution problems.

inversión de temperatura: **(pág. 292)** aumento de temperatura que ocurre al aumentar la altitud en alguna capa de la atmósfera; invierte la relación entre la altitud y la temperatura y puede empeorar los problemas de contaminación del aire.

temperature profile: **(p. 418)** plots changing ocean water temperatures against depth, which varies, depending on location and season.

perfil de temperatura: **(pág. 418)** diagramas que muestran cómo cambia la temperatura del océano con la profundidad; varía según la ubicación y la temporada.

tephra: **(p. 512)** rock fragments, classified by size, that are thrown into the air during a volcanic eruption and fall to the ground.

tefrita: **(pág. 512)** fragmentos rocosos que se clasifican por tamaño; son lanzados al aire durante una erupción volcánica y luego caen al suelo.

terrestrial planet: **(p. 804)** one of the rocky-surfaced, relatively small, dense inner planets closest to the Sun—Mercury, Venus, Earth, and Mars.

planetas terrestres: **(pág. 804)** planetas internos, densos, relativamente pequeños, con superficie rocosa y cercanos al Sol: Mercurio, Venus, la Tierra, y Marte.

tetrahedron: **(p. 96)** a geometric solid having four sides that are equilateral triangles

tetraedro: **(pág. 96)** sólido geométrico que tiene cuatro lados con forma de triángulo equilátero.

texture: **(p. 119)** the size, shape, and distribution of the crystals or grains that make up a rock.

textura: **(pág. 119)** tamaño, forma y distribución de los granos o cristales que forman una roca.

thermocline: **(p. 418)** transitional ocean layer that lies between the relatively warm, sunlit surface layer and the colder, dark, dense bottom layer and is characterized by temperatures that decrease rapidly with depth.

termoclina: **(pág. 418)** capa de transición del océano que se halla entre la capa superficial iluminada por el Sol, que tiene una temperatura relativamente tibia, y la capa inferior, que es densa, oscura y fría; se caracteriza por tener temperaturas que disminuyen rápidamente con la profundidad.

thermometer: **(p. 324)** instrument used to measure temperature using either the Faherenheit or Celsius scale.

termómetro: **(pág. 324)** instrumento que sirve para medir la temperatura en grados Fahrenheit o Celsius.

thermosphere: **(p. 284)** layer of Earth's atmosphere that is located above the mesopause; oxygen atoms absorb solar radiation causing the temperature to increase in this layer.

termosfera: **(pág. 284)** capa de la atmósfera terrestre ubicada por encima de la mesopausa; los átomos de oxígeno absorben radiación solar, haciendo que la temperatura aumente en esta capa.

tide: **(p. 423)** periodic rise and fall of sea level caused by the gravitational attraction among Earth, the Moon, and the Sun.

marea: **(pág. 423)** ascenso y descenso periódicos del nivel del mar causados por la atracción gravitacional entre la Tierra, la Luna y el Sol.

***TOPEX/Poseidon* satellite:** **(p. 42)** data-gathering satellite that uses radar to map features on the ocean floor.

satélite *TOPEX/Poseidon:* **(pág. 42)** satélite de recopilación de datos que usa un radar para trazar el relieve del fondo del océano.

topographic map: **(p. 36)** map that uses contour lines, symbols, and color to show changes in the elevation of Earth's surface and features such as mountains, bridges, and rivers.

topography: **(p. 562)** the change in elevation of the crust.

tornado: **(p. 352)** violent, whirling column of air in contact with the ground that forms when wind direction and speed suddenly change with height, is often associated with a supercell, and can be extremely damaging.

trace fossil: **(p. 608)** the only indirect fossil evidence of an organism; traces of worm trails, footprints, and tunneling burrows.

trade winds: **(p. 320)** two global wind systems that flow between 30° north and south latitudes, where air sinks, warms, and returns to the equator in a westerly direction.

transform boundary: **(p. 484)** place where two tectonic plates slide horizontally past each another; is characterized by long faults and shallow earthquakes.

transgression: **(p. 649)** occurs when sea level rises and causes the shoreline to move inland, resulting in deeper-water deposits overlying shallower-water deposits.

transported soil: **(p. 177)** soil that has been moved away from its parent material by water, wind, or a glacier.

tropical cyclone: **(p. 355)** large, low-pressure, rotating tropical storm that gets its energy from the evaporation of warm ocean water and the release of heat.

tropics: **(p. 378)** area of Earth that receives the most solar radiation, is generally warm year-round, and extends between 23.5° south and 23.5° north of the equator.

troposphere: **(p. 284)** layer of the atmosphere closest to Earth's surface, where most of the mass of the atmos-phere is found and in which most weather takes place and air pollution collects.

trough: **(p. 421)** lowest point of a wave.

tsunami (soo NAH mee): **(p. 548)** large, powerful ocean wave generated by the vertical motions of the seafloor during an earthquake; in shallow water, can form huge, fast-moving breakers exceeding 30 m in height that can damage coastal areas.

mapa topográfico: **(pág. 36)** mapa que usa curvas de nivel, símbolos y colores para mostrar los cambios en la elevación de la superficie terrestre, e incluye rasgos como las montañas, los puentes y los ríos.

topografía: **(pág. 562)** el cambio en la elevación de la corteza.

tornado: **(pág. 352)** violenta columna giratoria de aire en contacto con el suelo; se forma cuando la dirección y la velocidad del viento cambian repentinamente con la altura; a menudo está asociada con una supercelda y puede ser extremadamente dañina.

fósiles traza: **(pág. 608)** pruebas fósiles indirectas de un organismo: incluye rastros de gusanos, huellas de pasos y madrigueras.

vientos alisios: **(pág. 320)** dos sistemas globales de vientos que se desplazan entre los 30° de latitud norte y sur, donde el aire desciende, se calienta y regresa al ecuador con dirección oeste.

límite transformante: **(pág. 484)** lugar donde dos placas tectónicas se deslizan horizontalmente, una al lado de la otra y en sentidos opuestos; se caracteriza por presentar grandes fallas y terremotos superficiales.

transgresión: **(pág. 649)** ocurre cuando el nivel del mar aumenta y hace que el litoral retroceda hacia el interior; ocasiona depósitos de agua más profunda que cubren depósitos de agua menos profunda.

suelo transportado: **(pág. 177)** suelo que ha sido transportado lejos de su roca madre por el agua, el viento o un glaciar.

ciclón tropical: **(pág. 355)** gran tormenta giratoria de baja presión que obtiene su energía de la evaporación de las tibias aguas del mar y la liberación de calor.

trópicos: **(pág. 378)** área de la Tierra que recibe la mayor cantidad de radiación solar, generalmente es caliente todo el año y se extiende entre 23.5° sur y 23.5° norte del ecuador.

troposfera: **(pág. 284)** capa de la atmósfera más cercana a la superficie terrestre; en ella se halla la mayoría de la masa atmosférica, ocurren la mayoría de los fenómenos meteorológicos y se concentran la mayoría de los contaminantes.

seno: **(pág. 421)** punto más bajo de una onda.

tsunami: **(pág. 548)** enorme y poderosa ola marina generada por los movimientos verticales del fondo del mar durante un sismo; en aguas superficiales, puede formar inmensas olas muy rápidas de mas de 30 m de altura que pueden causar daños en las áreas costeras.

turbidity current: **(p. 448)** rapidly flowing ocean current that can cut deep-sea canyons in continental slopes and deposit the sediments in the form of a continental rise.

corriente de turbidez: **(pág. 448)** corriente oceánica de flujo rápido que puede formar cañones en los taludes continentales y depositar los sedimentos para formar el pie del talud continental.

U

unconformity: **(p. 598)** gap in the rock record caused by erosion or weathering.

disconformidad: **(pág. 598)** discontinuidad en el registro geológico causada por la erosión o la meteorización.

uniformitarianism: **(p. 595)** the theory that geologic processes occurring today have been occurring since Earth formed.

uniformitarianismo: **(pág. 595)** este principio establece que los procesos geológicos que ocurren actualmente han estado ocurriendo desde que la Tierra se formó.

uplifted mountain: **(p. 573)** mountain that forms when large regions of Earth are forced slowly upward without much deformation.

levantamiento montañoso: **(pág. 573)** montañas que se forman cuando grandes regiones de la Tierra son levantadas lentamente sin que ocurra mucha deformación.

upwelling: **(p. 426)** upward movement of ocean water that occurs when winds push surface water aside and it is replaced with cold, deeper waters that originate below the thermocline.

corriente resurgente: **(pág. 426)** movimiento ascendente de las aguas del océano que ocurre cuando los vientos remueven las aguas superficiales, causando que sean reemplazadas por aguas más frías y profundas prove-nientes de profundidades mayores que la termoclina.

V

valley glacier: **(p. 208)** glacier that forms in a valley in a mountainous area and widens V-shaped stream valleys into U-shaped glacial valleys as it moves downslope.

glaciar de valle: **(pág. 208)** glaciar que se forma en un valle de un área montañosa; al deslizarse cuesta abajo, ensancha los valles de corrientes con forma en V y los convierte en valles glaciales con forma de U.

variable star: **(p. 862)** star in the giant branch of the Hertzsprung-Russell diagram that pulsates in brightness due to its outer layers expanding and contracting.

estrella variable: **(pág. 862)** estrella en la rama de las gigantes del diagrama Hertzsprung-Russell, cuya luminosidad presenta pulsaciones debidas a la expansión y contracción de sus capas exteriores.

varve: **(p. 605)** alternating light-colored and dark-colored sedimentary layer of sand, clay, and silt deposited in a lake that can be used to date cyclic events and changes in the environment.

varve: **(pág. 605)** estratos sedimentarios de colores claros y oscuros alternados, compuestos de arena, arcilla y limo, depositados en un lago, que sirven para datar acontecimientos cíclicos y cambios en el ambiente.

vent: **(p. 505)** opening in Earth's crust through which lava erupts and flows out onto the surface.

chimenea: **(pág. 505)** abertura en la corteza terrestre por la cual fluye lava hacia la superficie.

ventifact: **(p. 203)** rock shaped by windblown sediments.

ventifacto: **(pág. 203)** roca moldeada por sedimentos arrastrados por el viento.

vesicular texture: **(p. 120)** a spongy-looking rock; lava whose gas bubbles do not escape.

textura vesicular: **(pág. 120)** roca de aspecto esponjoso; lava cuyas burbujas de gas no se escapan.

viscosity: **(p. 509)** a substance's internal resistance to flow.

viscosidad: **(pág. 509)** resistencia interna a fluir de una sustancia.

volcanism: **(p. 500)** describes all the processes associated with the discharge of magma, hot water, and steam.

vulcanismo: **(pág. 500)** describe todos los procesos asociados con la descarga de magma, agua caliente y vapor.

W

watershed: (p. 227) land area drained by a stream system.

cuenca: (pág. 227) área de terreno drenada por un sistema de corrientes de agua.

water table: (p. 254) upper boundary of the zone of saturation that rises during wet seasons and drops during dry periods.

capa freática: (pág. 254) límite superior de la zona de saturación; aumenta durante la temporada de lluvias y disminuye durante los períodos de sequía.

wave: (p. 421) rhythmic movement that carries energy through matter or space and, in oceans, is generated mainly by wind moving over the surface of the water.

onda (ola): (pág. 421) movimiento rítmico que transporta energía a través de la materia o el espacio; en los océanos, es generado principalmente por el movimiento del viento sobre la superficie del agua.

wave refraction: (p. 439) process in which waves advancing toward shore slow when they encounter shallower water, causing the initially straight wave crests to bend toward the headlands.

refracción de onda: (pág. 439) proceso en que las olas avanzan hacia la costa y reducen su velocidad, cuando llegan a aguas menos profundas, ocasionando que las crestas de las olas, inicialmente rectas, se inclinen hacia los promontorios.

weather: (p. 314) short-term variations in atmosphere phenomena that interact and affect the environment and life on Earth.

tiempo: (pág. 314) variaciones a corto plazo en los fenómenos que suceden en la atmósfera, que interactúan y afectan el entorno de la vida en la Tierra.

weathering: (p. 164) chemical or mechanical process that breaks down and changes rocks on or near Earth's surface and whose rate is influenced by factors such as precipitation and temperature.

meteorización: (pág. 164) proceso químico o mecánico que rompe y modifica las rocas que se hallan sobre o cerca de la superficie terrestre; su velocidad se ve influida por factores como la precipitación y la temperatura.

well: (p. 263) deep hole drilled or dug into the ground to reach a reservoir of groundwater.

pozo: (pág. 263) hoyo profundo perforado o excavado en el suelo para alcanzar un depósito de agua subterránea.

wetland: (p. 240) any land area, such as a bog or marsh, that is covered in water a large part of the year and supports specific plant species.

humedal: (pág. 240) toda área, como un pantano o una ciénaga, que se encuentra cubierta de agua gran parte del año y que alberga especies específicas de plantas.

windchill index: (p. 365) measures the windchill factor, by estimating the heat loss from human skin caused by a combination of wind and cold air.

índice de sensación térmica: (pág. 365) índice que toma en cuenta el efecto del viento en la sensación térmica, al estimar la pérdida de calor de la piel humana causada por la combinación de viento y aire frío.

Z

zircon: (p. 620) very stable and common mineral that scientists often use to age-date old rocks.

circón: (pág. 620) mineral sumamente estable que los científicos usan para datar rocas antiguas.

zone: (p. 812) high, cool, light-colored cloud that rises and flows rapidly in the Jovian atmosphere.

zona: (pág. 812) nubes altas, relativamente frías y de color claro, que se elevan y desplazan con rapidez en la atmósfera joviana.

zone of aeration (p. 254) region above the water table where materials are moist, but pores contain mostly air.

zona de aeración: (pág. 254) región sobre el manto freático en que los materiales están húmedos, pero los poros contienen principalmente aire.

zone of saturation: (p. 254) region below Earth's surface where all the pores of a material are completely filled with groundwater.

zona de saturación: (pág. 254) región profunda bajo la superficie terrestre donde todos los poros del material están completamente llenos con agua subterránea.

Glossary/
Glosario

Index

Index Key

Italic numbers = illustration/photo **Bold numbers = vocabulary term**
act = activity

Index

D

E

F

Index

I

Index

Index

Index

N

O

Index

Index

S

Index

Index

U

W

Credits

Photo Credits

Cover (t c)Getty Images, (b)Paul Chesley/Getty Images; **iv** (t to b)Courtesy of Francisco Borrero, (2)Courtesy of Frances Scelsi Hess, (3)Courtesy of Juno Hsu, (4)Courtesy of Gerhard Kunze, (5)Courtesy of Stephen Leslie; **v** (t to b)Courtesy of Stephen Letro, (2)Courtesy of Michael Manga, (3)Courtesy of Theodore Snow, (4)Courtesy of Dinah Zike; **xii** David Young-Wolff/PhotoEdit; **xx** David Doubilet/ National Geographic Image Collection; **xxi** George Seinmetz National Geographic Image Collection; **2–3** Stephen Alvarez/National Geographic Image Collection; **4** (tl)Eureka Slide/SuperStock, (tr)Gavriel Jecan/CORBIS, (bl)Stockbyte/SuperStock, (br)Bob O'Connor/Getty Images, (bkgd)Science VU/GSFC/Visuals Unlimited; **6** Roger Ressmeyer/CORBIS; **7** Alexis Rosenfeld/Photo Researchers; **10** (l)David Hay Jones/Photo Researchers, (r)Dwayne Newton/PhotoEdit; **11** David Wasserman/Brand X/CORBIS; **14** (l)SPL/Photo Researchers, (r)SSPL/The Image Works; **15** NASA/epa/CORBIS; **16** (inset)David Scharf/Photo Researchers, Royalty-free/CORBIS; **17** Royalty-free/CORBIS; **19** The British Library/HIP/The Image Works; **20** Jordon R. Beesley/U.S. Navy via Getty Images; **23** Roger Ressmeyer/ CORBIS; **24** Bill Varie/CORBIS; **28** Archivo Iconografico, S.A./CORBIS; **37** USGS; **41** produced by the U.S. Geological Survey; **42** (t)JPL/NASA, (bl)Gianni Dagli Orti/ CORBIS, (br)The Art Archive/Pharaonic Village Cairo/Dagli Orti; **43** (t)Boris Schulze, L-3 Communications ELAC Nautik GmbH, (b)CORBIS; **46** (tl tcl tcr tr)USGS; **46 47 49** USGS; **53** NASA/JPL/NOAA/The Cooperative Institute for Research in the Atmosphere (CIRA), Kathy Powell, SAIC and NASA Langley Research Center; **56–57** David Boyer/National Geographic Image Collection; **58** (tl br)Royalty-free/CORBIS, (cr)Doug Wilson/CORBIS, (bkgd)Momatiuk - Eastcott/CORBIS; **59** Doug Martin; **66** Stephen Frisch/Stock Boston; **70** Tim Courlas; **71** (tr)Gregor Schuster/zefa/CORBIS, (bl to br)Studiohio, (2)Mark Burnett, (3 6)Studiohio, (4)Matt Meadows, (5)Amanita Pictures, (7)Aaron Haupt; **73** (tl)Biophoto Associates/Photo Researchers, (tr)Charles D. Winters/Photo Researchers, (br)Mark A. Schneider/Photo Researchers, (bkgd)Doug Martin/Photo Researchers; **74** (t)John Evans, (b)SOHO/NASA; **75** PhotoAlto/SuperStock; **76** (l)Hugh Threlfall/Alamy Images, (r)Michael W. Davidson/Photo Researchers; **77** Matt Meadows; **84** (t)Richard Thom/Visuals Unlimited, (c)David Lazenby/ Animals Animals, (b)Dave Bunnell/Under Earth Images, (bkgd)David Muench/ CORBIS; **85** Holt Studios International Ltd/Alamy; **86** (l)Martin Bond/Photo Researchers, (r)Mark A. Schneider/Visuals Unlimited; **87** (t)Biophoto Associates/ Photo Researchers, (bl)GC Minerals/Alamy Images, (br)Lawrence Lawry/Photo Researchers; **88** (l to r)Piotr & Irena Kolasa/Alamy Images, (2)Jeff Weissman/ Photographic Guide to Mineral Species, (3)Vaughan Fleming/Photo Researchers, (4)Dr. Marli Miller/Visuals Unlimited; **89** (t)Albert Copley/Visuals Unlimited, (c)Scientifica/Visuals Unlimited, (bl)John Elk III/Getty Images, (br) Scientifica/ Visuals Unlimited; **90** (l)Andrew J. Martinez/Photo Researchers, (r)E.R. Degginger/ Animals Animals; **91**Matt Meadows; **92** (l)E.R. Degginger/Animals Animals, (c)Runk/Schoenberg/Grant Heilman Photography, (r)Science Museum/SSPL/The Image Works; **93** (t)Fundamental Photographs, (bl)Nikreates/Alamy Images, (bcl)Richard Carlton/Visuals Unlimited, (bcr)Mark A. Schneider/Photo Researchers, (br)E. R. Degginger/Photo Researchers; **94** (l to r)Mark A. Schneider/Visuals Unlimited, (2)Dr. Marli Miller/Visuals Unlimited, (3)Scientifica/Visuals Unlimited, (4)Charles George/Visuals Unlimited, (5)Mark A. Schneider/Visuals Unlimited; **95** Herve Berthoule/Photo Researchers; **98** (tl)Charles D. Winters/Photo Researchers, (tr)Scientifica/Visuals Unlimited, (c)SuperStock, Inc./SuperStock, (bl)Tom Bean/ CORBIS, (br)image used on pg. 85; **99** (tl)Wally Eberhart/Visuals Unlimited, (tr)Mark A. Schneider/Visuals Unlimited, (br)Royalty-free/CORBIS, (br)Sheila Terry/ Photo Researchers; **100** Ric Feld/AP Images; **101** (t)E. R. Degginger/Photo Researchers, (b)ImageState/Alamy Images; **102** Richard D. Fisher; **104** Albert Copley/Visuals Unlimited; **105** Matt Meadows; **106** Martin Bond/Photo Researchers; **110** (t)Albert J Copley/Visuals Unlimited, (t)Scientifica/Visuals Unlimited, (b)Breck P. Kent/Animals Animals, (bkgd)Nik Wheeler/CORBIS; **113** Lowell Georgia/CORBIS; **115** (t)Marli Miller/Visuals Unlimited, (b)Wally Eberhart/Visuals Unlimited; **116** Breck P. Kent/Animals Animals; **117** Douglas P. Wilson; Frank Lane Picture Agency/CORBIS; **118** (l)Wally Eberhart/Visuals Unlimited, (c)Scientifica/Visuals Unlimited on pg. 118 (c), (r)Albert Copley/Visuals Unlimited; **119** (l)Wally Eberhart/Visuals Unlimited, (c r)Breck P. Kent/Animals Animals; **120** (t)Albert J. Copley/Visuals Unlimited, (c)Jerome Wyckoff/Animals Animals, (b)Breck P. Kent/Animals Animals; **121** (tl)E. R. Degginger/Photo Researchers, (tr)Alfred Pasieka/Photo Researchers, (b)Breck P. Kent/Animals Animals; **122** (l)W. K. Fletcher/Photo Researchers, (r)Dave Bartruff/CORBIS; **123** (t)Koos van der Lende/age fotoStock, (b)Ken Lucas/Visuals Unlimited; **124** Roger Ressmeyer/CORBIS; **125** Matt Meadows; **126** (t)Marli Miller/Visuals Unlimited, (b)Albert J. Copley/Visuals Unlimited; **127** Breck P. Kent/Animals Animals; **128** (t)Breck P. Kent/Animals Animals, (b)Jerome Wyckoff/Animals Animals; **132** (t)Comstock Images/Alamy Images, (b)S.J. Krasemann/Peter Arnold, Inc., (bkgd)Joseph Sohm/ChromoSohm Inc./CORBIS; **133** John Cancalosi/Peter Arnold, Inc.; **134** Adrienne Gibson/Animals Animals; **135** (tl bl)Marli Miller/Visuals Unlimited, (tr)Julio Lopez Saguar/Getty Images, (br)Taylor S. Kennedy/National Geographic Image Collection; **136** (l)George Diebold Photography/Getty Images, (r)Eastcott Momatiuk/Getty Images; **137** Albert J Copley/Getty Images; **138** (t b)Doug Sokell/Visuals Unlimited; **140** (t)Rick Poley/Visuals Unlimited, (b)E. R. Degginger/Photo Researchers; **141** (l r)Breck P. Kent/Animals Animals; **143** (t)James Steinberg/Photo Researchers, (b)Mary Rhodes/Animals Animals; **144** (t)Lynn Stone/Animals Animals, (b)Albert Copley/Visuals Unlimited; **145** Tony Waltham/Robert Harding World Imagery/CORBIS; **146** (tl)George Whitely/Photo Researchers, (tr)Harry Taylor/Getty Images, (bl)Biophoto Associates/Photo Researchers, (br)Scientifica/Visuals Unlimited; **147** (tl tcl)Breck P. Kent/Animals Animals, (tcr)Bernard Photo Productions/Animals Animals, (tr)Andrew J. Martinez/ Photo Researchers, (bl)COLOR-PIC/Animals Animals, (bc)Joyce Photographics/ Photo Researchers, (br)Arthur Hill/Visuals Unlimited; **149** Ken Lucas/Visuals Unlimited; **150** (l)Altrendo Travel/Getty Images, (r)Pixtal/SuperStock; **152** (t)Rob Kim/Landov, (bl)Index Stock/Alamy Images, (br)Sandra Baker/Alamy Images; **153** Matt Meadows; **155** Mark A. Schneider/Visuals Unlimited; **157** Dr. Marli Miller/Visuals Unlimited; **160–161**, Galen Rowell/CORBIS; **162** (tl)Luther Linkhart/Visuals Unlimited, (cl bkgd) Gerald & Buff Corsi/Visuals Unlimited, (bl)Walt Anderson/Visuals Unlimited; **163** Matt Meadows; **164** Larry Stepanowicz/ Visuals Unlimited; **165** (tr)John Serrao/Visuals Unlimited, (b)Bruce Hayes/Photo Researchers; **166** Adam Hart-Davis/Photo Researchers; **167** Rob & Ann Simpson/ Visuals Unlimited; **169** (l)Mark Skalny/Visuals Unlimited, (r)Charles & Josette Lenars/CORBIS; **171** John Anderson/Animals Animals; **172** (t)William Banaszewski/Visuals Unlimited, (b)Inga Spence/Visuals Unlimited; **173** (t)Annie Griffiths Belt/National Geographic Image Collection, (b)Larry Cameron/Photo Researchers; **174** (t)William Manning/CORBIS, (b)David R. Frazier/Photo Researchers; **175** Robert Llewellyn/zefa/CORBIS; **177** William D. Bachman/Photo Researchers; **178** (tl tr)Photo courtesy of USDA Natural Resources Conservation Service; **180 181** State Soil Geographic Database (STATSGO)/NRCS/USDA; **183** The McGraw Hill Company; **185** Matt Meadows; **187** Royalty-free/CORBIS; **188** (l)Creatas/SuperStock, (r)Jim Reed/Photo Researchers; **189** (l)Annie Griffiths Belt/National Geographic Image Collection, (r)Photo courtesy of Dr. Ray Bryant of Cornell (now USDA-ARS in University Park, PA), Will Hanna (USDA-NRCS NY State Soil Scientist, retired), Dr. John Galbraith of Cornell (now Virginia Tech); **192** (t)Steve McCutcheon/Visuals Unlimited, (b)Bernhard Edmaier/Science Photo Library, (bkgd)Gregory Dimijian/Photo Researchers; **194** Dr. Marli Miller/Visuals Unlimited; **195** (t)David McNew/Getty Images, (b)Ralph Lee Hopkins/Photo Researchers; **196** (t)Steve Raymer/National Geographic Image Collection, (b)Gene Blevins/LA Daily News/CORBIS; **197** (t)Handout/Malacanang/Reuters/CORBIS, (b)Lloyd Cluff/CORBIS; **198** (t)Dr. Marli Miller/Visuals Unlimited, (b)Mauritius/ SuperStock; **199** (t)Ted Soqui/CORBIS, (b)Yann Arthus-Bertrand/CORBIS; **200** Michael Habicht/Animals Animals; **202** (t)Jerome Wyckoff/Animals Animals, (b)Remi Benali/CORBIS; **203** (l)Robert Barber/Visuals Unlimited, (r)David Nunuk/ Photo Researchers; **204** Thane/Animals Animals; **205** (b)ABPL Library/Photo Researchers, (others)George Steinmetz/CORBIS; **209** (l)Dr. Marli Miller/Visuals Unlimited, (c)Karl Weatherly/CORBIS, (r)Adam Jones/Visuals Unlimited; **210** E. R. Degginger/Photo Researchers; **211** (l)R.B. Colton/USGS, (c)Tom Bean/CORBIS, (r)Gustav Verderber/Visuals Unlimited; **212** Thomas & Pat Leeson/Photo Researchers; **213** David McNew/Getty Images; **214** USGS; **217** Gabe Palmer/ CORBIS; **218** Bill Kamin/Visuals Unlimited; **219** (l)Philip James Corwin/CORBIS, (r)USGS; **222** (l)Carl & Ann Purcell/CORBIS, (r)Elliott Kaufman/Beateworks/ CORBIS, (bkgd)Martin Garwood/Photo Researchers; **228** (tl)Salvatore Vasapolli/ Animals Animals, (bl)Lloyd Cluff/CORBIS, (br)Anthony Cooper/Ecoscene/CORBIS; **229** Jerry Grayson/Helifilms Australia PTY Ltd/Getty Images; **230** Barrie Rokeach/ Getty Images; **231** USGS; **233** (l)Mike Norton/Animals Animals, (r)Tom Bean/ CORBIS; **234** S.J. Krasemann/Peter Arnold, Inc.; **236** (t)Michael Andrews/Animals Animals, (bl br)USGS; **237** Louie Psihoyos/CORBIS; **238** Phil Schermeister/CORBIS; **239** (t)Michael Gadomski/Animals Animals, (b)Niall Benvie/CORBIS; **242** Guy Motil/CORBIS; **245** Yann Arthus-Bertrand/CORBIS; **246** (l)Dominique Braud/ Animals Animals, (r)Staffan Widstrand/CORBIS; **247** Michael Gadomski/Animals Animals; **250** Yvain Genevay/Geologos/CORBIS; **251** Doug Martin; **256** Jon Turk/ Visuals Unlimited; **259** Michele Burgess/Index Stock; **260** (l)Fritz Polking/Visuals Unlimited, (r)Adam Jones/Visuals Unlimited; **261** (t)Lloyd Homer/GNS Science, (b)Albert J. Copley/Visuals Unlimited; **262** Sheila Terry/Photo Researchers; **269** Kevin Fleming/CORBIS; **271** USGS; **278–279**, Douglas Faulkner/Photo Researchers; **275** Jon Turk/Visuals Unlimited; **280** (inset)Breck P. Kent/Animals Animals, (bkgd)Craig Tuttle/CORBIS; **281** Matt Meadows; **287** Michael Newman/ PhotoEdit; **289** David Hays Jones/Photo Researchers; **292** J Silver/SuperStock; **293** Royalty-free/CORBIS; **297** Fred Whitehead/Animals Animals; **301** Joyce Photographics/Photo Researchers; **302** (l)NCAR/Tom Stack & Associates, (r)Jim Reed/Photo Researchers; **305** Matt Meadows; **308** Pekka Parviainen/Photo Researchers; **312** (t)Tom Bean/CORBIS, (c)Royalty-free/CORBIS, (b)Marc Epstein/ Visuals Unlimited, (bkgd)Getty Images; **314** (l)Les David Manevitz/SuperStock; **321**NASA/CORBIS; **324** (l)Greg Vaughn/Tom Stack & Associates, (c)Stephen St. John/Getty Images, (b)Leonard Lessin, FBPA/Photo Researchers; **325** (tl)Aaron Haupt, (tr)Casella CEL Ltd, (b)Martin Bond/Photo Researchers; **326** United Nations; **327** (t b)NOAA Photo Library, NOAA Central Library, OAR/ERL/National Severe Storms Laboratory (NSSL); **328** (t b)NOAA; **331** Dwayne Newton/ PhotoEdit; **332** NASA/The Visible Earth/http:/visibleearth.nasa.gov/; **333** NOAA; **341** United Nations; **342** (t)Jim Reed/Photo Researchers, (cr)Radhika Chalasani/

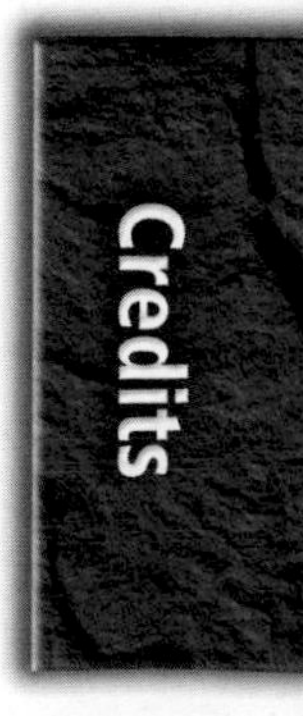

Getty Images, (b)Jim Reed/CORBIS, (bkgd)Scientifica/NOAA/Visuals Unlimited; **345** Royalty-free/CORBIS; **349** (l)G. Grob/zefa/CORBIS, (r)Mark A. Schneider/Visuals Unlimited; **350** Gene & Karen Rhoden/Visuals Unlimited; **351** (t)Jim Reed/Photo Researchers, (b)David Gray/Reuters/CORBIS; **353** (l)H. Baker/Weatherstock, (c)Keith Brewster/Weatherstock, (r)W. Balzer/Weatherstock; **354** Peter Guttman/CORBIS; **356** NASA/Photo Researchers; **358** (l)Reuters/CORBIS, (r)Bettmann/CORBIS; **359** (t)Eduardo Verdugo/AP Images, (c)Jim Reed/Photo Researchers, (b)NASA/Photo Researchers; **360** Paul J. Richards/AFP/Getty Images; **361** CORBIS SYGMA; **362** Don Smetzer/PhotoEdit; **364** David Pollack/CORBIS; **366** Mike Berger/Jim Reed Photography/Photo Researchers; **370** (l)NASA/Photo Researchers, (r)AP Images; **374** (t)Graham French/Masterfile, (c)Carol Polich/Getty Images, (b)Peter Griffith/Masterfile, (bkgd)Boston University and NASA Goddard Space Flight Center; **376** (l)Kelly-Mooney Photography/CORBIS, (r)Charles Bennett/AP Images; **379** (l)Mike Severns/Getty Images, (r)Bill Ross/CORBIS; **380** Galen Rowell/CORBIS; **382** (t)John E Marriott/Alamy Images, (tr)Theo Allofs/zefa/CORBIS, (b)Michael Lewis/CORBIS; **383** Wolfgang Kaehler/Alamy Images; **384** (t)age fotostock/SuperStock, (b)Eric Nguyen/Jim Reed Photography/Photo Researchers; **385** Frank Krahmer/Masterfile; **386** (l r)SVS/Goddard Space Flight Center/NASA; **392** Robert M. Carey/NOAA/Photo Researchers; **395** Joel W. Rogers/CORBIS; **396** Gabriel Bouys/AFP/Getty Images; **404** (t)Stuart Westmorland/CORBIS, (b)age fotostock/SuperStock, (bkgd)Philip James Corwin/CORBIS; **406** (l)Bettmann/CORBIS, (r)Archival Photography by Steve Nicklas/NOS/NGS; **407** (t)Donna C. Rona/Bruce Coleman, Inc., (bl)Jeffrey L. Rotman/CORBIS, (br)W.H.F.Smith/D.T. Sandwell/NOAA/NGDC; **408** (l)David Nunuk/Photo Researchers, (r)Ken Lucas/Visuals Unlimited; **411** Maria Stenzel/National Geographic Image Collection; **415** (l)Conrad Zobel/CORBIS, (r)Dr. Morley Read/Photo Researchers; **416** (t)Tony Hamblin/Frank Lane Picture Agency/CORBIS, (b)Gary Meszaros/Photo Researchers; **417** O.S.F./Animals Animals; **422** Royalty-free/CORBIS; **436** (t)Reinhard Dirscherl/Visuals Unlimited, (b)Gary Bell/zefa/CORBIS, (bkgd)B. S.P.I./CORBIS; **440** Elio Ciol/CORBIS; **442** Dr. Frank M. Hanna/Visuals Unlimited; **443** (tl tr)Joseph Melanson/Aero Photo Inc/www.skypic.com, (bl bc)T. Michot/USGS; **444** (t)Sexto Sol/Getty Images, (b)Tom Bean/CORBIS; **445** Anders Blomqvist/Lonely Planet Images; **446** Dr. Marli Miller/Visuals Unlimited; **449** Jerome Neufeld/The Experimental Nonlinear Physics Group/The University of Toronto; **450** Marie Tharp; **451** (t)Marie Tharp, (b)Science VU/NGDC/Visuals Unlimited; **452** (l)B. Murton/Southampton Oceanography Centre/Photo Researchers, (r)NOAA/www.oceanexplorer.noaa.gov; **453** Steve Gschmeissner/Photo Researchers; **454** Inst. of Oceanographic Sciences/NERC/Photo Researchers; **455** Ralph White/CORBIS; **456** Davis Barber/PhotoEdit; **458** Joseph Melanson/Aero Photo Inc/www.skypic.com; **464–465**, Krafft/Photo Researchers; **466** Kevin Schafer/CORBIS; **468** University of California, Berkeley; **470** John Cancalosi/Peter Arnold, Inc., (inset)Martin Land/Photo Researchers; **471** Australian Government Antarctic Division © Commonwealth of Australia; **472** Alfred Wegener Institute; **473** John F. Williams/U.S. Navy/Getty Images; **477** National Geophysical Data Center/NOAA/NGDC; **479** S. Jonasson/FLPA; **481** Altitude/Peter Arnold, Inc.; **482** (t)Joyce Photographics/Photo Researchers, (b)Andrew J. Martinez/Photo Researchers; **483** (tl)NASA/Photo Researchers, (tr)Kevin Schafer/Peter Arnold, Inc., (cl)Jeff Schmaltz/NASA, (cr)Ed Viggiani/Getty Images, (bl)Firstlight/Getty Images, (br)Woodfall/WWI/Peter Arnold, Inc.; **484** Marie Tharp; **485** Albert Copley/Visuals Unlimited; **486** Richard Megna/Fundamental Photographs; **489** (l r)courtesy of Vailulu'u 2005 Exploration/NOAA-OE; **491** National Geophysical Data Center/NOAA/NGDC; **494** Richard Megna/Fundamental Photographs; **495** USGS; **498** (t)Douglas Peebles/CORBIS, (c)Roger Ressmeyer/CORBIS, (b)Stephen & Donna O'Meara/Volcano Watch Int'l/Photo Researchers, (bkgd)George Steinmetz/CORBIS; **499** Matt Meadows; **502** (t)Science VU/NURP/Visuals Unlimited, (bl)Courtesy of the University of Texas Libraries, The University of Texas at Austin, (br)Roger Ressmeyer/CORBIS; **503**Ho/Reuters/CORBIS; **504** Michael T. Sedam/CORBIS; **505** David Muench/CORBIS; **506** (t)Roger Ressmeyer/CORBIS, (c)Kevin Schafer/CORBIS, (b)Steve Kaufman/Accent Alaska; **508** (l)Reuters/CORBIS, (r)Doug Beghtel/The Oregonian/CORBIS; **510** (t, b)Roger Ressmeyer/CORBIS, (c)Luis Magana/AP Images; **511** (l)Paul A. Souders/CORBIS, (c)Robert Hessler/Planet Earth Pictures, (r)Game McGimsey/epa/CORBIS; **512** (tl)Dr. John D. Cunningham/Visuals Unlimited, (tr)Jeremy Horner/CORBIS, (b)StockTrek/Getty Images; **513** (l)Bullit Marquez/AP Images, (r)Morris J. Elsing/National Geographic Image Collection; **515** (t)Farley Lewis/Photo Researchers, (c)CORBIS, (b)Breck P. Kent/Animals Animals; **516** (br)Marli Miller/Visuals Unlimited, (tr)Jess Alford/Getty Images, (bl)Jerome Wyckoff/Animals Animals; **517** Royalty-free/CORBIS; **518** Carsten Peter/National Geographic Image Collection; **519** Roger Ressmeyer/CORBIS; **526** (t b)Roger Ressmeyer/CORBIS, (c)Reuters/CORBIS, (bkgd)Bernhard Edmaier/Photo Researchers; **527** Bob Daemmrich; **530** (t)Karen Kasmauski/CORBIS, (b)Reuters/CORBIS; **531** (l)Wolfgang Langenstrassen/epa/CORBIS, (r)Paul Chesley/National Geographic Image Collection; **539** Zoriah/The Image Works; **540** Clay Mclachlan/Reuters; **545** R. Kachadoorian/USGS; **546** (t)Rong Shoujun/Xinhua Press/CORBIS, (b)Nik Wheeler/CORBIS; **547** CORBIS; **548** Benjamin Lowy/CORBIS; **550** Gary Kazanjian/AP Images; **552** Bettmann/CORBIS; **557** Joanne Huemoeller/Animals Animals; **560** (c)Royalty-free/CORBIS, (b)George H. H. Huey/CORBIS, (t)Ron Watts/CORBIS, (bkgd)Mark Burnett/Photo Researchers; **565** Adam Jones/Getty Images; **566** Galen Rowell/CORBIS; **568** (t)Jacques Descloitres/MODIS Rapid Response Team/NASA/GSFC, (b)Bob Krist/CORBIS; **569** Sinclair Stammer/Photo Researchers; **570** Worldsat International/Photo Researchers; **571** (l) E. R. Degginger/Photo Researchers, (r)Scott Camazine/Alamy Images; **573** Doug Sokell/Visuals Unlimited; **575** (t)Jon Arnold Images/SuperStock, (b)Tony Freeman/PhotoEdit; **576** Dr. Marli Miller/Visuals Unlimited; **577** Phil Schermeister/CORBIS; **578** Royalty-free/CORBIS; **579** USGS; **586–587**, Ira Block/National Geographic Image Collection; **588** (t)Tom Bean/CORBIS, (b)Richard T. Nowitz/CORBIS, (bkgd)David R. Frazier/Photo Researchers; **590** (l) Richard Hamilton Smith/CORBIS, (r)Royalty-free/CORBIS; **592** (t)Ken Lucas/Visuals Unlimited, (bc)SPL/Photo Reasearchers, Inc., (br)George H. H. Huey/CORBIS; **593** (t) O. Louis Mazzatenta/Getty Images, (bl)Jonathan Blair/CORBIS, (br)Ho New/Reuters; **594** James L. Amos/CORBIS; **595** John Lemker/Animals Animals; **597** (t)David Cavagnaro/Visuals Unlimited, (b)Dr. Marli Miller/Visuals Unlimited; **598** (tl)David Turner (Craven & Pendle Geological Society), (cl)Albert Copley/Visuals Unlimited, (bl)Dr. Marli Miller/Visuals Unlimited; **604** Vin Morgan/AFP/Getty Images; **605** (l)Damien Simonis/Lonely Planet Images, (t)Kevin Schafer/Peter Arnold; **606** Dr Dennis Kunkel/Getty Images; **607** (t)Alfred Pasieka/Photo Researchers, (bl)Tom Bean/CORBIS, (br)Ed Strauss; **608** Dick Roberts/Visuals Unlimited; **610** AP Images; **611** Bruce Heinemann/Getty Images; **618** (t)OSF/Kathy Atkinson/Animals Animals, (c)Layne Kennedy/CORBIS, (b)Dr. Tony Brian/Photo Researchers, (bkgd)Christopher Groenhout/Lonely Planet Images; **619** Charles D. Winters/Photo Researchers; **621** (br)Chris Butler/Photo Researchers, (others)Don Dixon/Cosmographica.com; **628** Gary Braasch/CORBIS; **629** Marli Miller/Visuals Unlimited; **630** (l)Dr. Marli Miller/Visuals Unlimited, (r)Jacques Jangoux/Getty Images; **631** Jack Dykinga; **632** ESA/DLR/FU Berlin (G. Neukum)/epa/CORBIS; **633** Bettmann/CORBIS; **634** (l)Joe Drivas/Getty Images, (c)B. Murton/Southampton Oceanography Centre/Photo Researchers, (r)Jerry Lodriguss/Photo Researchers; **635** Ralph White/CORBIS; **636** (t)Maria Stenzl/National Geographic Image Collection, (b)Chase Studio/Photo Researchers; **637** Hal Beral/Visuals Unlimited; **638** (t)NASA/Photo Researchers, (inset)Ames Research Center/NASA; **642** OSF/Kathy Atkinson/Animals Animals; **643** Jonathan Blair/CORBIS; **646** (t)John Koivula/Photo Researchers, (b)Breck P. Kent/Animals Animals, (bkgd)David Wall/Lonely Planet; **647** Doug Martin; **650** (t)B.S.P.I./CORBIS, (b)Kansas Geological Survey/kgs.ku.edu, (br)Zsolt Schléder and Janos L. Urai/Department of Geoscience, RWTH Aachen University, Germany; **654** Ludek Pesek/Photo Researchers; **656** Jeff Schmaltz, MODIS Rapid Response Team, NASA/GSFC; **658** (tl)Steve Gschmeissner/Photo Researchers, (tr)Ric Ergenbright/CORBIS, (b)Joe Tucciarone/Photo Researchers; **661** Michael T. Sedam/CORBIS; **662** Layne Kennedy/CORBIS; **663** Airphoto-Jim Wark/AirPhotoNA.com; **664** Planetary Visions Ltd/Photo Researchers; **665** Photo Researchers; **666** O. Louis Mazzatenta/National Geographic Image Collection; **667** (t)Ken Lucas/Visuals Unlimited, (b)DK Limited/CORBIS; **669** Ric Ergenbright/CORBIS; **671** Royalty-free/CORBIS; **674–675**, Gabe Palmer/CORBIS; **676** (t)Jim Cornfield/CORBIS, (b)Jim Vecchi/CORBIS, (bkgd)Susan Van Etten/PhotoEdit; **676–677** Victoria Pearson/PictureArts/CORBIS; **678** Clive Helm/CORBIS; **679** (t)Steven Mark Needham/Jupiter Images, (c)Ingram Publishing/SuperStock, (b)Royalty-free/Alamy Images, (bkgd)Royalty-free/CORBIS; **680** (tl)José Manuel Sanchis Calvete/CORBIS, (tr)George Whitely/Photo Researchers, (bl)Scientifica/Visuals Unlimited, (br)Walter Geiersperger/age fotostock; **685** (tc)Marli Miller/Visuals Unlimted, (t)Dr David Waters, Department of Earth Sciences, University of Oxford, (bc)John Cancalosi/Peter Arnold, Inc., (b)David Butow/CORBIS; **686** Julia Cheng/AP Images; **690** (t)U. S. Geological Survey/photo by J.D. Griggs, (b)Peter Essick/Aurora/Getty Images; **692** Will & Deni McIntyre/CORBIS; **694** Richard Hamilton Smith/CORBIS; **695** Du Huaju/XINHUA/CORBIS; **696** Glenn Fuentes/AP Images; **697** Juan José Pascual/age fotostock; **698** Larry Lee Photography/CORBIS; **703** Nick Hawkes, Ecoscene/CORBIS; **706** (l)Stuart Gregory/Getty Images, (r)U.S. Department of Energy/Photo Researchers, (bkgd)Discovery Science Center; **708** David Young-Wolff/PhotoEdit; **709** (l)imagebroker/Alamy Images, (r)Enzo & Paolo Ragazzini/CORBIS; **710** Dr. John D. Cunningham/Visuals Unlimited; **711** (t)Steve McCutcheon/Visuals Unlimited, (bl br)Mark A. Schneider/Visuals Unlimited; **713** U.S. Dept. of Energy/Photo Researchers; **715** (t)John Wilkinson, Ecoscene/CORBIS, (b)Gunter Marx Photography/CORBIS; **716** (bl)Jim Zuckerman/CORBIS, (t)AP Images, (br)Monty Fresco/Topical Press Agency/Getty Images; **717** (t)Roger Ressmeyer/CORBIS, (b)Simon Fraser/Photo Researchers; **719** Jim Richardson/CORBIS; **723** Paul Rapson/Photo Researchers; **724** Royalty-free/CORBIS; **726** (l)Steve McCutcheon/Visuals Unlimited, (c r)Mark A. Schneider/Visuals Unlimited; **727** Enzo & Paolo Ragazzini/CORBIS; **732** (t)Art Wolfe/Photo Researchers, (b)National Geographic Society, (bkgd)Craig Lovell/CORBIS; **734** Larry Lee Photography/CORBIS; **736** Casey Christie/The Bakersfield Californian; **737** Stephanie Maze/CORBIS; **738** (t)©1995 Hallmark Cards, Inc. Photography by John Perryman/courtesy of THE WILDS, (b)Charles E. Rotkin/CORBIS; **741** (tl)Envision/CORBIS, (tr)CORBIS, (b)Matt Meadows/Peter Arnold, Inc.; **742** Kayte M. Deioma/PhotoEdit; **747** Science VU/Visuals Unlimited; **748** Bob Rowan/Progressive Image/CORBIS; **749** (t)Michael St. Maur Sheil/CORBIS, (b)Jon Hicks/CORBIS; **760–761**, NASA/JSC Digital Image Collection; **762** (t)NASA/JPL-Caltech/CORBIS, (b) NASA/Photo Researchers, (bkgd)Craig Aurness/CORBIS; **764** (l)George Diebold, (c)Royalty-free/CORBIS, (r)Michael Nichols/National Geographic Image Collection; **765** (l)NOAO/AURA/NSF, (c)Roger Ressmeyer/CORBIS, (r)Paul Shambroom/Photo Researchers; **766** (l)Russell Croman/Photo Researchers, (r)Hemera Technologies/Alamy; **767** (t)Roger Ressmeyer/CORBIS, (bl)Gustavo Tomsich/CORBIS, (br)Science Museum/

SSPL/The Image Works; **768** NASA; **769** NASA/Photo Researchers; **770** NASA/Science Source; **771** (tl br) NASA Photo Researchers, (tr)NASA, (bl)NASA/CORBIS, (c)Frank Zullo/Photo Researchers; **772** Science VU/Visuals Unlimited; **773** (l)NASA/Photo Researchers, (r)Russell Croman/Science Photo Library; **775** age fotostock/SuperStock; **779** (cw from top)Jason Ware/Photo Researchers, (2)John Chumack/Photo Researchers, (3 5 7)John W. Bova/Photo Researchers, (4)John Sanford/Photo Researchers, (6)Frank Zullo/Photo Researchers, (bl)Chris Cook/Photo Researchers, (br)Eyebyte/Alamy; **781** George Post/Science Photo Library/Photo Researchers; **783** Fred Espenak/Photo Researchers; **784** Dennis di Cicco; **785 787** NASA; **794** (t)NASA/JPL-Caltech, (c)Amy Simon/Reta Beebe/Heidi Hammel/NASA, (b)John Chumack/Photo Researchers, (bkgd)Astrofoto/Peter Arnold, Inc.; **796** NASA/ESA/The Hubble Heritage Team (STScI/AURA); **799** Tuná Tezel; **802** NASA; **805** (t)NASA/JPL-Caltech, (b)NASA/JPL/Northwestern University; **806** JPL/NASA; **807** NASA/Roger Ressmeyer/CORBIS; **808** (l)CORBIS, (r)StockTrek/Getty Images; **809** (tl)USGS/Photo Researchers, (t)NASA/JPL/Cornell, (b)European Space Agency/DLR/FU Berlin/G. Neukum/Photo Researchers; **810** Phil James/Todd Clancy/Steve Lee/NASA; **811** (l)StockTrek/Getty Images, (r)NASA/JPL-Caltech; **812** (t)NASA/Photo Researchers, (b)StockTrek/Getty Images; **813** (t)NASA/JPL/Space Science Institute/Photo Researchers, (b)NASA/ESA/STScI/Photo Researchers; **814** California Association for Research in Astronomy/Photo Researchers; **815** (t)NASA/Photo Researchers, (b)CORBIS; **816** NASA/ESA/J. Parker/P. Thomas/L. McFadden/M. Mutchler/Z. Levay; **817** NASA/ESA/A. Feild; **818** NASA/Photo Researchers; **819** Dan Schechter/Photo Researchers; **820** NASA/JPL/Space Science Institute; **821** (l)NASA, (c)NASA/Mark Marten/Science Source/Photo Researchers, (r)USGS/Science Photo Library/Photo Researchers; **824** JPL/NASA; **828** (t c)STScI/NASA/Science Source, (b)Mark Garlick/Photo Researchers, (bkgd)NASA/ESA/J. Hester/A. Loll; **831** (tl)Kent Wood/Photo Researchers, (t)SOHO (ESA & NASA), (b)Fred Espenak/Photo Researchers; **832** (t)Hinrich Bósemann/dpa/CORBIS, (c)NASA/Photo Researchers, (b)John Chumack/Photo Researchers; **833** (t c)SOHO (ESA & NASA), (b)Detlev van Ravenswaay/Photo Researchers; **838** (tl)Chris Cook/Photo Researchers, (tr)John Chumack/Photo Researchers, (b)NASA/H.E. Bond/E. Nelan/M. Barstow/M. Burleigh/J.B. Holberg; **839** (tl)Jason T. Ware/Photo Researchers, (tr)L. Dodd/Photo Researchers, (c)Stephen & Donna O'Meara/Photo Researchers, (bl)John Chumac/Photo Researchers, (br)SPL/Photo Researchers; **843** (l)Spectral Images created by R. Pogge (OSU)/Spectra from JaCoby, GH, Hunter, DA, & Christian, CA 1984; (r)Matt Meadows; **848** NASA/Photo Researchers; **849** NASA/Andrew Fruchter/ERO Team/Sylvia Baggett (STScI)/Richard Hook (ST-ECF)/Zoltan Levay (STScI); **851** (t b)David Malin/Anglo-Australian Observatory; **852 856** SOHO (ESA & NASA); **860** (t)NASA/ESA/S. Beckwith (STScI)/The Hubble Heritage Team (STScI/AURA), (c)NASA/Holland Ford (JHU)/ACS Science Team/ESA, (b)NASA/ESA/The Hubble Heritage Team (STScI/AURA), (bkgd)NOAO/AURA/NSF/Photo Researchers; **862** (l r)NASA; **863** (t)Jerry Schad/Photo Researchers, (b)Ronald Royer/Science Photo Library/Photo Researchers; **864** NOAO/Photo Researchers; **865** NOAO/SPL/Photo Researchers; **867** European Southern Observatory/Photo Researchers; **869** John Chumack/Photo Researchers; **871** (t)Jason Ware/Photo Researchers, (cr)National Optical Astronomy Observatories/Photo Researchers, (tr)2MASS Image Gallery, (br)NASA/ESA/STScI/Photo Researchers; **872** (t)Luke Dodd/Photo Researchers, (b)Celestial Image Co./Science Photo Library/Photo Researchers; **873** STScI/NASA/CORBIS; **875** (t)AFP/CORBIS, (b)Atlas Photo Bank/Photo Researchers; **877** Chandra X-Ray Observatory/NASA/Photo Researchers; **880** (t)Bettmann/CORBIS, (b)NASA/WMAP Science Team; **881** NASA/Reuters/CORBIS; **882** NASA/ESA/A. M. Koekemoer/M. Dickinson/The GOODS Team; **883** Jean-Charles Cuillandre/Canada-France-Hawaii Telescope/Photo Researchers; **890** David Doubilet/National Geographic Image Collection; **891** George Steinmetz/National Geographic Image Collection; **892** NG Maps/National Geographic Image Collection; **893** Peter Ragg Conservation Air Patrol; **894–897** (c)George Steinmetz/National Geographic Image Collection; **898** Frans Lanting/National Geographic Image Collection; **899** NG Maps/National Geographic Image Collection; **900** (tl)Melissa Farlow/National Geographic Image Collection, (tr)Frans Lanting/National Geographic Image Collection, (b)Adriel Heisey/National Geographic Image Collection; **901** (t)Adriel Heisey/National Geographic Image Collection, (c b)Frans Lanting/National Geographic Image Collection; **902** (t b)Adriel Heisey/National Geographic Image Collection; **903** Frans Lanting/National Geographic Image Collection; **904** David Doubilet/National Geographic Image Collection; **905** August 1994 Landsat image by Thomas Gumbricht; NOAA AVHRR image pair by Philip Frost, Thomas Gumbricht, Jenny M. McCarthy and Frank Seidel/NG Maps/National Geographic Image Collection; **906** David Doubilet/National Geographic Image Collection; **907** Jennifer S. Hayes/National Geographic Image Collection; **908** David Doubilet/National Geographic Image Collection; **910** NASA; **911**Tim Loomis/NOAA Environmental Visualization Program/NG Maps/National Geographic Image Collection; **913** (all)NGM ART/National Geographic Image Collection; **914** Dr. Lynn 'Nick' Shay/Rosenstiel School of Marine & Atmospheric Science; **915** David Burnett/National Geographic Image Collection; **916** Peter Essick/National Geographic Image Collection; **917** (l)Photo courtesy of Bancroft Library/University of California,Berkeley, (r)Photo courtesy of U.S. Geological Survey; **918** Art by Charles Floyd/NGM ART/National Geographic Image Collection; **919 920** Peter Essick/National Geographic Image Collection; **922–923** Art by Lars Grant-West/National Geographic Image Collection; **924** (t)Mark Leong/National Geological Museum, Bejing/National Geographic Image Collection, (b)NG Maps/National Geographic Image Collection; **925** (t bl br)O. Louis Mazzatenta/National Geographic Image Collection, (cl cr)Art by Mark Dubeau/National Geographic Image Collection; **926** (l)Mark Leong/National Geographic Image Collection, (r)Art by Mark Dubeau/National Geographic Image Collection; **927** (l)Mark Leong/National Geographic Image Collection, (r)Art by Mark Dubeau/National Geographic Image Collection; **928–933** George Grall/National Geographic Image Collection; **934** NASA/JPL/CALTECH/Lori Allen and Joseph Hora, Harvard Smithsonian Center for Astrophysics; **935** Art by Bruce Morser/National Geographic Image Collection; **936** NASA/JPL/CALTECH/Oliver Krause, University of Arizona; **937** Bruce Morser/National Geographic Image Collection; **938** (tl)NASA/JPL/CALTECH/Kate Su, University of Arizona, (tr)NASA/JPL/CALTECH, (bl)NASA/JPL/CALTECH/Robert Hurt; **939** (t)NASA/Space Telescope Science Institute/Hubble Heritage Team, (r)NASA/Space Telescope Science Institute/Hubble Heritage Team and NASA/JPL/CALTECH/Robert Kennicutt, University of Arizona and University of Cambridge; **940** CORBIS; **941** (l)Albert Copley/Visuals Unlimited, (r)Charles D. Winters/Photo Researchers; **942** Mike Hoover for Deep Blue Productions; **947** Jose Pelaez/CORBIS; **949** Frances Roberts/Alamy Images.

GEOLOGIC TIME SCALE

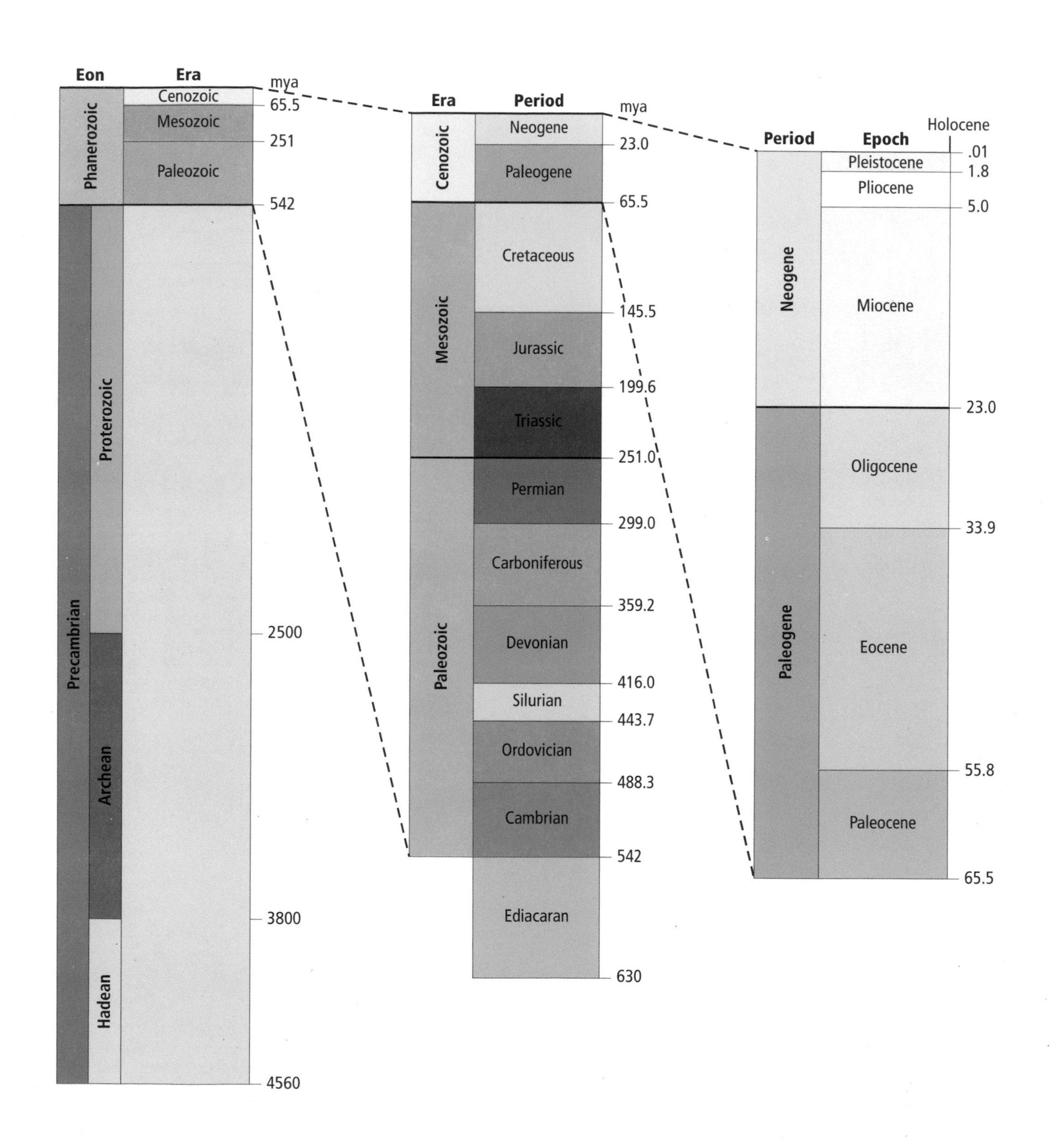

AN EARTH SCIENTIST'S GUIDE TO THE PERIODIC TABLE

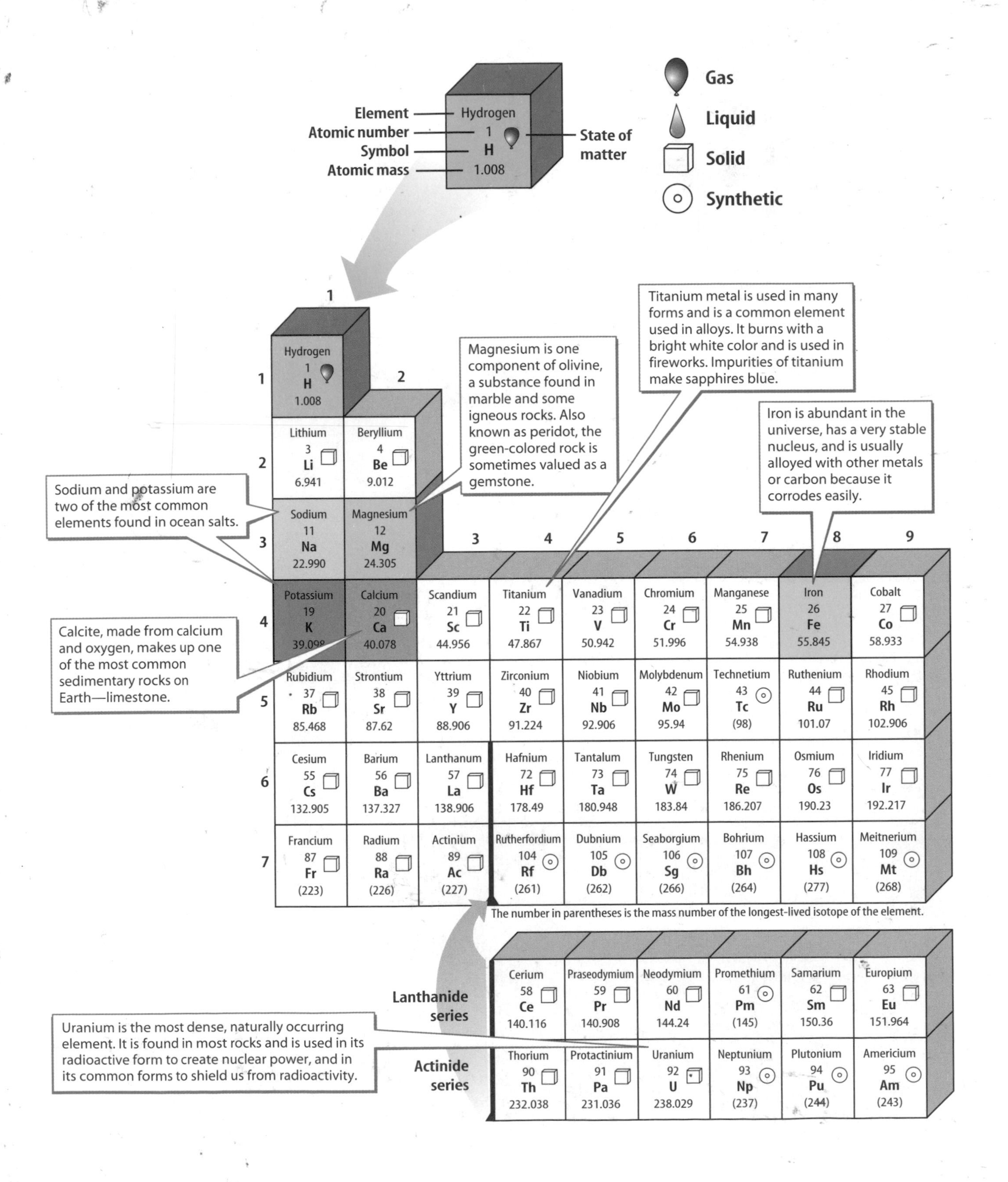